This book examines the theory of boundary value problems for elliptic systems of partial differential equations, a theory which has many applications in mathematics and the physical sciences. The aim is to simplify and to 'algebraize' the index theory by means of pseudo-differential operators and new methods in the spectral theory of matrix polynomials. This latter theory provides important tools that will enable the reader to work efficiently with the principal symbols of the elliptic and boundary operators. It also leads to important simplifications and unifications in the proofs of basic theorems such as the reformulation of the Lopatinskii condition in various equivalent forms, homotopy lifting theorems, the reduction of a system with boundary conditions to a system on the boundary, and the index formula for systems in the plane

This book is suitable for use in graduate level courses on partial differential equations, elliptic systems, pseudo-differential operators, and matrix algebra. All the theorems are proved in detail, and the methods are well illustrated through numerous examples and exercises.

BOUNDARY VALUE PROBLEMS FOR ELLIPTIC SYSTEMS

BOUNDARY VALUE PROBLEMS FOR ELLIPTIC SYSTEMS

J. T. WLOKA
Universität Kiel

B. ROWLEY
Champlain College

B. LAWRUK
McGill University

CAMBRIDGE
UNIVERSITY PRESS

CAMBRIDGE UNIVERSITY PRESS
Cambridge, New York, Melbourne, Madrid, Cape Town, Singapore, São Paulo

Cambridge University Press
The Edinburgh Building, Cambridge CB2 8RU, UK

Published in the United States of America by Cambridge University Press, New York

www.cambridge.org
Information on this title: www.cambridge.org/9780521430111

First published 1995
This digitally printed version 2008

A catalogue record for this publication is available from the British Library

Library of Congress Cataloguing in Publication data

Wloka, Joseph.
Boundary value problems for elliptic systems / J. T. Wloka, B. Rowley, B. Lawruk.
p. cm.
ISBN 0-521-43011-9
1. Boundary value problems. 2. Differential equations, Elliptic.
I. Rowley, B. II. Lawruk, B. III. Title.
QA379.W58 1995
515′.353—dc20 94-34827
CIP

ISBN 978-0-521-43011-1 hardback
ISBN 978-0-521-06143-8 paperback

Contents

Preface

This book examines the theory of boundary value problems for elliptic systems of partial differential equations, a theory which has many applications in mathematics and the physical sciences. The aim is to simplify and to algebraize the index theory by means of pseudo-differential operators and new methods in the spectral theory of matrix polynomials. This latter theory provides important tools that will enable the reader to work efficiently with the principal symbols of the elliptic and boundary operators. It also leads to important simplifications and unifications in the proofs of basic theorems such as the reformulation of the Lopatinskii condition in various equivalent forms, homotopy lifting theorems, the reduction of a system with boundary conditions to a system on the boundary, and the index formula for systems in the plane.

The book is suitable for use in graduate level courses on partial differential equations, elliptic systems, pseudo-differential operators, and matrix algebra. All the theorems are proved in detail, and the methods are well illustrated through numerous examples and exercises.

There are five parts to the book. Part I develops methods in the spectral theory of matrix polynomials which are used throughout the book; it could also be used independently as a text for a course in matrix algebra.

In Part II, there is a concise introduction to manifolds, vector bundles and differential forms. For the convenience of the reader, the development is mostly self-contained; however, it would be helpful for the reader to have had some previous acquaintance with manifold theory, and we recommend [Sp 1] for further background in the basic concepts in both the classical and modern contexts.

In Part III, pseudo-differential operators on $\mathbb{R}^n$ and on a compact manifold are studied. Chapter 7 develops the theory of pseudo-differential operators in $\mathbb{R}^n$ needed to define such operators on a manifold. Essentially, pseudo-differential operators on a manifold M are linear operators on $C^\infty(M)$ that are p.d.o.'s in local coordinates in $\mathbb{R}^n$ and satisfy a quasi-locality property. As it turns out, these operators have a main symbol (or principal symbol) defined modulo lower-order terms, and the aim of Chapter 8 is twofold: to develop the algebra of main and principal symbols, and to develop the Fredholm theory of elliptic operators in vector bundles, including the existence of a parametrix. It is beyond the scope of this book to discuss the Atiyah–Singer formula for the index of an elliptic operator on a compact manifold without boundary, but we develop, essentially, all the analytic properties required for the proof.

In the last chapter of Part III, one finds the main theorem for elliptic boundary value problems on bounded domains in $\mathbb{R}^n$: a necessary and sufficient condition for the Fredholm property to hold. An elliptic system with boundary conditions defines a Fredholm operator in appropriate Sobolev spaces if and only if the boundary operator satisfies a certain L-condition. It is assumed in the proof that the reader is familiar with the definition and basic properties of Sobolev spaces as outlined in the appendix to Chapter 7.

In Part IV, we make full use of the matrix algebra developed in Part I, and there are three important aims in this part.

First of all, a new version of the L-condition is formulated, which we refer to as the Δ-condition. This Δ-condition is expressed in terms of the boundary operator and a spectral pair for the matrix polynomial associated with the elliptic operator. In Chapter 10, we use the Δ-condition to give an elementary proof (using only matrix algebra) of the equivalence of the various formulations of the L-condition, i.e. the Lopatinskii condition, the complementing condition of Agmon, Douglis, and Nirenberg, and other conditions. Furthermore, the Δ-condition leads to an easy proof of some results of Agranovič and Dynin type, and, in addition, a homotopy lifting theorem for elliptic operators whereby a homotopy of elliptic operators is lifted to a homotopy of elliptic boundary problems. The second aim of Part IV is to show how to deform an elliptic boundary problem on Ω to a simpler form having the property that it is equivalent to an elliptic system on the double, $\tilde{\Omega}$, of Ω. The double is a compact manifold, without boundary, to which the Atiyah–Singer theory can be applied. Finally, in Part IV there is a broad discussion of the transmission property for pseudo-differential operators, and another proof of the main theorem for elliptic boundary problems. This second proof uses the Calderón operator to construct a parametrix, i.e. an operator which inverts the boundary value problem, modulo an integral operator with C^∞ kernel. We follow [Hö 3] in the construction of a parametrix by a method inspired by the classical integral representation of solutions of a boundary value problem for the Laplace operator in terms of single- and double-layer potentials (see the introduction to Chapter 14).

Part V is devoted to elliptic boundary problems on bounded domains in the plane. The aim of this part is to prove the index formula, i.e. that the topological index is equal to the analytical index. We study in further detail the L-condition for differential operators in the plane, then define the topological index of an elliptic boundary problem. The proof of the index formula relies on the homotopy lifting theorem mentioned above, in order to reduce an elliptic boundary value problem to a type of Riemann–Hilbert problem for which case there is a well-known formula for the index. With the exception of §16.5, we consider only differential operators in Part V, and the proofs here do not require the use of pseudo-differential operators. In the last chapter, a homotopy classification of 2×2 systems in the plane is accomplished and several examples are studied in detail.

Index of Notation

Transpose of a matrix, $[a_{ij}]^T = [a_{ji}]$
Transpose of a linear map, ${}^t f$
Inner product of column vectors in $\mathbb{R}^n$, $(a, b)_{\mathbb{R}^n} = b^T a = \sum a_i b_i$
Hermitian adjoint of a complex matrix, $[a_{ij}]^h = [\bar{a}_{ji}]$
Inner product of column vectors in $\mathbb{C}^n$, $(a, b)_{\mathbb{C}^n} = b^h a = \sum a_i \bar{b}_i$
Kernel of a linear map, ker A
Image of a linear map, im A

Part I

Part II

Parts III to V

Miscellaneous Notes

In Part II, finite dimensional vector spaces are denoted by bold letters **E**, **F**, etc., and the space of linear maps $\mathbf{E} \to \mathbf{F}$ is denoted by $L(\mathbf{E}, \mathbf{F})$. Elsewhere in the book, finite dimensional vector spaces are denoted by German letters $\mathfrak{M}$, $\mathfrak{N}$, etc., and the space of linear maps $\mathfrak{M} \to \mathfrak{N}$ is denoted by $\mathscr{L}(\mathfrak{M}, \mathfrak{N})$.

Throughout the book, ℓ denotes the degree of the matrix polynomial $L(\lambda) = \sum_{j=0}^{l} A_j \lambda^j$. The letter ℓ is also used occasionally to label the Sobolev spaces, for instance, as in Theorem 9.32.

Part I

A Spectral Theory of Matrix Polynomials

A matrix polynomial is a polynomial function of a complex variable of the form $L(\lambda) = \sum_{j=0}^{l} A_j\lambda^j$ where the coefficients, A_j, are $p \times p$ matrices of complex numbers. Basic matrix theory (including the spectral theory, Jordan form, etc.) is a theory of matrix polynomials $I\lambda - A$ of first degree. The purpose of Part I is to develop a spectral theory for matrix polynomials of arbitrary degree.

We always assume that $L(\lambda)$ is a *regular* matrix polynomial, i.e. $\det L(\lambda) \not\equiv 0$. The spectrum of $L(\lambda)$ is the set of all complex numbers λ such that $L(\lambda)$ is not invertible,

$$\mathrm{sp}(L) = \{\lambda; \det L(\lambda) = 0\}.$$

For a matrix polynomial of degree one, $I\lambda - A$, we write $\mathrm{sp}(A)$ rather than $\mathrm{sp}(I\lambda - A)$ in order to be consistent with the usual definition of the spectrum of the matrix A, that is,

$$\mathrm{sp}(A) = \{\lambda; \det(I\lambda - A) = 0\}.$$

Chapters 1 and 2 develop the spectral theory for the general case of any regular matrix polynomial $L(\lambda)$. Chapter 3 treats the *monic* case, where the leading coefficient, A_l, is the identity matrix. Chapter 4 is needed for technical reasons in Chapter 15.

The spectral triples introduced in Chapter 2 lead to a method for understanding the Lopatinskii condition for elliptic boundary problems, and for reformulating it in various ways as needed (see §10.1). The application of this spectral theory to partial differential equations is new, so we have made the presentation of Part I essentially self-contained, depending only on a knowledge of basic matrix theory and complex analysis. There is, of course, a close relationship between the spectral theory of matrix polynomials, $L(\lambda)$, and the solution space of the equation

$$L(d/dt)u = 0.$$

Due to our interest in differential equations, we do not hesitate to take advantage of this relationship in order to motivate or simplify proofs whenever possible.

1

Matrix polynomials

Let α denote the degree of det $L(\lambda)$. In this chapter we show how the spectral data for the matrix polynomial

$$L(\lambda) = \sum_{j=0}^{l} A_j \lambda^j$$

can be organized in a pair of matrices (X, J), where J is an $\alpha \times \alpha$ Jordan matrix and the columns of the $p \times \alpha$ matrix X are made up of various Jordan chains for each eigenvalue of $L(\lambda)$. There may of course be more than one Jordan chain corresponding to the same eigenvalue.

The idea is as follows. Let $\lambda_0 \in \text{sp}(L)$ be an eigenvalue, and let μ denote the multiplicity of λ_0 as a root of det $L(\lambda) = 0$. Suppose we have found a set of m Jordan chains (see §1.2 for definitions)

$$x_0^{(i)}, \ldots, x_{\mu_i - 1}^{(i)} \qquad i = 1, \ldots, m$$

for $L(\lambda)$ corresponding to the eigenvalue λ_0, such that the eigenvectors $x_0^{(1)}, \ldots, x_0^{(m)}$ (the first vector in each Jordan chain) are linearly independent and such that $\Sigma\mu_i = \mu$. Then introduce a $p \times \mu$ matrix X_0 and a $\mu \times \mu$ block diagonal Jordan matrix J_0 as follows:

$$X_0 = [x_0^{(1)} \cdots x_{\mu_1 - 1}^{(1)} | \cdots | x_0^{(m)} \cdots x_{\mu_m - 1}^{(m)}], \qquad J_0 = \text{diag}(J_0^{(i)})_{i=1}^{m}.$$

The columns of X_0 are the vectors in the given Jordan chains, and the $J_0^{(i)}$ are Jordan blocks of size $\mu_i \times \mu_i$ with eigenvalue λ_0. It can be shown that the pair (X_0, J_0) satisfies the properties

$$\text{(a)} \ \sum_{j=0}^{l} A_j X_0 J_0^j = 0, \qquad \text{(b)} \ \ker \begin{pmatrix} X_0 \\ X_0 J_0 \\ \vdots \\ X_0 J_0^{l-1} \end{pmatrix} = 0,$$

and we call (X_0, J_0) a *Jordan pair* of $L(\lambda)$ corresponding to the eigenvalue λ_0. Now, if we have a set of Jordan pairs $(X_1, J_1), \ldots, (X_s, J_s)$ corresponding to the distinct eigenvalues $\lambda_1, \ldots, \lambda_s$ of $L(\lambda)$. we let

$$X = [X_1 \cdots X_s], \qquad J = \text{diag}(J_k)_{k=1}^{s},$$

and it can be shown that the pair (X, J) also has the properties (a), (b). We call (X, J) a *finite Jordan pair* for $L(\lambda)$. Note that X is a $p \times \alpha$ matrix and J is $\alpha \times \alpha$, where α is the degree of det $L(\lambda)$.

By virtue of the properties (a), (b) we shall see that *every* solution of the equation $L(d/dt)u = 0$ can be written in the form

$$u(t) = Xe^{tJ}c$$

for a unique $c \in \mathbb{C}^\alpha$.

It turns out that the columns of the matrix X form a *canonical* set of Jordan chains for $L(\lambda)$ in the sense of Definitions 1.24, 1.27. The main result of this chapter is the proof that a canonical set of Jordan chains does exist.

For technical reasons, however, the order of development of the theory in Chapter 1 is somewhat different from what was indicated above. In the first section, §1.1, we show by row and column operations that any matrix polynomial $L(\lambda)$ can be transformed to a diagonal form $D(\lambda)$. In §1.2 it is proved that the length of a Jordan chain is preserved in the transformation from $L(\lambda)$ to $D(\lambda)$, which provides valuable information since the structure of the Jordan chains for a diagonal matrix polynomial is easy to establish. In §1.3 we define the notion of a *partial spectral pair* (X, T) – an admissible pair satisfying properties like (a), (b) but with T not necessarily a Jordan matrix and with no condition on the spectrum of T – and then prove some preliminary results which follow formally from this definition. In §1.4 the existence of a canonical set of Jordan chains is proved.

1.1 Smith canonical form

We begin by showing that any matrix polynomial $L(\lambda)$ can be reduced to upper triangular form (zeros below the diagonal) by means of elementary row operations. The types of operations permitted are the following:

(a) the interchange of two rows: $R_i \leftrightarrow R_j$,
(b) the addition to some row of another row multiplied by any polynomial in λ: $p(\lambda)R_j + R_i$,
(c) the multiplication of a row by a nonzero constant: cR_i.

A matrix polynomial obtained from the identity matrix I by means of one such row operation is referred to as an *elementary* matrix polynomial. Note that the determinant of such a matrix polynomial is a nonzero constant.

Two matrix polynomials, $L(\lambda)$ and $M(\lambda)$, are said to be *row equivalent* if there exists a finite sequence of row operations which transform $L(\lambda)$ to $M(\lambda)$ (or vice versa), i.e.

$$M(\lambda) = E(\lambda)L(\lambda),$$

where $E(\lambda)$ is a product of elementary matrix polynomials (and hence det $E(\lambda)$ is a nonzero constant). Note that row equivalence is an equivalence relation between matrix polynomials.

Proposition 1.1 *Every $p \times p$ matrix polynomial $L(\lambda)$ is row equivalent to a matrix polynomial in upper triangular form.*

Proof We may assume that the first column of $L(\lambda)$ is not identically 0. Let $a(\lambda)$ be the monic polynomial of least degree among all first-column entries of all matrix polynomials row equivalent to $L(\lambda)$, and let $M(\lambda) = [m_{ij}(\lambda)]$ be a matrix polynomial row equivalent to $L(\lambda)$ with $m_{11}(\lambda) = a(\lambda)$. The claim is that each entry in the first column of $M(\lambda)$ is necessarily divisible by $a(\lambda)$. Indeed, consider any entry $b(\lambda)$ in the first column of $M(\lambda)$, say in the $(i, 1)$ position. By the Euclidean algorithm for scalar polynomials $b(\lambda) = q(\lambda)a(\lambda) + r(\lambda)$, where $r(\lambda) \equiv 0$ or $\deg r(\lambda) < \deg a(\lambda)$. Now, the row operation $-q(\lambda)R_1 + R_i$ leads to a matrix polynomial row equivalent to $L(\lambda)$ having $r(\lambda)$ in the $(i, 1)$ position; therefore, by the way $a(\lambda)$ was chosen, it must be true that $r(\lambda) \equiv 0$. Performing the row operations $-b(\lambda)/a(\lambda)R_1 + R_i$, we obtain that $L(\lambda)$ is row equivalent to a matrix polynomial of the form

$$\left(\begin{array}{c|c} a(\lambda) & * \\ \hline 0 & L_1(\lambda) \end{array}\right),$$

where $L_1(\lambda)$ is a $(p-1) \times (p-1)$ matrix polynomial. Repeated application of this method leads to a proof of the proposition by induction on the dimension p of the matrix polynomial.

Corollary *Let $L(\lambda)$ be a $p \times p$ matrix polynomial with* $\det L(\lambda) \not\equiv 0$ *and let α denote the degree of $\det L(\lambda)$. Then the dimension of the solution space of the homogeneous differential equation*

$$L\left(\frac{d}{dt}\right)u(t) = 0 \tag{1}$$

is equal to α. Here $u(t) = [u_1(t) \cdots u_p(t)]^T$ is a $\mathbb{C}^p$-valued function.

Proof Let $M(\lambda) = [m_{ij}(\lambda)]$ be an upper triangular matrix polynomial obtained from $L(\lambda)$ by elementary row operations. The new system of equations

$$M\left(\frac{d}{dt}\right)u(t) = 0 \tag{2}$$

is equivalent to the original, since each such operation is reversible. Let l_i be the degree of $m_{ii}(\lambda)$ $(i = 1, \ldots, p)$. The pth equation is a scalar equation $m_{pp}(d/dt)u_p(t) = 0$ with l_p-dimensional solution space. Then by induction on the number p of equations, it is easily seen that the solution space of (2) (and hence (1)) has dimension $\sum_{i=1}^{p} l_i = \deg \det L(\lambda)$.

We will also need the fact that any matrix polynomial can be transformed by row and column operations to *diagonal* form. The column operations that are permitted are of the same three types as stated earlier for row operations, and there is a corresponding definition of column equivalence.

Also, we shall say that $L(\lambda)$ and $M(\lambda)$ are *equivalent under row and column operations* if there is a finite sequence of row or column operations that transforms $M(\lambda)$ to $L(\lambda)$, i.e.

$$M(\lambda) = E(\lambda)L(\lambda)F(\lambda),$$

where $E(\lambda)$ and $F(\lambda)$ are products of elementary matrix polynomials.

Proposition 1.2 *Let $L(\lambda)$ be a $p \times p$ matrix polynomial. Then $L(\lambda)$ is equivalent under row and column operations to a matrix polynomial of the form*

$$\tilde{L}(\lambda) = \left(\begin{array}{c|c} a(\lambda) & 0 \cdots 0 \\ \hline 0 & L_1(\lambda) \end{array}\right), \tag{3}$$

where $L_1(\lambda)$ is a $(p-1) \times (p-1)$ matrix polynomial and each entry of $L_1(\lambda)$ is divisible by $a(\lambda)$.

Proof We may assume that $L(\lambda) \not\equiv 0$. Let $a(\lambda)$ be the monic polynomial of least degree among all entries of all matrix polynomials row equivalent to $L(\lambda)$, and let $M(\lambda) = [m_{ij}(\lambda)]$ be a matrix polynomial row equivalent to $L(\lambda)$ with $m_{11}(\lambda) = a(\lambda)$. As in the proof of Proposition 1.1, each entry in the first column of $M(\lambda)$ is necessarily divisible by $a(\lambda)$. Since we are now permitting column operations, the same is true of each entry in the first row of $M(\lambda)$. By row and column operations, we then obtain that $L(\lambda)$ is equivalent to a matrix polynomial of the form (3). The claim is that each entry of $L_1(\lambda)$ must be divisible by $a(\lambda)$. Indeed, consider any entry $b(\lambda)$ of $L_1(\lambda)$, say in the (i, j) position of $\tilde{L}(\lambda)$ $(i > 1, j > 1)$. Then $b(\lambda) = q(\lambda)a(\lambda) + r(\lambda)$, where $r(\lambda) \equiv 0$ or $\deg r(\lambda) < \deg a(\lambda)$. By performing the row operation $R_i + R_1$ and the column operation $-q(\lambda)C_1 + C_j$ on $\tilde{L}(\lambda)$, we obtain a matrix polynomial that has $r(\lambda)$ in the $(1, j)$ position. Since this matrix polynomial is equivalent to $L(\lambda)$, then, by the way $a(\lambda)$ was chosen, we must have $r(\lambda) \equiv 0$. Hence $b(\lambda)$ is divisible by $a(\lambda)$. This completes the proof of the proposition.

Corollary *Repeated application of Proposition 1.2 leads to the following result: Every matrix polynomial $L(\lambda)$ is equivalent to a diagonal matrix polynomial*

$$D(\lambda) = \begin{pmatrix} d_1(\lambda) & & & & & \\ & \ddots & & & & \\ & & d_s(\lambda) & & & \\ & & & 0 & & \\ & & & & \ddots & \\ & & & & & 0 \end{pmatrix},$$

with monic scalar polynomials $d_i(\lambda)$ such that $d_i(\lambda)$ is divisible by $d_{i-1}(\lambda)$.

The scalar polynomials $d_i(\lambda)$ in the preceding corollary are uniquely determined by $L(\lambda)$ (see Theorem 1.4); $D(\lambda)$ is called the *Smith canonical form* of $L(\lambda)$. Note that if $\det L(\lambda) \not\equiv 0$ then $D(\lambda)$ has no zero entries on the diagonal, that is, $s = p$.

Let $L(\lambda) = [a_{ij}(\lambda)]$ be a $p \times p$ matrix polynomial, and choose k rows $1 \leqslant i_1 < \cdots < i_k \leqslant p$ and k columns $1 \leqslant j_1 < \cdots < j_k \leqslant p$ from $L(\lambda)$. Then the determinant

$$\det[a_{i_m j_n}(\lambda)]_{m,n=1}^{k}$$

is called a $k \times k$ subdeterminant, or minor, of $L(\lambda)$. We let $\phi_k(\lambda)$ denote the monic greatest common divisor (gcd) of all nonzero $k \times k$ subdeterminants of $L(\lambda)$ and let $\phi_k(\lambda) \equiv 0$ if all the subdeterminants of order k are zero. Also let $\phi_0(\lambda) = 1$.

Lemma 1.3 *The polynomials $\phi_k(\lambda)$ are invariant under row and column operations. That is, if $L(\lambda)$ and $M(\lambda)$ are equivalent matrix polynomials, then $\phi_k^M(\lambda) = \phi_k^L(\lambda)$, where $\phi_k^M(\lambda)$, $\phi_k^L(\lambda)$ denote the monic gcd of $k \times k$ subdeterminants of $M(\lambda)$, $L(\lambda)$, respectively.*

Proof First we prove the lemma under the assumption that $M(\lambda)$ is obtained from $L(\lambda)$ by a single row operation. The rows of $M(\lambda)$ are linear combinations with coefficients in $\mathbb{C}[\lambda]$ of the rows of $L(\lambda)$. Due to the multilinearity (i.e. linearity in each row) of the determinant function, it follows that each $k \times k$ subdeterminant of $M(\lambda)$ is a linear combination of the $k \times k$ subdeterminants of $L(\lambda)$. Consequently, $\phi_k^M(\lambda)$ is divisible by $\phi_k^L(\lambda)$. But if we consider the inverse row operation transforming $L(\lambda)$ to $M(\lambda)$, it follows that $\phi_k^L(\lambda)$ is divisible by $\phi_k^M(\lambda)$, and since both are monic, that $\phi_k^M(\lambda) = \phi_k^L(\lambda)$. The same proof holds if $M(\lambda)$ is obtained from $L(\lambda)$ by a single column operation, and then it clearly holds for any number of row and column operations.

Theorem 1.4 *Every matrix polynomial is equivalent under row and column operations to a unique diagonal matrix polynomial*

$$D(\lambda) = \begin{pmatrix} d_1(\lambda) & & & & & \\ & \ddots & & & & \\ & & d_s(\lambda) & & & \\ & & & 0 & & \\ & & & & \ddots & \\ & & & & & 0 \end{pmatrix}, \tag{4}$$

with monic scalar polynomials $d_i(\lambda)$ such that $d_i(\lambda)$ is divisible by $d_{i-1}(\lambda)$. The diagonal entries $d_i(\lambda)$ are given by the following formulas:

$$d_i(\lambda) = \frac{\phi_i(\lambda)}{\phi_{i-1}(\lambda)} \qquad i = 1, \ldots, s \tag{5}$$

Proof The existence of the diagonal form has been proved in Proposition 1.2. Suppose now that $L(\lambda)$ is equivalent to a diagonal matrix polynomial (4) where the $d_i(\lambda)$ are monic scalar polynomials such that $d_i(\lambda)$ is divisible by $d_{i-1}(\lambda)$. By Lemma 1.3, it follows that

$$\begin{aligned} \phi_k(\lambda) &= \text{gcd of } k \times k \text{ subdeterminants of } D(\lambda) \\ &= d_1(\lambda) \cdots d_k(\lambda), \end{aligned}$$

and the formulas (5) follow immediately, proving uniqueness of the Smith canonical form.

1.2 Eigenvectors and Jordan chains

A non-zero column vector $x_0 \in \mathbb{C}^p$ such that $L(\lambda_0)x_0 = 0$ is said to be an *eigenvector* of the matrix polynomial $L(\lambda)$ corresponding to the eigenvalue λ_0. It follows that $\det L(\lambda_0) = 0$. The *spectrum* of $L(\lambda)$, denoted $\mathrm{sp}(L)$, is the set of $\lambda \in \mathbb{C}$ such that $\det L(\lambda) = 0$. From now on, it is always assumed that $\det L(\lambda) \not\equiv 0$, so that $\mathrm{sp}(L)$ consists of a finite number of eigenvalues.

We seek solutions of (1) in the form of vector-valued functions

$$u(t) = \left[\frac{t^{k-1}}{(k-1)!}x_0 + \frac{t^{k-2}}{(k-2)!}x_1 + \cdots + x_{k-1}\right]e^{\lambda_0 t} \tag{6}$$

where $x_j \in \mathbb{C}^p$ and $x_0 \neq 0$. It is not hard to show that x_0 must be an eigenvector of $L(\lambda)$ corresponding to λ_0. In fact we have the following proposition.

Proposition 1.5 *The vector-valued function $u(t)$ given by* (6) *is a solution of* (1) *if and only if*

$$\sum_{j=0}^{i} \frac{1}{j!} L^{(j)}(\lambda_0) x_{i-j} = 0 \qquad i = 0, \ldots, k-1 \tag{7}$$

Proof The Taylor series for $L(\lambda)$ about λ_0 is

$$L(\lambda) = L(\lambda_0) + L'(\lambda_0)\cdot(\lambda - \lambda_0) + \cdots + \frac{1}{l!}L^{(l)}(\lambda_0)\cdot(\lambda - \lambda_0)^l$$

Then, replacing λ by d/dt, we obtain

$$L\left(\frac{d}{dt}\right)u(t) = L(\lambda_0)u(t) + L'(\lambda_0)\cdot\left(\frac{d}{dt} - \lambda_0\right)u(t) + \cdots + \frac{1}{l!}L^{(l)}(\lambda_0)$$
$$\times \left(\frac{d}{dt} - \lambda_0\right)^l u(t) \tag{8}$$

Computation shows that

$$\left(\frac{d}{dt} - \lambda_0\right)^j u(t) = \left[\frac{t^{k-j-1}}{(k-j-1)!}x_0 + \frac{t^{k-j-2}}{(k-j-2)!}x_1 + \cdots + x_{k-j-1}\right]e^{\lambda_0 t}$$

for $j = 0, \ldots, k-1$ and $(d/dt - \lambda_0)^j u(t) = 0$ for $j = k, k+1, \ldots$, and by substitution of these results in the formula (8), the equalities (7) are obtained.

A sequence of p-dimensional column vectors $x_0, x_1, \ldots, x_{k-1}$, where $x_0 \neq 0$, for which equalities (7) hold is called a *Jordan chain of length k for $L(\lambda)$* corresponding to the eigenvalue λ_0. The vectors $x_1, \ldots, x_{k-1}$ are sometimes known as generalized eigenvectors.

We shall say that an eigenvector x_0 of $L(\lambda)$ corresponding to λ_0 is of *rank*

k if the maximal length of a Jordan chain of $L(\lambda)$ corresponding to λ_0 with x_0 as eigenvector is k. Note that if $x_0, x_1, \ldots, x_{k-1}$ is a Jordan chain for $L(\lambda)$, then so is $x_0, x_1, \ldots, x_{s-1}$ for any $s \leqslant k$.

The definition of a Jordan chain of a matrix polynomial is a generalization of the usual idea of Jordan chains for a matrix T. Indeed, let $v_0, v_1, \ldots, v_{k-1}$ be a Jordan chain for T, that is, $Tv_0 = \lambda_0 v_0$, $Tv_1 = \lambda_0 v_1 + v_0, \ldots, Tv_{k-1} = \lambda_0 v_{k-1} + v_{k-2}$. Then these equalities mean exactly that $v_0, v_1, \ldots, v_{k-1}$ is a Jordan chain for the matrix polynomial $I\lambda - T$.

For a matrix polynomial of degree one, $I\lambda - T$, the vectors in a Jordan chain are linearly independent. This is not true for a matrix polynomial of degree greater than one. The following examples show that even the zero vector is admissible as a generalized eigenvector.

Example 1.6

$$L(\lambda) = \begin{pmatrix} (\lambda-2)^2 & -\lambda+2 \\ 0 & (\lambda-2)^2 \end{pmatrix}$$

Since $\det L(\lambda) = (\lambda - 2)^4$, there is one eigenvalue, namely, $\lambda_0 = 2$. Every non-zero vector in $\mathbb{C}^2$ is an eigenvector of $L(\lambda)$ corresponding to λ_0. It is easy to show that all Jordan chains of $L(\lambda)$ can be described as follows:

(1) Jordan chains of length 1 are $x_0 = \begin{pmatrix} a \\ b \end{pmatrix}$, where $a, b \in \mathbb{C}$ are not both zero.

(2) Jordan chains of length 2 are $x_0 = \begin{pmatrix} a \\ b \end{pmatrix}$, x_1, where $a \neq 0$ and $x_1 \in \mathbb{C}^2$ is arbitrary.

(3) Jordan chains of length 3 are $x_0 = \begin{pmatrix} a \\ 0 \end{pmatrix}$, $x_1 = \begin{pmatrix} b \\ a \end{pmatrix}$, x_2, where $a \neq 0$ and $b \in \mathbb{C}$, $x_2 \in \mathbb{C}^2$ are arbitrary.

There are no Jordan chains of length $\geqslant 4$.

Example 1.7 Let $L(\lambda)$ be a scalar polynomial ($p = 1$). Let $L(\lambda_0) = 0$, and let k be the multiplicity of the root λ_0. Then any sequence of complex numbers $x_0, x_1, \ldots, x_{k-1}$, where $x_0 \neq 0$, is a Jordan chain of maximal length of $L(\lambda)$ corresponding to λ_0. There are no Jordan chains of length $> k$ since $L^{(j)}(\lambda_0) = 0$ for $j = 0, \ldots, k-1$ but $L^{(k)}(\lambda_0) \neq 0$.

Example 1.8 Let $L(\lambda) = \operatorname{diag}(d_i(\lambda))_{i=1}^p$ be a diagonal matrix polynomial, where the $d_i(\lambda)$ are scalar polynomials. Let $\det L(\lambda_0) = 0$ and let κ_i be the multiplicity of λ_0 as a root of $d_i(\lambda) = 0$. Also, denote by e_i the ith coordinate vector in $\mathbb{C}^p$. Then it is easily verified that the following are Jordan chains of maximal length for $L(\lambda)$ for any $i = 1, \ldots, p$ such that $\kappa_i \neq 0$:

$$c_0 e_i, \ldots, c_{\kappa_i - 1} e_i,$$

where $c_j \in \mathbb{C}$ and $c_0 \neq 0$. Hence the eigenvector e_i has rank κ_i. More generally, if $x_0 = [x_0^{(i)}]_{i=1}^p$ is any eigenvector of $L(\lambda)$ corresponding to λ_0, then the rank of x_0 is equal to $\min\{\kappa_i; x_0^{(i)} \neq 0\}$.

The next goal will be to characterize the lengths of Jordan chains of maximal length (Proposition 1.10), and then to construct a canonical set of Jordan chains (§1.4).

Lemma 1.9 *Let $M(\lambda)$ be a $p \times p$ matrix polynomial and let $E(\lambda)$ and $F(\lambda)$ be $p \times p$ matrix polynomials with constant nonzero determinants. Then $x_0, \ldots, x_{k-1}$ is a Jordan chain of the matrix polynomial $L(\lambda) := E(\lambda)M(\lambda)F(\lambda)$ corresponding to some $\lambda_0 \in \mathrm{sp}(L)$ if and only if the vectors*

$$y_j = \sum_{i=0}^{j} \frac{1}{i!} F^{(i)}(\lambda_0) x_{j-i} \qquad j = 0, \ldots, k-1$$

form a Jordan chain of $M(\lambda)$ corresponding to λ_0.

Proof Let $x_0, \ldots, x_{k-1}$ *be a Jordan chain of* $L(\lambda)$ corresponding to λ_0 and define $u(t)$ as in (6). Then, as in Proposition 1.5, we have $L(d/dt)u(t) = 0$ so that

$$E\left(\frac{d}{dt}\right)M\left(\frac{d}{dt}\right)F\left(\frac{d}{dt}\right)u(t) = 0 \tag{9}$$

Since $E^{-1}(\lambda)$ is also a matrix polynomial, we can operate with $E^{-1}(d/dt)$ on both sides of (9) to obtain

$$M\left(\frac{d}{dt}\right)v(t) = 0, \tag{10}$$

where $v(t) = F(d/dt)u(t)$. Then, writing

$$F(\lambda) = \sum_{j=0}^{\deg F} \frac{1}{j!} F^{(j)}(\lambda_0)(\lambda - \lambda_0)^j,$$

it follows that

$$v(t) = \left[\frac{t^{k-1}}{(k-1)!} y_0 + \frac{t^{k-2}}{(k-2)!} y_1 + \cdots + y_{k-1}\right] e^{\lambda_0 t} \tag{11}$$

where

$$y_i = \sum_{j=0}^{i} \frac{1}{j!} F^{(j)}(\lambda_0) x_{i-j} \qquad i = 0, \ldots, k-1. \tag{12}$$

Due to (10), (11) and Proposition 1.5, we see that $y_0, \ldots, y_{k-1}$ is a Jordan chain of $M(\lambda)$ corresponding to λ_0. Note that $y_0 \neq 0$ since $\det F(\lambda_0) \neq 0$. The converse is obviously true since $M(\lambda) = E^{-1}(\lambda)L(\lambda)F^{-1}(\lambda)$, where $E^{-1}(\lambda)$ and $F^{-1}(\lambda)$ are matrix polynomials with constant non-zero determinants.

Let $D(\lambda) = \mathrm{diag}(d_i(\lambda))_{i=1}^p$ denote the Smith canonical form of $L(\lambda)$. If $\det L(\lambda_0) = 0$, let κ_i denote the multiplicity of λ_0 as a zero of $d_i(\lambda)$. Then $\kappa_1 \leqslant \cdots \leqslant \kappa_p$ are known as the *partial multiplicities* of $L(\lambda)$ at λ_0. Note that

$$\sum_{i=1}^{p} \kappa_i = \alpha_0,$$

the multiplicity of λ_0 as a zero of det $L(\lambda)$.

Proposition 1.10 *The lengths of Jordan chains of $L(\lambda)$ of maximal length are exactly the non-zero partial multiplicities of $L(\lambda)$ corresponding to λ_0.*

Proof By Theorem 1.4, there exist matrix polynomials $E(\lambda)$ and $F(\lambda)$ with constant non-zero determinant such that $L(\lambda) = E(\lambda)D(\lambda)F(\lambda)$, where $D(\lambda)$ is the Smith canonical form of $L(\lambda)$. Then the equations (12) of Lemma 1.9 define a one-to-one correspondence between Jordan chains of $L(\lambda)$ and Jordan chains of $D(\lambda)$ corresponding to λ_0 – the inverse to (12) is found by using the coefficients of the matrix polynomial $F^{-1}(\lambda)$ – that preserves the length of Jordan chains. Since $D(\lambda)$ is a diagonal matrix, then the lengths of Jordan chains of $D(\lambda)$ – and hence $L(\lambda)$ – of maximal length are simply the non-zero partial multiplicities of $L(\lambda)$ corresponding to λ_0 (see Example 1.8).

Remark It follows from Proposition 1.10 that the rank of any eigenvector corresponding to λ_0 does not exceed the multiplicity of λ_0 as a zero of det $L(\lambda)$.

The next proposition introduces a convenient way to summarize the information given in a Jordan chain.

Proposition 1.11 *The vectors $x_0, x_1, \ldots, x_{k-1}$ form a Jordan chain corresponding to λ_0 for the $p \times p$ matrix polynomial $L(\lambda) = \sum_{j=0}^{l} A_j\lambda^j$ if and only if $x_0 \neq 0$ and*

$$A_0X_0 + A_1X_0J_0 + \cdots + A_lX_0J_0^l = 0 \tag{13}$$

where $X_0 = [x_0 \cdots x_{k-1}]$ is a $p \times k$ matrix, and

$$J_0 = \begin{pmatrix} \lambda_0 & 1 & & & \\ & \lambda_0 & \cdot & & \\ & & \cdot & \cdot & \\ & & & \cdot & 1 \\ & & & & \lambda_0 \end{pmatrix} \tag{14}$$

is the $k \times k$ Jordan block with eigenvalue λ_0.

Proof Since

$$e^{tJ_0} = \begin{pmatrix} 1 & t & \cdots & \dfrac{t^{k-1}}{(k-1)!} \\ & 1 & \cdot & \\ & & \cdot & \cdot \\ & & & t \\ & & & 1 \end{pmatrix} e^{t\lambda_0} \tag{15}$$

the functions (6) can be written $u(t) = X_0\, e^{tJ_0}\, c$, where $c = [0 \cdots 0\ 1]^T$. Thus $L(d/dt)u = 0$ if and only if

$$(A_0X_0 + A_1X_0J_0 + \cdots + A_lX_0J_0^l)c = 0 \tag{16}$$

We can get this equation for all c by considering the functions $u_j = (d/dt - \lambda_0)^j u$ for $j = 0, 1, \ldots$ Since $u_j = X_0\, e^{tJ_0}\,(J_0 - \lambda_0 I)^j = X_0\, e^{tJ_0}\, c'$, where $c' = [0 \cdots 1 \cdots 0]^T$ with 1 in the $(k-j)$th position and 0's elsewhere, it follows that $L(d/dt) = u = 0$ if and only if (16) holds for all $c \in \mathbb{C}^k$.

1.3 Partial spectral pairs

The pair (X_0, J_0) as defined in Proposition 1.11 is called a partial Jordan pair of $L(\lambda)$ corresponding to the eigenvalue λ_0. The discussion of Jordan pairs will be continued later.

We turn now to some definitions in order to generalize the idea of a Jordan pair, making it more applicable as a theoretical tool. As usual, if $\mathfrak{M}$ and $\mathfrak{N}$ are finite dimensional vector spaces, $\mathscr{L}(\mathfrak{M}, \mathfrak{N})$ denotes the set of linear operators from $\mathfrak{M}$ to $\mathfrak{N}$.

Definition 1.12 *A pair of operators (X, T) is called an admissible pair if $X \in \mathscr{L}(\mathfrak{M}, \mathbb{C}^p)$ and $T \in \mathscr{L}(\mathfrak{M}) = \mathscr{L}(\mathfrak{M}, \mathfrak{M})$. The spaces $\mathfrak{M}$ and $\mathbb{C}^p$ are called the base space and target space, respectively, of the admissible pair. Two admissible pairs, (X, T) and (X', T'), are called similar if there exists an invertible operator $M \in \mathscr{L}(\mathfrak{M}', \mathfrak{M})$ such that $X' = XM$ and $T' = M^{-1}TM$.*

In the sequel, we will alternate between the matrix and operator viewpoints without comment, using whichever viewpoint is appropriate for the context. Note that if (X, T) is an admissible pair of matrices, the matrix T need not be a Jordan matrix. If $S_j \in \mathscr{L}(\mathfrak{M}, \mathfrak{N}_j)$, $j = 1, \ldots, n$, we define

$$\operatorname{col}(S_j)_{j=1}^n = \begin{pmatrix} S_1 \\ \vdots \\ S_n \end{pmatrix} \in \mathscr{L}(\mathfrak{M}, \mathfrak{N}_1 \oplus \cdots \oplus \mathfrak{N}_n)$$

Similarly, if $T_j \in \mathscr{L}(\mathfrak{M}_j, \mathfrak{N})$, $j = 1, \ldots, n$, we define

$$\operatorname{row}(T_j)_{j=1}^n = [T_1 \cdots T_n] \in \mathscr{L}(\mathfrak{M}_1 \oplus \cdots \oplus \mathfrak{M}_n, \mathfrak{N}).$$

Also, if $M_j \in \mathfrak{L}(\mathfrak{M}_j, \mathfrak{N}_j)$, $j = 1, \ldots, n$, we define

$$\operatorname{diag}(M_j)_{j=1}^n = \begin{pmatrix} M_1 & & \\ & \ddots & \\ & & M_n \end{pmatrix} \in \mathscr{L}(\mathfrak{M}_1 \oplus \cdots \oplus \mathfrak{M}_n, \mathfrak{N}_1 \oplus \cdots \oplus \mathfrak{N}_n).$$

Definition 1.13 *An admissible pair (X, T) is said to be a partial spectral pair for the matrix polynomial $L(\lambda) = \sum_{j=0}^l A_j\lambda^j$ if the following conditions hold:*

(a) $\sum_{j=0}^l A_jXT^j = 0$,
(b) $\operatorname{col}(XT^j)_{j=0}^{l-1}$ *is injective.*

Recall that for a matrix polynomial $L(\lambda)$ the spectrum, sp(L), is the set of $\lambda \in \mathbb{C}$ such that $\det L(\lambda) = 0$, while (in order to be consistent with the standard definition in linear algebra) for a *matrix* T the spectrum, sp(T), is the set of $\lambda \in \mathbb{C}$ such that $\det(I\lambda - T) = 0$. Now if (X, T) is a partial spectral pair for $L(\lambda)$ we claim that $\mathrm{sp}(T) \subset \mathrm{sp}(L)$. Indeed, if $Tv_0 = \lambda_0 v_0$, where $v_0 \neq 0$, then $Xv_0 \neq 0$ (otherwise $XT^j v_0 = 0$ for all $j \geqslant 0$, in contradiction with (b)) and

$$0 = \sum A_j X T^j v_0 = \sum A_j \lambda_0^j X v_0 = L(\lambda_0) X v_0.$$

In fact more than this is true as we show in the next proposition.

Proposition 1.14 *Let (X, T) be a partial spectral pair for $L(\lambda)$. If $v_0, \ldots, v_{k-1}$ is a Jordan chain of $I\lambda - T$ corresponding to λ_0 then $Xv_0, \ldots, Xv_{k-1}$ is a Jordan chain of $L(\lambda)$ corresponding to λ_0.*

Proof Let $u(t) = X\, e^{tT} v_{k-1}$. Since $\sum A_j X T^j = 0$, we have $L(d/dt)u(t) = 0$. In view of Proposition 1.5, it suffices to show that

$$u(t) = \left[\frac{t^{k-1}}{(k-1)!} X v_0 + \frac{t^{k-2}}{(k-2)!} X v_1 + \cdots + X v_{k-1} \right] e^{\lambda_0 t} \qquad (17)$$

Consider the subspace $\mathfrak{L}$ spanned by $v_0, \ldots, v_{k-1}$. Relative to the basis $v_0, \ldots, v_{k-1}$, the operator $T|_{\mathfrak{L}}$ may be represented by the $k \times k$ Jordan block (14) with eigenvalue λ_0. Also v_{k-1} is represented by the column vector $[0 \cdots 0\ 1]^T$ and $X|_{\mathfrak{L}}$ is represented by the matrix $[Xv_0 \cdots Xv_{k-1}]$. Then (17) follows immediately from (15).

Proposition 1.15 *Let (X, T) be a partial spectral pair for $L(\lambda)$ with base space $\mathfrak{M}$. Then* $\dim \mathfrak{M} \leqslant \alpha$, *where α is the degree of* $\det L(\lambda)$. *Moreover, if* $\dim \mathfrak{M} = \alpha$ *then every solution of $L(d/dt)u(t) = 0$ can be written in the form $u(t) = Xe^{tT}c$ for a unique $c \in \mathfrak{M}$.*

Proof Consider the linear map $\Phi\colon \mathfrak{M} \to C^\infty(\mathbb{R}, \mathbb{C}^p)$ defined by

$$c \mapsto u(t) = Xe^{tT}c.$$

If $\Phi c = 0$ it follows upon differentiation that $XT^j c = 0$ for all $j = 0, 1, \ldots,$ hence (b) of Definition 1.13 implies $c = 0$. Thus Φ is injective. Since (a) of Definition 1.13 implies that the image of Φ is a subspace of the solution space, $\mathfrak{M}_L$, of the homogeneous equation

$$L\left(\frac{d}{dt}\right)u(t) = 0,$$

it follows that $\dim \mathfrak{M} \leqslant \alpha$ (the dimension of $\mathfrak{M}_L$). Moreover, if $\dim \mathfrak{M} = \alpha$ then Φ is a bijection from $\mathfrak{M}$ to $\mathfrak{M}_L$.

Definition 1.16 *If (X, T) and (X', T') are admissible pairs then (X', T') is called a restriction of (X, T) if there exists a T-invariant subspace $\mathfrak{L}$ of the base space $\mathfrak{M}$ of (X, T) such that (X', T') is similar to $(X|_{\mathfrak{L}}, T|_{\mathfrak{L}})$.*

Let (X, T) be a partial spectral pair for $L(\lambda)$. As shown in the proof of Proposition 1.15 the base space $\mathfrak{M}$ must satisfy $\dim \mathfrak{M} \leqslant \alpha$. The following proposition shows that the partial spectral pairs which are maximal with respect to the relation defined by restriction of admissible pairs are exactly those whose base space has dimension α.

Proposition 1.17 *Let (X, T) be a partial spectral pair for $L(\lambda)$ and let $\mathfrak{M}$ denote its base space. Then the following are equivalent:*

(*a*) $\dim \mathfrak{M} = \alpha$
(*b*) *Every partial spectral pair for $L(\lambda)$ is a restriction of (X, T).*

Proof (a) $\Rightarrow$ (b) Assume that $\dim \mathfrak{M} = \alpha$ and let (X', T') be any partial spectral pair for $L(\lambda)$, with base space $\mathfrak{M}'$. Let $c' \in \mathfrak{M}'$ and consider the function

$$u(t) = X'e^{tT'}c'.$$

Since $u \in \mathfrak{M}_L$ it follows from Proposition 1.15 that $u(t) = Xe^{tT}c$ for a unique $c \in \mathfrak{M}$. Then upon differentiating the equation $X'e^{tT'}c' \equiv Xe^{tT}c$ with respect to t and setting $t = 0$, we obtain $X'(T')^i c' = XT^i c$ for all $i = 0, 1, \ldots$. Let $M\colon \mathfrak{M}' \to \mathfrak{M}$ be defined by $Mc' = c$, then we have

$$X'(T')^i = XT^iM \qquad \text{for all } i = 0, 1, \ldots \tag{18}$$

Note that M is injective due to the injectivity of $\operatorname{col}(X'(T')^i)_{i=0}^{l-1}$. We now claim that $X' = XM$ and $TM = MT'$. The first equation is simply (18) with $i = 0$; for the second equation observe that

$$\begin{aligned}
\operatorname{col}(XT^i)_{i=0}^{l-1}\, TM &= \operatorname{col}(XT^{i+1}M)_{i=0}^{l-1} \\
&= \operatorname{col}(X'(T')^{i+1})_{i=0}^{l-1} \\
&= \operatorname{col}(X'(T')^i)_{i=0}^{l-1}\, T' \\
&= \operatorname{col}(XT^i)_{i=0}^{l-1}\, MT'
\end{aligned}$$

Hence $TM = MT'$ follows from the injectivity of $\operatorname{col}(XT^i)_{i=0}^{l-1}$. Hence $\mathfrak{L} := \operatorname{im} M$ is T-invariant and (X', T') is similar to $(X|_{\mathfrak{L}}, T|_{\mathfrak{L}})$.

(b) $\Rightarrow$ (a) Suppose that every partial spectral pair is a restriction of (X, T). By Lemma 1.23 and the remark following Definition 1.27 (see below), there exists a partial spectral pair (in fact, a Jordan pair) of $L(\lambda)$ with α-dimensional base space. If this pair is to be a restriction of (X, T) then – in view of the fact that $\dim \mathfrak{M} \leqslant \alpha$ – it must be true that $\dim \mathfrak{M} = \alpha$.

The next lemma is needed in §1.4.

Lemma 1.18 *Let $L(\lambda) = \sum_{j=0}^{l} A_j\lambda^j$ be a matrix polynomial of degree l and define*

$$L_j(\lambda) = A_j + A_{j+1}\lambda + \cdots + A_l\lambda^{l-j} \qquad j = 0, \ldots, l.$$

If (X, T) is an admissible pair such that $\sum_{j=0}^{l} A_j X T^j = 0$ then

$$L(\lambda)\cdot X = \sum_{j=0}^{l-1} L_{j+1}(\lambda)\cdot X T^j (I\lambda - T) \tag{19}$$

Proof Since $\lambda L_{j+1}(\lambda) = L_j(\lambda) - A_j$, then

$$\begin{aligned}\sum_{j=0}^{l-1} L_{j+1}(\lambda) X T^j (I\lambda - T) &= \sum_{j=0}^{l-1} (L_j(\lambda) - A_j) X T^j - \sum_{j=0}^{l-1} L_{j+1}(\lambda) X T^{j+1} \\ &= L_0(\lambda) X - \sum_{j=0}^{l} A_j X T^j \\ &= L(\lambda) X\end{aligned}$$

Proposition 1.19 *Let (X, T) be an admissible pair such that $\sum_{j=0}^{l} A_j X T^j = 0$. Then the following identity holds:*

$$\operatorname{col}(X T^j)_{j=0}^{l-1}\cdot T = C_1\cdot \operatorname{col}(X T^j)_{j=0}^{l-1} \tag{20}$$

where C_1 is the $pl \times pl$ matrix

$$C_1 = \frac{1}{2\pi i}\int_\Gamma \begin{pmatrix} I \\ \vdots \\ \lambda^{l-1} I \end{pmatrix} \lambda L^{-1}(\lambda)[L_1(\lambda)\cdots L_l(\lambda)]\, d\lambda$$

and Γ is a simple, closed rectifiable contour such that $\operatorname{sp}(L)$ *is contained inside Γ.*

Proof First of all,

$$X T^j = \frac{1}{2\pi i}\int_\Gamma \lambda^j X (I\lambda - T)^{-1}\, d\lambda \tag{21}$$

for $j = 0, 1, \ldots$. Indeed, let Γ be a circle of sufficiently large radius and observe that

$$(I\lambda - T)^{-1} = \lambda^{-1}(I + T/\lambda + T^2/\lambda^2 + \cdots),$$

the series converging absolutely when $|\lambda| > \|T\|$. Now integrate term by term to obtain (21), then use (19) to obtain

$$X T^j = \frac{1}{2\pi i}\int_\Gamma \lambda^j L^{-1}(\lambda)\left(\sum_{k=0}^{l-1} L_{k+1}(\lambda) X T^k\right) d\lambda \tag{22}$$

for $j = 0, 1, \ldots$, from which (20) follows immediately.

Note that (22) implies that if $X T^j c = 0$ for $j = 0, 1, \ldots, l-1$, then $X T^j c = 0$ for all $j \geqslant 0$. Thus we obtain the following corollary.

Corollary 1.20 *Let (X, T) be an admissible pair such that $\sum_{j=0}^{l} A_j X T^j = 0$. If $\operatorname{col}(X T^j)_{j=0}^{k}$ is injective for some k then it is injective when $k = l - 1$.*

Remark 1.21 The equation (20) implies that if (X, T) satisfies Definition 1.13(a) then the kernel of $\operatorname{col}(XT^j)_{j=0}^{l-1}$ is T-invariant. This implies that 1.13(b) can be imposed without loss of generality, for if $\mathfrak{N}$ denotes a complement to this kernel in the base space of (X, T) then $(X|_{\mathfrak{N}}, T|_{\mathfrak{N}})$ satisfies both (a) and (b).

1.4 Canonical set of Jordan chains

We now continue with the discussion of Jordan pairs.

Lemma 1.22 *Let $x_0^{(i)}, \ldots, x_{\mu_i-1}^{(i)}$ $(i = 1, \ldots, m)$ be a set of Jordan chains of $L(\lambda)$ corresponding to an eigenvalue λ_0 such that the eigenvectors $x_0^{(1)}, \ldots, x_0^{(m)}$ are linearly independent (where $m \leqslant \dim\ker L(\lambda_0)$). Let $\mu = \sum_{i=1}^m \mu_i$ and define a $p \times \mu$ matrix X_0 with the vectors of the Jordan chains as its columns:*

$$X_0 = [x_0^{(1)} \cdots x_{\mu_1-1}^{(1)} | \cdots | x_0^{(m)} \cdots x_{\mu_m-1}^{(m)}]$$

and a $\mu \times \mu$ block diagonal Jordan matrix:

$$J_0 = \operatorname{diag}(J_0^{(i)})_{i=1}^m$$

where the $J_0^{(i)}$ are Jordan blocks of size $\mu_i \times \mu_i$ $(i = 1, \ldots, m)$ with eigenvalue λ_0. Then

$$\ker \operatorname{col}(X_0 J_0^i)_{i=0}^{l-1} = \{0\}.$$

Note: There are two different uses of superscripts: J_0^i denotes the ith power of J_0 while $J_0^{(i)}$ denotes the ith Jordan block of J_0.

Proof Let $\mathfrak{L} = \ker \operatorname{col}(X_0 J_0^i)_{i=0}^{l-1}$. First we show that $\mathfrak{L}$ is a J_0-invariant subspace. By applying Proposition 1.11 to each Jordan block, we see that the admissible pair (X_0, J_0) satisfies (a) of Definition 1.13. In view of Proposition 1.19,

$$\operatorname{col}(X_0 J_0^j)_{j=0}^{l-1} \cdot J_0 = C_1 \cdot \operatorname{col}(X_0 J_0^j)_{j=0}^{l-1}$$

so that $c \in \mathfrak{L}$ implies $J_0 c \in \mathfrak{L}$.

Suppose now that $\mathfrak{L}$ contains a vector $c \neq 0$. Write $\mathbb{C}^\mu$ in the obvious way as a direct sum $\mathbb{C}^{\mu_1} \oplus \cdots \oplus \mathbb{C}^{\mu_m}$ so that $J_0^{(i)}$ is the restriction of J_0 to $\mathbb{C}^{\mu_i}$, and let $e_j^{(i)}$ be the standard basis consisting of coordinate unit vectors, where $j = 0, \ldots, \mu_i - 1$ and $i = 1, \ldots, m$. Then we write c in the form

$$c = [c_0^{(1)} \cdots c_{\mu_1-1}^{(1)} | \cdots | c_0^{(m)} \cdots c_{\mu_m-1}^{(m)}] = \sum_{i=1}^{m} \sum_{j=0}^{\mu_i-1} c_j^{(i)} e_j^{(i)},$$

where at least one $c_j^{(i)} \neq 0$. Without loss of generality, some $c_{\mu_i-1}^{(i)} \neq 0$. Now let

$$\nu = \max\{\mu_i;\ c_{\mu_i-1}^{(i)} \neq 0,\ i = 1, \ldots, m\}$$

and let $S = \{i;\ \mu_i = \nu\}$. Then

$$(J_0 - \lambda_0 I)^{\nu-1} c = \sum_{i \in S} c_{\mu_i-1}^{(i)} e_0^{(i)}$$

Since $c \in \mathfrak{L}$ and $\mathfrak{L}$ is J_0-invariant, it follows that $(J_0 - \lambda_0 I)^{\nu-1}c \in \mathfrak{L}$; in particular, $X_0(J_0 - \lambda_0 I)^{\nu-1}c = 0$, that is, $\sum_{i\in S} c^{(i)}_{\mu_i-1}x^{(i)}_0 = 0$. By assumption, the eigenvectors $x^{(i)}_0$ are linearly independent. Hence $c^{(i)}_{\mu_i-1} = 0$ for all $i \in S$, which contradicts the definition of S. Hence $\mathfrak{L} = \{0\}$, and condition (b) holds.

Let (X_0, J_0) be an admissible pair as defined in Lemma 1.22. Then $(X_0. J_0)$ will be referred to as a *partial* Jordan pair of $L(\lambda)$ corresponding to the eigenvalue λ_0.

Lemma 1.23 *Let $L(\lambda) = \sum_{j=0}^{l} A_j\lambda^j$ be a $p \times p$ matrix polynomial of degree l and let $\lambda_1, \ldots, \lambda_s$ be the distinct eigenvalues of $L(\lambda)$. Let (X_i, J_i) be a partial Jordan pair of $L(\lambda)$ corresponding to λ_i, and let*

$$X = [X_1 \cdots X_s], \qquad J = \operatorname{diag}(J_i)_{i=1}^{s}$$

Then (X, J) is a partial spectral pair of $L(\lambda)$.

Proof Condition (a) of Definition 1.13 holds because we can apply Proposition 1.11 to each Jordan block. Let $\mathfrak{L}$ denote the kernel of $\operatorname{col}(XJ^j)_{j=0}^{l-1}$. We must show that $\mathfrak{L} = \{0\}$. First, as in the proof of Lemma 1.22, one sees that $\mathfrak{L}$ is a J-invariant subspace. Then according to a standard result of spectral theory (see [GLR], p. 359), it follows that

$$\mathfrak{L} = (\mathfrak{L} \cap \mathfrak{W}_1) \oplus \cdots \oplus (\mathfrak{L} \cap \mathfrak{W}_S),$$

where

$$\mathfrak{W}_i = \ker(J - \lambda_i I)^{\alpha} \qquad i = 1, \ldots, s$$

is the root subspace of J corresponding to the eigenvalue λ_i. Now

$$\mathfrak{L} \cap \mathfrak{W}_i = \ker \operatorname{col}(X_i J_i^k)_{k=0}^{l-1} \qquad i = 1, \ldots, s$$

and by Lemma 1.22 we have $\mathfrak{L} \cap \mathfrak{W}_i = \{0\}$ for all i. Hence $\mathfrak{L} = \{0\}$. The proof of the lemma is complete.

We now define a canonical set of Jordan chains corresponding to an eigenvalue λ_0. The definition proceeds by an inductive selection of eigenvectors with maximal rank.

Definition 1.24 *Let λ_0 be an eigenvalue of $L(\lambda)$. Choose an eigenvector $x^{(1)}_0$ of maximal rank μ_1. If eigenvectors $x^{(1)}_0, \ldots, x^{(i-1)}_0$ are already chosen then, among all eigenvectors not in the span of $x^{(1)}_0, \ldots, x^{(i-1)}_0$, choose an eigenvector $x^{(i)}_0$ of maximal rank μ_i. Continue this process until the subspace* $\ker L(\lambda_0)$ *of all eigenvectors corresponding to λ_0 is exhausted. Then for each of the eigenvectors selected, we have Jordan chains*

$$x^{(1)}_0, \ldots, x^{(1)}_{\mu_1-1}, \ldots, x^{(m)}_0, \ldots, x^{(m)}_{\mu_m-1} \tag{23}$$

where $m =$ dim ker *$L(\lambda_0) \leqslant p$. Such a set of Jordan chains is said to form a canonical set of Jordan chains of $L(\lambda)$ corresponding to λ_0.*

Proposition 1.25 *The lengths of the Jordan chains in a canonical set of Jordan chains of $L(\lambda)$ corresponding to λ_0 are exactly the non-zero partial multiplicities of $L(\lambda)$ at λ_0.*

Proof This is the same argument as in the proof of Proposition 1.10, but with one refinement: we have to verify that the one-to-one correspondence between Jordan chains of $L(\lambda)$ and $D(\lambda)$ defined by (12) is also a one-to-one correspondence between *canonical* sets of Jordan chains. But this is clear from the fact that if the vectors $x_0^{(1)}, \ldots, x_0^{(m)}$ in $\ker L(\lambda_0)$ are linearly independent then so are the vectors $y_0^{(1)}, \ldots, y_0^{(m)}$ in $\ker D(\lambda_0)$, since the matrix $F(\lambda_0)$ is nonsingular.

Corollary 1.26 *The sum $\sum_{i=1}^m \mu_i$ of the lengths of Jordan chains in a canonical set corresponding to an eigenvalue λ_0 of the matrix polynomial $L(\lambda)$ coincides with the multiplicity of λ_0 as a zero of* $\det L(\lambda)$.

Let $\lambda_0 \in \mathrm{sp}(L)$, and let α_0 denote the multiplicity of λ_0 as a zero of $\det L(\lambda)$. Choose a canonical set of Jordan chains as in (23). Then, in the same manner as in Lemma 1.22, define a $p \times \alpha_0$ matrix X_0 with these vectors as its columns and an $\alpha_0 \times \alpha_0$ block diagonal Jordan matrix J_0. Then we say that (X_0, J_0) is a *Jordan pair of $L(\lambda)$ corresponding to the eigenvalue λ_0*.

Now, we define the notion of a *finite Jordan pair for $L(\lambda)$*.

Definition 1.27 *Let $\lambda_1, \ldots, \lambda_s$ be the eigenvalues of $L(\lambda)$ with multiplicities $\alpha_1, \ldots, \alpha_s$, and let $\alpha = \sum \alpha_i$ denote the degree of* $\det L(\lambda)$. *A pair of matrices (X, J), where X is a $p \times \alpha$ matrix and J is an $\alpha \times \alpha$ matrix, is called a finite Jordan pair for $L(\lambda)$ if*

$$X = [X_1 \cdots X_s], \qquad J = \mathrm{diag}(J_i)_{i=1}^s$$

where (X_i, J_i) is a Jordan pair of $L(\lambda)$ corresponding to λ_i, $i = 1, \ldots, s$.

In principle, Definition 1.24 enables us to construct a Jordan pair for $L(\lambda)$ corresponding to an eigenvalue λ_0; doing so for each of the distinct eigenvalues of $L(\lambda)$ we obtain a finite Jordan pair for $L(\lambda)$. But Definition 1.24 is not convenient for determining whether a given admissible pair of matrices $(\tilde{X}_0, \tilde{J}_0)$ is in fact a Jordan pair of $L(\lambda)$ corresponding to λ_0. The next theorem states the necessary and sufficient conditions that the pair $(\tilde{X}_0, \tilde{J}_0)$ must satisfy in order for it to be a Jordan pair corresponding to λ_0.

Theorem 1.28 *Let $(\tilde{X}_0, \tilde{J}_0)$ be a pair of matrices, where $\tilde{X}_0$ is a $p \times \mu$ matrix and $\tilde{J}_0$ is a $\mu \times \mu$ block diagonal Jordan matrix with unique eigenvalue λ_0. Then the following conditions are necessary and sufficient in order that $(\tilde{X}_0, \tilde{J}_0)$ be a Jordan pair of $L(\lambda) = \sum_{j=0}^l A_j \lambda^j$ corresponding to λ_0:*

(*i*) $\det L(\lambda)$ *has a zero λ_0 of multiplicity μ,*
(*ii*) $\ker \mathrm{col}(\tilde{X}_0 \tilde{J}_0^j)_{j=0}^{l-1} = \{0\}$,
(*iii*) $A_0 \tilde{X}_0 + A_1 \tilde{X}_0 \tilde{J}_0 + \cdots + A_l \tilde{X}_0 \tilde{J}_0^l = 0$.

Proof Suppose that $(\tilde{X}_0, \tilde{J}_0)$ is a Jordan pair of $L(\lambda)$ corresponding to λ_0. Then (*i*) follows from Corollary 1.26, (*ii*) follows from Lemma 1.22, and (*iii*) follows from Proposition 1.11.

Conversely, suppose that (*i*)–(*iii*) hold. Let $\tilde{X}_0^{(i)} = [x_0^{(i)} \cdots x_{\mu_i - 1}^{(i)}]$ denote the part of $\tilde{X}_0$ corresponding to each Jordan block $\tilde{J}_0^{(i)}$ in $\tilde{J}_0$ ($i = 1, \ldots, m$).

By Proposition 1.11, the columns of $\tilde{X}_0^{(i)}$ form a Jordan chain of $L(\lambda)$ corresponding to λ_0. Condition (*ii*) ensures that the eigenvectors $x_0^{(1)}, \ldots, x_0^{(m)}$ of these Jordan chains are linearly independent. Indeed, suppose not; then

$$\sum_{i=1}^{m} c_i x_0^{(i)} = 0,$$

where the $c_i \in \mathbb{C}$ are not all zero. Then also $\sum_{i=1}^{m} c_i \lambda_0^j x_0^{(i)} = 0$ for $j = 1, \ldots, l-1$, so that the columns of the matrix

$$\operatorname{col}(\lambda_0^j x_0^{(1)}, \ldots, \lambda_0^j x_0^{(m)})_{j=0}^{l-1}. \tag{24}$$

are linearly dependent. But this contradicts (*ii*), since the columns of (24) are part of $\operatorname{col}(\tilde{X}_0 \tilde{J}_0^j)_{j=0}^{l-1}$ (the leftmost part corresponding to each Jordan block).

To conclude the proof we must show that the lengths of these Jordan chains coincide with the lengths of the Jordan chains in a canonical set. To this end, let $\lambda_0, \ldots, \lambda_s$ be the distinct eigenvalues of $L(\lambda)$ with multiplicities $\alpha_0, \ldots, \alpha_s$, and let $(X_0, J_0), \ldots, (X_s, J_s)$ be the corresponding Jordan pairs. Then we define the finite Jordan pair (X, J) as in Definition 1.27:

$$X = [X_0 \cdots X_s], \qquad J = \operatorname{diag}(J_i)_{i=0}^{s}$$

and an analogous definition for $(\tilde{X}, \tilde{J})$ with X_0 and J_0 replaced by $\tilde{X}_0$ and $\tilde{J}_0$, respectively. In view of Proposition 1.11, Lemma 1.22 and Proposition 1.17, the pairs $(\tilde{X}, \tilde{J})$ and (X, J) are restrictions of each other – both are partial spectral pairs with α-dimensional base spaces – and hence they are similar, that is, $\tilde{X} = XM$ and $\tilde{J} = M^{-1}JM$ for some invertible $\alpha \times \alpha$ matrix M. Now write $M = [M_{ij}]$ where M_{ij} is an $\alpha_j \times \alpha_j$ matrix (i and $j = 0, \ldots, s$). From the fact that $M\tilde{J} = JM$ we obtain

$$M_{i0}\tilde{J}_0 = J_i M_{i0} \qquad \text{for } i = 0, \ldots, s$$

$$M_{ij}\tilde{J}_j = J_i M_{ij} \qquad \text{for } i = 0, \ldots, s \text{ and } j = 1, \ldots, s$$

By Lemma 1.29 (see below), $M_{ij} = 0$ for $i \neq j$. Hence M_{00} is invertible and $\tilde{J}_0$ is similar to J_0, which implies that $\tilde{J}_0$ and J_0 have the same size Jordan blocks. In other words, the lengths of Jordan chains given in the matrix $\tilde{X}_0$ coincide with those in the canonical set given in X_0. This completes the proof of the theorem.

Lemma 1.29 *Let A and B be square matrices of dimension $m \times n$ and $n \times n$, respectively, such that* $\operatorname{sp}(B) = \{\lambda_0\}$ *and* $\lambda_0 \notin \operatorname{sp}(A)$. *Then the only solution of the matrix equation $AM = MB$ is the trivial solution $M = 0$.*

Proof Let M be an $m \times n$ matrix such that $AM = MB$. By hypothesis, B has a single eigenvalue λ_0, so that we have $(B - \lambda_0 I)^k = 0$ for some k. Then $(A - \lambda_0 I)^k M = M(B - \lambda_0 I)^k = 0$. Since $\lambda_0 \notin \operatorname{sp}(A)$, it follows that $M = 0$.

Remark 1.30 As in Lemma 1.22, let $x_0^{(i)}, \ldots, x_{\mu_i - 1}^{(i)}$ $(i = 1, \ldots, m)$ be a set of Jordan chains of $L(\lambda)$ corresponding to an eigenvalue λ_0 such that the eigenvectors $x_0^{(1)}, \ldots, x_0^{(m)}$ are linearly independent. Then by the same method of proof as in Theorem 1.28 it follows that $\sum_{i=1}^{m} \mu_i \leqslant \alpha_0$, and equality holds if and only if the set is canonical.

Example 1.31 (continuation of Example 1.6) Let

$$L(\lambda) = \begin{pmatrix} (\lambda-2)^2 & -\lambda+2 \\ 0 & (\lambda-2)^2 \end{pmatrix}.$$

Let us construct a canonical set of Jordan chains corresponding to the eigenvalue 2. First note that dim ker $L(2) = 2$, so that there exist two linearly independent eigenvectors. Choose an eigenvector of maximal rank 3, say $\begin{pmatrix} 1 \\ 0 \end{pmatrix}$, and a corresponding Jordan chain of length 3,

$$x_0^{(1)} = \begin{pmatrix} 1 \\ 0 \end{pmatrix}, \quad x_1^{(1)} = \begin{pmatrix} 0 \\ 1 \end{pmatrix}, \quad x_2^{(1)} = \begin{pmatrix} 0 \\ 0 \end{pmatrix}$$

Then choose an eigenvector of maximal rank among eigenvectors not in the span of $x_0^{(1)}$, say $x_0^{(2)} = \begin{pmatrix} 0 \\ 1 \end{pmatrix}$. This gives us the finite Jordan pair (X, J), where

$$X = \left(\begin{array}{ccc|c} 1 & 0 & 0 & 0 \\ 0 & 1 & 0 & 1 \end{array}\right), \quad J = \left(\begin{array}{ccc|c} 2 & 1 & & \\ & 2 & 1 & \\ & & 2 & \\ \hline & & & 2 \end{array}\right).$$

Example 1.32 Let

$$L(\lambda) = \begin{pmatrix} \lambda^2 & \lambda^2-\lambda \\ 4\lambda & \lambda^2 \end{pmatrix}.$$

The determinant is $\det L(\lambda) = \lambda^2(\lambda-2)^2$. Let us construct canonical sets of Jordan chains corresponding to the eigenvalues 0 and 2. Note that dim $L(0) = 2$, so that there exist two linearly independent eigenvectors corresponding to 0. For instance, we can take $x_0^{(1)} = \begin{pmatrix} 1 \\ 0 \end{pmatrix}$, $x_0^{(2)} = \begin{pmatrix} 0 \\ 1 \end{pmatrix}$. In view of Remark 1.30, this set is canonical; thus,

$$X_0 = \left(\begin{array}{c|c} 1 & 0 \\ 0 & 1 \end{array}\right), \quad J_0 = \left(\begin{array}{c|c} 0 & \\ \hline & 0 \end{array}\right)$$

is a Jordan pair of $L(\lambda)$ corresponding to the eigenvalue 0. Consider now the eigenvalue 2. We have $L(2) = \begin{pmatrix} 4 & 2 \\ 8 & 4 \end{pmatrix}$ so that dim $L(2) = 1$ and there is one linearly independent eigenvector, $x_0^{(1)} = \begin{pmatrix} 1 \\ -2 \end{pmatrix}$. To find the first generalized eigenvector, $x_1^{(1)}$, we solve the equation $L'(2)x_0^{(1)} + L(2)x_1^{(1)} = 0$, that is,

$$\begin{pmatrix} 4 & 3 \\ 4 & 4 \end{pmatrix}\begin{pmatrix} 1 \\ -2 \end{pmatrix} + \begin{pmatrix} 4 & 2 \\ 8 & 4 \end{pmatrix} x_1^{(1)} = \begin{pmatrix} 0 \\ 0 \end{pmatrix}. \tag{25}$$

As the first generalized eigenvector, we can take any solution of this equation, say $x_1^{(1)} = \begin{pmatrix} 0 \\ -1 \end{pmatrix}$. There are no Jordan chains of length > 2, as one can verify directly or by referring to Remark 1.30. Hence

$$X_2 = \begin{pmatrix} 1 & 0 \\ -2 & -1 \end{pmatrix}, \qquad J_2 = \begin{pmatrix} 2 & 1 \\ & 2 \end{pmatrix}$$

is a Jordan pair of $L(\lambda)$ corresponding to 2, and

$$X = \begin{pmatrix} 1 & 0 & 1 & 0 \\ 0 & 1 & -2 & -1 \end{pmatrix}, \qquad J = \begin{pmatrix} 0 & & & \\ & 0 & & \\ & & 2 & 1 \\ & & & 2 \end{pmatrix}$$

is a finite Jordan pair of $L(\lambda)$.

Example 1.33 Let $L(\lambda) = (\lambda - \lambda_1)^{\alpha_1} \cdots (\lambda - \lambda_k)^{\alpha_k}$ be a scalar polynomial where $\lambda_i \neq \lambda_j$ for $i \neq j$. Define a $1 \times \alpha$ matrix X and an $\alpha \times \alpha$ matrix J as follows:

$$X = [X_1 \cdots X_k], \qquad J = \operatorname{diag}(J_i)_{i=1}^k$$

where $X_i = [1\ 0 \cdots 0]$ is a $1 \times \alpha_j$ matrix and J_i is the $\alpha_i \times \alpha_i$ Jordan block with eigenvalue λ_i. Then (X_i, J_i) is a Jordan pair of $L(\lambda)$ corresponding to λ_i, (X, J) is a finite Jordan pair of $L(\lambda)$, and the $\alpha \times \alpha$ matrix $\operatorname{col}(XJ^i)_{i=0}^{\alpha-1}$ is invertible. If $L(\lambda)$ has simple roots ($\alpha_j = 1$), this is the usual Vandermonde matrix.

Addendum Let (X, J) be a finite Jordan pair of $L(\lambda)$. If u is a solution of $L(d/dt)u = 0$ then $u(t) = Xe^{tJ}c$ for a unique $c \in \mathbb{C}^\alpha$. It follows that

$$u(t) = \frac{1}{2\pi i}\int_\Gamma e^{t\lambda} X(I\lambda - J)^{-1} c \, d\lambda$$

and so, by Lemma 1.18,

$$u(t) = \frac{1}{2\pi i}\int_\Gamma e^{t\lambda} L^{-1}(\lambda)\left(\sum_{k=0}^{l-1} L_{k+1}(\lambda) XJ^k\right) c \, d\lambda$$

Since $u^{(k)}(0) = XJ^k c$ for $k = 0, 1, \ldots$, we have proved:

Every solution of $L(d/dt)u = 0$ admits the representation

$$u(t) = \frac{1}{2\pi i}\int_\Gamma e^{t\lambda} L^{-1}(\lambda) \sum_{k=0}^{l-1} L_{k+1}(\lambda) \cdot u^{(k)}(0) \, d\lambda$$

where Γ is a contour having all of $\operatorname{sp}(L)$ *in its interior.*

It follows that if $L(d/dt)u(t) = 0$ and $u^{(k)}(0) = 0$ for $k = 0, \ldots, l-1$, then $u \equiv 0$. (This fact is nontrivial in the case that the leading coefficient, A_l, of $L(\lambda)$ is not invertible.)

For more information on the Jordan theory of matrix polynomials and related topics, one should consult the books [GLR], [BGR].

2

Spectral triples for matrix polynomials

Let $L(\lambda) = \sum A_j \lambda^j$ be a $p \times p$ matrix polynomial with $\det L(\lambda) \not\equiv 0$ and γ a simple, closed contour not intersecting $\mathrm{sp}(L)$. Let r denote the number of roots of $\det L(\lambda) = 0$ inside γ with multiplicities counted. If in Definition 1.27 we take just the eigenvectors, generalized eigenvectors and Jordan blocks corresponding to the eigenvalues inside γ, then we obtain a pair of matrices (X_+, J_+) called a *γ-spectral Jordan pair*. Thus, X_+ is a $p \times r$ matrix whose columns are various Jordan chains of $L(\lambda)$ and J_+ is an $r \times r$ Jordan matrix. One of the aims of this chapter is to show that there exists an $r \times p$ matrix Y_+ such that

$$L^{-1}(\lambda) - X_+(I\lambda - J_+)^{-1} Y_+$$

has an analytic continuation inside γ. It turns out that the matrix Y_+ is uniquely determined by this condition and the *rows* of the matrix Y_+ are *left* Jordan chains. Of course, the left Jordan chains that make up the rows of Y_+ cannot be chosen arbitrarily, and it is difficult to give a direct proof of their existence. In this chapter we take a more abstract approach from which the existence of Y_+ follows as an easy consequence.

In §2.1 we introduce triples of operators (X_+, T_+, Y_+) which contain all the spectral information about a matrix polynomial at the eigenvalues inside γ. These triples are required to satisfy two properties: $\mathrm{sp}(T_+)$ lies inside γ and $L^{-1}(\lambda) - X_+(I\lambda - T_+)^{-1} Y_+$ has an analytic continuation inside γ. There are two other conditions they must satisfy (see Definition 2.1) which can be imposed without loss of generality and have the effect of making γ-spectral triples unique up to similarity. Such a triple

$$(X_+, T_+, Y_+)$$

is called a *γ-spectral triple* for $L(\lambda)$, and an admissible pair of operators

$$(X_+, T_+)$$

is called a *γ-spectral pair* for $L(\lambda)$ if there exists Y_+ such that (X_+, T_+, Y_+) is a γ-spectral triple. The existence of γ-spectral triples is proved in §2.1, and in §2.2 we develop the properties of these triples which are needed later on in Parts IV and V of the book.

For technical reasons, we also need the following definition. An admissible pair (X, T) is said to be a *partial γ-spectral pair* of $L(\lambda)$ if it is a partial spectral pair (Definition 1.13) and $\mathrm{sp}(T) \subset G$. Now we define an order relation in the class of admissible pairs as follows:

$$(X, T) \leqslant (X', T') \qquad \text{if } (X, T) \text{ is a restriction of } (X', T').$$

The terminology "partial γ-spectral pair" is justified by Proposition 2.12 which shows that a partial γ-spectral pair (X_+, T_+) is a γ-spectral pair if and only if it is maximal with respect to the relation $\leqslant$. Once this has been established it is easily shown that any γ-spectral Jordan pair is a γ-spectral pair.

Having introduced Jordan pairs, however, it should be mentioned that we use them in only one place in the theory for this chapter, in the proof of Proposition 2.5. The reason for developing the Jordan theory in Chapter 1 was to motivate the use of admissible pairs of matrices as a means of organizing the spectral data for a matrix polynomial. The proofs of §§2.1, 2.2 do not rely on the specific use of the Jordan theory, but they are of course related to it because they make use of the solution space of the homogeneous equation $L(d/dt)u = 0$. Moreover, it turns out that γ-spectral pairs are unique up to similarity (see Corollary 2.13), so the existence of a Jordan pair provides an algorithm for computing the γ-spectral pairs (X_+, T_+), and this pair can then be completed uniquely to form a γ-spectral triple (X_+, T_+, Y_+).

As an illustration of the use of spectral triples, we apply them in §2.3 to obtain a representation for the singular part of the resolvent $L^{-1}(\lambda)$ in terms of right and left eigenvectors and generalized eigenvectors of $L(\lambda)$. Finally, in §2.4, we look at the behaviour of spectral triples under a Möbius transformation of the matrix polynomial $L(\lambda)$.

2.1 Spectral triples

A triple of operators (X, T, Y) is called an admissible triple if $X \in \mathscr{L}(\mathfrak{M}, \mathbb{C}^p)$, $T \in \mathscr{L}(\mathfrak{M})$ and $Y \in \mathscr{L}(\mathbb{C}^p, \mathfrak{M})$. Two admissible triples, (X, T, Y) and (X', T', Y'), are called *similar* if there exists an invertible operator $M \in \mathscr{L}(\mathfrak{M}', \mathfrak{M})$ such that $X' = XM$, $T' = M^{-1}TM$ and $Y' = M^{-1}Y$. The admissible pair (X, T) is sometimes referred to as a *right* admissible pair, while (T, Y) is called a *left* admissible pair.

Let $L(\lambda) = \sum_{j=0}^{l} A_j\lambda^j$ be a $p \times p$ matrix polynomial and let γ be a simple, closed (rectifiable) contour not intersecting $\mathrm{sp}(L)$. Also let G denote the region inside γ.

Definition 2.1 *A γ-spectral triple for $L(\lambda)$ is defined to be an admissible triple (X_+, T_+, Y_+) with the following properties:*

(*i*) $\mathrm{sp}(T_+) \subset G$ (*i.e. inside* γ),
(*ii*) $L^{-1}(\lambda) - X_+(I\lambda - T_+)^{-1}Y_+$ *has an analytic continuation in* G,
(*iii*) $\mathrm{col}(X_+T_+^j)_{j=0}^{l-1}$ *is injective,*
(*iv*) $\mathrm{row}(T_+^j Y_+)_{j=0}^{l-1}$ *is surjective.*

Also, we say that (X_+, T_+) is a (right) γ-spectral pair for $L(\lambda)$ and (T_+, Y_+) is a left γ-spectral pair for $L(\lambda)$.

If the contour γ is chosen so that all of $\mathrm{sp}(L)$ lies inside γ then we obtain a triple (X, T, Y) which we call a *finite* spectral triple (i.e. for the finite complex plane). If γ is chosen so that it surrounds just one eigenvalue $\lambda_0 \in \mathrm{sp}(L)$ then we obtain a triple (X_0, T_0, Y_0) which we call a *spectral triple at* λ_0. If $T_0 = J_0$ is a Jordan matrix then we refer to (X_0, J_0, Y_0) as a *Jordan triple at* λ_0. There is also a spectral triple at ∞, defined as a spectral triple at 0 for the matrix polynomial $\tilde{L}(\lambda) = \lambda^l L(\lambda^{-1})$, but discussion of this is postponed until Chapter 4. (Often we deal with matrix polynomials with invertible leading coefficient, A_l, in which case there is no spectrum at ∞.)

Proposition 2.2 *Property (ii) of Definition* 2.1 *can be replaced by*

$$(ii') \qquad \frac{1}{2\pi i}\int_\gamma \lambda^j L^{-1}(\lambda)\, d\lambda = X_+ T_+^j Y_+, \qquad j = 0, 1, \ldots$$

Proof The property (*ii*) holds if and only if

$$\frac{1}{2\pi i}\int_\gamma \lambda^j \{L^{-1}(\lambda) - X_+(I\lambda - T_+)^{-1} Y_+\}\, d\lambda = 0 \qquad j = 0, 1, \ldots$$

Since $\mathrm{sp}(T_+) \subset G$, this is equivalent to (*ii'*).

Proposition 2.3 *If* (X_+, T_+) *is a* γ*-spectral pair for* $L(\lambda)$ *then*

$$\sum_{j=0}^{l} A_j X_+ T_+^j = 0.$$

Proof By definition, there exists an operator Y_+ such that (X_+, T_+, Y_+) is a γ-spectral triple for $L(\lambda)$. Then property (*ii'*) implies that

$$\sum_{j=0}^{l} A_j X_+ T_+^j \cdot T_+^k Y_+ = \frac{1}{2\pi i}\int_\gamma \sum_{j=0}^{l} A_j \lambda^j \cdot \lambda^k L^{-1}(\lambda)\, d\lambda = 0$$

for $k = 0, 1, \ldots$ and, by property (*iv*), it follows that $\sum_{j=0}^{l} A_j X_+ T_+^j = 0$.

Remark 2.4 If (X_+, T_+, Y_+) is a γ-spectral triple for $L(\lambda)$, then (Y_+^T, T_+^T, X_+^T) is a γ-spectral triple for $L^T(\lambda)$, where $L^T(\lambda) = \sum A_j^T \lambda^j$ is the matrix polynomial obtained by transposing the coefficients of $L(\lambda)$. Analogous to Proposition 2.3 one can show that if (T_+, Y_+) is a left γ-spectral pair for $L(\lambda)$ then $\sum_{j=0}^{l} T_+^j Y_+ A_j = 0$.

Let $\mathfrak{M}_L$ denote the set of solutions $u \in C^\infty(\mathbb{R}, \mathbb{C}^p)$ of the equation $L(d/dt)u(t) = 0$, and let (X, J) be a finite Jordan pair of $L(\lambda)$ (Definition 1.27). As shown in the addendum to Chapter 1, every $u \in \mathfrak{M}_L$ has a representation $u(t) = Xe^{tJ}c$ for a unique $c \in \mathbb{C}^\alpha$, and thus can be written in the form

$$u(t) = \sum p_i(t)\, e^{t\lambda_i}, \qquad (1)$$

where the p_i are $\mathbb{C}^p$-valued polynomials. The complex numbers λ_i are eigenvalues of J, so they are roots of $\det L(\lambda) = 0$. By virtue of the corollary to Proposition 1.1 the dimension of $\mathfrak{M}_L$ is equal to the degree, α, of $\det L(\lambda)$.

Now let γ be a simple, closed contour not intersecting sp(L) and let r denote the number of roots of det $L(\lambda) = 0$ inside γ (multiplicities counted). If we write $X = [X_+ \; X_-]$, $J = \begin{pmatrix} J_+ & \\ & J_- \end{pmatrix}$, where the eigenvalues of J_+ and J_- lie inside and outside γ, respectively, then every $u \in \mathfrak{M}_L$ has a representation

$$u(t) = X_+ e^{tJ_+} c_+ + X_- e^{tJ_-} c_-$$

for unique $c_+ \in \mathbb{C}^r$ and $c_- \in \mathbb{C}^{\alpha - r}$. Functions of the form $X_\pm e^{tJ_\pm} c_\pm$ belong to $\mathfrak{M}_L^\pm$, where $\mathfrak{M}_L^+$ (resp. $\mathfrak{M}_L^-$) denotes the subspace of $\mathfrak{M}_L$ consisting of solutions of the form (1) such that the eigenvalues λ_i lie inside (resp. outside) γ. Since $\operatorname{col}(X_\pm J_\pm^j)_{j=0}^{l-1}$ is injective, we have dim $\mathfrak{M}_L^+ \geqslant r$ and dim $\mathfrak{M}_L^- \geqslant \alpha - r$. Now, the dimension of $\mathfrak{M}_L$ equals α, and it is easily verified that $\mathfrak{M}_L^+ \cap \mathfrak{M}_L^- = 0$, hence

$$\dim \mathfrak{M}_L^+ = r, \qquad \dim \mathfrak{M}_L^+ = \alpha - r$$

and it follows that *every* $u \in \mathfrak{M}_L^\pm$ can be represented in the form $u(t) = X_\pm e^{tJ_\pm} c_\pm$ (for unique $c_\pm$).

Proposition 2.5 *Every $u \in \mathfrak{M}_L^+$ has the representation*

$$u(t) = \frac{1}{2\pi i} \int_\gamma e^{t\lambda} L^{-1}(\lambda) \sum_{j=0}^{l-1} L_{j+1}(\lambda) u^{(j)}(0) \, d\lambda \tag{2}$$

where $L_j(\lambda) = A_j + A_{j+1}\lambda + \cdots + A_l \lambda^{l-j}$, $j = 0, \ldots, l$. This expresses u in terms of its Cauchy data at $t = 0$.

Proof If $u \in \mathfrak{M}_L^+$ then in view of the above remarks $u(t) = X_+ e^{tJ_+} c$, where $c \in \mathbb{C}^r$. By Lemma 1.18, we obtain

$$\begin{aligned} u(t) &= \frac{1}{2\pi i} \int_\gamma e^{t\lambda} X_+ (I\lambda - J_+)^{-1} c \, d\lambda \\ &= \frac{1}{2\pi i} \int_\gamma e^{t\lambda} L^{-1}(\lambda) \sum_{j=0}^{l-1} L_{j+1}(\lambda) X_+ J_+^j c \, d\lambda \end{aligned}$$

The formula (2) now follows immediately since $u^{(j)}(0) = X_+ J_+^j c$ for $j = 0, 1, \ldots$

In anticipation of ideas which will be further developed in Chapter 3, we make the following remark. Let G denote the region inside γ. If f is a $\mathbb{C}^p$-valued function which is continuous in $\bar{G}$ and analytic in G then the function

$$u(t) = \frac{1}{2\pi i} \int_\gamma e^{t\lambda} L^{-1}(\lambda) f(\lambda) \, d\lambda \tag{3}$$

belongs to $\mathfrak{M}_L^+$. In view of (2) we see that (3) then holds with $f(\lambda)$ replaced by $p(\lambda) = \sum_{j=0}^{l-1} L_{j+1}(\lambda) u^{(j)}(0)$, which is a polynomial of degree $\leqslant l - 1$. Hence the difference $L^{-1}(\lambda) f(\lambda) - L^{-1}(\lambda) p(\lambda)$ has an analytic continuation in G. In other words, for every such f there exists an analytic quotient $q(\lambda)$ and a polynomial remainder $p(\lambda)$ of degree $\leqslant l-1$ such that $f(\lambda) = L(\lambda)q(\lambda) + p(\lambda)$. In §3.4 we consider a Euclidean algorithm for matrix polynomials which is based on the same idea.

Now we prove existence of γ-spectral triples. They are also unique up to similarity as we show later on in Corollary 2.13.

Theorem 2.6 *Let γ be a simple, closed contour not intersecting* sp(L). *Then there exists a γ-spectral triple for $L(\lambda)$.*

Proof We define an admissible triple (X_+, T_+, Y_+) with base space $\mathfrak{M}_L^+$ as follows:

$$X_+(u) = u(0)$$

$$T_+u = \frac{du}{dt}$$

$$(Y_+x)(t) = \frac{1}{2\pi i}\int_\gamma e^{t\lambda}L^{-1}(\lambda)x\,d\lambda$$

(Note that if $u \in \mathfrak{M}_L^+$ then $du/dt \in \mathfrak{M}_L^+$). We will show that (X_+, T_+, Y_+) satisfies properties (*i*), (*ii′*), (*iii*) and (*iv*) of Definition 2.1.

Let $\lambda_0 \notin \bar{G}$. Then define $S \in \mathscr{L}(\mathfrak{M}_L^+)$ as follows:

$$(Su)(t) = \frac{1}{2\pi i}\int_\gamma e^{t\lambda}L^{-1}(\lambda)f(\lambda)(\lambda_0 - \lambda)^{-1}\,d\lambda, \tag{4}$$

where f determines u as in (3). Note that if $\int_\gamma e^{t\lambda}L^{-1}(\lambda)f(\lambda)\,d\lambda = 0$ for all t then $L^{-1}(\lambda)f(\lambda)$ has an analytic continuation inside γ; consequently, the expression defined by the right-hand side of (4) is also 0, and S is well-defined. One can now verify immediately that $(I\lambda_0 - T_+)Su = u$ and $S(I\lambda_0 - T_+)u = u$ for all $u \in \mathfrak{M}_L^+$. Hence $(I\lambda_0 - T_+)^{-1}$ exists and sp$(T_+) \subset \bar{G}$. In fact, sp$(T_+) \subset G$, since there exists a contour γ' contained in G such that the part of sp(L) lying inside γ' is the same as that inside γ, and we can replace γ and γ' in the formula (4). This proves (*i*).

The fact that (*ii′*) of Proposition 2.2 holds is an immediate consequence of the definition of (X_+, T_+, Y_+).

Now recall that for every $u \in \mathfrak{M}_L^+$ the identity (2) holds. Since $u^{(j)}(0) = X_+T_+^ju$, the injectivity of $\mathrm{col}(X_+T_+^j)_{j=0}^{l-1}$ follows from (2). This proves (*iii*).

Finally, to prove (*iv*), note that (2) can be written in the form

$$u(t) = \frac{1}{2\pi i}\int_\gamma e^{t\lambda}L^{-1}(\lambda)[I \cdots \lambda^{l-1}I]\,d\lambda \cdot \mathscr{X} \cdot \mathscr{U}$$

where $\mathscr{U} = \mathrm{col}(u^{(j)}(0))_{j=0}^{l-1}$, the Cauchy data of u at $t = 0$, and

$$\mathscr{X} = \begin{pmatrix} A_1 & A_2 & \cdots & A_l \\ A_2 & & \cdot & \\ \vdots & \cdot & & 0 \\ A_l & & & \end{pmatrix} \tag{5}$$

a $pl \times pl$ matrix. By the definition of T_+ and Y_+ above, we obtain

$$u = \text{row}(T_+^j Y_+)_{j=0}^{l-1} \cdot \mathscr{X} \cdot \text{col}(X_+ T_+^k)_{k=0}^{l-1} u$$

so that $\text{row}(T_+^j Y_+)_{j=0}^{l-1}$ is surjective. This completes the proof of the theorem.

Proposition 2.7 *If* (X_+, T_+, Y_+) *is a* γ*-spectral triple for* $L(\lambda)$ *then* $\text{sp}(T_+) = \text{sp}(L) \cap G$.

Proof By definition, we have $\text{sp}(T_+) \subset G$. Now if $\lambda_0 \in \text{sp}(T_+)$, there exists $v_0 \neq 0$ such that $T_+ v_0 = \lambda_0 v_0$. Then, by Proposition 2.3,

$$0 = \sum_{j=0}^{l} A_j X_+ T_+^j v_0 = \sum_{j=0}^{l} A_j \lambda_0^j X_+ v_0 = L(\lambda_0) X_+ v_0.$$

Note that $X_+ v_0 \neq 0$; otherwise if $X_+ v_0 = 0$ then $X_+ T_+^j v_0 = \lambda_0^j X_+ v_0 = 0$, for all $j = 0, 1, \ldots,$ and, due to (*iii*) of Definition 2.1, we would have $v_0 = 0$. Hence $\lambda_0 \in \text{sp}(L)$ and it follows that $\text{sp}(T_+) \subset \text{sp}(L) \cap G$.

Next we show that if $\lambda \in G \backslash \text{sp}(L)$ then $\lambda \notin \text{sp}(T_+)$. Since

$$M(\lambda) := L^{-1}(\lambda) - X_+(I\lambda - T_+)^{-1} Y_+$$

is analytic in G, it follows that $L^{-1}(\lambda)$ has an analytic continuation to $G \backslash \text{sp}(T_+)$, which is $N(\lambda) := M(\lambda) + X_+(I\lambda - T_+)^{-1} Y_+$. If $\lambda \in G \backslash \text{sp}(L)$ then

$$\begin{aligned} I &= L(\lambda) N(\lambda) \\ &= L(\lambda) M(\lambda) + L(\lambda) X_+ (I\lambda - T_+)^{-1} Y_+ \\ &= L(\lambda) M(\lambda) + \sum_{j=0}^{l-1} L_{j+1}(\lambda) X_+ T_+^j Y_+, \end{aligned} \tag{6}$$

where the latter equality holds by Lemma 1.18. But G is connected, so (6) holds for all $\lambda \in G$. Another application of Lemma 1.18 implies $I = L(\lambda) N(\lambda)$ for all $\lambda \in G \backslash \text{sp}(T_+)$. Similarly, $N(\lambda) L(\lambda) = I$ there. Hence $(G \backslash \text{sp}(T_+)) \cap \text{sp}(L) = \varnothing$, and it follows that $\text{sp}(T_+) = \text{sp}(L) \cap G$.

2.2 Properties of spectral triples and the Calderón projector

For any $u \in \mathfrak{M}_L$, the column vector $\mathscr{U} \in \mathbb{C}^{pl}$ defined by

$$\mathscr{U} = \text{col}(u^{(j)}(0))_{j=0}^{l-1}$$

is the Cauchy data (or initial conditions) of u at $t = 0$. Recall that every $u \in \mathfrak{M}_L^+$ has the representation

$$u(t) = \frac{1}{2\pi i} \int_\gamma e^{t\lambda} L^{-1}(\lambda) [L_1(\lambda) \cdots L_l(\lambda)] \, d\lambda \cdot \mathscr{U} \tag{2'}$$

(see Proposition 2.5), then by taking initial conditions on the left-hand side of this equation we obtain $\mathscr{U} = P_\gamma \cdot \mathscr{U}$, where

$$P_\gamma = \frac{1}{2\pi i} \int_\gamma \begin{pmatrix} I \\ \vdots \\ \lambda^{l-1} I \end{pmatrix} L^{-1}(\lambda) \cdot [L_1(\lambda) \cdots L_l(\lambda)]\, d\lambda \tag{7}$$

The following theorem shows that P_γ is a projector, which we call the Calderón projector because of the reference to Calderón in [Se 3].

Theorem 2.8 *P_γ is a projector in $\mathbb{C}^{pl}$. The image of P_γ is equal to the set of all Cauchy data $\mathscr{U}$ of functions $u \in \mathfrak{M}_L^+$.*

Proof In view of the equation $\mathscr{U} = P_\gamma \cdot \mathscr{U}$, the set of Cauchy data of functions $u \in \mathfrak{M}_L^+$ is contained in the image of P_γ. On the other hand, the image of P_γ is contained in the set of Cauchy data, for if $c = [c_0, \ldots, c_{l-1}] \in \mathbb{C}^{pl}$ let

$$u(t) = (2\pi i)^{-1} \int_\gamma e^{t\lambda} L^{-1}(\lambda) \sum_{j=0}^{l-1} L_{j+1}(\lambda) c_j \, d\lambda$$

Then $u \in \mathfrak{M}_L^+$ and its Cauchy data vector is $\mathscr{U} = \operatorname{col}(u^{(j)}(0))_{j=0}^{l-1} = P_\gamma c$. The fact that P_γ is a projector is now clear, for the equation $\mathscr{U} = P_\gamma \cdot \mathscr{U}$ implies that P_γ is the identity on its image.

Corollary 2.9 *Let (X_+, T_+) be a γ-spectral pair of $L(\lambda)$ (Definition 2.1). Then P_γ and $\operatorname{col}(X_+ T_+^j)_{j=0}^{l-1}$ have the same image. Hence every $u \in \mathfrak{M}_L^+$ has a representation*

$$u(t) = X_+ e^{tT_+} c$$

for a unique c in the base space of (X_+, T_+).

Proof Let (X_+, T_+, Y_+) be a γ-spectral triple for $L(\lambda)$ with base space denoted $\mathfrak{M}_+$. By virtue of (*ii′*), we have

$$P_\gamma = \operatorname{col}(X_+ T_+^j)_{j=0}^{l-1} \cdot \operatorname{row}(T_+^j Y_+)_{j=0}^{l-1} \cdot \mathscr{L} \tag{8}$$

where $\mathscr{L}$ is defined by (5). Hence the image of P_γ is contained in that of $\operatorname{col}(X_+ T_+^j)_{j=0}^{l-1}$. On the other hand, for any $c \in \mathfrak{M}_+$ the Cauchy data vector, $\mathscr{U}$, of the function $u \in \mathfrak{M}_L^+$ defined by

$$u(t) = X_+ e^{tT_+} c \tag{*}$$

satisfies $\mathscr{U} = \operatorname{col}(X_+ T_+^j)_{j=0}^{l-1} c$. This proves the first part of the corollary, i.e. P_γ and $\operatorname{col}(X_+ T_+^j)_{j=0}^{l-1}$ have the same image.

Functions of the form (*) certainly belong to $\mathfrak{M}_L^+$ by virtue of Proposition 2.3 and the fact that $\operatorname{sp}(T_+) \subset$ inside γ. Conversely, if $u \in \mathfrak{M}_L^+$ then by Theorem 2.8 its Cauchy data vector, $\mathscr{U}$, is equal to $P_\gamma c$ for some $c \in \mathbb{C}^{pl}$. Since u and the function $\tilde{u}(t) = X_+ e^{tT_+} c$ have the same initial conditions,

$$u^{(j)}(0) = \tilde{u}^{(j)}(0) \qquad \text{for } j = 0, 1, \ldots, l-1,$$

it follows from (2′) that $u = \tilde{u}$. Hence every $u \in \mathfrak{M}_L^+$ has a representation (*), and c is unique due to the injectivity of $\mathrm{col}(X_+T_+^j)_{j=0}^{l-1}$.

Corollary 2.10 *The base space of any γ-spectral pair of $L(\lambda)$ has dimension r.*

Proof This follows from the uniqueness of c in Corollary 2.9 and the fact that r is the dimension of $\mathfrak{M}_L^+$.

The next corollary will be used frequently in later chapters.

Corollary 2.11 *If (X_+, T_+, Y_+) is a γ-spectral triple of $L(\lambda)$ then*

$$\mathrm{row}(T_+^j Y_+)_{j=0}^{l-1} \cdot \mathscr{X} \cdot \mathrm{col}(X_+T_+^j)_{j=0}^{l-1} = I_{\mathfrak{M}_+}.$$

Thus, $\mathrm{row}(T_+^j Y_+)_{j=0}^{l-1} \cdot \mathscr{X}$ *is a left inverse of* $\mathrm{col}(X_+T_+^j)_{j=0}^{l-1}$*, or, stated another way,* $\mathscr{X} \cdot \mathrm{col}(X_+T_+^j)_{j=0}^{l-1}$ *is a right inverse of* $\mathrm{row}(T_+^j Y_+)_{j=0}^{l-1}$.

Proof Since $P_\gamma \cdot \mathrm{col}(X_+T_+^j)_{j=0}^{l-1} = \mathrm{col}(X_+T_+^j)_{j=0}^{l-1}$, the result follows by virtue of (8) and the injectivity of $\mathrm{col}(X_+T_+^j)_{j=0}^{l-1}$.

We define an order relation in the class of admissible pairs as follows: $(X, T) \leqslant (X', T')$ if (X, T) is a restriction of (X', T'), see Definition 1.16. An admissible pair (X, T) is said to be a *partial γ-spectral pair* of $L(\lambda)$ if it is a partial spectral pair (Definition 1.13) and $\mathrm{sp}(T) \subset G$.

The following proposition shows that γ-spectral pairs are maximal with respect to the relation $\leqslant$.

Proposition 2.12 *Let (X_+, T_+) be a γ-spectral pair for $L(\lambda)$. Any partial γ-spectral pair (X, T) for $L(\lambda)$ is a restriction of (X_+, T_+). In fact, $X = X_+M$ and $MT = T_+M$, where $M = $* $\mathrm{row}(T_+^j Y_+)_{j=0}^{l-1} \cdot \mathscr{X} \cdot \mathrm{col}(XT^k)_{k=0}^{l-1}$.

Proof Since $\mathrm{sp}(T) \subset G$ we have

$$XT^k = \frac{1}{2\pi i}\int_\gamma \lambda^k X(I\lambda - T)^{-1}\, d\lambda$$
$$= \frac{1}{2\pi i}\int_\gamma \lambda^k L^{-1}(\lambda) \sum_{j=0}^{l-1} L_{j+1}(\lambda) X T^j\, d\lambda$$

for $k = 0, 1, \ldots$ (the second equality holds because of Lemma 1.18). Therefore $\mathrm{col}(XT^k)_{k=0}^{l-1} = P_\gamma \cdot \mathrm{col}(XT^k)_{k=0}^{l-1}$ and then substitution of the formula (8) for P_γ gives

$$\mathrm{col}(XT^k)_{k=0}^{l-1} = \mathrm{col}(X_+T_+^j)_{j=0}^{l-1} \cdot M \tag{9}$$

where $M \in \mathscr{L}(\mathfrak{M}, \mathfrak{M}_+)$ is defined in the statement of the proposition. In particular $X = X_+M$, and it remains to show that $MT = T_+M$. It follows from (9) that if $c \in \mathfrak{M}$ (the base space of (X, T)) then $Xe^{tT}c \equiv X_+e^{tT_+}Mc$, since both functions are in $\mathfrak{M}_L^+$ and have the same initial conditions. Differentiating this equation with respect to t and letting $t = 0$, we obtain

$XT^k = X_+ T_+^k M$ for all $k = 0, 1, \ldots$. Thus

$$\mathrm{col}(X_+ T_+^k)_{k=0}^{l-1} \cdot MT = \mathrm{col}(X_+ T_+^k)_{k=0}^{l-1} \cdot T_+ M,$$

and due to injectivity of $\mathrm{col}(X_+ T_+^k)_{k=0}^{l-1}$ we obtain $MT = T_+ M$. In particular the subspace $\mathfrak{L} = \mathrm{im}\, M$ is T_+-invariant. Also note that M is injective due to (9) and the injectivity of $\mathrm{col}(XT^k)_{k=0}^{l-1}$. Hence (X, T) is similar to $(X_+|_{\mathfrak{L}}, T_+|_{\mathfrak{L}})$.

Since any γ-spectral pair is also a partial γ-spectral pair we obtain the following corollary.

Corollary 2.13 *Any two γ-spectral triples for $L(\lambda)$ are similar.*

Proof Let (X_+, T_+, Y_+) and $(\tilde{X}_+, \tilde{T}_+, \tilde{Y}_+)$ be γ-spectral triples for $L(\lambda)$. The γ-spectral pairs (X_+, T_+) and $(\tilde{X}_+, \tilde{T}_+)$ are restrictions of each other, so they must be similar, and we obtain $\tilde{X}_+ = X_+ M$, $\tilde{T}_+ = M^{-1} T_+ M$ for some isomorphism M between the base spaces. Also $\tilde{Y}_+ = M^{-1} Y_+$ due to the fact that $\tilde{Y}_+$ is uniquely determined given $(\tilde{X}_+, \tilde{T}_+)$ (see properties (*ii'*) and (*iii*)).

Another way to state the conclusion of Proposition 2.12 is as follows. If (X, T) is a partial γ-spectral pair then there exists an operator Y such that (X, T, Y) is a γ-spectral triple if and only if (X, T) is maximal, or, in other words, its base space has dimension r.

Corollary 2.14 *Let (X, T) be a partial γ-spectral pair for $L(\lambda)$. (X, T) is a γ-spectral pair for $L(\lambda)$ if and only if its base space has dimension r.*

Proof See Corollary 2.10.

We now give the definitions for *left* admissible pairs. Note that the analogue of Lemma 1.18 holds: if (T, Y) is a left admissible pair such that $\sum_{j=0}^{l} T^j Y A_j = 0$, then

$$Y \cdot L(\lambda) = (I\lambda - T) \cdot \sum_{j=0}^{l-1} T^j Y L_{j+1}(\lambda).$$

By analogy with Definition 1.13, a left admissible pair (T, Y) is said to be a left partial spectral pair of $L(\lambda)$ if

(*a*) $\sum_{j=0}^{l} T^j Y A_j = 0$,
(*b*) $\mathrm{row}(T^j Y)_{j=0}^{l-1}$ is surjective.

In view of Definition 2.1, (T_+, Y_+) is a left γ-spectral pair of $L(\lambda)$ if and only if (Y_+^T, T_+^T) is a (usual) γ-spectral pair of $L^T(\lambda)$. Thus, every statement about γ-spectral pairs has a dual statement for left γ-spectral pairs. For example, the "left" Calderón projector is

$$P'_\gamma = \frac{1}{2\pi i}\int_\gamma \begin{pmatrix} L_1(\lambda) \\ \vdots \\ L_l(\lambda) \end{pmatrix} L^{-1}(\lambda)[I \cdots \lambda^{l-1}I]\, d\lambda$$

Theorem 2.15 *P'_γ is a projector in $\mathbb{C}^{pl}$. Also, if (T_+, Y_+) is any left γ-spectral pair of $L(\lambda)$ then* $\ker P'_\gamma = \ker \operatorname{row}(T^j_+ Y_+)^{l-1}_{j=0}$.

Proof Note that

$$P'_\gamma = \mathscr{X} \cdot \begin{pmatrix} X_+ \\ \vdots \\ X_+ T_+^{l-1} \end{pmatrix} \cdot [Y_+ \cdots T_+^{l-1} Y_+] \tag{10}$$

and by Corollary 2.11 it follows that $P'_\gamma \cdot P'_\gamma = P'_\gamma$. The inclusion

$$\ker[Y_+ \cdots T_+^{l-1} Y_+] \subset \ker P'_\gamma$$

is obvious. Conversely, if $P'_\gamma c = 0$ then $[Y_+ \cdots T_+^{l-1} Y_+] \cdot P'_\gamma c = 0$, and using (10) and Corollary 2.11 we get $[Y_+ \cdots T_+^{l-1} Y_+]c = 0$. Hence

$$\ker P'_\gamma \subset \ker[Y_+ \cdots T_+^{l-1} Y_+],$$

so equality holds.

There is also the concept of "restriction" of left admissible pairs which is dual to that for right admissible pairs. If $\mathfrak{L}$ is a subspace of a vector space $\mathfrak{M}$ let π denote the canonical projection $\mathfrak{M} \to \mathfrak{M}/\mathfrak{L}$. If $T \in \mathscr{L}(\mathfrak{M})$ and $\mathfrak{L}$ is T-invariant, let $\bar{T} \in \mathscr{L}(\mathfrak{M}/\mathfrak{L})$ be the operator defined by $\pi T = \bar{T}\pi$.

Definition 2.16 *If (T, Y) and (T', Y') are left admissible pairs then (T', Y') is called a restriction of (T, Y) if there exists a T-invariant subspace $\mathfrak{L}$ of the base space of (T, Y) such that (T', Y') is similar to $(\bar{T}, \pi Y)$.*

Analogous to Proposition 2.12 there is the following result: If (T_+, Y_+) is a left γ-spectral pair for $L(\lambda)$, then any partial left γ-spectral pair (T, Y) for $L(\lambda)$ is a restriction of (T_+, Y_+). The proof is left as an exercise.

Our aim now is to show the connection between the Calderón projector for a matrix polynomial and the Riesz projector for an operator. (See the third corollary below.)

Theorem 2.17 *Let γ and Γ be simple, closed contours not intersecting* $\operatorname{sp}(L)$ *such that γ is contained in the interior of Γ. Let (X, T, Y) be a Γ-spectral triple of $L(\lambda)$ and let*

$$Q_\gamma = \frac{1}{2\pi i}\int_\gamma (I\lambda - T)^{-1}\, d\lambda,$$

the Riesz spectral projector. Let $\mathfrak{M}^+ = \operatorname{im} Q_\gamma$. *Then* $(X_+, T_+, Y_+) := (X|_{\mathfrak{M}^+}, T|_{\mathfrak{M}^+}, Q_\gamma Y)$ *is a* γ*-spectral triple of* $L(\lambda)$ *with base space* $\mathfrak{M}^+$.

Proof From Proposition 2.7, sp(T) consists of the eigenvalues of $L(\lambda)$ inside Γ. In particular, $\operatorname{sp}(T) \cap \gamma = \varnothing$, so that Q_γ is well-defined. To prove that (X_+, T_+, Y_+) is a γ-spectral triple of $L(\lambda)$, we verify conditions (*i*) to (*iv*) of Definition 2.1. Since im Q_γ is the invariant subspace of T corresponding to the eigenvalues of T inside γ, it follows that sp(T_+) lies inside γ. This proves (*i*). To prove (*iii*) note that $\operatorname{col}(X_+ T_+^j)_{j=0}^{l-1} = \operatorname{col}(XT^j)_{j=0}^{l-1}|_{\mathfrak{M}^+}$. Hence the injectivity of $\operatorname{col}(X_+ T_+^j)_{j=0}^{l-1}$ follows from that of $\operatorname{col}(XT^j)_{j=0}^{l-1}$. The proof of (*iv*) is similar. Finally, observe that

$$\begin{aligned} X_+ T_+^j Y_+ &= XT^j Q_\gamma Y \\ &= XT^j \frac{1}{2\pi i} \int_\gamma (I\lambda - T)^{-1}\, d\lambda\, Y \\ &= \frac{1}{2\pi i} \int_\gamma \lambda^j X(I\lambda - T)^{-1} Y\, d\lambda \\ &= \frac{1}{2\pi i} \int_\gamma \lambda^j L^{-1}(\lambda)\, d\lambda \qquad j = 0, 1, \ldots \end{aligned}$$

where the last equality holds since $L^{-1}(\lambda) - X(I\lambda - T)^{-1}Y$ has an analytic continuation inside Γ and, therefore, inside γ. Thus condition (*ii′*) and hence (*ii*) holds.

Corollary *Suppose that the portion of* sp(L) *inside* Γ *is divided into two parts by contours* γ^+ *and* γ^-. *Let* $(X_\pm, T_\pm, Y_\pm)$ *be* $\gamma^\pm$*-spectral triples of* $L(\lambda)$, *respectively. Then*

$$X = [X_+\ X_-], \qquad T = \begin{pmatrix} T_+ & \\ & T_- \end{pmatrix}, \qquad Y = \begin{pmatrix} Y_+ \\ Y_- \end{pmatrix} \tag{11}$$

is a Γ*-spectral triple of* $L(\lambda)$.

Proof If (X, T, Y) is any Γ-spectral triple of $L(\lambda)$ (not necessarily of the form (11)) with base space $\mathfrak{M}$, let

$$Q_{\gamma\pm} = \frac{1}{2\pi i} \int_{\gamma^\pm} (I\lambda - T)^{-1}\, d\lambda$$

be the Riesz projectors. Then $\mathfrak{M} = \mathfrak{M}^+ \oplus \mathfrak{M}^-$, where $\mathfrak{M}^\pm = \operatorname{im} Q_{\gamma^\pm}$, and in view of Theorem 2.17, it follows that relative to this decomposition the triple (X, T, Y) may be represented in the form (11) where $(X_\pm, T_\pm, Y_\pm) := (X|_{\mathfrak{M}^\pm}, T|_{\mathfrak{M}^\pm}, Q_{\gamma^\pm} Y)$. Due to the fact that any two γ^+-spectral (resp. γ^--spectral) triples of $L(\lambda)$ are similar, the corollary is proved.

Corollary *Suppose that the portion of* sp(L) *inside* Γ *is divided into two disjoint parts by contours* γ^+ *and* γ^-. *Define Calderón projectors* P_{γ^+} *and* P_{γ^-} *as in* (7) *relative to* γ^+ *and* γ^-, *respectively. Then* $P_{\gamma^+} P_{\gamma^-} = P_{\gamma^-} P_{\gamma^+} = 0$.

Proof As in the preceding corollary, let

$$X = [X_+ \; X_-], \qquad T = \begin{pmatrix} T_+ & \\ & T_- \end{pmatrix}, \qquad Y = \begin{pmatrix} Y_+ \\ Y_- \end{pmatrix}$$

be a Γ-spectral triple of $L(\lambda)$, where $(X_\pm, T_\pm, Y_\pm)$ are $\gamma^\pm$-spectral triples of $L(\lambda)$, respectively. By Corollary 2.11, $\mathrm{row}(T^j Y)_{j=0}^{l-1} \cdot \mathscr{Z}$ is a left inverse to $\mathrm{col}(XT^j)_{j=0}^{l-1}$, that is,

$$\begin{pmatrix} Y_+ \cdots T_+^{l-1} Y_+ \\ Y_- \cdots T_-^{l-1} Y_- \end{pmatrix} \cdot \mathscr{Z} \cdot \begin{pmatrix} X_+ & X_- \\ \vdots & \vdots \\ X_+ T_+^{l-1} & X_- T_-^{l-1} \end{pmatrix} = \begin{pmatrix} I_{\mathfrak{M}^+} & 0 \\ 0 & I_{\mathfrak{M}^-} \end{pmatrix}$$

where $I_{\mathfrak{M}^\pm}$ denote the identity operators on the base spaces, $\mathfrak{M}^\pm$, of $(X_\pm, T_\pm, Y_\pm)$, respectively. Hence

$$\left.\begin{aligned} \mathrm{row}(T_+^j Y_+)_{j=0}^{l-1} \cdot \mathscr{Z} \cdot \mathrm{col}(X_- T_-^k)_{k=0}^{l-1} = 0 \\ \text{and} \quad \mathrm{row}(T_-^j Y_-)_{j=0}^{l-1} \cdot \mathscr{Z} \cdot \mathrm{col}(X_+ T_+^k)_{k=0}^{l-1} = 0 \end{aligned}\right\} \tag{12}$$

Now, by (8), we have

$$P_{\gamma^\pm} = \mathrm{col}(X_\pm T_\pm^j)_{j=0}^{l-1} \cdot \mathrm{row}(T_\pm^j Y_\pm)_{j=0}^{l-1} \cdot \mathscr{Z}$$

and it follows from (12) that $P_{\gamma^+} P_{\gamma^-} = P_{\gamma^-} P_{\gamma^+} = 0$.

Corollary *Suppose that the portion of* $\mathrm{sp}(L)$ *inside* Γ *is divided into two disjoint parts by contours* γ^+ *and* γ^-. *Let* (X, T) *be a* Γ*-spectral pair of* $L(\lambda)$. *Then*

$$\mathrm{col}(XT^j)_{j=0}^{l-1} \cdot Q_{\gamma\pm} = P_{\gamma\pm} \cdot \mathrm{col}(XT^j)_{j=0}^{l-1}$$

where $Q_{\gamma^\pm} = 1/2\pi i \int_{\gamma^\pm} (I\lambda - T)^{-1}\, d\lambda$, *the Riesz projectors, and* $P_{\gamma^\pm}$ *are the Calderón projectors* (7) *relative to* $\gamma^\pm$, *respectively.*

Proof Let $\mathfrak{M}$ denote the base space of (X, T). Then $\mathfrak{M} = \mathfrak{M}^+ \oplus \mathfrak{M}^-$, where $\mathfrak{M}^\pm = \mathrm{im}\, Q_{\gamma^\pm}$, and corresponding to this decomposition of $\mathfrak{M}$ we may write

$$X = [X_+ \; X_-] \qquad T = \begin{pmatrix} T_+ & \\ & T_- \end{pmatrix}$$

where $(X_\pm, T_\pm)$ are $\gamma^\pm$-spectral pairs of $L(\lambda)$, respectively. Then since $P_{\gamma^+} P_{\gamma^-} = P_{\gamma^-} P_{\gamma^+} = 0$ and $\mathrm{im}\, P_{\gamma^\pm} = \mathrm{im}\, \mathrm{col}(X_\pm T_\pm^j)_{j=0}^{l-1}$ we obtain

$$\begin{aligned} \mathrm{col}(XT^j)_{j=0}^{l-1} \cdot Q_{\gamma^+} &= [\mathrm{col}(X_+ T_+^j)_{j=0}^{l-1} \quad 0] \\ &= P_{\gamma^+} \cdot [\mathrm{col}(X_+ T_+^j)_{j=0}^{l-1} \quad \mathrm{col}(X_- T_-^j)_{j=0}^{l-1}] \\ &= P_{\gamma^+} \cdot \mathrm{col}(XT^j)_{j=0}^{l-1} \end{aligned}$$

and similarly for Q_{γ^-} and P_{γ^-}.

The following corollary will be needed in Chapter 10. As usual α denotes the degree of det $L(\lambda)$ and r is the number of zeros of det $L(\lambda)$ inside γ^+. In the statement of the corollary, $(X_\pm, T_\pm, Y_\pm)$ are $\gamma^\pm$-spectral triples of $L(\lambda)$ consisting of matrices. This means that X_+ is an $r \times p$ matrix, T_+ is an $r \times r$ matrix and Y_+ is a $p \times r$ matrix, and similarly for X_-, T_-, Y_- with r replaced by $\alpha - r$.

Matrix polynomials such as $B(\lambda)$ below, which have *rectangular* matrix coefficients, arise from boundary operators for elliptic systems (see Chapter 10).

Corollary *Suppose that the portion of* sp(L) *inside* Γ *is divided into two disjoint parts by contours* γ^+ *and* γ^- *and let* $(X_\pm, T_\pm, Y_\pm)$ *be* $\gamma^\pm$*-spectral triples of* $L(\lambda)$*, consisting of matrices. Then for any* $r \times r$ *matrix* M_+ *and* $r \times (\alpha - r)$ *matrix* M_- *there exists an* $r \times p$ *matrix polynomial* $B(\lambda) = \sum_{j=0}^{l-1} B_j \lambda^j$ *such that*

$$\sum_{j=0}^{l-1} B_j X_+ T_+^j = M_+ \quad \textit{and} \quad \sum_{j=0}^{l-1} B_j X_- T_-^j = M_- \tag{13}$$

and its coefficients are given by

$$B_j = M_+ \sum_{k=0}^{l-j-1} T_+^k Y_+ A_{j+k+1} + M_- \sum_{k=0}^{l-j-1} T_-^k Y_- A_{j+k+1} \tag{14}$$

Proof Let $X = [X_+ \ X_-]$, $T = \begin{pmatrix} T_+ & \\ & T_- \end{pmatrix}$ and $Y = \begin{pmatrix} Y_+ \\ Y_- \end{pmatrix}$. By the first corollary above, (X, T, Y) is a finite spectral triple of $L(\lambda)$, that is, a Γ-spectral triple where Γ is a contour having all of sp(L) in its interior. Since the equations (13) can be written in the form $\sum_{j=0}^{l-1} B_j X T^j = [M_+ \ M_-]$, or

$$[B_0 \cdots B_{l-1}] \operatorname{col}(XT^j)_{j=0}^{l-1} = [M_+ \ M_-]$$

then, in view of the fact that $[Y \cdots T^{l-1}Y]\mathscr{Z}$ is a left inverse of $\operatorname{col}(XT^j)_{j=0}^{l-1}$, there is the solution $[B_0 \cdots B_{l-1}] = [M_+ \ M_-] \cdot [Y \cdots T^{l-1} Y]\mathscr{Z}$, that is,

$$B_j = [M_+ \ M_-] \sum_{k=0}^{l-j-1} T^k Y A_{j+k+1},$$

which is equal to (14).

Remark If $L(\lambda)$ has invertible leading coefficient then B_j are unique, since $\operatorname{col}(XT^j)_{j=0}^{l-1}$ is invertible (see §3.1). For the case when the leading coefficient is not invertible see §4.2.

2.3 Left Jordan chains and a representation for the resolvent $L^{-1}(\lambda)$

If in Definition 1.27 we take just the eigenvectors, generalized eigenvectors and Jordan blocks corresponding to the eigenvalues inside γ, then we obtain a pair of matrices (X_+, J_+) which we call a *γ-spectral Jordan pair*. Note that

X_+ is a $p \times r$ matrix and J_+ is $r \times r$, where r is the number of zeros of $\det L(\lambda)$ inside γ.

Theorem 2.18 *A γ-spectral Jordan pair (X_+, J_+) is a γ-spectral pair, i.e. there exists a $p \times r$ matrix Y_+ such that (X_+, J_+, Y_+) is a γ-spectral triple.*

Proof See Corollary 2.14 and Lemma 1.23.

A sequence of row vectors $y_0, \ldots, y_{k-1}$, where $y_0 \neq 0$, is said to be a *left* Jordan chain of $L(\lambda)$ corresponding to λ if

$$\sum_{j=0}^{i} \frac{1}{j!} y_{i-j} L^{(j)}(\lambda_0) = 0 \qquad i = 0, \ldots, k-1$$

Note that $y_0, \ldots, y_{k-1}$ is a left Jordan chain of $L(\lambda)$ if and only if $y_0^T, \ldots, y_{k-1}^T$ is a (usual) Jordan chain of the transposed matrix polynomial $L^T(\lambda) = \sum_{j=0}^{l} A_j^T \lambda^j$.

Let (X_0, J_0, Y_0) be a Jordan triple for $L(\lambda)$ at λ_0, that is, a γ_0-spectral triple of $L(\lambda)$, where γ_0 is a small circle separating λ_0 from the other eigenvalues of $L(\lambda)$ and J_0 is a Jordan matrix with unique eigenvalue λ_0. The columns of X_0, when partitioned into blocks consistent with the partition of J_0 into Jordan blocks, form a canonical set of Jordan chains for $L(\lambda)$ corresponding to λ_0. The rows of Y_0 have a dual meaning as we show in the following lemma.

Lemma *Let (X_0, J_0, Y_0) be a Jordan triple for $L(\lambda) = \sum_{j=0}^{l} A_j \lambda^j$ corresponding to the eigenvalue λ_0. Then the rows of Y_0, when partitioned into blocks consistent with the partition of J_0 into Jordan blocks, and taken in each block in the reverse order, form a canonical set of left Jordan chains for $L(\lambda)$.*

Proof It follows from Definition 2.1(*iv*) and Remark 2.4 that $\operatorname{col}(Y_0^T (J_0^T)^i)_{i=0}^{l-1}$ is injective and $\sum_{i=0}^{l} A_i^T Y_0^T (J_0^T)^i = 0$. Since the base space of (Y_0^T, T_0^T) has dimension α_0 (the multiplicity of λ_0 as a zero of $\det L^T(\lambda)$), it follows by Corollary 2.14 that (Y_0^T, T_0^T) is a γ_0-spectral pair for $L^T(\lambda) = \sum_{j=0}^{l} A_j^T \lambda^j$. Write $J_0 = \operatorname{diag}(J_0^{(i)})_{i=1}^m$ where $J_0^{(i)}$ is a Jordan block of size $\mu_i \times \mu_i$. Then $P J_0^T P = J_0$, where $P = \operatorname{diag}(P^{(i)})_{i=1}^m$ and

$$P^{(i)} = \begin{pmatrix} & & 1 \\ & \cdot^{\cdot^{\cdot}} & \\ 1 & & \end{pmatrix}, \qquad \text{a } \mu_i \times \mu_i \text{ permutation matrix.}$$

Hence (Y_0^T, J_0^T) is similar to the Jordan pair $(Y_0^T P, J_0)$. Thus the columns of $Y_0^T P$, when partitioned into blocks consistent with the partition of J_0 into Jordan blocks, form a canonical set of Jordan chains for $L^T(\lambda)$. Noting that the rows of $Y_0^T P$ in each block are the same as those of Y_0, but in reverse order, the proof of the lemma is complete.

In a neighbourhood of $\lambda_0 \in \mathrm{sp}(L)$, we have the Laurent expansion

$$L^{-1}(\lambda) = \sum_{j=-\nu}^{\infty} B_j(\lambda - \lambda_0)^j,$$

and the singular part of $L^{-1}(\lambda)$ near $\lambda = \lambda_0$ is

$$\Xi(L^{-1}(\lambda)) := B_{-\nu}(\lambda - \lambda_0)^{-\nu} + \cdots + B_{-1}(\lambda - \lambda_0)^{-1}.$$

Theorem 2.19 *Let $\lambda_0 \in \mathrm{sp}(L)$. Then for every canonical set $x_0^{(i)}, \ldots, x_{\mu_i-1}^{(i)}$ $(i = 1, \ldots, m)$ of Jordan chains of $L(\lambda)$ corresponding to λ_0, there exists a canonical set $y_0^{(i)}, \ldots, y_{\mu_i-1}^{(i)}$ $(i = 1, \ldots, m)$ of left Jordan chains of $L(\lambda)$ corresponding to λ_0 such that the singular part of $L^{-1}(\lambda)$ near $\lambda = \lambda_0$ has the representation*

$$\Xi(L^{-1}(\lambda)) = \sum_{i=1}^{m} \sum_{j=1}^{\mu_i} (\lambda - \lambda_0)^{-j} \cdot \sum_{k=0}^{\mu_i - j} x_{\mu_i-j-k}^{(i)} \cdot y_k^{(i)} \tag{15}$$

Proof Let (X_0, J_0) be the Jordan pair for $L(\lambda)$ corresponding to λ_0 formed in the usual way from the given canonical set. By Corollary 2.14, (X_0, J_0) is a γ_0-spectral pair for $L(\lambda)$, so that there exists an $\alpha_0 \times p$ matrix Y_0 such that (X_0, J_0, Y_0) is a γ_0-spectral triple. Since, by definition,

$$L^{-1}(\lambda) - X_0(I\lambda - T_0)^{-1}Y_0$$

has an analytic continuation inside γ_0, then

$$\begin{aligned}\Xi(L^{-1}(\lambda)) &= X_0(I\lambda - J_0)^{-1}Y_0 \\ &= \sum_{j=0}^{\infty} X_0(J_0 - \lambda_0 I)^j Y_0 (\lambda - \lambda_0)^{-(j+1)}\end{aligned} \tag{16}$$

If we write

$$X_0 = \mathrm{row}(X_i)_{i=1}^m, \qquad \text{where } X_i = [x_0^{(i)}, \ldots, x_{\mu_i-1}^{(i)}]$$

$$J_0 = \mathrm{diag}(J_i)_{i=1}^m$$

$$Y_0 = \mathrm{col}(Y_i)_{i=1}^m, \qquad \text{where } Y_i = \begin{pmatrix} y_{\mu_i-1}^{(i)} \\ \vdots \\ y_0^{(i)} \end{pmatrix}$$

then (16) is equal to

$$\Xi(L^{-1}(\lambda)) = \sum_{i=1}^{m} \sum_{j=0}^{\mu_i-1} X_i J_i^j Y_i (\lambda - \lambda_0)^{-(j+1)}$$

and an easy computation completes the proof of the theorem.

The sum (16) is actually a finite series since $J_0 - \lambda_0 I$ is nilpotent, i.e. there exists $\mu \geqslant 0$ such that $(J_0 - \lambda_0 I)^\mu = 0$. Note that the order of the pole of $L^{-1}(\lambda)$ at λ_0 is equal to the minimal such μ (the length of the longest Jordan chain). To see this, write

$$\Xi(L^{-1}(\lambda)) = \sum_{j=0}^{\infty} B_j(\lambda - \lambda_0)^{-(j+1)}$$

where $B_j = X_0(J_0 - \lambda_0 I)^j Y_0$. Now if $(J_0 - \lambda_0 I)^\mu = 0$ then $B_j = 0$ for $j \geqslant \mu$. Conversely if $B_j = 0$ for $j \geqslant \mu$, then

$$\operatorname{col}(X_0 J_0^i)_{i=0}^{l-1}.\ (J_0 - \lambda_0 I)^\mu.\ \operatorname{row}(T_0^i Y_0)_{i=0}^{l-1} = 0$$

and hence $(J_0 - \lambda_0 I)^\mu = 0$ because $\operatorname{col}(X_0 J_0^i)_{i=0}^{l-1}$ is injective and $\operatorname{row}(T_0^i Y_0)_{i=0}^{l-1}$ is surjective.

Example 2.20 (continuation of Example 1.31)

$$L(\lambda) = \begin{pmatrix} (\lambda-2)^2 & -\lambda+2 \\ 0 & (\lambda-2)^2 \end{pmatrix}$$

From Example 1.31, we have the following Jordan pair corresponding to $\lambda_0 = 2$:

$$X_0 = \left(\begin{array}{ccc|c} 1 & 0 & 0 & 0 \\ 0 & 1 & 0 & 1 \end{array}\right), \qquad J_0 = \left(\begin{array}{ccc|c} 2 & 1 & & \\ & 2 & 1 & \\ & & 2 & \\ \hline & & & 2 \end{array}\right).$$

Let us find a 4×2 matrix Y_0 such that (X_0, J_0, Y_0) is a Jordan triple; since $l = 2$, it suffices to find Y_0 such that

$$X_0 Y_0 = \frac{1}{2\pi i}\int_{\gamma_0} L^{-1}(\lambda)\, d\lambda = \begin{pmatrix} 0 & 0 \\ 0 & 0 \end{pmatrix}$$

and

$$X_0 J_0 Y_0 = \frac{1}{2\pi i}\int_{\gamma_0} \lambda L^{-1}(\lambda)\, d\lambda = \begin{pmatrix} 1 & 0 \\ 0 & 1 \end{pmatrix}.$$

The unique solution is

$$Y_0 = \left(\begin{array}{cc} 0 & 0 \\ 1 & 0 \\ 0 & 1 \\ \hline -1 & 0 \end{array}\right)$$

Thus, we have

$$x_0^{(1)} = \begin{pmatrix} 1 \\ 0 \end{pmatrix}, \quad x_1^{(1)} = \begin{pmatrix} 0 \\ 1 \end{pmatrix}, \quad x_2^{(1)} = \begin{pmatrix} 0 \\ 0 \end{pmatrix}, \quad x_0^{(2)} = \begin{pmatrix} 0 \\ 1 \end{pmatrix}$$

$$y_0^{(1)} = [0\ 1], \quad y_1^{(1)} = [1\ 0], \quad y_2^{(1)} = [0\ 0], \quad y_0^{(2)} = [-1\ 0]$$

By Theorem 2.19, we have

$$\Xi(L^{-1}(\lambda)) = \sum_{j=1}^{3} (\lambda-2)^{-k} \sum_{k=0}^{3-j} x_{3-j-k}^{(1)} \cdot y_k^{(1)} + (\lambda-2)^{-1} x_0^{(2)} \cdot y_0^{(2)}$$
$$= \begin{pmatrix} (\lambda-2)^2 & (\lambda-2)^{-3} \\ 0 & (\lambda-2)^2 \end{pmatrix},$$

as expected.

2.4 Transformations of matrix polynomials

Let $\mathbb{C}_\infty = \mathbb{C} \cup \{\infty\}$ denote the extended complex plane. An automorphism of $\mathbb{C}_\infty$ is a meromorphic bijection $\varphi: \mathbb{C}_\infty \to \mathbb{C}_\infty$. Let $GL_2(\mathbb{C})$ denote the group of 2×2 invertible matrices. It is well known that $\mathrm{Aut}(\mathbb{C}_\infty)$, the group of automorphisms of $\mathbb{C}_\infty$, consists of the fractional linear transformations

$$\varphi(\lambda) = \frac{a\lambda + b}{c\lambda + d} \tag{17}$$

where $a, b, c, d \in \mathbb{C}$, $ad - bc \neq 0$. This defines a group homomorphism $\alpha \mapsto \varphi_\alpha$ from $GL_2(\mathbb{C})$ to $\mathrm{Aut}(\mathbb{C}_\infty)$, that is,

$$\varphi_{\alpha\beta} = \varphi_\alpha \circ \varphi_\beta \qquad \text{for any } \alpha, \beta \in GL_2(\mathbb{C})$$

and $\varphi_I = id$, where I is the 2×2 identity matrix and id is the identity transformation. The kernel of the homomorphism consists of the matrices kI, $k \neq 0$.

As usual, let $L(\lambda)$ be a $p \times p$ matrix polynomial of degree l. Then the matrix polynomial

$$\tilde{L}(\lambda) = \sum_{j=0}^{l} A_j(d\lambda - b)^j(-c\lambda + a)^{l-j} \tag{18}$$

is called the *transformation of* $L(\lambda)$ *under* $\varphi = \varphi_\alpha$, $\alpha = \begin{pmatrix} a & b \\ c & d \end{pmatrix}$. Note that

$$\tilde{L}(\lambda) = (-c\lambda + a)^l L(\varphi^{-1}(\lambda)).$$

Hence $\det \tilde{L}(\lambda) = (\det L(\lambda))^\sim$ is true only when $\det L(\lambda)$ has degree pl, and if the degree of $\det L(\lambda)$ is less than pl then $(\det L(\lambda))^\sim$ is a proper factor of $\det \tilde{L}(\lambda)$ and $\mathrm{sp}(\tilde{L}) = \varphi(\mathrm{sp}(L) \cup \infty)$.

As usual we let γ be a simple, closed contour not intersecting $\mathrm{sp}(L)$. Let $\tilde{\gamma} = \varphi(\gamma)$ denote the image of γ under φ.

Theorem 2.21 *If* (X_+, T_+, Y_+) *is a* γ*-spectral triple of* $L(\lambda)$ *and* φ *is the transformation* (17), *with* $-d/c$ *in the exterior of* γ, *then*

$$\tilde{X}_+ = X_+,$$
$$\tilde{T}_+ = \varphi(T_+) = (aT_+ + bI)(cT_+ + dI)^{-1},$$
$$\tilde{Y}_+ = (ad - bc)^{-(l-1)}(cT_+ + dI)^{l-2} Y_+$$

is a $\tilde{\gamma}$*-spectral triple of* $\tilde{L}(\lambda)$.

Proof Since $-d/c$ lies in the exterior of γ and $\varphi(-d/c) = \infty$, then φ maps the exterior (resp. interior) of γ to the exterior (resp. interior) of $\tilde{\gamma}$. By the spectral mapping theorem, see [DS 1, p. 569], it follows that $\mathrm{sp}(\tilde{T}_+)$ lies inside $\tilde{\gamma}$. Also, a/c lies in the exterior of $\tilde{\gamma}$ because $\varphi(\infty) = a/c$, hence $\det \tilde{L}(\lambda)$ has the same number of zeros inside $\tilde{\gamma}$ (counting multiplicities) as $\det L(\lambda)$ has inside γ, namely, r zeros.

To continue we need a formula expressing the coefficients of $\tilde{L}(\lambda)$ in terms of those of $L(\lambda)$. If we multiply the numbers a, b, c, and d by any $k \neq 0$ the effect is to replace $\tilde{L}(\lambda)$ by $k^l \cdot \tilde{L}(\lambda)$. Thus we may assume without loss of generality that $ad - bc = 1$. Let

$$(d\lambda - b)^j(-c\lambda + a)^{l-j} = \sum_{k=0}^{l} h_{jk}\lambda^k \tag{19}$$

for $j = 0, \ldots, l$. Since $ad - bc \neq 0$, the matrix $[h_{jk}]_{j,k=0}^{l}$ is invertible (see Lemma 2.23). Then, in view of (18), the coefficients of $\tilde{L}(\lambda) = \sum_{k=0}^{l} \tilde{A}_k \lambda^k$ are given by

$$[\tilde{A}_0 \cdots \tilde{A}_l] = [A_0 \cdots A_l] \cdot \mathscr{H} \tag{20}$$

where $\mathscr{H} = [h_{jk} I]_{j,k=0}^{l}$, and $\mathscr{H}$ is invertible (I is the $p \times p$ identity). Replacing λ by $\varphi(\lambda)$ in (19) we have

$$\lambda^j(c\lambda + d)^{-l} = \sum_{k=0}^{l} h_{jk}(\varphi(\lambda))^k \qquad j = 0, \ldots, l,$$

and then substituting T_+ for λ and multiplying by $X_+ = \tilde{X}_+$ gives us

$$X_+ T_+^j \cdot (cT_+ + dI)^{-l} = \sum_{k=0}^{l} h_{jk} \tilde{X}_+ \tilde{T}_+^k \qquad j = 0, \ldots, l.$$

Hence

$$\mathrm{col}(X_+ T_+^j)_{j=0}^{l} \cdot (cT_+ + dI)^{-l} = \mathscr{H} \cdot \mathrm{col}(\tilde{X}_+ \tilde{T}_+^k)_{k=0}^{l} \tag{21}$$

The next step is to show that $(\tilde{X}_+, \tilde{T}_+)$ is a $\tilde{\gamma}$-spectral pair of $\tilde{L}(\lambda)$. First we verify conditions (a) and (b) of Definition 1.13. By (20), (21) it follows that

$$\begin{aligned}\sum_{j=0}^{l} \tilde{A}_j \tilde{X}_+ \tilde{T}_+^j &= [\tilde{A}_0 \cdots \tilde{A}_l] \cdot \mathrm{col}(\tilde{X}_+ \tilde{T}_+^j)_{j=0}^{l} \\ &= [A_0 \cdots A_l] \cdot \mathrm{col}(X_+ T_+^j)_{j=0}^{l} \cdot (cT_+ + dI)^{-l} \\ &= \sum_{j=0}^{l} A_j X_+ T_+^j \cdot (cT_+ + dI)^{-l} = 0.\end{aligned}$$

Also, in view of (21), the injectivity of $\mathrm{col}(X_+ T_+^j)_{j=0}^{l}$ implies that of $\mathrm{col}(\tilde{X}_+ \tilde{T}_+^j)_{j=0}^{l}$. By virtue of Corollary 1.20, $\mathrm{col}(\tilde{X}_+ \tilde{T}_+^j)_{j=0}^{l-1}$ is also injective. Hence $(\tilde{X}_+, \tilde{T}_+)$ is a partial spectral pair of $\tilde{L}(\lambda)$. Since $\mathrm{sp}(\tilde{T}_+)$ lies inside $\tilde{\gamma}$ it follows that $(\tilde{X}_+, \tilde{T}_+)$ is a partial $\tilde{\gamma}$-spectral pair for $L(\lambda)$. The base space of $(\tilde{X}_+, \tilde{T}_+)$ has dimension r, so it follows by Corollary 2.14 that $(\tilde{X}_+, \tilde{T}_+)$ is a $\tilde{\gamma}$-spectral pair of $\tilde{L}(\lambda)$.

Hence there exists a matrix $\tilde{Y}'_+$ such that $(\tilde{X}_+, \tilde{T}_+, \tilde{Y}'_+)$ is a $\tilde{\gamma}$-spectral triple for $\tilde{L}(\lambda)$, and to complete the proof it remains to verify property (*ii'*) of Proposition 2.2 for the triple $(\tilde{X}_+, \tilde{T}_+, \tilde{Y}_+)$. (Due to the injectivity of $\mathrm{col}(\tilde{X}_+\tilde{T}^j_+)_{j=0}^{l-1}$, $\tilde{Y}_+$ is uniquely determined by the equations (*ii'*), whence $\tilde{Y}_+ = \tilde{Y}'_+$.) First we show that

$$X_+T^j_+(cT_+ + dI)^{-1}Y_+ = \frac{1}{2\pi i}\int_\gamma \lambda^j(c\lambda + d)^{-1}L^{-1}(\lambda)\,d\lambda \tag{22}$$

for $j = 0, 1, \ldots$. If $c = 0$ then $d \neq 0$ and (22) follows from property (*ii'*). If $c \neq 0$ then (22) can be written in the form

$$X_+T^j_+(T_+ - zI)^{-1}Y_+ = \frac{1}{2\pi i}\int_\gamma \lambda^j(\lambda - z)^{-1}L^{-1}(\lambda)\,d\lambda \tag{23}$$

where $z = -d/c$ lies in the exterior of γ. Since both sides are analytic functions of z in the exterior of γ, it suffices to prove (23) for large $|z|$. Let $R = \max(\|T_+\|, \sup_{\lambda\in\gamma}|\lambda|)$, then

$$(T_+ - zI)^{-1} = -z^{-1}(I + T_+/z + T_+^2/z^2 + \cdots)$$

and

$$(\lambda - z)^{-1} = -z^{-1}(1 + \lambda/z + \lambda^2/z^2 + \cdots),$$

with both series converging absolutely for $|z| > R$. Substitution of these series into left- and right-hand sides of (23) establishes the formula (again by property (*ii'*) for triple (X_+, T_+, Y_+)).

Now differentiate both sides of (22) with respect to d to obtain

$$X_+T^j_+(cT_+ + dI)^{-k}Y_+ = \frac{1}{2\pi i}\int_\gamma \lambda^j(c\lambda + d)^{-k}L^{-1}(\lambda)\,d\lambda,$$

j and $k = 0, 1, \ldots$, and hence

$$X_+T^j_+(\varphi(T_+))^kY_+ = \frac{1}{2\pi i}\int_\gamma \lambda^j(\varphi(\lambda))^kL^{-1}(\lambda)\,d\lambda,$$

j and $k = 0, 1, \ldots$. Now, making the substitution $\lambda = \varphi(w)$, we see that

$$\begin{aligned}\frac{1}{2\pi i}\int_{\tilde{\gamma}} \lambda^j\tilde{L}^{-1}(\lambda)\,d\lambda &= \frac{1}{2\pi i}\int_{\tilde{\gamma}} \lambda^j(-c\lambda + a)^{-l}L^{-1}(\varphi^{-1}(\lambda))\,d\lambda\\
&= \frac{1}{2\pi i}\int_\gamma (\varphi(w))^j(cw + d)^{l-2}L^{-1}(w)\,dw\\
&= X_+(\varphi(T_+))^j(cT_+ + dI)^{l-2}Y_+\\
&= \tilde{X}_+\tilde{T}^j_+\tilde{Y}_+\end{aligned}$$

for $j = 0, 1, \ldots$, which verifies the condition (*ii'*) of Proposition 2.2. Hence $(\tilde{X}_+, \tilde{T}_+, \tilde{Y}_+)$ is a $\tilde{\gamma}$-spectral of $\tilde{L}(\lambda)$.

We will also need to consider the transformation of a matrix polynomial with *rectangular* coefficients. (Such polynomials arise from boundary operators for elliptic systems.) If $B(\lambda) = \sum_{j=0}^m B_j\lambda^j$ is an $r \times p$ matrix polynomial of degree m, the transformation of $B(\lambda)$ under φ is defined just as in (18):

$$\tilde{B}(\lambda) = \sum_{j=0}^{m} B_j(d\lambda - b)^j(-c\lambda + a)^{m-j} = \sum_{j=0}^{m} \tilde{B}_j \lambda^j,$$

and the coefficients of $\tilde{B}(\lambda)$ can be expressed in terms of those of $B(\lambda)$ as follows:

$$[\tilde{B}_0 \cdots \tilde{B}_m] = [B_0 \cdots B_m] \cdot \mathscr{H}'$$

where

$$\mathscr{H}' = [h'_{jk} I]_{j,k=0}^{l} \quad \text{and} \quad (d\lambda - b)^j(-c\lambda + a)^{m-j} = \sum_{k=0}^{m} h'_{jk} \lambda^k$$

for $j = 0, \ldots, m$.

The following lemma can be proved in the same way that (20) was proved in Theorem 2.21. (As before we can suppose that $ad - bc = 1$.)

Lemma 2.22

$$\sum_{j=0}^{m} \tilde{B}_j \tilde{X}_+ \tilde{T}_+^j = (ad - bc)^m \sum_{j=0}^{m} B_j X_+ T_+^j \cdot (cT_+ + dI)^{-m}$$

The following lemma will sometimes be useful. The proof is left as an exercise.

Lemma 2.23 *The equations*

$$(a\lambda + b)^j(c\lambda + d)^{l-j} = \sum_{k=0}^{l} h_{jk} \lambda^k \qquad j = 0, \ldots, l$$

define a homomorphism $\begin{pmatrix} a & b \\ c & d \end{pmatrix} \mapsto [h_{jk}]_{j,k=0}^{l}$ *from* $GL_2(\mathbb{C})$ *to* $GL_{l+1}(\mathbb{C})$. *Moreover*

$$\det[h_{jk}]_{j,k=0}^{l} = (ad - bc)^{l(l+1)/2}.$$

3

Monic matrix polynomials

A matrix polynomial $L(\lambda) = I\lambda^l + \sum_{j=0}^{l-1} A_j\lambda^j$ in which the leading coefficient is the $p \times p$ identity matrix I is said to be *monic*. It is important to put special emphasis on this case for two reasons. First, the matrix polynomials which occur in analysis and applications are frequently already in this form. Second, the spectral theory is particularly simple in the monic case. In this case the degree of det $L(\lambda)$ is $\alpha = pl$, so $L(\lambda)$ has pl eigenvalues. If (X, T) is a finite spectral pair for $L(\lambda)$ then $\mathrm{col}(XT^j)_{j=0}^{l-1}$ is not just injective, but invertible since it is a square matrix.

3.1 Standard pairs and triples

Definition 3.1 *Let $L(\lambda)$ be a monic $p \times p$ matrix polynomial of degree l. A pair of matrices (X, T), where X is $p \times pl$ and T is $pl \times pl$ is called a standard pair of $L(\lambda)$ if the following conditions are satisfied:*

(*i*) $\mathrm{col}(XT^i)_{i=0}^{l-1}$ *is invertible*
(*ii*) $\sum_{j=0}^{l-1} A_jXT^j + XT^l = 0.$

A standard pair of $L(\lambda)$ is simply a finite spectral pair. It is important, however, to have a new terminology here in order to single out the monic case. Strictly speaking, a standard pair refers only to a pair of *matrices*, not operators.

Theorem 3.2 *Let $L(\lambda)$ be a monic $p \times p$ matrix polynomial, and let (X, T, Y) be a triple of matrices, where X is $p \times pl$, T is $pl \times pl$ and Y is $pl \times p$. Then the following are equivalent:*

(*a*) *(X, T, Y) is a finite spectral triple of $L(\lambda)$.*
(*b*) $L^{-1}(\lambda) = X(I\lambda - T)^{-1}Y$
(*c*) *(X, T) is a standard pair for $L(\lambda)$ and*

$$Y = (\mathrm{col}(XT^i)_{i=0}^{l-1})^{-1} \cdot \begin{pmatrix} 0 \\ \vdots \\ 0 \\ I \end{pmatrix}. \qquad (1)$$

We call (X, T, Y) a standard triple for $L(\lambda)$ if any (hence all) of these properties hold.

Proof (a) $\Rightarrow$ (b) Let (X, T, Y) be a finite spectral triple of $L(\lambda)$. By definition, $L^{-1}(\lambda) - X(I\lambda - T)^{-1}Y$ has an analytic continuation to the finite complex plane. Since both $L^{-1}(\lambda)$ and $X(I\lambda - T)^{-1}Y$ approach 0 as $|\lambda| \to \infty$ (the former due to the fact that $L(\lambda)$ is monic), it follows by Liouville's theorem that

$$L^{-1}(\lambda) - X(I\lambda - T)^{-1}Y \equiv 0$$

(b) $\Rightarrow$ (c) Let (b) hold and note that for $|\lambda|$ sufficiently large, $L^{-1}(\lambda)$ can be developed into a power series

$$L^{-1}(\lambda) = \lambda^{-l}I + \lambda^{-l-1}Z_1 + \lambda^{-l-2}Z_2 + \cdots$$

for some matrices $Z_1, Z_2, \ldots$. Consequently, if Γ is a circle having $\mathrm{sp}(L)$ in its interior, then

$$\frac{1}{2\pi i}\int_\Gamma \lambda^j L^{-1}(\lambda)\,d\lambda = \begin{cases} 0 & \text{if } j = 0, 1, \ldots, l-2 \\ I & \text{if } j = l-1 \end{cases} \tag{2}$$

We may also choose Γ large enough so that

$$\frac{1}{2\pi i}\int_\Gamma \lambda^j (I\lambda - T)^{-1}\,d\lambda = T^j \qquad j = 0, 1, \ldots \tag{3}$$

and then it follows from (b) that

$$\begin{pmatrix} X \\ \vdots \\ XT^{l-1} \end{pmatrix} \cdot [Y \cdots T^{l-1}Y] = \frac{1}{2\pi i}\int_\Gamma \begin{pmatrix} I \\ \vdots \\ \lambda^{l-1}I \end{pmatrix} L^{-1}(\lambda)[I \cdots \lambda^{l-1}I]\,d\lambda$$

$$= \begin{pmatrix} 0 & 0 & \cdots & I \\ \vdots & & \cdot^{\cdot^{\cdot}} & * \\ 0 & I & * & \\ I & * & & * \end{pmatrix}. \tag{4}$$

Since the matrix (4) is invertible, both $\mathrm{col}(XT^j)_{j=0}^{l-1}$ and $\mathrm{row}(T^jY)_{j=0}^{l-1}$ are invertible. Now we observe that for $j = 0, 1, \ldots, l-1$,

$$\begin{aligned} 0 &= \frac{1}{2\pi i}\int_\Gamma \lambda^j L(\lambda)L^{-1}(\lambda)\,d\lambda \\ &= \frac{1}{2\pi i}\int_\Gamma \lambda^j L(\lambda)X(I\lambda - T)^{-1}Y\,d\lambda \\ &= (A_0X + A_1XT + \cdots + A_{l-1}XT^{l-1} + XT^l)T^jY, \end{aligned}$$

and, since $\mathrm{row}(T^jY)_{j=0}^{l-1}$ is invertible, we obtain

$$A_0X + A_1XT + \cdots + A_{l-1}XT^{l-1} + XT^l = 0.$$

Hence (X, T) is a standard pair for $L(\lambda)$. Also, in view of (4), the formula (1) holds.

(c) ⇒ (a) Let (c) hold. Then we have

$$\begin{pmatrix} X \\ \vdots \\ XT^{l-1} \end{pmatrix} \cdot [Y \cdots T^{l-1}Y] = \begin{pmatrix} 0 & 0 & \cdots & I \\ \vdots & & \cdot^{\cdot^{\cdot}} & * \\ 0 & I & * & \\ I & * & & * \end{pmatrix},$$

so that $\operatorname{row}(T^jY)_{j=0}^{l-1}$ is invertible. Let Γ be a circle of sufficiently large radius having $\operatorname{sp}(L)$ and $\operatorname{sp}(T)$ in its interior such that (2) and (3) hold. In view of (1) and (2), we have

$$\frac{1}{2\pi i}\int_\Gamma \lambda^j L^{-1}(\lambda)\,d\lambda = XT^jY \qquad \text{for } j = 0, 1, \ldots, l-1 \tag{5}$$

It now follows that (5) holds for *all* $j = 0, 1, \ldots$. Indeed, suppose that (5) holds for $j = 0, 1, \ldots, v$ where $v \geqslant l - 1$. Then, since

$$\frac{1}{2\pi i}\int_\Gamma \lambda^{v+1-l}L(\lambda)L^{-1}(\lambda)\,d\lambda = 0,$$

we have

$$\begin{aligned}\frac{1}{2\pi i}\int_\Gamma \lambda^{v+1}L^{-1}(\lambda)\,d\lambda &= -\sum_{j=0}^{l-1} A_j \frac{1}{2\pi i}\int_\Gamma \lambda^{v+1-l+j}L^{-1}(\lambda)\,d\lambda \\ &= -\sum_{j=0}^{l-1} A_jXT^{v+1-l+j}Y \\ &= XT^l \cdot T^{v+1-l}Y \\ &= XT^{v+1}Y\end{aligned}$$

where in the third equality we used condition (*ii*) of Definition 3.1, i.e. $XT^l = -\sum_{j=0}^{l-1} A_jXT^j$. By induction, it follows that (5) holds for all $j = 0, 1, \ldots$. In view of (5) and (3),

$$L^{-1}(\lambda) - X(I\lambda - T)^{-1}Y$$

has an analytic continuation to the finite complex plane. We have shown that the triple (X, T, Y) satisfies conditions (*i*) to (*iv*) of Definition 2.1 relative to the contour Γ. Hence (X, T, Y) is a finite spectral triple of $L(\lambda)$.

Remark 3.3 For future reference note that by Corollary 2.11 we have

$$(\operatorname{col}(XT^i)_{i=0}^{l-1})^{-1} = \operatorname{row}(T^jY)_{j=0}^{l-1} \cdot \mathscr{Z}, \tag{6}$$

where $\mathscr{Z}$ is the matrix as defined by (5) in §2.1 (here $A_l = I$):

$$\mathscr{L} = \begin{pmatrix} A_1 & A_2 & \cdots & A_{l-1} & I \\ A_2 & & & I & \\ \vdots & \vdots & \cdot^{\cdot^{\cdot}} & & \\ A_{l-1} & I & & 0 & \\ I & & & & \end{pmatrix} \tag{7}$$

As a consequence of (6), we see that the matrix (4) is equal to $\mathscr{L}^{-1}$.

The next theorem shows how a standard triple of a product of matrix polynomials can be obtained from a standard triple of each factor.

Theorem 3.4 *Let $L_1(\lambda)$ and $L_2(\lambda)$ be monic matrix polynomials with standard triples (X_1, T_1, Y_1) and (X_2, T_2, Y_2), respectively. Then the product $L(\lambda) = L_2(\lambda)L_1(\lambda)$ has standard triple*

$$x = [X_1\ 0], \qquad T = \begin{pmatrix} T_1 & Y_1X_2 \\ 0 & T_2 \end{pmatrix}, \qquad Y = \begin{pmatrix} 0 \\ Y_2 \end{pmatrix}. \tag{8}$$

Proof By definition, a standard triple of $L(\lambda)$ is a triple of matrices (X, T, Y) satisfying condition (b) of Theorem 3.2. Using Theorem 3.2 for the matrix polynomials $L_1(\lambda)$ and $L_2(\lambda)$, we find that

$$\begin{aligned} L^{-1}(\lambda) &= L_1^{-1}(\lambda)L_2^{-1}(\lambda) \\ &= X_1(I\lambda - T_1)^{-1}Y_1 \cdot X_2(I\lambda - T_2)^{-1}Y_2 \end{aligned}$$

But it is easily verified that

$$(I\lambda - T)^{-1} = \begin{pmatrix} (I\lambda - T_1)^{-1} & (I\lambda - T_1)^{-1}Y_1X_2(I\lambda - T_2)^{-1} \\ 0 & (I\lambda - T_2)^{-1} \end{pmatrix},$$

and hence $L^{-1}(\lambda) = X(I\lambda - T)^{-1}Y$. Thus (X, T, Y) is a standard triple of $L(\lambda)$.

Analogous to Definition 3.1, there is the definition of *left* standard pair. A pair of matrices (T, Y), where Y is $pl \times p$ and T is $pl \times pl$, is called a left standard pair of $L(\lambda)$ if

(*ii*) $\text{row}(T^jY)_{j=0}^{l-1}$ is invertible
(*ii*) $\sum_{j=0}^{l-1} T^jYA_j + T^lY = 0$.

It is easily seen from the proof of Theorem 3.2 that (T, Y) is a left standard pair for $L(\lambda)$ if and only if there exists a $p \times pl$ matrix X such that (X, T, Y) is a standard triple for $L(\lambda)$. Moreover, if (X, T, Y) is a standard triple for $L(\lambda)$ then

$$X = [0 \cdots 0\ I] \cdot (\text{row}(T^jY)_{j=0}^{l-1})^{-1}$$

As a rule we prove only the results for (usual) standard pairs and then mention briefly the dual result for left standard pairs.

3.2 Linearization

Let $L(\lambda) = \sum_{j=0}^{l} A_j \lambda^j$ be a $p \times p$ matrix polynomial. A $pl \times pl$ linear matrix polynomial $B_0 + B_1\lambda$ is called a *linearization* of $L(\lambda)$ if

$$\begin{pmatrix} L(\lambda) & 0 \\ 0 & I_{p(l-1)} \end{pmatrix} = E(\lambda)(B_0 + B_1\lambda)F(\lambda)$$

for some $pl \times pl$ matrix polynomials $E(\lambda)$ and $F(\lambda)$ with constant nonzero determinants. This means that $\begin{pmatrix} L(\lambda) & \\ & I \end{pmatrix}$ can be obtained from $B_0 + B_1\lambda$ by a finite sequence of elementary row and column operations. (Hence both matrix polynomials have the same Smith form, or, equivalently, their Jordan chains may be put in one-to-one correspondence with each other.)

The *companion polynomial* of $L(\lambda)$ is the $pl \times pl$ linear matrix polynomial

$$C_L(\lambda) = \begin{pmatrix} I & & & & \\ & I & & & \\ & & \ddots & & \\ & & & I & \\ & & & & A_l \end{pmatrix}\lambda + \begin{pmatrix} 0 & -I & & & \\ & 0 & -I & & \\ & & \ddots & \ddots & \\ & & & 0 & -I \\ A_0 & A_1 & \cdots & & A_{l-1} \end{pmatrix} \tag{9}$$

Also define the matrix polynomials $L_j(\lambda)$ as in Lemma 1.18. Note that $L_j(\lambda) = A_j + \lambda \cdot L_{j+1}(\lambda)$, $j = 0, \ldots, l-1$, and $L_0(\lambda) = L(\lambda)$.

Theorem 3.5 *$C_L(\lambda)$ is a linearization of $L(\lambda)$. In fact,*

$$E(\lambda)C_L(\lambda)F(\lambda) = \begin{pmatrix} L(\lambda) & 0 \\ 0 & I_{p(l-1)} \end{pmatrix},$$

where

$$E(\lambda) = \begin{pmatrix} L_1(\lambda) & & \cdots & L_{l-1}(\lambda) & I \\ -I & 0 & & & \\ & -I & \ddots & & \\ & & \ddots & & \\ & & & -I & 0 \end{pmatrix}$$

and

$$F(\lambda) = \begin{pmatrix} I & & & & \\ \lambda I & I & & & \\ \vdots & \lambda I & \ddots & & \\ & \vdots & & I & \\ \lambda^{l-1}I & \lambda^{l-2}I & \cdots & \lambda I & I \end{pmatrix}.$$

Proof A direct computation shows that

$$E(\lambda)\cdot\begin{pmatrix} I\lambda & -I & & & \\ & I\lambda & -I & & \\ & & \ddots & \ddots & \\ & & & & -I \\ A_0 & A_1 & & \cdots & A_l\lambda + A_{l-1} \end{pmatrix} = \begin{pmatrix} L(\lambda) & & & & \\ -I\lambda & I & & & \\ & -I\lambda & \ddots & & \\ & & \ddots & \ddots & \\ & & & -I\lambda & I \end{pmatrix},$$

so that

$$E(\lambda)\cdot C_L(\lambda) = \begin{pmatrix} L(\lambda) & \\ & I \end{pmatrix}\cdot F^{-1}(\lambda),$$

where

$$F^{-1}(\lambda) = \begin{pmatrix} I & & & \\ -I\lambda & I & & \\ & -I\lambda & \ddots & \\ & & \ddots & \ddots & \\ & & & -I\lambda & I \end{pmatrix}.$$

The differential equation

$$L(d/dt)u = \sum_{j=0}^{l} A_j(d/dt)^j u = 0 \tag{10}$$

can be reduced to a first-order system in the usual way; we let

$$u_0 = u, \quad u_1 = \frac{d}{dt}u_0, \quad \ldots, \quad u_{l-1} = \frac{d}{dt}u_{l-2} \tag{11}$$

and then (11), (10) take the form

$$C_L(d/dt)U = 0,$$

where $U = [u_0\, u_1 \cdots u_{l-1}]^T$. Of course, if $L(\lambda)$ is monic ($A_l = I$) then $C_L(\lambda) = I\lambda - C_1$ so that

$$\frac{dU}{dt} = C_1 U, \qquad \text{where } C_1 = \begin{pmatrix} 0 & I & & & \\ & 0 & & & \\ \vdots & & \ddots & & \\ & & & 0 & I \\ -A_0 & -A_1 & \cdots & & -A_{l-1} \end{pmatrix} \tag{12}$$

and C_1 is called the *$pl \times pl$ companion matrix* of $L(\lambda)$. The solutions of (12) have the form $U(t) = e^{tC_1}\eta$, for some $\eta \in \mathbb{C}^{pl}$, so that the solutions of the original equation $L(d/dt)u = 0$ have the form $u(t) = P_1 e^{tC_1}\eta$, $\eta \in \mathbb{C}^{pl}$, where $P_1 = [I\,0 \cdots 0]$, a $p \times pl$ matrix. The pair (P_1, C_1) is called the (first) *companion pair* of $L(\lambda)$.

Theorem 3.6 *Let $L(\lambda) = I\lambda^l + \sum_{j=0}^{l-1} A_j\lambda^j$ be a monic matrix polynomial and let the companion matrix C_1 be defined as above. Then*

$$X = [I\,0\cdots 0], \qquad T = C_1, \qquad Y = \begin{pmatrix} 0 \\ \vdots \\ 0 \\ I \end{pmatrix}$$

is a standard triple for $L(\lambda)$, called the first companion triple.

Proof We first verify that $(X, T) = (P_1, C_1)$ is a standard pair for $L(\lambda)$. It is clear that

$$P_1 C_1^j = [0\cdots I\cdots 0] \qquad j = 0, \ldots, l-1$$

with I in the $(j + 1)$s position and 0's elsewhere, and

$$P_1 C_1^l = [-A_0 \quad -A_1 \quad \cdots \quad -A_{l-1}].$$

Hence $\operatorname{col}(P_1 C_1^j)_{j=0}^{l-1} = I$, and condition (*i*) of Definition 3.1 is satisfied. Also,

$$\begin{aligned}\sum_{j=0}^{l-1} A_j P_1 C_1^j + P_1 C_1^l &= \sum_{j=0}^{l-1} A_j[0\cdots I\cdots 0] + [-A_0 \; -A_1 \cdots -A_{l-1}] \\ &= 0\end{aligned}$$

so that (*ii*) is satisfied. Hence (X, T) is a standard pair for $L(\lambda)$. The fact that (X, T, Y) is a standard triple is now clear from (c) of Theorem 3.2.

The second companion matrix of $L(\lambda)$ is defined as follows:

$$C_2 = \begin{pmatrix} 0 & & \cdots & -A_0 \\ I & 0 & & -A_1 \\ & & \ddots & \vdots \\ & & I & -A_{l-1} \end{pmatrix},$$

and the second companion triple of $L(\lambda)$ is:

$$X' = [0\cdots 0\, I], \qquad T' = C_2, \qquad Y' = \begin{pmatrix} I \\ 0 \\ \vdots \\ 0 \end{pmatrix},$$

It is easily verified that the triple (X', T', Y') is similar to the one in

Theorem 3.6:

$$X' = [I\,0\cdots 0]\mathscr{X}^{-1}, \qquad C_2 = \mathscr{X}C_1\mathscr{X}^{-1}, \qquad Y' = \mathscr{X}\begin{pmatrix}0\\ \vdots\\ 0\\ I\end{pmatrix},$$

where $\mathscr{X}$ is given by (7). Thus (X', T', Y') is also a standard triple for $L(\lambda)$.

Proposition 3.7 *Any two standard pairs (X, T) and (X', T') of a monic matrix polynomial $L(\lambda)$ are similar, i.e., there exists an invertible $pl \times pl$ matrix M such that $X' = XM$, $T' = M^{-1}TM$. The matrix M is defined uniquely by (X, T) and (X', T') and is given by the formula*

$$M = (\mathrm{col}(XT^i)_{i=0}^{l-1})^{-1}\cdot \mathrm{col}(X'(T')^i)_{i=0}^{l-1} \tag{13}$$

Proof This is an immediate consequence of Proposition 1.13. However, we will give an alternative proof here using the companion matrix C_1. Observe that the following equalities hold:

$$C_1\cdot \mathrm{col}(XT^i)_{i=0}^{l-1} = \mathrm{col}(XT^i)_{i=0}^{l-1}\cdot T \tag{14}$$

$$C_1\cdot \mathrm{col}(X'(T')^i)_{i=0}^{l-1} = \mathrm{col}(X'(T')^i)_{i=0}^{l-1}\cdot T' \tag{14'}$$

Indeed, the equality in all but the last block row in (14) or (14′) is evident; the equality in the last block-row follows from (ii) of Definition 3.1. Comparison of the equalities (14), (14′) gives $T' = M^{-1}TM$, with M defined by (13). Also, by the definition of the inverse matrix, we have

$$X\cdot(\mathrm{col}(XT^i)_{i=0}^{l-1})^{-1} = [I\,0\cdots 0],$$

so that $XM = [I\,0\cdots 0]\cdot \mathrm{col}(X'(T')^i)_{i=0}^{l-1} = X'$. Hence (X', T') is similar to (X, T). Uniqueness of M follows from the fact that if (X, T) and (X', T') are related by the equations $X' = XM$ and $T' = M^{-1}TM$, then M satisfies the equation

$$\mathrm{col}(X'(T')^i)_{i=0}^{l-1} = \mathrm{col}(XT^i)_{i=0}^{l-1}\cdot M.$$

Remark An equality like (14) was encountered once before in Proposition 1.19. In fact assuming that $L(\lambda)$ is monic we can now show the following:

$$C_1 = \frac{1}{2\pi i}\int_\Gamma \begin{pmatrix} I\\ \vdots\\ \lambda^{l-1}I\end{pmatrix}\lambda L^{-1}(\lambda)[L_1(\lambda)\cdots L_l(\lambda)]\,d\lambda, \tag{15}$$

where Γ is a simple, closed contour having all of $\mathrm{sp}(L)$ in its interior. To prove this, let (X, T) be a standard pair of $L(\lambda)$, i.e. a Γ-spectral pair for the monic $L(\lambda)$. Consider the projector P_Γ as defined in Theorem 2.8. Then $\mathrm{im}\, P_\Gamma = \mathrm{im}\,\mathrm{col}(XT^i)_{i=0}^{l-1} = \mathbb{C}^{pl}$, so that $P_\Gamma = I_{pl}$, the $pl \times pl$ identity matrix.

This means that

$$\frac{1}{2\pi i}\int_{\Gamma}\lambda^j L^{-1}(\lambda)L_{k+1}(\lambda)\,d\lambda = \delta_{jk}I \tag{16}$$

for j and $k = 0, 1, \ldots, l-1$. The formula (15) is now evident in the first $l-1$ block-rows. For the last block, note that

$$\frac{1}{2\pi i}\int_{\Gamma}L(\lambda)L^{-1}(\lambda)L_{k+1}(\lambda)\,d\lambda = 0$$

for $k = 0, \ldots, l-1$. Since $L(\lambda) = I\lambda^l + \sum_{j=0}^{l-1} A_j\lambda^j$, we obtain

$$\begin{aligned}\frac{1}{2\pi i}\int_{\Gamma}\lambda^l L^{-1}(\lambda)L_{k+1}(\lambda)\,d\lambda &= -\sum_{j=0}^{l-1} A_j \frac{1}{2\pi i}\int_{\Gamma}\lambda^j L^{-1}(\lambda)L_{k+1}(\lambda)\,d\lambda \\ &= -\sum_{j=0}^{l-1} A_j\delta_{jk} \\ &= -A_k\end{aligned}$$

for $k = 0, \ldots, l-1$.

3.3 Representation of a monic matrix polynomial in terms of a standard pair

The next theorem shows that a monic matrix polynomial is uniquely determined by a standard pair.

Theorem 3.8 *Let $L(\lambda)$ be a monic matrix polynomial with standard pair (X, T). Then $L(\lambda)$ admits the following representation:*

$$L(\lambda) = I\lambda^l - XT^l(V_1 + V_2\lambda + \cdots + V_l\lambda^{l-1}) \tag{17}$$

where V_i are $pl \times p$ matrices such that

$$(\mathrm{col}(XT^i)_{i=0}^{l-1})^{-1} = [V_1 \cdots V_l]$$

Proof Since $A_0X + A_1XT + \cdots + A_{l-1}XT^{l-1} + XT^l = 0$, we obtain

$$[A_0 \cdots A_{l-1}]\cdot\mathrm{col}(XT^i)_{i=0}^{l-1} = -XT^l,$$

that is,

$$[A_0 \cdots A_{l-1}] = -XT^l\cdot(\mathrm{col}(XT^i)_{i=0}^{l-1})^{-1}.$$

Hence $A_j = -XT^lV_{j+1}$ for $j = 0, \ldots, l-1$.

Remark Note that (17) is independent of the choice of standard pair (X, T), since any two such pairs are similar.

Corollary *Let (X, T) be a pair of matrices, where X is $p \times pl$ and T is $pl \times pl$, such that $\mathrm{col}(XT^i)_{i=0}^{l-1}$ is invertible. Then there exists matrix polynomial $L(\lambda)$ having (X, T) as standard pair.*

Proof The uniqueness follows from Theorem 3.8. As for existence, define a matrix polynomial $L(\lambda)$ by means of the formula (17). Then

$$\begin{aligned} A_0X + A_1XT + \cdots + A_{l-1}XT^{l-1} + XT^l \\ &= -XT^l[V_1 \cdots V_l]\cdot \operatorname{col}(XT^i)_{i=0}^{l-1} + XT^l \\ &= -XT^l + XT^l \\ &= 0. \end{aligned}$$

By Definition 3.1, (X, T) is a standard pair for $L(\lambda)$.

Since (T, Y) is a left standard pair of $L(\lambda)$ if and only if (Y^T, T^T) is a (usual) standard pair of $L^T(\lambda)$, every statement on standard pairs has a dual statement for left standard pairs. For example, $L^T(\lambda)$ admits a representation of the form (17) in terms of the standard pair (Y^T, T^T). Transposing this result, we find that $L(\lambda)$ admits a representation in terms of the left standard pair (T, Y):

$$L(\lambda) = I\lambda^l - (W_1 + W_2\lambda + \cdots + W_l\lambda^{l-1})T^lY$$

where W_i are $p \times pl$ matrices such that

$$(\operatorname{row}(T^jY)_{j=0}^{l-1})^{-1} = \begin{pmatrix} W_1 \\ \vdots \\ W_l \end{pmatrix}.$$

Of course, this representation could be derived directly in a manner similar to the proof of Theorem 3.8.

3.4 Euclidean algorithm in terms of a standard pair

In this section we derive a formula for the division of an arbitrary matrix polynomial $L(\lambda)$ (not necessarily monic) by a monic matrix polynomial $L_1(\lambda)$ in terms of a standard triple of the divisor. This result is stated in Theorem 3.11 below.

But first some remarks concerning the division of matrix polynomials in general are in order. When we speak of the *right* division of $L(\lambda)$ by $L_1(\lambda)$ we mean a representation of the form

$$L(\lambda) = Q(\lambda)L_1(\lambda) + R(\lambda), \tag{18}$$

where the quotient $Q(\lambda)$ and the remainder $R(\lambda)$ are matrix polynomials, with either $R(\lambda) \equiv 0$ or $\deg R(\lambda) < \deg L_1(\lambda)$. If $R(\lambda) \equiv 0$ then $L_1(\lambda)$ is said to be a right divisor of $L(\lambda)$. Similarly, by the *left* division of $L(\lambda)$ by $L_1(\lambda)$, we mean a representation of the form

$$L(\lambda) = L_1(\lambda)Q(\lambda) + R(\lambda), \tag{19}$$

with either $R(\lambda) \equiv 0$ or $\deg R(\lambda) < \deg L_1(\lambda)$. If $R(\lambda) \equiv 0$ then $L(\lambda)$ is said to be a left divisor of $L(\lambda)$.

It is important to note that, in general, division of matrix polynomials in the form (18) or (19) may not be possible. However, in the case that the divisor $L_1(\lambda)$ is monic, division is always possible as we prove in the following proposition. The proof is the same as for the case of scalar polynomials.

Proposition 3.9 *If the divisor $L_1(\lambda)$ is monic, then the right (or left) division is always possible, and the right (or left) quotient and remainder are unique.*

Proof It suffices to prove the proposition for right division; the proof for left division is similar. Let $L(\lambda) = \sum_{j=0}^{l} A_j\lambda^j$ and $L_1(\lambda) = I\lambda^k + \sum_{j=0}^{k-1} B_j\lambda^j$. We can suppose that $l \geqslant k$ (otherwise let $Q(\lambda) = 0$ and $R(\lambda) = L(\lambda)$), so that (18) has the form

$$\sum_{j=0}^{l} A_j\lambda^j = \left(\sum_{j=0}^{l-k} Q_j\lambda^j\right)\cdot\left(I\lambda^k + \sum_{j=0}^{k-1} B_j\lambda^j\right) + \sum_{j=0}^{k-1} R_j\lambda^k, \tag{20}$$

where Q_j and R_j are to be determined. Now compare the coefficients of λ^l, $\lambda^{l-1}, \ldots, \lambda^k$ on both sides:

$$\begin{aligned} A_l &= Q_{l-k} \\ A_{l-1} &= Q_{l-k-1} + Q_{l-k}B_{k-1} \\ &\vdots \\ A_k &= Q_0 + \sum_{j=0}^{k-1} Q_{k-j}B_j \end{aligned}$$

By forward substitution, we find the coefficients $Q_{l-k}, Q_{l-k-1}, \ldots, Q_0$ of $Q(\lambda)$. Then from (20), we obtain

$$R_i = A_i - \sum_{j=0}^{i} Q_jB_{i-j} \qquad \text{for } i = 0, \ldots, k-1$$

Reversing these steps proves existence of the quotient and remainder. To prove uniqueness, suppose

$$L(\lambda) = Q(\lambda)L_1(\lambda) + R(\lambda) = \tilde{Q}(\lambda)L_1(\lambda) + \tilde{R}(\lambda)$$

where the degrees of $R(\lambda)$ and $\tilde{R}(\lambda)$ do not exceed $k-1$. Then

$$(Q(\lambda) - \tilde{Q}(\lambda))L_1(\lambda) = \tilde{R}(\lambda) - R(\lambda),$$

and, since $L_1(\lambda)$ is monic of degree k, it follows that the left-hand side of this equation has degree $\geqslant k$ if $Q(\lambda) - \tilde{Q}(\lambda) \neq 0$. Hence $Q(\lambda) - \tilde{Q}(\lambda) = 0$ and $\tilde{R}(\lambda) - R(\lambda) = 0$.

Lemma 3.10 *Let $L(\lambda) = \sum_{j=0}^{l} A_j\lambda^j$ and $L_1(\lambda)$ be matrix polynomials of degrees l and k, respectively. Suppose that $L_1(\lambda)$ is monic with standard triple (X_1, T_1, Y_1) and suppose that $Q(\lambda) := L(\lambda)L_1^{-1}(\lambda)$ has no poles in the finite complex plane. Then $Q(\lambda)$ is a polynomial of degree $\leqslant l-k$ and admits the* *following representation:*

$$Q(\lambda) = \sum_{j=0}^{l-k} \lambda^j \cdot \left(\sum_{i=j+1}^{l} A_i X_1 T_1^{i-j-1} Y_1 \right) \tag{21}$$

Proof The fact that $L(\lambda)L_1^{-1}(\lambda)$ has no poles implies that

$$\sum_{i=0}^{l} A_i X_1 T_1^i = 0 \tag{22}$$

Indeed, since (X_1, T_1, Y_1) is a finite spectral triple of $L_1(\lambda)$ we may apply Proposition 2.2 to obtain

$$\begin{aligned}\sum_{i=0}^{l} A_i X_1 T_1^i \cdot T_1^j Y_1 &= \frac{1}{2\pi i} \int_\Gamma \sum_{i=0}^{l} A_i \lambda^i \cdot \lambda^j L_1^{-1}(\lambda)\, d\lambda \\ &= \frac{1}{2\pi i} \int_\Gamma \lambda^j L(\lambda) L_1^{-1}(\lambda)\, d\lambda = 0\end{aligned}$$

for $j = 0, 1, \ldots$, where Γ is a circle containing $\mathrm{sp}(L_1)$ in its interior, and then (22) must hold because $\mathrm{row}(T_1^j Y_1)_{j=0}^{k-1}$ is invertible. Consequently, using (b) of Theorem 3.2 and Lemma 1.18, we obtain

$$\begin{aligned}Q(\lambda) = L(\lambda)L_1^{-1}(\lambda) &= L(\lambda)X_1(I\lambda - T_1)^{-1}Y_1 \\ &= \sum_{j=0}^{l-1} (A_{j+1} + A_{j+2}\lambda + \cdots + A_l \lambda^{l-j-1}) X_1 T_1^j Y_1 \\ &= \sum_{j=0}^{l-1} \sum_{i=j+1}^{l} A_i X_1 T_1^{i-j-1} Y_1 \lambda^j \end{aligned} \tag{23}$$

Since $L_1(\lambda)$ is monic of degree k, it follows that $X_1 T_1^i Y_1 = 0$ for $i = 0, \ldots, k-2$ (see (c) of Theorem 3.2). Hence the coefficients of λ^j in (23) vanish for $j > l-k$. This completes the proof of the lemma.

Theorem 3.11 *Let* $L(\lambda) = \sum_{j=0}^{l} A_j \lambda^j$ *and* $L_1(\lambda)$ *be matrix polynomials of degrees* l *and* k, *respectively. Suppose that* $L_1(\lambda)$ *is monic with standard pair* (X_1, T_1), *and let*

$$(\mathrm{col}(X_1 T_1^j)_{j=0}^{k-1})^{-1} = [V_1 \cdots V_k], \tag{24}$$

where V_j *are* $pk \times p$ *matrices. Then*

$$L(\lambda) = Q(\lambda) L_1(\lambda) + R(\lambda),$$

where

$$Q(\lambda) = \sum_{j=0}^{l-k} \lambda^j \cdot \left(\sum_{i=j+1}^{l} A_i X_1 T_1^{i-j-1} \right) V_k \tag{25}$$

and

$$R(\lambda) = \sum_{j=1}^{k} \lambda^{j-1} \cdot \left(\sum_{i=0}^{l} A_i X_1 T_1^i \right) V_j \tag{26}$$

Proof The method of proof will be to find a matrix polynomial $R(\lambda)$ of degree less than k such that $Q(\lambda) := (L(\lambda) - R(\lambda))L_1^{-1}(\lambda)$ has no poles in the finite complex plane, i.e. for which

$$\frac{1}{2\pi i}\int_\Gamma \lambda^j Q(\lambda)\, d\lambda = 0, \qquad j = 0, 1, \ldots$$

or, in other words,

$$\frac{1}{2\pi i}\int_\Gamma \lambda^j L(\lambda) L_1^{-1}(\lambda)\, d\lambda = \frac{1}{2\pi i}\int_\Gamma \lambda^j R(\lambda) L_1^{-1}(\lambda)\, d\lambda \tag{27}$$

where Γ is a circle containing $\mathrm{sp}(L_1)$ in its interior.

Let (X_1, T_1, Y_1) be a standard triple of $L_1(\lambda)$. Then

$$\begin{aligned}\frac{1}{2\pi i}\int_\Gamma \lambda^j L(\lambda) L_1^{-1}(\lambda)\, d\lambda &= \frac{1}{2\pi i}\sum_{i=0}^{l} A_i \int_\Gamma \lambda^{i+j} L_1^{-1}(\lambda)\, d\lambda \\ &= \sum_{i=0}^{l} A_i X_1 T_1^i \cdot T_1^j Y_1,\end{aligned}$$

for $j = 0, 1, \ldots$, and using the fact that $T_1^j = 1/2\pi i \int_\Gamma \lambda^j (I\lambda - T_1)^{-1}\, d\lambda$, we see that (27) holds, where

$$R(\lambda) = \sum_{i=0}^{l} A_i X_1 T_1^i \cdot (I\lambda - T_1)^{-1} Y_1 L_1(\lambda) \tag{28}$$

However, it remains to show that $R(\lambda)$ defined by (28) is a matrix *polynomial* of degree less than k; we shall do this by showing that $R(\lambda)$ is given by formula (26). Consider the following:

$$\begin{aligned}\mathrm{col}(X_1 T_1^i)_{i=0}^{k-1} \cdot (I\lambda - T_1)^{-1} Y_1 &= \mathrm{col}(X_1 T_1^i \cdot (I\lambda - T_1)^{-1} Y_1)_{i=0}^{k-1} \\ &= \mathrm{col}(\lambda^i X_1 (I\lambda - T_1)^{-1} Y_1)_{i=0}^{k-1} \\ &= \mathrm{col}(\lambda^i L_1^{-1}(\lambda))_{i=0}^{k-1}\end{aligned} \tag{29}$$

For the second equality in (29), we used the formulas

$$X_1 T_1^i \cdot (I\lambda - T_1)^{-1} Y_1 = \lambda^i X_1 (I\lambda - T_1)^{-1} Y_1$$

for $i = 0, 1, \ldots, k-1$ (proved by expanding $(I\lambda - T_1)^{-1}$ in a power series valid for $|\lambda| > \|T_1\|$ and using the fact that $X_1 T_1^i Y_1 = 0$ for $i = 0, 1, \ldots, k-2$ since $L_1(\lambda)$ is monic). In view of (24), we see that (29) takes the form

$$(I\lambda - T_1)^{-1} Y_1 = \sum_{j=1}^{k} \lambda^{j-1} V_j L_1^{-1}(\lambda),$$

and substituting this formula in (28) we obtain (26).

Since $Q(\lambda)$ has no poles in the finite complex plane, we may apply Lemma

3.10 (with $L(\lambda)$ replaced by $L(\lambda) - R(\lambda)$) to obtain

$$Q(\lambda) = \sum_{j=0}^{l-k} \lambda^j \cdot \left(\sum_{i=j+1}^{l} A_i X_1 T_1^{i-j-1} Y_1 \right) \tag{30}$$

Note that the formula (30) for $Q(\lambda)$ is not affected by the values of A_i for $i = 0, \ldots, k-1$ since $X_1 T_1^j Y_1 = 0$ for $j = 0, \ldots, k-2$. Finally, recall from (c) of Theorem 3.2 that

$$Y_1 = (\mathrm{col}(X_1 T_1^i)_{i=0}^{l-1})^{-1} \cdot \begin{pmatrix} 0 \\ \vdots \\ 0 \\ I \end{pmatrix} = V_k,$$

so that (25) follows from (30).

Corollary *Let $L_1(\lambda)$ and $L(\lambda)$ be as in Theorem* 3.11. *Then $L_1(\lambda)$ is a right divisor of $L(\lambda)$ if and only if the equality*

$$A_0 X_1 + A_1 X_1 T_1 + \cdots + A_l X_1 T_1^l = 0 \tag{31}$$

holds.

Proof Let $L(\lambda) = Q(\lambda) L_1(\lambda) + R(\lambda)$, where $R(\lambda)$ is given by (26). If (31) holds then $R(\lambda) \equiv 0$ and $L_1(\lambda)$ is a right divisor of $L(\lambda)$. Conversely, if $R(\lambda) \equiv 0$ then by (26)

$$\left(\sum_{i=0}^{l} A_i X_1 T_1^i \right) V_j \equiv 0 \qquad \text{for } j = 1, \ldots, k$$

Since $[V_1 \cdots V_k]$ is invertible, this means that (31) holds.

Remark Since any solution of $L_1(d/dt)u(t) = 0$ can be written in the form $u(t) = X_1 e^{tT_1} c$, for some c in the base space of (X_1, T_1), another way to view (31) is the following:

> $L_1(\lambda)$ is a right divisor of $L(\lambda)$ if and only if every solution of $L_1(d/dt)u(t) = 0$ is also a solution of $L(d/dt)u(t) = 0$,

or, in other words, *every Jordan chain of $L_1(\lambda)$ is a Jordan chain of $L(\lambda)$.*

There is a dual version of Theorem 3.11 for left division of matrix polynomials. This can be proved by an analogous argument, or inferred by duality since the left division $L(\lambda) = L_1(\lambda) Q(\lambda) + R(\lambda)$ is equivalent under transposition to the right division $L^T(\lambda) = Q^T(\lambda) L_1^T(\lambda) + R^T(\lambda)$. Since (Y_1^T, T_1^T) is a standard pair of $L_1^T(\lambda)$ whenever (T_1, Y_1) is a left standard pair of $L_1(\lambda)$, we obtain the following theorem.

Theorem 3.12 *Let $L(\lambda) = \sum_{j=0}^{l} A_j \lambda^j$ and $L_1(\lambda)$ be matrix polynomials of degrees l and k, respectively. Suppose that $L_1(\lambda)$ is monic with left standard* *pair* (T_1, Y_1) *and let*

$$(\operatorname{row}(T_1^j Y_1)_{j=0}^{k-1})^{-1} = \begin{pmatrix} W_1 \\ \vdots \\ W_k \end{pmatrix}$$

where W_j *are* $p \times pk$ *matrices. Then*

$$L(\lambda) = L_1(\lambda)Q(\lambda) + R(\lambda)$$

where

$$Q(\lambda) = \sum_{j=0}^{l-k} \lambda^j \left(\sum_{i=j+1}^{l} W_k \cdot T_1^{i-j-1} Y_1 A_i \right)$$

and

$$R(\lambda) = \sum_{j=1}^{k} \lambda^{j-1} \left(W_j \cdot \sum_{i=0}^{l} T_1^i Y_1 A_i \right)$$

Corollary *Let* $L_1(\lambda)$ *and* $L(\lambda)$ *be as in Theorem* 3.12. *Then* $L_1(\lambda)$ *is a left divisor of* $L(\lambda)$ *if and only if the equality*

$$Y_1 A_0 + T_1 Y_1 A_1 + \cdots + T_1^l Y_1 A_l = 0$$

holds.

3.5 Monic divisors

Let $L(\lambda)$ be a $p \times p$ matrix polynomial with $\det L(\lambda) \not\equiv 0$ (not necessarily monic). If $L(\lambda) = L_2(\lambda)L_1(\lambda)$ we say that $L_1(\lambda)$ is a *right divisor* of $L(\lambda)$ and $L_2(\lambda)$ is a left quotient (or $L_2(\lambda)$ is a *left divisor* of $L(\lambda)$ and $L_1(\lambda)$ is a right quotient). In this section there is no restriction on the spectra of L_1 and L_2, but in §3.6 we consider in more detail the special case which is most important to us, where the spectra are disjoint.

Let (X, T) be a finite spectral pair of $L(\lambda)$. The following theorem establishes a one-to-one correspondence between *monic* right divisors $L_1(\lambda)$ of degree k and certain pk-dimensional subspaces $\mathfrak{L}$ of the base space of (X, T).

The subspace $\mathfrak{L}$ corresponding to a right divisor $L_1(\lambda)$ of $L(\lambda)$ (by formula (32) below) is called the *supporting subspace of* $L_1(\lambda)$. Of course, the supporting subspace depends on the choice of (X, T), but once (X, T) has been fixed the supporting subspace depends only on the right divisor $L_1(\lambda)$.

Theorem 3.13 *Let* $L(\lambda)$ *be a* $p \times p$ *matrix polynomial, and choose a finite spectral pair* (X, T) *of* $L(\lambda)$. *Then for every* pk*-dimensional* T*-invariant subspace* $\mathfrak{L}$ *such that the restriction* $\operatorname{col}(X|_{\mathfrak{L}}(T|_{\mathfrak{L}})^j)_{j=0}^{k-1}$ *is invertible, there exists a unique monic right divisor* $L_1(\lambda)$ *of* $L(\lambda)$ *of degree* k *with standard pair similar to* $(X|_{\mathfrak{L}}, T|_{\mathfrak{L}})$, *given by*

$$L_1(\lambda) = I\lambda^k - X|_{\mathfrak{L}}(T|_{\mathfrak{L}})^k(V_1 + V_2\lambda + \cdots + V_k\lambda^{k-1}), \tag{32}$$

where $[V_1 \cdots V_k] = (\operatorname{col}(X|_{\mathfrak{L}}(T|_{\mathfrak{L}})^j)_{j=0}^{k-1})^{-1}$.

Conversely, any monic right divisor $L_1(\lambda)$ *of* $L(\lambda)$ *of degree k has standard pair similar to* $(X|_{\mathfrak{L}}, T|_{\mathfrak{L}})$ *for a unique pk-dimensional T-invariant subspace* $\mathfrak{L}$ *(and then* $\mathrm{col}(X|_{\mathfrak{L}}(T|_{\mathfrak{L}})^j)_{j=0}^{k-1}$ *is invertible and the formula* (32) *holds). This subspace is given by*

$$\mathfrak{L} = \mathrm{im}(\mathrm{col}(XT^j)_{j=0}^{l-1})^{-1}\,\mathrm{col}(X_1T_1^j)_{j=0}^{l-1}, \tag{33}$$

where (X_1, T_1) *is any standard pair of* $L_1(\lambda)$.

Proof Suppose that $\mathfrak{L}$ is a T-invariant subspace such that $\mathrm{col}(X|_{\mathfrak{L}}(T|_{\mathfrak{L}})^j)_{j=0}^{k-1}$ is invertible for some k. In view of Theorem 3.8 and its corollary, there is a unique monic matrix polynomial $L_1(\lambda)$ having a standard pair which is similar to $(X|_{\mathfrak{L}}, T|_{\mathfrak{L}})$, and $L_1(\lambda)$ is given by the formula (32). Also, since $\sum_{j=0}^{l} A_jXT^j = 0$, it follows that

$$\sum_{j=0}^{l} A_jX|_{\mathfrak{L}}(T|_{\mathfrak{L}})^j = 0.$$

Hence, in view of the corollary to Theorem 3.11, $L_1(\lambda)$ is a right divisor of $L(\lambda)$.

Conversely, let $L_1(\lambda)$ be a monic right divisor of $L(\lambda)$. Let (X_1, T_1) be a standard pair of $L(\lambda)$, then clearly (X_1, T_1) satisfies the following conditionsy:

(a) $\mathrm{sp}(T_1) = \mathrm{sp}(L_1) \subset \mathrm{sp}(L)$,
(b) $\sum_{j=0}^{l} A_jX_1T_1^j = 0$,
(c) $\mathrm{col}(X_1T_1^j)_{j=0}^{l-1}$ is injective.

Consequently, (X_1, T_1) is partial finite spectral pair of $L(\lambda)$, so that, by Proposition 2.12, (X_1, T_1) is a restriction of (X, T), i.e. (X_1, T_1) is similar to $(X|_{\mathfrak{L}}, T|_{\mathfrak{L}})$ for some T-invariant subspace $\mathfrak{L}$. The fact that $\dim \mathfrak{L} = pk$ follows since $\mathrm{col}(X_1T_1^j)_{j=0}^{k-1}$ is invertible. $\mathfrak{L}$ is unique due to the injectivity of $\mathrm{col}(XT^j)_{j=0}^{l-1}$. Also, due to Theorem 3.8, the formula (32) holds, and the formula (33) is easily verified since $(X|_{\mathfrak{L}}, T|_{\mathfrak{L}})$ is similar to (X_1, T_1). This completes the proof of the theorem.

3.6 Monic spectral divisors

Let $L(\lambda)$ be any matrix polynomial (not necessarily monic) and let γ be a simple, closed contour not intersecting $\mathrm{sp}(L)$. Then a matrix polynomial $L_1(\lambda)$ is said to be a *γ-spectral right divisor* of $L(\lambda)$ if $L(\lambda) = L_2(\lambda)L_1(\lambda)$ and $\mathrm{sp}(L_1)$, $\mathrm{sp}(L_2)$ are inside and outside γ, respectively. The next theorem gives necessary and sufficient conditions for the existence of *monic* γ-spectral right divisors of $L(\lambda)$. Note that one necessary condition is evident: $\det L(\lambda)$ must have exactly pk zeros inside γ, counting multiplicities.

Theorem 3.14 *Let* $L(\lambda)$ *be* $p \times p$ *matrix polynomial of degree l and let* γ *be a simple, closed contour not intersecting* $\mathrm{sp}(L)$. *Suppose that there are exactly pk zeros of* $\det L(\lambda)$ *inside* γ*, counting multiplicities. Then* $L(\lambda)$ *has a monic* γ*-spectral right divisor of degree k if and only if the* $pk \times pl$ *matrix*

$$M_{kl} = \frac{1}{2\pi i}\int_\gamma \begin{pmatrix} I \\ \vdots \\ \lambda^{k-1}I \end{pmatrix} L^{-1}(\lambda)[I \cdots \lambda^{l-1}I]\, d\lambda$$

has rank pk. If this condition is satisfied, then the monic γ-spectral divisor $L_1(\lambda)$ is unique and is given by the formula $L_1(\lambda) = I\lambda^k + \sum_{j=0}^{k-1} B_j\lambda^j$, where

$$[B_0 \cdots B_{k-1}] = -\frac{1}{2\pi i}\int_\gamma \lambda^k L^{-1}(\lambda)[I \cdots \lambda^{l-1}I]\, d\lambda \cdot M_{kl}^I, \tag{34}$$

and M_{kl}^I is any right inverse of M_{kl}.

Proof Let (X_+, T_+, Y_+) be a γ-spectral triple for $L(\lambda)$ and let $\mathfrak{M}^+$ denote its base space. Since there are pk zeros of $L(\lambda)$ inside γ, it follows that $\dim \mathfrak{M}^+ = pk$. By choosing a basis in $\mathfrak{M}^+$, we may assume that the triple consists of matrices, that is, X_+ is a $p \times pk$ matrix, T_+ is $pk \times pk$ and Y_+ is $pk \times p$. In view of Proposition 2.2, we have

$$M_{kl} = \mathrm{col}(X_+T_+^i)_{i=0}^{k-1} \cdot \mathrm{row}(T_+^j Y_+)_{j=0}^{l-1} \tag{35}$$

and, since $\mathrm{row}(T_+^j Y_+)_{j=0}^{l-1}$ is surjective, M_{kl} has rank pk if and only if $\mathrm{col}(X_+T_+^i)_{i=0}^{k-1}$ is an invertible matrix. Thus Theorem 3.13 implies that $L(\lambda)$ has a monic right divisor $L_1(\lambda)$ of degree k with standard pair (X_+, T_+) if and only if M_{kl} has rank pk. The right divisor $L_1(\lambda)$, if it exists, is necessarily a γ-*spectral* right divisor. In fact, by Proposition 2.7 we have $\mathrm{sp}(L_1) = \mathrm{sp}(T_+)$, so that all pk eigenvalues of $L_1(\lambda)$ lie inside γ. In view of the equation $\det L(\lambda) = \det L_2(\lambda) \det L_1(\lambda)$, we see that $\det L_2(\lambda)$ cannot have any zeros inside γ (otherwise $\det L(\lambda)$ would have more than pk zeros inside γ, contrary to assumption). Hence the eigenvalues of $L_2(\lambda)$ lie outside γ, which means that $L_1(\lambda)$ is a γ-spectral right divisor of $L(\lambda)$.

Now, if $\mathrm{col}(X_+T_+^i)_{i=0}^{k-1}$ is invertible, let $L_1(\lambda)$ be the monic matrix polynomial of degree k with standard pair (X_+, T_+), i.e. $L_1(\lambda) = I\lambda^k + \sum_{j=0}^{k-1} B_j\lambda^j$ where the coefficients B_j are given by (see (32))

$$\begin{aligned}[B_0 \cdots B_{k-1}] &= -V_+T_+^k[V_1 \cdots V_k] \\ &= -X_+T_+^k(\mathrm{col}(X_+T_+^j)_{j=0}^{k-1})^{-1}.\end{aligned} \tag{36}$$

This proves uniqueness of $L_1(\lambda)$. Further, from (35) and the equation $M_{kl}M_{kl}^I = I$ it follows that

$$(\mathrm{col}(X_+T_+^j)_{j=0}^{k-1})^{-1} = \mathrm{row}(T_+^j Y_+)_{j=0}^{l-1} \cdot M_{kl}^I$$

Substitution of this formula in (36) and use of Proposition 2.2 proves (34).

A matrix polynomial $L_1(\lambda)$ is said to be a γ-spectral *left* divisor of $L(\lambda)$ if $L(\lambda) = L_1(\lambda)L_2(\lambda)$ and $\mathrm{sp}(L_1)$, $\mathrm{sp}(L_2)$ are inside and outside γ, respectively.

Theorem 3.15 *Let the hypotheses be as in Theorem 3.14. Then $L(\lambda)$ has a monic γ-spectral left divisor of degree k if and only if the $pl \times pk$ matrix*

$$M_{lk} = \frac{1}{2\pi i}\int_\gamma \begin{pmatrix} I \\ \vdots \\ \lambda^{l-1}I \end{pmatrix} L^{-1}(\lambda)[I \cdots \lambda^{k-1}I]\, d\lambda$$

has rank pk. If this condition is satisfied, then the monic γ-spectral divisor $L_1(\lambda)$ is unique and is given by the formula $L_1(\lambda) = I\lambda^k + \sum_{j=0}^{k-1} B_j\lambda^j$, where

$$\begin{pmatrix} B_0 \\ \vdots \\ B_{k-1} \end{pmatrix} = -M_{lk}^I \cdot \frac{1}{2\pi i}\int_\gamma \begin{pmatrix} I \\ \vdots \\ \lambda^{l-1}I \end{pmatrix} \lambda^k L^{-1}(\lambda)\, d\lambda$$

and M_{lk}^I is any left inverse of M_{lk}.

Proof The proof is analogous to that of Theorem 3.14. Or, simply observe that $L_1(\lambda)$ is a γ-spectral left divisor of $L(\lambda)$ if and only if $L_1^T(\lambda)$ is a γ-spectral right divisor of $L^T(\lambda)$.

Theorem 3.16 *Let the hypotheses be as in Theorem 3.14. Then $L(\lambda)$ has both a monic γ-spectral left divisor and a monic γ-spectral right divisor of degree k if and only if the $pk \times pk$ matrix*

$$M_{kk} = \int_\gamma \begin{pmatrix} I \\ \vdots \\ \lambda^{k-1}I \end{pmatrix} L^{-1}(\lambda)[I \cdots \lambda^{k-1}I]\, d\lambda$$

is invertible. If this condition is satisfied, then the monic γ-spectral right (resp. left) divisor $L_1(\lambda) = I\lambda^k + \sum_{j=0}^{k-1} B_j\lambda^j$ is given by the formula

$$[B_0 \cdots B_{k-1}] = -\frac{1}{2\pi i}\int_\gamma \lambda^k L^{-1}(\lambda)[I \cdots \lambda^{k-1}I]\, d\lambda \cdot M_{kk}^{-1}$$

$$\text{resp.} \begin{pmatrix} B_0 \\ \vdots \\ B_{k-1} \end{pmatrix} = -M_{kk}^{-1} \cdot \frac{1}{2\pi i}\int_\gamma \begin{pmatrix} I \\ \vdots \\ \lambda^{k-1}I \end{pmatrix} \lambda^k L^{-1}(\lambda)\, d\lambda$$

Proof Let (X_+, T_+, Y_+) be a γ-spectral triple for $L(\lambda)$. Then

$$M_{kk} = \begin{pmatrix} X_+ \\ \vdots \\ X_+T_+^{k-1} \end{pmatrix} \cdot [Y_+ \cdots T_+^{k-1}Y_+] \tag{37}$$

Since the base space of (X_+, T_+, Y_+) has dimension pk, the invertibility of

M_{kk} is equivalent to the invertibility of each factor in (37). But, from the proof of Theorem 3.14, we see that the invertibility of $\mathrm{col}(X_+T_+^j)_{j=0}^{k-1}$ is equivalent to the existence of a γ-spectral right divisor; similarly, the invertibility of $\mathrm{row}(T_+^j Y_+)_{j=0}^{k-1}$ is equivalent to the existence of a γ-spectral left divisor. The formulas for the γ-spectral divisors are verified in the same way as in the proof of Theorem 3.14.

3.7 Second degree matrix polynomials and examples

Let $L(\lambda) = I\lambda^2 + A_1\lambda + A_0$ be a monic $p \times p$ matrix polynomial of degree two. Consider the right division of $L(\lambda)$ by a matrix polynomial $I\lambda - S$ of degree one:

$$L(\lambda) = Q(\lambda)(I\lambda - S) + R,$$

where the remainder R is a constant matrix. A standard pair for $I\lambda - S$ is given by $(X_1, T_1) = (I, S)$, so that Theorem 3.11 implies the following formulas for the quotient and remainder:

$$Q(\lambda) = I\lambda + (A_1 + S),$$

$$R = S^2 + A_1 S + A_0,$$

which are easy to verify directly. Hence $I\lambda - S$ is a right divisor of $L(\lambda)$ if and only if $S^2 + A_1 S + A_0 = 0$.

Let γ be a contour not intersecting $\mathrm{sp}(L)$ and suppose there are exactly p zeros of $\det L(\lambda)$ inside γ, counting multiplicities. Let (X_+, T_+) be a γ-spectral pair for $L(\lambda)$, where X_+ and T_+ are $p \times p$ matrices. From the proof of Theorem 3.14, we see that the existence of a γ-spectral right divisor is equivalent to the invertibility of X_+. Indeed, if X_+ is invertible then from the equation $X_+T_+^2 + A_1X_+T_+ + A_0X_+ = 0$ we obtain

$$S^2 + A_1 S + A_0 = 0,$$

where $S = X_+T_+X_+^{-1}$; hence $I\lambda - S$ is a γ-spectral right divisor of $L(\lambda)$. Conversely, if $I\lambda - S$ is a γ-spectral right divisor of $L(\lambda)$ then $(X_+, T_+) = (I, S)$ is a γ-spectral pair for $L(\lambda)$, and $X_+ = I$ is invertible.

Moreover, if Y_+ is a $p \times p$ matrix such that (X_+, T_+, Y_+) is a γ-spectral triple of $L(\lambda)$ then

$$M_{12} = \frac{1}{2\pi i}\int_\gamma L^{-1}(\lambda)[I \quad \lambda I]\, d\lambda = X_+ \cdot [Y_+ \quad T_+Y_+]$$

If M_{12} has rank p (i.e. X_+ is invertible), then we have

$$X_+^{-1} = [Y_+ \quad T_+Y_+] \cdot M_{12}^I,$$

where M_{12}^I is a right inverse of M_{12}. Thus $L(\lambda)$ has γ-spectral right divisor $I\lambda - S$, where

$$\begin{aligned} S &= X_+ T_+ X_+^{-1} \\ &= X_+ T_+ [Y_+ \quad T_+ Y_+] M_{12}^I \\ &= \frac{1}{2\pi i} \int_\gamma L^{-1}(\lambda)[I \quad \lambda I]\, d\lambda \cdot M_{12}^I \end{aligned}$$

This is, of course, the formula as proved in the corollary to Theorem 3.14; again we see that the γ-spectral right divisor $I\lambda - S$ does not depend on the choice of right inverse M_{12}^I (since $S = X_+ T_+ X_+^{-1}$).

Similar results are true for left divisors since $I\lambda - \tilde{S}$ is a γ-spectral left divisor of $L(\lambda)$ if and only if $I\lambda - \tilde{S}^T$ is a γ-spectral right divisor of $L^T(\lambda)$. The details are omitted.

Instead, we will demonstrate a fact which is implicit in Theorem 3.16; namely, that if $L(\lambda)$ has a γ-spectral left divisor $I\lambda - \tilde{S}$, i.e.

$$L(\lambda) = (I\lambda - \tilde{S})(I\lambda - \tilde{R}),$$

where $\mathrm{sp}(\tilde{S})$ is inside γ and $\mathrm{sp}(\tilde{R})$ is outside γ, then the rank of M_{12} is equal to the rank of the $p \times p$ matrix $M := 1/(2\pi i) \int_\gamma L^{-1}(\lambda)\, d\lambda$. Indeed, note that

$$\begin{aligned} \int_\gamma L^{-1}(\lambda)\lambda I\, d\lambda - \int_\gamma L^{-1}(\lambda)\tilde{S}\, d\lambda &= \int_\gamma L^{-1}(\lambda)(\lambda I - \tilde{S})\, d\lambda \\ &= \int_\gamma (I\lambda - \tilde{R})^{-1}\, d\lambda \\ &= 0, \end{aligned} \tag{38}$$

the last equality holding since $\mathrm{sp}(\tilde{R})$ is outside γ. Thus we obtain

$$M_{12} = \frac{1}{2\pi i}\left[\int_\gamma L^{-1}(\lambda)\, d\lambda \quad \int_\gamma L^{-1}(\lambda)\tilde{S}\, d\lambda\right].$$

Recall now the basic fact that if M is a $p \times p$ matrix and x is a $p \times 1$ vector then Mx is a linear combination of the columns of M. Let $\tilde{S}_j$ $(j = 1, \ldots, p)$ denote the columns of $\tilde{S}$. Then the columns of $M\tilde{S}$ are $M\tilde{S}_j$; therefore the columns of $M\tilde{S}$ are linear combinations of the columns of M. Hence $\operatorname{rank} M_{12} = \operatorname{rank}[M \; M\tilde{S}] = \operatorname{rank} M$.

This equality of ranks explains why it is true that if $L(\lambda)$ has both a γ-spectral right divisor $I\lambda - S$ (i.e. M_{12} has rank p) and a γ-spectral left divisor $I\lambda - \tilde{S}$ (i.e. the rank of M_{12} is equal to the rank of M) then the matrix $M = 1/(2\pi i) \int_\gamma L^{-1}(\lambda)\, d\lambda$ is invertible, as stated in Theorem 3.16.

By a similar calculation as done in (38), it is easy to see that if $I\lambda - S$ is a γ-spectral right divisor of $L(\lambda)$ then

$$\int_\gamma L^{-1}(\lambda)\lambda I\, d\lambda = S \cdot \int_\gamma L^{-1}(\lambda)\, d\lambda,$$

which leads to the following formula

$$S = \int_\gamma \lambda L^{-1}(\lambda)\, d\lambda \cdot \left(\int_\gamma L^{-1}(\lambda)\, d\lambda \right)^{-1}, \tag{39}$$

again as stated in Theorem 3.16.

Example 3.17 Let $\varphi = \varphi_1 + i\varphi_2$ be a complex number such that $|\varphi| \neq 1$ and let $L(\lambda) = A\lambda^2 + B\lambda + C$, where

$$A = \begin{pmatrix} \varphi_1 & \varphi_2 + 1 \\ \varphi_2 - 1 & -\varphi_1 \end{pmatrix}, \quad B = -2I, \ C = A^T$$

Then $\det A = 1 - |\varphi|^2 \neq 0$, so that $L(\lambda)$ has invertible leading coefficient and $A^{-1}L(\lambda)$ is a monic matrix polynomial of degree 2. Also $\det L(\lambda) = (1 + \lambda^2)^2(1 - |\varphi|^2)$; thus $L(\lambda)$ has eigenvalues i and $-i$, each with multiplicity 2. Let γ^+ be a circle in the upper half-plane, $\operatorname{Im} \lambda > 0$, containing i. It is easy to verify that $\begin{pmatrix} 1 \\ -i \end{pmatrix}, \begin{pmatrix} -i\varphi_2 \\ i\varphi_1 \end{pmatrix}$ is a Jordan chain of $L(\lambda)$ corresponding to the eigenvalue i. Hence we have the following γ^+-spectral Jordan pair for $L(\lambda)$:

$$X_+ = \begin{pmatrix} 1 & -i\varphi_2 \\ -i & i\varphi_1 \end{pmatrix}, \quad J_+ = \begin{pmatrix} i & 1 \\ 0 & i \end{pmatrix}$$

This pair may be extended to a γ^+-spectral Jordan triple (X_+, J_+, Y_+) by finding the unique 2×2 matrix Y_+ such that

$$X_+ J_+^j Y_+ = \frac{1}{2\pi i} \int_{\gamma^+} \lambda^j L^{-1}(\lambda)\, d\lambda \qquad j = 0, 1 \tag{40}$$

To simplify the calculation of Y_+, assume for a moment that $\varphi \neq 0$. Then $\det X_+ = \bar{\varphi} \neq 0$, and

$$X_+^{-1} = \frac{1}{\bar{\varphi}} \begin{pmatrix} \varphi_1 & \varphi_2 \\ 1 & -i \end{pmatrix}.$$

By computing $L^{-1}(\lambda)$, then evaluating residues, one obtains

$$\frac{1}{2\pi i} \int_{\gamma^+} L^{-1}(\lambda)\, d\lambda = \frac{1}{1 - |\varphi|^2} \cdot \frac{i}{2} \cdot \begin{pmatrix} \varphi_1 & \varphi_2 \\ \varphi_2 & -\varphi_1 \end{pmatrix},$$

so that

$$\begin{aligned} Y_+ &= X_+^{-1} \cdot \frac{1}{2\pi i} \int_{\gamma^+} L^{-1}(\lambda)\, d\lambda \\ &= \frac{i}{2\bar{\varphi}} \cdot \frac{1}{1 - |\varphi|^2} \cdot \begin{pmatrix} |\varphi|^2 & 0 \\ \bar{\varphi} & i\bar{\varphi} \end{pmatrix} \\ &= \frac{i}{2(1 - |\varphi|^2)} \begin{pmatrix} \varphi & 0 \\ 1 & i \end{pmatrix}. \end{aligned}$$

Then (40) holds when $\varphi \neq 0$, and, by letting $\varphi \to 0$, we see that (40) also holds if $\varphi = 0$. Thus (X_+, J_+, Y_+) is a γ^+-spectral triple for $L(\lambda)$. Since $\det X_+ = \bar{\varphi}$ and $\det Y_+ = -\varphi/2(1 - |\varphi|^2)$, we see that X_+ and Y_+ are invertible if and only if $\varphi \neq 0$. Hence

(*i*) If $\varphi = 0$, $L(\lambda)$ has neither monic γ-spectral right divisor nor left divisor.
(*ii*) If $\varphi \neq 0$, $L(\lambda)$ has a γ-spectral right divisor $I\lambda - S$, where

$$S = X_+ J_+ X_+^{-1} = \frac{i}{\bar{\varphi}}\begin{pmatrix} \bar{\varphi} - i & -1 \\ -1 & \bar{\varphi} + i \end{pmatrix}$$

and a γ-spectral left divisor $I\lambda - \tilde{S}$, where

$$\tilde{S} = Y_+^{-1} J_+ Y_+ = \frac{i}{\varphi}\begin{pmatrix} \varphi - i & 1 \\ 1 & \varphi + i \end{pmatrix}$$

Example 3.18 Let a, b, c and d distinct complex numbers and let

$$L(\lambda) = \begin{pmatrix} (\lambda - a)(\lambda - b) & \lambda - a \\ 0 & (\lambda - c)(\lambda - d) \end{pmatrix}$$

Let γ be a contour with a and b inside γ, and c and d outside γ. Then $\det L(\lambda)$ has $p = 2$ zeros inside γ. Since

$$M_{12} = \frac{1}{2\pi i}\int_\gamma L^{-1}(\lambda)[I \quad \lambda I]\, d\lambda = \begin{pmatrix} 0 & \delta & 1 & b\delta \\ 0 & 0 & 0 & 0 \end{pmatrix}$$

and

$$M_{21} = \frac{1}{2\pi i}\int_\gamma \begin{pmatrix} I \\ \lambda I \end{pmatrix} L^{-1}(\lambda)\, d\lambda = \begin{pmatrix} 0 & \delta \\ 0 & 0 \\ 1 & b\delta \\ 0 & 0 \end{pmatrix},$$

where $\delta = -(b - c)^{-1}(b - d)^{-1}$, we see that M_{12} has rank 1 and M_{21} has rank 2. Hence there exists a monic γ-spectral left divisor, but no such right divisor. A γ-spectral triple for $L(\lambda)$ is given by

$$X_+ = \begin{pmatrix} 1 & 1 \\ 0 & 0 \end{pmatrix}, \quad J_+ = \begin{pmatrix} a & \\ & b \end{pmatrix}, \quad Y_+ = \frac{1}{a - b}\begin{pmatrix} 1 & 0 \\ -1 & (a - b)\delta \end{pmatrix},$$

and observe that X_+ is not invertible, while Y_+ is invertible.

3.8 Changing from complex to real matrix coefficients

A polynomial with coefficients that are matrices with real entries shall be referred to as a *real* matrix polynomial. In other words, $L(\lambda) = \sum A_j \lambda^j$ is a real matrix polynomial if $\bar{A}_j = A_j$. Real matrix polynomials are also characterized by the following condition: $\overline{L(\lambda)} = L(\bar{\lambda})$ for all $\lambda \in \mathbb{C}$.

Throughout this section we assume that $\det L(\lambda) \neq 0$ for real λ. Also we let $\gamma^{\pm}$ be simple, closed contours containing the eigenvalues of $L(\lambda)$ in the upper and lower half-planes, respectively. Note that if $L(\lambda)$ is a real matrix polynomial and if (X_+, T_+, Y_+) is a γ^+-spectral triple of $L(\lambda)$, then $(\bar{X}_+, \bar{T}_+, \bar{Y}_+)$ is a $\bar{\gamma}$-spectral triple of $L(\lambda)$.

In general, if $L(\lambda) = \sum A_j \lambda^j$ is a complex $p \times p$ matrix polynomial then $\overline{L(\lambda)} = \bar{L}(\bar{\lambda})$ where $\bar{L}(\lambda) = \sum \bar{A}_j \lambda^j$ and $\bar{A}_j$ denotes the complex conjugate of A_j. Let $L_1(\lambda)$ and $L_2(\lambda)$ be the real and imaginary parts of $L(\lambda)$, i.e. the unique real $p \times p$ matrices such that $L(\lambda) = L_1(\lambda) + iL_2(\lambda)$. We then associate with $L(\lambda)$ the following real $2p \times 2p$ matrix polynomial:

$$L_{\mathbb{R}}(\lambda) = \begin{pmatrix} L_1(\lambda) & -L_2(\lambda) \\ L_2(\lambda) & L_1(\lambda) \end{pmatrix}$$

There is a close connection between the spectral data of $L(\lambda)$ and that of $L_{\mathbb{R}}(\lambda)$ (see Theorem 3.20). This can be anticipated by examining the differential equation $L(d/dt)u(t) = 0$. If we write $u = u_1 + iu_2$, then

$$L_{\mathbb{R}}(d/dt)u_{\mathbb{R}}(t) = 0, \qquad \text{where } u_{\mathbb{R}} = \begin{pmatrix} u_1 \\ u_2 \end{pmatrix}.$$

For example, if $L(\lambda_0)x_0 = 0$ then the function $u(t) = e^{\lambda_0 t}x_0$ is a solution of $L(d/dt)u = 0$ and if we write $\lambda_0 = a + ib$ and $x_0 = x_1 + ix_2$ then the corresponding real function

$$\begin{aligned} u_{\mathbb{R}}(t) &= e^{at}\begin{pmatrix} x_1 \cos bt - x_2 \sin bt \\ x_1 \sin bt + x_2 \cos bt \end{pmatrix} \\ &= \frac{e^{\lambda_0 t} + e^{\bar{\lambda}_0 t}}{2}\begin{pmatrix} x_1 \\ x_2 \end{pmatrix} + \frac{e^{\lambda_0 t} - e^{\bar{\lambda}_0 t}}{2i}\begin{pmatrix} -x_1 \\ x_2 \end{pmatrix} \\ &= \tfrac{1}{2} e^{\lambda_0 t}\begin{pmatrix} I \\ -iI \end{pmatrix} x_0 + \tfrac{1}{2} e^{\bar{\lambda}_0 t}\begin{pmatrix} I \\ iI \end{pmatrix} \bar{x}_0 \end{aligned} \tag{41}$$

satisfies $L_{\mathbb{R}}(d/dt)u_{\mathbb{R}}(t) = 0$.

This implies $L_{\mathbb{R}}(\lambda_0)x_{0,\mathbb{R}} = 0$ and $L_{\mathbb{R}}(\bar{\lambda}_0)\bar{x}_{0,\mathbb{R}} = 0$, where $x_{0,\mathbb{R}} = \begin{pmatrix} I \\ -iI \end{pmatrix} x_0$, which is easy to verify directly since

$$L_{\mathbb{R}}(\lambda_0)x_{0,\mathbb{R}} = \begin{pmatrix} L(\lambda_0)x_0 \\ -iL(\lambda_0)x_0 \end{pmatrix} = \begin{pmatrix} 0 \\ 0 \end{pmatrix}.$$

Thus, to each eigenvalue λ_0 of $L(\lambda)$ with eigenvector x_0, there corresponds two eigenvalues, λ_0 and $\bar{\lambda}_0$, of $L_{\mathbb{R}}(\lambda)$ with eigenvectors $\begin{pmatrix} I \\ -iI \end{pmatrix} x_0$ and $\begin{pmatrix} I \\ iI \end{pmatrix} \bar{x}_0$, respectively.

Lemma 3.19 *Let A and B be $p \times p$ real matrices. Then*

$$\begin{pmatrix} A & -B \\ B & A \end{pmatrix}\begin{pmatrix} I & I \\ -iI & iI \end{pmatrix} = \begin{pmatrix} I & I \\ -iI & iI \end{pmatrix}\begin{pmatrix} A+iB & \\ & A-iB \end{pmatrix}$$

and

$$\det\begin{pmatrix} A & -B \\ B & A \end{pmatrix} = |\det(A+iB)|^2$$

Proof The first formula is proved by multiplying out both sides and comparing entries. The second formula follows immediately from the first.

By virtue of the lemma we have the following representation:

$$L_{\mathbb{R}}(\lambda) = \begin{pmatrix} I & I \\ -iI & iI \end{pmatrix}\begin{pmatrix} L(\lambda) & \\ & \bar{L}(\lambda) \end{pmatrix}\begin{pmatrix} I & I \\ -iI & iI \end{pmatrix}^{-1}, \tag{42}$$

Since $\det L(\lambda) \neq 0$ for real λ then $\det L_{\mathbb{R}}(\lambda) \neq 0$ for real λ. If r (resp. $\alpha - r$) denotes the number of eigenvalues of $L(\lambda)$ in the upper (resp. lower) half-plane, counting multiplicities, then $L_{\mathbb{R}}(\lambda)$ has $r + (\alpha - r) = \alpha$ eigenvalues in the upper half-plane and also α eigenvalues in the lower half-plane.

Theorem 3.20 *Let $L(\lambda)$ be a $p \times p$ matrix polynomial such that* $\det L(\lambda) \neq 0$ *for real λ. Let $\gamma^{\pm}$ be a simple, closed contour containing the eigenvalues of $L(\lambda)$ in the upper and lower half-planes, respectively. Let $(X_{\pm}, T_{\pm}, Y_{\pm})$ be a $\gamma^{\pm}$-spectral triple of $L(\lambda)$ consisting of matrices, i.e. X_+ is a $p \times r$ matrix, T_+ is $r \times r$, Y_+ is $r \times p$, X_- is $p \times (\alpha - r)$, T_- is $(\alpha - r) \times (\alpha - r)$ and Y_- is $(\alpha - r) \times p$. Then*

$$X_{\mathbb{R}}^{+} = \begin{pmatrix} I & I \\ -iI & iI \end{pmatrix}\begin{pmatrix} X_+ & \\ & \bar{X}_- \end{pmatrix}, \quad 2p \times \alpha \text{ matrix},$$

$$T_{\mathbb{R}}^{+} = \begin{pmatrix} T_+ & \\ & \bar{T}_- \end{pmatrix}, \quad \alpha \times \alpha \text{ matrix},$$

$$Y_{\mathbb{R}}^{+} = \begin{pmatrix} Y_+ & \\ & \bar{Y}_- \end{pmatrix}\begin{pmatrix} I & I \\ -iI & iI \end{pmatrix}^{-1}, \quad \alpha \times 2p \text{ matrix},$$

is a γ^+-spectral triple of $L_{\mathbb{R}}(\lambda)$. (I denotes the $p \times p$ identity matrix.)

Proof We shall verify (*i*), (*ii′*), (*iii*) and (*iv*) of the definition of spectral triples (see Definition 2.1 and Lemma 2.2).

(*i*) The fact that $\mathrm{sp}(T_{\mathbb{R}}^{+})$ lies inside γ^+ is clear from the definition of $T_{\mathbb{R}}^{+}$.

(*ii′*) Since $X_{\pm} T_{\pm}^{j} Y_{\pm} = 1/(2\pi i) \int_{\gamma^{\pm}} \lambda^j L^{-1}(\lambda)\, d\lambda$ for $j = 0, 1, \ldots$ then

$$\begin{aligned} \bar{X}_- \bar{T}_-^{j} \bar{Y}_- &= \overline{\frac{1}{2\pi i}\int_{\gamma^-} \lambda^j L^{-1}(\lambda)\, d\lambda} \\ &= -\frac{1}{2\pi i}\int_{\gamma^-} \bar{\lambda}^j \bar{L}^{-1}(\lambda)\, d\bar{\lambda} \\ &= \frac{1}{2\pi i}\int_{\gamma^+} \lambda^j \bar{L}^{-1}(\lambda)\, d\lambda \end{aligned}$$

Note: If γ^- is traversed in the counterclockwise sense, then $\bar{\gamma}^-$ is traversed clockwise. In view of (42) we have

$$\begin{aligned}\int_{\gamma^+} \lambda^j L_{\mathbb{R}}^{-1}(\lambda)\, d\lambda &= \begin{pmatrix} I & I \\ -iI & iI \end{pmatrix} \int_{\gamma^+} \lambda^j \begin{pmatrix} L^{-1}(\lambda) & \\ & \bar{L}^{-1}(\lambda) \end{pmatrix} d\lambda \begin{pmatrix} I & I \\ -iI & iI \end{pmatrix}^{-1} \\ &= \begin{pmatrix} I & I \\ -iI & iI \end{pmatrix} \begin{pmatrix} X_+ T_+^j Y_+ & \\ & \bar{X}_- \bar{T}_-^j \bar{Y}_- \end{pmatrix} \begin{pmatrix} I & I \\ -iI & iI \end{pmatrix}^{-1} \\ &= X_{\mathbb{R}}^+ (T_{\mathbb{R}}^+)^j Y_{\mathbb{R}}^+ \end{aligned}$$

for $j = 0, 1, \ldots$.

(*iii*) From the definition of $X_{\mathbb{R}}^+$ and $T_{\mathbb{R}}^+$, one obtains the following:

$$\operatorname{col}\{X_{\mathbb{R}}^+ (T_{\mathbb{R}}^+)^j\}_{j=0}^{l-1} = \begin{bmatrix} X_+ & \bar{X}_- \\ -iX_+ & i\bar{X}_- \\ X_+ T_+ & \bar{X}_- \bar{T}_- \\ -iX_+ T_+ & i\bar{X}_- \bar{T}_- \\ \vdots & \vdots \\ X_+ T_+^{l-1} & \bar{X}_- \bar{T}_-^{l-1} \\ -iX_+ T_+^{l-1} & i\bar{X}_- \bar{T}_-^{l-1} \end{bmatrix} \tag{43}$$

By means of block-row operations (i times the first block of p rows, added to the second block of p rows, etc.), the matrix (43) is transformed to

$$\begin{bmatrix} X_+ & \bar{X}_- \\ 0 & 2i\bar{X}_- \\ X_+ T_+ & \bar{X}_- \bar{T}_- \\ 0 & 2i\bar{X}_- \bar{T}_- \\ \vdots & \vdots \\ X_+ T_+^{l-1} & \bar{X}_- \bar{T}_-^{l-1} \\ 0 & 2i\bar{X}_- \bar{T}_-^{l-1} \end{bmatrix} \tag{44}$$

Since $\operatorname{col}(X_+ T_+^j)_{j=0}^{l-1}$ has r linearly independent columns and $\operatorname{col}(\bar{X}_- \bar{T}_-^j)_{j=0}^{l-1}$ has $\alpha - r$ linearly independent columns, it is clear from the form of (44) that $\operatorname{col}\{X_{\mathbb{R}}^+ (T_{\mathbb{R}}^+)^j\}_{j=0}^{l-1}$ has α linearly independent columns.

(*iv*) Similarly, one proves that $\operatorname{row}\{Y_{\mathbb{R}}^+ (T_{\mathbb{R}}^+)^j\}_{j=0}^{l-1}$ has α linearly independent rows.

Before stating the next results, some notation is required. For $k = 1, 2, \ldots$ and $m = 1, 2, \ldots$, let (as in §3.5)

$$M_{km} = \frac{1}{2\pi i}\int_{\gamma^+}\begin{pmatrix} I \\ \vdots \\ \lambda^{k-1}I \end{pmatrix} L^{-1}(\lambda)[I \cdots \lambda^{m-1}I]\, d\lambda, \qquad pk \times pm \text{ matrix}$$

and

$$M^{\mathbb{R}}_{km} = \frac{1}{2\pi i}\int_{\gamma^+}\begin{pmatrix} I \\ \vdots \\ \lambda^{k-1}I \end{pmatrix} L_{\mathbb{R}}^{-1}(\lambda)[I \cdots \lambda^{m-1}I]\, d\lambda, \qquad 2pk \times 2pm \text{ matrix.}$$

Also define $\tilde{M}_{km}$ just as M_{km} but with L replaced by $\bar{L}$. In the definitions of M_{km} and $\tilde{M}_{km}$, the matrix I is the $p \times p$ identity while in $M^{\mathbb{R}}_{km}$ it denotes the $2p \times 2p$ identity.

In view of (42), we have

$$\frac{1}{2\pi i}\int_{\gamma^+}\lambda^j L_{\mathbb{R}}^{-1}(\lambda)\, d\lambda\begin{pmatrix} I & I \\ -iI & iI \end{pmatrix} = \begin{pmatrix} I & I \\ -iI & iI \end{pmatrix}\begin{pmatrix} C_j & \\ & \tilde{C}_j \end{pmatrix}$$

for $j = 0, 1, \ldots$ where

$$C_j = \frac{1}{2\pi i}\int_{\gamma^+}\lambda^j L^{-1}(\lambda)\, d\lambda, \qquad \tilde{C}_j = \frac{1}{2\pi i}\int_{\gamma^+}\lambda^j \bar{L}^{-1}(\lambda)\, d\lambda$$

Hence it follows that

$$M^{\mathbb{R}}_{km} Q_m = Q_k\begin{pmatrix} M_{km} & \\ & \tilde{M}_{km} \end{pmatrix} \tag{*}$$

where

$$Q_k = \left[\begin{array}{cccc|cccc} I & & & & I & & & \\ -iI & & & & iI & & & \\ & I & & & & I & & \\ & -iI & & & & iI & & \\ & & \ddots & & & & \ddots & \\ & & & I & & & & I \\ & & & -iI & & & & iI \end{array}\right]$$
$$\underbrace{\qquad\qquad\qquad}_{k \text{ blocks}} \qquad \underbrace{\qquad\qquad\qquad}_{k \text{ blocks}}$$

a $2pk \times 2pk$ matrix. Here I is the $p \times p$ identity and there are k blocks of $\begin{pmatrix} I \\ -iI \end{pmatrix}$ and k blocks of $\begin{pmatrix} I \\ iI \end{pmatrix}$.

In general, $\tilde{M}_{km} \neq \bar{M}_{km}$ (e.g. if $L(\lambda) = (\lambda - i)^2$ then $\tilde{M}_{12} \neq \bar{M}_{12}$); however, we claim that $\tilde{M}_{km} = \bar{M}_{km}$ whenever $L(\lambda)$ is monic of degree l and $k + m \leqslant l$.

To prove this, first note that if $k+m \leqslant l$, then the blocks of $M_{km} = [C_{i+j}]_{i=0,\,j=0}^{k-1,\,m-1}$ have the form

$$C_j = \frac{1}{2\pi i}\int_{\gamma^+} \lambda^j L^{-1}(\lambda)\,d\lambda, \qquad \text{where } j \leqslant l-2.$$

If $L(\lambda)$ is monic of degree l then $\lambda^j L^{-1}(\lambda)$ is of order no greater than λ^{-2} as $|\lambda| \to \infty$, and it follows that

$$\int_{\gamma^+} \lambda^j L^{-1}(\lambda)\,d\lambda = \int_{-\infty}^{\infty} \lambda^j L^{-1}(\lambda)\,d\lambda$$

and

$$\int_{\gamma^+} \lambda^j \bar{L}^{-1}(\lambda)\,d\lambda = \overline{\int_{-\infty}^{\infty} \lambda^j L^{-1}(\lambda)\,d\lambda}$$

for $j = 0, \ldots, l-2$. The claim now follows immediately.

Lemma 3.21 *Suppose that $L(\lambda)$ is monic. Then the following identity holds for any positive integer s such that $2s \leqslant l$:*

$$M_{ss}^{\mathbb{R}} = Q_s \begin{pmatrix} M_{ss} & \\ & \bar{M}_{ss} \end{pmatrix} Q_s^{-1}$$

Therefore $\det M_{ss}^{\mathbb{R}} = |\det M_{ss}|^2$, *and* $\det M_{ss}^{\mathbb{R}} \neq 0$ *if and only if* $\det M_{ss} \neq 0$.

Proof The blocks in the matrix M_{ss} have the form

$$\frac{1}{2\pi i}\int_{\gamma^+} \lambda^j L^{-1}(\lambda)\,d\lambda,$$

where $j \leqslant 2s-2 \leqslant l-2$. Hence, by the remarks above, it follows that $\tilde{M}_{ss} = \bar{M}_{ss}$. The corollary now follows immediately from (*).

Let $L(\lambda)$ be monic and suppose $L(\lambda) = L_2(\lambda)\cdot L_1(\lambda)$ where the right divisor $L_1(\lambda)$ has invertible leading coefficient. Then the left divisor $L_2(\lambda)$ must also have invertible leading coefficient. Combined with Theorems 3.14 and 3.16 this observation leads to the following theorem.

Theorem 3.22 *Let $L(\lambda)$ be a monic $p \times p$ matrix polynomial of degree $l = 2s$ such that* $\det L(\lambda) \neq 0$ *for real λ, and suppose that* $\det L(\lambda)$ *has exactly ps zeros in the upper half-plane, counting multiplicities. Let γ^+ be, as usual, a contour in the upper half-plane containing the zeros of* $\det L(\lambda)$ *there. Then the following statements are equivalent:*

(*a*) *$L(\lambda)$ admits both a right and a left monic γ^+-spectral divisor of degree s;*
(*b*) $\det M_{ss} \neq 0$;
(*c*) *Both $M_{s,2s}$ and $\tilde{M}_{s,2s}$ have rank ps;*
(*d*) *$L_{\mathbb{R}}(\lambda)$ admits a monic γ^+-spectral right divisor of degree s.*

Proof The fact that (a) $\Leftrightarrow$ (b) is known from Theorem 3.16.

(a) $\Leftrightarrow$ (c). This follows from Theorem 3.14, since by complex conjugation it is clear that $L(\lambda)$ has a monic γ^+-spectral *right* (left) factorization if and only if $\bar{L}(\lambda)$ has a γ^+-spectral *left* (right) factorization.

(b) $\Leftrightarrow$ (d). Since $L_{\mathbb{R}}(\lambda)$ is a *real* matrix polynomial, it follows by complex conjugation that $L_{\mathbb{R}}(\lambda)$ has a monic γ^+-spectral right divisor of degree s if and only it has a monic γ^+-spectral left divisor of degree s. Therefore, by Theorem 3.16, (d) holds if and only if $\det M_{ss}^{\mathbb{R}} \neq 0$. (In view of (42), $L_{\mathbb{R}}(\lambda)$ has exactly $2ps$ eigenvalues in the upper half-plane, counting multiplicities).) The fact that (b) and (d) are equivalent is now clear from Lemma 3.21.

Corollary *If $L(\lambda)$ is a real matrix polynomial satisfying the hypotheses of Theorem* 3.22, *then $M_{s,2s}$ has rank ps if and only if* $\det M_{ss} \neq 0$.

Proof By Theorem 3.14, $M_{s,2s}$ has rank ps if and only if $L(\lambda)$ has a monic γ^+-spectral right divisor of degree s, i.e. $L(\lambda) = L_2(\lambda)L_1(\lambda)$ where $\mathrm{sp}(L_1)$ lies in the upper half-plane and $\mathrm{sp}(L_2)$ lies in the lower half-plane. However, by complex conjugation it is clear that this holds if and only if

$$L(\lambda) = \bar{L}_2(\lambda)\bar{L}_1(\lambda)$$

where $\mathrm{sp}(\bar{L}_2) = \overline{\mathrm{sp}(L_2)}$ lies in the upper half-plane and $\mathrm{sp}(\bar{L}_1) = \overline{\mathrm{sp}(L_1)}$ lies in the lower half-plane. This means that $L(\lambda)$ has a monic γ^+-spectral right divisor of degree s if and only if it has a monic γ^+-spectral left divisor of degree s. Hence, the corollary follows from Theorem 3.16.

We conclude this section by showing in detail the connection between γ^+-spectral factorizations of $L(\lambda)$ and $L_{\mathbb{R}}(\lambda)$.

Suppose that $L(\lambda)$ admits both a right and a left monic γ^+-spectral factorization, i.e. $L(\lambda) = L^-(\lambda)\cdot L^+(\lambda) = M^+(\lambda)\cdot M^-(\lambda)$, where L^+ and M^+ (resp. L^- and M^-) are monic of degree s with spectrum contained in the upper (resp. lower) half-plane. In view of (42)

$$L_{\mathbb{R}}(\lambda) = \begin{pmatrix} I & I \\ -iI & iI \end{pmatrix}\begin{pmatrix} L^-(\lambda) & \\ & \bar{M}^+(\lambda) \end{pmatrix}\begin{pmatrix} L^+(\lambda) & \\ & \bar{M}^-(\lambda) \end{pmatrix}\begin{pmatrix} I & I \\ -iI & iI \end{pmatrix}^{-1} \tag{45}$$

so we obtain the monic γ^+-spectral right factorization

$$L_{\mathbb{R}}(\lambda) = \begin{pmatrix} A^-(\lambda) & -B^-(\lambda) \\ B^-(\lambda) & A^-(\lambda) \end{pmatrix}\begin{pmatrix} A^+(\lambda) & -B^+(\lambda) \\ B^+(\lambda) & A^+(\lambda) \end{pmatrix}, \tag{46}$$

where the matrix polynomials $A^{\pm}(\lambda)$ and $B^{\pm}(\lambda)$ are defined such that

$$L^+(\lambda) = A^+(\lambda) + iB^+(\lambda), \qquad \bar{M}^-(\lambda) = A^+(\lambda) - iB^+(\lambda),$$
$$L^-(\lambda) = A^-(\lambda) + iB^-(\lambda), \qquad \bar{M}^+(\lambda) = A^-(\lambda) - iB^-(\lambda)$$

To justify this, note that (46) is obtained from (45) by means of Lemma 3.19, and since

$$\det\begin{pmatrix} A^+(\lambda) & -B^+(\lambda) \\ B^+(\lambda) & A^+(\lambda) \end{pmatrix} = \det L^+(\lambda)\det \bar{M}^-(\lambda),$$

with a similar result for the other factor of (46), it follows that the factorization (46) is a γ^+-spectral right factorization of $L_{\mathbb{R}}(\lambda)$. Also, in view of the fact that $L^{\pm}(\lambda)$ and $M^{\pm}(\lambda)$ are monic of degree s, $A^{\pm}(\lambda)$ is monic of degree s and $B^{\pm}(\lambda)$ has degree *less* than s. Hence both factors in (46) are monic of degree s.

These steps are also reversible, i.e. given a γ^+-spectral factorization of $L_{\mathbb{R}}(\lambda)$ of form (46) one obtains right and left γ^+-spectral factorizations of $L(\lambda)$.

4
Further results

We consider the spectrum at infinity and related results in §§4.1 and 4.2. The main result is Theorem 4.3 which shows how the finite and infinite spectral pairs together form a generalization of the concept of a standard pair for the case of non-monic matrix polynomials. This result is not needed in the sequel, except very briefly in §10.4, but it is included here for the sake of completeness.

In §4.4 we consider a product of matrix polynomials,

$$L(\lambda) = L_2(\lambda)L_1(\lambda)$$

and show how to obtain a γ-spectral triple for it in terms of γ-spectral triples of each factor. The main result is proved in Theorem 4.17. Let (X_j^+, T_j^+, Y_j^+) be a γ-spectral triple of $L_j(\lambda)$ and denote the base space by $\mathfrak{M}_j^+$ $(j = 1, 2)$. Then the product $L(\lambda) = L_2(\lambda)L_1(\lambda)$ has the following γ-spectral triple:

$$X_+ = [X_1^+ \; A], \qquad T_+ = \begin{pmatrix} T_1^+ & Y_1^+ \cdot X_2^+ \\ 0 & T_2^+ \end{pmatrix}, \qquad Y_+ = \begin{pmatrix} B \\ Y_2^+ \end{pmatrix},$$

where

$$A = \frac{1}{2\pi i}\int_\gamma L_1^{-1}(\lambda)X_2^+(I\lambda - T_2^+)^{-1}\, d\lambda$$

$$B = \frac{1}{2\pi i}\int_\gamma (I\lambda - T_1^+)^{-1}Y_1^+\{L_2^{-1}(\lambda) - X_2^+(I\lambda - T_2^+)^{-1}Y_2^+\}\, d\lambda.$$

This theorem is proved using the operators Q and S which are defined in the statement of Lemmas 4.12 and 4.15. In §4.5 we look at how these operators Q and S are affected by a Möbius transformation of the matrix polynomials. Section 4.3 is essentially a continuation of §2.2 and is needed for §§4.4 and 4.5.

In Part V of the book we will prove an index formula for elliptic systems in the plane based on the spectral theory of matrix polynomials which has been developed in Chapters 1 to 3. Sections 4.4 and 4.5 develop further tools which are important for a full understanding of the index formula.

4.1 The inhomogeneous equation $L(d/dt)u = f$

Let $L(\lambda) = \sum_{j=0}^{l} A_j \lambda^j$ be a $p \times p$ matrix polynomial with $\det L(\lambda) \neq 0$. The Laurent series for the resolvent

$$L^{-1}(\lambda) = \sum_{j=-\infty}^{\kappa} C_j \lambda^j$$

converges when $|\lambda|$ is large, where

$$C_j = \frac{1}{2\pi i} \int_\Gamma \lambda^{-j-1} L^{-1}(\lambda)\, d\lambda \qquad j = 0, \pm 1, \pm 2, \dots \tag{1}$$

and Γ is a large circle having all of $\mathrm{sp}(L)$ in its interior. Also we assume $C_j = 0$ when $j > \kappa$ and κ is the smallest non-negative integer with this property, i.e. κ is the order of the pole at infinity.

For $j \leqslant -1$ the coefficients C_j can be expressed in terms of a *finite* spectral triple of $L(\lambda)$, i.e. $C_j = XT^{-j-1}Y$ where (X, T, Y) is a Γ-spectral triple of $L(\lambda)$. By analytic continuation we have

$$L^{-1}(\lambda) = X(I\lambda - T)^{-1}Y + \sum_{j=0}^{\kappa} C_j \lambda^j \tag{2}$$

since both sides have the same Laurent expansion when $|\lambda|$ is large.

(To compute the coefficients when $j \geqslant 0$ we need to introduce spectral triples at infinity. This will be taken up in §4.2.)

The representation (2) of the resolvent $L^{-1}(\lambda)$ can be used to obtain a particular solution of the equation $L(d/dt)u = f$.

Theorem 4.1 *Let $L(\lambda)$ be a $p \times p$ matrix polynomial with* $\det L(\lambda) \not\equiv 0$. *Then for any $f \in C^\infty(\mathbb{R}, \mathbb{C}^p)$, the differential equation $L(d/dt)u = f$ has the particular solution*

$$u(t) = \frac{1}{2\pi i} \int_0^t \left(\int_\Gamma e^{(t-s)\lambda} L^{-1}(\lambda)\, d\lambda \right) f(s)\, ds + \sum_{j=0}^{\kappa} C_j f^{(j)}(t)$$

where Γ is a simple, closed contour having all of $\mathrm{sp}(L)$ *in its interior, and C_j are as defined above.*

Proof Let (X, T, Y) be a finite spectral triple for $L(\lambda)$, and let

$$\begin{aligned} u_0(t) &:= \frac{1}{2\pi i} \int_0^t \left(\int_\Gamma e^{(t-s)\lambda} L^{-1}(\lambda)\, d\lambda \right) f(s)\, ds \\ &= \int_0^t X\, e^{(t-s)T}\, Y f(s)\, ds \end{aligned}$$

We have to show that

$$L(d/dt)\left(u_0(t) + \sum_{j=0}^{\kappa} C_j f^{(j)}(t) \right) = f(t) \tag{3}$$

First of all, it follows by induction that

$$\frac{d^i}{dt^i} u_0 = XYf^{(i-1)}(t) + \cdots + XT^{i-1}Yf(t) + \int_0^t XT^i e^{(t-s)T} Yf(s)\, ds$$

Then,

$$\begin{aligned} L(d/dt)u_0 &= \sum_{i=0}^{l} A_i \frac{d^i}{dt^i} u_0 \\ &= \sum_{i=1}^{l} A_i(XYf^{(i-1)}(t) + \cdots + XT^{i-1}Yf(t)) \\ &\quad + \int_0^t \sum_{i=0}^{l} A_i XT^i e^{(t-s)T} Yf(s)\, ds \\ &= \sum_{j=0}^{l-1} (A_{j+1}XY + \cdots + A_l XT^{l-j-1}Y) \frac{d^j}{dt^j} f(t), \end{aligned}$$

where, in the last equality, we used the fact that $\sum_{i=0}^{l} A_i XT^i = 0$. By Lemma 4.2 (see below), it follows that

$$L(d/dt)u_0 = \sum_{j=0}^{l-1} L_{j+1}(d/dt)XT^jYf(t) \tag{4}$$

Now multiply both sides of (2) by $L(\lambda)$ on the left to obtain

$$I = L(\lambda)X(I\lambda - T)^{-1}Y + L(\lambda) \cdot \sum_{j=0}^{\kappa} C_j\lambda^j,$$

and then it follows from Lemma 1.18 that

$$\sum_{j=0}^{l-1} L_{j+1}(\lambda)XT^jY = I - L(\lambda) \cdot \sum_{j=0}^{\kappa} C_j\lambda^j \tag{5}$$

Finally, replacing λ by d/dt in (5) and using (4), we obtain (3).

Lemma 4.2 *Let $L(\lambda)$ be a $p \times p$ matrix polynomial and define matrix polynomials $L_j(\lambda)$ $(j = 1, \ldots, l)$ as in Lemma 1.18. Then for any admissible pair (X, T), the following identity holds:*

$$\sum_{j=0}^{l-1} L_{j+1}(\lambda)XT^j = \sum_{j=0}^{l-1} (A_{j+1}X + \cdots + A_l XT^{l-j-1})\lambda^j.$$

Proof Let $\mathscr{X}$ be the matrix 2.1(5). Then

$$\begin{aligned} \sum_{j=0}^{l-1} L_{j+1}(\lambda)XT^j &= [L_1(\lambda) \cdots L_l(\lambda)] \cdot \operatorname{col}(XT^j)_{j=0}^{l-1} \\ &= [I \cdots \lambda^{l-1}I] \cdot \mathscr{X} \cdot \operatorname{col}(XT^j)_{j=0}^{l-1} \\ &= \sum_{j=0}^{l-1} (A_{j+1}X + \cdots + A_l XT^{l-j-1})\lambda^j. \end{aligned}$$

4.2 Infinite spectral triples

Let $L(\lambda) = \sum_{j=0}^{l} A_j\lambda^j$ be a $p \times p$ matrix polynomial with $\det L(\lambda) \not\equiv 0$, and let α denote the degree of $\det L(\lambda)$. We have $\alpha \leqslant pl$, and $\alpha = pl$ only when leading coefficient, A_l, of $L(\lambda)$ is invertible. Now consider the matrix polynomial

$$\tilde{L}(\lambda) = \lambda^l L(\lambda^{-1}) = \sum_{j=0}^{l} A_{l-j}\lambda^j.$$

If $\alpha < pl$ then $\det \tilde{L}(\lambda)$ has a zero of multiplicity $pl - \alpha$ at the origin. An *infinite* spectral triple for $L(\lambda)$ is by definition a spectral triple for $\tilde{L}(\lambda)$ corresponding to the eigenvalue 0. The base space of an infinite spectral triple $(X_\infty, T_\infty, Y_\infty)$ for $L(\lambda)$ thus has dimension $pl - \alpha$. By making the substitution $\lambda \to \lambda^{-1}$ we have

$$\begin{aligned} X_\infty T_\infty^j Y_\infty &= \frac{1}{2\pi i}\int_{\tilde{\gamma}} \lambda^j \tilde{L}^{-1}(\lambda)\, d\lambda \\ &= \frac{1}{2\pi i}\int_{\Gamma} \lambda^{l-2-j} L^{-1}(\lambda)\, d\lambda \qquad j = 0, 1, \ldots \end{aligned} \tag{6}$$

where $\tilde{\gamma}$ is a small circle centered at the origin not containing any other zeros of $\det \tilde{L}(\lambda)$, and Γ is a large circle having all of $\mathrm{sp}(L)$ in its interior. In view of (1) it follows that

$$C_j = X_\infty T_\infty^{l-1+j} Y_\infty \qquad \text{when } j \geqslant 0. \tag{7}$$

Since T_∞ is nilpotent (eigenvalue 0), $(I - \lambda T_\infty)^{-1} = I + \lambda T_\infty + \lambda^2 T_\infty^2 + \cdots$ and we have therefore proved

$$L^{-1}(\lambda) = X(I\lambda - T)^{-1}Y + X_\infty T_\infty^{l-1}(I - \lambda T_\infty)^{-1} Y_\infty \tag{8}$$

Remark Since $C_j = 0$ when $j > \kappa$ then $T_\infty^{l+\kappa} = 0$. This follows from (7) due to injectivity of $\mathrm{col}(X_\infty T_\infty^j)_{j=0}^{l-1}$ and surjectivity of $\mathrm{row}(T_\infty^j\, Y_\infty)_{j=0}^{l-1}$.

We know from §2.2 that $\mathrm{col}(XT^j)_{j=0}^{l-1}$ has left inverse $\mathrm{row}(T^j Y)_{j=0}^{l-1} \cdot \mathscr{X}$. The following theorem shows more, namely that $\mathrm{col}(XT^j)_{j=0}^{l-1}$ is part of a larger matrix which is invertible.

Theorem 4.3 *Let (X, T) and (X_∞, T_∞) be finite and infinite spectral pairs for $L(\lambda)$, respectively. Then the matrix*

$$S_{l-1} = \begin{pmatrix} X & X_\infty T_\infty^{l-1} \\ XT & X_\infty T_\infty^{l-2} \\ \vdots & \vdots \\ XT^{l-1} & X_\infty \end{pmatrix}$$

is invertible.

Proof Note that S_{l-1} is a square $pl \times pl$ matrix since the base space of (X, T) has dimension α while the base space of (X_∞, T_∞) has dimension $pl - \alpha$.

Suppose first that $0 \notin \mathrm{sp}(L)$. That means $\det A_0 \neq 0$ so the matrix polynomial $\tilde{L}(\lambda) = \lambda^l L(\lambda^{-1})$ has pl eigenvalues (counting multiplicities). We claim that

$$\tilde{X} = [X \quad X_\infty], \qquad \tilde{T} = \begin{pmatrix} T^{-1} & \\ & T_\infty \end{pmatrix}$$

is a standard pair for $\tilde{L}(\lambda)$. In fact, Theorem 2.21 implies that (X, T^{-1}) is a spectral pair for $\tilde{L}$ corresponding to the eigenvalues in $\mathrm{sp}(\tilde{L})\backslash 0$, and by definition (X_∞, T_∞) is a spectral pair for $\tilde{L}$ at 0, so the claim follows immediately from Lemma 1.23. Hence $\mathrm{col}(\tilde{X}\tilde{T}^j)_{j=0}^{l-1}$ is invertible. Since

$$\mathrm{col}(\tilde{X}\tilde{T}^{l-1-j})_{j=0}^{l-1} = S_{l-1} \cdot T^{-(l-1)}$$

then S_{l-1} is also invertible.

In the general case we choose $a \in \mathbb{C}$ such that the matrix polynomial $\hat{L}(\lambda) = L(\lambda - a)$ is invertible at $\lambda = 0$. By virtue of Theorem 2.21, $\hat{L}(\lambda)$ has a finite spectral pair

$$(X, \hat{T}), \qquad \text{where } \hat{T} = T + aI$$

and an infinite spectral pair

$$(X_\infty, \hat{T}_\infty), \qquad \text{where } \hat{T}_\infty = T_\infty(I + aT_\infty)^{-1}.$$

Thus $\hat{S}_{l-1} = \mathrm{col}(X\hat{T}^j \ X_\infty \hat{T}_\infty^{l-1-j})_{j=0}^{l-1}$ is invertible. Finally observe that

$$\hat{S}_{l-1}\begin{pmatrix} I & \\ & (I + aT_\infty)^{-(l-1)} \end{pmatrix} = \mathrm{col}(X(T + aI)^j \ X_\infty T_\infty^{l-1-j}(I + aT_\infty)^j)_{j=0}^{l-1}$$
$$= M \cdot S_{l-1}$$

where $M = [M_{jk}]$ and $M_{jk} = 0,\ j < k,\ M_{jk} = \binom{j}{k} a^{j-k} I,\ j \geqslant k$ when $j, k = 0, \ldots, l-1$. Hence S_{l-1} is invertible.

There exist operators Y and Y_∞ such that (X, T, Y) and $(X_\infty, T_\infty, Y_\infty)$ are finite and infinite spectral triples of $L(\lambda)$. We can give an alternative proof of Theorem 4.3 by finding an explicit formula for the inverse of S_{l-1}.

Theorem 4.3′

$$S_{l-1}^{-1} = \begin{pmatrix} [Y \cdots T^{l-1}Y]\mathscr{Z} \\ [T_\infty^{l-1}Y_\infty \cdots Y_\infty]\mathscr{Z}' \end{pmatrix} \tag{9}$$

where $\mathscr{Z}$ is defined by equation (5) in §2.1 and the matrix $\mathscr{Z}'$ is defined by equation (10) below.

Proof Let

$$L_j(T, Y) := \sum_{k=j}^{l} T^{k-j} Y A_k, \qquad j = 1, \ldots, l$$

(note the correspondence with the notation $L_j(\lambda)$ used in Lemma 1.18). Then we have

$$XT^\nu \cdot L_j(T, Y) = \frac{1}{2\pi i} \int_\Gamma \lambda^{\nu-j} L^{-1}(\lambda) \sum_{k=j}^{l} A_k \lambda^k \, d\lambda,$$

$\nu = 0, \ldots, l-1$ and $j = 1, \ldots, l$. Also let

$$\tilde{L}_j(T_\infty, Y_\infty) := \sum_{k=j}^{l} T_\infty^{k-j} Y_\infty A_{l-k}$$

(A_{l-k} are the coefficients of $\tilde{L}(\lambda)$). Then we have

$$\begin{aligned} X_\infty T_\infty^\nu \cdot \tilde{L}_j(T_\infty, Y_\infty) &= \frac{1}{2\pi i} \int_{\tilde{\gamma}} \lambda^{\nu-j} \tilde{L}^{-1}(\lambda) \sum_{k=j}^{l} A_{l-k} \lambda^k \, d\lambda \\ &= \frac{1}{2\pi i} \int_\Gamma \lambda^{-\nu+j-2} L^{-1}(\lambda) \sum_{k=0}^{l-j} A_k \lambda^k \, d\lambda \end{aligned}$$

(for the second equality see (6)). Hence

$$\begin{aligned} &XT^\nu \cdot L_{j+1}(T, Y) + X_\infty T_\infty^{l-1-\nu} \cdot \tilde{L}_{l-j}(T_\infty, Y_\infty) \\ &\quad = \frac{1}{2\pi i} \int_\Gamma \lambda^{\nu-j-1} L^{-1}(\lambda) \sum_{k=j+1}^{l} A_k \lambda^k \, d\lambda + \frac{1}{2\pi i} \int_\Gamma \lambda^{\nu-j-1} L^{-1}(\lambda) \sum_{k=0}^{j} A_k \lambda^k \, d\lambda \\ &\quad = \frac{1}{2\pi i} \int_\Gamma \lambda^{\nu-j-1} I \, d\lambda \\ &\quad = \delta_{\nu j} I, \qquad \nu = 0, \ldots, l-1 \quad \text{and} \quad j = 0, \ldots, l-1 \end{aligned}$$

Thus

$$S_{l-1} \begin{pmatrix} L_1(T, Y) \cdots L_l(T, Y) \\ \tilde{L}_l(T_\infty, Y_\infty) \cdots \tilde{L}_1(T_\infty, Y_\infty) \end{pmatrix} = I_{pl},$$

and this proves (9) once we note that

$$[\tilde{L}_l(T_\infty, Y_\infty) \cdots \tilde{L}_1(T_\infty, Y_\infty)] = [T_\infty^{l-1} Y_\infty \cdots Y_\infty] \mathscr{L}',$$

where

$$\mathscr{L}' = \sigma \tilde{\mathscr{L}} \sigma^{-1} = \begin{pmatrix} 0 & & A_0 \\ & \cdot^{\cdot^{\cdot}} & \vdots \\ A_0 & \cdots & A_{l-1} \end{pmatrix} \tag{10}$$

Corollary 4.4 *Let*

$$P_\Gamma = \begin{pmatrix} X \\ \vdots \\ XT^{l-1} \end{pmatrix} \cdot [Y \cdots T^{l-1}Y]\mathscr{Z}$$

and

$$P_\infty = \begin{pmatrix} X_\infty T_\infty^{l-1} \\ \vdots \\ X_\infty \end{pmatrix} \cdot [T_\infty^{l-1}Y_\infty \cdots Y_\infty]\mathscr{Z}'$$

where $\mathscr{Z}'$ *is defined by equation* (10). *Then* P_Γ *and* P_∞ *are projectors and we have*

(*a*) $P_\Gamma + P_\infty = I_{pl}$
(*b*) $P_\Gamma P_\infty = P_\infty P_\Gamma = 0$

Proof That P_Γ is a projector is known from §2.2. That P_∞ is a projector follows from the equation $P_\infty = \sigma \tilde{P}_{\tilde{\gamma}} \sigma^{-1}$, where $\tilde{P}_{\tilde{\gamma}}$ is the projector for the matrix polynomial $\tilde{L}(\lambda)$ corresponding to the contour $\tilde{\gamma}$ and σ is the permutation of $p \times p$ blocks in $\mathbb{C}^{pl}$ that maps the jth block to the $(l-j+1)$th block, $j = 1, \ldots, l$. (Note that $\mathscr{Z}' = \sigma \tilde{\mathscr{Z}} \sigma^{-1}$, where $\tilde{\mathscr{Z}}$ is the matrix 2.2(5) for the matrix polynomial $\tilde{L}(\lambda)$.) The fact that $P_\Gamma + P_\infty = I_{pl}$ is simply another way of expressing the equation (9). To prove (*b*) note that (9) also implies

$$\begin{pmatrix} [Y \cdots T^{l-1}Y]\mathscr{Z} \\ [T_\infty^{l-1}Y_\infty \cdots Y_\infty]\mathscr{Z}' \end{pmatrix} \begin{pmatrix} X & X_\infty T_\infty^{l-1} \\ \vdots & \vdots \\ XT^{l-1} & X_\infty \end{pmatrix} = \begin{pmatrix} I_{\mathfrak{M}} & 0 \\ 0 & I_{\mathfrak{M}_\infty} \end{pmatrix}.$$

where $\mathfrak{M}$ and $\mathfrak{M}_\infty$ are the base spaces of (X, T, Y) and $(X_\infty, T_\infty, Y_\infty)$, respectively. Hence

$$[T_\infty^{l-1}Y_\infty \cdots Y_\infty]\mathscr{Z}' \begin{pmatrix} X \\ \vdots \\ XT^{l-1} \end{pmatrix} = 0 \tag{11}$$

and

$$[Y \cdots T^{l-1}Y]\mathscr{Z} \begin{pmatrix} X_\infty T_\infty^{l-1} \\ \vdots \\ X_\infty \end{pmatrix} = 0. \tag{12}$$

From (12) it follows that $P_\Gamma P_\infty = 0$ and (11) implies $P_\infty P_\Gamma = 0$.

The following proposition will be of use in Chapter 10. In the statement of the proposition it is assumed that (X, T, Y) and $(X_\infty, T_\infty, Y_\infty)$ are matrices (rather than operators).

Proposition 4.5 *Let r be a positive integer. Then for any $r \times \alpha$ matrix, M, and $r \times (pl - \alpha)$ matrix, M_∞, there is a unique $r \times p$ matrix polynomial $B(\lambda) = \sum_{j=0}^{l-1} B_j \lambda^j$ such that*

$$\sum_{j=0}^{l-1} B_j X T^j = M \tag{13}$$

and

$$\sum_{j=0}^{l-1} B_j X_\infty T_\infty^{l-j-1} = M_\infty, \tag{14}$$

given by

$$B_j = M \cdot \sum_{k=0}^{l-j-1} T^k Y A_{j+k+1} + M_\infty \cdot \sum_{k=0}^{j} T_\infty^k Y_\infty A_{j-k} \tag{15}$$

for $j = 0, \ldots, l-1$.

Proof The equations (14), (13) can be written in the form

$$[B_0 \cdots B_{l-1}] S_{l-1} = [M \; M_\infty],$$

which has the unique solution

$$\begin{aligned} [B_0 \cdots B_{l-1}] &= [M \; M_\infty] S_{l-1}^{-1} \\ &= [M \; M_\infty] \begin{pmatrix} [T \cdots T^{l-1} Y] \mathscr{Z} \\ [T_\infty^{l-1} Y_\infty \cdots Y_\infty] \mathscr{Z}' \end{pmatrix} \end{aligned}$$

Upon substituting the definitions of $\mathscr{Z}$ and $\mathscr{Z}'$ we obtain the formulas (15) for B_j.

Remark Since $[Y \cdots T^{l-1} Y] \mathscr{Z}$ is a left inverse of $\operatorname{col}(XT^j)_{j=0}^{l-1}$, one solution of (13) is given by $B_j = M \cdot \sum_{k=0}^{l-j-1} T^k Y A_{j+k+1}$, $j = 0, \ldots, l-1$. This corresponds to choosing $M_\infty = 0$.

Addendum: infinite spectral triples and linearization

We now show that a linearization of a matrix polynomial $L(\lambda)$ (see §3.2) may be obtained from the spectral data as given in finite and infinite spectral pairs of $L(\lambda)$. This result is then used to prove a generalization of Theorem 3.2 for the non-monic case (Theorem 4.7).

In the statement of the following theorem, S_{l-1} denotes the $pl \times pl$ matrix defined in Theorem 4.3, and S_{l-2} denotes the following $p(l-1) \times pl$ sub-matrix

$$S_{l-2} = \begin{pmatrix} X & X_\infty T_\infty^{l-2} \\ \vdots & \vdots \\ XT^{l-2} & X_\infty \end{pmatrix}$$

Also recall the definition of the companion polynomial $C_L(\lambda)$ from §3.1:

$$C_L(\lambda) = \begin{pmatrix} I & & & & \\ & I & & & \\ & & \ddots & & \\ & & & I & \\ & & & & A_l \end{pmatrix} \lambda + \begin{pmatrix} 0 & -I & & & \\ & 0 & -I & & \\ & & \ddots & \ddots & \\ & & & 0 & -I \\ A_0 & A_1 & & \cdots & A_{l-1} \end{pmatrix} \tag{16}$$

Theorem 4.6 *Let (X, T) and (X_∞, T_∞) be finite and infinite spectral pairs of $L(\lambda)$. Then*

$$C_L(\lambda) S_{l-1} = \begin{pmatrix} S_{l-2} \\ W \end{pmatrix} \begin{pmatrix} I\lambda - T & \\ & T_\infty \lambda - I \end{pmatrix}, \tag{17}$$

where $W = [A_l X T^{l-1} \quad -\sum_{i=0}^{l-1} A_i X_\infty T_\infty^{l-1-i}]$.

Proof By equating like coefficients of λ on both sides of (17), we see that (17) is equivalent to the following pair of equalities:

$$\begin{pmatrix} I & & & \\ & \ddots & & \\ & & I & \\ & & & A_l \end{pmatrix} S_{l-1} = \begin{pmatrix} S_{l-2} \\ W \end{pmatrix} \begin{pmatrix} I & \\ & T_\infty \end{pmatrix} \tag{18}$$

and

$$\begin{pmatrix} 0 & -I & & & \\ & 0 & -I & & \\ & & \ddots & \ddots & \\ & & & 0 & -I \\ A_0 & A_1 & & \cdots & A_{l-1} \end{pmatrix} S_{l-1} = -\begin{pmatrix} S_{l-2} \\ W \end{pmatrix} \begin{pmatrix} T & \\ & I \end{pmatrix}, \tag{19}$$

But, equality in the first $l-1$ block-rows of (18) or (19) clearly holds for any admissible pairs (X, T) and (X_∞, T_∞), and in the last block-row it holds since (in fact, if and only if)

$$\sum_{j=0}^{l} A_j X T^j = 0, \qquad \sum_{j=0}^{l} A_j X_\infty T_\infty^{l-j} = 0. \tag{20}$$

Corollary *The matrix* $\begin{pmatrix} S_{l-2} \\ W \end{pmatrix}$ *is invertible.*

Proof In view of Theorem 3.5, $\mathrm{sp}(C_L) = \mathrm{sp}(L)$. Also $\mathrm{sp}(T) = \mathrm{sp}(L)$. If we choose $\lambda \notin \mathrm{sp}(L)$ then the equation (17) implies that $\begin{pmatrix} S_{l-2} \\ W \end{pmatrix}$ is invertible since $C_L(\lambda)$, $\begin{pmatrix} I\lambda - T & \\ & T_\infty\lambda - I \end{pmatrix}$ and S_{l-1} are invertible.

Recall that a triple of operators (X, T, Y) is called an *admissible triple* if $X \in \mathscr{L}(\mathbb{C}^p, \mathfrak{M})$, $T \in \mathscr{L}(\mathfrak{M})$ and $Y \in \mathscr{L}(\mathfrak{M}, \mathbb{C}^p)$ for some finite dimensional vector space $\mathfrak{M}$. If $\dim \mathfrak{M} = \mu$, we say that (X, T, Y) is an admissible triple of *order* μ.

Theorem 4.7 *Let $L(\lambda) = \sum_{j=0}^{l} A_j\lambda^j$ be a $p \times p$ matrix polynomial of degree l. Let (X, T, Y) and $(X_\infty, T_\infty, Y_\infty)$ be admissible triples of order α and $pl - \alpha$, respectively, where α is the degree of* $\det L(\lambda)$. *Then the following are equivalent*:

(*a*) *(X, T, Y) is a finite spectral triple of $L(\lambda)$ and $(X_\infty, T_\infty, Y_\infty)$ is an infinite spectral triple of $L(\lambda)$,*

(*b*) $L^{-1}(\lambda) = X(I\lambda - T)^{-1}Y + X_\infty T_\infty^{l-1}(I - \lambda T_\infty)^{-1}Y_\infty$, $\mathrm{sp}(T_\infty) = \{0\}$ *and* $XT^jY = X_\infty T_\infty^{l-2-j}Y_\infty$ *when* $j = 0, 1, \ldots, l-2$,

(*c*) *(X, T) is a finite spectral pair of $L(\lambda)$, (X_∞, T_∞) is an infinite spectral pair of $L(\lambda)$, and*

$$\begin{pmatrix} Y \\ -Y_\infty \end{pmatrix} = \begin{pmatrix} S_{l-2} \\ W \end{pmatrix}^{-1} [0 \cdots 0\, I]^T,$$

where W is the matrix defined in Theorem 4.6.

Proof (a) $\Rightarrow$ (b) This was proved in the discussion at the beginning of §4.1.

(b) $\Rightarrow$ (a) Now suppose that (b) holds. Since $\mathrm{sp}(T_\infty) = \{0\}$, then $T_\infty^k = 0$ for some k. Hence

$$(I - \lambda T_\infty)^{-1} = I + \lambda T_\infty + \lambda^2 T_\infty^2 + \cdots, \tag{21}$$

which is a polynomial in λ, and it follows that $L^{-1}(\lambda) - X(I\lambda - T)^{-1}Y$ is analytic in the finite complex plane. Hence

$$\begin{aligned} \frac{1}{2\pi i}\int_\Gamma \lambda^j L^{-1}(\lambda)\, d\lambda &= \frac{1}{2\pi i}\int_\Gamma \lambda^j X(I\lambda - T)^{-1}Y\, d\lambda \\ &= XT^jY \end{aligned} \tag{22}$$

when $j = 0, 1, \ldots,$ and it follows that

$$\begin{aligned} \mathrm{col}(XT^j)_{j=0}^{l-1}\, \mathrm{row}(T^jY)_{j=0}^{l-1}\, \mathscr{Z} &= \frac{1}{2\pi i}\int_\Gamma \begin{pmatrix} I \\ \vdots \\ \lambda^{l-1}I \end{pmatrix} L^{-1}(\lambda)[I \cdots \lambda^{l-1}I]\,\mathscr{Z}\, d\lambda \\ &= P_\Gamma. \end{aligned}$$

Since P_Γ has rank α and the base space of (X, T, Y) has dimension α, it follows that $\operatorname{col}(XT^j)_{j=0}^{l-1}$ and $\operatorname{row}(T^jY)_{j=0}^{l-1}$ must be injective and surjective, respectively. Hence (X, T, Y) is a finite spectral triple of $L(\lambda)$.

We now show that $\tilde{L}^{-1}(\lambda) - X_\infty(I\lambda - T_\infty)^{-1}Y_\infty$ has an analytic continuation near $\lambda = 0$, where $\tilde{L}(\lambda) = \lambda^l L(\lambda^{-1})$. Indeed, by making the substitution $w = \lambda^{-1}$, we obtain

$$\begin{aligned}\frac{1}{2\pi i}\int_{\tilde{\gamma}} \lambda^j \tilde{L}^{-1}(\lambda)\,d\lambda &= \frac{1}{2\pi i}\int_{\tilde{\gamma}} \lambda^{j-l}L^{-1}(\lambda^{-1})\,d\lambda \\ &= \frac{1}{2\pi i}\int_{\Gamma} w^{l-j-2}L^{-1}(w)\,dw \\ &= XT^{l-2-j}Y \\ &= X_\infty T_\infty^j Y_\infty\end{aligned}$$

when $j = 0, \ldots, l-2$, while for $j = l-1+k$ $(k \geqslant 0)$ we have

$$\begin{aligned}\frac{1}{2\pi i}\int_{\tilde{\gamma}} \lambda^j \tilde{L}^{-1}(\lambda)\,d\lambda &= \frac{1}{2\pi i}\int_{\tilde{\gamma}} \lambda^{k-1}L^{-1}(\lambda^{-1})\,d\lambda \\ &= \frac{1}{2\pi i}\int_{\tilde{\gamma}} \lambda^k X(I - \lambda T)^{-1}Y\,d\lambda \\ &\quad + \frac{1}{2\pi i}\int_{\tilde{\gamma}} \lambda^k X_\infty T_\infty^{l-1}(I\lambda - T_\infty)^{-1}Y_\infty\,d\lambda \\ &= 0 + X_\infty T_\infty^{l-1+k}Y_\infty \\ &= X_\infty T_\infty^j Y_\infty\end{aligned}$$

As usual, $\tilde{\gamma}$ is a small circle centered at the origin, of radius $< \|T\|^{-1}$. In the above calculation, we used the fact that $(I - \lambda T)^{-1}$ is analytic when $|\lambda| < \|T\|^{-1}$ and $(I\lambda - T_\infty)^{-1}$ is analytic when $\lambda \neq 0$.

Thus $\tilde{L}^{-1}(\lambda) - X_\infty(I\lambda - T_\infty)^{-1}Y_\infty$ has an analytic continuation near $\lambda = 0$. By the same method as shown for the finite spectral triple, it follows that $(X_\infty, T_\infty, Y_\infty)$ is a spectral triple of $\tilde{L}(\lambda)$ corresponding to $\lambda = 0$, and hence an infinite spectral triple of $L(\lambda)$.

(a) $\Rightarrow$ (c) Suppose that (a) holds. Then trivially (X, T) and (X_∞, T_∞) are finite and infinite spectral pairs of $L(\lambda)$, respectively. In view of (6) we have $X_\infty T_\infty^j Y_\infty = XT^{l-2-j}Y$ when $j = 0, \ldots, l-2$ and

$$\begin{aligned}A_l XT^{l-1}Y + \sum_{j=0}^{l-1} A_j X_\infty T_\infty^{l-1-j}Y_\infty &= \frac{1}{2\pi i}A_l\int_\Gamma \lambda^{l-1}L^{-1}(\lambda)\,d\lambda \\ &\quad + \frac{1}{2\pi i}\sum_{j=0}^{l-1} A_j \int_\Gamma \lambda^{j-1}L^{-1}(\lambda)\,d\lambda \\ &= \frac{1}{2\pi i}\int_\Gamma \lambda^{-1}\left(\sum_{j=0}^{l} A_j\lambda^j\right)L^{-1}(\lambda)\,d\lambda \\ &= \frac{1}{2\pi i}\int_\Gamma \lambda^{-1}I\,d\lambda = I.\end{aligned}$$

It follows that

$$\begin{pmatrix} S_{l-2} \\ W \end{pmatrix} \begin{pmatrix} Y \\ -Y_\infty \end{pmatrix} = \begin{pmatrix} 0 \\ \vdots \\ 0 \\ I \end{pmatrix} \tag{23}$$

(c) $\Rightarrow$ (a) Suppose that (c) holds. Then there exists an operator Y' such that (X, T, Y') is a finite spectral triple of $L(\lambda)$ and an operator Y'_∞ such that $(X_\infty, T_\infty, Y'_\infty)$ is an infinite spectral triple of $L(\lambda)$. Now since (23) holds for Y' and Y'_∞ then

$$\begin{pmatrix} Y \\ -Y_\infty \end{pmatrix} = \begin{pmatrix} S_{l-2} \\ W \end{pmatrix}^{-1} \begin{pmatrix} 0 \\ \vdots \\ I \end{pmatrix} = \begin{pmatrix} Y' \\ -Y'_\infty \end{pmatrix},$$

so that $Y = Y'$ and $Y_\infty = Y'_\infty$, and (X, T, Y) and $(X_\infty, T_\infty, Y_\infty)$ are finite and infinite spectral triples of $L(\lambda)$, respectively.

4.3 More on restrictions of spectral pairs

This section is essentially a continuation of §2.2. In the following theorem $L_1(\lambda)$ and $L(\lambda)$ are any $p \times p$ matrix polynomials and γ is a simple, closed contour not intersecting $\mathrm{sp}(L_1)$ or $\mathrm{sp}(L)$.

Theorem 4.8 *Let* $L(\lambda) = \sum_{j=0}^{l} A_j \lambda^j$. *If* (X_+, T_+) *and* (X_1^+, T_1^+) *are* γ*-spectral pairs of* $L(\lambda)$ *and* $L_1(\lambda)$*, respectively, the following are equivalent:*

(*a*) *The rational matrix function* $L(\lambda)L_1^{-1}(\lambda)$ *has an analytic continuation inside* γ;
(*b*) $\sum_{j=0}^{l} A_j X_1^+ (T_1^+)^j = 0$;
(*c*) (X_1^+, T_1^+) *is a restriction of* (X_+, T_+).

Proof Let Y_1^+ be the unique operator such that (X_1^+, T_1^+, Y_1^+) is a γ-spectral triple of $L_1(\lambda)$.

(a) $\Leftrightarrow$ (b) $L(\lambda)L_1^{-1}(\lambda)$ has an analytic continuation inside γ if and only if the integrals $1/2\pi i \int_\gamma \lambda^k L(\lambda) L_1^{-1}(\lambda)\, d\lambda$ vanish for $k = 0, 1, \ldots$. In view of property (*ii'*) of spectral triples these integrals are equal to $\sum_{j=0}^{l} A_j X_1^+ (T_1^+)^j \cdot (T_1^+)^k Y_1^+$ and due to property (*iv*) they vanish exactly when $\sum_{j=0}^{l} A_j X_1^+ (T_1^+)^j = 0$.

(b) $\Leftrightarrow$ (c) The equation $\sum_{j=0}^{l} A_j X_1^+ (T_1^+)^j = 0$ holds exactly when (X_1^+, T_1^+) is a partial γ-spectral pair of $L(\lambda)$, which in turn holds if and only if (X_1^+, T_1^+) is a restriction of (X_+, T_+).

Corollary 4.9 $L(\lambda)L_1^{-1}(\lambda)$ *is analytic and invertible inside* γ *if and only if the* γ*-spectral pairs of* $L(\lambda)$ *and* $L_1(\lambda)$ *coincide.*

Proof Applying Theorem 4.8 to the rational matrix functions $Q(\lambda) = L(\lambda)L_1^{-1}(\lambda)$ and $Q^{-1}(\lambda) = L_1(\lambda)L^{-1}(\lambda)$ we see that (X_1^+, T_1^+) is a restriction of (X_+, T_+) and vice versa. It follows that (X_1^+, T_1^+) and (X_+, T_+) are similar.

Recall that (c) means that there is a T_+-invariant subspace $\mathfrak{L}$ such that $(X_+|_{\mathfrak{L}}, T_+|_{\mathfrak{L}})$ is a γ-spectral pair of $L_1(\lambda)$. This holds if and only if every $u \in \mathfrak{M}_{L_1}^+$ has a representation $u(t) = X_+ e^{tT_+} c$ for some c. Hence (a), (b) and (c) are also equivalent to the following condition: $\mathfrak{M}_{L_1}^+ \subset \mathfrak{M}_L^+$.

Note that $\mathfrak{L}$ is unique due to the injectivity of $\operatorname{col}(X_+ T_+^j)_{j=0}^{l-1}$.

Corollary 4.10 *Let (X_+, T_+) be a γ-spectral pair of $L(\lambda)$, and denote its base space by $\mathfrak{M}^+$. If $L(\lambda)L_1^{-1}(\lambda)$ has an analytic continuation inside γ then the unique T_+-invariant subspace $\mathfrak{L}$ such that $(X_+|_{\mathfrak{L}}, T_+|_{\mathfrak{L}})$ is a γ-spectral pair of $L_1(\lambda)$ is*

$$\mathfrak{L} = \{c;\, c \in \mathfrak{M}^+,\, L_1(d/dt) X_+ e^{tT_+} c \equiv 0\}, \tag{24}$$

Proof The inclusion $\subset$ is clear. Conversely, if we have $c \in \mathfrak{M}^+$ such that $L_1(d/dt)X_+ e^{tT_+} c \equiv 0$ then the function $u(t) = X_+ e^{tT_+} c$ belongs to $\mathfrak{M}_{L_1}^+$. But $(X_+|_{\mathfrak{L}}, T_+|_{\mathfrak{L}})$ is a γ-spectral pair of $L_1(\lambda)$, so

$$u(t) = X_+|_{\mathfrak{L}}\, e^{tT_+|_{\mathfrak{L}}} c',$$

for some $c' \in \mathfrak{L}$, and thus $X_+ e^{tT_+} c = X_+ e^{tT_+} c'$ for all $t \in \mathbb{R}$. By the injectivity of $\operatorname{col}(X_+ T_+^j)_{j=0}^{l-1}$ it follows that $c = c' \in \mathfrak{L}$.

Remark We can also describe $\mathfrak{L}$ as the set of all $c \in \mathfrak{M}^+$ such that

$$\sum_{j=0}^{l_1} A_j^{(1)} X_+ T_+^j \cdot T_+^k c = 0 \qquad \text{for all } k = 0, 1, \ldots$$

where $A_j^{(1)}$, $j = 0, 1, \ldots, l_1$, denote the coefficients of $L_1(\lambda)$.

We mention without proof the analogue of Theorem 4.8 for left γ-spectral pairs, where γ is a simple, closed contour not intersecting $\operatorname{sp}(L)$ or $\operatorname{sp}(L_2)$.

Theorem 4.11 *Let $L(\lambda) = \sum_{j=0}^{l} A_j \lambda^j$. If (T_2^+, Y_2^+) and (T_+, Y_+) are left γ-spectral pairs of $L_2(\lambda)$ and $L(\lambda)$, respectively, the following are equivalent:*

(1) *The rational matrix function $L_2^{-1}(\lambda)L(\lambda)$ is analytic inside γ;*
(2) $\sum_{j=0}^{l} (T_2^+)^j Y_2^+ A_j = 0$;
(3) *(T_2^+, Y_2^+) is a restriction of (T_+, Y_+) (see Definition* 2.16).

Corollary *$L_2^{-1}(\lambda)L(\lambda)$ is analytic and invertible inside γ if and only if the left γ-spectral pairs of $L(\lambda)$ and $L_2(\lambda)$ coincide.*

4.4 Spectral triples of products

In this section we construct a γ-spectral triple for a product $L(\lambda) = L_2(\lambda)L_1(\lambda)$ of matrix polynomials given a γ-spectral triple for each factor. We do this using the operators Q and S which are defined below in Lemmas 4.12 and 4.15.

Lemma 4.12 *Let $L(\lambda) = L_2(\lambda)L_1(\lambda)$ be a product of matrix polynomials. If (X_+, T_+), (X_2^+, T_2^+) are γ-spectral pairs of $L(\lambda)$, $L_2(\lambda)$, respectively, then there is a (unique) surjection $Q \in \mathscr{L}(\mathfrak{M}^+, \mathfrak{M}_2^+)$ such that*

$$L_1(d/dt)X_+ e^{tT_+} = X_2^+ e^{tT_2^+} \cdot Q \tag{25}$$

Further, it follows from (25) that $\ker Q = \mathfrak{L}$, *where $\mathfrak{L}$ is the unique T_+-invariant subspace (24) such that $(X_+|_{\mathfrak{L}}, T_+|_{\mathfrak{L}})$ is a γ-spectral pair of $L_1(\lambda)$.*

Proof Let the base spaces of (X_+, T_+), (X_2^+, T_2^+) be denoted $\mathfrak{M}^+$, $\mathfrak{M}_2^+$, respectively. For any $c \in \mathfrak{M}^+$ the function $v = L_1(d/dt)X_+ e^{tT_+} c$ satisfies the equation $L_2(d/dt)v = 0$ and hence belongs to $\mathfrak{M}_{L_2}^+$. It follows that for every $c \in \mathfrak{M}^+$ there exists a unique $c_2 \in \mathfrak{M}_2^+$ such that

$$L_1(d/dt)X_+ e^{tT_+} c \equiv X_2^+ e^{tT_2^+} c_2 \tag{26}$$

Now define $Qc = c_2$; then Q is a map from $\mathfrak{M}^+$ to $\mathfrak{M}_2^+$ such that (25) holds. To show that Q is surjective let $c_2 \in \mathfrak{M}_2^+$. By virtue of Proposition 2.5 the function $v = X_2^+ e^{tT_2^+} c_2$ has the form

$$v = \frac{1}{2\pi i}\int_\gamma e^{t\lambda} L_2^{-1}(\lambda)p(\lambda)\, d\lambda$$

for some polynomial $p(\lambda)$ whence

$$v = L_1(d/dt)u \qquad \text{where } u = \frac{1}{2\pi i}\int_\gamma e^{t\lambda} L^{-1}(\lambda)p(\lambda)\, d\lambda$$

Now $u \in \mathfrak{M}_L^+$, so there exists $c \in \mathfrak{M}^+$ such that $u = X_+ e^{tT_+} c$. This means that (26) holds and Q is surjective.

Due to the injectivity of $\operatorname{col}(X_2^+(T_2^+)^j)_{j=0}^{l_2-1}$ we see that $Qc = 0$ if and only if $L_1(d/dt)X_+ e^{tT_+} c = 0$, which, in view of (24), holds if and only if $c \in \mathfrak{L}$. Hence $\ker Q = \mathfrak{L}$.

Lemma 4.13 *In addition to the hypotheses of Lemma 4.12, let Y_+ and Y_2^+ be the unique operators such that (X_+, T_+, Y_+) and (X_2^+, T_2^+, Y_2^+) are γ-spectral triples of $L(\lambda)$ and $L_2(\lambda)$, respectively. Then*

$$QT_+ = T_2^+Q \quad \textit{and} \quad QY_+ = Y_2^+.$$

Proof Differentiating both sides of (25) with respect to t gives

$$\begin{aligned} X_2^+ e^{tT_2^+} T_2^+ Q &= L_1(d/dt)X_+ e^{tT_+} T_+ \\ &= X_2^+ e^{tT_2^+} QT_+. \end{aligned}$$

Since $\operatorname{col}(X_2^+(T_2^+)^j)_{j=0}^{l_2-1}$ is injective we obtain $T_2^+Q = QT_+$. Similarly,

$$\begin{aligned} X_2^+ e^{tT_2^+} QY_+ &= L_1(d/dt)X_+ e^{tT_+} Y_+ \\ &= L_1(d/dt)\frac{1}{2\pi i}\int_\gamma e^{t\lambda} L^{-1}(\lambda)\, d\lambda \\ &= \frac{1}{2\pi i}\int_\gamma e^{t\lambda} L_2^{-1}(\lambda)\, d\lambda \\ &= X_2^+ e^{tT_2^+} Y_2^+, \end{aligned}$$

so that $QY_+ = Y_2^+$.

Remark 4.14 Let π denote the canonical projection $\mathfrak{M}^+ \to \mathfrak{M}^+/\mathfrak{L}$. Since $\mathfrak{L} = \ker Q$ there exists an operator $\bar{Q} \in \mathscr{L}(\mathfrak{M}^+/\mathfrak{L}, \mathfrak{M}_2^+)$ such that $Q = \bar{Q}\pi$. Also let $\bar{T}_+$, $\bar{Y}_+$ be the operators induced by π from T_+, Y_+, respectively. In view of the preceding lemma we have

$$\bar{Q}\bar{T}_+ = T_2^+\bar{Q} \quad \text{and} \quad \bar{Q}\bar{Y}_+ = Y_2^+.$$

Since $\bar{Q}$ is invertible this means that (T_2^+, Y_2^+) is a restriction of (T_+, Y_+) (see Definition 2.16).

Remark Lemmas 4.12, 4.13 are essentially a further development of the remark made after Definition 2.16.

As we showed in Lemma 4.12 the map Q is surjective. The next lemma shows that there is a natural way to define a section of Q, that is, an operator $S \in \mathscr{L}(\mathfrak{M}_2^+, \mathfrak{M}^+)$ such that $QS = I$ (the identity operator on $\mathfrak{M}_2^+$).

Lemma 4.15 *Let the hypotheses and notation be as in Lemma 4.12. There exists a unique map $S \in \mathscr{L}(\mathfrak{M}_2^+, \mathfrak{M}^+)$ such that*

$$X_+ e^{tT_+} S = \frac{1}{2\pi i}\int_\gamma e^{t\lambda} L_1^{-1}(\lambda) X_2^+ (I\lambda - T_2^+)^{-1}\, d\lambda, \tag{27}$$

for all $t \in \mathbb{R}$, and the equation (27) implies that $QS = I$.

Proof Let $c_2 \in \mathfrak{M}_2^+$ and consider the function

$$v(t) = \frac{1}{2\pi i}\int_\gamma e^{t\lambda} L_1^{-1}(\lambda) X_2^+ (I\lambda - T_2^+)^{-1} c_2\, d\lambda.$$

Since $L_2(\lambda)X_2^+(I\lambda - T_2^+)^{-1}c_2$ is a polynomial in λ due to Lemma 1.18, we see that $L(d/dt)v = 0$, so $v = \mathfrak{M}_L^+$, and therefore $v(t) = X_+ e^{tT_+} c$ for a unique $c \in \mathfrak{M}^+$. Now define $Sc_2 = c$; then S is a map from $\mathfrak{M}_2^+$ to $\mathfrak{M}^+$ such that (27) holds. The uniqueness of the map S follows from the injectivity of $\operatorname{col}(X_+ T_+^j)_{j=0}^{l-1}$.

To show that S is a section of Q, observe that differentiation of both sides of (27) with respect to t yields

$$\begin{aligned} L_1(d/dt) X_+ e^{tT_+} S &= \frac{1}{2\pi i}\int_\gamma e^{t\lambda} X_2^+ (I\lambda - T_2^+)^{-1}\, d\lambda \\ &= X_2^+ e^{tT_2^+} \end{aligned}$$

and then (25) implies

$$X_2^+ e^{tT_2^+} QS = X_2^+ e^{tT_2^+} \qquad \text{for all } t \in \mathbb{R}.$$

Hence QS must be the identity on $\mathfrak{M}_2^+$.

We now turn to the problem of constructing a γ-spectral triple for a product $L(\lambda) = L_2(\lambda)L_1(\lambda)$ of matrix polynomials given a γ-spectral triple for each factor. As shown in §4.3 there is a unique subspace $\mathfrak{L}$ such that

$(X_1^+, T_1^+) = (X_+|_{\mathfrak{L}}, T_+|_{\mathfrak{L}})$ is a γ-spectral pair for $L_1(\lambda)$. Let $Y_1^+ \in \mathfrak{L}(\mathbb{C}^p, \mathfrak{L})$ be the unique operator such that (X_1^+, T_1^+, Y_1^+) is a γ-spectral triple of $L_1(\lambda)$.

We claim that the operator S from Lemma 4.15 satisfies the equations

$$X_+ S = \frac{1}{2\pi i} \int_\gamma L_1^{-1}(\lambda) X_2^+ (I\lambda - T_2^+)^{-1} \, d\lambda \tag{28}$$

and

$$T_+ S = S T_2^+ + Y_1^+ X_2^+, \tag{29}$$

Of course (28) follows from (27) by letting $t = 0$. Also (29) is equivalent to

$$X_+ e^{tT_+} T_+ S = X_+ e^{tT_+} S T_2^+ + X_1^+ e^{tT_1^+} Y_1^+ X_2^+ \tag{30}$$

due to the injectivity of $\operatorname{col}(X_+ T_+^j)_{j=0}^{l-1}$. However, in view of (27) and the fact that

$$X_1^+ e^{tT_1^+} Y_1^+ = \frac{1}{2\pi i} \int_\gamma e^{t\lambda} X_1^+ (I\lambda - T_1^+)^{-1} Y_1^+ \, d\lambda = \frac{1}{2\pi i} \int_\gamma e^{t\lambda} L_1^{-1}(\lambda) \, d\lambda,$$

we see that the right-hand side of (30) is equal to

$$\begin{aligned}
\frac{1}{2\pi i} \int_\gamma & e^{t\lambda} L_1^{-1}(\lambda) X_2^+ (I\lambda - T_2^+)^{-1} T_2^+ \, d\lambda + \frac{1}{2\pi i} \int_\gamma e^{t\lambda} L_1^{-1}(\lambda) X_2^+ \, d\lambda \\
&= \frac{1}{2\pi i} \int_\gamma e^{t\lambda} L_1^{-1}(\lambda) X_2^+ \{(I\lambda - T_2^+)^{-1} T_2^+ + I\} \, d\lambda \\
&= \frac{1}{2\pi i} \int_\gamma e^{t\lambda} L_1^{-1}(\lambda) \lambda X_2^+ (I\lambda - T_2^+)^{-1} \, d\lambda \\
&= X_+ e^{tT_+} T_+ S,
\end{aligned}$$

which is the left-hand side of (30). This proves (30) and hence (29).

Lemma 4.16 *Suppose that* $S \in \mathscr{L}(\mathfrak{M}_2^+, \mathfrak{M}^+)$ *and* $Q \in \mathscr{L}(\mathfrak{M}^+, \mathfrak{M}_2^+)$ *such that* $QS = I$ *(the identity operator on* $\mathfrak{M}_2^+$*), and let* $\mathfrak{M}_1^+ = \ker Q$*. Then the operator* $M \in \mathscr{L}(\mathfrak{M}_1^+ \times \mathfrak{M}_2^+, \mathfrak{M}_+)$ *defined by*

$$M\begin{pmatrix} c_1 \\ c_2 \end{pmatrix} = c_1 + Sc_2, \qquad c_1 \in \mathfrak{M}_1^+, c_2 \in \mathfrak{M}_2^+,$$

is an isomorphism with inverse $M^{-1} = \begin{pmatrix} P \\ Q \end{pmatrix}$*, where* $P = I - SQ \in \mathscr{L}(\mathfrak{M}^+)$ *is a projector with* $\operatorname{im} P = \mathfrak{M}_1^+$ *and* $\ker P = \operatorname{im} S$.

The proof is left as an exercise.

In the following theorem, $L_1(\lambda)$ and $L_2(\lambda)$ are $p \times p$ matrix polynomials and γ is a simple, closed contour not intersecting $\operatorname{sp}(L_1)$ or $\operatorname{sp}(L_2)$.

Theorem 4.17 *Let (X_j^+, T_j^+, Y_j^+) be a γ-spectral triple of $L_j(\lambda)$ and denote the base space by $\mathfrak{M}_j^+$ $(j = 1, 2)$. Then the product $L(\lambda) = L_2(\lambda)L_1(\lambda)$ has the following γ-spectral triple:*

$$X_+ = [X_1^+ \ A], \qquad T_+ = \begin{pmatrix} T_1^+ & Y_1^+ \cdot X_2^+ \\ 0 & T_2^+ \end{pmatrix}, \qquad Y_+ = \begin{pmatrix} B \\ Y_2^+ \end{pmatrix}, \tag{31}$$

where

$$A = \frac{1}{2\pi i}\int_\gamma L_1^{-1}(\lambda)X_2^+(I\lambda - T_2^+)^{-1}\,d\lambda$$

$$B = \frac{1}{2\pi i}\int_\gamma (I\lambda - T_1^+)^{-1}Y_1^+\{L_2^{-1}(\lambda) - X_2^+(I\lambda - T_2^+)^{-1}Y_2^+\}\,d\lambda.$$

Proof We begin by letting (X_+, T_+, Y_+) be any γ-spectral triple of $L(\lambda)$, not necessarily of the form (31), with base space denoted $\mathfrak{M}^+$. Without loss of generality $(X_1^+, T_1^+) = (X_+|_{\mathfrak{L}}, T_+|_{\mathfrak{L}})$. In view of Lemma 4.16, the operator $M \in \mathscr{L}(\mathfrak{M}_1^+ \times \mathfrak{M}_2^+, \mathfrak{M}^+)$ defined by $M\begin{pmatrix} c_1 \\ c_2 \end{pmatrix} = c_1 + Sc_2$ is an isomorphism. We claim that

$$X_+M = [X_1^+ \ A], \qquad M^{-1}T_+M = \begin{pmatrix} T_1^+ & Y_1^+ \cdot X_2^+ \\ 0 & T_2^+ \end{pmatrix}, \qquad M^{-1}Y_+ = \begin{pmatrix} B \\ Y_2^+ \end{pmatrix} \tag{31'}$$

(This is sufficient to prove the theorem since an admissible triple that is similar to a γ-spectral triple of $L(\lambda)$ is itself a γ-spectral triple.)

By the definition of M one verifies easily that, for all $c_1 \in \mathfrak{M}_1^+$, $c_2 \in \mathfrak{M}_2^+$,

$$X_+M\begin{pmatrix} c_1 \\ c_2 \end{pmatrix} = X_+c_1 + X_+Sc_2 = [X_1^+ \ X_+S]\begin{pmatrix} c_1 \\ c_2 \end{pmatrix},$$

and by (29)

$$T_+M\begin{pmatrix} c_1 \\ c_2 \end{pmatrix} = T_+c_1 + T_+Sc_2 = M\begin{pmatrix} T_1^+ & Y_1^+ \cdot X_2^+ \\ 0 & T_2^+ \end{pmatrix}\begin{pmatrix} c_1 \\ c_2 \end{pmatrix}.$$

In view of (28), the first two equations in (31′) hold.

Further, note that the image of $Y_+ - SY_2^+$ is contained in $\mathfrak{M}_1^+$. Indeed, by Lemma 4.13, we have $QY_+ = Y_2^+$ so that

$$Y_+ - SY_2^+ = (I - SQ)Y_+ = PY_+.$$

Since the image of P is equal to $\mathfrak{M}_1^+$ (Lemma 4.16), the image of $Y_+ - SY_2^+$ is a subspace of $\mathfrak{M}_1^+$ and then

$$M\begin{pmatrix} Y_+ - SY_2^+ \\ Y_2^+ \end{pmatrix} = Y_+ - SY_2^+ + SY_2^+ = Y_+.$$

Hence the last equation in (31′) holds provided we show that $Y_+ - SY_2^+ = B$. But consider the fact that

$$X_+ e^{tT_+}(Y_+ - SY_2^+) = X_+ e^{tT_+} Y_+ - X_+ e^{tT_+} SY_2^+,$$

so that by virtue of (27) we obtain

$$\begin{aligned} X_+ e^{tT_+}(Y_+ - SY_2^+) &= \frac{1}{2\pi i}\int_\gamma e^{t\lambda} L_1^{-1}(\lambda)\{L_2^{-1}(\lambda) - X_2^+(I\lambda - T_2^+)^{-1}Y_2^+\}\, d\lambda \\ &= \frac{1}{2\pi i}\int_\gamma e^{t\lambda} X_1^+(I\lambda - T_1^+)^{-1}Y_1^+ \\ &\quad \times \{L_2^{-1}(\lambda) - X_2^+(I\lambda - T_2^+)^{-1}Y_2^+\}\, d\lambda \\ &= X_1^+ e^{tT_1^+} B \\ &= X_+ e^{tT^+} B. \end{aligned}$$

Since $\operatorname{col}(X_+T_+^j)_{j=0}^{l-1}$ is injective, it follows that $Y_+ - SY_2^+ = B$.

Remark Theorem 4.17 generalizes the result of Theorem 3.2 for if we assume that $L_1(\lambda)$ and $L_2(\lambda)$ are monic and consider standard triples for $L_1(\lambda)$ and $L_2(\lambda)$ (i.e. let γ be a contour having all of $\operatorname{sp}(L_1)$ and $\operatorname{sp}(L_2)$ in its interior), it is easily seen that $A = 0$ and $B = 0$.

It is convenient for some purposes to state Theorem 4.17 in a modified form. Recall that we have let $\mathfrak{L}$ denote the unique subspace such that $(X_+|_\mathfrak{L}, T_+|_\mathfrak{L})$ is a γ-spectral pair of $L_1(\lambda)$.

Theorem 4.17′ *Let (X_+, T_+) be a γ-spectral pair of the product $L(\lambda) = L_2(\lambda)L_1(\lambda)$. Also let (X_j^+, T_j^+) be a γ-spectral pair of $L_j(\lambda)$ $(j = 1, 2)$. Then there exists a unique isomorphism $M \in \mathscr{L}(\mathfrak{M}_1^+ \times \mathfrak{M}_2^+, \mathfrak{M}^+)$ such that*

$$X_+M = [X_1^+ \; A], \qquad T_+M = M\begin{pmatrix} T_1^+ & Y_1^+ \cdot X_2^+ \\ 0 & T_2^+ \end{pmatrix},$$

where A is defined in the statement of Theorem 4.17 and $Y_1^+ \in \mathscr{L}(\mathbb{C}^p, \mathfrak{M}_1^+)$ is the unique operator such that (X_1^+, T_1^+, Y_1^+) is a γ-spectral triple of $L_1(\lambda)$. Further, if $(X_1^+, T_1^+) = (X_+|_\mathfrak{L}, T_+|_\mathfrak{L})$, then M has the form $M\begin{pmatrix} c_1 \\ c_2 \end{pmatrix} = c_1 + Sc_2$, where $S \in \mathscr{L}(\mathfrak{M}_2^+, \mathfrak{M}^+)$ is the unique operator satisfying the equations (28), (29).

The existence of M follows from the proof of Theorem 4.17 and uniqueness is a consequence of the injectivity of $\operatorname{col}(X_+T_+^j)_{j=0}^{l-1}$. The fact that there is only one operator S that satisfies (28), (29) also follows from the injectivity of $\operatorname{col}(X_+T_+^j)_{j=0}^{l-1}$ for if both S and S' satisfy (28), (29) then $X_+(S - S') = 0$ and $T_+ \cdot (S - S') = (S - S') \cdot T_2^+$ which implies

$$X_+T_+^j(S - S') = 0 \qquad j = 0, 1, \ldots$$

whence $S - S' = 0$.

Remark If we choose the first two operators in (31) as the γ-spectral pair (X_+, T_+) then $M = I$. It follows in this case that the operators S and Q in Lemmas 4.12 and 4.15 are

$$Sc_2 = \begin{pmatrix} 0 \\ c_2 \end{pmatrix}, \qquad Q\begin{pmatrix} c_1 \\ c_2 \end{pmatrix} = c_2, \qquad c_1 \in \mathfrak{M}_1^+, c_2 \in \mathfrak{M}_2^+$$

We now introduce rectangular matrix polynomials (these occur in boundary problems for elliptic systems):

$$B_i(\lambda) = \sum_{j=0}^{m_i} B_j^{(i)}\lambda^j, \qquad i = 1, 2$$

of degree m_i where the coefficients $B_j^{(i)}$ are $r_i \times p$ matrices, and then let

$$B(\lambda) = \begin{pmatrix} B_1(\lambda) \\ B_2(\lambda)L_1(\lambda) \end{pmatrix} = \sum_{j=0}^{m} \beta_j\lambda^j$$

which is a matrix polynomial of degree $\leqslant m = \max(m_1, m_2 + l_1)$ where the coefficients β_j are $(r_1 + r_2) \times p$ matrices.

Corollary 4.18 *Let* $\Delta_i = \sum_{j=0}^{m_i} B_j^{(i)} X_i^+ (T_i^+)^j$, $i = 1, 2$, *and* $\Delta = \sum_{j=0}^{m} \beta_j X_+ T_+^j$. *Then* $\Delta \cdot M = \begin{pmatrix} \Delta_1 & Z \\ 0 & \Delta_2 \end{pmatrix}$, *where M denotes the isomorphism in Theorem* 4.17′ *and*

$$Z = \frac{1}{2\pi i}\int_\gamma B_1(\lambda)L_1^{-1}(\lambda)X_2^+(I\lambda - T_2^+)^{-1}\, d\lambda$$

Proof The isomorphism $M: \mathfrak{M}_1^+ \times \mathfrak{M}_2^+ \to \mathfrak{M}^+$ is defined by $M\begin{pmatrix} c_1 \\ c_2 \end{pmatrix} = c_1 + Sc_2$. Note that $\Delta = B(d/dt)X_+ e^{tT_+}|_{t=0}$, so that if $c_1 \in \mathfrak{M}_1^+$ then

$$\begin{aligned}
\Delta c_1 &= B(d/dt)X_1^+ e^{tT_1^+} c_1|_{t=0} \\
&= \begin{pmatrix} B_1(d/dt) \\ B_2(d/dt)L_1(d/dt) \end{pmatrix} X_1^+ e^{tT_1^+}|_{t=0}\, c_1 \\
&= \begin{pmatrix} \Delta_1 c_1 \\ 0 \end{pmatrix}
\end{aligned}$$

where we used the fact that $L_1(d/dt)X_1^+ e^{tT_1^+} \equiv 0$ since (X_1^+, T_1^+) is a γ-spectral pair of $L_1(\lambda)$. Next we use (27) to obtain

$$\begin{aligned}
\Delta \cdot S &= B(d/dt)X_+ e^{tT_+} S|_{t=0} \\
&= \begin{pmatrix} B_1(d/dt) \\ B_2(d/dt)L_1(d/dt) \end{pmatrix} \frac{1}{2\pi i}\int_\gamma e^{t\lambda} L_1^{-1}(\lambda)X_2^+(I\lambda - T_2^+)^{-1}\, d\lambda|_{t=0} \\
&= \begin{pmatrix} Z \\ \dfrac{1}{2\pi i}\displaystyle\int_\gamma B_2(\lambda)X_2^+(I\lambda - T_2^+)^{-1}\, d\lambda \end{pmatrix} \\
&= \begin{pmatrix} Z \\ \Delta_2 \end{pmatrix}.
\end{aligned}$$

The proof is complete.

4.5 Transformations of products

Let $L(\lambda)$ and $L_1(\lambda)$ be $p \times p$ matrix polynomials such that $L(\lambda)L_1^{-1}(\lambda)$ has an analytic continuation inside γ. Let (X_+, T_+) be a γ-spectral pair of $L(\lambda)$. As shown in §4.3 the subspace

$$\mathfrak{L} = \{\eta;\, \eta \in \mathfrak{M}^+,\, L_1(d/dt)X_+\, e^{tT_+}\, \eta \equiv 0\}$$

is the unique T_+-invariant subspace of $\mathfrak{M}^+$ such that $(X_+|_{\mathfrak{L}}, T_+|_{\mathfrak{L}})$ is a γ-spectral pair of $L_1(\lambda)$.

Let $\varphi(\lambda) = (a\lambda + b)/(c\lambda + d)$ be a Möbius transformation as in §2.4 with $-d/c$ outside γ. The transformed matrix polynomial

$$\tilde{L}(\lambda) = \sum_{j=0}^{l} A_j(d\lambda - b)^j(-c\lambda + a)^{l-j} = (-c\lambda + a)^l L(\varphi^{-1}(\lambda))$$

has the $\tilde{\gamma}$-spectral pair $(\tilde{X}_+, \tilde{T}_+)$ which is defined in Theorem 2.21. Ordinarily we choose l to be the degree of $L(\lambda)$, i.e. the minimal integer such that $A_j = 0$ for $j > l$, but it will be important for the proof of the lemma below to be able to increase l arbitrarily by adding on zero coefficients. This has the effect of multiplying $\tilde{L}(\lambda)$ by powers of $-c\lambda + a$, but since a/c lies outside $\tilde{\gamma}$ this does not change the spectrum of $\tilde{L}(\lambda)$ inside $\tilde{\gamma}$, nor does it affect the $\tilde{\gamma}$-spectral pairs (see Corollary 4.9).

The following lemma implies that the subspace $\mathfrak{L}$ is unchanged by the transformation $\sim$, that is, $\tilde{\mathfrak{L}} = \mathfrak{L}$.

Lemma 4.19 $(\tilde{X}_+|_{\mathfrak{L}}, \tilde{T}_+|_{\mathfrak{L}})$ *is a* $\tilde{\gamma}$*-spectral pair of* $\tilde{L}_1(\lambda)$.

Proof Since a/c lies outside $\tilde{\gamma}$, the number of zeros of $\det L_1(\lambda)$ inside γ is equal to the number of zeros of $\det \tilde{L}_1(\lambda)$ inside $\tilde{\gamma}$. Also

$$\tilde{L}(\lambda)\tilde{L}_1^{-1}(\lambda) = (-cd + a)^{l-l_1}L(\varphi^{-1}(\lambda))L_1^{-1}(\varphi^{-1}(\lambda))$$

is analytic inside $\tilde{\gamma}$ so there is a unique $\tilde{T}_+$-invariant subspace $\tilde{\mathfrak{L}}$ such that $(\tilde{X}_+|_{\tilde{\mathfrak{L}}}, \tilde{T}_+|_{\tilde{\mathfrak{L}}})$ is a $\tilde{\gamma}$-spectral pair of $\tilde{L}_1(\lambda)$. To prove the lemma it suffices to show that

$$\tilde{L}_1(d/dt)\tilde{X}_+\, e^{t\tilde{T}_+}\, \eta \equiv 0 \qquad \text{for all } \eta \in \mathfrak{L} \tag{32}$$

This would imply that $\mathfrak{L} \subset \tilde{\mathfrak{L}}$, and since $\dim \tilde{\mathfrak{L}} = \dim \mathfrak{L} = r_1$ (the number of zeros of $\det L_1(\lambda)$ inside γ, or $\det \tilde{L}_1(\lambda)$ inside $\tilde{\gamma}$) then $\tilde{\mathfrak{L}} = \mathfrak{L}$ must hold.

Without loss of generality, by adding on zero coefficients to either $L(\lambda)$ or $L_1(\lambda)$, we may assume that the same "degree" is used in the definitions of $\tilde{L}(\lambda)$ and $\tilde{L}_1(\lambda)$, that is, $l = l_1$.

We will use formulas (20), (21) from Theorem 2.21. As in the proof of that theorem we assume without loss of generality that $ad - bc = 1$. Let the coefficients of $L_1(\lambda)$, $\tilde{L}_1(\lambda)$ be denoted $A_j^{(1)}$, $\tilde{A}_j^{(1)}$, respectively, $j = 0, \ldots, l$. By 2.4(20) we have

$$[\tilde{A}_0^{(1)} \cdots \tilde{A}_l^{(1)}] = [A_0^{(1)} \cdots A_l^{(1)}] \cdot \mathscr{H},$$

and by making use of 2.4(21)

$$\begin{aligned}\sum_{j=0}^{l} \tilde{A}_j^{(1)}\tilde{X}_+\tilde{T}_+^j &= [\tilde{A}_0^{(1)}\cdots\tilde{A}_l^{(1)}]\cdot\operatorname{col}(\tilde{X}_+\tilde{T}_+^j)_{j=0}^l \\ &= [A_0^{(1)}\cdots A_l^{(1)}]\cdot\operatorname{col}(X_+T_+^j)_{j=0}^l(cT_+ + dI)^{-l} \\ &= \sum_{j=0}^{l} A_j^{(1)}X_+T_+^j(cT_+ + dI)^{-l}.\end{aligned}$$

Thus

$$\tilde{L}_1(d/dt)\tilde{X}_+ \mathrm{e}^{t\tilde{T}_+}\eta = \sum_{j=0}^{l} A_j^{(1)}X_+T_+^j\tilde{\eta}, \tag{33}$$

where $\tilde{\eta} = (cT_+ + dI)^{-l}e^{t\tilde{T}_+}\eta$ (for a fixed $t \in \mathbb{R}$). Now let $\eta \in \mathfrak{L}$. Since $\mathfrak{L}$ is invariant under T_+ it is also invariant under $cT_+ + dI$. But $-d/c$ lies outside γ so $cT_+ + dI$ is invertible and $\mathfrak{L}$ is therefore invariant under the inverse operator $(cT_+ + dI)^{-1}$. Since $\tilde{T}_+ = (aT_+ + bI)(cT_+ + dI)^{-1}$ then $\tilde{T}_+(\mathfrak{L}) \subset \mathfrak{L}$. Hence $\tilde{\eta} \in \mathfrak{L}$ and, in view of (33), it follows that (32) holds.

We now consider the effect of the transformation $\tilde{}$ on a product of matrix polynomials, $L(\lambda) = L_2(\lambda)L_1(\lambda)$, and on the operators Q and S associated with this product as in §4.4.

Let l_1, l_2 and l denote the degrees of $L_1(\lambda)$, $L_2(\lambda)$ and $L(\lambda)$. Then $l \leqslant l_1 + l_2$ and we have $(-c\lambda + a)^{l_1+l_2-l}\tilde{L}(\lambda) = \tilde{L}_2(\lambda)\tilde{L}_1(\lambda)$. By adding zero coefficients to $L(\lambda)$ (see the remarks preceding Lemma 4.19) we may assume that $l = l_1 + l_2$ and then $\tilde{L}(\lambda) = \tilde{L}_2(\lambda)\tilde{L}_1(\lambda)$.

In §4.4 the operators Q, S enabled us to construct γ-spectral triples of a product in terms of γ-spectral triples of each factor. The question we now consider is this: what is the relationship between the operators $\tilde{Q}$, $\tilde{S}$ defined with respect to the product $\tilde{L}_2(\lambda)\tilde{L}_1(\lambda)$ and the operators Q, S of the original product $L_2(\lambda)L_1(\lambda)$? This question is answered by Proposition 4.22 below.

As usual, (X_+, T_+) is a γ-spectral pair of $L(\lambda)$ with base space $\mathfrak{M}^+$, and $(\tilde{X}_+, \tilde{T}_+)$ denotes its transformation under φ as in Theorem 2.21. Also (X_i^+, T_i^+) is a γ-spectral pair of $L_i(\lambda)$ $(i = 1, 2)$ with base space $\mathfrak{M}_i^+$, with $(\tilde{X}_i^+, \tilde{T}_i^+)$ the corresponding transformation under φ. Without loss of generality we may assume that $(X_1^+, Y_1^+) = (X_+|_{\mathfrak{L}}, T_+|_{\mathfrak{L}})$, where $\mathfrak{L}$ is the subspace (24). In view of Lemma 4.19, $(\tilde{X}_1^+, \tilde{T}_1^+) = (\tilde{X}_+|_{\mathfrak{L}}, \tilde{T}_+|_{\mathfrak{L}})$ for the same subspace $\mathfrak{L}$. Also we let Y_1^+ be the unique operator such that (X_1^+, T_1^+, Y_1^+) is a γ-spectral triple of $L_1(\lambda)$; then $(\tilde{X}_1^+, \tilde{T}_1^+, \tilde{Y}_1^+)$, its transformation under φ, is a $\tilde{\gamma}$-spectral triple of $\tilde{L}_1(\lambda)$.

Lemma 4.20 *Under the hypotheses and notation as stated above, we have*

$$\tilde{T}_+S = S\tilde{T}_2^+ + (-c\tilde{T}_1^+ + aI)Y_1^+\cdot X_2^+(cT_2^+ + dI)^{-1} \tag{34}$$

Proof By Theorem 4.17′

$$X_+ = [X_1^+ \; A]M^{-1}, \qquad T_+ = M\begin{pmatrix} T_1^+ & Y_1^+\cdot X_2^+ \\ 0 & T_2^+ \end{pmatrix}M^{-1},$$

where $M\begin{pmatrix} x \\ y \end{pmatrix} = x + Sy$. Then

$$\tilde{T}_+ = (aT_+ + bI)(cT_+ + dI)^{-1}$$
$$= M\begin{pmatrix} aT_1^+ + bI & aY_1^+ \cdot X_2^+ \\ 0 & aT_2^+ + bI \end{pmatrix}\begin{pmatrix} cT_1^+ + dI & cY_1^+ \cdot X_2^+ \\ 0 & cT_2^+ + dI \end{pmatrix}^{-1} M^{-1}$$

and, using the fact that for an upper triangular block matrix of the form $\begin{pmatrix} A & C \\ 0 & B \end{pmatrix}$ the inverse is given by $\begin{pmatrix} A^{-1} & -A^{-1}CB^{-1} \\ 0 & B^{-1} \end{pmatrix}$, it follows that

$$\tilde{T}_+ M = M\begin{pmatrix} \tilde{T}_1^+ & (-c\tilde{T}_1^+ + aI)\, Y_1^+ X_2^+ (cT_2^+ + dI)^{-1} \\ 0 & \tilde{T}_2^+ \end{pmatrix} \tag{35}$$

Now (34) is an immediate consequence of (35) since $M\begin{pmatrix} 0 \\ y \end{pmatrix} = Sy$.

Lemma 4.21 *We have*

$$\tilde{X}_+ \tilde{S} = (ad - bc)^{-l_1} X_+ (cT_+ + dI)^{l_1 - 1} S (cT_2^+ + dI) \tag{36}$$

Proof In view of (28) in §4.4 we have

$$\tilde{X}_+ \tilde{S} = \frac{1}{2\pi i} \int_{\tilde{\gamma}} \tilde{L}_1^{-1}(\lambda) \tilde{X}_2^+ (I\lambda - \tilde{T}_2^+)^{-1}\, d\lambda$$
$$= \frac{1}{2\pi i} \int_{\tilde{\gamma}} (-c\lambda + a)^{-l_1} L_1^{-1}(\varphi^{-1}(\lambda)) X_2^+ (I\lambda - \varphi(T_2^+))^{-1}\, d\lambda$$

and, with the substitution $\lambda = \varphi(w) = (aw + b)/(cw + d)$, it follows that

$$\tilde{X}_+ \tilde{S} = \frac{1}{2\pi i} \int_{\gamma} (-c\varphi(w) + a)^{-l_1} L_1^{-1}(w) X_2^+ (I\varphi(w) - \varphi(T_2^+))^{-1} \frac{ad - bc}{(cw + d)^2}\, dw.$$

Since $-c\varphi(w) + a = (ad - bc)/(cw + d)$ and

$$I\varphi(w) - \varphi(T_2^+) = \frac{ad - bc}{cw + d}(Iw - T_2^+)(cT_2^+ + dI)^{-1},$$

we obtain

$$\tilde{X}_+ \tilde{S} = (ad - bc)^{-l_1} \frac{1}{2\pi i}$$
$$\times \int_{\tilde{\gamma}} (cw + d)^{l_1 - 1} L_1^{-1}(w) X_2^+ (Iw - T_2^+)^{-1}\, dw\, (cT_2^+ + dI)$$
$$= (ad - bc)^{-l_1} X_+ (cT_+ + dI)^{l_1 - 1} S (cT_2^+ + dI).$$

Proposition 4.22 *We have also*

$$\tilde{S} = (ad - bc)^{-l_1}(cT_+ + dI)^{l_1 - 1}S(cT_2^+ + dI) \tag{37}$$

$$\tilde{Q} = (ad - bc)^{l_1}(cT_2^+ + dI)^{-l_1}Q \tag{38}$$

and

$$\tilde{M} = M\begin{pmatrix} I & Z \\ 0 & (ad - bc)^{-l_1}(cT_2^+ + dI)^{l_1} \end{pmatrix}, \tag{39}$$

where I is the identity operator on $\mathfrak{M}_1^+$ *and Z is the operator from* $\mathfrak{M}_2^+$ *to* $\mathfrak{M}_1^+$ *given by*

$$Z = c(ad - bc)^{-l_1} \sum_{j+k=l_1-2} (cT_1^+ + dI)^j Y_1^+ \cdot X_2^+(cT_2^+ + dI)^{k+1} \tag{40}$$

Proof Let $\tilde{S}'$ denote the right-hand side of (37). In view of Lemma 4.21, we have $\tilde{X}_+\tilde{S} = \tilde{X}_+\tilde{S}'$, and hence

$$\tilde{X}_+(\tilde{S} - \tilde{S}') = 0. \tag{41}$$

Also, by Lemma 4.20, we have

$$\tilde{T}_+S = S\tilde{T}_2^+ + (c\tilde{T}_1^+ + aI)Y_1^+ \cdot X_2^+(cT_2^+ + dI)^{-1}$$

Then, since $\tilde{X}_2^+ = X_2^+$, $\tilde{Y}_1^+ = (ad - bc)^{-(l_1-1)}(cT_1^+ + dI)^{l_1-2}Y_1^+$ and $-c\tilde{T}_1 + aI = (ad - bc)(cT_1^+ + dI)^{-1}$, it follows that

$$\tilde{T}_+S = S\tilde{T}_2^+ + (ad - bc)^{l_1}(cT_1^+ + dI)^{-(l_1-1)}\tilde{Y}_1^+\tilde{X}_2^+(cT_2^+ + dI)^{-1}.$$

Upon multiplying this equation by $(ad - bc)^{-l_1}(cT_+ + dI)^{l_1-1}$ on the left and $cT_2^+ + dI$ on the right, we obtain

$$\tilde{T}_+\tilde{S}' = \tilde{S}'\tilde{T}_2^+ + \tilde{Y}_1^+\tilde{X}_2^+. \tag{42}$$

However, due to (28) and the definition of $\tilde{S}$, we also have $\tilde{T}_+\tilde{S} = \tilde{S}\tilde{T}_2^+ + \tilde{Y}_1^+\tilde{X}_2^+$ and subtraction of (42) from this equation yields

$$\tilde{T}_+(\tilde{S} - \tilde{S}') = (\tilde{S} - \tilde{S}')T_2^+ \tag{43}$$

Now (41) and (43) imply that

$$\tilde{X}_+\tilde{T}_+^j(\tilde{S} - \tilde{S}') = 0 \qquad \text{for all } j = 0, 1, \ldots.$$

Hence $\tilde{S} - \tilde{S}' = 0$ since $\operatorname{col}(\tilde{X}_+\tilde{T}_+^j)_{j=0}^{l-1}$ is injective. Thus $\tilde{S}' = \tilde{S}$ and (37) holds.

Next we prove (39). By the corollary to Lemma 4.23 (see below), it follows from (37) that

$$\tilde{S} = (ad - bc)^{-l_1}S(cT_2^+ + dI)^{l_1} + Z,$$

where Z is given by (40). Then, since $M\begin{pmatrix} x \\ y \end{pmatrix} = x + Sy$ and $\tilde{M}\begin{pmatrix} x \\ y \end{pmatrix} = x + \tilde{S}y$ for all $x \in \mathfrak{M}_1^+$ and $y \in \mathfrak{M}_2^+$, it follows that (39) holds.

To prove (38), recall from Lemma 4.16 that $M^{-1} = \begin{pmatrix} P \\ Q \end{pmatrix}$. Thus $Q = [0\ I]M^{-1}$, where I is the identity operator on $\mathfrak{M}_2^+$; similarly, $\tilde{Q} = [0\ I]\tilde{M}^{-1}$. Then

$$\begin{aligned}\tilde{Q} &= [0\ I]\tilde{M}^{-1}\\ &= [0\ I]\begin{pmatrix} I & * \\ 0 & (ad-bc)^{l_1}(cT_2^+ + dI)^{-l_1}\end{pmatrix}M^{-1}\\ &= (ad-bc)^{l_1}(cT_2^+ + dI)^{-l_1}Q.\end{aligned}$$

The asterisk * denotes the operator $-Z(ad-bc)^{l_1}(cT_2^+ + dI)^{-l_1}$, and is of no significance in the derivation of the formula (38). This completes the proof of the theorem.

Remark If the numbers a, b, c and d are multiplied by $k \neq 0$ then the Möbius transformation φ is unchanged but $\tilde{L}_1(\lambda)$ is replaced by $k^{l_1}\tilde{L}_1(\lambda)$ and then $\tilde{S}$ is replaced by $k^{-l_1}\tilde{S}$ and $\tilde{Q}$ is replaced by $k^{l_1}Q$. This is the reason for the factors $ad - bc$ in the formulas (37) and (38). Note that by choosing $k = (ad-bc)^{-1}$, we may assume without loss of generality that $ad - bc = 1$ in all formulas in this section.

Remark If the formulas (37) and (38) are inverted, we obtain the following:

$$\begin{aligned}S &= (-c\tilde{T}_+ + aI)^{l_1-1}\tilde{S}(-c\tilde{T}_2^+ + aI),\\ Q &= (-c\tilde{T}_2^+ + aI)^{-l_1}\tilde{Q},\end{aligned}$$

and the analogues of (34) and (36) are

$$T_+\tilde{S} = \tilde{S}T_2^+ + (cT_1^+ + dI)\tilde{Y}_1^+ \cdot \tilde{X}_2^+(-c\tilde{T}_2^+ + aI)^{-1}$$

and

$$X_+S = \tilde{X}_+(-c\tilde{T}_+ + aI)^{l_1-1}\tilde{S}(-c\tilde{T}_2^+ + aI).$$

Lemma 4.23

$$T_+^m S = S(T_2^+)^m + \sum_{j+k=m-1} (T_1^+)^j Y_1^+ \cdot X_2^+(T_2^+)^k \tag{44}$$

Proof By (29) of §4.4 we have

$$T_+S = ST_2^+ + Y_1^+ \cdot X_2^+,$$

so that the lemma holds if $m = 1$, and then we may proceed by induction on m.

Corollary *For any real numbers c and d,*

$$(cT_+ + dI)^m S = S(cT_2^+ + dI)^m + c\sum_{j+k=m-1}(cT_1^+ + dI)^j Y_1^+ \cdot X_2^+(cT_2^+ + dI)^k$$

Proof The formula (44) holds if $T_+ + \alpha I$ is substituted for T_+ (since (29) remains true). Now if $c \neq 0$, simply let $\alpha = d/c$.

In §16.2 we will need a generalized version of Corollary 4.18.

Proposition 4.24 *In addition to the hypotheses and notation of Corollary* 4.18, *let* $\varphi(\lambda) = (a\lambda + b)/(c\lambda + d)$ *be a Möbius transformation with* $-d/c$ *in the exterior of* γ. *Let* $\tilde{B}_i(\lambda) = \sum_{j=0}^{m_i} \tilde{B}_j^{(i)}\lambda^j$ *denote the transformation of* $B_i(\lambda)$, $i = 1, 2$. *Also let*

$$\tilde{B}(\lambda) = \begin{pmatrix} \tilde{B}_1(\lambda) \\ \tilde{B}_2(\lambda)\tilde{L}_1(\lambda) \end{pmatrix} = \sum_{j=0}^{m} \tilde{\beta}_j \lambda^j.$$

Without loss of generality, $ad - bc = 1$, *and then the following formula holds:*

$$\tilde{\Delta} \cdot M = \begin{pmatrix} \tilde{\Delta}_1 & * \\ 0 & \tilde{\Delta}_2 \cdot (cT_2^+ + dI)^{-l_1} \end{pmatrix},$$

where

$$\tilde{\Delta} = \sum_{j=0}^{m} \tilde{\beta}_j \tilde{X}_+ \tilde{T}_+^j, \qquad \tilde{\Delta}_i = \sum_{j=0}^{m} \tilde{B}_j^{(i)} \tilde{X}_i^+ (\tilde{T}_i^+)^j, \qquad i = 1, 2$$

and $(\tilde{X}_+, \tilde{T}_+)$ *and* $(\tilde{X}_i^+, \tilde{T}_i^+)$, $i = 1, 2$, *are the transformations of* (X_+, T_+) *and* (X_i^+, T_i^+), *respectively. The asterisk* * *denotes an operator from* $\mathfrak{M}_2^+$ *to* $\mathbb{C}^{r_1}$ *for which an explicit formula is given below.*

Remark $\tilde{B}(\lambda)$ itself is not the transformation of $B(\lambda)$, unless $m_1 = m_2 + l_1$. Also, the transformation of $L_2(\lambda)L_1(\lambda)$ is not equal to $\tilde{L}_2(\lambda)\tilde{L}_1(\lambda)$ unless $l = l_1 + l_2$; see the discussion preceding Lemma 4.19.

Proof We first apply Proposition 4.18 to the product $\tilde{L}_2(\lambda)\tilde{L}_1(\lambda)$ and the matrix polynomials $\tilde{B}_1(\lambda)$, $\tilde{B}_2(\lambda)$ and $\tilde{B}(\lambda)$ to obtain

$$\tilde{\Delta} \cdot \tilde{M} = \begin{pmatrix} \tilde{\Delta}_1 & \tilde{Z} \\ 0 & \tilde{\Delta}_2 \end{pmatrix},$$

where

$$\tilde{Z} = \frac{1}{2\pi i} \int_\gamma \tilde{B}_1(\lambda) \tilde{L}_1^{-1}(\lambda) \tilde{X}_2^+ (I\lambda - \tilde{T}_2^+)^{-1} \, d\lambda.$$

Also, by Proposition 4.22 we have

$$\tilde{M} = M \begin{pmatrix} I & W \\ 0 & (cT_2^+ + dI)^{l_1} \end{pmatrix},$$

where

$$W = c \sum_{j+k=l_1-2} (cT_1^+ + dI)^j Y_1^+ X_2^+ (cT_2^+ + dI)^{k+1}.$$

Therefore

$$\begin{aligned}
\tilde{\Delta}\cdot M &= \tilde{\Delta}\tilde{M}\cdot(M^{-1}\tilde{M})^{-1} \\
&= \begin{pmatrix} \tilde{\Delta}_1 & \tilde{Z} \\ 0 & \tilde{\Delta}_2 \end{pmatrix} \begin{pmatrix} I & W \\ 0 & (cT_2^+ + dI)^{-l_1} \end{pmatrix}^{-1} \\
&= \begin{pmatrix} \tilde{\Delta}_1 & \tilde{Z} \\ 0 & \tilde{\Delta}_2 \end{pmatrix} \begin{pmatrix} I & -W(cT_2^+ + dI)^{-l_1} \\ 0 & (cT_2^+ + dI)^{-l_1} \end{pmatrix} \\
&= \begin{pmatrix} \tilde{\Delta}_1 & * \\ 0 & \tilde{\Delta}_2\cdot(cT_2^+ + dI)^{-l_1} \end{pmatrix}
\end{aligned}$$

where the asterisk * denotes the operator $(-\tilde{\Delta}_1 W + \tilde{Z})(cT_2^+ + dI)^{-l_1}$.

Part II

Manifolds and Vector Bundles

5
Manifolds and vector bundles

5.1 Background and notation

We begin with a brief review of differential calculus in finite dimensional spaces. Let E and F be finite dimensional spaces over $\mathbb{R}$ with norms denoted by $|\ |$. (All norms are equivalent on finite dimensional vector spaces.) A function $f: U \to F$ where U is an open subset of E is *differentiable* at $x \in U$ if there is an element $Df(x) \in L(E, F)$ such that

$$|f(x + h) - f(x) - Df(x)h| = o(|h|), \qquad h \to 0$$

Here $L(E, F)$ is the space of linear transformations from E to F, which we endow with the norm $\|T\| = \sup_{|x| \leqslant 1} |Tx|$. The notation $o(t)$ denotes a real-valued function of a real variable such that $\lim_{t \to 0} o(t)/t = 0$.

We denote by $C^1(U, F)$ the set of continuously differentiable functions from U to F, that is, the set of functions $f: U \to F$ which are differentiable at every point and for which $U \ni x \mapsto Df(x) \in L(E, F)$ is continuous.

We will often need to use the chain rule which states that if $f: U \to V$ is differentiable at x_0 (U open in E, V open in F) and if $g: V \to G$ is differentiable at $f(x_0)$, then $g \circ f$ is differentiable at x_0 and

$$D(g \circ f)(x_0) = Dg(f(x_0)) \circ Df(x_0) \tag{1}$$

The formula shows that if f and g are of class C^1 then so is $g \circ f$.

Let $f: U \subset E \to F$ be of class C^1, that is, the map

$$Df: U \to L(E, F)$$

is continuous. Since $L(E, F)$ is itself a vector space we can consider the second derivative $D^2f = D(Df)$ which, if it exists, is a map $U \to L(E, L(E, F))$. The space $L(E, L(E, F))$ can be identified with $L^2(E, F)$, the set of bilinear maps $E \times E \to F$, so we have

$$D^2f: U \to L^2(E, F),$$

and we say that the *value* of $D^2f(x)$ at $(h, v) \in E \times E$ is $D^2f(x)(h, v) := (D^2f(x) \cdot h) \cdot v$.

Higher-order derivatives are defined inductively. Let $E^{(r)} = E \times \cdots \times E$ and let $L^r(E, F)$ denote the space of r-multilinear maps from E to F, that is, the set of maps $E^{(r)} \to F$ which are linear in each argument separately. We endow the vector space $L^r(E, F)$ with the norm $\|T\| = \sup_{|x_i| \leqslant 1} |T(x_1, \ldots, x_r)|$. Note that $L(E, L^{r-1}(E, F))$ can be identified with $L^r(E, F)$; the identification is given by the equation $T(x_1, \ldots, x_r) = T(x_1)(x_2, \ldots, x_r)$ and is norm-preserving. The rth derivative $D^r f$ is defined as $D(D^{r-1}f)$ and is a map of U into $L(E, L^{r-1}(E, F))$, so

$$D^r f\colon U \to L^r(E, F)$$

A function $f\colon U \to F$ is said to be of *class* C^r if all derivatives $D^k f\colon U \to L^k(E, F)$, $1 \leqslant k \leqslant r$, are defined and continuous. We denote the set of all such maps by $C^r(U, F)$. Note that if $F = \mathbb{R}$ we write $L^r(E) = L^r(E, \mathbb{R})$ and $C^r(U) = C^r(U, \mathbb{R})$.

If f is differentiable at $x \in U$ we say that $Df(x)\cdot v$ is the *directional derivative* of f at x in the direction $v \in E$, and it follows from the chain rule that the directional derivative is given by

$$Df(x)\cdot v = \frac{d}{dt} f(x + tv)|_{t=0} \tag{2}$$

By repeated application of this result we obtain the values of $D^r f(x)$,

$$D^r f(x)\cdot(v_1, \ldots, v_r) = \frac{d}{dt_r}\cdots\frac{d}{dt_1} f(x + t_1 v_1 + \cdots + t_r v_r)|_{t_1 = \cdots = t_r = 0}, \tag{3}$$

a formula which is often convenient for calculation of higher-order derivatives.

Example

(i) If $f \in L(E, F)$, a linear map $E \to F$, then $Df(x) = f$ for all $x \in E$. Since Df is constant (independent of x), it follows that $D^2 f = 0$.

(ii) If $f \in L^2(E, F)$, a bilinear map $E \times E \to F$, then

$$|f(x_1 + h_1, x_2 + h_2) - f(x_1, x_2) - [f(h_1, x_2) + f(x_1, h_2)]| = |f(h_1, h_2)|,$$

and the right-hand side is $o(|(h_1, h_2)|)$, where $|(h_1, h_2)| = (|h_1|^2 + |h_2|^2)^{1/2}$. It follows that the derivative $Df(x_1, x_2)\colon E \times E \to F$ is given by

$$Df(x_1, x_2)\cdot(h_1, h_2) = f(h_1, x_2) + f(x_1, h_2)$$

Differentiating once more we obtain

$$D^2 f(x_1, x_2)\cdot((h_1, h_2), (k_1, k_2)) = f(h_1, k_2) + f(k_1, h_2)$$

Hence the value of $D^2 f$ is constant (independent of $x = (x_1, x_2)$), and therefore $D^3 f = 0$.

(iii) If $f \in L^r(E, F)$, a multilinear map $E^{(r)} \to F$, then

$$Df(x_1, \ldots, x_r)\cdot(h_1, \ldots, h_r) = f(h_1, x_2, \ldots, x_r) + f(x_1, h_2, \ldots, x_r) + \cdots + f(x_1, \ldots, x_{r-1}, h_r)$$

Proceeding inductively, it is clear that f is C^∞ and $D^k f = 0$ when $k > r$.

Proposition 5.1 *If* $f: U \to V$ and $g: V \to G$ *are of class* C^r, *then* $g \circ f$ *is also class* C^r.

Proof The map $\varphi: L(F, G) \times L(E, F) \to L(E, G)$ defined by $(A, B) \mapsto A \circ B$ is bilinear, whence C^∞. By the chain rule (1), $D(g \circ f)$ is the composition of two maps:

$$U \to L(F, G) \times L(E, F) \xrightarrow{\varphi} L(E, G)$$

$$x \mapsto (Dg(f(x)), Df(x)) \mapsto Dg(f(x)) \circ Df(x),$$

where the first map is C^{r-1} since f and g are C^r, and the second map is C^∞. Thus $D(g \circ f)$ is C^{r-1}, hence $g \circ f$ is C^r.

Let $L(E) = L(E, E)$ and let $GL(E)$ denote the set of invertible maps (linear isomorphisms) in $L(E)$. Note that $GL(E)$ is an open set in $L(E)$ for by choosing a basis in E we can assume that $E = \mathbb{R}^n$, and an $n \times n$ matrix is invertible if and only if its determinant does not vanish. (Also see Exercise 5.)

Proposition 5.2 *Let* $\mathscr{I}: GL(E) \to GL(E)$ *be defined by* $A \mapsto A^{-1}$. *Then the map* $\mathscr{I}$ *is of class* C^∞ *and for the first derivative* $D\mathscr{I}(A): L(E) \to L(E)$ *at* $A \in GL(E)$, *we have*

$$D\mathscr{I}(A)B = -A^{-1} \circ B \circ A^{-1}$$

Proof If we knew already that the map $\mathscr{I}$ is differentiable then, by differentiating the equation $A \circ A^{-1} = \text{id}$, the formula for the derivative $D\mathscr{I}$ would follow easily. In fact it is clear that $\mathscr{I}$ is C^∞ because we can assume $E = \mathbb{R}^n$ and then use the formula $A^{-1} = A^{co}/\det A$ where A^{co} is the cofactor matrix of the $n \times n$ matrix A.

We will give another proof, however, which has the advantage that it holds when E is any Banach space. Let $A \in GL(E)$. Observe that

$$\begin{aligned}(A + H)^{-1} - A^{-1} + A^{-1} \circ H \circ A^{-1} &\\ = (A + H)^{-1} \circ [A - (A + H) + (A + H) \circ A^{-1} \circ H] \circ A^{-1} &\\ = (A + H)^{-1} \circ H \circ A^{-1} \circ H \circ A^{-1} &\end{aligned}$$

and the last expression is $o(\|H\|)$. Hence $\mathscr{I}$ is differentiable and $D\mathscr{I}(A)H = -A^{-1}HA^{-1}$. To prove that $\mathscr{I}$ is C^∞ we proceed by induction. The derivative $D\mathscr{I}$ can be expressed as a composite of two maps

$$D\mathscr{I}: GL(E) \to GL(E) \times GL(E) \xrightarrow{\varphi} L(L(E), L(E))$$

$$A \mapsto (A^{-1}, A^{-1}) \mapsto -A^{-1}(\cdot)A^{-1}$$

The second map, φ, is the restriction of the bilinear map $L(E) \times L(E) \to L(L(E), L(E))$ given by $(A, B) \mapsto -A(\cdot)A$, and is hence C^∞. If we assume as the inductive step that $\mathscr{I}$ is C^r then $D\mathscr{I}$ is the composition of a C^r map and a C^∞ map. Thus $D\mathscr{I}$ is C^r, hence $\mathscr{I}$ is C^{r+1}.

Let $f: U \to F$ where U is an open set in $E_1 \times E_2$ and E_1, E_2 are any finite dimensional vector spaces. Let $x = (x_1, x_2) \in U$. The derivatives of the mappings $y_1 \mapsto f(y_1, x_2)$ and $y_2 \mapsto f(x_1, y_2)$, if they exist, are called *partial derivatives* of f at the point x and are denoted by $D_1 f(x) \in L(E_1, F)$ and $D_2 f(x) \in L(E_2, F)$, respectively. Then we have the following proposition.

Proposition 5.3 *Let $U \subset E_1 \times E_2$ be open and $f\colon U \to F$. If f is differentiable, then the partial derivatives exist and we have*

$$Df(x)\cdot(h_1, h_2) = D_1 f(x)\cdot h_1 + D_2 f(x)\cdot h_2$$

Moreover, f is of class C^r if and only if $D_i f\colon U \to L(E_i, F)$, $i = 1, 2$, $i = 1, 2$, both exist and are of class C^{r-1}.

Let $L_s^r(E, F)$ denote the space of *symmetric* r-multilinear maps, that is, maps in $L^r(E, F)$ whose value at $(v_1, \dots, v_r)$ is not changed by a permutation of $v_1, \dots, v_r$.

Proposition 5.4 *If $f\colon U \subset E \to F$ is C^r then $D^r f(x) \in L_s^r(E, F)$, i.e. $D^r f(x)$ is symmetric.*

The notation for partial derivatives of a function defined on open sets in $\mathbb{R}^n$ is as follows. If $f\colon U \subset \mathbb{R}^n \to \mathbb{R}^m$ is a vector-valued function we write out f in component form

$$f(x_1, \dots, x_n) = [f_1(x_1, \dots, x_n), \dots, f_m(x_1, \dots, x_n)]^T$$

where each $f_i\colon U \to \mathbb{R}$ is a real-valued function. The superscript T indicates transpose of a matrix, that is, we regard the values of f as column vectors in $\mathbb{R}^m$. The total derivative $Df(x)\colon \mathbb{R}^m \to \mathbb{R}^n$ can be regarded as an $m \times n$ matrix

$$Df(x) = \left[\frac{\partial f_i}{\partial x_j}\right]_{m\times n}$$

where the rows are indicated by the index $i = 1, \dots, m$ and the columns by $j = 1, \dots, n$ and $\partial f_i/\partial x_j$ is the derivative of f_i with respect to x_j, keeping the other variables $x_1, \dots, x_{j-1}, x_{j+1}, \dots, x_n$ fixed. In view of (3) the components of $D^r f(x)$ in terms of the standard basis $e_1, \dots, e_n$ in $\mathbb{R}^n$ are

$$D^r f(e_{i_1}, \dots, e_{i_r}) = \frac{\partial}{\partial x_{i_1}} \cdots \frac{\partial f}{\partial x_{i_r}} \tag{4}$$

Proposition 5.3 implies that f is of class C^s if and only if the partial derivatives (4) are defined and continuous for all $r \leqslant s$.

By Proposition 5.4 we can write (4) in the form

$$\partial_1^{\alpha_1} \cdots \partial_n^{\alpha_n} f \qquad (\partial_j = \partial/\partial x_j)$$

where $\alpha = (\alpha_1, \dots, \alpha_n)$ is a multi-index, i.e. an n-tuple of non-negative integers. For short we write $\partial^\alpha = \partial_1^{\alpha_1} \cdots \partial_n^{\alpha_n}$ and let $|\alpha| = \sum \alpha_j = r$ denote the order of differentiation. By (3) we have

$$D^r f(x)\cdot(h,\ldots,h) = \sum_{|\alpha|=r} \binom{r}{\alpha} \partial^\alpha f(x) h^\alpha \tag{5}$$

where $\binom{r}{\alpha} = r!/\alpha!$, $\alpha! = \alpha_1!\cdots\alpha_n!$ and $h^\alpha = h_1^{\alpha_1}\cdots h_n^{\alpha_n}$ where $h = (h_1,\ldots,h_n) \in \mathbb{R}^n$.

Taylor's formula

Let $[x, y]$ denote the line segment $\{x + t(y - x); 0 \leqslant t \leqslant 1\}$ between two points $x, y \in E$. If $f \in C^1(U, F)$ where U is a neighbourhood of the line segment $[x, y]$ then we have $d/dt\, f(x + t(y - x)) = Df(x + t(y - x))\cdot(y - x)$, and integrating this equation with respect to t we have

$$f(y) - f(x) = \int_0^1 Df(x + t)y - x))\cdot(y - x)\,dt \tag{6}$$

In the statement of Taylor's formula below we use the notation $h^{(k)} = (h,\ldots,h)$, where h is repeated k times. The notation $D^k f(x)\cdot h^{(k)}$ means, therefore, the value $D^k f(x)(h,\ldots,h)$.

Theorem 5.5 *Let $x, h \in E$ and let $f \in C^N(U, F)$ where U is a neighbourhood of the line segment $[x, x + h]$. Then*

$$f(x + h) = f(x) + Df(x)\cdot h + \frac{1}{2!} D^2 f(x)\cdot h^{(2)} + \cdots +$$

$$\frac{1}{(N-1)!} D^{N-1} f(x)\cdot h^{(N-1)} + R_N(x, h)$$

where

$$R_N(x, h) = \int_0^1 \frac{(1 - t)^{N-1}}{(N-1)!} D^N f(x + th)\cdot h^{(N)}\,dt$$

Proof For $N = 1$ this is just (6). The theorem can be proved by induction on N. If the formula holds for some N then it holds for $N + 1$ by integration by parts on the integral $R_N(x, h)$ since

$$\frac{d}{dt} D^N f(x + th)\cdot h^{(N)} = D^{N+1} f(x + th)\cdot h^{(N+1)}$$

which follows from (2).

Remark 5.6 If $f\colon U \subset \mathbb{R}^n \to \mathbb{R}^m$ then, due to (5), Taylor's formula takes the form

$$f(x + h) = \sum_{|\alpha|<N} \frac{1}{\alpha!} \partial^\alpha f(x) h^\alpha + R_N(x, h)$$

where

$$R_N(x, h) = \sum_{|\alpha|=N} N \int_0^1 (1-t)^{N-1} \frac{1}{\alpha!} \partial^\alpha f(x+th)h^\alpha \, dt$$

Homogeneous polynomial maps

If $T \in L^r(E, F)$ then the map p: $E \to F$ defined by $p(x) = T(x, \ldots, x)$ is called a *homogeneous polynomial map* of degree r from E to F. The set of all such maps is denoted $P^{(r)}(E, F)$. We also define $P^{(0)}(E, F) = F$, the space of constant polynomials.

Each permutation $\sigma \in S_r$ defines a map σ: $L^r(E, F) \to L^r(E, F)$ by

$$(\sigma T)(v_1, \ldots, v_r) = T(v_{\sigma(1)}, \ldots, v_{\sigma(r)})$$

Thus by definition $T \in L_s^r(E, F)$ (i.e. T is a symmetric r-multilinear map) if $\sigma T = T$ for all $\sigma \in S_r$. Define the linear map Sym: $L^r(E, F) \to L_s^r(E, F)$ by

$$\operatorname{Sym} T = \frac{1}{r!} \sum_{\sigma \in S_r} \sigma T$$

We have $\operatorname{Sym} T = T$ if and only if $T \in L_s^r(E, F)$, and the operator Sym is a projector, i.e. $\operatorname{Sym} \circ \operatorname{Sym} = \operatorname{Sym}$.

Proposition 5.7 *Let $p \in P^{(r)}(E, F)$, i.e. $p(x) = T(x, \ldots, x)$ for some map $T \in L^r(E, F)$. Then for all $x \in E$ we have*

$$\frac{1}{r!} D^r p(x) = \operatorname{Sym} \mathrm{T} \tag{7}$$

Hence for any $p \in P^{(r)}(E, F)$ we have $p(x) = D^r p(0) \cdot (x, \ldots, x)/r!$. The map $L_s^r(E, F) \to P^{(r)}(E, F)$ defined by $T \mapsto p$ is an isomorphism; its inverse is $p \mapsto D^r p(0)/r!$.

Proof By virtue of (3), we have

$$D^r p(x) \cdot (v_1, \ldots, v_r) = \frac{d}{dt_r} \cdots \frac{d}{dt_1} p\left(x + \sum t_i v_i\right)\Big|_{t_1 = \cdots = t_r = 0} \tag{8}$$

Since $p(v) = T(v, \ldots, v)$, where T is multilinear, the expression $p(x + \Sigma\, t_i v_i)$ is polynomial in the variables $t_1, \ldots, t_r$. It is clear that the only terms that remain after substitution in (8) are those that contain the product $t_1 \cdots t_r$. In particular, it follows that (8) is independent of x. So let $x = 0$. Then we have

$$\begin{aligned} p\left(\sum t_i v_i\right) &= T\left(\sum t_i v_i, \ldots, \sum t_i v_i\right) \\ &= \sum t_{i_1} \cdots t_{i_r} T(v_{i_1}, \ldots, v_{i_r}) \qquad \text{by multilinearity} \end{aligned}$$

and the terms in this sum that contain the product $t_1 \cdots t_r$ correspond to permutations $(i_1, \dots, i_r) \in S_r$. Hence, substituting in (8), we get

$$D^r p(x)(v_1, \dots, v_r) = \sum T(v_{i_1}, \dots, v_{i_r})$$

where the sum is taken over all permutations $\sigma = (i_1, \dots, i_r) \in S_r$. This proves the first statement (7). The second statement also follows immediately. To complete the proof observe that if $p(x) = T(x, \dots, x)$ then we can replace T by Sym T, so without loss of generality T is symmetric, in which case (7) gives $D^r p(x)/r! = T$ so we can recover $T \in L_s^r(E, F)$ from p.

Inverse function theorem

A map $f: U \to V$, where U is open in E and V is open in F, is said to be a C^r *diffeomorphism* ($r \geqslant 1$) if it is of class C^r, is bijective, and has an inverse $g: V \to U$ which is also of class C^r. By virtue of the chain rule it is clear that if f is a diffeomorphism then the linear map $Df(x): E \to F$ must be invertible at each $x \in U$; in fact, by differentiating the equations $f \circ f^{-1} = \text{id}$ and $f^{-1} \circ f = \text{id}$, we infer from (1) that

$$Df^{-1}(y) = [Df(f^{-1}(y))]^{-1} \tag{9}$$

For the converse we have the Inverse Function Theorem, the proof of which can be found in any textbook on real analysis, see [Ca] or [La 1].

Theorem 5.8 *Let U be an open subset of E and let $f: U \to F$ be of class C^r. Assume that for some point $x_0 \in U$, the derivative $Df(x_0): E \to F$ is an invertible map. Then f is a C^r diffeomorphism of some neighbourhood of x_0 onto some neighbourhood of $f(x_0)$.*

The main part of the proof is to show that f is a homeomorphism from some neighbourhood U of x_0 onto some neighbourhood V of $f(x_0)$. Then one shows that (9) holds in a smaller neighbourhood $V' \subset V$ of $f(x_0)$ and $f^{-1}: V' \to U' \subset U$ is the inverse in that neighbourhood. Since (9) can be written in the form

$$Df^{-1} = \mathscr{I} \circ Df \circ f^{-1} \tag{10}$$

it follows from Proposition 5.2 that Df^{-1} is continuous, whence f^{-1} is C^1. In that way one shows that $f: U' \to V'$ is a C^1 diffeomorphism. By induction using (10) one then shows that $f^{-1}: V' \to U'$ is of class C^r.

Exterior product of a vector space

The exterior product of a vector space is needed later when we define differential forms on manifolds. For the convenience of the reader we write down the definitions and basic properties of such products. For the proofs, see [La 2].

Let E_i, $i = 1, \dots, r$ and F be finite-dimensional vector spaces over $\mathbb{R}$. A map $f: E_1 \times \cdots \times E_r \to F$ is said to be *multilinear* if it is linear in each variable separately. We denote the set of such maps by $L(E_1, \dots, E_r; F)$. If

$E_i = E$ for all i we write $E^{(r)} = E \times \cdots \times E$ and denote as before the set of r-multilinear maps $E^{(r)} \to F$ by $L^r(E; F)$.

An r-multilinear map A: $E^{(r)} \to F$ is said to be *alternating* if $A(v_1, \ldots, v_r) = 0$ when $v_i = v_j$, $i \neq j$. This condition can be stated in another way: the sign of $A(v_1, \ldots, v_r)$ changes if any two of its arguments are switched. The set of alternating maps $E^{(r)} \to F$ is denoted $L_a^r(E, F)$.

For each positive integer r there exists a vector space, $\bigwedge^r E$, and an r-multilinear, alternating map, $E^{(r)} \to \bigwedge^r E$, denoted by

$$(v_1, \ldots, v_r) \mapsto v_1 \wedge \cdots \wedge v_r,$$

which is universal with respect to r-multilinear alternating maps on E. By "universal" we mean that if g: $E^{(r)} \to G$ is an r-multilinear alternating map, then there exists a unique linear map g_*: $\bigwedge^r(E) \to G$ such that the following diagram is commutative:

$$\begin{array}{ccc} E^{(r)} & \longrightarrow & \bigwedge^r(E) \\ & \searrow^{g} & \downarrow g_* \\ & & G \end{array}$$

The universal property characterizes $\bigwedge^r E$ up to a unique isomorphism, and it is called the rth *exterior product* of E. An element of $\bigwedge^r E$ of the form $v_1 \wedge \cdots \wedge v_r$ is said to be *decomposable*. Every member of $\bigwedge^r E$ is a linear combination of such decomposable elements.

Note that $\bigwedge^r E = 0$ for $r > n$, where $n = \dim E$. Now let $1 \leqslant r \leqslant n$. It is a consequence of the universal property that if $e_1, \ldots, e_n$ is a basis of E then the elements

$$e_{i_1} \wedge \cdots \wedge e_{i_r}, \qquad i_1 < \cdots < i_r$$

form a basis of $\bigwedge^r E$. Thus, the dimension of $\bigwedge^r E$ is $\binom{n}{r}$. For $r = 0$ we let $\bigwedge^0 E = \mathbb{R}$.

For each pair of positive integers r, s there exists a unique product (bilinear map)

$$\textstyle\bigwedge^r E \times \bigwedge^s E \to \bigwedge^{r+s} E$$

such that if $v_1, \ldots, v_r, w_1, \ldots, w_s \in E$ then

$$(v_1 \wedge \cdots \wedge v_r) \times (w_1 \wedge \cdots \wedge w_s) \mapsto v_1 \wedge \cdots \wedge v_r \wedge w_1 \wedge \cdots \wedge w_s$$

This product is associative. The *exterior algebra*, $\bigwedge(E)$, is the direct sum $\bigoplus_{r=0}^{\infty} \bigwedge^r E$, that is, $\bigwedge(E) = \mathbb{R} \oplus \bigwedge^1 E \oplus \bigwedge^2 E \oplus \cdots \oplus \bigwedge^n E$, with multiplication defined by

$$\left(\sum_{r=0}^{\infty} \alpha_r\right) \wedge \left(\sum_{s=0}^{\infty} \beta_s\right) = \sum_{k=0}^{\infty} \left(\sum_{r+s=k} \alpha_r \wedge \beta_s\right) \tag{11}$$

It is easy to verify that $\bigwedge(E)$ is an algebra over $\mathbb{R}$ in the sense that it is a vector space over $\mathbb{R}$ and the multiplication $\wedge$ is a bilinear map. The bilinearity of $\wedge$ means that for all $c \in \mathbb{R}$ and α, β and ω in $\bigwedge(E)$,

$$(c\alpha) \wedge \beta = \alpha \wedge (c\beta) = c(\alpha \wedge \beta)$$
$$(\alpha + \beta) \wedge \omega = \alpha \wedge \omega + \beta \wedge \omega$$
$$\omega \wedge (\alpha + \beta) = \omega \wedge \alpha + \omega \wedge \beta$$

Further, the multiplication is associative, that is, $\alpha \wedge (\beta \wedge \omega) = (\alpha \wedge \beta) \wedge \omega$ for all α, β and ω in $\bigwedge(E)$, so $\bigwedge(E)$ is an associative algebra over $\mathbb{R}$. Also, note that if $\alpha_r \in \bigwedge^r(E)$ and $\beta_s \in \bigwedge^s(E)$ then

$$\beta_s \wedge \alpha_r = (-1)^{rs}\alpha_r \wedge \beta_s \tag{12}$$

It follows from (12) that if β is a sum of terms of *even* degree then α commutes with any $\beta \in \bigwedge(E)$.

Remark As a vector space over $\mathbb{R}$, $\bigwedge(E)$ has dimension

$$1 + \binom{n}{1} + \binom{n}{2} + \cdots + \binom{n}{n} = 2^n.$$

If $f \in L(E, F)$ is a linear map, then, due to the universal property of exterior products, there is a unique linear map $\bigwedge^r(f) \in L(\bigwedge^r E, \bigwedge^r F)$ for $r = 1, 2, \ldots,$ that maps

$$v_1 \wedge \cdots \wedge v_r \mapsto f(v_1) \wedge \cdots \wedge f(v_r)$$

Also, we let $\bigwedge^0(f) \in L(\mathbb{R}, \mathbb{R})$ be the identity. Note that $\bigwedge^{r+s}(f)(\alpha \wedge \beta) = \bigwedge^r(f) \wedge \bigwedge^s(f)(\beta)$ for all $\alpha \in \bigwedge^r E$ and $\beta \in \bigwedge^s E$. Each map $\bigwedge^r(f)$ is linear so there is an induced linear map $\bigwedge(f)\colon \bigwedge(E) \to \bigwedge(F)$ on the direct sums, and it is evident that $\bigwedge(f)$ is a homomorphism of algebras, that is,

$$\textstyle\bigwedge(f)(\alpha \wedge \beta) = \bigwedge(f)(\alpha) \wedge \bigwedge(f)(\beta) \tag{13}$$

for all α and $\beta \in \bigwedge E$, where the multiplication $\wedge$ is defined by (11).

If E and F are vector spaces over $\mathbb{R}$, there is a natural isomorphism of algebras over $\mathbb{R}$

$$\textstyle\bigwedge(E \oplus F) \simeq \bigwedge(E) \bigoplus_{\mathbb{R}} \bigwedge(F)$$

such that the vector space $\bigwedge^k(E \oplus F)$ is isomorphic to $\bigoplus_{i+j=k} \bigwedge^i E \oplus \bigwedge^j F$ for all $k = 0, 1, 2, \ldots$

Later on we will be working with the exterior product $\bigwedge^r(E^*)$ of the dual space E^*. Sometimes it is convenient to identify it with $L_a^r(E, \mathbb{R})$, the space of alternating r-multilinear forms on E.

Theorem 5.9 *There is an isomorphism* $\bigwedge^r(E^*) \to L_a^r(E, \mathbb{R})$ *which maps* $\xi^1 \wedge \cdots \wedge \xi^r \in \bigwedge^r(E^*)$ *to the r-form* $(v_1, \ldots, v_r) \mapsto \det[\xi^i(v_j)]_{i,j=1}^r$.

Sketch of proof If $\xi^1, \ldots, \xi^r \in E^*$ and $v_1, \ldots, v_r \in E$ then the value $\det[\xi^i(v_j)]_{i,j=1}^r$ is an alternating, multilinear map as a function of both

$(\xi^1, \ldots, \xi^r)$ and $(v_1, \ldots, v_r)$, hence defines an alternating r-multilinear map $E^{*(r)} \to L_a^r(E, \mathbb{R})$. By the universal property of the exterior product (with $G = L_a^r(E, \mathbb{R})$) there is a uniquely determined linear map

$$\bigwedge^r(E^*) \to L_a^r(E, \mathbb{R}) \tag{*}$$

Choosing a basis $e_1, \ldots, e_n$ in E and the dual basis $\lambda^1, \ldots, \lambda^n$ in E^*, it is not hard to verify that the basis vectors $\lambda^{i_1} \wedge \cdots \wedge \lambda^{i_r}$ of $\bigwedge^r(E^*)$ are mapped to basis vectors in $L_a^r(E, \mathbb{R})$. Hence the map (*) is an isomorphism.

From now on we identify the element $\xi_1 \wedge \cdots \wedge \xi_r \in \bigwedge^r(E^*)$ with its image under this isomorphism, i.e.

$$\xi_1 \wedge \cdots \wedge \xi_r(v_1, \ldots, v_r) = \det[\xi_i(v_j)]_{i,j=1}^r \tag{14}$$

Realization and complexification

A complex vector space E can be viewed as a real vector space together with a linear map $i: E \to E$ such that $i^2 = -I$. The underlying real vector space is called the *realization* of E. Let $A: E \to E$ be a $\mathbb{C}$-linear map. The realization of A, denoted $A_{\mathbb{R}}$, is the same map, but now regarded as an $\mathbb{R}$-linear map.

Example The realization of $\mathbb{C}^p$ is $\mathbb{R}^{2p}$, where for each $v_1, v_2 \in \mathbb{R}^p$ we identify $v_1 + iv_2 \in \mathbb{C}^p$ with $(v_1, v_2) \in \mathbb{R}^{2p}$. Given a linear map $A: \mathbb{C}^p \to \mathbb{C}^p$, i.e. a $p \times p$ matrix with complex entries, we can write $A = A_1 + iA_2$ where A_1 and A_2 are $p \times p$ matrices with real entries. Since

$$(A_1 + iA_2)(v_1 + iv_2) = A_1 v_1 - A_2 v_2 + iA_2 v_1 + iA_1 v_2,$$

the realization of A is the block operator

$$A_{\mathbb{R}} = \begin{pmatrix} A_1 & -A_2 \\ A_2 & A_1 \end{pmatrix}$$

If we take A to be multiplication by i, then $i_{\mathbb{R}} = \begin{pmatrix} 0 & -I \\ I & 0 \end{pmatrix}$.

Now let E be a vector space over $\mathbb{R}$. Then $E \otimes_{\mathbb{R}} \mathbb{C}$ is a complex vector space in a natural way: $c(v \otimes z) = v \otimes cz$ for $c \in \mathbb{C}$, and is called the *complexification* of E. Since $\mathbb{C} = \mathbb{R} \oplus i\mathbb{R}$, then every element of $E \otimes \mathbb{C}$ can be written uniquely in the form $v_1 \otimes 1 + v_2 \otimes i$ where $v_1, v_2 \in E$. As a real vector space (i.e. the realization) we have $E \otimes \mathbb{C} \simeq E \oplus E$, where $v_1 \otimes 1 + v_2 \otimes i$ is identified with (v_1, v_2). Multiplication by i is then an $\mathbb{R}$-linear map $E \oplus E \to E \oplus E$ having the block operator form $\begin{pmatrix} 0 & -I \\ I & 0 \end{pmatrix}$, where I is the identity operator on E.

An inner product $\langle , \rangle$ on the real vector space E gives rise to a natural inner product on the direct sum $E \otimes E$ by

$$\langle (v_1, v_2), (w_1, w_2) \rangle := \langle v_1, w_1 \rangle + \langle v_2, w_2 \rangle$$

and to a Hermitian inner product $(\,,)$ on the complex vector space $E \otimes \mathbb{C}$ by

$$(v \otimes c, w \otimes d) := \langle v, w\rangle \cdot c\bar{d} \qquad v, v' \in E, c, d \in \mathbb{C}$$

There is a relation between the two inner products: if we substitute $v = v_1 \otimes 1 + v_2 \otimes i$ and $w = w_1 \otimes 1 + w_2 \otimes i$ into the Hermitian inner product on $E \otimes \mathbb{C}$, we obtain a bilinear form

$$((v_1, v_2), (w_1, w_2)) = \langle v_1, w_1\rangle + \langle v_2, w_2\rangle + i\langle v_2, w_1\rangle - i\langle v_1, w_2\rangle$$

on the underlying real vector space, $E \oplus E$. We see that the real part of the Hermitian inner product on $E \otimes \mathbb{C}$ is nothing but the inner product on the real vector space $E \oplus E$ induced by $\langle\ ,\ \rangle$, that is, $\mathrm{Re}((v_1, v_2), (w_1, w_2)) = \langle v_1, w_1\rangle + \langle v_2, w_2\rangle$.

If $A: E \otimes \mathbb{C} \to E \otimes \mathbb{C}$ is a $\mathbb{C}$-linear map, we let $A_{\mathbb{R}}$ denote the same map A, but viewed as an $\mathbb{R}$-linear map $E \oplus E \to E \oplus E$. We call $A_{\mathbb{R}}$ the *realization* of A. Analogous to the matrix example given above, we have the following result.

Lemma 5.10 *Let E and F be vector spaces over $\mathbb{R}$. If $A: E \otimes \mathbb{C} \to F \otimes \mathbb{C}$ is a $\mathbb{C}$-linear map then*

$$A_{\mathbb{R}} = \begin{pmatrix} A_1 & -A_2 \\ A_2 & A_1 \end{pmatrix}$$

for unique $\mathbb{R}$-linear maps $A_i: E \to F$, $i = 1, 2$. Further, given inner products on E and F, let ${}^tA_i$ denote the transpose of A_i, and let ${}^hA$ denote the adjoint of A with respect to the corresponding Hermitian inner products on $E \otimes \mathbb{C}$ and $F \otimes \mathbb{C}$. Then the realization of ${}^hA: F \otimes \mathbb{C} \to E \otimes \mathbb{C}$ is

$$({}^hA)_{\mathbb{R}} = \begin{pmatrix} {}^tA_1 & {}^tA_2 \\ -{}^tA_2 & {}^tA_1 \end{pmatrix}.$$

Exercises

1. Prove Lemma 5.10. Also, show that the correspondence $A \mapsto A_{\mathbb{R}}$ is a homomorphism from $\mathbb{C}$-linear maps to $\mathbb{R}$-linear maps, $(AB)_{\mathbb{R}} = A_{\mathbb{R}}B_{\mathbb{R}}$.

Let E be a finite dimensional vector space over $\mathbb{R}$. The dual space, E^*, is the set of linear functionals on E, i.e. $E^* = L(E, \mathbb{R})$. Recall that if $e_1, \ldots, e_n$ is a basis of E, and we define $\lambda^1, \ldots, \lambda^n \in E^*$ by

$$\langle \lambda^j, e_i\rangle = \delta_{ij}$$

then $\lambda^1, \ldots, \lambda^n$ is the dual basis of E^* corresponding to $e_1, \ldots, e_n$.

2. Let $f: E \to F$ be linear. The transpose map ${}^t\!f: F^* \to E^*$ is defined by ${}^t\!f(\lambda) = \lambda \circ f$, that is, $\langle {}^t\!f(\lambda), v\rangle = \langle \lambda, f(v)\rangle$. Show that if f has matrix $[a_{ij}]$ with respect to bases for E and F, then ${}^t\!f$ has matrix $[a_{ij}]^T = [a_{ji}]$ with respect to the corresponding dual bases in F^* and E^*.

3. Suppose that $V = E \oplus F$, where $\dim F = 1$. *Let* $\omega \in \bigwedge^p(V)$, and let $v \in F$ such that $v \neq 0$. Show that $\omega = \alpha + v \wedge \beta$ for unique $\alpha \in \bigwedge^p(E)$ and $\beta \in \bigwedge^{p-1}(E)$.

4. Let E, E' and F, F' be arbitrary finite dimensional vector spaces over $\mathbb{R}$. Show that the map $L(E, E') \times L(F, F') \to L(E \otimes F, E' \otimes F')$ defined by $(f, g) \mapsto f \otimes g$ is smooth. By $f \otimes g$ we mean the map such that $(f \otimes g)(v \otimes w) = f(v) \otimes g(w)$. Hint: The map is bilinear.

5. If A is an $n \times n$ matrix and $\|A\| < 1$, show that $I - A$ is invertible and has inverse given by the Neumann series

$$(I - A)^{-1} = I + A + A^2 + \cdots$$

Use this result to give another proof of the fact that $GL(\mathbb{R}^n)$ is open in $L(\mathbb{R}^n)$.

6. (a) Let $A = A(t)$ be an $n \times n$ matrix where each entry $a_{ij} = a_{ij}(t)$ is a differentiable function of a single variable t. Show that $\det A(t)$ is a differentiable function of t and

$$\frac{d}{dt} \det A(t) = \sum_{i=1}^{n} A_i'(t)$$

where $A_i'(t)$ denotes the determinant of the $n \times n$ matrix obtained from $A(t)$ by replacing each entry $a_{ij}(t)$ of row i by its derivative $(d/dt)(a_{ij}(t))$.
(b) Suppose we regard $L(\mathbb{R}^n)$ as $\mathbb{R}^{n^2}$ with points denoted $x = [x_{ij}]$, then $\det\colon GL(\mathbb{R}^n) \to \mathbb{R}$ is a map defined on an open set in $\mathbb{R}^{n^2}$. Find $\partial(\det)/\partial x_{ij}$. Show that det is a C^∞ function.

7. Let E and F be finite dimensional real vector spaces. Also let $\varphi\colon E \to \mathbb{R}^n$ and $\psi\colon F \to \mathbb{R}^m$ be linear isomorphisms, obtained from a basis in E and in F, respectively. Show that $f\colon U \subset E \to F$ is of class C^r if and only if $\psi \circ f \circ \varphi^{-1}\colon \varphi(U) \subset \mathbb{R}^n \to \mathbb{R}^m$ is of class C^r.

8. Let $f_i\colon U \subset E \to F_i$, $1 \leqslant i \leqslant p$, and define $f = f_1 \times \cdots \times f_p$ by $f(x) = (f_1(x), \ldots, f_p(x))$. Show that f is differentiable at x if and only if each f_i is differentiable at x, and then

$$Df(x) = Df_1(x) \times \cdots \times Df_p(x)$$

9. (Leibniz rule) Let $U \subset E$ be an open set. Let $f\colon U \to F_1$ and $g\colon U \to F_2$ be of class C^2. If $B\colon F_1 \times F_2 \to G$ is a bilinear map show that

$$D[B(f, g)](x) \cdot v = B(Df(x) \cdot v, g(x)) + B(f(x), Dg(x) \cdot v)$$

where $B(f, g)\colon U \to G$ is defined by $x \mapsto B(f(x), g(x))$. (The usual Leibniz rule is obtained when $F_1 = F_2 = G = \mathbb{R}$ and B is multiplication.)
Hint: $B(f, g) = B \circ (f \times g)$ where $(f \times g)(x) = (f(x), g(x))$ as in Exercise 4. Now use the chain rule.

10. Let $\langle , \rangle$ be an inner product (or any symmetric bilinear form) on E. Define the polynomial map $f\colon E \to \mathbb{R}$ by $f(x) = \langle x, x \rangle$. Show that $\frac{1}{2}D^2 f(x) = \langle , \rangle$.

11. Let f be a map $E \to F$. Let $e_1, \ldots, e_n$ be a basis and let $\varphi\colon E \to \mathbb{R}^n$ be the linear isomorphism defined by $\varphi(\sum x_i e_i) = (x_1, \ldots, x_n)$. Show that $f \in P^{(r)}(E, F)$ if and only if $\tilde{f} = f \circ \varphi^{-1}$ is a polynomial in the variables $x_1, \ldots, x_n$ (with coefficients in F).

12. A map $f: E \to F$ is *homogeneous* of degree $k \geqslant 0$ if $f(tx) = t^k f(x)$ for all $t \in \mathbb{R}$ and $x \in E$. Show that if f is C^k then it is necessarily a polynomial map, i.e. $f \in P^{(k)}(E, F)$ (and hence f is C^∞).
Hint: By Taylor's formula we get

$$f(h) = \frac{1}{k!} D^k f(0) \cdot h^{(k)} + o(|h|^k)$$

since $D^j f(0) = 0$ for $j < k$. Replace h by $t \cdot h$, divide both sides by t^k and let $t \to 0$.

13. Define the function $f: \mathbb{R}^2 \to \mathbb{R}$ by

$$f(x, y) = \begin{cases} \dfrac{xy(x^2 - y^2)}{x^2 + y^2} & \text{if } (x, y) \neq (0, 0) \\ 0 & \text{if } (x, y) = 0, 0) \end{cases}$$

Show that f is of class C^1. Conclude from Exercise 12 that f is *not* of class C^2.

A *linear differential operator* $\mathscr{A}$ of degree k on an open set U in $\mathbb{R}^n$ has the form

$$\mathscr{A}(x, \partial) = \sum_{|\alpha| \leqslant k} a_\alpha(x) \frac{\partial^{\alpha_1}}{\partial x_1^{\alpha_1}} \cdots \frac{\partial^{\alpha_n}}{\partial x_n^{\alpha_n}}$$

where A_α are C^∞ real-valued (or matrix) functions on U. For short we write $\mathscr{A} = \sum_{|\alpha| \leqslant k} a_\alpha(x)\, \partial^\alpha$, and the function

$$\mathscr{A}(x. \xi) = \sum_{|\alpha| \leqslant k} a_\alpha(x) \xi^\alpha, \qquad \text{a polynomial in } \xi \in \mathbb{R}^n,$$

is called the *complete symbol* of $\mathscr{A}$. Note that $\mathscr{A}(x, \xi) = e^{-x \cdot \xi} \mathscr{A}(x, \partial)\, e^{x \cdot \xi}$ where $\cdot$ denotes the usual inner product of vectors in $\mathbb{R}^n$.

14. Let $\mathscr{A}(x, \partial)$ be a linear differential operator on $U \subset \mathbb{R}^n$. Prove Hörmander's generalized form of the Leibniz rule:

$$\mathscr{A}(x, \partial)(uv) = \sum (\mathscr{A}^{(\alpha)}(x, \partial) u)\, \partial^\alpha v(x), \qquad u, v \in C^\infty(U)$$

where $\mathscr{A}^{(\alpha)}(x, \xi) = \partial_\xi^\alpha \mathscr{A}(x, \xi)$.
Hint: If we fix v then each side of this formula can be regarded as a differential operator in u; show that both have the same complete symbol (use Taylor's formula).

5.2 Manifolds

A locally Euclidean space M of dimension n is a topological space M for which each point has a neighbourhood homeomorphic to an open subset of $\mathbb{R}^n$. A *coordinate map* is a pair (U, κ) where $U \subset M$ is open and κ is a homeomorphism of U onto an open set in $\mathbb{R}^n$. (Often it is convenient to assume that U is connected.)

When dealing with more than one coordinate map we use the notation $\kappa\colon U_\kappa \to V_\kappa$ to indicate the domain U_κ and range $V_\kappa \subset \mathbb{R}^n$ of the coordinate map.

Definition 5.11 *Let r be an integer $\geqslant 1$ or ∞. A C^r structure on a locally Euclidean space M is a family $\mathfrak{F}$ of coordinate maps $\kappa\colon U_\kappa \to V_\kappa$ such that*

(*i*) *The domains U_κ cover M: $\bigcup_{\kappa\in\mathfrak{F}} U_\kappa = M$.*
(*ii*) *If $\kappa, \kappa' \in \mathfrak{F}$ and $U_\kappa \cap U_{\kappa'} \neq \varnothing$ then the overlap map*

$$\kappa'\circ\kappa^{-1}\colon \kappa(U_\kappa \cap U_{\kappa'}) \to \kappa'(U_\kappa \cap U_{\kappa'}) \qquad \text{is } C^r$$

(*iii*) *The family $\mathfrak{F}$ is maximal with respect to (ii): if κ_0 is a coordinate map such that $\kappa_0\circ\kappa^{-1}$ and $\kappa\circ\kappa_0^{-1}$ are C^r for all $\kappa\in\mathfrak{F}$ then $\kappa_0\in\mathfrak{F}$.*

Note: Since κ is a homeomorphism the set $\kappa(U_\kappa \cap U_{\kappa'})$ is open in $\mathbb{R}^n$. Also, if the overlap map $\kappa'\circ\kappa^{-1}$ is C^r then so is its inverse $\kappa\circ\kappa'^{-1}$ by Theorem 5.8.

If $\mathscr{A} = \{(U_i, \kappa_i)\}$ (i ranging in some index set) is any collection of coordinate maps satisfying (*i*) and (*ii*) then there is a unique C^r structure $\mathfrak{F}$ containing $\mathfrak{A}$; namely, let $\mathfrak{F} = \{\kappa;\ \kappa$ is a coordinate map and $\kappa_i\circ\kappa^{-1}$ is C^r for all $i\}$. Then $\mathfrak{F} \supset \mathfrak{A}$, and it is easily checked that it satisfies (*i*) and (*ii*). $\mathfrak{F}$ is maximal with respect to (*ii*) by construction.

A C^r structure can thus be defined by an arbitrary family $\mathfrak{A}$ satisfying (*i*) and (*ii*) only. Such a family is called a C^r *atlas* and two atlases are called *equivalent* if they define the same C^r structure. Clearly, two atlases $\mathfrak{A}_1$ and $\mathfrak{A}_2$ are equivalent if and only if $\mathfrak{A}_1 \cup \mathfrak{A}_2$ is an atlas.

Definition 5.12 *A C^r manifold is a pair $(M, \mathfrak{F})$ where M is a locally Euclidean, Hausdorff, second countable space (i.e. there is a countable basis for the topology), and $\mathfrak{F}$ is a C^r structure on M.*

Usually we refer to the underlying space M as the manifold. The maps in $\mathfrak{F}$ are called admissible coordinate maps, or *charts*. If κ is a chart with domain $U \ni x$, we say that (U, κ) is a chart at x.

It is known that any C^r atlas ($r \geqslant 1$) on a set contains a C^∞ atlas; see [Hi]. Thus, it is no loss of generality to assume that $r = \infty$, which we do from now on.

Examples

(1) An n-sphere is the set $S^n = \{x \in \mathbb{R}^{n+1};\ |x| = 1\}$ with the relative topology as a subset of $\mathbb{R}^{n+1}$. Let $N = (0, \ldots, 0, 1)$ and $S = (0, \ldots, 0, -1)$ denote the north and south poles. Then a C^∞ structure on S^n is defined by the atlas $\{\kappa_1, \kappa_2\}$ where $\kappa_1\colon S^n\backslash N \to \mathbb{R}^n$ is defined by

$$(x_1, \ldots, x_{n+1}) \mapsto (x_1/(1 - x_{n+1}), \ldots, x_n/(1 - x_{n+1}))$$

and $\kappa_2\colon S^n\backslash S \to \mathbb{R}^n$ is defined by

$$(x_1, \ldots, x_{n+1}) \mapsto (x_1/(1 + x_{n+1}), \ldots, x_n/(1 + x_{n+1}))$$

The overlap map is $\kappa_2 \circ \kappa_1^{-1}\colon \mathbb{R}^n\backslash 0 \to \mathbb{R}^n\backslash 0$ and is given by $\kappa_2 \circ \kappa_1^{-1}(y) = y/|y|^2$, $y \in \mathbb{R}^n\backslash 0$ which together with its inverse is C^∞ so condition (*ii*) is satisfied.

Remark The charts κ_1 and κ_2 are essentially "stereographic projection" from the north and south poles, respectively. For example, the stereographic

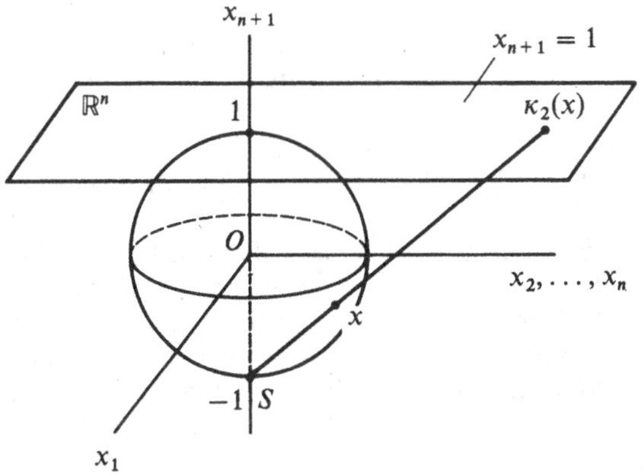

Fig. 1. Stereographic projection from the south pole.

projection of the point $x = (x_1, \ldots, x_{n+1})$ from the south pole to the plane $x_{n+1} = 1$ is the point with coordinates $(2x_1/(1 + x_{n+1}), \ldots, 2x_n/(1 + x_{n+1}), 1)$.

(2) $\mathbb{R}^n$ is a manifold with an atlas formed by a single chart ($\mathbb{R}^n$, identity). This is called the standard C^∞ structure on $\mathbb{R}^n$. Further, if E is a finite dimensional vector space then, by virtue of a linear isomorphism $\kappa\colon E \to \mathbb{R}^n$ defined by any basis of E, there is an atlas $\{\kappa\}$ for a C^∞ structure on E. Any two such linear isomorphisms κ, κ' define the same C^∞ structure because the overlap map $\kappa' \circ \kappa^{-1}\colon \mathbb{R}^n \to \mathbb{R}^n$ is a constant matrix, hence is C^∞.

(3) It is possible for a locally Euclidean space to possess non-equivalent atlases. For instance consider $M = \mathbb{R}$ and the following coordinate maps on $\mathbb{R}$: $\kappa(t) = t$ and $\omega(t) = t^3$. Then $\mathfrak{A}_1 = \{\kappa\}$ and $\mathfrak{A}_2 = \{\omega\}$ are atlases on $\mathbb{R}$ but $\omega^{-1} \circ \kappa$ is not differentiable at the origin. In other words $\mathfrak{A}_1$ and $\mathfrak{A}_2$ define distinct C^∞ structures on $\mathbb{R}$ (but they are diffeomorphic; see Exercise 3 at the end of this section.

(4) An open subset U of a manifold M with C^∞ structure $\mathfrak{F}$ is itself a manifold with C^∞ structure $\mathfrak{F}_U = \{\kappa \in \mathfrak{F};$ domain of $\kappa \subset U\}$. Note that if $\kappa \in \mathfrak{F}$ and the domain of κ intersects U, then its restriction to U also belongs to $\mathfrak{F}$ by virtue of the maximality condition (iii).

Next we define the concept of a smooth map between manifolds and the local representation of a map.

Definition 5.13 *Let M and N be manifolds. We say that $f: M \to N$ is of class C^∞ (or smooth) if*

(i) *$f: M \to N$ is continuous*
(ii) *for each chart (U, φ) of M and chart (V, ψ) of N such that $f(U) \subset V$, the local representation, $f_{\varphi\psi} = \psi \circ f \circ \varphi^{-1}$, is of class C^∞.*

The set of all smooth maps $f: M \to N$ is denoted $C^\infty(M, N)$.

Proposition 5.14 *Let $\mathfrak{A}_M$ be an atlas on M and $\mathfrak{A}_N$ an atlas on N. Let $f: M \to N$ be continuous. Then f is of class C^∞ if and only if*

for each $m \in M$, there exist charts $(U, \varphi) \in \mathfrak{A}_M$ and $(V, \psi) \in \mathfrak{A}_N$ such that $m \in U$, $f(U) \subset V$, and the local representation $f_{\varphi\psi}$ is C^∞.

Proof The necessity is clear. Conversely, if this condition holds, let (U, φ) and (V, ψ) be *any* charts of M and N, respectively, such that $f(U) \subset V$. We must show that the local representation $f_{\psi\varphi}$ is C^∞; it suffices to show this in a neighbourhood of each $m \in U$. Now, by hypothesis, there exist charts $(U', \varphi') \in \mathfrak{A}_M$ and $(V', \psi') \in \mathfrak{A}_N$ such that $m \in U'$, $f(U') \subset V'$, and the local representation $f_{\varphi'\psi'}$ is C^∞. By restricting the domains of φ, ψ, φ' and ψ' we may assume that $U = U'$ and $V = V'$. Then $f_{\psi\varphi} = (\psi \circ \psi'^{-1}) \circ f_{\psi'\varphi'} \circ (\varphi \circ \varphi'^{-1})^{-1}$ is C^∞ since it is a composition of C^∞ functions defined on open sets in $\mathbb{R}^n$ (Proposition 5.1).

If we recall that the composition of C^∞ maps on open sets in Euclidean space is C^∞ we obtain the following proposition.

Proposition 5.15 *If $f \in C^\infty(M, N)$ and $g \in C^\infty(N, P)$ then $g \circ f \in C^\infty(M, P)$.*

A map $f: M \to N$ where M and N are manifolds is called a *diffeomorphism* if f is C^∞, bijective, and $f^{-1}: N \to M$ is C^∞. If a diffeomorphism exists between two manifolds, they are called diffeomorphic.

The *product* of two manifolds is defined as follows. If M_1 and M_2 are manifolds with C^∞ structures $\mathfrak{F}_1$ and $\mathfrak{F}_2$ then $M_1 \times M_2$ is a manifold with atlas

$$\{\kappa = \kappa_1 \times \kappa_2; \kappa_j \in \mathfrak{F}_j, j = 1, 2\}$$

where if $\kappa_j: U_{\kappa_j} \to V_{\kappa_j} \subset \mathbb{R}^{n_j}$ then $\kappa_1 \times \kappa_2: U_{\kappa_1} \times U_{\kappa_2} \to V_{\kappa_1} \times V_{\kappa_2} \subset \mathbb{R}^{n_1+n_2}$ is the product map. The C^∞ structure for $M_1 \times M_2$ is the maximal family containing this atlas.

An example of a product manifold is the n-torus $T^n = S^1 \times \cdots \times S^1$ (n times).

We conclude this section with a study of the axioms for an atlas. Often it happens that an atlas is constructed on a set S which is then used to define a topology on S.

By an *atlas on a set S* we mean a collection of pairs $\mathfrak{A} = \{(U_i, \kappa_i)\}$ (i ranging in some index set) satisfying the following conditions:

AT 1 *The U_i's cover S*

AT 2 *Each κ_i is a bijection of U_i onto on open set $\kappa_i(U_i)$ in $\mathbb{R}^n$, and for any i, j, $\kappa_i(U_i \cap U_j)$ is open in $\mathbb{R}^n$.*

AT 3 *If $U_i \cap U_j \neq \varnothing$, the overlap map*

$$\kappa_i \circ \kappa_j^{-1}: \kappa_j(U_j \cap U_i) \to \kappa_i(U_j \cap U_i) \qquad \text{is } C^\infty$$

Conditions AT 1 and AT 3 are the same as (i), (ii) previously. It is necessary to add condition AT 2 since S is just a set so that the κ_i are not (*a priori*) homeomorphisms.

Proposition 5.16 *Let $\mathfrak{A} = \{(U_i, \kappa_i)\}$ be an atlas on a set S. Then the collection of subsets of S,*

$$\mathfrak{S} = \{W; \exists \kappa_j \in \mathfrak{A} \text{ such that } W \subset U_j \text{ and } \kappa_j(W) \text{ is open in } \mathbb{R}^n\},$$

forms a basis for a topology on S. This topology makes S into a locally Euclidean space and the family $\mathfrak{A}$ is an atlas for a C^∞ structure on this space.

Proof The key point in the proof is the following:

$$\text{if } W \in \mathfrak{S}, \text{ then } \kappa_i(U_i \cap W) \text{ is open in } \mathbb{R}^n \text{ for } \textit{every } \kappa_i \in \mathfrak{A} \tag{15}$$

By definition there exists $\kappa \in \mathfrak{A}$ such that $W \subset U_\kappa$ and $\kappa(W)$ is open in $\mathbb{R}^n$. Then, for another $\kappa_i \in \mathfrak{A}$, we have

$$\kappa_i(U_i \cap W) = \kappa_i \circ \kappa^{-1}(\kappa(U_i \cap W))$$

which is open in $\mathbb{R}^n$ by virtue of the Inverse Function Theorem since $\kappa_i \circ \kappa^{-1}$ is a diffeomorphism and $\kappa(U_i \cap W) = \kappa(U_i \cap U_\kappa) \cap \kappa(W)$ is open in $\mathbb{R}^n$.

Let $W, W' \in \mathfrak{S}$. Then for some $\kappa_i, \kappa_j \in \mathfrak{A}$ we have $W \subset U_i$, $W' \subset U_j$ and $\kappa_i(W)$ and $\kappa_j(W')$ are open in $\mathbb{R}^n$. If we can show that $W \cap W' \in \mathfrak{S}$, then $\mathfrak{S}$ is a basis for a topology on S. But this follows since $W \cap W' \subset U_j$ and

$$\kappa_j(W \cap W') = \kappa_j(U_i \cap U_j) \cap \kappa_j(W') \cap \kappa_j(U_j \cap W)$$

which is an intersection of three open sets in $\mathbb{R}^n$ (see (15) for the third set) and hence is open itself.

We claim that with this topology each $\kappa \in \mathfrak{A}$ becomes a homeomorphism. By the definition of $\mathfrak{S}$ it is immediately obvious that κ is continuous. It remains to show that $\kappa(U)$ is open in $\mathbb{R}^n$ for every open set U in the domain of κ. But since U is a union of sets in $\mathfrak{S}$ it follows that

$$\kappa(U) = \bigcup \{\kappa(W); W \in \mathfrak{S}, W \subset U\}$$

is open since it is a union of sets $\kappa(W)$ which are open by (15). Hence κ is a homeomorphism. Thus every point in S has a neighbourhood homeomorphic to an open set in $\mathbb{R}^n$.

Remark The set S with the topology in Lemma 5.16 is not necessarily second countable, or even Hausdorff (see Exercise 8 below). In other words, we cannot conclude that S is a manifold in the sense of Definition 5.12.

Exercises

1. Show that the upper hemisphere $S^2_+ = \{x \in \mathbb{R}^3;\ |x| = 1,\ x_3 > 0\}$ of the 2-sphere is diffeomorphic to the open disc $D = \{z \in \mathbb{R}^2;\ |z| < 1\}$, where S^2_+ and D are given the C^∞ structures as open sets in S^2 and $\mathbb{R}^2$, respectively.

2. Let $U_i^\pm = \{x \in S^n;\ \pm x_i > 0\}$ and define $\kappa_i^\pm\colon U_i^\pm \to \{x' \in \mathbb{R}^n;\ |x'| < 1\}$ by

$$x \mapsto x' = (x_1, \ldots, x_{i-1}, x_{i+1}, \ldots, x_{n+1})$$

Show that $\mathfrak{A} = \{(U_i^\pm, \kappa_i^\pm)\}_{i=1,\ldots,n+1}$ is an atlas of $2n + 2$ charts for S^n, and that the C^∞ structure defined by $\mathfrak{A}$ is the same as that defined by the atlas in Example (1).

3. Consider the two C^∞ structures on $\mathbb{R}$ in Example (3). Show that the map $t \to t^{1/3}$ is a diffeomorphism from $(\mathbb{R}, \mathfrak{F}_1)$ to $(\mathbb{R}, \mathfrak{F}_2)$.

4. Let $B_r(x_0)$ denote the open ball in $\mathbb{R}^n$ of radius r and center x_0. Show that the map $x \mapsto x\cdot(1 - |x|^2)^{-1/2}$ is a diffeomorphism from $B_1(0)$ to $\mathbb{R}^n$. Conclude that any manifold has an atlas $\{(U_i, \kappa_i)\}$ for which all $\kappa_i(U_i) = \mathbb{R}^n$.

5. (a) For $f_i\colon M \to N_i$, $i = 1, 2$, define $f\colon M \to N_1 \times N_2$ by

$$f(m) = (f_1(m), f_2(m)).$$

Show that $f \in C^\infty(M, N_1 \times N_2)$ if and only if both $f_i \in C^\infty(M, N_i)$, $i = 1, 2$.
(b) For $g_i\colon M_i \to N_i$, $i = 1, 2$, define $g_1 \times g_2\colon M_1 \times M_2 \to N_1 \times N_2$ in the obvious way. Show that $g_1 \times g_2$ is of class C^∞ if and only if g_i are of class C^∞, $i = 1, 2$.

Quotient topology

An equivalence relation on a set S is a binary relation $\sim$ such that for all x, y and $z \in S$:

(i) $x \sim x$
(ii) $x \sim y$ and $y \sim x$
(iii) $x \sim y$ and $y \sim z$ implies $x \sim z$

The equivalence class containing x is $[x] = \{y \in S;\ x \sim y\}$. The set of equivalence classes is denoted $S/\!\sim$ and the map $\pi\colon S \to S/\!\sim$ defined by $x \mapsto [x]$ is called the canonical projection. If S is a topological space, the collection of sets

$$\{U \subset S/\!\sim;\ \pi^{-1}(U) \text{ is open in } S\}$$

satisfies the axioms for a topology, called the *quotient topology* on $S/\!\sim$. Also see the text at the end of §5.4 for a discussion of quotient manifolds.

6. Let $\sim$ be an equivalence relation on a topological space S and $\pi\colon S \to S/\!\sim$ the canonical projection. Let T be another topological space. Prove that a map $\varphi\colon S/\!\sim\ \to T$ is continuous if and only if $\varphi \circ \pi\colon S \to T$ is continuous.

7. The set $\Gamma = \{(x, x'); x \sim x'\} \subset S \times S$ is called the *graph* of the equivalence relation $\sim$. Note that $\Gamma = (\pi \times \pi)^{-1}(\Delta)$ where Δ is the diagonal of $(S/\sim) \times (S/\sim)$.
(a) If $S/\sim$ is a Hausdorff space, show that Γ is closed in $S \times S$.
(b) Conversely, if Γ is closed in $S \times S$ and π is an open map (i.e. the image of any open set in $S/\sim$ is open in S), show that $S/\sim$ is Hausdorff.

8. (line with two origins) Let $S = (\mathbb{R} \times 0) \cup (\mathbb{R} \times 1)$ and define the equivalence relation $\sim$ on S which identifies $(t, 0)$ with $(t, 1)$ for $t \neq 0$. Let κ_i: $\mathbb{R} \times i \to \mathbb{R}$ be defined as $\kappa_i(t, i) = t$ for $i = 0, 1$.
(a) Show that $\mathfrak{A} = \{(\mathbb{R} \times i, \kappa_i)\}_{i=0,1}$ is an atlas on $S/\sim$.
(b) Show that the induced topology on $S/\sim$ (by Proposition 5.16) is locally Euclidean but non-Hausdorff.
(c) Show that the topology in (b) is the quotient topology.

9. The *real projective space* $\mathbb{R}P^n$ is defined as S^n (the unit sphere in $\mathbb{R}^{n+1}$) with antipodal points x and $-x$ identified for all $x \in S^n$. The canonical map π: $S^n \to \mathbb{R}P^n$ is given by $x \mapsto [x] = \{x, -x\}$.
(a) Show that $\mathbb{R}P^n$ with the quotient topology is a compact, connected and Hausdorff space.
(b) Show that the restriction of the canonical map π to any open hemisphere in S^n is a homeomorphism onto an open set of $\mathbb{R}P^n$. Conclude that $\mathbb{R}P^n$ is locally Euclidean.
(c) Show that there is a unique C^∞ structure on $\mathbb{R}P^n$ such that π restricted to any open hemisphere is a diffeomorphism, and with this structure $\mathbb{R}P^n$ is a manifold.

5.3 The tangent bundle

Let M be a manifold and $m \in M$. In this section we define the tangent space, T_mM, at the point m and show that the union, TM, of all these tangent spaces itself forms a manifold, called the tangent bundle. Then we show that a smooth map between manifolds induces in a natural way a map between their tangent bundles.

Let I denote an open internal in $\mathbb{R}$ with $0 \in I$. A curve at m is a C^∞ map $c: I \to M$ such that $c(0) = m$. We say that two curves c_1 and c_2 at m are *tangent at m* if they have identical tangent vectors in every chart, i.e.

$$(\kappa \circ c_1)'(0) = (\kappa \circ c_2)'(0) \tag{16}$$

for every chart (U, κ) with $m \in U$ (we denote the derivative of a function $g: I \to \mathbb{R}^n$ at $t = 0$ by $g'(0) \in \mathbb{R}^n$). The intervals I_j on which c_j are defined are of course taken sufficiently small so that $c_j(I_j)$ lies in the domain of $\kappa, j = 1, 2$.

Tangency of curves at $m \in M$ is a concept that is independent of the chart used, i.e. it suffices that (16) holds for one chart κ at m. Indeed, let $\tilde{\kappa}$ be another chart at m. Then if we multiply (16) by the matrix $D(\tilde{\kappa} \circ \kappa^{-1})(\kappa(m))$, it follows by the chain rule that $(\tilde{\kappa} \circ c_1)'(0) = (\tilde{\kappa} \circ c_2)'(0)$ also.

Tangency at $m \in M$ is an equivalence relation among curves at m. An equivalence class of such curves is denoted $[c]_m$, where c is a representative of the class.

Definition 5.17 *The tangent space to M at the point m is the set of equivalence classes of curves at m*

$$T_mM = \{[c]_m;\ c \text{ is a curve at } m\}$$

The disjoint union of the tangent spaces is the *tangent bundle*

$$TM = \bigcup_{m \in M} T_mM$$

To emphasize that this is a disjoint union we often write elements of T_mM in the form (m, t) where $t = [c]_m$. The map $\pi: TM \to M$ defined by $\pi([c]_m) = m$ is the tangent bundle *projection*. If U is an open subset of M, it inherits a C^∞ structure and we identify $T_mU = T_mM$ for $m \in U$. Then $TU = \pi^{-1}U = \bigcup_{m \in U} T_mM$ with projection $\pi_U: TU \to U$.

The set of smooth real-valued functions $f: M \to \mathbb{R}$ is denoted $C^\infty(M, \mathbb{R})$, or just $C^\infty(M)$ for simplicity. If $f \in C^\infty(M)$ we define the *differential* of f at $m \in M$ to be the map

$$df(m): T_mM \to \mathbb{R} \qquad \text{defined by } [c]_m \mapsto (f \circ c)'(0)$$

which is well-defined, independent of the representative c, by virtue of the chain rule. Then we define the map $df: TM \to \mathbb{R}$ such that its restriction to T_mM is $df(m)$, that is, $df|_{T_mM} = df(m)$. We will often write $df|_m$ instead of $df(m)$, using whichever notation is convenient.

Remark

(i) If U is open in M then $d(f|_U) = df|_{TU}$.

(ii) Leibniz' rule holds for $f, g \in C^\infty(M)$

$$d(f \cdot g) = f \cdot dg + g \cdot df,$$

that is, $d(f \cdot g)(m) = f(m) \cdot dg(m) + g(m) \cdot df(m)$ for all $m \in M$.

If $f: U \to \mathbb{R}^k$ is vector-valued, then df is defined in the same way as above. Note that if we write $f = (f_1, \ldots, f_k)$ then $df = (df_1, \ldots, df_k): TU \to \mathbb{R}^k$.

Theorem 5.18 *T_mM has a unique vector space structure such that, for any function $f \in C^\infty(U, \mathbb{R}^k)$, where U is an open neighbourhood of m, the map $df: T_mM \to \mathbb{R}^k$ is linear.*
(Note: When no confusion is possible we write df instead of $df(m)$.)

Proof If (U, κ) is a chart at $m \in U$, then $\kappa: U \to \mathbb{R}^n$ and we have a map $d\kappa: T_mM \to \mathbb{R}^n$. By the definition of $d\kappa$ and of tangency of curves it is clear that $d\kappa$ is injective. We claim that $d\kappa$ is bijective. For $w \in \mathbb{R}^n$ define the curve $c(t) = \kappa^{-1}(\kappa(m) + tw)$, $t \in (-\delta, \delta)$; δ is sufficiently small. Then $d\kappa[c]_m = (\kappa \circ c)'(0) = w$ which shows that $d\kappa$ is surjective. Since $d\kappa$ is bijective we can use it to impose a vector space structure on T_mM; in other words such that

$$\left.\begin{aligned} d\kappa([c_1]_m \oplus [c_2]_m) &= d\kappa[c_1]_m + d\kappa[c_2]_m \\ d\kappa(\lambda \odot [c]_m) &= \lambda \cdot d\kappa[c]_m, \lambda \in \mathbb{R} \end{aligned}\right\} \tag{17}$$

This definition of the vector space operations $\oplus$ and $\odot$ in T_mM is independent of the choice of chart at m, for if (V, φ) is another chart at m then multiplication of the equations (17) by the map $D(\varphi \circ \kappa^{-1})(\kappa(m))$: $\mathbb{R}^n \to \mathbb{R}^n$, which is *linear*, shows via the chain rule that

$$\left.\begin{aligned} d\varphi([c_1]_m \oplus [c_2]_m) &= d\varphi[c_1]_m + d\varphi[c_2]_m \\ d\varphi(\lambda \odot [c]_m) &= \lambda \cdot d\varphi[c]_m, \lambda \in \mathbb{R} \end{aligned}\right\} \tag{17'}$$

Thus the equations (17) hold if and only if (17′) holds and the vector space operations on T_mM are well-defined.

Now let f: $U \to \mathbb{R}^k$ be a smooth map. We wish to show that df: $T_mM \to \mathbb{R}^k$ is linear. Let $[c_j]_m \in T_mM$, $j = 1, 2$ and let $[c]_m = [c_1]_m \oplus [c_2]_m$. Then by definition of df and applying the chain rule it follows that

$$df([c_1]_m \oplus [c_2]_m) = (f \circ c)'(0) = D(f \circ \kappa^{-1})(\kappa(m)) \cdot (\kappa \circ c)'(0).$$

By virtue of the definition (17) of $[c]_m = [c_1]_m \oplus [c_2]_m$, we have $(\kappa \circ c)'(0) = (\kappa \circ c_1)'(0) + (\kappa \circ c_2)'(0)$ for any chart κ at m, hence the linearity of $D(f \circ \kappa^{-1})(\kappa(m))$ implies

$$\begin{aligned} df([c_1]_m \oplus [c_2]_m) & \\ &= D(f \circ \kappa^{-1})(\kappa(m)) \cdot (\kappa \circ c_1)'(0) + D(f \circ \kappa^{-1})(\kappa(m)) \cdot (\kappa \circ c_2)'(0) \\ &= (f \circ c_1)'(0) + (f \circ c_2)'(0) \\ &= df[c_1]_m + df[c_2]_m. \end{aligned}$$

Similarly, one shows $df(\lambda \odot [c]_m) = \lambda \cdot (f \circ c)'(0) = \lambda \cdot df[c]_m$ for all $\lambda \in \mathbb{R}$. Hence df: $T_mM \to \mathbb{R}^k$ is a linear map.

The differentials, $d\kappa$, make it possible to define a topology and a C^∞ structure on the tangent bundle. If (U, κ) is a chart, we have the map $T\kappa$: $TU \to \kappa(U) \times \mathbb{R}^n$ given by

$$[c]_m \mapsto (\kappa(m), (\kappa \circ c)'(0))$$

that is, $T\kappa = (\kappa \circ \pi, d\kappa)$ where π: $TU \to U$ is the tangent bundle projection. As shown above, $d\kappa(m)$: $T_mM \to \mathbb{R}^n$ is a bijection for each m, whence $T\kappa$ is a bijection from TU to $\kappa(U) \times \mathbb{R}^n$. Now we can use Proposition 5.16 to prove the following theorem.

Theorem 5.19 *The family*

$$\mathfrak{A} = \{(TU, T\kappa); (U, \kappa) \text{ is a chart on } M\}$$

is an atlas for a C^∞ structure on TM, making TM into a manifold. With this structure, then for any C^∞ map f: $M \to \mathbb{R}^k$, the map df: $TM \to \mathbb{R}^k$ is also C^∞. The tangent bundle projection π: $TM \to M$ is a C^∞ map. Also $M \subset TM$

if we identify $m \in M$ with the zero vector $0_m \in T_m M$, and this makes M into a C^∞ submanifold of TM.

Proof The conditions AT 1 and AT 2 obviously hold (Proposition 5.16). To verify AT 3, let (U, κ) and (V, φ) be two charts for M. If $x \in \kappa(U \cap V)$, $w \in \mathbb{R}^n$, we have $(T\kappa)^{-1}(x, w) = [c]_m$ whence c is a curve at $m = \kappa^{-1}(x)$ such that $(\kappa \circ c)'(0) = w$. Then

$$\begin{aligned} T\varphi \circ (T\kappa)^{-1}(x, w) &= T\varphi([c]_m) \\ &= (\varphi(m), (\varphi \circ c)'(0)) \\ &= (\varphi(\kappa^{-1}(x)), D(\varphi \circ \kappa^{-1})(x) \cdot w) \end{aligned}$$

which is a C^∞ map from $\kappa(U \cap V) \times \mathbb{R}^n$ to $\varphi(U \cap V) \times \mathbb{R}^n$. This shows that $\mathfrak{A}$ is an atlas for a C^∞ structure on TM. It is not hard to show – with the topology on TM as defined in Proposition 5.16 – that TM is Hausdorff and has a countable basis since this is true of M and $\mathbb{R}^n$. The proofs of the last three statements in the theorem are left to the reader.

Remarks

(i) If U is an open subset of a manifold M then it is itself a manifold in the obvious way, and $T_m U$ and $T_m M$ can be identified for each $m \in U$.

(ii) There is an identification $T\mathbb{R}^k \simeq \mathbb{R}^k \times \mathbb{R}^k$ given by the collection of maps from $T_x\mathbb{R}^k \to \{x\} \times \mathbb{R}^k$ defined by $[c]_x \mapsto (x, c'(0))$.

Definition 5.20 *Let $f: M \to N$ be a C^∞ map. We define $Tf: TM \to TN$ by*

$$Tf([c]_m) = [f \circ c]_{f(m)}$$

Tf is called the tangent of f. The notation $T_m f$ is also used for the restriction of Tf to the tangent space $T_m M$; we regard $T_m f$ as a map $T_m M \to T_{f(m)} N$.

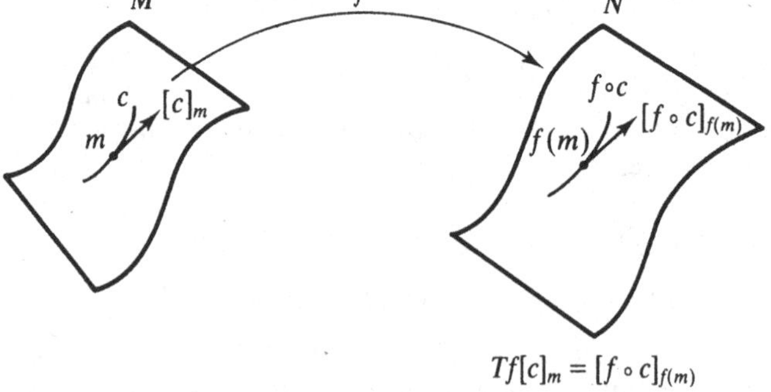

Fig. 2. The tangent map Tf.

Note: df is defined only if f is real-valued (or vector-valued) on M, whereas the tangent map, Tf, is defined for any C^∞ map f between two manifolds and Tf is a map between the tangent bundles.

By virtue of Remark (ii) above we see that if $f: M \to \mathbb{R}^k$ is a C^∞ map then

$$Tf([c]_m) = (f(m), df[c]_m) \qquad [c]_m \in T_m M \tag{*}$$

Locally, Tf has the form (*) for any $f: M \to N$. Indeed, if (U, κ) and (V, φ) are charts on M and N, respectively, such that $f(U) \subset V$ then $T\varphi \circ Tf \circ (T\kappa)^{-1}$ is given by

$$\kappa(U) \times \mathbb{R}^n \to \varphi(V) \times \mathbb{R}^d \qquad (n = \dim M,\, d = \dim N)$$

$$(x, w) \mapsto ((\varphi \circ f \circ \kappa^{-1})(x), D(\varphi \circ f \circ \kappa^{-1})(x) \cdot w), \tag{18}$$

for if c is a curve at $m = \kappa^{-1}(x)$ such that $(\kappa \circ c)'(0) = w$ then

$$\begin{aligned} T\varphi \circ Tf \circ (T\kappa)^{-1}(x, w) = T\varphi \circ Tf[c]_m &= T\varphi[f \circ c]_{f(m)} \\ &= (\varphi(f(m)), (\varphi \circ f \circ c)'(0)) \\ &= ((\varphi \circ f \circ \kappa^{-1})(x), D(\varphi \circ f \circ \kappa^{-1})(x) \cdot w) \end{aligned}$$

Theorem 5.21

(*i*) *If $f: M \to N$ is a C^∞ map then $Tf: TM \to TN$ is also C^∞.*

(*ii*) *Suppose $f: M \to N$ and $g: N \to P$ are C^∞ maps of manifolds. Then $g \circ f: M \to P$ is C^∞ and*

$$T(g \circ f) = Tg \circ Tf$$

(*iii*) *If $\mathscr{I}: M \to M$ is the identity map, then $T\mathscr{I}: TM \to TM$ is also the identity.*

(*iv*) *If $f: M \to N$ is a diffeomorphism, then $Tf: TM \to TN$ is a bijection and $(Tf)^{-1} = T(f^{-1})$.*

Proof The statement (*i*) follows from (18) since $\varphi \circ f \circ \kappa^{-1}$ is smooth by definition (see Definition 5.13). By using local representatives it is easily seen that $g \circ f$ is C^∞ if f and g are C^∞ (see Proposition 5.15). Moreover,

$$T(g \circ f)[c]_m = [g \circ f \circ c]_{g \circ f(m)}$$

and

$$(Tg \circ Tf)[c]_m = Tg([f \circ c]_{f(m)}) = [g \circ f \circ c]_{g \circ f(m)},$$

whence $T(g \circ f) = Tg \circ Tf$.

(*iii*) is obvious, and (*iv*) follows from (*ii*) and (*iii*).

Remark The following chain rule holds for smooth functions $g: N \to \mathbb{R}$ and $f: M \to N$,

$$d(g \circ f) = dg \circ Tf \tag{19}$$

Exercise

Let M_1 and M_2 be manifolds and $p_1: M_1 \times M_2 \to M_1$ and $p_2: M_1 \times M_2 \to M_2$ the canonical projections. Show that the map (Tp_1, Tp_2) is a diffeomorphism of the tangent bundle $T(M_1 \times M_2)$ with the product manifold $TM_1 \times TM_2$. With this identification, show that if $g_1: N \to M_1$ and $g_2: N \to M_2$ are smooth maps then

$$T(g_1 \times g_2) = Tg_1 \times Tg_2.$$

5.4 Submanifolds

Let us write $\mathbb{R}^n = \mathbb{R}^{n-k} \times \mathbb{R}^k$ and denote points of $\mathbb{R}^n$ by $x = (x', x'')$ where $x' \in \mathbb{R}^{n-k}$, $x'' \in \mathbb{R}^k$. The notation $\mathbb{R}^{n-k} \times 0$ indicates the subspace of $\mathbb{R}^n$ consisting of points $(x', 0, \dots, 0)$ (k zeros).

An $(n - k)$-dimensional *submanifold* of an n-dimensional manifold M is a subset $Y \subset M$ endowed with the topology induced from M and with the property that for each $y \in Y$ there is a chart (U, κ) of M with $y \in U$ such that

$$\text{SM} \qquad \kappa(U \cap Y) = \kappa(U) \cap (\mathbb{R}^{n-k} \times 0)$$

(the submanifold property for κ). The number k is called the *codimension* of Y in M.

An open subset of M is a submanifold of M of codimension 0. A submanifold of codimension 1 is called a hypersurface of M.

Proposition 5.22 *Let Y be a submanifold of a manifold M, and give Y the relative topology induced from M. Let $\{(U_i, \kappa_i)\}_{i\in I}$ be a family of charts satisfying the* SM *property and such that the sets $U_i \cap Y$ cover Y. If we identify $\mathbb{R}^{n-k}$ with the subspace $\mathbb{R}^{n-k} \times 0$, then the family of pairs*

$$\{(U_i \cap Y, \kappa_i|_{U_i \cap Y})\}_{i\in I}$$

is an atlas for a C^∞ structure on Y, making Y into an $(n - k)$-dimensional manifold. The inclusion map $\iota: Y \to M$ is C^∞.

Proof Any coordinate map $\kappa: U \to \kappa(U)$ (i.e. a homeomorphism to an open set in $\mathbb{R}^n$) on M restricts to a homeomorphism $\kappa|_{U\cap Y}: U \cap Y \to \kappa(U \cap Y)$ and, further, if κ satisfies property SM then $\kappa|_{U\cap Y}$ is a coordinate map on Y (i.e. a homeomorphism to an open set in $\mathbb{R}^{n-k}$). We may let $\kappa = \kappa_i$ for any $i \in I$, and the open sets $U_i \cap Y$ cover Y, thus it follows that Y is locally Euclidean.

To complete the proof we must show that the overlap of two maps,

$$(U_i \cap Y, \kappa_i|_{U_i\cap Y}) \quad \text{and} \quad (U_j \cap Y, \kappa_j|_{U_j\cap Y}),$$

is C^∞. But this is clear since $\kappa_j \circ \kappa_i^{-1}$ is a C^∞ function defined on an open set in $\mathbb{R}^n$ and

$$\kappa_j|_{U_j\cap Y} \circ (\kappa_i|_{U_i\cap Y})^{-1} = \kappa_j \circ \kappa_i^{-1}|\kappa_i(U_i \cap Y)$$

is a restriction of this C^∞ function to an open set in $\mathbb{R}^{n-k}$. Hence (ii) of Definition 5.11 holds and $\mathfrak{A}$ is an atlas for Y. Also Y is Hausdorff and has a countable basis since this is true for M.

The inclusion map ι is C^∞ for if (U, κ) is a chart on M satisfying the SM property then the local representation of ι with respect to charts κ and $\kappa|_{U \cap Y}$ is a restriction of the inclusion $\mathbb{R}^{n-k} \to \mathbb{R}^n$ and is hence smooth.

Example Consider the set $Y \subset \mathbb{R}^2$ as in the figure below:

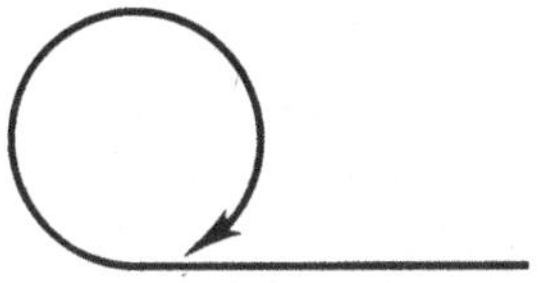

Fig. 3. A manifold which is *not* a submanifold of $\mathbb{R}^2$.

The arrow means that the line approaches itself arbitrarily close at p, without touching. It is clear that there is a bijection $\mathbb{R} \to Y$. By means of this bijection one can define a topology and C^∞ structure on Y; however, Y is *not* a submanifold of $\mathbb{R}^2$ because the topology on Y is not the relative topology induced from $\mathbb{R}^2$. (To show this, examine neighbourhoods of the point p.)

Submanifolds have the following universal property.

Proposition 5.23 *Let Y be a submanifold of M. Given another manifold N, and a C^∞ map $f: N \to M$ such that $f(N) \subset Y$, let $f_1: N \to Y$ denote the induced map. Then $f_1 \in C^\infty(N, Y)$.*

Proof Note that in the simplest case when $Y = \mathbb{R}^{n-k} \times 0$ and $M = \mathbb{R}^n$, the proposition is obvious (see §5.2, Exercise 5). The same is true when $Y = W \cap (\mathbb{R}^{n-k} \times 0)$ and $M = W$ is some open set in $\mathbb{R}^n$. The SM property enables us to reduce the general case to this case.

Suppose that f is C^∞. Then f is continuous, and since Y has the relative topology in M it follows that f_1 is continuous. To show that f_1 is C^∞ it suffices to show it is smooth in a neighbourhood of each $x \in N$. By property SM there is a chart (U, κ) at $y = f(x)$ such that $\kappa(U \cap Y) = \kappa(U) \cap (\mathbb{R}^{n-k} \times 0)$. Then there is a neighbourhood $V \ni x$ such that $f_1(V) \subset U \cap Y$. The function $\kappa \circ f_1|_V$ maps into $\kappa(U) \cap (\mathbb{R}^{n-k} \times 0)$, so it is smooth since $\kappa \circ f|_V$ is smooth.

Corollary *The universal property in Proposition* 5.23 *uniquely characterizes the C^∞ structure on Y.*

Proof If $\mathfrak{F}_1$ and $\mathfrak{F}_2$ are any two such C^∞ structures on Y then this property implies that the (set-theoretic) identity map $Y \to Y$ is C^∞ from $(Y, \mathfrak{F}_1)$ to $(Y, \mathfrak{F}_2)$. Hence $\mathfrak{F}_2 \subset \mathfrak{F}_1$, for by composing a chart in $\mathfrak{F}_2$ with the identity it becomes a chart in $\mathfrak{F}_1$. Similarly, $\mathfrak{F}_1 \subset \mathfrak{F}_2$, so that $\mathfrak{F}_1 = \mathfrak{F}_2$. The structures are the *same*, not just diffeomorphic.

For examples of submanifolds, see the Submersion Theorem below.

Let us now consider the tangent space to a submanifold $Y \subset M$. The inclusion map $\imath: Y \to M$ is smooth, and we claim that the tangent map

$$T_y \imath: T_y Y \to T_y M, \qquad y \in Y,$$

is injective; in this way the tangent space, $T_y Y$, can be identified with a subspace of $T_y M$.

To prove that the tangent map is injective it suffices to work in local coordinates. Let (U, κ) be a chart having the SM property. With respect to charts $\kappa|_{U \cap Y}$ and κ, the local representation of $\imath$ is $\kappa \circ \imath \circ (\kappa|_{U \cap Y})^{-1}$ and is just a restriction of the inclusion $\mathbb{R}^{n-k} \to \mathbb{R}^n$. By Theorem 5.21 we see that the local representation of the tangent map $T\imath: TY \to TM$ is

$$T\kappa \circ T\imath \circ (T\kappa|_{U \cap Y})^{-1}$$

and is a restriction of the inclusion $\mathbb{R}^{n-k} \times \mathbb{R}^{n-k} \to \mathbb{R}^n \times \mathbb{R}^n$ (also see (18)). It follows that the tangent map $T_y \imath: T_y Y \to T_y M$ is an inclusion for each $y \in Y$.

Remark With this identification (of $T_y Y$ as a subspace of $T_y M$), we have

$$T_y Y = (T_y \kappa)^{-1}(\mathbb{R}^{n-k} \times 0)$$

for any chart (U, κ) at y satisfying the SM property.

In the following lemma, E and F denote finite dimensional vector spaces.

Lemma 5.24 *Let $f: U \to F$ be of class C^1 where $U \subset E$ is open. Suppose that, for some $u_0 \in U$, $Df(u_0)$ is surjective. Let $E_1 = \ker Df(u_0)$. Then there exists an open subset U' of U containing u_0 and a C^1 diffeomorphism $\psi: W_1 \times W_2 \to U'$, where $W_1 \subset E_1$ and $W_2 \subset F$ are open, such that*

$$(f \circ \psi)(x_1, x_2) = x_2 \qquad \textit{for all } (x_1, x_2) \in W_1 \times W_2$$

Proof Let $E_1 = \ker Df(u_0)$ and choose a complement E_2 such that $E = E_1 \oplus E_2$. Then the partial derivative $D_2 f(u_0): E_2 \to F$ is a linear isomorphism. Thus, we may assume that $E_2 = F$ and $E = E_1 \times E_2$.

Define the map $g: U \to E_1 \times E_2$ by $g(x_1, x_2) = (x_1, f(x_1, x_2))$ and note that the derivative of g at u_0 is the linear map $E_1 \times E_2 \to E_1 \times E_2$ given by

$$Dg(u_0) = \begin{bmatrix} I_{E_1} & 0 \\ D_1 f(u_0) & D_2 f(u_0) \end{bmatrix}$$

which is an isomorphism. By the Inverse Function Theorem, g is a diffeomorphism of some open set U' containing u_0 onto an open set $W_1 \times W_2$ containing $g(u_0)$. Let $\psi = (g|_{U'})^{-1}$. If $(x_1, x_2) \in W_1 \times W_2$ we have $(x_1, x_2) = g(\psi(x_1, x_2))$ and hence

$$(x_1, x_2) = (x_1, f(\psi(x_1, x_2)))$$

Thus $f(\psi(x_1, x_2)) = x_2$.

Let $f \in C^\infty(M, N)$. A point $y \in N$ is called a *regular value* if for each $x \in f^{-1}(y)$, the tangent map $T_x f$ is surjective. A *critical point* is a point $x \in M$ where $T_x f$ is not surjective. If $T_x f$ is surjective for each x in a set S, we say that f is a *submersion* on S.

Theorem 5.25 (*Submersion Theorem*) *Let $f: M \to N$ be of class C^∞ and let $y_0 \in N$ be a regular value. Assume that the level set $f^{-1}(y_0) = \{x; x \in M, f(x) = y_0\}$ is non-empty. Then $f^{-1}(y_0)$ is a submanifold of M with tangent space at $x \in f^{-1}(y_0)$ given by* $\ker T_x f$.

Proof Let $x_0 \in f^{-1}(y_0)$. Let (V, φ) be a chart at y_0 and (U, κ) a chart at x_0 such that $f(U) \subset V$. Consider the local representation $f_{\varphi\kappa} = \varphi \circ f \circ \kappa^{-1}$. We have that $T(f_{\varphi\kappa}) = T\varphi \circ Tf \circ (T\kappa)^{-1}$ is the map $\kappa(U) \times \mathbb{R}^n \to \varphi(V) \times \mathbb{R}^d$ given by

$$(x, w) \mapsto (f_{\varphi\kappa}(x), D(f_{\varphi\kappa})(x) \cdot w)$$

(see (18)). Since $T_{x_0} f$ is surjective, it follows that the derivative $D(f_{\varphi\kappa})(x_0)$ is a surjective map. Hence we can apply Lemma 5.24. Let $n = \dim M$ and $d = \dim N$. By shrinking U and V, if necessary (and replacing κ by $\psi^{-1} \circ \kappa$), we may assume that $\kappa(U) = W_1 \times W_2$, $\varphi(V) = W_2$ where $W_1 \subset \mathbb{R}^{n-d}$ and $W_2 \subset \mathbb{R}^d$ are open, such that

$$f_{\varphi\kappa} \text{ is the projection } W_1 \times W_2 \to W_2 \text{ on the second factor.}$$

We also assume that $\kappa(x_0) = 0$. With this choice of charts, we have $\kappa(U \cap f^{-1}(y_0)) = (f_{\varphi\kappa})^{-1}(0) = W_1 \times 0$, hence

$$\kappa(U \cap f^{-1}(y_0)) = \kappa(U) \cap (\mathbb{R}^{n-d} \times 0),$$

which is the submanifold (SM) property. Hence $f^{-1}(y_0)$ is a submanifold of M of codimension $d = \dim N$.

We also have for $u \in W_1 \times 0$

$$T_u(f_{\varphi\kappa}^{-1}(0)) = \mathbb{R}^{n-d} \times 0 = \ker(T_u f_{\varphi\kappa})$$

which in view of $T_u(f_{\varphi\kappa}) = T_{f(x)}\varphi \circ T_x f \circ (T_x\kappa)^{-1}$, where $\kappa(x) = u$, implies that

$$T_x(f^{-1}(y_0)) = (T_x\kappa)^{-1}(\mathbb{R}^{n-d} \times 0) = \ker T_x f.$$

Example Let $f: \mathbb{R}^{n+1} \to \mathbb{R}$ be defined by $f(x) = |x|^2$. By identifying $T(\mathbb{R}^{n+1}) = \mathbb{R}^{n+1} \times \mathbb{R}^{n+1}$, we get

$$Tf(x, v) = (|x|^2, 2\langle x, v\rangle)$$

where $\langle\, ,\rangle$ is the usual scalar product of vectors in $\mathbb{R}^{n+1}$. Let $x \in S^n$, i.e. $f(x) = 1$. Then $T_x f: T_x\mathbb{R}^{n+1} \to T_1\mathbb{R}$ is the map $v \mapsto 2\langle x, v\rangle$ and is surjective since $x \neq 0$. Hence 1 is a regular value of f, so the n-sphere $S^n = f^{-1}(1)$ is a submanifold of $\mathbb{R}^{n+1}$, and for any $x \in S^n$ we have

$$T_x(S^n) = \{v; v \in \mathbb{R}^{n+1}, \langle x, v\rangle = 0\}.$$

We conclude this section with a study of quotient manifolds (see also §5.2, Exercises 6 to 9).

Definition 5.26 *An equivalence relation* $\sim$ *on a manifold is called* regular *if the quotient space* $M/\sim$ *has a manifold structure such that the canonical projection* $\pi: M \to M/\sim$ *is a submersion.*

If $\sim$ is a regular equivalence relation, then $M/\sim$ is called the *quotient manifold* of M by $\sim$.

Proposition 5.27 *Let* $\sim$ *be a regular equivalence relation on* M.

(i) *A map* $f: M/\sim \to N$ *is* C^∞ *if and only if* $f \circ \pi: M \to N$ *is* C^∞.
(ii) *Any* C^∞ *map* $g: M \to N$ *which factors through the equivalence relation* $\sim$, *that is,*

$$x \sim y \Rightarrow g(x) = g(y),$$

defines a unique C^∞ *map* $\hat{g}: M/\sim \to N$ *such that* $\hat{g} \circ \pi = g$.

Proof (i) If f is C^∞, then so is $f \circ \pi$ by Proposition 5.15. Conversely, let $f \circ \pi$ be C^∞. Since π is a submersion, it can be locally expressed as a projection, i.e. (as shown above) there exist charts (U, κ) at $m \in M$ and (V, φ) at $\pi(m) \in M/\sim$ such that $\kappa(U) = W_1 \times W_2$, $\varphi(V) = W_2$ where $W_1 \subset \mathbb{R}^{n-d}$ and $W_2 \subset \mathbb{R}^d$ are open, and

$$\pi_{\varphi\kappa}: W_1 \times W_2 \to W_2 \text{ is the projection on the second factor.}$$

Hence if (V', ψ) is a chart at $f \circ \pi(m)$ in N satisfying $f \circ \pi(U) \subset V'$, then

$$f_{\psi\varphi} = \psi \circ f \circ \varphi^{-1} = \psi \circ (f \circ \pi) \circ \kappa^{-1}|_{\{0\} \times W_2}.$$

Therefore $f_{\psi\varphi}$ is C^∞, being the restriction to $\{0\} \times W_2$ of the C^∞ function $(f \circ \pi)_{\psi\kappa} = \psi \circ (f \circ \pi) \circ \kappa^{-1}$.
(ii) It is evident that g is uniquely determined; and it is C^∞ by (i).

Corollary 5.28 *The manifold structure of* $M/\sim$ *is unique.*

Proof Let $(M/\sim)_1$ and $(M/\sim)_2$ be two manifold structures on $M/\sim$ having π as a submersion. If we apply Proposition 5.27(ii) to the map $\pi: M \to (M/\sim)_2$, it follows that the set-theoretic identity map $(M/\sim)_1 \to (M/\sim)_2$ is C^∞. But of course the roles of $(M/\sim)_1$ and $(M/\sim)_2$ can be reversed, so the identity map induces a diffeomorphism.

There is in fact a bijective correspondence between surjective submersions and quotient manifolds; see Exercise 6.

Exercises

1. (a) Show that if Y_i is a submanifold of M_i, $i = 1, 2$, then $Y_1 \times Y_2$ is a submanifold of $M_1 \times M_2$.
(b) Show that if $f: M \to N$ is a diffeomorphism and $Y \subset M$ is a submanifold then $f(Y)$ is a submanifold of N.

2. Let $h: M \to N$ be C^∞. Show that if Y is a submanifold of M then the restriction $h|_Y: Y \to N$ is C^∞. Hint: The inclusion map $\iota: Y \to M$ is C^∞.

3. Let Y be a submanifold of M. Show that $f \in C^\infty(Y, N)$ if and only if for every $y \in Y$ there is an open set U in M and a function $\tilde{f} \in C^\infty(U, N)$ such that $y \in U$ and

$$\tilde{f}|_{U \cap Y} = f|_{U \cap Y}$$

4. Verify that the C^∞ structure on S^n defined in the example above (by means of Proposition 5.22) is the same as the one defined by stereographic projection.

5. Let $f \in C^\infty(M, N)$. The graph of f is defined to be $\text{graph}(f) = \{(x, f(x));\ x \in M\}$.
(a) Show that $\text{graph}(f)$ is a submanifold of $M \times N$ of codimension $k = \dim N$.
(b) Show that the inclusion map, $\text{graph}(f) \to M \times N$, induces a diffeomorphism $T_{(m, f(m))}(\text{graph}(f)) \simeq \text{graph}(T_m f)$.
Hint for (a). First consider the case $N = \mathbb{R}^k$. Define the map $\varphi: M \times N \to M \times N$ by $\varphi(x, y) = (x, y - f(x))$ and use the Inverse Function Theorem to show that φ is a diffeomorphism of a neighbourhood of $(x_0, f(x_0))$ onto a neighbourhood of $(x_0, 0)$.

6. Let $f: M \to N$ be a submersion and let $\sim$ be the equivalence relation defined by f, i.e.

$$x \sim y \Leftrightarrow f(x) = f(y)$$

Show that $\sim$ is regular, $M/\sim$ is diffeomorphic to $f(M)$, and $f(M)$ is open in N.

7. Let M and N be manifolds, and R and Q regular equivalence relations on M and N, respectively. If $f: M \to N$ is a C^∞ map compatible with R and Q, that is, if

$$xRy \Rightarrow f(x)Qf(y),$$

then f induces a unique C^∞ map $\bar{f}: M/R \to N/Q$ such that $\pi_N \circ f = \bar{f} \circ \pi_M$.

A C^∞ map $f: M \to N$ is called an *immersion at* m if the linear map $T_m f: T_m M \to T_m N$ is injective. If f is an immersion at each m, we just say that f is an *immersion*. A C^∞ map $f: M \to N$ is called an *embedding* if it is an immersion that is a homeomorphism of M onto the image $f(M)$ with the relative topology induced from N.

8. Let $f: M \to N$ be a C^∞ map. Show that the following are equivalent:
(a) f is an immersion at m.
(b) there are charts (U, κ) and (V, ψ) on M and N, respectively, with $f(U) \subset V$, $m \in U$, $\kappa(m) = 0$, $\kappa: U \to U'$ and $\psi: V \to U' \times V'$ such that $\psi \circ f \circ \kappa^{-1}: U' \to U' \times V'$ is the inclusion $x \mapsto (x, 0)$.
(c) there is a neighbourhood U of m such that $f(U)$ is a submanifold in N and f restricted to U is a diffeomorphism of U onto $f(U)$.

9. Find an example of an injective immersion which is *not* an embedding.

10. Let N be a manifold. Show that a subset $S \subset N$ is a submanifold if and only if S is the image of an embedding.

5.5 Vector fields

A *vector field* on a manifold M is a C^∞ map $X: M \to TM$ such that $X_m \in T_m M$ for all $m \in M$ (i.e. a smooth section of the tangent bundle TM). The set of vector fields on M is denoted by $\mathscr{X}(M)$. $C^\infty(M)$ is a commutative ring under pointwise addition and multiplication of functions, and $\mathscr{X}(M)$ can be given a $C^\infty(M)$-module structure as follows: if $X, Y \in \mathscr{X}(M)$ and $f \in C^\infty(M)$ then $X + Y \in \mathscr{X}(M)$ and $fX \in \mathscr{X}(M)$ are defined by

$$(X + Y)(m) = X(m) + Y(m) \quad \text{and} \quad (fX)(m) = f(m)\cdot X(m)$$

To facilitate calculations in local coordinates, we introduce some notation. Let $\lambda^i: \mathbb{R}^n \to \mathbb{R}$ denote the projection on the ith component. If (U, κ) is a chart for M, we write

$$\kappa = (x_1, \ldots, x_n),$$

that is, $x_i = \lambda^i \circ \kappa: U \to \mathbb{R}$ are the component functions of κ. Note that x_i is C^∞ since λ^i and κ are C^∞. Hence $dx_i: TU \to \mathbb{R}$ is C^∞, and, by Theorem 5.18, $dx_i|_m \in (T_m M)^*$, a linear function on $T_m M$ for each $m \in U$. We have $d\kappa = (dx_1, \ldots, dx_n)$ so the definition of $T\kappa$ takes the form

$$T\kappa(m, v) = (\kappa(m), dx_1(v), \ldots, dx_n(v))$$

for $v \in T_m M$. Let $e_1, \ldots, e_n$ denote the standard basis of $\mathbb{R}^n$ and let $\partial/\partial x_j|_m$ denote the unique vector in $T_m M$ such that $d\kappa|_m(\partial/\partial x_j|_m) = e_j$, that is,

$$dx_i|_m(\partial/\partial x_j|_m) = \delta_{ij} \qquad i, j = 1, \ldots, n$$

For each $m \in U$, the vectors $\partial/\partial x_1|_m, \ldots, \partial/\partial x_n|_m$ form a basis of $T_m M$ and $dx_1|_m, \ldots, dx_n|_m$ is the corresponding dual basis in $(T_m M)^*$. Note that $\partial/\partial x_1, \ldots, \partial/\partial x_n$ are vector fields on U since

$$T\kappa \circ \partial/\partial x_j \circ \kappa^{-1}(x) = (x, e_j), \qquad x \in \kappa(U), \tag{20}$$

is C^∞ on $\kappa(U)$, whence $\partial/\partial x_j: U \to TU$ is smooth.

Proposition 5.29 *Let $(U, \kappa) = (U, x_1, \ldots, x_n)$ be a chart on M. Then $\mathscr{X}(U)$ is generated by $\partial/\partial x_1, \ldots, \partial/\partial x_n$ as $C^\infty(U)$-module, i.e., a map $X: U \to TU$ is a vector field on U if and only if it has the form*

$$X = X^1 \frac{\partial}{\partial x_1} + \cdots + X^n \frac{\partial}{\partial x_n}$$

where $X^i \in C^\infty(U)$, $i = 1, \ldots, n$.

Proof Let X be a vector field on U. Since $X(m) \in T_m M$, the definition of

$\partial/\partial x_j|_m$ implies that

$$X(m) = X^1(m)\frac{\partial}{\partial x_1}\bigg|_m + \cdots + X^n(m)\frac{\partial}{\partial x_n}\bigg|_m \tag{21}$$

where $X^i(m) = dx_i(X(m))$. The X^i are smooth since $X^i = dx_i \circ X$ where $v\colon U \to TU$ and $dx_i\colon TU \to \mathbb{R}$ are smooth. Conversely, any map $X\colon U \to TU$ of this form is obviously smooth and $X(m) \in T_m M$ for all $m \in U$; thus it is a vector field on U.

Let $X \in \mathscr{X}(M)$, a vector field on M. The *directional derivative* of $f \in C^\infty(M)$ along X is defined to be $Xf := df \circ X$, that is,

$$(Xf)(m) = df(m)(X(m)).$$

Since df and X are smooth then so is Xf. See Exercise 2 at the end of this section for properties of the directional derivative.

Now let $(U, \kappa) = (U, x_1, \ldots, x_n)$ be a chart on M. For the vector fields $\partial/\partial x_j \in \mathscr{X}(U)$, it follows by the chain rule that the directional (partial) derivative $(\partial/\partial x_j)f = df \circ \partial/\partial x_j$ is given by

$$\frac{\partial f}{\partial x_j}(m) = \partial_j(f \circ \kappa^{-1})(\kappa(m)), \qquad m \in U,$$

where $f \in C^\infty(U)$ and ∂_j is the jth partial differentiation operator for functions defined on open sets in $\mathbb{R}^n$. Then we have

$$df(m) = \frac{\partial f}{\partial x_1}(m)\, dx_1(m) + \cdots + \frac{\partial f}{\partial x_n}(m)\, dx_n(m), \qquad m \in U \tag{22}$$

Now we turn to the definition of pull-back and push-forward for functions and vector fields.

If $\varphi\colon M \to N$ is a diffeomorphism and $X \in \mathscr{X}(M)$, a vector field on M, the *push-forward* of X by φ is defined by

$$\varphi_* X = T\varphi \circ X \circ \varphi^{-1} \in \mathscr{X}(N) \tag{23}$$

It is the vector field that makes the diagram

$$\begin{array}{ccc} T(M) & \xrightarrow{T\varphi} & T(N) \\ {\scriptstyle X}\uparrow & & \uparrow{\scriptstyle \varphi_* X} \\ M & \xrightarrow{\varphi} & N \end{array}$$

commutative. Note: If $x \in N$ then $(\varphi_* X)|_x \in T_x N$, so the push-forward of X is indeed a vector field on N.

We also define a *pull-back* map φ^* for vector fields by replacing φ by φ^{-1}, that is, $\varphi^* X = T\varphi^{-1} \circ X \circ \varphi$.

For functions, the pull-back map $\varphi^*\colon C^\infty(N) \to C^\infty(M)$ is defined by

$$\varphi^*(f) = f \circ \varphi, \qquad f \in C^\infty(N)$$

Similarly, the push-forward for functions is defined by $\varphi_*(f) = f \circ \varphi^{-1}$.

Remark

(i) We have $\varphi^* = (\varphi^{-1})_*$ for both functions and vector fields.

(ii) The pull-back of functions, $\varphi^*(f) = f \circ \varphi$, is defined for *any* smooth map φ, but the pull-back of vector fields makes sense only for diffeomorphisms (but see Exercise 4).

The push-forward and pull-back maps are $\mathbb{R}$-linear on both functions and vector fields. Also, on functions, the pull-back and push-forward maps are algebra homomorphisms, that is,

$$\varphi^*(fg) = \varphi^*(f)\cdot\varphi^*(g) \qquad f, g \in C^\infty(N)$$

and

$$\varphi_*(fg) = \varphi_*(f)\cdot\varphi_*(g) \qquad f, g \in C^\infty(M)$$

The *local representation* of a function f on M with respect to a chart (U, κ) is

$$\tilde{f} = f \circ \kappa^{-1}$$

The *local representation* of a vector field X on M with respect to the charts (U, κ) on M and $(TU, T\kappa)$ on TM is

$$\tilde{X} = T\kappa \circ X \circ \kappa^{-1}$$

(compare with (20)).

Integral curves

Recall that we say $c: I \to M$ is a curve at m if I is an open interval, $0 \in I$, and $c(0) = m$. Since I is an open set in $\mathbb{R}$ we can identify $T_tI \simeq \mathbb{R}$ for each $t \in I$ in a canonical manner, and the tangent map $T_tc: \mathbb{R} \to T_{c(t)}M$ enables us to assign a tangent vector to the curve at each point $c(t)$ by

$$c'(t) = T_tc(1)$$

Note that if $\varphi: M \to N$ is a smooth map, then $\varphi \circ c$ is a curve on N; by the chain rule (Theorem 5.21) we have $(\varphi \circ c)'(t) = T_{c(t)}\varphi(c'(t))$.

An *integral curve of a vector field* $X \in \mathscr{X}(M)$ is a curve $c: I \to M$ such that

$$c'(t) = X(c(t))$$

for all $t \in I$; and if it also has the initial value $c(0) = m$, we say that c is an *integral curve of X at m*. The local representation of c with respect to a chart (U, κ) at m is the curve $\tilde{c} = \kappa \circ c$ in $\mathbb{R}^n$. Applying the map $T_{c(t)}\kappa$ to the equation $c'(t) = X(c(t))$ we obtain

$$\tilde{c}'(t) = \tilde{X}(\tilde{c}(t)) \tag{24}$$

where $\tilde{X} = T\kappa \circ X \circ \kappa^{-1}$ is the local representation of the vector field X with respect to the charts (U, κ) on M and $(TU, T\kappa)$ on TM. Then equation (24) is a system of ordinary differential equations in n unknowns

$$\frac{dc^1}{dt} = \tilde{X}^1(c^1(t), \ldots, c^n(t))$$
$$\vdots \qquad \vdots$$
$$\frac{dc^n}{dt} = \tilde{X}^n(c^1(t), \ldots, c^n(t))$$

where $\tilde{c} = (c^1, \ldots, c^n)$ and $\tilde{X} = (\tilde{X}^1, \ldots, \tilde{X}^n)$.

Thus, for existence of integral curves we may use the following well-known result on systems of ordinary differential equations. The most difficult part of the proof is the smooth dependence of the solutions c on the initial conditions $c(0)$; the proof can be found in [AMR].

Theorem 5.30 *Let U be an open set in $\mathbb{R}^n$ and let $X: U \to \mathbb{R}^n$ be of class C^∞. For each $x_0 \in U$, there is a curve $c: I \to U$ at x_0 such that*

$$dc/dt = X(c(t)) \qquad (*)$$

for all $t \in I$. Any two such curves are equal on the intersection of their domains. Furthermore, there is a neighbourhood U_0 of x_0, a real number $a > 0$ or $a = \infty$, and a C^∞ mapping

$$F: U_0 \times I_a \to \mathbb{R}^n,$$

where I_a is the open interval $(-a, a)$, such that $F(x, 0) = x$ and the curve $I_a \to \mathbb{R}^n$ given by $t \mapsto F(x, t)$ satisfies the differential equation (*).

This theorem is local in nature; it can be applied, essentially without any further proof required, to manifolds.

Theorem 5.31 *Let M be a manifold and $X \in \mathscr{X}(M)$. For each $m_0 \in M$ there exists a neighbourhood U_0 of m_0, a real number $a > 0$ or $a = \infty$, and a C^∞ map*

$$F: U_0 \times I_a \to M$$

such that $F(m, 0) = m$ and the curve $I_a \to M$ given by $t \mapsto F(m, t)$ is an integral curve of X. If $F': U'_0 \times I_{a'} \to M$ is another map with the same properties then F and F' are equal on $(U_0 \cap U'_0) \times (I_a \cap I_{a'})$.

Proof Let (U, κ) be a chart at x_0, and let $\tilde{X}: \kappa(U) \to \mathbb{R}^n$ be the local representative of X, given by $\tilde{X} = T\kappa \circ X \circ \kappa^{-1}$. By Theorem 5.30 there exists a neighbourhood $\tilde{U}_0 \subset \kappa(U)$ and a smooth function $\tilde{F}: \tilde{U}_0 \times I \to \mathbb{R}^n$ such that $\tilde{F}(x, 0) = x$ and the curve $t \mapsto \tilde{F}(x, t)$ satisfies the equation (24). By shrinking $\tilde{U}_0$ and I we may assume that $\tilde{F}(\tilde{U}_0 \times I) \subset U$. Now let $U_0 = \kappa^{-1}(\tilde{U}_0)$ and define $F: U_0 \times I \to M$ by $F(m, t) = \kappa^{-1}(\tilde{F}(\kappa(m), t))$. Since c is an integral curve of X if and only if $\kappa \circ c$ is an integral curve of $\tilde{X}$, it is clear that F satisfies the required properties.

Now let F' be another function with the same properties as F. Due to the uniqueness of integral curves proved below in Theorem 5.32, for each $x \in U_0 \cap U_0'$ we have $F|_{\{x\} \times I} = F'|_{\{x\} \times I}$ where $I = I_a \cap I_{a'}$. Hence $F = F'$ on $(U_0 \cap U_0') \times I$.

Observe that solutions $c: I \to M$ of $c'(t) = X(c(t))$ remain solutions under translations; that is, if $t_0 \in I$ then $\alpha(t) = c(t_0 + t)$ is defined in a neighbourhood of $t = 0$ and

$$\alpha'(t) = c'(t_0 + t) = X(c(t_0 + t)) = X(\alpha(t))$$

This fact allows us to extend the uniqueness part of Theorem 5.31 to get a global uniqueness theorem on manifolds. It is important here that we have required that all manifolds be Hausdorff.

Theorem 5.32 *If c_1 and c_2 are integral curves of X at $m \in M$ then $c_1 = c_2$ on the intersection of their domains.*

Proof It is not sufficient just to refer to Theorem 5.30, since we do not assume that the domains of $c_1: I_1 \to M$ and $c_2: I_2 \to M$ lie in a single chart. However, observe that the intersection, $I = I_1 \cap I_2$, of the domains of c_1 and c_2 is a connected set. The set $J = \{s \in I; c_1(s) = c_2(s)\}$ is certainly closed since M is Hausdorff. On the other hand, J is also open, for if $s \in J$ then the curves $t \mapsto c_1(t + s)$ and $t \mapsto c_2(t + s)$ are integral curves of X passing through $c_1(s) = c_2(s)$ when $t = 0$, so by virtue of Theorem 5.30 they coincide in some neighbourhood of $t = 0$. Hence $c_1 = c_2$ on a neighbourhood of s. Thus J is both open and closed in I, which implies $J = I$ since I is connected.

Corollary *Let $c_1: I_1 \to M$ and $c_2: I_2 \to M$ be integral curves of X and suppose that $c_1(t_1) = c_2(0)$ for some $t_1 \in I_1$. Then the curve $t \mapsto c_1(t + t_1)$ is defined on the open interval $I_1 - t_1$ (which contains 0) and must coincide with c_2 on the intersection $(I_1 - t_1) \cap I_2$ since both are integral curves for X at $c_2(0)$. Hence*

$$c(t) = \begin{cases} c_1(t), & t \in I_1 \\ c_2(t - t_1), & t \in t_1 + I_2 \end{cases}$$

is also an integral curve of X.

The proof of the corollary is obvious, but the fact that any two integral curves of X meeting at some point can be pieced together is an important idea to keep in mind when proving the global results in Theorem 5.33.

In view of Theorems 5.31 and 5.32 there exists an integral curve at m for X with maximal domain. We denote this integral curve by c_m and, for emphasis, we mention once more that $c_m'(t) = X(c_m(t))$, the domain of c_m is an open interval containing 0, and we have $c_m(0) = m$. The right- and left-hand end-points of the domain of c_m are denoted by $t^+(m)$ and $t^-(m)$, respectively.

Let $\mathscr{D}_X$ be the subset of $M \times \mathbb{R}$ consisting of the points (m, t) such that

$$t^-(m) < t < t^+(m)$$

and define the map $F\colon \mathscr{D}_X \to M$ by $F(m, t) = c_m(t)$. We call F the (global) *flow* of X.

Note that $\mathscr{D}_X$ is the set of $(m, t) \in M \times \mathbb{R}$ such that X has an integral curve $c\colon I \to M$ at m with $t \in I$. In particular $\mathscr{D}_X \supset M \times 0$.

The vector field X is said to be *complete* if $\mathscr{D}_X = M \times \mathbb{R}$. Thus, X is complete if and only if each integral curve can be extended so that its domain becomes $(-\infty, \infty)$; that is, $t^+(m) = \infty$ and $t^-(m) = -\infty$ for all $m \in M$.

Examples

(1) On $M = \mathbb{R}^2$, let X be the constant vector field, $X(m) = (0, 1)$. The integral curve through $m = (x, y)$ is $t \mapsto (x, y + t)$. Each integral curve of X is defined for all $t \in \mathbb{R}$, so X is complete.

(2) On $M = \mathbb{R}^2 \backslash 0$, let X be the same vector field. The integral curves are the same as before, except for the fact that any curve that passed through $(0, 0)$ in Example (1) cannot now be extended to infinity; thus X is not complete. For example, $t^-(0, 1) = -1$ and $t^+(0, 1) = \infty$.

(3) For $M = \mathbb{R}$ let $X \in \mathscr{X}(\mathbb{R})$ be defined by $X(x) = -x^2$. The integral curves of X are the solutions of $dc/dt = -[c(t)]^2$. Solving this differential equation gives $c(t) = (t + c(0)^{-1})^{-1}$ if $c(0) \neq 0$, and $c(t) \equiv 0$ if $c(0) = 0$. X is not complete; for example, $t^+(1) = \infty$ but $t^-(1) = -1$.

Theorem 5.33 *Let M be a manifold and $X \in \mathscr{X}(M)$. Then*

(i) *$\mathscr{D}_X$ is open in $M \times \mathbb{R}$ and contains $M \times 0$.*
(ii) *The map $F\colon \mathscr{D}_X \to M$ is smooth.*
(iii) *If $(m, t) \in \mathscr{D}_X$ then $(F(m, t), s) \in \mathscr{D}_X$ iff $(m, t + s) \in \mathscr{D}_X$; in this case*

$$F(m, t + s) = F(F(m, t), s)$$

Proof Let $(m, t) \in \mathscr{D}_X$. Both $s \mapsto c_m(t + s)$ and $s \mapsto c_{F(m,t)}(s)$ are integral curves of X with the same initial condition at $s = 0$. Hence they must coincide and have the same (maximal) domain of definition. Taking $s = t'$, (iii) follows immediately.

It remains to show that $\mathscr{D}_X$ is open in $M \times \mathbb{R}$ and that F is smooth on $\mathscr{D}_X$. It is clear from Theorem 5.31 that F is smooth in a neighbourhood of $M \times 0$, but an additional argument is required for the other points of $\mathscr{D}_X$. Let $m_0 \in M$, and let J be the set of points t in the domain of c_{m_0} such that F is C^∞ at (m_0, t), that is, for which there exists a number $b > 0$ and a neighbourhood U of m_0 such that the product $U \times (t - b, t + b)$ is contained in $\mathscr{D}_X$ and the restriction of the flow F to this product is a C^∞ map. To show that $\mathscr{D}_X$ is open and F is C^∞ everywhere on $\mathscr{D}_X$ we must show that, for any $m_0 \in M$, every point in the domain of c_{m_0} belongs to J.

We do this by a connectedness argument. It is clear that J is open. It remains to show that J is closed in the domain of c_{m_0}. Let s be in the closure of J; then by Theorem 5.31 there is a neighbourhood V of $c_{m_0}(s)$ and a number $a > 0$ such that

$$\text{the flow } F \text{ is smooth on } V \times (-a, a). \tag{25}$$

Since c_{m_0} is smooth (hence continuous) at s there exists $t_1 \in J$ such that $c_{m_0}(t_1)$ lies in $\bar{V}_1 \subset V$. Since $t_1 \in J$, the flow F is C^∞ at (m_0, t_1). By the continuity of F there is a number $b > 0$ and neighbourhood U of m_0 such that

$$F \text{ maps the product } U \times (t_1 - b, t_1 + b) \text{ into } V \tag{26}$$

and F is C^∞ on this product. We can also suppose that $|t_1 - s| < a$.

Thus for any $m \in U$ we have integral curves for X,

$$c_m: (t_1 - b, t_1 + b) \to M \quad \text{and} \quad c_{F(m,t_1)}: (-a, a) \to M,$$

and, in view of the corollary to Theorem 5.32, every $t \in t_1 + (-a, a)$ lies in the domain of the integral curve c_m when $m \in U$ and we have $c_m(t) = c_{F(m,t_1)}(t - t_1)$. In other words, the product $U \times (t_1 - a, t_1 + a)$ is contained in $\mathscr{D}_X$ and

$$F(m, t) = F(F(m, t_1), t - t_1)$$

for all (m, t) in this product. Hence in view of (25) and (26) the restriction of F to this product, $U \times (t_1 - a, t_1 + a)$, is a composition of two C^∞ functions, so it is C^∞ itself. Since $U \times (t_1 - a, t_1 + a)$ is a neighbourhood of (m_0, s) we have $s \in J$.

Thus J is both open and closed. By connectedness, J is the whole domain of c_{m_0}.

From now on we write the domain of the flow as $\mathscr{D}$ instead of $\mathscr{D}_X$. Let $\mathscr{D}_t \subset M$ be the set of points $m \in M$ such that $(m, t) \in \mathscr{D} = \mathscr{D}_X$. Since $\mathscr{D}$ is open in $M \times \mathbb{R}$ it is evident that $\mathscr{D}_t$ is open in M. The restriction of F to $\mathscr{D}_t$ defines a map $F_t: \mathscr{D}_t \to M$ by $F_t(m) = F(m, t)$. Note that (iii) of Theorem 5.33 says that $F_s \circ F_t = F_{s+t}$ wherever it is defined.

Corollary 5.34 *Let X be a vector field on M and $F: \mathscr{D} \to M$ its flow. Then $\mathscr{D}_t$ is open in M for each $t \in M$, and F_t is a diffeomorphism of $\mathscr{D}_t$ onto an open set in M. In fact, $F_t(\mathscr{D}_t) = \mathscr{D}_{-t}$ and $F_t^{-1} = F_{-t}$.*

Proof The fact that $\mathscr{D}_t$ is open has already been proved. Now let $m \in \mathscr{D}_t$. Then $(m, t) \in \mathscr{D}$ and (iii) implies $(F_t(m), -t) \in \mathscr{D}$, since $(m, 0) \in \mathscr{D}$ always holds. Hence $F_t(m) \in \mathscr{D}_{-t}$ and

$$F_{-t}(F_t(m)) = m \qquad (m \in \mathscr{D}_t)$$

Similarly, if $m \in \mathscr{D}_{-t}$ then $F_t(F_{-t}(m)) = m$.

Proposition 5.35 *Let $X \in \mathscr{X}(M)$. Let c be an integral curve of X with a maximal domain of definition. If for every finite open interval (a, b) in its domain, $c((a, b))$ lies in a compact subset of M, then c is defined for all $t \in \mathbb{R}$.*

Proof It suffices to show that the end-points a and b of the interval (a, b) lie in the domain of c. Let $t_n \in (a, b)$ with $t_n \to b$. By passing to a subsequence we can assume that $c(t_n)$ converges to a point $x \in M$. Now, since the domain of the flow contains a neighbourhood of $(x, 0)$, there is a neighbourhood U of x and a number $\tau > 0$ such that integral curves starting at points in U persist for a time longer than τ. If we choose n such that $c(t_n) \in U$ and $b - t_n < \tau$, we can extend c to a time greater than b. Hence b is in the domain of c, since c has maximal domain. A similar argument holds for the other endpoint, a.

The *support* of a vector field X defined on a manifold M is the closure of the set $\{m \in M;\ X(m) \neq 0\}$.

Corollary 5.36 *A vector field with compact support on a manifold M is complete. In particular, if M is a compact manifold then all vector fields on M are complete.*

Note that when the vector field X is complete, the maps F_t are diffeomorphisms $M \to M$ for each $t \in \mathbb{R}$ and $F_t \circ F_s = F_{t+s}$ for all $s, t \in \mathbb{R}$. The set $\{F_t;\ t \in \mathbb{R}\}$ is called a *one-parameter group of diffeomorphisms.*

Exercises

1. Let c be a curve on M. Show that

$$c'(t_0)f = \frac{d}{dt} f(c(t))|_{t=t_0} \qquad \text{for all } f \in C^\infty(M)$$

2. Let $X \in \mathscr{X}(M)$, a vector field on M. Show that the directional derivative $X: C^\infty(M) \to C^\infty(M)$ defined by $Xf = df \circ X$ has the following properties for $f, g \in C^\infty(M)$ and $\lambda \in \mathbb{R}$:

$$X(f+g) = Xf + Xg, \qquad X(\lambda g) = \lambda \cdot Xg, \quad \text{and} \quad X(f \cdot g) = f \cdot Xg + g \cdot Xf.$$

3. Let X be a vector field on M, and let F denote its flow. Show that $Xf = (d/dt)F_t^*(f)|_{t=0}$, that is, for all $m \in M$,

$$(Xf)(m) = \frac{d}{dt} f(F_t(m))|_{t=0} \qquad (f \in C^\infty(M))$$

4. Let $\varphi: M \to N$ be a smooth map of manifolds. A vector field $X \in \mathscr{X}(M)$ is said to be *φ-related* to $Y \in \mathscr{X}(N)$ if

$$T\varphi \circ X = Y \circ \varphi.$$

For smooth maps $\varphi: M \to N$ and $\psi: N \to P$ show that if X is φ-related to Y and Y is φ-related to Z then X is $(\psi \circ \varphi)$-related to Z.

Even when the vector field $X \in \mathscr{X}(M)$ is not complete, the flows, F_t, are local diffeomorphisms in the sense of the following definition (see [AMR]).

Definition *A flow box of X at $m \in M$ is a triple (U_0, a, F) such that*

(*i*) *$U_0 \subset M$ is open, $m \in U_0$, and $a \in \mathbb{R}$, where $a > 0$ or $a = +\infty$ and F is a smooth map $U_0 \times I_a \to M$ where $I_a = (-a, a)$.*
(*ii*) *for each $x \in U_0$, the curve $I_a \to M$ given by $t \mapsto F(x, t)$ is an integral curve of X at the point x.*
(*iii*) *if F_t: $U_0 \to M$ is defined by $F_t(x) = F(x, t)$, then*

$F_t(U_0)$ is open, and
F_t is a diffeomorphism onto its image,

for all $t \in I_a$.

For each $m \in M$, there exists a flow box of X at m, as the reader is asked to show in the next exercise.

5. Let X be a vector field on the manifold M. If the triple (U_0, a, F) satisfies (*i*) and (*ii*) in the definition above, choose an open set $V_0 \subset U_0$ and a positive real number $b < a$ such that $F_t(V_0) \subset U_0$ for all $t \in (-b, b)$. Show that $(V_0, b, F|_{V_0 \times (-b,b)})$ is a flow box at m.

6. Let φ: $M \to N$ be a C^∞ mapping, and let $X \in \mathscr{X}(M)$ and $Y \in \mathscr{X}(N)$ be vector fields on M and N, with flows denoted F_t^X and F_t^Y, respectively.

(a) Show that X is φ-related to Y if and only if $\varphi \circ F_t^X = F_t^Y \circ \varphi$.
(b) If φ is a diffeomorphism, show that Y is the push-forward of X (i.e. $Y = \varphi_* X$) if and only if the flow of Y is $\varphi \circ F_t^X \circ \varphi^{-1}$.

Hint for (a). For any curve c on M, we have $(\varphi \circ c)'(t) = T_{c(t)}\varphi(c'(t))$.

5.6 Partitions of unity

Before we can prove the existence of a partition of unity on a manifold M, we need the following lemma in $\mathbb{R}^n$ which guarantees a plentiful supply of C^∞ functions.

Lemma 5.37 *Let $r_1 < r_2$. There exists a C^∞ function h: $\mathbb{R}^n \to [0, 1]$ such that $h(x) = 1$ if $|x| \leqslant r_1$, $0 < h(x) < 1$ if $r_1 < |x| < r_2$ and $h(x) = 0$ if $|x| \geqslant r_2$.*

Proof By a scaling transformation we can assume $r_1 = 1$ and $r_2 = 3$. The function φ: $\mathbb{R} \to [0, 1]$ defined by

$$\varphi(r) = \begin{cases} \exp(-1/(1 - r^2)), & |r| < 1 \\ 0, & |r| \geqslant 1 \end{cases}$$

is easily shown to be C^∞. Now let $k = \int_{-1}^{1} \varphi(r)\, dr > 0$ and

$$h(x) = \frac{1}{k} \int_{-1}^{2-|x|} \varphi(r)\, dr, \qquad x \in \mathbb{R}^n.$$

Then $h(x)$ equals 0 if $|x| \geqslant 3$ and equals 1 if $|x| \leqslant 1$, and h is strictly between 0 and 1 if $1 < |x| < 3$. Clearly h is C^∞ when $x \neq 0$. But h is constant near $x = 0$ so it is C^∞ everywhere.

Corollary 5.38 *Let M be a manifold and $m \in M$. Then given any neighbourhood V of m there is a function $\hbar \in C^\infty(M)$, $0 \leqslant \hbar \leqslant 1$, such that $\hbar = 1$ in a neighbourhood of m and $\hbar = 0$ outside V.* (We call $\hbar$ a "bump function".)

Proof There exists a chart (U, κ) at m such that $U \subset V$, $\kappa(m) = 0$, and $\kappa(U)$ is the ball $|x| < 3$. By the lemma there exists a function $h \in C^\infty(\mathbb{R}^n)$, $0 \leqslant h \leqslant 1$, equal to 1 on $|x| \leqslant 1$ and vanishing when $|x| > 2$. Let

$$\hbar = \begin{cases} h \circ \kappa & \text{on } U \\ 0 & \text{on } \complement U \end{cases}$$

Since the support of h is a compact subset of $\kappa(U)$, it follows that $\hbar \in C^\infty(M)$.

Let M be a manifold. Recall that the *support* of a function $f \in C^\infty(M)$ is the closure of the set of points $x \in M$ such that $f(x) \neq 0$. A C^∞ *partition of unity* on M consists of a system of smooth real-valued functions $\rho_j \in C^\infty(M)$, $j \in J$, satisfying the following conditions:

(a) for all $x \in M$ we have $\rho_j(x) \geqslant 0$
(b) every point has a neighbourhood in which $\Sigma_j \rho_j$ is a finite sum
(c) for each point $x \in M$ we have $\sum_j \rho_j(x) = 1$

If $\{V_j\}_{j \in J}$ is an open covering of M we say that the partition of unity $\{\rho_j\}$ is *subordinate* to $\{V_j\}$ if supp $\rho_j \subset V_j$ for all $j \in J$.

A collection $\mathscr{C}$ of subsets of a manifold M (or any topological space) is said to be *locally finite* if each point $m \in M$ has a neighbourhood U such that $U \cap C = \varnothing$ except for finitely many $C \in \mathscr{C}$. Condition (b) says that the supports of ρ_j are locally finite.

It turns out that there exists a partition of unity subordinate to any given open cover. The open cover $\mathscr{V} = \{V_j\}$ is not assumed to be locally finite; however, the first step in the proof of existence of a partition of unity is to find a *refinement* of $\mathscr{V}$ which *is* locally finite. By a refinement of $\mathscr{V}$ we mean another open cover $\{U_i\}$ such that for every $i \in I$, $U_i \subset V_j$ for some j.

If the manifold M is compact one can just choose a finite subcover, but in the non-compact case the following lemma is needed. Recall that a set in a topological space is said to be *σ-compact* if it is a countable union of compact sets.

Lemma 5.39 *Any manifold M is σ-compact. In fact, there exists a sequence of open sets $G_1, G_2, \ldots$ such that*

$$\bar{G}_n \text{ is compact, } \bar{G}_n \subset G_{n+1}, \ \bigcup_{n=1}^{\infty} G_n = M.$$

Proof In a countable basis for the topology of M, select those sets which have compact closure. The sequence $U_1, U_2, \ldots$ of open sets obtained in this way covers M, by local compactness of M. The G_n's can now be defined inductively as a union of these sets. Let $G_1 = U_1$, and suppose

$$G_n = U_1 \cup U_2 \cup \cdots \cup U_{j_n}$$

has already been defined such that $\bar{G}_n$ is compact for $n \leqslant N$ and $\bar{G}_n \subset G_{n+1}$ for $n < N$. Since $\bar{G}_N$ is compact it is covered by a finite number of the sets $U_1, U_2, \ldots$. Then we let $G_{N+1} = U_1 \cup \cdots \cup U_{j_{N+1}}$, where j_{N+1} is the smallest positive integer greater than j_N such that $U_1 \cup \cdots \cup U_{j_{N+1}} \supset \bar{G}_N$. The sequence of open sets $G_1, G_2, \ldots$ defined in this way covers M, since $j_n \to \infty$.

A topological space is said to be *paracompact* if it is Hausdorff and every open cover has a locally finite refinement.

Corollary 5.40 *Any manifold is paracompact.*

Proof Let $G_1, G_2, \ldots$ be a sequence of open sets with properties as in the lemma. Also let $G_i = \varnothing$ when $i \leqslant 0$. We have $M = \bigcup_{n=1}^{\infty} A_n$, where $A_n = \bar{G}_n \backslash G_{n-1}$ (intuitively, each A_n is a closed "annulus").

Let $\{V_j\}_{j \in J}$ be any open cover of M. Since $\{V_j \cap (G_{n+1} \backslash \bar{G}_{n-2})\}_{j \in J}$ is an open cover of A_n, and A_n is compact, there is a finite set $J_n \subset J$ such that $A_n \subset \bigcup_{j \in J_n} V_j \cap (G_{n+1} \backslash \bar{G}_{n-2})$. In this way we obtain an open cover

$$\{V_j \cap (G_{n+1} \backslash \bar{G}_{n-2})\}_{n \in \mathbb{N},\, j \in J_n}$$

of M, which refines $\{V_j\}_{j \in J}$, and is locally finite since a point in G_i is not in $G_{n+1} \backslash \bar{G}_{n-2}$ when $n \geqslant 2 + i$.

In the statement of the next theorem, $B(0, r)$ denotes the open ball in $\mathbb{R}^n$ with center at the origin and radius r.

Theorem 5.41 *Let M be a manifold. Any open cover of M has a locally finite refinement consisting of a countable number of charts (U_i, κ_i), $i \in \mathbb{N}$, such that $\kappa_i(U_i) = B(0, 3)$ and such that the open sets $\kappa_i^{-1}(B(0, 1))$ cover M. There is a partition of unity $\{\rho_i\}$ subordinate to $\{U_i\}$ with* supp ρ_i *a* compact *subset of U_i.*

Proof Let $\{V_j\}$ be a given open cover of M. For each $x \in M$ we can find an *arbitrarily* small chart (U_x, κ_x) at x such that $\kappa_x(U_x) = B(0, 3)$. The proof of Corollary 5.40 shows that there is a countable number of charts (U_i, κ_i) such that

$\{U_i\}$ is a locally finite refinement of $\{V_j\}$,
$\kappa_i(U_i) = B(0, 1)$ for each i, and the open sets $W_i = \kappa_i^{-1}(B(0, 1))$ cover M.

Indeed, for each point $x \in A_n$, we can find a chart (U_x, κ_x) at x such that $\kappa_x(U_x) = B(0, 3)$, $U_x \subset G_{n+1} \backslash \bar{G}_{n-2}$ and $U_x \subset V_{j_x}$ for some j_x. By compactness of A_n, finitely many of the sets $W_x = \kappa_x^{-1}(B(0, 1))$ cover A_n. We let $\{(U_i, \kappa_i)\}$ be the collection of charts obtained in this way for $n = 1, 2, \ldots$

By Lemma 5.37 there is a bump function h with $h \geqslant 0$ everywhere, $h = 1$ on $B(0, 1)$ and supp $h \subset B(0, 3)$. Transporting h to the manifold

$$\psi_i(x) = \begin{cases} h(\kappa_i(x)), & x \in U_i \\ 0, & x \notin U_i \end{cases}$$

we obtain a smooth function $\psi_i \in C^\infty(M)$, $0 \leqslant \psi_i \leqslant 1$ everywhere, such that

$\psi_i = 1$ in $\kappa_i^{-1}(B(0, 1))$ and supp $\psi_i \subset U_i$ is compact. Now we let

$$\rho_i = \frac{\psi_i}{\sum_i \psi_i}$$

and obtain a partition of unity subordinate to $\{U_i\}$ such that supp ρ_i is a compact subset of U_i.

In general, an arbitrary open cover of M does *not* have a subordinate partition of unity consisting of functions with compact support. (For example, consider the real line $\mathbb{R}$ with the open cover consisting of precisely one open set, namely $\mathbb{R}$ itself.) However, we do have the following result.

Corollary 5.42 *Any open cover has a subordinate partition of unity.*

Proof Let $\{V_j\}$ be an open cover of M. There is a locally finite refinement, $\{U_i\}$, of $\{V_j\}$ with the properties stated in the theorem, and a subordinate partition of unity $\{\rho_i\}$. For each i, we have $U_i \subset V_{j(i)}$ for some $j(i)$. We let $\varphi_j = \sum_{\{i;\, j(i) = j\}} \rho_i$, where it is to be understood that $\varphi_j = 0$ if $j(i) \neq j$ for all i. Then $\sum \varphi_j = \sum \rho_i = 1$ and

$$\operatorname{supp} \varphi_j \subset \bigcup_{\{i;\, j(i) = j\}} \operatorname{supp} \rho_i \subset V_j \qquad \text{for all } j$$

(For the first inclusion, recall that if $\{A_i\}$ is a locally finite collection of closed sets then $\bigcup A_i$ is closed). The supports of φ_j are locally finite since any open set which intersects infinitely many of these supports would *a priori* intersect infinitely many of the supports of ρ_i.

Corollary 5.43 *Let U be open in M and let F be closed in M with $F \subset U$. Then there exists a C^∞ function φ: $M \to [0, 1]$ with $\varphi(x) = 1$ for all $x \in F$ and* supp $\varphi \subset U$.

Proof By Corollary 5.42, the cover $\{U, M \backslash F\}$ has a partition of unity $\{\varphi, \psi\}$ subordinate to it with supp $\varphi \subset U$ and supp $\psi \subset M \backslash F$. Then φ is the desired function.

We will need the following in §5.11. It says that we can "shrink" a locally finite open cover of a manifold (this result is true for any paracompact space).

Lemma 5.44 *If $\{U_i\}$ is a locally finite open cover of M, then there exists a locally finite open covering $\{V_i\}$ such that $\bar{V}_i \subset U_i$.*

Proof Since M has a countable basis, any locally finite open covering is necessarily countable. So let us suppose that the cover $\{U_i\}$ is indexed by the natural numbers. Then

$$A_1 = M \backslash (U_2 \cup U_3 \cup \cdots)$$

is a closed set contained in U_1. Because M is paracompact, and hence a normal space, there is an open set V_1 with $A_1 \subset V_1 \subset \bar{V}_1 \subset U_1$. Then $M = V_1 \cup U_2 \cup U_3 \cup \cdots$ and

$$A_2 = M \backslash (V_1 \cup U_3 \cup U_4 \cup \cdots)$$

is a closed set contained in U_2. Let V_2 be an open set with $A_2 \subset V_2 \subset \bar{V}_2 \subset U_2$. Then we have $M = V_1 \cup V_2 \cup U_3 \cup U_4 \cup \cdots$ and continuing in this way it is not hard to establish by induction the following result: There exist open sets $V_1, V_2, \ldots$ with $\bar{V}_i \subset U_i$ and such that for any j

the open sets $V_1, V_2, \ldots, V_j, U_{j+1}, U_{j+2}, \ldots$ cover M.

Now, for any $x \in M$, there is a largest j with $m \in U_j$, since $\{U_i\}$ is locally finite. It follows that $x \in V_1 \cup V_2 \cup \cdots$ because replacing $U_{j+1}, U_{j+2}, \ldots$ by $V_{j+1}, V_{j+2}, \ldots$ cannot possibly eliminate x. Hence $\{V_i\}$ is an open cover of M and we are done.

5.7 Vector bundles

Let M be any manifold. Let $\{U_i\}$ be an open cover of M consisting of domains of charts (U_i, κ_i), and define the maps $\beta_i: \pi^{-1}(U_i) \to U_i \times \mathbb{R}^n$ by

$$[c]_x \mapsto (x, d\kappa_i([c]_x)).$$

The pair (TM, π), where $\pi: TM \to M$ is the tangent bundle projection $\pi([c]_x) = x$, satisfies the following properties: (i) Each fibre $\pi^{-1}(x) = T_xM$ is a vector space. (ii) Each β_i is a diffeomorphism, and the restriction to a fibre over any $x \in U_i$ is a vector space isomorphism.

This means that (TM, π) is a *real vector bundle over M* in the sense of the following definition.

Definition 5.45 *Let M be a manifold. A triple (E, p, M) where E is a manifold and $p: E \to M$ is a C^∞ map is called a real vector bundle over M if the following conditions are satisfied:*

(i) *for each $x \in M$ the fibre $E_x = p^{-1}(x)$ is a vector space over $\mathbb{R}$;*
(ii) *there is an open cover $\{U_i\}_{i \in I}$ of M and, for each $i \in I$, a positive integer r and a diffeomorphism*

$$\beta_i: p^{-1}(U_i) \to U_i \times \mathbb{R}^r$$

such that for all $x \in U_i$, $\beta_i(E_x) = \{x\} \times \mathbb{R}^r$ and the map $\beta_{ix}: E_x \to \mathbb{R}^r$ induced by β_i on the fibres is a vector space isomorphism.

Complex vector bundles are defined in the same way if we replace $\mathbb{R}$ by $\mathbb{C}$.

The maps β_i are called *trivializing maps.* Sometimes it is convenient to assume that the U_i's are chart domains on M and then to extend the diffeomorphism β_i by the coordinate map $\kappa_i: U_i \to V_i \subset \mathbb{R}^n$ to get coordinate maps $\chi_i = (\kappa_i \times id) \circ \beta_i$ for the bundle E, that is,

$$\chi_i: p^{-1}(U_i) \xrightarrow{\beta_i} U_i \times \mathbb{R}^r \xrightarrow{\kappa_i \times id} V_i \times \mathbb{R}^r$$

where $V_i = \kappa_i(U_i) \subset \mathbb{R}^n$.

The number r may depend on i, but it is constant on each connected component of M. We will usually assume without mention that r is constant on all of M. (E, p, M) is then called a vector bundle of *rank* r. We can replace $\mathbb{R}^r$ by an arbitrary r-dimensional vector space $\mathbf{E}$, in which case we say that the *vector bundle is modelled on* $\mathbf{E}$. This does not really enlarge the class of vector bundles; any trivializing map $\beta: p^{-1}(U) \to U \times \mathbf{E}$ can be composed with a linear isomorphism $\mathbf{E} \to \mathbb{R}^r$ bringing us back to Definition 5.45.

E is called the *total space*, and M is the *base space*. The map p is referred to as the vector bundle *projection*. Usually we refer to the total space E as the vector bundle, unless there is some reason to emphasize the role of the projection.

Let E, F be vector bundles over the same manifold M. A map $f: E \to F$ is said to be *fibre-preserving* if $p'f = p$, where p and p' denote the projections of E and F on M, respectively. In other words, f is fibre-preserving if $f(E_x) \subset F_x$ for all $x \in M$. In such a case we let f_x denote the restriction of f to the fibre over x

$$f_x: E_x \to F_x$$

and f is said to be "linear on each fibre" if f_x is a linear map for each x. A C^∞ map $f: E \to F$ that is fibre-preserving and linear on each fibre is called a *homomorphism* from E to F. The set of vector bundle homomorphisms from E to F is denoted $\mathrm{HOM}(E, F)$.

A vector bundle *isomorphism* is a homomorphism $f \in \mathrm{HOM}(E, F)$ for which there exists a homomorphism $g \in \mathrm{HOM}(F, E)$ such that $f \circ g = \mathrm{id}_F$ and $g \circ f = \mathrm{id}_E$. If there exists an isomorphism from E to F we write $E \simeq F$.

The vector bundle $(M \times \mathbb{R}^r, p, M)$, where $p: M \times \mathbb{R}^r \to M$ is the projection on the first factor, is referred to as a *product bundle*. More generally, a vector bundle E over M is called *trivial* if it is isomorphic to a product bundle, i.e. $E \simeq M \times \mathbb{R}^r$ for some r.

Lemma *A map $f: M \times \mathbb{R}^r \to M \times \mathbb{R}^d$ is fibre-preserving and linear on each fibre if and only if it has the form*

$$f(x, v) = (x, \tilde{f}(x)v) \tag{*}$$

for some map $\tilde{f}: M \to L(\mathbb{R}^r, \mathbb{R}^d)$. If (*) holds then f is smooth if and only if $\tilde{f}$ is smooth.

Proof The first statement is obvious. For the second statement, the "if" part is clear. Conversely, suppose that f is smooth. Let pr denote the projection $M \times \mathbb{R}^d \to \mathbb{R}^d$ on the second factor. This is a linear map, hence C^∞. Let $e_1, \ldots, e_r$ be the standard basis of $\mathbb{R}^r$. Taking $v = e_i$ we find that

$$\tilde{f}(x)e_i = pr \circ f(x, e_i)$$

is smooth in x. But $\tilde{f}(x)e_i$ is just the ith column of the matrix $\tilde{f}(x)$, $i = 1, \ldots, r$.

Remark As usual we identify $L(\mathbb{R}^r, \mathbb{R}^d)$ with the space of $d \times r$ real matrices.

If $f \in \mathrm{HOM}(E, F)$ and (U, β) and (U, ψ) are local trivializations of E and F, respectively, then, since f is fibre-preserving, the map

$$f_{\psi\beta} = \psi \circ f \circ \beta^{-1}: U \times \mathbb{R}^r \to U \times \mathbb{R}^d$$

(d = dimension of the fibres of F) has the form

$$(x, v) \mapsto (x, \tilde{f}(x)v),$$

where $\tilde{f}: U \to L(\mathbb{R}^r, \mathbb{R}^d)$ is smooth. We refer to $\tilde{f}$ as the *local representation* of f with respect to given trivializing maps. The following properties hold: (i) The local representation of the identity map $f = \mathrm{id}_E \in \mathrm{HOM}(E, E)$ is the $r \times r$ identity matrix; (ii) if $f \in \mathrm{HOM}(E, F)$ and $g \in \mathrm{HOM}(F, G)$ and (U, β), (U, ψ) and (U, φ) are local trivializations of E, F and G, respectively, then $(g \circ f)_{\varphi\beta} = g_{\varphi\psi} \circ f_{\psi\beta}$. This means that the local representation of $g \circ f$ is the map $x \mapsto \tilde{g}(x) \cdot \tilde{f}(x)$ (matrix multiplication).

Lemma 5.46 *Let E, F be vector bundles over M. If $f \in \mathrm{HOM}(E, F)$ and is bijective, then f is a vector bundle isomorphism.*

Proof Since f is fibre-preserving we have $f(E_x) \subset F_x$ for all $x \in M$. But f is also bijective so it must be that $f(E_x) = F_x$. The set-theoretic inverse $g: F \to E$ is therefore fibre-preserving and it is linear on each fibre since f is. To prove that g is smooth we can work locally. The local representation of f has the form $\tilde{f}: U \to GL(\mathbb{R}^r)$, and then the local representation of g is the map $U \to GL(\mathbb{R}^r)$ given by $\tilde{g}(x) = [\tilde{f}(x)]^{-1}$. In the notation of Lemma 5.2 this means that $g = \mathscr{I} \circ \tilde{f}$, so $\tilde{g}$ is C^∞ and hence g is C^∞.

Let E be a vector bundle over M, and choose an open cover $\{U_i\}_{i \in I}$ of M and trivializing maps β_i as in Definition 5.45. The maps $h_{ij}: U_i \cap U_j \to GL(\mathbb{R}^r)$ defined by

$$h_{ij}(x) = (\beta_i \beta_j^{-1})_x$$

are called the *transition matrices* (or functions) for E, and we have

$$h_{ji} = (h_{ij})^{-1} \qquad \text{in } U_i \cap U_j$$

$$h_{ij} \cdot h_{jk} = h_{ik} \qquad \text{in } U_i \cap U_j \cap U_k,$$

these properties being known as the *cocycle conditions*. Given a system $\{h_{ij}\}$ of $r \times r$ matrix functions with C^∞ entries satisfying the cocycle conditions, one can define a vector bundle having h_{ij} as its transition matrices by forming the disjoint union of all $U_i \times \mathbb{R}^r$, $i \in I$,

$$Q = \bigcup_{i \in I} \{i\} \times U_i \times \mathbb{R}^r,$$

and by identifying

$$(i, x, v) \in U_i \times \mathbb{R}^r \sim (j, x, v') \in U_j \times \mathbb{R}^r$$

if $v' = h_{ji}(x) \cdot v$. It follows from the cocycle conditions that this is an equivalence relation $\sim$ on Q, and it follows from Proposition 5.47 below that the quotient set $S = Q/\sim$ is a vector bundle with projection $p: S \to M$ defined by $[(i, x, v)] \mapsto x$ and trivialization $\beta_i: p^{-1}(U_i) \to U_i \times \mathbb{R}^r$ given by $[(i, x, v)] \mapsto (x, v)$. Also (S, p, M) is isomorphic to E if h_{ij} were obtained from local trivializations of E as explained above.

The following proposition states the axioms by which a collection of trivializing maps on an arbitrary set S define a vector bundle structure on S.

Proposition 5.47 *Let M be a manifold, and $p: S \to M$ a mapping from some set S into M. Let $\{U_i\}$ be an open cover of M and, for each i, suppose we are given a bijection $\beta_i: p^{-1}(U_i) \to U_i \times \mathbb{R}^r$ satisfying the following properties:*

VB 1 *The map β_i is fibre-preserving, i.e. the diagram*

$$\begin{array}{ccc} p^{-1}(U_i) & \xrightarrow{\beta_i} & U_i \times \mathbb{R}^r \\ & \searrow \quad U_i \quad \swarrow & \end{array}$$

commutes. Consequently, there is an induced map on the fibres, which we denote by $\beta_{ix}: p^{-1}(x) \to \mathbb{R}^r$.

VB 2 *If U_i and U_j are two members of the cover such that $U_i \cap U_j \neq \varnothing$, then for each $x \in U_i \cap U_j$, the map $(\beta_i\, \beta_j^{-1})_x \in L(\mathbb{R}^r)$ is a linear isomorphism, and the map of $U_i \cap U_j$ into $GL(\mathbb{R}^r)$ given by*

$$x \mapsto (\beta_i \beta_j^{-1})_x$$

is C^∞.

Then there is a unique topology and C^∞ structure on S, and a unique vector space structure on each fibre $p^{-1}(x)$, such that β_{ix} is a linear isomorphism and (S, p, M) is a vector bundle with β_i as trivializing maps.

Proof For $x \in U_i \cap U_j$, let $h_{ij}(x) = (\beta_i\, \beta_j^{-1})_x$. By hypothesis, h_{ij} is a smooth map $U_i \cap U_j \to GL(\mathbb{R}^r)$. It follows that the overlap maps

$$\beta_i \beta_j^{-1}: (U_i \cap U_j) \times \mathbb{R}^r \to (U_i \cap U_j) \times \mathbb{R}^r$$

are given by $(x, v) \mapsto (x, h_{ij}(x)v)$ and are therefore smooth. These maps are in fact diffeomorphisms since $h_{ij}^{-1} = h_{ji}$ is also smooth. We claim that by Proposition 5.16 there is a unique C^∞ structure on S such that the β_i's are diffeomorphisms. The collection $\{(p^{-1}(U_i), \beta_i)\}$ is not quite an atlas for S since the image of β_i is not a subset of Euclidean space. But we can argue as follows: each U_i is a union of chart domains so we may assume without loss of generality that we are given an open cover $\{U_i\}$ with charts (U_i, κ_i) for M. We then let $\chi_i = (\kappa_i \times \mathrm{id}) \circ \beta_i$ to obtain a bijection

$$\chi_i: p^{-1}(U_i) \to \kappa_i(U_i) \times \mathbb{R}^r \subset \mathbb{R}^n \times \mathbb{R}^r$$

which satisfies AT 1, AT 2 and AT 3 of Proposition 5.16. Hence there is a unique C^∞ structure on S such that the χ_i's (and hence the β_i's) are diffeomorphisms.

It is clear that p is smooth, since $p: S \to M$ is locally obtained as a composite of smooth maps, i.e., $p|_{U_i} = p_i \circ \beta_i$ where $p_i: U_i \times \mathbb{R}^r \to U_i$ is the projection on the first factor. If $x \in U_i$ we can transport the vector space structure of $\mathbb{R}^r$ to the fibre $p^{-1}(x)$ by means of β_{ix}. The result is independent of the choice of U_i since $(\beta_i \, \beta_j^{-1})_x$ is a linear isomorphism.

Remark When applying Proposition 5.47 it often happens that the fibres $p^{-1}(x)$ already possess a vector space structure such that β_{ix} is linear. The vector space structure defined in the proof will agree with the given structure.

Let (E, p, M) be a vector bundle over M. A smooth map $v: M \to E$ is called a *section* of E if $p \circ v$ is the identity on M; in other words,

$$v(x) \in E_x = p^{-1}(x) \qquad \text{for all } x \in M.$$

The space of all sections of E is denoted $C^\infty(M, E)$. If U is an open set in M we refer to a section $s \in C^\infty(U, E|_U)$ as a section of E over U.

Let $u \in C^\infty(M, E)$ be a section of E. If we have a covering $\{U_i\}$ of M as above with local trivializations $\beta_i: E|_{U_i} \to U_i \times \mathbb{R}^r$, then $u_i = \beta_i \circ u \in C^\infty(U_i, \mathbb{R}^r)$ and we have

$$u_i = h_{ij} \cdot u_j \qquad \text{in } U_i \cap U_j, \tag{27}$$

where h_{ij} are the transition matrices introduced above. Conversely, any system $u_i \in C^\infty(U_i, \mathbb{R}^r)$ satisfying (27) defines a section of the vector bundle.

Lemma 5.48 *A collection of sections $v_1, \ldots, v_r \in C^\infty(U, E)$ is called a* local basis *for E if for every point x in U, the vectors $v_1(x), \ldots, v_r(x)$ form a basis in the vector space E_x. E has a local basis over U if and only if $E|_U$ is trivial.*

Proof If $\beta: E|_U \to U \times \mathbb{R}^r$ is a trivializing map and e_i denote the standard unit vectors in $\mathbb{R}^r$ then the sections

$$v_i(x) = (\beta_x)^{-1}(e_i), \qquad i = 1, \ldots, r$$

are a local basis on U. Conversely, if $v_1, \ldots, v_r$ is a local basis on U, then the map φ from $U \times \mathbb{R}^r$ to $E|_U$ defined by

$$(x, \xi) \mapsto \sum_{i=1}^{r} \xi_i \cdot v_i(x) \tag{28}$$

is a vector bundle isomorphism (it has a smooth inverse by Lemma 5.46). Hence $\beta = \varphi^{-1}$ is a trivializing map.

The following corollary is the analogue of Proposition 5.29.

Corollary 5.49 *Let E be a vector bundle over M, and let $v_1, \ldots, v_r$ be a local basis on U. If v is a section of E over U, that is, $v \in C^\infty(U, E)$ then we have*

$$v(x) = \sum_{i=1}^{r} f_i(x) \cdot v_i(x), \qquad x \in U$$

for unique functions $f_i \in C^\infty(U)$, $i = 1, \ldots, r$. Conversely, any map $v: U \to E$ defined in this way is a section of E over U.

Proof Let $\beta\colon E|_U \to U \times \mathbb{R}^r$ denote the trivializing map defined as the inverse of the map (28). Then

$$\beta\left(\sum_{i=1}^{r} \xi_i \cdot v_i(x)\right) = (x, \xi_1, \ldots, \xi_r) \qquad x \in U,\ \xi \in \mathbb{R}^r$$

Now, $\beta \circ v$ is a section of the trivial bundle $U \times \mathbb{R}^r$, so we have

$$\beta(v(x)) = (x, f(x))$$

for a unique function $f \in C^\infty(U, \mathbb{R}^r)$. This proves the first statement since $f(x) = \sum_{i=1}^{r} f_i(x) \cdot e_i$ and $v_i(x) = (\beta_x)^{-1}(e_i)$. The second statement is obvious.

If $E \simeq M \times \mathbb{R}^r$ is a trivial bundle the space of sections of E can be identified with the smooth vector-valued maps $M \to \mathbb{R}^r$ and we write $C^\infty(M, \mathbb{R}^r)$ rather than $C^\infty(M, E)$. If $r = 1$ we write $C^\infty(M)$.

Note that E is trivial if and only if there exists a global basis, that is, a basis defined on all M.

Examples

(1) Let S^1 denote the unit circle $|x| = 1$ in $\mathbb{R}^2$. Since S^1 is a submanifold of $\mathbb{R}^2$ the tangent space to S^1 at any point is identified with a subspace of $\mathbb{R}^2$; at the point $x = (x_1, x_2)$, we have

$$T_x(S^1) = \{\xi \in \mathbb{R}^2;\ x_1\xi_1 + x_2\xi_2 = 0\}$$

We say that $T(S^1)$ is a *line bundle* because its fibers are one-dimensional. Note that $T(S^1)$ has a non-vanishing section: the section $v\colon S^1 \to T(S^1)$ defined by $(x_1, x_2) \mapsto (-x_2, x_1)$. Hence $T(S^1)$ is trivial because there is the vector bundle isomorphism $S^1 \times \mathbb{R} \to T(S^1)$ defined by $(x, \lambda) \mapsto \lambda \cdot v(x)$.

(2) Let $\mathbb{T}^n = S^1 \times \cdots \times S^1$ (n copies) be the n-torus. Then

$$T(\mathbb{T}^n) \simeq T(S^1) \times \cdots \times T(S^1)$$

and, since $T(S^1) \simeq S^1 \times \mathbb{R}$, we have $T(\mathbb{T}^n) \simeq \mathbb{T}^n \times \mathbb{R}^n$. Thus the tangent bundle to the n-torus is trivial.

(3) Let S^2 denote the unit sphere $|x| = 1$ in $\mathbb{R}^3$. At any point $x = (x_1, x_2, x_3)$ we have

$$T_x(S^2) = \{\xi \in \mathbb{R}^3;\ x_1\xi_1 + x_2\xi_2 + x_3\xi_3 = 0\}$$

Using degree theory it is possible to prove that $T(S^2)$ does not have a non-vanishing section (see [AMR], §7.5). This implies that $T(S^2)$ is non-trivial.

(4) (Möbius band) Next we give an example of a non-trivial vector bundle $\mathcal{M}$ over S^1. On the product manifold $\mathbb{R} \times \mathbb{R}$, consider the equivalence relation defined by

$$(t, v) \sim (t', v') \Leftrightarrow t' = t + k,\ v' = (-1)^k v \qquad \text{for some } k \in \mathbb{Z}$$

and denote by $\pi\colon \mathbb{R} \times \mathbb{R} \to \mathcal{M}$ the quotient topological space. $\mathcal{M}$ is called the *Möbius band*. By means of Exercise 7 in §5.2 it is not hard to show that it is a Hausdorff space.

Let the points in $\mathscr{M}$ be denoted $[t, v] = \pi(t, v)$. There is a continuous map

$$p: \mathscr{M} \to S^1$$
$$[t, v] \mapsto e^{2\pi i t}$$

and we claim that $(\mathscr{M}, p)$ has a vector bundle structure. First, note that each fibre inherits a vector space structure from $\mathbb{R}$:

$$[t, v_1] + [t, v_2] = [t, v_1 + v_2]$$
$$\lambda \cdot [t, v] = [t, \lambda v]$$

(The definitions of $+$ and $\cdot$ are independent of the choice of representatives in the equivalence classes since the map $v \mapsto -v$ is linear.) Let $U_1 = S^1 \backslash \{1\}$ and $U_2 = S^1 \backslash \{-1\}$ and define maps

$$\beta_j: p^{-1}(U_j) \to U_j \times \mathbb{R}$$
$$[t, v] \mapsto (e^{2\pi i t}, v)$$

for $j = 1, 2$ where it is to be understood that $0 < t < 1$ when $j = 1$ and $-\frac{1}{2} < t < \frac{1}{2}$ when $j = 2$. Then $U_1 \cap U_2 = S^1 \backslash \{1, -1\}$ is a union of two connected parts, U^+ and U^- (above and below the real axis, respectively). If $x \in U^+$ then $(\beta_2 \circ \beta_1^{-1})(x, v) = (x, v)$. On the other hand, if $x \in U^-$ then $x = e^{2\pi i t} = e^{2\pi i (t-1)}$ where $\frac{1}{2} < t < 1$. Since $[t, v] = [t - 1, -v]$ it follows that

$$(\beta_2 \circ \beta_1^{-1})(x, v) = (x, -v) \qquad \text{if } x \in U^-$$

Hence the maps β_j satisfy VB 1 and VB 2 so that $(\mathscr{M}, p)$ is a vector bundle over S^1.

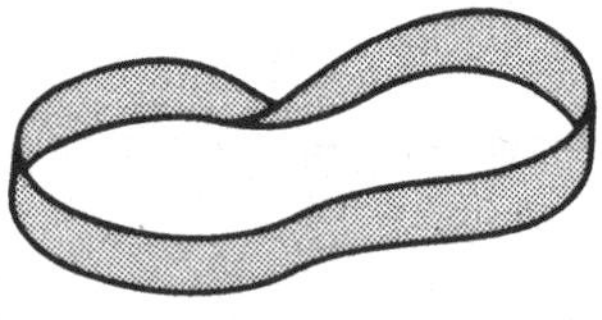

Fig. 4. The Möbius band.

To prove that $\mathscr{M}$ is non-trivial, we will show that any section $s: S^1 \to \mathscr{M}$ must vanish at some point, that is, $s(x) = 0_x$ for some $x \in S^1$. (Also note that $0_x = [t, 0]$ where $x = e^{2\pi i t}$). Intuitively, this is clear from the above diagram. The fibres are represented by vertical line segments I (an open interval diffeomorphic to $\mathbb{R}$) and the base S^1 is regarded as the interval $[0, 1]$ with 0 and 1 identified. We get the Möbius band from the rectangle $[0, 1] \times I$ by identifying $0 \times I$ and $1 \times I$ with a twist. The curved line in the diagram represents a section $s: S^1 \to \mathscr{M}$. Note that for s to be defined on all S^1 the end points must match under the twist, and hence s must cross the 0 section at some point.

Now we verify this fact analytically. Suppose that there is a non-vanishing section $s: S^1 \to \mathcal{M}$. Let c be the closed curve in $\mathcal{M}$ defined by $c(t) = s(e^{2\pi i t})$, $0 \leqslant t \leqslant 1$. Recall that β_1 defines a trivialization of $\mathcal{M}|_{U_1}$ and let $f: (0, 1) \to \mathbb{R}$ be defined by

$$f(t) = \text{second component of } \beta_1(c(t)).$$

A simple argument shows that f can be continuously extended to $[0, 1]$. In fact, near $t = 0$, $f(t)$ is equal to the second component of $\beta_2(c(t))$, and near $t = 1$ it is equal to *minus* the second component of $\beta_2(c(t))$. Now let $\tilde{c}: [0, 1] \to \mathbb{R} \times \mathbb{R}$ be defined by $\tilde{c}(t) = (t, f(t))$. Then $\tilde{c}$ is a "lifting" of c in the sense that $c = \pi \circ \tilde{c}$, that is,

$$c(t) = [t, f(t)], \; 0 \leqslant t \leqslant 1$$

Since $c(0) = c(1)$ we find that $[0, f(0)] = [1, f(1)]$. Hence $f(1) = -f(0)$. The continuity of f and the fact that $[0, 1]$ is connected implies that $f(t_0) = 0$ for some t_0. Then $c(t_0) = [t_0, 0] = 0_{t_0}$ which contradicts the assumption that s is non-vanishing.

Sub-bundles

Let (E, p, M) be a vector bundle over M. A subset $F \subset E$ is called a *sub-bundle* if for each $x \in M$ there is an open set U containing x and a local trivialization $\beta: p^{-1}(U) \to U \times \mathbb{R}^r$ such that

SB $$\beta(p^{-1}(U) \cap F) = U \times (\mathbb{R}^{r-k} \times 0).$$

Proposition 5.50 *Let (E, p, M) be a vector bundle over M. If F is a sub-bundle of E, then $(F, p|_F, M)$ is a vector bundle over M such that the inclusion map $F \to E$ is a vector bundle homomorphism.*

Proof Let $\{U_i\}$ be an open cover of M such that there exist local trivializations $\beta_i: p^{-1}(U_i) \to U_i \times \mathbb{R}^r$ satisfying the SB property relative to F. The conditions VB 1 and VB 2 of Proposition 5.47 are easily verified so $(F, p|_F, M)$ has a vector bundle structure. For instance, note that if $h_{ij}(x) = (\beta_i \, \beta_j^{-1})_x$ are the transition functions for E, then F has transition functions $h'_{ij}(x) = h_{ij}(x)|_{\mathbb{R}^{r-k} \times 0}$ which are smooth from $U_i \cap U_j$ to $GL(\mathbb{R}^{r-k})$.

The inclusion map $F \to E$ is obviously smooth because if β is a trivializing map satisfying the SB property then its local representation with respect to $\beta|_F$ and β is the inclusion $U \times (\mathbb{R}^{r-k} \times 0) \to U \times \mathbb{R}^r$.

Sub-bundles have the following universal property.

Proposition 5.51 *Let F be a sub-bundle of E, and G another vector bundle over M. Let $f: G \to E$ be a vector bundle homomorphism such that $f(G) \subset F$. Then the induced map $f_1: G \to F$ is a vector bundle homomorphism.*

The proof is similar to Proposition 5.23. Also note that the vector bundle structure on F is uniquely determined by the requirement that the inclusion map $F \to E$ be a vector bundle homomorphism.

Proposition 5.52 *Let E and F be vector bundles over M. If $f: E \to F$ is a vector bundle homomorphism let $f_x: E_x \to F_x$ be the restriction of f to the fibre over $x \in M$, and define the kernel and image of f by*

$$\ker(f) = \bigcup_{x \in M} \ker(f_x) \quad \text{and} \quad \mathrm{im}(f) = \bigcup_{x \in M} \mathrm{im}(f_x)$$

If the rank of f_x is locally constant then $\ker(f)$ *is a sub-bundle of E and* $\mathrm{im}(f)$ *is a sub-bundle of F.*

Proof Let U be an open set in M such that there exist trivializing maps $\beta: E_U \to U \times \mathbb{R}^r$ and $\psi: F_U \to U \times \mathbb{R}^d$. The local representation

$$f_{\psi\beta} = \psi \circ f \circ \beta^{-1}: U \times \mathbb{R}^r \to U \times \mathbb{R}^d$$

has the form $(x, v) \mapsto (x, \tilde{f}(x)v)$, where $\tilde{f}: U \to L(\mathbb{R}^r, \mathbb{R}^d)$ is smooth.

Fix $x_0 \in U$. If we let $\mathbb{R}^r = \mathscr{V}_1 \oplus \mathscr{V}_2$ where $\mathscr{V}_1$ is the kernel of $\tilde{f}(x_0)$ and $\mathbb{R}^d = \mathscr{W}_1 \oplus \mathscr{W}_2$ where $\mathscr{W}_1$ is the image of $\tilde{f}(x_0)$, then $\tilde{f}(x)$ has the block matrix representation

$$\tilde{f}(x) = \begin{bmatrix} a_{11}(x) & a_{12}(x) \\ a_{21}(x) & a_{22}(x) \end{bmatrix} \in L(\mathscr{V}_1 \oplus \mathscr{V}_2, \mathscr{W}_1 \oplus \mathscr{W}_2)$$

where $a_{ij}: U \to L(\mathscr{V}_j, \mathscr{W}_i)$ is smooth. The choice of $\mathscr{V}_2$ and $\mathscr{W}_1$ implies that $a_{12}(x_0)$ is an isomorphism. Therefore, $a_{12}(x)$ is an isomorphism for all x in a neighbourhood of x_0; by shrinking U we can assume that this holds for all $x \in U$. Note also that $a_{11}(x_0) = 0$, $a_{21}(x_0) = 0$ and $a_{22}(x_0) = 0$. Since the rank of $\tilde{f}(x)$ is locally constant we can shrink U further, if necessary, so that $\tilde{f}(x)$ has constant rank for all $x \in U$. Since $a_{12}(x)$ is an isomorphism and the rank of $\tilde{f}(x)$ is equal to $\dim \mathscr{W}_1$, it follows that

$$a_{21}(x) = 0 \quad \text{and} \quad a_{22}(x) = 0 \qquad \text{for all } x \in U$$

Now if we let

$$A(x) = \begin{bmatrix} I & 0 \\ -a_{12}(x)^{-1} \cdot a_{11}(x) & I \end{bmatrix} \in GL(\mathscr{V}_1 \oplus \mathscr{V}_2, \mathscr{V}_1 \oplus \mathscr{V}_2)$$

then

$$\tilde{f}(x) \cdot A(x) = \begin{bmatrix} 0 & a_{12}(x) \\ 0 & 0 \end{bmatrix} \tag{29}$$

Replacing the trivializing map β by $A^{-1}\beta$, we may assume that $\tilde{f}(x)$ is equal to the matrix on the right-hand side of (29). Then we have $\ker \tilde{f}(x) = \mathscr{V}_1$ and $\mathrm{im}\, \tilde{f}(x) = \mathscr{W}_1$ for all $x \in U$. Finally, by a linear isomorphism (change of basis) in $\mathbb{R}^r$ and $\mathbb{R}^d$ we can assume that $\mathscr{V}_1 = \mathbb{R}^{r-k} \times 0$ and $\mathscr{W}_1 = \mathbb{R}^k \times 0$, where k is the rank of $\tilde{f}(x)$. Now we have

$$\beta(E_U \cap \ker(f)) = U \times (\mathbb{R}^{r-k} \times 0) \quad \text{and} \quad \psi(F_U \cap \mathrm{im}(f)) = U \times (\mathbb{R}^k \times 0)$$

which is the SB property for $\ker(f)$ and $\mathrm{im}(f)$.

Remark From the proof of the proposition we see that if $f: E \to F$ is any homomorphism then $\operatorname{rank}(f_x) \geqslant \operatorname{rank}(f_{x_0})$ for all x near x_0. Thus, $\operatorname{rank}(f_x)$ is an upper semi-continuous function of x.

Definition 5.53 *A projection operator P for a vector bundle E is a homomorphism $P: E \to E$ with $P^2 = P$.*

If P is a projection operator for E, then $\operatorname{im}(P_x) \oplus \operatorname{im}(I - P_x) = E_x$ and

$$\operatorname{rank}(P_x) + \operatorname{rank}(I - P_x) = \dim E_x.$$

Since both $\operatorname{rank}(P_x)$ and $\operatorname{rank}(I - P_x)$ are upper semi-continuous functions of x, they are locally constant. Thus by Proposition 5.52 both $\ker(P) = \operatorname{im}(I - P)$ and $\operatorname{im}(P)$ are sub-bundles of E and

$$E = \operatorname{im}(P) \oplus \operatorname{im}(I - P).$$

Proposition 5.54 *Let (E, p, M) be a vector bundle over M, and $F \subset E$ a sub-bundle. Let E/F be the (disjoint) union of the quotient vector spaces E_x/F_x, $x \in M$. Then E/F has a unique vector bundle structure over M for which the canonical map $\pi: E \to E/F$ is a vector bundle homomorphism.*

Proof Let $\bar{p}: E/F \to M$ be the map induced from $p: E \to M$. By the SB property, there is an open cover $\{U_i\}$ of M and local trivializations $\beta_i: E|_{U_i} \to U_i \times \mathbb{R}^r$ such that $\beta_i(F|_{U_i}) = U_i \times (\mathbb{R}^{r-k} \times 0)$. Then β_i induces a unique map

$$\bar{\beta}_i: \bar{p}^{-1}(U_i) \to U_i \times (0 \times \mathbb{R}^k)$$

by the condition $\bar{\beta}_i \circ \pi = \beta_i|_{\beta_i^{-1}(U_i \times (0 \times \mathbb{R}^k))}$. Note that $\bar{\beta}_i$ is a bijection and property VB 1 holds. To verify VB 2, let h denote the transition function $x \mapsto (\beta_i \circ \beta_j^{-1})_x$ for E. We claim that the transition function $\bar{h}$ for E/F given by $x \mapsto (\bar{\beta}_i \circ \bar{\beta}_j^{-1})_x$ is also smooth. Writing h as a block matrix with respect to the direct sum $\mathbb{R}^r = (\mathbb{R}^{r-k} \times 0) \oplus (0 \times \mathbb{R}^k)$ we have

$$h(x) = \begin{pmatrix} h_{11}(x) & h_{12}(x) \\ h_{21}(x) & h_{22}(x) \end{pmatrix}$$

where h_{11}, h_{12}, h_{21} and h_{22} are smooth matrix functions on $U_i \cap U_j$ of dimensions $(r-k) \times (r-k)$, $(r-k) \times k$, $k \times (r-k)$ and $k \times k$, respectively. Since $\mathbb{R}^{r-k} \times 0$ has to be carried into itself by h, we have $h_{21} = 0$. Since $h(x)$ is invertible, then so is $h_{22}(x)$ for each x. The proof is complete since $\bar{h} = h_{22}$. By Proposition 5.47, $(E/F, \bar{p}, M)$ is a vector bundle. It is evident from the local representations that the canonical map π is a vector bundle homomorphism (i.e. smooth).

Exercises

1. Let $Y \subset M$ be a submanifold.

(a) Show that TY is a sub-bundle of $TM|_Y$.

(b) The *normal bundle* of Y is defined to be the quotient bundle $\nu(Y) = (TM|_Y)/TY$. Let k denote the codimension of Y. Show that $\nu(Y)$ is trivial iff there are smooth maps $X_i\colon Y \to TM$, $i = 1, \ldots, k$, such that $X_i(y) \in T_yM$ and $X_1(y), \ldots, X_k(y)$ span a subspace V_y satisfying $T_yM = T_yY \oplus V_y$ for all $y \in Y$.

2. Let $S^n \subset \mathbb{R}^{n+1}$ be the sphere. Show that the normal bundle of S^n is trivial.

Let $\mathbf{E}$ be a finite dimensional vector space and let $\mathbf{F} \subset \mathbf{E}$ be a subspace. The annihilator of $\mathbf{F}$, denoted $\mathbf{F}^0$, is the set of $\lambda \in \mathbf{E}^*$ such that

$$\langle \lambda, v \rangle = 0 \qquad \text{for all } v \in \mathbf{F}.$$

There is a natural isomorphism $\mathbf{F}^0 \simeq (\mathbf{E}/\mathbf{F})^*$.

3. Let $Y \subset M$ be a submanifold. The *conormal bundle*, $\mu(Y)$, is defined to be the union of the annihilators $(T_yY)^0 \subset T_y^*M$, $y \in Y$, i.e.

$$\mu(Y) = \{\xi \in T_y^*M;\ \langle \xi, v \rangle = 0 \text{ for all } v \in T_yY,\ y \in Y\}$$

(a) Show that $\mu(Y)$ is a sub-bundle of $T^*M|_Y$.
(b) Show that $\mu(Y) \simeq \nu(Y)^*$, i.e. the conormal bundle is isomorphic to the dual of the normal bundle.

5.8 Operations on vector bundles

Let Y be a submanifold of M. If E is a vector bundle on M we can restrict it to Y

$$E|_Y = \bigcup_{y \in Y} E_y \qquad (= p^{-1}(Y))$$

For the sake of simplifying the notation we also write $E_Y = E|_Y$. Each fibre of E_Y is a vector space (being a fibre of E) and we would like to show that it is a vector bundle. The crucial step is to show first that E_Y is a *submanifold* of E.

Now, for each $y \in Y$ there exists an open set U containing y and a local trivialization $\beta\colon p^{-1}(U) \to U \times \mathbb{R}^r$. By shrinking U we may assume that it is the domain of a coordinate map κ satisfying the submanifold property SM:

$$\kappa(U \cap Y) = \kappa(U) \cap (\mathbb{R}^{n-k} \times 0) \qquad (k = \text{codimension of } Y \text{ in } M)$$

Then the composite map $\chi = (\kappa \times \mathrm{id}) \circ \beta$, that is,

$$\chi\colon p^{-1}(U) \xrightarrow{\ \beta\ } U \times \mathbb{R}^r \xrightarrow{\ \kappa \times id\ } \kappa(U) \times \mathbb{R}^r \subset \mathbb{R}^{n+r},$$

satisfies the following condition:

$$\chi(p^{-1}(U) \cap E_Y) = \chi(p^{-1}(U)) \cap (\mathbb{R}^{n-k} \times 0 \times \mathbb{R}^r)$$

Thus $(p^{-1}(U), \chi)$ is a chart on E that satisfies the SM property for E_Y in E; this shows that E_Y is a submanifold (also of codimension k) in E.

Proposition 5.55 *Let E be a vector bundle with projection $p\colon E \to M$, and let Y be a submanifold of M. Then*

$$E_Y = \bigcup_{y \in Y} E_y$$

is a submanifold of E and a vector bundle on Y with projection $p|_{E_Y}\colon E_Y \to Y$.

Proof We have already shown that E_Y is a submanifold of E. Now, $p|_{E_Y}$ is a C^∞ map since it is the restriction of a C^∞ map to a submanifold (see Corollary 5.23). For the same reason, if we have a collection of local trivializations of E,

$$p^{-1}(U_i) \to U_i \times \mathbb{R}^r,$$

such that $\{U_i\}$ cover M, and restrict them to E_Y, we obtain a collection of local trivializations of E_Y,

$$p^{-1}(U_i) \cap E_Y \to (U_i \cap Y) \times \mathbb{R}^r,$$

such that the sets $U_i \cap Y$ cover Y. (The restriction is C^∞ and the same holds for the restriction of the inverse since $(U_i \cap Y) \times \mathbb{R}^r$ is a submanifold of $U_i \times \mathbb{R}^r$.)

Let N and M be manifolds and (E, p, M) a vector bundle over M. Any smooth map $f\colon N \to M$ induces a vector bundle $f^{-1}E$ on N, called the *pull-back of E by f*. The pull-back $f^{-1}E$ is the subset of $N \times E$ consisting of pairs $(n, e) \in N \times E$ with $p(e) = f(n)$; it is the unique maximal subset of $N \times E$ that makes the following diagram commutative:

$$\begin{array}{ccc} f^{-1}E & \longrightarrow & E \\ {\scriptstyle f^{-1}p}\downarrow & & \downarrow{\scriptstyle p} \\ N & \xrightarrow{\;f\;} & M \end{array}$$

where the vertical map $f^{-1}p$ on the left is given by $(n, e) \mapsto n$ and the top horizontal map is $(n, e) \mapsto e$. The fibre of $f^{-1}E$ over a point $n \in N$ is

$$(f^{-1}E)_n = \{n\} \times E_{f(n)},$$

which we identify with $E_{f(n)}$, and thus has a vector space structure. We also give $f^{-1}E$ the relative topology as a subset of $N \times E$.

Example Suppose that $E = M \times \mathbb{R}^r$, and p the projection on the first factor. Then if $f\colon N \to M$ we have

$$f^{-1}(M \times \mathbb{R}^r) = \{(n, f(n), v);\, n \in N, v \in \mathbb{R}^r\} = \Gamma \times \mathbb{R}^r$$

where Γ is the graph of f. By virtue of Exercise 5 in §5.4, $f^{-1}(M \times \mathbb{R}^r)$ is a submanifold of $N \times (M \times \mathbb{R}^r)$.

Suppose E is trivial, that is, there is an isomorphism $\varphi\colon M \times \mathbb{R}^r \to E$. Then the diffeomorphism $\mathrm{id}_N \times \varphi\colon N \times (M \times \mathbb{R}^r) \to N \times E$ maps $\not{f}^{-1}(M \times \mathbb{R}^r)$ to $\not{f}^{-1}(E)$. Hence $\not{f}^{-1}(E)$ is a submanifold of $N \times E$ (see §5.4, Exercise 1(b)).

Since any vector bundle is locally trivial, it follows that $\not{f}^{-1}E$ is a submanifold of $N \times E$ when E is any vector bundle over M. Indeed, let U be an open set in M such that E_U is trivial. Then $V = \not{f}^{-1}(U)$ is open in N, and the pull-back of E_U under the map $\not{f}|_V\colon V \to U$ is a submanifold of $V \times E_U$. Note also that this pull-back is equal to

$$\not{f}^{-1}E|_V = (V \times E_U) \cap \not{f}^{-1}E$$

which is open in $\not{f}^{-1}E$ since $V \times E_U$ is open in $N \times E$. Thus, every point of $\not{f}^{-1}E$ belongs to an open set (in $\not{f}^{-1}E$) which is a submanifold of $N \times E$, and, consequently, $\not{f}^{-1}E$ is itself a submanifold.

Proposition 5.56 *Let $\not{f}\colon N \to M$ be a smooth map and E a vector bundle with projection $p\colon E \to M$. Then $\not{f}^{-1}E$ is a submanifold of $N \times E$, and is a vector bundle over N with projection $\not{f}^{-1}p\colon \not{f}^{-1}E \to N$.*

Proof In fact, if $\{U_i\}$ is an open cover of M such that $\beta_i\colon E|_{U_i} \to U_i \times \mathbb{R}^r$ are trivializing maps, let β_{i2} denote the second component of β_i (the composition of β_i with the map $U_i \times \mathbb{R}^r \to \mathbb{R}^r$). Then the sets $V_i = \not{f}^{-1}(U_i)$ are open and cover N, and the maps

$$\begin{aligned} \varphi_i\colon \not{f}^{-1}E|_{V_i} &\mapsto V_i \times \mathbb{R}^r \\ (n, e) &\mapsto (n, \beta_{i2}(e)) \end{aligned}$$

are trivializing maps for $\not{f}^{-1}E$. Each map φ_i is C^∞ since it is the restriction of a smooth map on $N \times E$ to a submanifold of $N \times E$. The inverse of φ_i is also smooth due to Lemma 5.46.

Natural operations on vector spaces carry over to vector bundles. In this section it is convenient to extend slightly the generality of Proposition 5.47. We can replace $\mathbb{R}^r$ by any r-dimensional vector space $\mathbf{E}$. The bijections $\beta_i\colon p^{-1}(U_i) \to U_i \times \mathbf{E}$ satisfying VB 1 and VB 2 then define a vector bundle structure on S modelled on $\mathbf{E}$. The transition functions h_{ij} are smooth maps $U_i \cap U_j \to GL(\mathbf{E})$.

As the first example of an operation on vector bundles, we consider the direct sum. If E and F are vector bundles over M modelled on vector spaces $\mathbf{E}$ and $\mathbf{F}$, then

$$E \oplus F = \bigcup_{x \in M} E_x \oplus F_x$$

has a vector bundle structure over M modelled on $\mathbf{E} \oplus \mathbf{F}$. In fact, there exists an open cover $\{U_i\}$ of M and trivializing maps

$$\left.\begin{aligned} \varphi_i\colon E_{u_i} &\to U_i \times \mathbf{E} \\ \varphi_i'\colon F_{u_i} &\to U_i \times \mathbf{F} \end{aligned}\right\} \qquad (30)$$

so that we have a bijection

$$\beta_i\colon (E \oplus F)_{U_i} \to U_i \times (\mathbf{E} \oplus \mathbf{F})$$

and, for each $x \in M$, $\beta_{ix} = \varphi_{ix} \oplus \varphi'_{ix}$ is a linear map $E_x \oplus F_x \to \mathbf{E} \oplus \mathbf{F}$. Thus VB 1 and the first part of VB 2 are satisfied; the map $p\colon E \oplus F \to M$ is the obvious one defined by $p^{-1}(x) = E_x \oplus F_x$ for each $x \in M$. As for the smoothness requirement, let $h_{ij}(x) = (\varphi_i \, \varphi_j^{-1})_x \in GL(\mathbf{E})$ and $h'_{ij}(x) = (\varphi'_i \, \varphi'^{-1}_j)_x \in GL(\mathbf{F})$ be the transition functions for E and F, respectively. Then $E \oplus F$ has transition functions $U_i \cap U_j \to GL(\mathbf{E} \oplus \mathbf{F})$ given by

$$x \mapsto \begin{pmatrix} h_{ij}(x) & 0 \\ 0 & h'_{ij}(x) \end{pmatrix}$$

which are smooth since h_{ij} and h'_{ij} are smooth. By Proposition 5.47, $E \oplus F$ is a vector bundle modelled on $\mathbf{E} \oplus \mathbf{F}$. If E is rank r and F is rank d then $E \oplus F$ is rank $r + d$ (since $\mathbb{R}^r \oplus \mathbb{R}^d \simeq \mathbb{R}^{r+d}$).

Similarly, we can define other natural operations on vector bundles.

For the tensor product $E \otimes F = \bigcup E_x \otimes F_x$, given the trivializing maps (30) we have the bijection

$$\beta_i\colon (E \otimes F)_{U_i} \to U_i \times (\mathbf{E} \otimes \mathbf{F}), \tag{31}$$

and, for each $x \in M$, it is a linear map $E_x \otimes F_x \to \mathbf{E} \otimes \mathbf{F}$ defined by $\beta_{ix} = \varphi_{ix} \otimes \varphi'_{ix}$. The transition functions $x \mapsto h_{ij}(x) \otimes h'_{ij}(x) \in GL(E \otimes F)$ are again smooth, so by Proposition 5.47, $\mathbf{E} \otimes \mathbf{F}$ is a vector bundle modelled on $\mathbf{E} \otimes \mathbf{F}$. If E has rank r and F has rank d then $E \otimes F$ has rank $r \cdot d$ (since $\mathbb{R}^r \otimes \mathbb{R}^d \simeq \mathbb{R}^{r \cdot d}$).

For the dual bundle $E^* = \bigcup (E_x)^*$, the transition functions are not quite so immediate. Recall that for a real vector space $\mathbf{E}$, the dual space is $\mathbf{E}^* = L(\mathbf{E}, \mathbb{R})$, the real-valued linear functionals on $\mathbf{E}$. A linear map $f\colon \mathbf{E} \to \mathbf{F}$ induces the transpose map ${}^t f\colon \mathbf{F}^* \to \mathbf{E}^*$ defined by

$$\langle {}^t f(\alpha), v \rangle = \langle \alpha, f(v) \rangle, \qquad \alpha \in \mathbf{F}^*, v \in \mathbf{E},$$

that is, ${}^t f(\alpha) = \alpha \circ f$ $(\alpha \in \mathbf{F}^*)$. Note that the transpose operator t is a linear map $L(\mathbf{E}, \mathbf{F}) \to L(\mathbf{F}^*, \mathbf{E}^*)$. Now, if $\varphi_i\colon E_{U_i} \to U_i \times \mathbf{E}^r$ is a trivializing map for E, then

$$\beta_i\colon (E^*)_{U_i} \to U_i \times \mathbf{E}^* \qquad \text{where } \beta_{ix} = ({}^t \varphi_{ix})^{-1}$$

is a bijection which has transition functions

$$x \mapsto (\beta_i \, \beta_j^{-1})_x = [{}^t h_{ij}(x)]^{-1} \in GL(\mathbf{E}^*)$$

where $h_{ij}(x) = (\varphi_i \, \varphi_j^{-1})_x \in GL(\mathbf{E})$. Since the transpose operator t is linear, hence smooth, these transition functions are smooth in view of Proposition 5.2. Thus E^* is a vector bundle over M modelled on $\mathbf{E}^*$. If E is rank r then E^* is also rank r since $(\mathbb{R}^r)^* \simeq \mathbb{R}^r$.

These types of constructions of vector bundles can be unified with the language of category theory. Proposition 5.57 and the series of exercises following it are meant to illustrate this idea.

Recall that a *category* $\mathfrak{A}$ consists of a class of objects and for any two objects A and B, a set $\mathrm{Hom}(A, B)$ of *morphisms* from A to B, satisfying the following properties. If f is a morphism from A to B and g a morphism from B to C, then the composite morphism $g \circ f$ from A to C is defined; furthermore, the composition operation is required to be associative and to have an identity 1_A in $\mathrm{Hom}(A, A)$ for every object A. Some examples of categories are:

The class of all topological spaces, with morphisms being the continuous maps.
The class of finite dimensional vector spaces, with morphisms being the linear maps.
The class of all groups, with morphisms being the group homomorphisms.

For a concise introduction to categories, see [Do].

A *covariant functor* τ from a category $\mathfrak{A}$ to a category $\mathfrak{B}$ is a rule which to each object A in $\mathfrak{A}$ associates an object $\tau(A)$ in $\mathfrak{B}$ and with each morphism $f: A \to B$ a morphism $\tau(f): \tau(A) \to \tau(B)$ such that

$$\tau(f \circ g) = \tau(f) \circ \tau(g) \tag{32}$$

whenever f and g are morphisms such that $f \circ g$ is defined, and

$$\tau(1_A) = 1_{\tau(A)} \tag{33}$$

A *contravariant* functor τ is defined by reversing the arrows in (32) in the following sense: τ is a rule which to each object A in $\mathfrak{A}$ associates an object $\tau(A)$ in $\mathfrak{B}$ and with each morphism $f: A \to B$ a morphism $\tau(f): \tau(B) \to \tau(A)$ (going in the opposite direction) such that (33) holds and

$$\tau(f \circ g) = \tau(g) \circ \tau(f) \tag{32'}$$

whenever f and g are morphisms such that $f \circ g$ is defined.

From now on, we always work with the category $\mathfrak{A}$ of finite dimensional vector spaces over $\mathbb{R}$, with morphisms being the linear maps. Let τ be a covariant functor on $\mathfrak{A}$. Then the functor τ defines a map

$$\tau: L(\mathbf{E}, \mathbf{F}) \to L(\tau(\mathbf{E}), \tau(\mathbf{F}));$$

we say that τ is a *smooth functor* if this map is smooth for any real finite dimensional vector spaces $\mathbf{E}$ and $\mathbf{F}$.

Let $VB(M, \mathfrak{A})$ denote the category of vector bundles over M modelled on vector spaces in $\mathfrak{A}$, with morphisms taken to be the vector bundle homomorphisms. The next proposition shows that a smooth functor on $\mathfrak{A}$ can be extended to $VB(M, \mathfrak{A})$.

Proposition 5.57 *Suppose that τ is a smooth covariant functor on $\mathfrak{A}$, the category of finite dimensional real vector spaces. If E is a vector bundle over M, define the family of vector spaces,*

$$\tau(E) = \bigcup_x \tau(E_x) \qquad \textit{(disjoint union)}$$

If $f: E \to F$ is a vector bundle homomorphism, define the map $\tau(f): \tau(E) \to \tau(F)$ such that its value on the fibre over x is given by $\tau(f_x): \tau(E_x) \to \tau(F_x)$. Then, for each $E \in VB(M, \mathfrak{A})$, there is a topology and C^∞ structure on $\tau(E)$ such that $\tau(E) \in VB(M, \mathfrak{A})$ and the following properties hold:

(*a*) *If $f: E \to F$ is a vector bundle homomorphism then $\tau(f)$ is a vector bundle homomorphism $\tau(E) \to \tau(F)$.*
(*b*) *τ is a functor: $\tau(f \circ g) = \tau(f) \circ \tau(g)$ whenever $f \circ g$ is defined, and $\tau(id_E) = id_{\tau(E)}$.*
(*c*) *If $E = M \times \mathbf{E}$ is a trivial vector bundle then $\tau(E) = M \times \tau(\mathbf{E})$ is a trivial vector bundle.*
(*d*) *If $f: N \to M$ is a smooth map between manifolds then $\tau(f^{-1}E) = f^{-1}\tau(E)$.*

(To be precise in (d) we should write $\tau_N(f^{-1}E) = f^{-1}\tau_M(E)$; that is, there is a distinct functor $\tau = \tau_M$ for each manifold M. But (d) shows that there is no harm in dropping the reference to the manifold.)

Proof As a preliminary remark, note that it follows from (32), (33) that if $f: \mathbf{E} \to \mathbf{F}$ is an isomorphism then so is $\tau(f): \tau(\mathbf{E}) \to \tau(\mathbf{F})$, and in fact $\tau(f)^{-1} = \tau(f^{-1})$.

We assume that M is connected, so that each fibre is isomorphic to a fixed vector space $\mathbf{E}$. Let $\{U_i\}$ be an open cover of M with trivializing maps for E

$$\beta_i: E|_{U_i} \to U_i \times \mathbf{E}$$

Then we have a bijection

$$\tau(\beta_i): \tau(E|_{U_i}) \to U_i \times \tau(\mathbf{E})$$

defined on the fibers over $x \in U_i$ by the linear isomorphism $\tau(\beta_{ix}): \tau(E_x) \to \tau(\mathbf{E})$. We must show that these bijections satisfy the conditions of Proposition 5.47. There is an obvious map $\pi: \tau(E) \to M$ such that VB 1 holds for $(\tau(E), \pi, M)$. To verify VB 2 note that the transition functions for $\tau(E)$ are

$$x \mapsto \tau(\beta_{ix}) \circ \tau(\beta_{jx})^{-1}$$

The expression on the right is equal to $\tau((\beta_i \circ \beta_j^{-1})_x) = \tau(h_{ij}(x))$ where h_{ij} are the transition functions for E. Since h_{ij} is smooth from $U_i \cap U_j$ to $GL(\mathbf{E})$ then $\tau \circ h_{ij}$ is smooth from $U_i \cap U_j$ to $GL(L(\mathbf{E}))$ due to the smoothness property of the functor τ. Thus there is a unique topology and C^∞ structure on $\tau(E)$ such that $\tau(\beta_{ix})$ is a linear isomorphism on each fibre $\pi^{-1}(x) = \tau(E_x)$ and $(\tau(E), \pi, M)$ is a vector bundle with $\tau(\beta_i)$ as trivializing maps.

Now, for each vector bundle homomorphism $f: E \to F$ we have the map $\tau(f): \tau(E) \to \tau(F)$ which is obviously a vector bundle homomorphism, provided we can show that it is C^∞. To do so we can argue locally. We have the commutative diagram

$$\begin{array}{ccc} E|_{U_i} & \xrightarrow{\;f\;} & F|_{U_i} \\ \big\downarrow{\scriptstyle \beta_i} & & \big\downarrow{\scriptstyle \psi_i} \\ U_i \times \mathbf{E} & \longrightarrow & U_i \times \mathbf{F} \end{array}$$

where the map in the bottom row is $(x, v) \mapsto (x, \tilde{f}(x)v)$, and by the functorial property of τ we obtain another commutative diagram

$$\begin{array}{ccc} \tau(E|_{U_i}) & \xrightarrow{\tau(f)} & \tau(F|_{U_i}) \\ \Big\downarrow{\scriptstyle \tau(\beta_i)} & & \Big\downarrow{\scriptstyle \tau(\psi_i)} \\ U_i \times \tau(\mathbf{E}) & \longrightarrow & U_i \times \tau(\mathbf{F}) \end{array}$$

where the map in the bottom row is $(x, w) \mapsto (x, \tau(\tilde{f}(x))w)$. The map $\tilde{f}: U_i \to L(\mathbf{E}, \mathbf{F})$ is the local representation of f and is smooth, so that by the smoothness property of τ, the map $x \mapsto \tau(\tilde{f}(x))$ is also smooth. Hence $\tau(f)$ is smooth.

Conditions (b) and (c) are obviously satisfied. We leave the proof of (d) as an exercise.

Examples

(1) Let $\tau(\mathbf{E}) = \bigwedge^k(\mathbf{E})$, the kth exterior product (see §5.1). If $f \in L(\mathbf{E}, \mathbf{F})$ there is a unique linear map $\tau(f) = \bigwedge^k f \in L(\bigwedge^k \mathbf{E}, \bigwedge^k \mathbf{F})$ such that

$$v_1 \wedge \cdots \wedge v_k \mapsto f(v_1) \wedge \cdots \wedge f(v_k) \tag{34}$$

Then τ is a covariant functor on the category of finite dimensional real vector spaces, that is, $\bigwedge^k(f \circ g) = \bigwedge^k f \circ \bigwedge^k g$ whenever $f \circ g$ is defined and $\bigwedge^k(\mathrm{id}_{\mathbf{E}}) = \mathrm{id}_{\bigwedge^k \mathbf{E}}$.

(2) Let $\tau(\mathbf{E}) = \mathbf{E}^*$ and for each linear map $f \in L(\mathbf{E}, \mathbf{F})$ define the linear map $\tau(f) = {}^t f \in L(\mathbf{F}^*, \mathbf{E}^*)$. Then τ is a *contravariant* functor on the category of finite dimensional real vector spaces.

One can also define functors of several variables $\tau(\mathbf{E}, \mathbf{F}, \ldots)$, covariant in some variables, contravariant in others. For instance, a functor τ in two variables is said to be contravariant in the first variable and covariant in the second if

$$\tau(f_1 \circ f_2, g) = \tau(f_2, g) \circ \tau(f_1, g)$$

$$\tau(f, g_1 \circ g_2) = \tau(f, g_1) \circ \tau(f, g_2)$$

and $\tau(\mathrm{id}_{\mathbf{E}}, \mathrm{id}_{\mathbf{F}}) = \mathrm{id}_{\tau(\mathbf{E}, \mathbf{F})}$. Further, τ is said to be *smooth* if the map

$$L(\mathbf{E}', \mathbf{E}) \times L(\mathbf{F}, \mathbf{F}') \to L(\tau(\mathbf{E}, \mathbf{F}), \tau(\mathbf{E}', \mathbf{F}'))$$

defined by $(f, g) \mapsto \tau(f, g)$ is smooth. See Exercise 7 below.

Exercises

1. Let Y be a submanifold of M, and $\imath: Y \to M$ the inclusion map. Show that $\imath^{-1}E \simeq E|_Y$. Hint: The isomorphism is given by $(y, e) \mapsto e$ where $y \in Y$, $e \in E$, with inverse $e \mapsto (p(e), e)$.

2. (a) Let $f: N \to M$ and $g: Z \to N$ be smooth maps. Show that $(f \circ g)^{-1}E \simeq g^{-1}(f^{-1}E)$.
(b) Show that isomorphic vector bundles have isomorphic pull-backs.

3. Show that $\bigwedge^k$ is a smooth covariant functor in the category $\mathfrak{A}$ of finite dimensional real vector spaces. That is, if $f \in L(E, F)$ then there is the induced linear map $\bigwedge^k f \in L(\bigwedge^k \mathbf{E}, \bigwedge^k \mathbf{F})$ defined by

$$v_1 \wedge \cdots \wedge v_k \mapsto f(v_1) \wedge \cdots \wedge f(v_k) \tag{34}$$

and it has the following properties:

(i) If $g \in L(\mathbf{F}, \mathbf{G})$ and $f \in L(\mathbf{E}, \mathbf{F})$ then $\bigwedge^k(g \circ f) = \bigwedge^k g \circ \bigwedge^k f$.
(ii) If $\mathrm{id}_{\mathbf{E}} \in L(\mathbf{E}, \mathbf{E})$ denotes the identity on $\mathbf{E}$ then $\bigwedge^k(\mathrm{id}_{\mathbf{E}}) = \mathrm{id}_{\bigwedge^k \mathbf{E}}$, and the smoothness property:
(iii) The map $\bigwedge^k: L(\mathbf{E}, \mathbf{F}) \to L(\bigwedge^k \mathbf{E}, \bigwedge^k \mathbf{F})$ is C^∞.

Hint for (iii). If bases are selected for the vector spaces, then $\bigwedge^k$ becomes a matrix function with polynomial entries.

Let τ be a smooth covariant functor on $\mathfrak{A}$. Let E be a vector bundle over M, modelled on $\mathbf{E}$, and let $s \in C^\infty(M, \tau(E))$ be a section, that is, a smooth map $s: M \to \tau(E)$ such that $s(x) \in \tau(E_x)$ for all $x \in M$. If $\beta: E|_U \to U \times \mathbf{E}$ is a trivializing map then we obtain an isomorphism $\beta_x: E_x \to \mathbf{E}$, hence an isomorphism $\tau(\beta_x): \tau(E_x) \to \tau(\mathbf{E})$ for each x. Thus, s has a *local representation*

$$\tilde{s}: U \to \tau(\mathbf{E}) \tag{*}$$

where $\tilde{s}$ is smooth.

4. Consider once again the functor $\bigwedge^k$. Write out in detail what the local representation (*) of a section $s \in C^\infty(M, \bigwedge^k(E))$ looks like in this case, given a basis $v_1, \ldots, v_r$ in $\mathbf{E}$.

5. Show that Proposition 5.57 holds for smooth *contravariant* functors, where in (a) the arrow is reversed. (A contravariant functor gives a map $L(\mathbf{E}, \mathbf{F}) \to L(\tau(\mathbf{F}), \tau(\mathbf{E}))$ and, as before, τ is said to be a smooth functor if this map is smooth for all $\mathbf{E}$ and $\mathbf{F}$.)

6. Let $\tau(\mathbf{E}) = E^*$, and for each $f \in L(\mathbf{E}, \mathbf{F})$ let $\tau(f) = {}^t f \in L(\mathbf{F}^*, \mathbf{E}^*)$ be the transpose of f, i.e. $\tau(f)h = h \circ f$. Show that τ is a smooth contravariant functor on the category of finite dimensional real vector spaces.

The functors considered in Proposition 5.57 and Exercises 3 to 6 are functors of a single variable. Similarly, one can define functors of many variables, contravariant in some variables and covariant in others.

7. Let τ be a smooth functor of two variables in the category $\mathfrak{A}$, contravariant in the first variable and covariant in the second. If E and F are vector bundles over M, define

$$\tau(E, F) = \bigcup_x \tau(E_x, F_x),$$

and, for each pair of vector bundle homomorphisms $f: E' \to E$ and

$g: F \to F'$, define the map

$$\tau(f, g): \tau(E, F) \to \tau(E', F')$$

whose value on the fibres over x is given by $\tau(f_x, g_x): \tau(E_x, F_x) \to \tau(E'_x, F'_x)$. Show that τ is a functor in the category of vector bundles with properties analogous to (a)–(d) of Proposition 5.57.

8. Let Hom or L be the functor which associates to finite dimensional vector spaces $\mathbf{E}$ and $\mathbf{F}$ the finite dimensional vector space

$$\mathrm{Hom}(\mathbf{E}, \mathbf{F}) = L(\mathbf{E}, \mathbf{F})$$

(the space of linear maps from $\mathbf{E}$ to $\mathbf{F}$), and associates to each pair of linear maps $f: \mathbf{E}' \to \mathbf{E}$ and $g: \mathbf{F} \to \mathbf{F}'$ the linear map $\mathrm{Hom}(\mathbf{E}, \mathbf{F}) \to \mathrm{Hom}(\mathbf{E}', \mathbf{F}')$ defined by $h \mapsto g \circ h \circ f$. Show that τ is a smooth functor, contravariant in the first variable and covariant in the second. (Note: We can conclude from Exercise 7 that for each pair of vector bundles E, F over M,

$$\mathrm{Hom}(E, F) = L(E, F) = \bigcup_x L(E_x, F_x)$$

is a vector bundle over M. However, this vector bundle should be distinguished from $\mathrm{HOM}(E, F)$. $\mathrm{HOM}(E, F)$ is the space of sections of $\mathrm{Hom}(E, F)$; see the discussion preceding Lemma 5.59.)

9. If $\mathbf{E}$ and $\mathbf{F}$ are vector spaces then $\mathbf{E}^* \otimes \mathbf{F} \simeq L(\mathbf{E}, \mathbf{F})$ by a canonical isomorphism such that $\lambda \otimes w \in \mathbf{E}^* \otimes \mathbf{F}$ corresponds to the linear map $\mathbf{E} \to \mathbf{F}$ given by $v \mapsto \langle \lambda, v \rangle \cdot w$. Show that this induces an isomorphism

$$L(E, F) \simeq E^* \otimes F$$

for any two vector bundles E and F over M.

10. Formulate the definition of a smooth functor τ of two variables which is

(i) covariant in both variables
(ii) contravariant in both variables.

Show that in each case τ can be extended to the category of vector bundles with properties analogous to (a)–(d) of Proposition 5.57.

11. Let $L^2(\mathbf{E})$ denote the set of bilinear maps $\mathbf{E} \times \mathbf{E} \to \mathbb{R}$.

(a) If E is any vector bundle over M, show that $L^2(E) = \bigcup_x L^2(E_x)$ has a natural vector bundle structure with properties analogous to Proposition 5.57 (a)–(d).
(b) Show that the canonical isomorphism $L^2(\mathbf{E}) \simeq \mathbf{E}^* \otimes \mathbf{E}^*$ of vector spaces induces an isomorphism

$$L^2(E) \simeq E^* \otimes E^*$$

of vector bundles over M.

12. Let E be a vector bundle over M. Let $v_1, \ldots, v_r$ be a local basis for E on U:

(a) At each point $x \in U$ define the dual basis $\lambda^1(x), \ldots, \lambda^r(x) \in E_x^*$ by

$$\langle \lambda^j(x), v_i(x) \rangle = \delta_{ij}$$

Show that λ^j is smooth and hence a section of $E^*|_U$. Hint: Consider the isomorphism φ defined by (28). What is the image of $\lambda^j(x)$ under the transpose map ${}^t(\varphi_x)$?

(b) As in (b) of the preceding exercise, we identify $L^2(E)$ with $E^* \otimes E^*$. Show that every section g of $L^2(E)$ can be written locally in the form

$$g_x = \sum g_{ij}(x)\lambda^i(x) \otimes \lambda^j(x), \qquad x \in U$$

where $g_{ij} \in C^\infty(U)$ are unique.

5.9 Homotopy property for vector bundles

Let E be a vector bundle over M. A map $s\colon A \to E$ is said to be a C^∞ section defined on the closed set $A \subset M$ if for every $x \in A$ there is an open set $U = U_x$ containing x and a C^∞ section $s_1\colon U \to E$ such that $s_1|_{U \cap A} = s|_{U \cap A}$. The following lemma, a smooth version of the Tietze extension theorem, shows that s can be smoothly extended to a section $\tilde{s}\colon M \to E$ defined on all M.

Lemma 5.58 (*Smooth Tietze Extension Theorem*) *Let E be a vector bundle over M. Suppose that $s\colon A \to E$ is a smooth section defined on the closed set A, that is, s is smooth in the sense defined above and $p \circ s$ is the identity on A. Then s can be smoothly extended to a section of E.*

Proof Consider the open cover $\{U_x; x \in A\}$ of A which is given by the definition of smoothness of A. If we include also the open set $M \setminus A$, we obtain an open cover of M. Let $\{V_i\}$ be a locally finite refinement of this cover, and let $\{\rho_i\}$ be a partition of unity subordinate to $\{V_i\}$ with supp ρ_i compact in V_i. Let s_i denote a C^∞ section $V_i \to E|_{V_i}$ so that s_i and s coincide on $A \cap V_i$ (where $s_i \equiv 0$ if $A \cap V_i = \varnothing$). Then we get a C^∞ section $\tilde{s}_i$ of E by defining

$$\tilde{s}_i(x) = \begin{cases} \rho_i(x)s_i(x) & \text{if } x \in V_i \\ 0 & \text{if } x \notin V_i \end{cases}$$

Then $\tilde{s} = \sum_i \tilde{s}_i$ is a C^∞ section of E, and it is an extension of s, for if $x \in A$ then

$$\tilde{s}(x) = \sum_i \rho_i(x)s_i(x) = \sum_i \rho_i(x)s(x) = s(x).$$

Remark In particular, the lemma can be applied when E is a trivial bundle $M \times \mathbb{R}^r$. Thus if A is a closed set in a manifold M and $f\colon A \to \mathbb{R}^r$ is smooth, then f has an extension to a smooth function $\tilde{f}\colon U \to \mathbb{R}^r$ where $U \supset A$ is open.

For any two vector bundles E and F over M, we let $\mathrm{Hom}(E, F)$ be the vector bundle for which the fibre over $x \in M$ is the vector space of linear maps from E_x to F_x,

$$\operatorname{Hom}(E, F) = \bigcup_x L(E_x, F_x)$$

(see §5.8, Exercise 8). If $f: M \to \operatorname{Hom}(E, F)$ is a section, then for each $x \in M$ we have

$$f_x \in \operatorname{Hom}(E_x, F_x) = L(E_x, F_x)$$

and f has a local representation of the form $\tilde{f}: U \to L(\mathbb{R}^r, \mathbb{R}^d)$ where $\tilde{f}$ is smooth (r and d are the fibre dimensions of E and F, respectively). The smoothness requirement on f is the same, therefore, whether it is regarded as a section $M \to L(E, F)$ or as a homomorphism $E \to F$. Hence

$$C^\infty(M, \operatorname{Hom}(E, F)) = \operatorname{HOM}(E, F),$$

that is, the *sections* of $\operatorname{Hom}(E, F)$ are precisely the homomorphisms from E to F.

Lemma 5.59 *Let Y be a submanifold of M, and E and F two vector bundles over M. Then any isomorphism $\varphi: E|_Y \to F|_Y$ extends to an isomorphism $E|_U \to F|_U$ for some open set U containing Y.*

Proof Note that φ is a section of $\operatorname{Hom}(E, F)|_Y$. Applying Lemma 5.58 we can extend φ to a section g of $\operatorname{Hom}(E, F)$. Let U be the subset of M consisting of points x for which g_x is an isomorphism. Since $GL(\mathbb{R}^n)$ is open in $L(\mathbb{R}^n)$, U is open and contains Y.

Throughout the rest of this section, $I = (-\varepsilon, 1 + \varepsilon)$ denotes an *open* interval containing the unit interval $[0, 1]$.

Definition 5.60 *Two smooth maps $f_0: N \to M$ and $f_1: N \to M$ are said to be (smoothly) homotopic if there exists a smooth map $f: N \times I \to M$ such that $f(x, 0) = f_0(x)$ and $f(x, 1) = f_1(x)$ for all $x \in N$.*

Ordinarily, a homotopy is defined as a *continuous* map $f: N \times [0, 1] \to M$. However, if two smooth maps f_0 and f_1 are continuously homotopic then they are also smoothly homotopic (see §10.6) so there is no loss of generality in restricting attention to smooth homotopies.

Example Let $g: M \to M$ and $h: M \to \mathbb{R}$ be smooth maps. Then $(g, h): M \to M \times \mathbb{R}$ is homotopic to the map $(g, 0)$. The homotopy is $f(x, t) = (g(x), th(x))$, $x \in M$, $t \in I$.

In Lemma 5.61 and Theorem 5.62 we assume that the manifold M is compact. This is sufficient for our purposes, but these two results in fact hold for any (paracompact) manifold. See [Hi] or [Sp 2] for the proof in general.

For a fixed $t \in I$, $M \times \{t\}$ is a submanifold of $M \times I$, so it makes sense to consider the restriction $E|_{M \times \{t\}}$, and we regard it as a vector bundle over M.

Lemma 5.61 *Let M be a (compact) manifold. Then for any vector bundle E over $M \times I$*

$$E|_{M \times 0} \simeq E|_{M \times 1}$$

where $\simeq$ indicates an isomorphism as vector bundles over M.

Proof Fix $t_0 \in I$ and define $\imath: M \times I \to M \times I$ by $\imath(x, t) = (x, t_0)$. Also let $\iota: M \times t_0 \to M \times I$ denote the inclusion map. The restrictions

$$\imath^{-1}E|_{M \times t_0} \quad \text{and} \quad E|_{M \times t_0} \text{ are isomorphic}$$

since they are isomorphic to $\iota^{-1}\imath^{-1}E$ and $\iota^{-1}E$, respectively, and $\iota^{-1}\imath^{-1}E \simeq (\imath \circ \iota)^{-1}E = \iota^{-1}E$ since $\imath \circ \iota = \iota$. By Lemma 5.59 this isomorphism can be extended to an isomorphism

$$\imath^{-1}E|_U \simeq E|_U \tag{35}$$

for some open set $U \supset M \times t_0$. Since we are assuming that M is compact, then $U \supset M \times (t_0 - \varepsilon, t_0 + \varepsilon)$ for some $\varepsilon > 0$, and restricting (35) to the submanifolds $M \times t$ we obtain

$$E|_{M \times t} \simeq E|_{M \times t_0} \qquad \text{when } |t - t_0| < \varepsilon$$

(regarded as vector bundles over M). This implies that the set $I_0 = \{t \in I; E|_{M \times t} \simeq E|_{M \times 0}\}$ is both open and closed in I. Since I is connected we have $I_0 = I$.

Theorem 5.62 *Let N and M be (compact) manifolds and let $f_i: N \to M$, $i = 0, 1$, be smoothly homotopic maps. Then $f_1^{-1}E \simeq f_0^{-1}E$ for any vector bundle E over M.*

Proof Let $f: N \times I \to M$ be a homotopy connecting f_1 and f_0. Since $f_t = f \circ \iota_t$, where $\iota_t: N \times t \to N \times I$ denotes the inclusion map, we have

$$f_t^{-1}E \simeq f^{-1}E|_{N \times t} \qquad t = 0, 1$$

and, by virtue of Lemma 5.61, the restrictions $f^{-1}E|_{N \times 1}$ and $f^{-1}E|_{N \times 0}$ are isomorphic vector bundles over N.

A manifold M is said to be *contractible* if there exists a point $x_0 \in M$ and a smooth homotopy $f: M \times I \to M$ such that $f(x, 1) = x$ and $f(x, 0) = x_0$ for all $x \in M$. In other words, the identity map on M and the constant map are homotopic.

Corollary 5.63 *If M is contractible, then any vector bundle over M is trivial.*

For example, any star-shaped open set in $\mathbb{R}^n$ is contractible. Recall that $U \subset \mathbb{R}^n$ is star-shaped if there is a point $x_0 \in U$ such that, for any $x \in U$, the line segment from x to x_0 lies in U.

5.10 Riemannian and Hermitian metrics

Let E be a vector bundle over M. We say that a *Riemannian metric* has been defined on E if we have an inner product $\langle\,,\,\rangle_x = \langle\,,\,\rangle_{E_x}$ on each fibre E_x which is smooth in the sense that for any sections s and $t \in C^\infty(M, E)$ the function

$$x \mapsto \langle s(x), t(x)\rangle_x$$

belongs to $C^\infty(M)$. Note that $\langle\,,\,\rangle_x$ satisfies the smoothness property if and only if $g_{ij}(x) = \langle s_i(x), s_j(x)\rangle_x$ belongs to $C^\infty(U)$ for any local basis $s_1, \ldots, s_r$ for E on U. (This is a consequence of Corollary 5.49.)

Example The product bundle $M \times \mathbb{R}^r$ has the Riemannian metric

$$\langle v, w\rangle = \sum_{k=1}^{r} v_k w_k \qquad v, w \in \mathbb{R}^r$$

for each $x \in M$.

It follows that any trivial bundle E has a Riemannian metric. Indeed, if β is a trivializing map $E \to M \times \mathbb{R}^r$ then by transporting the metric on $M \times \mathbb{R}^r$ we get a metric on E:

$$\langle v, w\rangle_x = \langle \beta_x v, \beta_x w\rangle \qquad v, w \in E_x$$

where β_x: $E_x \to \mathbb{R}^r$ is the map induced on the fibres by β. It is clear that $\langle\,,\,\rangle_x$ is an inner product since β_x is a linear isomorphism, and, further, the smoothness property holds.

Proposition 5.64 *Every vector bundle E has a Riemannian metric.*

Proof Let $\{U_i, \rho_i\}$ be a partition of unity such that $E|_{U_i}$ is trivial and ρ_i has compact support in U_i. Introduce any Riemannian metric $\langle\,,\,\rangle^{(i)}$ into $E|_{U_i}$ and then let

$$\langle v, w\rangle_x = \sum_i \rho_i(x)\cdot\langle v, w\rangle_x^{(i)} \qquad v, w \in E_x$$

where the ith term is taken to be 0 if $x \notin U_i$. The sum is well-defined since it is a finite sum for each x. Thus $\langle\,,\,\rangle_x$ is bilinear and $\langle v, v\rangle_x \geqslant 0$ since $\rho_i(x) \geqslant 0$ for all i. Also $\rho_i(x) > 0$ for some i, hence $\langle v, v\rangle_x > 0$ if $v \neq 0$. The smoothness property holds for $\langle\,,\,\rangle_x$ because it holds for each term in the sum and every point has a neighbourhood in which the sum is finite.

Let E be a vector bundle over M. In view of §5.8, Exercise 11,

$$L^2(E) = \bigcup_{x\in M} L^2(E_x)$$

is a vector bundle over M. A section g: $M \to L^2(E)$ of this vector bundle has a local representation of the form $\tilde{g}$: $U \to L^2(\mathbb{R}^r)$ where $\tilde{g}$ is smooth (see §5.8, Exercise 12). Since $L^2(\mathbb{R}^r)$, the set of bilinear maps on $\mathbb{R}^r$, can be identified by a linear isomorphism with the set of $r \times r$ matrices, we find that the local representation of g has the form

$$\tilde{g}(x) = [g_{ij}(x)] \qquad \text{where } g_{ij} \in C^\infty(U).$$

It follows that the set of Riemannian metrics on E is the set of sections $g_x = \langle\,,\,\rangle_x$ of $L^2(E)$ such that g_x is symmetric and positive definite on E_x for all $x \in M$.

An inner product $\langle\,,\,\rangle$ in a vector space $\mathbf{E}$ induces an isomorphism between $\mathbf{E}$ and its dual, $\mathbf{E}^*$. We call the map $\mathbf{E} \to \mathbf{E}^*$ defined by $v \mapsto \langle v, \cdot\rangle$ the "index-lowering" operator. Its inverse, $\mathbf{E}^* \to \mathbf{E}$, is called the "index-raising" operator. We denote the index lowering and raising operators by the symbols $_\#$ (subscript) and $^\#$ (superscript), respectively. That is,

$$X_\# = \langle X, \cdot\rangle \quad \text{and} \quad \xi = \langle \xi^\#, \cdot\rangle$$

The reason for this terminology comes from the Einstein summation convention where the components of vectors $X \in \mathbf{E}$ with respect to a basis are written as superscripts X^i, while components of covectors $\xi \in \mathbf{E}^*$ are written as subscripts ξ_i. Let $[g_{ij}]$ denote the matrix of $\langle\,,\,\rangle$ with respect to a basis $v_1, \ldots, v_r$ of $\mathbf{E}$, that is, $g_{ij} = \langle v_i, v_j\rangle$. The index-lowering operator maps the vector $X \in \mathbf{E}$ with components X^i to the covector $X_\# \in \mathbf{E}^*$ with components $X_i = \sum g_{ij}X^j$,

$$\left(\sum X^i v_i\right)_\# = \sum X_i \lambda^i$$

where $\lambda^1, \ldots, \lambda^r$ is the dual basis corresponding to $v_1, \ldots, v_r$. Now, let $[g^{ij}] = [g_{ij}]^{-1}$ denote the inverse matrix. The index-raising operator maps the covector $\xi \in \mathbf{E}^*$ with components ξ_i to the vector $\xi^\# \in \mathbf{E}$ with components $\xi^i = \sum g^{ij}\xi_j$,

$$\left(\sum \xi_i \lambda^i\right)^\# = \sum \xi^i v_i$$

The inner product on $\mathbf{E}$ induces an inner product on $\mathbf{E}^*$ by $\langle \xi, \eta\rangle = \langle \xi^\#, \eta^\#\rangle$; it is not hard to verify that it has matrix $[g^{ij}]$ with respect to the dual basis $\lambda^1, \ldots, \lambda^n$, that is, $g^{ij} = \langle \lambda^i, \lambda^j\rangle$. Thus,

$$\|\xi\|^2 = \langle \xi, \xi\rangle = \sum g^{ij}\xi_i\xi_j = \sum \xi^j \xi_j$$

is the length of ξ with respect to the induced inner product on $\mathbf{E}^*$.

Definition 5.65 *A Riemannian manifold is a manifold M with a Riemannian metric g on the tangent bundle TM.*

By virtue of Proposition 5.64, Riemannian metrics exist on any manifold. A Riemannian metric on TM leads to the important subject of Riemannian geometry, with geodesics, curvature, and so on. However, we omit such topics since we will have no occasion to apply them; we need Riemannian metrics for various technical reasons and in order to study the Laplace-de Rham operator on M.

In view of the index-lowering operator $TM \to T^*M$ defined above we have a one-to-one correspondence between vector fields $X \in C^\infty(M, TM)$

and covector fields (or 1-forms) $X_\# \in C^\infty(M, T^*M)$, which in local coordinates takes the form

$$X = \sum X^i \frac{\partial}{\partial x_i} \mapsto X_\# = \sum X_i \, dx^i,$$

where $X_i = \sum g_{ij} X^j$ and $g_{ij}(x) = \langle \partial/\partial x_i, \partial/\partial x_j \rangle_x$. The inverse of this correspondence is of course given by the index-raising operator, i.e. $X^j = \sum g^{ij} X_i$.

Now let E be a complex vector bundle over M, i.e. the fibres are vector spaces over $\mathbb{C}$. A C^∞ *inner product* (or Hermitian metric) on E is a $\mathbb{C}$-inner product

$$(\ , \ )_{E_x} \tag{36}$$

in each fibre E_x such that the map $x \mapsto (s(x), t(x))_{E_x}$ belongs to $C^\infty(M, \mathbb{C})$ for all sections $s, t \in C^\infty(M, E)$.

Every complex vector bundle has a C^∞ inner product. This is proved just as for the real case: we start with the usual inner product in $\mathbb{C}^r$,

$$\langle v, w \rangle = \sum_{k=1}^{r} v_k \bar{w}_k \qquad v, w \in \mathbb{C}^r,$$

then make use of local trivializations $E|_{U_i}$ of E to get a C^∞ inner product on $E|_{U_i}$ and finally piece them together with a partition of unity.

A *Hermitian bundle* is a complex vector bundle E with a given C^∞ inner product. The reason that we will need Hermitian bundles is the following. Given a measure $d\mu$ on M (see §6.5), one can define an inner product on $C_0^\infty(M, E)$, the set of smooth sections of E with compact support, by

$$(s, t)_{L_2(M,E)} = \int_M (s(x), t(x))_{E_x} \, d\mu(x), \qquad s, t \in C_0^\infty(M, E)$$

Then $L_2(M, E)$ is the completion of $C_0^\infty(M, E)$ with respect to this inner product.

The Gram-Schmidt orthonormalization procedure can be applied in Hermitian bundles. Let $s_1, \ldots, s_r \in C^\infty(U, E)$ be a local basis for E on the open set U. Since $s_1(x), \ldots, s_r(x) \in E_x$ are linearly independent at each $x \in U$, we can orthonormalize them in the usual manner: we define $u_1 = s_1/\|s_1\|$ and if $u_1, \ldots, u_{j-1}$ have already been defined we let

$$u_j' = s_j - \sum_{i=1}^{j-1} (u_i, s_i) s_i, \qquad u_j = u_j'/\|u_j'\| \tag{37}$$

Then for each $j = 1, \ldots, r$ and each $x \in U$, the vectors $u_1(x), \ldots, u_j(x) \in E_x$ are orthonormal and span the same subspace as do $s_1(x), \ldots, s_j(x)$. In particular,

$$u_1(x), \ldots, u_r(x) \qquad \text{is an orthonormal basis for } E_x,$$

for all $x \in U$. Moreover, it is immediate from (37) and the assumed C^∞

property of (36) that the sections $u_1, \ldots, u_r$ are smooth, i.e. belong to $C^\infty(U, E)$.

In this way we have essentially proved the following lemma.

Lemma 5.66 *For any Hermitian bundle E over M and any open set $U \subset M$ such that $E|_U$ is trivial, there exists a trivialization*

$$\beta: E|_U \to U \times \mathbb{C}^r$$

such that for all $x \in U$ we have $(v, w)_{E_x} = (\beta v, \beta w)_{\mathbb{C}^r}$ for all $v, w \in E_x$.

Proof Let $u_1, \ldots, u_r \in \mathbb{C}^\infty(U, E)$ be the sections that we constructed above by the Gram-Schmidt process, and consider the map $U \times \mathbb{C}^r \to E|_U$ defined by

$$(x, c_1, \ldots, c_r) \mapsto \sum_{j=1}^{r} c_j \cdot u_j(x).$$

This map is a vector bundle homomorphism and it is bijective, hence by Lemma 5.47 it has a smooth inverse. Then we let β be this inverse; that β has the required property is due to the fact that $u_1(x), \ldots, u_r(x)$ is an orthonormal basis of E_x for each $x \in U$.

Exercises

1. Let E be a real vector bundle over M with Riemannian metric $g_x = \langle\, , \rangle_x$.

(a) The unit ball, $B(E_x)$, in the fibre over x is the set of $v \in E_x$ such that $|v|_x < 1$. Show that the unit ball bundle

$$B(E) = \bigcup_x B(E_x)$$

is an open subset of E.

(b) The unit sphere, $S(E_x)$, in the fibre over x is the set of $v \in E_x$ such that $|v|_x = 1$. Show that the unit sphere bundle

$$S(E) = \bigcup_x S(E_x)$$

is a submanifold of E of codimension 1.

Hint: If U is an open set in M such that $E|_U$ is trivial, use the Submersion Theorem 5.25 to show that $S(E)|_U$ is a submanifold of $E|_U$.

2. Let E be a real vector bundle over M with Riemannian metric $g_x = \langle\, , \rangle_x$. Show that the map $v \mapsto v/(1 + |v|_x^2)^{1/2}$ is a diffeomorphism of E onto its unit ball bundle $B(E)$. (It is not hard to find an explicit formula for the inverse.)

3. Let E be a vector bundle over M with a given Riemannian metric. Show that the map $E \to E^*$ defined by the index-lowering operator $E_x \to E_x^*$ on each fibre is a vector bundle isomorphism.

4. Let E be a vector bundle over M with a given Riemannian metric. Let F be a sub-bundle of E and set

$$F^{\perp} = \bigcup_{x \in M} (F_x)^{\perp},$$

where $(F_x)^{\perp}$ is the orthogonal space of F_x in E_x. Show that $F^{\perp}$ is a sub-bundle of E, and $E = F \oplus F^{\perp}$.

5. Let F and V sub-bundles of E such that $E = F \oplus V$. Show that there exists a Riemannian metric on E such that $V = F^{\perp}$.

6. Let $Y \subset M$ be a submanifold, and let the tangent bundle TM be given some Riemannian metric. Show that the normal bundle $\nu(Y)$ (see §5.7, Exercise 1) is isomorphic to the sub-bundle $(TY)^{\perp} = \bigcup_{y \in Y} (T_y Y)^{\perp}$.

5.11 Manifolds with boundary

Let $\bar{\mathbb{R}}^n_+$ denote the upper half-space $\{x \in \mathbb{R}^n \mid x_n \geqslant 0\}$ with the topology induced from $\mathbb{R}^n$, that is, $U \subset \bar{\mathbb{R}}^n_+$ is open if and only if $U = U_1 \cap \bar{\mathbb{R}}^n_+$ for some open set U_1 in $\mathbb{R}^n$. We call $\operatorname{Int} U = U \cap \{x \in \mathbb{R}^n \mid x_n > 0\}$ the *interior* of U and

$$\partial U = U \cap \{x \in \mathbb{R}^n \mid x_n = 0\}$$

the *boundary* of U (this is not the topological boundary of U in $\mathbb{R}^n$, only the part which meets $\partial \bar{\mathbb{R}}^n_+ = \mathbb{R}^{n-1} \times 0$).

A locally Euclidean space with boundary is obtained by piecing together sets which are homeomorphic to open sets in $\mathbb{R}^n$ or in $\bar{\mathbb{R}}^n_+$. To carry this definition over to manifolds we need a definition of local smoothness in $\bar{\mathbb{R}}^n_+$, to be used for overlap maps of charts.

Definition 5.67 *Let $U \subset \bar{\mathbb{R}}^n_+$ be an open set. A map $f: U \to \mathbb{R}^k$ is called smooth if for each $x \in U$ there exists an open neighbourhood U_1 of x in $\mathbb{R}^n$ and a smooth map $f_1: U_1 \to \mathbb{R}^k$ such that*

$$f_1|_{U \cap U_1} = f|_{U \cap U_1}$$

For each $x \in U$, we define the derivative of f to be

$$Df(x) = Df_1(x)$$

(Of course, if $\partial U = \varnothing$ the present definition of smoothness is the same as the one used previously.)

We must show that the definition of Df is independent of the choice of f_1. This is obvious if $x \in \operatorname{Int} U$, while if $x \in \partial U$ it amounts to showing that if $\psi: U_1 \to \mathbb{R}^k$ is a smooth map with U_1 open in $\mathbb{R}^n$ such that $\psi|_{U_1 \cap \bar{\mathbb{R}}^n_+} = 0$, then $D\psi(x) = 0$ for all $x \in U_1 \cap \partial \bar{\mathbb{R}}^n_+$. But this follows by taking a sequence of points $x_n \in \operatorname{Int}(U_1 \cap \bar{\mathbb{R}}^n_+)$ such that $x_n \to x$.

A map $f: U \to V$ is a *diffeomorphism* between open sets U and V in $\bar{\mathbb{R}}^n_+$ if there is a smooth map $g: V \to U$ which is an inverse of f.

It is clear that the open sets in half-spaces together with the smooth maps between them form a category in the sense that the composite of smooth maps is associative when defined, and for any open set U in $\bar{\mathbb{R}}^n_+$ the identity map $\iota: U \to U$ is smooth. The diffeomorphisms are the isomorphisms in this category.

Lemma 5.68 *Let U be open in $\bar{\mathbb{R}}^n_+$ and $\varphi: U \to \bar{\mathbb{R}}^n_+$ a smooth map. Assume that for some $x_0 \in U$, we have $\varphi(x_0) \in \partial\bar{\mathbb{R}}^n_+$. If we write $\varphi = (\varphi', \varphi_n)$ where $\varphi': U \to \mathbb{R}^{n-1}$ and $\varphi_n: U \to \mathbb{R}_+$ then the derivative of φ at x_0 has the form*

$$D\varphi(x_0) = \begin{pmatrix} D'\varphi'(x_0) & * \\ 0\cdots 0 & D_n\varphi_n(x_0) \end{pmatrix}, \qquad \text{where } D_n\varphi_n(x_0) \geqslant 0, \qquad (38)$$

and D' denotes the derivative in the first $n-1$ variables and $D_n = \partial/\partial x_n$. Further, if $x_0 \in \operatorname{Int} U$ then also $D_n\varphi_n(x_0) = 0$. (Hence $D\varphi(x_0)$ is not invertible.)

Proof Since φ maps into $\bar{\mathbb{R}}^n_+$ we have $\varphi_n(x) \geqslant 0$ for all $x \in U$. By hypothesis $\varphi_n(x_0) = 0$, so the partial derivatives of φ_n are

$$\begin{aligned} D_j\varphi_n(x_0) &= \lim_{h\to 0} (\varphi_n(x_0 + he_j) - \varphi_n(x_0))/h, \qquad j = 1, \ldots, n \\ &= \lim_{h \to 0} \varphi_n(x_0 + he_j)/h. \end{aligned}$$

Now, if $x_0 \in \operatorname{Int} U$ we can let $h \to 0^+$ and $h \to 0^-$ from which it follows that $D_j\varphi_n(x_0) \geqslant 0$ and $D_j\varphi_n(x_0) \leqslant 0$, whence $D_j\varphi_n(x_0) = 0$ for all $j = 1, \ldots, n$. If we have $x_0 \in \partial U$ then this holds only for $j = 1, \ldots, n-1$, and we have $D_n\varphi_n(x_0) \geqslant 0$ by letting $h \to 0^+$ only.

Remark The second statement in the lemma says that if φ preserves upper half-spaces and maps an interior point to the boundary then the derivative must be zero in the normal direction.

The derivative operator D satisfies a functorial property, namely, if $\varphi: U \to V$ and $f: V \to W$ are smooth then

$$D(f \circ \varphi)(x) = Df(\varphi(x)) \circ D\varphi(x), \qquad x \in U \qquad (39)$$

For $x \in \operatorname{Int} U$ this formula is just the chain rule, while if $x \in \partial U$ it follows by taking a limit $x_n \to x$ where $x_n \in \operatorname{Int} U$.

Proposition 5.69 *Let U be open in $\bar{\mathbb{R}}^n_+$, V open in $\bar{\mathbb{R}}^n_+$ and $\varphi: U \to V$ a diffeomorphism. Then φ restricts to diffeomorphisms $\operatorname{Int} U \to \operatorname{Int} V$ and $\partial U \to \partial V$ (between open sets in $\mathbb{R}^n$ and between open sets in $\mathbb{R}^{n-1}$, respectively).*

Proof Since φ is a diffeomorphism, $D\varphi(x)$ is invertible for all $x \in U$ (this follows from (39) with $f = \varphi^{-1}$). If $x \in \operatorname{Int} U$, then $\varphi(x) \in \operatorname{Int} V$ by Lemma 5.68. Hence $\varphi(\operatorname{Int} U) \subset \operatorname{Int} V$. By considering the inverse $\varphi^{-1}: \operatorname{Int} V \to \operatorname{Int} U$ it also follows that $\varphi(\operatorname{Int} U) \supset \operatorname{Int} V$. Hence $\varphi(\operatorname{Int} U) = \operatorname{Int} V$. But then $\varphi(\partial U) = \partial V$ as well. Thus φ restricts to smooth maps $\operatorname{Int} U \to \operatorname{Int} V$ and

$\partial U \to \partial V$. Since the same is true of the inverse φ^{-1}, it follows that these restrictions are diffeomorphisms.

This proposition shows that the boundary and interior of an open set in $\bar{\mathbb{R}}^n_+$ are differentiable invariants, so we can define a manifold with boundary just as in Definition 5.11 except that the chart domains are now homeomorphic to open sets in $\mathbb{R}^n$ or in $\bar{\mathbb{R}}^n_+$ (and the smoothness of the overlap maps in Definition 5.11(ii) is understood in the sense of Definition 5.67). The *boundary* and *interior* of M are defined to be

$$\partial M = \bigcup \{\kappa^{-1}(\partial(\kappa(U))); (U, \kappa) \text{ is a chart on } M\}$$

and

$$\text{Int } M = \bigcup \{\kappa^{-1}(\text{Int}(\kappa(U))); (U, \kappa) \text{ is a chart on } M\},$$

respectively. In other words, $x \in \partial M$ if and only if $\kappa(x) \in \partial(\kappa(U))$ for some (and hence all) chart (U, κ) at x, and similarly for $x \in \text{Int } M$ (see Fig. 5).

From now on, we use the term "manifold" to refer to the extended definition. If M is a manifold in the original sense then $\partial M = \varnothing$, and we say that M is a manifold without boundary.

The next theorem follows at once from Proposition 5.69.

Theorem 5.70 *If M is a manifold of dimension n, then its interior* Int M *and its boundary ∂M are manifolds without boundary of dimensions n and $n - 1$, respectively. If $\varphi: M \to N$ is a diffeomorphism, N being another manifold, then φ restricts to diffeomorphisms* Int $M \to$ Int N *and* $\partial M \to \partial N$.

Most of the concepts introduced previously go over without modification to manifolds with boundary. But there are some differences. For example, products $M \times N$ do not exist in this category; the product of two manifolds with boundary would be a "manifold with corner" which we do not wish to introduce here. This is the reason why in the definition of a smooth homotopy in §5.8 we required the interval I to be open. If M is a manifold with boundary ∂M, then $M \times I$ is also a manifold with boundary $\partial M \times I$, and the homotopy results hold true as before (with the same proofs).

The smoothness of a map $f: M \to N$ is defined just as in Definition 5.13, keeping in mind that the smoothness of local representatives at boundary points now refers to Definition 5.67. A map $c: I \to M$, where I is an open, half-open or closed interval in $\mathbb{R}$, is said to be a curve at m if it is smooth, $0 \in I$ and $c(0) = m$. Two curves c_1 and c_2 at m are tangent at m if $(\kappa \circ c_1)'(0) = (\kappa \circ c_2)'(0)$ for some (hence all) chart (U, κ) at m.

As before, a tangent vector at $m \in M$ is an equivalence class $[c]_m$ of curves c at m (two curves at m being equivalent if they are tangent at m). The tangent space

$$T_m M = \{[c]_m; c \text{ is a curve at } m\}$$

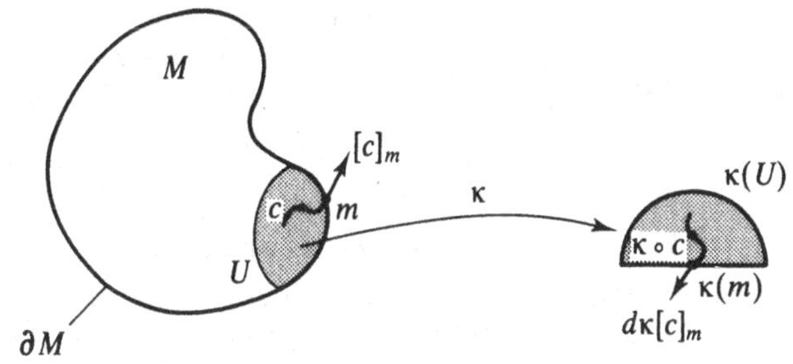

Fig. 5. An outward-pointing tangent vector at a point on the boundary.

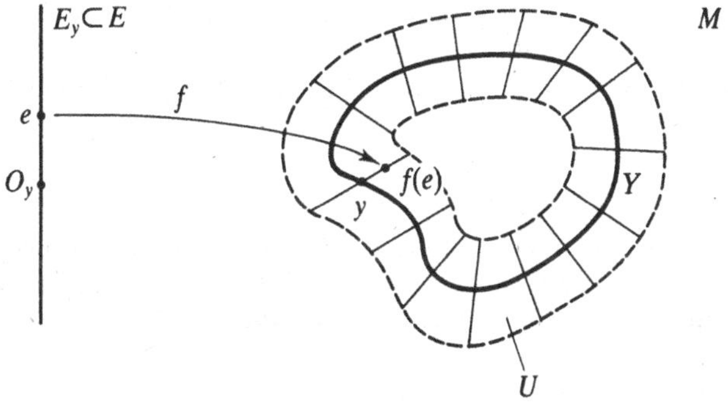

Fig. 6. A partial tubular neighbourhood.

is once again an n-dimensional vector space, even for points $m \in \partial M$ (see Fig. 5), and the tangent bundle $TM = \bigcup_m T_m M$ is a vector bundle over M.

For every $m \in \partial M$, the subspace $T_m(\partial M)$ of vectors tangent to ∂M has codimension one in $T_m M$. Hence $T_m M$ is the union of two half-spaces containing $T_m(\partial M)$. An *outward-pointing* vector $v = [c]_m \in T_m M$ is one for which in any chart (U, κ) on M at m the representative of v is a vector $\tilde{v} = (\kappa \circ c)'(0) \in \mathbb{R}^n$ with $\tilde{v}_n < 0$. Similarly, v is *inward-pointing* if $\tilde{v}_n > 0$. The choice of a chart does not affect the property of a vector being outward- or inward-pointing. To see this suppose that κ_1 is another chart, then we have by the chain rule

$$(\kappa_1 \circ c)'(0) = D(\kappa_1 \circ \kappa^{-1})(\kappa(m)) \cdot \tilde{v},$$

and by Lemma 5.68 the matrix $D(\kappa_1 \circ \kappa^{-1})(\kappa(m))$ has the form (38), i.e. $D_j(\kappa_1 \circ \kappa^{-1})(\kappa(m)) = 0$ for $j = 1, \ldots, n-1$ and $\geqslant 0$ if $j = n$. But since $\kappa_1 \circ \kappa^{-1}$ is a diffeomorphism we must in fact have $D_n(\kappa_1 \circ \kappa^{-1})(\kappa(m)) > 0$, and it follows that $(\kappa_1 \circ c)'(0)$ is outward (inward) pointing if $\tilde{v} = (\kappa \circ c)'(0)$ is outward (inward) pointing.

Let $m \in \partial M$ and let c be a curve at m. In Fig. 5 we have drawn a curve c at m with domain of c being the half-open interval $I = (-\delta, 0]$; in this case the tangent vector $[c]_m \in T_m M$ is outward-pointing at m. Similarly, $[c]_m$ is inward-pointing at m if $I = [0, \delta)$. Finally, if the domain is an *open* interval

$I = (-\delta, \delta)$ then $[c]_m$ is tangent to ∂M, that is, $[c]_m \in T_m(\partial M)$.

Having defined the tangent bundle, all our definitions and constructions given in previous sections carry over directly to manifolds with boundary, with the same proofs, as the reader can easily verify.

5.12 Tubular neighbourhoods and collars

Let $Y \subset M$ be a submanifold, and assume that Y and M are manifolds without boundary. A *tubular neighbourhood* of Y in M is a pair (E, f), where E is a vector bundle over Y with projection $p: E \to Y$, and f is a diffeomorphism

$$f: E \to U$$

of E onto an open set U in M containing Y, such that f is the identity on Y (where Y is identified with the zero section of E). If we denote the elements of E as pairs (y, v) where $v \in E_y$, the latter condition on f can be written in the form

$$f(y, 0) = y \qquad (y \in Y)$$

We call f the *tubular map*.

A slightly more general concept is that of a *partial tubular neighbourhood* of Y. This means a triple (E, Z, f) where E is a vector bundle over Y, Z is an open neighbourhood of the zero section in E, and

$$f: Z \to U$$

is a diffeomorphism of Z onto an open set U in M containing Y such that $f(y, 0) = y$ for all $y \in Y$. See Fig. 6.

Proposition 5.71 *Let $p: E \to Y$ be a vector bundle over Y and let $Z \subset E$ be an open neighbourhood of the zero section. Then there exists a diffeomorphism*

$$h: E \to Z_1$$

of E onto an open subset Z_1 of Z containing the zero section such that:

(i) h preserves the fibres of E, that is, $p \circ h = p$.
(ii) h is the identity on the zero section, that is, $h(y, 0) = (y, 0)$ for all $y \in Y$.

Proof Choose a Riemannian metric $\langle\,,\rangle_x$ on E. Given a positive function $\sigma \in C^\infty(Y)$, let $E(\sigma)$ denote the union of the open balls in E_x centred at the origin of radius $\sigma(x)$,

$$E(\sigma) = \bigcup_{x \in Y} \{v;\, v \in E_x, |v|_x < \sigma(x)\}.$$

We now construct a function σ such that $E(\sigma) \subset Z$. For each $x \in Y$, there exists an open neighbourhood V_x and a number $a_x > 0$ such that the vectors in $E|_{V_x}$ which are of length $< a_x$ lie in Z. Choose a partition of unity $\{(U_i, \rho_i)\}$ such that each U_i is contained in some $V_{x(i)}$, and let

$$\sigma = \sum a_{x(i)} \rho_i$$

Note that $\sigma(y) > 0$ for all $y \in Y$. Now let $v \in E_y$ such that $|v|_y < \sigma(y)$. Note that $\sigma(y) < a_{x(j)}$ where $a_{x(j)}$ denotes the maximum among the numbers $a_{x(i)}$ such that $\rho_i(y) \neq 0$ (there are only finitely many such indices i). Hence $|v|_y < a_{x(j)}$ and, since $y \in V_{x(j)}$, it follows that $v \in Z$. Thus $E(\sigma) \subset Z$.

In view of the diffeomorphism of E onto its unit ball bundle $B(E)$ given in §5.10, Exercise 2, we obtain a diffeomorphism of E onto $E(\sigma)$ given by

$$v \mapsto \sigma(p(v)) \cdot \frac{v}{(1 + |v|^2)^{1/2}}.$$

This completes the proof of the proposition since we can take $Z_1 = E(\sigma)$.

Corollary 5.72 *Any partial tubular neighbourhood (E, Z, f) of Y in M contains a tubular neighbourhood in the following sense: there is a tubular neighbourhood (E, g) of Y in M such that $g(E) \subset f(Z)$.*

Proof Let $h\colon E \to Z_1$ be a diffeomorphism having the properties as stated in the proposition above. Then let $g = f \circ h$.

Thus it suffices to prove the existence of a partial tubular neighbourhood. We prove existence only for the case when Y has codimension 1 in M and there exists a vector field $X = \mathscr{X}(M)$ such that $X(y) \notin T_y Y$ for all $y \in Y$. The basic idea of the proof is to use the integral curves of X starting at the points of the submanifold Y to define the tubular neighbourhood.

The existence of X is equivalent to assuming that the normal bundle of TY in $TM|_Y$ is trivial (see §5.6, Exercise 3).

It is true that *any* submanifold of M has a tubular neighbourhood, but we will have no occasion to use the more general result. See [La 2] or [Hi] for the proof in general.

Theorem 5.73 *Let Y be a submanifold of M of codimension 1 such that $\partial Y = \partial M = \varnothing$. Then there exists a tubular neighbourhood of Y in M.*

Proof As mentioned above we prove the theorem only in the case when there exists a vector field $X \in \mathscr{X}(M)$ such that $X(y) \notin T_y Y$ for all $y \in Y$. Consider the integral curves c_m of X through the points $m \in M$. The flow F of X,

$$F\colon (m, t) \mapsto c_m(t),$$

is a smooth map of $\mathscr{D}$ into M, where $\mathscr{D}$ is an open set in $M \times \mathbb{R}$ that contains $M \times 0$ (see §5.5). Now consider the restriction $F|_{\mathscr{D} \cap (Y \times \mathbb{R})}$ of the flow, in other words, the flow determined by the integral curves of X that begin at the points of Y. We claim that for some open set Z in $Y \times \mathbb{R}$ such that $Z \supset Y \times 0$, *the flow F is a diffeomorphism of Z onto an open set in M.* Since $F(y, 0) = y$ for all $y \in Y$ this will prove the existence of a partial tubular neighbourhood of Y, namely

$$(Y \times \mathbb{R}, Z, F|_Z).$$

First we show that $F|_{Y\times\mathbb{R}}$ is a local diffeomorphism at each $(y, 0) \in Y \times 0$. Let (U, κ) be a chart at y with the submanifold property, i.e. $\kappa(U \cap Y) = \kappa(U) \cap (\mathbb{R}^{n-1} \times 0)$. The local representative of X is $T\kappa \circ X \circ \kappa^{-1}$ and has the form $x \mapsto (x, \tilde{X}(x))$ where $\tilde{X}: \kappa(U) \to \mathbb{R}^n$ is a vector function. The flow $\tilde{F}$ of $\tilde{X}$ is a smooth map from an open set in $\kappa(U) \times \mathbb{R}$ to $\kappa(U)$ and it satisfies the equations

$$\frac{d}{dt}\tilde{F}(x, t) = \tilde{X}(F(x, t)), \qquad \tilde{F}(x, 0) = x \tag{40}$$

$\tilde{F}$ is just the local representative of F, i.e. $\tilde{F}(x, t) = \kappa(F(\kappa^{-1}(x), t))$. Thus, it suffices to show that the local representative $\tilde{F}|_{\kappa(U\cap Y)\times\mathbb{R}}$ is a local diffeomorphism at each point $((x', 0), 0) \in \kappa(U \cap Y) \times 0$. The derivative of $\tilde{F}|_{\kappa(U\cap Y)\times\mathbb{R}}$ as a map from an open set in $\mathbb{R}^{n-1} \times \mathbb{R}$ to $\mathbb{R}^n$ is

$$(D_{x'}\tilde{F}, D_t\tilde{F}) \tag{41}$$

and at the point $((x', 0), 0)$ these partial derivatives are equal to (see (40))

$$D_{x'}\tilde{F} = \begin{pmatrix} I_{n-1} \\ 0 \end{pmatrix}, \qquad D_t\tilde{F} = \tilde{X}(x', 0)$$

As X is not tangent to Y, we have $\tilde{X}^n(x', 0) \neq 0$, hence the derivative (41) is an invertible map at the point $((x', 0), 0)$. By virtue of the Inverse Function Theorem we obtain the following result (after transporting $\tilde{F}$ back to the manifold Y by means of the coordinate map):

> for each $y \in \partial\Omega$, there is an open set $Z \subset Y \times \mathbb{R}$ containing $(y, 0)$ and an open set U in M containing y such that $F|_Z: Z \to U$ is a diffeomorphism.

By shrinking U, if necessary, we can assume the additional property: if $x \in U \cap Y$ then $(x, 0) \in Z$.

It remains to show how these local inverses can be pieced together to form a global inverse of F defined on some neighbourhood of Y in M. We can find a locally finite open covering of Y by open sets U_i in M such that for each i there are smooth maps

$$f_i: Z_i \to U_i \quad \text{and} \quad g_i: U_i \to Z_i$$

where Z_i is an open set in $Y \times \mathbb{R}$ and $f_i = F|_{Z_i}$, g_i and f_i are inverses of each other, and f_i and g_i are the identity on the points of Y in their respective domains. (Y is viewed both as a subset of M and as the zero section in $Y \times \mathbb{R}$.) The last condition on g_i implies that if $x \in U_i \cap Y$ then $(x, 0) \in Z_i$. Now shrink $\{U_i\}$ to obtain a cover $\{V_i\}$ of Y by open sets of M such that $\bar{V}_i \subset U_i$ (by Lemma 5.44), and let $V = \bigcup V_i$. Let

$$W = \{x \in V;\ \text{if } x \in \bar{V}_i \cap \bar{V}_j, \text{ then } g_i(x) = g_j(x)\}$$

Then W certainly contains Y because g_i is the identity on $U_i \cap Y$. We claim that W is a neighbourhood of Y in M, i.e. contains an open set in M containing Y.

Let $y \in Y$. There exists an open neighbourhood W_y of y in M which meets only a finite number of $\bar{V}_i$, say $\bar{V}_{i_1}, \ldots, \bar{V}_{i_r}$. We can assume that y belongs to each one of these sets, for if $y \notin \bar{V}_{i_k}$, simply replace W_y by its intersection with the complement $\complement \bar{V}_{i_k}$ and repeat for each such k. Next, replacing W_y by its intersection with $U_{i_1}, \ldots, U_{i_r}$, we can assume that

$$y \in W_y \subset U_{i_1} \cap \cdots \cap U_{i_r}$$

Shrinking W_y further, if necessary, we find that $g_{i_1}, \ldots, g_{i_r}$ must agree on W_y since the respective inverses $f_{i_1}, \ldots, f_{i_r}$ are restrictions of F. Indeed, we have $g_{i_j}(y) = (y, 0) \in Z_{i_j}$ for all $j = 1, \ldots, r$ and hence, if W_y is chosen small enough, the image

$$g_{i_j}(W_y) \subset Z_{i_1} \cap \cdots \cap Z_{i_r} \qquad \text{for all } j = 1, \ldots, r$$

Since $F(g_{i_j}(x)) = x$ for all $x \in W_y$ and the restriction of F to $Z_{i_1} \cap \cdots \cap Z_{i_r}$ is one-to-one it follows that $g_{i_1}(x) = \cdots = g_{i_r}(x)$ when $x \in W_y$. Since W_y meets $\bar{V}_i$ when $i = i_1, \ldots, i_r$, but not for any other value of i, it follows that $W_y \subset W$.

Let W' be the union of the W_y's for each $y \in Y$, and Z the union of the open sets $g_i(W' \cap V_i)$. Note that $W' \supset Y$ and $Z \supset Y \times 0$. Since $W' \subset W$ we can define a smooth map

$$G\colon W' \to Z$$

by taking G equal to g_i on $W' \cap V_i$. Also $F(G(x)) = x$ when $x \in W'$, and if $u = g_i(x)$ for some $x \in W' \cap U_i$ then $G(F(u)) = G(x) = g_i(x) = u$. Hence G and $F|_Z\colon Z \to W'$ are inverses of each other.

Collar of a boundary

The boundary ∂M of a manifold cannot have a tubular neighbourhood in M. However, it has a "half-tubular" neighbourhood called a collar. A *collar* of ∂M is a diffeomorphism $f\colon \partial M \times [0, \infty) \to V$ onto an open set V in M containing ∂M such that $f(y, 0) = y$ for all $y \in \partial M$.

In the proof of the following theorem we make the assumption that M is a submanifold of a manifold $\hat{M}$ *without boundary*. This is done so that we can easily apply Theorem 5.73.

The existence of $\hat{M}$ is no restriction for us since later in the book we will assume that $M = \bar{\Omega}$ where Ω is an open set in $\mathbb{R}^n$ with smooth boundary; in this case we can take $\hat{M} = \mathbb{R}^n$. We leave it to the interested reader to work through the details of the proof of Theorem 5.74 without the *a priori* assumption that $\hat{M}$ exists.

Theorem 5.74 *Let M be a manifold with boundary. Then ∂M has a collar.*

Proof Assume that M is embedded as a submanifold of a manifold, $\hat{M}$, without boundary. Choose a vector field $X\colon M \to TM$ such that $X(y)$ is inward-pointing for all $y \in \partial M$ (constructed via a partition of unity). Then extend X to a vector field on $\hat{M}$. This is possible since local extensions exist by definition, so Lemma 5.58 guarantees existence of a global extension.

Since ∂M is a submanifold of $\hat{M}$, and both are manifolds without boundary, we can use Theorem 5.73. It follows that there is a tubular neighbourhood (E, f) of ∂M in $\hat{M}$. As in the proof of Theorem 5.73, E is a trivial vector bundle $\partial M \times \mathbb{R}$. Thus, we have a diffeomorphism

$$F: \partial M \times \mathbb{R} \to \hat{U}$$

such that $F(y, 0) = y$ for all $y \in \partial M$, and $\hat{U}$ is an open set in $\hat{M}$. It follows that $F(\partial M \times [0, \infty))$ is an open set in M, and, since M is a submanifold of $\hat{M}$, the restriction

$$f = F|_{\partial M \times [0, \infty)}$$

is a diffeomorphism of $\partial M \times [0, \infty)$ onto $F(\partial M \times [0, \infty))$. The map f is a collar of ∂M.

If we had chosen X to be *outward*-pointing at each point of ∂M the collar would have been a diffeomorphism $f: \partial M \times (-\infty, 0] \to U$.

Theorem 5.75 *Let M be a manifold with boundary ($\partial M \neq \varnothing$) embedded as a submanifold of a manifold $\hat{M}$ without boundary ($\partial \hat{M} = \varnothing$). Then there exists an open set U in $\hat{M}$ containing ∂M, and a diffeomorphism $f: \partial M \times (-1, 1) \to U$ such that*

$$f(y, 0) = y \qquad \textit{for all } y \in \partial M,$$

$f(\partial M \times (-1, 0])$ is an open set in M containing ∂M, and

$$f(\partial M \times (0, 1)) \cap M = \varnothing.$$

Proof In the proof of Theorem 5.74, take the vector field X to be outward-pointing at each point of ∂M and then just compose F with a diffeomorphism $\mathbb{R} \to (-1, 1)$, for instance, $x \mapsto x/(1 + x^2)^{1/2}$.

Later on when we use this theorem, the diffeomorphism f will be fixed once and for all. Thus, points in M near the boundary ∂M can be identified with (y, t), $y \in \partial M$, $-1 < t \leqslant 0$ and the points in $\hat{M} \backslash M$ near the boundary are identified with (y, t), $y \in \partial M$, $0 < t < 1$. Also ∂M is identified with $\partial M \times 0$, of course. We will usually refer to $\partial M \times (-1, 1)$ itself as the *tubular neighbourhood* and $\partial M \times (-1, 0]$ as the *collar*.

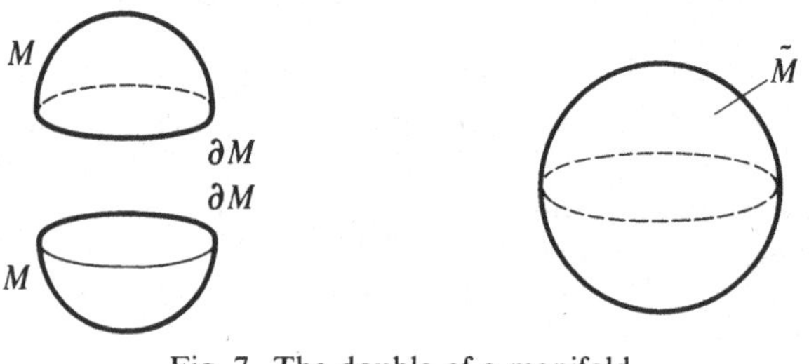

Fig. 7. The double of a manifold.

The boundaryless double of a manifold

Let M be a manifold with $\partial M \neq \varnothing$. Two copies of M can be glued along the boundary ∂M to form a manifold without boundary, called the *double* of M (see Fig. 7). We will need the double of a manifold in Chapter 11.

Let the two copies of M be denoted $M \times 0$ and $M \times 1$. The (disjoint) union $(M \times 0) \cup (M \times 1)$ is given the obvious topology where a set is open if its intersection with $M \times 0$ and with $M \times 1$ is open in $M \times 0$ and in $M \times 1$, respectively. The *double* of M, denoted $\tilde{M}$, is the union of $M \times 0$ and $M \times 1$ with the points $(x, 0)$ and $(x, 1)$ identified whenever $x \in \partial M$. To be precise, we have the equivalence relation on $(M \times 0) \cup (M \times 1)$ defined by

$$(x, i) \sim (y, j) \qquad \text{if either } y = x \in \partial M, \text{ or } y = x \in M \text{ and } i = j$$

and $\tilde{M}$ is the quotient space. We use the notation $[(x, i)]$ for the points of $\tilde{M}$. If $x \in \operatorname{Int} M$ we write just (x, i) (since there is a unique representative in the equivalence class).

Let $\pi: (M \times 0) \cup (M \times 1) \to \tilde{M}$ be the canonical map $(x, i) \mapsto [(x, i)]$. We give $\tilde{M}$ the quotient topology: a set $V \subset \tilde{M}$ is open if $\pi^{-1}(V)$ is open in $(M \times 0) \cup (M \times 1)$.

Lemma 5.76 *The double $\tilde{M}$ is a locally Euclidean Hausdorff space without boundary.*

Proof The graph, Γ, of the equivalence relation is equal to the union of three closed sets: the diagonal in $(M \times 0) \cup (M \times 0)$, the diagonal in $(M \times 1) \cup (M \times 1)$ and the set $\{((x, 0), (x, 1)); x \in \partial M\}$. Hence Γ is closed. It is evident that π is an open map, i.e. $\pi^{-1}(\pi(W))$ is open in $(M \times 0) \cup (M \times 1)$ for any open set W in $(M \times 0) \cup (M \times 1)$, whence $\tilde{M}$ is Hausdorff by §5.2, Exercise 7.

For every point $x \in \partial M$ there is an open set $U \ni x$ and a diffeomorphism $\kappa: U \to V$ where V is an open set in the half-space $\bar{\mathbb{R}}^n_+$. Let $\imath: \mathbb{R}^n \to \mathbb{R}^n$ be the reflection in the plane $x_n = 0$, that is, $\imath(x', x_n) = (x', -x_n)$, and then define the map $\tilde{\kappa}: \tilde{U} \to \tilde{V}$ by

$$\tilde{\kappa}(\tilde{x}) = \begin{cases} \kappa(x) & \text{if } \tilde{x} = [(x, 0)] \\ \imath \circ \kappa(x) & \text{if } \tilde{x} = [(x, 1)] \end{cases}$$

where $\tilde{U} = \pi((U \times 0) \cup (U \times 1))$ is an open set in $\tilde{M}$ and $\tilde{V} = V \cup \imath(V)$ is an open set in $\mathbb{R}^n$. It is not hard to verify that $\tilde{\kappa}$ is a homeomorphism. Hence $\tilde{M}$ is locally Euclidean, and has no boundary (each point has a neighbourhood homeomorphic to an open set in $\mathbb{R}^n$).

Remark If we define $\varphi: \mathbb{R}^n \to \mathbb{R}^n$ by $\varphi(x', x_n) = (x', x_n^2)$ if $x_n \geqslant 0$ and $\varphi(x', x_n) = (x', -x_n^2)$ if $x_n < 0$, then the restriction of φ to each half-space $\bar{\mathbb{R}}^n_\pm$ is of class C^∞, but φ itself is not C^∞. A chart (U, κ) on M induces a coordinate map $\tilde{\kappa}$ on $\tilde{M}$ as in the proof above, and in the same way the chart $(U, \varphi \circ \kappa)$ on M induces a coordinate map $\varphi \circ \tilde{\kappa}$ on $\tilde{M}$, but *the overlap map* $(\varphi \circ \tilde{\kappa}) \circ \tilde{\kappa}^{-1} = \varphi$ *is not of class* C^∞.

To define a C^∞ structure on $\tilde{M}$ for an arbitrary manifold, therefore, takes a little more work due to the necessity of verifying the smoothness of overlap maps.

In order to define a C^∞ structure on $\tilde{M}$ we use a collar of ∂M in M. The existence of a collar means that we have a diffeomorphism $\alpha\colon W \to \partial M \times [0, 1)$ such that $\alpha(x) = (x, 0)$ for all $x \in \partial M$, where W is an open set in M containing ∂M. This induces a map $\tilde{\alpha}\colon \tilde{W} \to \partial M \times (-1, 1)$ defined by

$$\tilde{\alpha}(\tilde{x}) = \begin{cases} \alpha(x) & \text{if } \tilde{x} = [(x, 0)] \\ \imath(\alpha(x)) & \text{if } \tilde{x} = [(x, 1)] \end{cases}$$

and $\tilde{W} = \pi((W \times 0) \cup (W \times 1))$ is open in $\tilde{M}$.

Theorem 5.77 *Let M be a manifold with $\partial M \neq \varnothing$, and W an open set in M containing ∂M on which there is a diffeomorphism $\alpha\colon W \to \partial M \times [0, 1)$ such that $\alpha(x) = (x, 0)$ for all $x \in \partial M$. Then there is a unique C^∞ structure on the double $\tilde{M}$ such that both inclusions $M \to \tilde{M}$ are C^∞, and such that the map $\tilde{\alpha}\colon \tilde{W} \to \partial M \times (-1, 1)$ induced by α is a diffeomorphism.*

Proof We take as coordinate maps on $\tilde{M}$ the following:

$(U \times 0, \kappa)$ and $(U \times 1, \kappa)$, if (U, κ) is a chart on M with $U \subset \operatorname{Int} M$

$(\tilde{\alpha}^{-1}(U' \times (-1, 1)), (\kappa' \times \mathrm{id}) \circ \tilde{\alpha})$, if (U', κ') is a chart on ∂M

where id denotes the identity map $(-1, 1) \to (-1, 1)$. It is straightforward to verify that the overlap of two such maps is of class C^∞. The domains of these maps cover $\tilde{M}$ so we have (an atlas for) a C^∞ structure on $\tilde{M}$.

Exercises

1. Suppose that $\alpha_i\colon W_i \to \partial M \times (-1, 1)$, $i = 1, 2$ are two diffeomorphisms satisfying the conditions of Theorem 5.77.

(a) Show that the C^∞ structures on $\tilde{M}$ determined by α_1 and α_2 are *diffeomorphic*.

(b) Under what conditions are the two C^∞ structures the same?

2. Let M and N be manifolds and $f\colon M \to N$ and $g\colon M \to N$ continuous maps such that $f(x) = g(x)$ for all $x \in \partial M$. Then there is an induced map $h\colon \tilde{M} \to \tilde{N}$ given by $h([(x, 0)]) = [(f(x), 0)]$ and $h([(x, 1)]) = [(g(x), 1)]$. Show that h is continuous. If f and g are of class C^r, what additional conditions are required in order that h also be of class C^r?

3. Let M be a manifold with $\partial M \neq \varnothing$. Show that there exists a vector field $X \in \mathscr{X}(M)$ that is inward-pointing when restricted to ∂M.

4. Let (X, d) be a compact metric space and $A \subset X$ a closed subset. Let $f\colon X \to S$ be a local homeomorphism such that $f|_A$ is one to one.

(a) Let K be the set of points in $X \times X$ such that $x \neq y$ and $f(x) = f(y)$. Show that K is compact.

(b) Show that there is a neighbourhood U of A such that $f|_U$ is one to one. (Hint: Define $\delta: K \to \mathbb{R}$ by $\delta(x, y) = d(x, A) + d(y, A)$; since K is compact, there is an $\varepsilon > 0$ such that $\delta \geqslant \varepsilon$ on K.)

An open set Ω in $\mathbb{R}^n$ is said to have C^∞ *boundary* if for each $x_0 \in \partial\Omega$ there is a C^∞ function φ defined on an open set W (in $\mathbb{R}^n$) containing x_0 such that

$$\varphi(x_0) = 0, \text{ grad } \varphi(x_0) \neq 0, \text{ and } \Omega \cap W = \{x \in W; \varphi(x) > 0\}$$

5. Let Ω be an open set in $\mathscr{R}^n$ with C^∞ boundary. Show that $\bar{\Omega}$ is a submanifold of $\mathbb{R}^n$, with boundary $\partial\Omega$.

6. Let Ω be an open set in $\mathbb{R}^n$ with C^∞ boundary. Note that if $v: \bar{\Omega} \to \mathbb{R}^n$ is a vector field on $\bar{\Omega}$ then it has flow $F(x, t) = x + t \cdot v(x)$. Suppose now that $\partial\Omega$ is compact, and use Exercise 4 to give a short proof of Theorem 5.73 in this context, i.e. that $\partial\Omega$ has a tubular neighbourhood in $\mathbb{R}^n$.

6
Differential forms

6.1 Differential forms

Before introducing differential forms we need the following result, which is just a restatement of Proposition 5.57 for the functor $\bigwedge^k$. Recall that $\bigwedge^k$ is a smooth functor by virtue of §5.8, Exercise 3.

Proposition 6.1 *The functor $\bigwedge^k$ extends to the category of vector bundles over M. For any vector bundle E over M let $\bigwedge^k E = \bigcup_{x \in M} \bigwedge^k(E_x)$. If $f\colon E \to F$ is a vector bundle homomorphisms define the map $\bigwedge^k f\colon \bigwedge^k E \to \bigwedge^k F$ such that its value on the fibre over x is given by the map $\bigwedge^k(f_x)\colon \bigwedge^k(E_x) \to \bigwedge^k(F_x)$. Then $\bigwedge^k E$ has a natural topology and C^∞ structure making it into a vector bundle with the following properties:*

(a) If $f\colon E \to F$ is a vector bundle homomorphism then $\bigwedge^k(f)$ is a vector bundle homomorphism $\bigwedge^k(E) \to \bigwedge^k(F)$.
(b) $\bigwedge^k(f \circ g) = \bigwedge^k(f) \circ \bigwedge^k(g)$ whenever $f \circ g$ is defined and $\bigwedge^k(id_E) = id_{\bigwedge^k(E)}$.
(c) If $E = M \times \mathbf{E}$ is a trivial vector bundle then $\bigwedge^k(E) = M \times \bigwedge^k(\mathbf{E})$ is a trivial vector bundle.
(d) If $\not{f}\colon Y \to M$ is a smooth map between manifolds then $\bigwedge^k(\not{f}^{-1}E) = \not{f}^{-1}\bigwedge^k(E)$.

If E is a vector bundle over M, there is a natural map $E^{(k)} \xrightarrow{\wedge} \bigwedge^k E$ defined on the fibres over x, which is easily seen to be smooth. Indeed, locally on an open set U in M where $E_{|U}$ is trivial, it is expressed as the map $U \times \mathbf{E}^{(k)} \to U \times \bigwedge^k \mathbf{E}$ given by $(x, v_1, \ldots, v_k) \mapsto (x, v_1 \wedge \cdots \wedge v_k)$, and the map $\mathbf{E}^{(k)} \xrightarrow{\wedge} \bigwedge^k \mathbf{E}$ is C^∞ because it is multilinear.

If $v_1, \ldots, v_k$ is a local basis for E on U, then the sections

$$x \mapsto v_{i_1}(x) \wedge \cdots \wedge v_{i_k}(x), \qquad i_i < \cdots < i_k,$$

form a local basis for $\bigwedge^k E$ on U; these sections are smooth because they are a composition of smooth maps. In view of Corollary 5.49 applied to the bundle $\bigwedge^k E$, any section $s \in C^\infty(U, \bigwedge^k E)$ can therefore be written in the form

$$s(x) = \sum s_{i_1 \cdots i_k}(x) \cdot v_{i_k}(x) \wedge \cdots \wedge v_{i_k}(x)$$

where $s_{i_1 \cdots i_k} \in C^\infty(U)$ and the sum is taken over all indices $i_1 < \cdots < i_k$.

Let M be a manifold and consider the tangent bundle TM. We know from the general considerations in §5.8 that the dual bundle $T^*M = \bigcup_{x\in M} (T_xM)^*$ is a vector bundle over M, which we call the *cotangent bundle*.

Let us write out explicitly what the trivializing maps look like for T^*M. Let $(U, \kappa) = (U, x_1, \ldots, x_n)$ be a chart for M, then $TU = \bigcup_{x\in U} T_xM$ has the trivializing map β: $TU \to U \times \mathbb{R}^n$ defined by $v \mapsto (x, d\kappa(v))$. Now let $\lambda^1, \ldots, \lambda^n$ be the dual basis corresponding to the standard basis in $\mathbb{R}^n$, i.e. λ^i: $\mathbb{R}^n \to \mathbb{R}$ is the projection on the ith coordinate. Since ${}^t(\beta_x)(\lambda^i) = \lambda^i \circ \beta_x = dx_{i|x}$ we see that

$$dx_1, \ldots, dx_n \text{ is a local basis for } T^*M \text{ on } U$$

and is the dual basis corresponding to $\partial/\partial x_1, \ldots, \partial/\partial x_n$. Thus T^*M has the trivializing map $T^*U \to U \times \mathbb{R}^n$ given by

$$\sum_{i=1}^{n} \xi_i \, dx_{i|x} \mapsto (x, \xi_1, \ldots, \xi_n)$$

A C^∞ section of T^*M is called a *covector field* on M. For example, if $f \in C^\infty(M)$ then the differential df can be regarded as a section $M \to T^*M$ defined by $df(x) = df_{|T_xM} \in (T_xM)^*$. It is clear that $df \in C^\infty(M, T^*M)$ because locally in the coordinate chart $(U, x_1, \ldots, x_n)$ we have (see the text following Proposition 5.29)

$$df = \frac{\partial f}{\partial x_1} dx_1 + \cdots + \frac{\partial f}{\partial x_n} dx_n$$

and $\partial f/\partial x_i \in C^\infty(U)$. Thus df is a covector field.

Definition 6.2 *We define the set of differential forms of degree r on M to be*

$$\Omega^r(M) = C^\infty(M, \textstyle\bigwedge^r(T^*M)),$$

the C^∞ sections of the vector bundle $\bigwedge^r(T^*M)$. In other words an r-form $\omega \in \Omega^r(M)$ is a C^∞ map

$$\omega: M \to \textstyle\bigwedge^r(T^*M) \qquad \text{such that} \qquad \omega(x) \in \bigwedge^r(T_x^*M)$$

for each $x \in M$.

Note that $\bigwedge^0(T^*M)$ is the trivial bundle $M \times \mathbb{R}$, so a 0-form is just a C^∞ function on M. Also, the 1-forms are the covector fields since $\bigwedge^1(T^*M) = T^*M$.

We can also define differential forms on any open set by replacing M with U. In view of the comments following Proposition 6.1, if $(U, x_1, \ldots, x_n)$ is a chart for M then $\bigwedge^r(T^*M)$ has the local basis $dx_{i_1} \wedge \cdots \wedge dx_{i_r}$, $i_1 < \cdots < i_r$, on U. Thus a differential k-form on U can be written uniquely in the form

$$\omega(x) = \sum f_{i_1 \cdots i_r}(x) \, dx_{i_1} \wedge \cdots \wedge dx_{i_r} \tag{1}$$

where $f_{i_1 \cdots i_r} \in C^\infty(U)$ and the sum is taken over indices $i_1 < \cdots < i_r$. For short we often write $\omega = \sum f_I \, dx_I$.

By identifying the differential form dx_i with the corresponding form dx_i on the open set $\kappa(U) \subset \mathbb{R}^n$, we see that (1) is the *local representation* of $\omega \in C^\infty(M, \bigwedge^r(T^*M))$ as defined in Exercise 4 of §5.8.

The set $\Omega^r(M)$ of r-forms is a vector space over $\mathbb{R}$, with addition and scalar multiplication defined pointwise, and in fact it is a $C^\infty(M)$-module. The product of differential forms is also defined pointwise: if α and η are forms of degree r and s, respectively, on the manifold M (or any open set) then the *wedge product* $\alpha \wedge \eta$ is defined by

$$(\alpha \wedge \eta)(x) = \alpha(x) \wedge \eta(x) \qquad x \in M$$

Note that $\alpha \wedge \eta$ is smooth, which is clear from the local representations of α and η, whence $\alpha \wedge \eta \in \Omega^{r+s}(M)$. If f is a form of degree 0 (a real-valued function) and α is a form of degree r then

$$f \wedge \alpha = f\alpha,$$

where $f\alpha$ is the r-form on M given by $(f\alpha)(x) = f(x)\cdot\alpha(x)$. The operation $\wedge$ is bilinear over $\mathbb{R}$ and associative since this holds pointwise. In fact $\wedge$ is bilinear over $C^\infty(M)$.

Let $\Omega(M)$ denote the direct sum of $\Omega^r(M)$, $r = 0, 1, \ldots, n$ together with its structure as a module over $C^\infty(M)$, and with multiplication $\wedge$ extended componentwise to $\Omega(M)$. We call $\Omega(M)$ the *algebra of differential forms on* M, and it is an associative algebra over $C^\infty(M)$.

6.2 The exterior derivative d

We begin with the definition of differential forms on open sets in $\mathbb{R}^n$ and define the exterior derivative of such forms. The exterior derivative satisfies certain properties which characterize it uniquely, and make it possible to define this operator on manifolds.

We identify $T_m\mathbb{R}^n \simeq \mathbb{R}^n$ by the canonical isomorphism $[c]_m \mapsto c'(0)$. Then $dx_1, \ldots, dx_n$ is the dual basis in $(\mathbb{R}^n)^*$ corresponding to the standard basis $e_1, \ldots, e_n$ in $\mathbb{R}^n$. The canonical isomorphism $T_m\mathbb{R}^n \simeq \mathbb{R}^n$ means that we can also identify $\bigwedge^r(T_m^*\mathbb{R}^n) \simeq \bigwedge^r((\mathbb{R}^n)^*)$ which we do from now on. Recall from §5.1 that $\bigwedge((\mathbb{R}^n)^*)$ is the algebra over $\mathbb{R}$ generated by $dx_1, \ldots, dx_n$ with the relations

$$dx_i \wedge dx_j = -dx_j \wedge dx_i.$$

As a vector space over $\mathbb{R}$, $\bigwedge((\mathbb{R}^n)^*)$ has basis

$$1,\ dx_i,\ \underset{i<j}{dx_i \wedge dx_j},\ \underset{i<j<k}{dx_i \wedge dx_j \wedge dx_k},\ \ldots,\ dx_1 \wedge \cdots \wedge dx_n$$

Every $\xi \in \bigwedge^r((\mathbb{R}^n)^*)$ can be uniquely written as

$$\xi = \sum \xi_{i_1 \cdots i_r} dx_{i_1} \wedge \cdots \wedge dx_{i_r}$$

where the sum is taken over indices $i_1 < \cdots < i_r$. The real numbers

$\xi_{i_1 \cdots i_r}$ are the coordinates of ξ with respect to the $\binom{n}{r}$-dimensional basis $\{d_{i_1} \wedge \cdots \wedge dx_{i_r};\ i_1 < \cdots < i_r\}$ of $\bigwedge^r((\mathbb{R}^n)^*)$.

The C^∞ differential forms of degree r on an open set $U \subset \mathbb{R}^n$ are the elements of the $C^\infty(U)$-module

$$\Omega^r(U) = C^\infty(U) \underset{\mathbb{R}}{\otimes} \textstyle\bigwedge^r((\mathbb{R}^n)^*) \qquad \text{(tensor product of vector spaces over } \mathbb{R})$$

If ω is such a form it can be written uniquely as

$$\omega = \sum f_{i_1 \cdots i_r} dx_{i_1} \wedge \cdots \wedge dx_{i_r}$$

where the coefficients $f_{i_1 \cdots i_r}$ are C^∞ functions on U. For short we also write $\omega = \sum f_I\, dx_I$ where I is a multi-index $(i_1, \ldots, i_r)$ with $i_1 < \cdots < i_r$. Note that $\Omega^0(U) = C^\infty(U)$, and $\Omega^r(U) = 0$ if $r > n$.

The *wedge product* of two differential forms is defined by

$$\alpha \wedge \eta = \sum f_I g_J\, dx_I \wedge dx_J \in \Omega^{r+s}(U)$$

if $\alpha = \sum f_I\, dx_I \in \Omega^r(U)$ and $\eta = \sum g_J\, dx_J \in \Omega^s(U)$. We have $\alpha \wedge \eta = (-1)^{rs} \eta \wedge \alpha$ and

$$f\alpha \wedge \eta = \alpha \wedge f\eta, \qquad f \in C^\infty(U) \tag{2}$$

Then $\Omega(U) = \Omega^0(U) \oplus \Omega^1(U) \oplus \cdots \oplus \Omega^n(U)$ is an algebra with product $\wedge$ defined in the obvious way (see §5.1).

Remark In the definition of the wedge product $\wedge$ one has to take care of the ordering: I and J are ordered, but $\sum_{I,J}$ is not. For instance, if $\alpha = \sum a_i dx_i \in \Omega^1(U)$ and $\eta = \sum b_j dx_j \in \Omega^1(U)$ then

$$\alpha \wedge \eta = \sum_{i<j} (a_i b_j - a_j b_i)\, dx_i \wedge dx_j.$$

For 0-forms (functions) $f \in \Omega^0(U) = C^\infty(U)$ we define

$$df = \sum \partial f / \partial x_i\, dx_i \tag{3}$$

so that df is a one-form on U, $df \in \Omega^1(U)$. (This agrees with the definition in §5.3.) The operator d can be extended to forms of degree > 1 as we show in the next proposition.

Proposition 6.3 *Let U be open in $\mathbb{R}^n$. The operator d extends uniquely to an operator $d\colon \Omega^r(U) \to \Omega^{r+1}(U)$ for each $r = 0, 1, \ldots, n$, satisfying the following properties*:

(*i*) *d is an anti-derivation with respect to $\wedge$, that is, d is $\mathbb{R}$-linear and*

$$d(\alpha \wedge \eta) = d\alpha \wedge \eta + (-1)^r \alpha \wedge d\eta$$

for $\alpha \in \Omega^r(U)$ and $\eta \in \Omega^s(U)$.

(*ii*) $d^2 = d \circ d = 0.$

Proof For uniqueness, note that if $\alpha = \sum f_I \, dx_I \in \Omega^r(U)$ then property (i) gives

$$d\alpha = \sum df_I \wedge dx_I + \sum f_I \, d(dx_I)$$

By properties (i) and (ii) we have $d(dx_I) = d(dx_{i_1} \wedge \cdots \wedge dx_{i_r}) = 0$, so in view of (3) it follows that

$$d\alpha = \sum \frac{\partial f_I}{\partial x_j} dx_j \wedge dx_I \tag{4}$$

To prove existence, if $\alpha = \sum f_I \, dx_I$ is an r-form, $r \geqslant 1$, we simply define $d\alpha$ by the formula (4) and show that d satisfies (i) and (ii). Let $\eta = \sum g_J \, dx_J$, then

$$\alpha \wedge \eta = \sum_{I,J} f_I g_J \, dx_I \wedge dx_J$$

and, by definition of d and properties of $\wedge$,

$$\begin{aligned} d(\alpha \wedge \eta) &= \sum d(f_I \cdot g_J) \wedge dx_I \wedge dx_J \\ &= \sum (g_J \cdot df_I + f_I \cdot dg_J) \wedge dx_I \wedge dx_J \\ &= \sum df_I \wedge dx_I \wedge \sum g_J \, dx_J + (-1)^r \sum f_I \, dx_I \wedge \sum dg_J \wedge dx_J \\ &= d\alpha \wedge \eta + (-1)^r \alpha \wedge d\eta \end{aligned}$$

Note that in the third equation we used (2) and the fact that $dg_J \wedge dx_I = (-1)^r \, dx_I \wedge dg_J$, since $dx_I = dx_{i_1} \wedge \cdots \wedge dx_{i_r}$ is an r-form.

To verify (ii) we first show that if $f \in C^\infty(U)$ then $d(df) = 0$. This follows since

$$d(df) = d\left(\sum \frac{\partial f}{\partial x_i} dx_i\right) = \sum \frac{\partial^2 f}{\partial x_j \partial x_i} dx_j \wedge dx_i = 0$$

by symmetry of the second partial derivatives and the fact that $dx_j \wedge dx_i = -dx_i \wedge dx_j$. Now, if $\alpha = \sum f_I \, dx_I$ is an r-form, then (ii) follows from property (i) since $f_I \in C^\infty(U)$ and $x_{i_1}, \ldots, x_{i_r} \in C^\infty(U)$ so that $d(df_I) = 0$ and $d(dx_{i_j}) = 0$.

The operator d is called *exterior differentiation*. The uniqueness of d in Proposition 6.3 implies that its definition is independent of the coordinate system. This makes it possible to define the exterior derivative operator on manifolds.

Let $\omega \in \Omega^r(M)$ where M is a manifold. If (U, κ) is a chart on M, then ω has the local representation $\omega_{|(U,\kappa)} = \sum f_I \, dx_{i_1} \wedge \cdots \wedge dx_{i_r}$, where $f_I \in C^\infty(U)$, and we define

$$d\omega|_{(U,\kappa)} = \sum \frac{\partial f_I}{\partial x_i} dx_i \wedge dx_{i_1} \wedge \cdots \wedge dx_{i_r}, \tag{5}$$

the sum being taken over all indices $i = 1, \ldots, n$ and $i_1 < \cdots < i_r$. Now, if (U', κ') is another chart, then, by virtue of the uniqueness stated in Proposition 6.3, we have

$$d\omega|_{(U \cap U', \kappa)} = d\omega|_{(U \cap U', \kappa')}$$

Definition 6.4 *For each $\omega \in \Omega^r(M)$ we define the exterior derivative on M by*

$$d\omega|_U = d\omega|_{(U,\kappa)}$$

for each open set U that is the domain of a chart (U, κ) on M.

As shown above, the exterior derivative $d\omega$ is well-defined (independent of the coordinates κ). The local representations (5) show that $d\omega: M \to \bigwedge^{r+1}(T^*M)$ is a C^∞ section, so that $d\omega \in \Omega^{r+1}(M)$. In this way we have defined an $\mathbb{R}$-linear operator $d: \Omega^r(M) \to \Omega^{r+1}(M)$ for each $r = 0, 1, \ldots, n$.

Theorem 6.5 *For each $r = 0, 1, \ldots, n$ there exists an operator $d: \Omega^r(M) \to \Omega^{r+1}(M)$, satisfying the following properties:*

(*i*) *d is an anti-derivation with respect to $\wedge$, that is, d is $\mathbb{R}$-linear and*

$$d(\alpha \wedge \eta) = d\alpha \wedge \eta + (-1)^r \alpha \wedge d\eta$$

for $\alpha \in \Omega^r(M)$ and $\eta \in \Omega^s(M)$.

(*ii*) $d^2 = d \circ d = 0$

By the manner in which d was defined, no additional proof is required here, since these properties hold in a collection of open sets that cover M.

Since $d: \Omega^r(M) \to \Omega^{r+1}(M)$ is $\mathbb{R}$-linear for $r = 0, 1, \ldots$, it induces a linear mapping, $d: \Omega(M) \to \Omega(M)$, on the direct sums by

$$d(\omega_0 + \cdots + \omega_n) = d\omega_0 + \cdots + d\omega_n \qquad (\omega_r \in \Omega^r(M),\ r = 0, 1, \ldots, n)$$

Let $\varphi: M \to N$ be a smooth map. We would like to define the pull-back of a differential form α on N by the map φ. The tangent of φ is a vector bundle homomorphism, $T\varphi: TM \to TN$, and hence we have a vector bundle homomorphism ${}^tT\varphi: T^*N \to T^*M$ given by the transpose map in each fibre, and then a homomorphism $\bigwedge^r({}^tT\varphi): \bigwedge^r(T^*N) \to \bigwedge^r(T^*M)$ (see Proposition 6.1).

Definition 6.6 *Let $\varphi: M \to N$ be a smooth map, and let $\alpha \in \Omega^r(N)$. We define the pull-back of α by φ as follows:*

$$\varphi^*(\alpha) = \textstyle\bigwedge^r({}^tT\varphi) \circ \alpha \circ \varphi \tag{6}$$

(For the case $r = 0$, we have $\Omega^0(N) = C^\infty(N)$ and $\varphi^*(\alpha) = \alpha \circ \varphi$, coinciding with the definition in §5.5 for the pull-back of C^∞ functions)

Note that $\varphi^*(\alpha)$ is a smooth map $M \to \bigwedge^r(T^*M)$ (it is a composition of smooth maps) and $\varphi^*(\alpha)_y \in \bigwedge^r(T_y^*M)$ for each $y \in M$. Hence the pull-back, $\varphi^*(\alpha)$, is a differential form of degree r on M, i.e. $\varphi^*(\alpha) \in \Omega^r(M)$.

Let us examine what $\varphi^*(\alpha)$ looks like in local coordinates. Let $(U, x_1, \ldots, x_n)$ be a chart for N and $(V, y_1, \ldots, y_p)$ a chart for M such that $\varphi(V) \subset U$. We claim that

$$\varphi^*(\textstyle\sum f_I\, dx_{i_1} \wedge \cdots \wedge dx_{i_r}) = \sum (f_I \circ \varphi)\, d\varphi_{i_1} \wedge \cdots \wedge d\varphi_{i_r} \tag{7}$$

where $\varphi_i = x_i \circ \varphi$ is the ith component of the function $\varphi|_V$.

First, let us show that ${}^t(T_y\varphi)(dx_i) = d\varphi_i|_y$. This follows directly from the definition of the tangent map (Definition 5.20) and the differential of a function:

$$\begin{aligned} {}^t(T_y\varphi)(dx_i)[c]_y &= dx_i\, T_y\varphi[c]_y \qquad \text{where } [c]_y \in T_yM \\ &= dx_i[\varphi \circ c]_{\varphi(y)} \\ &= (x_i \circ \varphi \circ c)'\,(0) \\ &= d\varphi_i[c]_y \end{aligned}$$

Now, to verify (7), it suffices to consider a monomial term $\alpha = f_I\, dx_{i_1} \wedge \cdots \wedge dx_{i_r}$. For $y \in V$, we have by definition (6)

$$\begin{aligned} \varphi^*(\alpha)_y &= \textstyle\bigwedge^r({}^tT\varphi)(f_I(\varphi(y))\cdot dx_{i_1}|_{\varphi(y)} \wedge \cdots \wedge dx_{i_r}|_{\varphi(y)}) \\ &= f_I(\varphi(y))\cdot {}^t(T_y\varphi)(dx_{i_1}) \wedge \cdots \wedge {}^t(T_y\varphi)(dx_{i_r}) \end{aligned}$$

and we get the expression (7) since ${}^t(T_y\varphi)(dx_i) = d\varphi_i|_y$.

Theorem 6.7 *Let $\varphi\colon M \to N$ be a C^∞ mapping of manifolds. Then the pull-back map $\varphi^*\colon \Omega^r(N) \to \Omega^r(M)$ has the following properties:*

(i) $(\varphi \circ \psi)^* = \psi^* \circ \varphi^*$ whenever $\varphi \circ \psi$ is defined
(ii) If $\iota\colon M \to M$ is the identity map then $\iota^*\colon \Omega^r(M) \to \Omega^r(M)$ is also the identity.

Proof If we recall that $T(\varphi \circ \psi) = T\varphi \circ T\psi$ from Theorem 5.21, and the functorial property of $\bigwedge^r$, Proposition 6.1, then

$$\begin{aligned} (\varphi \circ \psi)^*(\alpha) &= \textstyle\bigwedge^r({}^tT(\varphi \circ \psi)) \circ \alpha \circ \varphi \circ \psi \\ &= \textstyle\bigwedge^r({}^tT\psi \circ {}^tT\varphi) \circ \alpha \circ \varphi \circ \psi \\ &= \textstyle\bigwedge^r({}^tT\psi) \circ \bigwedge^r({}^tT\varphi) \circ \alpha \circ \varphi \circ \psi \\ &= \psi^*(\varphi^*(\alpha)). \end{aligned}$$

(ii) is obvious.

Let $\varphi\colon M \to N$ be a C^∞ map. Since $\varphi^*\colon \Omega^r(N) \to \Omega^r(M)$ is $\mathbb{R}$-linear for $r = 0, 1, \ldots$, it induces a linear mapping on the direct sums, $\varphi^*\colon \Omega(N) \to \Omega(M)$.

Theorem 6.8 *If $\varphi\colon M \to N$ is a C^∞ map then $\varphi^*\colon \Omega(N) \to \Omega(M)$ is a homomorphism of differential algebras, that is,*

(i) $\varphi^*(\alpha \wedge \eta) = \varphi^*(\alpha) \wedge \varphi^*(\eta)$ for $\alpha \in \Omega^r(N)$ and $\eta \in \Omega^s(N)$.
(ii) $\varphi^*(d\alpha) = d(\varphi^*\alpha)$ for $\alpha \in \Omega^r(N)$.

Proof Property (*i*) follows from the corresponding pointwise result (see (13) in Chapter 5). Property (*ii*) follows from the result proved below which characterizes the pull-back map locally.

Proposition 6.9 *Let U and V be open in $\mathbb{R}^n$ and $\varphi\colon V \to U$ a smooth map. The pull-back map, φ^*, on C^∞ functions (defined by $\varphi^*(f) = f \circ \varphi$ where*

$f \in C^\infty(U)$) extends uniquely to an operator $\varphi^*\colon \Omega^r(U) \to \Omega^r(V)$, $r = 0, 1, \ldots, n$ satisfying the following properties:

(i) $\varphi^*(\alpha \wedge \eta) = \varphi^*\alpha \wedge \varphi^*\eta$ *for* $\alpha \in \Omega^r(U)$ *and* $\eta \in \Omega^s(U)$
(ii) $d \circ \varphi^* = \varphi^* \circ d$

Proof Clearly if φ^* satisfies (i), (ii) then

$$\varphi^*\left(\sum f_I\, dx_{i_1} \wedge \cdots \wedge dx_{i_r}\right) = \sum (f_I \circ \varphi)\, d\varphi_{i_1} \wedge \cdots \wedge d\varphi_{i_r} \tag{8}$$

where $\varphi_i = x_i \circ \varphi$ is the ith component of the function φ. This proves uniqueness. For existence, we define $\varphi^*\colon \Omega^r(V) \to \Omega^r(U)$ by the formula (8). The property (i) is evident. Since φ^* is linear, it suffices to verify (ii) on monomial terms, $\alpha = f_I\, dx_{i_1} \wedge \cdots \wedge dx_{i_r}$. Now

$$\begin{aligned} d\varphi^*\alpha &= d((f_I \circ \varphi)\, d\varphi_{i_1} \wedge \cdots \wedge d\varphi_{i_r}) \\ &= d(f_I \circ \varphi) \wedge d\varphi_{i_1} \wedge \cdots \wedge d\varphi_{i_r} \end{aligned}$$

and

$$\begin{aligned} \varphi^* d\alpha &= \varphi^*\left(\sum \frac{\partial f_I}{\partial x_j}\, dx_j \wedge dx_{i_1} \wedge \cdots \wedge dx_{i_r}\right) \\ &= \sum \left(\frac{\partial f_I}{\partial x_j} \circ \varphi\right) d\varphi_j \wedge d\varphi_{i_1} \wedge \cdots \wedge d\varphi_{i_r} \\ &= d(f_I \circ \varphi) \wedge d\varphi_{i_1} \wedge \cdots \wedge d\varphi_{i_r} \end{aligned}$$

Remark If $\varphi\colon V \to U$ is a diffeomorphism then $y_i = x_i \circ \varphi$ ($i = 1, 2, \ldots, n$) are new coordinates on U. It is property (ii) which shows that the definition of d is independent of the coordinate system.

We conclude this section with the definition of the contraction of a vector field with a differential form.

But first let us consider the case of a vector $X \in \mathbf{E}$ and a p-form $\alpha \in \bigwedge^p(\mathbf{E}^*)$, where $\mathbf{E}$ is a finite dimensional vector space over $\mathbb{R}$. Due to the universal property of exterior products there exists a linear map $i_X\colon \bigwedge^p(\mathbf{E}^*) \to \bigwedge^{p-1}(\mathbf{E}^*)$ such that its value on decomposable p-forms is

$$i_X(\alpha^1 \wedge \cdots \wedge \alpha^p) = \sum_{i=1}^{p} (-1)^{i+1} \alpha^i(X) \alpha^1 \wedge \cdots \wedge \widehat{\alpha^i} \wedge \cdots \wedge \alpha^p \tag{9}$$

since the right-hand side is an alternating function of $\alpha^1, \ldots, \alpha^p$. (The symbol $\wedge$ above the element α^i means that it is omitted.) We define $i_X c = 0$ if $c \in \bigwedge^0(\mathbf{E}^*) = \mathbb{R}$, and extend i_X by linearity to the direct sum $\bigwedge(\mathbf{E}^*)$.

Proposition 6.10 *Let* $\mathbf{E}$ *be a finite dimensional real vector space and let* $X \in \mathbf{E}$. *There is a unique linear operator* $i_X\colon \bigwedge(\mathbf{E}^*) \to \bigwedge(\mathbf{E}^*)$ *which sends* $\bigwedge^p(\mathbf{E}^*)$ *to* $\bigwedge^{p-1}(\mathbf{E}^*)$, $p = 1, 2, \ldots$, *and satisfies the following properties:*

(i) $i_X c = 0$ *if* $c \in \mathbb{R}$
(ii) $i_X \lambda = \lambda(X)$ *if* $\lambda \in \mathbf{E}^*$

(iii) i_X *is an anti-derivation with respect to* $\wedge$, *that is,* i_X *is linear and*

$$i_X(\alpha \wedge \eta) = i_X\alpha \wedge \eta + (-1)^p \alpha \wedge i_X\eta$$

for all $\alpha \in \bigwedge^p(\mathbf{E}^*)$ *and* $\eta \in \bigwedge^q(\mathbf{E}^*)$.

Proof The uniqueness of i_X is evident, since every element of $\bigwedge^p(\mathbf{E}^*)$ is a sum of decomposable elements. The contraction operator i_X defined above satisfies (i) and is linear. Also, when $p = 1$ the definition (9) is just (ii). It suffices to prove (iii) when $\alpha = \alpha^1 \wedge \cdots \wedge \alpha^p$ and $\eta = \eta^1 \wedge \cdots \wedge \eta^q$, where α^i and $\eta^j \in \mathbf{E}^*$, because such elements span $\bigwedge^p(\mathbf{E}^*)$ and $\bigwedge^q(\mathbf{E}^*)$, respectively. Now, let $\alpha^{p+i} = \eta^i$ for $i = 1, \ldots, q$. Then

$$i_X(\alpha \wedge \eta) = \sum_{i=1}^{p+q} (-1)^{i+1} \alpha^i(X) \alpha^1 \wedge \cdots \wedge \widehat{\alpha^i} \wedge \cdots \wedge \alpha^{p+q}$$

If we write the sum in two parts, $\sum_{i=1}^{p} + \sum_{i=p+1}^{p+q}$, and in the second sum change the index of summation to $i' = i - p$, we get (iii) because $(-1)^{i+1} = (-1)^p \cdot (-1)^{i'+1}$.

Remark Properties (i)–(iii) imply that $i_X \circ i_X = 0$. Also note that the map $(\alpha, X) \mapsto i_X\alpha$ is $\mathbb{R}$-bilinear.

For a vector field $X \in \mathscr{X}(M)$ and a differential form $\alpha \in \Omega^p(M)$, $i_X\alpha$ is defined pointwise:

$$(i_X\alpha)(m) = i_{X(m)}\alpha(m) \qquad m \in M$$

We claim that $i_X\alpha$ is smooth when X and α are smooth. In view of (1), it suffices to show this when $\alpha = dx_{i_1} \wedge \cdots \wedge dx_{i_p}$ in local coordinates $(U, x_1, \ldots, x_n)$. By definition we have

$$i_X(dx_{i_1} \wedge \cdots \wedge dx_{i_p}) = \sum_{r=1}^{p} (-1)^{r+1} X^{i_r} dx_{i_1} \wedge \cdots \wedge \widehat{dx_{i_r}} \wedge \cdots \wedge dx_{i_p},$$

which belongs to $\Omega^p(U) = C^\infty(U, \bigwedge^p(T^*U))$ because $X^{i_r} = dx_{i_r}(X) \in C^\infty(U)$.

We call $i_X\alpha$ the *contraction* or *interior product* of the vector field X with the differential form α. Sometimes it is more convenient to use the notation

$$\alpha \lrcorner X,$$

instead of $i_X\alpha$.

The interior product has the following properties since they hold pointwise by Proposition 6.10.

Theorem 6.11 *Let* $X \in \mathscr{X}(M)$, *a vector field on* M. *The operator* $i_X\colon \Omega(M) \to \Omega(M)$ *sends* $\Omega^p(M)$ *to* $\Omega^{p-1}(M)$, $p = 1, 2, \ldots,$ *and satisfies the following properties*:

(i) $i_X f = 0$ *if* $f \in \Omega^0(M) = C^\infty(M)$
(ii) $i_X\omega = \omega(X)$ *if* $\omega \in \Omega^1(M)$. *By* $\omega(X)$ *we mean the function in* $C^\infty(M)$ *defined by* $\omega(X)(m) = \omega(m)(X(m))$.
(iii) i_X *is an anti-derivation with respect to* $\wedge$, *that is,* i_X *is* $\mathbb{R}$*-linear and*

$$i_X(\alpha \wedge \eta) = i_X\alpha \wedge \eta + (-1)^p \alpha \wedge i_X\eta$$

for all $\alpha \in \Omega^p(M)$ *and* $\eta \in \Omega^q(M)$.

Exercises

1. Let $\mathbf{E}$ be a vector space over $\mathbb{R}$. We showed in Theorem 5.9 that there is an isomorphism $\bigwedge^k(\mathbf{E}^*) \simeq L_a^k(\mathbf{E}, \mathbb{R})$ which maps each decomposable element $\alpha^1 \wedge \cdots \wedge \alpha^k$ to the alternating k-form $(v_1, \ldots, v_k) \mapsto \det\langle \alpha^i, v_j \rangle$. Show that if $\alpha \in \bigwedge^p(\mathbf{E}^*)$, and $i_X \in \bigwedge^{p-1}(\mathbf{E}^*)$ is identified with its image in $L_a^{p-1}(\mathbf{E}, \mathbb{R})$, then

$$i_X\alpha(v_1, \ldots, v_{p-1}) = \alpha(X, v_1, \ldots, v_{p-1})$$

Hint: First verify this when α is a decomposable element, then extend by linearity.

2. Let $\varphi\colon M \to N$ be a C^∞ mapping. If $X \in \mathscr{X}(M)$ and $Y \in \mathscr{X}(N)$ are φ-related, i.e. $T\varphi \circ X = Y \circ \varphi$ (see §5.5, Exercise 4), show that

$$i_X \circ \varphi^* = \varphi^* \circ i_Y.$$

Hint: It suffices to verify this for differential forms of degree 0 or 1 (because we can work locally).

3. Suppose that $\varphi\colon M \to N$ is a diffeomorphism. Then $i_{\varphi^*Y} \circ \varphi^* = \varphi^* \circ i_Y$ where $\varphi^*Y \in \mathscr{X}(N)$ is the pull-back of $Y \in \mathscr{X}(M)$ (see §5.5).

6.3 The Poincaré lemma

A differential form ω is said to be *closed* if $d\omega = 0$ and *exact* if $\omega = d\alpha$ for some α. Since $d \circ d = 0$, every exact form is closed. The converse is not true as is clear from the elementary example

$$\omega = \frac{-y}{x^2 + y^2}\,dx + \frac{x}{x^2 + y^2}\,dy \qquad \text{on } M = \mathbb{R}^2\backslash 0$$

In this section we show that if M is *contractible* then every closed form on M is exact. Recall that a manifold M is contractible to a point $m_0 \in M$ if there is a C^∞ function $H\colon M \times I \to M$ such that $H(m, 1) = m$ and $H(m, 0) = m_0$ for all $m \in M$, where $I = (-\varepsilon, 1 + \varepsilon)$ is an open interval containing $[0, 1]$. For $t \in [0, 1]$, we define $\imath_t\colon M \to M \times I$ by

$$\imath_t(m) = (m, t)$$

and we claim that is ω if a form on $M \times I$ with $d\omega = 0$, then

$$\imath_1^*\omega - \imath_0^*\omega \qquad \text{is exact} \tag{10}$$

Assuming that (10) has been proved we can easily prove the Poincaré Lemma.

Theorem 6.12 *Let M be contractible. Then any closed form on M is exact.*

Proof Let $\alpha \in \Omega^r(M)$. Then $H^*\alpha$ is a form on $M \times I$ and in view of (10) we have

$$\imath_1^*H^*\alpha - \imath_0^*H^*\alpha = d\eta$$

for some $\eta \in \Omega^r(M)$. But observe that $\imath_1^*H^* = (H \circ \imath_1)^*$ is the identity on M

and $i_0^*H^* = (H \circ i_0)^* = 0$ since $H \circ i_0$ is a constant map, so we obtain

$$\alpha - 0 = d\eta$$

Before proving (10) we need to develop a few preliminaries concerning the manifold $M \times I$. There is a global coordinate function, t, on $M \times I$ (namely, the projection $M \times I \to I$ on the second factor). If $(U, x_1, \ldots, x_n)$ is a chart on M and π is the projection $M \times I \to M$ on the second factor then

$$(U \times I, x_1 \circ \pi, \ldots, x_n \circ \pi, t)$$

is a chart on $M \times I$. We will denote $x_i \circ \pi$ by $\bar{x}_i$, and more generally if f is any C^∞ function on M, we denote the pull-back $f \circ \pi$ by $\bar{f}$. The dual basis of $\partial/\partial t, \partial/\partial \bar{x}_1, \ldots, \partial/\partial \bar{x}_n$ is given by $dt, d\bar{x}_1, \ldots, d\bar{x}_n$.

Let $\hat{\Omega}^r(M \times I)$ denote the set of r-forms $\omega \in \Omega^r(M \times I)$ such that locally in each chart we have

$$\omega_{|U \times I} = \sum f_{i_1 \cdots i_r} \, d\bar{x}_{i_1} \wedge \cdots \wedge d\bar{x}_{i_r}$$

where the f's are C^∞ functions on $U \times I$ and the sum is taken over all multi-indices $I = (i_1, \ldots, i_r)$ with $i_1 < \cdots < i_r$.

Note that if the f's were independent of t then ω would be the pull-back by π of a differential form on M.

We need an intrinsic characterization of $\hat{\Omega}^r(M \times I)$. First note that there is a canonical isomorphism

$$T_{(m,t)}(M \times I) \simeq T_m M \times T_t I,$$

so that $T_m M$ can be identified with the subspace $T_m M \times 0_t$. Explicitly, this identification is obtained as follows: The tangent to the map $i_t: M \to M \times I$ is a linear map $T_m M \to T_m M \times T_t I$ with image $T_m M \times 0_t$ and it sends the vector $\partial/\partial x_i|_m \in T_m M$ to the vector $\partial/\partial \bar{x}_i|_{(m,t)} \in T_{(m,t)}(M \times I)$. Similarly, we can identify $T_t I$ with the subspace $0_m \times T_t I$ by mapping $\partial/\partial t|_t \mapsto \partial/\partial t|_{(m,t)}$. With this understanding, we have, for each point $(m, t) \in M \times I$, a direct sum of subspaces

$$T_{(m,t)}(M \times I) \simeq (T_m M \times 0_t) \oplus (0_m \times T_t I)$$

Upon taking the dual of this isomorphism we obtain

$$T^*_{(m,t)}(M \times I) \simeq (T_m M \times 0_t)^\circ \oplus (0_m \times T_t I)^\circ, \tag{11}$$

a direct sum of annihilators of $T_m M$ and $T_t I$ in $T^*_{(m,t)}(M \times I)$. It is not hard to verify (if one recalls the definition of a dual basis) that

$(T_m M \times 0_t)^\circ$ has basis $dt|_{(m,t)}$

$(0_m \times T_t I)^\circ$ has basis $d\bar{x}_1|_{(m,t)}, \ldots, d\bar{x}_n|_{(m,t)}$

Thus we see that $\hat{\Omega}^r(M \times I)$ consists of the C^∞ sections of the vector bundle $\bigcup_{(m,t)} \bigwedge^r((0_m \times T_t I)^\circ)$, that is,

$$\hat{\Omega}^r(M \times I) = C^\infty(M \times I, \bigcup_{(m,t)} \bigwedge^r((0_m \times T_t I)^\circ)).$$

Lemma 6.13 *Let $\pi: M \times I \to M$ be the projection on the first factor. If $\omega \in \Omega^r(M \times I)$ then we have*

$$\omega = \omega_1 + dt \wedge \eta$$

for unique $\omega_1 \in \hat{\Omega}^r(M \times I)$ and $\eta \in \hat{\Omega}^{r-1}(M \times I)$.

Proof There are really two issues to verify here. First of all we have to show that for any $(m, t) \in M \times I$ we have

$$\omega|_{(m,t)} = \omega_1 + dt|_{(m,t)} \wedge \eta$$

for unique $\omega_1 \in \bigwedge^r((0_m \times T_t I)^\circ)$ and $\eta \in \bigwedge^{r-1}((0_m \times T_t I)^\circ)$. Secondly, we must show that these forms ω_1 and η which are uniquely defined at each point do indeed define smooth forms on $M \times I$. But, the existence and uniqueness of ω_1 and η at each point is clear from (11) and Exercise 3 in §5.1. Locally, the existence and uniqueness of ω_1 and η is also evident, because then ω is a sum of terms of the form $f\, d\bar{x}_{i_1} \wedge \cdots \wedge d\bar{x}_{i_r}$ and $f\, dt \wedge d\bar{x}_{i_1} \wedge \cdots \wedge d\bar{x}_{i_r}$ and $f\, dt \wedge d\bar{x}_{i_1} \wedge \cdots \wedge d\bar{x}_{i_{r-1}}$ and we can separate off the terms containing the factor dt. The local expressions show that ω_1 and η are smooth.

Remark Using the properties of the interior product in Theorem 6.11, we see that the unique η in Lemma 6.13 is given by $\eta = i_{\partial/\partial t}\omega$.

Now we define the operator $K: \Omega^r(M \times I) \to \Omega^{r-1}(M)$ as follows. If $\omega \in \Omega^r(M \times I)$ we have $\omega = \omega_1 + dt \wedge \eta$ where $\eta \in \hat{\Omega}^{r-1}(M \times I)$; for fixed $m \in M$, the map

$$t \mapsto \eta|_{(m,t)} = (\iota_t^* \eta)|_m$$

is a smooth function on I with values in the finite dimensional vector space $\bigwedge^{r-1}(T_m^* M)$ so its integral is well-defined. We define

$$K\omega|_m = \int_0^1 \eta|_{(m,t)}\, dt \tag{12}$$

It is evident that $K\omega|_m \in \bigwedge^{r-1}(T_m^* M)$ for each $m \in M$, and we must show that $K\omega$ is smooth. But it is clear from the local representation of elements of $\hat{\Omega}^{r-1}(M \times I)$ that for all $m \in U$ (a chart domain), $K\omega|_m$ is a sum of terms of the form

$$\left(\int_0^1 f(m, t)\, dt\right) \cdot dx_{i_1} \wedge \cdots \wedge dx_{i_{r-1}},$$

where $f \in C^\infty(U \times I)$. Hence $K\omega$ is smooth, so $K\omega \in \Omega^{r-1}(M)$.

Theorem 6.14 *We have on $\Omega^r(M \times I)$*

$$\iota_1^*\omega - \iota_0^*\omega = d(K\omega) + K(d\omega)$$

Consequently, if $\omega \in \Omega^r(M \times I)$ and $d\omega = 0$ then $\iota_1^*\omega - \iota_0^*\omega = d(K\omega)$, so that (10) *holds.*

Proof The definition (12) is an intrinsic definition, independent of coordinates; to prove the theorem we can just as well work in a coordinate system $(U \times I, \bar{x}_1, \ldots, \bar{x}_n, t)$. The operator K is clearly linear, so we need only consider two cases:

(1) $\omega = f\, d\bar{x}_{i_1} \wedge \cdots \wedge d\bar{x}_{i_r} = f\, d\bar{x}_I$. Then

$$d\omega = \frac{\partial f}{\partial t}\, dt \wedge d\bar{x}_I + \text{other terms}$$

so that by the definition of K

$$\begin{aligned} K(d\omega)|_m &= \left(\int_0^1 \frac{\partial f}{\partial t}(m, t)\, dt \right) dx_I|_m \\ &= [f(m, 1) - f(m, 0)]\, dx_I|_m \\ &= \iota_1^* \omega|_m - \iota_0^* \omega|_m \end{aligned}$$

Since $K\omega = 0$, the theorem is proved in this case.

(2) $\omega = f\, dt \wedge d\bar{x}_{i_1} \wedge \cdots \wedge d\bar{x}_{i_{r-1}} = f\, dt \wedge d\bar{x}_I$. Then

$$d\omega = \sum_{j=1}^{n} \frac{\partial f}{\partial x_j}\, dx_j \wedge dt \wedge d\bar{x}_I$$

so that

$$K(d\omega)|_m = -\sum_{j=1}^{n} \left(\int_0^1 \frac{\partial f}{\partial x_j}(m, t)\, dt \right) dx_j \wedge dx_I$$

(the negative sign arises because $dx_j \wedge dt = -dt \wedge dx_j$) and

$$\begin{aligned} d(K\omega) &= d\left(\int_0^1 f(m, t)\, dt \right) \wedge dx_I \\ &= \sum_{j=1}^{n} \frac{\partial}{\partial x_j} \left(\int_0^1 f(m, t)\, dt \right) dx_j \wedge dx_I \end{aligned}$$

Clearly $K(d\omega) + d(K\omega) = 0$ because we can differentiate under the integral sign. Since $\iota_1^*\omega = 0$ and $\iota_0^*\omega = 0$, the theorem is proved.

Exercise

Let $f, g: M \to N$ be C^∞ mappings which are homotopic, and let α be a closed form on N. Prove that $g^*\alpha - f^*\alpha$ is an exact form on M.

6.4 Orientation of a vector bundle

Let $\mathbf{E}$ be a finite dimensional vector space over $\mathbb{R}$. Two ordered bases $v_1, \ldots, v_n$ and $v_1', \ldots, v_n'$ for $\mathbf{E}$ determine an isomorphism $\varphi: \mathbf{E} \to \mathbf{E}$ by

$\varphi(v_i) = v_i'$. The matrix $[a_{ij}]$ of φ is given by the equations

$$v_j' = \sum_{i=1}^{n} a_{ij} v_i$$

We say that $v_1, \ldots, v_n$ and $v_1', \ldots, v_n'$ have the *same orientation* if $\det [a_{ij}] > 0$ and the *opposite orientation* if $\det[a_{ij}] < 0$. The relation of being equally oriented is clearly an equivalence relation, dividing the collection of all ordered bases of **E** into just two equivalence classes. Either of these two equivalence classes is called an *orientation* for **E**. The class to which $v_1, \ldots, v_n$ belongs will be denoted by μ. We also say that the ordered basis $v_1, \ldots, v_n$ "belongs to the orientation", or is "positively oriented" with respect to μ.

An *oriented vector space* is a pair $(\mathbf{E}, \mu)$ where **E** is a vector space and μ is an orientation on **E**. If μ is one orientation on **E**, the other orientation is denoted by $-\mu$. If $(\mathbf{E}, \mu)$ and $(\mathbf{F}, \nu)$ are two oriented vector spaces, an isomorphism $\varphi: \mathbf{E} \to \mathbf{F}$ is called *orientation-preserving* if it maps a positively oriented basis of $(\mathbf{E}, \mu)$ to a positively oriented basis of $(\mathbf{F}, \nu)$. Of course, if this condition holds for one positively oriented basis of **E**, then it holds for all.

We give $\mathbb{R}^n$ the standard orientation, that is, the orientation which contains the standard basis $e_1, \ldots, e_n$.

Definition 6.15 *Let E be a vector bundle over M. An orientation μ of E is a collection of orientations μ_x for each fibre E_x which satisfy the following compatibility condition:*

> *for any local trivialization $\beta: E|_U \to U \times \mathbb{R}^n$ where U is an open, connected set $\subset M$, the maps $\beta_x: E_x \to \mathbb{R}^n$ are either orientation-preserving for all $x \in U$ or orientation-reversing for all $x \in U$.*

If β and ψ are two local trivializations $E|_U \to U \times \mathbb{R}^n$, then the overlap map $\psi \circ \beta^{-1}: U \times \mathbb{R}^n \to U \times \mathbb{R}^n$ has the form

$$\psi \circ \beta^{-1}(x, e_j) = \sum_{i=1}^{n} a_{ij}(x) \cdot (x, e_i)$$

where $\det[a_{ij}(x)] \neq 0$ for all $x \in U$. Since U is connected this means that either the determinant is positive for all $x \in U$, or negative for all $x \in U$. Thus we can put Def. 6.15 in a more convenient form.

Definition 6.15′ *An orientation μ of E is a collection of orientations μ_x for each fibre E_x which satisfy the following compatibility condition:*

> *there is a collection $\{U_i\}$ of open, connected sets which cover M, and local trivializations $\beta_i: E|_{U_i} \to U_i \times \mathbb{R}^n$ such that $\beta_{ix}: E_x \to \mathbb{R}^n$ is orientation-preserving for all $x \in U_i$.*

For example, we can choose the U_i's such that they are the domains of charts in an atlas on M.

A vector bundle is said to be *orientable* if it has an orientation. Note that if E is an orientable vector bundle over M, and M is connected, then E has just two orientations; if $\mu = \{\mu_x\}$ is one orientation, the other is $-\mu = \{-\mu_x\}$.

Examples

(1) Every trivial bundle is orientable. Indeed, let $\beta: E \to M \times \mathbb{R}^n$ be a trivialization. For the trivial bundle $M \times \mathbb{R}^n$ we can put the standard orientation on each fiber $\{x\} \times \mathbb{R}^n$, then choose the orientation μ_x on E_x to be such that β_x is orientation-preserving for all $x \in M$.

(2) The Möbius band $\mathscr{M}$ (see §5.7, Example (4)) is *not* orientable. Indeed, for each $x = e^{2\pi i t} \in S^1$ we have just two vectors of "unit length" in the fibre $\mathscr{M}_x$, namely, $[t, 1]$ and $[t, -1]$. If $\mathscr{M}$ were orientable then it would be possible to define a smooth nonvanishing section $s: S^1 \to \mathscr{M}$ by letting $s(x)$ be the unique vector of unit one-dimensional length such that $s(x) \in \mathscr{M}_x$ is a positively oriented basis of the one-dimensional space $\mathscr{M}_x$. But we have already seen that every smooth section of $\mathscr{M}$ vanishes at some point.

It is convenient to reformulate the definition of orientation using exterior products. Let $n = \dim \mathbf{E}$. A nonzero element, ω, of the one-dimensional space $\bigwedge^n(\mathbf{E}^*)$ is called a *volume element*. By §5.1 (14), ω is identified with an alternating map $\mathbf{E}^{(n)} \to \mathbb{R}$ and then we have

$$\omega(\varphi(e_1), \ldots, \varphi(e_n)) = \det\varphi \cdot \omega(e_1, \ldots, e_n) \tag{13}$$

for all $\varphi \in \mathrm{L}(\mathbf{E}, \mathbf{E})$ and $e_i \in \mathbf{E}$. It follows that two ordered bases $v_1, \ldots, v_n$ and $v_1', \ldots, v_n'$ have the same orientation if and only if $\omega(v_1, \ldots, v_n)$ and $\omega(v_1', \ldots, v_n')$ have the same sign.

The following proposition is a consequence of (13).

Proposition 6.16 *Let* $\mathbf{E}$ *be a vector space over* $\mathbb{R}$ *and let* ω *be a volume element on* $\mathbf{E}$. *Then there is a unique orientation for* $\mathbf{E}$ *such that the basis* $v_1, \ldots, v_n$ *is positively oriented if and only if* $\omega(v_1, \ldots, v_n) > 0$.

Two volume elements ω_1 and ω_2 are said to be *equivalent* if there is a number $c > 0$ such that $\omega_1 = c\omega_2$. Note that the orientation for $\mathbf{E}$ defined by a volume element depends only on the equivalence class of the volume element. Thus, an equivalence class $[\omega]$ of volume elements defines an orientation on $\mathbf{E}$, the opposite orientation being $[-\omega]$.

If $v_1, \ldots, v_n$ is a positively oriented basis with respect to a given orientation of $\mathbf{E}$, then that orientation is the one induced by $\omega = \lambda^1 \wedge \cdots \wedge \lambda^n$, where $\lambda^1, \ldots, \lambda^n$ is the dual basis corresponding to $v_1, \ldots, v_n$.

Volume elements are determined only up to a positive constant. To specify a volume element uniquely, we need additional structure such as an inner product on $\mathbf{E}$.

Proposition 6.17 *Let* $\mathbf{E}$ *be an oriented vector space over* $\mathbb{R}$ *and let* $\langle , \rangle$ *be an inner product on* $\mathbf{E}$. *Then there exists a unique volume element* $\mu \in \bigwedge^n(\mathbf{E}^*)$ *such that*

$$\mu(e_1, \ldots, e_n) = 1$$

for all positively oriented orthonormal bases $e_1, \ldots, e_n$ *of* $\mathbf{E}$. *In fact, if* $\mu^1, \ldots, \mu^n$ *is the dual basis then* $\mu = \mu^1 \wedge \cdots \wedge \mu^n$. *More generally, if* $v_1, \ldots, v_n$ *is any positively oriented basis of* $\mathbf{E}$ *and* $\lambda^1, \ldots, \lambda^n$ *is the dual basis then*

$$\mu = (\det[\langle v_i, v_j \rangle])^{1/2} \lambda^1 \wedge \cdots \wedge \lambda^n \tag{14}$$

Proof Clearly the condition $\mu(e_1, \ldots, e_n) = 1$ uniquely determines μ by multilinearity. If $\mu^1, \ldots, \mu^n$ is the dual basis in $\mathbf{E}^*$, then $e_1, \ldots, e_n$ is positively oriented with respect to $\mu = \mu^1 \wedge \cdots \wedge \mu^n$ since by virtue of §5.1 (14) we have $\mu(e_1, \ldots, e_n) = \det[\delta_{ij}] = 1$, which is > 0. Now let $v_1, \ldots, v_n$ be any other positively oriented basis of $\mathbf{E}$ (not necessarily orthonormal) and consider the map $\varphi \in GL(\mathbf{E})$ defined by $\varphi(e_i) = v_i$. We have $v_j = \sum a_{ij} e_i$ for some $n \times n$ matrix $A = [a_{ij}]$. Then

$$\langle v_j, v_k \rangle = \sum a_{ij} a_{lk} \cdot \langle e_i, e_l \rangle = \sum a_{ij} a_{ik}$$

and hence

$$\det[\langle v_j, v_k \rangle] = \det A^T A = (\det A)^2$$

Therefore, $(\det[\langle v_j, v_k \rangle])^{1/2} = |\det A|$, and since $\det A > 0$ we have

$$(\det[\langle v_j, v_k \rangle])^{1/2} = \det A.$$

Now, since $\bigwedge^n(\mathbf{E}^*)$ is a one-dimensional space, we have $\mu = c \cdot \lambda^1 \wedge \cdots \wedge \lambda^n$ for some $c \in \mathbb{R}$, and then by §5.1 (14) we get

$$c = \mu(v_1, \ldots, v_n) = \det[\langle \mu^i, v_j \rangle] = \det[a_{ij}].$$

Thus, $c = (\det[\langle v_i, v_j \rangle])^{1/2}$ which proves (14).

In summary, corresponding to each inner product $\langle , \rangle$ on $\mathbf{E}$ there is a unique n-form μ such that

$$\mu(v_1, \ldots, v_n) = (\det[\langle v_i, v_j \rangle])^{1/2}$$

for each positively oriented basis $v_1, \ldots, v_n$ of $\mathbf{E}$. We call μ the *canonical volume element* determined by the given inner product and orientation on $\mathbf{E}$.

The orientability of a vector bundle can now be expressed in a manner analogous to Proposition 6.26, by using the functor $\bigwedge^k$.

Let E be a vector bundle over M of rank n (of course, the rank need not be the same as the dimension of M). If there exists $\omega \in C^\infty(M, \bigwedge^n(E^*))$ such that $\omega(x) \neq 0$ everywhere, then for each $x \in M$ the nonzero $\omega(x) \in \bigwedge^n(E_x^*)$ determines an orientation μ_x of E_x by Proposition 6.16. It is easy to see that the collection of orientations $\{\mu_x\}$ satisfies the "compatibility condition" of Definition 6.15, thus $\{\mu_x\}$ is an orientation of E.

It follows that E is orientable whenever the vector bundle $\bigwedge^n(E^*)$ has a section which is nowhere 0. The converse also holds, as we show in the following proposition.

Proposition 6.18 *Let E be a vector bundle over M of rank n. Then E is orientable if and only if the vector bundle $\bigwedge^n(E^*)$ has a section which is nowhere* 0.

Proof Suppose there exists $\omega \in C^\infty(M, \bigwedge^n(E^*))$ such that $\omega(x) \neq 0$ everywhere. Then for each $x \in M$ the nonzero $\omega(x) \in \bigwedge^n(E_x^*)$ determines an orientation μ_x of E_x by Proposition 6.16. We claim that these orientations

$\{\mu_x\}$ satisfy the "compatibility condition" of Definition 6.15′. Let $x_0 \in M$ and choose an open, connected set $U \ni x_0$ on which there exists a trivialization $\beta: E|_U \to U \times \mathbb{R}^n$. The sections

$$v_i(x) = (\beta_x)^{-1}(e_i), \qquad i = 1, \ldots, n$$

form a local basis of E on U, and without loss of generality the basis $\{v_1(x_0), \ldots, v_n(x_0)\}$ for E_{x_0} belongs to the orientation μ_{x_0}, that is $\omega_{x_0}(v_1(x_0), \ldots, v_n(x_0)) > 0$ when $x = x_0$. By continuity we then have

$$\omega_x(v_1(x), \ldots, v_n(x)) > 0 \qquad \text{for all } x \in U_0$$

where $U_0 \subset U$ is an open set containing x_0. This means $\beta_x: E_x \to \mathbb{R}^n$ is orientation-preserving for all $x \in U_0$, and we have thus verified Definition 6.15′, i.e. $\{\mu_x\}$ is an orientation of E.

It remains to prove the "only if" part. Suppose E is orientable, and let U be an open, connected set in M for which there is a trivialization $\beta: E|_U \to U \times \mathbb{R}^n$ such that the maps β_x are orientation-preserving for all $x \in U$. Let $e_1, \ldots, e_n$ be the standard basis in $\mathbb{R}^n$. Then $\beta_x^{-1}(e_1), \ldots, \beta_x^{-1}(e_n)$ is a positively oriented basis in E_x, and we let $\lambda_x^1, \ldots, \lambda_x^n$ be the corresponding dual basis in E_x^*. It is easy to see that $\lambda^i \in C^\infty(U, E^*)$ (see §5.8, Exercise 11) and, if we define $\omega_U \in C^\infty(U, \bigwedge^n E)$ by $\omega_U(x) = \lambda_x^1 \wedge \cdots \wedge \lambda_x^n$, $x \in U$, then $v_1, \ldots, v_n$ is a positively oriented basis of E_x if and only if $\omega_U(v_1, \ldots, v_n) > 0$.

Therefore, we can cover M by a collection of open, connected sets U_i on which there exists $\omega_i \in C^\infty(U_i, \bigwedge^n(E^*))$ such that, for all $x \in U_i$,

$$\omega_i(v_1, \ldots, v_n) > 0 \Leftrightarrow v_1, \ldots, v_n \text{ is a positively oriented basis of } E_x.$$

Let ρ_i be a partition of unity subordinate to $\{U_i\}$, and set

$$\omega = \sum \rho_i \omega_i.$$

Because the supports of ρ_i are locally finite, this sum is finite in some neighbourhood of each point in M, whence $\omega \in C^\infty(M, \bigwedge^n(E^*))$. (As usual $\rho_i\omega_i$ is defined to be 0 outside U_i, hence it is smoothly defined on all M.)

At each point $x \in M$, if $v_1, \ldots, v_n$ is a positively oriented basis of E_x, we have

$$(\rho_i\omega_i)(x)(v_1, \ldots, v_n) \geqslant 0 \qquad \text{for all } i$$

and $\rho_i(x) > 0$ for at least one index i. Thus, $\omega(x) \neq 0$.

Since the fibres of the vector bundle $\bigwedge^n(E^*)$ are one-dimensional we can also write the proposition in the following way: *E is orientable if and only if $\bigwedge^n(E^*)$ is a trivial vector bundle.*

Exercises

1. Let E and F be vector bundles over M. Show that if two of E, F and $E \oplus F$ are orientable, so is the third.

2. Let $f: N \to M$ be a smooth map, and let E be a vector bundle over M. Show that if E is orientable, then the pull-back bundle $f^{-1}E$ is also orientable. Find an example showing that the converse is not true.

3. Show that if E is any vector bundle (orientable or not), then $E \oplus E$ is orientable. Hint: For any vector space $\mathbf{E}$ there is a "natural" orientation on $\mathbf{E} \oplus \mathbf{E}$.

6.5 Orientation of a manifold

A manifold M is said to be orientable if its tangent bundle TM is an orientable vector bundle. It follows from Proposition 6.18 that M is orientable if and only if there is an n-form $\mu \in \Omega^n(M)$ such that $\mu(x) \neq 0$ for all $x \in M$.

Any two nowhere vanishing n-forms μ, $\mu' \in \Omega^n(M)$ on an orientable manifold M of dimension n differ by a nowhere vanishing function: $\mu = f\mu'$. If M is connected, then f is either everywhere positive or everywhere negative. We say that μ and μ' are *equivalent* if f is positive. Thus on a connected orientable manifold M the nowhere vanishing n-forms fall into two equivalence classes. Either class defines an orientation on M by Proposition 6.16, and these are the only two orientations on M.

Examples

(1) The unit sphere $S^n \subset \mathbb{R}^{n+1}$ is orientable. Let $\mathscr{n}: S^n \to \mathbb{R}^{n+1}$ be the unit normal vector field on S^n, i.e. $\mathscr{n}(x) = x$. The orientation on the tangent spaces $T_x(S^n)$ is defined as follows. A basis $v_2, \ldots, v_n$ in $T_x(S^n)$ is positively oriented if and only if $\mathscr{n}(x), v_2, \ldots, v_n$ is positively oriented in $\mathbb{R}^{n+1}$ with the standard orientation. We leave it as an exercise to show that this defines an orientation on the tangent bundle $T(S^n)$ (i.e. satisfies the compatibility condition).

(2) The real projective space $\mathbb{R}P^n$ (see §5.2, Exercise 9) is *orientable if n is odd, but not orientable if n is even.* Let $\pi: S^n \to \mathbb{R}P^n$ be the canonical projection $x \mapsto [x]$. Since π is a local diffeomorphism, the tangent map $T_x\pi: T_x(S^n) \to T_{[x]}(\mathbb{R}P^n)$ is an isomorphism for each $x \in S^n$. Regarding $T(S^n)$ as a sub-bundle of the trivial bundle $T(\mathbb{R}^{n+1}) = \mathbb{R}^{n+1} \times \mathbb{R}^{n+1}$, we have the map

$$(T_{-x}\pi)^{-1} \circ T_x\pi \quad \text{given by} \quad (x, v) \mapsto (-x, -v) \tag{15}$$

from $T_x(S^n)$ to $T_{-x}(S^n)$. In view of the orientation defined on S^n in Example (1), we see that this map is orientation-preserving if n is odd, but orientation-reversing if n is even (since $n + 1$ is even and odd, respectively). In fact, this holds for any orientation on S^n since S^n is connected and there are only two orientations on S^n, the standard orientation and the opposite orientation.

If n is odd, we can define an orientation on each tangent space $T_{[x]}\mathbb{R}P^n$ by requiring that $T_x\pi$ be orientation-preserving. This definition is independent of the choice of the representative x in the class $[x]$, and it clearly satisfies the compatibility condition of Definition 6.15 since π is a local diffeomorphism.

If n is even, then $\mathbb{R}P^n$ cannot be endowed with an orientation for if this were possible then we could define an orientation on S^n by requiring that $T_x\pi$ be orientation-preserving for each $x \in S^n$. But then $(T_{-x}\pi)^{-1} \circ T_x\pi$ would be orientation-preserving which contradicts (15).

From the point of view of integration in §6.6 we will need to recast the definition of orientability in a more convenient form.

A diffeomorphism $\varphi: V \to W$ between open sets in $\mathbb{R}^n$ is said to be *orientation-preserving* if

$$\det D\varphi(x) > 0 \quad \text{for all} \quad x \in V,$$

i.e., the derivative map $D\varphi(x): \mathbb{R}^n \to \mathbb{R}^n$ is orientation-preserving for all $x \in V$.

Theorem 6.19 *Let M be a manifold. The following are equivalent:*

(*i*) *The tangent bundle TM is orientable.*
(*ii*) *There is an atlas $\{(U_i, \kappa_i)\}$ on M such that all the overlap maps $\kappa_i \circ \kappa_j^{-1}$ are orientation-preserving diffeomorphisms, i.e.* $\det D(\kappa_i \circ \kappa_j^{-1}) > 0$ *everywhere on $\kappa_j(U_j \bigcap U_i)$.*

Proof (ii) → (i) Suppose M has an oriented atlas $\{(U_i, \kappa_i)\}$. We define an orientation μ_x on T_xM, $x \in U_i$, as follows: A basis $v_1, \ldots, v_n$ of T_xM is "positively oriented" if $d\kappa_i|_x(v_1), \ldots, d\kappa_i|_x(v_n)$ is a positively oriented basis in $\mathbb{R}^n$. Since the overlap maps are orientation-preserving, it follows that the definition of μ_x is well-defined for all $x \in M$. The collection of orientations $\{\mu_x\}$ obviously satisfies the compatibility condition of Definition 6.15, so TM is orientable.

(i) → (ii) Conversely, suppose TM is orientable. Choose an atlas $\{(U_i, \kappa_i)\}$ of M where the domains U_i are connected. Then the maps $d\kappa_i|_x: T_xM \to \mathbb{R}^n$ are either orientation-preserving or orientation-reversing for all $x \in U_i$. In the latter case we can interchange any two component functions of κ_i to make it orientation-preserving. Thus, we can assume that the $d\kappa_i$'s are orientation-preserving everywhere for all i. It follows that the overlap maps, $\kappa_i \circ \kappa_j^{-1}$, are orientation-preserving.

We say that an atlas $\{(U_i, \kappa_i)\}$ satisfying (ii) of Theorem 6.19 is an *oriented atlas*. An *orientation* on M is a maximal oriented atlas $\mathcal{O}$. Any chart (U, κ) that belongs to $\mathcal{O}$ is said to be *positively oriented* with respect to the orientation.

We conclude this section by showing how a Riemannian metric on an oriented manifold determines a unique volume element.

Let M be a Riemannian manifold (see Definition 5.65), and assume that M is oriented. Then the inner product $g_x = \langle\, , \rangle_x$ determines a canonical volume element μ_x on T_xM, and we obtain a map $x \in M \mapsto \mu_x \in \bigwedge^n(T_x^*M)$. This defines a differential form $\mu \in \Omega^n(M)$ provided we can show that μ is smooth, and to do so we can work locally. Let $(U, x_1, \ldots, x_n)$ be a chart on M, then in view of Proposition 6.17 we have

$$\mu = \rho^{1/2}\, dx_1 \wedge \cdots \wedge dx_n \quad \text{in} \quad U$$

where $\rho := \det[g_{ij}]$ and $g_{ij}(x) = \langle \partial/\partial x_i, \partial/\partial x_j \rangle_x$. Since each $g_{ij} \in C^\infty(U)$ then $g \in C^\infty(U)$ so that the local representation of μ is smooth.

We call μ the *canonical volume form* on M determined by the Riemannian metric and the orientation.

If X is a vector field on M, we define the *divergence* of X to be the function $\operatorname{div} X \in C^\infty(M)$ defined by

$$\operatorname{div} X \cdot \mu = d(\mu \lrcorner X)$$

($\operatorname{div} X$ is well-defined since $\mu_x \neq 0$, and $\bigwedge^n(T_x^* M)$ is a one-dimensional space for each $x \in M$.) The smoothness of $\operatorname{div} X$ is a consequence of the local formula in the next proposition.

Proposition 6.20 *In a positively oriented chart $(U, x_1, \ldots, x_n)$, we have*

$$\operatorname{div} X = \rho^{-1/2} \sum \frac{\partial}{\partial x_i} (\rho^{1/2} X^i)$$

where $X = \sum X^i \partial/\partial x_i$.

Proof The Riemannian metric and volume form have the local representations $[g_{ij}]$ and $\mu = \rho^{1/2}\, dx_1 \wedge \cdots \wedge dx_n$, where $\rho = \det[g_{ij}]$. Due to (9) we have

$$\mu \lrcorner X = \rho^{1/2} \sum (-1)^{i+1} X^i\, dx_1 \wedge \cdots \wedge \widehat{dx_i} \wedge \cdots \wedge dx_n$$

so that

$$d(\mu \lrcorner X) = \sum \frac{\partial}{\partial x_i} (\rho^{1/2} X^i)\, dx_1 \wedge \cdots \wedge dx_n$$

$$= \rho^{-1/2} \sum \frac{\partial}{\partial x_i} (\rho^{1/2} X^i) \cdot \mu$$

Since $\operatorname{div} X \cdot \mu = d(\mu \lrcorner X)$, the proposition follows.

Let $f \in C^\infty(M)$. The *gradient* of f is defined to be the vector field $\operatorname{grad} f = X \in \mathscr{X}(M)$ defined by

$$df(Y) = g(X, Y), \quad Y \in \mathscr{X}(M)$$

In other words, $\operatorname{grad} f$ is the image of the 1-form df by the index-raising operator (see §5.10). In local coordinates, we have $\operatorname{grad} f = \sum (\operatorname{grad} f)^i\, \partial/\partial x_i$ where $(\operatorname{grad} f)^i = \sum g^{ij}\, \partial f/\partial x_j$.

The *Laplacian* of f is defined to be the divergence of $\operatorname{grad} f$, $\Delta f = \operatorname{div}(\operatorname{grad} f) \in C^\infty(M)$, and we have the local expression

$$\Delta f = \rho^{-1/2} \sum \frac{\partial}{\partial x_i} (\rho^{1/2} g^{ij}\, \partial f/\partial x_j) \quad \text{in} \quad U$$

Exercises

1. Let M and N be manifolds, and $\pi_M\colon M \times N \to M$ and $\pi_N\colon M \times N \to N$ the projections onto the first and second factors, respectively. Show that

(a) $T(M \times N) \simeq \pi_M^{-1}(TM) \oplus \pi_N^{-1}(TN)$.
(b) $M \times N$ is orientable if and only if both M and N are orientable.

2. Show that the torus $S^1 \times S^1$ is an orientable manifold.

3. Let the Möbius band $\mathcal{M}$ be defined as in §5.7, Example (4), and let $\pi\colon \mathbb{R}^2 \to \mathcal{M}$ be the canonical projection. Show that $\mathcal{M}$ is *not* an orientable manifold. Hint: If $\mu \in \Omega^2(\mathcal{M})$ define $f \in C^\infty(\mathbb{R}^2)$ by $\pi^*\mu = f \cdot (dx \wedge dy)$. Show that $f(x+1, -y) = -f(x, y)$, and conclude that f must vanish at some point.

4. Show that the sphere S^n is orientable by using Proposition 6.19 (ii) and the charts given in §5.2. Example (1). Show that the orientation on S^n defined in this way coincides with the orientation defined in Example (1) above.

5. Let M be a Riemannian manifold; denote the Riemannian metric by $g_x = \langle\, ,\rangle_x$ for each $x \in M$. Let $f \in C^\infty(M)$ and suppose $k \in \mathbb{R}$ is a regular value of f. Show that the vector field grad f is perpendicular to the manifold $f^{-1}(k)$, i.e. for all $x \in f^{-1}(k)$

$$\langle \operatorname{grad} f(x), v\rangle_x = 0 \quad \text{for all} \quad v \in T_x(f^{-1}(k))$$

Hint: v is the tangent vector to a curve c such that $f(c(t)) = k$ for all t.

6. Suppose M is an orientable manifold, $f \in C^\infty(M)$ and $k \in \mathbb{R}$ is a regular value of f. Show that $f^{-1}(k)$ is an orientable manifold (if it is nonempty). Hint: Endow M with a Riemannian metric and let μ be the canonical volume form on M. Then let $X = \operatorname{grad} f$ in Lemma 6.21 (see below) to show that there exists a nowhere vanishing $(n-1)$-form on $f^{-1}(k)$.

7. Show that the sphere S^n is an orientable manifold by constructing an explicit n-form which is nowhere 0. Hint: Consider the volume form $\mu = dx_1 \wedge \cdots \wedge dx_{n+1}$ in $\mathbb{R}^{n+1}$ and see the preceding exercise.

8. Let M be an oriented manifold with oriented atlas $\{(U_i, \kappa_i)\}$ and let E be an oriented vector bundle over M with orientation $\{(U_i, \beta_i)\}$. Show that E can be made into an oriented *manifold* with orientation given by the atlas $\{(E_{U_i}, \chi_i)\}$ where

$$\chi_i = (\kappa_i \times id) \circ \beta_i\colon E_{U_i} \xrightarrow{\ \beta_i\ } U_i \times \mathbb{R}^r \xrightarrow{\ \kappa_i \times id\ } \kappa_i(U_i) \times \mathbb{R}^r \subset \mathbb{R}^n \times \mathbb{R}^r$$

This is called the *local product orientation* on E.

Addendum

In this addendum we prove a lemma which is useful for providing an orientation for a submanifold of an oriented manifold.

Let $\mathbf{E}$ be a subspace of a vector space $\mathbf{V}$. The transpose of the inclusion map $\iota\colon \mathbf{E} \to \mathbf{V}$ is the restriction map ${}^t\iota\colon \mathbf{V}^* \to \mathbf{E}^*$ given by $\lambda \mapsto \lambda|_{\mathbf{E}}$. Applying the functor $\bigwedge^r$ to the restriction map we get a map $\bigwedge^r({}^t\iota)\colon \bigwedge^r(\mathbf{V}^*) \to \bigwedge^r(\mathbf{E}^*)$ which we also denote by the symbol $|_{\mathbf{E}}$. By virtue of (31) in Chapter 5, we have

$$(\lambda^1 \wedge \cdots \wedge \lambda^r)|_{\mathbf{E}} = \lambda^1|_{\mathbf{E}} \wedge \cdots \wedge \lambda^r|_{\mathbf{E}} \quad \text{for all } \lambda^i \in \mathbf{V}^*.$$

Lemma 6.21 *Let μ be a volume element on a vector space* **V**. *Let* **E** *be a subspace of codimension* 1, *and* X a vector in **V** *which is not in* **E**. *Then* $(\mu \lrcorner X)|_{\mathbf{E}} = 0$ *if and only if* $X \in \mathbf{E}$.

Proof Let X be a vector in **V** which does not belong to **E**. Choose a basis $v_1, \ldots, v_n$ in **V** such that $v_1 = X$ and $v_2, \ldots, v_n \in \mathbf{E}$, and let $\lambda^1, \ldots, \lambda^n$ be the corresponding dual basis. Since $0 \neq \mu \in \bigwedge^n(\mathbf{V}^*)$, and $\bigwedge^n(\mathbf{V}^*)$ is a one-dimensional space we may assume that $\mu = \lambda^1 \wedge \cdots \wedge \lambda^n$. By definition (9) of the interior product, we have

$$\mu \lrcorner X = (-1)^{1+1} \lambda^2 \wedge \cdots \wedge \lambda^n,$$

because $\lambda^1(X) = 1$ and $\lambda^i(X) = 0$ for $i = 2, \ldots, n$. Since $\lambda^2|_{\mathbf{E}}, \ldots, \lambda^n|_{\mathbf{E}}$ is a basis of $\mathbf{E}^*$, it follows that $(\mu \lrcorner X)|_{\mathbf{E}} \neq 0$. Conversely, if $X \in \mathbf{E}$, a similar method shows that $(\mu \lrcorner X)|_{\mathbf{E}} = 0$.

It follows that if $X \notin \mathbf{E}$ then $(\mu \lrcorner X)|_{\mathbf{E}} \in \bigwedge^{n-1}(\mathbf{E})$ is a volume element on **E**.

If **V** is given an inner product, then, by working with orthonormal bases, we obtain the following result.

Corollary 6.22 *In addition to the hypotheses of the Lemma, let* **V** *be given some inner product* $\langle\,,\rangle$, *and choose an orientation on* **V** *and on* **E**. *If μ (resp. σ) denotes the canonical volume element on* **V** *(resp.* **E***) determined by* $\langle\,,\rangle$ *(resp. the restriction of* $\langle\,,\rangle$ *to* **E**), *then*

$$\sigma = (\mu \lrcorner n)|_{\mathbf{E}}$$

for a unique unit normal vector $n \in \mathbf{E}^{\perp}$.

6.6 Integration on manifolds

The support of a differential form $\alpha \in \Omega^k(M)$ is defined just as for functions, i.e. supp α is the closure of the set of points $x \in M$ such that $\alpha(x) \neq 0$. We denote by $\Omega_c^k(M)$ the set of k-forms α with compact support in M. If $\partial M \neq \emptyset$ we of course allow the possibility that supp α meets ∂M.

Although we did not mention it earlier, everything we have done in §§6.1 to 6.5 holds for manifolds with boundary. For points on the boundary, we have to use coordinate neighbourhoods which are diffeomorphic to open sets in the half-space $\bar{\mathbb{R}}^n_+$, but apart from that all the proofs remain the same (see the discussion in §5.11). The same holds for the definitions and proofs in this section (which is important for §6.7 where we prove Stokes' theorem on a manifold with boundary), but to focus on the essential ideas we continue to write out the proofs under the assumption that the boundary of M is empty.

The Change of Variables Theorem in $\mathbb{R}^n$ will enable us to define the integral of any differential form $\alpha \in \Omega_c^n(M)$ ($n = \dim M$) when M is an orientable manifold.

Theorem 6.23 (*Change of Variables*) *Let U and V be open in $\mathbb{R}^n$ and let $\varphi\colon V \to U$ be a C^1 diffeomorphism. Then for any integrable function $f \in L_1(U)$ we have $f \circ \varphi\, |J_\varphi| \in L_1(V)$ and*

$$\int_U f(x)\, dx = \int_V f(\varphi(y))\, |J_\varphi(y)|\, dy$$

where $J_\varphi(y) = \det D\varphi(y) = \partial x/\partial y$, the determinant of the Jacobian matrix.

Let U be an open set in $\mathbb{R}^n$. If $\alpha \in \Omega_c^n(U)$ then $\alpha = f\, dx_1 \wedge \cdots \wedge dx_n$ for some function $f \in C^\infty(U)$ with compact support. We let

$$\int_U \alpha = \int_U f(x_1, \ldots, x_n)\, dx$$

The dx in the integral on the right-hand side denotes Lebesgue measure on $\mathbb{R}^n$. The order of $dx_1, \ldots, dx_n$ in the Lebesgue integral makes no difference but it does in the differential form; for example in $\mathbb{R}^2$ we have

$$\int f\, dx_2 \wedge dx_1 = -\int f\, dx_1 \wedge dx_2.$$

Now let $\varphi\colon V \to U$ be an orientation-preserving diffeomorphism between open sets in $\mathbb{R}^n$. Let $x_1, \ldots, x_n$ and $y_1, \ldots, y_n$ be the standard coordinate functions on U and V, respectively. Then $\varphi_i = x_i \circ \varphi$ are the component functions of φ and we have

$$\begin{aligned} \varphi^*(f\, dx_1 \wedge \cdots \wedge dx_n) &= f(\varphi(y))\, d\varphi_1 \wedge \cdots \wedge d\varphi_n \\ &= f(\varphi(y)) \det D\varphi(y)\, dy_1 \wedge \cdots \wedge dy_n, \end{aligned}$$

the second equality holding because of the alternating property of the wedge product. For any $\alpha \in \Omega^n(U)$ with compact support we have $\varphi^*(\alpha) \in \Omega^n(V)$ with compact support equal to $\varphi^{-1}(\operatorname{supp} \alpha)$, and so by the Change of Variables Theorem 6.23 we have

$$\int_V \varphi^*\alpha = \int_U \alpha$$

If (U, κ) is a chart on a manifold M, we let $\kappa_* = (\kappa^{-1})^*$ denote the *push-forward* of a differential form defined on $U \subset M$ to a differential form defined on $\kappa(U) \subset \mathbb{R}^n$. Note that if $\alpha \in \Omega_c^n(U)$ then $\kappa_*\alpha \in \Omega_c^n(\kappa(U))$.

Corollary 6.24 *Let (U, κ) and (V, ψ) be charts on a manifold M such that the overlap map $\psi \circ \kappa^{-1}\colon \kappa(U \cap V) \to \psi(U \cap V)$ is orientation-preserving (in $\mathbb{R}^n$). If $\alpha \in \Omega^n(M)$ has compact support in $U \cap V$ then*

$$\int_{\kappa(U)} \kappa_*(\alpha_{|U}) = \int_{\psi(V)} \psi_*(\alpha_{|V})$$

Now assume that M is an orientable manifold, and that some orientation has been selected. Let (U, κ) be a positively oriented chart. If $\alpha \in \Omega^n(M)$ has compact support inside U we define

$$\int_M \alpha := \int_{\kappa(U)} \kappa_*(\alpha_{|U})$$

Due to Corollary 6.24, the definition of $\int_M \alpha$ does not depend on the choice of coordinate map κ. Given an n-form $\alpha \in \Omega_c^n(M)$ with arbitrary compact support we then define its integral by

$$\int_M \alpha := \sum_i \int_{U_i} \rho_i \alpha$$

where $\{\rho_i\}$ is a partition of unity subordinate to an oriented atlas $\{(U_i, \kappa_i)\}$.

Proposition 6.25 *The definition of $\int_M \alpha$ is independent of the oriented atlas $\{(U_i, \kappa_i)\}$ and the partition of unity $\{\rho_i\}$.*

Proof Let $\{V_j\}$ be another oriented atlas of M (with respect to the same orientation!) and $\{\chi_j\}$ a partition of unity subordinate to $\{V_j\}$. Then

$$\begin{aligned}\sum_i \int_{U_i} \rho_i \alpha &= \sum_{i,j} \int_{U_i} \rho_i \chi_j \alpha && \text{since } \Sigma_j \chi_j = 1\\ &= \sum_{i,j} \int_{V_j} \rho_i \chi_j \alpha && \text{since } \rho_i \chi_j \alpha \text{ has support in } U_i \cap V_j\\ &= \sum_j \int_{V_j} \chi_j \alpha && \text{since } \Sigma_i \rho_i = 1\end{aligned}$$

Remark If M is connected then there are just two orientations on M and changing the orientation reverses the sign of the integral $\int_M \alpha$.

Densities

We assumed above that the manifold is oriented so that we could use the calculus of differential forms. But even on nonoriented manifolds it is possible to integrate quantities known as "densities".

To see how a density should be defined, let us investigate the behaviour of a linear functional λ on $C_c(M)$ with respect to a change of coordinates. If (U, κ) is a chart then λ defines a linear functional λ_κ on $C_c(\kappa(U))$ by

$$\langle \lambda_\kappa, f \rangle = \langle \lambda, f \circ \kappa \rangle, \quad f \in C_c(\kappa(U))$$

($f \circ \kappa$ is defined as 0 outside U). Now, if (U', κ') is another chart we have

$$\begin{aligned}\langle \lambda_{\kappa'}, f \rangle &= \langle \lambda, f \circ \kappa' \rangle\\ &= \langle \lambda_\kappa, f \circ (\kappa' \circ \kappa^{-1}) \rangle, \quad f \in C_c(\kappa'(U' \cap U))\end{aligned}$$

For the sake of the argument let us suppose that $\lambda_\kappa = \tilde{s}_\kappa(x)\, dx$ for some

$\tilde{s}_\kappa \in C(\kappa(U))$; then we obtain

$$\int_{V'} f(x')\tilde{s}_{\kappa'}(x')\,dx' = \int_V f(\varphi(x))\tilde{s}_\kappa(x)\,dx, \quad f \in C_c(V'),$$

where $V = \kappa(U \cap U')$ and $V' = \kappa'(U \cap U')$ are open sets in $\mathbb{R}^n$ and $\varphi = \kappa' \circ \kappa^{-1}$ is a diffeomorphism between them. Now, by substituting $x' = \varphi(x)$ in the first integral, we see by the change of variables theorem that $\tilde{s}_\kappa(x) = \tilde{s}_{\kappa'}(\varphi(x)). \,|\det D\varphi(x)|$, or, expressed another way, the system of functions $s_\kappa = \tilde{s}_\kappa \circ \kappa \in C(U_\kappa, \mathbb{R}^1)$ (where $U_\kappa = U$ is the domain of κ) satisfies the transformation property

$$s_\kappa = |\det D(\kappa' \circ \kappa^{-1}) \circ \kappa| \cdot s_{\kappa'} \quad \text{in} \quad U \cap U' \tag{16}$$

Now, in view of Proposition 5.47 (the text preceding it), there exists a line bundle having 1×1 transition matrices

$$h_{\kappa\kappa'} = |\det D(\kappa' \circ \kappa^{-1}) \circ \kappa| \in GL(\mathbb{R}^1) \quad \text{in} \quad U \cap U'.$$

We denote this bundle by $|\Lambda|$ and call it the *density line bundle*. In classical notation we can also write $h_{\kappa'\kappa} = |\partial x'/\partial x|$, where $\partial x'/\partial x$ is the Jacobian determinant for the coordinate transformation $x \to x'$. A system of functions $s_\kappa \in C(U_\kappa, \mathbb{R}^1)$ satisfying (16), i.e.

$$s_\kappa = h_{\kappa\kappa'} \cdot s_{\kappa'},$$

is equivalent to having a section $s \in C(M, |\Lambda|)$, which we call a *density*. If $s \in C^\infty(M, |\Lambda|)$ we say that s is a C^∞ *density*.

The condition (16) can be written in an invariant form by pulling it back to the density bundle. Let $(U, \kappa) = (U, x_1, \ldots, x_n)$ be a chart with trivialization $\beta_U: |\Lambda|_U \to U \times \mathbb{R}$. The constant function $1 \in C^\infty(U, \mathbb{R})$ gives us the density $\beta_U^{-1} 1 \in C^\infty(U, |\Lambda|)$ and we denote it by the symbol $dx_1 \cdots dx_n$. Pulling back $s_\kappa \in C(U, \mathbb{R})$ by β_U^{-1} we obtain, therefore, the density $\beta_U^{-1} s_\kappa = s_\kappa\, dx_1 \cdots dx_n \in C(U, |\Lambda|)$. Now let $(U', x_1', \ldots, x_n')$ be another chart with a trivialization $\beta_{U'}: |\Lambda|_{U'} \to U' \times \mathbb{R}$. Then (16) can be written $s_\kappa = h_{\kappa\kappa'} s_{\kappa'}$ where $h_{\kappa\kappa'} = \beta_U \beta_{U'}^{-1}$. Since $dx_1' \cdots dx_n' := \beta_{U'}^{-1}\, 1$ we have

$$dx_1' \cdots dx_n' = \beta_{U'}^{-1}\, \beta_U\, dx_1 \cdots dx_n.$$

It is not hard to see that the map $\beta_{U'}^{-1}\, \beta_U: U \cap U' \to \mathrm{Hom}(|\Lambda|_{U \cap U'})$ is given by multiplication by the scalar function $h_{\kappa'\kappa} = (h_{\kappa\kappa'})^{-1}$, thus

$$dx_1' \cdots dx_n' = h_{\kappa'\kappa}\, dx_1 \cdots dx_n \quad \text{in} \quad U \cap U',$$

which means that (16) can be written equivalently as

$$s_\kappa\, dx_1 \cdots dx_n = s_{\kappa'}\, dx_1' \cdots dx_n' \quad \text{in} \quad U \cap U'.$$

The integral of a density is defined as follows. Choose an atlas $\{(U_i, \kappa_i)\}$ for M and a subordinate partition of unity $\{\psi_i\}$. If $s \in C_c(M, |\Lambda|)$ is a density with compact support, we set

$$\int_M s := \sum_i \int_{\mathbb{R}^n} (\psi_i \circ \kappa_i^{-1}) \cdot \tilde{s}_{\kappa_i}\, dx, \tag{17}$$

where dx is the Lebesgue measure on $\mathbb{R}^n$. Just as for n-forms it is not hard to show that the integral does not depend on the choice of atlas or partition of unity.

The measure associated with a Riemannian metric

Now let M be a Riemannian manifold. The Riemannian metric g defines a C^∞ density. For every chart $(U, x_1, \ldots, x_n)$ we let

$$\rho = \rho_U = (\det[g_{ij}])^{1/2} \quad \text{in} \quad U$$

where $[g_{ij}] = [g(\partial/\partial x_i, \partial/\partial x_j)]$ is the local representative of the metric on the chart. This system of functions does indeed satisfy the transformation property for if $(U', x'_1, \ldots, x'_n)$ is another chart then

$$\rho_U^{1/2} = \rho_{U'}^{1/2}\left|\frac{\partial x'}{\partial x}\right| \quad \text{in} \quad U \cap U'$$

by virtue of the proof of Proposition 6.17 since $\partial/\partial x_i = \sum_j a_{ij}\, \partial/\partial x'_j$ where $a_{ij} = \partial x'_j/\partial x_i$. We let $d\mu \in C^\infty(M, |\Lambda|)$ denote the corresponding C^∞ density, i.e.

$$d\mu = \rho_U^{1/2}\, dx_1 \cdots dx_n \quad \text{in} \quad U \tag{18}$$

If $f \in C_c(M)$ has compact support then $f \cdot d\mu \in C_c(M, |\Lambda|)$ and we call $\int_M f \cdot d\mu$ the integral of f with respect to $d\mu$. In view of (17) and (18) we have

$$\int_M f \cdot d\mu = \sum_i \int_{\mathbb{R}^n} (\psi_i \cdot f \cdot \rho_{U_i}^{1/2}) \circ \kappa_i^{-1}\, dx. \tag{19}$$

Let X be a locally compact Hausdorff space. A linear map $\lambda\colon C_c(X) \to \mathbb{C}$ is said to be a *positive* functional if $\lambda f \geqslant 0$ whenever f is real and $f \geqslant 0$. By the Riesz Representation Theorem (see [Ru 1]) there is a unique positive measure m defined on the Borel sets in X such that

$$\lambda f = \int f\, dm, \quad f \in C_c(X),$$

and such that $m(K) < \infty$ for every compact set $K \subset X$. Furthermore, if X is σ-compact then m is a regular measure.

In particular, a manifold M is a locally compact and σ-compact Hausdorff space to which we can apply this theorem. Let $d\mu \in C(M, |\Lambda|)$ be a density on M. It is not hard to see that the map $f \mapsto \int f \cdot d\mu$ is a positive linear functional on $C_c(M)$. By the Riesz Representation Theorem there is a unique regular Borel measure, m_μ, such that

$$\int f\, d\mu = \int f\, dm_\mu \quad \text{when} \quad f \in C_c(M)$$

The space $L_p(M, m_\mu)$, $p \in \mathbb{R}$, consists of all measurable functions f such that $|f|^p$ is integrable with respect to m_μ. For $p \geqslant 1$, the norm

$$\|f\| = \left(\int |f|^p\, dm_\mu\right)^{1/p}$$

makes $L_p(M, m_\mu)$ into a Banach space, where functions that differ only on a set of measure zero are identified. For $p = 2$, the space $L_2(M, m_\mu)$ is a Hilbert space with inner product

$$(f, g)_{L_2(M)} = \int_M f\bar{g}\, dm_\mu$$

From now on we always write $d\mu$ rather than dm_μ.

Exercise

1. Let $d\mu \in C(M, |\Lambda|)$ be a density and $f \in C_c(M)$. Verify that if U is an open set containing the support of f then $\int_M f\, d\mu = \int_U f\, d\mu$. ($U$ is not necessarily a chart domain.)

2. Let $d\mu \in C(M, |\Lambda|)$ be a density. Verify that the map $f \mapsto \int_M f \cdot d\mu$ is a positive linear functional on $C_c(M)$.

6.7 Stokes' theorem

Let M be an oriented manifold with boundary ∂M. Before we can state Stokes' theorem we need to verify that the boundary of an oriented manifold M is orientable, and then to choose the appropriate orientation on ∂M.

Recall from Proposition 5.69 that if φ: $U \to V$ is a diffeomorphism between open sets in $\bar{\mathbb{R}}^n_+$, then φ restricts to a diffeomorphism φ': $\partial U \to \partial V$. Furthermore, if $x = (x', 0) \in \partial U$ we have by virtue of (38) of Lemma 5.68 that

$$D\varphi(x) = \begin{pmatrix} D'\varphi'(x) & * \\ 0 & \partial_n\varphi_n(x) \end{pmatrix}.$$

Now, if the Jacobian determinant of φ is everywhere positive,

$$\det D\varphi(x) > 0 \quad \text{for all} \quad x \in U,$$

it follows that

$$\det D'\varphi'(x) > 0 \quad \text{for all} \quad x \in \partial U,$$

i.e. φ' also has positive Jacobian determinant everywhere.

We claim that *if M is orientable then its boundary ∂M is also orientable.* To see this let $\{(U_i, \kappa_i)\}$ be an oriented atlas on M, and restrict it to get an atlas on ∂M

$$U_i' = U_i \cap \partial M, \quad \kappa_i' = \kappa_i|_{U_i \cap \partial M}.$$

Since the overlap maps for the atlas $\{(U_i, \kappa_i)\}$ have positive Jacobian determinant everywhere, the overlap maps for the atlas $\{(U_i, \kappa_i)\}$ on ∂M also have positive Jacobian determinant everywhere.

Now, let the upper half-space $\bar{\mathbb{R}}^n_+ = \{x; x_n \geqslant 0\}$ in $\mathbb{R}^n$ be given the standard orientation $dx_1 \wedge \cdots \wedge dx_n$. The *boundary orientation on* $\partial\bar{\mathbb{R}}^n_+ = \mathbb{R}^{n-1} = \{x; x_n = 0\}$ is, by definition, the equivalence class of the volume form

$$\sigma = (-1)^n\, dx_i \wedge \cdots \wedge dx_{n-1} \tag{20}$$

The factor $(-1)^n$ is needed to make Stokes' theorem come out sign-free.

Let M be an oriented manifold, and let $\{(U_i, \kappa_i)\}$ be an oriented atlas on M. The *boundary orientation* on ∂M is defined by the following requirement: if κ is an orientation-preserving diffeomorphism of some open set U in M into the upper half-space $\bar{\mathbb{R}}^n_+$ then the orientation on $\partial U = \partial M \cap U$ is the equivalence class of $\kappa^*\sigma$, that is, if $m \in \partial U$ then the basis $v_1, \ldots, v_{n-1} \in T_m(\partial M)$ is positively oriented if the local representatives $\tilde{v}_j = d\kappa|_m(v_j)$, $j = 1, \ldots, n-1$, satisfy

$$\sigma(\tilde{v}_1, \ldots, \tilde{v}_{n-1}) = (-1)^n \det[\langle dx_i, v_j \rangle]_{i,j=1}^{n-1} > 0 \tag{21}$$

By the discussion above, this orientation does not depend on the choice of local coordinates. Moreover, it is also clear from (21) (by continuity) that the "compatibility condition" of Definition 6.15′ is satisfied, so we have indeed defined an orientation on $T(\partial M)$.

Let $v_1 \in T_m M$ be any outward-pointing vector at $m \in \partial M$. It is not hard to verify the following: *a basis $v_2, \ldots, v_n$ of $T_m(\partial M)$ is positively oriented if and only if $v_1, v_2, \ldots, v_n$ is a positively oriented basis in $T_m M$.*

Example Let $e_1, e_2, \ldots, e_n$ be the standard basis on $\mathbb{R}^n$. If M is the half-space $\{x \in \mathbb{R}^n;\ x_1 \leqslant 0\}$, then $e_2, \ldots, e_n$ is positively oriented in the boundary orientation on ∂M.

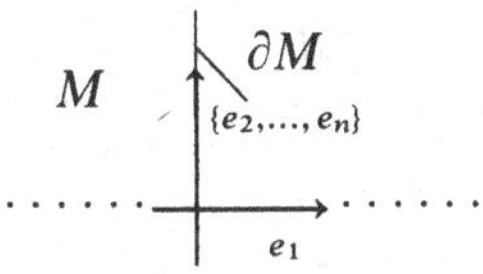

We also have the following continuation of Theorem 5.70.

Theorem 6.26 *If M and N are oriented manifolds and $\varphi: M \to N$ is an orientation-preserving diffeomorphism then the induced map $\partial M \to \partial N$ is also an orientation-preserving diffeomorphism with respect to the boundary orientations on ∂M and ∂N.*

Now we are ready to prove Stokes' theorem.

Theorem 6.27 *Let M be an oriented manifold and $\alpha \in \Omega_c^{n-1}(M)$ have compact support. Let $\iota: \partial M \to M$ denote the inclusion map. Then*

$$\int_M d\alpha = \int_{\partial M} \iota^*\alpha$$

or, for short,

$$\int_M d\alpha = \int_{\partial M} \alpha \tag{22}$$

If M has no boundary then $\int_M d\alpha = 0$.

Proof *Case* 1. M is diffeomorphic to an open set U in $\mathbb{R}^n$. By definition of the integral we may assume $M = U$. Write

$$\alpha = \sum_{i=1}^{n} f_i \, dx_1 \wedge \cdots \wedge \widehat{dx_i} \wedge \cdots \wedge dx_n$$

where the notation $\widehat{dx_i}$ indicates that dx_i is omitted. Then

$$d\alpha = \sum_{i=1}^{n} (-1)^{i-1} \frac{\partial f_i}{\partial x_i} dx_1 \wedge \cdots \wedge dx_n$$

so that

$$\int_U d\alpha = \sum_{i=1}^{n} \int_{\mathbb{R}^n} (-1)^{i-1} \frac{\partial f_i}{\partial x_i} dx_1 \cdots dx_n.$$

Now, $\int_{-\infty}^{\infty} \partial f_i/\partial x_i \, dx_i = 0$ since f_i has compact support; thus

$$\int_U d\alpha = 0.$$

Case 2. M is diffeomorphic to an open set U in $\bar{\mathbb{R}}^n_+$, and $\partial U \neq \varnothing$. Again we may assume $M = U$. The same result holds as in Case 1 except for the last term where we get by the fundamental theorem of calculus

$$\int_{\mathbb{R}^{n-1}} \int_0^{\infty} \frac{\partial f_n}{\partial x_n} dx_n \, dx_1 \cdots dx_{n-1} = -\int_{\mathbb{R}^{n-1}} f_n(x', 0) \, dx_1 \cdots dx_{n-1}$$

since f_n has compact support; thus

$$\int_U d\alpha = -\int_{\mathbb{R}^{n-1}} (-1)^{n-1} f_n(x', 0) \, dx_1 \cdots dx_{n-1}$$

On the other hand, since $\imath^*(dx_n) = d(x_n \circ \imath) = 0$ on $\partial \bar{\mathbb{R}}^n_+$, we have

$$\int_{\partial U} \imath^* \alpha = \int_{\partial \bar{\mathbb{R}}^n_+} \imath^* \alpha = \int_{\partial \bar{\mathbb{R}}^n_+} f_n(x', 0) \, dx_1 \wedge \cdots \wedge dx_{n-1}$$

which is equal to $\int_U d\alpha$ by virtue of (20).

Case 3. M is any manifold. If $\rho \in C^\infty(M)$ has its support contained inside some chart domain U, then the support of $\rho \cdot \alpha$ is a compact subset of U. Then

$$\int_M d(\rho\alpha) = \int_U d(\rho\alpha) = \int_{\partial U} \rho\alpha = \int_{\partial M} \rho\alpha$$

where the first and last equalities hold because of the support of $\rho \cdot \alpha$, and the middle equality holds because of Case 1 or 2. Now if we choose an oriented atlas on M and a subordinate partition of unity, then Stokes' formula (22) follows at once since both sides are linear in α.

Remark There is also a version of Stokes' theorem on nonoriented manifolds which involves the integration of densities. See [BT].

Let M be an oriented Riemannian manifold. The Riemannian metric and orientation on the tangent bundle TM determine a volume form $\mu_M \in \Omega^n(M)$. Also, the restriction of the Riemannian metric to $T(\partial M)$, together with the boundary orientation on ∂M, determine a volume form $\mu_{\partial M} \in \Omega^{n-1}(\partial M)$. Let m_M and $m_{\partial M}$ denote the measures associated with μ_M and $\mu_{\partial M}$.

In the next theorem we apply Stokes' theorem to a Riemannian manifold and thus obtain the Gauss (or Divergence) Theorem.

Theorem 6.28 *Let M be an oriented Riemannian manifold. Then for any vector field $X \in \mathscr{X}(M)$ with compact support we have*

$$\int_M (\operatorname{div} X)\, dm_M = \int_{\partial M} \langle X, n \rangle\, dm_{\partial M}$$

where n is the outward-pointing unit normal vector along ∂M.

Proof We have $\operatorname{div} X \cdot \mu = d(\mu \lrcorner X)$, hence Stokes' theorem implies

$$\int_M \operatorname{div} X \cdot \mu = \int_{\partial M} i^*(\mu \lrcorner X)$$

where i is the inclusion map $\partial M \to M$. For the right-hand side observe that $X = (X \cdot n)n + v$, where v is tangent along ∂M. In view of Lemma 6.21 and its corollary, we see that

$$i^*(\mu \lrcorner X) = (X \cdot n)\mu_{\partial M} + 0$$

and Gauss' Theorem follows at once.

Remark We see from the proof that if X is an outward-pointing vector then the boundary orientation on ∂M is the equivalence class of the volume element

$$i^*(\mu \lrcorner X) \in \Omega^{n-1}(T^*(\partial M)).$$

6.8 Differential operators in vector bundles

Let M be a manifold without boundary. Let E and F be complex vector bundles on M, and let $A: C^\infty(M, E) \to C^\infty(M, F)$ be a linear operator. We say that A is a *local operator* if $Au(x_0) = 0$ whenever u vanishes in a neighbourhood of x_0. This implies that if $u = 0$ in an open set U then $Au = 0$ in U, so that

$$\operatorname{supp} Au \subset \operatorname{supp} u \quad \text{for all } u \in C^\infty(M, E)$$

We let $C_0^\infty(M, E)$ denote the sections $u \in C^\infty(M, E)$ with compact support.

Lemma 6.29 *If $A: C^\infty(M, E) \to C^\infty(M, F)$ is a local operator then for any open set $U \subset M$ there is a unique local operator $A_U: C^\infty(U, E_U) \to C^\infty(U, F_U)$ such that*

$$A_U(f|_U) = (Af)|_U \tag{23}$$

for all $f \in C^\infty(M, E)$.

Proof Any $f \in C_0^\infty(U, E_U)$ with compact support in U can be extended to a section $f \in C^\infty(M, E)$ by setting it equal to 0 outside U. In this way the equation (23) defines an operator $C_0^\infty(U, E_U) \to C^\infty(U, F_U)$ which for convenience we denote by the same symbol A. Now, if $f \in C^\infty(U, E_U)$ we define

$$A_U f(x) = A(\psi f)(x), \quad x \in U,$$

where $\psi \in C_0^\infty(U, \mathbb{R})$ is a bump function equal to 1 on a neighbourhood of x. Since A is local, the value of $A(\psi f)(x)$ does not depend on the choice of ψ. It is not hard to show that A_U is local and satisfies (23). Uniqueness is also clear.

We summarize the equation (23) by saying that "A_U is equal to A on U". The same kind of argument shows that A_U is equal to A_V on $U \cap V$. Thus it makes sense to denote the restrictions A_U by the same symbol A. We always do this from now on.

So let $A: C^\infty(M, E) \to C^\infty(M, F)$ be a local operator and let $(U, \kappa) = (U, x_1, \dots, x_n)$ be local coordinates on M, and $e_1, \dots, e_p$ and $e'_1, \dots, e'_q$ local bases of E and F on U, respectively. If $u = \sum_{j=1}^{p} u_j e_j$ is a section of $E|_U$, then $Au = \sum_{i=1}^{q} (Au)_i e'_i$ is a section of $F|_U$, and we have

$$(Au)_i = \sum_{j=1}^{p} A_{ij} u_j, \quad i = 1, \dots, q \tag{24}$$

where $A_{ij}: C^\infty(U) \to C^\infty(U)$ are linear operators.

Before stating the definition of a differential operator we need the following notion. Let (U, κ) be a chart at x_0. The partial differentiation operator $\partial^\alpha = \partial_{x_1}^{\alpha_1} \cdots \partial_{x_n}^{\alpha_n}$, where $\partial_{x_j} = \partial/\partial x_j$, is defined on functions $f \in C^\infty(U, \mathbb{C})$ by transporting the partial derivatives of functions in $C^\infty(\kappa(U), \mathbb{C})$ using the chart κ, that is, $\partial^\alpha f := (\partial^\alpha(f \circ \kappa^{-1})) \circ \kappa$ (the definition is chart-dependent of course). We say that $f \in C^\infty(M, \mathbb{C})$ vanishes to kth order at x_0 if

$$\partial^\alpha f(x_0) = 0 \quad \text{when} \quad |\alpha| \leqslant k \quad \text{(for any chart at } x_0\text{)},$$

that is, all derivatives of order $\leqslant k$ vanish at x_0. By the chain rule this latter condition makes intrinsic sense, independent of the chart at x_0.

If $f \in C^\infty(M, \mathbb{C}^p)$ is a vector-valued function then it is said to vanish to kth order at x_0 if each of its component functions vanishes to kth order at x_0.

Finally, if E is a complex vector bundle over M and $f \in C^\infty(M, E)$, we say that f vanishes to kth order at x_0 if it does so in terms of local coordinates at x_0, i.e. for each neighbourhood U of x_0 such that $E|_U$ is trivial:

the local representative $\tilde{f}: U \to \mathbb{C}^p$ vanishes to kth order at x_0.

This condition makes intrinsic sense, independent of the local trivialization of E (and chart at x_0). Recall that the local representative $\tilde{f}: U \to \mathbb{C}^p$ is the vector function $(f_1, \dots, f_p)$ where $f_j \in C^\infty(U, \mathbb{C})$ are defined by $f = \sum f_j e_j$.

Definition 6.30 *A kth order linear differential operator from E to F is a linear map $A: C^\infty(M, E) \to C^\infty(M, F)$ such that for all $x_0 \in M$:*

$$Au(x_0) = 0 \quad \textit{whenever } u \in C^\infty(M, E) \textit{ vanishes to } k\textit{th order at } x_0.$$

We denote by $\mathrm{Diff}_k(E, F)$ the set of kth order differential operators from E to F. Note that if $A \in \mathrm{Diff}_k(E, F)$ then also $A \in \mathrm{Diff}_{k+1}(E, F)$. Evidently, a differential operator is a local operator.

Proposition 6.31 *Let A be a local operator $C^\infty(M, E) \to C^\infty(M, F)$. A necessary and sufficient condition for A to be a kth order differential operator,*

$$A \in \mathrm{Diff}_k(E, F),$$

is the following: for each chart $(U, x_1, \ldots, x_n)$ on M the operators A_{ij} in equation (24) are kth order differential operators in the usual sense in the coordinates $x_1, \ldots, x_n$,

$$A_{ij}(x, \partial) = \sum_{|\alpha| \leqslant k} a_{ij}^{(\alpha)}(x) \frac{\partial^{\alpha_1}}{\partial x_1^{\alpha_1}} \cdots \frac{\partial^{\alpha_n}}{\partial x_n^{\alpha_n}}$$

Proof The sufficiency of this condition is obvious. To prove necessity, we use Taylor's formula. By virtue of (24) we may assume that $A \in \mathrm{Diff}_k(U \times \mathbb{C}, U \times \mathbb{C})$, i.e. a scalar operator.

Let $x_0 \in U$ and $\kappa(x_0) = 0$. If $f \in C^\infty(U, \mathbb{C})$ then by transporting Taylor's formula for $f \circ \kappa^{-1}$ by the chart κ we obtain

$$f(x) = \sum_{|\alpha| \leqslant k} \frac{1}{\alpha!} \partial^\alpha f(x_0) x^\alpha + R_{k+1}(x), \quad x \in U \tag{25}$$

where

$$R_{k+1}(x) = \sum_{|\alpha| \leqslant k+1} x^\alpha \cdot g_{k,\alpha}(x) \quad \text{and} \quad g_{k,\alpha} \in C^\infty(U, \mathbb{C}).$$

Note: By x^α we mean the function in $C^\infty(U, \mathbb{R})$ defined by $x_1^{\alpha_1} \cdots x_n^{\alpha_n}$ where $x_i \in C^\infty(U, \mathbb{R})$ are the component functions of the chart κ. Since $\kappa(x_0) = 0$ we have $x_i(x_0) = 0$ for $i = 1, \ldots, n$. Then x^α, $|\alpha| = k + 1$, is a product of $k + 1$ functions vanishing at x_0, so R_{k+1} vanishes to kth order at x_0. Since $A \in \mathrm{Diff}_k(U \times \mathbb{C}, U \times \mathbb{C})$ then by applying A to (25) we obtain that

$$Af(x_0) = \sum_{|\alpha| \leqslant k} \frac{1}{\alpha!} \partial^\alpha f(x_0) \cdot A(x^\alpha)(x_0) + 0.$$

This holds for any $x_0 \in U$; thus $A = \sum_{|\alpha| \leqslant k} a_\alpha \partial^\alpha$ where $a_\alpha = A(x^\alpha)/\alpha!$.

Next, we define the principal symbol of a differential operator. We let $p\colon T^*M \to M$ denote the cotangent bundle projection and then $p^{-1}E$ and $p^{-1}F$ are the pull-backs of E and F to T^*M. Recall that the set of vector bundle homomorphisms from $p^{-1}E$ to $p^{-1}F$ is the set of C^∞ sections of $\mathrm{Hom}(p^{-1}E, p^{-1}F)$, that is,

$$\mathrm{HOM}(p^{-1}E, p^{-1}F) = C^\infty(T^*M, \mathrm{Hom}(p^{-1}E, p^{-1}F)).$$

Definition 6.32 *Let $A \in \mathrm{Diff}_k(E, F)$. The* principal symbol *of order k is the vector bundle homomorphism $\pi_k A \in C^\infty(T^*M, \mathrm{Hom}(p^{-1}E, p^{-1}F))$ defined as* *follows: for any* $(x, \xi) \in T^*M$ *and* $c \in E_x$ *we let*

$$\pi_k A(x, \xi)c = A\left(\frac{1}{k!}\varphi^k \cdot u\right)(x) \tag{26}$$

where $\varphi \in C^\infty(M)$ *and* $u \in C^\infty(M, E)$ *are chosen such that* $\varphi(x) = 0$, $d_x\varphi = \xi$ *and* $u(x) = c$. *The map* $\pi_k\colon \mathrm{Diff}_k(E, F) \to C^\infty(T^*M, \mathrm{Hom}(p^{-1}E, p^{-1}F))$ *is called the principal symbol map of order k.*

Of course we must show that the definition of $\pi_k A$ is independent of the choice of φ and u. Let $(U, x_1, \ldots, x_n)$ be local coordinates on M such that E and F have local bases on U. The elements $c \in E_x$ and $\pi_k A(x, \xi)c \in F_x$ can then be regarded as column vectors in $\mathbb{C}^p$ and $\mathbb{C}^q$, respectively, and we have

$$\pi_k A(x, \xi)c = \sum_{|\alpha| = k} a_\alpha(x)\xi^\alpha \cdot c \tag{26'}$$

where $a_\alpha = [a_{ij}^{(\alpha)}]_{q \times p}$ is a smooth matrix function on U, $\xi^\alpha = \xi_1^{\alpha_1} \cdots \xi_n^{\alpha_n}$ and the ξ_i's are the coordinates of $\xi = \Sigma \xi_i\, dx_i$. Since (26′) does not depend on the choice of φ and u, then neither does (26). At the same time, we have shown that the local definition (26′) is independent of the choice of coordinates and local bases in E and F. Furthermore, (26′) shows that $\pi_k A(x, \xi)$ is a linear map $E_x \to F_x$, and, since the expression on the right-hand side depends smoothly on x and ξ, we have defined a vector bundle homomorphism $\pi_k A\colon p^{-1}E \to p^{-1}F$, i.e. a C^∞ section from T^*M to $\mathrm{Hom}(p^{-1}E, p^{-1}F)$.

Remark The representation (26′) corresponds to the local representation

$$\tilde{A} = [A_{ij}] = \sum_{|\alpha| \leq k} a_\alpha(x)\, \partial^\alpha, \tag{27}$$

in Proposition 6.31.

We say that $s \in C^\infty(T^*M, \mathrm{Hom}(p^{-1}E, p^{-1}F))$ is a kth *order homogeneous symbol* if

$$s(x, \lambda\xi) = \lambda^k \cdot s(x, \xi) \quad \text{for all } \lambda > 0.$$

The set of all such symbols is denoted $\mathrm{Symbl}_k(E, F)$ and is a linear subspace of $C^\infty(T^*M, \mathrm{Hom}(p^{-1}E, p^{-1}F))$. If $A \in \mathrm{Diff}_k(E, F)$ then (26′) implies that $\pi_k A \in \mathrm{Symbl}_k(E, F)$. Not every element of $\mathrm{Symbl}_k(E, F)$, however, is the principal symbol of a differential operator.

Let $\mathrm{DSymbl}_k(E, F)$ denote the set of homomorphisms $s \in C^\infty(T^*M, \mathrm{Hom}(p^{-1}E, p^{-1}F))$ such that for all $x \in M$, $s_x\colon T_x^*M \to L(E_x, F_x)$ is a homogeneous polynomial map as defined in §5.1, that is, $s_x \in P^{(k)}(T_x^*M, L(E_x, F_x))$. It is clear that π_k maps $\mathrm{Diff}_k(E, F)$ into $\mathrm{DSymbl}_k(E, F)$.

Recall that a sequence of linear maps $\mathbf{E}_1 \xrightarrow{f_1} \mathbf{E}_2 \xrightarrow{f_2} \cdots \mathbf{E}_n \xrightarrow{f_n} \mathbf{E}_{n+1}$ is called *exact* if the image of f_i is equal to the kernel of f_{i+1} for $i = 1, 2, \ldots, n-1$.

Theorem 6.33 *The following sequence of complex vector spaces and linear maps is exact:*

$$0 \to \mathrm{Diff}_{k-1}(E, F) \to \mathrm{Diff}_k(E, F) \xrightarrow{\pi_k} \mathrm{DSymbl}_k(E, F) \to 0$$

Proof The sequence is obviously exact at the first arrow since the inclusion map $\mathrm{Diff}_{k-1}(E, F) \to \mathrm{Diff}_k(E, F)$ is injective. For the second arrow, if $A \in \mathrm{Diff}_{k-1}(E, F)$ then $\pi_k A = 0$; in fact by (26) we have $\pi_k A(x, \xi)c = 0$ since $\varphi^k u$ vanishes to $(k-1)$th order at x. Conversely, if $A \in \mathrm{Diff}_k(E, F)$ then locally A has the representation

$$Au = \sum_{i=1}^{q} \sum_{j=1}^{p} A_{ij} u_j e'_i \quad \text{where} \quad A_{ij} = \sum_{|\alpha| \leqslant k} a_{ij}^{(\alpha)} \partial^\alpha.$$

If $\pi_k A = 0$, then (26′) implies that $a_{ij}^{(\alpha)} = 0$ when $|\alpha| = k$, and hence $A \in \mathrm{Diff}_{k-1}(E, F)$ so we have exactness at the second arrow.

To complete the proof we must show that the principal symbol map π_k is surjective. Let $s \in \mathrm{DSymbl}_k(E, F)$. We cover M with charts (U_i, κ_i) such that E_{U_i} and F_{U_i} are trivial. In local coordinates on U_i we have $T^*(U_i) \simeq U_i \times \mathbb{R}^n$, $E_{U_i} \simeq U_i \times \mathbb{C}^p$ and $F_{U_i} \simeq U_i \times \mathbb{C}^q$ and the local representation of s is a kth order homogeneous polynomial of $\xi \in \mathbb{R}^n$ with $q \times p$ matrix coefficients. It follows that there exists $A_i \in \mathrm{Diff}_k(E_{U_i}, F_{U_i})$ such that $\pi_k A_i = s$ on the fibres of $p^{-1}E$ over U_i. Now let $1 = \sum \psi_i$ be a partition of unity subordinate to the cover U_i and set $A = \sum \psi_i A_i$. Every point $x \in M$ has a neighbourhood that intersects only finitely many of the supports of ψ_i, so it follows that $A \in \mathrm{Diff}_k(E, F)$ and

$$\pi_k A(x, \xi) = \sum_i \psi_i(x) \cdot \pi_k A_i(x, \xi) = \sum \psi_i(x) \cdot s(x, \xi) = s(x, \xi).$$

Definition 6.34 *A differential operator* $A \in \mathrm{Diff}_k(E, F)$ *is elliptic at* $x \in M$ *if*

$$\pi_k A(x, \xi) \textit{ is an isomorphism } E_x \to F_x \textit{ for all } (x, \xi) \in T^*M \setminus 0.$$

The operator A is said to be elliptic if it is elliptic at each $x \in M$. From now on we write $\pi = \pi_k$ if no confusion is possible.

Let the tangent bundle TM be given some Riemannian metric $g = \langle\ ,\ \rangle$. This metric induces an "index lowering" operator $T_x M \to T_x^* M$, defined by $v \mapsto \langle v, \cdot \rangle_x$ on each fibre, which gives rise to a Riemannian metric on the cotangent bundle T^*M. Thus, we can consider the *unit cotangent bundle*,

$$ST^*M = \bigcup_{x \in M} \{\xi \in T_x^* M; \langle \xi, \xi \rangle_x = 1\},$$

which is a submanifold of T^*M (see §5.10, Exercise 1). Due to the homogeneity of the principal symbol, the differential operator A is elliptic if $\pi A(x, \xi)$ *is an isomorphism for all* $(x, \xi) \in ST^*M$.

Let $(U, x_1, \ldots, x_n)$ be a chart on M. With respect to the local basis $\partial/\partial x_1, \ldots, \partial/\partial x_n$ of TM, the Riemannian metric on TM has matrix $[g_{ij}]$ where $g_{ij} \in C^\infty(U)$ are given by

$$g_{ij}(x) = \langle \partial/\partial x_i, \partial/\partial x_j \rangle_x,$$

and, with respect to the local basis $dx_1, \ldots, dx_n$ of T^*M, the induced Riemannian metric on T^*M has matrix $[g^{ij}] = [g_{ij}]^{-1}$. We have

$$ST^*M|_U = \bigcup_{x \in U} \{\xi \in T_x^*(M); \sum g^{ij}(x)\xi_i\xi_j = 1, \quad \text{where } \xi = \sum \xi_i \, dx_i|_x\}$$

Let E be a Hermitian bundle, i.e. a $\mathbb{C}$-vector bundle with a C^∞ inner product $(,)_{E_x}$ on the fibers (see §5.10). Let $d\mu$ be the measure on M induced by the Riemannian metric g (see §6.5). We define an inner product on $C_0^\infty(M, E)$ by

$$(u, v)_{L_2(M,E)} = \int_M (u(x), v(x))_{E_x} \, d\mu(x), \quad u, v \in C_0^\infty(M, E).$$

$L_2(M, E)$ is defined to be the completion of $C_0^\infty(M, E)$ with respect to this inner product.

If E and F are Hermitian bundles and $A: C^\infty(M, E) \to C^\infty(M, F)$ is a differential operator, then the (formal) *adjoint* A^* is the linear operator $C_0^\infty(M, F) \to L_2(M, E)$ defined by

$$(u, A^*v)_{L_2(M,E)} = (Au, v)_{L_2(M,F)} \tag{28}$$

for all $u \in C_0^\infty(M, E)$ and $v \in C_0^\infty(M, F)$.

We claim that the adjoint A^* is also a differential operator. Since A^* is a local operator (see Exercise 7), it suffices to show this in a chart $(U, \kappa) = (U, x_1, \ldots, x_n)$ on M for which E_U and F_U are trivial. By virtue of the Gram-Schmidt orthonormalization procedure, E and F have smooth local bases that are orthonormal for all $x \in U$:

$$(e_i(x), e_j(x))_{E_x} = \delta_{ij} \quad \text{and} \quad (f_i(x), f_j(x))_{F_x} = \delta_{ij} \tag{29}$$

Now, if A has the local representation (27), then from (28) applied to $u \in C_0^\infty(U, E_U)$ and $v \in C_0^\infty(U, E_U)$ it follows that the local representation of A^* is

$$\tilde{A}^*u = \sum_{|\alpha| \leq k} (-1)^{|\alpha|} \frac{1}{\rho} \partial^\alpha(\rho \, a_\alpha^h \, u), \quad u \in C^\infty(U, \mathbb{C}^q), \tag{30}$$

where $\rho = \det[g_{ij}]$ is the Riemannian volume density and a_α^h is the Hermitian adjoint (conjugate transpose) of the matrix a_α. Thus, $A^*u \in C^\infty(M, F)$, and since A^* is local it extends in a unique way to an operator $A^*: C^\infty(M, F) \to C^\infty(M, E)$. Hence A^* is a differential operator of order $\leq k$.

By considering just the highest order terms in (27), (30) we find that locally the principal symbols of A and A^* are related by

$$\pi_k \tilde{A}^*(x, \xi) = (-1)^k \sum_{|\alpha| = k} a_\alpha^h(x)\xi^\alpha = (-1)^k [\pi_k \tilde{A}(x, \xi)]^h,$$

and we have proved the first part of the following proposition, since to verify it, it suffices to do so locally.

Proposition 6.35 *Let E and F be Hermitian bundles over M. If $A \in \mathrm{Diff}_k(E, F)$ then its adjoint $A^* \in \mathrm{Diff}_k(F, E)$ has principal symbol*

$$\pi_k A^*(x, \xi) = (-1)^k[\pi_k A(x, \xi)]^h \tag{31}$$

where the superscript ^{h} indicates the Hermitian adjoint $F_x \to E_x$ of a linear map $E_x \to F_x$. Also, $A^{**} = A$ and if $B \in \mathrm{Diff}_{k'}(E', E)$ then $(A \circ B)^* = B^* \circ A^*$.

The proof of the second part is left as an exercise.

Exercises

1. Let $\{U_i\}$ be an open cover of M and let E and F be complex vector bundles over M. Given local operators

$$A_i\colon C^\infty(U_i, E_{U_i}) \to C^\infty(U_i, F_{U_i}) \quad \text{with} \quad A_{U_i} = A_{U_j} \quad \text{on} \quad U_i \cap U_j$$

show that there exists a unique local operator $A\colon C^\infty(M, E) \to C^\infty(M, F)$ such that $A_i(f|_{U_i}) = (Af)|_{U_i}$ for all $f \in C^\infty(E)$. If all A_i's are kth order differential operators show that the same is true of A.

2. Show that a linear map $A\colon C^\infty(M, E) \to C^\infty(M, F)$ is a first-order differential operator if and only if the following holds: $A(fg)(x_0) = 0$ whenever $f \in C^\infty(M, \mathbb{R})$ and $g \in C^\infty(M, E)$ satisfy $f(x_0) = 0$ and $g(x_0) = 0$.

3. Let X be a vector field on M. Then we have the derivation (or directional derivative)

$$D_X\colon C^\infty(M, \mathbb{R}) \to C^\infty(M, \mathbb{R})$$

defined by $D_X f = Xf := df \circ X$. Show that $D_X \in \mathrm{Diff}_1(M \times \mathbb{R}, M \times \mathbb{R})$ and find the principal symbol $\pi_1(D_X)$.

4. Let M be an oriented Riemannian manifold. The Riemannian metric g on TM induces a Riemannian metric on the cotangent bundle T^*M by the index lowering operator defined in §5.10. Show that the Laplacian $\Delta\colon C^\infty(M) \to C^\infty(M)$ is a second-order differential operator on M with principal symbol $\|\xi\|^2 = g(\xi, \xi)$.

5. Show that if $A \in \mathrm{Diff}_r(E, F)$ and $B \in \mathrm{Diff}_s(E', E)$ then $A \circ B \in \mathrm{Diff}_{r+s}(E', F)$. Prove that the principal symbol is multiplicative, that is, $\pi_{r+s}(A \circ B) = \pi_r A \circ \pi_s B$.

6. Let $A\colon C^\infty(M, E) \to C^\infty(M, F)$ be a differential operator of order k. Fix $(x, \xi) \in T^*M$. Show that for any $\varphi \in C^\infty(M)$ and $u \in C^\infty(M, E)$ such that $d\varphi_x = \xi$ and $u(x) = c$ we have

$$\pi A(x, \xi)c = \lim_{t \to \infty} t^{-k} e^{-\iota\varphi} A(e^{\iota\varphi} u)(x), \quad c \in E_x$$

7. Let E and F be Hermitian vector bundles over M. Show that if $A\colon C^\infty(M, E) \to C^\infty(M, F)$ is a differential operator then its adjoint A^* defined by (28) is a local operator. If A has the local representation $[A_{ij}]_{q \times p}$

(see (24)) with respect to the local bases (29), show that A^* has the local representation $[A^*_{ji}]_{p \times q}$.

8. Verify the local formula (30) for the adjoint of a differential operator.

9. Prove Proposition 6.35.

10. Recall that the operators div and grad were defined at the end of §6.5.
(a) Show that div: $C^\infty(M, TM) \to C^\infty(M)$ and grad: $C^\infty(M) \to C^\infty(M, TM)$ are first-order differential operators and $(\text{div})^* = -\text{grad}$.
(b) The operators div and grad are operators on sections of the real vector bundles TM and $M \times \mathbb{R}$. If we complexify the bundles to obtain Hermitian vector bundles $TM \otimes \mathbb{C}$ and $M \times \mathbb{C}$ (see §5.1) then we have induced operators

$$\text{div}: C^\infty(M, TM \otimes \mathbb{C}) \to C^\infty(M, \mathbb{C})$$

and

$$\text{grad}: C^\infty(M, \mathbb{C}) \to C^\infty(M, TM \otimes \mathbb{C}).$$

Show that the formula $(\text{div})^* = -\text{grad}$ still holds.

11. If E is a Hermitian bundle over M then we say that $A \in \text{Diff}_{2k}(E, E)$ is *strongly elliptic* of order $2k$ if, for all $(x, \xi) \in T^*M\backslash 0$, we have

$$((-1)^k \pi_{2k} A(x, \xi)v, v)_E > 0 \quad \text{for all } v \in E_x, v \neq 0.$$

(a) Let $A \in \text{Diff}_k(E, F)$. Show that $A^*A \in \text{Diff}_{2k}(E, E)$ is strongly elliptic if and only if $\pi_k A(x, \xi)$: $E_x \to F_x$ is injective for all $(x, \xi) \in T^*M\backslash 0$.
(b) If E and F have the same rank (fibre dimension) show that A is kth order elliptic if and only if A^*A is $2k$th order strongly elliptic.

If I and J are ideals in a commutative ring R then $I \cdot J$ denotes the ideal generated by all products fg, $f \in I$, $g \in J$. Clearly, $I \cdot J$ is the set of all finite sums $\sum f_i g_i$ where $f_i \in I$ and $g_j \in J$. Since $I(JK) = (IJ)K$ for any three ideals I, J and K in R, it makes sense to define $I^k = I \cdots I$, and then I^k is the ideal in R generated by products of k elements of I.

Note that $C^\infty(M, \mathbb{C})$ is a commutative ring, with addition and multiplication of functions defined pointwise.

12. Fix $x_0 \in M$, and let I_{x_0} denote the ideal in $C^\infty(M, \mathbb{C})$ consisting of the functions $f \in C^\infty(M, \mathbb{C})$ vanishing at x_0, i.e. $f(x_0) = 0$. Then, f vanishes to kth order at x_0 if and only if $f \in I^{k+1}_{x_0}$, that is,

$$f \text{ is a finite sum of products of } k+1 \text{ functions in } I_{x_0}.$$

(Hint: The sufficiency of this condition is clear. To prove necessity, use Taylor's formula in local coordinates near x_0.)

For the following exercise, recall that if $\mathcal{M}$ is a module over R and I is an ideal in R then $I \cdot \mathcal{M}$ denotes the submodule generated by the products $a \cdot f$, $a \in I$, $f \in \mathcal{M}$; it consists of all finite sums $\sum a_i f_i$ where $a_i \in I$ and $f_i \in \mathcal{M}$.

13. If E is a complex vector bundle over M, then $C^\infty(M, E)$ is a module over the ring $C^\infty(M, \mathbb{C})$, with addition and multiplication defined pointwise.

Fix a point $x \in M$, and let Z_x^k denote the submodule of $C^\infty(M, E)$ consisting of sections vanishing to kth order at x. Show that

$$Z_x^k = I_k^{k+1} \cdot C^\infty(M, E).$$

14. Let E be a complex vector bundle over M. Define the complex vector space of "k-jets" at x by

$$J^k(E)_x = C^\infty(M, E)/Z_x^k$$

where Z_x^k is defined in Exercise 13. The canonical linear map $C^\infty(M, E) \to J^k(E)_x$ is denoted by $f \mapsto j_k(f)_x = [f]_x$. Show that the family of complex vector spaces

$$J^k(E) = \bigcup_x J^k(E)_x$$

has a unique vector bundle structure such that $j_k(f) \in C^\infty(M, J^k(E))$ for all $f \in C^\infty(M, E)$. Hint: For any trivialization $\varphi\colon E|_U \to U \times \mathbb{C}^p$, define the map

$$J^k(E)_{|U} \to U \times {}_{|\alpha| \leqslant k}\mathbb{C}^p$$
$$[f]_x \mapsto (x, (\partial^\alpha \tilde{f}(x))_{|\alpha| \leqslant k}), \quad x \in U,$$

where $\tilde{f}$ is the local representative of f. Now apply Proposition 5.47.

6.9 The Hodge star operator and the Laplace–de Rham operator

Let M be a manifold with some Riemannian metric g on TM. The goal of this section is to define the Laplace–de Rham operator on differential forms. But first it is necessary to define the Hodge star operator, and in order to do so we need to verify that the inner product $\langle\,,\rangle_x$ on T_xM induces in a natural way an inner product on the pth exterior product $\bigwedge^p(T_x^*M)$.

It is convenient to consider this problem on the level of abstract inner product spaces, so that we can use orthonormal bases rather than being confined to the coordinate vectors ∂/x_i.

Thus let $\mathbf{E}$ be a vector space over $\mathbb{R}$ and let $\langle\,,\rangle$ be an inner product in $\mathbf{E}$. From §5.10 we have the index-lowering operator $_\#\colon \mathbf{E} \to \mathbf{E}^*$ given by $v \mapsto \langle v, \cdot\,\rangle$. Due to the functoriality of $\bigwedge^p$, there is an induced map $\bigwedge^p(_\#)\colon \bigwedge^p(\mathbf{E}) \to \bigwedge^p(\mathbf{E}^*)$ such that

$$\xi = w_1 \wedge \cdots \wedge w_p \mapsto \xi_\# = (w_1)_\# \wedge \cdots \wedge (w_p)_\#, \tag{32}$$

which we also refer to as the index-lowering operator. Let us examine how this map looks in terms of a basis for E. Let $v_1, \ldots, v_n$ be any basis for $\mathbf{E}$. If $\xi \in \bigwedge^p\mathbf{E}$ then

$$\xi = \sum_{i_1 < \cdots < i_p} \xi^{i_1 \ldots i_p} v_{i_p} \wedge \cdots \wedge v_{i_p} \tag{33}$$

for unique real numbers $\xi^{i_1 \ldots i_p}$. We can extend the definition of the coordinates $\xi^{i_1 \ldots i_p}$, so that they are defined for all $i_1, \ldots, i_p$ and alternating (anti-symmetric) in those indices. Then

$$\xi = \frac{1}{p!} \sum \xi^{i_1 \ldots i_p} v_{i_p} \wedge \cdots \wedge v_{i_p} \quad \text{(sum over all } i_1, \ldots, i_p). \tag{33'}$$

Note that *distinct* indices $i_1, \ldots, i_p$ can be ordered in $p!$ ways and the terms in the sum are a product of two alternating terms, hence symmetric in $i_1, \ldots, i_p$.

Lemma 6.36 *As usual, let $g_{ij} = \langle v_i, v_j \rangle$ and let $\lambda^1, \ldots, \lambda^n$ be the dual basis in* $\mathbf{E}^*$ *corresponding to the given basis in* $\mathbf{E}$. *Then $\xi_\# \in \bigwedge^p(\mathbf{E}^*)$ is given by*

$$\xi_\# = \sum_{j_1 < \cdots < j_p} \xi_{j_1 \ldots j_p} \lambda^{j_1} \wedge \cdots \wedge \lambda^{j_p}, \tag{34}$$

where

$$\xi_{j_1 \ldots j_p} = \sum \xi^{i_1 \ldots i_p} g_{i_1 j_1} \cdots g_{i_p j_p} \quad (\textit{sum over } \text{all } i_1, \ldots, i_p). \tag{35}$$

Proof The index-lowering operator $\mathbf{E} \to \mathbf{E}^*$ is given by

$$\sum X^i v_i \mapsto \sum X_i \lambda^i, \quad \text{where } X_i = \sum g_{ij} X^j.$$

In particular, the basis vector $v_i \in \mathbf{E}$ is mapped to $\sum g_{ij} \lambda^j \in \mathbf{E}^*$, so the action of (32) on the corresponding basis vectors in $\bigwedge^p \mathbf{E}$ is given by

$$v_{i_1} \wedge \cdots \wedge v_{i_p} \mapsto \sum g_{i_1 j_1} \cdots g_{i_p j_p} \lambda^{j_1} \wedge \cdots \wedge \lambda^{j_p},$$

where the sum is taken over *all* $j_1, \ldots, j_p$. Due to linearity, ξ in (33′) is mapped to

$$\xi_\# = \frac{1}{p!} \sum \sum g_{i_1 j_1} \cdots g_{i_p j_p} \xi^{i_1 \ldots i_p} \lambda^{j_1} \wedge \cdots \wedge \lambda^{j_p}$$

$$= \sum_{j_1 < \cdots < j_p} \sum_{\substack{\text{all} \\ i_1 \ldots i_p}} g_{i_1 j_1} \cdots g_{i_p j_p} \xi^{i_1 \ldots i_p} \lambda^{j_1} \wedge \cdots \wedge \lambda^{j_p}$$

This completes the proof.

Remark Note that the $\xi_{j_1 \ldots j_p}$'s as defined by (35) are alternating in the indices $j_1, \ldots, j_p$ so that we can also write (34) in the form

$$\xi_\# = \frac{1}{p!} \sum \xi_{j_1 \ldots j_p} \lambda^{j_1} \wedge \cdots \wedge \lambda^{j_p}. \tag{34'}$$

We now wish to show that an inner product on $\mathbf{E}$ induces a natural inner product on $\bigwedge^p \mathbf{E}$, the pth exterior product.

Theorem 6.37 *Let $\langle , \rangle$ be an inner product in the n-dimensional space* $\mathbf{E}$. *There exists a unique inner product $\langle , \rangle_{(p)}$ in $\bigwedge^p \mathbf{E}$ such that*

$$\langle v_1 \wedge \cdots \wedge v_p, w_1 \wedge \cdots \wedge w_p \rangle_{(p)} = \det[\langle v_i, w_j \rangle]_{i,j=1}^p \tag{36}$$

for all v_i and $w_j \in \mathbf{E}$. This inner product is also uniquely determined by the fact that for any orthonormal basis $e_1, \ldots, e_n$ in $\mathbf{E}$, *the basis* $\{e_{i_1} \wedge \cdots \wedge e_{i_p};\ i_1 < \cdots < i_p\}$ *is orthonormal in* $\bigwedge^p \mathbf{E}$.

Proof The uniqueness of $\langle , \rangle_{(p)}$ is clear. The existence of a bilinear map $\langle , \rangle_{(p)}$ satisfying (36) follows from the universal property for exterior products. Indeed,

$$\det[\langle v_i, w_j \rangle]_{i,j=1}^{p}$$

is alternating and multilinear as a function of both $(v_1, \ldots, v_p)$ and $(w_1, \ldots, w_p)$, so it defines an alternating p-multilinear map $\mathbf{E}^{(p)} \to L_a^p(\mathbf{E}, \mathbb{R})$, and by the universal property there is a uniquely determined linear map $\bigwedge^p(\mathbf{E}) \to L_a^p(\mathbf{E}, \mathbb{R})$. This map can be viewed as an alternating p-multilinear map $\mathbf{E}^{(p)} \to (\bigwedge^p \mathbf{E})^*$ and hence there is a uniquely determined linear map $\bigwedge^p \mathbf{E} \to (\bigwedge^p \mathbf{E})^*$, i.e. a bilinear map $\bigwedge^p \mathbf{E} \times \bigwedge^p \mathbf{E} \to \mathbb{R}$ such that

$$(v_1 \wedge \cdots \wedge v_p, w_1 \wedge \cdots \wedge w_p) \mapsto \det[\langle v_i, w_j \rangle]_{i,j=1}^{p}$$

Now let us prove the second statement in the theorem concerning the orthonormal basis, namely

$$\langle e_{i_1} \wedge \cdots \wedge e_{i_p}, e_{j_1} \wedge \cdots \wedge e_{j_p} \rangle_{(p)} = \delta_{i_1 j_1} \cdots \delta_{i_p j_p} \tag{37}$$

By definition we have

$$\langle e_{i_1} \wedge \cdots \wedge e_{i_p}, e_{j_1} \wedge \cdots \wedge e_{j_p} \rangle_{(p)} = \det[\delta_{i_a j_b}]_{a,b=1}^{p}$$

Keeping in mind that $i_1 < \cdots < i_p$ and $j_1 < \cdots < j_p$, we see that if $i_p \neq j_p$ then either $j_p > i_p$ so there is a column of 0's, or $i_p > j_p$ in which case the pth row is 0. In either case, $\det[\delta_{i_a j_b}]_{a,b=1}^{p} = 0$ if $i_p \neq j_p$. If $i_p = j_p$ then expanding by cofactors about the last column or row we get

$$\det[\delta_{i_a j_b}]_{a,b=1}^{p} = \det[\delta_{i_a j_b}]_{a,b=1}^{p-1}$$

so it is evident that (37) follows by induction on p.

To show that $\langle , \rangle_{(p)}$ is positive definite, let $\alpha \in \bigwedge^p \mathbf{E}$ and express it in terms of the orthonormal basis, i.e.

$$\alpha = \sum \alpha^{i_1 \ldots i_p} e_{i_1} \wedge \cdots \wedge e_{i_p}$$

where the sum is taken over all indices $i_1 < \cdots < i_p$. By bilinearity, we have

$$\begin{aligned}\langle \alpha, \alpha \rangle_{(p)} &= \sum \alpha^{i_1 \ldots i_p} \alpha^{j_1 \ldots j_p} \delta_{i_1 j_1} \cdots \delta_{i_p j_p} \\ &= \sum (\alpha^{i_1 \ldots i_p})^2 \geqslant 0,\end{aligned}$$

whence $\langle \alpha, \alpha \rangle_{(p)} = 0$ if and only if $\alpha^{i_1 \ldots i_p} = 0$ for all indices $i_1 < \cdots < i_p$. This proves that $\langle , \rangle_{(p)}$ is an inner product on $\bigwedge^p \mathbf{E}$.

Corollary 6.38 *Let $v_1, \ldots, v_n$ be any basis for* **E**, *and let*

$$\xi = \frac{1}{p!} \sum \xi^{i_1 \ldots i_p} v_{i_1} \wedge \cdots \wedge v_{i_p}$$

$$\eta = \frac{1}{p!} \sum \eta^{i_1 \ldots i_p} v_{i_1} \wedge \cdots \wedge v_{i_p}$$

be elements of $\bigwedge^p \mathbf{E}$, *with coordinates* $\xi^{i_1 \ldots i_p}$ *and* $\eta^{i_1 \ldots i_p}$ *extended so that they are defined for all* $i_1, \ldots, i_p$ *and alternating in those indices. Then the value of the inner product in Theorem 6.37 is*

$$\langle \xi, \eta \rangle_{(p)} = \frac{1}{p!} \sum g_{i_1 j_1} \cdots g_{i_p j_p} \xi^{i_1 \ldots i_p} \eta^{j_1 \ldots j_p}. \tag{38}$$

Proof By virtue of bilinearity and the definition of $\langle\ ,\ \rangle_{(p)}$ we have

$$\langle \xi, \eta \rangle_{(p)} = \frac{1}{p!} \frac{1}{p!} \sum \sum \xi^{i_1 \ldots i_p} \eta^{j_1 \ldots j_p} \det[g_{i_a j_b}]_{a,b=1}^{p}$$

because $g_{ij} = \langle v_i, v_j \rangle$. By expanding the determinant we get

$$\langle \xi, \eta \rangle_{(p)} = \frac{1}{p!} \frac{1}{p!} \sum \sum \sum_{\sigma \in S_p} \xi^{i_1 \ldots i_p} \eta^{j_1 \ldots j_p} \operatorname{sgn} \sigma \cdot g_{i_{\sigma(1)}, j_1} \cdots g_{i_{\sigma(p)}, j_p}$$

where S_p is the group of permutations of $\{1, \ldots, p\}$. Hence

$$\begin{aligned} \langle \xi, \eta \rangle_{(p)} &= \frac{1}{p!} \frac{1}{p!} \sum_{i_1, \cdots, i_p} \sum_{\sigma \in S_p} \xi^{i_1 \cdots i_p} \eta_{i_{\sigma(1)} \cdots i_{\sigma(p)}} \operatorname{sgn} \sigma \\ &= \frac{1}{p!} \frac{1}{p!} \sum_{i_1, \cdots, i_p} \sum_{\sigma \in S_p} \xi^{i_1 \cdots i_p} \eta_{i_1 \cdots i_p} \\ &= \frac{1}{p!} \sum \xi^{i_1 \cdots i_p} \eta_{i_1 \cdots i_p} \end{aligned}$$

which is equal to the right-hand side of (38).

Remark From now on we will denote the induced inner product on $\bigwedge^p \mathbf{E}$ by $\langle , \rangle$ rather than $\langle , \rangle_{(p)}$.

Now, consider the exterior product $\bigwedge^p (\mathbf{E}^*)$. An inner product $\langle , \rangle$ on **E** induces an inner product on $\mathbf{E}^*$ by means of the index-lowering operator $v \mapsto \langle v, \cdot \rangle$, and we denote it by the same symbol $\langle , \rangle$. By applying Theorem 6.37 to this inner product on $\mathbf{E}^*$, we get the induced inner product $\langle , \rangle = \langle , \rangle_{(p)}$ in $\bigwedge^p(\mathbf{E}^*)$, $p \geqslant 1$. One could easily write down the analogue of (38) for this inner product, but the important point to remember here is that for any orthonormal basis $\mu_1, \ldots, \mu_n$ in $\mathbf{E}^*$, the basis

$$\mu^{i_1} \wedge \cdots \wedge \mu^{i_p}, \quad i_1 < \cdots < i_p,$$

is orthonormal in $\bigwedge^p(\mathbf{E}^*)$. If $p = 0$ we define $\langle\,,\rangle$ to be the obvious inner product on $\bigwedge^0(\mathbf{E}^*) = \mathbb{R}$.

Now we define the Hodge star operator.

Proposition 6.39 *Let* **E** *be an oriented n-dimensional vector space with inner product* $\langle\,,\rangle$ *and let* $\mu \in \bigwedge^n(\mathbf{E}^*)$ *denote the volume element on* **E**. There is a unique isomorphism $*: \bigwedge^p(\mathbf{E}^*) \to \bigwedge^{n-p}(\mathbf{E}^*)$ *satisfying*

$$\alpha \wedge *\eta = \langle \alpha, \eta \rangle \mu \quad \text{for} \quad \alpha, \eta \in \textstyle\bigwedge^p(\mathbf{E}^*) \tag{39}$$

Proof First we prove uniqueness of $*$. Let $e_1, \ldots, e_n$ be an *oriented* orthonormal basis of **E** and $\mu^1, \ldots, \mu^n$ the corresponding dual basis of $\mathbf{E}^*$. Then $\bigwedge^p(\mathbf{E}^*)$ has the orthonormal basis

$$\mu^{j_1} \wedge \cdots \wedge \mu^{j_p}, \quad j_1 < \cdots < j_p, \tag{40}$$

and similarly for $\bigwedge^{n-p}(\mathbf{E}^*)$. The volume element for the given orientation is $\mu = \mu^1 \wedge \cdots \wedge \mu^n$.

Let $\eta = \mu^{i_1} \wedge \cdots \wedge \mu^{i_p}$ where $i_1 < \cdots < i_p$ and $\alpha = \mu^{j_1} \wedge \cdots \wedge \mu^{j_p}$ where $j_1 < \cdots < j_p$. Since $\langle \alpha, \eta \rangle = 0$ unless $(i_1, \ldots, i_p) = (j_1, \ldots, j_p)$ it follows from (39) that

$$*\eta = c \cdot \mu^{i_{p+1}} \wedge \cdots \wedge \mu^{i_n} \qquad \text{for some constant } c$$

where $i_1 < \cdots < i_p$ and $i_{p+1} < \cdots < i_n$ are complementary sets of indices, i.e. the map $i: a \mapsto i_a$ is a permutation of $\{1, \ldots, n\}$. But then $\eta \wedge *\eta = c \cdot \operatorname{sgn}(i)\mu$ where $\operatorname{sgn}(i)$ denotes the sign of the permutation i. Since $\langle \eta, \eta \rangle = 1$ it follows from (39) that $c \cdot \operatorname{sgn}(i) = 1$, thus $c = \operatorname{sgn}(i)$, so that $*$ is unique.

Now, define the operator $*: \bigwedge^p(\mathbf{E}^*) \to \bigwedge^{n-p}(\mathbf{E}^*)$ by

$$*(\mu^{i_1} \wedge \cdots \wedge \mu^{i_p}) = \operatorname{sgn}(i) \cdot \mu^{i_{p+1}} \wedge \cdots \wedge \mu^{i_n} \tag{41}$$

where $i: a \to i_a$ is a permutation of $\{1, \ldots, n\}$ such that $i_1 < \cdots < i_p$ and $i_{p+1} < \cdots < i_n$. Then (39) holds because it holds on the basis elements (40), and the operator $*$ is an isomorphism because it maps a basis of $\bigwedge^p(\mathbf{E}^*)$ to a basis of $\bigwedge^{n-p}(\mathbf{E}^*)$.

Remark The operator $*$ is characterized by the fact that $*1 = \mu$, $*\mu = 1$, and for *any* positively oriented orthonormal basis $\mu^1, \ldots, \mu^n$ of $\mathbf{E}^*$

$$*(\mu^1 \wedge \cdots \wedge \mu^p) = \mu^{p+1} \wedge \cdots \wedge \mu^n, \qquad 0 < p < n$$

The operator $*$ is called the *Hodge star operator*. By virtue of the symmetry of $\langle\,,\rangle$ we have for all $\alpha, \eta \in \bigwedge^p(\mathbf{E}^*)$

$$\alpha \wedge *\eta = \langle \alpha, \eta \rangle \mu = \eta \wedge *\alpha \tag{42}$$

Proposition 6.40 *The Hodge star operator* $*$ *satisfies the following properties*:

$$**\alpha = (-1)^{p(n-p)}\alpha, \qquad \alpha \in \textstyle\bigwedge^p(\mathbf{E}^*)$$

$$\langle \alpha, \eta \rangle = \langle *\alpha, *\eta \rangle, \qquad \alpha, \eta \in \textstyle\bigwedge^p(\mathbf{E}^*)$$

Proof Let $i: a \to i_a$ by a permutation of $\{1, \ldots, n\}$ such that $i_1 < \cdots < i_p$ and $i_{p+1} < \cdots < i_n$. By (41) we have

$$*(\mu^{i_{p+1}} \wedge \cdots \wedge \mu^{i_n}) = c \cdot \mu^{i_1} \wedge \cdots \wedge \mu^{i_p}$$

where c is the sign of the permutation

$$\begin{pmatrix} 1 & \cdots & n-p & n-p+1 & \cdots & n \\ i_{p+1} & \cdots & i_n & i_1 & \cdots & i_p \end{pmatrix}$$

Thus, $c = (-1)^{p(n-p)} \cdot \operatorname{sgn}(i)$ and the first equation follows by using (41) again. For the second equation we have

$$\langle *\alpha, *\eta \rangle \mu = *\alpha \wedge **\eta = (-1)^{p(n-p)} *\alpha \wedge \eta$$

By virtue of the anti-commutativity of the wedge product $\wedge$, the right-hand side of this equation is equal to $\eta \wedge *\alpha$, or $\langle \alpha, \eta \rangle \mu$ using (42). Hence $\langle \alpha, \eta \rangle = \langle *\alpha, *\eta \rangle$.

Example Let $e_1 = (1, 0,0)$, $e_2 = (0, 1, 0)$ and $e_3 = (0, 0, 1)$ be the standard basis of $\mathbb{R}^3$ and dx, dy and dz the dual basis. Then the Hodge star operator on $\mathbb{R}^3$ (with the standard inner product and orientation) is given by

$$*dx = dy \wedge dz, \; *dy = -dx \wedge dz, \; *dz = dx \wedge dy$$

This follows from the remark above.

The next proposition shows that the transpose of the interior multiplication operator $i_X: \bigwedge^p(\mathbf{E}^*) \to \bigwedge^{p-1}(\mathbf{E}^*)$ is exterior multiplication by $X_\#$.

Proposition 6.41 *Let* $\mathbf{E}$ *be a vector space with inner product* $\langle \, , \rangle$. *Denote the induced inner product on each exterior product* $\bigwedge^p(\mathbf{E}^*)$ *by the same symbol* $\langle \, , \rangle$. *If* $X \in \mathbf{E}$ *then*

$$\langle i_X \alpha, \eta \rangle = \langle \alpha, X_\# \wedge \eta \rangle \tag{43}$$

for all $\alpha \in \bigwedge^p(\mathbf{E}^*)$ *and* $\eta \in \bigwedge^{p-1}(\mathbf{E}^*)$.

Proof Let $e_1, \ldots, e_n$ be an orthonormal basis of $\mathbf{E}$ and let $\mu^1, \ldots, \mu^n$ denote the (orthonormal) dual basis in $\mathbf{E}^*$. By linearity it suffices to verify (43) when $X = e_i$ and α and η belong to the corresponding orthonormal bases in $\bigwedge^p(\mathbf{E}^*)$ and $\bigwedge^{p-1}(\mathbf{E}^*)$, i.e.

$$\alpha = \mu^{i_1} \wedge \cdots \wedge \mu^{i_p} \quad \text{and} \quad \eta = \mu^{j_1} \wedge \cdots \wedge \mu^{j_{p-1}},$$

where $i_1 < \cdots < i_p$ and $j_1 < \cdots < j_{p-1}$. By definition of the interior product (see (9)), we have

$$i_X \alpha = i_{e_i} \alpha = \begin{cases} 0 & \text{if } i \notin \{i_1, \ldots, i_p\} \\ (-1)^{r+1} \mu^{i_1} \wedge \cdots \wedge \widehat{\mu^{i_r}} \wedge \cdots \wedge \mu^{i_p} & \text{if } i = i_r \end{cases}$$

Since $X = e_i$ then $X_\# = \mu^i$ and (43) is now easy to verify.

Remark The inner product can be extended to the direct sum $\bigwedge(\mathbf{E}^*) = \bigoplus_{1 \leqslant p \leqslant n} \bigwedge^p(\mathbf{E}^*)$ as follows:

$$\left\langle \sum_{p=1}^{n} \alpha_p, \sum_{p=1}^{n} \eta_p \right\rangle = \sum_{p=1}^{n} \langle \alpha_p, \eta_p \rangle, \qquad \alpha_p \in \textstyle\bigwedge^p(\mathbf{E}^*), \eta_p \in \bigwedge^p(\mathbf{E}^*),$$

i.e. the inner product is extended by linearity such that $\bigwedge^p\mathbf{E}$ is orthogonal to $\bigwedge^q\mathbf{E}$ if $p \neq q$. Then we have

$$\langle i_X\alpha, \eta \rangle = \langle \alpha, X_\# \wedge \eta \rangle$$

for all $\alpha \in \bigwedge(\mathbf{E}^*)$ and $\eta \in \bigwedge(\mathbf{E}^*)$.

Given a linear operator $A: E \to E$, where $(E, \langle\, , \rangle)$ is an inner product space, then there is a transpose map ${}^tA: E \to E$. If we take the complexification $E \otimes \mathbb{C}$ then we get a $\mathbb{C}$-linear operator $A: E \otimes \mathbb{C} \to E \otimes \mathbb{C}$ defined by

$$A(v_1 + iv_2) = Av_1 + iAv_2,$$

and with respect to the induced inner product on $E \otimes \mathbb{C}$ (see §5.1) the adjoint map ${}^hA: E \otimes \mathbb{C} \to E \otimes \mathbb{C}$ is just the analogous $\mathbb{C}$-extension of ${}^tA$, that is,

$${}^hA(v_1 + iv_2) = {}^tAv_1 + i\,{}^tAv_2.$$

It follows that the spaces on which i_X acts can be complexified without affecting Proposition 6.41. We complexify the real vector space $\bigwedge^p(\mathbf{E}^*)$ by taking $\bigwedge^p(\mathbf{E}^*) \otimes \mathbb{C}$, i.e. the space of "complex-valued" p-forms $\eta = \eta_1 + i\eta_2$ where $\eta_1, \eta_2 \in \bigwedge^p(\mathbf{E}^*)$ are "real-valued" p-forms. The inner product $\langle\, , \rangle$ on $\bigwedge^p(\mathbf{E}^*)$ is extended by $\mathbb{C}$-linearity to obtain a Hermitian inner product

$$\langle \alpha_1 + i\alpha_2, \eta_1 + i\eta_2 \rangle = \langle \alpha_1, \eta_1 \rangle + \langle \alpha_2, \eta_2 \rangle + i(\langle \alpha_2, \eta_1 \rangle - \langle \alpha_1, \eta_2 \rangle) \quad (44)$$

on the complex-valued forms. The interior multiplication operator i_X, $X \in \mathbf{E}$, extends by $\mathbb{C}$-linearity to an operator $i_X: \bigwedge^p(\mathbf{E}^*) \otimes \mathbb{C} \to \bigwedge^{p-1}(\mathbf{E}^*) \otimes \mathbb{C}$, and its adjoint is exterior multiplication by $X_\#$ as before.

We also extend the Hodge star operator $*: \bigwedge^p(\mathbf{E}^*) \otimes \mathbb{C} \to \bigwedge^p(\mathbf{E}^*) \otimes \mathbb{C}$ by conjugate-linearity, i.e.

$$*(\eta_1 + i\eta_2) = *\eta_1 - i *\eta_2,$$

so that (39) holds with the inner product (44). The properties in Proposition 6.40 continue to hold.

The codifferential and the Laplace-de Rham operator

Let M be an oriented manifold, without boundary. We consider the following vector bundles over M:

$$\textstyle\bigwedge^p(T^*M),\ p = 0, \ldots, n \quad \text{and} \quad \bigwedge(T^*M) = \bigoplus_{p=0}^{n} \bigwedge^p(T^*M),$$

with the corresponding spaces of sections

$$C^\infty(M, \textstyle\bigwedge^p(T^*M)),\ p = 0, \ldots, n, \quad \text{and} \quad C^\infty(M, \bigwedge(T^*M)),$$

i.e. the smooth differential forms on M (p is the degree of the form). For brevity we will usually denote these spaces simply $C^\infty(M, \bigwedge^p)$ and $C^\infty(M, \bigwedge)$, and in fact we will be dealing with the complexified spaces $C^\infty(M, \bigwedge^p \otimes \mathbb{C})$ and $C^\infty(M, \bigwedge \otimes \mathbb{C})$. The exterior derivative

$$d: C^\infty(M, \textstyle\bigwedge^p \otimes \mathbb{C}) \to C^\infty(M, \textstyle\bigwedge^{p+1} \otimes \mathbb{C})$$

on complex-valued forms is defined by $d(\eta_1 + i\eta_2) = d\eta_1 + id\eta_2$.

Let g be a Riemannian metric on TM, with local representation $[g_{ij}]$ with respect to some coordinate chart $(U, x_1, \ldots, x_n)$. By virtue of the index-lowering operator $T_xM \to T_x^*M$, there is an induced inner product on T_x^*M with local representation $[g^{ij}] = [g_{ij}]^{-1}$, the inverse matrix. By Theorem 6.37 and Corollary 6.38 we obtain an induced Riemannian metric on $\bigwedge^p(T^*M)$ defined as follows: if

$$\xi = \frac{1}{p!} \sum \xi_{i_1 \cdots i_p} dx_{i_1} \wedge \cdots \wedge dx_{i_p},$$

$$\eta = \frac{1}{p!} \sum \eta_{j_1 \cdots j_p} dx_{j_1} \wedge \cdots \wedge dx_{j_p},$$

are elements of $\bigwedge^p(T_x^*M)$ then

$$\langle \xi, \eta \rangle_x = \frac{1}{p!} \sum g^{i_1 j_1} \cdots g^{i_p j_p} \xi_{i_1 \cdots i_p} \eta_{j_1 \cdots j_p} \tag{45}$$

The inner product (45) can be complexified as in (44) to obtain an inner product on $\bigwedge^p(T_x^*M) \otimes \mathbb{C}$. With this inner product, we then define the Hodge star operator

$$*: C^\infty(M, \textstyle\bigwedge^p \otimes \mathbb{C}) \to C^\infty(M, \textstyle\bigwedge^{n-p} \otimes \mathbb{C})$$

pointwise at each point $x \in M$, that is, $(*\alpha)(x) = *\alpha(x)$. The operator $*$ has the properties stated in Propositions 6.39 and 6.40; in particular

$$\alpha \wedge *\eta = \langle \alpha, \eta \rangle \mu \qquad \alpha, \eta \in C^\infty(M, \textstyle\bigwedge^p \otimes \mathbb{C})$$

where μ is the volume form on M induced by the Riemannian metric.

The inner product on $\bigwedge^p(T^*M) \otimes \mathbb{C}$ induces an inner product on the sections $C^\infty(M, \bigwedge^p(T^*M) \otimes \mathbb{C})$ in the usual way (see §6.8),

$$\begin{aligned}(\alpha, \eta) &= \int_M \alpha \wedge *\beta \\ &= \int_M \langle \alpha(x), \beta(x) \rangle \, d\mu(x).\end{aligned}$$

Definition 6.42 *The codifferential* $\delta: C^\infty(M, \bigwedge^{p+1} \otimes \mathbb{C}) \to C^\infty(M, \bigwedge^p \otimes \mathbb{C})$ *is defined by*

$$\delta\eta = (-1)^{np+1} *d*\eta \qquad \text{if } \eta \in C^\infty(M, \textstyle\bigwedge^{p+1} \otimes \mathbb{C}),$$

and $\delta f = 0$ *if* $f \in C^\infty(M, \bigwedge^0 \otimes \mathbb{C}) = C^\infty(M, \mathbb{C})$.

Since $d^2 = 0$ and $**$ is a constant multiple of the identity it follows that

$$\delta^2 = 0.$$

The *Laplace–de Rham* operator $\Delta\colon C^\infty(M, \bigwedge^p \otimes \mathbb{C}) \to C^\infty(M, \bigwedge^p \otimes \mathbb{C})$ is defined by

$$\Delta = (d + \delta)^2 = d\delta + \delta d,$$

and is a second-order differential operator since d and δ are first-order operators. (This is clear from the local representations.) Also note that Δ preserves the degree of differential forms.

Theorem 6.43 *The operators d and δ are adjoints of each other. Further, at each point $(x, \xi) \in T_x^*M$, we have*

the principal symbol of d is exterior multiplication $c \mapsto \xi \wedge c$
the principal symbol of δ is interior multiplication $-i_{\xi^\#}$
the principal symbol of Δ is multiplication by the scalar $-\|\xi\|^2$

In particular, it follows that the Laplace operator Δ is strongly elliptic.

Proof By the anti-derivation property of d, and the fact that $*\delta = (-1)^{p+1} d*$, we have

$$\begin{aligned} d(\alpha \wedge *\eta) &= d\alpha \wedge *\eta + (-1)^p \alpha \wedge d*\eta \\ &= d\alpha \wedge *\eta + (-1)^p(-1)^{p+1}\alpha \wedge *\delta\eta \\ &= d\alpha \wedge *\eta - \alpha \wedge *\delta\eta \end{aligned}$$

whence by Stokes' theorem (since ∂M is empty) we have

$$\int_M d\alpha \wedge *\eta - \int_M \alpha \wedge *\delta\eta = 0,$$

so that $(d\alpha, \eta) = (\alpha, \delta\eta)$. This proves that d and δ are adjoints of each other.

Since d is a first-order differential operator, the principal symbol of d is the map $\pi d(x, \xi)\colon \bigwedge^p(T_x^*M) \otimes \mathbb{C} \to \bigwedge^{p+1}(T_x^*M) \otimes \mathbb{C}$ given by

$$\begin{aligned} \pi d(x, \xi)c &= d(\varphi \cdot \omega)(x) \\ &= d_x\varphi \wedge \omega(x) + \varphi(x) d_x\omega \\ &= \xi \wedge c \end{aligned}$$

where $\varphi \in C^\infty(M)$ and $\omega \in C^\infty(M, \bigwedge^p(T^*M) \otimes \mathbb{C})$ are chosen so that $\varphi(x) = 0$, $d_x\varphi = \xi$ and $\omega(x) = c$. Since δ is the adjoint of d, then

$$\eta\delta(x, \xi) = -[\pi d(x, \xi)]^h \qquad \text{(see (31))}$$

which by Proposition 6.41 is equal to $-i_{\xi^\#}$.

Lastly, we determine the principal symbol of Δ. Since $\Delta = d \circ \delta + \delta \circ d$, then

$$\pi\Delta(x, \xi) = \pi d(x, \xi) \circ \pi\delta(x, \xi) + \pi\delta(x, \xi) \circ \pi d(x, \xi),$$

whence $\pi\Delta(x, \xi)$: $\bigwedge^p(T_x^*M) \otimes \mathbb{C} \to \bigwedge^p(T_x^*M) \otimes \mathbb{C}$ is the negative of the map

$$c \mapsto \xi \wedge i_{\xi^\#}c + i_{\xi^\#}(\xi \wedge c),$$

and by Lemma 6.44 it follows that $\pi\Delta(x, \xi)c = -\|\xi\|^2 \cdot c$.

Remark Since the principal symbol of Δ is $-\|\xi\|^2$, then in local coordinates this means that if

$$\omega = \sum \omega_{i_1 \cdots i_p} dx_{i_1} \wedge \cdots \wedge dx_{i_p}$$

then

$$\Delta\omega = \sum (\Delta\omega)_{i_1 \cdots i_p} dx_{i_1} \wedge \cdots \wedge dx_{i_p},$$

where

$$(\Delta\omega)_{i_1 \cdots i_p} = -\sum_{i,j} g^{ij} \frac{\partial^2 \omega_{i_1 \cdots i_p}}{\partial x_i \, \partial x_j} + \text{lower-order terms}$$

To complete the proof of Theorem 6.43, it remains to prove the following lemma.

Lemma 6.44 *Let* **E** *be a vector space with inner product* $\langle\,,\rangle$. *Denote the induced inner product on* $\mathbf{E}^*$ *by the same symbol* $\langle\,,\rangle$. *Let* $\xi \in \mathbf{E}^*$ *and let* $\xi^\# \in \mathbf{E}$ *be its image by the index-raising operator. Then*

$$\xi \wedge i_{\xi^\#}\alpha + i_{\xi^\#}(\xi \wedge \alpha) = \|\xi\|^2 \cdot \alpha, \qquad \alpha \in \textstyle\bigwedge^p(\mathbf{E}^*) \otimes \mathbb{C} \tag{46}$$

where $\|\xi\|^2 = \langle \xi, \xi \rangle$.

Proof Let $v_1, \ldots, v_n$ be a basis of **E** and let $\lambda^1, \ldots, \lambda^n$ denote the corresponding dual basis. If $\xi \in \mathbf{E}^*$ we have $\xi = \sum \xi_i \lambda^i$ for some $\xi_i \in \mathbb{R}$. Then $\xi^\# = \sum \xi^i v_i$ where $\xi^i = \sum g^{ij}\xi_j$. It suffices to prove (46) when α is a basis element, $\alpha = \lambda^{i_1} \wedge \cdots \wedge \lambda^{i_p}$, $i_1 < \cdots < i_p$, since both sides of (46) are linear in α (note that a basis for a real vector space E is still a basis over $\mathbb{C}$ for the complexified space $E \otimes \mathbb{C}$). In view of (9) we then have

$$\xi \wedge i_{\xi^\#}\alpha = \xi \wedge \sum_{r=1}^{p} (-1)^{r+1} \xi^{i_r} \lambda^{i_1} \wedge \cdots \wedge \widehat{\lambda^{i_r}} \wedge \cdots \wedge \lambda^{i_p}$$

and

$$i_{\xi^\#}(\xi \wedge \alpha) = \sum_i \sum_j \xi_i \xi^j i_{v_j}(\lambda^i \wedge \alpha).$$

Also, by using the anti-derivation property from Theorem 6.11, $i_{v_j}(\lambda^i \wedge \alpha) = i_{v_j}(\lambda^i)\alpha - \lambda^i \wedge i_{v_j}\alpha$, and keeping in mind that $\lambda^i(v_j) = \delta_{ij}$ we obtain

$$\begin{aligned} i_{\xi^\#}(\xi \wedge \alpha) &= \sum_i \xi_i \xi^i \alpha + \sum_i \sum_r \xi_i \xi^{i_r} (-1)^r \lambda^i \wedge \lambda^{i_1} \wedge \cdots \wedge \widehat{\lambda^{i_r}} \wedge \cdots \wedge \lambda^{i_p} \\ &= \|\xi\|^2 \alpha - \xi \wedge i_{\xi^\#}\alpha, \end{aligned}$$

which completes the proof of the lemma.

Example 6.45 Let $(U, x_1, \ldots, x_n)$ be a chart on M. Let us determine the formula in local coordinates for the codifferential on (real-valued) 1-forms,

$$\delta: C^\infty(M, T^*M) \to C^\infty(M)$$

and the divergence of vector fields (see §6.5)

$$\text{div}: C^\infty(M, TM) \to C^\infty(M).$$

We claim that

$$\delta\omega = -\rho^{-1/2} \sum \frac{\partial}{\partial x_i}(\rho^{1/2} g^{ij} \omega_j), \qquad \omega = \sum \omega_j \, dx_j \tag{47}$$

and

$$\text{div}\, X = -\delta X_\#, \qquad X = \sum X^j \, \partial/\partial x_j.$$

Note that $X_\# = \sum X^j \, dx_j$ is the image of X by the index-lowering operator $TM \to T^*M$.

By definition, with $p = 1$, we have

$$\delta\omega = -{*d*}\omega \tag{49}$$

Since $*\omega = \sum \omega_j {*dx_j}$ we first need to work out the value of the Hodge star operator on the 1-forms dx_j. Now, $*dx_j$ is an $(n-1)$-form, so

$$*dx_j = \sum c_{ij} \, dx_1 \wedge \cdots \wedge \widehat{dx_i} \wedge \cdots \wedge dx_n$$

for some functions c_{ij} on U. By definition of the $*$ operator, see (42), we have

$$dx_i \wedge {*dx_j} = \langle dx_i, dx_j \rangle \cdot \mu.$$

Since $\langle dx_i, dx_j \rangle = g^{ij}$ and $\mu = \rho^{1/2} \, dx_1 \wedge \cdots \wedge dx_n$, it follows that $c_{ij} = (-1)^i g^{ij} \cdot \rho^{1/2}$, thus

$$*dx_j = \sum (-1)^i g^{ij} \cdot \rho^{1/2} \, dx_1 \wedge \cdots \wedge \widehat{dx_i} \wedge \cdots \wedge dx_n.$$

Since $*\omega = \sum \omega_j {*dx_j}$ then, by applying the exterior derivative operator d, we obtain

$$d{*\omega} = \sum_{i,j} \frac{\partial}{\partial x_i}(g^{ij} \rho^{1/2} \omega_j) \, dx_1 \wedge \cdots \wedge dx_n.$$

Now $*\mu = 1$, so $*(dx_1 \wedge \cdots \wedge dx_n) = \rho^{-1/2}$, and (47) follows at once from (49). The formula (48) is now evident if we refer to the local expression for div X in Proposition 6.20.

Example 6.46 Using Example 6.45 we can determine the value of Δ when it operates on *functions* $f \in C^\infty(M)$, i.e. $p = 0$. Since $\delta f = 0$ then

$$\Delta f = d\delta f + \delta df = \delta df.$$

Because df is a 1-form then by (48) we have

$$\Delta f = -\operatorname{div}(df)^{\#} = -\operatorname{div}(\operatorname{grad} f).$$

This is usually called the Laplace-Beltrami operator, see §6.5.

Example 6.47 Now we consider the Laplace-de Rham operator on p-forms in $\mathbb{R}^n$. Let

$$\omega = f\, dx_1 \wedge \cdots \wedge dx_p.$$

With the usual inner product and orientation on $\mathbb{R}^n$, we have

$$\delta\omega = \sum_{j=1}^{p} (-1)^j \frac{\partial f}{\partial x_j} dx_1 \wedge \cdots \wedge \widehat{dx_j} \wedge \cdots \wedge dx_p$$

and then taking the exterior derivative

$$\begin{aligned} d\delta\omega = &-\sum_{j=1}^{p} \frac{\partial^2 f}{\partial x_j^2} dx_1 \wedge \cdots \wedge dx_p \\ &+ \sum_{j=1}^{p} \sum_{k=p+1}^{n} (-1)^{j+p+1} \frac{\partial^2 f}{\partial x_k \, \partial x_j} dx_1 \wedge \cdots \wedge \widehat{dx_j} \wedge \cdots \wedge dx_p \wedge dx_k \end{aligned} \tag{50}$$

Further,

$$\begin{aligned} d\omega &= \sum_{k=p+1}^{n} \frac{\partial f}{\partial x_k} dx_k \wedge dx_1 \wedge \cdots \wedge dx_p \\ &= \sum_{k=p+1}^{n} (-1)^p \frac{\partial f}{\partial x_k} dx_1 \wedge \cdots \wedge dx_p \wedge dx_k \end{aligned}$$

so that

$$*d\omega = \sum_{k=p+1}^{n} (-1)^{k+1} \frac{\partial f}{\partial x_k} dx_{p+1} \wedge \cdots \wedge \widehat{dx_k} \wedge \cdots \wedge dx_n$$

and then

$$\begin{aligned} d{*}d\omega = &-\sum_{k=p+1}^{n} \frac{\partial^2 f}{\partial x_k^2} dx_{p+1} \wedge \cdots \wedge dx_n \\ &+ \sum_{j=1}^{p} \sum_{k=p+1}^{n} (-1)^{k+1} \frac{\partial^2 f}{\partial x_j \, \partial x_k} dx_j \wedge dx_{p+1} \wedge \cdots \wedge \widehat{dx_k} \wedge \cdots \wedge dx_n \end{aligned}$$

whence

$$\begin{aligned} \delta d\omega &= (-1)^{np+1} {*}d{*}d\omega \\ &= -\sum_{k=p+1}^{n} \frac{\partial^2 f}{\partial x_k^2} dx_1 \wedge \cdots \wedge dx_p \\ &\quad + \sum_{j=1}^{p} \sum_{k=p+1}^{n} (-1)^{j+p} \frac{\partial^2 f}{\partial x_j \, \partial x_k} dx_1 \wedge \cdots \wedge \widehat{dx_j} \wedge \cdots \wedge dx_p \wedge dx_k \end{aligned} \tag{51}$$

By adding (50) and (51) it follows that

$$\Delta\omega = -\left(\sum_{k=1}^{n} \frac{\partial^2 f}{\partial x_k^2}\right) dx_1 \wedge \cdots \wedge dx_p.$$

Exercise

Let

$$F = \sum_{j=1}^{3} F^j \frac{\partial}{\partial x_j}$$

be a vector field in $\mathbb{R}^3$, and let $F_{\#} = \sum_{j=1}^{3} F^j \, dx_j$ denote its image by the index-lowering operator which maps $\partial/\partial x_j \to dx_j$. Show that the divergence and curl of this vector field are given by

$$\text{div } F = {*d*}(F_{\#}), \qquad (\text{curl } F)_{\#} = {*d}(F_{\#}).$$

Using the properties of * and d, verify the following vector identities:

(a) curl grad $F = 0$
(b) div curl $F = 0$
(c) curl curl $F = -\nabla^2 F + \text{grad div } F$.

Part III

Pseudo-Differential Operators and Elliptic Boundary Value Problems

7
Pseudo-differential operators on $\mathbb{R}^n$

7.1 Some remarks about generalizing integrals

While working with pseudo-differential operators our main tool will be the Fourier transformation $\mathfrak{F}$,

$$\hat{u}(\xi) = \int e^{-i(x,\xi)} u(x)\, dx, \tag{1}$$

with the inverse $\mathfrak{F}^{-1}$,

$$u(x) = (2\pi)^{-n} \int e^{i(x,\xi)} \hat{u}(\xi)\, d\xi.$$

The Fourier transformation is well-defined on the L. Schwartz space, $\mathscr{S}(\mathbb{R}^n)$, i.e. the set of infinitely differentiable functions $u \in C^\infty(\mathbb{R}^n)$ such that for all multi-indices α and β,

$$\sup_{x \in \mathbb{R}^n} |x^\alpha (D^\beta u)(x)| < \infty.$$

Taking the maximum of these quantities for $|\alpha| + |\beta| \leqslant k$ gives us a semi-norm on $\mathscr{S}$ which we denote by $|u|_{k,\mathscr{S}}$. These semi-norms for $k = 0, 1, 2, \ldots$ define a topology on $\mathscr{S}(\mathbb{R}^n)$, making it into a Fréchet space. It is well-known that the Fourier transformation defines a linear, topological isomorphism $\mathfrak{F}\colon \mathscr{S}(\mathbb{R}^n) \to \mathscr{S}(\mathbb{R}^n)$, which can be extended to the dual space, $\mathfrak{F}\colon \mathscr{S}'(\mathbb{R}^n) \to \mathscr{S}'(\mathbb{R}^n)$. If we denote by $\langle\, ,\rangle$ the duality between $\mathscr{S}$ and $\mathscr{S}'$, we have $\langle \mathfrak{F}\Phi, u\rangle = \langle \Phi, \mathfrak{F}u\rangle$ which is the definition of $\mathfrak{F}$ on $\mathscr{S}'$.

We wish to give a meaning to the integral in (1), and also to other similar integrals, in the case that the continuous function $u(x)$ satisfies for some $C > 0$ and $m > 0$ the condition

$$|u(x)| \leqslant C(1 + |x|^2)^m. \tag{2}$$

We do this by introducing a simple class of oscillatory integrals. For $k > m + n/2$, let

$$\int e^{-i(x,\xi)} u(x)\, dx := (1 - \Delta_\xi)^k \int e^{-i(x,\xi)} \frac{u(x)}{(1 + |x|^2)^k}\, dx, \tag{3}$$

where the derivatives in ξ are taken in the distributional sense. This definition is motivated by the formula

$$e^{-i(x,\xi)} = (1 + |x|^2)^{-k}(1 - \Delta_\xi)^k e^{-i(x,\xi)}$$

Since $k > m + n/2$, the integral on the right-hand side of (3) is absolutely convergent.

To show that the definition (3) is independent of the choice of k, we look at another way to define the oscillatory integral, using the duality of $\mathscr{S}$ and $\mathscr{S}'$. Choose $\chi \in \mathscr{S}$ such that $\chi(0) = 1$ and set $\chi_\varepsilon(x) = \chi(\varepsilon x)$, $\varepsilon > 0$. If u satisfies the condition (2) then $u \in \mathscr{S}'$. Since $\chi_\varepsilon \cdot u \to u$ in $\mathscr{S}'$ as $\varepsilon \to 0$, it follows that $(\chi_\varepsilon u)\hat{} \to \hat{u}$ in $\mathscr{S}'$, i.e. for $\psi \in \mathscr{S}$ and $k > m + n/2$, we have

$$\begin{aligned}\langle \hat{u}, \psi \rangle &= \lim_{\varepsilon \to 0} \int (\chi_\varepsilon u)\hat{}(\xi)\psi(\xi)\, d\xi, \\ &= \lim_{\varepsilon \to 0} \iint e^{-i(x,\xi)}\chi(\varepsilon x)u(x)\psi(\xi)\, dx\, d\xi \\ &= \iint e^{-i(x,\xi)} \frac{u(x)}{(1 + |x|^2)^k}(1 - \Delta_\xi)^k\psi(\xi)\, dx\, d\xi\end{aligned}$$

In the last step we integrated by parts, then let $\varepsilon \to 0$ which is justified by the Lebesgue dominated convergence theorem since the integrand is absolutely convergent. This shows that $\hat{u}$ is equal to the oscillatory integral $\int e^{-i(x,\xi)} u(x)\, dx$ defined in (3). In particular, (3) does not depend on the choice of k.

The second method of definition suggests quite generally that one may operate on oscillatory integrals just as with standard integrals, differentiation can be performed under the integral sign, orders of integration can be interchanged and so on (these operations will be verified when the need for them arises).

Example If u satisfies the condition (2), let us calculate $\mathfrak{F}^{-1}\mathfrak{F}u$ in the sense of an oscillatory integral. Formally, we have

$$(\mathfrak{F}^{-1}\mathfrak{F}u)(x) = (2\pi)^{-n} \iint e^{i(x-y,\xi)}u(y)\, dy\, d\xi.$$

Now, consider $e^{i(x-y,\xi)}u(y)$ as a function of y, ξ (with x as a parameter) and regard $\iint \cdots dy\, d\xi$ as an oscillatory integral. Choose $\chi \in C_0^\infty(\mathbb{R}^n)$ such that $\chi(0) = 1$. Then

$$\begin{aligned}\iint e^{i(x-y,\xi)}u(y)\, dy\, d\xi &:= \lim_{\varepsilon \to 0} (2\pi)^{-n} \iint \chi(\varepsilon y)\chi(\varepsilon\xi)\, e^{i(x-y,\xi)}u(y)\, dy\, d\xi \\ &= \lim_{\varepsilon \to 0} \int \chi(\varepsilon y)u(y)(\mathfrak{F}^{-1}\chi)((x - y)/\varepsilon)\, dy \\ &= \lim_{\varepsilon \to 0} \int \chi(\varepsilon(x - \varepsilon z))u(x - \varepsilon z)(\mathfrak{F}^{-1}\chi)(z)\, dz \\ &= u(x),\end{aligned}$$

as expected.

The purpose of this section has *not* been to prove any specific results; rather it has been to illustrate a method for generalizing integrals which will be used in §7.2. The following proposition is typical of the kind of oscillatory integral considered there. For further details on oscillatory integrals, see [Hö 1] and [Shu].

Let $\mathcal{O}^m$ denote the set of all $a \in C^\infty(\mathbb{R}^n \times \mathbb{R}^n)$ such that $|D_\xi^\alpha D_x^\beta a(x, \xi)| \leqslant C_{\alpha,\beta}(1 + |\xi|)^m$ for all multi-indices α, β, and introduce the semi-norms

$$|a|_{k,\mathcal{O}^m} = \max_{|\alpha+\beta|\leqslant k} \sup_{(x,\xi)} |D_\xi^\alpha D_x^\beta a(x, \xi)| \cdot (1 + |\xi|)^{-m}$$

$k = 0, 1, 2, \ldots$.

Proposition *Let $a \in \mathcal{O}^m$. Then the value of the oscillatory integral*

$$I = \iint e^{i(x,\xi)} a(x, \xi)\, dx\, d\xi := \lim_{\varepsilon \to 0} \iint e^{i(x,\xi)} \chi(\varepsilon x, \varepsilon\xi) a(x, \xi)\, dx\, d\xi,$$

is independent of the choice of $\chi \in C_0^\infty(\mathbb{R}^n \times \mathbb{R}^n)$ satisfying $\chi(0, 0) = 1$. Further, if M, N are even natural numbers such that $M > n$ and $N > m + n$, then the oscillatory integral is equal to

$$I = \iint e^{i(x,\xi)} \langle x\rangle^{-M} \langle D_\xi\rangle^M [\langle\xi\rangle^{-M} \langle D_x\rangle^N a(x, \xi)]\, dx\, d\xi, \tag{4}$$

where the integrand is in $L_1(\mathbb{R}^n \times \mathbb{R}^n)$. Further, $|I| \leqslant C \cdot |a|_{M+N,\mathcal{O}^m}$, where the constant C depends only on M and N and not on the function "a".

Proof Let $\langle x\rangle^2 = 1 + |x|^2$ and $\langle D\rangle^2 = 1 - \Delta$, where Δ is the Laplace operator. Using the identities

$$\langle x\rangle^{-M} \langle D_\xi\rangle^M e^{i(x,\xi)} = e^{i(x,\xi)} = \langle\xi\rangle^{-N} \langle D_x\rangle^N e^{i(x,\xi)},$$

where M, N are even natural numbers, we can integrate by parts since $\chi(\varepsilon x, \varepsilon\xi) \in C_0^\infty$ to obtain

$$\begin{aligned} I_\varepsilon &= \iint e^{i(x,\xi)}\, \chi(\varepsilon x, \varepsilon\xi) a(x, \xi)\, dx\, d\xi \\ &= \iint e^{i(x,\xi)} \langle x\rangle^{-M} \langle D_\xi\rangle^M \{\langle\xi\rangle^{-N} \langle D_x\rangle^N [\chi(\varepsilon x, \varepsilon\xi) a(x, \xi)]\}\, dx\, d\xi. \end{aligned}$$

It is not hard to see that the derivatives of $\chi(\varepsilon x, \varepsilon\xi)$ are bounded by a constant independent of ε, and for any real number s, each derivative of $\langle\xi\rangle^s$ is certainly bounded by a constant times itself, so it follows that we can estimate the integrand by

$$|\langle x\rangle^{-M} \langle D_\xi\rangle^M \{\langle\xi\rangle^{-N} \langle D_x\rangle^N [\chi(\varepsilon x, \varepsilon\xi) a(x, \xi)]\}| \leqslant \text{const}\, |a|_{m+n} \langle\xi\rangle^{m-N} \langle x\rangle^{-M}$$

(and the constant depends only on M and N). If we choose $M > n$ and $N > m + n$ then the integrand belongs to $L_1(\mathbb{R}^n \times \mathbb{R}^n)$. By dominated convergence, $I = \lim_{\varepsilon \to 0} I_\varepsilon$ exists and is equal to (4).

Remark When $a \in L_1(\mathbb{R}^n \times \mathbb{R}^n)$, the oscillatory integral coincides with the usual value $\iint e^{i(x,\xi)} a(x, \xi)\, dx\, d\xi$.

7.2 The classes S^m

Consider a differential operator

$$A(x, D) = \sum_{|\alpha| \leqslant m} a_\alpha(x) D^\alpha, \qquad x \in \mathbb{R}^n. \tag{5}$$

Here $\alpha = (\alpha_1, \ldots, \alpha_n)$ is a multi-index, $D^\alpha = D_1^{\alpha_1} \cdots D_n^{\alpha_n}$ where $D_j = i^{-1}\, \partial/\partial x_j$, and $|\alpha| = \alpha_1 + \cdots + \alpha_n$ is the order of D^α. The Fourier transformation yields

$$\begin{aligned} A(x, D)u(x) &= (2\pi)^{-n} \sum_\alpha a_\alpha(x) \int e^{i(x,\xi)}\, \xi^\alpha \hat{u}(\xi)\, d\xi \\ &= (2\pi)^{-n} \int e^{i(x,\xi)}\, A(x, \xi) \hat{u}(\xi)\, d\xi \end{aligned}$$

where the polynomial $A(x, \xi) = \sum_{|\alpha| \leqslant m} a_\alpha(x) \xi^\alpha$ is called the symbol of $A(x, \xi)$. Recall that differential operators are local, i.e. $A(x, D)u(x) = 0$ in every open set in which $u = 0$.

To define pseudo-differential operators we shall use the same formula

$$Au(x) = (2\pi)^{-n} \int e^{i(x,\xi)}\, A(x, \xi) \hat{u}(\xi)\, d\xi$$

with a larger class of symbols $A(x, \xi)$ suitably chosen so that the operator A still has a pseudo-local property, namely, $Au \in C^\infty$ in every open set in which $u \in C^\infty$. Not every function $A(x, \xi)$ in $\mathcal{O}^m$ leads to a pseudo-differential operator; for instance if $A(x, \xi) = e^{i(x_0, \xi)}$, x_0 fixed, then A is a translation operator and hence not pseudo-local. The appropriate class of symbols is stated in the following definition. The pseudo-local property is proved later after we have developed basic properties of the operators $A(x, D)$ (see Corollary 7.14).

The symbol $A(x, \xi)$ of the differential operator (5) belongs to S^m, $m \in \mathbb{N}$, if the coefficients a_α are C^∞ functions with bounded derivatives of all orders, i.e. $a_\alpha \in C_b^\infty(\mathbb{R}^n)$.

Definition 7.1 *If m is a real number, $m \in \mathcal{R}$, then $S^m = S^m(\mathbb{R}^n \times \mathbb{R}^n)$ is the set of all $A \in C^\infty(\mathbb{R}^n \times \mathbb{R}^n)$ such that for all α, β*

$$|D_\xi^\alpha D_x^\beta A(x, \xi)| \leqslant C_{\alpha,\beta}(1 + |\xi|)^{m - |\alpha|}, \qquad \text{for all } x, \xi \in \mathbb{R}^n. \tag{6}$$

S^m is called the space of symbols of order m. We write $S^{-\infty} = \bigcap S^m$, $S^\infty = \bigcup S^m$.

It is clear that S^m is a Fréchet space with semi-norms given by the smallest constants which can be used in (6), and we let $|A|_k^{(m)}$ denote the maximum of these constants $C_{\alpha,\beta}$ for $|\alpha| + |\beta| \leqslant k$, that is,

$$|A|_k^{(m)} = \max_{|\alpha|+|\beta|\leqslant k} \sup_{(x,\xi)} |D_\xi^\alpha D_x^\beta A(x, \xi)| \cdot (1 + |\xi|)^{-m+|\alpha|}$$

for $k = 0, 1, 2, \ldots$.

The following relations are obvious or easily verified: $S^{m_1} \subset S^{m_2}$ for $m_1 \leqslant m_2$; if $A \in S^m$ then $D_\xi^\alpha D_x^\beta A \in S^{m-|\alpha|}$; and if $A \in S^m$, $B \in S^{m'}$ then $A \cdot B \in S^{m+m'}$. Also, note that if $\varphi \in \mathscr{S}$ then $A(x, \xi) := \varphi(\xi) \in S^{-\infty}$.

Proposition 7.2 *If* $A \in S^m$ *and* $u \in \mathscr{S}$ *then*

$$A(x, D)u(x) = \frac{1}{(2\pi)^n} \int e^{i(x,\xi)} A(x, \xi)\hat{u}(\xi)\, d\xi, \tag{7}$$

defines a function $A(x, D)u(x) \in \mathscr{S}$, *and the map* $A(x, D)\colon \mathscr{S} \to \mathscr{S}$ *is continuous. In fact, the bilinear map* $S^m \times \mathscr{S} \ni (A, u) \mapsto A(x, D)u \in \mathscr{S}$ *is continuous.*

Remark One calls $A(x, D)$ a *pseudo-differential operator (p.d.o.) of order m* and one writes $A(x, D) \in OS^m$. The operator $A(x, D)$ is sometimes denoted $\mathrm{Op}(A(x, \xi))$.

Proof Let $u \in \mathscr{S}$. To prove that $A(x, D)u \in \mathscr{S}$ and the continuity of the bilinear map $(A, u) \mapsto A(x, D)u$ we need only prove that, for any two multi-indices α and β, the quantity

$$\sup_{x\in\mathbb{R}^n} |x^\alpha D_x^\beta(A(x, D)u)(x)| \tag{8}$$

is bounded by a product of a semi-norm of A in S^m and a semi-norm of u in $\mathscr{S}$.

The integral (7) is well-defined in the usual Lebesgue sense since $\hat{u}(\xi)$ is rapidly decreasing as $|\xi| \to \infty$. The derivatives of $\hat{u}$ are also rapidly decreasing, so integrating by parts and using Leibniz' formula gives us

$$\begin{aligned} x^\alpha D_x^\beta(Au)(x) &= \frac{x^\alpha}{(2\pi)^n} \int D_x^\beta\{e^{i(x,\xi)}\, A(x, \xi)\}\hat{u}(\xi)\, d\xi \\ &= \frac{(-1)^{|\alpha|}}{(2\pi)^n} \int \sum_{\gamma\leqslant\beta} \sum_{\delta\leqslant\alpha} \binom{\beta}{\gamma}\binom{\alpha}{\delta} e^{i(x,\xi)}(D_\xi^{\alpha-\delta} D_x^{\beta-\gamma} A(x, \xi)) \\ &\quad \times D_\xi^\delta(\xi^\gamma \hat{u}(\xi))\, d\xi. \end{aligned}$$

Using the estimate (6), we obtain for $|\alpha| + |\beta| \leqslant k$,

$$|x^\alpha D_x^\beta(Au)(x)| \leqslant C_k |A|_k^{(m)} \cdot \sum_{\substack{\gamma\leqslant\beta \\ \delta\leqslant\alpha}} \int (1 + |\xi|)^{m-|\alpha|+|\delta|} \cdot |D_\xi^\delta(\xi^\gamma \hat{u}(\xi))|\, d\xi \tag{9}$$

Now, since $\hat{u} \in \mathscr{S}$, we can always achieve that

$$\int |D_\xi^\delta(\xi^\gamma \hat{u}(\xi))| \, d\xi \leqslant C(1 + |\xi|)^{-(m+n+1)} \qquad \text{for all } \delta \leqslant \alpha, \gamma \leqslant \beta$$

whence the "constant" C is bounded by a semi-norm of u in $\mathscr{S}$; by substituting this inequality in (9) we obtain the desired result.

We will also need the relation between the symbol $A(x, \xi)$ and the distributional kernel K of the map $A(x, D)$. In general, the distributional kernel (or Schwartz kernel) of any map $A\colon \mathscr{S} \to \mathscr{S}'$ is a distribution $K \in \mathscr{S}'(\mathbb{R}^n \times \mathbb{R}^n)$ such that $\langle Au, v\rangle = \langle K, u \otimes v\rangle$, $u, v \in \mathscr{S}$. This is a variant of the Schwartz kernel theorem which is usually stated for the spaces $\mathscr{D}$ and $\mathscr{D}'$ (see [Hö 1], Chap. 5). Formally, we write

$$A(x, D)u(x) = \int K(x, y)u(y) \, dy.$$

For the case of an operator $A = A(x, D) \in OS^m$, the Schwartz kernel is given by

$$K(x, y) = \frac{1}{(2\pi)^n} \int e^{i(x-y,\xi)} A(x, \xi) \, d\xi, \tag{10a}$$

which exists as an oscillatory integral, and by Fourier inversion we have

$$A(x, \xi) = \int K(x, y) \, e^{-i(x-y,\xi)} \, dy. \tag{10b}$$

The following lemma is often useful in order to regularize symbols. Let $A \in S^m$, choose χ as in Lemma 7.3 and set $A_\varepsilon(x, \xi) = \chi(\varepsilon\xi) \cdot A(x, \xi)$. Then $A_\varepsilon \in S^{-\infty}$, and it follows from the lemma that $\{A_\varepsilon\}$ is a bounded set in S^m and for any $m' > m$ we have $A_\varepsilon \to A$ in $S^{m'}$ as $\varepsilon \to 0$.

Lemma 7.3 *Let $\chi \in C_0^\infty(\mathbb{R}^n)$ be equal to 1 in a neighbourhood of the origin, and let $\chi_\varepsilon(\xi) = \chi(\varepsilon\xi)$. Then χ_ε is bounded in S^0, $0 \leqslant \varepsilon \leqslant 1$, and $\chi_\varepsilon \to 1$ in S^m for every $m > 0$ where $\varepsilon \to 0$.*

Proof First of all, it is obvious that $\chi \in S^0$. Then

$$\begin{aligned} |D_\xi^\alpha \chi_\varepsilon(\xi)| &= C_\alpha \varepsilon^{|\alpha|}(1 + |\varepsilon\xi|)^{-|\alpha|} \\ &\leqslant C_\alpha(1 + |\xi|)^{-|\alpha|} \qquad \forall\alpha, \text{ since } \varepsilon^{-1} \geqslant 1, \end{aligned}$$

which shows that χ_ε is bounded in S^0. By hypothesis, $\chi(\xi) = 1$ when $|\xi| \leqslant \delta$ for some $\delta > 0$, hence $\varepsilon|\xi| \geqslant \delta$ on the support of $\chi_\varepsilon - 1$. It follows that on the support of $\chi_\varepsilon - 1$ we have $\delta \leqslant \varepsilon(1 + |\xi|)$, so

$$|D_\xi^\alpha(\chi_\varepsilon(\xi) - 1)| \leqslant C'_\alpha \delta^{-m} \varepsilon^m (1 + |\xi|)^{m-|\alpha|}, \qquad \forall\alpha. \tag{11}$$

Hence $\chi_\varepsilon \to 1$ in S^m for every $m > 0$ when $\varepsilon \to 0$.

It is not essential that there be as many x variables as ξ variables. For some purposes it is useful to have a wider class of symbols available, namely, the so-called general symbols or amplitudes. Let m be a real number. We say that $a(x, y, \xi) \in S^m(\mathbb{R}^n \times \mathbb{R}^n \times \mathbb{R}^n) = S^m$ if $a \in C^\infty(\mathbb{R}^{3n})$ and for all α, β, γ

$$|D_\xi^\alpha D_x^\beta D_y^\gamma a(x, y, \xi)| \leqslant C_{\alpha\beta\gamma}(1 + |\xi|)^{m-|\alpha|}, \qquad x, y, \xi \in \mathbb{R}^n. \tag{12}$$

If $a(x, y, \xi) \in S^m$ then as before we let $|a|_k^{(m)}$, $k = 0, 1, 2, \ldots$, denote the semi-norm which is the maximum of the smallest constants $C_{\alpha\beta\gamma}$ which can be used in (12) for $|\alpha| + |\beta| + |\gamma| \leqslant k$. With each general symbol $a(x, y, \xi)$ we associate an operator

$$Au(x) = \frac{1}{(2\pi)^n} \iint a(x, y, \xi) u(y)\, e^{i(x-y, \xi)}\, dy, d\xi, \qquad u \in \mathscr{S}. \tag{13}$$

We sometimes denote the operator A by $\mathrm{Op}(a(x, y, \xi))$ or just $\mathrm{Op}(a)$ for short.

Let $\Lambda = (1 + |\xi|^2)^{1/2}$. Then $\Lambda^m \in S^m$, $m \in \mathbb{R}$. We have already used the pseudo-differential operator $\Lambda^{2k}(D)$, that is,

$$(1 + |D|^2)^k = (1 - \Delta)^k, \qquad \text{when } k \text{ is an integer.}$$

See the proposition at the end of §7.1.

Proposition 7.4 *If we define the integral in* (13) *as an oscillatory integral, the operator* A *gives a continuous map* $A\colon \mathscr{S} \to \mathscr{S}$.

Proof For short we write $\langle x \rangle^2 = 1 + |x|^2$ and $\langle D \rangle^2 = 1 - \Delta$. Using the identities

$$\langle x - y \rangle^{-M} \langle D_\xi \rangle^M e^{i(x-y,\xi)} = e^{i(x-y,\xi)} = \langle \xi \rangle^{-N} \langle D_y \rangle^N e^{i(x-y,\xi)},$$

where M, N are even natural numbers, we define the oscillatory integral in (13) as

$$Au(x) = \frac{1}{(2\pi)^n} \iint e^{i(x-y,\xi)} \langle x - y \rangle^{-M} \langle D_\xi \rangle^M \langle D_y \rangle^N \times [\langle \xi \rangle^{-N} a(x, y, \xi) u(y)]\, dy\, d\xi, \tag{14}$$

with x regarded as a parameter. If the amplitude $a \in S^m$ satisfies (12) and if $u \in \mathscr{S}$, then clearly for $M > n$ and $N > m + n$ the integral in (14) becomes absolutely convergent, defining a continuous function of $x \in \mathbb{R}^n$. Increasing M and N, we will obtain integrals which are also convergent after differentiation with respect to x. Using the inequality

$$(1 + |x|)^k \leqslant (1 + |y|)^k (1 + |x - y|)^k, \qquad k \geqslant 0,$$

we see from (14) that

$$(1 + |x|)^k |Au(x)| \leqslant C_k$$

for any $k > 0$, and a similar estimate holds if we replace $Au(x)$ by $D_x^\alpha(Au(x))$.

This means that $Au \in \mathscr{S}$. A closer look at this result would give an estimate of the semi-norms guaranteeing the continuity of the map $A: \mathscr{S} \to \mathscr{S}$.

Let $A(x, \xi) \in S^m(\mathbb{R}^n \times \mathbb{R}^n)$. Considering it as constant in y, $a(x, y, \xi) := A(x, \xi)$, we see that $a(x, y, \xi) \in S^m(\mathbb{R}^n \times \mathbb{R}^n \times \mathbb{R}^n)$. For this amplitude, definition (13) becomes

$$\begin{aligned} Au(x) &= (2\pi)^{-n} \iint A(x, \xi) u(y)\, e^{-i(y,\xi) + i(x,\xi)}\, dy\, d\xi \\ &= (2\pi)^{-n} \int A(x, \xi) \hat{u}(\xi)\, e^{i(x,\xi)}\, d\xi, \end{aligned}$$

which is (7); hence both definitions agree, and the name "general symbol" is justified.

Our next aim is to show that the operator A defined by (13) belongs to OS^m. To do so we must determine how to associate to a general symbol, $a(x, y, \xi)$, an ordinary symbol $A(x, \xi)$.

If $a(x, y, \xi) \in S^m$ is a general symbol, the Schwartz kernel for the corresponding operator (13) is given by $K(x, y) = (2\pi)^{-n} \int e^{i(x-y,\xi)} a(x, y, \xi)\, d\xi$. This formula and (10b) provide the motivation for (15) below. Also see (31) in the remark that follows Theorem 7.10.

Note that if $a(x, y, \xi)$ does not depend on y, then (15) is the Fourier inversion formula (substitute $y - x = z$) as it should be.

Proposition 7.5 *Let $a(x, y, \xi) \in S^m$ be a general symbol. If we define the ordinary symbol by*

$$A(x, \xi) := \frac{1}{(2\pi)^n} \iint a(x, y, \theta)\, e^{i(x-y,\theta)}\, e^{-i(x-y,\xi)}\, dy\, d\theta, \tag{15}$$

(the integral is oscillatory!), then $A(x, \xi) \in S^m$, with semi-norms $|A|_k^{(m)}$ bounded by a constant multiplied by $|a|_{2k+2(n+1)+|m|}^{(m)}$. Further, both definitions (7) and (13) of the operator Au agree: $\mathrm{Op}(A(x, \xi)) = \mathrm{Op}(a(x, y, \xi))$, *that is,*

$$\int e^{i(x,\xi)} A(x, \xi) \hat{u}(\xi)\, d\xi = \iint a(x, y, \xi) u(y)\, e^{i(x-y,\xi)}\, dy\, d\xi, \qquad u \in \mathscr{S}, \tag{16}$$

Proof Substituting $z = y - x$, $\eta = \theta - \xi$ and regularizing as before, we rewrite (15) in the form

$$\begin{aligned} A(x, \xi) = \frac{1}{(2\pi)^n} \iint & e^{-i(z,\eta)} \langle z \rangle^{-M} \langle D_\eta \rangle^M \langle D_z \rangle^N \\ & \times [\langle \eta \rangle^{-N} \cdot a(x, x + z, \xi + \eta)]\, dz\, d\eta \end{aligned}$$

where M, N are even and non-negative. Taking into account the inequality $\langle \xi + \eta \rangle^{\pm 1} \leqslant 2\langle \xi \rangle^{\pm 1} \langle \eta \rangle$, it follows that $\langle \xi + \eta \rangle^s \leqslant 2^{|s|} \langle \xi \rangle^s \langle \eta \rangle^{|s|}$ for all

$s \in \mathbb{R}$, so we obtain when $|\alpha| + |\beta| \leqslant k$ that

$$\begin{aligned} |D_\xi^\alpha D_x^\beta A(x, \xi)| &\leqslant \text{const } |a|_{k+M+N}^{(m)} \cdot \iint \langle z \rangle^{-M} \langle \xi + \eta \rangle^{m-|\alpha|} \langle \eta \rangle^{-N} \, dz \, d\eta \\ &\leqslant \text{const } |a|_{k+M+N}^{(m)} \cdot \langle \xi \rangle^{m-|\alpha|} \cdot \iint \langle z \rangle^{-M} \cdot \langle \eta \rangle^{|m-|\alpha||-N} \, dz \, d\eta \\ &\leqslant \text{const } |a|_{k+M+N}^{(m)} \cdot (1 + |\xi|)^{m-|\alpha|} \end{aligned}$$

if we take $M > n$ and $N > n + |m - |\alpha||$. It follows that $A(x, \xi) \in S^m$, and the estimates for the semi-norms $|A|_k^{(m)}$ are evident.

It remains to verify that both definitions (7) and (13) of the operator Au agree. To this end, choose $\chi \in C_0^\infty$ with $\chi(0) = 1$, then set

$$a_\varepsilon(x, y, \xi) = a(x, y, \xi)\chi(\varepsilon y)\chi(\varepsilon \xi)$$

and define A_ε as in (15) with a replaced by a_ε. By virtue of Lemma 7.3 we know that $\chi(\varepsilon\xi)$ is bounded in S^0 and $\chi(\varepsilon y)$ has bounded derivatives, independently of $\varepsilon > 0$, so it follows that a_ε is bounded in S^m, i.e. all semi-norms $|a_\varepsilon|_k^{(m)}$ are bounded independently of $\varepsilon > 0$. The estimates that we established above shows that the same is true for $A_\varepsilon \in S^m$. Hence by dominated convergence it suffices to verify (16) when $a = a_\varepsilon$ and $A = A_\varepsilon$. But since a and A now have compact support in y and ξ all integrals exist in the ordinary Lebesgue sense and we can interchange the order of integration to get

$$\begin{aligned} \text{Op}(a)u &= \iint a(x, y, \theta) u(y) \, e^{i(x-y, \theta)} \, dy \, d\theta \\ &= (2\pi)^{-n} \iiint a(x, y, \theta) \, e^{i(x-y, \theta)} \, e^{i(y, \xi)} \, \hat{u}(\xi) \, d\xi \, dy \, d\theta \\ &= (2\pi)^{-n} \int e^{i(x, \xi)} \left[\iint a(x, y, \theta) \, e^{i(x-y, \theta)} \, e^{i(y, \xi)} \, e^{-i(x, \xi)} \, \hat{u}(\xi) \, dy \, d\theta \right] d\xi \\ &= (2\pi)^{-n} \int e^{i(x, \xi)} \, A(x, \xi) \hat{u}(\xi) \, d\xi \\ &= A(x, D)u. \end{aligned}$$

Remark The symbol $A(x, \xi)$ of the operator $A(x, D)$ is uniquely determined; see (31) below. But the general symbol $a(x, y, \xi)$ is *not* uniquely determined by $\text{Op}(a(x, y, \xi))$. For example, let $\varphi \in C_0^\infty(\mathbb{R}^n)$, $\varphi \not\equiv 0$. Then $a(x, y, \xi) = \xi_1 \varphi(y) - \varphi(x)\xi_1 - D_{x_1}\varphi(x)$ defines the zero operator since $D_{x_1}(\varphi u) = \varphi \cdot D_{x_1} u + D_{x_1}\varphi \cdot u$ for all $u \in \mathcal{S}$.

An operator A of class $OS^{-\infty}$ can be characterized by means of a C^∞ kernel representation.

In the statement of the next theorem we make a change of variable in the Schwartz kernel and write (formally)

$$Au(x) = \int \tilde{K}(x, x - y)u(y)\, dy.$$

Then the relations (10a), (10b) with $z = x - y$ become

$$\tilde{K}(x, z) = \frac{1}{(2\pi)^n} \int e^{i(z,\xi)} A(x, \xi)\, d\xi, \tag{17a}$$

and (Fourier inversion)

$$A(x, \xi) = \int \tilde{K}(x, z)\, e^{-i(z,\xi)}\, dz. \tag{17b}$$

Theorem 7.6 *Given $A = A(x, D) \in OS^{-\infty}$ we define the Schwartz kernel $\tilde{K}$ as in (17a). Then $\tilde{K}(x, z) \in C^\infty(\mathbb{R}^n \times \mathbb{R}^n)$ and it satisfies*

$$\sup_{x,z} \{(1 + |z|)^l \cdot |\partial_x^\beta \partial_z^\gamma \tilde{K}(x, z)|\} < \infty \tag{*}$$

for any β, γ, and $l \geqslant 0$, and we can write

$$Au(x) = \int \tilde{K}(x, x - y)u(y)\, dy \qquad \textit{for } u \in \mathcal{S}.$$

Conversely, if $\tilde{K}(x, z) \in C^\infty(\mathbb{R}^n \times \mathbb{R}^n)$ satisfies $(*)$ then the operator A defined in this way (classically) is an operator of class $OS^{-\infty}$ with symbol $A(x, \xi) \in S^{-\infty}$ given by (17b).

Proof The relations (17a) and (17b) between $A(x, \xi)$ and $\tilde{K}(x, z)$ are clear. The equivalence of $A \in S^{-\infty}$ and $(*)$ can be derived from

$$z^\alpha \partial_x^\beta \partial_z^\gamma \tilde{K}(x, z) = \frac{1}{(2\pi)^n} \int e^{i(z,\xi)} (i\partial_\xi)^\alpha \{(i\xi)^\gamma \partial_x^\beta A(x, \xi)\}\, d\xi$$

and

$$\xi^\alpha \partial_x^\beta \partial_\xi^\gamma A(x, \xi) = \int e^{-i(z,\xi)} (-i\partial_z)^\alpha \{(-iz)^\gamma \partial_x^\beta \tilde{K}(x, z)\}\, dz.$$

Here we integrated by parts in ξ and in z using the identities $z^\alpha e^{i(z,\xi)} = (-i\partial_\xi)^\alpha e^{i(z,\xi)}$ and $\xi^\alpha e^{-i(z,\xi)} = (i\partial_z)^\alpha e^{-i(z,\xi)}$, respectively.

Remark For the Schwartz kernel in (10a), the condition $(*)$ becomes that $K \in C^\infty(\mathbb{R}^n \times \mathbb{R}^n)$ and

$$\sup_{x,y} \{(1 + |x - y|)^l \cdot |\partial_x^\beta \partial_y^\gamma K(x, y)|\} < \infty$$

for any β, γ and l.

7.3 Pseudo-differential operator algebra and asymptotics

An important notion in the theory of pseudo-differential operators (p.d.o.'s) is the asymptotic expansion of a symbol. Let $A_j(x, \xi) \in S^{m_j}$, where $m = m_1 > m_2 \cdots > m_j \to -\infty$ as $j \to +\infty$, and let $A(x, \xi) \in C^\infty(\mathbb{R}^n \times \mathbb{R}^n)$. We will write

$$A \sim \sum_{j=1}^{\infty} A_j,$$

if for any integer $r \geqslant 2$ we have

$$A - \sum_{j=1}^{r-1} A_j \in S^{m_r}. \tag{18}$$

From this it follows in particular that $A(x, \xi) \in S^m$.

Proposition 7.7 *For any given sequence* $A_j \in S^{m_j}$, $j = 1, 2, \ldots$, *with* $m_{j-1} > m_j \to -\infty$ *as* $j \to +\infty$, *we can always find a function* $A \in S^m$ *(where* $m = m_1$*) such that*

$$A \sim \sum_{j=1}^{\infty} A_j.$$

The function A is uniquely determined modulo $S^{-\infty}$.

Proof The uniqueness follows immediately from (18). To prove existence, we use Lemma 7.3. Choose a function $\chi \in C_0^\infty$ which is equal to 1 in a neighbourhood of the origin, define $\chi_\varepsilon(\xi) = \chi(\varepsilon\xi)$, then put

$$A = \sum_{j=1}^{\infty} (1 - \chi_{\varepsilon_j}) \cdot A_j, \tag{19}$$

where $0 < \varepsilon_j \leqslant 1$ approaches 0 so rapidly as $j \to \infty$ that

$$|D_\xi^\alpha D_x^\beta[(1 - \chi(\varepsilon_j \xi)) A_j(x, \xi)]| \leqslant 2^{-j}(1 + |\xi|)^{m_j + 1 - |\alpha|}, \tag{20}$$

for all x, ξ and $|\alpha| + |\beta| \leqslant j$. Let us show that this is always possible. Since $A_j \in S^{m_j}$ and $1 - \chi(\varepsilon\xi)$ satisfies the estimates (11) in Lemma 7.3 (with $m = 1$) we obtain by Leibniz' formula that

$$\begin{aligned} |D_\xi^\alpha D_x^\beta[(1 - \chi(\varepsilon\xi)) A_j(x, \xi)]| &= |D_\xi^\alpha[(1 - \chi(\varepsilon\xi)) D_x^\beta A_j(x, \xi)]| \\ &\leqslant C_{j\alpha\beta}\varepsilon(1 + |\xi|)^{m_j + 1 - |\alpha|} \end{aligned}$$

Choosing $\varepsilon = \varepsilon_j$ such that

$$C_{j\alpha\beta}\varepsilon_j \leqslant 2^{-j} \qquad \text{for } |\alpha| + |\beta| \leqslant j,$$

we get (20), as desired. The sum in (19) is locally finite; hence $A \in C^\infty$. Also, observe that

$$A - \sum_{j=1}^{N-1} A_j = \sum_{j=N}^{\infty} (1 - \chi_{\varepsilon_j}) A_j - \sum_{j=1}^{N-1} A_j \cdot \chi_{\varepsilon_j} \tag{21}$$

and $A_j \cdot \chi_{\varepsilon_j} \in S^{-\infty}$ because $\chi_{\varepsilon_j} \in C_0^\infty$. For any fixed α, β and r, we can choose N so large that $N \geqslant |\alpha| + |\beta|$ and $m_N + 1 \leqslant m_r$. Since m_j is a decreasing sequence, we obtain from (20) by summing a geometric series:

$$\left| D_\xi^\alpha D_x^\beta \left[\sum_{j=N}^{\infty} (1 - \chi(\varepsilon_j \xi)) A_j(x, \xi) \right] \right| \leqslant (1 + |\xi|)^{m_r - |\alpha|},$$

which means that the first sum on the right-hand side of (21) belongs to S^{m_r}, hence $A - \sum_{j=1}^{N-1} A_j \in S^{m_r}$. But $A_j \in S^{m_r}$ for $j \geqslant r$, so it follows that

$$A - \sum_{j=1}^{r-1} A_j \in S^{m_r}$$

as required.

It is useful to restate Proposition 7.7 in the language of operators instead of symbols. Let $A_j(x, D) \in OS^{m_j}$, where $m_1 > m_2 \cdots > m_j \to -\infty$ as $j \to +\infty$, and let $A(x, D) \in OS^{m_1}$. We will write

$$A(x, D) \sim \sum_{j=1}^{\infty} A_j(x, D),$$

if for any integer $r \geqslant 2$ we have

$$A(x, D) - \sum_{j=1}^{r-1} A_j(x, D) \in OS^{m_r}. \tag{18'}$$

Proposition 7.7′ *For any given sequence $A_j(x, D) \in OS^{m_j}$, $j = 1, 2, \ldots$, with $m_{j-1} > m_j \to -\infty$ as $j \to \infty$, we can always find a p.d.o. $A(x, D) \in OS^{m_1}$ such that*

$$A(x, D) \sim \sum_{j=1}^{\infty} A_j(x, D).$$

$A(x, D)$ is uniquely determined modulo $OS^{-\infty}$.

Proof The uniqueness follows immediately from (18′). To prove existence we use Proposition 7.7. Since the p.d.o.'s $A_j(x, D) \in OS^{m_j}$ have symbols $A_j(x, \xi) \in S^{m_j}$ we can find a symbol $A(x, \xi) \in S^{m_1}$ such that (18) holds. But then (18′) is also satisfied for $A(x, D)$.

Proposition 7.5 gives the connection between the amplitude $a(x, y, \xi)$ and the ordinary symbol $A(x, \xi)$ of a p.d.o., see (15). We complete this proposition by giving an asymptotic expansion.

Proposition 7.8 *Let the amplitude $a(x, y, \xi) \in S^m$. Then the ordinary symbol $A(x, \xi)$ given by* (15) *has the asymptotic expansion*

$$A(x, \xi) \sim \sum_\alpha \frac{i^{|\alpha|}}{\alpha!} D_\xi^\alpha D_y^\alpha a(x, y, \xi)|_{y=x}$$

Proof Obviously $D_\xi^\alpha D_y^\alpha a(x, y, \xi)|_{y=x} \in S^{m-|\alpha|}$ and the asymptotic sum is therefore meaningful. Since $i^{|\alpha|}D_y^\alpha = \partial_y^\alpha$, what we have to show is that

$$A(x, \xi) \sim \sum_\alpha \frac{1}{\alpha!} D_\xi^\alpha \partial_y^\alpha a(x, x, \xi) \tag{22}$$

Using the notation $\mathrm{Op}(A(x, \xi))$ for the operator $A(x, D)$, we may reformulate this as an asymptotic expansion of operators:

$$A(x, D) = \mathrm{Op}(A(x, \xi)) \sim \sum_\alpha \frac{1}{\alpha!} \mathrm{Op}(D_\xi^\alpha \partial_y^\alpha a(x, x, \xi)). \tag{22'}$$

By Proposition 7.7, there exists $A_1(x, \xi) \in S^m$ with an asymptotic expansion given by the right-hand side of (22), so to complete the proof it suffices to show that

$$A(x, D) - A_1(x, D) \in OS^{-\infty} = \bigcap_k OS^k.$$

Our aim now is to apply Theorem 7.6. To verify the conditions of this theorem, we use Taylor's formula (Remark 5.6) in the y variable in order to expand $a(x, y, \xi)$ in powers of $y - x$ about $y = x$. This gives, after an integration by parts using the identity $(-D_\xi)^\alpha e^{i(x-y,\xi)} = (y - x)^\alpha e^{i(x-y,\xi)}$, that the operator (13) (which by Proposition 7.5 coincides with $A(x, D)$) is equal to

$$Au = (2\pi)^{-n} \iint e^{i(x-y,\xi)} \sum_{|\alpha| \leqslant N-1} \frac{1}{\alpha!} D_\xi^\alpha \partial_y^\alpha a(x, x, \xi) u(y)\, dy\, d\xi + R_N u,$$

where

$$R_N u = \int K_N(x, y) u(y)\, dy$$

and

$$K_N(x, y) = (2\pi)^{-n} \int e^{i(x-y,\xi)} \sum_{|\alpha|=N} \frac{N}{\alpha!}$$

$$\times \int_0^1 (1 - t)^{N-1} D_\xi^\alpha \partial_y^\alpha a(x, x + t(y - x), \xi)\, dt\, d\xi$$

Now, the symbol estimates for "a" imply that if $N > m + n + k$, then $K_N \in C^k$ and

$$|D_x^\beta D_y^\gamma K_N(x, y)| \leqslant C_{\beta\gamma} \qquad \text{when } |\beta| + |\gamma| \leqslant k.$$

In fact, integrating by parts in ξ, gives us for $N > m + n + k$

$$|(x - y)^\delta D_x^\beta D_y^\gamma K_N(x, y)| \leqslant C_{\delta\beta\gamma} \qquad \text{when } |\beta| + |\gamma| \leqslant k. \tag{23}$$

The difference of operators, $A - A_1$, can now be written in the form

$$A(x, D) - A_1(x, D) = R_N - r_N(x, D), \qquad \forall N,$$

where

$$r_N(x, \xi) = A_1(x, \xi) - \sum_{|\alpha| \leqslant N-1} \frac{1}{\alpha!} D_\xi^\alpha \partial_y^\alpha a(x, x, \xi)$$

is the Nth remainder in the asymptotic expansion for A_1. As we have shown, for each k, R_N has a C^k kernel satisfying (23) for $N > m + n + k$. The same is true for r_N since $r_N \in S^{m-N}$ by definition of the asymptotic expansion. Since N is arbitrary then by virtue of Theorem 7.6 it follows that $A - A_1 \in OS^{-\infty}$.

The following proposition facilitates the verification of $A \sim \sum_{j=1}^{\infty} A_j$. It shows that little attention has to be paid to the derivatives.

Proposition 7.9 *Let $A_j \in S^{m_j}$, $j = 1, 2, \ldots$ and assume that $m_{j-1} > m_j \to -\infty$ when $j \to \infty$. Let $A(x, \xi) \in C^{\infty}(\mathbb{R}^n \times \mathbb{R}^n)$ and assume that for all α, β we have for some $\mu = \mu(\alpha, \beta)$ and $C = C(\alpha, \beta)$ depending on α and β*

$$|D_\xi^\alpha D_x^\beta A(x, \xi)| \leqslant C(1 + |\xi|)^\mu, \qquad x, \xi \in \mathbb{R}^n. \tag{24}$$

If there is a sequence $\mu_k \to -\infty$ such that

$$\left| A(x, \xi) - \sum_{j=1}^{k-1} A_j(x, \xi) \right| \leqslant C_k(1 + |\xi|)^{\mu_k}, \tag{25}$$

it follows that $A \in S^{m_1}$ and that $A \sim \sum A_j$.

Proof By Proposition 7.7, we can find an $\tilde{A} \in S^{m_1}$ such that $\tilde{A} \sim \sum A_j$. It remains to show that $A - \tilde{A} \in S^{-\infty}$. Note that $A - \tilde{A}$ is rapidly decreasing since

$$\begin{aligned} |A(x, \xi) - \tilde{A}(x, \xi)| &\leqslant \left| A(x, \xi) - \sum_{j=1}^{k-1} A_j \right| + \left| \tilde{A}(x, \xi) - \sum_{j=1}^{k-1} A_j \right| \\ &\leqslant C_k(1 + |\xi|)^{\mu_k} + C_k'(1 + |\xi|)^{m_k} \\ &\leqslant C_N(1 + |\xi|)^{-N} \end{aligned} \tag{26}$$

for k such that $\mu_k, m_k \leqslant -N$. We must verify that the same is true for all the derivatives of $A - \tilde{A}$; it suffices to do so for first-order derivatives and iterate the conclusion. If η is a unit vector and $P = A - \tilde{A}$, then using Taylor's formula, and the fact that $\tilde{A} \in S^{m_1}$ and (24), we obtain

$$|P(x, \xi + \varepsilon\eta) - P(x, \xi) - (\operatorname{grad}_\xi P(x, \xi), \varepsilon\eta)| \leqslant C\varepsilon^2(1 + |\xi|)^{\tilde{\mu}},$$

where $0 < \varepsilon < 1$ and $\tilde{\mu} = \max(m_1, \mu)$. Hence dividing by ε we obtain

$$|(\operatorname{grad}_\xi P(x, \xi), \eta)| \leqslant C\varepsilon(1 + |\xi|)^{\tilde{\mu}} + |P(x, \xi) - P(x, \xi + \varepsilon\eta)| \cdot \varepsilon^{-1},$$

which gives

$$|\operatorname{grad}_\xi(A(x, \xi) - \tilde{A}(x, \xi))| \leqslant C_N'(1 + |\xi|)^{\tilde{\mu} - N},$$

if we take $\varepsilon = (1 + |\xi|)^{-N}$ and $2N$ instead of N in (26). The derivatives with respect to x can be handled in the same way, and this completes the proof.

We now use Proposition 7.9 to give another proof of Proposition 7.8 using a Taylor expansion in the ξ variable rather than y.

Alternative proof of Proposition 7.8 Since $A(x, \xi) \in S^m$ by Proposition 7.5, then an estimate of type (24) holds for $A(x, \xi)$. We change the variables $z = y - x$, $\eta = \theta - \xi$ in (15) and obtain

$$A(x, \xi) = \frac{1}{(2\pi)^n} \iint e^{-i(z,\eta)}\, a(x, x + z, \xi + \eta)\, dz\, d\eta.$$

Using Taylor's formula to expand $a(x, x + z, \xi + \eta)$ in powers of η about $\eta = 0$, we find that

$$A(x, \xi) = (2\pi)^{-n} \iint e^{-i(z,\eta)} \sum_{|\alpha| \leqslant N-1} \partial_\xi^\alpha a(x, x + z, \xi) \frac{\eta^\alpha}{\alpha!}\, dz\, d\eta + R_N(x, \xi)$$

where

$$R_N(x, \xi) = (2\pi)^{-n} \iint e^{-i(z,\eta)} \sum_{|\alpha| = N} N \frac{\eta^\alpha}{\alpha!}$$

$$\times \int_0^1 (1 - t)^{N-1}\, \partial_\xi^\alpha a(x, x + z, \xi + t\eta)\, dt\, dz\, d\eta$$

By Fourier's inversion formula

$$\frac{1}{(2\pi)^n} \iint \partial_\xi^\alpha a(x, x + z, \xi) \eta^\alpha e^{-i(z,\eta)}\, dz\, d\eta = \partial_\xi^\alpha D_z^\alpha(x, x + z, \xi)|_{z=0}$$

which gives the finite terms in the asymptotic expansion (22).

To establish an estimate of type (25) we must estimate the remainder $R_N(x, \xi)$. Regularizing in z and η using the identities

$$\langle z \rangle^{-n-1} \langle D_\eta \rangle^{n+1} e^{-i(z,\eta)} = e^{-i(z,\eta)} = \langle \eta \rangle^{-k} \langle D_z \rangle^k e^{-i(z,\eta)},$$

then R_N becomes a sum of integrals with respect to t of functions

$$R_{\alpha\beta\gamma t}(x, \xi) = f(t) \cdot \iint \frac{e^{-i(z,\eta)}}{\langle z \rangle^{n+1} \langle \eta \rangle^k} \partial_\xi^{\alpha+\gamma} D_z^{\alpha+\beta} a(x, x + z, \xi + t\eta)\, dz\, d\eta$$

where $|\alpha| = N$, $|\beta| \leqslant k$ and $|\gamma| \leqslant n + 1$, and $f(t)$ is bounded on $[0, 1]$. We must then estimate $R_{\alpha\beta\gamma t}$, uniformly in $t \in [0, 1]$. Since $a \in S^m$ we have the estimate

$$|R_{\alpha\beta\gamma t}(x, \xi)| \leqslant C_{k,N} \iint \langle z \rangle^{-n-1} \langle \eta \rangle^{-k} (1 + |\xi + t\eta|)^{m-N}\, dz\, d\eta. \qquad (27)$$

Let us decompose the region of integration into two parts, one part over the set $\{(z, \eta); |\eta| \leqslant \frac{1}{2}|\xi|\}$, and the other over its complement. If $|\eta| \leqslant \frac{1}{2}|\xi|$, then $\frac{1}{2}|\xi| \leqslant |\xi + t\eta| \leqslant \frac{3}{2}|\xi|$, so we obtain that the integral, I_1, over the first region is bounded by

$$|I_1| \leqslant C'_{k,N}(1 + |\xi|)^{m-N} \int_{|\eta| \leqslant 1/2|\xi|} d\eta \int \langle z \rangle^{-n-1}\, dz$$

$$\leqslant C''_{k,N}(1 + |\xi|)^{m+n-N},$$

where the second inequality holds because the volume of the domain of integration in η does not exceed $C|\xi|^n$. On the other hand, for $\frac{1}{2}|\xi| \leqslant |\eta|$, we obtain $(1 + |\xi + t\eta|)^{m-N} \leqslant C(1 + |\eta|)^p$ where $p = \max(m - N, 0)$. Hence the integral, I_2, over the second region is bounded by

$$|I_2| \leqslant C_{k,N} \int_{|\eta| \geqslant 1/2|\xi|} \int \langle z \rangle^{-n-1} \langle \eta \rangle^{p-k} \, dz \, d\eta$$

and if $p - k + n + 1 \leqslant 0$ it follows that

$$|I_2| \leqslant C'_{k,N}(1 + |\xi|)^{p-k+n+1} \iint \langle z \rangle^{-n-1} \langle \eta \rangle^{-n-1} \, dz \, d\eta$$
$$\leqslant C''_{k,N}(1 + |\xi|)^{p+n+1-k}.$$

Thus, by substituting the estimates for $|I_1|$ and $|I_2|$ in (27) and by choosing k sufficiently large we obtain

$$|R_N(x, \xi)| \leqslant C_N(1 + |\xi|)^{m+n-N},$$

which ensures the applicability of Proposition 7.8, so the proof is finished.

We shall now determine the adjoint $A^*(x, D)$ of the p.d.o. $A(x, D)$ with respect to the L_2 scalar product

$$(u, v)_{L_2} = \int u(x)\overline{v(x)} \, dx, \qquad u, v \in \mathcal{S}.$$

The adjoint A^* is defined by the equation $(A(x, D)u, v)_{L_2} = (u, A^*(x, D)v)_{L_2}$.

Theorem 7.10 *Let $A(x, D) \in OS^m$. Then its adjoint is a p.d.o., $A^*(x, D) \in OS^m$, which has the representation*

$$A^*(x, D)v = \frac{1}{(2\pi)^n} \iint \overline{A(y, \xi)} \, e^{i(x-y,\xi)} \, v(y) \, dy \, d\xi. \tag{28}$$

The symbol of $A^*(x, D)$ is given by

$$A^*(x, \xi) = \frac{1}{(2\pi)^n} \iint \overline{A(y, \theta} \, e^{i(x-y,\theta-\xi)} \, dy \, d\theta, \tag{29}$$

each semi-norm of $A^*(x, \xi)$ in S^m is bounded by a semi-norm of $A(x, \xi)$ in S^m, and we have the asymptotic expansion

$$A^*(x, \xi) \sim \sum_\alpha \frac{i^{|\alpha|}}{\alpha!} D_\xi^\alpha D_x^\alpha \overline{A(x, \xi)} \tag{30}$$

Proof We have

$$(A(x, D)u, v)_{L_2} = \frac{1}{(2\pi)^n} \int \overline{v(y)} \iint e^{i(y-x,\xi)} A(y, \xi) u(x) \, dx \, d\xi \, dy$$
$$= \frac{1}{(2\pi)^n} \int u(x) \cdot \overline{\iint e^{i(x-y,\xi)} \overline{A(y, \xi)} v(y) \, dy \, d\xi} \, dx$$
$$= (u, A^*(x, D)v)_{L_2},$$

hence

$$A^*(x, D)v = \frac{1}{(2\pi)^n} \iint \overline{A(y, \xi)}\, e^{i(x-y,\xi)}\, v(y)\, dy\, d\xi,$$

which proves (28). Comparing this with (13), we see that A^* is given by an amplitude representation, with amplitude $a(x, y, \xi) = \overline{A(y, \xi)}$. Since $\overline{A(y, \xi)} \in S^m$, then $A^*(x, D) \in OS^m$ by Proposition 7.5. Also, (29) is formula (15) in Proposition 7.5 and the assertion on the semi-norms of A^* is evident. Finally, (30) is an application of Proposition 7.8.

Remark Theorem 7.10 is also true for the dual ${}^tA$ of A, defined by $\langle Au, v\rangle = \langle u, {}^tAv\rangle$, $u, v \in \mathscr{S}$, where $\langle u, v\rangle = \int u(x)v(x)\, dx$. Since ${}^tA$ maps continuously ${}^tA\colon \mathscr{S} \to \mathscr{S}$, we are able to define

$$A\colon \mathscr{S}' \to \mathscr{S}'$$

by duality:

$$\langle Au, v\rangle = \langle u, {}^tAv\rangle, \qquad u \in \mathscr{S}', v \in \mathscr{S}.$$

Now, because the function $e^{i(x,\xi)} \in \mathscr{S}'$ (for fixed ξ), it is possible to recover the symbol $A(x, \xi)$ from the operator $A(x, D)$:

$$A(x, \xi) = e^{-i(x,\xi)}\, A(x, D)\, e^{i(x,\xi)}. \tag{31}$$

To prove this formula, we observe that if $A \in S^m$ then a direct computation gives

$$A(x, D)(e^{i(x,\xi)}\, u) = e^{i(x,\xi)}\, A(x, D + \xi)u, \qquad u \in \mathscr{S}', \xi \in \mathbb{R}^n.$$

If we take $u(x) = v(\varepsilon x)$ where $\hat{v} \in C_0^\infty$ and $v(0) = 1$, it follows when $\varepsilon \to 0$ that

$$A(x, D)\, e^{i(x,\xi)} = A(x, \xi)\, e^{i(x,\xi)}$$

since $e^{i(x,\xi)}u \to e^{i(x,\xi)}$ in $\mathscr{S}'$ and $A(x, D + \xi)u = (2\pi)^{-n} \int A(x, \varepsilon\eta + \xi)\hat{v}(\eta)\, d\eta \to A(x, \xi)$.

Corollary 7.11 *If* $A(x, \xi) \in S^m$, *then*

$$A^*(x, D) - \overline{A(x, D)} \in S^{m-1}.$$

Proof Look at (30).

Next, we study the composition of pseudo-differential operators.

Theorem 7.12 *Let* $A(x, \xi) \in S^{m_1}$ *and* $B(x, \xi) \in S^{m_2}$. *Then the composition of* $A(x, D)$ *and* $B(x, D)$ *is once more a p.d.o.,* $A(x, D) \circ B(x, D) \in OS^{m_1+m_2}$, *and it has the representation*

$$A(x, D) \circ B(x, D)u = \frac{1}{(2\pi)^n} \iint e^{i(x-y,\xi)}\, A(x, \xi)\overline{B^*(y, \xi)}u(y)\, dy\, d\xi$$

The symbol of $C(x, D) := A(x, D) \circ B(x, D)$ *is given by*

$$C(x, \xi) = \frac{1}{(2\pi)^n} \iint A(x,\theta)\overline{B^*(y, \theta)}\, e^{i(x-y,\theta-\xi)}\, dy\, d\theta, \tag{32}$$

and each semi-norm of $C(x, \xi)$ *in* $S^{m_1+m_2}$ *is bounded by a product of a semi-norm of* $A(x, \xi)$ *in* S^{m_1} *and a semi-norm of* $B(x, \xi)$ *in* S^{m_2}. *Further, we have the asymptotic expansion*

$$C(x, \xi) \sim \sum_\alpha \frac{i^{|\alpha|}}{\alpha!} D_\xi^\alpha A(x, \xi) D_x^\alpha B(x, \xi). \tag{33}$$

Proof First, we apply Theorem 7.10 to the operator $B^*(x, D)$ to obtain

$$B(x, D)u = B^{**}(x, D)u = \frac{1}{(2\pi)^n} \iint \overline{B^*(y, \xi)}\, e^{i(x-y,\xi)}\, u(y)\, dy\, d\xi.$$

This implies

$$(B(x, D)u)\hat{}\,(\xi) = \frac{1}{(2\pi)^n} \int e^{-i(y,\xi)}\, \overline{B^*(y, \xi)}\, u(y)\, dy,$$

thus

$$A(x, D) \circ B(x, D)u = \int e^{i(x,\xi)}\, A(x, \xi) B(x, D)u(\xi)\, d\xi$$

$$= \frac{1}{(2\pi)^n} \iint e^{i(x-y,\xi)}\, A(x, \xi)\overline{B^*(y, \xi)}u(y)\, dy\, d\xi \tag{34}$$

which is the representation (13), with amplitude $a(x, y, \xi) = A(x, \xi)\overline{B^*(y, \xi)}$. By Theorem 7.10, $B^* \in S^{m_2}$, hence $a(x, y, \xi) \in S^{m_1+m_2}$ and we conclude that $A(x, D) \circ B(x, D) \in OS^{m_1+m_2}$ by virtue of Proposition 7.5. Now, (32) is formula (15) in Proposition 7.5, and the assertion on the semi-norms of $C(x, \xi)$ is evident from the form of $a(x, y, \xi)$. It remains to show the asymptotic expansion. Applying Proposition 7.8 to the amplitude $A(x, \xi)\overline{B^*(y, \xi)}$ and Theorem 7.10 to $B^*(y, \xi)$, we obtain

$$C(x, \xi) \sim \sum \frac{i^{|\alpha|}}{\alpha!} D_\xi^\alpha D_y^\alpha (A(x, \xi)\overline{B^*(y, \xi)})|_{y=x}$$

$$\sim \sum \frac{i^{|\alpha|-|\beta|}}{\alpha! \cdot \beta!} D_\xi^\alpha A(x, \xi) D_\xi^\beta D_y^{\alpha+\beta} B(y, \xi)|_{y=x}.$$

If one expands this, numerous terms cancel and (33) is produced (see [Shu] p. 28).

Corollary 7.13 *Suppose that* $A \in S^{m_1}$ *and* $B \in S^{m_2}$. *Then the commutator*

$$[A, B] := A(x, D) \circ B(x, D) - B(x, D) \circ A(x, D) \in OS^{m_1+m_2-1}. \tag{35}$$

If $A(x, D) * B(x, D)$ *denotes the p.d.o. with the symbol* $A(x, \xi) \cdot B(x, \xi)$, *we also have* $A(x, D) \circ B(x, D) - A(x, D) * B(x, D) \in OS^{m_1+m_2-1}$.

Proof Look at the asymptotic expansions.

We end this section by showing that pseudo-differential operators are quasi-local.

Corollary 7.14 *Let $\varphi, \psi \in C_b^\infty(\mathbb{R}^n)$ and let $A \in S^m$. If the supports of φ and ψ are disjoint, i.e.* supp $\varphi \cap$ supp $\psi = \varnothing$, *we have*

$$\varphi A(x, D)\psi \in OS^{-\infty}. \tag{36}$$

Proof Recall that $C_b^\infty(\mathbb{R}^n)$ is the set of C^∞ functions $\varphi(x)$ with bounded derivatives of all orders. Since $C_b^\infty(\mathbb{R}^n) \subset S^0(\mathbb{R}^n \times \mathbb{R}^n)$, the composition $\varphi \circ A \circ \psi$ is well-defined and belongs to S^m. Now let us apply (34) to the operator

$$C(x, D) = (\varphi \circ A(x, D)) \circ \psi.$$

The symbol of $\varphi \circ A(x, D)$ being $\varphi(x) \cdot A(x, \xi)$, we obtain by the support property of φ, ψ that

$$\begin{aligned} C(x, \xi) &\sim \sum \frac{i^{|\alpha|}}{\alpha!} \varphi(x) D_\xi^\alpha A(x, \xi) \cdot D_x^\alpha \psi(x) \\ &= \sum \frac{i^{|\alpha|}}{\alpha!} \varphi(x) \cdot D_x^\alpha \psi(x) \cdot D_\xi^\alpha A(x, \xi) = 0, \end{aligned}$$

which means that $C(x, \xi) \in S^{-\infty}$.

Remark The Schwartz kernel $K(x, y)$ of a p.d.o. is C^∞ when $x \neq y$. Indeed, the kernel of $\varphi A \psi$ is $\varphi(x) K(x, y) \psi(y)$, and so it follows from Corollary 7.14 that $K(x, y)$ is C^∞ when $x \neq y$.

Another way to verify this remark is from the equation (10a). The following proposition will be used in the proof of Theorem 7.16.

Proposition 7.15 *Let the pseudo-differential operator A be defined by* (13) *with general symbol $a(x, y, \xi) \in S^m$, and let $K(x, y) = (2\pi)^{-n} \int e^{i(x-y, \xi)} a(x, y, \xi)\, d\xi$ be its distributional (Schwartz) kernel. Then $K \in C^\infty(\mathbb{R}^n \times \mathbb{R}^n \backslash D)$, where D is the diagonal in $\mathbb{R}^n \times \mathbb{R}^n$. Further, if $a(x, y, \xi) = 0$ for $|x - y| \leqslant \delta$, for some $\delta > 0$, then $K(x, y) \in C^\infty$.*

Proof The Schwartz kernel for the p.d.o. (13) defined by the amplitude function a is

$$K(x, y) = (2\pi)^{-n} \int e^{i(x-y, \xi)} a(x, y, \xi)\, d\xi,$$

and then regularizing this oscillatory integral we get for $x \neq y$

$$K(x, y) = (2\pi)^{-n} |x - y|^{-2N} \int e^{i(x-y, \xi)} (-\Delta_\xi)^N a(x, y, \xi)\, d\xi. \tag{37}$$

Since $a \in S^m$ then

$$|(-\Delta_\xi)^N a(x, y, \xi)| \leqslant C_N(1 + |\xi|)^{m-2N},$$

thus $K(x, y) \in C^k$ when $x \neq y$ provided that $m - 2N + k < -n$. But N can be chosen to be arbitrarily large, so it follows that $K \in C^\infty$ when $x \neq y$. The second statement in the lemma is also clear because in that case we have $|x - y| \geqslant \delta$ on the support of K.

7.4 Transformations of p.d.o.'s under a diffeomorphism

To study p.d.o.'s on a manifold, we have to discuss the effect of a change of variables on OS^m. Let $\kappa: \mathbb{R}^n \to \mathbb{R}^n$ be a C^∞ diffeomorphism (coordinate transformation) and suppose that the following conditions hold:

$$\text{(a)} \qquad |D^\alpha \kappa(x)| \leqslant C_\alpha, \qquad \text{for } |\alpha| \geqslant 1, x \in \mathbb{R}^n$$

$$\text{(b)} \qquad 0 < c_0 \leqslant |\det d\kappa(x)| \leqslant c_1 < \infty, \qquad x \in \mathbb{R}^n, \tag{38}$$

where $d\kappa$ denotes the Jacobian matrix of κ. By the Inverse Function Theorem we see that (a) and (b) also hold for the inverse mapping κ^{-1}:

$$\left.\begin{array}{lll}\text{(a)} & |D^\alpha \kappa^{-1}(y)| \leqslant C'_\alpha, & \text{for } |\alpha| \geqslant 1, y \in \mathbb{R}^n, \\ \text{(b)} & 0 < c'_0 \leqslant |\det d\kappa^{-1}(y)| \leqslant c'_1 < \infty, & y \in \mathbb{R}^n.\end{array}\right\} \tag{38'}$$

We consider the induced mapping κ^* on functions, $\kappa^* u = u \circ \kappa$, that is,

$$\kappa^* u(x) = u(\kappa(x)),$$

which is called the *pull-back* operator. It is easy to see that $u \in \mathscr{S}(\mathbb{R}^n_y)$ implies $\kappa^* u \in \mathscr{S}(\mathbb{R}^n_x)$. Indeed, by the chain rule, it follows that $|D^\alpha_x(u \circ \kappa)(x)|$ is bounded by a constant multiplied by $\max_{\beta \leqslant \alpha} |(D^\beta_y u)(\kappa(x))|$, so if $y = \kappa(x)$ then for any k we have

$$|x|^k |D^\alpha_x(\kappa^* u)(x)| \leqslant C_k(|y|^k + 1) \max_{\beta \leqslant \alpha} |D^\beta u(y)|$$

since it follows from the boundedness of the derivatives of κ^{-1} that $|x| \leqslant c(|y| + 1)$. Thus every semi-norm of $\kappa^* u$ in $\mathscr{S}(\mathbb{R}^n_x)$ is bounded by some semi-norm of u in $\mathscr{S}(\mathbb{R}^n_y)$, whence

$$\kappa^*: \mathscr{S}(\mathbb{R}^n_y) \to \mathscr{S}(\mathbb{R}^n_x)$$

is continuous. Since it has the inverse $(\kappa^{-1})^*$ which is also continuous, we see that the pull-back operator κ^* is an isomorphism.

Let A be a p.d.o. from OS^m, $m \in \mathbb{R}$. We define the transformed operator A_κ by

$$A_\kappa u := [A(u \circ \kappa)] \circ \kappa^{-1}, \qquad u \in \mathscr{S}(\mathbb{R}^n_y), \tag{39}$$

that is, $A_\kappa = (\kappa^*)^{-1} \circ A \circ \kappa^*$, or, with help of the commutative diagram

$$\begin{array}{ccc} \mathscr{S}(\mathbb{R}^n_x) & \xrightarrow{A} & \mathscr{S}(\mathbb{R}^n_x) \\ \kappa^* \uparrow & & \downarrow \kappa^{*-1} \\ \mathscr{S}(\mathbb{R}^n_y) & \xrightarrow{A_\kappa} & \mathscr{S}(\mathbb{R}^n_y). \end{array}$$

In this section we will show that if $A \in OS^m$ then $A_\kappa \in OS^m$. We begin with the simplest case, i.e. operators in $OS^{-\infty} = \bigcap_m OS^m$.

Lemma *Let the coordinate transformation* $\kappa: \mathbb{R}^n \to \mathbb{R}^n$ *satisfy* (38), (38′). *If* $A \in OS^{-\infty}$ *then also* $A_\kappa \in OS^{-\infty}$.

Proof By Theorem 7.6, A has the representation

$$Au(x) = \int K(x, x')u(x')\,dx',$$

where $K \in C^\infty(\mathbb{R}^{2n})$ and $|\partial_x^\alpha \partial_{x'}^\beta K(x, x')| \leqslant C_{\alpha\beta N}(1 + |x - x'|)^{-N}$ for all α, β and N. The transformed operator (39) is then given by

$$\begin{aligned} A_\kappa u(y) &= \int K(\kappa^{-1}(y), x')u(\kappa(x'))\,dx' \\ &= \int K(\kappa^{-1})(y), \kappa^{-1}(y'))u(y')\det[d\kappa^{-1}(y')]\,dy', \end{aligned}$$

where in the second integral we made the change of variables $y' = \kappa(x')$. By virtue of the boundedness of the derivatives of κ, we have $|\kappa^{-1}(y) - \kappa^{-1}(y')| \geqslant \delta|y - y'|$ for some constant $\delta > 0$. Hence A_k has the representation

$$A_\kappa u(y) = \int K_\kappa(y, y')u(y')\,dy',$$

where the kernel $K_\kappa(y, y') = K(\kappa^{-1}(y), \kappa^{-1}(y'))u(y')\det[d\kappa^{-1}(y')]$ belongs to $C^\infty(\mathbb{R}^{2n})$ and satisfies the condition $|\partial_y^\alpha \partial_{y'}^\beta K_\kappa(y, y')| \leqslant C_{\alpha\beta N}(1 + |y - y'|)^{-N}$ for all α, β and N. By Theorem 7.6, we conclude that $A_\kappa \in OS^{-\infty}$.

Theorem 7.16 *Let the coordinate transformation* $\kappa: \mathbb{R}^n \to \mathbb{R}^n$ *satisfy* (38), (38′). *If* $A \in OS^m$, $m \in \mathbb{R}$, *then* $A_\kappa \in OS^m$. *Denoting by* $A_\kappa(y, \eta)$ *the symbol of* A_κ, *we have*

$$A_\kappa(y, \eta) - A(\kappa^{-1}(y), {}^t d\kappa(\kappa^{-1}(y))\cdot\eta) \in S^{m-1}, \tag{40}$$

where ${}^t d\kappa$ *denotes the transposed Jacobian matrix of* κ. *Each semi-norm of* A_κ *in* S^m *is bounded by a semi-norm of* A *in* S^m. *Further,* $A_\kappa(\kappa(x), \eta) = e^{-i\langle\kappa(x), \eta\rangle} A(x, D)\, e^{i\langle\kappa(x), \eta\rangle}$.

In the proof we give also an asymptotic expansion of A_κ in terms of the original symbol $A(x, \xi)$.

Proof The operator A may be expressed in terms of an oscillatory integral, see (13),

$$Au(x) = \frac{1}{(2\pi)^n} \iint e^{i(x-x', \xi)} A(x, \xi) u(x')\, dx'\, d\xi. \tag{41}$$

By definition (39) we have, for $u \in \mathscr{S}(\mathbb{R}^n_y)$,

$$\begin{aligned} A_\kappa u(y) &= \frac{1}{(2\pi)^n} \iint e^{i(\kappa^{-1}(y) - x, \xi)} A(\kappa^{-1}(y), \xi) u(\kappa(x))\, dx\, d\xi \\ &= \frac{1}{(2\pi)^n} \iint e^{i(\kappa^{-1}(y) - \kappa^{-1}(z), \xi)} A(\kappa^{-1}(y), \xi) u(z) \det[d\kappa^{-1}(z)]\, dz\, d\xi \end{aligned}$$

Now we use Hadamard's Lemma:

$$\kappa^{-1}(y) - \kappa^{-1}(z) = \Phi(y, z)(y - z),$$

where $\Phi(y, z) = \int_0^1 d\kappa^{-1}(z + t(y - z))\, dt$. Since $\Phi(y, y) = d\kappa^{-1}(y)$, the matrix $\Phi(y, z)$ is invertible near the diagonal in $\mathbb{R}^n \times \mathbb{R}^n$, say for $|\kappa^{-1}(y) - \kappa^{-1}(z)| \leqslant 2\varepsilon$, where $\varepsilon > 0$. We have

$$A_\kappa u(y) = \frac{1}{(2\pi)^n} \iint e^{i(y-z, {}^t\Phi(y, z)\xi)} A(\kappa^{-1}(y), \xi) u(z) \det[d\kappa^{-1}(z)]\, dz\, d\xi \tag{42}$$

and we would like to make the change of variable $\xi \to {}^t\Phi(y, z)\xi$ in order to put the exponential term in the form $e^{i(y-z, \xi)}$ so that we can recognize a pseudo-differential operator, but first we must take into account the region where ${}^t\Phi(y, z)$ is not invertible.

Let $h \in C_0^\infty(\mathbb{R})$ be such that $h(t) = 1$ for $|t| \leqslant 1$ and $h(t) = 0$ for $|t| \geqslant 2$. Then we decompose the original operator (41) as $A = A^1 + A^2$, where

$$A^1 u(x) = (2\pi)^{-n} \iint A(x, \xi) h(\varepsilon^{-1}|x - x'|) u(x')\, e^{i(x-x', \xi)}\, dx'\, d\xi \tag{43}$$

$$A^2 u(x) = (2\pi)^{-n} \iint A(x, \xi)[1 - h(\varepsilon^{-1}|x - x'|)] u(x')\, e^{i(x-x', \xi)}\, dx'\, d\xi \tag{44}$$

We claim that $A^2 \in OS^{-\infty}$. Indeed, the operator A^2 has the form of an integral operator

$$A^2 u(x) = \int K(x, y) u(y)\, dy,$$

where

$$K(x, y) = \int A(x, \xi)[1 - h(\varepsilon^{-1}|x - y|)]\, e^{i(x-y, \xi)}\, d\xi.$$

Thus A^2 is a p.d.o. with amplitude representation of the form (13) where $a(x, y, \xi) = A(x, \xi)[1 - h(\varepsilon^{-1}|x - y|)]$. Since $a(x, y, \xi) = 0$ for $|x - y| \leqslant \varepsilon$, then $K \in C^\infty$ by Proposition 7.15. By regularizing the oscillatory integral for K as in (37), we obtain

$$|D_x^\alpha D_y^\beta K(x, y)| \leqslant C_{\alpha\beta N}|x - y|^{-2N} \qquad \forall\alpha, \beta \text{ and } N \geqslant 0,$$

and then if we use the inequality $1 + |x - y| \leqslant \varepsilon^{-1}(1 + \varepsilon)|x - y|$ for $|x - y| \geqslant \varepsilon$, it follows that

$$|D_x^\alpha D_y^\beta K(x, y)| \leqslant C_{\alpha\beta N\varepsilon}(1 + |x - y|)^{-2N} \qquad \forall\alpha, \beta \text{ and } N \geqslant 0.$$

Hence $A^2 \in OS^{-\infty}$ by virtue of Theorem 7.6.

From (43), (44) we have $A_\kappa = A_\kappa^1 + A_\kappa^2$. Then $A_\kappa^2 \in OS^{-\infty}$, by virtue of the lemma proved above, so it remains to show that $A_\kappa^1 \in OS^m$ and that (40) holds for A_κ^1. Representations like (41), (42) hold for A^1 and A_κ^1, and then by the change of variable $\xi \to {}^t\Phi(y, z)\xi$ we have

$$A_\kappa^1 u(y) = \frac{1}{(2\pi)^n} \iint e^{i(y-z,\xi)}\, A(\kappa^{-1}(y), \psi(y, z)\xi)D(y, z)u(z)\, dz\, d\xi \tag{46}$$

where $\psi(y, z) = {}^t\Phi^{-1}(y, z)$ and

$$D(y, z) := \det[d\kappa^{-1}(z)] \cdot \det[\psi(y, z)] \cdot h(\varepsilon^{-1}|\kappa^{-1}(y) - \kappa^{-1}(z)|).$$

Now, (46) is a familiar representation of the p.d.o. A_κ^1 in the form (13) with the amplitude

$$a(y, z, \xi) = A(\kappa^{-1}(y), \psi(y, z)\xi) \cdot D(y, z). \tag{47}$$

Note that $\psi(y, z)$ is well-defined and smooth on the support of $D(y, z)$, i.e. when $|\kappa^{-1}(y) - \kappa^{-1}(z)| \leqslant 2\varepsilon$. Hence $a \in C^\infty$. A straightforward application of the chain rule shows that by (38′) we have

$$|D_y^\gamma D_z^\beta D_\xi^\alpha a(y, z, \xi)| \leqslant C_{\alpha\beta\gamma}(1 + |\xi|)^{m-|\alpha|},$$

which proves that $A_\kappa^1 \in OS^m$. In view of Propositions 7.5 and 7.8 applied to the amplitude (47), we obtain for the symbol of A_κ the asymptotic expansion

$$A_\kappa(y, \eta) \sim \sum_\alpha \frac{i^{|\alpha|}}{\alpha!} D_\eta^\alpha D_z^\alpha A(\kappa^{-1}(y), \psi(y, z)\eta)D(y, z)|_{z=y} \tag{48}$$

Now, $\psi(y, y) = {}^t d\kappa(\kappa^{-1}(y))$ and $D(y, y) = 1$, so by taking the first term in this expansion we obtain (40).

The assertion about the semi-norms of A_κ is clear from the estimates above and in Proposition 7.5.

Finally, we know from (31) that the symbol of the operator A_κ is given by $A_\kappa(y, \eta) = e^{i(y,\eta)}\, A_\kappa(y, D)\, e^{i(y,\eta)}$. By continuity, the formula (39) holds for $u(y) = e^{i(y,\xi)} \in \mathscr{S}'$, hence

$$A_\kappa(\kappa(x), \eta) = e^{-i(\kappa(x),\eta)}\, A(x, D)\, e^{i(\kappa(x),\eta)}.$$

Remark By making a more careful analysis of (48), one can rearrange the terms, simplify them, and arrive at the following expansion

$$A_\kappa(\kappa(x), \eta) \sim \sum_\alpha \frac{1}{\alpha!} \varphi_\alpha(x, \eta)\, \partial_\xi^\alpha A(x, {}^t d\kappa(x)\eta),$$

where $\varphi_\alpha(x, \eta) := D_z^\alpha \exp i(\kappa(z) - \kappa(x) - d\kappa(x)(z - x), \eta)|_{z=x}$ is a polynomial in η of degree $\leqslant |\alpha|/2$. For details we refer the reader to [Hö 3] and [Shu].

We also need another version of Theorem 7.16. Let X and Y be open subsets of $\mathbb{R}^n$ and $\kappa: X \to Y$ a diffeomorphism. If $A \in OS^m$ and the Schwartz kernel of A has compact support in $X \times X$, then there exists a function $\psi \in C_0^\infty(X)$ such that $A = A(\psi\ \cdot)$. The operator $A: \mathcal{S} \to \mathcal{S}$ can therefore be extended to $C^\infty(X) \to \mathcal{S}$ by letting $Au := A(\psi \cdot u)$. We then define the transformed operator

$$A_\kappa: \mathcal{S}(\mathbb{R}^n) \to \mathcal{S}(\mathbb{R}^n)$$

by

$$A_\kappa u(y) = \begin{cases} [A(u \circ \kappa)](\kappa^{-1}(y)) & \text{if } y \in Y \\ 0 & \text{if } y \notin Y \end{cases}$$

Theorem 7.17 *Let X and Y be open subsets of $\mathbb{R}^n$ and $\kappa: X \to Y$ a diffeomorphism. If $A \in OS^m$ and the kernel of A has compact support in $X \times X$ then $A_\kappa \in OS^m$, and the kernel of A_κ has compact support in $Y \times Y$. Denoting by $A_\kappa(y, \eta)$ the symbol of A_κ, we have*

$$A_\kappa(y, \eta) - A(\kappa^{-1}(y), {}^t d\kappa(\kappa^{-1}(y)) \cdot \eta) \in S^{m-1},$$

and $A(x, \xi) = 0$ for x outside a compact subset of X, see (10b), and $A_\kappa(y, \eta) = 0$ for y outside a compact subset of Y. Each semi-norm of A_κ in S^m is bounded by a semi-norm of A in S^m. Further, $A_\kappa(\kappa(x), \eta) = e^{-i(\kappa(x), \eta)} A(x, D)\, e^{i(\kappa(x), \eta)}$.

Proof The proof of Theorem 7.16 carries with obvious modification. Note that if the support of the kernel of A is contained inside $F \times F$, where $F \subset X$ is compact, it follows from (10b) that $A(x, \xi) = 0$ for $x \in X \backslash F$. Then we have $A_\kappa(y, \eta) = 0$ for $y \in Y \backslash \kappa(F)$.

For future reference, observe that if $\varphi, \psi \in C_0^\infty(X)$ then for any $A \in OS^m$ the kernel of $\varphi(x) A(x, D)[\psi(x) \cdot]$ has compact support in $X \times X$. In fact, if $K(x, y)$ denotes the kernel of A, then $\varphi(x) \cdot K(x, y) \cdot \psi(y)$ is the kernel of $\varphi A \psi$.

7.5 Classical symbols

We first define the classes J^m, consisting of homogeneous functions.

Definition 7.18 *Let $m \in \mathbb{R}$. A function $H(x, \xi)$ belongs to $J^m = J^m(\mathbb{R}^n \times \mathbb{R}^n)$ if it has the following properties:*

(1) *$H(x, \xi)$ is (positively) homogeneous of order m,*

$$H(x, c\xi) = c^m \cdot H(x, \xi), \qquad c > 0, \xi \neq 0$$

(2) *$H \in C^\infty(\mathbb{R}^n \times (\mathbb{R}^n \backslash 0))$ (the point $\xi = 0$ is excluded!)*

(3) *All derivatives $D_\xi^\alpha D_x^\beta H(x, \xi)$ are bounded on $\mathbb{R}^n \times S^{n-1}$,*

$$H \in C_b^\infty(\mathbb{R}^n \times S^{n-1})$$

where S^{n-1} is the unit sphere $|\xi| = 1$ in $\mathbb{R}^n$.

A function $\chi \in C^\infty(\mathbb{R}^n)$, $0 \leqslant \chi \leqslant 1$, is called a *cut-off function* if $\chi(\xi)$ is identically 0 near the origin and identically 1 for large $|\xi|$. Multiplication of a function $H \in J^m$ by χ cuts off the singularity of H at $\xi = 0$, that is, $\chi(\xi) \cdot H(x, \xi) \in C^\infty(\mathbb{R}^n \times \mathbb{R}^n)$. Clearly, every cut-off function can be written in the form

$$\chi(\xi) = 1 - \varphi(\xi)$$

where $\varphi \in C_0^\infty(\mathbb{R}^n)$, $0 \leqslant \varphi \leqslant 1$, and $\varphi(\xi) = 1$ for $|\xi| \leqslant \varepsilon$. We call φ a *patch function.*

Proposition 7.19

(a) If $H \in J^m$, then $\chi(\xi)H \in S^m$ where χ is a cut-off function.
(b) If χ_1 and χ_2 are two cut-off functions, then $\chi_1 H - \chi_2 H \in S^{-\infty}$ for $H \in J^m$.

This proposition says that we can associate with each homogeneous function H a pseudo-differential operator

$$(\chi \cdot H)(x, D), \tag{49}$$

which depends on χ only modulo $S^{-\infty}$. For convenience we often write $\chi(D)H(x, D)$ instead of (49), and we call $H(x, \xi)$ the "symbol" of this p.d.o., thus slightly abusing our terminology.

Proof (a) The derivatives $D_\xi^\alpha D_x^\beta H(x, \xi)$ are homogeneous functions of degree $m - |\alpha|$; we have for $\xi \neq 0$

$$D_\xi^\alpha D_x^\beta H(x, \xi) = |\xi|^{m-|\alpha|} D_\xi^\alpha D_x^\beta H(x, \xi/|\xi|).$$

From condition (3) in Definition 7.18 we obtain for each $\varepsilon > 0$ and $|\xi| \geqslant \varepsilon$

$$D_\xi^\alpha D_x^\beta H(x, \xi) \leqslant C_{\alpha\beta}|\xi|^{m-|\alpha|} \leqslant C_{\alpha\beta\varepsilon}(1 + |\xi|)^{m-|\alpha|}.$$

The Leibniz rule finishes the proof of (a).

(b) Since $A(x, \xi) := \chi_1 H - \chi_2 H = (\chi_1 - \chi_2)H$ is equal 0 in a neighbourhood of $\xi = 0$ and is also 0 for $|\xi| \geqslant R$, and bounded on $\mathbb{R}^n \times \{\chi; \varepsilon \leqslant |\xi| \leqslant R\}$, we can estimate

$$|D_\xi^\alpha D_\xi^\beta A(x, \xi)| \leqslant C_{\alpha\beta N}(1 + |\xi|)^{-N}, \qquad \text{for all } \alpha, \beta \text{ and } N \geqslant 0$$

which means that $A \in S^{-\infty}$.

Remark 7.20 When $m \geqslant 0$ we have

$$|D_x^\beta H(x, \xi)| \leqslant C_\beta(1 + |\xi|)^m, \qquad \xi \neq 0, \text{ for all } \beta, \tag{50}$$

which implies that these functions are locally bounded. Since the Lebesgue measure μ_{2n} of the singularities of $D_x^\beta H$ is $\mu_{2n}(\mathbb{R}^n \times \{0\}) = 0$, it follows that the functions $D_x^\beta H$ are locally integrable. We shall use this remark later.

Now we turn to the definition of classical symbols.

Definition 7.21 *We say that $A(x, \xi) \in S^m$ is a (-1) classical symbol if there is a representation*

$$A(x, \xi) = \chi(\xi)H_m(x, \xi) + A_{m-1}(x, \xi), \tag{51}$$

where χ is a cut-off function, $H_m \in J^m$ and $A_{m-1} \in S^{m-1}$. Obviously the existence of the representation does not depend on the choice of χ. We denote the set of (-1) classical symbols by $Cl^{-1}S^m$.

Remark It is also possible to define $(-k)$ classical symbols by $A(x, \xi) = \sum_{j=0}^{k-1} \chi(\xi)H_{m-j} + A_{m-k}$, where $H_{m-j} \in J^{m-j}$, $A_{m-k} \in S^{m-k}$, and to define $(-\infty)$ classical symbols by an asymptotic expansion

$$A(x, \xi) \sim \sum_{j=0}^{\infty} \chi(\xi)H_{m-j}, \qquad H_{m-j} \in J^{m-j}.$$

Classical symbols $A \in Cl^{-1}S^m$ have a *principal part* defined by

$$\pi A(x, \xi) := H_m(x, \xi). \tag{52}$$

In general, the principal part πA of $A \in S^m$ is understood to be the equivalence class $[A]$ in S^m/S^{m-1}, but for classical symbols we can be more concrete:

Proposition 7.22 *If A is a (-1) classical symbol then the principal part $H_m(x, \xi) \in J^m$ is uniquely determined by A and we have*

$$H_m(x, \xi) = \lim_{t \to +\infty} \frac{A(x, t\xi)}{t^m}, \qquad \xi \neq 0.$$

Further, in the equivalence class $[A] \in S^m/S^{m-1}$, there is only one function $H_m \in J^m$ such that $[A] = [\chi(\xi)H_m]$, where χ is a cut-off function. Hence it makes sense to identify $[A]$ with the homogeneous function H_m.

Proof Let $R > 0$ such that $\varphi(\xi) = 0$ for $|\xi| \geqslant R$. By definition we have $\chi = 1 - \varphi$ and $A - [1 - \varphi(\xi)]H_m \in S^{m-1}$, so that

$$|A(x, t\xi) - [1 - \varphi(t\xi)]H_m(x, t\xi)| \leqslant C(1 + t|\xi|)^{m-1} \leqslant C_\xi t^{m-1}.$$

Dividing by t^m, we get for $t \geqslant R/|\xi|$

$$\left|\frac{A(x, t\xi)}{t^m} - H_m(x, \xi)\right| \leqslant C_\xi t^{-1},$$

since $\varphi(t\xi) = 0$ and H_m is homogeneous of degree m. This proves the first part of the proposition. Suppose now that there are two functions $H_1, H_2 \in J^m$ with $(1 - \varphi_1(\xi))H_1 - (1 - \varphi_2(\xi))H_2$ belonging to S^{m-1}. Once more we have the estimate

$$|(1 - \varphi_1(t\xi))H_1(x, t\xi) - (1 - \varphi_2(t\xi))H_2(x, t\xi)| \leqslant C_\xi t^{m-1}.$$

Dividing by t^m and letting $t \geqslant R/|\xi|$, we obtain

$$|H_1(x, \xi) - H_2(x, \xi)| \leqslant C_\xi t^{-1},$$

since $\varphi_1(t\xi) = \varphi_2(t\xi) = 0$. This means that $H_1 \equiv H_2$.

Corollary *Let* $A \in Cl^{-1}S^m$. *Then* $\pi A = 0$ *if and only if* $A \in S^{m-1}$.

In connection with this corollary, note that $S^{m-1} \subset Cl^{-1}S^m$, that is, *any* symbol of order $m - 1$ is (-1) classical of order m.

The set of all (-1) classical operators forms a $*$ algebra, and the principal symbol map π is a $*$ algebra homomorphism. This is the subject of the next proposition; we leave the proof as an exercise.

Proposition 7.23

(*a*) *If* $A \in OCl^{-1}S^{m_1}$, $B \in OCl^{-1}S^{m_2}$ *then* $A \circ B \in OCl^{-1}S^{m_1+m_2}$ *and*

$$\pi(A \circ B) = \pi A \cdot \pi B,$$

(*b*) *If* $A \in OCl^{-1}S^m$ *then* $A^* \in OCl^{-1}S^m$ *and*

$$\pi A^* = \overline{\pi A}.$$

Next, we examine the behaviour of classical symbols under a coordinate transformation.

Theorem 7.16′ *The statement of Theorem* 7.16 *holds for* (-1) *classical operators, that is, if* $A \in OCl^{-1}S^m$ *then* $A_\kappa \in OCl^{-1}S^m$. *Moreover, for the principal parts we have*

$$\pi A_\kappa(y, \eta) = \pi A(\kappa^{-1}(y), {}^t d\kappa(\kappa^{-1}(y)) \cdot \eta). \tag{53}$$

This means that the principal part of a pseudo-differential operator $A \in OCl^{-1}S^m$ is invariantly defined on the cotangent space $T^*(\mathbb{R}^n)\backslash 0$.

Proof If we write $\pi A = H_m$ and apply the coordinate transformation κ to $A = \chi \cdot H_m + A_{m-1}$, where χ is a cut-off function and $A_{m-1} \in S^{m-1}$, we get

$$A_\kappa = (\chi H_m)_\kappa + (A_{m-1})_\kappa,$$

and $(A_{m-1})_\kappa \in S^{m-1}$ and $(\chi\, H_m)_\kappa \in S^m$ by Theorem 7.16.

Further, by the same theorem,

$$(\chi H_m)_\kappa = \chi({}^t d\kappa(\kappa^{-1}(y))\cdot\eta)H_m(\kappa^{-1}(y), {}^t d\kappa(\kappa^{-1}(y))\cdot\eta) + A'_{m-1}(y,\eta)$$

where $A'_{m-1} \in S^{m-1}$, whence

$$A_\kappa(y,\eta) = \chi({}^t d\kappa(\kappa^{-1}(y))\cdot\eta)H_m(\kappa^{-1}(y), {}^t d\kappa(\kappa^{-1}(y))\cdot\eta) + A''_{m-1}(y,\eta) \quad (54)$$

where $A''_{m-1} \in S^{m-1}$. The conditions (38), (38′) imply

$$c'|\eta| \leqslant |{}^t d\kappa(x)\cdot\eta| \leqslant c|\eta|, \qquad \text{for all } x \in \mathbb{R}^n,\ \eta \in \mathbb{R}^n,$$

where $c > 0$, $c' > 0$. Now let $\tilde{\chi}(\eta)$ be a cut-off function, i.e. $\tilde{\chi}(\eta) = 0$ near $\eta = 0$ and $\tilde{\chi}(\eta) = 1$ for large $|\eta|$. The function

$$\psi(y,\eta) = \tilde{\chi}(\eta) - \chi({}^t d\kappa(\kappa^{-1}(y))\cdot\eta)$$

is C^∞ and satisfies $\psi(y,\eta) = 0$ for $|\eta| \leqslant \varepsilon$ and for $|\eta| \geqslant R$ (where $\varepsilon > 0$ and $R > 0$ are independent of y), hence the symbol $\psi(y,\eta)\cdot H_m(\kappa^{-1}(y), {}^t d\kappa(\kappa^{-1}(y))\cdot\eta)$ belongs to $S^{-\infty}$, and we obtain from (54) that

$$A_\kappa(y,\eta) = \tilde{\chi}(\eta)H_m(\kappa^{-1}(y), {}^t d\kappa(\kappa^{-1}(y))\cdot\eta) + \tilde{A}_{m-1}(y,\eta) \quad (55)$$

where $\tilde{A}_{m-1} \in S^{m-1}$. Thus $A_\kappa \in OCl^{-1}S^m$ and Proposition 7.21 now gives us the assertion (53) of the theorem.

There is also a version of Theorem 7.17 that holds for (-1) classical symbols. This is important for Chapter 8.

Theorem 7.17′ *Let X and Y be open subsets of $\mathbb{R}^n$ and $\kappa: X \to Y$ a diffeomorphism. If $A \in OCl^{-1}S^m$ is a (-1) classical operator and the kernel of A has compact support in $X \times X$ then*

$$A_\kappa \in OCl^{-1}S^m,$$

and the kernel of A_κ has compact support in $Y \times Y$. Also.

$$\pi A_\kappa(y,\eta) = \pi A(\kappa^{-1}(y), {}^t d\kappa(\kappa^{-1}(y))\cdot\eta), \qquad y \in Y,\ \eta \in \mathbb{R}^n \quad (56)$$

where $A(x,\xi) = 0$ for x outside subset of X, and $A_\kappa(y,\eta) = 0$ for y outside a compact subset of Y.

7.6 Continuity in Sobolev spaces

In this section we follow Hörmander's proof [Hö 3] that any pseudo-differential operator $A(x, D) \in OS^m$ is a continuous operator

$$A(x,D)\colon W_2^l(\mathbb{R}^n) \to W_2^{l-m}(\mathbb{R}^n),$$

for every $l \in \mathbb{R}$. The Sobolev spaces

$$W_2^l(\mathbb{R}^n) = \left\{ u \in L^2_{\text{loc}};\ \|u\|_l^2 = \int |\hat{u}(\xi)|^2(1 + |\xi|^2)^l\, d\xi < \infty \right\}$$

are endowed with the Fourier norm $\|u\|_l$.

A summary of definitions and results on Sobolev spaces can be found in the appendix at the end of this chapter.

Proposition 7.24 *Let* $A(x, \xi) \in S^m$. *The commutators of* $A(x, D)$ *with* D_j *and multiplication by* x_j *are*

$$[A(x, D), D_j] = i \frac{\partial A}{\partial x_j}(x, D),$$

$$[A(x, D), x_j] = \frac{1}{i} \frac{\partial A}{\partial \xi_j}(x, D).$$

Proof The first relation states that

$$D_j(A(x, D)u(x)) = A(x, D)D_ju(x) - i \frac{\partial A}{\partial x_j}(x, D) \cdot u(x).$$

which follows by differentiation of

$$A(x, D)u(x) = (2\pi)^{-n} \int e^{i(x, \xi)} A(x, \xi)\hat{u}(\xi)\, d\xi$$

under the integral sign since $\xi_j \hat{u}(\xi)$ is the Fourier transform of $D_j u$. The Fourier transform of $x_j \cdot u(x)$ is $-D_j\hat{u}(\xi)$, so an integration by parts gives the second relation.

Proposition 7.25 *If* $A_1, \ldots, A_k \in S^0$ *and* $F \in C^\infty(\mathbb{C}^k)$ *then* $F(A_1, \ldots, A_k) \in S^0$.

Proof Since $\operatorname{Re} A_v$, $\operatorname{Im} A_v \in S^0$ we may assume that A_v are real valued and that $F \in C^\infty(\mathbb{R}^k)$. We use the fact that for $A_v \in S^0$ the values $A(x, \xi)$ belong to $\bar{B}(0, R)$ for some R and that C^∞-functions F are bounded on compact sets $\bar{B}$. This implies that $F(A) = F(A_1, \ldots, A_k)$ is bounded, and for the derivatives we have

$$\frac{\partial F(A)}{\partial x_j} = \sum_{v=1}^{k} \frac{\partial F}{\partial A_v} \frac{\partial A_v}{\partial x_j} \qquad \text{is bounded}$$

$$\frac{\partial F(A)}{\partial \xi_j} = \sum_{v=1}^{k} \frac{\partial F}{\partial A_v} \frac{\partial A_v}{\partial \xi_j} \qquad \text{is } O((1 + |\xi|)^{-1}),$$

since $\partial A_v/\partial x_j \in S^0$, $\partial A_v/\partial \xi_j \in S^{-1}$ and $\partial F(A)/\partial A_v$ is bounded. The proof is finished by induction with respect to the order $|\alpha| + |\beta|$ of the derivatives $D_\xi^\alpha D_x^\beta F$.

We need also the classical lemma of Schur.

Lemma 7.26 *If $K(x, y)$ is a continuous function in $\mathbb{R}^n \times \mathbb{R}^n$ and*

$$\sup_y \int |K(x, y)|\, dx \leqslant C, \qquad \sup_x \int |K(x, y)|\, dy \leqslant C, \tag{57}$$

then the integral operator $Ku := \int K(x, y)u(y)\, dy$ has norm $\leqslant C$ in $L_2(\mathbb{R}^n)$.

Proof The Cauchy-Schwarz inequality gives

$$|(Ku)(x)|^2 \leqslant \int |K(x, y)|\, |u(y)|^2\, dy \cdot \int |K(x, y)|\, dy.$$

If the last integral is estimated by C, an integration with respect to x gives

$$\int |(Ku)(x)|^2\, dx \leqslant C \iint |K(x, y)| \cdot |u(y)|^2\, dx\, dy \leqslant C^2 \int |u(y)|^2\, dy.$$

Theorem 7.27 *If $A \in S^0$ then the p.d.o. $A(x, D)$ is a bounded operator*

$$A(x, D)\colon L_2(\mathbb{R}^n) \to L_2(\mathbb{R}^n).$$

Proof Assume first that $A \in S^{-n-1}$. Then the kernel K of $A(x, D)$ is continuous (estimate the integral in (10a)) and

$$|K(x, y)| \leqslant (2\pi)^{-n} \int |A(x, \xi)|\, d\xi \leqslant C' \int \frac{d\xi}{(1 + |\xi|)^{n+1}} < \infty.$$

Now $(x - y)^\alpha K(x, y)$ is the kernel of the commutator

$$[x_1, [x_1, \ldots, [x_n, A(x, D)]]] = i^{|\alpha|}(\partial_\xi^\alpha A)(x, D) \in OS^{-n-1-|\alpha|}$$

by Proposition 7.24, so this is also a bounded function. Hence

$$(1 + |x - y|)^{n+1}|K(x, y)| \leqslant \tilde{C},$$

from which we obtain the estimates (57), and the L_2-continuity of $A(x, D)$ follows from Schur's Lemma 7.26.

Next we prove by induction that $A(x, D)$ is L_2-continuous if $A \in S^k$ and $k \leqslant -1$. To do so we observe that for $u \in \mathscr{S}$ (which is dense in $L_2(\mathbb{R}^n)$)

$$\|A(x, D)u\|^2 = (A(x, D)u, A(x, D)u) = (B(x, D)u, u),$$

where $B(x, D) = A^*(x, D) \circ A(x, D) \in OS^{2k}$. The continuity of $A(x, D)$ is therefore a consequence of that of $B(x, D)$ since

$$\|A(x, D)u\|^2 \leqslant \|B(x, D)u\| \cdot \|u\| \leqslant \|B(x, D)\| \cdot \|u\|^2.$$

From the first part of the proof the continuity of $A(x, D)$ for all $A \in S^k$ now follows successively for $k \leqslant -(n + 1)/2$, $k \leqslant -(n + 1)/4, \ldots,$ hence for $k \leqslant -1$ after a finite number of steps.

Assume now that $A \in S^0$ and choose $M > 2 \cdot \sup_{x,\xi} |A(x, \xi)|^2$. Then

$$C(x, \xi) := [M - |A(x, \xi)|^2]^{1/2} \in S^0,$$

by Proposition 7.25 since $M/2 \leqslant M - |A(x, \xi)|^2$ and we can choose

$F \in C^\infty(\mathbb{R})$ with $F(t) = t^{1/2}$ when $t \geqslant M/2$. Corollary 7.11 and Theorem 7.12 imply that

$$C^*(x, D) \circ C(x, D) = M - A^*(x, D) \circ A(x, D) + T(x, D),$$

where $T(x, D) \in OS^{-1}$ (look at the symbols!). Hence we have

$$\begin{aligned}(A(x, D)u, A(x, D)u) &= -(C(x, D)u, C(x, D)u) + M(u, u) + (T(x, D)u, u)\\ &\leqslant M(u, u) + (T(x, D)u, u),\end{aligned}$$

or

$$\|A(x, D)u\|^2 \leqslant M\|u\|^2 + \|T(x, D)u\| \cdot \|u\|,$$

which completes the proof since $T \in OS^{-1}$ is already known to be L_2 continuous.

By referring to the Fourier norm of W_2^l, it is obvious that the p.d.o. Λ^m with symbol $(1 + |\xi|^2)^{m/2}$ is an isometry, Λ^m: $W_2^l \to W_2^{l-m}$, and we have

$$\Lambda^{m_1} \circ \Lambda^{m_2} = \Lambda^{m_1 + m_2}, \qquad m_1, m_2 \in \mathbb{R}.$$

Using these isometries we can prove the following theorem.

Theorem 7.28 *If $A \in S^m$ then $A(x, D)$ is a continuous operator,*

$$A(x, D)\colon W_2^l(\mathbb{R}^n) \to W_2^{l-m}(\mathbb{R}^n) \qquad \textit{for every } l \in \mathbb{R}.$$

Proof We write $A(x, D) = A(x, D)\Lambda^{-m}\Lambda^m$, where $A(x, D) \circ \Lambda^{-m} \in OS^0$ by Theorem 7.12. In view of Theorem 7.27, the operator $A(x, D) \circ \Lambda^{-m}$ is continuous $W_2^0 \to W_2^0$, hence

$$A(x, D)\colon W_2^m \xrightarrow{\Lambda^m} W_2^0 \xrightarrow{A\Lambda^{-m}} W_2^0$$

is continuous, where $W_2^0 = L_2$. This proves the statement in the theorem if $l = m$. Now, for any $l \in \mathbb{R}$, we use the fact that Λ^{l-m} is isometric to obtain

$$\|Au\|_{l-m} = \|\Lambda^{l-m}Au\|_0 \leqslant C\|u\|_l,$$

by virtue of the result just proved, since $\Lambda^{l-m} \circ A(x, D) \in OS^l$.

Remark The proofs of Theorems 7.27 and 7.28 involve a finite number of adjoints and compositions of operators derived from the symbol $A(x, \xi)$. It follows from Theorems 7.10 and 7.12 that the norm of $A(x, D)$: $W_2^l \to W_2^{l-m}$ is bounded by a semi-norm of A in S^m.

The symbol spaces $\tilde{S}^m$

If the symbol $H(x, \xi) \in J^m$, $m \geqslant 0$, is independent of x for large x then it is possible to define operators $H(x, D)$ directly without a cut-off function, and to show the continuity of

$$H(x, D)\colon W_2^l(\mathbb{R}^n) \to W_2^{l-m}(\mathbb{R}^n), \qquad m \geqslant 0,$$

without introducing any ξ-derivatives. This is important for some applications. For a homogeneous symbol $H(x, \xi) \in J^m$, differentiation with respect to ξ of $H(x, \xi)$ would lead to serious singularities near $\xi = 0$ because for $|\alpha| > m$,

$$D_\xi^\alpha H(x, \xi) \sim c|\xi|^{m-|\alpha|}.$$

Let $\tilde{J}^m = \tilde{J}^m(\mathbb{R}^n \times \mathbb{R}^n)$, $m \in \mathbb{R}$, denote the set of functions satisfying (1) and (2) of Definition 7.18, but with condition (3) replaced by the following:

(3)$^\sim$ *$H(x, \xi)$ is independent of x for large $|x| \geqslant R$.*

It is obvious that $\tilde{J}^m \subset J^m$.

The new symbol spaces are defined as follows.

Definition 7.29 *We say that $A(x, \xi) \in \tilde{S}^m(\mathbb{R}^n \times \mathbb{R}^n)$, $m \in \mathbb{R}$, if*

(1) *all x-derivatives $D_x^\beta A(x, \xi)$ exist, are locally integrable and satisfy the estimate*

$$|D_x^\beta A(x, \xi)| \leqslant C_\beta(1 + |\xi|)^m, \tag{58}$$

(2) *$A(x, \xi)$ is independent of x for large $|x| \geqslant R$.*

Using Remark 7.20, we see that

$$\tilde{J}^m \subset \tilde{S}^m \qquad \text{for } m \geqslant 0. \tag{59}$$

Theorem 7.28′ *Let $A(x, \xi) \in \tilde{S}^m$, $m \in \mathbb{R}$. We define the operator $A(x, D)$ as before (see (7))*

$$A(x, D)u(x) := (2\pi)^{-n} \int A(x, \xi)\hat{u}(\xi)\, e^{i(x,\xi)}\, d\xi, \qquad u \in C_0^\infty(\mathbb{R}^n) \text{ or } u \in \mathscr{S} \tag{60}$$

Then $A(x, D)$ is continuous

$$A(x, D)\colon W_2^l(\mathbb{R}^n) \to W_2^{l-m}(\mathbb{R}^n) \tag{61}$$

for every $l \in \mathbb{R}$.

Proof The Fourier transform of functions $u(x) \in C_0^\infty(\mathbb{R}^n)$ (or $\mathscr{S}(\mathbb{R}^n)$) satisfies

$$|\hat{u}(\xi)| \leqslant C_N(1 + |\xi|)^{-N}, \qquad \text{for all } N \geqslant 0.$$

Putting $N = n + 1 + |m|$ and using estimate (58), we see that the integral in (60) is well-defined.

Let $A(\infty, \xi)$ be the value of $A(x, \xi)$ for large $|x| \geqslant R$, and set

$$A'(x, \xi) = A(x, \xi) - A(\infty, \xi).$$

We have

$$\left.\begin{array}{ll} \text{(a)} & |A(\infty, \xi)| \leqslant C(1 + |\xi|)^m \\ \text{(b)} & |D_x^\beta A'(x, \xi)| \leqslant C_\beta(1 + |\xi|)^m \\ \text{(c)} & A'(x, \xi) = 0 \quad \text{for} \quad |x| \geqslant R. \end{array}\right\} \tag{62}$$

By a Fourier transformation of the symbol $A'(x, \xi)$ in x we obtain from (62) parts (b) and (c) the estimate

$$(1 + |\eta|)^N |\hat{A}'(\eta, \xi)| \leqslant C_N(1 + |\xi|)^m, \qquad \text{for all } N \geqslant 0. \tag{63}$$

To prove the continuity of (61), we decompose the symbol

$$A(x, \xi) = A(\infty, \xi) + A'(x, \xi).$$

The action of $A(\infty, D)$ is multiplication by $A(\infty, \xi)$ in the Fourier norm,

$$A(\infty, D)u = \mathfrak{F}^{-1}(A(\infty, \xi)\hat{u}(\xi)),$$

and therefore the continuity of $A(\infty, D)$: $W_2^l \to W_2^{l-m}$ follows at once from the estimate (62) part (a). For the operator $A'(x, D)$, we Fourier transform the function $v(x) = A'(x, D)u$ to obtain, see (60),

$$\hat{v}(\eta) = (2\pi)^{-n} \int \hat{A}'(\eta - \xi, \xi)\hat{u}(\xi)\, d\xi$$

and then use the estimate (63) to get

$$|\hat{v}(\eta)| \leqslant \int \frac{C_N(1 + |\xi|)^m |\hat{u}(\xi)|}{(1 + |\eta - \xi|)^N}\, d\xi.$$

By multiplying this estimate by $(1 + |\eta|)^{l-m}$ and applying the inequality

$$(1 + |\eta|)^t \leqslant (1 + |\xi|)^t (1 + |\xi - \eta|)^{|t|}, \qquad t \in \mathbb{R},$$

(with $t = l - m$) we get

$$(1 + |\eta|)^{l-m} |\hat{v}(\eta)| \leqslant C_N \int \frac{(1 + |\xi|)^l |\hat{u}(\xi)|}{(1 + |\eta - \xi|)^{N - |l-m|}}\, d\xi.$$

Now, set $N = n + 1 + |l - m|$, apply Schur's Lemma 7.26 to the kernel $K(\eta, \xi) = (1 + |\eta - \xi|)^{-n-1}$ and recall that $(1 + |\xi|)^l \cdot \hat{u}(\xi) \in L_2$ since $u \in W_2^l$. By squaring and integrating, we get

$$\|A'(x, D)u\|_{l-m} = \|v\|_{l-m} \leqslant C_N' \|u\|_l,$$

thus, $A'(x, D)$ is a bounded operator $W_2^l \to W_2^{l-m}$. The proof is complete.

Corollary 7.30 *Let $m \geqslant 0$ and $H(x, \xi) \in \tilde{J}^m$. The operator*

$$H(x, D): W_2^l(\mathbb{R}^n) \to W_2^{l-m}(\mathbb{R}^n)$$

is bounded for all $l \in \mathbb{R}$. Also, we have

$$H(x, \xi) - \chi(\xi)H(x, \xi) \in \tilde{S}^{-\infty} = \bigcap_{l \in \mathbb{R}} \tilde{S}^l,$$

where χ is an arbitrary cut-off function.

Proof Since $\tilde{J}^m \subset \tilde{S}^m$ for $m \geqslant 0$, the first statement follows from Theorem 7.28′. Now $\tilde{J}^m \subset J^m$, hence $\chi \cdot H(x, D)$ is a p.d.o. in the former sense, and the second statement means that the difference between a new and an old p.d.o. is infinitely smoothing.

Now for the short proof. If $\varphi = 1 - \chi \in C_0^\infty(\mathbb{R}^n)$ is any patch function then from (50) we obtain the estimate

$$|D_x^\beta \varphi(\xi) H(x, \xi)| = |\varphi(\xi) D_x^\beta H(x, \xi)| = C_\beta |\varphi(\xi)| \cdot (1 + |\xi|)^m$$
$$\leqslant C_{\beta N}(1 + |\xi|)^{-N}, \qquad \forall \beta \text{ and } N \geqslant 0$$

because φ has compact support. Hence $\varphi(\xi) \cdot H(x, \xi) \in \tilde{S}^{-\infty}$, as was to be shown.

Remark 7.31 If $A \in Cl^{-1}S^m$ is a (-1) classical symbol with $m \geqslant 0$ and $\pi A = H_m(x, \xi)$ is independent of x for large $|x|$, then by Corollary 7.30 we can omit the cut-off function in the definition of $A(x, D)$. Indeed, by Definition 7.21 we have $A = \chi H_m + A_{m-1}$ where $A_{m-1} \in S^{m-1}$, so that

$$A(x, D) = H_m(x, D) - \varphi(D) H_m(x, D) + A_{m-1}(x, D),$$

where $\varphi H_m \in \tilde{S}^{-\infty}$. Therefore, the definition of the operator $A(x, D)$ is only affected modulo $\tilde{S}^{-\infty}$ if we omit the term $\varphi(D) H_m(x, D)$.

We know from equation (31) that the symbol $A(x, \xi)$ of a p.d.o. is uniquely determined by the operator $A(x, D)$. Further, if A is (-1) classical then the principal part will then be uniquely determined by the operator. The following lemma gives more information on this point.

Lemma 7.32 *Given $x_0 \in \mathbb{R}^n$ and $\xi_0 \in \mathbb{R}^n$, $|\xi_0| = 1$, there is a sequence $\{\phi_k\}$ in $C_0^\infty(\mathbb{R}^n)$ such that*

(*i*) $\phi_k(x) = 0$ *if* $|x - x_0| > 1/k$;
(*ii*) $\|\phi_k\|_0 = 1$ *and, for* $s < 0$, *we have* $\lim_{k \to \infty} \|\phi_k\|_s = 0$;
(*iii*) *if* $A \in OCl^{-1}S^0$ *then* $\lim_{k \to \infty} \|A(x, D)\phi_k - \pi A(x_0, \xi_0)\phi_k\|_0 = 0$.

It also follows from (i) and (ii) and dominated convergence that $\phi_k \to 0$ weakly in L_2, i.e. $\int \phi_k \psi \, dx \to 0$ for all $\psi \in L_2$.

Proof Choose $\phi \in C_0^\infty$ with $\|\phi\|_0 = 1$ and $\operatorname{supp} \phi \subset \{x \in \mathbb{R}^n; |x - x_0| < 1\}$. Then let

$$\phi_k(x) = k^{n/2}\phi((x - x_0)\cdot k)\, e^{i(x, k^2\xi_0)};$$

by a translation we may assume that $x_0 = 0$. Property (i) is obvious, and $\|\phi_k\|_0 = 1$ follows by a change of variables. Further, we have

$$\hat{\phi}_k(x) = k^{-n/2}\hat{\phi}((\xi - k^2\xi_0)/k) \tag{64}$$

hence for $s < 0$ the norm

$$\|\phi_k\|_s^2 = \int (1 + |k\xi + k^2\xi_0|^2)^s |\hat{\phi}(\xi)|^2\, d\xi \to 0 \qquad \text{as } k \to \infty$$

by the Lebesgue dominated convergence theorem.

Let s_k be the isometry of L_2 given by $f(x) \mapsto k^{n/2} f(kx)$, and note that $\phi_k = s_k\, [e^{ik(\cdot,\, \xi_0)}\phi]$. Since s_k and multiplication by $e^{ik(\cdot,\, \xi_0)}$ are isometries of L_2, we may establish (iii) by showing that $e^{-ik(\cdot,\, \xi_0)}\, s_k^{-1} A(x, D)\phi_k =: \psi_k$ approaches $\pi A(x_0, \xi_0)\phi$ in L_2. In view of (64) and the definition of $A(x, D)$, we have

$$\psi_k(x) = k^{-n}(2\pi)^{-n} \int e^{i(x/k,\, \xi - k^2\xi_0)}\, A(x/k, \xi)\hat{\phi}((\xi - k^2\xi_0)/k)\, d\xi$$

$$= (2\pi)^{-n} \int e^{i(x, \xi)}\, A(x/k, k\xi + k^2\xi_0)\hat{\phi}(\xi)\, d\xi$$

Since A is a (-1) classic symbol we have $A = \chi\cdot\pi A + A_{-1}$ where $\chi(\xi)$ is a cut-off function and $A_{-1} \in S^{-1}$. Then using the homogeneity of πA we obtain by dominated convergence that, for each $x \in \mathbb{R}^n$,

$$\psi_k(x) \to \psi(x) := (2\pi)^{-n} \int e^{i(x, \xi)}\, \pi A(0, \xi_0)\hat{\phi}(\xi)\, d\xi = \pi A(0, \xi_0)\phi(x)$$

We have shown that $\psi_k \to \psi$ pointwise and it remains to show that the convergence holds in L_2. We claim that $x^\alpha \psi_k(x)$ is bounded on $\mathbb{R}^n$. An integration by parts gives us

$$x^\alpha \psi_k(x) = (2\pi)^{-n} \int e^{i(x, \xi)}\, D_\xi^\alpha [A(x/k, k\xi + k^2\xi_0)\hat{\phi}(\xi)]\, d\xi,$$

hence by the Leibniz rule $|x^\alpha\, \psi_k(x)|$ is bounded by a sum of terms of the form

$$C_\alpha \cdot \int |\xi + k\xi_0|^{-|\beta|} D_\xi^\gamma \hat{\phi}(\xi)\, d\xi, \qquad \gamma + \beta = \alpha,$$

which are bounded by a constant independent of k. Thus, $|\psi_k(x)| \leqslant C_n(1 + |x|)^{-(n+1)/2} \in L_2$ for some constant C_n, so the dominated convergence theorem gives us $\int |\psi_k - \psi|^2\, dx \to 0$.

Remark Let $OP^m(\mathbb{R}^n)$ denote the set of linear maps $A: \mathscr{S} \to \mathscr{S}$ of order m, i.e. which satisfy

$$\|Au\|_{l-m} \leqslant C_l \|u\|_l, \qquad \forall l \in \mathbb{R}.$$

Theorem 7.28 shows that $OS^m \subset OP^m$. On the other hand, it follows from Lemma 7.32 that if $A \in OCl^{-1}S^m$ and $\pi A(x_0, \xi_0) \neq 0$ for some x_0, ξ_0 then $A \notin OP^{m'}$ for any $m' < m$.

7.7 Elliptic operators on $\mathbb{R}^n$

In this section we prove the existence of a parametrix for an elliptic pseudo-differential operator. We show also that ellipticity is a necessary condition for existence of a parametrix.

Let $A = A(x, D) \in OS^m(\mathbb{R}^n)$. A *parametrix* for A is an operator $B \in OS^{-m}(\mathbb{R}^n)$, if it exists, such that

$$\left.\begin{aligned} &\text{(i)} \qquad A(x, D)B(x, D) - I \in OS^{-\infty} \\ &\text{(ii)} \qquad B(x, D)A(x, D) - I \in OS^{-\infty} \end{aligned}\right\} \tag{65}$$

It is important to note that existence of a (two-sided) parametrix follows from the existence of a right parametrix and of a left parametrix as the following lemma implies.

Lemma 7.33 *If for some $B', B'' \in OS^{-m}$ we have $AB' - I \in OS^{-\infty}$ and $B''A - I \in OS^{-\infty}$ then $B' - B'' \in OS^{-\infty}$.*

Proof This follows at once from the equation

$$B'' - B' = B''(I - AB') + (B''A - I)B'$$

by virtue of Theorem 7.12.

Thus B' is not only a right parametrix but also a left parametrix since

$$B'A - I = (B' - B'')A + (B''A - I) \in OS^{-\infty}$$

The same is true of B''.

Remark The lemma also proves the uniqueness of a parametrix, modulo $OS^{-\infty}$.

Proposition 7.34 *Every operator of the form $A(x, D) = I - T$, where $T \in OS^{-1}$, has a parametrix of the same form*

$$B(x, D) = I - T', \qquad T' \in OS^{-1}. \tag{66}$$

Proof We denote $T^k = T \circ \cdots \circ T$ and define B by the asymptotic series

$$B \sim I + T + T^2 + \cdots$$

By Theorem 7.12 we have

$$T^k \in OS^{-k}, \qquad k = 1, 2, \ldots,$$

so B is well defined (see Proposition 7.7′) and clearly (66) is satisfied. Setting

$$B = I + T + \cdots + T^N + T'_{N+1} \tag{67}$$

where $T'_{N+1} \in OS^{-N-1}$, we obtain

$$\begin{aligned}(I - T) \circ B - I &= (I - T)(I + T + \cdots + T^N + T'_{N+1}) - I \\ &= (I - T) \circ T'_{N+1} - T^{N+1}\end{aligned}$$

Thus $(I - T) \circ B - I \in OS^{-N-1}$, for $N = 1, 2, \ldots$, which means that

$$(I - T) \circ B - I \in OS^{-\infty}.$$

Similarly we get $B \circ (I - T) - I \in OS^{-\infty}$, hence B is a parametrix for $I - T$.

We now consider p.d.o.'s of arbitrary degree m.

Theorem 7.35 *An operator $A(x, D) \in OS^m(\mathbb{R}^n)$ has a parametrix if and only if for some positive constants c and R*

$$|A(x, \xi)| > c|\xi|^m \qquad \text{if } |\xi| > R \tag{68}$$

Proof If A has a parametrix $B \in OS^{-m}$ then $A(x, D)B(x, D) - I \in OS^{-\infty}$ and the asymptotic expansion (33) implies that

$$A(x, \xi)B(x, \xi) - 1 \in S^{-1}.$$

Hence $|A(x, \xi) \cdot B(x, \xi) - 1| < \frac{1}{2}$ for $|\xi|$ large, so that

$$\tfrac{1}{2} < |A(x, \xi)B(x, \xi)| < C'|A(x, \xi)||\xi|^{-m} \qquad \text{for large } |\xi|,$$

which proves (68). (The second inequality holds because $B(x, \xi) \in S^{-m}$.)

Conversely suppose that (68) holds. Assume first that $m = 0$. If we choose $F \in C^\infty(\mathbb{C})$ so that $F(z) = 1/z$ when $|z| > c$, then $B = F(A) \in S^0$ (see Proposition 7.25) and due to (68) with $m = 0$ we have

$$A(x, \xi) \cdot B(x, \xi) = 1 \qquad \text{when } |\xi| > R \tag{69}$$

For arbitrary m we can introduce $A(x, \xi) \cdot (1 + |\xi|^2)^{-m/2}$ and $B(x, \xi) \cdot (1 + |\xi|^2)^{m/2}$ in place of A and B, so that (69) again holds with $B \in S^{-m}$. Therefore,

$$A(x, D)B(x, D) = I - T(x, D), \qquad T \in OS^{-1} \tag{70}$$

and we have constructed an operator which inverts $A(x, D)$ modulo OS^{-1}. But our aim is to prove the existence of $B(x, D) \in OS^{-m}$ which inverts $A(x, D)$ modulo $OS^{-\infty}$, i.e. such that (65) holds. By Proposition 7.34, $I - T(x, D)$ has a parametrix $Q(x, D) \in OS^0$ and by Theorem 7.12 we can write

$$B(x, D)Q(x, D) = B'(x, D), \qquad \text{where } B' \in S^{-m}.$$

Since $(I - T(x, D))Q(x, D) \in OS^{-\infty}$, it follows from (70) that $B'(x, D)$ is a right parametrix of $A(x, D)$, that is, 65(i) holds with B replaced by B'. In the same way one can find $B'' \in S^{-m}$ such that (ii) is fulfilled with B replaced by B''. The proof is completed by Lemma 7.33.

Definition 7.36 *An operator $A \in OS^m(\mathbb{R}^n)$ is said to be elliptic if there exists $B \in OS^{-m}(\mathbb{R}^n)$ which inverts A modulo OS^{-1}, i.e. such that*

$$A(x, D)B(x, D) - I \in OS^{-1} \quad \textit{and} \quad B(x, D)A(x, D) - I \in OS^{-1}.$$

We have shown in Theorem 7.35 that ellipticity is equivalent to (68), and therefore equivalent to existence of a parametrix.

If A is (-1) classical, $A \in OCl^{-1}S^m$, then (68) is of course equivalent to ellipticity of the principal part πA,

$$\pi A(x, \xi) \neq 0, \qquad \xi \neq 0, x \in \mathbb{R}^n.$$

In Chapter 14 we will need a version of Theorem 7.35 in which the condition (68) holds only for x in some open subset of $\mathbb{R}^n$.

Proposition 7.37 *Let $A(x, D) \in OS^m(\mathbb{R}^n)$. Suppose that for some open set $U \subset \mathbb{R}^n$ we have*

$$|A(x, \xi)| > c|\xi|^m \qquad \textit{if } |\xi| > R, x \in U$$

Let W be an open set with $\bar{W} \subset U$. Then there exists $B(x, D) \in OS^{-m}(\mathbb{R}^n)$ such that, for all $\varphi \in C_0^\infty(W)$, we have

$$\varphi \cdot (AB - I) \in OS^{-\infty} \quad \textit{and} \quad (BA - I) \cdot \varphi \in OS^{-\infty} \tag{71}$$

Proof By the same method as in the proof of Theorem 7.35 there exists $B \in S^{-m}$ such that

$$A(x, \xi) \cdot B(x, \xi) = 1 \qquad \text{when } |\xi| > R, x \in U. \tag{72}$$

By choosing $\rho \in C_0^\infty(U)$ such that $\rho = 1$ on W, we then have

$$\rho \cdot (A(x, D)B(x, D) - I) = T(x, D), \qquad T \in OS^{-1}(\mathbb{R}^n),$$

and in view of the fact that $\rho \cdot \varphi = \varphi$, we obtain

$$\varphi \cdot A(x, D)B(x, D) = \varphi \cdot (I + T(x, D))$$

for all $\varphi \in C_0^\infty(W)$.

We can now complete the proof as in Theorem 7.35. Let $Q \in OS^0$ be a parametrix of $I + T$, and set $B'(x, D) = B(x, D)Q(x, D) \in OS^m$. It follows that

$$\varphi \cdot (A(x, D)B'(x, D) - I) \in OS^{-\infty} \tag{73}$$

for all $\varphi \in C_0^\infty(W)$. Similarly there exists $B'' \in OS^{-\infty}$ such that

$$(B''(x, D)A(x, D) - I) \cdot \varphi \in OS^{-\infty} \tag{74}$$

for all $\varphi \in C_0^\infty(W)$. Then as in Lemma 7.33 we have

$$\varphi(B'' - B')\psi = \varphi B''(I - AB')\cdot\psi + \varphi\cdot(B''A - I)B'\psi \in OS^{-\infty} \qquad (75)$$

for all $\varphi, \psi \in C_0^\infty(W)$, since $(I - AB')\cdot\psi$ and $\varphi\cdot(B''A - I)$ belong to $OS^{-\infty}$ by virtue of the lemma below. Now choose $\psi \in C_0^\infty(W)$ such that $\psi = 1$ on supp φ. It follows that $(B''\psi A - I)\varphi \in OS^{-\infty}$, since this expression differs from (74) by the term $B''\circ(1 - \psi)A\varphi$ which belongs to $OS^{-\infty}$ by quasi-locality of A. If we choose $\zeta \in C_0^\infty(W)$ such that $\zeta = 1$ on supp ψ, then by quasi-locality of B'' we have

$$(\zeta B''\psi A - I)\varphi \in OS^{-\infty} \qquad (76)$$

and then (75) implies

$$(\zeta B'\psi A - I)\varphi \in OS^{-\infty}.$$

Reversing the steps which led from (74) to (76) (this time with B' instead of B'') we also find that

$$(B'(x, D)A(x, D) - I)\varphi \in OS^{-\infty} \qquad (77)$$

Thus, (73) and (77) demonstrate the existence of an operator B satisfying (71).

Lemma *Let $T \in OS^m(\mathbb{R}^n)$, and let U be an open set in $\mathbb{R}^n$. If $\psi\cdot T \in OS^{-\infty}$ for all $\psi \in C_0^\infty(U)$ then also $T\cdot\psi \in OS^{-\infty}$ for all $\psi \in C_0^\infty(U)$.*

Proof Choose $\psi' \in C_0^\infty(U)$ such that $\psi' = 1$ on the support of ψ. Then the supports of $1 - \psi'$ and of ψ are disjoint, hence the lemma follows from the equation

$$T\psi = \psi' T\psi + (1 - \psi')T\psi$$

since $(1 - \psi')T\psi \in OS^{-\infty}$ by quasi-locality and $\psi' T \in OS^{-\infty}$ by hypothesis.

Remark This lemma implies that we may add the following to the conclusion (71) in Proposition 7.37:

$$(AB - I)\cdot\varphi \in OS^{-\infty} \quad \text{and} \quad \varphi\cdot(BA - I) \in OS^{-\infty} \qquad (71')$$

7.8 Gårding's inequality and some results on the relation between the operator norm of a p.d.o. and the norm of its symbol

Gårding's inequality for strongly elliptic p.d.o.'s is one of the fundamental results of the p.d.o. theory. It was first proved by Gårding for differential operators and by Calderón and Zygmund [CZ] for singular integral operators. The proofs in this section are based on [KN].

Theorem 7.38 *Let $A(x, D) \in OS^0(\mathbb{R}^n)$ and suppose that*

$$\operatorname{Re} A(x, \xi) \geqslant 0. \qquad (78)$$

Then for any $\varepsilon > 0$ there is a constant $C = C(\varepsilon)$ such that

$$\mathrm{Re}(A(x, D)u, u) + \varepsilon\|u\|_0^2 \geqslant -C\|u\|_{-1/2}^2 \tag{79}$$

for all $u \in C_0^\infty$.

Proof Since $\mathrm{Re}\, A(x, \xi) \geqslant 0$ the symbol $B(x, \xi) := [\mathrm{Re}\, A(x, \xi) + \varepsilon]^{1/2}$ belongs to S^0 by Proposition 7.25. Then by Corollary 7.11 and Theorem 7.12 we get

$$\tfrac{1}{2}(A(x, D) + A^*(x, D)) + \varepsilon - B^*(x, D) \circ B(x, D) = T(x, D)$$

where $T(x, D) \in OS^{-1}$. Thus, $\mathrm{Re}(Au, u) + \varepsilon(u, u) = (Bu, Bu) + \mathrm{Re}(Tu, u)$, so we have $\mathrm{Re}(Au, u) + \varepsilon(u, u) \geqslant \mathrm{Re}(Tu, u)$. Now, T is an operator of order -1, so $|(Tu, u)| \leqslant \|Tu\|_{1/2} \cdot \|u\|_{-1/2} \leqslant \text{const}\, \|u\|_{-1/2}\|u\|_{-1/2}$ and (79) follows.

Remark There is a sharp form of Gårding's inequality which holds with $\varepsilon = 0$ and C some constant. For various improvements and for best constants see [Eg], [Hö 3] and [Tay].

We showed in Theorem 7.27 that a p.d.o. $A(x, D) \in OS^0$ defines a bounded operator on $L^2(\mathbb{R}^n)$. For some purposes it is important to have a sharper form which gives some information about its L^2 norm. The following theorem can be viewed as a corollary to Gårding's inequality.

Theorem 7.39 *Let $A(x, \xi) \in S^0$ and let*

$$\gamma = \varlimsup_{|\xi| \to \infty} \sup_x |A(x, \xi)|$$

Then for every $\varepsilon > 0$ there is a constant $C = C(\varepsilon)$ such that

$$\|A(x, D)u\|_0 \leqslant (\gamma + \varepsilon)\|u\|_0 + C\|u\|_{-1/2} \tag{80}$$

for all $u \in C_0^\infty$.

Proof Since $A \in S^0$, then γ is finite, and by definition of $\varlimsup$ for every $\varepsilon > 0$ there exists $R \geqslant 0$ such that

$$\sup_x |A(x, \xi)| \leqslant \gamma + \tfrac{1}{2}\varepsilon \qquad \text{for } |\xi| \geqslant R.$$

We take a patch function $\varphi \in C_0^\infty(\mathbb{R}^n)$, $0 \leqslant \varphi \leqslant 1$, with $\varphi(\xi) = 1$ for $|\xi| \leqslant 1$, and decompose

$$A(x, \xi) = A(x, \xi)\left(1 - \varphi\left(\frac{\xi}{R}\right)\right) + A(x, \xi) \cdot \varphi\left(\frac{\xi}{R}\right).$$

Then $A_2(x, \xi) = A(x, \xi) \cdot \varphi(\xi/R) \in S^{-\infty}$, hence obviously $\|A_2(x, D)u\|_0 \leqslant C\|u\|_{-1/2}$. It remains to estimate the p.d.o. with symbol $A_1(x, \xi) = A(x, \xi)(1 - \varphi(\xi/R))$. By definition of $\varphi(\xi/R)$, we have $A_1(x, \xi) = 0$ for $|\xi| \leqslant R$, whence

$$|A_1(x, \xi)| \leqslant \gamma + \tfrac{1}{2}\varepsilon, \qquad \text{for all } x, \xi \in \mathbb{R}^n.$$

By applying Gårding's inequality (79) to the operator $(\gamma + \varepsilon)^2 - A_1^* \circ A_1$ we obtain (80) since ε can be chosen arbitrarily small.

Gårding's inequality can be extended to the case of a p.d.o. of any order m, if we make use of the operators Λ^k with symbol $(1 + |\xi|^2)^{k/2}$.

Theorem 7.38′ *Let* $A(x, D) \in OS^m(\mathbb{R}^n)$ *and suppose that for some constant* c'

$$\operatorname{Re} A(x, \xi) \geqslant c'|\xi|^m \qquad \textit{for } |\xi| \textit{ large.} \tag{81}$$

Then for any $\varepsilon > 0$ *there is a constant* $C = C(\varepsilon, m)$ *such that*

$$\operatorname{Re}(A(x, D)u,u) \geqslant (c' - \varepsilon)\|u\|_{m/2}^2 - C\|u\|_{(m-1)/2}^2 \tag{82}$$

for all $u \in C_0^\infty$.

Proof Replacing $A(x, D)$ by $\Lambda^{-m/2}A(x, D)\Lambda^{-m/2}$, we can assume without loss of generality that $m = 0$. By consideration of $A - c'$ we may then assume that $c' = 0$. Also, we may modify A on a set of compact support in ξ so as to make it satisfy

$$\operatorname{Re} A(x, \xi) \geqslant 0 \qquad \text{for all } \xi.$$

(see the proof of Theorem 7.39). The new operator differs from the original one by an operator of order -1, which does not affect the form of (82) as we saw with the operator T in the proof of Theorem 7.38.

Remark When A is a (-1) classical operator the condition (81) is also necessary for Gårding's inequality. We can assume more that $m = 0$ and $c' = 0$, then let ϕ_k be the sequence from Lemma 7.32. By substituting $u = \phi_k$ in (79) and letting $k \to \infty$ we find that $\operatorname{Re} \pi A(x_0, \xi_0) + \varepsilon \geqslant 0$. Since $\varepsilon > 0$ is arbitrary, then $\operatorname{Re} \pi A(x, \xi) \geqslant 0$ for all x, ξ, and hence $\operatorname{Re} A(x, \xi) \geqslant 0$ for large $|\xi|$.

The next theorem is the analogue of Theorem 7.39 for pseudo-differential operators of arbitrary order.

Theorem 7.39′ *Let* $A(x, \xi) \in S^m$ *and set*

$$\gamma = \varlimsup_{|\xi| \to \infty} \sup_x \frac{|A(x, \xi)|}{(1 + |\xi|^2)^{m/2}}$$

Then for every $\varepsilon > 0$ *and* $l \in \mathbb{R}$ *there is a constant* $C = C(\varepsilon, l)$ *such that*

$$\|A(x, D)u\|_{l-m} \leqslant (\gamma + \varepsilon)\|u\|_l + C\|u\|_{l-1/2} \tag{80′}$$

for all $u \in C_0^\infty$.

Proof The case $m = 0$ follows from Theorem 7.39 since we can replace u by $\Lambda^l u$ in the estimate (80′) to reduce it to (80). Now, if $A(x, D) \in OS^m$ we consider the operator $\Lambda^{-m} \circ A(x, D) \in OS^0$, and by using the case just proved we obtain

$$\|\Lambda^{-m}\circ A(x,D)u\|_l \leqslant (\gamma+\varepsilon)\|u\|_l + C\|u\|_{l-1/2},$$

which proves (80′). Note: The definition of γ is not affected by this substitution since the asymptotic expansion for the symbol of $\Lambda^{-m}\circ A(x,D)$ is equal to $(1+|\xi|^2)^{-m/2}A(x,\xi)$ plus a term of order -1 which does not affect the limit.

The inequality in Theorem 7.39′ suggests the following question: how is the norm of A, as a map from W_2^l to W_2^{l-m}, related to the symbol? It would seem to be difficult to deal with this question in general. The next theorem, however, does give a positive result in this direction which has important consequences (see §8.10).

In the proof we use the fact that

$$e^{-i(\cdot,t\xi)}A(x,D)(u\,e^{i(\cdot,t\xi)}) = A(x,D+t\xi)u, \tag{83}$$

for all $u\in C_0^\infty$, which follows from (31).

The theorem is formulated merely for the case $l=m=0$.

Theorem 7.40 *If $A\in OCl^{-1}S^0$ is a (-1) classical symbol of order* 0, *then as a map of L_2 into L_2 it satisfies*

$$\inf_T \|A+T\| = \sup_{x,\xi}|\pi A(x,\xi)|$$

where $\|\ \|$ denotes the norm of the operator as a map of L_2 into L_2, and the infimum is taken over all operators $T\in OS^{-1}$. Here πA is the principal part of A; see §7.5.

Proof It follows by homogeneity of πA that the number γ in the statement of Theorem 7.39 is equal to $\sup_{x,\xi}|\pi A(x,\xi)|$. Thus, we must show that

$$\inf_T\|A+T\| = \gamma$$

We start by showing $\|A+T\|\geqslant\gamma$ for any operator $T\in OS^{-1}$; since $A+T$ has the same principal part as A we can assume that $T=0$. We claim that for all $u,v\in C_0^\infty(\mathbb{R}^n)$ and fixed $\xi_0\in\mathbb{R}^n\backslash 0$

$$(A(x,D)(u\,e^{i(\cdot,t\xi_0)}),\ v\,e^{i(\cdot,t\xi_0)})_{L^2}\to(\pi A(\cdot,\xi_0)u,v)_{L^2} \tag{84}$$

as $t\to\infty$; once this has been proved it follows that

$$|(\pi A(\cdot,\xi_0)u,v)|\leqslant\|A\|\,\|u\|\,\|v\| \qquad \text{for any } \xi_0,$$

and hence $\gamma=\sup_{x,\xi}|\pi A(x,\xi)|\leqslant\|A\|$. By definition of a (-1) classical symbol of order 0 we have

$$A=\chi\cdot\pi A + A'$$

where χ is a cut-off function, πA is homogeneous of degree 0, and $A'\in S^{-1}$.

In view of (83), the left-hand side of (84) is equal to

$$
\begin{aligned}
(A(x, D + t\xi_0)u, v) &= (2\pi)^{-n} \iint e^{i(x,\xi)} A(x, \xi + t\xi_0)\hat{u}(\xi)\bar{v}(x)\, d\xi\, dx \\
&\to (2\pi)^{-n} \iint e^{i(x,\xi)} \pi A(x, \xi_0)\hat{u}(\xi)\bar{v}(x)\, d\xi\, dx \\
&= (\pi A(\cdot, \xi_0)u, v),
\end{aligned}
$$

where we used the dominated convergence theorem and the fact that $(\chi \cdot \pi A)(x, \xi + t\xi_0) \to \pi A(x, \xi_0)$ (by homogeneity) and $A'(x, \xi + t\xi_0) \to 0$ (since $A' \in S^{-1}$) as $t \to \infty$.

To complete the proof we must show that $\inf_T \|A + T\| \leqslant \gamma$. To this end let $\psi(\xi)$ be a C_0^∞ symbol, $0 \leqslant \psi \leqslant 1$, which is equal to 1 on a large ball $|\xi| \leqslant R$. Then according to (80) we have

$$
\begin{aligned}
\|A(I - \psi(D))u\|_0 &\leqslant (\gamma + \varepsilon)\|(I - \psi(D))u\|_0 + C(\varepsilon)\|(I - \psi(D))u\|_{-1/2} \\
&\leqslant (\gamma + \varepsilon)\|u\|_0 + C(\varepsilon)\|(I - \psi(D))u\|_{-1/2}
\end{aligned}
$$

If we choose R very large then just as in the proof of Lemma 7.41 below we have $\|(I - \psi(D))u\|_{-1/2} \leqslant \delta\|u\|_0$ where δ is arbitrarily small. Hence

$$
\|Au - A \circ \psi(D)u\|_0 \leqslant (\gamma + 2\varepsilon)\|u\|_0,
$$

and since $T = -A \circ \psi(D)$ is of order $-\infty$, it follows that $\inf_T \|A + T\| \leqslant \gamma + 2\varepsilon$. The proof of the theorem is complete.

Now we prove a lemma which shows that the symbol $A(x, \xi)$ can be approximated locally by $A(x_0, \xi)$, a symbol with x "frozen" at the point x_0, i.e. independent of x. This is important for §9.4.

Lemma 7.41 (*Freezing Lemma*) *Let $A(x, \xi) \in S^m$. Then for every $x_0 \in \mathbb{R}^n$ and $\varepsilon > 0$ we can find a neighbourhood U of x_0 and p.d.o.'s $K(x, D)$ and $T(x, D)$ such that for all $\varphi \in C_0^\infty(U)$ we have the decomposition*

$$
\varphi(x)A(x, D) = \varphi(x)[A(x_0, D) + K(x, D) + T(x, D)], \tag{85}
$$

where $K(x, D) \in OS^m$ has a small norm

$$
\|K(x, D)u\|_{l-m} \leqslant \varepsilon\|u\|_l, \qquad u \in W_2^l(\mathbb{R}^n)
$$

and $T(x, D) \in OS^{-\infty}$.

Proof By the mean value theorem we have

$$
|A(x, \xi) - A(x_0, \xi)| \leqslant |x - x_0| \cdot |D_x A(x', \xi)|
$$

and since $D_x A(x, \xi)$ also belongs to S^m, see (6), we can find a ball $B(x_0, \delta)$, $\delta > 0$, such that

$$
\frac{|A(x, \xi) - A(x_0, \xi)|}{(1 + |\xi|^2)^{m/2}} \leqslant \frac{\varepsilon}{2} \qquad \text{for } x \in B(x_0, \delta) \text{ and all } \xi \in \mathbb{R}^n \tag{86}
$$

Now, we take an open set $U \ni x_0$ with closure lying inside $B(x_0, \delta)$ and a function $\psi_0 \in C_0^\infty(B(x_0, \delta))$ with $0 \leqslant \psi_0 \leqslant 1$ and $\psi_0 = 1$ on U. For the purpose of establishing (85) there is no loss of generality in assuming that (86) holds for all $x \in \mathbb{R}^n$ because we can replace $A(x, \xi)$ by $\psi_0(x) \cdot A(x, \xi)$ which has no effect on (85) because $\varphi(x) = \varphi(x)\psi_0(x)$ for all $\varphi \in C_0^\infty(U)$. Now we set

$$K(x, \xi) := A(x, \xi)\left(1 - \psi\left(\frac{\xi}{R}\right)\right), \qquad T(x, \xi) := A(x, \xi)\psi\left(\frac{\xi}{R}\right),$$

where $\psi \in C_0^\infty(\mathbb{R}^n)$ is a patch function, i.e. $0 \leqslant \psi \leqslant 1$ and $\psi(\xi) = 1$ for $|\xi| \leqslant 1$. The expansion (85) holds and we must verify the properties of K and T. Since (86) holds for all x and ξ in $\mathbb{R}^n$ then $\gamma \leqslant \varepsilon/2$ in Theorem 7.39′, so we obtain

$$\|A(x, D)u\|_{l-m} \leqslant \frac{2\varepsilon}{3}\|u\|_l + C_\varepsilon\|u\|_{l-1/2} \tag{87}$$

(the constant C_ε also depends on l). Obviously $T(x, \xi) \in S^{-\infty}$, and it remains to consider the operator K. From (87) we obtain

$$\begin{aligned}\|K(x, D)u\|_{l-m} &\leqslant \frac{2\varepsilon}{3}\left\|\left(1 - \psi\left(\frac{D}{R}\right)\right)u\right\|_l + C_\varepsilon\left\|\left(1 - \psi\left(\frac{D}{R}\right)\right)u\right\|_{l-1/2} \\ &\leqslant \frac{2\varepsilon}{3}\|u\|_l + C_\varepsilon\left\|\left(1 - \psi\left(\frac{D}{R}\right)\right)u\right\|_{l-1/2},\end{aligned} \tag{88}$$

where we used $0 \leqslant \psi(\xi/R) \leqslant 1$ to get $\|(1 - \psi(D/R))u\|_l \leqslant \|u\|_l$. Now by the definition of the norm in $W_2^{l-1/2}$ we get

$$\begin{aligned}\left\|\left(1 - \psi\left(\frac{\xi}{R}\right)\right)u\right\|_{l-1/2}^2 &= \int \frac{(1 - \psi(\xi/R))^2}{(1 + |\xi|^2)^{1/2}}|\hat{u}(\xi)|^2(1 + |\xi|^2)^l\, d\xi \\ &\leqslant (1 + R^2)^{-1/2}\|u\|_l^2\end{aligned} \tag{89}$$

and we choose R so large that $C_\varepsilon(1 + R^2)^{-1/4} \leqslant \varepsilon/3$. By this choice of R we obtain from (88) and (89) the inequality

$$\|K(x, D)u|_{l-m} \leqslant \varepsilon\|u\|_l, \qquad u \in W_2^l(\mathbb{R}^n).$$

The Freezing Lemma is also true for p.d.o.'s from $\tilde{J}^m$ (see §7.6) when $m \geqslant 0$.

Lemma 7.41′ *Let $H(x, \xi) \in \tilde{J}^m$, $m \geqslant 0$. Then for every x_0 and $\varepsilon > 0$ we can find a neighbourhood U of x_0 and operators $K(x, D) \in OS^m$ and $T(x, D) \in O\tilde{S}^{-\infty}$ such that for all $\varphi \in C_0^\infty(U)$ we have the decomposition*

$$\varphi(x) \cdot H(x, D) = \varphi(x) \cdot [H(x_0, D) + K(x, D) + T(x, D)]$$

with K small: $\|Ku\|_{l-m} \leqslant \varepsilon\|u\|_l$ for all $u \in W_2^l(\mathbb{R}^n)$.

Proof With Corollary 7.30 the proof is very simple. We take a cut-off function $\chi(\xi)$ and apply Lemma 7.41 to $\chi \cdot H$:

$$\begin{aligned}\varphi(x)\cdot H(x, D) &= \varphi(x)[\chi H + T_1] \\ &= \varphi(x)[(\chi H)(x_0, D) + T_2 + K + T_1] \\ &= \varphi(x)[H(x_0, D) + T_3 + T_2 + K + T_1],\end{aligned}$$

where $T_1, T_3 \in \tilde{S}^{-\infty}$ by Remark 7.31 and $T_2 \in S^{-\infty}$. Clearly we may also take $T_2 \in \tilde{S}^{-\infty}$, i.e. independent of x for large x. Then $T_1 + T_2 + T_3 = T \in \tilde{S}^{-\infty}$.

Exercises

1. The convolution of a pair of functions φ, ψ on $\mathbb{R}^n$ is the function

$$(\varphi * \psi)(x) = \int_{\mathbb{R}^n} \varphi(x - y)\psi(y)\, dy$$

Show that if $\varphi, \psi \in \mathscr{S}$ then $\varphi * \psi \in \mathscr{S}$ and $\mathfrak{F}(\varphi * \psi) = \mathfrak{F}\varphi \cdot \mathfrak{F}\psi$.

2. Show that $C_0^\infty(\mathbb{R}^n)$ is dense in $\mathscr{S}(\mathbb{R}^n)$.

3. Show that the Fourier transform of $f \in L_1(\mathbb{R}^n)$ is a bounded, continuous function on $\mathbb{R}^n$.

4. Let f be measurable on $\mathbb{R}^n$ and $|f(x)| \leqslant C(1 + |x|)^{-n-\varepsilon-m}$, where $\varepsilon > 0$ and m is a non-negative integer. Show that the Fourier transform $\mathfrak{F}f$ has bounded and continuous derivatives up to order $\leqslant m$.

5. The convolution $f * \varphi$ of $f \in \mathscr{S}'$ and $\varphi \in \mathscr{S}$ is defined by

$$(f * \varphi)(x) = \langle f(y), \varphi(x - y)\rangle,$$

where $\langle\, ,\rangle$ is the duality bracket between $\mathscr{S}'$ and $\mathscr{S}$. Show that the function $f * \varphi$ is infinitely differentiable and $|D^\alpha(f * \varphi)(x)| \leqslant C_\alpha(1 + |x|)^N$ for some number N independent of α.

6. If $f \in \mathscr{S}'$ and $\varphi \in \mathscr{S}$, show that $\mathfrak{F}(f * \varphi) = \mathfrak{F}f \cdot \mathfrak{F}\varphi$.

7. Let $a(\xi) \in S^m$ (independent of x). The operator $a(D) \in OS^m$ is then defined by $a(D)u = \mathfrak{F}^{-1}(a(\xi)\hat{u}(\xi))$; in particular if $m < -n$ we have

$$(a(D)u)(x) = \int_{\mathbb{R}^n} b(x - y)u(y)\, dy, \qquad u \in \mathscr{S}$$

where the function $b = \mathfrak{F}^{-1}(a(\xi))$ is bounded and continuous by Exercise 3. Differentiability properties of b follow from Exercise 4, for instance, if $a \in S^{-\infty}$ show that $b \in C_b^\infty(\mathbb{R}^n)$.

8. Let $a(\xi) \in S^m$ where m is arbitrary. Show that there exists a number l and a bounded, continuous function b such that

$$(a(D)u)(x) = \int_{\mathbb{R}^n} b(x - y)(-\Delta_y + 1)^l u(y)\, dy, \qquad u \in \mathscr{S}$$

9. Let $K(x, \xi) \in J^{-n}$ and let $\int_{|\xi|=1} K(x, \xi)\, d\sigma_\xi = 0$ ($d\sigma_\xi$ denotes the measure on the sphere $|\xi| = 1$). Show that for $u \in \mathscr{S}$ the following limit exists

$$(Au)(x) := \lim_{\varepsilon \to 0} \int_{|x-y| \geqslant \varepsilon} K(x, x-y) u(y)\, dy,$$

and defines a p.d.o. of order 0, that is, $A \in OS^0$. Hint: Consult the book [Es].

Appendix: summary of definitions and theorems for Sobolev spaces

In this appendix we state the main definitions and theorems of the theory of Sobolev spaces. For the proofs one should consult the books [Wl], [Tri], [Ne], [Ad]. We restrict ourselves to the spaces $W_2^l(\Omega)$ where $l \in \mathbb{R}_+$ ($l \geqslant 0$) in which the norms are based on L_2 norms of the derivatives.

Definition 7A.1 *Let Ω be an open set in $\mathbb{R}^n$. Suppose first that l is integral, that is, $l = 0, 1, 2, \ldots$. The Sobolev space $W_2^l(\Omega)$ is defined as the set of all functions $u \in L_2(\Omega)$ for which the distributional derivatives $D^s u$ (or weak derivatives) belong to $L_2(\Omega)$ for $|s| \leqslant l$:*

$$W_2^l(\Omega) = \{u \in L_2(\Omega);\ D^s u \in L_2(\Omega) \text{ for } |s| \leqslant l\}.$$

We introduce a scalar product on $W_2^l(\Omega)$ by means of

$$(u, v)_l := \sum_{|s| \leqslant l} \int_\Omega D^s u(x) \cdot \overline{D^s v(x)}\, dx,$$

where, as usual, functions which are equal almost everywhere are identified.

Now suppose that l is not integral, that is, $l = [l] + \lambda$, $0 < \lambda < 1$. As scalar product we take

$$(u, v)_l := \sum_{|s| \leqslant l} \int_\Omega D^s u(x) \cdot \overline{D^s v(x)}\, dx + \sum_{|s| \leqslant l} \iint_{\Omega \times \Omega} \frac{(D^s u(x) - D^s u(y))\overline{(D^s u(x) - D^s v(y))}}{|x-y|^{n+2\lambda}}\, dx \cdot dy$$

and let

$$W_2^l(\Omega) = \{u \in L_2(\Omega);\ D^s u \in L_2(\Omega) \text{ with } (u, u)_l < \infty\}.$$

The Sobolev space $W_2^l(\Omega)$, $l \in \mathbb{R}_+$, is a separable Hilbert space and hence has a countable basis. The norm is denoted $\|u\|_l^2 = (u, u)_l$, but for convenience we often use the following variations in the notation:

$$\|u\|_l = \|u\|_{W_2^l(\Omega)} = \|u\|_{l, \Omega}.$$

The *support* of a function u, denoted by supp u, is the complement of the largest open set on which $u = 0$. We denote by $C_0^\infty(\Omega)$ the space of all C^∞ functions on $\mathbb{R}^n$ whose support is a compact subset of Ω. The subspace $\overset{\circ}{W}{}_2^l(\Omega)$

is defined as the closure of $C_0^\infty(\Omega)$ in the $W_2^l(\Omega)$ norm:

$$\mathring{W}_2^l(\Omega) := \overline{C_0^\infty(\Omega)}^{W_2^l}$$

Note that if $u \in W_2^l(\Omega)$ has compact support in Ω, then $u \in \mathring{W}_2^l(\Omega)$. This can be proved by the usual regularization technique. We choose a function $\varphi \in C_0^\infty(\mathbb{R})$ with support in the unit ball $|x| \leqslant 1$ and $\int \varphi(x)\,dx = 1$. Then $\varphi_\varepsilon(x) = \varepsilon^{-n}\varphi(x/\varepsilon)$ has support in $|x| \leqslant \varepsilon$, $\int \varphi_\varepsilon(x)\,dx = 1$, and we set $u_\varepsilon = \varphi_\varepsilon * u$, i.e.

$$u_\varepsilon(x) = \int \varphi_\varepsilon(x-y)u(y)\,dy = \int \varphi(y)u(x-\varepsilon y)\,dy,$$

where $\varepsilon < d(\operatorname{supp} u, \complement\Omega)$. Now $u_\varepsilon \in C_0^\infty(\Omega)$ and it can be shown that $u_\varepsilon \to u$ in $W_2^l(\Omega)$ as $\varepsilon \to 0$.

It is also true that functions in $W_2^l(\Omega)$ with bounded support (but not necessarily compact in Ω) are dense in $W_2^l(\Omega)$.

In the case $\Omega = \mathbb{R}^n$, where bounded support implies compact support, it follows that $\mathring{W}_2^l(\mathbb{R}^n) = W_2^l(\mathbb{R}^n)$; or, in other words, $C_0^\infty(\mathbb{R}^n)$ is dense in $W_2^l(\mathbb{R}^n)$.

Throughout this book, $\Omega \subset \mathbb{R}^n$ will denote an open set with a smooth boundary $\partial\Omega \in C^\infty$. In particular, Ω has the segment property so the following theorem holds, see [Wl].

Theorem 7A.2 *Let Ω be an open set in $\mathbb{R}^n$ with smooth boundary. The restrictions to Ω of functions from $C_0^\infty(\mathbb{R}^n)$ form a dense subset of $W_2^l(\Omega)$, $l \in \mathbb{R}_+$.*

By this theorem it is clear how to work in Sobolev spaces. We first derive inequalities and equalities for smooth functions ($\in C_0^\infty(\mathbb{R}^n)$) and continuous operators, and then argue by density to obtain them for the whole space $W_2^l(\Omega)$.

For Sobolev spaces on $\mathbb{R}^n$ and $\mathbb{R}_+^n$, we need the equivalent Fourier norms. The Fourier transform of a function $u \in L_1(\mathbb{R}^n)$ is defined by

$$\hat{u}(\xi) = \mathfrak{F}u(\xi) = \int_{\mathbb{R}^n} e^{-i(x,\xi)}\,u(x)\,dx$$

As is well known, the Fourier transformation can be extended to the space of tempered distributions, see [Hö 1] and [Wl], with good properties.

Consider now the weighted Hilbert space L_2^l with the norm

$$\|u\|_{L_2^l}^2 := \int_{\mathbb{R}^n} |u(x)|^2(1+|x|^2)^l\,dx.$$

We have the following theorem.

Theorem 7A.3 *For $l \in \mathbb{R}_+$ $(l \geqslant 0)$ the Fourier transformation is a (linear, topological) isomorphism $\mathfrak{F}: W_2^l(\mathbb{R}^n) \to L_2^l$, that is, the norm on $W_2^l(\mathbb{R}^n)$ is equivalent to the "Fourier norm",*

$$\|u\|_l^2 = \|u\|_{W_2^l}^2 \simeq \int_{\mathbb{R}^n} |\hat{u}(\xi)|^2(1 + |\xi|^2)^l \, d\xi$$

We also need the Sobolev space $W_2^l(\mathbb{R}_+^n)$ on the half-space $\mathbb{R}_+^n$, where l is a non-negative integer, $l = 0, 1, 2, \ldots$. The points in $\mathbb{R}_+^n$ are denoted by $x = (x', t)$ where $x' = (x_1, \ldots, x_{n-1}) \in \mathbb{R}^{n-1}$ and $t = x_n > 0$, and by Definition 7A.1 the norm in $W_2^l(\mathbb{R}_+^n)$ is

$$\|u\|_l^2 = \sum_{\nu=0}^{l} \int_{\mathbb{R}^{n-1}} \sum_{|s| \leqslant l-\nu} \int_0^\infty \left| \frac{\partial^\nu}{\partial t^\nu} D_{x'}^s \, u(x', t) \right|^2 dt \, dx'.$$

The partial Fourier transformation $\mathfrak{F}'$ of a function $u(x', t)$ on $\mathbb{R}_+^n$. is defined by

$$\tilde{u}(\xi', t) = \mathfrak{F}'u(\xi', t) = \int e^{-i(x', \xi')} u(x', t) \, dx', \qquad \xi' = (\xi_1, \ldots, \xi_{n-1}).$$

Theorem 7A.4 *On $W_2^l(\mathbb{R}_+^n)$, $l = 0, 1, 2, \ldots$ we have the equivalent "Fourier norm"*

$$\|u\|_l^2 = \|u\|_{l, \mathbb{R}_+^n}^2 \simeq \sum_{\nu=0}^{l} \int_{\mathbb{R}^{n-1}} (1 + |\xi'|^2)^{l-\nu} \int_0^\infty \left| \frac{d^\nu}{dt^\nu} \tilde{u}(\xi', t) \right|^2 dt \, d\xi'.$$

We need also the Sobolev spaces on a compact C^∞ manifold M (for instance, on the boundary $\partial\Omega$ of a bounded open set Ω).

Definition 7A.5 *Let M be a compact C^∞ manifold. Let $l \geqslant 0$, a non-negative real number. The Sobolev space $W_2^l(M)$ is the set of functions $u: M \to \mathbb{C}$ such that*

$$(\varphi \cdot u) \circ \kappa^{-1} \in W_2^l(\mathbb{R}^n)$$

for every coordinate map $\kappa: U \to V$ and for every $\varphi \in C_0^\infty(U)$. The topology on $W_2^l(M)$ is defined by the semi-norms

$$u \mapsto \|(\varphi \cdot u) \circ \kappa^{-1}\|_l,$$

i.e. the weakest topology for which the semi-norms are continuous.

Let $\{U_j, \kappa_j\}_{j=1,\ldots,N}$ be an atlas on M and $\{\varphi_j\}$ a subordinate partition of unity, $\operatorname{supp} \varphi_j \subset U_j$. For $u \in W_2^l(M)$ we let $u_j := (\varphi_j \cdot u) \circ \kappa_j^{-1}$ and we make $W_2^l(M)$ into a Hilbert space by defining the scalar product

$$(u, v)_{W_2^l(M)} = \sum_{j=1}^{N} (u_j, v_j)_l.$$

It is possible to show that the topology defined by the norm $\|u\|^2 = (u, u)_{W_2^l(M)}$ coincides with the topology in Definition 7A.5, in other words, two equivalent atlases and two corresponding subordinate partitions of unity yield equivalent norms.

If $u \in W_2^l(M)$ has support in a coordinate domain (U, κ) then it can be approximated by a function in $C_0^\infty(U)$ (by convolution with a smooth function as mentioned above). By means of a partition of unity it follows that

$$C^\infty(M) \text{ is dense in } W_2^l(M).$$

Let Ω be an arbitrary open set in $\mathbb{R}^n$. Elements from $W_2^l(\Omega)$ are distributions; therefore the restriction operator $\imath = \imath_{\Omega_1,\Omega}$ to smaller sets $\Omega_1 \subset \Omega$ is always defined, and it agrees on functions with the usual restriction. The continuity of the operator $\imath$: $W_2^l(\Omega) \to W_2^l(\Omega_1)$,

$$\|\imath u\|_{l,\Omega_1} \leqslant \|u\|_{l,\Omega},$$

follows because the norms are expressed in terms of integrals.

Definition 7A.6 *By an extension operator $\beta = \beta_{\Omega_1,\Omega}$ from an open set Ω_1 to a larger Ω, $\Omega_1 \subset \Omega$, we mean a continuous, linear operator*

$$\beta\colon W_2^l(\Omega_1) \to W_2^l(\Omega), \qquad l \in \mathbb{R}_+$$

such that $\imath \circ \beta$ is the identity on $W_2^l(\Omega_1)$.

The existence of extension operators is asserted in the next two theorems.

Theorem 7A.7 *For each $l \geqslant 0$ there exists a continuous linear extension operator*

$$\beta = \beta_l\colon W_2^l(\mathbb{R}_+^n) \to W_2^l(\mathbb{R}^n)$$

that has the property that if supp u *is bounded then* supp $\beta_l u$ *is also bounded. Furthermore, let $0 < L < \infty$ be given, then for each $0 \leqslant l \leqslant L$ the extension operator β_l can be chosen to be independent of l.*

By independent of l, we mean that if $0 \leqslant l \leqslant L$ and $0 \leqslant l' \leqslant L$ then $\beta_l u = \beta_{l'} u$ for all $u \in W_2^l(\mathbb{R}_+^n) \cap W_2^{l'}(\mathbb{R}_+^n)$.

The next theorem states the corresponding result for any bounded open set Ω with smooth boundary.

Theorem 7A.8 *Let Ω be bounded, with C^∞ boundary, and let $\Omega_1 \supset \bar{\Omega}$. Then for each $L \geqslant 0$ there exists a continuous linear extension operator $\beta = \beta_{\Omega_1,\Omega}$ which is a continuous, linear operator*

$$\beta\colon W_2^l(\Omega) \to \mathring{W}_2^l(\Omega_1),$$

and which is independent of l in the range $0 \leqslant l \leqslant L$.

Let $\imath'$ denote the restriction map from $\mathbb{R}^n$ to $\mathbb{R}^{n-1}$,

$$(\imath' u)(x') = u(x', 0), \qquad u \in C_0^\infty(\mathbb{R}^n).$$

By density of $C_0^\infty(\mathbb{R}^n)$ in $W_2^l(\mathbb{R}^n)$, the following lemma shows that $\imath'$ extends uniquely to a continuous, linear map from $W_2^l(\mathbb{R}^n)$ to $W_2^{l-1/2}(\mathbb{R}^{n-1})$ for any $l > \frac{1}{2}$.

Lemma *If* $l > \frac{1}{2}$ *then* $\|\imath' u\|_{l-1/2} \leqslant C_l \|u\|_l$ *for all* $u \in C_0^\infty(\mathbb{R}^n)$.

Proof If $u \in \mathscr{S}$ then

$$\mathfrak{F}'u(\xi', x_n) = (2\pi)^{-1} \int e^{ix_n\xi_n} \hat{u}(\xi)\, d\xi_n$$

where $\hat{u}$ is the Fourier transform in $\mathbb{R}^n$. Now let $v = \imath' u$, then $\hat{v}(\xi') = \mathfrak{F}'u(\xi', 0)$ and we have

$$\begin{aligned}|\hat{v}(\xi')|^2 &= |(2\pi)^{-1} \int \hat{u}(\xi)\, d\xi_n|^2 \\ &\leqslant (2\pi)^{-2} \int |\hat{u}(\xi)|^2 (1 + |\xi|^2)^l\, d\xi_n \cdot \int (1 + |\xi|^2)^{-l}\, d\xi_n,\end{aligned}$$

by the Cauchy-Schwarz inequality. By making the change of variable $t = \xi_n/(1 + |\xi'|^2)^{1/2}$ it follows that $\int (1 + |\xi|^2)^{-l}\, d\xi_n \leqslant C_l^2 (1 + |\xi'|^2)^{1/2 - l}$. (It is essential for convergence that $l > \frac{1}{2}$.) Hence

$$|\hat{v}(\xi')|^2 \leqslant C_l^2 \int |\hat{u}(\xi)|^2 (1 + |\xi|^2)^l (1 + |\xi'|^2)^{1/2 - l}\, d\xi_n.$$

If we multiply by $(1 + |\xi'|^2)^{l - 1/2}$ and integrate, it follows that

$$\|v\|_{l-1/2} \leqslant C_l \|u\|_l.$$

Now let us consider the restriction of functions to $\partial\Omega$. The trace operator $\imath'$ is defined on smooth functions $u \in C_0^\infty(\mathbb{R}^n)$ by

$$\imath' u = u|_{\partial\Omega} \in C^\infty(\partial\Omega).$$

By means of local coordinates and a partition of unity, it follows that $\|\imath' u\|_{l-1/2} \leqslant C_l' \|u\|_l$, $l > \frac{1}{2}$, so $\imath'$ extends uniquely to a continuous, linear map $\imath'\colon W_2^l(\Omega) \to W_2^{l-1/2}(\partial\Omega)$.

We also consider traces of the normal derivatives. Let $\nu = (\nu_j)\colon \partial\Omega \to \mathbb{R}^n$ be an inward pointing normal vector field along $\partial\Omega$, and consider the map $F\colon \partial\Omega \times \mathbb{R} \to \mathbb{R}^n$ defined by

$$(x', t) \mapsto x' + t \cdot \nu(x').$$

By virtue of §5.12, Exercise 6 (or Theorem 5.73), there is a neighbourhood $\partial\Omega \times (-\varepsilon, \varepsilon) \supset \partial\Omega \times 0$ on which F is a diffeomorphism; by a scaling which affects the length of ν only, we may assume that $\varepsilon = 1$. Then the image $F(\partial\Omega \times (-1, 1))$ is a tubular neighbourhood of $\partial\Omega$ in $\mathbb{R}^n$; we regard F as an identification, i.e. $(x', t) \in \partial\Omega \times (-1, 1)$ will denote the point $x' + t\nu(x') \in \mathbb{R}^n$.

The normal derivative along $\partial\Omega$ of a function $u \in C_0^\infty(\mathbb{R}^n)$ is defined by

$$\frac{\partial u}{\partial n} = \nu \cdot \operatorname{grad} u = \sum_{j=1}^{n} \nu_j \frac{\partial u}{\partial x_j},$$

and we have $\partial u/\partial n \in C^\infty(\partial\Omega)$. Note that $\partial u/\partial n = \partial(u \circ F)/\partial t$, that is,

$$\frac{\partial u}{\partial n}(x') = \lim_{t \to 0^+} \frac{\partial}{\partial t}[u(x' + t\nu(x'))], \qquad x' \in \partial\Omega.$$

With F regarded as an identification, the normal derivative $\partial/\partial n$ is just the partial derivative $\partial/\partial t$, and this gives us an extension of the normal derivative throughout the whole tubular neighbourhood:

$$\frac{\partial u}{\partial n}(x' + t \cdot \nu(x')) = \nu(x') \cdot \operatorname{grad} u(x' + t\nu(x')), \qquad x' \in \partial\Omega, \ -1 < t < 1.$$

We define the trace operator T_m by

$$T_m u = \left(\iota' u, \iota' \frac{\partial u}{\partial n}, \ldots, \iota' \frac{\partial^m u}{\partial n^m}\right) = \lim_{t \to 0^+} \left(u(x', t), \ldots, \frac{\partial^m u}{\partial t^m}(x', t)\right), \qquad (*)$$

when $u \in C_0^\infty(\mathbb{R}^n)$, and then extend its definition by continuity using the lemma above.

First we consider the case $\Omega = \mathbb{R}^n$, $\partial\Omega = \mathbb{R}^{n-1}$ and $\partial/\partial n = \partial/\partial x_n$.

Theorem 7A.9 *Let $l - m > \frac{1}{2}$, $m \in \mathbb{N}$, $l \in \mathbb{R}_+$. The operator (*) extends to a continuous linear map*

$$T_m \colon W_2^l(\mathbb{R}^n) \to \mathop{\times}_{j=0}^{m} W_2^{l-j-1/2}(\mathbb{R}^{n-1}),$$

called the trace operator of order m, and there exists a right inverse Z_m,

$$T_m \circ Z_m = I \qquad (\textit{the identity}),$$

which acts continuously and linearly

$$Z_m \colon \mathop{\times}_{j=0}^{m} W_2^{l-j-1/2}(\mathbb{R}^{n-1}) \to W_2^l(\mathbb{R}^n).$$

The extension operator Z_m solves the inverse problem, i.e. given functions $g_j \in W_2^{l-j-1/2}(\mathbb{R}^{n-1})$ then $u = Z_m(g_0, \ldots, g_m)$ satisfies

$$\iota' \partial^j u/\partial n^j = g_j \qquad \text{for } j = 0, 1, \ldots, m.$$

Remark To shorten the notation we usually write $\partial^j u/\partial n^j = g_j$, or sometimes $\partial^j u/\partial n^j|_{\partial\Omega} = g_j$.

By means of local coordinates and a partition of unity, one can obtain the corresponding result for any bounded open set Ω in $\mathbb{R}^n$ with smooth boundary.

Theorem 7A.10 *Suppose that Ω is bounded, with C^∞ boundary, and let $l - m > 1/2$, $m \in \mathbb{N}$, $l \in \mathbb{R}_+$, Then there exists a continuous, linear trace operator*

$$T_m: W_2^l(\Omega) \to \mathop{\times}_{j=0}^{m} W_2^{l-j-1/2}(\partial\Omega),$$

with the property (*). The operator T_m possesses a right inverse Z_m,

$$T_m \circ Z_m = I \text{ (the identity)},$$

which acts continuously and linearly

$$Z_m: \mathop{\times}_{j=0}^{m} W_2^{l-j-1/2}(\partial\Omega) \to W_2^l(\Omega).$$

We need also an important lemma on Sobolev spaces.

Lemma 7A.11 *Consider the trace operator T_m in Theorem 7A.9. Let $u, v \in W_2^{m+1}(\mathbb{R}^n)$. If $T_m u = T_m v$ then the function w defined by $w = u$, for $x_n > 0$, and $w = v$, for $x_n < 0$, is an element of $W_2^{m+1}(\mathbb{R}^n)$.*

Proof By virtue of a density argument, see Theorem 7A.2, we may suppose first that $u', v' \in C_0^\infty(\mathbb{R}^n)$ and define w' in the same way as w. For any $\varphi \in C_0^\infty(\mathbb{R}^n)$ we have for $|\alpha| \leqslant m + 1$

$$\int w' D^\alpha \varphi \, dx = (-1)^{|\alpha|}\left[\int_{x_n > 0} \varphi D^\alpha u' \, dx + \int_{x_n < 0} \varphi D^\alpha v' \, dx\right]$$
$$+ \text{ boundary terms involving } T_m u' \text{ and } T_m v'.$$

(There are no boundary terms in the variables $x_1, \ldots, x_{n-1}$ due to the compact supports.) Letting $u' \to u$ and $v' \to v$ converge in $W_2^{m+1}(\mathbb{R}^n)$, the sum of the boundary terms converges to 0 since T_m is a continuous operator and $T_m u = T_m v$. Thus

$$\left|\int w D^\alpha \varphi \, dx\right| \leqslant \text{const } \|\varphi\|_{L_2}, \qquad |\alpha| \leqslant m + 1$$

and so $w \in W_2^{m+1}(\mathbb{R}^n)$.

Remark If the statement of Lemma 7A.11 is modified so that $u \in W_2^{m+1}(\mathbb{R}^n_+)$ and $v \in W_2^{m+1}(\mathbb{R}^n_-)$, then the conclusion still holds, for we can extend u and v to elements of $W_2^{m+1}(\mathbb{R}^n)$ by Theorem 7A.7.

Let $\partial\Omega \times (-1, 1)$ be the tubular neighbourhood of $\partial\Omega$ introduced above. Now we form the boundaryless double $\tilde{\Omega}$ of $\bar{\Omega}$ consisting of two copies $\bar{\Omega}_1$ and $\bar{\Omega}_2$ of $\bar{\Omega}$ identified on $\partial\Omega$, as discussed at the end of §5.12. The C^∞ structure of $\tilde{\Omega}$ is defined as follows: the collar of $\partial\Omega$ in $\bar{\Omega}_1$ is identified with $\partial\Omega \times [0, 1)$, while that in $\bar{\Omega}_2$ is identified with $\partial\Omega \times (-1, 0]$ by changing the sign of the x_n-coordinate. Then we have

$$T_m u_1 = \lim_{t \to 0^+} \left(u_1, \frac{\partial u_1}{\partial t}, \ldots, \frac{\partial^m}{\partial t^m} u_1 \right),$$

$$\text{resp.} \quad T_m u_2 = \lim_{t \to 0^-} \left(u_2, \frac{\partial u_2}{\partial t}, \ldots, \frac{\partial^m}{\partial t^m} u_2 \right)$$

if $u_1 \in C^\infty(\bar{\Omega}_1)$, resp. $u_2 \in C^\infty(\bar{\Omega}_2)$. (Note the modification in the definition of $T_m u_2$).

The following lemma is easily proved using a partition of unity on the compact, closed manifold $\tilde{\Omega}$ and applying Lemma 7A.11 in the chart domains which intersect $\partial\Omega$.

Lemma 7A.12 *Let $\tilde{\Omega}$ be the double of $\bar{\Omega}$ consisting of two copies $\bar{\Omega}_1$ and $\bar{\Omega}_2$ of $\bar{\Omega}$ identified on $\partial\Omega$. If the functions $u_1 \in W_2^{m+1}(\Omega_1)$, $u_2 \in W_2^{m+1}(\Omega_2)$, are given with $T_m u_1 = T_m u_2$ on $\partial\Omega$, then there exists a function $w \in W_2^{m+1}(\tilde{\Omega})$ with $u = u_1$ on Ω_1 and $u = u_2$ in Ω_2.*

For an arbitrary open set $\Omega \subset \mathbb{R}^n$, the Sobolev spaces form a chain, that is, we have continuous embeddings

$$W_2^{l_1}(\Omega) \hookrightarrow W_2^{l_2}(\Omega) \qquad \text{for } l_1 \geqslant l_2$$

where $l_1, l_2 \in \mathbb{N}$ are non-negative integers.

Theorem 7A.13 *The following embeddings are continuous for $l_1 \geqslant l_2$:*

(*a*) $\mathring{W}_2^{l_1}(\Omega) \hookrightarrow \mathring{W}_2^{l_2}(\Omega)$, $\quad l_1, l_2 \in \mathbb{R}_+$ *arbitrary*, Ω *arbitrary*
(*b*) $W_2^{l_1}(\Omega) \hookrightarrow W_2^{l_2}(\Omega)$, $\quad l_1, l_2 \in \mathbb{N}$, Ω *arbitrary*,
(*c*) $W_2^{l_1}(\Omega) \hookrightarrow W_2^{l_2}(\Omega)$, $\quad l_1, l_2 \in \mathbb{R}_+$ *arbitrary, and* Ω *has the extension property, that is, there exist continuous extension operators* β_{l_i}: $W_2^{l_i}(\Omega) \to W_2^{l_i}(\mathbb{R}^n)$, $i = 1, 2$.
(*d*) $W_2^{l_1}(M) \hookrightarrow W_2^{l_2}(M)$, $\quad l_1, l_2 \in \mathbb{R}_+$ *arbitrary, and* M *is a compact manifold without boundary.*

If l is sufficiently large then $W_2^l(\Omega)$ consists only of continuous functions, indeed continuously differentiable functions. This is the content of Sobolev's Lemma. If Ω is an open set in $\mathbb{R}^n$ and l is an integer, we denote by $C^l(\bar{\Omega})$ the space of all functions $u \in C^l(\Omega)$ such that u and the derivatives $D^s u$, $|s| \leqslant l$, are bounded and extend continuously to the closure $\bar{\Omega}$; the norm is

$$\|u\|_{C^l} := \sup_{x \in \Omega, |s| \leqslant l} |D^s u(x)|.$$

If l is not an integer, i.e. $l = [l] + \lambda$, $0 < \lambda < 1$, we let $C^l(\bar{\Omega})$ be the corresponding Hölder space with norm

$$\|u\|_{C^l} := \sup_{\substack{x \in \Omega \\ |s| < l}} |D^s u(x)| + \sup_{\substack{x, y \in \Omega \\ |s| = [l]}} \frac{|D^s u(x) - D^s u(y)|}{|x - y|^\lambda}.$$

Theorem 7A.14 (*Sobolev's Lemma*) *The following embeddings are continuous for $l_1 - l_2 > n/2$.*

(*a*) $\mathring{W}_2^{l_1}(\Omega) \hookrightarrow \mathring{C}^{l_2}(\bar{\Omega})$, *$\Omega$ arbitrary, where $\mathring{C}^{l_2}(\bar{\Omega})$ is the sub-space of $C^{l_2}(\bar{\Omega})$ consisting of all functions which are zero on the boundary $\partial\Omega$ along with their derivatives of order $\leqslant l_2$ (in other words, $\mathring{C}^{l_2}(\bar{\Omega})$ is the closure of $C_0^\infty(\Omega)$ in the C^{l_2} norm).*

(*b*) $W_2^{l_1}(\Omega) \hookrightarrow C^{l_2}(\bar{\Omega})$, *where, either Ω is bounded with smooth boundary, or Ω satisfies some geometric condition, such as the cone condition, the star condition, etc. (without necessarily being bounded).*

(*c*) $W_2^{l_1}(M) \hookrightarrow C^{l_2}(M)$, *where M is a compact manifold without boundary.*

For partial differential equations, compact embeddings play a decisive role. They are true for bounded Ω, but the boundedness hypothesis cannot essentially be weakened.

Theorem 7A.15

(*a*) *Let Ω be open and bounded and let $l_1 > l_2$, where $l_1, l_2 \in \mathbb{R}_+$. Then the embedding*

$$\mathring{W}_2^{l_1}(\Omega) \hookrightarrow \mathring{W}_2^{l_2}(\Omega) \qquad \textit{is compact.}$$

(*b*) *Let Ω be bounded and with smooth boundary (or satisfying some geometric condition), and let $l_1 > l_2$, $l_1, l_2 \in \mathbb{R}_+$. Then the embedding*

$$W_2^{l_1}(\Omega) \hookrightarrow W_2^{l_2}(\Omega) \qquad \textit{is compact.}$$

(*c*) $W_2^{l_1}(M) \hookrightarrow W_2^{l_2}(M)$ *is compact for $l_1 > l_2$, where M is a compact manifold without boundary.*

From the compact embedding, Ehrling's Lemma follows:

Theorem 7A.16 *Let Ω be bounded, and let $0 \leqslant l_2 < l_1$, whence $l_1, l_2 \in \mathbb{R}_+$. Then for each $\varepsilon > 0$ there exists a constant $c(\varepsilon)$ such that for all $u \in \mathring{W}_2^{l_1}(\Omega)$*

$$\|u\|_{l_2} \leqslant \varepsilon \|u\|_{l_1} + c(\varepsilon)\|u\|_0 \qquad (**)$$

Further, (**) is also true for all $u \in W_2^{l_1}(\Omega)$ if Ω is bounded with smooth boundary, or all $u \in W_2^{l_1}(M)$ if M is a compact manifold without boundary.

Ehrling's Lemma is also true for the Sobolev spaces on $\mathbb{R}^n$, as one sees by Fourier transformation.

8
Pseudo-differential operators on a compact manifold

In this chapter we define the class $OS^m(M)$ of pseudo-differential operators of order m on a compact manifold M. Essentially, they are linear operators on $C^\infty(M)$ which are p.d.o.'s in local coordinates and satisfy a quasi-locality property.

The symbol of an operator $A \in OS^m(\mathbb{R}^n)$ in Euclidean space is uniquely determined by the formula (31) of Chapter 7, that is,

$$A(x, \xi) = e^{-i(x,\xi)}\, A(x, D)\, e^{i(x,\xi)}.$$

For an operator $A \in OS^m(M)$ there still exists a symbol – defined locally – but it is not unique due to the effect of coordinate transformations. It is possible, however, to define a symbol modulo lower-order terms, which we call a *main symbol*. The main symbol isomorphism

$$OS^m(M)/OS^{m-1}(M) \simeq S^m(T^*(M))/S^{m-1}(T^*(M))$$

is also an algebra isomorphism. It should be noted that other books use the term "principal symbol" where we use "main symbol". We reserve the former term for the special case of classic operators defined in §8.4 which are (-1) classic operators in local coordinates (see §7.5). Such operators A have a uniquely determined homogeneous principal part πA which is therefore well-defined on the cotangent bundle, and $\pi A \in C^\infty(T^*(M)\backslash 0)$.

This chapter comprises two parts. In the first part, §§8.1 to 8.6, the p.d.o. algebra is developed, and in addition to the scalar p.d.o.'s which were indicated above, we also consider p.d.o.'s acting on sections of vector bundles. In the second part of the chapter, §§8.7 to 8.10, the Fredholm theory of elliptic p.d.o.'s in vector bundles is developed, including the existence of a parametrix. For classic p.d.o.'s, ellipticity means invertibility of the values of the principal symbol. The index of an elliptic symbol is defined and we prove the invariance of the index under homotopies of various kinds, both for general p.d.o.'s and for classic p.d.o.'s that have a homogeneous principal part.

It is beyond the scope of this book to discuss the Atiyah–Singer formula for the index of elliptic p.d.o.'s (or elliptic symbols) but we develop, essentially, all the analytic properties which are required for its proof. As an illustration, we prove a special case, namely, Noether's formula for the index of elliptic p.d.o.'s of order zero on the unit circle.

8.1 Background and notation

Let M be a compact C^∞ manifold without boundary. A Riemannian metric on the tangent bundle $T(M)$ gives rise to a density $d\mu$ on M as discussed in §6.6. If (U, κ) is a chart on M then, locally, this density is given by $d\mu = g_U(x)\,dx$, where $g_U = \det[g_{ij}]$ is the Riemannian volume density. We use this density $d\mu$ to define the Hilbert space of square-integrable functions on M. For functions $u, v \in C^\infty(M)$ the scalar product is defined by

$$(u, v)_M := \int_M u\bar{v}\,d\mu,$$

and then $L_2(M) = L_2(M, \mathbb{C})$ is the completion of $C^\infty(M)$ with respect to this scalar product.

If X is an open set in M, we let $C_0^\infty(X) = C_0^\infty(X, \mathbb{C})$ denote the space of C^∞ functions on X with compact support. Since M is compact then $C_0^\infty(M) = C^\infty(M)$. We need also the Sobolev spaces. They are defined here for complex-valued functions and then later on in §8.5 for sections of a vector bundle.

Definition 8.1 *Let M be a compact C^∞ manifold. Let $l \geqslant 0$, a non-negative real number. The Sobolev space $W_2^l(M)$ is the set of functions $u\colon M \to \mathbb{C}$ such that*

$$(\varphi \circ u) \circ \kappa^{-1} \in W_2^l(\mathbb{R}^n)$$

for every coordinate map $\kappa\colon U \to V$ and for every $\varphi \in C_0^\infty(U)$. The topology on $W_2^l(M)$ is defined by the semi-norms

$$u \mapsto \|(\varphi \cdot u) \circ \kappa^{-1}\|_l,$$

i.e. the weakest topology for which the semi-norms are continuous.

Let $\{U_j, \kappa_j\}_{j=1,\ldots,N}$ be an atlas on M and $\{\varphi_j\}$ a subordinate partition of unity. For $u \in W_2^l(M)$ we let $u_j := (\varphi_j \cdot u) \circ \kappa_j^{-1}$, and make $W_2^l(M)$ into a Hilbert space by defining the scalar product

$$(u, v)_{W_2^l(M)} = \sum_{j=1}^{N} (u_j, v_j)_l.$$

It can be shown that the topology defined by the norm $\|u\|^2 = (u, u)_{W_2^l(M)}$ coincides with the topology in Definition 8.1, in other words, two equivalent atlases and two corresponding subordinate partitions of unity yield equivalent norms.

Note that $W_2^0(M)$ can be identified with $L_2(M)$. For $l < 0$ we define $W_2^l(M)$ as the *anti-dual* space

$$W_2^l(M) = (W_2^{-l}(M))^*,$$

or, alternatively, we could use the definition above with distributions u instead of functions (both definitions agree, giving equivalent norms).

We use the scalar product $(\,,\,)_M$ to identify $L_2(M) = W_2^0(M)$ with its anti-dual. The space $C^\infty(M)$ is dense in $W_2^l(M)$ for all $l \geqslant 0$. For $k \geqslant l \geqslant 0$

we therefore have a dense inclusion $W_2^k(M) \subset W_2^l(M)$, and by transposing it we obtain a dense inclusion $W_2^{-l}(M) \subset W_2^{-k}(M)$ for $-k \leqslant -l \leqslant 0$. If $k < 0$ and $l > 0$ we consider $W_2^k(M)$ as embedded in $W_2^l(M)$ by the inclusions $W_2^k(M) \subset W_2^0(M) \subset W_2^l(M)$. In this way we regard $W_2^k(M)$ as embedded in $W_2^l(M)$ for all $k < l$, and we say that $\{W_2^l(M)\}_{l \in \mathbb{R}}$ forms a *scale of Hilbert spaces* (see [Pa] for further details).

Recall that $C^\infty(M)$ is a Fréchet space where convergence of a sequence $\{u_j\}$ is defined to be uniform convergence of all derivatives of u_j on compact subsets of chart domains. By Sobolev's Lemma we have

$$C^\infty(M) = \bigcap_l W_2^l(M),$$

with the inverse limit topology, i.e. a sequence $\{u_j\} \subset C^\infty(M)$ converges to $u \in C^\infty(M)$ if and only if it converges in $W_2^l(M)$ for all $l \in \mathbb{R}$.

The scalar product $(u, v)_M$ defined for u, v in $L_2(M)$ extends to a pairing between $W_2^l(M)$ and $W_2^{-l}(M)$ for all l: if $l \geqslant 0$ and u is (an anti-linear functional) in $W_2^{-l}(M)$ then

$$(u, v)_M = \langle u, v \rangle \qquad \text{for all } v \in W_2^l(M);$$

and if u is in $W_2^l(M)$, $l \geqslant 0$, then

$$(u, v)_M = \overline{\langle v, u \rangle} \qquad \text{for all } v \in W_2^{-l}(M).$$

Definition 8.2 *A (continuous) linear mapping $B: C^\infty(M) \to C^\infty(M)$ is an operator of order m if for each $l \in \mathbb{R}$ it extends to a continuous mapping*

$$B: W_2^l(M) \to W_2^{l-m}(M) \tag{1}$$

The set of all such operators is denoted by $OP^m(M)$. This is a local space of operators, that is, $B \in OP^m(M)$ if and only if $\varphi B(\psi \cdot) \in OP^m(M)$ for all functions $\varphi, \psi \in C^\infty(M)$.

If $B: C^\infty(M) \to C^\infty(M)$ is a linear mapping then its formal adjoint B^* is defined by means of the scalar product $(\,,\,)_M$

$$(Bu, v)_M = (u, B^*v)_M, \qquad u, v \in C^\infty(M) \tag{2}$$

Note that the left-hand side of (2) is not necessarily a bounded operator of $u \in L_2(M)$, so the existence of B^*v is not assured. However, if

$$B \in OP^m(M)$$

we let $B_l: W_2^l(M) \to W_2^{l-m}(M)$ denote the maps of Definition 8.2 (1) which extend the operator B, and then the existence of the adjoint B^* is proved by considering the duality adjoints $B_l^*: W_2^{m-l}(M) \to W_2^{-l}(M)$ of the operators B_l. The formal adjoint B^* is the restriction of B_l^* to $C^\infty(M)$. We have $B^*: C^\infty(M) \to C^\infty(M)$ and

$$B^* \in OP^m(M).$$

The set of operators of order $-\infty$ is denoted

$$OP^{-\infty}(M) = \bigcap_{m \in \mathbb{R}} OP^m(M),$$

that is, B belongs to $OP^{-\infty}(M)$ if it extends to a continuous map $B: W_2^{l_1}(M) \to W_2^{l_2}(M)$ for all $l_1, l_2 \in \mathbb{R}$.

Proposition 8.3 *Let M be a compact, closed manifold. An operator $B: C^\infty(M) \to C^\infty(M)$ belongs to $OP^{-\infty}(M)$ if and only if it is an integral operator with a C^∞ kernel, that is,*

$$(Bf)(x) = \int_M K(x, y) f(y)\, d\mu(y), \qquad K \in C^\infty(M \times M). \tag{3}$$

Sketch of Proof Let $\mathscr{D}'(M)$ (resp. $\mathscr{E}'(M)$) be the space of all distributions (resp. with compact support) on M. Since M is compact then ([Tri])

$$\mathscr{E}'(M) = \mathscr{D}'(M) = \bigcup_l W_2^l(M),$$

with the inductive limit topology, i.e. a sequence $\{u_j\} \subset \mathscr{E}'(M)$ converges if and only if $\{u_j\} \subset W_2^l(M)$ converges for some $l \in \mathbb{R}$. It follows that if $B \in OP^{-\infty}(M)$ then it is also continuous as an operator $B: \mathscr{E}'(M) \to C^\infty(M)$. Let $\{(U_j, \kappa_j)\}$ be an atlas on M and $\{\psi_j\}$ a subordinate partition of unity. Then we can write

$$Bf = \sum_{i,j} \psi_i B \psi_j f, \tag{4}$$

and the operator $\psi_i B(\psi_j \cdot)$ is continuous from $\mathscr{E}'(U_j)$ to $C^\infty(U_i)$. The points in U_j and in U_i can be identified with their coordinates in $\mathbb{R}^n$ (by the maps κ_j and κ_i, respectively), and then by L. Schwartz's "Théoreme de noyaux" (see [Hö 1] or [Tri]), $\psi_i B(\psi_j \cdot)$ is an integral operator with a C^∞ kernel $K_{ij} \in C^\infty(U_i \times U_j)$, i.e. $(\psi_i B \psi_j f)(x) = \int_{U_j} K_{ij}(x, y) f(y)\, dy$. Now, if we set

$$K(x, y) = \sum_{i,j} \psi_i(x) \cdot K_{ij}(x, y) \cdot \psi_j(y) [g_{U_j}(y)]^{-1},$$

the representation (3) follows.

Conversely, if B has a C^∞ kernel K, we can localize as in (4), and it follows upon differentiating the kernel of $\psi_i B \psi_j$ that B acts continuously $B: W_2^{l_1}(M) \to W_2^{l_2}(M)$ for all $l_1, l_2 \in \mathbb{R}$ which means that $B \in OP^{-\infty}(M)$.

We write $B \in OC^\infty(M)$ if it is an integral operator with C^∞ kernel $K(x, y)$, i.e.

$$(Bf)(x) = \int_M K(x, y) f(y)\, \mathrm{d}\mu(y), \qquad K \in C^\infty(M \times M);$$

thus Proposition 8.3 implies that

$$OP^{-\infty}(M) = OC^\infty(M).$$

Now, we define the support of an operator and then the transfer (or push-forward) of an operator under a coordinate map.

Definition 8.4 *If* $B\colon C^\infty(M) \to C^\infty(M)$ *is a linear map we define the* support *of* B *to be the complement of the largest open set* $\mathcal{O} \subset M$ *such that*

(1) $B\varphi(x) = 0$ *if* $x \in \mathcal{O}$,
(2) $B\varphi = 0$ *if* $\operatorname{supp}\varphi \subset \mathcal{O}$

(*We also make the same definition for a linear map* $B\colon C_0^\infty(\mathbb{R}^n) \to C^\infty(\mathbb{R}^n)$.)

If $\varphi \in C^\infty(M)$ has its support disjoint from the support of B then clearly

$$\varphi B = B\varphi = 0.$$

If $\varphi \in C^\infty(M)$ is identically 1 on a neighbourhood of $\operatorname{supp} B$ then $(1 - \varphi)B = 0$, so $B = \varphi B$ and similarly $B = B\varphi$. Clearly, if $\varphi_1, \varphi_2 \in C^\infty(M)$ then for any B we have

$$\operatorname{supp}(\varphi_1 B \varphi_2) \subset \operatorname{supp}\varphi_1 \cup \operatorname{supp}\varphi_2. \tag{5}$$

Let $B\colon C^\infty(M) \to C^\infty(M)$ be a linear operator, and let $\kappa\colon U \to V \subset \mathbb{R}^n$ some chart for M. Let $\kappa^* f = f \circ \kappa$ denote the pull-back of $f \in C^\infty(V)$ and $\kappa^{*-1} g = g \circ \kappa^{-1}$ the push-forward of $g \in C^\infty(U)$. By considering $\imath \circ B \circ \imath$ where $\imath$ is the natural embedding $\imath\colon C_0^\infty(U) \to C^\infty(M)$ and $\imath\colon C^\infty(M) \to C^\infty(U)$ is the restriction operator, we can view B as an operator $B\colon C_0^\infty(U) \to C^\infty(U)$, and then the *push-forward* operator $B_\kappa\colon C_0^\infty(V) \to C^\infty(V)$ is just B in local coordinates:

$$B_\kappa = \kappa^{*-1} \circ B \circ \kappa^*, \tag{6}$$

i.e. $B_\kappa = B(f \circ \kappa) \circ \kappa^{-1}$, $f \in C_0^\infty(V)$, so the following diagram commutes:

$$\begin{array}{ccc} C_0^\infty(U) & \xrightarrow{\;B\;} & C^\infty(U) \\ {\scriptstyle \kappa^*}\big\uparrow & & \big\downarrow{\scriptstyle (\kappa^*)^{-1}} \\ C_0^\infty(V) & \xrightarrow{\;B_\kappa\;} & C^\infty(V) \end{array}$$

The push-forward map takes an operator acting on $U \subset M$ to one acting on $V \subset \mathbb{R}^n$.

For the purpose of §8.2 and subsequent sections, we would like to regard the push-forward as a map $B_\kappa\colon C_0^\infty(\mathbb{R}^n) \to C_0^\infty(\mathbb{R}^n)$ so that we can use the theory of p.d.o.'s in $\mathbb{R}^n$. This is possible if the support of B lies in the chart domain U, as we show in the next definition.

Definition 8.5 *Let* (U, κ) *be a chart for* M, *and* $B\colon C^\infty(M) \to C^\infty(M)$ *a linear map with* $\operatorname{supp} B \subset U$. *Let* $\zeta \in C_0^\infty(U)$ *be identically* 1 *on a neighbourhood of* $\operatorname{supp} B$. *We define a linear map* $B_\kappa = T(B, \kappa)\colon C_0^\infty(\mathbb{R}^n) \to C_0^\infty(\mathbb{R}^n)$, *called the transfer* (*or push-forward*) *of* B *by* κ, *as follows*:

$$T(B, \kappa)f = [\zeta \cdot B(\zeta \cdot (f \circ \kappa))] \circ \kappa^{-1}, \qquad f \in C_0^\infty(\mathbb{R}^n), \tag{7}$$

and extended to be zero outside $\kappa(U) = V$. *We can also write* (7) *in the form* $T(B, \kappa)f = \kappa^{*-1} \circ \zeta \cdot B \circ \kappa^* \tilde{\zeta} \cdot f$, *where* $\tilde{\zeta} = \zeta \circ \kappa^{-1}$.

The definition of $T(B, \kappa)$ is, of course, independent of the choice of ζ. In §8.2 we will be considering the transfer of an operator $B = \varphi A \psi$, where φ, $\psi \in C_0^\infty(U)$ are regarded as multiplication operators on $C^\infty(M)$. In that case there is really no need for the function ζ and we can write simply

$$T(\varphi A \psi, \kappa) = \kappa^{*-1}(\varphi A \psi)\kappa^*.$$

For future reference, note that

$$T(\varphi A \psi, \kappa) = (\varphi A \psi)_\kappa = \varphi_\kappa A_\kappa \psi_\kappa \tag{8}$$

where $\varphi_\kappa = \varphi \circ \kappa^{-1}$ and $\psi_\kappa = \psi \circ \kappa^{-1}$.

The support of $T(B, \kappa)$ is $\kappa(\operatorname{supp} B)$, a compact subset of $\kappa(U) = V$. Conversely, if $S: C_0^\infty(\mathbb{R}^n) \to C_0^\infty(\mathbb{R}^n)$ has compact support in $\kappa(U)$, then $S = T(B, \kappa)$ for some $B: C^\infty(M) \to C^\infty(M)$ with support in U. In fact, we define the transfer $B = T(S, \kappa^{-1})$: $C^\infty(M) \to C^\infty(M)$ having support in U by

$$T(S, \kappa^{-1})f = [\tilde{\zeta} \cdot S(\tilde{\zeta} \cdot (f \circ \kappa^{-1}))] \circ \kappa = \kappa^* \circ \tilde{\zeta} \cdot S \circ \kappa^{*-1} \zeta \cdot f.$$

The transfers $B \mapsto T(B, \kappa)$ and $S \mapsto T(S, \kappa^{-1})$, also called *push-forward* and *pull-back*, respectively, are mutually inverse one-to-one maps between linear maps $B: C^\infty(M) \to C^\infty(M)$ with support in U and linear maps $S: C_0^\infty(\mathbb{R}^n) \to C_0^\infty(\mathbb{R}^n)$ with compact support in $\kappa(U)$.

Analogous to Definition 8.2 we say that an operator $C_0^\infty(\mathbb{R}^n) \to C^\infty(\mathbb{R}^n)$ belongs to $OP^m(\mathbb{R}^n)$ if for each l it extends to a continuous mapping $W_2^l(\mathbb{R}^n) \to W_2^{l-m}(\mathbb{R}^n)$.

Theorem 8.6 *Let (U, κ) be a chart on M. Then a map $B: C^\infty(M) \to C^\infty(M)$ with* supp $B \subset U$ *belongs to $OP^m(M)$ if and only if the transferred map $T(B, \kappa)$ (having its compact support in $\kappa(U) = V$) belongs to $OP^m(\mathbb{R}^n)$.*

Proof Let $\zeta \in C_0^\infty(U)$ be identically 1 on some neighbourhood of supp B in U. Then we can write (see (7))

$$T(B, \kappa) = \imath \circ \nu \circ B \circ \mu \circ \lambda, \tag{9}$$

where $\imath: C_0^\infty(V) \to C_0^\infty(\mathbb{R}^n)$ is the inclusion, $\nu: C^\infty(M) \to C_0^\infty(V)$ is the map $f \mapsto (\zeta f) \circ \kappa^{-1}$, $\mu: C_0^\infty(V) \to C^\infty(M)$ is the map $f \mapsto f \circ \kappa$ (where $f \circ \kappa$ is extended to be zero outside U), and $\lambda: C_0^\infty(\mathbb{R}^n) \to C_0^\infty(V)$ is the map $f \mapsto \tilde{\zeta} \cdot f$. We define the space $\mathring{W}_2^l(V)$, $l \in \mathbb{R}$, as the completion of $C_0^\infty(V)$ in the $W_2^l(\mathbb{R}^n)$ norm. The inclusion $\imath: \mathring{W}_2^l(V) \hookrightarrow W_2^l(\mathbb{R}^n)$ is then continuous for all $l \in \mathbb{R}$. By the definition of the $W_2^l(M)$ topology (Definition 8.1) the maps ν and μ extend continuously to

$$\nu: W_2^l(M) \to \mathring{W}_2^l(V), \qquad \mu: \mathring{W}_2^l(V) \to W_2^l(M), \qquad \forall l \in \mathbb{R},$$

and, further, it is not hard to verify that multiplication by the function $\tilde{\zeta} \in C_0^\infty(V)$ is a continuous map

$$\lambda: W_2^l(\mathbb{R}^n) \to \mathring{W}_2^l(V), \qquad \forall l \in \mathbb{R}.$$

Now, if $B \in OP^m(M)$, i.e. $B: W_2^l(M) \to W_2^{l-m}(M)$ is continuous for all $l \in \mathbb{R}$, then all five operators in the chain (9) for $T(B, \kappa)$,

$$W_2^l(\mathbb{R}^n) \xrightarrow[\lambda]{} \mathring{W}_2^l(V) \xrightarrow[\mu]{} W_2^l(M) \xrightarrow[B]{} W_2^{l-m}(M) \xrightarrow[\nu]{} \mathring{W}_2^{l-m}(V) \xrightarrow[\iota]{} W_2^{l-m}(\mathbb{R}^n),$$

are continuous. Thus, $T(B, \kappa)$: $W_2^l(\mathbb{R}^n) \to W_2^{l-m}(\mathbb{R}^n)$ is continuous for all $l \in \mathbb{R}$ and we have proved that $T(B, \kappa) \in OP^m(\mathbb{R}^n)$.

To prove sufficiency, we use the pull-back $T(S, \kappa^{-1})$ where $S \in OP^m(\mathbb{R}^n)$. We have $T(S, \kappa^{-1}) = \mu \circ \lambda \circ S \circ \iota \circ \nu$, from which it follows that $T(S, \kappa^{-1}) \in OP^m(M)$.

8.2 Pseudo-differential operators on M

In this section we define pseudo-differential operators on a C^∞ manifold M. For simplicity we assume that the manifold is compact.

Definition 8.7 *A linear operator $A: C^\infty(M) \to C^\infty(M)$ is called a pseudo-differential operator on M of order $m \in \mathbb{R}$, and we write $A \in OS^m(M)$, if for every coordinate patch (U, κ) on M, and for all $\varphi, \psi \in C_0^\infty(U)$, the push-forward operator* (*see* §8.1),

$$T(\varphi A \psi, \kappa) = \kappa^{*-1}(\varphi A \psi)\kappa^*,$$

belongs to $OS^m(\mathbb{R}^n)$.

Thus, pseudo-differential operators on M are operators which are p.d.o.'s in local coordinates. Note that Definition 8.7 must hold for every coordinate patch, even one in which the domain is not connected. This is important for the next proposition. If (U, κ) and (U', κ') are two coordinate patches with $U \cap U' = \varnothing$ then $(U \cup U', \kappa \cup \kappa')$ is another coordinate patch, where $\kappa \cup \kappa'$ denotes the map

$$(\kappa \cup \kappa')(x) = \begin{cases} \kappa(x) & \text{if } x \in U \\ \kappa'(x) & \text{if } x \in U' \end{cases}$$

and without loss of generality we take κ and κ' to have disjoint image so that $\kappa \cup \kappa'$ has an inverse.

Proposition 8.8 *Operators $A \in OS(M)$ are quasi-local, that is, for any functions $\varphi, \psi \in C^\infty(M)$ with* $\operatorname{supp} \varphi \cap \operatorname{supp} \psi = \varnothing$, *we have*

$$\varphi A \psi \in OC^\infty(M) \qquad (\textit{an operator with } C^\infty \textit{ kernel, see } §8.1).$$

Proof Because the supports of φ and ψ are compact and disjoint from each other, we have

$$\operatorname{supp} \varphi \subset \mathscr{U} = U_1 \cup \cdots \cup U_p, \qquad \operatorname{supp} \psi \subset \mathscr{V} = U_{p+1} \cup \cdots \cup U_N$$

where $\mathscr{U} \cap \mathscr{V} = \varnothing$ and (U_j, κ_j), $j = 1, \ldots, N$, are some coordinate charts on M. Choose an open set $\mathscr{W}$ such that

$$\operatorname{supp} \varphi \cup \operatorname{supp} \psi \subset \mathscr{W} \subset \overline{\mathscr{W}} \subset \mathscr{U} \cup \mathscr{V}.$$

Then $U_1, \ldots, U_N$, $\complement\overline{\mathscr{W}}$ is an open cover of M, and we choose a subordinate partition of unity, that is, $\sum_{j=1}^{N+1} \varphi_j = 1$, supp $\varphi_j \subset U_j$ for $j = 1, \ldots, N$ and supp $\varphi_{N+1} \cap \mathscr{W} = \varnothing$. Then $\varphi A \psi = \sum' \varphi_j \varphi A \psi \varphi_k$, where the sum is taken over the indices $j = 1, \ldots, p$ and $k = p + 1, \ldots, N$, since these are the only indices for which $\varphi_j \varphi \neq 0$ and $\varphi_k \psi \neq 0$. If we choose functions $\psi_j \in C_0^\infty(U_j)$ equal to 1 on supp φ_j, then

$$\varphi A \psi = \sum{}' \psi_j \varphi(\varphi_j A \varphi_k) \psi \psi_k. \tag{10}$$

Joining the two coordinate patches (U_j, κ_j) and (U_k, κ_k) as above, we obtain a coordinate patch $(U_j \cup U_k, \kappa_j \cup \kappa_k)$, and then by Definition 8.7 the transferred map, $T(\varphi_j A \varphi_k, \kappa_j \cup \kappa_k)$, belongs to $OS^m(\mathbb{R}^n)$. Now, since $\operatorname{supp}(\psi_j \varphi) \cap \operatorname{supp}(\psi \psi_k) = \varnothing$, Corollary 7.14 implies that the push-forward,

$$T(\psi_j \varphi(\varphi_j A \varphi_k) \psi \psi_k, \kappa_j \cup \kappa_k),$$

of each term in the sum (10) belongs to $OS^{-\infty}(\mathbb{R}^n)$ (also see (8)). By Theorem 7.6 it therefore has a C^∞ kernel representation, and the support of this kernel is a compact subset of $\kappa_j(U_j) \times \kappa_k(U_k) \subset \mathbb{R}^n \times \mathbb{R}^n$. Pulling back to the manifold M, it follows that each term in the sum belongs to $OC^\infty(M)$, whence $\varphi A \psi \in OC^\infty(M)$.

In the next theorem, given a linear operator $A\colon C^\infty(M) \to C^\infty(M)$ which is already known to be quasi-local, we prove that $A \in OS^m(M)$ if and only if the condition of Definition 8.7 holds for all charts in some atlas on M.

Theorem 8.9 *Let $\mathfrak{A} = \{(U_j, \kappa_j)\}$ be an atlas on M. A linear operator $A\colon C^\infty(M) \to C^\infty(M)$ belongs to $OS^m(M)$ if and only if:*

(*i*) *For any chart $(U_j, \kappa_j) \in \mathfrak{A}$ and for all $\varphi, \psi \in C_0^\infty(U_j)$, we have*

$$T(\varphi A \psi, \kappa_j) = (\varphi A \psi)_{\kappa_j} \in OS^m(\mathbb{R}^n);$$

(*ii*) (*quasi-locality*) *For any functions $\varphi, \psi \in C^\infty(M)$ with* supp $\varphi \cap$ supp $\psi = \varnothing$, *we have*

$$\varphi A \psi \in OC^\infty(M).$$

Proof The necessity of (i) and (ii) is obvious from Definition 8.7 and Proposition 8.8. To prove sufficiency, let the operator $A\colon C^\infty(M) \to C^\infty(M)$ satisfy conditions (i) and (ii). Upon choosing a partition of unity $\{\varphi_j\}$ subordinate to the atlas $\mathfrak{A}$ and corresponding functions $\psi_j \in C_0^\infty(U_j)$ such that $\psi_j = 1$ on the support of φ_j, we have

$$A \cdot = \sum_j \varphi_j A \cdot = \sum_j \varphi_j A[\psi_j \cdot] + \sum_j \varphi_j A[(1 - \psi_j) \cdot] \tag{11}$$

Let $\varphi, \psi \in C_0^\infty(U)$. By quasi-locality, the terms $\varphi_j A[(1 - \psi_j) \cdot]$ belong to $OC^\infty(M)$. After composing them with the multiplication operators, φ and ψ, the push-forward of such terms are elements of $OS^{-\infty}(\mathbb{R}^n)$ because they have a C^∞ kernel with compact support in $\mathbb{R}^n \times \mathbb{R}^n$, and so we can invoke Theorem 7.6.

It remains to show that each term $\varphi_j A[\psi_j \cdot]$ belongs to $OS^m(M)$. Let (U, κ) be a chart on M; as in §8.1 we denote by $B_\kappa = T(B, \kappa)$ the transfer (push-forward) of an operator B. We must show for any $\varphi, \psi \in C_0^\infty(U)$ that the transferred operator $(\varphi\varphi_j A\psi_j\psi)_\kappa$ belongs to $OS^m(\mathbb{R}^n)$. By choosing functions $\chi_j \in C_0^\infty(U_j)$ equal to 1 on the support of ψ_j and using (8) we have

$$(\varphi\varphi_j A\psi_j\psi)_\kappa = [(\varphi\chi_j)_{\kappa_j}(\varphi_j A\psi_j)_{\kappa_j}(\chi_j\psi)_{\kappa_j}]_{\kappa\circ\kappa_j^{-1}} \tag{12}$$

By condition (i), we have $(\varphi_j A\psi_j)_{\kappa_j} \in OS^m(\mathbb{R}^n)$. Hence

$$(\varphi\chi_j)_{\kappa_j}(\varphi_j A\psi_j)_{\kappa_j}(\chi_j\psi)_{\kappa_j} \in OS^m(\mathbb{R}^n)$$

by Theorem 7.12, and the Schwartz kernel of this operator has compact support in $\kappa_j(U \cap U_j) \times \kappa_j(U \cap U_j) \subset \mathbb{R}^n \times \mathbb{R}^n$. By virtue of Theorem 7.17, the operator (12) belongs to $OS^m(\mathbb{R}^n)$, as was to be shown.

For some purposes it is useful to have another version of Theorem 8.9. By Lemma 5.44, we may shrink the open cover $\{(U_j, \kappa_j)\}_{j=1}^N$ to obtain an atlas $\mathfrak{A}' = \{(W_j, \kappa_j)\}_{j=1}^N$ where $\bar{W}_j \subset U_j$ for $j = 1, \ldots, N$. Then condition (i) in Theorem 8.9 can be replaced by the following:

(i′) *There exist p.d.o.'s* $\tilde{A}_j(x, D) \in OS^m(\mathbb{R}^n)$, $j = 1, \ldots, N$, *such that*

$$T(\varphi A\psi, \kappa_j) = \varphi_{\kappa_j}\tilde{A}_j(x, D)\psi_{\kappa_j},$$

i.e. pulling back by κ_j^{-1} *we have the representation*

$$\varphi A\psi = \varphi(\tilde{A}_j(x, D))_{\kappa_j^{-1}}\psi,$$

for all $\varphi, \psi \in C_0^\infty(W_j)$.

Theorem 8.10 *A linear operator* $A: C^\infty(M) \to C^\infty(M)$ *belongs to* $OS^m(M)$ *if and only if conditions* (i′) *and* (ii) *hold.*

Proof The sufficiency of conditions (i′) and (ii) is clear since they imply the conditions of Theorem 8.9. Now choose functions $\chi_j \in C_0^\infty(U_j)$ equal to 1 on $\bar{W}_j$. If $A \in OS^m(M)$ we have $\varphi A\psi = \varphi\chi_j A\chi_j\psi = \varphi(\tilde{A}_j(x, D))_{\kappa_j^{-1}}\psi$ where $\tilde{A}_j(x, D) = (\chi_j A\chi_j)_{\kappa_j}$, thereby proving the necessity of (i′).

Remark The proof shows that we can choose the operators $\tilde{A}_j(x, D)$ so that they have compact support in an arbitrarily small neighbourhood of $\kappa_j(\bar{W}_j)$ (in $\kappa_j(U_j)$). See Definition 8.4 for the definition of the support of an operator.

The following lemma shows how to construct pseudo-differential operators on a manifold. It will be of use in the proof of Proposition 8.16 and of Theorem 8.23.

Lemma 8.11 *Let* (U, κ) *be a chart on* M, *and let the operator* $\tilde{A}: C_0^\infty(\mathbb{R}^n) \to C_0^\infty(\mathbb{R}^n)$ *have compact support in* $\kappa(U) = V \subset \mathbb{R}^n$. *Then the pull-back operator* $A: C^\infty(M) \to C^\infty(M)$ *defined by*

$$A := T(\tilde{A}, \kappa^{-1})$$

is in $OS^m(M)$ *if and only if* $\tilde{A} \in OS^m(\mathbb{R}^n)$.

Proof Note that supp $A = \kappa^{-1}(\text{supp}\,\tilde{A}) \subseteq U$. Thus, $A = \varphi A \varphi$ if $\varphi \in C_0^\infty(U)$ is equal to 1 on supp A and hence $\tilde{A} \in OS^m(\mathbb{R}^n)$ by Definition 8.7 if $A \in OS^m(M)$. This proves the necessity part of the lemma. To prove sufficiency, suppose that $\tilde{A} \in OS^m(\mathbb{R}^n)$. There exists an atlas $\mathfrak{A} = \{(U_j, \kappa_j)\}$ for M such that $(U_{j_0}, \kappa_{j_0}) = (U, \kappa)$ and supp $A \cap U_j = \varnothing$ when $j \neq j_0$. Since $A = (\tilde{A}(x, D))_{\kappa^{-1}}$, then we obtain for any $\varphi, \psi \in C_0^\infty(U_j)$

$$(\varphi A \psi)_\kappa = \varphi_\kappa \tilde{A}(x, D)\psi_k \in OS^m(\mathbb{R}^n) \qquad \text{when } j = j_0,$$

and
$$(\varphi A \psi)_{\kappa_j} = 0 \in OS^m(\mathbb{R}^n) \qquad \text{when } j \neq j_0,$$

so A satisfies condition (i) of Theorem 8.9 for this atlas. The verification of condition (ii) is straightforward since p.d.o.'s in $\mathbb{R}^n$ are quasi-local.

Remark 8.12 If $\tilde{A}(x, D) \in OS^m(\mathbb{R}^n)$ is *any* p.d.o. in $\mathbb{R}^n$, and (U, κ) is a chart on M, then Lemma 8.11 applies to the operator $\tilde{\varphi}\tilde{A}(x, D)\tilde{\psi}$ for every $\tilde{\varphi}, \tilde{\psi} \in C_0^\infty(\kappa(U))$ since by (5) we have

$$\text{supp}\,\tilde{\varphi} A \tilde{\psi} \subset \text{supp}\,\tilde{\varphi} \cup \text{supp}\,\tilde{\psi} \subset \kappa(U),$$

and supp $\tilde{\varphi} \cup$ supp $\tilde{\psi}$ is compact in $\kappa(U)$.

The equation (11) in the proof of Theorem 8.9 constitutes a convenient formula for a p.d.o. on a manifold M, and enables us to localize the operator A so that we can use Theorem 8.6 to prove continuity in the Sobolev spaces on M.

Theorem 8.13 *Let $A \in OS^m(M)$. Then for each l the mapping*

$$A\colon W_2^l(M) \to W_2^{l-m}(M)$$

is continuous.

Proof Let $\mathfrak{A} = \{(U_j, \kappa_j)\}$ be an atlas on M, let $\{\varphi_j\}$ be a partition of unity subordinate to the cover $\{U_j\}$ and let $\psi_j \in C_0^\infty(U_j)$ be functions which are equal to 1 on supp φ_j. We have

$$A = \sum_j \varphi_j A \psi_j + \sum_j \varphi_j A(1 - \psi_j),$$

where the terms in the second sum belong to $OP^{-\infty}(M)$ by quasi-locality, so it remains to prove that the terms in the first sum belong to $OP^m(M)$. By Definition 8.7 we have $T(\varphi_j A \psi_j, \kappa_j) \in OS^m(\mathbb{R}^n)$. Then using Theorem 7.28 we get

$$T(\varphi_j A \psi_j, \kappa_j) \in OP^m(\mathbb{R}^n),$$

whence Theorem 8.6 implies that $\varphi_j A \psi_j \in OP^m(M)$. (The support property in Theorem 8.6 is obtained from (5), i.e. supp $\varphi_j A \psi_j \subset$ supp $\varphi_j \cup$ supp $\psi_j \subset U_j$.)

Recall that differential operators are local, i.e. if A is a differential operator then supp $Au \subset$ supp u for all $u \in W_2^l(M)$. Pseudo-differential operators have a quasi-local property that we saw in Proposition 8.8. For $u \in W_2^l(M)$ we define the *singular support* of u to be the complement of the largest open set on which u is C^∞,

$$\text{sing supp}\, u = M\backslash\{x \in M;\, u \in C^\infty(U) \text{ for some neighbourhood } U \text{ of } x\}$$

Then we have the following corollary of Proposition 8.8 and Theorem 8.13.

Corollary 8.14 *Let $A \in OS^m(M)$. Then for all $u \in W_2^l(M)$*

$$\text{sing supp}\, Au \subset \text{sing supp}\, u$$

Proof Let $x \notin \text{sing supp}\, u$. Then there exists a neighbourhood U of x such that $u \in C^\infty(U)$. Take $\varphi,\, \psi \in C_0^\infty(U)$ such that $\psi = 1$ on a neighbourhood of supp φ, and write

$$\varphi Au = \varphi A(\psi u) + \varphi A(1 - \psi)u$$

The first term on the right-hand side belongs to $C^\infty(M)$ since $\psi u \in C_0^\infty(U) \subset \bigcap_l W_2^l(M)$. Also, the second term belongs to $C^\infty(M)$ since the supports of φ and $1 - \psi$ are disjoint. Hence $\varphi Au \in C^\infty(M)$, and by taking $\varphi = 1$ near x, it follows that Au is C^∞ near x. Hence $x \notin \text{sing supp}\, Au$.

The next result is a continuation of Proposition 8.3. We let

$$OS^{-\infty}(M) = \bigcap_m OS^m(M).$$

Proposition 8.15 *We have $OP^{-\infty}(M) = OC^\infty(M) = OS^{-\infty}(M)$.*

Proof The proof of the first equality has already been indicated in §8.1. Obviously $OS^{-\infty}(M) \subset OP^{-\infty}(M)$ by Theorem 8.13, so it remains to prove $OC^\infty(M) \subset OS^{-\infty}(M)$. Let $A \in OC^\infty(M)$. Clearly A satisfies condition (ii) of Theorem 8.9. Now, let (U, κ) be a chart on M, and let $\varphi,\, \psi \in C_0^\infty(U)$. Then $\varphi A\psi \in OC^\infty(M)$ has compactly supported C^∞ kernel K with

$$\text{supp}\, K \subset U \times U.$$

The push-forward operator $(\varphi A\psi)_\kappa$ is also an integral operator with a compactly supported C^∞ kernel, $\tilde{K}$, with

$$\text{supp}\, \tilde{K} \subset \kappa(U) \times \kappa(U) \subset \mathbb{R}^n \times \mathbb{R}^n$$

Hence condition (*) of Theorem 7.6 is satisfied and we obtain

$$(\varphi A\psi)_\kappa \in OS^{-\infty}(\mathbb{R}^n),$$

which establishes condition (i) of Theorem 8.9 for any $m \in \mathbb{R}$.

We end this section by defining the notion of asymptotic expansions of pseudo-differential *operators* on a manifold M. Let $A_d \in OS^{m_d}(M)$, $d = 1, 2, \ldots$ with $m_{d-1} > m_d \to -\infty$ as $d \to \infty$ and let $A \in OS^{m_1}(M)$. We will write

$$A \sim \sum_{d=1}^{\infty} A_d$$

if for any integer $r \geqslant 2$ we have

$$A - \sum_{d=1}^{r-1} A_d \in OS^{m_r}(M) \tag{13}$$

Proposition 8.16 *For any given sequence* $A_d \in OS^{m_d}(M)$, $d = 1, 2, \dots$ *with* $m_{d-1} > m_d \to -\infty$ *as* $d \to \infty$, *there exists* $A \in OS^{m_1}(M)$ *such that* $A \sim \sum_{d=1}^{\infty} A_d$. *The p.d.o. A is uniquely determined modulo* $OS^{-\infty}(M)$.

Proof The uniqueness follows immediately from (13). To prove existence let $\mathfrak{A}' = \{(W_j, \kappa_j)\}$ be the atlas on M that was chosen just before Theorem 8.9′, and let $\{\varphi_j\}$ be a subordinate partition of unity. Also choose functions $\psi_j \in C_0^\infty(U_j)$ such that $\psi_j = 1$ on supp φ_j.

The condition (i′) in Theorem 8.10 for each operator $A_d \in OS^{m_d}(M)$ gives us the following: For every j there are operators $\tilde{A}_{d,j} \in OS^{m_d}(\mathbb{R}^n)$ such that

$$\varphi A_d \psi = \varphi(\tilde{A}_{d,j}(x, D))_{\kappa_j^{-1}} \psi \qquad \text{for all } \varphi, \psi \in C_0^\infty(W_j). \tag{14}$$

By Proposition 7.7, there exists a p.d.o. $\tilde{A}_j \in OS^{m_1}(\mathbb{R}^n)$ such that $\tilde{A}_j - \sum_{d=1}^{r-1} \tilde{A}_{d,j} \in OS^{m_r}(\mathbb{R}^n)$, $r = 2, 3, \dots$, and it follows from (14) that

$$\varphi_j(\tilde{A}_j)_{\kappa_j^{-1}} \psi_j - \varphi_j\left(\sum_{d=1}^{r-1} A_d\right)\psi_j \in OS^{m_r}(M). \tag{15}$$

Now, define

$$S = \sum_j \varphi_j(\tilde{A}_j)_{\kappa_j^{-1}} \psi_j,$$

and note that $A \in OS^{m_1}(M)$ by Lemma 8.11 (see Remark 8.12). Also, $\varphi_j(\sum_{d=1}^{r-1} A_d)(1 - \psi_j) \in OS^{-\infty}(M)$ because p.d.o.'s are quasi-local. Summing over j, using the fact that $\sum_j \varphi_j = 1$, we see that (15) implies that

$$A - \sum_{d=1}^{r-1} A_d = \sum_j \left[\varphi_j(\tilde{A}_j)_{\kappa_j^{-1}} \psi_j - \varphi_j\left(\sum_{d=1}^{r-1} A_d\right)\psi_j - \varphi_j\left(\sum_{d=1}^{r-1} A_d\right)(1 - \psi_j)\right]$$

belongs to $OS^{m_r}(M)$ for $r = 2, 3, \dots$.

8.3 Main symbols and p.d.o. algebra

First we discuss symbols on open sets in $\mathbb{R}^n$. If $X \subset \mathbb{R}^n$ is open we define $S^m_{\mathrm{loc}}(X \times \mathbb{R}^n)$ as the set of all $a \in C^\infty(X \times \mathbb{R}^n)$ such that $\varphi(x) \cdot a(x, \xi) \in S^m(\mathbb{R}^n \times \mathbb{R}^n)$ for every $\varphi \in C_0^\infty(X)$. This means that for every compact set $K \subset X$ we have

$$|D_\xi^\alpha D_x^\beta a(x, \xi)| \leqslant C_{\alpha\beta K}(1 + |\xi|)^{m - |\alpha|}, \qquad x \in K,\ \xi \in \mathbb{R}^n$$

for some constant $C_{\alpha\beta K}$. Note that if $X = \mathbb{R}^n$ then $S^m \subset S^m_{\mathrm{loc}}$ but equality does *not* hold.

We can define an operator $a(x, D)$: $\mathscr{S} \to C^\infty(X)$ by the formula,

$$a(x, D)u = (2\pi)^{-n} \int e^{i(x,\xi)}\, a(x, \xi)\hat{u}(\xi)\, d\xi. \tag{16}$$

In fact, if $\varphi \in C_0^\infty(X)$ then $\varphi(x)a(x, \xi) \in S^m$ and the corresponding operator $\varphi(x) \cdot a(x, D)$: $\mathscr{S} \to \mathscr{S}$ has already been discussed in Chapter 7. Restricting $a(x, D)$ to $C_0^\infty(X) \subset \mathscr{S}$ we obtain an operator

$$a(x, D)\colon C_0^\infty(X) \to C^\infty(X)$$

defined by (16).

The following lemma will be needed shortly. It says that symbols behave well under a change of variables.

Lemma 8.17 *Let X_1 and X_2 be open sets in $\mathbb{R}^n$ and let $\phi\colon X_1 \to X_2$ and $\Phi\colon X_1 \to GL(n, \mathbb{R})$ be C^∞ maps. Then*

$$a_1(x, \xi) = a_2(\phi(x), \Phi(x)\xi)$$

is in $S^m_{\mathrm{loc}}(X_1 \times \mathbb{R}^n)$ for every $a_2 \in S^m_{\mathrm{loc}}(X_2 \times \mathbb{R}^n)$.

Proof The functions ϕ and Φ and all their derivatives are bounded on compact subsets $K \subset X_1$ and

$$c|\xi| \leqslant |\Phi(x)\xi| \leqslant C|\xi|$$

when $x \in K$ (where $c > 0$ and $C > 0$ depend on K). The lemma now follows easily by repeated use of the chain rule.

For a pseudo-differential operator in $OS^m(\mathbb{R}^n)$, it is clear from (31) in Chapter 7 that the symbol in S^m is uniquely determined by the operator. But on open sets X in $\mathbb{R}^n$ the symbol is only determined modulo $S^{-\infty}_{\mathrm{loc}}(X \times \mathbb{R}^n)$. This is the subject of the next proposition which we take from [Hö 3].

Proposition 8.18 *Let $X \subset \mathbb{R}^n$ be an open set. If $A\colon C_0^\infty(X) \to C^\infty(X)$ is a linear map such that for all $\varphi, \psi \in C_0^\infty(X)$ the operator $\mathscr{S} \to \mathscr{S}$ given by*

$$u \mapsto \varphi A \psi u$$

is in OS^m, then one can find $a \in S^m_{\mathrm{loc}}(X \times \mathbb{R}^n)$ such that

$$A = a(x, D) + A_0$$

where the kernel of A_0 is in $C^\infty(X \times X)$, and a is uniquely determined modulo $S^{-\infty}_{\mathrm{loc}}(X \times \mathbb{R}^n)$.

Remark We call the equivalence class $[a] \in S^m_{\mathrm{loc}}/S^{-\infty}_{\mathrm{loc}}$ the complete symbol of the operator A.

Proof Let $\sum \psi_j = 1$, $\psi_j \in C_0^\infty(X)$, be a locally finite partition of unity in X. Then $\psi_j A \psi_k u = \tilde{A}_{jk}(x, D)u$, $u \in \mathscr{S}$, where $\tilde{A}_{jk} \in S^m$ and $\tilde{A}_{jk}(x, \xi) = 0$ when $x \notin \operatorname{supp} \psi_j$. Set

$$a(x, \xi) = \sum{}' \tilde{A}_{jk}(x, \xi)$$

where we sum over all j and k for which $\operatorname{supp} \psi_j \cap \operatorname{supp} \psi_k \neq \varnothing$. The sum is locally finite since any compact subset of X meets only finitely many $\operatorname{supp} \psi_j$ and they in turn meet only finitely many $\operatorname{supp} \psi_k$. Hence

$a \in S^m_{\text{loc}}(X \times \mathbb{R}^n)$. If $K(x, y)$ is the Schwartz kernel of A then the kernel of $A - a(x, D)$ is the sum,

$$\sum{}'' \psi_j(x)K(x, y)\psi_k(y),$$

taken over the indices for which $\operatorname{supp} \psi_j \cap \operatorname{supp} \psi_k = \varnothing$. It is in $C^\infty(X \times X)$ since the sum is locally finite and the terms are in C^∞ by Corollary 7.14.

To prove the uniqueness, we must show that if $a \in S^m_{\text{loc}}(X \times \mathbb{R}^n)$ and the kernel of $a(x, D)$ is in $C^\infty(X \times X)$ then $a \in S^{-\infty}_{\text{loc}}(X \times \mathbb{R}^n)$. Taking $\varphi, \psi \in C_0^\infty(X)$ with $\varphi = \psi = 1$ near any given point in X, the operator $\tilde{B}(x, D)u = \varphi a(x, D)\psi u$ belongs to OS^m and its Schwartz kernel is in $C_0^\infty(\mathbb{R}^n \times \mathbb{R}^n)$. Hence the symbol

$$\tilde{B}(x, \xi) = e^{-i(x, \xi)}\, \tilde{B}(x, D)\, e^{i(x, \xi)}$$

is rapidly decreasing when $\xi \to \infty$ (see the proof of Theorem 7.6), so in fact $\tilde{B}(x, D) \in OS^{-\infty}$. Choosing ψ equal to 1 in a neighbourhood of $\operatorname{supp} \varphi$, it follows that

$$\varphi a(x, D) = \varphi a(x, D)\psi + \varphi a(x, D)(1 - \psi)$$

is in $OS^{-\infty}$, by virtue of Corollary 7.14.

We now define the symbol space $S^m_{\text{loc}}(T^*(X))$ when X is an open set in the manifold M. Let (U, κ) be a chart on M with $U \subset X$. For each $x \in U$, consider the differential $d\kappa(x)\colon T_x(M) \to \mathbb{R}^n$ (see §5.3) and the transpose map

$${}^t d\kappa(x)\colon \mathbb{R}^n \to T_x^*(M).$$

Recall that $T^*(U) = \kappa(U) \times \mathbb{R}^n$. The push-forward of a function $a \in C^\infty(T^*(X))$ is the function in $C^\infty(T^*(U))$ defined by

$$a(\kappa^{-1}(y), {}^t d\kappa(\kappa^{-1}(y))\cdot\eta), \qquad y \in \kappa(U), \eta \in \mathbb{R}^n. \tag{17}$$

More formally, we define the map $t(\cdot, \kappa)\colon C^\infty(T^*(X)) \to C^\infty(T^*(U))$ by

$$t(a, \kappa)(y, \eta) = a|_{T^*U}(\kappa^{-1}(y), {}^t d\kappa(\kappa^{-1}(y)))\cdot\eta).$$

Then $t(a, \kappa) \in C^\infty(T^*(U))$ is called the *transfer* or *push-forward* of a function $a \in C^\infty(T^*(X))$ by the coordinate map κ.

Definition 8.19 *Let X be an open set in M. We define $S^m_{\text{loc}}(T^*(X))$ to be the set of all $a \in C^\infty(T^*(X))$ such that the push-forward function* (17) *belongs to S^m_{loc}, that is,*

$$t(a, \kappa) \in S^m_{\text{loc}}(\kappa(U) \times \mathbb{R}^n),$$

for every chart (U, κ) on X. If $X = M$, we write $S^m(T^*(M))$ instead of $S^m_{\text{loc}}(T^*(M))$.

By virtue of Lemma 8.17, it is enough to require that this condition hold for all charts (U, κ) in an atlas on X. In particular, the definition of $S^m_{\text{loc}}(T^*(X))$ agrees with the earlier definition when X is an open set in $\mathbb{R}^n$, on which there is the atlas consisting of the single chart $(X, \text{identity})$.

If $A \in OS^m(M)$ then the push-forward operator $A_\kappa\colon C_0^\infty(\kappa(U)) \to C^\infty(\kappa(U))$ in $\mathbb{R}^n$ satisfies the conditions of Proposition 8.18 with $X = \kappa(U)$ (see (8) and Definition 8.7), so it defines a complete symbol in $S^m_{\text{loc}}/S^{-\infty}_{\text{loc}}$. Letting $\tilde{a}_\kappa \in S^m_{\text{loc}}(\kappa(U) \times \mathbb{R}^n)$ denote a representative (see (20) below), and then pulling it back to U,

$$a_\kappa(x, \xi) = \tilde{a}_\kappa(\kappa(x), {}^t d\kappa(x)^{-1} \cdot \xi), \qquad x \in U, \xi \in T^*_x M \tag{18}$$

we get a function $a_\kappa \in S^m_{\text{loc}}(T^*(U))$ which is called a *complete symbol of A on the coordinate chart* (U, κ), and it is unique modulo $S^{-\infty}_{\text{loc}}(T^*(U))$. It is important to take note of an inconsistency in our use of notation: A_κ is the push-forward to $\mathbb{R}^n$ of the operator A, but $a_\kappa = t(\tilde{a}_\kappa, \kappa^{-1})$ is the pull-back to $U \subset M$ of the complete symbol $\tilde{a}_\kappa$ of A_κ, and therefore involves a "double transfer".

Lemma 8.20 *Let $A \in OS^m(M)$. Let (U, κ) and (U', κ') be charts on M and let $a_\kappa \in S^m_{\text{loc}}(T^*(U))$ and $a_{\kappa'} \in S^m_{\text{loc}}(T^*(U'))$ be the corresponding complete symbols of A on each chart, as defined above. If $U \cap U' \neq \varnothing$ then*

$$a_\kappa - a_{\kappa'} \in S^{m-1}_{\text{loc}}(T^*(U \cap U')). \tag{19}$$

Proof By definition, we have

$$A_\kappa = \tilde{a}_\kappa(y, D) + A_0, \qquad A_{\kappa'} = \tilde{a}_{\kappa'}(y, D) + A'_0 \tag{20}$$

where the kernel of A_0 is in $C^\infty(\kappa(U) \times \kappa(U))$ and the kernel of A'_0 is in $C^\infty(\kappa'(U') \times \kappa'(U'))$. Let $\tilde{\varphi} \in C^\infty_0(\kappa(U \cap U'))$ and choose $\tilde{\psi} \in C^\infty_0(\kappa(U \cap U'))$ such that $\tilde{\psi} = 1$ on supp $\tilde{\varphi}$. The operator

$$\tilde{\varphi} A_\kappa(y, D) \tilde{\psi} \in OS^m$$

has a kernel which is compactly supported in $\kappa(U \cap U') \times \kappa(U \cap U')$. By virtue of Theorem 7.17, with the coordinate transformation $\omega = \kappa' \circ \kappa^{-1}$, it follows that $(\tilde{\varphi} A_\kappa \tilde{\psi})_\omega \in OS^m$ and

$$(\tilde{\varphi} A_\kappa \tilde{\psi})(\omega^{-1}(y), {}^t d\omega(\omega^{-1}(y)) \cdot \eta) - (\tilde{\varphi} A_\kappa \tilde{\psi})_\omega(y, \eta) \in S^{m-1},$$

for all $y \in \kappa'(U \cap U')$, $\eta \in \mathbb{R}^n$. Then, by Corollary 7.13 (i.e. by commuting $\tilde{\varphi} A_\kappa$ and $\tilde{\psi}$) and the fact that $(\tilde{\varphi} A_\kappa \tilde{\psi})_\omega = \tilde{\varphi} \circ \omega^{-1} A_{\kappa'} \tilde{\psi} \circ \omega^{-1}$ and using (20), we obtain

$$\tilde{\varphi}(\omega^{-1}(y)) \cdot \tilde{a}_\kappa(\omega^{-1}(y), {}^t d\omega(\omega^{-1}(y)) \cdot \eta) - \tilde{\varphi}(\omega^{-1}(y)) \cdot \tilde{a}_{\kappa'}(y, \eta) \in S^{m-1}.$$

Substituting $\kappa'^{-1}(y) = x$, ${}^t d\kappa'(x) \cdot \eta = \xi$ and $\varphi = \tilde{\varphi} \circ \kappa$, and then using Lemma 8.17, we find that

$$\varphi(x) \cdot (\tilde{a}_\kappa(\kappa(x), {}^t d\kappa(x)^{-1} \xi) - \tilde{a}_{\kappa'}(\kappa'(x), {}^t d\kappa'(x)^{-1} \xi)) \in S^{m-1},$$

$$x \in U \cap U', \xi \in T^*_x M \tag{21}$$

for every $\varphi \in C^\infty_0(U \cap U')$. By definition, a_κ and $a_{\kappa'}$ are the pull-backs of $\tilde{a}_\kappa$ and $\tilde{a}_{\kappa'}$ by the maps κ and κ', respectively, and therefore (19) is a consequence of (21).

Our aim now is to define the symbol of a p.d.o. on a manifold, and to investigate its properties. Lemma 8.20 indicates that it is possible to define such symbols only modulo $S^{m-1}_{\text{loc}}(T^*(M))$.

Definition 8.21 *Let $A \in OS^m(M)$. Then $a \in S^m(T^*(M))$ is said to be a main symbol of A if*

$$a - a_\kappa \in S^{m-1}_{\text{loc}}(T^*(U)) \qquad \text{for all charts } (U, \kappa) \text{ on } M, \tag{22}$$

where a_κ is the complete symbol of A on the chart κ (see (18)).

Note: *By Definition* 8.19, *the condition* (22) *just means that the push-forward of $\varphi \cdot (a - a_\kappa)$ belongs to S^{m-1} for every $\varphi \in C_0^\infty(U)$, that is,*

$$\tilde{\varphi}(y) \cdot (a(\kappa^{-1}(y), {}^t d\kappa(\kappa^{-1}(y)) \cdot \eta) - \tilde{a}_\kappa(y, \eta)) \in S^{m-1}(\mathbb{R}^n \times \mathbb{R}^n)$$

for every $\tilde{\varphi} \in C_0^\infty(\kappa(U))$.

When checking that a is a main symbol of A, it suffices to verify (22) for all charts in an atlas on M. Indeed, let $\{\varphi_j\}$ be a partition of unity subordinate to a given atlas $\mathfrak{A} = \{(U_j, \kappa_j)\}$, i.e. supp $\varphi_j \subset U_j$, and let (U, κ) be any chart for M. Then, for every $\varphi \in C_0^\infty(U)$, the push-forward of

$$\varphi(a - a_\kappa) = \sum \varphi\varphi_j(a - a_{\kappa_j}) + \sum \varphi\varphi_j(a_{\kappa_j} - a_\kappa)$$

by the map κ is an element of $S^{m-1}(\mathbb{R}^n \times \mathbb{R}^n)$. The terms in the first sum are in S^{m-1} by hypothesis, while the terms in the second sum are in S^{m-1} by Lemma 8.20.

It remains to be shown in Theorem 8.23 that every p.d.o. $A \in OS^m(M)$ has a main symbol. But three facts are clear at this point: (1) If a and a' are both main symbols of A then $a - a' \in S^{m-1}(T^*(M))$; (2) If $A \in OS^{m-1}(M)$, then the 0 function is a main symbol of A (when it is regarded as an operator in $OS^m(M)$); and (3) If A has main symbol a, then any other function $a' \in C^\infty(T^*(M))$ such that $a'(x, \xi) = a(x, \xi)$ when $(x, \xi) \in T^*(M) \backslash K$, K a compact set in $T^*(M)$, is also a main symbol for A.

Lemma 8.22 *Let (U, κ) be a chart on M, and let $\tilde{A}(y, D) \in OS^m(\mathbb{R}^n)$ have compact support in $\kappa(U)$. Also, let $\tilde{A}(y, \eta)$ be the symbol of $\tilde{A}$. Then the pull-back operator*

$$A = T(\tilde{A}, \kappa^{-1}) = (\tilde{A}(y, D))_{\kappa^{-1}} \in OS^m(M)$$

(see Lemma 8.11) *has a main symbol $a \in C^\infty(T^*(M))$ given by*

$$a(x, \xi) = \tilde{A}(\kappa(x), {}^t d\kappa(x)^{-1}\xi) \quad \text{if } x \in U, \qquad a(x, \xi) = 0, \quad \text{if } x \in M \backslash U, \tag{23}$$

i.e. the main symbol of A is the pull-back, $t(\tilde{A}, \kappa^{-1})$, of the symbol $\tilde{A}$ to $T^*(U)$, extended by 0 *off $T^*(U)$.*

Proof Formula 7.31 for the symbol of a p.d.o. in $\mathbb{R}^n$ shows that $\tilde{A}(y, \eta) = 0$ when $y \in \kappa(U) \backslash \text{supp}\, \tilde{A}$ (see condition (1) of Definition 8.4). Hence

$$a(x, \xi) = 0 \qquad \text{when } x \in U \backslash \text{supp}\, A, \tag{24}$$

and it extends by zero to a C^∞ function on $T^*(M)$.

As mentioned above, to prove that a is a main symbol of A it suffices to verify (22) of Definition 8.21 for all charts in an atlas on M. Let $\mathfrak{A} = \{(U_j, \kappa_j)\}$ be the atlas that was used in the proof of Lemma 8.11. For $j \neq j_0$ we can choose $\tilde{a}_{\kappa_j} = 0$ since $A_{\kappa_j} = 0$ (supp A does not meet U_j); therefore $a_{\kappa_j} = 0$ on $T^*(U_j)$, and by (24) we also have $a = 0$ on $T^*(U_j)$. On the other hand, when $j = j_0$, we choose $\tilde{a}_\kappa = \tilde{A}$ and then it is evident by definition, see (18), that $a = a_\kappa$ on $T^*(U)$. Hence $a = a_{\kappa_j}$ for all j, and a satisfies (22) for all charts in this atlas.

Suppose that we are given an operator $A \in OS^m(M)$ with supp $A \subset U$. Then the push-forward operator $T(A, \kappa) = \tilde{A}: C_0^\infty(\mathbb{R}^n) \to C_0^\infty(\mathbb{R}^n)$ belongs to $OS^m(\mathbb{R}^n)$. In fact, this is immediate by Definition 8.7 if we observe that $A = \varphi A \varphi$ when $\varphi \in C_0^\infty(U)$ is equal to 1 on supp A. Therefore, Lemma 8.22 can be phrased in a different form:

Lemma 8.22′ *Let (U, κ) be a chart on M, and let $A \in OS^m(M)$ have support in U. Then A has a main symbol $a \in C^\infty(T^*(M))$ given by*

$$a(x, \xi) = \tilde{A}(\kappa(x), {}^t d\kappa(x)^{-1}\xi) \quad \text{if } x \in U, \qquad a(x, \xi) = 0, \quad \text{if } x \in M \setminus U,$$

where $\tilde{A}(y, \eta)$ is the symbol of the push-forward operator $T(A, \kappa) = \tilde{A}: C_0^\infty(\mathbb{R}^n) \to C_0^\infty(\mathbb{R}^n)$.

Lemmas 8.22 and 8.22′ are the basis for computing the main symbol of a pseudo-differential operator which is defined via a partition of unity.

Theorem 8.23 *Every $A \in OS^m(M)$ has a main symbol $a \in S^m(T^*(M))$ (i.e. satisfying Definition* 8.21). *We have a main symbol isomorphism*

$$p: OS^m(M)/OS^{m-1}(M) \simeq S^m(T^*(M))/S^{m-1}(T^*(M))$$

which is the linear map defined by $[A] \mapsto [a]$. We write also $pA = a$, to avoid being too meticulous.

Proof Choose an atlas $\{(U_j, \kappa_j)\}_{j=1}^N$ on M, and a subordinate partition of unity $\{\varphi_j\}$. Let $\chi_j \in C_0^\infty(U_j)$ be equal to 1 on supp φ_j, then write

$$A = \sum \varphi_j A \chi_j + \sum \varphi_j A(1 - \chi_j).$$

Because of quasi-locality the second sum is in $OS^{-\infty}(M)$ and thus has a main symbol of 0. By Lemma 8.22′, the operator $\varphi_j A \chi_j$, with support in U_j, has a main symbol $a_j \in S^m(T^*(M))$. Then it is easily verified that $a := \sum a_j \in S^m(T^*(M))$ is a main symbol of $\sum \varphi_j A \chi_j$, and hence it is a main symbol for A.

The main symbol map $[A] \mapsto [a]$ is well-defined because $a \in S^{m-1}(T^*(M))$ when $A \in OS^{m-1}(M)$. To prove injectivity, we must show that if $A \in OS^m(M)$ and has a main symbol $a \in S^{m-1}(T^*(M))$, then $A \in OS^{m-1}(M)$. But from (20) we have for every $\varphi, \psi \in C_0^\infty(U)$,

$$\begin{aligned}(\varphi A \psi)_\kappa &= \varphi_\kappa A_\kappa \psi_\kappa \\ &= \varphi_\kappa \tilde{a}_\kappa(y, D)\psi_\kappa + \varphi_\kappa A_0 \psi_\kappa; \end{aligned} \tag{25}$$

the second term belongs to $OS^{-\infty}(\mathbb{R}^n)$ while the first term belongs to $OS^{m-1}(\mathbb{R}^n)$ if $a \in S^{m-1}(T^*(M))$. Hence $A \in OS^{m-1}(M)$ by Definition 8.7. Finally, to prove surjectivity let $a \in S^m(T^*(M))$. Then choose $\psi_j \in C_0^\infty(U_j)$

such that $\sum \psi_j^2 = 1$ and set

$$A_j u = \psi_j (\tilde{a}_j(x, D))_{\kappa_j^{-1}} \psi_j u, \qquad u \in C^\infty(M),$$

where $\tilde{a}_j$ is the push-forward of a to $\kappa_j(U_j) \times \mathbb{R}^n$, see (17). By Lemma 8.11 it follows that $A_j \in OS^m(M)$ and then Lemma 8.22 shows that A_j has a main symbol $\psi_j^2 a$, whence $A = \sum A_j$ has a main symbol $\sum \psi_j^2 a = a$.

Our next aim is to develop the p.d.o. algebra on M, i.e. the composition and adjoints of operators. If $A \in OS^m(M)$, the adjoint operator A^* is taken with respect to some inner product $(,)_M$ on $L_2(M)$, see (2) in §8.1. The definition of A^* depends on the choice of inner product, but (b) of the following theorem shows that its main symbol does *not* depend on this choice.

Let $\{U_j, \kappa_j\}_{j=1,\dots,N}$ be an atlas on M and $\{\varphi_j\}$ a subordinate partition of unity. Also let $d\mu = g_{U_j}(x)\, dx$ be the density on U_j. We write down for later use the formulas

$$\int_M u\bar{v}\, d\mu = \sum_{j=1}^N \int_{U_j} \varphi_j(x) u(x) \overline{v(x)} g_{U_j}(x)\, dx, \tag{26}$$

and when u and v have their support in U_j,

$$\int_M u\bar{v}\, d\mu = \int_{U_j} u(x)\overline{v(x)} g_{U_j}(x)\, dx, \tag{27}$$

where dx is of course Lebesgue measure. (To simplify the notation we have identified points in U_i with their coordinates in $\mathbb{R}^n$.)

Theorem 8.24

(*a*) *If* $A \in OS^{m_1}(M)$ *and* $B \in OS^{m_2}(M)$ *then* $A \circ B \in OS^{m_1+m_2}(M)$. *Further, if* a *and* b *are main symbols of A and B, respectively, then* ab *is a main symbol of AB*:

$$\not p(A \circ B) = \not p A \cdot \not p B$$

(*b*) *If* $A \in OS^m(M)$ *then the adjoint operator,* A^*, *is once more a p.d.o. belonging to* $OS^m(M)$. *If* a *is a main symbol of A, then* $\bar{a}$ *is a main symbol of* A^*:

$$\not p A^* = \overline{\not p A}$$

Proof (a) Let $\varphi, \psi \in C_0^\infty(U)$ where (U, κ) is any coordinate chart on M, and choose $\chi \in C_0^\infty(U)$ equal to 1 in a neighbourhood of supp ψ. Then

$$\varphi AB\psi = (\varphi A\chi)(\chi B\psi) + \varphi A(1 - \chi^2)B\psi.$$

The second term on the right-hand side has a C^∞ kernel. In fact $(1 - \chi^2)B\psi \in OC^\infty(M) = OP^{-\infty}(M)$ since the supports of $1 - \chi^2$ and ψ are disjoint. It follows that $\varphi A(1 - \chi^2)B\psi \in OP^{-\infty}(M)$, so it has a C^∞ kernel by Proposition 8.3, and therefore belongs to $OS^{-\infty}(M)$ by Proposition 8.15. As for the first term on the right-hand side, the push-forward operator

$$(\varphi A\chi)_\kappa \circ (\chi B\psi)_\kappa$$

belongs to $OS^{m_1+m_2}(\mathbb{R}^n)$ by Theorem 7.12. The condition of Definition 8.7 therefore holds with $m = m_1 + m_2$, and we have proved that $A \circ B \in OS^{m_1+m_2}(M)$.

Next we prove that the main symbol of AB is the product of those of A and of B. It suffices to assume that A is the operator $\sum A_j$ with main symbol a that was constructed in the proof of Theorem 8.23, and B is an analogous operator $\sum B_j$, because $A - \sum A_j \in OS^{m_1-1}(M)$ and $B - \sum B_j \in OS^{m_2-1}(M)$, and then

$$AB - \sum A_j \cdot \sum B_k = \left(A - \sum A_j\right)B + \sum A_j\left(B - \sum B_k\right) \in OS^{m_1+m_2-1}(M).$$

Thus, we assume that

$$A = \sum \psi_j(\tilde{a}_j(x, D))_{\kappa_j^{-1}}\psi_j, \qquad B = \sum \psi_k(\tilde{b}_k(x, D))_{\kappa_k^{-1}}\psi_k,$$

where $\tilde{a}_j \in S^{m_1}$ and $\tilde{b}_k \in S^{m_2}$. Then we have

$$AB = \sum_{j,k}\sum \psi_j(\tilde{a}_j(x, D))_{\kappa_j^{-1}}\psi_j \cdot \psi_k(\tilde{b}_k(x, D))_{\kappa_k^{-1}}\psi_k,$$

and the (j, k) term in this sum has main symbol $\psi_j^2\psi_k^2 a \cdot b$ by Lemma 8.22. By linearity of the main symbol map, we obtain that AB has main symbol

$$\sum_{j,k}\sum \psi_j^2\psi_k^2 a \cdot b = a \cdot b \qquad \left(\text{since } \sum \psi_j^2 = 1\right).$$

Now to the proof of (b). Note first that the adjoint A^* exists and belongs to $OP^m(M)$; see §8.1. We will show that A^* satisfies the conditions (i′) and (ii) of Theorem 8.10.

For condition (ii), observe that if $B \in OC^\infty(M)$ with kernel $K(x, y)$, then $B^* \in OC^\infty(M)$ with C^∞ kernel $\overline{K(y, x)}$. Therefore, if φ and ψ have disjoint supports, it follows that $\varphi A^*\psi = (\bar{\psi}A\bar{\varphi})^* \in OC^\infty(M)$.

To verify (i′) we use the atlas $\{(W_j, \kappa_j)\}_{j=1}^N$ introduced just before Theorem 8.10. Let $\tilde{A}_j(x, D) \in OS^m(\mathbb{R}^n)$, $j = 1, \ldots, N$, be any p.d.o.'s such that the representation

$$\varphi A\psi = \varphi(\tilde{A}_j(x, D))_{\kappa_j^{-1}}\psi \tag{28}$$

holds for all $\varphi, \psi \in C_0^\infty(W_j)$, $j = 1, \ldots, N$. We must obtain a similar representation for A^*. To simplify the notation, let us identify points in U_j with their coordinates in $\kappa_j(U_j) \subset \mathbb{R}^n$. Now, if we recall $\bar{W}_j \subset U_j$, we obtain for $\varphi, \psi \in C^\infty(W_j)$ that

$$\begin{aligned}(\varphi A^*\psi u, v)_M = (\psi u, A\bar{\varphi}v)_M &= \int_{W_j} \psi u \cdot \overline{A(\bar{\varphi}v)} g_{W_j}\, dx \qquad \text{(see (27))}\\ &= \int_{\mathbb{R}^n} \psi u \cdot \overline{\tilde{A}_j(x, D)\bar{\varphi}v\tilde{g}_j}\, dx\\ &= \int_{\mathbb{R}^n} \varphi\tilde{g}_j^{-1}\tilde{A}_j^*(x, D)(\tilde{g}_j\psi u) \cdot \bar{v}\tilde{g}_j\, dx\end{aligned}$$

where g_{W_j} and $(g_{W_j})^{-1}$ are extended to functions $\tilde{g}_j$ and $\tilde{g}_j^{-1}$ in $C_b^\infty(\mathbb{R}^n)$, respectively, such that $\tilde{g}_j^{-1}\tilde{g}_j = 1$ on the support of the operator $\tilde{A}_j(x, D)$ (see Definition 8.4 for the definition of support). Hence

$$\varphi A^*\psi = \varphi(\tilde{g}_j^{-1}\tilde{A}_j^*(x, D)\tilde{g}_j)_{\kappa_j^{-1}}\psi \tag{29}$$

for every $\varphi, \psi \in C_0^\infty(W_j)$. In other words, A^* satisfies condition (i′) of Theorem 8.9′ with the role of the p.d.o.'s $\tilde{A}_j$ being played by

$$\tilde{g}_j^{-1}\tilde{A}_j^*(x, D)\tilde{g}_j \in OS^m(\mathbb{R}^n), \qquad j = 1, \ldots, N,$$

where we used Theorems 7.10 and 7.12. (Theorem 7.10 also shows us how to get $\tilde{A}_j^*(x, \xi)$ from $\tilde{A}_j(x, \xi)$.)

It remains to show that $\bar{a}$ is a main symbol of A^*. Let $\{\varphi_j\}$ be a partition of unity subordinate to the cover $\{W_j\}$, and choose functions $\psi_j \in C_0^\infty(W_j)$ which are equal to 1 on supp φ_j. By Lemma 8.22, the operator $\varphi_j A\psi_j$ has a main symbol $a_j \in S^m(T^*(M))$ given by

$$a_j(x, \xi) = \varphi_j(x)\tilde{A}_j(\kappa_j(x), {}^t d\kappa_j(x)^{-1}\xi)\psi_j(x) \qquad (= 0 \text{ when } x \notin W_j)$$

and by virtue of (29) and Theorem 7.10, and Lemma 8.22 once again, $\bar{a}_j$ is a main symbol of $\varphi_j A^*\psi_j$. Hence $a := \sum a_j$ is a main symbol of $A = \sum \varphi_j A\psi_j + \sum \varphi_j A(1 - \psi_j)$, while $\sum \bar{a}_j = \bar{a}$ is a main symbol of $A^* = \sum \varphi_j A^*\psi_j + \sum \varphi_j A^*(1 - \psi_j)$.

Corollary 8.25 *If $A \in OS^{m_1}(M)$ and $B \in OS^{m_2}(M)$ then*

$$AB - BA \in OS^{m_1 + m_2 - 1}(M).$$

Proof The operator $AB - BA \in OS^{m_1 + m_2}(M)$ has a main symbol $ab - ba = 0$, and therefore by the main symbol isomorphism, i.e. Theorem 8.23, we get $AB - BA \in OS^{m_1 + m_2 - 1}(M)$.

The next corollary follows at once from Theorem 8.24.

Corollary 8.26 *If $A \in OS^m(M)$ has main symbol a, then for every φ, $\psi \in C^\infty(M)$, we have $\varphi \circ A \circ \psi \in OS^m(M)$ with main symbol $\varphi(x)\psi(x)a(x, \xi)$, and $(\varphi \circ A \circ \psi)^* = \bar{\psi}A^*\bar{\varphi} \in OS^m(M)$ with main symbol $\bar{\varphi}(x)\bar{\psi}(x)\bar{a}(x, \xi)$.*

It follows that $OS^m(M)$ is a *local space* of operators, i.e. $A \in OS^m(M)$ if and only if $\varphi A\psi \in OS^m(M)$ for every pair of C^∞ functions φ, ψ on M.

The definitions and results of this section extend to matrix p.d.o.'s with the obvious modifications. A matrix pseudo-differential operator is a $p \times q$ matrix,

$$A = [A_{ij}]_{p \times q} \in OS^m(M, p \times q)$$

with entries, $A_{ij} \in OS^m(M)$, that are scalar p.d.o.'s of order m. For any open set X in M, let $S^m_{\text{loc}}(T^*X, p \times q)$ be the set of $p \times q$ matrix functions whose entries belong to $S^m_{\text{loc}}(T^*X)$. Then Definition 8.21 of the main symbol extends to matrix operators. It is easy to see that the matrix operator A has a main symbol

$$a = [a_{ij}] \in S^m(T^*M, p \times q),$$

where $a_{ij} \in S^m(T^*M)$ is a main symbol of A_{ij}. Once again, we have a main symbol isomorphism $\not{h}$ and Theorem 8.24 also holds with obvious modifications. For instance, if $A \in OS^m(M, p \times q)$ has a main symbol a then the adjoint operator $A^* \in OS^m(M, q \times p)$ has main symbol, ${}^h a$, the Hermitian adjoint (conjugate transpose) of a.

8.4 Classic operators on M

An operator $A \in OS^m(M)$ is said to be (-1) *classic*, and we write

$$A \in OClS^m(M),$$

if for any chart (U, κ) on M and for all $\varphi, \psi \in C_0^\infty(U)$ the push-forward

$$(\varphi A \psi)_\kappa \in OS^m(\mathbb{R}^n) \text{ is a } (-1) \text{ classic operator}$$

(Definition 7.21). Naturally it suffices to verify this condition for all charts in an atlas on M (see Theorem 7.17′).

Let $\mathfrak{A} = \{(U_j, \kappa_j)\}_{j=1}^N$ be an atlas on M, and shrink it to obtain an atlas $\mathfrak{A}' = \{(W_j, \kappa_j)\}_{j=1}^N$ with $\bar{W}_j \subset U_j$. Then we have (-1) classic operators

$$\tilde{A}_j(x, D) \in OS^m(\mathbb{R}^n), \qquad j = 1, \dots, N,$$

such that condition (i′) in Theorem 8.10 holds. The operators $\tilde{A}_j$ have a principal symbol, homogeneous of degree m, defined by

$$\pi \tilde{A}_j(y, \eta) = \lim_{t \to \infty} \frac{\tilde{A}_j(y, t\eta)}{t^m} \qquad \text{(see Proposition 7.22)}$$

for $y \in \kappa_j(W_j)$ and $\eta \neq 0$, and if $W_j \cap W_k \neq \varnothing$ then

$$\pi \tilde{A}_k(y, \eta) = \pi \tilde{A}_j(\omega^{-1}(y), {}^t d\omega(\omega^{-1}(y)) \cdot \eta), \qquad \omega = \kappa_k \circ \kappa_j^{-1},$$

for all $y \in \kappa_k(W_j \cap W_k)$ and $\eta \in \mathbb{R}^n \backslash 0$. Consequently, we can define $\pi A \in C^\infty(T^*(M) \backslash 0)$ by letting

$$\pi A(x, \xi) = \pi \tilde{A}_j(\kappa_j(x), {}^t d\kappa_j^{-1}(x) \cdot \xi) \qquad \text{when } x \in W_j,\ \xi \in T_x^* M \backslash 0. \quad (30)$$

This definition does not depend on the choice of atlas because the union of two atlases is once again an atlas.

Theorem 8.27 *Let $A \in OClS^m(M)$ be a (-1) classic operator. Then the function $\pi A \in C^\infty(T^*(M) \backslash 0)$ defined by* (30) *satisfies the following properties*:

(*a*) *πA is positively homogeneous of degree m in ξ,*

$$\pi A(x, c\xi) = c^m \cdot \pi A(x, \xi), \qquad c > 0,$$

(*b*) *for any $\chi \in C^\infty(T^*(M))$ which vanishes near the* 0 *section of $T^*(M)$ and equals* 1 *outside a compact subset of $T^*(M)$, the function $\chi \cdot \pi A$ is a main symbol of A (in the sense of Definition* 8.21).

Moreover, πA is the unique function in $C^\infty(T^*(M)\backslash 0)$ having properties (a) and (b).

Proof The first property (a) is obvious. We prove that $a = \chi\cdot\pi A$ is a main symbol of A and leave the proof of uniqueness as an exercise. From the definition of $\tilde{A}_j$ in the proof of Theorem 8.9′ it is evident that we can choose

$$\tilde{a}_{\kappa_j}(y, \eta) = \tilde{A}_j(y, \eta), \qquad y \in \kappa_j(W_j)$$

as the representative in the symbol class $S^m_{\text{loc}}/S^{-\infty}_{\text{loc}}$ of the operator

$$A_{\kappa_j}\colon C_0^\infty(\kappa_j(W_j)) \to C^\infty(\kappa_j(W_j))$$

on the open set $\kappa_j(W_j) \subset \mathbb{R}^n$. Then a_{κ_j} is defined by (18), and we must show that $a - a_{\kappa_j} \in S^{m-1}_{\text{loc}}(T^*(W_j))$ for all j, that is,

$$\tilde{\varphi}(y)\cdot(a(\kappa_j^{-1}(y), {}^t d\kappa_j(\kappa_j^{-1}(y))\cdot\eta) - \tilde{a}_{\kappa_j}(y, \eta)) \in S^{m-1} \quad \text{(see Definition 8.21)}$$

for all $\tilde{\varphi} \in C_0^\infty(\kappa_j(W_j))$, or, since $a = \chi\cdot\pi A$, what we must show is that

$$\tilde{\varphi}(y)\cdot(\tilde{\chi}(\eta)\cdot\pi\tilde{A}_j(y, \eta) - \tilde{A}_j(y, \eta)) \in S^{m-1},$$

where $\tilde{\chi}$ is a cut-off function (see the definition of πA in (30)). But this follows at once from the definition of (-1) classic symbols in §7.5.

We call πA the principal part (or principal symbol) of A. Clearly $OS^{m-1}(M) \subset OClS^m(M)$, i.e. every p.d.o. of order $m-1$ is a (-1) classic operator of order m (with principal part 0). To be precise then we should denote the principal part of A by $\pi_m A$ but normally we omit the subscript m. If $A \in OClS^m(M)$ then $\pi_m A = 0$ if and only if $A \in OS^{m-1}(M)$.

Let $\text{Symbl}^m(M)$ denote the class of all functions $h(x, \xi) \in C^\infty(T^*(M)\backslash 0)$ which are homogeneous of order m in ξ. The next theorem shows that $\text{Symbl}^m(M)$ is the class of all principal parts of operators in $OClS^m(M)$.

Theorem 8.28 *If $h \in \text{Symbl}^m(M)$ then there exists a (-1) classic operator $A \in OClS^m(M)$ with principal part $\pi A = h$.*

Proof If we multiply h by a function $\chi \in C^\infty(T^*(M))$ which vanishes near the 0 section of $T^*(M)$ and equals 1 outside a compact subset of $T^*(M)$ then we obtain a function

$$a = \chi\cdot h \in C^\infty(T^*(M)).$$

By Theorem 8.23 there exists a p.d.o. $A \in OS^m(M)$ with main symbol $\chi\cdot h$ (in the sense of Definition 8.21). It is not hard to verify that A is a (-1) classic operator and $\pi A = h$.

Corollary 8.29 *The following sequence is exact:*

$$0 \longrightarrow OS^{m-1}(M) \xrightarrow{i} OClS^m(M) \xrightarrow{\pi_m} \text{Symbl}^m(M) \longrightarrow 0$$

Remarks 8.30

(i) If $A', A \in OClS^m(M)$ are p.d.o.'s with the same principal part, $\pi A' = \pi A$, then $A' - A \in OS^{m-1}(M)$ by Theorem 8.23.

(ii) $OS^{-\infty}(M) = OClS^{-\infty}(M)$, i.e. operators of order $-\infty$ are always (-1) classic.

(iii) If $A \sim \sum_{d=1}^{\infty} A_d$ is an asymptotic expansion (see Proposition 8.16) and $m_2 \leqslant m_1 - 1$, then A is (-1) classic if and only if A_1 is (-1) classic, and we have

$$\pi A = \pi A_1.$$

To prove this, we need only observe that relation (13) gives $A - A_1 \in OS^{m_2}(M) \subset OS^{m_1-1}(M)$.

One easily verifies that if A and B are (-1) classic operators then so are the operators $A \circ B$ and A^*. Thus, Theorem 8.24 implies the following:

Theorem 8.31

(*a*) *If $A \in OClS^{m_1}(M)$ and $B \in OClS^{m_2}(M)$ then $A \circ B \in OClS^{m_1+m_2}(M)$ and we have*

$$\pi(A \circ B) = \pi A \cdot \pi B$$

(*b*) *The adjoint of $A \in OClS^m(M)$ is an operator $A^* \in OClS^m(M)$ and we have*

$$\pi(A^*) = \overline{\pi A}$$

8.5 Definitions for operators in vector bundles

The purpose of this section is to generalize the concepts in the introductory section §8.1 to operators acting between sections of vector bundles. Then we define pseudo-differential operators in bundles in §8.6.

Two vector bundles (E_1, π_1, M_1) and (E_2, π_2, M_2) are said to be *isomorphic* if there is a diffeomorphism $\kappa: M_1 \to M_2$ and a smooth map $\chi: E_1 \to E_2$ which is linear and invertible on each fibre and such that $\pi_2 \circ \chi = \kappa \circ \pi_1$, i.e. such that the diagram

$$\begin{array}{ccc} E_1 & \xrightarrow{\chi} & E_2 \\ {\scriptstyle \pi_1}\big\downarrow & & \big\downarrow{\scriptstyle \pi_2} \\ M_1 & \xrightarrow{\kappa} & M_2 \end{array}$$

commutes. We say that χ is an *isomorphism over* κ. There is then a push-forward operation which maps sections $f: M_1 \to E_1$ to sections $\kappa_* f: M_2 \to E_2$, defined by

$$\chi_* f = \chi \circ f \circ \kappa^{-1} \tag{31}$$

(the sections are not assumed to be smooth). Of course, there is also a pull-back operation χ^* defined by $\chi^* f = \chi^{-1} \circ f \circ \kappa$ which maps sections of E_2 to sections of E_1. In the case of trivial bundles $E_1 = M_1 \times \mathbb{C}^p$ and $E_2 = M_2 \times \mathbb{C}^p$ and a smooth map $\kappa: M_1 \to M_2$, the push-forward and pull-back operations become simply

$$\kappa_* f = f \circ \kappa^{-1} \quad \text{and} \quad \kappa^* f = f \circ \kappa.$$

The Sobolev space $W_2^l(M, \mathbb{C}^p)$ for the trivial bundle $M \times \mathbb{C}^p$ is defined as

$$W_2^l(M, \mathbb{C}^p) = \bigoplus_{j=1}^{p} W_2^l(M).$$

To define the Sobolev spaces for a general vector bundle E, we will use the push-forward operation when χ is a local trivialization of E over a chart κ on M. The basic idea is that $u \in W_2^l(M, E)$ provided that whenever $x_1, \ldots, x_n$ are local coordinates on $U \subset M$ and $e_1, \ldots, e_p$ is a basis for E_U then for all $\varphi \in C_0^\infty(U)$ we have

$$\varphi \cdot u = \sum u_i(x) e_i,$$

where $u_i \in W_2^l$ (in the local coordinates). It is then clear that all the definitions given above can be generalized to vector bundles with no essential difficulty.

To formalize this definition (see (32) below), let $\kappa: U \to V$ be a chart for M on which there is a trivialization $\beta: E_U \to U \times \mathbb{C}^p$. Then the vector bundle isomorphism

$$\chi: E_U \to V \times \mathbb{C}^p$$

over κ is defined by $\kappa = (\kappa, l_p) \circ \beta$. The push-forward (31) of a section $f \in C^\infty(U, E_U)$ by the trivialization χ is a section of the trivial bundle $V \times \mathbb{C}^p$, that is, $\chi_* f = \chi \circ f \circ \kappa^{-1} \in C^\infty(V, \mathbb{C}^p)$. There is also a push-forward $\beta_* f = \beta \circ f \in C^\infty(U, \mathbb{C}^p)$, but in this case we omit the asterisk $*$ and denote it simply by βf.

We call (U, κ, χ) a *trivialization-chart* and (U, β) a *trivializing patch* for E. The letter β is meant to suggest a "basis" for the vector bundle, while κ and χ denote "coordinates".

Definition 8.32 *Let $l \geqslant 0$, a non-negative real number. The Sobolev space $W_2^l(M, E)$ is the set of sections $u: M \to E$ such that*

$$\chi_*(\varphi u) = \chi \circ (\varphi \cdot u) \circ \kappa^{-1} \in W_2^l(\mathbb{R}^n, \mathbb{C}^p) \tag{32}$$

for every coordinate map $\kappa: U \to V$ and trivialization $\chi: E_U \to V \times \mathbb{C}^p$ over κ, and for every $\varphi \in C_0^\infty(U)$. The topology on $W_2^l(M, E)$ is defined by the semi-norms

$$u \mapsto \|\chi \circ (\varphi \cdot u) \circ \kappa^{-1}\|_l,$$

i.e. the weakest topology for which the semi-norms are continuous.

Let $\{U_j, \kappa_j\}_{j=1,\ldots,N}$ be an atlas on M and $\chi_j: E_{U_j} \to V_j \times \mathbb{C}^p$ trivializations over κ_j, and choose a subordinate partition of unity $\{\varphi_j\}$. For $u \in W_2^l(M, E)$ we let $u_j = \chi_j \circ (\varphi_j \cdot u) \circ \kappa_j^{-1}$, and we make $W_2^l(M, E)$ into a Hilbert space by defining the scalar product

$$(u, v)_{W_2^l(M,E)} = \sum_{j=1}^{N} (u_j, v_j)_l.$$

It is possible to show that the topology defined by the norm $\|u\|^2 = (u, u)_{W_2^l(M,E)}$ coincides with the topology in Definition 8.32, in other words, two equivalent atlases and two corresponding subordinate partitions of unity yield equivalent norms.

For $l < 0$ we define $W_2^l(M, E)$ as the *anti-dual* space

$$W_2^l(M, E) = (W_2^l(M, E))^*,$$

or, alternatively, we could use Definition 8.32 with distributional sections.

Now we endow E with a C^∞ inner product $(,)_{E_x}$, $x \in M$, making it into a Hermitian bundle, see §5.10. For sections $u, v \in C^\infty(M, E)$ we then define a scalar product by

$$(u, v)_E := \int_M (u(x), v(x))_{E_x} \, d\mu(x), \tag{33}$$

and $L_2(M, E)$ is defined to be the completion of $C^\infty(M, E)$ with respect to this scalar product ($d\mu$ is the invariant volume element on M introduced in §8.1). We have $W_2^0(M, E) \simeq L_2(M, E)$ as before, and the general comments about the scale of Sobolev spaces following Definition 8.1 hold once again.

Let E and F be complex vector bundles over M. Just as in Definition 8.2, we say that a linear mapping $B: C^\infty(M, E) \to C^\infty(M, F)$ is an operator of order m if for each $l \in \mathbb{R}$ it extends to a continuous mapping

$$B: W_2^l(M, E) \to W_2^{l-m}(M, F).$$

The set of all such operators is denoted by $OP^m(E, F)$, and it is a local space of operators.

If $B: C^\infty(M, E) \to C^\infty(M, F)$ is a linear mapping then its formal adjoint B^* is defined by means of the scalar products $(,)_E$ and $(,)_F$,

$$(Bu, v)_E = (u, B^*v)_F, \qquad u \in C^\infty(M, E), v \in C^\infty(M, F)$$

If $B \in OP^m(E, F)$ then we have $B^* \in OP^m(F, E)$. (The existence of B^* is proved as before.)

The set of operators of order $-\infty$ is denoted

$$OP^{-\infty}(E, F) = \bigcap_{m \in \mathbb{R}} OP^m(E, F),$$

that is, B belongs to $OP^{-\infty}(E, F)$ if it extends to a continuous map

$$B: W_2^{l_1}(M, E) \to W_2^{l_2}(M, F),$$

for all $l_1, l_2 \in \mathbb{R}$. We say that $B \in OC^\infty(E, F)$ if it is an integral operator with a kernel K which is a C^∞ section of the bundle $\mathrm{Hom}(\pi_1^{-1}E, \pi_2^{-1}F)$ over $M \times M$, i.e.

$$Bf(x) = \int_M K(y, x) f(y) \, d\mu(y), \qquad K \in C^\infty(M \times M, \mathrm{Hom}(\pi_1^{-1}E, \pi_2^{-1}F)),$$

where $\pi_1: M \times M \to M$ and $\pi_2: M \times M \to M$ are the projections on the first and second components, so that $K(y, x)$ is a linear map from E_y to F_x. Once again the classes $OP^{-\infty}(E, F)$ and $OC^{\infty}(E, F)$ coincide; using local trivializations the generalization of Proposition 8.3 to vector bundles is straightforward.

Given a chart $\kappa: U \to V$ on M and local trivializations $\beta^E: E_U \to U \times \mathbb{C}^p$ and $\beta^F: F_U \to U \times \mathbb{C}^q$ of the bundles E and F, there is a notion of the transfer of an operator $B: C^{\infty}(M, E) \to C^{\infty}(M, F)$ similar to that for scalar operators. As in the discussion preceding Definition 8.32 we let $\chi^E = (\kappa, 1_p) \circ \beta^E$ and $\chi^F = (\kappa, 1_q) \circ \beta^F$ and we call $(U, \kappa, \chi^E, \chi^F)$ a *trivialization-chart* for the pair of bundles E, F. The push-forward operators B_β and B_χ are defined by the diagram

$$\begin{array}{ccc} C_0^{\infty}(U, E_U) & \xrightarrow{\ B\ } & C^{\infty}(U, F_U) \\ \uparrow{\scriptstyle (\beta^E)^{-1}} & & \downarrow{\scriptstyle \beta^F} \\ C_0^{\infty}(U, \mathbb{C}^p) & \xrightarrow{\ B_\chi\ } & C^{\infty}(U, \mathbb{C}^q) \\ \uparrow{\scriptstyle \kappa^*} & & \downarrow{\scriptstyle \kappa^{*-1}} \\ C_0^{\infty}(V, \mathbb{C}^p) & \xrightarrow{\ B_\chi\ } & C^{\infty}(V, \mathbb{C}^q) \end{array} \qquad (34)$$

that is, $B_\beta = \beta^F \circ B \circ (\beta^E)^{-1}$. Since $\chi^F_* = \kappa^{*-1} \circ \beta^F$, and similarly for χ^E_*, we have

$$\begin{aligned} B_\chi &= \chi^F_* \circ B \circ (\chi^E_*)^{-1} \\ &= \kappa^{*-1} \circ B_\beta \circ \kappa^*. \end{aligned}$$

The push-forward by β (resp. the push-forward by χ) takes an operator B acting in the bundles $E_U \to F_U$ to a matrix operator B_β over $U \subset M$ (resp. to a matrix operator B_χ over $V \subset \mathbb{R}^n$).

To be precise, the operator B in the first row of (34) is really $\imath_F \circ B \circ \imath_E$, where $\imath_E$ is the natural embedding $\imath_E: C_0^{\infty}(U, E_U) \to C^{\infty}(M, E)$ and $\imath_F: C^{\infty}(M, F) \to C^{\infty}(U, F_U)$ is the restriction operator.

The support of a linear map $B: C^{\infty}(M, E) \to C^{\infty}(M, F)$ is defined as in Definition 8.4. If the support of B lies in a chart domain U, we have an extended definition of the push-forwards B_β and B_χ.

Definition 8.33 *Let* $(U, \kappa, \chi^E, \chi^F)$ *and* (U, β^E, β^F) *be trivialization-charts for* E, F, *and let* B *be a linear operator*

$$B: C^{\infty}(M, E) \to C^{\infty}(M, F),$$

with supp $B \subset U$. *Also let* $\zeta \in C_0^{\infty}(U)$ *be such that* $\zeta = 1$ *on a neighbourhood of* supp B. *We define the following transfers (or push-forwards):*
(*i*) $B_\beta = T(B, \beta)$: $C^{\infty}(M, \mathbb{C}^p) \to C^{\infty}(M, \mathbb{C}^q)$ *is a matrix operator,*

$$T(B, \beta)f := \beta^F \circ \zeta \cdot B \circ (\beta^E)^{-1} \zeta \cdot f = [B_{t,s}]_{q \times p} f,$$

and we have supp $T(B, \beta) =$ supp $B \subset U$;
(*ii*) $B_\chi = T(B, \chi)$: $C_0^{\infty}(\mathbb{R}^n, \mathbb{C}^p) \to C_0^{\infty}(\mathbb{R}^n, \mathbb{C}^q)$ *is a matrix operator,*

$$T(B, \chi) := T(T(B, \beta), \kappa) = [T(B_{t,s}, \kappa)]_{q \times p},$$

where the transfer $T(\cdot, \kappa)$ *is from Definition* 8.5, *and we have* supp $T(B, \chi) = \kappa(\text{supp } B) \subset V$. *Obviously, we have*

$$\begin{aligned} T(B, \chi)g &= \kappa^{*-1}\beta^F \circ \zeta \cdot B \circ (\beta^E)^{-1}\kappa^*\tilde{\zeta} \cdot g \\ &= \chi_*^F \circ \zeta \cdot B \circ (\chi_*^E)^{-1}\tilde{\zeta} \cdot g, \end{aligned}$$

where $\tilde{\zeta} = \zeta \circ \kappa^{-1}$, *and we see the complete analogy with* (7).

The remarks mentioned after Definition 8.5 hold in this case as well; in particular, we point out once again that if $B = \varphi A \psi$ where $\varphi, \psi \in C_0^\infty(U)$, there is no need to introduce another function ζ.

Theorem 8.6 remains true with obvious changes in the proof.

Theorem 8.34 *Let the assumptions be as in Definition* 8.33. *Then the following are equivalent:*

(*i*) $B \in OP^m(E, F)$ *with* supp $B \subset U$,
(*ii*) $T(B, \beta) \in OP^m(M, q \times p)$ *and* supp $T(B, \beta) =$ supp $B \subset U$,
(*iii*) $T(B, \chi) \in OP^m(\mathbb{R}^n, q \times p)$ *and* supp $T(B, \chi)$ *is a compact subset of* V.

8.6 Pseudo-differential operators in vector bundles

The definition of the transfer of an operator introduced in §8.5 is crucial to the development of p.d.o.'s in vector bundles. Now that we have this definition, we can take over all theorems and all proofs from §§8.2 to 8.4 almost without change.

As usual, let $OS^m(M, q \times p)$ denote the set of $q \times p$ matrix operators with entries in $OS^m(M)$. The next definition stipulates that pseudo-differential operators in bundles be locally of this form.

Definition 8.35 *A linear operator* $A: C^\infty(M, E) \to C^\infty(M, F)$ *belongs to* $OS^m(E, F)$,

$$A \in OS^m(E, F),$$

if and only if for every trivializing patch (U, β^E, β^F) *and for all* $\varphi, \psi \in C_0^\infty(U)$, *the push-forward operator*

$$T(\varphi A \psi, \beta) = \beta^F(\varphi A \psi)\beta^{E-1} \tag{35}$$

belongs to $OS^m(M, q \times p)$.

Since $T(\varphi A \psi, \chi)$ is the κ-push-forward of $T(\varphi A \psi, \beta)$, then using Definition 8.7 we have the equivalent definition:

χ-Definition 8.35 *A linear operator* $A: C^\infty(M, E) \to C^\infty(M, F)$ *is called a pseudo-differential operator of order* $m \in \mathbb{R}$, *and we write* $A \in OS^m(E, F)$, *if for every coordinate patch* $(U, \kappa, \chi^E, \chi^F)$ *trivializing* E *and* F, *and for all* φ, $\psi \in C_0^\infty(U)$, *the push-forward operator*

$$T(\varphi A \psi, \chi) = \kappa^{*-1}\beta^F(\varphi A \psi)\beta^{E-1}\kappa^* \tag{36}$$

belongs to $OS^m(\mathbb{R}^n, q \times p)$.

If we were studying p.d.o.'s in vector bundles *ab initio*, it would be necessary to work with the χ-definition, which involves not just a change of basis β in the fibre but also a coordinate transformation κ on the manifold. We can rely, however, on the work already done in §8.2, so we need only consider β-transfers. We leave it as an exercise for the reader to verify that the results stated below in terms of β-transfers have their counterparts in terms of χ-transfers (and vice versa). As in Proposition 8.8, one proves that from χ-Definition 8.35 it follows that A is quasi-local: *if* $\varphi, \psi \in C^\infty(M)$ *have disjoint support,* $\operatorname{supp}\varphi \cap \operatorname{supp}\psi = \emptyset$, *then*

$$\varphi A \psi \in OP^{-\infty}(E, F).$$

Then we obtain the following theorem, the proof of which follows by substituting κ by χ in the proofs of Theorems 8.9 and 8.10 and then pulling back to the manifold.

Theorem 8.36 *Let* $\mathfrak{A} = \{(U_j, \beta_j^E, \beta_j^F)\}_{j=1}^N$ *be an "atlas" on* M *which trivializes* E *and* F. *A linear operator* $A: C^\infty(M, E) \to C^\infty(M, F)$ *belongs to* $OS^m(E, F)$ *if and only if:*

(*i*) *For any trivializing patch* $(U_j, \beta_j^E, \beta_j^F) \in \mathfrak{A}$ *and for all* $\varphi, \psi \in C_0^\infty(U_j)$ *we have*

$$T(\varphi A \psi, \beta_j) \in OS^m(M, q \times p).$$

(*ii*) *A is quasi-local.*

Let the "atlas" $\mathfrak{A}' = \{(W_j, \beta_j^E, \beta_j^F)\}_{j=1}^N$ *be obtained by shrinking* $\bar{W}_j \subset U_j$; *see Lemma 5.44. We can replace* (*i*) *by the following condition, obtaining thus another equivalent definition:*

(*i'*) *There exist matrix p.d.o.'s* $\tilde{A}_j \in OS^m(M, q \times p)$, $j = 1, \ldots, N$, *such that the following representation holds for all* $\varphi, \psi \in C_0^\infty(W_j)$:

$$T(\varphi A \psi, \beta_j) = \varphi \tilde{A}_j \psi$$

or using pull-backs by β_j^{-1}:

$$\varphi A \psi = \varphi(\tilde{A}_j)_{\beta_j^{-1}} \psi.$$

We can also choose the operators $\tilde{A}_j$ *so that they have compact support in* U_j.

Note that in the statement of Theorem 8.36, we do not assume that the sets U_j are chart domains, although in any application of it we will assume tacitly that the U_j's have been chosen this way.

The following lemma which is the counterpart of Lemma 8.11 shows how to construct p.d.o.'s in bundles.

χ-Lemma 8.37 *Let* $(U, \kappa, \chi^E, \chi^F)$ *be a trivialization-chart on* M, *and let the matrix operator* $\tilde{A}: C_0^\infty(\mathbb{R}^n, \mathbb{C}^p) \to C_0^\infty(\mathbb{R}^n, \mathbb{C}^q)$ *have compact support in* $\kappa(U) = V \subset \mathbb{R}^n$. *Then the pull-back operator* $A: C^\infty(M, E) \to C^\infty(M, F)$ *defined by*

$$A := T(\tilde{A}, \chi^{-1})$$

is in $OS^m(E, F)$ *if and only if* $\tilde{A} \in OS^m(\mathbb{R}^n, q \times p)$.

It is obvious that the corresponding β-Lemma 8.37 is also true: If $\tilde{A}$ has compact support in $U \subset M$, then

$$A = T(\tilde{A}, \beta^{-1}) \in OS^m(E, F)$$

if and only if $\tilde{A} \in OS^m(M, q \times p)$.

The proof of continuity in Sobolev spaces is analogous to Theorem 8.13.

Theorem 8.38 *Let $A \in OS^m(E, F)$. Then for each $l \in \mathbb{R}$ the mapping*

$$A: W_2^l(M, E) \to W_2^{l-m}(M, F)$$

is continuous.

Proof Let $\{\varphi_j\}$ be a partition of unity subordinate to the cover $\{U_j\}$ and let $\psi_j \in C_0^\infty(U_j)$ be functions which are equal to 1 on supp φ_j. We have

$$A = \sum_j \varphi_j A \psi_j + \sum_j \varphi_j A(1 - \psi_j),$$

where the second sum belongs to $OP^{-\infty}(E, F)$ by quasi-locality, so it remains to prove that the first sum belongs to $OP^m(E, F)$. By Definition 8.35(i) we have $T(\varphi_j A \psi_j, \chi_j) \in OS^m(\mathbb{R}^n, q \times p)$. Then using Theorem 7.28 we get

$$T(\varphi_j A \psi_j, \chi_j) \in OP^m(\mathbb{R}^n, q \times p),$$

whence Theorem 8.34 implies that

$$\varphi_j A \psi_j \in OP^m(E, F),$$

which finishes the proof. (The support property in Theorem 8.34 is obtained from (5), i.e. supp $\varphi_j A \psi_j \subset$ supp $\varphi_j \cup$ supp $\psi_j \subset U_j$.)

Now we define the main symbol of an operator $A \in OS^m(E, F)$. Once again, since we can rely on the work done in §8.3 we need only investigate the effect of a change of basis in the fibre, i.e. we look at β-transfers instead of χ-transfers.

Let π_* be the projection $\pi_*: T^*M \to M$, and as usual let $\mathrm{Hom}(\pi_*^{-1}E, \pi_*^{-1}F)$ denote the vector bundle over T^*M whose fibre over $(x, \xi) \in T^*M$ is the vector space of linear transformations $E_x \to F_x$. To obtain a main symbol for $A \in OS^m(E, F)$ the general idea is as follows. We push A forward by a trivializing patch (U, β^E, β^F) to get a matrix operator $A_U \in OS^m(M, q \times p)$, for which we have already defined the main symbol

$$\not{a} A_U \in C^\infty(T^*M, \mathrm{Hom}(\mathbb{C}^p, \mathbb{C}^q))$$

(see the last paragraph of §8.3), and it can now be pulled back by $(\beta^F)^{-1} \circ \cdots \circ \beta^E$ to obtain the section

$$a \in C^\infty(T^*M, \mathrm{Hom}(\pi_*^{-1}E, \pi_*^{-1}F))$$

as a main symbol for A.

To make this idea more precise, some definitions are required. Let (U, β^E, β^F) be a trivializing patch on M. The transfer of

$$a \in C^\infty(T^*M, \mathrm{Hom}(\pi_*^{-1}E, \pi_*^{-1}F))$$

is defined by

$$t(a, \beta)(x, \xi) := \beta_x^F \circ a(x, \xi) \circ (\beta_x^E)^{-1}, \qquad (x, \xi) \in T^*M|_U = T^*U. \quad (37)$$

Then $t(a, \beta) \in C^\infty(T^*M|_U, \mathrm{Hom}(\mathbb{C}^p, \mathbb{C}^q))$, i.e. it is a $q \times p$ matrix function on $T^*M|_U$.

For short we often use the abbreviation $\mathrm{Hom} := \mathrm{Hom}(\pi_*^{-1}E, \pi_*^{-1}F)$.

Definition 8.39 *Let X be an open set in M. We define $S_{\mathrm{loc}}^m(T^*X, \mathrm{Hom})$ to be the set of all $a \in C^\infty(T^*X, \mathrm{Hom})$ such that the transfer matrix* (37) *belongs to S_{loc}^m, that is,*

$$t(a, \beta) \in S_{\mathrm{loc}}^m(T^*X|_U, q \times p)$$

for every trivializing patch (U, β^E, β^F) on X. If $X = M$, we write $S^m(T^*M, \mathrm{Hom})$ instead of $S_{\mathrm{loc}}^m(T^*M, \mathrm{Hom})$.

It is sufficient to require that this condition hold for a collection of trivializing patches covering M (an "atlas") since the transformation from one basis (U, β^E, β^F) to another, (U', β'^E, β'^F), just involves multiplying by smooth matrix functions on $T^*(U \cap U')$, that is, $t(a, \beta') = g^F \circ t(a, \beta) \circ (g^E)^{-1}$, where $g^E = \beta'^E \circ (\beta^E)^{-1}$ and $g^F = \beta'^F \circ (\beta^F)^{-1}$ are the transition matrices. In particular, the definition of $S_{\mathrm{loc}}^m(T^*(X), \mathrm{Hom})$ agrees with the earlier definition when the bundles are trivial, $E = M \times \mathbb{C}^p$ and $F = M \times \mathbb{C}^q$, on which there is a trivialization given by $\beta^E = $ identity on E and $\beta^F = $ identity on F.

Let (U, β^E, β^F) be a trivializing patch, and let W be an open set with $\bar{W} \subset U$. Let $A \in OS^m(E, F)$. Then for all $\varphi, \psi \in C_0^\infty(W)$, the transfer $T(\varphi A \psi, \beta) = (\varphi A \psi)_\beta$ is given by

$$(\varphi A \psi)_\beta = \varphi A_U \psi, \quad (38)$$

for some $A_U \in OS^m(M, q \times p)$ with $\mathrm{supp}\, A_U \subset U$. If (38) holds for another representative $\bar{A}_U$ then $\varphi(A_U - \bar{A}_U) \in OS^{-\infty}$ for all $\varphi \in C_0^\infty(W)$, as follows if we take $\psi \in C_0^\infty(W)$ such that $\psi = 1$ on $\mathrm{supp}\, \varphi$. Passing to main symbols we obtain by Theorem 8.23 that $\varphi \cdot (\not{p} A_U - \not{p} \bar{A}_U) \in S_{\mathrm{loc}}^{m-1}(T^*M, q \times p)$, and therefore

$$\not{p} A_U - \not{p} \bar{A}_U \in S_{\mathrm{loc}}^{m-1}(T^*M|_W, q \times p).$$

For further use we introduce the notation

$$a_\beta := t(\not{p} A_U, \beta^{-1}), \quad (39)$$

and we call a_β a *symbol of A in the basis β*. This local symbol is therefore unique modulo $S_{\mathrm{loc}}^{m-1}(T^*M|_W, \mathrm{Hom}(\pi_*^{-1}E, \pi_*^{-1}F)|_W)$.

Definition 8.40 *Let $A \in OS^m(E, F)$. Then $a \in S^m(T^*M, \mathrm{Hom}(\pi_*^{-1}E, \pi_*^{-1}F))$ is said to be a main symbol of A if*

$$a - a_\beta \in S_{\mathrm{loc}}^{m-1}(T^*M|_W, \mathrm{Hom}(\pi_*^{-1}E, \pi_*^{-1}F)|_W) \quad (40)$$

for all trivializing patches (U, β^E, β^F) where $\bar{W} \subset U$. (By Definition 8.39 this is equivalent to $t(a, \beta) - \not{p}A_U \in S^{m-1}_{\text{loc}}(T^*M|_W, q \times p)$.)

It suffices to verify (40) for a collection of trivializing patches covering M (an "atlas"), as follows from the next lemma; see the text after Definition 8.21.

Lemma 8.41 *Let $A \in OS^m(E, F)$. Then let $(U, \beta^E_U, \beta^F_U)$ and $(U', \beta^E_{U'}, \beta^F_{U'})$ be two trivializing patches, and A_U and $A_{U'}$ the local representatives of A, see (38). If $\bar{W} \subset U$, $\bar{W}' \subset U'$ and $W \cap W' \neq \emptyset$, we have*

$$a_{\beta'} - a_\beta = t(\not{p}A_{U'}, \beta_{U'}^{-1}) - t(\not{p}A_U, \beta_U^{-1} \in S^{m-1}_{\text{loc}}(T^*M|_{W \cap W'}, \text{Hom}|_{W \cap W'}) \quad (41)$$

Proof If we denote by $g_{U'U} = \beta_{U'} \circ \beta_U^{-1}$ the transition matrix for a vector bundle, then we have for different transfers, $(\varphi A \psi)_\beta = \beta^F \circ \varphi A \psi \circ (\beta^E)^{-1}$, the following relation:

$$(\varphi A \psi)_{\beta_{U'}} = g^F_{U'U} \circ (\varphi A \psi)_{\beta_U} \circ (g^E_{U'U})^{-1}.$$

Hence for local representatives, see (38), we have

$$\varphi \cdot (A_{U'} - g^F_{U'U} \circ A_U \circ (g^E_{U'U})^{-1}) \cdot \psi = 0$$

for all $\varphi, \psi \in C_0^\infty(W \cap W')$ and then for main symbols

$$\not{p}A_{U'} - g^F_{U'U} \circ \not{p}A_U \circ (g^E_{U'U})^{-1} \in S^{m-1}_{\text{loc}}(T^*M|_{W \cap W'}, q \times p). \quad (42)$$

Applying the operator $t(\cdot, \beta_{U'}^{-1})$ to (42) and using the definition of $g_{U'U}$ we get (41).

Now we prove the lemma which is crucial for computing main symbols of p.d.o.'s in vector bundles.

Lemma 8.42 *Let (U, β^E, β^F) be a trivializing patch on M, let $A \in OS^m(E, F)$ have support in U, and consider the push-forward operator $\tilde{A} = T(A, \beta) \in OS^m(M, q \times p)$. Choose a main symbol $\tilde{a} = \not{p}\tilde{A}$ such that $\tilde{a}(x, \xi) = 0$ when $x \notin W$ where* supp $A \subset W \subset \bar{W} \subset U$ *(which can always be achieved by choosing any main symbol $\tilde{a}$ of $\tilde{A}$ and then replacing it by $\zeta \cdot \tilde{a}$ where $\zeta \in C_0^\infty(W)$ is equal to* 1 *on* supp A*). Then A has a main symbol*

$$a \in C^\infty(T^*M, \text{Hom}(\pi_*^{-1}E, \pi_*^{-1}F))$$

given by

$$a(x, \xi) = (\beta^F_x)^{-1} \circ \tilde{a}(x, \xi) \circ \beta^E_x \text{ if } x \in U, \qquad a(x, \xi) = 0 \text{ if } x \in M \setminus U \quad (43)$$

i.e., the pull-back $t(\tilde{a}, \beta^{-1})$ on T^*U, extended by O off T^*U.

Proof Choose an "atlas" on M which consists of trivializing patches $(U_j, \beta^E_j, \beta^F_j)$ such that

$$\text{supp } A \subset W \subset \bar{W} \subset U \quad \text{and} \quad (U_{j_0}, \beta^E_{j_0}, \beta^F_{j_0}) = (U, \beta^E, \beta^F)$$

and $$W \cap U_j = \emptyset \quad \text{when} \quad j \neq j_0.$$

We will check Definition 8.40 for this atlas. We choose $A_U = \tilde{A}$ and $\not{p}A_U = \tilde{a}$ when $j = j_0$, while for $j \neq j_0$ we can choose $A_{U_j} = 0$ and $\not{p}A_{U_j} = 0$. By (43) we have $a = a_\beta$ on T^*U, so (40) is satisfied when $j = j_0$. If $j \neq j_0$ we have $a_{\beta_j} = t(0, \beta_j^{-1}) = 0$ on $T^*(U_j)$ and by (43) we have $a = 0$ on $T^*(U_j)$ (since $\tilde{a}(x, \xi) \neq 0$ implies $x \notin U_j$), hence once more (40) is satisfied.

Theorem 8.43 *Every $A \in OS^m(E, F)$ has a main symbol*

$$a \in S^m(T^*M, \mathrm{Hom}(\pi_*^{-1}E, \pi_*^{-1}F))$$

(*i.e. satisfying Definition* 8.40). *We have a main symbol isomorphism*

$$\not{p}\colon \frac{OS^m(E, F)}{OS^{m-1}(E, F)} \simeq \frac{S^m(T^*M, \mathrm{Hom}(\pi_*^{-1}E, \pi_*^{-1}F))}{S^{m-1}(T^*M, \mathrm{Hom}(\pi_*^{-1}E, \pi_*^{-1}F))} \tag{44}$$

where $\not{p}$ is the linear map defined by $[A] \mapsto [a]$.

Proof Choose an atlas $\mathfrak{A}' = \{(W_j, \beta_j^E, \beta_j^F)\}$ as in Theorem 8.36 and a subordinate partition of unity $(\varphi_j\}$, i.e. supp $\varphi_j \subset W_j \subset \bar{W}_j \subset U_j$. As usual we let $\psi_j \in C_0^\infty(W_j)$ be equal to 1 on supp φ_j, then write

$$A = \sum \varphi_j A\psi_j + \sum \varphi_j A(1 - \psi_j).$$

Because of quasi-locality the second sum has a main symbol of 0. Let $a_j \in S^m(T^*M, \mathrm{Hom})$ be a main symbol of $\varphi_j A\psi_j$ (which exists by Lemma 8.42). Then $a := \sum a_j \in S^m(T^*M, \mathrm{Hom})$ is a main symbol of the first sum, and hence a main symbol for A.

The main symbol map $[A] \mapsto [a]$ is well-defined because

$$a \in S^{m-1}(T^*M, \mathrm{Hom}) \quad \text{when} \quad A \in OS^{m-1}(E, F).$$

To prove injectivity, we must show that if $A \in OS^m(E, F)$ and has a main symbol $a \in S^{m-1}(T^*M, \mathrm{Hom})$, then $A \in OS^{m-1}(E, F)$. To show this we use Definition 8.35. Let (U, β^E, β^F) be a trivializing patch and let $\bar{W} \subset U$. Since a is a main symbol of A then $a - a_\beta \in S_{\mathrm{loc}}^{m-1}(T^*M|_W, \mathrm{Hom})$; by Definition 8.39 we see that

$$\beta^F a \beta^{E-1} - \not{p}A_U \in S_{\mathrm{loc}}^{m-1}(T^*M|_W, q \times p),$$

and, since $a \in S^{m-1}(T^*M, \mathrm{Hom})$, then $\beta^F a \beta^{E-1} \in S_{\mathrm{loc}}^{m-1}(T^*M|_U, q \times p)$, whence

$$\not{p}A_U \in S_{\mathrm{loc}}^{m-1}(T^*M|_W, q \times p).$$

Thus, $\varphi \not{p}A_U\psi \in S^{m-1}(T^*M, q \times p)$ for every $\varphi, \psi \in C_0^\infty(W)$ and by the main symbol isomorphism, Theorem 8.23, it follows that

$$\varphi A_U \psi \in OS^{m-1}(M, q \times p).$$

In view of (38) we have $(\varphi A\psi)_\beta \in OS^{m-1}(M, q \times p)$ and this actually holds for all $\varphi, \psi \in C_0^\infty(U)$ because W is just an arbitrary open set such that $\bar{W} \subset U$. Hence $A \in OS^{m-1}(E, F)$ by Definition 8.35.

Finally, to prove surjectivity let $a \in S^m(T^*M, \mathrm{Hom})$. As above let $\mathfrak{A}'$ be an atlas and $\{\varphi_j\}$ a subordinate partition of unity. Consider the transfer $\beta_j^F a(\beta_j^E)^{-1} \in S_{\mathrm{loc}}^m(T^*M|_{U_j}, q \times p)$; by Theorem 8.23 there is an operator

$A_j \in OS^m(M, q \times p)$ with main symbol $a_j := \beta_j^F \varphi_j a(\beta_j^E)^{-1} \in S^m(T^*M, q \times p)$. By replacing A_j by $\psi_j A_j \psi_j$ where $\psi_j \in C_0^\infty(W_j)$ is equal to 1 on supp φ_j, we may assume that the support of the operator A_j lies in W_j. Now set

$$A = \sum (\beta_j^F)^{-1} A_j \beta_j^E;$$

then for the operator A we have

$$\text{main symbol} = \sum (\beta_j^F)^{-1} a_j \beta_j^E = \sum \varphi_j a = a,$$

since $\sum \varphi_j = 1$. This completes the proof.

The next two theorems show that $\not{p}$ is a *-algebra homomorphism.

Theorem 8.44 *If* $A \in OS^{m_1}(E, F)$ *and* $B \in OS^{m_2}(G, E)$ *then*

$$A \circ B \in OS^{m_1+m_2}(G, F).$$

Further, if a *and* b *are main symbols of* A *and* B, *respectively, then* $a \circ b$ *is a main symbol of* $A \circ B$:

$$\not{p}(A \circ B) = \not{p}A \circ \not{p}B$$

Note: We compose the linear mappings $a \in \mathrm{Hom}(\pi_*^{-1}E, \pi_*^{-1}F)$ *and* $b \in \mathrm{Hom}(\pi_*^{-1}G, \pi_*^{-1}E)$ *fibrewise to get the mapping* $a \circ b \in \mathrm{Hom}(\pi_*^{-1}G, \pi_*^{-1}F)$.

Proof The proof follows just as in Theorem 8.24 if we substitute κ by β.

Theorem 8.45 *We endow* M *with a volume density* μ *and* E, F *with Hermitian structures. Let* $A: C^\infty(M, E) \to C^\infty(M, F)$ *be a linear operator. Using the scalar products, see* (33),

$$(\ ,\ )_E = \int_M (\ ,\ )_{E_x}\, d\mu(x), \qquad (\ ,\ )_F = \int_M (\ ,\ )_{F_x}\, d\mu(x),$$

we define the adjoint operator $A^*: C^\infty(M, F) \to C^\infty(M, E)$ *of* A *by*

$$(A^*u, v)_E = (u, Av)_F, \qquad u \in C^\infty(M, F),\ v \in C^\infty(M, E).$$

Now, if $A \in OS^m(E, F)$ *then the adjoint operator,* A^*, *is once more a p.d.o., belonging to* $OS^m(F, E)$. *If* a *is a main symbol of* A, *then* ${}^h a$ *is a main symbol of* A^*:

$$\not{p}A^* = {}^h(\not{p}A).$$

Here ${}^h a(x, \xi): F_x \to E_x$ *denotes the Hermitian adjoint of the mapping* $a(x, \xi): E_x \to F_x$, *and since* $a \in C^\infty(T^*M, \mathrm{Hom}(\pi_*^{-1}E, \pi_*^{-1}F))$ *it follows that* ${}^h a \in C^\infty(T^*M, \mathrm{Hom}(\pi_*^{-1}F, \pi_*^{-1}E))$.

Proof Note first that the adjoint A^* exists and belongs to $OP^m(F; E)$; see §8.1. We will show that A^* satisfies the conditions of (i′) and (ii) of Theorem 8.36.

For condition (ii), observe that if $B \in OC^\infty(E, F)$ with kernel $K(x, y) \in C^\infty(M \times M, \mathrm{Hom}(\pi_1^{-1}E, \pi_2^{-1}F))$, see §8.1, then $B^* \in OC^\infty(F, E)$ with the (Hermitian transposed) kernel ${}^h K(y, x) \in C^\infty(M \times M, \mathrm{Hom}(\pi_2^{-1}F, \pi_1^{-1}E))$.

Therefore, if φ and ψ have disjoint supports, it follows that

$$\varphi A^*\psi = (\bar{\psi} A\bar{\varphi})^* \in OC^\infty(F, E).$$

To verify (i′) we use the atlas $\{(W_j, \beta_j^E, \beta_j^F)\}_{j=1}^N$ introduced in Theorem 8.36. Since over U_j the Hermitian bundle E is trivial, then by means of the Gram–Schmidt orthonormalization process, see §5.10, we can find an orthonormal basis, i.e. a trivialization mapping β_j^E such that

$$(f, g)_{E_x} = (\beta_j^E f, \beta_j^F g)_{\mathbb{C}^p}, \qquad x \in U_j,$$

and of course the same holds for F:

$$(u, v)_{F_x} = (\beta_j^E u, \beta_j^F v)_{\mathbb{C}^q}, \qquad x \in U_j.$$

Let $\tilde{A}_j \in OS^m(M, q \times p)$, $j = 1, \ldots, N$ be matrix p.d.o.'s such that the representation

$$\varphi A\psi = \varphi(\tilde{A}_j)_{\beta_j^{-1}}\psi, \qquad \operatorname{supp} \tilde{A}_j \subset U_j, \tag{45}$$

holds for all φ, $\psi \in C_0^\infty(W_j)$, $j = 1, \ldots, N$. We must obtain a similar representation for A^*. If we recall $\bar{W}_j \subset U_j$, we obtain for φ, $\psi \in C_0^\infty(W_j)$ that

$$(\varphi A^*\psi u, v)_E = (u, \bar{\psi} A\bar{\varphi} v)_F = \int_{U_j} (u, \bar{\psi} A\bar{\varphi} v)_{F_x}\, d\mu(x)$$

$$= \int_{U_j} (\beta_j^F u, \beta_j^F \bar{\psi} A\bar{\varphi} v)_{\mathbb{C}^q}\, d\mu$$

and, since $\bar{\psi} A\bar{\varphi} = (\beta_j^F)^{-1}\bar{\psi}\tilde{A}\bar{\varphi}\beta_j^E$, we get

$$= \int_{U_j} (\beta_j^F u, \bar{\psi}\tilde{A}\bar{\varphi}\beta_j^E v)_{\mathbb{C}^q}\, d\mu$$

$$= \int_{U_j} (\varphi\tilde{A}_j^*\psi\beta_j^F u, \beta_j^E v)_{\mathbb{C}^q}\, d\mu$$

and, since $\varphi(\tilde{A}_j^*)_{\beta_j^{-1}}\psi = \varphi(\beta_j^E)^{-1}\tilde{A}_j^*\beta_j^F\psi$, we obtain finally

$$= \int_{U_j} (\beta_j^E \varphi(\tilde{A}_j^*)_{\beta_j^{-1}}\psi u, \beta_j^E v)_{\mathbb{C}^q}\, d\mu$$

$$= (\varphi(\tilde{A}_j^*)_{\beta_j^{-1}}\psi u, v)_E,$$

where, given the matrix $\tilde{A}_j(x, D) = [\tilde{A}_{j,(s,t)}(x, D)]_{q \times p}$, we denoted by $\tilde{A}_j^*(x, D)$ the matrix

$$\tilde{A}_j^*(x, D) = [\tilde{A}_{j,(t,s)}^*(x, D)]_{q \times p}$$

Thus we have established that

$$\varphi A^*\psi = \varphi(\tilde{A}_j^*)_{\beta_j^{-1}}\psi \tag{46}$$

for every φ, $\psi \in C_0^\infty(W_j)$, where $\tilde{A}_j^* \in OS^m(M, q \times p)$. In other words, A^* satisfies condition (i′) of Theorem 8.36.

It remains to show that ${}^h\!a$ is a main symbol of A^*. By means of a partition of unity – see the proof of Theorem 8.24(b) – this reduces to the problem of finding a main symbol of (46) given the symbol a of A. Now, we have $\bar{\psi}A\bar{\varphi} = \bar{\psi}(\tilde{A}_j)_{\beta_j^{-1}}\bar{\varphi}$ for $\varphi, \psi \in C_0^\infty(W_j)$, and by Lemma 8.42 the main symbol of $\bar{\psi}\tilde{A}_j\bar{\varphi}$ is equal to the matrix $\bar{\psi}(\beta_j^F a \beta_j^{E^{-1}})\bar{\varphi}$. Using Theorem 8.24(b) and matrix algebra we see that the symbol of $\varphi\tilde{A}_j^*\psi$ is equal to $\varphi(\beta_j^E \circ {}^h\!a \circ \beta_j^{F^{-1}})\psi$. Applying once more Lemma 8.42 we get the symbol of (46):

$$\not{p}(\varphi A^*\psi) = \varphi\, {}^h\!a\psi.$$

Finally, we define classic p.d.o.'s in vector bundles.

Definition 8.46 *An operator $A \in OS^m(E, F)$ is said to be* (-1) *classic and we write*

$$A \in OClS^m(E, F)$$

if for any trivialization (U, β) and for all $\varphi, \psi \in C_0^\infty(U)$ the push-forward

$$(\varphi A\psi)_\beta \in OS^m(M, q \times p)$$

is a (-1) *classic matrix operator on M, see* §8.4.

Let $\mathfrak{A} = \{(U_j, \beta_j^E, \beta_j^F)\}$ be an atlas of trivializing patches on M, and shrink it to obtain an atlas $\mathfrak{A}' = \{(W_j, \beta_j^E, \beta_j^F)\}$ with $\bar{W}_j \subset U_j$. Then we have classic matrix operators

$$\tilde{A}_j(x, D) \in OClS^m(M, q \times p), \qquad \operatorname{supp} \tilde{A}_j \subset U_j,$$

such that condition (i′) in Theorem 8.36 holds. The operators $\tilde{A}_j = [\tilde{a}_{j,(s,t)}]_{q \times p}$ have a principal symbol, homogeneous of degree m, defined by

$$\pi\tilde{A}_j(x, \xi) := [\pi\tilde{a}_{j,(s,t)}]_{q \times p}$$

where $\pi\tilde{a}_{j,(s,t)} \in C^\infty(T^*M\backslash 0)$. Since $\operatorname{supp} \tilde{A}_j \subset U_j$, then we have $\pi\tilde{A}_j(x, \xi) = 0$ for x outside U_j, and from (42) it follows that on $W_j \cap W_k \neq \emptyset$ we have

$$\pi\tilde{A}_k(x, \xi) = g^F_{U_kU_j}(x)\pi\tilde{A}_j(x, \xi)(g^E_{U_kU_j}(x))^{-1}.$$

Consequently, by pulling $\pi\tilde{A}_j$ back to the manifold, i.e.

$$\pi A(x, \xi) = (\beta_j^F)^{-1}\pi\tilde{A}_j(x, \xi)\beta_j^E, \qquad \text{when } x \in W_j,\ \xi \in T_x^*M\backslash 0 \tag{47}$$

we have that $\pi A \in C^\infty(T^*M\backslash 0)$, $\operatorname{Hom}(\pi_*^{-1}E, \pi_*^{-1}F))$ is well-defined, and, further, its definition does not depend on the choice of atlas or of trivializing patches. We call πA the *principal part* or principal symbol of A, and the proof of the following theorem is now essentially the same as Theorem 8.27.

Theorem 8.47 *Let $A \in OClS^m(E, F)$ be a* (-1) *classic operator. Then the section $\pi A \in C^\infty(T^*M\backslash 0, \operatorname{Hom}(\pi_*^{-1}E, \pi_*^{-1}F))$ defined by (47) satisfies the following properties:*

(a) *πA is positively homogeneous of degree m in ξ;*
(b) *for any $\chi \in C^\infty(T^*M)$ which vanishes near the* 0 *section of T^*M and equals* 1 *outside a compact subset of T^*M, the section $\chi \cdot \pi A$ is a main symbol of A.*

We denote by

$$\mathrm{Symbl}^m(E, F)$$

the vector space of all C^∞ sections $\mathscr{h} \in C^\infty(T^*M\backslash 0), \mathrm{Hom}(\pi_*^{-1}E, \pi_*^{-1}F))$ which are homogeneous of degree m in ξ. By homogeneous of degree m in ξ, we mean that the maps $\mathscr{h}(x, \xi): E_x \to F_x$, $(x, \xi) \in T^*M\backslash 0$, satisfy the condition

$$\mathscr{h}(x, c\xi) = c^m \cdot \mathscr{h}(x, \xi) \qquad \text{for all } c > 0.$$

Clearly $OS^{m-1}(E, F) \subset OClS^m(E, R)$. As in Corollary 8.29 we have the following result.

Theorem 8.48 *The following sequence is exact:*

$$0 \longrightarrow OS^{m-1}(E, F) \xrightarrow{\iota} OClS^m(E, F) \xrightarrow{\pi m} \mathrm{Symbl}^m(E, F) \longrightarrow 0$$

(Exactness at the middle two arrows means that $\pi_m A = 0$ if and only if $A \in OS^{m-1}(E, F)$, while exactness at the last two arrows means that π_m is surjective.)

Further, the mapping π is a homomorphism:

Theorem 8.49
(a) If $A \in OClS^{m_1}(F, G)$ and $B \in OClS^{m_2}(E, F)$ then $A \circ B \in OClS^{m_1+m_2}(E, G)$ and we have

$$\pi(A \circ B) = \pi A \circ \pi B$$

(b) If the bundles E, F are Hermitian then the adjoint of $A \in OClS^m(E, F)$ is an operator $A^* \in OClS^m(F, E)$ and we have

$$\pi A^* = {}^h(\pi A)$$

The proofs of Theorems 8.48 and 8.49 follow from Theorems 8.44, 8.45 and 8.47(b).

Theorem 8.50
(a) For any given sequence $A_d \in OS^{m_d}(E, F)$, $d = 1, 2, \ldots$, with $m_{d-1} > m_d \to -\infty$ as $d \to +\infty$, there exists $A \in OS^{m_1}(E, F)$ such that $A \sim \sum_{d=1}^\infty A_d$ (see (13)). The p.d.o. is uniquely determined modulo $OS^{-\infty}(E, F)$.
(b) If $m_2 \leqslant m_1 - 1$ then A is (-1) classic if and only if A_1 is (-1) classic, and we have $\pi A = \pi A_1$.

The proof of (a) is obtained by localizing as in Proposition 8.16, while (b) is obvious since we have $A - A_1 \in OS^{m_2}(E, F) \subset OS^{m_1-1}(E, F)$.

We end this section with a lemma which gives some information on the relation between an operator and its principal symbol. By virtue of Lemma 8.67 (see §8.7) there is no loss of generality if we consider just p.d.o.'s of order 0.

Lemma 8.51 *Let E and F be Hermitian bundles. Given $x_0 \in M$, there exists a sequence $\{f_k\}$ in $C^\infty(M)$ with the following properties: for all sections $e \in C^\infty(M, E)$ with $e(x)$ of length 1 for x in a neighbourhood of x_0, and 1-forms $\xi \in C^\infty(M, T^*M)$ with $\xi(x_0) \neq 0$,*

(*i*) *the support of f_k converges to x_0,*
(*ii*) $\|f_k e\|_{L_2} = 1$,
(*iii*) *for each $A \in OClS^0(E, F)$, we have*

$$\|A(f_k e)(x) - \pi A(x, \xi(x)) \cdot f_k(x) e(x)\|_{L_2} \to 0 \qquad \text{as } k \to \infty.$$

Proof First suppose that $E = F = M \times \mathbb{C}$ are trivial bundles, then we may take $e(x) \equiv 1$. Consequently, it suffices to prove existence of $\{f_k\}$ satisfying (i), (ii) and

$$\|Af_k(x) - \pi A(x_0, \xi_0) \cdot f_k(x)\|_{L_2} \to 0 \tag{iii$'$}$$

since it would follow from (i) and (ii) that

$$\|\pi A(x, \xi(x)) \cdot f_k(x) - \pi A(x_0, \xi_0) \cdot f_k(x)\|_{L_2} \to 0$$

where $\xi_0 = \xi(x_0)$, and (iii) would then follow.

Let $\kappa: U \to V$ be a coordinate map in a neighbourhood U of x_0. Let $\psi \in C^\infty(M)$ equal 1 in a neighbourhood of x_0 with supp $\psi \subset U$; we have

$$\psi A \psi = \psi(\tilde{A}(x, D))_{\kappa^{-1}} \psi$$

for some p.d.o. $\tilde{A}(x, D) \in OS^0(\mathbb{R}^n)$. Letting $\tilde{x}_0 = \kappa(x_0)$ and $\tilde{\xi}_0 = {}^t d\kappa^{-1}(x_0)(\xi_0)$, and choosing $\tilde{\phi}_k \in C_0^\infty(\mathbb{R}^n)$ as in Lemma 7.32 for the point $(\tilde{x}_0, \tilde{\xi}_0) \in \mathbb{R}^n \times \mathbb{R}^n$, we have $\|\tilde{A}(x, D)\tilde{\phi}_k - \pi\tilde{A}(\tilde{x}_0, \xi) \cdot \tilde{\phi}_k\|_0 \to 0$. Since the support of $\tilde{\phi}_k$ converges to x_0, then for k sufficiently large, we may let $\phi_k = \tilde{\phi}_k \circ \kappa \in C^\infty(M)$ be the pull-back to the manifold. For k sufficiently large we also have $\psi^2 \phi_k = \phi_k$, so that $A\phi_k = \psi A \psi \phi_k + A_{-1}\phi_k$, where $A_{-1} \in OS^{-1}$. Now $A_{-1}\phi_k \to 0$ since $\|\phi_k\|_s \to 0$ when $s < 0$, and it follows that

$$\|A\phi_k - \pi\tilde{A}(\tilde{x}_0, \tilde{\xi}_0) \cdot \phi_k\|_{L_2} \to 0.$$

In view of (30) we have $\pi A(x_0, \xi_0) = \pi\tilde{A}(\tilde{x}_0, \tilde{\xi}_0)$, and since $\|\phi_k\|_{L_2}^2 = \int_{\kappa(U)} |\tilde{\phi}_k(x)|^2 g_U(x)\, dx \geqslant C > 0$, we can set $f_k = \phi_k / \|\phi_k\|_{L_2}$, thereby establishing (ii) and (iii$'$).

In the case where E and F are general bundles, choose trivializations $\beta^F: F_U \to U \times \mathbb{R}^q$ and $\beta^E: E_U \to U \times \mathbb{R}^q$ with $\beta^E \circ e = e_1$ (the first coordinate vector in $\mathbb{R}^n$). Then we have

$$\psi A \psi = \psi(\beta^F)^{-1}(\tilde{A}(x, D))_{\kappa^{-1}} \beta^E \psi$$

for some matrix operator $\tilde{A}(x, D) = [\tilde{A}_{ij}(x, D)] \in OS^m(\mathbb{R}^n, q \times p)$. Now choose $\tilde{\phi}_k$ as above. Since $\beta^F \circ e = e_1$ it follows that

$$\psi A \psi \phi_k e = \psi(\beta^F)^{-1}([\tilde{A}_{i1}(x, D)]_{i=1}^q)_{\kappa^{-1}} \phi_k \psi.$$

Then by Lemma 7.32 we have as before $\|\tilde{A}_{i1}(x, D)\tilde{\phi}_k - \pi\tilde{A}_{i1}(\tilde{x}_0, \tilde{\xi}_0) \cdot \tilde{\phi}_k\|_0 \to 0$, and the sequence $f_k = \phi_k / \|\phi_k\|_{L_2}$ again has the required properties, in view of (47).

Corollary 8.52 *It follows that* $\|A(f_k e)\|_{L_2} \to |\pi A(x_0, \xi_0)e(x_0)|_{F_{x_0}}$, *and hence, for any* $0 \neq \xi_0 \in T^*_{x_0}M$, *the section e in Corollary* 8.51 *can be chosen so that*

$$\|A(f_k e)\|_{L_2} \to \|\pi A(x_0, \xi_0)\| \qquad \text{as } k \to \infty. \tag{48}$$

Proof Let β^E and β^F be the trivializations introduced above, and in addition choose them so that they map orthonormal bases to orthonormal bases. It follows that

$$\|\pi A(x, \xi(x)) \cdot f_k(x)e(x)\|^2_{L_2} = \sum_{i=1}^{q} \int_M |\pi \tilde{A}_{i1}(\kappa(x), {}^t d\kappa^{-1}(x)(\xi(x)))|^2 |f_k(x)|^2 \, d\mu(x)$$

$$\to \sum_i |\pi \tilde{A}_{i1}(\kappa(x_0), {}^t d\kappa^{-1}(x_0)(\xi(x_0)))|^2$$

$$= |\pi A(x_0, \xi_0)e(x_0)|^2_{F_{x_0}}$$

since $\int_M |f_k|^2 \, d\mu = 1$ and the support of f_k converges to x_0. Then (48) follows from Corollary 8.51 if we choose the section $e \in C^\infty(M, E)$ so that

$$|\pi \mathscr{A}(x_0, \xi_0)e(x_0)|_{F_{x_0}} = \|\pi \mathscr{A}(x_0, \xi_0)\|$$

and $|e(x)|_{E_x} = 1$ in a neighbourhood of x_0.

8.7 Elliptic operators

An operator $A \in OS^m(E, F)$, with main symbol

$$a \in S^m(T^*M, \operatorname{Hom}(\pi_*^{-1}E, \pi_*^{-1}F)),$$

is said to be *elliptic* of order m if there exists some

$$b \in S^{-m}(T^*M, \operatorname{Hom}(\pi_*^{-1}F, \pi_*^{-1}E))$$

such that

$$\left.\begin{aligned} ab - 1_F \in S^{-1}(T^*M, \operatorname{Hom}(\pi_*^{-1}F, \pi_*^{-1}F)) \\ ba - 1_E \in S^{-1}(T^*M, \operatorname{Hom}(\pi_*^{-1}E, \pi_*^{-1}E)) \end{aligned}\right\} \tag{49}$$

where 1_E and 1_F denote the identity mappings $E \to E$ and $F \to F$. We will also refer to a as an *elliptic symbol.* Let

$$\operatorname{Ell}^m(E, F)$$

denote the set of all elliptic operators $A \in OS^m(E, F)$ of order m and $\operatorname{Ell} = \bigcup_m \operatorname{Ell}^m$ the set of elliptic operators of arbitrary order. Note that if $A \in \operatorname{Ell}$ then $A \in \operatorname{Ell}^m$ for a unique m (see Exercise 8). Similarly, the set of all elliptic symbols $a \in S^m(T^*(M), \operatorname{Hom}(\pi_*^{-1}E, \pi_*^{-1}F))$ of order m is denoted

$$\mathscr{ell}^m(E, F),$$

and $\mathscr{ell} = \bigcup_m \mathscr{ell}^m$ is the set of elliptic symbols of arbitrary order.

It can be shown that the existence of b satisfying both conditions in (49) implies that E and F have the same fibre dimension (see Exercise 19). If we had assumed this at the start, the two conditions would have been equivalent; see the addendum at the end of the chapter.

We begin by showing that elliptic p.d.o.'s are Fredholm operators.

Theorem 8.53 *Let $A \in OS^m(E, F)$. If A is elliptic then*

$$A\colon W_2^l(M, E) \to W_2^{l-m}(M, F) \tag{50}$$

is a Fredholm operator (for all $l \in \mathbb{R}$). The kernel, $\ker A$, *is contained in $C^\infty(M, E)$, and the image,* $\operatorname{im} A$, *is the orthogonal space of the kernel $\subset C^\infty(M, F)$ of the adjoint $A^* \in OS^m(F, E)$, i.e.*

$$\operatorname{im} A = \{f \in W_2^{l-m}(M, F);\ (f, g)_F = 0 \text{ for all } g \in \ker A^*\} \tag{51}$$

Consequently, the index is independent of l since $\operatorname{ind} A = \dim\ker A - \dim\ker A^*$. *If $A' \in OS^m(E, F)$ is another elliptic operator such that $A - A' \in OS^{m-1}(E, F)$ then* $\operatorname{ind} A = \operatorname{ind} A'$; *hence if two elliptic operators in $OS^m(E, F)$ have the same main symbol then they have the same index.*

Proof By Theorems 8.43 and 8.44, the condition (49) is equivalent to the existence of an operator $B \in OS^{-m}(F, E)$ such that

$$AB - I =: T_r \in OS^{-1} \quad \text{and} \quad BA - I =: T_l \in OS^{-1}. \tag{52}$$

T_r and T_l are operators of order -1 in the scale of Sobolev spaces by Theorem 8.13 and by the compact embedding theorem for Sobolev spaces, Theorem 7A.13, it follows that

$$T_r\colon W_2^{l-m}(M, F) \to W_2^{l-m+1}(M, F) \hookrightarrow W_2^{l-m}(M, F)$$

and

$$T_l\colon W_2^l(M, E) \to W_2^{l+1}(M, E) \hookrightarrow W_2^l(M, E)$$

are compact operators. Hence B is a right and a left regularizer, and by Theorem 9.6 the operator A is Fredholm.

If $u \in \ker A \subset W_2^l$ then $u = -T_l u \in W_2^{l+1}$, and by iterating this result we get

$$u \in \bigcap W_2^l(M, E) = C^\infty(M, E)$$

Similarly, the adjoint $A^* \in OS^m$ of A is an elliptic p.d.o. that maps $W_2^{m-l}(M, F)$ to $W_2^{-l}(M, E)$, and $\ker A^* \subset C^\infty(M, F)$. Since the image of A as an operator from $W_2^l(M, E)$ to $W_2^{l-m}(M, F)$ is closed, it is the orthogonal space of the kernel of the anti-dual operator which can be identified with the adjoint $A^*\colon W_2^{m-l}(M, F) \to W_2^{-l}(M, E)$ by means of the duality brackets, $(\,,\,)_E$ and $(\,,\,)_F$, see §8.5, and this proves (51). The index is independent of l because $\operatorname{ind} A = \dim\ker A - \dim\ker A^*$, and both $\ker A \subset C^\infty$ and $\ker A^* \subset C^\infty$ are independent of l. The last statement in the theorem is clear by virtue of Theorem 9.11 because the operator $A - A'\colon W_2^l(M, E) \to W_2^{l-m+1}(M, F) \hookrightarrow W_2^{l-m}(M, F)$ is compact.

Substituting $l = m$ we obtain the following.

Corollary 8.54 *If $A \in OS^m(E, F)$ is elliptic, we have the orthogonal decomposition*

$$L_2(M, F) = \text{im } A \oplus \ker A^*, \tag{53}$$

where both subspaces, im A *and* ker A^*, *are closed. By* im A *we mean the image of the operator* A: $W_2^m(M, E) \to L_2(M, F)$.

From (52) we obtain the following regularity theorem. The discussion of smoothable operators at the end of §9.1 is relevant here.

Corollary 8.55 (*Regularity Theorem*) *If* $u \in W_2^l(M, E)$ *and* $Au \in W_2^{l-m+1}(M, F)$ *then* $u \in W_2^{l+1}(M, E)$. *Hence if* $u \in \bigcup W_2^l(M, E)$ *and* $Au \in W_2^r(M, F)$ *then* $u \in W_2^{r+m}(M, E)$.

Proof Since $Au \in W_2^{l-m+1}(M, F)$ and B has order $-m$ then $BAu \in W_2^{l+1}(M, E)$. Now, by (52) we have $u = BAu - T_l u \in W_2^{l+1}(M, E)$ which proves the first statement. The second statement follows easily, for if $u \in W_2^l(M, E)$ with $l - m + 1 < r$ then $u \in W_2^{l+1}(M, E)$, and we can iterate this result until $l - m + 1 = r$.

There is also the following result which shows that $A: C^\infty(M, E) \to C^\infty(M, F)$ is Fredholm and has the same index as the operator (50). The topology on $C^\infty(M, E)$ is the Fréchet space topology in which a sequence $\{u_j\} \subset C^\infty(M, E)$ converges if all derivatives of u_j converge uniformly on compact subsets of chart domains.

Corollary 8.56 *If* $u \in W_2^l(M, E)$ *for some* l *and* $Au \in C^\infty(M, F)$, *then* $u \in C^\infty(M, E)$. (*This is known as Weyl's Lemma.*) *Hence*

$$C^\infty(M, F) = A(C^\infty(M, E)) \oplus \ker A^*, \tag{53'}$$

where both subspaces $A(C^\infty(M, E))$ *and* $\ker A^*$ *are closed in* $C^\infty(M, F)$. *Moreover,* $A(C^\infty(M, E))$ *consists of all* $f \in C^\infty(M, F)$ *for which* $(f, g)_F = 0$ *for all* $g \in \ker A^*$.

Proof If $u \in W_2^l(M, E)$ for some l, and $Au \in C^\infty(M, F)$, then $u \in \bigcap W_2^l(M, E) = C^\infty(M, E)$ by virtue of Corollary 8.55. Now, since $\ker A^* \subset C^\infty(M, F)$ it follows from (53) that

$$C^\infty(M, F) = (C^\infty(M, F) \cap \text{im } A) \oplus \ker A^*$$

which implies (53′) by Weyl's Lemma just proved. Once again it follows from Weyl's Lemma that $A(C^\infty(M, E))$ is closed in $C^\infty(M, F)$ since im A is closed in $L_2(M, F)$ and the embedding $C^\infty(M, F) \hookrightarrow L_2(M, F)$ is continuous. Obviously $\ker A^*$ is closed in $C^\infty(M, F)$ because it is a finite dimensional subspace. Finally, the last statement in the corollary follows from (51) and Weyl's Lemma.

We now turn to the problem of determining a parametrix for an operator $A \in OS^m(E, F)$.

Definition 8.57 *An operator $B \in OS^{-m}(F, E)$ is a parametrix for A if*

$$A \circ B - I \in OS^{-\infty}(F, F) \quad \text{and} \quad B \circ A - I \in OS^{-\infty}(E, E) \tag{54}$$

We show in the next theorem that ellipticity is the necessary and sufficient condition for existence of a parametrix. To prove this result two lemmas are required.

Lemma 8.58 *If $T \in OS^{-1}(E, E)$ then $I - T$ has a parametrix.*

Proof We let $T^k := T \circ \cdots \circ T$ and define Q by the asymptotic series

$$Q \sim I + T + T^2 + \cdots$$

Since by Theorem 8.44 we have $T^k \in OS^{-k}(E, E)$, then Q is well-defined (see Theorem 8.50). Setting $Q = I + T + \cdots + T^N + T'_{N+1}$ where $T'_{N+1} \in OS^{-N-1}(E, E)$ we obtain

$$\begin{aligned} (I - T) \circ Q - I &= (I - T)(I + T + \cdots + T^N + T'_{N+1}) - I \\ &= I - T^{N+1} + (I - T) \circ T'_{N+1} - I \\ &= (I - T) \circ T'_{N+1} - T^{N+1} \end{aligned}$$

which belongs to $OS^{-N-1}(E, E)$ for $N = 1, 2, \ldots$, whence

$$(I - T) \circ Q - I \in OS^{-\infty}(E, E).$$

Similarly, we get $Q \circ (I - T) - I \in OS^{-\infty}(E, E)$. Hence Q is a parametrix for $I - T$.

The following lemma shows that if A has a left parametrix and a right parametrix then any right or left parametrix of A is in fact a parametrix.

Lemma 8.59 *Let $A \in OS^m(E, F)$. If for some $B \in OS^{-m}(F, E)$ and $B' \in OP^{-m}(F, E)$ (i.e. an operator of order $-m$) we have*

$$A \circ B - I \in OS^{-\infty} \quad \text{and} \quad B' \circ A - I \in OS^{-\infty},$$

then $B - B' \in OS^{-\infty}$. (In particular, it follows that $B' \in OS^{-m}(F, E)$.)

Proof This follows at once from the equation

$$B' - B = B'(I - A \circ B) + (B' \circ A - I)B$$

by virtue of Theorems 8.44 and 8.38 and the fact that $OP^{-\infty}(E, F) = OS^{-\infty}(E, F)$.

Theorem 8.60 *Let $A \in OS^m(E, F)$. Then A has a parametrix if and if it is elliptic.*

Proof If A has a parametrix B satisfying (54) then certainly $AB - I \in OS^{-1}$ and $BA - I \in OS^{-1}$. If we let $\mathscr{b}$ denote a main symbol of B, it follows from Theorems 8.43 and 8.44 that (49) holds, i.e. A is elliptic.

Conversely, if A is elliptic then as in the proof of Theorem 8.53 there is an operator $B_1 \in OS^{-m}(F, E)$ that inverts A modulo OS^{-1}, i.e.

$$A \circ B_1 = I - T \qquad \text{where } T \in OS^{-1}(F, F).$$

By Lemma 8.58, $I - T$ has a parametrix $Q \in OS^0(F, F)$ and if we let $B := B_1 \circ Q \in OS^{-m}(F, E)$ it follows that

$$A \circ B - I = A \circ B_1 \circ Q - I = (I - T) \circ Q - I \in OS^{-\infty},$$

so B is a right parametrix. Similarly, there exists a left parametrix B', and by Lemma 8.59, we have $B - B' \in OS^{-\infty}$. Thus B is a two-sided parametrix of A.

We turn now to the case of classic pseudo-differential operators $A \in OClS^m(E, F)$. Classic operators have a principal part $\pi A \in \mathrm{Symbl}^m(E, F)$ which exists by Theorem 8.47.

Theorem 8.61 *Let $A \in OClS^m(E, F)$. Then A is elliptic if and only if*

$$\pi A(x, \xi)\colon E_x \to F_x \text{ is invertible} \tag{55}$$

for all $(x, \xi) \in T^*M \backslash 0$.

Proof The inverse $[\pi A(x, \xi)]^{-1}$ is taken pointwise for each $(x, \xi) \in T^*M \backslash 0$. Then $[\pi A]^{-1} \in C^\infty(T^*M \backslash 0, \mathrm{Hom}(\pi_*^{-1}F, \pi_*^{-1}E))$ follows from Proposition 5.2 in local coordinates because πA is C^∞. And since $\pi A(x, c\xi) = c^m \pi A(x, \xi)$ if $c > 0$, it follows that

$$[\pi A(x, c\xi)]^{-1} = [c^m \pi A(x, \xi)]^{-1} = c^{-m}[\pi A(x, \xi)]^{-1},$$

and therefore $[\pi A(x, \xi)]^{-1} \in \mathrm{Symbl}^{-m}(F, E)$. By Theorem 8.48 there exists $B \in OClS^{-m}(F, E)$ such that $\pi B = [\pi A]^{-1}$, whence $\pi(A \circ B - I_E) = 0$ and $\pi(B \circ A - I_F) = 0$ (by Theorem 8.49), i.e.

$$A \circ B - I_F \in OS^{-1}(F, F) \quad \text{and} \quad B \circ A - I_E \in OS^{-1}(E, E).$$

Applying the main symbol isomorphism, Theorem 8.45, we get $\mathscr{a} \circ \mathscr{b} - 1_F \in S^{-1}$ and $\mathscr{b} \circ \mathscr{a} - 1_E \in S^{-1}$, which is ellipticity in the sense of (49).

To prove the necessity of (55), choose some Riemannian metric on M, i.e. a smoothly defined set of inner products $\langle\, , \rangle_x$ on each tangent space T_xM. Let $|\xi| = \langle \xi, \xi \rangle_x^{1/2}$ where $\langle\, , \rangle_x$ is the induced inner product on T_x^*M. Also, choose Hermitian structures in E and in F.

Let $\mathscr{a} \in S^m$ be a main symbol for A such that $\mathscr{a} = \pi A$ for large $|\xi|$. If A is elliptic, there exists a symbol $\mathscr{b} \in S^{-m}$ such that (49) holds, and, since $\mathscr{b}\mathscr{a} - 1 \in S^{-1}$, then by working in local coordinates near a fixed point $x \in M$ it is easily established that

$$\|\mathscr{b}(x, \xi)\mathscr{a}(x, \xi) - 1\| < 1 \qquad \text{when } |\xi|_x \text{ is sufficiently large,}$$

where $\| \ \|$ denotes the operator norm for linear maps $E_x \to E_x$ (relative to the inner product on E_x). It follows that, when $|\xi|_x$ is sufficiently large, $b(x, \xi)a(x, \xi)$ is invertible, so $a(x, \xi)$ has a left inverse. Hence $\pi A(x, \xi)$ has a left inverse when $|\xi|_x$ is large; by homogeneity it follows that $\pi A(x, \xi)$ has a left inverse for all $(x, \xi) \in T^*M \backslash 0$. Similarly, by consideration of $ab - 1 \in S^{-1}$, we see that $\pi A(x, \xi)$ has a right inverse for all $(x, \xi) \in T^*M \backslash 0$. Hence $\pi A(x, \xi)$ is invertible for all $(x, \xi) \in T^*M \backslash 0$.

Corollary 8.62 *The operator $A \in OClS^m(E, E)$ is elliptic, i.e.* (55) *holds, if and only if it has a parametrix $P \in OClS^{-m}(E, E)$.*

Proof We know from Theorem 8.60 that A has a parametrix if and only if it is elliptic, and by Theorem 8.61 the ellipticity of A is equivalent to (55).

It remains to show that if A is a (-1) classic operator and elliptic, then it has a (-1) *classic* parametrix P. But the existence of P is proved just as in Theorem 8.60: we begin by choosing $B_1 \in OClS^{-m}$ such that $\pi B_1 = (\pi A)^{-1}$, then continue with $B = B_1 \circ Q$ where $Q \sim I + T + T^2 + \cdots$ is the parametrix of $T = A \circ B_1 - I \in OClS^{-1}$. Now from Theorem 8.50(b) we obtain that Q is (-1) classic because the first term, I, in the asymptotic series for Q is obviously (-1) classic. Hence the parametrix $P = B_1 \circ Q$ is (-1) classic by Theorem 8.49(a).

Before continuing with the study of elliptic operators, we use them to define the index of any continuous function $a\colon ST^*M \to \mathrm{Hom}(\pi_*^{-1}E, \pi_*^{-1}F)$ which has invertible values, i.e. $a\colon ST^*M \to \mathrm{Iso}(\pi_*^{-1}E, \pi_*^{-1}F)$. By the notation

$$\mathrm{Iso}(\pi_*^{-1}E, \pi_*^{-1}F)$$

we mean the bundle over T^*M whose fibre over $(x, \xi) \in T^*M$ is the vector space of linear isomorphisms $E_x \to F_x$; it is *not* a vector bundle, but rather an open subset of the vector bundle $\mathrm{Hom}(\pi_*^{-1}E, \pi_*^{-1}F)$.

Let $a \in \mathrm{Symbl}^0(E, F)$ have invertible values; for short we write $a \in \mathrm{Symbl}^0(E, F) \cap \mathit{ell}$. Then we define $\mathrm{ind}_s a$ to be the index of any elliptic p.d.o. $A \in OClS^0(E, F)$ with principal part $\pi A = a$. By virtue of Theorem 8.53, the map

$$\mathrm{ind}_s\colon \mathrm{Symbl}^0(E, F) \cap \mathit{ell} \to \mathbb{Z}$$

is well-defined. By restriction of symbols to ST^*M, there is a one-to-one correspondence

$$\mathrm{Symbl}^0(E, F) \cap \mathit{ell} \simeq C^\infty(ST^*M, \mathrm{Iso}(\pi_*^{-1}E, \pi_*^{-1}F))$$

given by $a \mapsto a|_{ST^*M}$, which we normally regard as an identification.

Note that $\mathrm{ind}_s a = \mathrm{ind}_s a_1$ if a is sufficiently near to a_1, i.e.

$$\sup_{x,\xi} \|a(x, \xi)^{-1} a_1(x, \xi) - 1\| < 1, \tag{56}$$

where $\| \ \|$ is the operator norm in the space of linear maps $E_x \to E_x$. To see this, choose $A, A_1 \in OClS^0$ with principal parts a, a_1, respectively. Then for $0 \leqslant t \leqslant 1$ the operator $A_t = (1 - t)A + tA_1$ is elliptic of order 0, so it defines a homotopy of Fredholm operators in $\mathscr{L}(L_2(M, E), L_2(M, F))$. By Theorem

9.12 it follows that ind A = ind A_1, whence $\mathrm{ind}_s\, a = \mathrm{ind}_s\, a_1$.

The following lemma is a consequence of Exercise 14 at the end of the chapter.

Lemma 8.63 *$C^\infty(ST^*M, \mathrm{Hom}(\pi_*^{-1}E, \pi_*^{-1}F))$ is dense in*

$$C(ST^*M, \mathrm{Hom}(\pi_*^{-1}E, \pi_*^{-1}F))$$

with the norm

$$|f| = \sup_{x,\xi} \|f(x, \xi)\|, \tag{57}$$

where $\|f(x, \xi)\|$ denotes the norm of the linear map $f(x, \xi)\colon E_x \to F_x$ relative to the inner products on E_x and F_x.

We can therefore extend the definition of ind_s to continuous symbols. If $a \in C(ST^*M, \mathrm{Hom}(\pi_*^{-1}E, \pi_*^{-1}F))$ has invertible values, i.e.

$$a \in C(ST^*M, \mathrm{Iso}(\pi_*^{-1}E, \pi_*^{-1}F)),$$

but not necessarily C^∞, we define

$$\mathrm{ind}_s\, a = \mathrm{ind}_s\, a_1,$$

where a_1 is any C^∞ elliptic symbol satisfying (56) (and since a_1 satisfies (56) it necessarily has invertible values). The definition is independent of the choice of a_1 since (56) implies that $\sup_{x,\xi}\|a_2(x, \xi)^{-1}a_1(x, \xi) - 1\| < 1$ for some $a_2 \in C^\infty$ arbitrarily close to a, hence $\mathrm{ind}_s\, a_1 = \mathrm{ind}_s\, a_2$.

For this extended definition of ind_s, it is still true that $\mathrm{ind}_s\, a = \mathrm{ind}_s\, a_1$ if a and a_1 are just continuous, elliptic and homogeneous symbols of degree 0 satisfying (56).

If $A \in OClS^0(E, F) \cap \mathrm{Ell}$, then, trivially, by definition of the symbol index, we have

$$\mathrm{ind}\, A = \mathrm{ind}_s(\pi A|_{ST^*M}).$$

This equation also holds if A has any order m, for if we choose an elliptic operator $\Lambda^{-m} \in OClS^{-m}(E, E)$ with principal part $|\xi|^{-m}1_E$ then $A \circ \Lambda^{-m}$ has order 0, ind $A = \mathrm{ind}(A \circ \Lambda^{-m})$ since ind $\Lambda^{-m} = 0$ (see Lemma 8.68 below), and the principal part of $A \circ \Lambda^{-m}$ is $\pi A(x, \xi)\cdot|\xi|^{-m}$ which equals $\pi A(x, \xi)$ when $|\xi| = 1$. It follows that for most questions concerning the Fredholm property and the index, it is no restriction to consider operators of order 0 (see §8.9, however, for an exception).

We now study the existence of just a one-sided parametrix. For instance, if $A \in OClS^m(E, F)$, where dim $E \neq$ dim F, then we may expect that A has a left parametrix if πA is injective and a right parametrix if πA is surjective. The basic idea in the proof of the next theorem is that πA is injective if and only if ${}^h\pi A \cdot \pi A$ is invertible so A has a left parametrix if and only if A^*A is elliptic.

Theorem 8.64 *Let $A \in OClS^m(E, F)$ be a (-1) classic matrix p.d.o. of order m. Then the following are equivalent (of course it is necessary that* $\dim F \geqslant \dim E$*):*

(i) *The operator $A\colon W_2^l(M, E) \to W_2^{l-m}(M, F)$ has finite dimensional kernel, and closed image (for any $l \in \mathbb{R}$).*
(ii) *The principal part $\pi\mathscr{A}(x, \xi)\colon E_x \to F_x$ is injective, that is, $v \in E_x$ and $\pi A(x, \xi)v = 0$ implies $v = 0$, for all $(x, \xi) \in T^*(M)\backslash 0$.*
(iii) *A has a left regularizer $B \in OClS^{-m}(F, E)$, that is, $B \circ A - I \in OS^{-1}$.*
(iv) *A has a left parametrix $B \in OClS^{-m}(F, E)$, that is, $B \circ A - I \in OS^{-\infty}$.*

Note: It follows from the equivalence of these conditions that if (i) holds for some $l \in \mathbb{R}$ then it holds for all $l \in \mathbb{R}$.

Proof The equivalence of (iii) and (iv) is proved just as in Theorem 8.60. (iv) $\Rightarrow$ (ii) If $B \circ A - I \in OS^{-\infty}$ then $\pi(B \circ A - I) = 0$, and since π is a homomorphism we get

$$\pi B(x, \xi) \cdot \pi A(x, \xi) = 1,$$

from which it follows that $\pi A(x, \xi)$ is injective.

(ii) $\Rightarrow$ (iv) Conversely, suppose that the principal part $\pi A(x, \xi)$ is injective for all $(x, \xi) \in T^*(M)\backslash 0$. By Theorem 8.49(b) the adjoint A^* belongs to $OClS^m$ and has principal part $[\pi A(x, \xi)]^h$. Injectivity of $\pi A(x, \xi)$ implies the invertibility of $[\pi A(x, \xi)]^h \circ \pi A(x, \xi)$, hence A^*A is elliptic and it has a parametrix $Q \in OClS^{-2m}$ (see Corollary 8.62). Then QA^* is a left parametrix of A.

By Theorem 9.6(a) it is clear that (iv) or (iii) implies (i). In the proof of (i) $\Rightarrow$ (ii) we need to make use of the fact that

$$t^{-s}\|u\, e^{i(\cdot, t\xi)}\|_s \to \|u\|\, |\xi|^s, \qquad u \in C_0^\infty(\mathbb{R}^n) \tag{58}$$

as $t \to \infty$, where $\|u\|$ is the L^2 norm of u. Since the Fourier transform of $u\, e^{i(\cdot, t\xi)}$ is $\hat{u}(\eta + t\xi)$, the left-hand side of (58) is equal to

$$\left((2\pi)^{-n} \int |\hat{u}(\eta)|^2 (t^{-2} + |\xi + \eta/t|^2)^s \, d\eta\right)^{1/2}$$

and the result follows by the Lebesgue dominated convergence theorem and the Plancherel theorem.

Now, if (i) holds then by Proposition 9.20 we have the a priori estimate

$$\|u\|_l \leqslant c[\|Au\|_{l-m} + \|u\|_{l-1}] \qquad \text{for all } u \in W_2^l(M, E). \tag{59}$$

It is no restriction to assume that $l = m$ since we can multiply A to the left by an elliptic operator of order $l - m$. Let (U, κ) be a local coordinate patch on which E_U and F_U are trivial; to simplify the notation we identify points in U with their local coordinates $x_1, \ldots, x_n$, and E_U and F_U with $U \times \mathbb{C}^p$ and $U \times \mathbb{C}^q$, respectively. Choose an open set W such that $\overline{W} \subset U$, a function $\psi \in C_0^\infty(W)$ and a function $\varphi \in C_0^\infty(U)$ equal to 1 on W. We have $A\psi = \varphi A\psi + (1 - \varphi)A\psi$, where $(1 - \varphi)A\psi \in OP^{-\infty}$ by quasi-locality since the supports of $1 - \varphi$ and ψ are disjoint. Hence applying (59) (with $l = m$)

we obtain

$$\|\psi u\|_m \leqslant c'[\|\varphi A\psi u\|_0 + \|\psi u\|_{m-1}], \tag{60}$$

for all $u \in C_0^\infty(\mathbb{R}^n, \mathbb{C}^p)$.

Let χ be a cut-off function, then $a = \chi \cdot \pi A$ is a main symbol of A. In terms of the local coordinates, we have $\varphi A\psi = \tilde{A}(x, D)$, where $\tilde{A} \in OClS^m(\mathbb{R}^n, q \times p)$ is (-1) classic and

$$\tilde{A}(x, \xi) - \psi(x)\cdot a(x, \xi) \in S^{m-1}(\mathbb{R}^n, q \times p). \tag{61}$$

Now, for $\xi \in \mathbb{R}^n$ and $h \in \mathbb{C}^p$, $|h| = 1$, we set

$$u_\xi(x) = \psi(x)\, e^{i(x,\xi)}\, h, \qquad \psi \in C_0^\infty(W),$$

and obtain $\varphi Au_\xi = \tilde{A}(x, \xi)\, e^{i(x,\xi)}\, h$; hence by substituting u_ξ in (60) we get

$$\|u_\xi\|_m \leqslant c'[\|\tilde{A}(\cdot, \xi)h\|_0 + \|u_\xi\|_{m-1}].$$

Now we replace ξ with $t\xi$, then multiply the inequality by t^{-m} and let $t \to \infty$, thus obtaining from (58) and (61) that

$$|\xi|^m \|\psi\|_0 \leqslant c' \|\psi \pi A(\cdot, \xi)h\|_0,$$

since $t^{-m}a(x, t\xi) = \pi A(x, \xi)$ for large t. Because $\psi \in C_0^\infty(W)$ is arbitrary, it follows that

$$|\xi|^m \leqslant c'|\pi A(x, \xi)h| \qquad \text{for all } x \in W$$

whence $\pi A(x, \xi)$ is injective for $x \in W$ and all $\xi \in \mathbb{R}^n \backslash 0$. This proves (ii) since M can be covered with such open sets W.

There is also a dual version of Theorem 8.64.

Theorem 8.65 *Let $A \in OClS^m(E, F)$ be a (-1) classic matrix p.d.o. of order m. Then the following are equivalent:*

(*i*) *The image of the operator $A: W_2^l(M, E) \to W_2^{l-m}(M, F)$ has finite codimension (for any $l \in \mathbb{R}$).*
(*ii*) *The principal part $\pi A(x, \xi)$ is surjective for all $(x, \xi) \in T^*(M)\backslash 0$, that is, for all $w \in F$, there exists $v \in E$, such that $\pi A(x, \xi)v = w$.*
(*iii*) *A has a right regularizer $B \in OClS^{-m}(F, E)$, that is, $A \circ B - I \in OS^{-1}$.*
(*iv*) *A has a right parametrix $B \in OClS^{-m}(F, E)$, that is, $A \circ B - I \in OS^{-\infty}$.*

Note: It follows from the equivalence of these conditions that if (i) holds for some $l \in \mathbb{R}$ then it holds for all $l \in \mathbb{R}$.

Proof By considering of the adjoint operator A^* the proof is reduced to that of Theorem 8.65.

Putting the two preceeding theorems together, we obtain the following result when $\dim E = \dim F$, i.e. when E and F have the same fibre dimension.

Theorem 8.66 *Let $A \in OClS^m(E, F)$ be a (-1) classic matrix p.d.o. of order m. Then the following are equivalent:*

(i) The operator $A: W_2^l(M, E) \to W_2^{l-m}(M, F)$ is Fredholm (for any $l \in \mathbb{R}$).
(ii) The principal part $\pi A(x, \xi)$ is invertible for all $(x, \xi) \in T^*(M)\backslash 0$.
(iii) A has a right and left regularizer $B \in OClS^{-m}(F, E)$.
(iv) A has a right and left parametrix $B \in OClS^{-m}(F, E)$.

The following lemma is useful in order to modify the order of a pseudo-differential operator.

Lemma 8.67 *Given any function $\lambda \in$ Symbl$^m(M)$, $m \neq 0$, such that $\lambda(x, \xi) > 0$ for all $(x, \xi) \in T^*M\backslash 0$, there exists an operator $\Lambda \in OClS^m(M)$ with principal part λ such that $\Lambda^* = \Lambda$, and $\Lambda: W_2^l(M) \to W_2^{l-m}(M)$ is an isomorphism for all $l \in \mathbb{R}$.*

Proof Assume first that $m > 0$. In view of Theorem 8.28 there exists an operator $B \in OClS^{m/2}(M)$ with principal part $\lambda^{1/2}$. We set $\Lambda = I + B^*B$ which belongs to $OClS^m(M)$ and has principal part $\lambda^{1/2} \cdot \lambda^{1/2} = \lambda$ by Theorem 8.31 because $m > 0$. Now, $(\Lambda u, u)_M \geqslant (u, u)_M$, so the operator $\Lambda: W_2^l(M) \to W_2^{l-m}(M)$ is one-to-one, and being self-adjoint its image is dense in $W_2^{l-m}(M)$. On the other hand, Λ is elliptic because $\lambda > 0$, so the image of Λ is closed. Thus Λ is bijective, hence an isomorphism by virtue of the Open Mapping Theorem.

Further, by Corollary 8.62, Λ has a (-1) classic parametrix. Since $\Lambda^{-1} \in OP^{-m}$ is certainly a parametrix (it is an inverse!) and then by Lemma 8.59 it follows that $\Lambda^{-1} \in OClS^{-m}(M)$. By Theorem 8.31, the equation $\Lambda \circ \Lambda^{-1} = I$ implies that the principal part of Λ^{-1} is λ^{-1}.

Thus, the case $m < 0$ may be handled by applying the result just proved to the function $\lambda^{-1} \in$ Symbl$^{-m}(M)$ since $-m > 0$.

Remark The same proof works even in bundles if $\lambda \in$ Symbl$^m(E, E)$ and $\lambda(x, \xi)$ is positive definite for all $(x, \xi) \in T^*M\backslash 0$.

But for most purposes the following weaker version is sufficient.

Lemma 8.68 *Let $\lambda \in$ Symbol$^m(M)$ be real-valued and suppose that $\lambda \neq 0$. If $\Lambda \in OClS^m(M)$ is any p.d.o. with principal symbol λ then* ind $\Lambda = 0$.

Proof Since λ is real-valued, then Λ and Λ^* have the same principal symbol. The difference between Λ and Λ^* is therefore in $OS^{m-1}(M)$, and by virtue of Theorem 8.53 we have ind Λ = ind $\Lambda^* = -$ind Λ, whence Λ has zero index.

We now consider briefly the case of $p \times p$ systems which are elliptic in the sense of Douglis–Nirenberg. Let the real numbers $s_1, \ldots, s_p, t_1, \ldots, t_p$ be given (we can take *any* real numbers without restriction). We denote by $OClS^{s+t}(M, p \times p)$ the set of all $p \times p$ matrix operators

$$A = [A_{ij}]_{p \times p}, \qquad A_{ij} \in OClS^{m_{ij}}(M)$$

such that each $A_{ij} \in OClS^{m_{ij}}(M)$ is (-1) classic of order $m_{ij} \leqslant s_i + t_j$. The DN principal part is the $p \times p$ matrix

$$\pi_D A(x, \xi) := [a_{ij}(x, \xi)]$$

where the entries are $a_{ij} = \pi_{s_i+t_j} A_{ij}$ if the order of A_{ij} is $s_i + t_j$, and 0 otherwise. Its elements a_{ij} are homogeneous functions in ξ of order $s_i + t_j$, so we have the homogeneity property

$$\pi_D A(x, c\xi) = S(c)\pi_D A(x, \xi)T(c), \qquad c > 0, \tag{62}$$

where $S(c)$, $T(c)$ are the diagonal matrices

$$S(c) = [\delta_{ij} c^{s_i}]_{p \times p}, \qquad T(c) = [\delta_{ij} c^{t_j}]_{p \times p}. \tag{63}$$

In general π_D is not an algebra homomorphism since the DN numbers of two operators A and B do not necessarily match up to give DN numbers for AB.

The matrix p.d.o. A is said to be DN *elliptic* if there exist real numbers $s_1, \ldots, s_p, t_1, \ldots, t_p$ such that $A \in OClS^{s+t}(M, p \times p)$ and

$$\det \pi_D A(x, \xi) \neq 0, \qquad \text{for all } (x, \xi) \in T^*(M)\backslash 0. \tag{64}$$

Theorem 8.69 *Let $A = [A_{ij}] \in OClS^{s+t}(M, p \times p)$ be a DN matrix p.d.o. with* DN *numbers $s_1, \ldots, s_p, t_1, \ldots, t_p$. Then the following are equivalent:*

(*i*) *The operator $A\colon \bigoplus_{j=1}^{p} W_2^{l+t_j}(M) \to \bigoplus_{i=1}^{p} W_2^{l-s_i}(M)$ is Fredholm (for any $l \in \mathbb{R}$)*

(*ii*) *A is* DN *elliptic, see* (64).

(*iii*) *A has a right and left regularizer $B = [B_{ij}] \in OClS^{-t-s}(M, p \times p)$ with* DN *numbers $-t_1, \ldots, -t_p, -s_1, \ldots, -s_p$.*

(*iv*) *A has a right and left parametrix $B = [B_{ij}] \in OClS^{-t-s}(M, p \times p)$ with* DN *numbers $-t_1, \ldots, -t_p, -s_1, \ldots, -s_p$.*

Note: It follows from the equivalence of these conditions that if (i) holds for some $l \in \mathbb{R}$ then it holds for all $l \in \mathbb{R}$.

Proof Choose any elliptic operators $\Lambda^k \in OClS^k(M)$, $k \in \mathbb{R}$, say with principal symbol $|\xi|^k$ for some Riemannian metric. (By Lemma 8.68 these operators have zero index but we do not need this fact here.) Let $P = [\delta_{ij}\Lambda^{-s_i}]$ and $Q = [\delta_{ij}\Lambda^{-t_j}]$ and note that P and Q are also elliptic.

(i) $\Rightarrow$ (ii) If A is a Fredholm operator, then $A' = PAQ \in OClS^0(M, p \times p)$ is a Fredholm operator in L_2. Hence $\det A'(x, \xi) \neq 0$ for all $(x, \xi) \in T^*M\backslash 0$ by Theorem 8.66. Now,

$$\pi A' = [\delta_{ij}|\xi|^{-s_i}] \cdot \pi_D A \cdot [\delta_{ij}|\xi|^{-t_j}],$$

so $\det \pi_D A(x, \xi) \neq 0$ for all $(x, \xi) \in T^*M\backslash 0$, i.e. A is DN elliptic.

(ii) $\Rightarrow$ (iv) If $\det \pi_D A(x, \xi) \neq 0$ for all $(x, \xi) \in T^*M\backslash 0$ then the same is true for $\det \pi A'(x, \xi)$. By Theorems 8.60 and 8.61, A' has a right and left classic regularizer $B' \in OClS^0(M, p \times p)$, i.e.

$$B'(PAQ) - I \in OS^{-\infty}$$

and

$$(PAQ)B' - I \in OS^{-\infty}.$$

Now, if we multiply the first expression by Q on the left and by a parametrix of Q on the right, we find that $QB'P$ is a right parametrix for A. Similarly, if we multiply the second expression by P on the right and by a parametrix of P on the left, we find that $QB'P$ is also a left parametrix for A.

(iii) $\Rightarrow$ (i) This follows as in Theorem 8.53.

(iii) $\Rightarrow$ (iv) This follows as in Theorem 8.60.

Remark The analogues of Theorems 8.64 and 8.65 also hold for rectangular $q \times p$ DN systems. This is accomplished by the simple device of left and right multiplications by diagonal matrices P and Q as in the proof of Theorem 8.69. Any questions concerning the index of A can be reduced to the index of the operator $A' = PAQ$ since ind P = ind $Q = 0$ and therefore ind A' = ind A.

For the sake of completeness we indicate a more explicit way to construct the left parametrix of $A \in OClS^{s+t}(M, q \times p)$ in the case when the principal part of πA is injective, i.e. $v \in \mathbb{C}^q$, $\pi A(x, \xi)v = 0 \Rightarrow v = 0$. Let $a = \pi A$ and $(x_0, \xi_0) \in ST^*(M)$. The injectivity of the $p \times q$ matrix $a(x_0, \xi_0)$ implies that it has a non-vanishing $q \times q$ minor. By continuity, the minor of $a(x, \xi)$ formed from the same rows and columns will not vanish for all (x, ξ) in some neighbourhood $V_0 \ni (x_0, \xi_0)$. Thus, there exists a $q \times p$ matrix function $b_0(x, \xi)$ with entries in $C^\infty(V_0)$ such that

$$b_0(x, \xi) \cdot a(x, \xi) = 1, \qquad (x, \xi) \in V_0.$$

By compactness $ST^*(M)$ can be covered with a finite number of such neighbourhoods, and piecing together these local constructions using a partition of unity we obtain a $q \times p$ matrix function with entries in $C^\infty(ST^*(M))$ such that

$$b(x, \xi) \cdot a(x, \xi) = 1, \qquad (x, \xi) \in ST^*(M). \tag{65}$$

Extending b by homogeneity we obtain

$$b = [b_{ij}], \qquad b_{ij} \text{ homogeneous of degree } -t_i - s_j$$

and (65) holding for all $(x, \xi) \in T^*(M)\backslash 0$. Now let $B = [B_{ij}] \in OClS^{-t-s}$ where B_{ij} is (-1) classic with principal part b_{ij}. In view of (65) we have

$$BA = I + T,$$

where each entry of T is in $OClS^{-1}(M)$. Since $I + T$ has a parametrix $Q \in OClS^0$, then A has left parametrix QB.

8.8 An illustration: the Hodge decomposition theorem

Let M be a compact, oriented manifold. As usual we assume that M has no boundary. We consider the following vector bundles over M

$$\textstyle\bigwedge^p(T^*M),\ p = 0, \ldots, n; \qquad \bigwedge(T^*M) = \bigoplus_{p=0}^{n} \bigwedge^p(T^*M),$$

and the spaces of sections $\Omega^p(M) = C^\infty(M, \bigwedge^p)$, $\Omega(M) = C^\infty(M, \bigwedge)$. Recall from §6.9 that the Laplace–de Rham operator, $\Delta = d\delta + \delta d$, is an elliptic second-order differential operator, $\Delta\colon C^\infty(M, \bigwedge^p) \to C^\infty(M, \bigwedge^p)$. A form α for which $\Delta\alpha = 0$ is called *harmonic*. Let

$$\mathscr{H}^p = \ker_p \Delta = \{\alpha \in C^\infty(M, \textstyle\bigwedge^p);\ \Delta\alpha = 0\}$$

denote the vector space of harmonic p-forms. In §6.9 we introduced the L_2 inner product in $C^\infty(M, \bigwedge^p)$ defined by

$$(\alpha, \eta) = \int_M \alpha \wedge {}^*\eta$$

and showed in Theorem 6.43 that d and δ are adjoints, i.e. $(d\alpha, \eta) = (\alpha, \delta\eta)$. Therefore, for $\varphi, \psi \in C^\infty(M, \bigwedge^p)$ we get

$$(\Delta\psi, \varphi) = (\delta d\psi, \varphi) + (d\delta\psi, \varphi) = (d\psi, d\varphi) + (\delta\psi, \delta\varphi) = (\psi, \Delta\varphi),$$

and it follows that $\Delta = \Delta^*$, i.e. Δ *is self-adjoint.*

Proposition 8.70 *If $\omega \in C^\infty(M, \bigwedge^p)$ then $\Delta\omega = 0$ if and only if $d\omega = 0$ and $\delta\omega = 0$. Thus harmonic forms are closed and co-closed.*

Proof If $d\omega = 0$ and $\delta\omega = 0$ then by definition $\Delta\omega = d\delta\omega + \delta d\omega = 0$. Conversely, if $d\omega = 0$ and $\delta\omega = 0$ then the previous computation shows that $0 = (\Delta\omega, \omega) = (d\omega, d\omega) + (\delta\omega, \delta\omega)$, hence $d\omega = 0$ and $\delta\omega = 0$.

Since Δ is elliptic then by Theorem 8.53 we have

$$\ker \Delta \subset C^\infty(M, \textstyle\bigwedge), \qquad \dim \ker \Delta < \infty.$$

Theorem 8.71 (*Hodge Decomposition Theorem*) *Let $\omega \in C^\infty(M, \bigwedge^p)$, $p = 0, \ldots, n$. Then there exist forms $\alpha \in C^\infty(M, \bigwedge^{p-1})$, $\beta \in C^\infty(M, \bigwedge^{p+1})$ and $\gamma \in C^\infty(M, \bigwedge^p)$ such that*

$$\omega = d\alpha + \delta\beta + \gamma \quad \textit{and} \quad \Delta\gamma = 0. \tag{66}$$

Further, this decomposition is orthogonal, hence unique, so we have the direct sum

$$C^\infty(M, \textstyle\bigwedge^p) = \ker_p \Delta \oplus \operatorname{im}_p d \oplus \operatorname{im}_p \delta \tag{67}$$

where $\operatorname{im}_p d$ is the image of $d\colon C^\infty(M, \bigwedge^{p-1}) \to C^\infty(M, \bigwedge^p)$ and $\operatorname{im}_p \delta$ is the image of $\delta\colon C^\infty(M, \bigwedge^{p+1}) \to C^\infty(M, \bigwedge^p)$.

Proof Since Δ is elliptic and $\Delta = \Delta^*$ then by Theorem 8.53 we have

$$\ker_p \Delta \subset C^\infty(M, \textstyle\bigwedge^p), \qquad \dim \ker_p \Delta < \infty, \tag{68}$$

and by Corollary 8.56 we have the orthogonal decomposition

$$C^\infty(M, \textstyle\bigwedge^p) = \Delta(C^\infty(M, \bigwedge^p)) \oplus \ker_p \Delta, \qquad p = 0, \ldots, n \tag{69}$$

Thus we can write a p-form $\omega \in C^\infty(M, \bigwedge^p)$ as $\omega = \Delta\eta + \gamma$ for some $\gamma \in \ker_p \Delta$ and $\eta \in C^\infty(M, \bigwedge^p)$. Therefore, $\omega = d\delta\eta + \delta d\eta + \gamma$, and to get (66) we can choose $\alpha = \delta\eta$ and $\beta = d\eta$. Hence we have established the sum

$$C^\infty(M, \wedge^p) = \ker_p \Delta + \operatorname{im}_d d + \operatorname{im}_d \delta.$$

If we show that the factors in this sum are orthogonal to each other, we are done because uniqueness of (66) follows from orthogonality.

(1) $\operatorname{im}_p d \perp \operatorname{im}_p \delta$. Let $u = d\alpha \in \operatorname{im}_p d$ and $v = \delta\beta \in \operatorname{im}_p d$. Then, since d is the adjoint of δ, we get $(u, v) = (d\alpha, \delta\beta) = (d^2\alpha, \beta) = 0$.
(2) $\operatorname{im}_p d \perp \ker_p \Delta$. Let $\Delta\omega = 0$ and $u = d\alpha \in \operatorname{im}_p d$. Then $\delta\omega = 0$ (see Proposition 8.70) and $(\omega, u) = (\omega, d\alpha) = (\delta\omega, \alpha) = 0$.
(3) $\operatorname{im}_p \delta \perp \ker_p \Delta$. Now, from $\Delta\omega = 0$ it follows that $d\omega = 0$, thus composing with $v = \delta\beta$ we get $(\omega, v) = (\omega, \delta\beta) = (d\omega, \beta) = 0$.

Definition 8.72 *A p-form α on a manifold M is called closed if $d\alpha = 0$; it is called exact if there is a $(p-1)$-form β such that $\alpha = d\beta$. Since $d^2 = 0$, every exact form is closed. We define the vector spaces*

$$Z^p(M) := \{\omega \in C^\infty(M, \wedge^p); d\omega = 0\}$$

$$B^p(M) := \{\omega \in Z^p(M); \omega = d\alpha, \alpha \in C^\infty(M, \wedge^{p-1})\}$$

and the quotient vector space

$$H^p(M) = Z^p(M)/B^p(M), \qquad p = 0, \ldots, n \tag{70}$$

is called the pth de Rham cohomology group of M. The points in $H^p(M)$, i.e. the classes in $Z^p(M)$ modulo $B^p(M)$, are called cohomology classes.

Theorem 8.73 *Each cohomology class in $Z^p(M)$ contains a unique harmonic representative from* $\ker_p \Delta$.

Proof Let $\omega \in Z^p(M)$. From the Hodge decomposition theorem 8.71 we have $\omega = d\alpha + \delta\beta + \gamma$, where $\gamma \in \ker_p \Delta$. Since γ is harmonic then $d\gamma = 0$ and, further, since $\omega \in Z^p(M)$ then $d\omega = 0$, so that $d\delta\beta = 0$ too. It follows that $(\delta\beta, \delta\beta) = (\beta, d\delta\beta) = 0$, so $\delta\beta = 0$. Thus $\omega = d\alpha + \gamma$, and the cohomology class $[\omega]$ contains the harmonic representative γ, i.e. $[\omega] = [\gamma]$.

If two harmonic forms α_1 and α_2 differ by an exact form $d\beta$, then we have

$$0 = d\beta + (\alpha_1 - \alpha_2). \tag{71}$$

But since α_1 and α_2 are harmonic then $\delta\alpha_1 = \delta\alpha_2 = 0$, and it follows that $d\beta$ and $(\alpha_1 - \alpha_2)$ are orthogonal because $(d\beta, \alpha_1 - \alpha_2) = (\beta, \delta\alpha_1 - \delta\alpha_2) = (\beta, 0) = 0$. Hence by uniqueness of orthogonal decompositions we have

$$d\beta = 0 \quad \text{and} \quad \alpha_1 - \alpha_2 = 0.$$

Thus, there is a unique harmonic form $\alpha_1 = \alpha_2$ in each de Rham cohomology class.

Using Theorem 8.73 we immediately obtain that the de Rham cohomology groups are finite dimensional.

Theorem 8.74 *The vector spaces*

$$H^p(M) \simeq \ker_p \Delta, \qquad p = 0, \ldots, n$$

are isomorphic; and since the vector spaces $\ker_p \Delta$ *are finite dimensional, we have established the finite dimensionality of* $H^p(M)$,

$$\dim H^p(M) < \infty.$$

Proof We map $H^p(M) \to \ker_p \Delta$ by $[\omega] \mapsto \gamma$ where γ is the unique harmonic form such that $\omega - \gamma$ is exact. It is easily verified that the map is linear, and it is evident that it is surjective (since harmonic forms are closed). It is also injective because if $\gamma = 0$ then ω is exact, whence $[\omega] = 0$. This establishes an isomorphism $H^p(M) \simeq \ker_p \Delta$.

8.9 Limits of pseudo-differential operators

In this section we look at limits of pseudo-differential operators in the operator norm. Much of the Fredholm theory of elliptic p.d.o.'s can be carried over to limits of such operators.

The primary reason why we introduce these limits is the following. If $A: C^\infty(M) \to C^\infty(M)$ is a pseudo-differential operator on M then we can define

$$A': C^\infty(M \times N) \to C^\infty(M \times N)$$

by letting A act on just the first variable. But A' need not be a p.d.o. because in local coordinates the estimates for a function $\tilde{a}'(x, y; \xi, \eta) = \tilde{a}(x, \xi)$, namely,

$$|D^\alpha_{(\xi,\eta)} D^\beta_{(x,y)} \tilde{a}(x, \xi)| \leqslant c(1 + |\xi| + |\eta|)^{m-|\alpha|},$$

cannot hold for $|\alpha| > m$ unless $D^\alpha_\xi D^\beta_x \tilde{a}$ vanishes identically (i.e. when $\tilde{a}$ is polynomial in ξ). Also, note that if $\tilde{a}(x, \xi)$ were defined and non-zero only for $|\xi| > C$ this would not affect the Fredholm property of the operator A because $\{\xi \in \mathbb{R}^n; |\xi| \leqslant C\}$ is compact. But for A' we must regard $\mathbb{R}^n$ as embedded in $\mathbb{R}^n \times \mathbb{R}^{n'}$; then $\{(\xi, \eta) \in \mathbb{R}^n \times \mathbb{R}^{n'}; |\xi| \leqslant C\}$ is not compact because it contains all the points in $0 \times \mathbb{R}^{n'}$.

When $m > 0$, however, it can be shown that $A' = A \otimes I$ is a limit in OP^m of pseudo-differential operators of order m (see Lemma 8.78) and then we may use Theorem 8.76. The basic idea here is that when $m > 0$ the principal part, $(\pi A \otimes 1)(x, y, \xi, \eta) = |\xi|^m \pi A(x, \xi/|\xi|)$, is continuous for $|\xi| + |\eta| > 0$ (including $\xi = 0$).

We begin with the following lemma.

Lemma 8.75 *Let* $A_j \in OClS^m(E, F)$ *be* (-1) *classic operators,* $j = 1, 2, \ldots,$ *and assume that* A_j *converges in the operator norm in* $\mathscr{L}(W^l_2, W^{l-m}_2)$ *for some* $l \in \mathbb{R}$. *Then the principal parts* πA_j *converge uniformly on compact subsets of* $T^*M\backslash 0$.

Proof If $A \in S^m(\mathbb{R}^n \times \mathbb{R}^n)$ has main symbol $\chi \cdot \pi A$ then

$$t^{-m}(A(x, D)(u\, e^{i(\cdot, t\xi)}), v\, e^{i(\cdot, t\xi)})_{L_2} = (t^{-m}A(x, D + t\xi)u, v)_{L_2} \to (\pi A(\cdot, \xi)u, v)$$

for $u, v \in C_0^\infty(\mathbb{R}^n)$; see the proof of Theorem 7.40. It follows that

$$|(\pi A(\cdot, \xi)u, v)_{L_2}| \leqslant \lim_{t \to \infty} t^{-m}|(A(x, D)(u\, e^{i(\cdot, t\xi)}), v\, e^{i(\cdot, t\xi)})|$$

and, by making use of (58),

$$|(\pi A(\cdot, \xi)u, v)_{L_2}| \leqslant M\|u\|\,\|v\|, \qquad |\xi| = 1,$$

where M is the norm of $A(x, D)$ as operator from W_2^l to W_2^{l-m} and $\|\ \|$ is the L_2 norm. Since $u, v \in C_0^\infty$ are arbitrary we have

$$\sup_{|\xi|=1} |\pi A(x, \xi)| \leqslant M.$$

If we apply this result to $\varphi(A_j - A_k)\psi$ with $\varphi, \psi \in C^\infty(M)$ having support in a coordinate patch on which E and F are trivial, it follows that $\pi A_j - \pi A_k$ converges to 0 uniformly on compact subsets of $T^*M\backslash 0$ as j and $k \to \infty$, so $\lim \pi A_j$ exists and is a continuous function, homogeneous of degree m.

Theorem 8.76 *Let A: $W_2^l(M, E) \to W_2^{l-m}(M, F)$ be bounded, and suppose that there exist $A_j \in OClS^m(E, F)$ such that $A_j \to A$ in $\mathscr{L}(W_2^l, W_2^{l-m})$ for some $l \in \mathbb{R}$. We define the principal part of A to be the function $\pi A := \lim_{j\to\infty} \pi A_j$. (The limit exists by Lemma* 8.75, *and defines a continuous function on $T^*M\backslash 0$ homogeneous of degree m.) If*

$$\pi A(x, \xi) \text{ is invertible for all } (x, \xi) \in ST^*M, \tag{72}$$

then A is a Fredholm operator, with index equal to $\operatorname{ind}_s \pi A$ (see §8.7, *text following Corollary* 8.62, *for the definition of* ind_s).

Proof Assume for the moment that $l = m = 0$. Since πA has invertible values and the homogeneous symbols πA_j converges uniformly on $T^*M\backslash 0$ by Lemma 8.75, there is a uniform bound for $(\pi A_j)^{-1}$ on $T^*M\backslash 0$. By Theorem 7.40 we can choose a pseudo-differential operator $B_j \in OClS^0$ with principal part $(\pi A_j)^{-1}$ such that the norm of B_j in $\mathscr{L}(L_2, L_2)$ is uniformly bounded. Now

$$B_jA = I + B_j(A - A_j) + (B_jA_j - I),$$

and $I + B_j(A - A_j)$ is invertible in $\mathscr{L}(L_2, L_2)$ for large j, and $B_jA_j - I$ is compact. Hence B_jA is a Fredholm operator for large j, so $\ker A$ is finite dimensional. For similar reasons AB_j is a Fredholm operator, whence A has closed image of finite codimension. Hence A is a Fredholm operator.

We have $\operatorname{ind} A_j = \operatorname{ind}_s \pi A_j$ for any j by definition of ind_s, and since both the analytical index, ind, and the symbol index, ind_s, are locally constant then by taking large j it follows that $\operatorname{ind} A = \operatorname{ind}_s \pi A$.

By Theorem 8.48 for each $k \in \mathbb{R}$ there exists $\Lambda_E^k \in OClS^k(E, E)$ with principal part $|\xi|^k \cdot 1_E$, and similarly for F. Then, for arbitrary l and m, it follows from what was just proved that $\Lambda_F^{l-m} \circ A \circ \Lambda_E^{-l} \in OClS^0(E, F)$ is a

Fredholm operator. If we multiply by Λ_F^{m-l} on the left and Λ_E^l on the right, we find that A is a compact perturbation of a Fredholm operator, hence A is also Fredholm. For large j the index of A is equal to that of A_j, thus equal to $\operatorname{ind}_s \pi A_j = \operatorname{ind}_s \pi A$.

We define a topology on $OP^m(E, F)$ as follows: Convergence holds in OP^m, i.e.

$$A_j \to A \qquad \text{in } OP^m$$

if and only if the convergence holds in operator norm $W_2^l(M, E) \to W_2^{l-m}(M, F)$ for all $l \in \mathbb{R}$. Also, let $\mathscr{P}^m$ denote the closure of $OClS^m(E, F)$ in $OP^m(E, F)$.

Since $OClS^m(E, F)$ and $OP^m(E, F)$ are local spaces of operators, it follows that $\mathscr{P}^m$ is also a local space, i.e. $A \in \mathscr{P}^m$ if and only if $\varphi A \psi \in \mathscr{P}^m$ for every pair of functions $\varphi, \psi \in C^\infty(M)$.

Corollary 8.77 *Suppose that $A \in \mathscr{P}^m$, that is, $A \in OP^m(E, F)$ and there exist $A_j \in OClS^m(E, F)$ such that $A_j \to A$ in OP^m. Then the hypothesis of Theorem 8.76 holds for* all $l \in \mathbb{R}$, *so if $\pi A(x, \xi)$ is invertible for all $(x, \xi) \in ST^*M$ then*

ker A *is a finite dimensional subspace of $C^\infty(M, E)$, and*

im A *is the orthogonal space of a finite dimensional subspace of $C^\infty(M, F)$.*

Proof The subspace $\ker A \subset W_2^l(M, E)$ decreases with l. Also the codimension of im A is equal to the dimension of $\ker A^* \subset W_2^{m-l}(M, F)$ which increases with l. Since the index is independent of l, it follows that ker A and ker A^* are independent of l. Thus,

$$\ker A \subset \bigcap W_2^l(M, E) = C^\infty(M, E)$$

and similarly $\ker A^* \subset C^\infty(M, F)$, which completes the proof since im A is closed in $W_2^{l-m}(M, F)$.

Remark Obviously it suffices that the hypotheses of Corollary 8.77 hold for all integers l (or just for some sequence $l_j \to \infty$).

The following lemma is often used to construct the approximating sequence in Theorem 8.76. To be able to apply this lemma, however, it is necessary to show the existence of the function χ. First we choose a function $\sigma \in C^\infty(\mathbb{R}^{n+n'})$, $0 \leqslant \sigma \leqslant 1$, which is homogeneous of degree 0 and

$$\rho(\xi, \eta) = \begin{cases} 0 & \text{for } |\eta| \leqslant |\xi| \\ 1 & \text{for } |\eta| \geqslant 2|\xi| \end{cases}$$

Also let $\phi \in C^\infty(\mathbb{R})$ such that $\phi(\lambda) = 0$ for $|\lambda| \leqslant \sqrt{2}$, $\varphi(\lambda) = 1$ for $|\lambda| \geqslant 2$, and put

$$\chi(\xi, \eta) = 1 - \phi((|\xi|^2 + |\eta|^2)^{1/2}) \cdot \sigma(\xi, \eta).$$

Then $\chi \in C^\infty(\mathbb{R}^{n+n'})$ and the appearance of this function is indicated below ($\chi = 1$ in the horizontally shaded region, and $\chi = 0$ in the vertically shaded region).

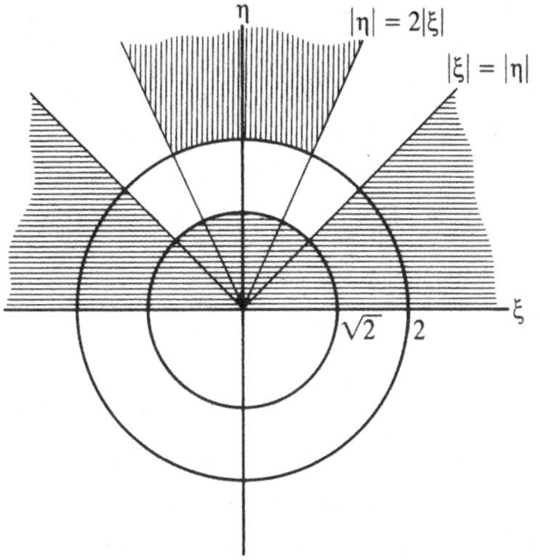

Fig. 8. Values of the function χ.

Lemma 8.78 *Let $A(x, y, \xi) \in S^m(\mathbb{R}^{n+n'} \times \mathbb{R}^n)$ be (-1) classic where $m > 0$. Choose $\chi \in C^\infty(\mathbb{R}^{n+n'})$ so that $\chi(\xi, \eta) = 1$ when $|\eta| \leqslant \max(1, |\xi|)$ and $\chi(\xi, \eta) = 0$ when $|\eta| \geqslant \max(2, 2|\xi|)$, while $\chi(\xi, \eta)$ is homogeneous of degree 0 when $|\eta| > 2$. Then*

$$A_\varepsilon(x, y, \xi, \eta) = A(x, y, \xi)\chi(\xi, \varepsilon\eta) \in ClS^m(\mathbb{R}^{n+n'} \times \mathbb{R}^{n+n'}),$$

the principal symbol πA_ε converges uniformly to πA on $|\xi|^2 + |\eta|^2 = 1$ as $\varepsilon \to 0$, and the norm of $A_\varepsilon(x, y, D_x, D_y) - A(x, y, D_x)$ as operator from $W_2^l(\mathbb{R}^{n+n'})$ to $W_2^{l-m}(\mathbb{R}^{n+n'})$ tends to 0 as $\varepsilon \to 0$ for every $l \in \mathbb{R}$. (In particular, it follows that $A(x, y, D_x) \in OP^m(\mathbb{R}^{n+n'})$.)

Proof Since χ is homogeneous of degree 0 when $|\xi|^2 + |\eta|^2 > 8$ (in fact >4 for the function constructed above) it follows that $\chi \in OClS^0(\mathbb{R}^{n+n'} \times \mathbb{R}^{n+n'})$. Also it is clear that $A_\varepsilon \in S^m$ in both variables ξ, η, that is,

$$|D_x^\alpha D_y^{\alpha'} D_\xi^\beta D_\eta^{\beta'} A_\varepsilon(x, y, \xi, \eta)| \leqslant \text{const}(1 + |\xi| + |\eta|)^{m-|\beta|-|\beta'|},$$

because $\chi(\xi, \eta) \neq 0$ only when $|\eta| < 2|\xi|$; and then it is not hard to see that $A_\varepsilon \in OClS^m$ with principal part $\pi A(x, y, \xi) \cdot \pi\chi(\xi, \varepsilon\eta)$. The magnitude of the difference between the principal parts πA_ε and πA is bounded by a constant times $|\xi|^m \cdot |\pi\chi(\xi, \varepsilon\eta) - 1|$ which tends to 0 uniformly on $|\xi|^2 + |\eta|^2 = 1$ as $\varepsilon \to 0$ because $m > 0$, $\pi\chi$ is bounded, and $\pi\chi(\xi, \varepsilon\eta) \to \pi\chi(\xi, 0) = 1$ uniformly when $|\xi| \geqslant \text{const} > 0$.

Now we must prove convergence of the corresponding p.d.o.'s in Sobolev norm. For fixed y, let $C(y)$ the norm of the operator $A(x, y, D_x)$: $W_2^m(\mathbb{R}^n) \to L_2(\mathbb{R}^n)$ and note that $C(y)$ is bounded by a semi-norm of A, thus $C(y)$ is bounded by a constant C independent of y. If $v \in \mathscr{S}(\mathbb{R}^n \times \mathbb{R}^{n'})$ we then have for fixed y

$$\int |A(x, y, D_x)(x, y)|^2 \, dx \leqslant C\|v(\cdot, y)\|_m^2,$$

and then integrating with respect to y,

$$\|A(x, y, D_x)v\|_0^2 \leqslant C \int \|v(\cdot, y)\|_m^2 \, dy = C \iint |\hat{v}(\xi, \eta)|^2 (1 + |\xi|^2)^m \, d\xi \, d\eta.$$

Now, since

$$A_\varepsilon(x, y, D_x, D_y)u - A(x, y, D_x)u = A(x, y, D_x)v_\varepsilon$$

where $v_\varepsilon = \chi(D_x, \varepsilon D_y)u - u$, it follows that

$$\begin{aligned} &\|A_\varepsilon(x, y, D_x, D_y)u - A(x, y, D_x)u\|_0^2 \\ &\qquad \leqslant C \iint |\hat{u}(\xi, \eta)|^2 (1 + |\xi|^2)^m |\chi(\xi, \varepsilon\eta) - 1|^2 \, d\xi \, d\eta \\ &\qquad \leqslant \|u\|_m^2 \cdot \sup C(1 + |\xi|^2)^m |\chi(\xi, \varepsilon\eta) - 1|^2 (1 + |\xi|^2 + |\eta|^2)^{-m} \qquad (73) \end{aligned}$$

When $\chi(\xi, \varepsilon\eta) \neq 1$ we have $1 + |\xi|^2 \leqslant 2|\varepsilon\eta|^2$, hence

$$1 + |\xi|^2 \leqslant 2\varepsilon^2(1 + |\xi|^2 + |\eta|^2).$$

This proves

$$\|(A_\varepsilon(x, y, D_x, D_y) - A(x, y, D_x))u\|_0 \leqslant \text{const } \varepsilon^m \|u\|_m, \qquad (74)$$

and next we shall prove that

$$\|(A_\varepsilon(x, y, D_x, D_y) - A(x, y, D_x))u\|_l \leqslant C_l \varepsilon^m \|u\|_{m+l} \qquad (74')$$

if l is a positive integer. To do so we observe that $\chi(D_x, \varepsilon D_y)$ commutes with differentiations while the commutator of $A(x, y, D_x)$ and a differentiation is of the same form as A (see Proposition 7.24). Hence

$$D^\alpha(A_\varepsilon(x, y, D_x, D_y) - A(x, y, D_x))u$$

is a sum of terms of the form

$$(B_\varepsilon(x, y, D_x, D_y) - B(x, y, D_x)) \cdot D^\beta u$$

with $|\beta| \leqslant |\alpha|$ and B of the same form as A. This gives (74) when l is a positive integer. If $l = -k$ is a negative integer and $u \in W_2^{m+l}$, we claim that it is possible to write u in the form

$$u = \sum_{|\alpha| \leqslant -l} u_\alpha, \qquad \text{where} \sum_{|\alpha| \leqslant -l} \|u_\alpha\|_m^2 \leqslant C\|u\|_{m+l}^2. \qquad (75)$$

If we commute the derivatives D^α with $A(x, y, D_x)$ as above and apply (74), we obtain (74′) also when l is a negative integer. By interpolation, (74′) holds for all $l \in \mathbb{R}$ (but $l \in \mathbb{Z}$ will suffice for the applications).

Remark It is easy to verify (75) when $l = -1$, for if $u \in W_2^{m-1}$, we write $u = \sum D_j u_j + u_0$, where $\hat{u}_j = \xi_j(1 + |\xi|^2)^{-1}\hat{u}(\xi)$ and $\hat{u}_0 = (1 + |\xi|^2)^{-1}\hat{u}$. The general case then follows by induction on l.

Let M and N be compact manifolds, E, F vector bundles over M and G a vector bundle over N. Let $E \boxtimes G$ be the vector bundle over $M \times N$ with fibres $E_x \otimes G_y$, $(x, y) \in M \times N$, which we call the exterior tensor product to distinguish it from the tensor product defined in §5.8. If $f \in C^\infty(M, E)$ and $g \in C^\infty(N, G)$ we define $fg \in C^\infty(M \times N, E \boxtimes G)$ by $(fg)(x, y) = f(x)g(y)$. Note that finite sums of the form

$$\sum f_j g_j, \qquad f_j \in C^\infty(M, E),\ g_j \in C^\infty(N, G)$$

are dense in $C^\infty(M \times N, E \boxtimes G)$; see Exercise 17.

Theorem 8.79 *Let M and N be compact manifolds, let E, F be vector bundles over M and G a vector bundle over N. Let $m > 0$, $A \in OClS^m(E, F)$. Then there exists a unique continuous, linear operator*

$$A'\colon C^\infty(M \times N, E \boxtimes G) \to C^\infty(M \times N, F \boxtimes G),$$

such that $A'(f \cdot g) = Af \cdot g$ for all $f \in C^\infty(M, E)$, $g \in C^\infty(N, G)$. Moreover, $A' \in \mathscr{P}^m$, i.e. there exist operators $A'_\varepsilon \in OClS^m(E \boxtimes G, F \boxtimes G)$ such that $A'_\varepsilon \to A'$ in OP^m with principal symbols converging to $\pi A \otimes 1$ uniformly on $ST^*(M \times N)$, where 1 is the identity on G.

Remark We also use the notation $A' = A \otimes I$, where I is the identity on $C^\infty(M, G)$, hence

$$\pi(A \otimes I) = \pi A \otimes I.$$

Proof Uniqueness of A' is clear since the action of A' on the dense subset of separated functions is

$$A'\left(\sum f_j \cdot g_j\right) = \sum (Af^j) \cdot g_j.$$

On the other hand, if we define A' in this way, and if we show the existence of operators $A'_\varepsilon \in OClS^m(E \boxtimes G, F \boxtimes G)$ such that

$$\|(A' - A'_\varepsilon)u\|_l \leqslant C_l \varepsilon^m \|u\|_{m+l}$$

for all $l \in \mathbb{R}$ and functions $u = \sum f_j g_j$, then it follows that $A' \in \mathscr{P}^m$; in particular A' can be extended continuously to a map $C^\infty(M \times N, E \boxtimes G) \to C^\infty(M \times N, F \boxtimes G)$ (also see Exercise 18).

There exists an atlas $\{(U_j, \kappa_j, \beta_j^E, \beta_j^F)\}$ on M in which the covering is so fine that $U_j \cup U_k$ is also contained in a coordinate patch whenever $U_j \cap U_k \neq \varnothing$ (see Exercise 16); then the same is true when $U_j \cup U_k = \varnothing$ because we can join the coordinate systems to get $(U_j \cup U_k, \kappa_j \cup \kappa_k)$ as in the remark preceding Proposition 8.8. By using a partition of unity on M subordinate to this atlas, and a similar one on N, we can write A' as a sum of operators of the form $\varphi\varphi_1 A' \psi\psi_1$ where $\varphi, \psi \in C_0^\infty(U)$, $U \subset M$ is a chart domain on which E, F are trivial, $\varphi_1, \psi_1 \in C_0^\infty(U_1)$ and $U_1 \subset N$ is a chart domain on which G is trivial, so it suffices to show that

$$\varphi\varphi_1 A' \psi\psi_1 \in \mathscr{P}^m. \tag{76}$$

Let us identify points in U and U_1 with their coordinates in $\mathbb{R}^n$ and $\mathbb{R}^{n'}$, respectively, and E, F and G with the trivial bundles $U \times \mathbb{C}^p$, $U \times \mathbb{C}^q$ and $U_1 \times \mathbb{C}^r$, respectively. By shrinking U we may assume that $\varphi A \psi = \varphi(\tilde{A}(x, D))\psi$ where $\tilde{A} \in S^m(\mathbb{R}^n, q \times p)$ (in terms of the local coordinates and trivializations) and we have

$$\varphi\varphi_1 A'\psi\psi_1 = \varphi\tilde{A}(x, y, D_x)\psi, \qquad \text{where } \tilde{A}(x, y, \xi) = \psi_1^2(y)\tilde{A}(x, \xi).$$

Since $\tilde{A}(x, y, \xi) \in ClS^m(\mathbb{R}^{n+n'} \times \mathbb{R}^n, qr \times pr)$, we find

$$\tilde{A}_\varepsilon \in S^m(\mathbb{R}^{n+n'} \times \mathbb{R}^{n+n'}, qr \times pr)$$

having the properties in Lemma 8.78. We let $A'_\varepsilon \in OS^m(E \boxtimes G, F \boxtimes G)$ be the pull-back of $\tilde{A}_\varepsilon$ to $M \times N$ (see Lemma 8.11), and it follows that $A'_\varepsilon \to \varphi\varphi_1 A'\psi\psi_1$ in OP^m with principal symbol converging uniformly to $\varphi\varphi_1 \pi A \otimes 1\psi\psi_1$, so (76) holds.

8.10 The index of elliptic symbols

In this section we define the index of any continuous section $a\colon T^*M \to \mathrm{Hom}(\pi_*^{-1}E, \pi_*^{-1}F)$ which has invertible values outside a compact subset of T^*M. Recall that if a is homogeneous, i.e. determined by its restriction to ST^*M, we have already defined the symbol index, $\mathrm{ind}_s\, a$, in the text following Corollary 8.62. By making use of homotopies, we will show that it is possible to extend the definition of $\mathrm{ind}_s\, a$ (see Theorems 8.89 and 8.90).

We first prove a few more results concerning the case of homogeneous symbols (i.e. for the symbols of classic operators), and then prove Noether's formula for the index of an elliptic p.d.o. on the unit circle (Theorem 8.83).

The following lemma implies a converse to Theorem 8.76, that is, (72) is also *necessary* for $A = \lim A_j$ to be Fredholm.

Lemma 8.80 *If $A \in OClS^0(E, F)$ then*

$$\sup_{x,\xi} \|\pi A(x, \xi)\| = \inf_K \|A + K\| \tag{77}$$

where the infimum is taken over all compact operators K from $L_2(M, E)$ to $L_2(M, F)$. ($\|\pi A(x, \xi)\|$ denotes the operator norm of $\pi A(x, \xi)\colon E_x \to F_x$ with respect to the Hermitian inner products on E_x and F_x.)

Remark We will use the notation $|\pi A| := \sup_{x,\xi} \|\pi A(x, \xi)\|$.

Proof The sequence $f_k e$ in Corollary 8.52 converges weakly to 0 in $L_2(M, E)$, so that $K(f_k e) \to 0$ in $L_2(M, F)$. Since $\|f_k e\|_{L_2} = 1$, it follows that

$$\|\pi A(x, \xi)\| \leqslant \|A + K\|$$

for any compact operator K.

The proof of the other inequality could be based on Theorem 7.40, but for the sake of interest we give another proof which is due to [Se 2]. Let us prove it first when $E = F$ and $A = A^*$ is self-adjoint. Recall that for any

Fredholm operator B, the image of B is the orthogonal complement of the kernel of B^*. In particular, when B is a self-adjoint Fredholm operator, B is invertible if and only if its kernel is trivial.

Now, suppose λ_0 is an eigenvalue of A and $|\lambda_0| > |\pi A|$. Then $\lambda_0 I - A$ is an elliptic p.d.o., so it has a finite dimensional kernel; the orthogonal projection P_0 on this kernel is therefore compact. Since λ_0 is real, the operator $\lambda_0 I - A - P_0$ is a self-adjoint Fredholm operator, and it has trivial kernel so it is invertible; and further there is a $\delta > 0$ such that $|\lambda_0| > |\pi A| + \delta$ and $\lambda I - A - P_0$ is invertible for $|\lambda - \lambda_0| < \delta$. We claim that if λ is a real number such that $0 < |\lambda - \lambda_0| < \delta$ then $\lambda \notin \mathrm{sp}(A)$. In fact, if $\lambda f - Af = 0$ then $(\lambda - \lambda_0)f = -\lambda_0 f + Af$ belongs to the image of $\lambda_0 I - A$ and is hence orthogonal to the image of P_0. Since $\lambda - \lambda_0 \neq 0$ it follows that f is orthogonal to the image of P_0, so that $P_0 f = 0$. Hence $\lambda f - Af - P_0 f = 0$, which implies $f = 0$ since $|\lambda - \lambda_0| < \delta$. Therefore, $\lambda I - A$ has trivial kernel, so it is invertible.

Thus we have shown that the eigenvalue of A satisfying $|\lambda| > |\pi A|$ are isolated points of finite multiplicity. Let $\varepsilon > 0$ be given. There are only finitely many eigenvalues λ of A such that $|\lambda| > |\pi A| + \varepsilon$; we denote them by $\lambda_1, \dots, \lambda_N$ and let $P_1, \dots, P_N$ be the projections on the corresponding eigenspaces. Then $\sum \lambda_j P_j$ is compact, and $\lambda I - (A - \sum \lambda_j P_j)$ is invertible for $|\lambda| > |\pi A| + \varepsilon$, that is,

$$\mathrm{sp}\left(A - \sum \lambda_j P_j\right) \subset \{\lambda; |\lambda| \leqslant |\pi A| + \varepsilon\},$$

hence

$$\left\| A - \sum \lambda_j P_j \right\| \leqslant |\pi A| + \varepsilon,$$

which proves the required inequality.

In the case where A is not self-adjoint, we can consider the operator on $L_2(M, E \oplus F) = L_2(M, E) \oplus L_2(M, F)$ defined by $\tilde{A}(e, f) = (A^* f, Ae)$. Here $E \oplus F$ denotes the orthogonal direct sum of the bundles E and F, and $L_2(M, E) \oplus L_2(M, F)$ is the orthogonal direct sum of the Hilbert spaces $L_2(M, E)$ and $L_2(M, F)$. It is not hard to verify that $\|\tilde{A}\| = \|A\|$ and $|\pi \tilde{A}| = |\pi A|$, and $\tilde{A}$ is a self-adjoint p.d.o. Further, $\tilde{A}$ can be represented in block operator form by the matrix

$$\begin{pmatrix} 0 & A^* \\ A & 0 \end{pmatrix}$$

and in view of what we proved above there is a matrix K of compact operators such that the operator

$$\begin{pmatrix} K_{11} & A^* + K_{12} \\ A + K_{21} & K_{22} \end{pmatrix}$$

has norm $\leqslant |\pi A| + \varepsilon$. Since the norm of each entry in such a matrix is dominated by the norm of the operator represented by the matrix, we have $\|A + K_{21}\| \leqslant |\pi A| + \varepsilon$, and the lemma is proved.

We denote the closure of $OClS^0(E, F)$ in the L_2 operator norm by

$$\mathscr{P} = \mathscr{P}(E, F).$$

Clearly $\mathscr{P} \supset \mathscr{K}$, the set of compact operators $L_2(M, E) \to L_2(M, F)$. In fact, operators of finite rank $Pf = \sum (f, g_i)_E h_i$, where $g_i \in C^\infty(M, E)$, $h_i \in C^\infty(M, F)$, are dense in $\mathscr{K}$, and such operators belong to $OClS^0(E, F)$ because they are operators of order $-\infty$.

Since we can identify $\mathrm{Symbl}^0(E, F)$ with $C^\infty(ST^*M, \mathrm{Hom}(\pi_*^{-1}E, \pi_*^{-1}F))$ by restriction to ST^*M, we have a principal symbol map

$$\pi = \pi|_{ST^*M}: OClS^0(E, F) \to C^\infty(ST^*M, \mathrm{Hom}(\pi_*^{-1}E, \pi_*^{-1}F)) \qquad (78)$$

which is surjective by Theorem 8.48. By virtue of Lemma 8.80, the map π induces an *isometry* $\bar{\pi}: \mathscr{P}/\mathscr{K} \to C(ST^*M, \mathrm{Hom}(\pi_*^{-1}E, \pi_*^{-1}F))$, which necessarily has closed image because $\mathscr{P}/\mathscr{K}$ is a Banach space (i.e. a complete space). Further, the image of $\bar{\pi}$ contains $C^\infty(ST^*M, \mathrm{Hom}(\pi_*^{-1}E, \pi_*^{-1}F))$ and these sections are dense in the continuous sections by Lemma 8.63, so $\bar{\pi}$ is in fact surjective. In other words, the principal symbol map (78) extends to a map

$$\pi: \mathscr{P}(E, F) \to C(ST^*M, \mathrm{Hom}(\pi_*^{-1}E, \pi_*^{-1}F)) \qquad (79)$$

which is surjective, mapping onto the space of continuous sections of $\mathrm{Hom}(\pi_*^{-1}E, \pi_*^{-1}F)$.

Theorem 8.81 *Let M be a compact manifold, and let $\mathscr{P} = \mathscr{P}(E, F)$ denote the closure of $OClS^0(E, F)$ in the L_2 operator norm. The principal symbol map (78) extends to a surjective map (79) which has the set $\mathscr{K} = \mathscr{K}(E, F)$ of compact operators $L_2(M, E) \to L_2(M, F)$ as kernel. The following properties hold:*

(*i*) *The induced map $\bar{\pi}: \mathscr{P}/\mathscr{K} \to C(ST^*M, \mathrm{Hom}(\pi_*^{-1}E, \pi_*^{-1}F))$ is bijective and an isometry;*
(*ii*) *$\mathscr{P}$ is closed under composition of operators and taking of adjoints, and the map π is a $*$ homomorphism.*

Moreover, an operator $A \in \mathscr{P}(E, F)$ is Fredholm if and only if $\pi A(x, \xi): E_x \to F_x$ is invertible for all $(x, \xi) \in ST^*M$. If $A \in \mathscr{P}(E, F)$ is Fredholm then it has a regularizer $B \in \mathscr{P}(F, E)$, and we have $\mathrm{ind}\, A = \mathrm{ind}_s\, \pi A$.

Proof The property (i) follows because (77) in Lemma 8.80 carries over to operators $A \in \mathscr{P}$ by taking limits, and it is immediate that $\pi A = 0$ if and only if $A \in \mathscr{K}$. Property (ii) is obtained by taking limits in Theorems 8.44 and 8.45.

We have already proved in Theorem 8.76 that invertibility of πA on ST^*M is sufficient for A to be Fredholm. In that case there is an operator $B \in \mathscr{P}(F, E)$ such that $\pi B = (\pi A)^{-1}$ and it follows that $\pi(AB - I) = 0$ and $\pi(BA - I) = 0$, hence $AB - I \in \mathscr{K}$ and $BA - I \in \mathscr{K}$, i.e. B is a regularizer for A.

To prove necessity, let $A \in \mathscr{P}(E, F)$ be a Fredholm operator. Then A^*A has index zero (since it is self-adjoint), so there is a compact K such that

$A^*A + K$ is invertible, thus,

$$\|(A^*A + K)u\| \geqslant c\|u\|, \qquad c > 0. \tag{80}$$

The conclusion (48) in Corollary 8.52 carries over to operators in $\mathscr{P}$, so by applying it to the operator A^*A and using (ii) we obtain

$$\|A^*A(f_k e)\| \to |\pi A^*(x_0, \xi_0) \circ \pi A(x_0, \xi_0) e(x_0)| \qquad \text{as } k \to \infty.$$

Also $f_k e \to 0$ weakly in $L_2(M, E)$, thus $K(f_k e) \to 0$ in $L_2(M, F)$, and by substituting $u = f_k e$ in (80) and taking the limit as $k \to \infty$ we find that

$$|\pi A^*(x_0, \xi_0) \circ \pi A(x_0, \xi_0) e(x_0)| \geqslant c > 0$$

for all $e(x_0) \in E_{x_0}$ of length 1. Therefore, $\pi A(x_0, \xi_0)$ is injective and an identical argument shows that $\pi(A^*) = {}^h\pi A$ is injective at any $(x_0, \xi_0) \in T^*M\backslash 0$, so πA has invertible values.

Let $\mathscr{P}_l^m(E, F)$ denote the closure of $OClS^m(E, F)$ in the operator norm $W_2^l(M, E) \to W_2^{l-m}(M, F)$. In Theorem 8.82 we will show that the index of $A \in OClS^m$ is independent of the homotopy class of $\pi A|_{ST^*M}$. By the usual device (see the proof of Theorem 8.76) it suffices to consider the class

$$\mathscr{P} = \mathscr{P}_0^0,$$

i.e. the closure of $OClS^0(E, F)$ in the L_2 operator norm.

Let $a_t \in \text{Symbl}^0(E, F)$, $0 \leqslant t \leqslant 1$, and suppose that $a_t(x, \xi)$ is invertible for all $(x, \xi) \in ST^*M$. We call $\{a_t\}$ *a homotopy of* a_0 *with* a_1 if the map

$$a\colon ST^*M \times [0, 1] \to \text{Iso}(\pi_*^{-1}E, \pi_*^{-1}F)$$

defined by $a_t(x, \xi; t) = a_t(x, \xi)$ is continuous.

Theorem 8.82 *Let $A \in OClS^m(E, F)$ and $B \in OClS^{m'}(E, F)$ be elliptic. If πA and πB are homotopic (i.e., their restrictions to ST^*M are homotopic) then*

$$\text{ind } A = \text{ind } B.$$

Proof Without loss of generality $m = m' = 0$, and then we may use Theorem 8.81. By hypothesis we have a homotopy $a\colon ST^*M \times [0,1] \to \text{Iso}(\pi_*^{-1}E, \pi_*^{-1}F)$ with $a_0 = \pi A|_{ST^*M}$ and $a_1 = \pi B|_{ST^*M}$. Using the isomorphism $\bar{\pi}$ in Theorem 8.81 we get a continuous map

$$F\colon [0, 1] \to \mathscr{P}/\mathscr{K} \qquad \text{defined by } F_t = \bar{\pi}^{-1} \circ a_t.$$

Since a_t has invertible values then by Theorem 8.81(iii) the representatives in the equivalence class F_t are Fredholm operators. Now if

$$\|B + \mathscr{K} - (A + \mathscr{K})\| < \varepsilon$$

then there is a compact K such that $\|B + K - A\| < \varepsilon$, so by Theorems 9.11, 9.12 we obtain that ind $A = \text{ind}(B + K) = \text{ind } B$. Hence ind F_t is locally constant in t, and by the usual connectedness argument, ind F_t is constant for all $t \in [0, 1]$. Thus, ind $A = \text{ind } F_0 = \text{ind } F_1 = \text{ind } B$.

As an illustration of how Theorem 8.82 may be applied, we use it to prove Noether's formula for the index of a pseudo-differential operator of order 0 on the unit circle.

We denote by $\mathbb{T}$ the unit circle $|z| = 1$ in $\mathbb{R}^2$, and let P be the operator which is defined on trigonometric polynomials by

$$P\left(\sum_{|k|<N} c_k z^k\right) = \sum_{k=0}^{N} c_k z^k,$$

the analytic part of the Fourier expansion of u. It is easily verified that $\|Pu\|_{L_2(\mathbb{T})} = \sum_{k\geqslant 0} |c_k|^2$, and, since the trigonometric polynomials are dense in $L_2(\mathbb{T})$, we obtain an operator P: $L_2(\mathbb{T}) \to \mathrm{L}^2(\mathbb{T})$ of norm 1.

Further, P restricts to an operator P: $C^\infty(\mathbb{T}) \to C^\infty(\mathbb{T})$ because the elements of $C^\infty(\mathbb{T})$ are characterized by the condition that $\sum_{-\infty}^{\infty} |c_k|^2(1 + k^2)^N < \infty$ for all N. We show in the next lemma that P is a pseudo-differential operator of order 0.

The definition of P can also be written in another form. If $u = \sum c_k\, e^{ikx}$ is a trigonometric polynomial we let $u_r = \sum c_k r^k\, e^{ikx}$, $r < 1$, then

$$Pu_r = \frac{1}{2\pi i}\int_{|z|=1} \frac{u(z)}{z - r\,e^{ix}}\,dz,$$

where the contour integral is taken along the unit circle in the counter-clockwise direction, and

$$Pu = \lim_{r\to 1^-} Pu_r = \lim_{r\to 1^-} \frac{1}{2\pi}\int_{\mathbb{T}} \frac{u(e^{iy})}{1 - r\,e^{i(x-y)}}\,dy, \tag{81}$$

the convergence holding in the L_2 sense if $u \in L_2(\mathbb{T})$.

The volume element on $\mathbb{T}$ induced from $\mathbb{R}^2$ and the counterclockwise orientation is $\tau(e^{ix}) = x_1\,dx_2 - x_2\,dx_1$, where $e^{ix} = x_1 + ix_2$. If ξ is a covector, we write $\xi > 0$ if ξ is a positive multiple of τ, i.e. pointing in the counterclockwise direction.

Lemma *$P \in OS^0(\mathbb{T})$ and is a (-1) classic p.d.o. with principal symbol*

$$\pi P(e^{ix}, \xi) = \begin{cases} 1 & \textit{for } \xi > 0 \\ 0 & \textit{for } \xi < 0 \end{cases} \tag{82}$$

Proof We have to verify the conditions of Theorem 8.9. To verify condition (ii), let φ and $\psi \in C^\infty(\mathbb{T})$ with $\operatorname{supp}\varphi \cap \operatorname{supp}\psi = \varnothing$. Then due to (81), and using the notation $u(y) := u(e^{iy})$, we have

$$\begin{aligned}(\psi P \varphi u)(x) &= \lim_{r\to 1^-} \psi(x)\cdot\frac{1}{2\pi}\int \frac{\varphi(y)u(y)}{1 - r\,e^{i(x-y)}}\,dy \\ &= \frac{1}{2\pi}\int K(x, y)u(y)\,dy\end{aligned}$$

where $K(x, y) = \psi(x)\varphi(y)/(1 - e^{i(x-y)}) \in C^\infty(\mathbb{T} \times \mathbb{T})$. Thus, the operator $\psi P \varphi$ has a C^∞ kernel, so $\psi P \varphi \in OP^{-\infty}(\mathbb{T})$.

To verify condition (i) we construct an atlas for $\mathbb{T}$. Let $V_1 = (\delta, 2\pi - \delta)$ and $V_2 = (-\pi + \delta, \pi - \delta)$ where δ is a small positive number, and let U_j be the image of V_j by the map $x \mapsto e^{ix}$. In this way we obtain an atlas $\mathfrak{A} = \{(U_j, \kappa_j)\}_{j=1,2}$ for $\mathbb{T}$ where κ_j: $U_j \to V_j \subset \mathbb{R}$ is defined by $\kappa_j(e^{ix}) = x$. The operator P in local coordinates,

$$P_{\kappa_2}: C_0^\infty(V_2) \to C^\infty(V_2),$$

acts on functions $u \in C_0^\infty(V_2)$ as follows:

$$P_{\kappa_2} u(x) = \lim_{r \to 1^-} \frac{1}{2\pi} \int_{-\infty}^{\infty} \frac{u(y)}{1 - r\, e^{i(x-y)}}\, dy, \qquad x \in V_2.$$

Writing $r = e^{-\varepsilon}$ and

$$\frac{1}{1 - e^{-\varepsilon}\, e^{ix}} = \frac{-1}{i(x + i\varepsilon)} + F(i(x + i\varepsilon)),$$

where $F(z) = 1/(1 - e^z) + 1/z$ is analytic if $z \neq \pm 2\pi i, \pm 4\pi i, \ldots$, and letting $\varepsilon \to 0^+$, we see that the limit in $\mathcal{S}'$ of $1/(1 - e^{-\varepsilon}\, e^{ix})$ is $i/(x + i0)$ plus a term which is in $C^\infty((-2\pi, 2\pi))$. Thus, for $u \in C_0^\infty(V_2)$ we have

$$P_{\kappa_2} u = u * \frac{i}{2\pi(x + i0)} + \int K(x, y) u(y)\, dy$$

where $K(x, y) = \chi(x - y) F(i(x - y)) \in C_0^\infty(\mathbb{R} \times \mathbb{R})$ and $\chi \in C_0^\infty((-\pi, \pi))$ is equal to 1 in a neighbourhood of $\bar{V}_2$. Since $i/2\pi(x + i0)$ is the inverse Fourier transform of the Heaviside function (see Exercise 3), this shows that the operator P_{κ_2} is the restriction to V_2 of a $(-\infty)$ classic p.d.o. belonging to $OS^0(\mathbb{R})$. Similarly for P_{κ_1}. Thus, $P \in OS^0(\mathbb{T})$ is a (-1) classic operator. By virtue of (30) the principal symbol πP is the pull-back of the Heaviside function, thus (82) follows.

Theorem 8.83 (*Noether's formula*) *Let $A \in OClS^0(\mathbb{T})$ with principal symbol*

$$\pi A(z, \xi) = a_+(z),\ \xi > 0, \quad \textit{and} \quad \pi A(z, \xi) = a_-(z),\ \xi < 0,$$

and suppose that $a_\pm(z) \neq 0$ for all $z = e^{ix} \in \mathbb{T}$. Then A: $L_2(\mathbb{T}) \to L_2(\mathbb{T})$ is a Fredholm operator with index

$$\text{ind}\, A = -\frac{1}{2\pi} [\arg a_+(e^{ix})]_{\mathbb{T}} + \frac{1}{2\pi} [\arg a_-(e^{ik})]_{\mathbb{T}}.$$

Proof The fact that A is Fredholm is clear since A is elliptic (Theorem 8.53). Now, multiplying A by the multiplication operator a_-^{-1} which has index zero (because it is invertible), the symbol becomes a_+/a_- when $\xi > 0$ and 1 when $\xi < 0$. If m denotes the winding number of a_+/a_- along $\mathbb{T}$ then

$$a_+(e^{ix})/a_-(e^{ix}) = e^{imx + \phi(x)}$$

where ϕ is periodic. Replacing ϕ by $t\phi$ for $0 \leqslant t \leqslant 1$ gives us a homotopy of Fredholm operators. Hence by Theorem 8.82 we can assume that $\phi = 0$. We can also assume that

$$Au = e^{imx} Pu + u - Pu$$

because this operator has the same principal symbol as the given operator and therefore the same index by Theorem 8.53. Let $u = \sum_{-\infty}^{\infty} c_k e^{ikx} \in L_2(\mathbb{T})$. Then $Au \in L_2(\mathbb{T})$ has the Fourier expansion

$$Au = \sum_{-\infty}^{-1} c_k e^{ikx} + e^{imx} \sum_{0}^{\infty} c_k e^{ikx}.$$

If $m = 0$ then A is the identity and ind $A = 0$ of course. If $m > 0$ it is evident that A has zero kernel and its image consists of all $f \in L_2(\mathbb{T})$ with kth Fourier coefficient equal to zero for $k = 0, \ldots, m - 1$. Hence ind $A = 0 - m = -m$. On the other hand, if $m < 0$, then the image of A is all of $L_2(\mathbb{T})$ and its kernel consists of all $u = \sum_{-\infty}^{\infty} c_k e^{ikx} \in L_2(\mathbb{T})$ with Fourier coefficients satisfying the conditions

$$c_k = 0 \text{ for } k \geqslant -m \text{ or } k < m, \qquad c_{k-m} + c_k = 0 \text{ for } k = m, \ldots, -1.$$

Hence $\ker A$ has dimension $|m|$ and $\operatorname{im} A$ has codimension 0, so ind $A = |m| - 0 = -m$. In any case we have ind $A = -m$ as was to be shown.

The next theorem is the matrix version of Theorem 8.83.

Theorem 8.84 *Let $A \in OClS^0(\mathbb{T}, p \times p)$ with principal symbol $\pi A(z, \xi)$ equal to $a_+(z)$, $\xi > 0$, and equal to $a_-(z)$, $\xi < 0$, and suppose that* $\det a_\pm(z) \neq 0$ *for all z. Then A: $L^2(\mathbb{T}, \mathbb{C}^p) \to L^2(\mathbb{T}, \mathbb{C}^p)$ is a Fredholm operator with index*

$$\operatorname{ind} A = -\frac{1}{2\pi} [\arg \det a_+(e^{ix})]_{\mathbb{T}} + \frac{1}{2\pi} [\arg \det a_-(e^{ix})]_{\mathbb{T}}.$$

Proof We can assume that

$$A = a_+ \cdot P + a_- \cdot (I - P)$$

since this operator has the same principal symbol as the given A. In view of Proposition 15.5 there exist homotopies $a_+^{(t)}$: $\mathbb{T} \to GL_p(\mathbb{C})$ and $a_+^{(t)}$: $\mathbb{T} \to GL_p(\mathbb{C})$, $0 \leqslant t \leqslant 1$, such that

$$a_\pm^{(0)} = a_\pm \quad \text{and} \quad a_\pm^{(1)} = \begin{pmatrix} \det a_\pm & & & \\ & 1 & & \\ & & \ddots & \\ & & & 1 \end{pmatrix}$$

Then $A^{(t)} = a_+^{(t)} \cdot P + a_-^{(t)} \cdot (I - P)$ is a homotopy of Fredholm operators with

$$A^{(0)} = A \quad \text{and} \quad A^{(1)} = \begin{pmatrix} B & & & \\ & 1 & & \\ & & \ddots & \\ & & & 1 \end{pmatrix}$$

where B denotes the 1×1 operator $\det a_+ \cdot P + \det a_- \cdot (I - P)$. Now by Theorem 8.82

$$\operatorname{ind} A = \operatorname{ind} A^{(1)} = \operatorname{ind} B$$

and then Theorem 8.83 finishes the proof.

Remark Theorem 8.82 is an important tool for calculating the index, and it was proved using the closure, $\mathscr{P}$, of $OClS^0$ in the L_2 operator norm. However, this theorem can be proved by other means as we show below in Remark 8.88.

Our aim now is to extend the definition of the symbol index, $\operatorname{ind}_s a$, to the case where a is any continuous section $T^*M \to \operatorname{Hom}(\pi_*^{-1}E, \pi_*^{-1}F)$ (not necessarily homogeneous) that is invertible outside a compact subset of T^*M. To do so, however, it is necessary to allow a wider class of homotopies of elliptic operators than was considered in Theorem 8.82.

But first we need a few preliminary definitions. By definition of the manifold structure on T^*M,

$$a \in C^\infty(T^*M)$$

if and only if for every coordinate chart (U, κ) on M, the push-forward function $a_\kappa = t(a, \kappa)$, see (17), belongs to $C^\infty(\kappa(U) \times \mathbb{R}^n)$. Then for each compact set $K \subset \kappa(U) \times \mathbb{R}^n$, consider the semi-norms

$$\sup_{\substack{(y,\eta) \in K \\ |\alpha| + |\beta| \leq d}} |D_\eta^\alpha D_y^\beta a_\kappa(y, \eta)|, \qquad d = 0, 1, 2, \ldots$$

(derivatives in local coordinates of the push-forward function). If we let (U, κ) vary over all charts, and K over all compact sets in $\kappa(U) \times \mathbb{R}^n$, we get a system of semi-norms defining a topology on $C^\infty(T^*M)$.

A sequence $a_n \in C^\infty(T^*M)$ is convergent if, for each chart (U, κ) on M, all derivatives of the push-forward functions converge on compact subsets of $\kappa(U) \times \mathbb{R}^n$. The limit function $a = \lim_{n \to \infty} a_n$ will then also be in $C^\infty(T^*M)$.

By using a locally finite partition of unity on T^*M, one could write down a countable family of separating semi-norms on $C^\infty(T^*M)$, so the topology is metrizable and hence a Fréchet space topology. For the case of $C^\infty(ST^*M)$, it would suffice to use a partition of unity on M due to the compactness of ST^*M.

The semi-norms in $S^m = S^m(\mathbb{R}^n \times \mathbb{R}^n)$ can also be transported to the manifold: for an arbitrary chart (U, κ) and compact set $K \subset \kappa(U)$ we have the semi-norm

$$\|a\|_{K,d}^{(m)} = \sup_{\substack{y \in K \\ |\alpha|+|\beta| \leqslant d}} |D_\eta^\alpha D_y^\beta a_\kappa(y, \eta)| \cdot (1 + |\eta|)^{|\alpha| - m}, \qquad d = 0, 1, 2, \ldots \quad (83)$$

where $a_\kappa = t(a, \kappa)$, and this system of semi-norms defines a topology on $S^m(T^*M)$, with continuous inclusion $S^m(T^*M) \hookrightarrow C^\infty(T^*M)$. Once again, using a partition of unity on M we obtain a countable collection of separating semi-norms for $S^m(T^*M)$, and the topology on $S^m(T^*M)$ is a Fréchet space topology.

A subset $\mathfrak{S} \subset S^m$ is said to be *bounded* if each semi-norm in S^∞ has bounded values on $\mathfrak{S}$. It is not hard to see by Ascoli's theorem that, on bounded sets in S^m, the following topologies are identical: the C^∞ topology, the topology of pointwise convergence, and the $S^{m'}$ topology for any $m' > m$.

Definition 8.85 *A sequence $a_n \in S^m(T^*M)$, $n = 1, 2, \ldots$ is said to converge* weakly *as $n \to \infty$ if*

(*i*) *$\{a_n\}_{n=1}^\infty$ is a bounded set in $S^m(T^*M)$, i.e. each semi-norm is bounded on this set;*
(*ii*) *$\{a_n\}_{n=1}^\infty$ converges in $C^\infty(T^*M)$.*

(As remarked above, condition (i) implies that (ii) is equivalent to pointwise convergence.)

It is clear that if a_n is weakly convergent then the limit function $a(x, \xi) = \lim_{n\to\infty} a_n$ is also in $S^m(T^*M)$.

If Z is any topological space, then we say that a map $S^m \to Z$ is *weakly continuous* if the restriction to any bounded set in S^m is continuous in the C^∞ topology.

In the proof of Theorem 8.23 we defined a right inverse to the main symbol isomorphism

$$S^m(T^*M) \ni a \mapsto A = \sum \psi_j(\tilde{a}_j(x, D))_{\kappa_j^{-1}} \psi_j \in OS^m(M), \qquad (84)$$

where $\tilde{a}_j$ is the push-forward of a to $\kappa_j(U_j) \times \mathbb{R}^n$, see (17), and it defines a continuous map $S^m(T^*M) \to \mathscr{L}(W_2^l, W_2^{l-m})$ when S^m has the semi-norm topology and $\mathscr{L}(W_2^l, W_2^{l-m})$ is given the operator-norm topology; see (86) below.

Lemma 8.86 *Let $a_j \to a$ weakly in $S^m = S^m(\mathbb{R}^n \times \mathbb{R}^n)$ as $j \to \infty$. Then we have*

$$a_j(x, D)u \to a(x, D)u \qquad \text{in } W_2^{l-m}(\mathbb{R}^n) \qquad (85)$$

for every $u \in W_2^l(\mathbb{R}^n)$.

Proof The operator norm of $a(x, D)$: $W_2^l \to W_2^{l-m}$ is bounded by a semi-norm of a in S^m (see the remark following Theorem 7.28), i.e.

$$\|a(x, D)u\|_{l-m} \leqslant \text{const}|a|_k^{(m)} \|u\|_l, \qquad \text{for some } k, \qquad (86)$$

where the constant depends only on l. Consequently, the operator norm of $a_j(x, D)$ is bounded by some constant, independently of j, so it suffices to verify (85) on the dense subset $u \in \mathscr{S}$. In fact, we will show that if $u \in \mathscr{S}$ then $a_j(x, D)u \to a(x, D)u$ also in $\mathscr{S}$ as $j \to \infty$.

Choose a function $\chi \in C_0^\infty(\mathbb{R}^n)$ equal to 1 in a neighbourhood of the origin and, for $b \in S^m$, set

$$b^\nu(x, \xi) = \chi(x/\nu)\chi(\xi/\nu)b(x, \xi), \qquad \nu \geqslant 1.$$

By Lemma 7.3 and the estimates for the semi-norms in Theorem 7.12, it follows that the symbols $\{b^\nu\}_{\nu \geqslant 1}$ form a bounded set in S^m, and $b^\nu \to b$ in S^{m+1} as $\nu \to \infty$. In fact, more is true: since any semi-norm of $b^\nu - b$ in S^{m+1} is bounded by a semi-norm of b in S^m and a semi-norm of $\chi(x/\nu)\chi(\xi/\nu) - 1$ in S^1, it follows that if $\mathfrak{S}$ is a bounded set in S^m then $b^\nu \to b$ in S^{m+1} as $\nu \to \infty$, uniform in $b \in \mathfrak{S}$. Hence $b^\nu(x, D)u \to b(x, D)u$ in $\mathscr{S}$, uniform in $b \in \mathfrak{S}$ (see the estimates in the proof of Proposition 7.2). To be precise, this means that if V is any neighbourhood of 0 in $\mathscr{S}$ then there for some ν_0 it follows that $b^\nu(x, D)u - b(x, D)u \in V$ for all $\nu \geqslant \nu_0$.

Now, if $a_j \to a$ weakly in S^m then we write

$$a_j(x, D)u - a(x, D)u = (a_j(x, D)u - a_j^\nu(x, D)u) + (a_j^\nu(x, D) - a^\nu(x, D))u$$
$$+ (a^\nu(x, D)u - a(x, D)u). \tag{87}$$

Since $\{a_j\}$ is a bounded set in S^m, then, for any neighbourhood V of 0 in $\mathscr{S}$, there is some ν such that $a_j(x, D)u - a_j^\nu(x, D)u \in \frac{1}{3}V$ for all j and $a(x, D)u - a^\nu(x, D)u \in \frac{1}{3}V$. It remains to show that with this (fixed) ν, the middle term on the right-hand side of (87) lies in $\frac{1}{3}V$ for sufficiently large j. Now, since $a_j \to a$ on compact subsets of $\mathbb{R}^{2n}$ as $j \to \infty$ and $a_j^\nu - a^\nu$ is equal to $a_j - a$ multiplied by $\chi(x/\nu)\chi(\xi/\nu)$, a function with compact support, it follows that $a_j^\nu \to a^\nu$ in S^m. Hence $a_j^\nu(x, D)u \to a^\nu(x, D)u$ in $\mathscr{S}$ as $j \to \infty$, that is, $a_j^\nu(x, D)u \to a^\nu(x, D)u \in \frac{1}{3}V$ for all $j \geqslant j_0$. We conclude that $a_j(x, D)u - a(x, D)u \in V$ for all $j \geqslant j_0$. Hence $a_j(x, D)u \to a(x, D)u$ in $\mathscr{S}$ as $j \to \infty$.

Lemma 8.86 implies that the map (84) is weakly continuous when $\mathscr{L}(W_2^l, W_2^{l-m})$ is given the strong operator topology. For this reason we can use Theorem 9.13 to prove the following theorem, which we take from [Hö 3].

Theorem 8.87 *Let I be the interval $[0, 1]$ in $\mathbb{R}$. Let $\mathscr{a}: I \to C^\infty(T^*M)$ and $\mathscr{b}: I \to C^\infty(T^*M)$ be continuous maps such that, for $t \in I$, $\mathscr{a}(t)$ is bounded in S^m, $\mathscr{b}(t)$ is bounded in S^{-m}, $\mathscr{a}(t)\mathscr{b}(t) - 1$ is bounded in S^{-1}, and $\mathscr{b}(t)\mathscr{a}(t) - 1$ is bounded in S^{-1}. If $A_0, A_1 \in OS^m(M)$ have main symbols $\mathscr{a}(0)$ and $\mathscr{a}(1)$, respectively, then*

$$\operatorname{ind} A_0 = \operatorname{ind} A_1.$$

Proof By composing $t \mapsto \mathscr{a}(t)$ with the map (84) we obtain a continuous map

$$I \ni t \mapsto A(t) \in \mathscr{L}(W_2^l(M), W_2^{l-m}(M))$$

with the strong operator topology, and similarly for $t \mapsto B(t)$. This verifies the first part of the hypotheses in Theorem 9.13.

Choose a covering of M by coordinate patches (U_j, κ_j) which is so fine that $U_j \cup U_k$ is contained in another coordinate patch (U_{jk}, κ_{jk}) whenever $U_j \cap U_k \neq \varnothing$ (see Exercise 16). By virtue of (84), the operators

$$A(t) = \sum \psi_j(\tilde{a}_j(t, x, D))_{\kappa_j^{-1}}\psi_j \quad \text{and} \quad B(t) = \sum \psi_j(\tilde{b}_j(t, x, D))_{\kappa_j^{-1}}\psi_j,$$

have principal symbols $a(t)$ and $b(t)$, respectively. We know that $A(t)B(t) - I$ is of order -1 (because it has main symbol $a(t)b(t) - 1 \in S^{-1}$), and we wish to show that its norm as an operator $W_2^l \to W_2^{l+1}$ is uniformly bounded in t (for any l).

Analogous to (84), let

$$P(t) = \sum \psi_j\psi_k((\tilde{a}_{jk}\tilde{b}_{jk})(t, x, D))_{\kappa_{jk}^{-1}}\psi_j\psi_k,$$

an operator with principal symbol $a(t)b(t)$, where $\tilde{a}_{jk}$ is the push-forward of a to $\kappa_{jk}(U_{jk}) \times \mathbb{R}^n$, see (17), and similarly for $\tilde{b}_{jk}$. Then

$$P(t) - I = \sum \psi_j\psi_k((\tilde{a}_{jk}\tilde{b}_{jk} - 1)(t, x, D))_{\kappa_{jk}^{-1}}\psi_j\psi_k \tag{88}$$

is uniformly bounded in t as an operator $W_2^l \to W_2^{l+1}$ by the hypothesis on $a(t)b(t) - 1$, so it remains to verify the same for the operator $A(t)B(t) - P(t)$. We have

$$\begin{aligned} A(t)B(t) &= \sum \psi_j(\tilde{a}_j(t, x, D))_{\kappa_j^{-1}}\psi_j\psi_k(\tilde{b}_k(t, x, D))_{\kappa_k^{-1}}\psi_k \\ &= \sum \psi_j\psi_k(\tilde{a}_j(t, x, D))_{\kappa_j^{-1}}(\tilde{b}_k(t, x, D))_{\kappa_k^{-1}}\psi_j\psi_k + T_1(t) \\ &= \sum \psi_j\psi_k(\tilde{a}_{jk}(t, x, D) \circ \tilde{b}_{jk}(t, x, D))_{\kappa_{jk}^{-1}}\psi_j\psi_k + T_2(t) + T_1(t), \end{aligned}$$

where the operator $T_1(t)$ is of order -1 and contains terms involving the commutators

$$[\psi_k, (\tilde{a}_j(t, x, D))_{\kappa_j^{-1}}] \in OS^{m-1} \quad \text{and} \quad [\psi_j, (\tilde{b}_k(t, x, D))_{\kappa_k^{-1}}] \in OS^{-m-1},$$

multiplied by $(\tilde{b}_k(x, t, D))_{\kappa_k^{-1}} \in OS^{-m}$ and $(\tilde{a}_j(x, t, D))_{\kappa_j^{-1}} \in OS^m$, respectively, while $T_2(t)$ is also of order -1, and contains terms that arise from the asymptotic expansion in Theorem 7.17. If one looks carefully at these terms and the definition of the Sobolev norm on M it is clear that the boundedness assumption on the main symbols $a(t)$ and $b(t)$ implies that the norms of $T_1(t)$ and $T_2(t)$ as operators $W_2^l \to W_2^{l+1}$ are uniformly bounded in t. Finally, by applying Theorem 7.12 to the composite operator $\tilde{a}_{jk}(t, x, D) \circ \tilde{b}_{jk}(t, x, D)$ we get

$$A(t)B(t) = P(t) + T_3(t) + T_2(t) + T_1(t) \tag{89}$$

where $T_3(t)$ is also of order -1 with uniformly bounded Sobolev norm.

In view of (88) and (89) we conclude that the norm of $A(t)B(t) - I$ as an operator $W_2^l \to W_2^{l+1}$ is uniformly bounded in t. Similarly for $B(t)A(t) - I$. All the hypotheses of Theorem 9.13 have been verified, and the proof is complete since by Theorem 8.53 we have

$$\operatorname{ind} A_0 = \operatorname{ind} A(0) = \operatorname{ind} A(1) = \operatorname{ind} A_1.$$

The preceding definitions of the semi-norm topologies on $C^\infty(T^*M)$ and $S^m(T^*M)$ extend to matrix functions, $C^\infty(T^*M, p \times p)$ and $S^m(T^*M, p \times p)$ in the obvious way, and then we define weak convergence of symbols just as in Definition 8.85. Moreover, since the definitions are stated locally we can even extend them to $S^m(T^*M, \mathrm{Hom}(\pi_*^{-1}E, \pi_*^{-1}F))$ if we first take push-forwards $\beta^F a(\beta^E)^{-1}$ of the symbols under local trivializations of the bundles E and F in order to get matrix methods.

As before, we say that a set $\mathfrak{S} \subset S^m(T^*M, \mathrm{Hom}(\pi_*^{-1}E, \pi_*^{-1}F))$ is *bounded* if each semi-norm is bounded on $\mathfrak{S}$. Theorem 8.87 may now be generalized as follows.

Theorem 8.87′ *Let* $a: I \to C^\infty((T^*M, \mathrm{Hom}(\pi_*^{-1}E, \pi_*^{-1}F))$ *and*

$$b: I \to C^\infty(T^*M, \mathrm{Hom}(\pi_*^{-1}F, \pi_*^{-1}E))$$

be continuous maps such that, for $t \in I$, $a(t)$ *is bounded in* S^m, $b(t)$ *is bounded in* S^{-m}, $a(t)b(t) - 1$ *is bounded in* S^{-1}, *and* $b(t)a(t) - 1$ *is bounded in* S^{-1}. *If* $A_0, A_1 \in OS^m(E, F)$ *have main symbols* $a(0)$ *and* $a(1)$, *respectively, then*

$$\operatorname{ind} A_0 = \operatorname{ind} A_1$$

Proof Let $\{(U_j, \kappa_j)\}$ be an atlas on M for which there exist trivializations

$$\beta_j^E: E_{U_j} \to U_j \times \mathbb{C}^p, \qquad \beta_j^F: F_{U_j} \to U_j \times \mathbb{C}^p$$

and as before choose functions $\psi_j \in C_0^\infty(U_j)$ with $\sum \psi_j^2 = 1$. If $a \in S^m(T^*M, \mathrm{Hom})$ then let $\tilde{a}_j$ be the push-forward of the matrix function $\beta_j^F a \beta_j^{E-1}$ to $\kappa_j(U_j) \times \mathbb{R}^n$, i.e.

$$\tilde{a}_j(y, \eta) = \beta_j^F a(\kappa_j^{-1}(y), {}^t d\kappa_j(\kappa_j^{-1}(y)) \cdot \eta) \beta_j^{E-1}, \qquad \text{see (17) and (37),}$$

which is a $p \times p$ matrix function with entries in $S^m_{\mathrm{loc}}(\kappa_j(U_j) \times \mathbb{R}^n)$. We may then define a right inverse to the main symbol isomorphism, $\not{p}$, by

$$S^m(T^*M, \mathrm{Hom}) \ni a \mapsto A = \sum \psi_j \beta_j^{F-1}(\tilde{a}_j(y, D))_{\kappa_j^{-1}} \beta_j^E \psi_j \in OS^m(E, F) \qquad (90)$$

The rest of the proof is essentially the same as for Theorem 8.87.

Remark 8.88 In view of (86) and the semi-norms (83), the map (90) is continuous when S^m has the semi-norm topology and $\mathscr{L}(W_2^l, W_2^{l-m})$ has the *operator-norm* topology. This leads to an alternative proof of Theorem 8.82 as follows: If πA and πB are homotopic then by Exercise 14 it can be shown that they are C^∞ homotopic (cf. proof of Lemma 10.22), i.e. there exists a C^∞ map

$$h: ST^*M \times I \to \mathrm{Hom}(\pi_*^{-1}E, \pi_*^{-1}F)$$

with $h_0 = \pi A|_{ST^*M}$ and $h_1 = \pi B|_{ST^*M}$. Now extend h by homogeneity so that it becomes a smooth map $T^*M \backslash 0 \times I \to \mathrm{Hom}(\pi_*^{-1}E, \pi_*^{-1}F)$, and let χ be a cut-off function as in Theorem 8.47. Then $a_t = \chi \cdot h_t$ defines a continuous map $a: I \to S^m(T^*M, \mathrm{Hom})$ with the semi-norm topology. By composition of this map with (90) we obtain a continuous map $I \to \mathscr{L}(W_2^l, W_2^{l-m})$ with the operator norm topology; denoting its values by $t \mapsto A(t)$, we obtain by

Theorem 9.12 that

$$\operatorname{ind} A = \operatorname{ind} A(0) = \operatorname{ind} A(1) = \operatorname{ind} B,$$

as required.

The next theorem shows that in the context of determining the index of elliptic operators, it suffices to consider operators which are (-1) classic. This is where we need the full strength of Theorem 8.87′.

Theorem 8.89 *Let $A \in OS^m(E, F)$ have main symbol a such that $a(x, \xi)$ is invertible when $|\xi| \geqslant C$. Then any operator with main symbol $a(x, r\cdot\xi/|\xi|)$ is elliptic and has the same index as A when $r \geqslant C$.*

Proof It suffices to prove the statement when C is an arbitrarily small positive number C_1 by virtue of the homotopy $a(x, r\xi)$, $1 \leqslant r \leqslant C/C_1$.

First assume that $m = 0$. Choose $b \in S^0(T^*M, \operatorname{Hom}(\pi_*^{-1}F, \pi_*^{-1}E))$ such that $a \circ b = 1$ and $b \circ a = 1$ when $|\xi| \geqslant C$. For $0 \leqslant \varepsilon \leqslant 1$, let

$$a_\varepsilon(x, \xi) = a(x, \xi\cdot\chi(\varepsilon|\xi|)), \qquad b_\varepsilon(x, \xi) = b(x, \xi\cdot\chi(\varepsilon|\xi|))$$

where χ is a decreasing C^∞ function on $\mathbb{R}$ with $\chi(t) = 1$ when $t < 1$ and $\chi(t) = 1/t$ when $t > 2$. Then we have

$$a_\varepsilon \circ b_\varepsilon - 1 = a \circ b - 1, \qquad b_\varepsilon \circ a_\varepsilon - 1 = b \circ a - 1$$

provided that $|\xi|\chi(\varepsilon|\xi|) \geqslant C$ when $\varepsilon|\xi| \geqslant 1$, that is,

$$\varepsilon C \leqslant \min_{t \geqslant 1} t\chi(t). \tag{91}$$

If we replace A by an operator with main symbol $a(x, r\xi)$, then (91) becomes $\varepsilon C/r \leqslant \min_{t\geqslant 1} t\chi(t)$, and by choosing a sufficiently large value of r, we may assume that (91) holds.

Since $a_0 = a$, our intention is to use Theorem 8.87′ to finish the proof; this is justified provided we can show that a_ε and b_ε are uniformly bounded in S^0 when $0 \leqslant \varepsilon \leqslant 1$. First of all, because a is a symbol of order 0, the functions

$$a(x, \theta/\varepsilon), \qquad 0 < \varepsilon \leqslant 1,$$

are uniformly bounded in C^∞ in the compact subset of $T^*M\backslash 0$ where $\frac{1}{2} \leqslant |\theta| \leqslant 2$. (This is verified by working in a finite number of local coordinate systems, taking notice of the fact that $|\theta|$ is bounded away from 0.) Substituting $\theta = \varepsilon\xi\cdot\chi(\varepsilon|\xi|)$, it follows that $a_\varepsilon(x, \xi) = a(x, \theta/\varepsilon)$ is uniformly bounded in C^∞ on the compact set where $1 \leqslant \varepsilon|\xi| \leqslant 2$. This is sufficient to conclude that a_ε is uniformly bounded in S^0, since $a_\varepsilon = a$ when $\varepsilon|\xi| \leqslant 1$ and a_ε is homogeneous of degree 0 when $\varepsilon|\xi| \geqslant 2$. For the same reasons, b_ε is also uniformly bounded in S^0.

It is now clear that a_ε and b_ε depend continuously on ε since $a_\varepsilon \to a$ and $b_\varepsilon \to b$ pointwise as $\varepsilon \to 0$. Thus, we have verified all the hypotheses of Theorem 8.87′.

It follows that $\operatorname{ind} A = \operatorname{ind} A_1$ where $A_1 \in OS^0$ has main symbol a_1. But $a_1(x, \xi) = a(x, \xi\cdot\chi(|\xi|))$ which is equal to $a(x, \xi/|\xi|)$ when $|\xi| > 2$, whence

A_1 has main symbol $a(x, \xi/|\xi|)$ (see the comments following Definition 8.21). The proof is now complete when $m = 0$. For the case $m \neq 0$, we choose a p.d.o. $\Lambda^{-m} \in OS^{-m}(M)$ with principal symbol $|\xi|^{-m}$; by Lemma 8.68, the index of Λ^{-m} is zero, so $A \circ \Lambda^{-m}$ has the same index as A. In view of Theorem 8.24(a) the operator $A \circ \Lambda^{-m} \in OS^0$ has main symbol $a(x, \xi) \cdot |\xi|^{-m}$, so it has the same index as any p.d.o. with main symbol

$$a(x, r\xi/|\xi|) \cdot \left|\frac{r\xi}{|\xi|}\right|^{-m} = a(x, r\xi/|\xi|) r^{-m}$$

for some sufficiently large r. This completes the proof.

This result enables us to make the final extension of the symbol index function.

Theorem 8.90 *There is a unique function,* ind_s, *defined on all*

$$a \in C(T^*M, \text{Hom}(\pi_*^{-1}E, \pi_*^{-1}F))$$

such that $a(x, \xi)$ *is invertible outside a compact set in* T^*M, *and such that*

(*i*) $\text{ind}_s\, a = \text{ind}\, A$ *if* $A \in OClS^0(E, F)$ *is a* (-1) *classic elliptic operator*;
(*ii*) $\text{ind}_s\, a_t$ *is independent of* t *if* a_t *is a continuous function of* $t \in I$ *with values in* $C(T^*M, \text{Hom}(\pi_*^{-1}E, \pi_*^{-1}F))$ *such that* $a_t(x, \xi)$ *is invertible except when* (x, ξ) *is in a compact set.*

Proof We have already defined $\text{ind}_s\, a$ when $a \in C(T^*M, \text{Hom})$ is homogeneous of order 0. Now, even if a is not homogeneous, there is a constant C such that $\det a(x, \xi) \neq 0$ when $|\xi| \geqslant C$, so we define

$$\text{ind}_s\, a = \text{ind}_s\, a',$$

where $a'(x, \xi) = a(x, r\xi/|\xi|)$ is a homogeneous symbol of order 0. The definition is independent of the choice of $r \geqslant C$. When a is homogeneous of order 0 we have $a' = a$, so the property (i) holds. The property (ii) follows by Theorem 8.87′. The uniqueness of the index function, ind_s, is a consequence of the homotopy used in the proof of Theorem 8.89.

Remark In particular, the definition of ind_s does not depend on the choice of Riemannian metric $<, >$ on T^*M.

We review some basic facts about tensor products of finite dimensional vector spaces, then conclude with a multiplicative property of the index. Let E and F be finite dimensional vector spaces over $\mathbb{C}$. The tensor product $E \otimes F$ is characterized up to unique isomorphism by the following universal property: There is a bilinear map $\varphi: E \times F \to E \otimes F$ such that for any bilinear map $\psi: E \times F \to G$ there exists a unique linear map $\tilde{\psi}: E \otimes F \to G$

such that the diagram

$$\begin{array}{ccc} E \otimes F & & \\ \uparrow{\scriptstyle \varphi} & \searrow{\scriptstyle \bar{\varphi}} & \\ E \times F & \xrightarrow{\psi} & G \end{array}$$

commutes. We write $x \otimes y = \varphi(x, y)$ for $x \in E$, $y \in F$. It follows from the universal property that if $\{e_i\}$ and $\{f_j\}$ are bases of E and of F, respectively, then $\{e_i \otimes f_j\}$ is a basis of $E \otimes F$. If E and F have inner products then there is a unique inner product on $E \otimes F$ such that $(x \otimes y, x' \otimes y') = (x, x')_E \cdot (y, y')_F$. Then $\{e_i \otimes f_j\}$ is an orthonormal basis of $E \otimes F$ if $\{e_i\}$ and $\{f_j\}$ are orthonormal bases of E and F.

For $A \in \mathscr{L}(E_1, F_1)$ and $B \in \mathscr{L}(E_2, F_2)$ we define $A \otimes B$ to be the unique linear map in $\mathscr{L}(E_1 \otimes E_2, F_1 \otimes F_2)$ such that $(A \otimes B)(x \otimes y) = Ax \otimes By$.

Lemma 8.91 *Let E_1, E_2, F_1 and F_2 be finite dimensional vector spaces endowed with Hermitian inner products, and let $S \in \mathscr{L}(E_1, F_1)$ and $T \in \mathscr{L}(E_2, F_2)$, and define*

$$S \# T := \begin{pmatrix} S \otimes I_{E_2} & -I_{F_1} \otimes T^* \\ I_{E_1} \otimes T & S^* \otimes I_{F_2} \end{pmatrix} \in \mathscr{L}(E_1 \otimes E_2 \oplus F_1 \otimes F_2, F_1 \otimes E_2 \oplus E_1 \otimes F_2)$$

in block operator form. We have

$$\ker(S \# T) = \ker S \otimes \ker T \oplus \ker S^* \otimes \ker T^*, \tag{92}$$

$$\ker(S \# T)^* = \ker S^* \otimes \ker T \oplus \ker S \otimes \ker T^*. \tag{93}$$

Hence if S and T are isomorphisms then so is $S \# T$.

Proof First of all, note that the adjoint $(S \# T)^*$ is taken relative to the natural Hermitian inner products on $E_1 \otimes E_2 \oplus F_1 \otimes F_2$ and $F_1 \otimes E_2 \oplus E_1 \otimes F_2$ in which the summands are orthogonal, and is of the same form with S replaced by S^* and T by $-T$. Let $Z = S \# T$, then it follows that

$$\begin{aligned} Z^*Z &= \begin{pmatrix} S^* \otimes I & I \otimes T^* \\ I \otimes T & S \otimes I \end{pmatrix} \begin{pmatrix} S \otimes I & -I \otimes T^* \\ I \otimes T & S^* \otimes I \end{pmatrix} \\ &= \begin{pmatrix} S^*S \otimes I + I \otimes T^*T & 0 \\ 0 & I \otimes TT^* + SS^* \otimes I \end{pmatrix} \end{aligned} \tag{94}$$

Let $\{e_j\}$ be an orthonormal basis of E_1 of eigenvectors of S^*S with eigenvalues $\lambda_j \geqslant 0$. Let $\{f_j\}$, $\{e'_j\}$, $\{f'_j\}$ and μ_j, λ'_j, μ'_j be the corresponding objects for SS^*, T^*T and TT^*, respectively. Let $e_{ij} = (e_i \otimes e'_j, 0)$ and $f_{ij} = (0, f_i \otimes f'_j)$; then the set $\{e_{ij}, f_{kl}\}$ is an orthonormal basis of $E_1 \otimes E_2 \oplus F_1 \otimes F_2$ and

$$Z^*Ze_{ij} = (\lambda_i + \lambda'_j)e_{ij}, \qquad Z^*Zf_{ij} = (\mu_i + \mu'_j)f_{ij}.$$

Since all the eigenvalues are non-negative it follows that $(\lambda_i + \lambda'_j)e_{ij} = 0$ if and only if $\lambda_i = \lambda'_j = 0$ and $(\mu_i + \mu'_j)f_{ij} = 0$ if and only if $\mu_i = \mu'_j = 0$, and then it is easy to see that $\ker Z = \ker S \otimes \ker T \oplus \ker S^* \otimes \ker T^*$. Replacing S by S^* and T by $-T$ we also get the equation for $\ker Z^*$. Finally, if S and T are invertible then we obtain $\ker Z = 0$ and $\ker Z^* = 0$, so Z is also invertible.

Theorem 8.92 *Let M and N be compact manifolds, and let E_1, F_1 be vector bundles over M and E_2, F_2 vector bundles over N. Let $A \in OClS^m(E_1, F_1)$ and $B \in OClS^m(E_2, F_2)$ be elliptic operators of order $m > 0$, and define the operator*

$$A \# B = \begin{pmatrix} A \otimes I & -I \otimes B^* \\ I \otimes B & A^* \otimes I \end{pmatrix}: C^\infty(M \times N, E_1 \boxtimes E_2 \oplus F_1 \boxtimes F_2)$$
$$\to C^\infty(M \times N, F_1 \boxtimes E_2 \oplus E_1 \boxtimes F_2)$$

Then $A \# B$ is a Fredholm operator, and

$$\operatorname{ind}(A \# B) = \operatorname{ind} A \cdot \operatorname{ind} B.$$

Proof Applying Theorem 8.79 to each of the four operators in $A \# B$, it follows that

$$A \# B \in \mathscr{P}^m(E_1 \boxtimes E_2 \oplus F_1 \boxtimes F_2, F_1 \boxtimes E_2 \oplus E_1 \boxtimes F_2)$$

with $\pi(A \# B) = \pi A \# \pi B$. Now, by Lemma 8.91, the linear map

$$(\pi A \# \pi B)(x, y, \xi, \eta) = \pi A(x, \xi) \# \pi B(y, \eta)$$

is an isomorphism for each $(x, y, \xi, \eta) \in T^*(M \times N)\backslash 0$, thus, $\ker(A \# B) \subset C^\infty$ and $\ker(A \# B)^* \subset C^\infty$ are finite dimensional spaces by Corollary 8.77. The same calculation which gave (94) also gives

$$(A \# B)^*(A \# B) = \begin{pmatrix} A^*A \otimes I + I \otimes B^*B & 0 \\ 0 & I \otimes BB^* + AA^* \otimes I \end{pmatrix}$$

and a similar formula for $(A \# B)(A \# B)^*$ with A^*A replaced by AA^* and so on. It follows that (92), (93) remain true, so $\operatorname{ind}(A \# B) = \operatorname{ind} A \cdot \operatorname{ind} B$.

Corollary 8.93 *Suppose that*

$$a \in C(T^*M, \operatorname{Hom}(\pi_*^{-1}E_1, \pi_*^{-1}F_1)), \qquad b \in C(T^*M, \operatorname{Hom}(\pi_*^{-1}E_2, \pi_*^{-1}F_2))$$

are isomorphisms outside compact subsets $K_1 \subset M$ and $K_2 \subset N$. Then

$$a \# b \in C(T^*(M \times N), \operatorname{Hom}(\pi_*^{-1}E_1 \boxtimes \pi_*^{-1}E_2$$
$$\oplus \pi_*^{-1}F_1 \boxtimes \pi_*^{-1}F_2, \pi_*^{-1}F_1 \boxtimes \pi_*^{-1}E_2 \oplus \pi_*^{-1}E_1 \boxtimes \pi_*^{-1}F_2))$$

is an isomorphism outside $K_1 \times K_2$, and

$$\operatorname{ind}_s(a \# b) = \operatorname{ind}_s a \cdot \operatorname{ind}_s b. \tag{95}$$

Proof The fact that $a \# b$ is an isomorphism outside $K_1 \times K_2$ follows from Lemma 8.91. To prove (95) we may assume that a and b are homogeneous of degree 1 and C^∞ outside the zero section, for a and b are homotopic to such functions. Then $a \# b$ is also homogeneous of degree 1 and continuous outside the zero section, but will not in general be C^∞ when $\xi = 0$ or $\eta = 0$. If we choose classic pseudo-differential operators A and B with principal symbols a and b then it follows that

$$\operatorname{ind}_s(a \# b) = \operatorname{ind}(A \# B) = \operatorname{ind} A \cdot \operatorname{ind} B = \operatorname{ind}_s a \cdot \operatorname{ind}_s b.$$

Exercises

1. The Cauchy principal value of $1/x$ is the distribution defined by

$$\left\langle \mathrm{pv}\frac{1}{x}, \varphi \right\rangle = \lim_{\varepsilon \to 0^+} \int_{|x| \geqslant \varepsilon} \frac{\varphi(x)}{x}\, dx, \qquad \varphi \in \mathscr{S}(\mathbb{R})$$

Show that $\mathrm{pv}(1/x) \in \mathscr{S}'(\mathbb{R})$.

2. Show that $1(x \pm i0) \in \mathscr{S}'(\mathbb{R})$ where

$$\left\langle \frac{1}{x \pm i0}, \varphi \right\rangle := \lim_{\varepsilon \to 0^+} \int_{-\infty}^{\infty} \frac{\varphi(x)}{x \pm i\varepsilon}\, dx, \qquad \varphi \in \mathscr{S}(\mathbb{R})$$

Also show that $1/(x \pm i0) = \mathrm{pv}(1/x) \mp i\pi\delta$, where δ is the Dirac distribution.

3. Show that the Fourier transform of $i/2\pi(x + i0)$ is equal to the Heaviside function θ (where $\theta(\xi) = 1$ when $\xi > 0$, and $= 0$ when $\xi < 0$). Hint: Apply the residue theorem to evaluate the limit

$$\lim_{R \to \infty} \int_{-R}^{R} e^{-ix\xi} \frac{1}{x + i\varepsilon}\, dx.$$

4. Show that the Fourier transform of $\mathrm{pv}(1/x)$ is equal to $-\pi i \operatorname{sgn} \xi$, where $\operatorname{sgn} \xi = 1$ if $\xi > 0$ and $\operatorname{sgn} \xi = -1$ if $\xi < 0$.

5. (a) If $Q \in C^1(\mathbb{T}, p \times p)$ show that

$$\frac{1}{\det Q}\, d(\det Q) = \operatorname{tr}(Q^{-1}\, dQ)$$

where the operator d is exterior differentiation and $\operatorname{tr}(\cdot)$ denotes the trace of a matrix. (b) Show that the index formula of Theorem 8.84 can be written in the form

$$\operatorname{ind} A = -\frac{1}{2\pi} \int_{\mathbb{T}} \operatorname{tr}(a_+^{-1}\, da_+) + \frac{1}{2\pi} \int_{\mathbb{T}} \operatorname{tr}(a_-^{-1}\, da_-)$$

6. Demonstrate that the one-dimensional singular integral operator on a smooth closed (compact) curve $\Gamma \subset \mathbb{R}^2$, defined by

$$Au(t) := a(t)u(t) + \lim_{\varepsilon \to 0} \int_{|t-\tau| \geqslant \varepsilon} \frac{K(t, \tau)}{t - \tau} u(\tau), d\tau, \qquad t, \tau \in \Gamma,$$

for $a(t) \in C^\infty(\Gamma)$ and $K(t, \tau) \in C^\infty(\Gamma \times \Gamma)$, belongs to the class $OS^0(\Gamma)$. Also show that the operator is (-1) classic and compute the principal part.

7. Prove that for any sequence $a_d \in S^{m_d}(T^*M)$, $d = 1, 2, \ldots$ with $m_{d-1} > m_d \to -\infty$ as $d \to \infty$, there exists $a \in S^{m_1}(T^*M)$ such that $a \sim \sum_{d=1}^{\infty} a_d$, that is,

$$a - \sum_{d=1}^{r-1} a_d \in S^{m_r}(T^*M) \qquad \text{for any integer } r \geqslant 2.$$

The function a is uniquely determined modulo $S^{-\infty}(T^*M)$. Hint: See Proposition 8.16 and recall that every function in $S^m(T^*M)$ is the main symbol of an operator in $OS^m(M)$.

8. Show that $A \in OS^m(M)$ cannot be elliptic when it is regarded as an element of $OS^{m_1}(M)$ for $m_1 > m$.

9. Let $A \in OS^m(E, F)$ be an elliptic operator. Suppose that $\operatorname{ind} A = 0$ and $\ker A = 0$. Show that $A: C^\infty(M, E) \to C^\infty(M, F)$ is bijective and $A^{-1} \in OS^{-m}(F, E)$. Further, if $A \in OClS^m(E, F)$ then $A^{-1} \in OClS^{-m}(F, E)$. Hint: See Lemma 8.59.

10. Prove *Gårding's inequality*: Let $A \in OS^m(M)$, $m \in \mathbb{R}$, and suppose that $\operatorname{Re} A(x, \xi) \geqslant 0$ for large $|\xi|$. Then for any $\varepsilon > 0$, there is a constant $C = C(\varepsilon)$ such that

$$\operatorname{Re}(Au, u)_{L^2(M)} + \varepsilon \|u\|_{m/2}^2 \geqslant -C\|u\|_{(m-1)/2}^2$$

for all $u \in C^\infty(M)$. Hint: Choose $\Lambda^k \in OClS^k(M)$, $k \in \mathbb{R}$, with principal part $|\xi|^k$, such that $\Lambda^k: W_2^l(M) \to W_2^{l-k}(M)$ is an isomorphism for all $l \in \mathbb{R}$. Then replacing A by $\Lambda^{-m/2} \circ A \circ \Lambda^{-m/2}$, we may assume without loss of generality that $m = 0$. The rest of the proof is just as in the proof of Gårding's inequality in $\mathbb{R}^n$, Theorem 7.38.

11. Formulate, and prove, Gårding's inequality for $A \in OS^m(E, F)$, where E and F are Hermitian bundles.

12. Let $P \in OS^m(M)$, $m > 0$. Suppose P is self-adjoint and that P has a main symbol p with

$$\operatorname{Re} p(x, \xi) \geqslant C|\xi|^m, \qquad |\xi| \text{ large}.$$

Show that there exists a (formally) self-adjoint $P_+ \in OS^m(M)$ such that $P_+ - P \in OS^{-\infty}$ and $P_+ \geqslant cI > 0$.

13. Let $A \in OS^m(M)$ be elliptic and $m > 0$. If A is formally self-adjoint, i.e.

$$(Au, v)_M = (u, Av)_M, \qquad u, v \in C^\infty(M),$$

show that $\pm iI + A: W_2^m(M) \to L_2(M)$ are isomorphisms.

14. Let M be a compact manifold and E a Hermitian bundle over M. Show that $C^\infty(M, E)$ is dense in the Banach space, $C(M, E)$, of continuous sections $f\colon M \to E$ with the supremum norm, $|f| = \sup_{x\in M}\|f(x)\|$. Hint: Using a partition of unity one can reduce the proof to the case where $f \in C_0^\infty(\mathbb{R}^n, \mathbb{C}^p)$.

15. Let (X, d) be a compact metric space and $\{U_j\}$ an open covering of X. Show that there exists a positive number λ (called a Lebesgue number of the covering) such that each ball $B(x, \lambda)$ is contained in at least one U_j.

16. Let M be a compact manifold. Show that it is possible to choose a covering of M by coordinate patches $\{U_j\}$ which is so fine that $U_j \cup U_k$ is also contained in some coordinate patch whenever $U_j \cap U_k \neq \varnothing$. Hint: The topology on M is metrizable, hence one can use Exercise 15.

17. Let M and N be compact manifolds, and E, F vector bundles over M and G a vector bundle over N. Prove that finite sums $\sum f_j \otimes g_j$ with $f_j \in C^\infty(M, E)$, $g_j \in C^\infty(N, G)$ are dense in $C^\infty(M \times N, E \boxtimes G)$. Hint: If a function has compact support in $\{x; |x_i| < \pi, i = 1, \ldots, n\}$ then it can be viewed as a function on the torus $\mathbb{T}^n$ for which a multiple Fourier series argument can be applied.

18. If A is a continuous linear map $C^\infty(M, E) \to C^\infty(M, F)$ prove that there is a unique continuous linear map

$$A'\colon C^\infty(M \times N, E \boxtimes G) \to C^\infty(M \times N, F \boxtimes G)$$

such that

$$A'(f \cdot g) = Af \cdot g, \qquad f \in C^\infty(M, E),\ g \in C^\infty(N, G).$$

Let $m \geqslant 0$ be an integer. Show that if A has a continuous extension $A_l\colon W_2^l \to W_2^{l-m}$ for all integers l then A' has a continuous extension $A_l'\colon W_2^l \to W_2^{l-m}$ too, and

$$\|A_l'\| \leqslant \text{const} \max_{-l \leqslant s \leqslant l} \|A_s\| \tag{96}$$

for any integer l (the constant depends on l, of course). Hint: To verify (96), divide the proof into three cases: $l \geqslant m$, $l \leqslant 0$, and $0 < l < m$. If $l \geqslant m$, then in local coordinates we have if $|\alpha| + |\beta| \leqslant l - m$, $u \in C_0^\infty(\mathbb{R}^{n+n'})$, and if we integrate first with respect to x with fixed y:

$$\begin{aligned}\iint_{\mathbb{R}^{n'} \times \mathbb{R}^n} |D_x^\alpha D_y^\beta A'u|^2\, dx\, dy &= \int_{\mathbb{R}^{n'}} dy \int_{\mathbb{R}^n} |D_x^\alpha A' D_y^\beta u|^2\, dx \\ &\leqslant C_l \int dy \int \sum_{|\gamma| \leqslant |\alpha| + m} |D_x^\gamma D_y^\beta u|^2\, dx \\ &\leqslant C_l \iint \sum_{|\beta| + |\gamma| \leqslant l} |D_x^\gamma D_y^\beta u|^2\, dx\, dy,\end{aligned}$$

that is, $\|A'u\|_{l-m} \leqslant C_l \|u\|_l$, where the constant C_l depends only on the norms of $A\colon W_2^s(\mathbb{R}^n) \to W_2^{s-m}(\mathbb{R}^n)$ for $0 \leqslant s \leqslant l$. Passing to adjoints gives the corresponding estimate for $l \leqslant 0$. The estimates for $0 < l < m$ can be handled easily by using the Fourier norm in the Sobolev spaces.

19. Prove that if both conditions hold in (49), then $\dim E = \dim F$ (i.e. E and F have the same fibre dimension). Hint: see Theorem 8.94 in the addendum below.

20. Formulate, and prove, versions of Theorems 8.64, 8.65 and 8.66 for any operator $A \in OS^m(E, F)$, i.e. when A is not assumed to be a classic operator.

Addendum

The proof of (iii) $\Leftrightarrow$ (ii) in Theorem 8.64 can be generalized to the case where the operator A is not assumed to be a classic operator.

Theorem 8.94 *Let $A \in OS^m(E, F)$. The following are equivalent:*

(1) If $a \in S^m(T^*M, \mathrm{Hom}(\pi_*^{-1}E, \pi_*^{-1}F))$ *is a main symbol of A then a has a left inverse $b \in S^{-m}(T^*M, \mathrm{Hom}(\pi_*^{-1}F, \pi_*^{-1}E))$, i.e.*

$$b a - 1_E \in S^{-1}.$$

(2) *For each point $x_0 \in M$, let (U, κ) be a coordinate patch on which there exist trivializations $\chi_U^E\colon E_U \to U \times C^p$ and $\chi_U^F\colon F_U \to U \times \mathbb{C}^q$. Then*

$$\varliminf_{|\eta| \to \infty} \inf_{|v|=1} \inf_{y \in \kappa(W)} |\tilde{a}_\kappa(y, \eta)v| / (1 + |\eta|)^m > 0$$

where $\tilde{a}_\kappa \in S^m_{\mathrm{loc}}(\kappa(U) \times \mathbb{R}^n, q \times p)$ is a complete symbol of the matrix operator A_χ on $\kappa(U)$ (see (20)) and W is an open set such that $x_0 \in \overline{W} \subset U$.

Proof To simplify the notation, we assume that E and F are the trivial bundles $M \times \mathbb{C}^p$ and $M \times \mathbb{C}^q$, respectively. This does not alter the proof in any significant way; we leave the details for the reader.

(1) $\Rightarrow$ (2) We have $a - a_\kappa \in S^{m-1}_{\mathrm{loc}}(T^*U, \mathrm{Hom})$ by Definition 8.21, and multiplying by b gives $b(a - a_\kappa) \in S^{-1}_{\mathrm{loc}}(T^*U, \mathrm{Hom})$, whence

$$1 - b a_\kappa \in S^{-1}_{\mathrm{loc}}(T^*U, \mathrm{Hom}).$$

Choosing $\varphi \in C_0^\infty(U)$ such that $0 \leqslant \varphi \leqslant 1$ and $\varphi = 1$ on W, then pushing forward we obtain

$$\tilde{\varphi}(1 - \tilde{b} \cdot \tilde{a}_\kappa) \in S^{-1}(\mathbb{R}^n \times \mathbb{R}^n, q \times p)$$

where $\tilde{\varphi}(y) = \varphi(\kappa^{-1}(y))$ and $\tilde{b}(y, \eta) = b(\kappa^{-1}(y), {}^t d\kappa(\kappa^{-1}(y) \cdot \eta)$ is the push-forward of b (see (20)). Therefore,

$$\tilde{\varphi}(y) \| 1 - \tilde{b}(y, \eta) \cdot \tilde{a}_\kappa(y, \eta) \| < \tfrac{1}{2}$$

for $y \in \kappa(U)$ and $\eta \in \mathbb{R}^n$ with $|\eta|$ sufficiently large. Since $\tilde{\varphi}(y) \| \tilde{b}(y, \eta) \| < c(1 + |\eta|)^{-m}$ for large $|\eta|$, and $\tilde{\varphi}(y) = 1$ when $y \in \kappa(W)$, it follows that

$$|\tilde{a}_\kappa(y, \eta)v| \geqslant \varepsilon(1 + |\eta|)^m \qquad \text{for some } \varepsilon > 0, \tag{*}$$

for all $y \in \kappa(W)$, $v \in \mathbb{C}^p$ with $|v| = 1$, and $\eta \in \mathbb{R}^n$ such that $|\eta|$ is sufficiently large.

(2) ⇒ (1) The inequality (*) implies that

$$({}^h\tilde{a}_\kappa(y, \eta)\cdot \tilde{a}_\kappa(y, \eta)v, v)_{\mathbb{C}^p} \geqslant \varepsilon^2(1 + |\eta|)^{2m}|v|^2, \qquad y \in \kappa(W),\ v \in \mathbb{C}^p,$$

if $|\eta|$ is sufficiently large (the superscript ^{h} indicates the Hermitian adjoint). The Hermitian matrix ${}^h\tilde{a}_\kappa \cdot \tilde{a}_\kappa$ belongs to $S^{2m}_{\text{loc}}(\kappa(W) \times \mathbb{R}^n)$, and since the eigenvalues are $\geqslant \varepsilon^2(1 + |\eta|)^{2m}$, the determinant $\in S^{2mp}_{\text{loc}}$ is therefore bounded from below by $\varepsilon^{2p}(1 + |\eta|)^{2mp}$ so $\det({}^h\tilde{a}_\kappa \cdot \tilde{a}_\kappa)$ has an inverse in S^{-2mp}_{loc} by Theorem 7.35. The product q of this inverse by the cofactor matrix of ${}^h\tilde{a}_\kappa \cdot \tilde{a}_\kappa$ belongs to S^{-2m}_{loc} and satisfies $q{}^h\tilde{a}_\kappa \cdot \tilde{a}_\kappa = 1$. Letting $b = q{}^h\tilde{a}_\kappa$, we have therefore constructed a symbol $\tilde{b}_\kappa \in S^{-m}_{\text{loc}}$ such that

$$\tilde{b}_\kappa \cdot \tilde{a}_\kappa = 1. \tag{97}$$

Let a_κ and b_κ denote the pull-backs of $\tilde{a}_\kappa$ and $\tilde{b}_\kappa$, respectively, as defined by (18). If (U', κ') is another coordinate patch, then by Lemma 8.20 we have $a_\kappa - a_{\kappa'} \in S^{m-1}_{\text{loc}}(T^*(U \cap U'))$ and by the manner in which b_κ and $\tilde{b}_\kappa$ were constructed it is not hard to see that $b_\kappa - b_{\kappa'} \in S^{-m-1}_{\text{loc}}(T^*(U \cap U'), \text{Hom})$. Now we claim that there exists $b \in S^{-m}(T^*M, \text{Hom})$ such that

$$b - b_\kappa \in S^{-m-1}_{\text{loc}}(T^*U, \text{Hom}) \qquad \text{for all charts } (U, \kappa).$$

Indeed, let $\{(U_j, \kappa_j)\}$ be an atlas on M, and $\{\varphi_j\}$ a subordinate partition of unity and then define the function $b = \sum \varphi_j b_{\kappa_j} \in S^{-m}(T^*M, \text{Hom})$. For any $\varphi \in C_0^\infty(U)$ the push-forward of $\varphi \cdot (b - b_\kappa) = \sum \varphi\varphi_j(b_{\kappa_j} - b_\kappa)$ by the map κ is an element of $S^{-m-1}(\mathbb{R}^n \times \mathbb{R}^n, q \times p)$, whence $b - b_\kappa \in S^{-m-1}_{\text{loc}}(T^*U, \text{Hom})$.

If we take into account that (97) implies $b_\kappa \cdot a_\kappa - 1 = 0$ we obtain

$$\varphi \cdot (ba - 1) = \varphi \cdot (b(a - a_\kappa) + (b - b_\kappa)a_\kappa + 0) \in S^{-1}$$

for any chart (U, κ) and $\varphi \in C_0^\infty(U)$. Hence $ba - 1 \in S^{-1}$.

Corollary 8.95 *Let a and b denote main symbols of $A \in OS^m(E, F)$ and $B \in OS^{-m}(F, E)$ respectively. If* $\dim E = \dim F$ *(i.e. E and F have the same fibre dimension) then the two conditions in* (49) *are equivalent:*

$$ab - 1 \in S^{-1} \qquad \textit{if and only if } ba - 1 \in S^{-1}.$$

Proof In local coordinates (U, κ) where E_U and F_U are identified with $U \times \mathbb{C}^p$, when we have $\tilde{b}_\kappa \cdot \tilde{a}_\kappa = 1$ in the proof of Theorem 8.94 then also $\tilde{a}_\kappa \cdot \tilde{b}_\kappa = 1$, since these are square $p \times p$ matrices.

9
Elliptic systems on bounded domains in $\mathbb{R}^n$

Let Ω be a bounded domain in $\mathbb{R}^n$. In this chapter we consider boundary value problems for systems,

$$\mathscr{A}(x, D)u = f(x),\ x \in \Omega; \qquad \mathscr{B}(y, D)u = g(y),\ y \in \partial\Omega,$$

where $\mathscr{A}$ is a differential operator with coefficients which are smooth $p \times p$ matrix functions on $\bar{\Omega}$. The boundary operator $\mathscr{B}$ is polynomial in the normal derivative D_n, with coefficients which are permitted to be pseudo-differential in the tangential variables.

Ellipticity of the matrix operator $\mathscr{A}$ is defined in §9.2. There are two definitions of ellipticity, one due to Douglis and Nirenberg [DN] and another due to Volevič [Vo 2]. In fact, the two definitions are equivalent and we show this in §9.2 by using a result from linear programming. The boundary problem $\mathfrak{L} = (\mathscr{A}, \mathscr{B})$ is said to be *L-elliptic* if $\mathscr{A}$ is elliptic in $\bar{\Omega}$ and $(\mathscr{A}, \mathscr{B})$ satisfies the "L-condition" on $\partial\Omega$ as defined in §9.3. In §9.4 we prove that L-ellipticity is equivalent to the Fredholm property for the operator $\mathfrak{L}$ (in suitable Sobolev spaces) and also equivalent to two other properties: the existence of smoothing regularizers and the a priori estimate.

The proof of the main theorem will involve the use of p.d.o.'s in $\mathbb{R}^n$ and on $\partial\Omega$. A key ingredient in the proof is the Freezing Lemma of §7.8 which enables us, essentially, to reduce the general case to the case of operators with constant coefficients in the half-space $\Omega = \mathbb{R}^n_+$, $\partial\Omega = \mathbb{R}^{n-1}$.

9.1 Fredholm operators and some functional analysis

We begin with some definitions and preliminary results from functional analysis that are important for elliptic boundary value problems. This material should be quite familiar to the reader, so our presentation will be brief and serves mainly to establish the context.

Let X be a Banach space, i.e. a normed and complete vector space, and let its dual space be denoted by X'. For our purposes – differential operators in Hilbert spaces and their adjoints – it makes more sense to work with the anti-dual space X^* which consists of all anti-linear continuous functionals $f: X \to \mathbb{C}$. For $f \in X^*$, we write $f(x) = \langle f, x\rangle$, and then

$$\langle f, \lambda x + \mu y\rangle = \bar{\lambda}\langle f, x\rangle + \bar{\mu}\langle f, y\rangle, \qquad \text{for } x, y \in X, \lambda, \mu \in \mathbb{C},$$

and X^* is a Banach space with

$$\langle \lambda f_1 + \mu f_2, x \rangle = \lambda \langle f_1, x \rangle + \mu \langle f_2, x \rangle, \qquad f_1, f_2 \in X^*,$$

and norm defined by $\| f \| = \sup_{\|x\| \leqslant 1} |\langle f, x \rangle|$. The replacement of X' by X^* is not essential; it does, however, introduce a simplification for Hilbert spaces, H, since Riesz' theorem now states that H^* is isomorphic to H, and anti-dual operators, A^*, coincide with adjoint operators.

Let X and Y be Banach spaces, and let $\mathscr{L}(X, Y)$ denote the set of continuous, linear maps $A: X \to Y$. For $A \in \mathscr{L}(X, Y)$, we define the anti-dual map $A^*: Y^* \to X^*$ by $\langle A^* y^*, x \rangle_X = \langle y^*, Ax \rangle_Y$. It is well-known that $\|A^*\| = \|A\|$ and therefore A^* is a continuous linear map, i.e. $A^* \in \mathscr{L}(Y^*, X^*)$.

Definition 9.1 *We say that a continuous linear operator $A: X \to Y$ (where X, Y are Banach spaces) is a Fredholm operator if A has the following two properties:*

$$\alpha(A) = \dim \ker A < \infty,$$

where $\ker A = \{x \in X; Ax = 0\}$ *is the kernel of the operator A, and*

$$\beta(A) = \dim \operatorname{coker} A = \dim(Y/\operatorname{im} A) < \infty,$$

where $\operatorname{im} A = \{y \in Y; y = Ax, x \in X\}$ *is the image of A and* $\operatorname{coker} A = Y/\operatorname{im} A$ *is the cokernel.*

The numbers $\alpha(A)$, $\beta(A)$ are called the *d-numbers* (defect numbers) of A, and the index of A is the integer

$$\operatorname{ind} A = \alpha(A) - \beta(A).$$

We write $A \in \Phi(X, Y)$, i.e. $\Phi(X, Y)$ is the set of Fredholm operators from X to Y.

Proposition 9.2 *Let X and Y be Banach spaces and $A: X \to Y$ a continuous linear operator. If $\beta(A) < \infty$ then the image,* $\operatorname{im} A$, *is closed in Y.*

Proof Note that $X/\ker A$ is a Banach space, since $\ker A$ is closed by continuity of A. Let $\bar{A}: X/\ker A \to Y$ be the uniquely defined and continuous, linear, bijective map defined by $\bar{A}\bar{x} = Ax$ $(x \in \bar{x})$. Obviously the maps $\bar{A}$ and A have the same image.

Let W be a complementary subspace for $\operatorname{im} A$ in Y. By hypothesis, W is finite dimensional, and hence it is a Banach space. Consider now the map

$$S: (X/\ker A) \times W \to Y,$$

defined by $S(\bar{x}, w) := \bar{A}\bar{x} + w$. Since the Cartesian product of two Banach spaces is again a Banach space, then $(X/\ker A) \times W$ is a Banach space. The mapping S is continuous, linear and bijective, so it has a continuous inverse by the Open Mapping Theorem. In view of the fact that $(X/\ker A) \times 0$ is closed in $(X/\ker A) \times W$, we conclude that $\operatorname{im} A = S((X/\ker A) \times 0)$ is closed in Y.

Remark We can therefore add to the definition of a Fredholm operator $A: X \to Y$ the property that

$$\operatorname{im} A \text{ is closed in } Y,$$

without altering the class of Fredholm operators. This remark is important for the next proposition.

Proposition 9.3 *If A is a Fredholm operator, $A \in \Phi(X, Y)$, then the anti-dual A^* is also a Fredholm operator, i.e. $A^* \in \Phi(Y^*, X^*)$. The converse also holds. Further, for the d-numbers we have*

$$\alpha(A^*) = \beta(A) \quad \text{and} \quad \beta(A^*) = \alpha(A),$$

from which it follows that

$$\operatorname{ind} A^* = -\operatorname{ind} A.$$

Proof It is easy to see that

$$\ker A^* = (\operatorname{im} A)^{\perp} \cong (Y/\operatorname{im} A)^* = (\operatorname{coker} A)^*$$

and, since coker A is finite dimensional, then $\dim(\operatorname{coker} A)^* = \dim \operatorname{coker} A$, i.e. $\alpha(A^*) = \beta(A)$. To prove the second equality, we use the fact that im A is closed in Y, and apply the Closed Range Theorem of Banach (see [TL], p. 240), by which

$$\operatorname{im} A^* = (\ker A)^{\perp}.$$

This latter result is the key to the proof since we now obtain

$$\operatorname{coker} A^* = X^*/\operatorname{im} A^* = X^*/(\ker A)^{\perp} \cong (\ker A)^*,$$

from which it follows that

$$\beta(A^*) = \dim \operatorname{coker} A^* = \dim(\ker A)^* = \dim \ker A = \alpha(A).$$

The proof of the converse is similar.

Definition 9.4 *Let $A \in \mathscr{L}(X, Y)$. The operator $R_l \in \mathscr{L}(Y, X)$ (resp., the operator $R_r \in \mathscr{L}(Y, X)$) is called a left regularizer (resp., right regularizer) for A if the operator*

$$\mathscr{k}_1 := R_l \circ A - I_X \qquad (\textit{resp.}, \mathscr{k}_2 := A \circ R_r - I_Y),$$

is compact. Here I_X and I_Y denote the identity maps on the spaces X and Y.

Regularizers are unique modulo a compact operator, as the following lemma implies.

Lemma 9.5 *If for some R and R' we have $R \circ A - I \in \mathscr{K}$ and $A \circ R' - I \in \mathscr{K}$, where $\mathscr{K}$ denotes the class of compact operators then $R - R' \in \mathscr{K}$.*

The proof follows at once from the equation

$$R - R' = R(I - A \circ R') + (R \circ A - I)R'.$$

Theorem 9.6 *Let $A \in \mathscr{L}(X, Y)$.*

(*a*) *If A possesses a left regularizer, R_l, then $\alpha(A) < \infty$ and* im *A is closed.*
(*b*) *If A possesses a right regularizer, R_r, then $\beta(A) < \infty$ and* im *A is closed.*
(*c*) *A processes both left and right regularizers if and only if A is a Fredholm operator. (In this case one can take $R_t = R_l$.)*

Proof (a) Let A have a left regularizer $R_l \in \mathscr{L}(Y, X)$,

$$R_lA = I_X = \mathscr{k}_1, \qquad \text{where } \mathscr{k}_1 \in \mathscr{L}(X, X) \text{ is compact.}$$

If $x \in \ker A$, i.e. $Ax = 0$, then we get $x + \mathscr{k}_1 x = 0$, which means that x is an eigenvector of the compact operator $\mathscr{k}_1$ corresponding to the eigenvalue -1, and the finite dimensionality of $\ker A$ (i.e. $\alpha(A) < \infty$) follows from the Riesz-Schauder theory of compact operators (see [TL]). In order to show that im A is closed, one factors A and proceeds in a manner similar to the proof of Proposition 9.2. For the details see [Wl].

(b) Let A have a right regularizer $R_r \in \mathscr{L}(Y, X)$,

$$AR_r = I_Y + \mathscr{k}_2, \qquad \text{where } \mathscr{k}_2 \in \mathscr{L}(Y, Y) \text{ is compact.}$$

By duality we have $R_r^* \circ A^* = I_{Y^*} + \mathscr{k}_2^*$, where the operator $\mathscr{k}_2^*$ is compact. Hence $\beta(A) = \alpha(A^*) < \infty$, and then im A is closed by Proposition 9.2.

(c) It follows from (a) and (b) that existence of right and left regularizers is sufficient for A to be a Fredholm operator. Conversely, suppose that $A \in \Phi(X, Y)$. Then there exist closed subspaces Z and W in X and Y, respectively, such that

$$X = \ker A \oplus Z, \qquad Y = \operatorname{im} A \oplus W.$$

We have $\alpha = \dim \ker A$ linearly independent functionals $\ker A \to \mathbb{C}$, and they can be extended to elements $f_1, \ldots, f_\alpha$ of X^* by the Hahn-Banach Theorem. We then let Z be the intersection of the kernels of $f_1, \ldots, f_\alpha$. The existence of W is clear (see Proposition 9.2).

The restricted operator $B = A|_Z$ is continuous and bijective from Z to im A. Since Z and im A are Banach spaces then $B^{-1}\colon \operatorname{im} A \to Z$ is also continuous by the Open Mapping Theorem. We put $R = B^{-1} \circ P$, where P is the projection of Y onto im A, and have

$$RA - I = -\mathrm{proj}_{\ker A}, \qquad AR - I = -\mathrm{proj}_W.$$

This completes the proof of (c), because, with finite dimensional image, the continuous operators $\mathscr{k}_1 = -\mathrm{proj}_{\ker A}$ and $\mathscr{k}_2 = -\mathrm{proj}_W$ are compact.

Remark 9.7 From the proof we see that $\mathscr{k}_1$, $\mathscr{k}_2$ can be taken from the class of operators with finite dimensional image.

Theorem 9.8

(a) *If* $A_1 \in \Phi(X, Y)$ *and* $A_2 \in \Phi(Y, Z)$ *are Fredholm operators, then*

$$A := A_2 \circ A_1$$

is again a Fredholm operator and

$$\operatorname{ind}(A_2 \circ A_1) = \operatorname{ind} A_1 + \operatorname{ind} A_2.$$

(b) *If* $A_1 \in \Phi(X_1, Y_1)$ *and* $A_2 \in \Phi(X_2, Y_2)$ *then the direct sum* $A_1 \oplus A_2$: $X_1 \oplus X_2 \to Y_1 \oplus Y_2$ *is also a Fredholm operator and*

$$\operatorname{ind}(A_1 \oplus A_2) = \operatorname{ind} A_1 + \operatorname{ind} A_2.$$

Proof By part (c) of Theorem 9.6 there exist operators $R_1 \in \mathscr{L}(Y, X)$ and $R_2 \in \mathscr{L}(Z, Y)$ such that the operators $k_{11} := R_1 A_1 - I$, $k_{21} := A_1 R_1 - I$, $k_{12} := R_2 A_2 - I$, and $k_{22} := A_2 R_2 - I$, are compact. We put $R = R_1 \circ R_2$ and then

$$\begin{aligned} RA - I &= R_1 R_2 A_2 A_1 - I \\ &= R_1(I + k_{12})A_1 - I \\ &= k_{11} + R_1 k_{12} A_1 \end{aligned}$$

is a compact operator; similarly, $AR - I = k_{22} + A_2 k_{21} R_2$ is a compact operator. Therefore $A \in \Phi(X, Z)$ by Definition 9.4.

In order to prove the index formula we use Proposition 9.3 and proceed as follows Let $\{\psi_1, \ldots, \psi_{\beta(A_1)}\}$ be a basis of $\ker A_1^*$ and $\{\varphi_1, \ldots, \phi_{\alpha(A_2)}\}$ a basis of $\ker A_2$. If $A_2 \circ A_1 x = 0$ then $A_1 x = \sum_{k=1}^{\alpha(A_2)} c_k \varphi_k$ for some $c_k \in \mathbb{C}$; because $\operatorname{im} A_1 = (\ker A_1^*)^{\perp}$ it must be that

$$\sum_{k=1}^{\alpha(A_2)} c_k \langle \psi_j, \varphi_k \rangle = 0 \qquad \text{for } j = 1, \ldots, \beta(A_1).$$

These steps are reversible whence the equation $Ax = A_2 A_1 x = 0$ has exactly

$$\alpha(A) = \alpha(A_1) + (\alpha(A_2) - s)$$

linearly independent solutions, where s = rank of the matrix $\{\langle \psi_j, \varphi_k \rangle\}$ and $\alpha(A_2) - s$ is the dimension of the kernel of this $\beta(A_1) \times \alpha(A_2)$ matrix. If one proceeds similarly with the equation $A^* z = A_1^* A_2^* z = 0$, one deduces that it has

$$\alpha(A^*) = \alpha(A_2^*) + (\alpha(A_1^*) - s)$$

linearly independent solutions (where s is the rank of the transpose of the same matrix as before). Therefore we have

$$\begin{aligned} \operatorname{ind} A &= \alpha(A) - \alpha(A^*) \\ &= \alpha(A_1) + \alpha(A_2) - s - \alpha(A_1^*) - \alpha(A_2^*) + s \\ &= \operatorname{ind} A_1 + \operatorname{ind} A_2 \end{aligned}$$

Finally, to prove part (b) of the theorem we factor

$$A_1 \oplus A_2 = (I_{Y_1} \oplus A_2) \circ (A_1 \oplus I_{X_2}),$$

and observe that $\operatorname{ind}(I_{Y_1} \oplus A_2) = \operatorname{ind} A_2$ and $\operatorname{ind}(A_1 \oplus I_{X_2}) = \operatorname{ind} A_1$, and then apply part (a).

Corollary 9.9 *If $A \in \Phi(X, Y)$ is a Fredholm operator and $S_1 \in L(Y, Z)$, $S_2 \in L(V, X)$ are isomorphisms, then $S_1 \circ A \circ S_2 \colon V \to Z$ is again a Fredholm operator with the same index.*

In fact, because an isomorphism is Fredholm with index 0, the corollary follows from part (a) of Theorem 9.8.

Theorem 9.10 *If $A \in \Phi(X, Y)$ is a Fredholm operator and if R is a left regularizer for A, then R is also a Fredholm operator, $R \in \Phi(Y, X)$, and we have*

$$\operatorname{ind} R = -\operatorname{ind} A$$

By duality the same holds for right regularizers.

Proof We have $\dim\ker R \circ A = \dim\ker(I_X + \mathscr{k}) < \infty$, and from linear algebra

$$\dim\ker R \leqslant \dim\ker R \circ A + \operatorname{codim}\operatorname{im} A.$$

Since, by assumption, $\operatorname{codim}\operatorname{im} A = \dim\operatorname{coker} A < \infty$, it follows that $\alpha(R) = \dim\ker R < \infty$. Moreover, A is a right regularizer for R, so that by part (b) of Theorem 9.6 the dimension of the cokernel, $\beta(R)$, is also finite. Hence R is Fredholm. As for the index, we apply Theorem 9.8(a) to the equation

$$R \circ A = I_X + \mathscr{k},$$

obtaining

$$\operatorname{ind}(R \circ A) = \operatorname{ind} R + \operatorname{ind} A = \operatorname{ind}(I_X + \mathscr{k}),$$

which concludes the proof since $I_X + \mathscr{k}$ is a Riesz-Schauder operator and therefore has zero index (see [TL]).

Stability of the index under perturbations

The index of a Fredholm operator has certain stability properties which make it a homotopy invariant. We first prove the stability of the index under compact perturbations.

Theorem 9.11 *If $A \in \Phi(X, Y)$ is a Fredholm operator and $\mathscr{k} \in \mathscr{K}(X, Y)$ is compact, then $A + \mathscr{k} \in \Phi(X, Y)$ is again a Fredholm operator and we have*

$$\operatorname{ind}(A + \mathscr{k}) = \operatorname{ind} A.$$

Proof With $R_l \circ A - I_X$ and $A \circ R_r - I_Y$ being compact operators, the operators

$$R_l(A + \mathscr{k}) - I_X \quad \text{and} \quad (A + \mathscr{k})R_r - I_Y$$

are also compact, hence $A + \mathscr{k}$ is also both left and right regularizable with the same regularizers as for A. Then Theorem 9.6(c) yields

$$A + \mathscr{k} \in \Phi(X, Y)$$

and, by virtue of Theorem 9.10, we have $\operatorname{ind}(A + \mathscr{k}) = -\operatorname{ind} R_l = \operatorname{ind} A$.

We next study "continuous perturbations" of Fredholm operators.

Theorem 9.12 *Let $A \in \Phi(X, Y)$ be a Fredholm operator with a (left and right) regularizer R. If $S \in \mathscr{L}(X, Y)$ is any operator such that $\|S\| \cdot \|R\| < 1$, then $A + S \in \Phi(X, Y)$ is again a Fredholm operator and*

$$\operatorname{ind}(A + S) = \operatorname{ind} A.$$

It follows that $\Phi(X, Y)$ is an open set in $\mathscr{L}(X, Y)$, and that the index is homotopically stable in the operator-norm topology.

Proof We know from Theorem 9.6(c) that for any $A \in \Phi(X, Y)$ there exists a regularizer R which is simultaneously a right and left regularizer.

Assuming $\|S\| \cdot \|R\| < 1$, the Neumann series for $(I_X + R \circ S)^{-1}$ and $(I_Y + S \circ R)^{-1}$ converge, giving bounded linear operators. It is easily shown that

$$R_l = (I_X + R \circ S)^{-1} \circ R$$

is a left regularizer and

$$R_r = R \circ (I_Y + S \circ R)^{-1}$$

is a right regularizer for $A + S$, so that $A + S \in \Phi(X, Y)$. Since the equations $R_l y = 0$ and $Ry = 0$ are equivalent we have $\alpha(R_l) = \alpha(R)$, and passing to duals we get in the same manner $\beta(R_l) = \alpha(R_l^*) = \alpha(R^*) = \beta(R)$. We apply Theorem 9.10 and obtain

$$\operatorname{ind}(A + S) = -\operatorname{ind} R_l = -\operatorname{ind} R = \operatorname{ind} A.$$

To prove the last assertion of our theorem, consider a continuous path $\{A_t; 0 \leqslant t \leqslant 1\}$ in $\Phi(X, Y)$. Since a path, being a continuous image of $[0, 1]$, is connected and compact, we may apply the first part of our theorem finitely often, obtaining thus $\operatorname{ind} A_0 = \operatorname{ind} A_t = \operatorname{ind} A_1$.

The index has much stronger stability properties than stated in Theorem 9.12; this is the content of the following theorem of Hörmander [Hö 3].

Recall that an operator function $A_t \in \mathscr{L}(X_1, X_2)$, $t \in I$, is said to be strongly continuous if, for $t_n \to t_0$, the sequence $A_{t_n} f \to A_{t_0} f$ converges in X_2 for every $f \in X_1$.

Theorem 9.13 *Let X_1 and X_2 be two Banach spaces and let I be a compact space (i.e. $I = [0, 1]$). If $A_t \in \mathscr{L}(X_1, X_2)$ and $R_t \in \mathscr{L}(X_2, X_1)$ are strongly continuous as functions of $t \in I$, and if*

$$\mathscr{k}_{1t} = R_t A_t - I_1, \; \mathscr{k}_{2t} = A_t R_t - I_2$$

are uniformly compact in the sense that for $j = 1, 2$

$$M_j = \{\mathscr{k}_{jt} f; t \in I, f \in X_j, \|f\|_j \leqslant 1\}$$

is precompact, in X_j, *then* A_t *and* R_t *are Fredholm operators,* dim ker A_t *and* dim ker R_t *are upper semi-continuous and* ind $A_t = -$ind R_t *is locally constant, hence constant if* I *is connected.*

Remark 9.14 The hypotheses of Theorem 9.13 are very natural in the study of elliptic pseudo-differential operators A_t on a compact manifold M; the operators R_t are their parametrices (see §8.6).

For the proof we need the classical lemma of F. Riesz, which we state in a form appropriate to our purposes.

Lemma 9.15 *Let* L *be a closed subspace of the Banach space* X. *We have*

$$\dim L \geqslant k$$

if and only if there exist $f_j \in L$, $j = 1, \ldots, k$ *such that*

$$\|f_j\| = 1 \leqslant \left\| f_j + \sum_{i<j} a_i f_i \right\|,$$

for all $j = 1, \ldots, k$ *and all* $a_i \in \mathbb{C}$.

Proof If the last condition holds the f_j's must be linearly independent; hence $\dim L \geqslant k$. To prove the necessity, let us suppose that $f_1, \ldots, f_{j-1}$ have already been chosen, and denote by L_{j-1} the linear space spanned by these elements. If $j-1<k$ we choose $g \in L$ outside L_{j-1}, then there is a point $h \in L_{j-1}$ which minimizes $\|g-h\|$, because L_{j-1} is finite dimensional (hence locally compact) and the norm $\to \infty$ if $h \to \infty$. Then

$$\left\| g - h - \sum_{i<j} a_i f_i \right\| \geqslant \|g-h\| \qquad \text{for all } a_i \in \mathbb{C}$$

and $f_j = (g-h)/\|g-h\|$ has the required properties.

Now we prove Theorem 9.13. First of all, we claim that for any finite dimensional space $W \subset X_2$ the set

$$N_W = \{(t, f) \in I \times X_1;\ \|f\|_1 = 1,\ A_t f \in W\}$$

is compact in $I \times X_1$. In particular if we let $W = \{0\}$ it follows that

$$N = \{(t, f) \in I \times X_1;\ \|f\|_1 = 1,\ A_t f = 0\}.$$

is compact in $I \times X_1$.

To show first that N is closed let $(t_n, f_n) \in N$ and let $t_n \to t_0$ and $f_n \to f_0$. By the principle of uniform boundedness of Banach-Steinhaus (here we use strong continuity, see [TL]), we have $\|A_{t_n} - A_{t_0}\| \leqslant M < \infty$, whence

$$\|(A_{t_n} - A_{t_0})(f_n - f_0)\| \leqslant M \|f_n - f_0\| \to 0.$$

Since $A_{t_n} f_n = 0$ and

$$(A_{t_n} - A_{t_0})(f_n - f_0) = A_{t_n} f_n - A_{t_0} f_n - A_{t_n} f_0 + A_{t_0} f_0,$$

we obtain by letting $n \to \infty$,

$$0 = 0 - A_{t_0} f_0 - A_{t_0} f_0 + A_{t_0} f_0$$

Therefore $A_{t_0} f_0 = 0$, and $(t_0, f_0) \in N$, so N is closed. In the same way if we have $(t_n, f_n) \in N_W$ then $A_{t_n} f_n \in W$ and it follows that $A_{t_0} f_0 \in W$, so $(t_0, f_0) \in N_W$ and N_W is also closed.

The compactness of N now follows immediately from our assumptions. In fact N is a subset of $I \times M_1$ because $A_t f = 0$ implies $f = -\mathscr{k}_{1t} f \in M_1$; therefore N is closed in the compact space $I \times \bar{M}_1$ and is therefore compact itself. To prove compactness of N_W for any finite dimensional space W requires only a little more proof. When $(t, f) \in N_W$ we have $f = R_t \circ A_t f - \mathscr{k}_{1t} f$ where $\mathscr{k}_{1t} f \in M_1$ and

$$\| A_t f \|_2 \leqslant C \| f \|_1 = C$$

since A_t is strongly continuous in t (by Banach-Steinhaus). The compactness of N_W follows from the fact that the map $(t, g) \mapsto R_t g$ is continuous and the set $\{(t, g) \in I \times W; \|g\|_2 \leqslant C\}$ is compact.

The hypotheses of Theorem 9.13 imply that A_t and R_t are Fredholm operators by Theorems 9.6(c) and 9.10. Consider now the set

$$I_k = \{t \in I; \dim \ker A_t \geqslant k\}.$$

By applying Lemma 9.15 we get

$$I_k = \left\{ t \in I; \exists f_1, \ldots, f_k \text{ such that } (t, f_j) \in N \text{ and } \left\| f_j + \sum_{i<j} a_i f_i \right\| \geqslant 1 \right.$$
$$\left. j = 1, \ldots, k \text{ and all } a_i \in \mathbb{C} \right\}$$

and the set of $(f_1, \ldots, f_k)$'s considered here is closed in the cartesian product N^k, so I_k is the projection into I of a compact set in N^k. Hence I_k is compact; this means that $\{t; \dim \ker A_t < k\}$ is open. Hence $\dim \ker A_t$ is upper semi-continuous and the same is of course true for $\dim \ker R_t$ (since R_t satisfies the same hypotheses).

Since $\operatorname{ind} A_t + \operatorname{ind} R_t = 0$ (see Theorem 9.10), the proof of the theorem will be complete if we show that $\operatorname{ind} A_t$ (and therefore $\operatorname{ind} R_t$) is upper semi-continuous. Let $t_0 \in I$ and choose a closed subspace $Z \subset X_1$ such that $X_1 = Z \oplus \ker A_{t_0}$. Since the injection $Z \to X_1$ has index equal to $-\dim \ker A_{t_0}$, the index of the restriction of A_t to Z is by Theorem 9.8(a) equal to

$$\operatorname{ind} A_t|_Z = \operatorname{ind} A_t - \dim \ker A_{t_0}.$$

Hence it is sufficient to show that $\operatorname{ind} A_t|_Z$ is upper semi-continuous. (It is easily verified that the operators $A_t|_Z$ satisfy the same conditions as those imposed on A_t in the statement of the theorem.) Since $\dim \ker A_{t_0}|_Z = 0 < 1$, then, by the upper semi-continuity of $\dim \ker A_t|_Z$ proved above, we have $\dim \ker A_t|_Z < 1$ for all t in a neighbourhood of t_0. Hence $A_t|_Z$ is injective for all t near t_0, and

$$\text{ind } A_t|_Z = 0 - \text{codim } A_t Z.$$

To prove the upper semi-continuity of $\text{ind } A_t|_Z$ we must therefore show that

$$\text{codim } A_t Z \geqslant \text{codim } A_{t_0} Z = \text{codim } A_{t_0} X_1$$

for all t near t_0. Choose W with $\dim W = \dim \text{coker } A_{t_0} < \infty$ and $W \cap A_{t_0} X_1 = \{0\}$. From what was proved earlier we know that the set

$$N_W \cap I \times Z = \{(t, f) \in I \times Z;\ \|f\|_1 = 1,\ A_t f \in W\}$$

is compact in $I \times Z$ (this is just the set N_W with the closed subspace Z substituted for X_1); hence its projection into I,

$$I_W = \{t \in I;\ (t, f) \in N_W \text{ for some } f \in Z\},$$

is compact in I. The complement is therefore open in I, and it contains t_0 by choice of W. For all t in a neighbourhood of t_0 it follows that $A_t f \notin W$ for all $f \in Z$, thus

$$\text{codim } A_t Z \geqslant \dim W = \text{codim } A_{t_0} X_1,$$

as required.

Smoothable operators

We now turn to the definition of a smoothable operator. The definitions and results of this sub-section have already been used in §8.6 for elliptic operators on a compact manifold, and are used again in §9.4.

Definition 9.16 *Let X_1, X_2, Y_1 and Y_2 be four Banach spaces, with embeddings $X_2 \hookrightarrow X_1$ and $Y_2 \hookrightarrow Y_1$. Also let $A\colon X_1 \to Y_1$ be linear and continuous and consider the diagram*

$$\begin{array}{ccc} X_2 & \hookrightarrow & X_1 \\ & & \downarrow A \\ Y_2 & \hookrightarrow & Y_1 \end{array}$$

We say that A is smoothable if there exist operators R_l and $R_r \in \mathscr{L}(Y_1, X_1)$ such that

$$T_l := R_l \circ A - I_{X_1} \in \mathscr{L}(X_1, X_2),$$

$$T_r := A \circ R_r - I_{Y_1} \in \mathscr{L}(Y_1, Y_2),$$

and with the property that the restricted operators $R_l|_{Y_2}$ and $R_r|_{Y_2}$ belong to $\mathscr{L}(Y_2, X_2)$.

We call the operator T_l (resp. T_r) a left (resp. right) *smoothing operator*. The operator R_l (resp. R_r) is called a left (resp. right) *smoothing regularizer*.

Weyl's Lemma holds for smoothable operators.

Lemma 9.17 (*Weyl's Lemma*) *Let A be smoothable as defined above. Then from the equation $Ax = y$, $x \in X_1$, $y \in Y_2$ it follows that x is an element of X_2. In particular,* ker A *is contained in X_2.*

Proof We have $T_l = R_l A - I_{X_1}$, and so $x = R_l y - T_l x \in X_2$ by the defining properties.

Proposition 9.18 *Let A be smoothable. If the embeddings $X_2 \hookrightarrow X_1$ and $Y_2 \hookrightarrow Y_1$ are compact, then A is a Fredholm operator.*

Proof We extend the codomain of the smoothing operators

$$T_l\colon X_1 \to X_2 \hookrightarrow X_1, \qquad T_r\colon Y_1 \to Y_2 \hookrightarrow Y_1$$

to obtain compact operators and then apply Theorem 9.6(c).

Proposition 9.19 *Let $A \in \mathscr{L}(X_1, Y_1)$ be smoothable as defined above, and let the following additional conditions be satisfied: A restricts to an operator $A|_{X_2} \in \mathscr{L}(X_2, Y_2)$, the embeddings $X_2 \hookrightarrow X_1$ and $Y_2 \hookrightarrow Y_1$ are compact, and Y_2 is dense in Y_1. Then both*

$$A \text{ and } A|_{X_2} \text{ are Fredholm operators,}$$

and their d-numbers are equal, that is, $\alpha(A) = \alpha(A|_{X_2})$ and $\beta(A) = \beta(A|_{X_2})$, from which it follows that

$$\operatorname{ind} A = \operatorname{ind}(A|_{X_2}).$$

Proof The Fredholm property of A was proved in Proposition 9.18, and the equality of the kernels of A and of $A|_{X_2}$ follows from Lemma 9.17.

For the cokernel of A we have by definition

$$Y_1 = A(X_1) \oplus W,$$

where $\dim W = \beta(A) < \infty$; moreover, we can take $W \subset Y_2$ since Y_2 is dense in Y_1 (the Gohberg–Krein lemma; see [GK] or [Wl], p. 185). We claim that $Y_2 \cap A(X_1) = A(X_2)$, from which it follows that

$$Y_2 = A(X_2) \oplus W. \tag{*}$$

Now, by assumption A restricts to an operator $X_2 \to Y_2$ so we certainly have the inclusion $Y_2 \cap A(X_1) \supset A(X_2)$. The other inclusion $\subset$ is also clear because A is smoothable: if $y \in Y_2 \cap A(X_1)$ then by Lemma 9.17 we must have $y = Ax$ where $x \in X_2$. Therefore (*) holds, so we obtain $\beta(A|_{X_2}) = \dim W = \beta(A)$, whence $A|_{X_2}$ is Fredholm and $\operatorname{ind} A = \operatorname{ind}(A|_{X_2})$.

The following proposition gives an abstract a priori estimate.

Proposition 9.20 *Let X_1, X and Y be three Banach spaces, and consider the "Schauder scheme"*

$$\begin{array}{ccc} X & \xrightarrow{A} & Y \\ \uparrow & & \\ X_1 & & \end{array}$$

where the operator $A \in \mathscr{L}(X, Y)$ has finite dimensional kernel and closed image. Then there exists a constant $c > 0$ such that for all $x \in X$ we have

$$\|x\|_X \leqslant c[\|Ax\|_Y + \|x\|_{X_1}].$$

Proof Since $\dim \ker A < \infty$, we can decompose X into a direct sum,

$$X = \ker A \oplus Z, \tag{1}$$

where Z is a closed subspace. The restriction of A to Z is continuous and bijective with closed image, im A. Then, by the Open Mapping Theorem, the inverse $(A|_Z)^{-1}$ is also continuous, and we have the estimate

$$\|z\|_X \leqslant c\|Az\|_Y, \qquad \text{for all } z \in Z. \tag{2}$$

Because $X \subset X_1$, we can consider $\ker A$ as a subspace of X_1, and, since $\ker A$ is finite dimensional, all norms on $\ker A$ are equivalent and $\ker A$ is again closed in X_1. Thus we may assume that

$$X_1 = \ker A \oplus Z_1,$$

where Z_1 is a closed subspace of X_1 containing Z. Then the projection

$$P: X_1 \to \ker A \subset X$$

is continuous

$$\|Px\|_X \leqslant c'\|x\|_{X_1}, \qquad \text{for all } x \in X_1. \tag{3}$$

Also note that P agrees on X with the projection $X \to \ker A$ induced by the direct sum (1) since $Z \subset Z_1$.

Now let $x \in X$. By the direct sum decomposition (1) we have $x = Px + z$, where $Px \in \ker A$ and $z \in Z$, whence

$$\begin{aligned} \|x\|_X &\leqslant \|Px\|_X + \|z\|_X \\ &\leqslant c'\|x\|_{X_1} + c\|Az\|_Y \\ &\leqslant \text{const}\,[\|x\|_{X_1} + \|Ax\|_Y], \end{aligned}$$

where we have used (2), (3) and the fact that $Ax = Az$.

9.2 Elliptic systems of Douglis-Nirenberg type

Let Ω be a bounded domain in $\mathbb{R}^n$ (a domain being a connected, open set). We will consider systems of linear differential equations $\mathscr{A}(x, D)u(x) = f(x)$, $x \in \Omega$, where u and f are p-vector functions and

$$\mathscr{A}(x, D) = [A_{ij}(x, D)]_{p \times p}, \qquad x \in \Omega,$$

is a $p \times p$ matrix, such that the elements, A_{ij}, are linear differential operators

$$A_{ij}(x, D) = \sum_{|\alpha| \leqslant \alpha_{ij}} a_{ij}^{(\alpha)}(x) D^\alpha$$

with smooth coefficients, $a_{ij}^{(\alpha)} \in C^\infty(\bar{\Omega})$. Here we are using the usual multi-index notation: $D^\alpha = D_1^{\alpha_1} \cdots D_n^{\alpha_n}$, $|\alpha| = \alpha_1 + \cdots + \alpha_n$, and for convenience when operating with the Fourier transformation, the basic derivatives are equipped with the factor $1/i$, that is, $D_j = i^{-1}\partial/\partial x_j$, where $i = \sqrt{-1}$. The boundary $\partial\Omega$ of the domain Ω is also assumed to be C^∞. For weaker smoothness assumptions see [Be] or [Wl].

Remark For later use, we assume that the coefficients, $a_{ij}^{(\alpha)}$, have been extended from $\bar{\Omega}$ to $\mathbb{R}^n$ in such a way that they belong to $C_0^\infty(\mathbb{R}^n)$. By virtue of Lemma 5.58 this is always possible.

Let α_{ij} denote the order of A_{ij}. If $A_{ij} \equiv 0$, then we set $\alpha_{ij} = -\infty$. For any permutation $T = (j_1, \ldots, j_p)$ of the numbers $(1, \ldots, p)$, let $m(T)$ denote the sum of orders $\alpha_{ij_1} + \cdots + \alpha_{pj_p}$, and put

$$m = \max_T m(T). \tag{4}$$

Let the point $x \in \bar{\Omega}$ be fixed. Upon replacing the derivatives D_j with ξ_j, we obtain the polynomial $\det \mathscr{A}(x, \xi)$ of n real variables $\xi = (\xi_1, \ldots, \xi_n)$ and it has order $\leqslant m$. The *characteristic polynomial* of $\mathscr{A}$ is

$$\chi(\xi) = \pi_m \det \mathscr{A}(x, \xi),$$

i.e. the sum of the terms of order m in $\det \mathscr{A}(x, \xi)$. Note that it is possible that $\chi \equiv 0$ even if $\det \mathscr{A} \not\equiv 0$ (see Exercise 14).

Definition 9.21 *The differential operator $\mathscr{A}$ is called non-degenerate at $x \in \bar{\Omega}$ if the characteristic polynomial χ is not identically equal to* 0. *It is* elliptic *at x if*

$$\chi(\xi) = \pi_m \det \mathscr{A}(x, \xi) \neq 0 \qquad \text{for all } 0 \neq \xi \in \mathbb{R}^n.$$

This definition of ellipticity is due to Volevič [Vo 2]. Now we define ellipticity in the sense of Douglis-Nirenberg [DN] and show that it is equivalent to Volevič-ellipticity.

First we need a preliminary lemma, the proof of which is postponed until the appendix at the end of this section. Note that if $s_1, \ldots, s_p, t_1, \ldots, t_p$ are any integers such that $\alpha_{ij} \leqslant s_i + t_j$ for all i and j, then

$$\alpha_{1,j_1} + \cdots + \alpha_{p,j_p} \leqslant \sum_{i=1}^{p} (s_i + t_i) \tag{5}$$

for any permutation $j_1, \ldots, j_p$.

Lemma 9.22 *For any $p \times p$ matrix $\alpha = [\alpha_{ij}]$ whose elements are integers, or equal to $-\infty$, there exist integers $s_1, \ldots, s_p, t_1, \ldots, t_p$ such that*

$$\sum_{i=1}^{p} (s_i + t_i) = m$$

$$s_i + t_j \geqslant \alpha_{ij}$$

where, as before, m denotes the maximum of $\alpha_{1,j_1} + \cdots + \alpha_{p,j_p}$ *taken over all permutations* $(j_1, \ldots, j_p)$ *of the numbers* $(1, \ldots, p)$.

Let $m' = \min \sum_{i=1}^{p} (s_1 + t_i)$, where the minimum is taken over all integers $s_1, \ldots, s_p, t_1, \ldots, t_p$ such that $s_i + t_j \geqslant \alpha_{ij}$. In view of (4) and (5), we always have

$$m \leqslant m'.$$

The proof of Lemma 9.22 is equivalent to showing that $m = m'$. This is in fact true and is just an instance of the duality principle in linear programming, i.e. the maximum for the primal problem is equal to the minimum for the dual problem. But for the particular max-min problem that we are concerned with here, it is possible to give a simple, direct proof based on Lemma 9.27 in the appendix.

Remark A system of integers $s_1, \ldots, s_p, t_1, \ldots, t_p$ satisfying the conditions of Lemma 9.22 is said to be a set of *DN numbers* for the matrix $[\alpha_{ij}]$.

Thus, there always exists a set of DN numbers for the matrix $[\alpha_{ij}]$ of orders of the differential operator $\mathscr{A}$, i.e. there exist integers $s_1, \ldots, s_p$, $t_1, \ldots, t_p$ such that $\sum_{i=1}^{p} (s_i + t_i) = m$ and the operator A_{ij} has order $\alpha_{ij} \leqslant s_i + t_j$, where it is to be understood that $A_{ij} \equiv 0$ if $s_i + t_j < 0$. Clearly, any given s_i, t_j may be replaced by s_i + constant, t_j − same constant.

Then we let $A'_{ij}(x, D)$ denote the sum of terms in $A_{ij}(x, D)$ which are exactly of the order $s_i + t_j$, with lower-order terms replaced by zeros. For arbitrary real $\xi = (\xi_1, \ldots, \xi_n) \in \mathbb{R}^n$ we define the DN principal part $\pi_D\mathscr{A}(x, \xi)$ as the polynomial $p \times p$ matrix

$$\pi_D\mathscr{A}(x, \xi) := [A'_{ij}(x, \xi)]_{p \times p} = [\pi_{s_i + t_j} A_{ij}(x, \xi)]_{p \times p}. \tag{6}$$

Lemma 9.23 *The polynomial* det $\pi_D\mathscr{A}(x, \xi)$ *does not depend on the choice of DN numbers* s_i *and* t_j. *In fact,* $\det \pi_D\mathscr{A}(x, \xi) = \chi(\xi)$, *the characteristic polynomial of* $\mathscr{A}$.

Proof The terms in the polynomial $\det \pi_D\mathscr{A}(x, \xi)$ have order $\sum_{i=1}^{p} (s_i + t_i) = m$; hence they form a subset of the terms of $\chi = \pi_m \det \mathscr{A}$. On the other hand, terms in $\pi_D\mathscr{A}$ which are 0 are not counted in χ. Indeed, $\det \mathscr{A}$ is a sum of terms of the form $\pm A_{1,j_1} \cdots A_{p,j_p}$ and if $A'_{k,j_k} \equiv 0$ for some k then

$$\text{either } A_{k,j_k} \equiv 0 \quad \text{or} \quad \alpha_{k,j_k} < s_k + t_{j_k},$$

and in the latter case the order of the product $A_{1,j_1} \cdots A_{p,j_p}$ is less than m.

Definition 9.21′ *The operator* $\mathscr{A}(x, D)$ *is said to be* DN-elliptic *at* $x \in \bar{\Omega}$ *if there exist DN numbers* $s_1, \ldots, s_p$, $t_1, \ldots, t_p$ *such that the characteristic* *polynomial*

$$\chi(\xi) = \det \pi_D \mathscr{A}(x, \xi) \neq 0 \qquad \text{for each } 0 \neq \xi \in \mathbb{R}^n. \tag{7}$$

By virtue of Lemmas 9.22 and 9.23 we see that $\mathscr{A}$ is DN-elliptic if and only if it is Volevič-elliptic. From now on we say "elliptic" rather than DN-elliptic or Volevič-elliptic since the two notions are equivalent.

Note that the choice of DN numbers is not unique. Nevertheless, from now on we write $\pi\mathscr{A}$ instead of $\pi_D\mathscr{A}$ for the principal part. The characteristic polynomial χ is homogeneous of degree m,

$$\chi(c\xi) = \det \pi\mathscr{A}(x, c\xi) = c^m \det \pi\mathscr{A}(x, \xi) = c^m\chi(\xi). \tag{8}$$

Further, we have

$$\pi\mathscr{A}(x, c\xi) = S(c)\cdot\pi\mathscr{A}(x, \xi)\cdot T(c), \qquad c \neq 0, \tag{9}$$

where $S(c) = [\delta_{ij}c^{s_i}]_{p\times p}$, $T(c) = [\delta_{ij}\, c^{t_j}]_{p\times p}$ are diagonal matrices.

Example We illustrate the definition of DN numbers on a 2×2 matrix $\mathscr{A}(x, D)$ with the orders

$$\alpha = \begin{pmatrix} \alpha_{11} & \alpha_{12} \\ \alpha_{21} & \alpha_{22} \end{pmatrix};$$

and assume nontriviality, i.e. not all $\alpha_{ij} = -\infty$. There are three possibilities:

(i) $m = \alpha_{11} + \alpha_{22} = \alpha_{12} + \alpha_{21}$
(ii) $m = \alpha_{11} + \alpha_{22} > \alpha_{12} + \alpha_{21}$
(iii) $\alpha_{11} + \alpha_{22} < \alpha_{12} + \alpha_{21} = m$

In cases (ii) and (iii) we have $m \geqslant 0$, and degeneracy can only occur in case (i). Let us verify Lemma 9.22, i.e. the fact that DN numbers s_1, s_2, t_1 and t_2 always exist. Without loss of generality $t_1 = 0$.

In cases (i) and (ii) we seek integers s_i and t_j such that

$$m = \alpha_{11} + \alpha_{22} = s_1 + 0 + s_2 + t_2 \qquad \text{and } \alpha_{ij} \leqslant s_i + t_j \ \forall i, j. \tag{10}$$

Solve for s_1, s_2 and t_2 successively from the equations $s_1 = \alpha_{11}$, $s_1 + t_2 = \alpha_{12}$, $s_2 + t_2 = \alpha_{22}$. Since $\alpha_{11} + \alpha_{22} - \alpha_{12} \geqslant \alpha_{21}$, it follows that

$$\alpha_{21} \leqslant s_1 + s_2 + t_2 - s_1 - t_2 = s_2 + 0 = s_2 + t_1,$$

so (10) is satisfied.

In case (iii) we seek integers s_i and t_j such that

$$m = \alpha_{12} + \alpha_{21} = s_1 + 0 + s_2 + t_2 \qquad \text{and } \alpha_{ij} \leqslant s_i + t_j \ \forall i, j.$$

Solve for s_1, s_2 and t_2 from the equations $s_1 = \alpha_{11}$, $s_1 + t_2 = \alpha_{12}$, $s_2 = \alpha_{21}$. Since $\alpha_{22} < \alpha_{12} + \alpha_{21} - \alpha_{11}$, it follows that

$$\alpha_{22} < s_1 + t_2 + s_2 - s_1 = s_2 + t_2.$$

We say that the differential operator $\mathscr{A}(x, D)$ is *elliptic* if it is elliptic at each point $x \in \bar{\Omega}$ with a fixed set of DN numbers. If $\mathscr{A}(x, D)$ is elliptic then by taking the minimum of the continuous function $|\det \pi\mathscr{A}(x, \xi)| > 0$ over the compact set $\bar{\Omega} \times S^n$, we obtain $|\det \pi\mathscr{A}(x, \xi)| \geqslant c$, $\forall x \in \bar{\Omega}$, $|\xi| = 1$. Substituting $\xi/|\xi|$, $\xi \neq 0$, and using the homogeneity (8) of $\det \pi\mathscr{A}(x, \xi)$, we get

$$|\det \pi\mathscr{A}(x, \xi)| \geqslant c|\xi|^m, \qquad \forall x \in \bar{\Omega},\ \xi \in \mathbb{R}^n, \tag{11}$$

where c is independent of x and ξ, and m is the degree of the characteristic polynomial $\chi = \det \pi\mathscr{A}$. Ellipticity means that the inverse matrix $(\pi\mathscr{A})^{-1}(x, \xi)$ exists for $x \in \bar{\Omega}$, $0 \neq \xi \in \mathbb{R}^n$; in the proof of Theorems 9.29, 9.32 we shall need an estimate for the entries $I_{ij}(x, \xi)$ of this matrix. From (9) we obtain

$$\begin{aligned}\pi\mathscr{A}^{-1}(x, \xi) &= \pi\mathscr{A}^{-1}\left(x, |\xi| \cdot \frac{\xi}{|\xi|}\right)\\ &= T^{-1}(|\xi|) \cdot (\pi\mathscr{A})^{-1}\left(x, \frac{\xi}{|\xi|}\right) \cdot S^{-1}(|\xi|).\end{aligned}$$

In view of (11) the entries of the matrix $\pi\mathscr{A}^{-1}(x, \xi/|\xi|)$ are bounded on the compact set $\bar{\Omega} \times S^n \ni (x, \xi/|\xi|)$ by some $M < \infty$, whence the entries

$$I_{ij}(x, \xi) = |\xi|^{-t_i} \cdot I_{ij}(x, \xi/|\xi|) \cdot |\xi|^{-s_j}$$

of $\pi\mathscr{A}^{-1}(x, \xi)$ are bounded by $M|\xi|^{-(t_i + s_j)}$ for all $0 \neq \xi \in \mathbb{R}^n$. Hence for $\xi = (\xi', \tau)$ with $0 \neq \xi' \in \mathbb{R}^{n-1}$ we have the estimate

$$|I_{ij}(x, (\xi', \pi))| \leqslant \frac{M}{(|\xi'| + |\tau|)^{t_i + s_j}}, \ \forall \tau \in \mathbb{R}, \tag{12}$$

which shows that for $\xi' \neq 0$ there is no singularity at $\tau = 0$.

In the proof of Theorem 9.32, we will need to work in local coordinates. For a point lying inside a compact subset of Ω we use the standard coordinates in $\mathbb{R}^n$. For points near the boundary, $\partial\Omega$, we use admissible coordinates defined as follows. By Theorem 5.73, a tubular neighbourhood of $\partial\Omega$ in $\mathbb{R}^n$ exists and can be identified with $\partial\Omega \times (-1, 1)$. A chart (U', x') on $\partial\Omega$ gives rise to a chart

$$(U' \times (-1, 1); x', x_n), \qquad x' \in \mathbb{R}^{n-1},\ x_n \in (-1, 1),$$

on the tubular neighbourhood $\partial\Omega \times (-1, 1)$. Then (x', x_n) are coordinates on $\partial\Omega \times (-1, 1)$ such that the vector fields $\partial/\partial x_1, \ldots, \partial/\partial x_{n-1}$ are tangential to $\partial\Omega$, and we can assume that $\partial/\partial x_n$ is normal to $\partial\Omega$ (see text preceding Theorem 7A.9). We refer to (x', x_n) as *admissible coordinates*; note that x_n is a global coordinate on $\partial\Omega \times (-1, 1)$, while $x_1, \ldots, x_{n-1}$ are local coordinates.

Remark For future use, let us also note that the set $\hat{\Omega} = \Omega \cup (\partial\Omega \times (-1, 0))$ is an open set in $\mathbb{R}^n$ containing $\bar{\Omega}$.

Let $\nu: \partial\Omega \to T(\mathbb{R}^n)$ be the inward-pointing unit normal along $\partial\Omega$. For each $x \in \partial\Omega$ there is the direct sum decomposition $T_x\mathbb{R}^n = T_x(\partial\Omega) \oplus \text{span}(\nu_x)$, and by passing to the dual, we have $T_x^*\mathbb{R}^n = [\text{span}(\nu_x)]^0 \oplus [T_x(\partial\Omega)]^0$. By identifying the annihilator subspace $[\text{span}(\nu_x)]^0$ with the cotangent space $T_x^*(\partial\Omega)$ by means of the restriction map $\xi' \mapsto \xi'|_{T_x(\partial\Omega)}$ we have

$$T_x^*\mathbb{R}^n = T_x^*(\partial\Omega) \oplus [T_x(\partial\Omega)]^0$$
$$\xi = \xi' + \xi_n$$

where $\xi' \in T_x^*(\partial\Omega)$ is cotangent to $\partial\Omega$ and ξ_n is "conormal" to $\partial\Omega$, i.e. $\langle \xi_n, v \rangle = 0$ for all $v \in T_x(\partial\Omega)$.

Now let $n: \partial\Omega \to T^*(\mathbb{R}^n)$ be the image of ν by the index-lowering operator $T(\mathbb{R}^n) \to T^*(\mathbb{R}^n)$ that maps $\partial/\partial x_i$ to dx_i. Because the space of conormal vectors at a point $x \in \partial\Omega$ is one-dimensional and $n(x) \neq 0$ for each x, then every $\xi \in T_x^*(\mathbb{R}^n)$ can be written uniquely in the form

$$\xi = \xi' + \xi_n \cdot n(x), \qquad \text{where } \xi' \in T_x^*(\partial\Omega),\ \xi_n \in \mathbb{R}. \tag{13}$$

This defines a vector bundle isomorphism $T^*(\mathbb{R}^n) \simeq T^*(\partial\Omega) \oplus (\partial\Omega \times \mathbb{R})$ and we are justified in writing for each $\xi \in T_x^*\mathbb{R}^n$,

$$\xi = (\xi', \xi_n) \qquad \text{where } \xi' \in T_x^*(\partial\Omega),\ \xi_n \in \mathbb{R}. \tag{13'}$$

The definition of proper ellipticity at a point $x \in \partial\Omega$ on the boundary is as follows. When we substitute $\xi = (\xi', \lambda) = \xi' + \lambda \cdot n(x) \in T_x^*\mathbb{R}^n$ in the characteristic polynomial $\chi(\xi) = \det \pi\mathscr{A}(x, \xi)$, it is to be understood that the canonical isomorphism $T_x^*(\mathbb{R}^n) \simeq \mathbb{R}^n$ is taken into account. We may also permit λ to be a complex number and then $\xi \in T_x^*\mathbb{R}^n \otimes \mathbb{C} \subset \mathbb{C}^n$.

Definition 9.24 *Let* $n \geqslant 2$. *The elliptic operator* $\mathscr{A}(x, D)$ *is* proper *at* $x \in \partial\Omega$ *if for each* $0 \neq \xi' \in T_x^*(\partial\Omega)$ *the polynomial in* $\lambda \in \mathbb{C}$

$$P(\lambda) = \chi(\xi', \lambda) = \det \pi\mathscr{A}(x, (\xi', \lambda)) \tag{14}$$

has as many roots, r, in the upper half-plane $\text{Im}\,\lambda > 0$ *as in the lower half-plane* $\text{Im}\,\lambda < 0$, *counting multiplicities.*

Here $\chi = \det \pi\mathscr{A}$ is the characteristic polynomial (7). Since it is a polynomial of degree m in (ξ', λ) then

$$P(\lambda) = \det \pi\mathscr{A}(x, (\xi', \lambda)) = a_m(x)\lambda^m + \cdots,$$

with $a_m(x)$ independent of ξ'. By ellipticity with $\xi' = 0$, $\lambda = 1$ it follows that $a_m(x) \neq 0$, $x \in \partial\Omega$, whence m is the degree of $P(\lambda)$. Also because of ellipticity there are no roots on the real axis, so if $\mathscr{A}$ is properly elliptic we must have $m = 2r$. Hence m is even and the number r in Definition 9.24 is independent of $x \in \partial\Omega$ and $\xi' \neq 0$.

Using the fact that any pair of linearly independent vectors, $\xi_1, \xi_2 \in \mathbb{R}^n$, $n > 2$, is homotopic to the pair $(1, \ldots, 0)$, $(0, \ldots, 1)$ through pairs of linearly independent vectors, we obtain the equivalent definitions of proper ellipticity:

(i) $\det \pi\mathscr{A}(x, (\xi', \lambda))$ is proper (Definition 9.24);
(ii) If $\xi_1, \xi_2 \in \mathbb{R}^n$ are linearly independent, the polynomial in the complex variable λ, $\det \pi\mathscr{A}(x, \xi_1 + \xi_2)$, has as many roots in $\operatorname{Im} \lambda < 0$ as in $\operatorname{Im} \lambda > 0$;
(iii) The polynomial $\det \pi\mathscr{A}(x, (1, \ldots, 0) + \lambda(0, \ldots, 1))$ has r roots in $\operatorname{Im} \lambda < 0$ and r in $\operatorname{Im} \lambda > 0$.

For $n = 2$, either ξ_1, ξ_2 or $\xi_1, -\xi_2$ is homotopic to the pair $(1, 0), (0, 1)$ and then using (8) with $c = -1$, we again obtain the equivalence of conditions (i), (ii), and (iii). It is also well known that proper ellipticity differs from ellipticity only in the case $n = 2$.

Remark One could try to generalize proper ellipticity and propose that the number r of roots of $P(\lambda) = 0$ in the upper half-plane $\operatorname{Im} \lambda > 0$ be *independent* of $\xi' \in T_x^*(\partial\Omega)\backslash 0$. But by homogeneity of $\det \pi\mathscr{A}(x, \xi)$ with $c = -1$ we have

$$\det \pi\mathscr{A}(x, (-\xi', -\lambda)) = (-1)^m \det \pi\mathscr{A}(x, (\xi', \lambda)), \qquad x \in \partial\Omega,$$

from which it follows that the number, $m - r$, of roots in the lower half-plane $\operatorname{Im} \lambda < 0$ is equal to

$$m - r = r.$$

Thus we get proper ellipticity.

Let $\mathscr{A}(x, D)$ be elliptic on $\bar{\Omega}$. We say that $\mathscr{A}$ is *properly elliptic on* $\bar{\Omega}$ if condition (ii) holds for all $x \in \bar{\Omega}$. Since the domain $\bar{\Omega}$ is connected, it suffices that this condition hold for one point $x_0 \in \bar{\Omega}$ and one pair of linearly independent vectors $\xi_1, \xi_2 \in \mathbb{R}^n$. For instance, we can take $x_0 \in \partial\Omega$, a single cotangent vector $\xi_1 = \xi' \in T_{x_0}^*(\partial\Omega)$ and a conormal vector $\xi_2 = n(x_0)$ as in Definition 9.24.

Definition 9.25 *If in the definition of DN ellipticity (Definition 9.21′) we can put $s_1 = \cdots = s_p$ and $t_1 = \cdots = t_p$, then the operator $\mathscr{A}$ is said to be* homogeneously *elliptic. Because s_i and t_j can be replaced by s_i + constant, t_j − same constant, we can choose $t_1 = \cdots = t_p = 0$. (Another possible choice is of course $s_1 = \cdots = s_p = 0$.)*

If $t_1 = \cdots = t_p = 0$ and $s_1 = \cdots = s_p = l$, then writing the homogeneous operator $\mathscr{A}$ in the form

$$\mathscr{A}(x, D) = \sum_{|\alpha| \leqslant l} A_\alpha(x) D^\alpha$$

where the $A_\alpha(x)$ are $p \times p$ matrices, we have

$$\pi\mathscr{A}(x, \xi) = \sum_{|\alpha| = l} A_\alpha(x) \xi^\alpha,$$

and the characteristic polynomial $\chi(\xi) = \det \pi\mathscr{A}(x, \xi)$ has the (maximal) order $l \cdot p$. Often we have $l = 2s$.

Definition 9.26 *If in the definition of DN ellipticity we can put either $s_1 = \cdots = s_p$ or $t_1 = \cdots = t_p$, then $\mathscr{A}$ is said to be elliptic in the sense of* Petrovskii. *In the former case one can put $s_1 = \cdots = s_p = 0$ and have $t_j \geqslant 0$, while in the latter case one can put $t_1 = \cdots = t_p = 0$ and have $s_i \geqslant 0$.*

If $s_1 = \cdots = s_p = 0$, then formula (9) reduces to

$$\pi\mathscr{A}(x, c\xi) = \pi\mathscr{A}(x, \xi)\cdot T(c).$$

Examples (see [Vo 1])

(a) The system $\Delta u_1 = 0$, $\partial^m u_1/\partial x_1^m + \Delta u_2 = 0$, with the 2×2 matrix

$$\begin{bmatrix} \Delta & 0 \\ \partial^m/\partial x_1^m & \Delta \end{bmatrix}$$

has characteristic polynomial $\chi(\xi) = |\xi|^4$. It is DN elliptic with

$$s_1 = 2, s_2 = m, t_1 = 0, t_2 = 2 - m$$

and it is homogeneously elliptic if $m = 2$.

(b) The system

$$\frac{\partial^p u_1}{\partial x_1^p} - \frac{\partial^q u_2}{\partial x_2^q} = 0, \qquad \frac{\partial^r u_1}{\partial x_2^r} + \frac{\partial^s u_2}{\partial x_1^s} = 0$$

with the 2×2 matrix

$$\begin{bmatrix} \partial^p/\partial x_1^p & -\partial^q/\partial x_2^q \\ \partial^r/\partial x_2^r & \partial^s/\partial x_1^s \end{bmatrix}$$

has the characteristic polynomial $\chi(\xi) = \xi_1^{p+s} + \xi_2^{r+q}$. It is DN elliptic if $p + s = r + q =$ even, with the DN numbers

$$s_1 = p, s_2 = r, t_1 = 0, t_2 = q - p.$$

The system is elliptic in the sense of Petrovskii if either

$$p = r,\ q = s \text{ and } p + s \text{ is even} \qquad \text{or } p = q,\ r = s \text{ and } p + r \text{ is even}.$$

It is homogeneously elliptic if $p = q = r = s$.

(c) The linearized system of Navier-Stokes

$$\Delta u_i - \frac{\partial u_4}{\partial x_i} = 0,\ i = 1, 2, 3, \qquad \frac{\partial u_1}{\partial x_1} + \frac{\partial u_2}{\partial x_2} + \frac{\partial u_3}{\partial x_3} = 0$$

is DN elliptic with

$$s_1 = s_2 = s_3 = 1, s_4 = 0, t_1 = t_2 = t_3 = 1, t_4 = 0,$$

but not Petrovskii elliptic.

Addendum: the marriage lemma

A fundamental question in combinatorics is the "marriage problem". Let $X = [x_{ij}]_{p \times p}$ be a matrix of zeros and ones. Suppose that the p rows correspond to girls $g_1, \ldots, g_p$ and p columns correspond to boys $b_1, \ldots, b_p$. Then $x_{ij} = 1$ if g_i and b_j are willing to marry, and $x_{ij} = 0$ if not. A full set of p marriages (if it is possible) is called a *complete matching*. A complete matching means of course that $x_{i, j_i} = 1$, $i = 1, \ldots, p$, for some permutation $j_1, \ldots, j_p$ of the integers $1, \ldots, p$.

For example, consider the matrix

$$X = \begin{bmatrix} 1 & 1 & 1 & 1 \\ 0 & 0 & 0 & 1 \\ 0 & 1 & 0 & 1 \\ 0 & 1 & 0 & 0 \end{bmatrix}$$

It is evident that a complete matching does not exist for this matrix because the last three girls like only two boys. In other words, the lack of a complete matching is due to the existence of a 3×2 submatrix of zeros, namely, the submatrix consisting of rows 2, 3 and 4 and columns 1 and 3. One necessary condition for a complete matching is therefore evident, known as *Hall's condition*:

There is no $r \times s$ submatrix of zeros for which $r + s > p$

(otherwise, there would be r girls who like $\leqslant p - s < r$ boys). It is remarkable that this condition is also sufficient for the existence of a complete matching.

Lemma 9.27 *Let $X = [x_{ij}]_{p \times p}$ be a matrix of zeros and ones. Hall's condition is necessary and sufficient for existence of a complete matching.*

Proof If we have an $r \times s$ submatrix of zeros then the r girls corresponding to the rows of this submatrix like no more than $p - s$ boys, and vice versa. Consequently, Hall's condition can be stated another way:

Every set of r girls, $1 \leqslant r \leqslant p$, likes at least r boys

We will prove the sufficiency of this condition by induction on the order p of the matrix. For $p = 1$, it is obviously sufficient. Now suppose that Hall's condition holds for some p, and that it is sufficient for existence of a complete matching for any matrix of order $< p$.

There are two cases to consider:

Case 1. Every set of r girls, $r < p$, likes *more than* r boys. Then the first girl can marry anyone she likes, and Hall's condition holds for the remaining $p - 1$ girls and $p - 1$ boys. By the induction hypothesis the remaining $p - 1$ girls and boys can be completely matched.

Case 2. Some set of r girls, $r < p$, likes *exactly* r boys. This is the more difficult case. After reordering the rows and columns, the matrix X takes the form

$$X = \begin{bmatrix} A & C \\ 0 & B \end{bmatrix}$$

where 0 denotes an $r \times (p - r)$ matrix of zeros. The $r \times r$ matrix B corresponds to the r girls who like exactly r boys. Within this smaller matrix, Hall's condition still holds (it applied to every set of girls). Hence, by induction, r marriages can be arranged between the r girls and r boys in B.

To complete the proof we show that Hall's condition holds for the remaining $p - r$ girls and $p - r$ boys, i.e. for the matrix A. Consider any R girls in this remaining subset; we must show that they like at least R of the remaining (unmarried) boys. The trick is to consider these R girls together with the original r. By Hall's condition, this group likes at least $r + R$ boys, and since the r girls liked *only* the r boys they married, then the R girls necessarily like at least R of the unmarried boys. Hall's condition holds therefore for the matrix A and, by induction, $p - r$ marriages can be arranged between the girls and boys in A. In this way we have obtained a complete matching.

The proof of Lemma 9.27 was taken from the book [Str]. This book contains many interesting insights into the topics of applied mathematics, including a chapter on linear programming. For instance, the following lemma is just the strong duality principle for the "assignment problem".

Lemma 9.22 *For any $p \times p$ matrix $\alpha = [\alpha_{ij}]$ whose elements are integers, or equal to $-\infty$, there exist integers $s_1, \ldots, s_p, t_1, \ldots, t_p$ such that*

$$\sum_{i=1}^{p} (s_i + t_i) = m$$

$$s_i + t_j \geqslant \alpha_{ij}$$

where, as before, m denotes the maximum of $\alpha_{1,j_1} + \cdots + \alpha_{p,j_p}$ taken over all permutations $(j_1, \ldots, j_p)$ of the numbers $(1, \ldots, p)$.

Proof Let $s_1, \ldots, s_p$, $t_1, \ldots, t_p$ be integers such that $s_i + t_j \geqslant \alpha_{ij}$ and $\sum_{i=1}^{p} (s_i + t_i) = m'$ (the minimum of such sums). Define a $p \times p$ matrix $[x_{ij}]$ of zeros and ones as follows:

$$\begin{aligned} x_{ij} &= 1 \qquad \text{if } s_i + t_j = \alpha_{ij} \\ &= 0 \qquad \text{if } s_i + t_j > \alpha_{ij}. \end{aligned}$$

We claim that this matrix has a complete matching, i.e. there exists a permutation $(j_1, \ldots, j_p)$ such that $x_{i,j_i} = 1$. From this fact the lemma follows at once. Now, to prove the claim, we argue by contradiction. If such a permutation does not exist then by Lemma 9.27 there must be an $r \times s$ submatrix of zeros with $r + s > p$ that prevented a matching. Define new integers $s'_1, \ldots, s'_p, t'_1, \ldots, t'_p$ by subtracting 1 from the t_j's corresponding to the s columns of this submatrix and adding 1 to the s_i corresponding to the

$p - r$ rows *not* in this submatrix. Then the inequalities $s'_i + t'_j \geqslant \alpha_{ij}$ are still true, but

$$\sum_{i=1}^{p} (s'_i + t'_i) = \sum_{i=1}^{p} (s_i + t_i) + (p - r - s)$$

$$< \sum_{i=1}^{p} (s_i + t_i)$$

which contradicts the choice of the integers $s_1, \ldots, s_p, t_1, \ldots, t_p$.

Example The proof of Lemma 9.22 suggests a way to find DN numbers s_i and t_j for a given matrix α. For example let

$$\alpha = [\alpha_{ij}] = \begin{bmatrix} 5 & 4 & 0 \\ 9 & 7 & 3 \\ 3 & 2 & 5 \end{bmatrix}.$$

As a first guess, let all s_i be zero and let t_j be the largest of the elements α_{ij} in column j, that is, $t_1 = 9$, $t_2 = 7$ and $t_3 = 5$. That certainly achieves $s_i + t_j \geqslant \alpha_{ij}$. However, if we form the matrix $[x_{ij}]$ as in the proof above we obtain

$$[x_{ij}] = \begin{bmatrix} 0 & 0 & 0 \\ 1 & 1 & 0 \\ 0 & 0 & 1 \end{bmatrix}$$

which does *not* have a complete matching due to the first row, a 1×3 submatrix of zeros. Hence the sum $\sum_{i=1}^{3} (s_i + t_i) = 21$ is not yet minimized. While keeping $s_1 = 0$ (which is no loss of generality) we decrease t_1, t_2 and t_3 as much as possible by the same amount, i.e. we subtract 3 to obtain $t'_1 = 6$, $t'_2 = 4$ and $t'_3 = 2$ and then increase s_2 and s_3 by the same amount to obtain $s'_1 = 0$, $s'_2 = 3$ and $s'_3 = 3$. We still have $s'_i + t'_j \geqslant \alpha_{ij}$ and now the matrix $[x_{ij}]$ has the form

$$[x_{ij}] = \begin{bmatrix} 0 & 1 & 0 \\ 1 & 1 & 0 \\ 0 & 0 & 1 \end{bmatrix}$$

which has a complete matching $x_{12} = x_{21} = x_{33} = 1$. The sum

$$\sum_{i=1}^{3} (s'_i + t'_i) = 18$$

is minimized because it is equal to $m = \alpha_{12} + \alpha_{21} + \alpha_{33}$.

9.3 Boundary operators and the *L*-condition

We turn now to the formulation of boundary value problems for a (properly) elliptic operator $\mathscr{A}$, with DN numbers $s_1, \ldots, s_p, t_1, \ldots, t_p$. As usual, we let $D_n = i^{-1}\,\partial/\partial n$ where n is the inward pointing unit (co)normal vector field on $\partial\Omega$. Let D_n^κ be the normal trace operator $W_2^l(\Omega) \to W_2^{l-\kappa-1/2}(\partial\Omega)$, $l - \kappa > \frac{1}{2}$, from Theorem 7A.10.

Points on the boundary will be denoted by $y \in \partial\Omega$. Let r be the number of roots of the polynomial $P(\lambda) = \det \pi\mathscr{A}(y, (\xi', \lambda))$ in the upper half-plane $\operatorname{Im}\lambda > 0$, i.e. half the order of the characteristic polynomial χ. In addition to the p equations, $\mathscr{A}(x, D)u(x) = f(x)$, in Ω, we consider r boundary conditions

$$\sum_{j=1}^{p} B_{kj}(y, D)u_j(y) = g_k(y), \qquad k = 1, \ldots, r, \tag{15}$$

that is, $\mathscr{B}(y, D)u(y) = g(y)$, where $\mathscr{B}(y, D)$ is the matrix operator $[B_{kj}(y, D)]_{r\times p}$. The boundary operators B_{kj} are taken in the form

$$B_{kj}(y, D) = \sum_{\kappa=0}^{l_{kj}} b_{kj}^\kappa(y, D')D_n^\kappa, \; y \in \partial\Omega, \tag{16}$$

where $b_{kj}^\kappa(y, D') \in OClS^{\beta_{kj}^\kappa}(\partial\Omega)$ are (-1) classical pseudo-differential operators on the manifold $\partial\Omega$. The principal parts are denoted by $\pi b_{kj}^\kappa(y, \xi')$, $(y, \xi') \in T^*(\partial\Omega)\backslash 0$, and we also write $\beta_{kj}^\kappa = \operatorname{ord} b_{kj}^\kappa$.

The DN principal part of the boundary operator $\mathscr{B}$ is defined as follows. Let

$$m_{kj} := \max_\kappa \operatorname{ord}(b_{kj}^\kappa D_n^\kappa) = \max_\kappa (\beta_{kj}^\kappa + \kappa) \tag{17}$$

(the numbers m_{kj} can be negative and also non-integer, i.e. $m_{kj} \in \mathbb{R}$) and then let

$$m_k := \max_{1\leqslant j\leqslant p} (m_{kj} - t_j), \qquad k = 1, \ldots, r, \tag{18}$$

so that $m_{kj} \leqslant m_k + t_j$. The *DN principal part* of the boundary operator $\mathscr{B}(y, D)$ is defined as the $r \times p$ matrix

$$\pi_D\mathscr{B}(y, \xi) = [B'_{kj}(y, \xi)]_{r\times p}, \qquad B'_{kj}(y, \xi) = \sum_\kappa{}' \pi b_{kj}^\kappa(y, \xi')\xi_n^\kappa, \tag{19}$$

where $\sum_\kappa'$ denotes the sum over those terms with $\beta_{kj}^\kappa + \kappa = m_k + t_j$. In other words, $B'_{kj}(y, \xi)$ consists of the principal parts of the terms in B_{kj} which are just of order $m_k + t_j$, with the other terms replaced by 0. As usual, $\xi = (\xi', \xi_n)$, $\xi' \in T_y^*(\partial\Omega)\backslash 0$ and ξ_n is conormal at y.

Usually, we denote the DN principal part by $\pi\mathscr{B}$ rather than $\pi_D\mathscr{B}$; it is polynomial in ξ_n, and, as easily checked, the following homogeneity property holds:

$$\pi\mathscr{B}(y, c\xi) = M(c)\cdot\pi\mathscr{B}(y, \xi)\cdot T(c), \qquad c > 0, \tag{20}$$

where

$$M(c) = [\delta_{ik}\, c^{m_k}]_{r\times r}, \qquad T(c) = [\delta_{ij}\, c^{t_j}]_{p\times p}$$

are diagonal matrices.

The operators b^{κ}_{kj} can have negative order, $\beta^{\kappa}_{kj} < 0$; for some purposes, however, it is convenient to assume that all these orders are non-negative, and this can always be achieved as follows. Let $\Lambda^q \in OS^q(\partial\Omega)$ be a (-1) classical operator with principal symbol $|\xi'|^q$, then replace b^{κ}_{kj} by $\Lambda^q \circ b^{\kappa}_{kj}$ where q is chosen so that $q + \beta^{\kappa}_{kj} \geqslant 0$ for all k, j, that is, we choose

$$q = -\min_{\kappa, k, j} \beta^{\kappa}_{kj}. \tag{21}$$

Remark If $\mathscr{A}(x, D)$ is Petrovskii elliptic, Definition 9.26, say $t_1 = \cdots = t_p = 0$, then the principal part $\pi\mathscr{B}$ is obtained by computing in the kth row the maximal order m_k and then taking from (16) the principal parts of order m_k.

Our intention now is to formulate the L-condition for elliptic boundary value problem operators $(\mathscr{A}, \mathscr{B})$, but some more preparation is needed.

Choose a tubular neighbourhood of $\partial\Omega$ with admissible coordinates, and consider the decomposition (13′) of the cotangent space $T^*_{y_0}(\partial\Omega)$ at the boundary point $y_0 \in \partial\Omega$ where y_0 is fixed. We substitute ξ_n by $(1/i)(d/dt)$ and fix $\xi' \neq 0$ in the DN principal part of $\mathscr{A}$ to obtain the system of ordinary differential equations (with constant coefficients)

$$\pi\mathscr{A}\left(y_0, \left(\xi', \frac{1}{i}\frac{d}{dt}\right)\right)w(t) = 0, \qquad t > 0, \qquad \xi' \in T^*_{y_0}(\partial\Omega)\backslash 0. \tag{22}$$

It should be noted that there is a slight difference in our treatment here from that in Part I. In Chapters 1 to 4 we studied the connection between the spectrum of $L(\lambda)$ and the equation $L(d/dt)w = 0$, whereas here – because of the Fourier transformation – we are concerned with

$$L\left(\frac{1}{i}\frac{d}{dt}\right)w = 0, \qquad \text{where } L(\lambda) = \pi\mathscr{A}(y_0, (\xi', \lambda)).$$

The solutions of this equation are p-columns of exponential polynomials of the form $\sum p_j(t)\, e^{i\lambda_j t}$ where the p_j's are polynomials in t and the λ_j's are eigenvalues of $L(\lambda)$. The solution space $\mathfrak{M} = \mathfrak{M}_{L(1/i\, d/dt)}$ of (22) decomposes directly into

$$\mathfrak{M} = \mathfrak{M}^- \oplus \mathfrak{M}^+,$$

where $\mathfrak{M}^+$ consists of all solutions $w(t)$ with $\lim_{t\to\infty} w(t) = 0$. We have $\dim \mathfrak{M} = m$, $\dim \mathfrak{M}^+ = r$ and $\dim \mathfrak{M}^- = m - r$, and any $w \in \mathfrak{M}^+$ is a p-column of exponential polynomials such that $\operatorname{Im} \lambda_j > 0$ for all j. It is evident that if $w \in \mathfrak{M}^+$ then w and all its derivatives belong to $L^2(\mathbb{R}_+)$ since $|e^{i\lambda_j t}| \leqslant e^{-ct}$ for some constant $c > 0$.

Before continuing on, we make some observations about the weights $s_1, \ldots, s_p, t_1, \ldots, t_p$ in the definition of ellipticity. Note that

$$\max_i s_i \geqslant -t_j \qquad \text{for all } j \tag{24}$$

Otherwise, there would exist j_0 such that $s_i + t_{j_0} < 0$ for all i, so that $A_{ij_0}(x, D) \equiv 0$, in contradiction to ellipticity. Replacing the weights s_i and t_j with s_i − const and t_j + const we may assume that the weights are normalized as follows:

$$s_i \leqslant 0,\ t_j \geqslant 0, \qquad \text{for all } i, j, \tag{25}$$
$$\max s_i = 0.$$

Definition 9.28 *The pair of operators*

$$\mathscr{A}(y, D), \mathscr{B}(y, D), \qquad y \in \partial\Omega,$$

is said to fulfil the L-condition if one of the following three statements holds for all $y \in \partial\Omega$, $0 \neq \xi' \in T_y^*(\partial\Omega)$.

I. The zero initial value problem

$$\pi\mathscr{A}\left(y, \left(\xi', \frac{1}{i}\frac{d}{dt}\right)\right)w(t) = 0, \qquad t > 0,$$

$$\pi\mathscr{B}\left(y, \left(\xi', \frac{1}{i}\frac{d}{dt}\right)\right)w(t)|_{t=0} = 0,$$

has in $\mathfrak{M}^+$ *the unique solution* $w(t) = 0$.

II. The initial value problem

$$\pi\mathscr{A}\left(y, \left(\xi', \frac{1}{i}\frac{d}{dt}\right)\right)w(t) = 0, \qquad t > 0,$$

$$\pi\mathscr{B}\left(y, \left(\xi', \frac{1}{i}\frac{d}{dt}\right)\right)w(t)|_{t=0} = g,$$

has for every choice of $g \in \mathbb{C}^r$ *a unique solution* w *in* $\mathfrak{M}^+$.

III. Let

$$l \geqslant \max s_i = 0 \quad \textit{and} \quad l \geqslant \max m_k + q + 1 \ (\textit{for } q \textit{ see } (21)) \tag{26}$$

Then the operator

$$\left(\pi\mathscr{A}\left(y, \left(\xi', \frac{1}{i}\frac{d}{dt}\right)\right), \pi\mathscr{B}\left(y, \left(\xi', \frac{1}{i}\frac{d}{dt}\right)\right)\Big|_{t=0}\right): \mathscr{W}^{l+\mathbf{t}}(\mathbb{R}_+) \to \mathscr{W}^{l-\mathbf{s}}(\mathbb{R}_+) \times \mathbb{C}^r \tag{2}$$

is an algebraic and topological isomorphism.

In the statement of the condition III, we have used the spaces

$$\mathscr{W}^{l+\mathbf{t}}(\mathbb{R}_+) := W_2^{l+t_1}(\mathbb{R}_+) \times \cdots \times W_2^{l+t_p}(\mathbb{R}_+)$$
$$\mathscr{W}^{l-\mathbf{s}}(\mathbb{R}_+) := W_2^{l-s_1}(\mathbb{R}_+) \times \cdots \times W_2^{l-s_p}(\mathbb{R}_+)$$

with norms $\|u\|_{l+\mathbf{t}}^2 = \sum_{j=1}^p \|u\|_{l+t_j}^2$ and $\|f\|_{l-\mathbf{s}}^2 = \sum_{j=1}^p \|f_j\|_{l-s_j}^2$, respectively. It is important to recall that we defined the Sobolev spaces $W_2^l(\mathbb{R}_+)$ only for $l \geqslant 0$. In view of (24), the condition $l \geqslant \max s_i$ ensures that $l + t_j \geqslant 0$ for all j and hence all spaces $W_2^{l+t_j}(\mathbb{R}_+)$ and $W_2^{l-s_i}(\mathbb{R}_+)$ are well-defined. With the normalization (25), the first condition in (26) becomes $l \geqslant 0$. The second condition is necessary when taking boundary values.

Theorem 9.29 *Suppose that the operator $\mathscr{A}(y, D)$, $y \in \partial\Omega$, is properly elliptic (Definition 9.24). Then the statements I, II, III from Definition 9.28 are equivalent. If III holds we have in particular the estimate*

$$\|\varphi\|_{l+\mathbf{t}} \leqslant C(\xi') \cdot \left[\left\| \pi\mathscr{A}\left(y, \left(\xi', \frac{1}{i}\frac{d}{dt}\right)\right)\varphi \right\|_{l-\mathbf{s}} + \left\| \pi\mathscr{B}\left(y, \left(\xi', \frac{1}{i}\frac{d}{dt}\right)\right)\varphi(0) \right\|_{\mathbb{C}^r} \right], \tag{28}$$

where on compact sets $K \subset T^*(\partial\Omega)\backslash 0$, the constant $C(\xi')$ can be chosen independently of $\xi' \in K$.

Proof We have $\mathfrak{M}^+ \subset \mathscr{W}^{l+\mathbf{t}}(\mathbb{R}_+)$, and in fact it is not hard to show that

$$\mathfrak{M}^+ = \mathfrak{M} \cap \mathscr{W}^{l+\mathbf{t}}(\mathbb{R}_+),$$

so the implications III → II → I are obvious. To prove the implication I → II, we consider the map

$$\pi\mathscr{B}\left(\ldots, \frac{1}{i}\frac{d}{dt}\right)\bigg|_{t=0} : \mathfrak{M}^+ \to \mathbb{C}^r.$$

Supposing that I holds, this map acts injectively, and since

$$\dim \mathfrak{M}^+ = r = \dim \mathbb{C}^r,$$

then by linear algebra it acts also surjectively, and we have proved II.

Before proving the implication II → III, we must show the continuity of the map (27). For fixed y and $\xi' \neq 0$, the matrix differential operator $\pi\mathscr{A}(y, (\xi', (1/i)(d/dt))) = [A'_{ij}(y, (\xi', (1/i)(d/dt)))]$ has constant coefficients, and it follows that

$$\pi\mathscr{A}: \mathscr{W}^{l+\mathbf{t}}(\mathbb{R}_+) \to \mathscr{W}^{l-\mathbf{s}}(\mathbb{R}_+) \tag{29}$$

is continuous since the order of A'_{ij} is $\leqslant s_i + t_j$. Similarly, $(\pi\mathscr{B}(y, (\xi', (1/i) \times (d/dt))) = [B'_{kj}(y, (\xi', (1/i)(d/dt)))]_{r \times p}$ for fixed y, ξ' has constant coefficients and the order of B'_{kj} is $\leqslant m_k + t_j + q$ (see (18) and (21) for the definition of m_k and q), hence if we use the trace theorems for Sobolev spaces (Theorem

7A.9 or [Wl], p. 121), we see that $\pi\mathscr{B}$ acts continuously as follows:

$$\pi\mathscr{B}\colon \mathscr{W}^{l+\mathbf{t}}(\mathbb{R}_+) \to \mathbb{C}^r. \tag{30}$$

Thus (29) and (30) prove the continuity of (27).

Suppose now that II or I holds. Then the map (27) is injective since $\mathscr{W}^{l+\mathbf{t}}(\mathbb{R}_+) \cap \mathfrak{M} = \mathfrak{M}^+$, and once we prove surjectivity, then the Open Mapping Theorem gives the continuity of the inverse map, thus proving the implication II $\to$ III and simultaneously the estimate (28). Let $f \in \mathscr{W}^{l-\mathbf{s}}(\mathbb{R}_+)$ and $g \in \mathbb{C}^r$. We can extend f to the whole real axis $\mathbb{R}^1$, where for the extension $\tilde{f}$ we have

$$\tilde{f} \in \mathscr{W}^{l-\mathbf{s}}(\mathbb{R}^1), \qquad \|\tilde{f}\|_{l-\mathbf{s},\mathbb{R}^1} \leqslant c\|f\|_{l-\mathbf{s},\mathbb{R}_+},$$

with c independent of f (see Theorem 7A.7 or [Wl], p. 100), and then apply the one-dimensional Fourier transformation to the variable t ($\to \tau$) and put

$$\psi(\tau) := [\pi\mathscr{A}(y, (\xi', \tau))]^{-1}(\mathfrak{F}\tilde{f})(\tau), \qquad \xi' \neq 0.$$

By ellipticity (Definition 9.21′), $(\pi\mathscr{A})^{-1}$ exists and the estimates (12) show that multiplication by $(\pi\mathscr{A})^{-1}(y, (\xi', \tau))$ gives a map $L^{l-\mathbf{s}}(\mathbb{R}^1) \to L^{l+\mathbf{t}}(\mathbb{R}^1)$, where $L^{l-\mathbf{s}}(\mathbb{R}^1)$ denotes the space of p-vector functions $f(\tau) = [f_i(\tau)]$ such that $f_i(\tau)\cdot(1 + |\tau|)^{l-s_i} \in L_2(\mathbb{R}^1)$ for all $i = 1, \ldots, p$, and analogously for $L^{l+\mathbf{t}}(\mathbb{R}^1)$. Taking the inverse Fourier transformation, we find that $\varphi(t) := \mathfrak{F}^{-1}(\psi(\tau))$ belongs to $\mathscr{W}^{l+\mathbf{t}}(\mathbb{R}^1)$ and satisfies the equation

$$\pi\mathscr{A}\left(\ldots, \frac{1}{i}\frac{d}{dt}\right)\varphi(t) = \tilde{f}(t), \qquad \forall t \in \mathbb{R}^1.$$

Let $\varphi_1(t)$ be the restriction of $\varphi(t)$ to $\mathbb{R}_+$; then we have $\varphi_1 \in \mathscr{W}^{l+\mathbf{t}}(\mathbb{R}_+)$. In view of (30) we can calculate the boundary values

$$\pi\mathscr{B}\left(\ldots, \frac{1}{i}\frac{d}{dt}\right)\varphi_1(t)|_{t=0} = c.$$

By II there exists a solution $\varphi_2 \in \mathfrak{M}^+ \subset \mathscr{W}^{l+\mathbf{t}}(\mathbb{R}_+)$ of the equations

$$\pi\mathscr{A}\left(\ldots, \frac{1}{i}\frac{d}{dt}\right)\varphi_2(t) = 0, \qquad t \in \mathbb{R}_+,$$

$$\pi\mathscr{B}\left(\ldots, \frac{1}{i}\frac{d}{dt}\right)\varphi_2|_{t=0} = g - c,$$

and then $w = \varphi_1 + \varphi_2 \in \mathscr{W}^{l+\mathbf{t}}(\mathbb{R}_+)$ is a solution of

$$\pi\mathscr{A}\left(\ldots, \frac{1}{i}\frac{d}{dt}\right)w(t) = f(t), \quad t \in \mathbb{R}_+,$$

$$\pi\mathscr{B}\left(\ldots, \frac{1}{i}\frac{d}{dt}\right)w|_{t=0} = g,$$

thereby proving the surjectivity of (27).

To prove the last assertion (concerning the constant $C(\xi')$), it is sufficient to show that for each $\xi'_0 \neq 0$, there exists a ball $B(\xi'_0, \delta)$ and a constant $C(\xi'_0, \delta) > 0$ such that (28) holds for all $\xi' \in B(\xi'_0, \delta)$ with $C(\xi'_0, \delta)$ substituted for $C(\xi')$. We take an $\varepsilon > 0$ such that $1 - 2\varepsilon C(\xi'_0) > 0$ holds, then choose a $\delta > 0$ such that

$$\left.\begin{aligned} \left\| \pi\mathscr{A}\left(y, \left(\xi'_0, \frac{1}{i}\frac{d}{dt}\right)\right)\varphi - \pi\mathscr{A}\left(y, \left(\xi', \frac{1}{i}\frac{d}{dt}\right)\right)\varphi \right\|_{l-\mathbf{s}} &\leqslant \varepsilon \|\varphi\|_{l+\mathbf{t}} \\ \left\| \pi\mathscr{B}\left(y, \left(\xi'_0, \frac{1}{i}\frac{d}{dt}\right)\right)\varphi\Big|_{t=0} - \pi\mathscr{B}\left(y, \left(\xi', \frac{1}{i}\frac{d}{dt}\right)\right)\varphi\Big|_{t=0} \right\|_{\mathbb{C}^r} &\leqslant \varepsilon \|\varphi\|_{l+\mathbf{t}} \end{aligned}\right\} \tag{31}$$

for all $\xi' \in B(\xi'_0, \delta)$. This is always possible because of the continuous dependence of the coefficients of the expressions $\pi\mathscr{A}$, $\pi\mathscr{B}$ on ξ'. Combining (28) (with ξ'_0) and (31) gives us

$$\|\varphi\|_{l+\mathbf{t}} \leqslant C(\xi'_0) \cdot \left[\left\| \pi\mathscr{A}\left(y, \left(\xi', \frac{1}{i}\frac{d}{dt}\right)\right)\varphi \right\|_{l-\mathbf{s}} + \left\| \pi\mathscr{B}\left(y, \left(\xi', \frac{1}{i}\frac{d}{dt}\right)\right)\varphi\Big|_{t=0} \right\|_{\mathbb{C}^r} \right]$$
$$+ 2\varepsilon C(\xi'_0)\|\varphi\|_{l+\mathbf{t}}$$

for all $\xi' \in B(\xi'_0, \delta)$, or

$$\|\varphi\|_{l+\mathbf{t}} \leqslant C(\xi'_0, \delta) \cdot \left[\left\| \pi\mathscr{A}\left(y, \left(\xi', \frac{1}{i}\frac{d}{dt}\right)\right)\varphi \right\|_{l-\mathbf{s}} + \left\| \pi\mathscr{B}\left(y, \left(\xi', \frac{1}{i}\frac{d}{dt}\right)\varphi\Big|_{t=0} \right\|_{\mathbb{C}^r} \right]$$

where we have put $C(\xi'_0, \delta) := C(\xi'_0)/(1 - 2\varepsilon C(\xi'_0))$.

Now we state the definition of L-ellipticity of a boundary value problem.

Definition 9.30 *The boundary value problem*

$$\mathscr{A}(x, D)u(x) = f(x), \qquad x \in \Omega,$$
$$\mathscr{B}(y, D)u(y) = g(y), \qquad y \in \partial\Omega,$$

is said to be L-elliptic in $\bar{\Omega}$ if:

(i) the operator $\mathscr{A}(x, D)$ is elliptic for all $x \in \bar{\Omega}$, see Definition 9.21′;
(ii) the operator $\mathscr{A}(x, D)$ is properly elliptic for all $x \in \partial\Omega$, see Definition 9.24;
(iii) $\mathscr{A}(y, D)$, $\mathscr{B}(y, D)$ satisfies for all $y \in \partial\Omega$ the L-condition of Definition 9.28.

Remark 9.31 The second condition is actually superfluous, i.e. it follows from conditions (i) and (iii), for if the mapping

$$\pi\mathscr{B}\left(y, \left(\xi', \frac{1}{i}\frac{d}{dt}\right)\right)\Big|_{t=0} : \mathfrak{M}^+(\xi') \to \mathbb{C}^r,$$

is bijective for all $\xi' \neq 0$, then the dimension of $\mathfrak{M}^+(\xi')$ must be independent of ξ', i.e. the number of roots of $P(\lambda) = \det \pi\mathscr{A}(y, (\xi', \lambda)) = 0$ in $\operatorname{Im} \lambda > 0$ is independent of ξ' (equal to the number of boundary conditions) and proper ellipticity of $\mathscr{A}$ follows as in the remark after Definition 9.24.

9.4 The main theorem for elliptic boundary value problems

We consider the equation

$$\mathscr{A}(x, D)u(x) = f(x), \qquad x \in \Omega,$$

where $\mathscr{A}(x, D) = [A_{ij}(x, D)]_{p \times p}$ is a DN differential operator as in §9.2, with the numbers $s_1, \ldots, s_p, t_1, \ldots, t_p$ normalized by the condition $s_i \leqslant 0$, $t_j \geqslant 0$, $\max s_i = 0$ (see (25)), and take r boundary conditions, that is,

$$\mathscr{B}(y, D)u(y) = g(y), \qquad y \in \partial\Omega,$$

where $\mathscr{B}(y, D)$ is an $r \times p$ matrix. The type of boundary operator that we admit is set out in §9.3, i.e.

$$\mathscr{B}(y, D) = [B_{kj}(y, D)]_{r \times p}, \; B_{kj}(y, D) = \sum_{\kappa=0}^{l_{kj}} b_{kj}^{\kappa}(y, D')D_n^{\kappa}, \qquad y \in \partial\Omega, \quad (32)$$

where the D_n^{κ} are the normal trace operators and the coefficients $b_{kj}^{\kappa} \in OS^{\beta_{kj}^{\kappa}}(\partial\Omega)$ are (-1) classical p.d.o.'s on $\partial\Omega$ of arbitrary order, and $\mathscr{B}$ has DN numbers $m_1, \ldots, m_r$ and $t_1, \ldots, t_p$.

Let $\mathbf{s} = (s_1, \ldots, s_p)$, $\mathbf{t} = (t_1, \ldots, t_p)$ and $\mathbf{m} = (m_1, \ldots, m_r)$, and define the Hilbert spaces

$$\mathscr{W}^{l+\mathbf{t}}(\Omega) = W_2^{l+t_1}(\Omega) \times \cdots \times W_2^{l+t_p}(\Omega),$$
$$\mathscr{W}^{l-\mathbf{s}}(\Omega) = W_2^{l-s_1}(\Omega) \times \cdots \times W_2^{l-s_p}(\Omega),$$
$$\mathscr{W}^{l-\mathbf{m}-1/2}(\partial\Omega) = W_2^{l-m_1-1/2}(\partial\Omega) \times \cdots \times W_2^{l-m_r-1/2}(\partial\Omega)$$

with the natural inner products induced by the Cartesian products. For instance, the norm for $u \in \mathscr{W}^{l+\mathbf{t}}(\Omega)$ is $\|u\|_{l+\mathbf{t}}^2 = \sum_{j=1}^{p} \|u_j\|_{l+t_j}^2$ and similarly for $\mathscr{W}^{l-\mathbf{s}}(\Omega)$ and $\mathscr{W}^{l-\mathbf{m}-1/2}(\partial\Omega)$. Here l is an integer such that

$$l \geqslant l_0 := \max\left\{\max_k m_k + q + 1, 0\right\}, \qquad (33)$$

and $q = -\min_{\kappa,k,j} \beta_{kj}^{\kappa}$; see (21) and (26).

Corresponding to the boundary value problem (36) we have the operator

$$\mathfrak{L} = (\mathscr{A}(x, D), \mathscr{B}(y, D)): \mathscr{W}^{l+\mathbf{t}}(\Omega) \to \mathscr{W}^{l-\mathbf{s}}(\Omega) \times \mathscr{W}^{l-\mathbf{m}-1/2}(\partial\Omega) \quad (34)$$

which is a continuous linear operator,

$$\|\mathscr{A}u\|_{l-\mathbf{s}} + \|\mathscr{B}u\|_{l-\mathbf{m}-1/2} \leqslant \text{const}\, \|u\|_{l+\mathbf{t}}$$

since $\operatorname{ord} A_{ij} \leqslant s_i + t_j$ and $\operatorname{ord} B_{kj} \leqslant m_k + t_j$.

Remark Note that

$$l_0 + t_j \geqslant 1 \qquad \text{for all } j. \qquad (35)$$

This is clearly true for $\max_k m_k + q \geqslant 0$, so it remains to consider the case when $\max_k m_k + q < 0$. By definition of m_k and q, we then have

$$m_{kj} - t_j < \beta_{kj}^{\kappa}$$

for all $\kappa = 0, \ldots, l_{kj}$, $j = 1, \ldots, l$ and $k = l, \ldots, r$. The definition of m_{kj} implies that $\beta_{kj}^{\kappa} + \kappa - t_j < \beta_{kj}^{\kappa}$ for all j, k and κ. Hence $t_j > 0$, and t_j being an integer we have $t_j \geqslant 1$. This proves (35) since $l_0 \geqslant 0$.

The inequality (35) ensures that $\mathscr{W}^{l+\mathbf{t}-1}(\Omega) \subset L_2(\Omega) \times \cdots \times L_2(\Omega)$ when $l \geqslant l_0$.

Theorem 9.32 *Let Ω be a bounded domain in $\mathbb{R}^n$ with a C^∞ boundary $\partial\Omega$. For the boundary value problem*

$$\left.\begin{aligned} \mathscr{A}(x, D)u(x) &= f(x), \qquad x \in \Omega, \\ \mathscr{B}(y, D)u(y) &= g(y), \qquad y \in \partial\Omega, \end{aligned}\right\} \tag{36}$$

the following four conditions are equivalent:

(*a*) *The boundary value problem* (36) *is L-elliptic* (*Definition* 9.30).
(*b*) *The operator* $\mathfrak{L} = (\mathscr{A}, \mathscr{B})$, *see* (34),

$$\begin{array}{c} \mathfrak{L}\colon \mathscr{W}^{l+\mathbf{t}}(\Omega) \to \mathscr{W}^{l-\mathbf{s}}(\Omega) \times \mathscr{W}^{l-\mathbf{m}-1/2}(\partial\Omega) \\ \quad\circlearrowleft \qquad\qquad\qquad \circlearrowleft \\ \mathfrak{L}\colon \mathscr{W}^{l+1+\mathbf{t}}(\Omega) \to \mathscr{W}^{l+1-\mathbf{s}}(\Omega) \times \mathscr{W}^{l+1-\mathbf{m}-1/2}(\partial\Omega) \end{array} \tag{37}$$

is smoothable for every $l \geqslant l_0$ (see Definition 9.16).
(*c*) *The operator* $\mathfrak{L} = (\mathscr{A}, \mathscr{B})\colon \mathscr{W}^{l+\mathbf{t}}(\Omega) \to \mathscr{W}^{l-\mathbf{s}}(\Omega) \times \mathscr{W}^{l-\mathbf{m}-1/2}(\partial\Omega)$, $l \geqslant l_0$, *is Fredholm.*
(*d*) *For all $u \in \mathscr{W}^{l+\mathbf{t}}(\Omega)$, $l \geqslant l_0$, there is the* a priori *estimate*

$$\|u\|_{l+\mathbf{t}} \leqslant c[\|\mathscr{A}u\|_{l-\mathbf{s}} + \|\mathscr{B}u\|_{l-\mathbf{m}-1/2} + \|u\|_{l+\mathbf{t}-1}] \tag{38}$$

Remark It will follow from the proof that if (b) holds for just one $l \geqslant l_0$ then it holds for *every* $l \geqslant l_0$. The same is true for (c) and (d).

Further, if l lies in a bounded interval the smoothing regularizers in (b) can be chosen to be independent of l.

Before starting the proof of Theorem 9.32 we mention one of its consequences. Let $u \in \mathscr{W}^{l+\mathbf{t}}(\Omega)$ be a solution of $\mathfrak{L}u = F = (f, g)$. If $F \in \mathscr{W}^{l+1-\mathbf{s}}(\Omega) \times \mathscr{W}^{l+1-\mathbf{m}-1/2}(\partial\Omega)$, then by Lemma 9.17 applied to the statement (b) of Theorem 9.32 it follows that u must in fact be in $\mathscr{W}^{l+1+\mathbf{t}}(\Omega)$.

Consequently, we obtain the classical form of Weyl's Lemma if we let l take all values and then apply Sobolev's lemma (see Theorem 7A.14 or [Wl], p. 111).

Corollary 9.33 (*Weyl's Lemma*) *If $\mathfrak{L}u = F$ and $F \in C^\infty$, then $u \in C^\infty$.*

Moreover, using the compactness of the embeddings in (37) and applying Proposition 9.19 to (b) and (c), we obtain that the defect numbers α, β and the index of the operator $\mathfrak{L}$ are independent of l.

Remark 9.34 If any one of the statements (a), (b), (c) or (d) is true for $\mathfrak{L} = (\mathscr{A}, \mathscr{B})$, then it is also true for any operator $\mathfrak{L}'$ that has the same principal parts $\pi\mathscr{A}$ and $\pi\mathscr{B}$ as $\mathfrak{L}$; in other words,

each condition in Theorem 9.32 is independent of lower-order operators.

To verify this, write $\mathfrak{L}' = \mathfrak{L} + \tilde{\mathfrak{L}} = (\mathscr{A}, \mathscr{B}) + (\tilde{\mathscr{A}}, \tilde{\mathscr{B}})$, where in each entry the operators $\tilde{\mathscr{A}}$ and $\tilde{\mathscr{B}}$ are of order *less* than the order that is used to determine the principal parts, that is, $\tilde{\mathscr{A}} = [\tilde{A}_{ij}]$ and $\tilde{\mathscr{B}} = [\tilde{B}_{kj}]$ with ord $\tilde{A}_{ij} \leqslant s_i + t_j - 1$ and ord $\tilde{B}_{kj} \leqslant m_k + t_j - 1$.

The independence of (a) from these lower-order terms is obvious since by definition L-ellipticity is formulated in terms of principal parts of highest order.

According to Definition 9.16, the condition (b) means that there exist left and right (smoothing) regularizers R_l and R_r: $\mathscr{W}^{l-\mathbf{s}}(\Omega) \times \mathscr{W}^{l-\mathbf{m}-1/2}(\partial\Omega) \to \mathscr{W}^{l+\mathbf{t}}(\Omega)$ for $\mathfrak{L}$, that is,

$$R_l \circ \mathfrak{L} = I + T_l, \qquad \mathfrak{L} \circ R_r = I + T_r,$$

where

$$T_l\colon \mathscr{W}^{l+\mathbf{t}}(\Omega) \to \mathscr{W}^{l+1+\mathbf{t}}(\Omega),$$
$$T_r\colon \mathscr{W}^{l-\mathbf{s}}(\Omega) \times \mathscr{W}^{l-\mathbf{m}-1/2}(\partial\Omega) \to \mathscr{W}^{l+1-\mathbf{s}}(\Omega) \times \mathscr{W}^{l+1-\mathbf{m}-1/2}(\partial\Omega)$$

are *operators of order* -1. Then (b) does not depend on the lower-order terms, for

$$R_l \circ (\mathfrak{L} + \tilde{\mathfrak{L}}) = I + T_l + R_l \circ \tilde{\mathfrak{L}}, \qquad (\mathfrak{L} + \tilde{\mathfrak{L}}) \circ R_r = I + T_r + \tilde{\mathfrak{L}} \circ R_r,$$

where $R_l \circ \tilde{\mathfrak{L}}$ and $\tilde{\mathfrak{L}} \circ R_r$ are also operators of order -1 since the entries of $\tilde{\mathscr{L}}$ are of order less than the corresponding entries of the principal part of $\mathfrak{L}$.

The fact that the statement (c) does not depend on lower-order operators is due to Proposition 9.11 since

$$\tilde{\mathfrak{L}}\colon \mathscr{W}^{l+\mathbf{t}}(\Omega) \to \mathscr{W}^{l+1-\mathbf{s}}(\Omega) \times \mathscr{W}^{l+1-\mathbf{m}-1/2}(\partial\Omega)$$

and the embedding

$$\mathscr{W}^{l+1-\mathbf{s}}(\Omega) \times \mathscr{W}^{l+1-\mathbf{m}-1/2}(\partial\Omega) \hookrightarrow \mathscr{W}^{l-\mathbf{s}}(\Omega) \times \mathscr{W}^{l-\mathbf{m}-1/2}(\partial\Omega)$$

is a compact mapping by Theorem 7A.15. Finally, for (d), the independence from lower-order terms is obvious since

$$\|\tilde{\mathscr{A}}\varphi\|_{l-\mathbf{s}} + \|\tilde{\mathscr{B}}u\|_{l-\mathbf{m}-1/2} \leqslant \text{const}\, \|u\|_{l+\mathbf{t}-1}.$$

The proof of Theorem 9.32

Since the vertical maps in (37) are compact (by Theorem 7A.15), an application of Proposition 9.18 gives the implication (b) $\to$ (c). The implication (c) $\to$ (d) is obtained from Proposition 9.20.

Now we show the implication (d) $\to$ (a), i.e. if the a priori estimate (38) holds for some $l \geqslant l_0$ then $(\mathscr{A}, \mathscr{B})$ is L-elliptic. Since L-ellipticity is verified pointwise, we fix a point $x_0 \in \bar{\Omega}$. If $x_0 \in \Omega$ we must show that $\mathscr{A}(x_0, D)$ is elliptic, while if $x_0 \in \partial\Omega$ we must in addition verify that $(\mathscr{A}(x_0, D), \mathscr{B}(x_0, D))$ satisfies the L-condition.

Let us consider just the case $x_0 \in \partial\Omega$ (the first case is simpler). See the text following (13) for the definition of admissible coordinates. We choose admissible coordinates in a small neighbourhood U of x_0 in $\mathbb{R}^n$, thus $U' = U \cap \partial\Omega$ is a coordinate patch on $\partial\Omega$, and by identifying points in U with their coordinates $(x', x_n) \in \mathbb{R}^n$ we have

$$U = U' \times (-1, 1), \qquad \text{where } U' \subset \mathbb{R}^{n-1}.$$

We also write $U_+ := U \cap \Omega = U' \times (0, 1)$. Since the topology defined by the Sobolev norms on $\partial\Omega$ does not depend on the choice of atlas we may assume that the Sobolev norm in $W_2^k(\partial\Omega)$ of a function $v \in C_0^\infty(U')$ is equal to

$$\|v\|_k^2 = \int |\hat{v}(\xi')|^2(1 + |\xi'|^2)^k \, d\xi' \qquad \text{in } \mathbb{R}^{n-1}.$$

Now we choose a neighbourhood $W' \ni x_0$ in $\partial\Omega$ such that $\bar{W}' \subset U'$, and let W be the following neighbourhood of x_0,

$$W = W' \times (-\delta, \delta),$$

i.e. the set of points with coordinates (x', x_n) where $x' \in W'$ and $-\delta < x_n < \delta$. Then $\bar{W} \subset U$, and we also write $W_+ = W \cap \Omega$.

Our first aim is to remodel the a priori estimate (38) in the form (39) below. We accomplish this goal in two steps.

First step: We localize the operator $\mathscr{A}$ into U and the operator $\mathscr{B}$ into U'. Let $\varphi \in C_0^\infty(U)$ such that $\varphi = 1$ on a neighbourhood of W; by locality of the differential operator $\mathscr{A}$ we have $(1 - \varphi)\mathscr{A}u = 0$, that is, $\mathscr{A}u = \varphi\mathscr{A}u$ for all $u \in C_0^\infty(W)$. Also let $\zeta \in C_0^\infty(U')$ such that $\zeta = 1$ on a neighbourhood of W'. Then we have

$$\mathscr{B}u = \zeta\mathscr{B}u + (1 - \zeta)\mathscr{B}u,$$

and for $u \in C_0^\infty(W)$ the supports of $1 - \zeta$ and $u|_{\partial\Omega}$ are disjoint. By quasi-locality, the term $(1 - \zeta)\mathscr{B}u$ is therefore of order $-\infty$, in particular, $\|(1 - \zeta)\mathscr{B}u\|_{l-\mathbf{m}-1/2} \leqslant \text{const}\, \|u\|_{l+\mathbf{t}-1}$ so that the a priori estimate (11) becomes

$$\|u\|_{l+\mathbf{t}} \leqslant c'[\|\varphi\mathscr{A}u\|_{l-\mathbf{s}} + \|\zeta\mathscr{B}u\|_{l-\mathbf{m}-1/2} + \|u\|_{l+\mathbf{t}-1}], \qquad u \in C_0^\infty(W).$$

Here $\|\ \|_{l-\mathbf{m}-1/2}$ is the Sobolev norm on $W' \subset \mathbb{R}^{n-1}$ and the norms for the other terms are the Sobolev norms on $W_+ \subset \mathbb{R}^n_+$.

Second step: We introduce principal parts of $\mathscr{A}$ and $\mathscr{B}$ by absorbing the lower-order terms in the norm $\|\ \|_{l+\mathbf{t}-1}$. We set

$$A := \pi\mathscr{A}(x, D) \quad \text{and} \quad B := \chi(D')\pi\mathscr{B}(y, D')$$

where $\chi(\xi')$, $\xi' \in \mathbb{R}^{n-1}$, is a cut-off function and then have

$$\|u\|_{l+\mathbf{t}} \leqslant c'[\|\varphi Au\|_{l-\mathbf{s}} + \|\zeta Bu\|_{l-\mathbf{m}-1/2} + \|u\|_{l+\mathbf{t}-1}], \qquad u \in C_0^\infty(W). \tag{39}$$

The cut-off function χ is necessary only if some of the tangential coefficients,

b_{kj}^{κ}, of $\pi\mathscr{B}$ have negative order (i.e. $q > 0$). In view of (32) we have

$$\chi(D')\circ B = \left[\sum_{\kappa} \chi(D')\circ \pi b_{kj}^{\kappa}(y, D')D_n^{\kappa}\right]_{r\times p}$$

where the tangential coefficients $\chi(D')\circ\pi b_{kj}^{\kappa}(y, D')$ are p.d.o.'s on $\mathbb{R}^{n-1}$ and we may assume they have compact support in y (since they are multiplied in (39) by the function ζ).

Note that (39) simplifies for $u \in C_0^\infty(W_+)$ to

$$\|u\|_{l+\mathbf{t}} \leqslant c'[\|\varphi\cdot A(x, D)u\|_{l-\mathbf{s}} + \|u\|_{l+\mathbf{t}-1}] \tag{40}$$

We claim that the ellipticity of $\mathscr{A}$ follows from (40). Let $h \in \mathbb{C}^p$ and $\xi_0 \in \mathbb{R}^n$, $|\xi_0| = 1$, be fixed; we put

$$u_\lambda(x) = \psi(x)\cdot v_\lambda(x), \qquad v_\lambda = \lambda^{-l} e^{i\lambda(x,\xi_0)}\, T^{-1}(\lambda)h,$$

where $\psi \in C_0^\infty(W_+)$ and $T(\lambda)$ is the diagonal matrix from (9). The diagonal elements of $\lambda^{-l}T^{-1}(\lambda)$ are λ^{-l-t_i} so that by (58) in Chapter 8 we obtain

$$\|u_\lambda\|_{l+\mathbf{t}} \to \|\psi\|\,|h| \qquad \text{as } \lambda\to\infty, \tag{41}$$

where $\|\psi\|$ is the L^2 norm, and also

$$\|u_\lambda\|_{l+\mathbf{t}-1} = O(\lambda^{-1}) \qquad \text{as } \lambda\to\infty \tag{42}$$

Now, Leibniz's rule (or Corollary 7.13) gives us $A(x, D)u_\lambda = \psi\cdot A(x, D)v_\lambda + \tilde{A}(x, D)v_\lambda$, where $\tilde{A} = [\tilde{A}_{ij}]$ is of lower order in each entry (i.e. ord $\tilde{A}_{ij} \leqslant s_i + t_j - 1$), and by substituting u_λ in (40) we get

$$\|u_\lambda\|_{l+\mathbf{t}} \leqslant c'[\|\psi\cdot A(x, D)v_\lambda\|_{l-\mathbf{s}} + \|\varphi\cdot\tilde{A}(x, D)v_\lambda\|_{l-\mathbf{s}} + \|u_\lambda\|_{l+\mathbf{t}-1}] \tag{43}$$

since $\psi\varphi = \psi$. In view of the homogeneity property (9) of $A = \pi\mathscr{A}$ we have

$$A(x, D)v_\lambda = \lambda^{-l} e^{i\lambda(x,\xi_0)}\, S(\lambda)A(x, \xi_0)h,$$

and just as in (41), (42) it follows that $\|\psi\cdot A(x, D)v_\lambda\|_{l-\mathbf{s}} \to \|\psi\cdot A(x, \xi_0)h\|$ and $\|\varphi\cdot\tilde{A}(x, D)v_\lambda\|_{l-\mathbf{s}} = O(\lambda^{-1})$. Hence by letting $\lambda\to\infty$ in (43) we get

$$\|\psi\|\,|h| \leqslant c'\|\psi\cdot A(x, \xi_0)h\|$$

and since $\psi \in C_0^\infty(W)$ was arbitrary it follows that

$$|h| \leqslant c'|A(x, \xi_0)h| \qquad \text{for all } x\in W,$$

whence det $A(x_0, \xi_0) \neq 0$ and $\mathscr{A}$ is elliptic.

(We have been considering the case $x_0 \in \partial\Omega$. However, if $x_0\in\Omega$ then we can assume that U lies inside Ω, and by an analogous argument, without boundary operator, arrive again at the a priori estimate (43). The ellipticity of $\mathscr{A}$ at x_0 follows.)

To verify the L-ellipticity of $\mathfrak{L} = (\mathscr{A}, \mathscr{B})$ at $x_0 \in \partial\Omega$, it remains to show that Definition 9.28I is satisfied. Let $\xi_0' \in \mathbb{R}^{n-1}$, $|\xi_0'| = 1$, be fixed and suppose that $w \in \mathfrak{M}^+$, i.e. w is a vector function such that

$$\pi\mathscr{A}\left(x_0, \xi'_0, \frac{1}{i}\frac{d}{dx_n}\right)w(x_n) = 0, \qquad x_n > 0 \tag{44}$$

and $w(t) \to 0$ rapidly as $t \to \infty$. Let

$$v_\lambda(x', x_n) = \lambda^{1/2 - l}\, e^{i\lambda(x', \xi'_0)}\, T^{-1}(\lambda) w(\lambda x_n)$$

Let $\rho(x_n) \in C_0^\infty((-\delta, \delta))$ be equal to 1 near $x_n = 0$ and $\phi(x') \in C_0^\infty(W')$ and set

$$u_\lambda(x) = \psi(x)\cdot v_\lambda(x), \qquad \psi(x) = \phi(x')\cdot\rho(x_n),$$

where $x = (x', x_n)$. Then we have $\psi \in C_0^\infty(W)$ and $u_\lambda \in C_0^\infty(W)$. By making the change of variable $t = \lambda x_n$, $dt = \lambda\, dx_n$, it follows that

$$\|u_\lambda\|^2_{l+\mathbf{t}} \to \int_{\mathbb{R}^{n-1}} |\phi(x')|^2\, dx' \cdot \int_0^\infty |w(t)|^2\, dt \qquad \text{as } \lambda \to \infty \tag{45}$$

and $\|u_\lambda\|_{l+\mathbf{t}-1} = O(\lambda^{-1})$. As before $A(x, D)u_\lambda = \psi\cdot A(x, D)v_\lambda + \tilde{A}(x, D)v_\lambda$, where $\tilde{A}(x, D)$ is of lower order, and by Corollary 7.13 applied to the tangential coefficients of B we have similarly

$$B(x', D)u_\lambda = \phi\cdot B(x', D)v_\lambda + \tilde{B}(x', D)v_\lambda$$

where $\tilde{B} = [\tilde{B}_{kj}]$ is of lower order in each entry (ord $\tilde{B}_{kj} \leqslant m_k + t_j - 1$). Substitution of both expressions in (39) gives

$$\begin{aligned}\|u_\lambda\|_{l+\mathbf{t}} \leqslant c'[&\|\psi A v_\lambda\|_{l-\mathbf{s}} + \|\varphi\tilde{A}v_\lambda\|_{l-\mathbf{s}} \\ &+ \|\varphi B v_\lambda\|_{l-\mathbf{m}-1/2} + \|\zeta\tilde{B}v_\lambda\|_{l-\mathbf{m}-1/2} + \|u_\lambda\|_{l+\mathbf{t}-1}]\end{aligned} \tag{46}$$

since $\psi\varphi = \psi$ and $\zeta\phi = \phi$. In view of (44) and the homogeneity of $\pi\mathscr{A}$ we have $\|\psi A v_\lambda\|_{l-\mathbf{s}} = O(\lambda^{-1})$ as $\lambda \to \infty$, and as before $\|\varphi\tilde{A}v_\lambda\|_{l-\mathbf{s}} = O(\lambda^{-1})$ as $\lambda \to \infty$ since $\tilde{A}$ is of lower order. Further, by applying (7.31) to the tangential coefficients of $\pi\mathscr{B}$ and using the homogeneity property (20) of $\pi\mathscr{B}$ we obtain

$$B(x', D)v_\lambda = \lambda^{1/2 - l}\, e^{i\lambda(x', \xi'_0)}\, M(\lambda)\chi(\lambda\xi'_0)\pi\mathscr{B}\left(x', \xi'_0, \frac{1}{i}\frac{d}{dt}\right)w(t)|_{t=0}$$

where $t = \lambda x_n$. Since the diagonal elements of $\lambda^{1/2-l}M(\lambda)$ are $\lambda^{-l+m_k+1/2}$ we have $\|\phi\cdot Bv_\lambda\|_{l-\mathbf{m}-1/2} \to \|\phi\cdot\pi\mathscr{B}(x', \xi'_0, (1/i)(d/dt))w(t)|_{t=0}\|$, where $\|\ \|$ is the L^2 norm for $\mathbb{C}^r$-valued functions on $\mathbb{R}^{n-1}$, and $\|\zeta\cdot\tilde{B}v_\lambda\|_{l-\mathbf{m}-1/2} \to 0$ as $\lambda \to \infty$. Hence letting $\lambda \to \infty$ in (19) and using (45) we get from (46) that

$$\|\phi\|\cdot\left(\int_0^\infty |w(t)|^2\, dt\right)^{1/2} \leqslant c'\left\|\phi(x')\cdot\pi\mathscr{B}\left(x', \xi'_0, \frac{1}{i}\frac{d}{dt}\right)w(t)|_{t=0}\right\|$$

Since $\phi \in C_0^\infty(W')$ is arbitrary it follows that

$$\left(\int_0^\infty |w(t)|^2\, dt\right)^{1/2} \leqslant c'\left\|\pi\mathscr{B}\left(x', \xi'_0, \frac{1}{i}\frac{d}{dt}\right)w(t)|_{t=0}\right\|_{\mathbb{C}^r} \qquad \text{for all } x' \in W'.$$

In particular it follows that $\pi\mathscr{B}(x_0', \xi_0', (1/i)(d/dt))w(t)|_{t=0}$ is injective as a map $\mathfrak{M}^+ \to \mathbb{C}^r$, so we have verified that condition I of Definition 9.28 holds.

It remains to prove the implication (a) → (b), that is, if the boundary value problem (36) is L-elliptic then the operator $\mathfrak{L} = (\mathscr{A}(x, D), \mathscr{B}(y, D))$ is smoothable. We will do so first under the assumption that all terms b_{kj}^{κ} in the representation (32) have non-negative order

$$\operatorname{ord} b_{kj}^{\kappa}(y, D) \geqslant 0. \tag{R}$$

Later on we show how to get rid of this restriction.

Let $y_0 \in \partial\Omega$. We choose admissible coordinates in $U \ni y_0$, with $U' = U \cap \partial\Omega$ being a coordinate patch on $\partial\Omega$. As usual we identify the points in U with their coordinates in $\mathbb{R}^n$; in these local coordinates we have

$$\varphi\mathscr{B}(y, D)\psi = \varphi[\pi\mathscr{B}(y, D) + \tilde{\mathscr{B}}(y, D)]\psi, \qquad y \in U' \subset \mathbb{R}^{n-1}$$

for all $\varphi, \psi \in C_0^{\infty}(U)$, where

$$\left.\begin{aligned} \pi\mathscr{B}(y, D) = [B_{kj}'(y, D)]_{r\times p}, \qquad B_{kj}'(y, D) = \sum_{\kappa}{}' \pi b_{kj}^{\kappa}(y, D')D_n^{\kappa}, \\ \pi b_{kj}^{\kappa}(y, D')v = (2\pi)^{1-n}\int e^{i(y,\xi')}\,\pi b_{kj}^{\kappa}(y, \xi')\hat{v}(\xi')\,d\xi', \end{aligned}\right\} \tag{47}$$

and $\tilde{\mathscr{B}}$ consists of lower-order terms.

We may, of course, suppose that $\pi b_{kj}^{\kappa}(y, \xi')$ has compact support in $y \in U' \subset \mathbb{R}^{n-1}$, because the regularizers are eventually defined by a partition of unity. In view of the text after Theorem 7.28 concerning the symbol spaces $\tilde{S}^m$, we can therefore do *without* a cut-off function in ξ' in formula (48) for the purpose of establishing estimates in Sobolev norms.

In order to make the long proof of the implication (a) → (b) transparent, we first prove two lemmas.

The points in $\mathbb{R}_+^n$ are denoted by $x = (x', t)$, where

$$x' = (x_1, \ldots, x_{n-1}) \in \mathbb{R}^{n-1}$$

and $t = x_n \in \mathbb{R}_+^1$. If $u(x', t)$ is a function on $\mathbb{R}_+^n$, we make use of the partial Fourier transformation $\mathfrak{F}'$ defined by

$$\hat{u}(\xi', t) = \mathfrak{F}'u(\xi', t) = \int e^{-i(x',\xi')}u(x', t)\,dx', \qquad \xi' = (\xi_1, \ldots, \xi_{n-1}).$$

For integers $k = 0, 1, \ldots$, the norm on $W_2^k(\mathbb{R}_+^n)$ is equivalent to the Fourier norm

$$\|u\|_k^2 \cong \sum_{\nu=0}^{k}\int (1 + |\xi'|^2)^{k-\nu}\int_0^{\infty}\left|\frac{d^{\nu}}{dt^{\nu}}\hat{u}(\xi', t)\right|^2 dt\,d\xi', \qquad t = x_n$$

(see Theorem 7A.4), which in turn is equivalent to

$$\|u\|_k^2 = \sum_{\mu+\nu\leqslant k}\int |\xi'|^{2\mu}\int_0^{\infty}\left|\frac{d^{\nu}}{dt^{\nu}}\hat{u}(\xi', t)\right|^2 dt\,d\xi', \tag{48}$$

and this is a more convenient form to work with in the following lemma.

Lemma 9.35 *Fix a point $y_0 \in \partial\Omega$ on the boundary, let*

$$A(D) := \pi\mathscr{A}(y_0, D), \qquad B(D) := \pi\mathscr{B}(y_0, D), \tag{49}$$

and as above choose admissible coordinates in a neighbourhood of y_0. If $(A(D), B(D))$ is L-elliptic in $\mathbb{R}^n_+$ then we have the a priori estimate ($l \geqslant l_0$)

$$\|u\|_{l+\mathbf{t}} \leqslant c[\|A(D)u\|_{l-\mathbf{s}} + \|B(D)u\|_{l-\mathbf{m}-1/2} + \|u\|_{l+\mathbf{t}-1}] \tag{50}$$

for all $u \in \mathscr{W}^{l+\mathbf{t}}(\mathbb{R}^n_+)$.

Proof For $u \in \mathscr{W}^{l+\mathbf{t}}(\mathbb{R}^n_+)$ the norm is equivalent to

$$\|u\|^2_{l+\mathbf{t}} = \sum_{j=1}^{p} \|u\|^2_{l+t_j} = \sum_{j=1}^{p} \sum_{\mu+\nu \leqslant l+t_j} \int |\xi'|^{2\mu} \int_0^\infty \left| \frac{d^\nu}{dt^\nu} \hat{u}_j(\xi', t) \right|^2 dt\, d\xi'.$$

We use the notation $|\ |$ to indicate the terms of highest order $\mu + \nu = l + t_j$ in the sum, i.e.

$$|u|^2_{l+\mathbf{t}} = \sum_{j=1}^{p} \sum_{\nu=0}^{l+t_j} \int |\xi'|^{2(l+t_j-\nu)} \int_0^\infty \left| \frac{d^\nu}{dt^\nu} \hat{u}_j(\xi', t) \right|^2 dt\, d\xi'.$$

By the change of variables $t = x_n/|\xi'|$, $dt = dx_n/|\xi'|$, we get

$$\begin{aligned} |u|^2_{l+\mathbf{t}} &= \sum_{j=1}^{p} \sum_{\nu=0}^{l+t_j} \int |\xi'|^{2(l+t_j)-1} \int_0^\infty \left| \frac{d^\nu}{dx_n^\nu} \hat{u}_j(\xi', x_n/|\xi'|) \right|^2 dx_n\, d\xi' \\ &= \int |\xi'|^{2l-1} \|T(|\xi'|)\hat{u}(\xi', x_n/|\xi'|)\|^2_{l+\mathbf{t}, \mathbb{R}^1_+}\, d\xi', \end{aligned} \tag{51}$$

where $T(c)$ is the diagonal matrix from (9), and the norm on $\mathscr{W}^{l+\mathbf{t}}(\mathbb{R}^1_+)$ is given by

$$\|v(x_n)\|^2_{l+\mathbf{t}, \mathbb{R}^1_+} = \sum_{j=1}^{p} \sum_{\nu=0}^{l+t_j} \left| \int_0^\infty \frac{d^\nu}{dx_n^\nu} v_j(x_n) \right|^2 dx_n.$$

Recall that the space $\mathscr{W}^{l+\mathbf{t}}(\mathbb{R}^1_+)$ was defined after Definition 9.28.

There is of course an analogous expression for the norm on $\mathscr{W}^{l-\mathbf{s}}(\mathbb{R}^n_+)$, where the term $T(|\xi'|)$ is replaced by $S(|\xi'|^{-1})$, and an analogous expression for the norm on $\mathscr{W}^{l-\mathbf{s}}(\mathbb{R}^1_+)$.

Now we use the estimate (28) on the compact set $K = S^{n-1} \ni \xi'/|\xi'|$, which gives us

$$\begin{aligned} \|\varphi(x_n)\|^2_{l+\mathbf{t}, \mathbb{R}^1_+} \leqslant c\Bigg[&\left\| A\left(\frac{\xi'}{|\xi'|}, \frac{1}{i}\frac{d}{dx_n}\right)\varphi \right\|^2_{l-\mathbf{s}, \mathbb{R}^1_+} \\ &+ \left\| B\left(\frac{\xi'}{|\xi'|}, \frac{1}{i}\frac{d}{dx_n}\right)\varphi\Big|_{x_n=0} \right\|^2_{\mathbb{C}^r} \Bigg] \end{aligned} \tag{52}$$

where $c = c_K$. By making use of the homogeneity properties (9) and (20), we also have

$$A\left(\frac{\xi'}{|\xi'|}, \frac{1}{i}\frac{d}{dx_n}\right) = S(|\xi'|^{-1})A\left(\xi', \frac{1}{i}\frac{d}{d(x_n/|\xi'|)}\right)T(|\xi'|^{-1})$$

and

$$B\left(\frac{\xi'}{|\xi'|}, \frac{1}{i}\frac{d}{dx_n}\right) = M(|\xi'|^{-1})B\left(\xi', \frac{1}{i}\frac{d}{d(x_n/|\xi'|)}\right)T(|\xi'|^{-1}),$$

and if we substitute $\varphi(x_n) = T(|\xi'|)\hat{u}(\xi', x_n/|\xi'|)$ in (52) it follows that

$$\|T(|\xi'|)\hat{u}(\xi', x_n/|\xi'|)\|^2_{l+\mathbf{t},\mathbb{R}^1_+} \leqslant \text{const}\left[\left\|S(|\xi'|^{-1})A\left(\xi', \frac{1}{i}\frac{d}{d(x_n/|\xi'|)}\right)\hat{u}(\xi', x_n/|\xi'|)\right\|^2_{l-\mathbf{s},\mathbb{R}^1_+} + \left\|M(|\xi'|^{-1})B\left(\xi', \frac{1}{i}\frac{d}{dx_n}\right)\hat{u}(\xi', x_n)|_{x_n=0}\right\|^2_{\mathbb{C}^r}\right]$$

Multiplying this inequality by $|\xi'|^{2l-1}$, integrating over ξ', and taking into account that the diagonal elements of the matrix $|\xi'|^{(2l-1)/2}M(|\xi'|)^{-1}$ are $|\xi'|^{l-m_k-1/2}$, then in view of (51) we obtain

$$|u|^2_{l+\mathbf{t}} \leqslant c[|A(D)u|^2_{l-\mathbf{s}} + \|B(D)u\|^2_{l-\mathbf{m}-1/2}].$$

Now (50) follows since $\|u\|^2_{l+\mathbf{t}} = |u|^2_{l+\mathbf{t}} + \|u\|^2_{l+\mathbf{t}-1}$ and similarly for $\|A(D)\|^2_{l-\mathbf{s}}$.

As before let $A(D)$, $B(D)$ be given by (49). We denote by $\omega_j(\xi', t) \in \mathfrak{M}^+$ the solution of the initial value problem in Definition 9.28 II for $g = e_j = (0, \ldots, 0, 1, 0, \ldots, 0)$, and by $\omega(\xi', t)$ the rectangular $p \times r$ matrix whose columns are the vectors $\omega_j(\xi', t)$. Then we have

$$\left.\begin{aligned} A\left(\xi', \frac{1}{i}\frac{d}{dt}\right)\omega = 0, \qquad t > 0, \\ B\left(\xi', \frac{1}{i}\frac{d}{dt}\right)\omega|_{t=0} = 1_r, \end{aligned}\right\} \tag{53}$$

where 1_r denotes the $r \times r$ identity matrix. A matrix function $\omega(\xi', t)$ satisfying the equations (53) is called a *canonical matrix*. The solution of the initial value problem in Definition 9.28 II may be written in the form $w(\xi', t) = \omega(\xi', t)g$.

The existence of a unique canonical matrix at each fixed point ξ' follows from the presupposed L-ellipticity.

Lemma 9.36 *For the canonical matrix* $\omega(\xi', t)$, *we have the homogeneity relation*

$$T(c)\omega(c\xi', t/c)M(c) = \omega(\xi', t). \tag{54}$$

and it follows that

$$\int_0^\infty \left|\frac{d^\nu}{dt^\nu}\omega_{jk}(\xi', t)\right|^2 dt \leqslant C_\nu|\xi'|^{2\nu-1-2t_j-2m_k} \qquad \textit{for all } \nu = 0, 1, \ldots \tag{55}$$

Proof If we replace ξ' by $c\xi'$ and t by t/c in the system (53), and then use (9) and (20), we see that the matrix $T(c)\omega(c\xi', t/c)$ satisfies the equations

$$A\left(\xi', \frac{1}{i}\frac{d}{dt}\right)T(c)\omega(c\xi', t/c) = 0,$$

$$B\left(\xi', \frac{1}{i}\frac{d}{dt}\right)T(c)\omega(c\xi', 0) = M^{-1}(c).$$

By uniqueness it follows that $T(c)\omega(c\xi', t/c) = \omega(\xi', t)M^{-1}(c)$, which is (54). By compactness of S^{n-1} it can be shown that

$$\int_0^\infty \left|\frac{d^\nu}{dt^\nu}\omega_{jk}(\xi'/|\xi'|, t)\right|^2 dt \leqslant C_\nu < \infty$$

for all $\xi'/|\xi'| \in S^{n-1}$; see the remark at the end of §10.4. Then by substituting $t = |\xi'|x_n$, $dt = |\xi'|\,dx_n$ and using relation (54) with $c = |\xi'|^{-1}$ we obtain the result (55) at once.

To prove the implication (a) → (b), we must construct left and right regularizers R_l, R_r and smoothing operators T_l, T_r. By Remark 9.34, we may discard lower-order terms of the operators and consider just the principal parts.

Throughout this section, when we say regularizer it is understood to mean a "smoothing" regularizer; that is, R is a right (resp. left) smoothing regularizer of an operator A if $AR - I$ (resp. $RA - I$) is an operator of order -1. See also Definition 9.16.

The construction proceeds by the following steps:

(1) We freeze the coefficients of $\pi\mathscr{A}$ at a fixed point x_0 in $\bar{\Omega}$, and construct regularizers for the constant coefficient operator $A(D) = \pi\mathscr{A}(x_0, D)$ in $\mathbb{R}^n$.
(2) We fix $y_0 \in \partial\Omega$ and an admissible coordinate system around y_0 and construct regularizers for $(A(D), B(D))$ (see (49)) in the half-space $\mathbb{R}^n_+$.
(3) Applying the Freezing Lemma 7.41, we accomplish the transition from frozen x_0 to "local" right and left regularizers in a small neighbourhood $U \ni x_0$ for the original problem (36).
(4) We finish the construction by using a partition of unity.

Step 1 We consider the constant coefficient operator $A(D) = \pi\mathscr{A}(x_0, D)$, $x_0 \in \bar{\Omega}$ fixed, and define

$$Rf = \mathfrak{F}^{-1}|\xi|^\sigma(1 + |\xi|^\sigma)^{-1}A^{-1}(\xi)\mathfrak{F}f, \qquad 0 \neq \xi \in \mathbb{R}^n, \tag{56}$$

where $\mathfrak{F}$ is the Fourier transformation in $\mathbb{R}^n$ and $\sigma := \max(s_i + t_j) \geqslant 1$ (the trivial case $\sigma = 0$ of a multiplication operator is not considered here). Using the estimates (12), it is easy to verify that $R\colon \mathscr{W}^{l-\mathbf{s}}(\mathbb{R}^n) \to \mathscr{W}^{l+\mathbf{t}}(\mathbb{R}^n)$ is continuous. To see this, it is necessary to consider two cases, $|\xi| \leqslant 1$ and $|\xi| \geqslant 1$. Note that the choice of σ ensures that the matrix $|\xi|^\sigma(1 + |\xi|^\sigma)^{-1}A^{-1}(\xi)$ is a bounded function near $\xi = 0$ while for large $|\xi|$, it behaves like $A^{-1}(\xi)$. Now, we have

$$A \circ Rf = f + Tf, \qquad R \circ Au = u + Tu,$$

where $Tf = -\mathfrak{F}^{-1}(1 + |\xi|^\sigma)^{-1}\mathfrak{F}f$. Since $\sigma \geqslant 1$, it follows that T is smoothing, i.e. an operator of order -1,

$$T: \mathscr{W}^{l-\mathbf{s}}(\mathbb{R}^n) \to \mathscr{W}^{l+1-\mathbf{s}}(\mathbb{R}^n) \quad \text{and} \quad \mathscr{W}^{l+\mathbf{t}}(\mathbb{R}^n) \to \mathscr{W}^{l+1+\mathbf{t}}(\mathbb{R}^n),$$

whence R is both a left and a right regularizer.

Step 2 In the half-space $\mathbb{R}^n_+$ we consider the L-elliptic operator with constant coefficients, $\mathfrak{L}(D) = (\pi\mathscr{A}(y_0, D), \pi\mathscr{B}(y_0, D)) = (A(D), B(D))$, where $y_0 \in \partial\Omega$ is fixed and $A(D)$ is the same as in Step 1.

Let $\imath_+: W_2^m(\mathbb{R}^n) \to W_2^m(\mathbb{R}^n_+)$, $m \geqslant 0$, denote the restriction of a function on $\mathbb{R}^n$ to $\mathbb{R}^n_+$. Also let $\beta: W_2^m(\mathbb{R}^n_+) \to W_2^m(\mathbb{R}^n)$, $m \geqslant 0$, be a linear, continuous extension operator from $\mathbb{R}^n_+$ to $\mathbb{R}^n$. For all m in a bounded interval, it is possible to define β independently of m, i.e. the definition of βf does not depend on the particular space W_2^m to which f belongs (Theorem 7A.7).

As before, $\mathfrak{F}$ denotes the Fourier transform in $\mathbb{R}^n$ and $\mathfrak{F}'$ the partial Fourier transform in $\mathbb{R}^n_+$. We define the operators

$$R_0 f = \imath_+\mathfrak{F}^{-1}|\xi|^\sigma(1 + |\xi|^\sigma)^{-1}A^{-1}(\xi)\mathfrak{F}\beta f = \imath_+ R\beta f,$$
$$R_1 g = \mathfrak{F}'^{-1}\omega(\xi', x_n)|\xi'|^{\sigma'}(1 + |\xi'|^{\sigma'})^{-1}\mathfrak{F}' g, \qquad x_n > 0,$$

where R is from (56), $\omega(\xi', x_n)$ is the canonical matrix function satisfying (53) and $\sigma' := \max(m_k + t_j) + 1$. The continuity of $R_0 := \mathscr{W}^{l+\mathbf{t}}(\mathbb{R}^n_+) \to \mathscr{W}^{l-\mathbf{s}}(\mathbb{R}^n_+)$ follows from the continuity of R and the continuity of $\imath_+$ and β because

$$\|R_0 f\|_{l+\mathbf{t}, \mathbb{R}^n_+} \leqslant \|R\beta f\|_{l+\mathbf{t}, \mathbb{R}^n} \leqslant c_1\|\beta f\|_{l-\mathbf{s}, \mathbb{R}^n} \leqslant c_2\|f\|_{l-\mathbf{s}, \mathbb{R}^n_+}. \tag{57}$$

To show the continuity of $R_1: \mathscr{W}^{l-\mathbf{m}-1/2}(\mathbb{R}^{n-1}) \to \mathscr{W}^{l+\mathbf{t}}(\mathbb{R}^n_+)$, we use the estimates (55) of $\omega(\xi', x_n)$ to obtain

$$\begin{aligned}
\|R_1 g\|_{l+\mathbf{t}}^2 &\leqslant \sum_{j=1}^{p}\sum_{k=1}^{r}\int \sum_{\mu+\nu \leqslant l+t_j} |\xi'|^{2\mu} \int \left|\frac{d^\nu}{dt^\nu}\omega_{jk}(\xi', t)\right|^2 |\mathfrak{F}' g_k|^2 \, dt \, d\xi' \\
&\leqslant C\sum_{k=1}^{r}\int (1 + |\xi'|^{2l-2m_k-1})|\mathfrak{F}' g_k|^2 \, d\xi' \\
&\leqslant C'\|g\|_{l-\mathbf{m}-1/2}^2.
\end{aligned} \tag{58}$$

(Once again, one must consider two cases, $|\xi'| \leqslant 1$ and $|\xi'| \geqslant 1$.)

Now define

$$\mathfrak{R}(f, g) := R_0 f + R_1[g - B(D)R_0 f]. \tag{59}$$

We will show that the operator $\mathfrak{R}: \mathscr{W}^{l-\mathbf{s}}(\mathbb{R}^n_+) \times \mathscr{W}^{l-\mathbf{m}-1/2}(\mathbb{R}^{n-1}) \to \mathscr{W}^{l+\mathbf{t}}(\mathbb{R}^n_+)$ satisfies all the requirements imposed on a regularizer for $\mathfrak{L}$, namely, that it is continuous and satisfies the left and right regularizing properties.

First of all, it follows by (58), (57) and the continuity of $B(D)$ that

$$\begin{aligned}
\|R_1[g - B(D)R_0 f]\|_{l+\mathbf{t}} &\leqslant c\|g\|_{l-\mathbf{m}-1/2} + c\|B(D)R_0 f\|_{l-\mathbf{m}-1/2} \\
&\leqslant c\|g\|_{l-\mathbf{m}-1/2} + cc'\|R_0 f\|_{l+\mathbf{t}} \\
&\leqslant c\|g\|_{l-\mathbf{m}-1/2} + cc'c_2\|f\|_{l-\mathbf{s}},
\end{aligned}$$

which proves the continuity of $\mathfrak{R}$. Next, we show that $\mathfrak{R}$ is a right regularizer of $\mathfrak{L}$. Since $A(D)R_1 = 0$, then

$$A(D)\mathfrak{R}(f, g) = A(D)R_0 f = f + V_0 f,$$

where $V_0 f := -r_+\mathfrak{F}^{-1}(1 + |\xi|^\sigma)^{-1}\mathfrak{F}\beta f$. Also, $B(D)\mathfrak{R}(0, g) = B(D)R_1 g = g + V_1 g$, where $V_1 g := -\mathfrak{F}'^{-1}(1 + |\xi'|^{\sigma'})^{-1}\mathfrak{F}'g$, and then replacing g by $g - B(D)R_0 f$, we obtain

$$\begin{aligned} B(D)\circ\mathfrak{R}(f, g) &= B(D)\circ R_0 f + B(D)\circ R_1[g - B(D)R_0 f] \\ &= g + V_1[g - B(D)R_0 f]. \end{aligned}$$

We have shown, therefore, that

$$\mathfrak{L}(D)\circ\mathfrak{R}(f, g) = (f, g) + T_r(f, g), \tag{60}$$

where $T_r(f, g) = (V_0 f, V_1 g - V_1 B(D)R_0 f)$. Now, we claim that T_r is an operator of order -1. Indeed, $\|V_0 f\|_{l-\mathbf{s}+1} \leqslant c\|f\|_{l-\mathbf{s}}$ because $\sigma \geqslant 1$, and, similarly,

$$\begin{aligned} \|V_1[g - B(D)R_0 f]\|_{l+1-\mathbf{m}-1/2} &\leqslant c\|g - B(D)R_0 f\|_{l-\mathbf{m}-1/2} \\ &\leqslant c_1(\|g\|_{l-\mathbf{m}-1/2} + \|f\|_{l-\mathbf{s}}) \end{aligned}$$

since $\sigma' = \max(m_k + t_j) + 1 \geqslant 1$ (by virtue of the restriction (R)), which proves the continuity of

$$T_r\colon \mathscr{W}^{l-\mathbf{s}}(\mathbb{R}^n_+) \times \mathscr{W}^{l-\mathbf{m}-1/2}(\mathbb{R}^{n-1}) \to \mathscr{W}^{l+1-\mathbf{s}}(\mathbb{R}^n_+) \times \mathscr{W}^{l+1-\mathbf{m}-1/2}(\mathbb{R}^{n-1}), \tag{61}$$

Finally, to show that $\mathfrak{R}$ is a left regularizer, we define the operator T_l by

$$\mathfrak{R}\circ\mathfrak{L}(D)u = u + T_l u, \tag{62}$$

and must show that T_l is an operator of order -1, $T_l\colon \mathscr{W}^{l+\mathbf{t}}(\mathbb{R}^n_+) \to \mathscr{W}^{l+1+\mathbf{t}}(\mathbb{R}^n_+)$. First of all, it follows from the continuity of $\mathfrak{L}(D)$ and $\mathfrak{R}$ that T_l is continuous from $\mathscr{W}^{l+\mathbf{t}}(\mathbb{R}^n_+)$ to $\mathscr{W}^{l+\mathbf{t}}(\mathbb{R}^n_+)$. Further, in view of (60), (62), we have

$$\mathfrak{L}\circ\mathfrak{R}\circ\mathfrak{L}u = \mathfrak{L}(I + T_l)u = (I + T_r)\mathfrak{L}u,$$

so $\mathfrak{L}T_l u = T_r\mathfrak{L}u$, i.e.

$$(A(D)T_l u, B(D)T_l u) = T_r(A(D)u, B(D)u) \tag{63}$$

By virtue of the a priori estimate (50), we obtain

$$\|T_l u\|_{l+1+\mathbf{t}} \leqslant c[\|A(D)T_l u\|_{l+1-\mathbf{s}} + \|B(D)T_l u\|_{l+1-\mathbf{m}-1/2} + \|T_l u\|_{l+\mathbf{t}}]$$

and then by (63), (61)

$$\begin{aligned} \|T_l u\|_{l+1+\mathbf{t}} &\leqslant c'[\|A(D)u\|_{l-\mathbf{s}} + \|B(D)u\|_{l-\mathbf{m}-1/2} + \|T_l u\|_{l+\mathbf{t}}] \\ &\leqslant c''\|u\|_{l+\mathbf{t}}, \end{aligned}$$

by the continuity of $A(D)$, $B(D)$ and T_l. This proves the smoothing property of T_l.

Step 3 In this step we show that the original operator $\mathfrak{L} = (\mathscr{A}, \mathscr{B})$ in (34) has a "local" regularizing property in a neighbourhood of any point $x_0 \in \bar{\Omega}$. By Remark 9.34, we need to focus only on the principal parts

$$A(x, D) := \pi\mathscr{A}(x, D), \qquad B := \pi\mathscr{B}.$$

First let us fix $x_0 \in \Omega$. Since $\Omega \subset \mathbb{R}^n$ we use the standard coordinates in $\mathbb{R}^n$, and consider the operator $A(x_0, D)$ in $\mathbb{R}^n$; by Step 1 it has a regularizer $R(x_0)\colon \mathscr{W}^{l-\mathbf{s}}(\mathbb{R}^n) \to \mathscr{W}^{l+\mathbf{t}}(\mathbb{R}^n)$. We apply the Freezing Lemma 7.41 to the point x_0 with $\varepsilon = 1/(2\|R(x_0)\|)$ to obtain a neighbourhood $U \subset \Omega$ of x_0 such that for all $\varphi \in C_0^\infty(U)$ the decomposition

$$\varphi A(x, D) = \varphi[A(x_0, D) + K_{x_0} + T_{x_0}], \tag{64}$$

holds, where K_{x_0} is a matrix p.d.o. of the same order in each entry as $A(x, D)$ such that $\|K_{x_0}\| \leqslant \varepsilon$, and T_{x_0} is a matrix p.d.o. of lower order in each entry. Note that the norm of K_{x_0} refers to the norm of the operator $K_{x_0}\colon \mathscr{W}^{l+\mathbf{s}}(\mathbb{R}^n) \to \mathscr{W}^{l-\mathbf{t}}(\mathbb{R}^n)$, of course. Then we let

$$R_l(x_0) := (I + R(x_0) \circ K_{x_0})^{-1} \circ R(x_0),$$
$$R_r(x_0) := R(x_0) \circ (I + K_{x_0} \circ R(x_0))^{-1},$$

and observe that the inverse operators exist and are bounded because $\|R(x_0) \circ K_{x_0}\| < 1$ and $\|K_{x_0} \circ R(x_0)\| < 1$ by choice of ε. It can be verified that $(R_l(x_0), 0)$ and $(R_r(x_0), 0)$ are "local" left and right regularizers of $\mathfrak{L}$ in the sense that for all $\varphi, \psi \in C_0^\infty(U)$,

$$\varphi R_l(x_0) \circ \mathfrak{L}\psi = \varphi I\psi + \varphi T_l \psi$$
$$\varphi \mathfrak{L} \circ R_r(x_0)\psi = \varphi I\psi + \varphi T_r \psi$$

where $T_l \in OP^{-1}$ and $T_r \in OP^{-1}$ (i.e. operators of order -1). The verification is similar (but easier) to that of (67) and (68) below. The support of φ and of ψ do not meet $\partial\Omega$, so we have $\varphi(f, g)\psi = \varphi(f, 0)\psi$ for all $f \in \mathscr{W}^{l+\mathbf{s}}(\Omega)$ and $g \in \mathscr{W}^{l-\mathbf{m}-1/2}(\partial\Omega)$.

Now, let us fix a point on the boundary $y_0 \in \partial\Omega$. By choosing admissible coordinates in a neighbourhood of y_0, and by considering the push-forward operators for A and B, we may assume that $\Omega = \mathbb{R}^n_+$, and $\partial\Omega = \mathbb{R}^{n-1}$. Since we will finish the construction of the regularizers in Step 4 using a partition of unity we may assume that the coefficients of the differential operator $A(x, D)$ have compact support in $x \in \mathbb{R}^n$, and that the boundary operator $B = \pi\mathscr{B}$ has the form (47) where the normal coefficients, $\pi b^\kappa_{kj}(y, \xi')$, of B have compact support in $y \in \mathbb{R}^{n-1}$.

From Step 2 we know that the operator

$$\mathfrak{L}(y_0, D) = (A(y_0, D), B(y_0, D))$$

has a regularizer $R(y_0)\colon \mathscr{W}^{l-\mathbf{s}}(\mathbb{R}^n_+) \times \mathscr{W}^{l-\mathbf{m}-1/2}(\mathbb{R}^{n-1}) \to \mathscr{W}^{l+\mathbf{t}}(\mathbb{R}^n_+)$. The Freezing Lemma 7.41 applied to A and the Freezing Lemma 7.41′ applied to each normal coefficient, $\pi b^\kappa_{kj}(y, \xi')$, of B gives us for $\varepsilon = 1/(2\|R(y_0)\|)$ an open neighbourhood U in $\mathbb{R}^n$ of y_0 such that for all $\varphi \in C_0^\infty(U)$

$$\varphi\mathfrak{L} = \varphi[(A(y_0, D), B(y_0, D)) + K_{y_0} + T_{y_0}], \tag{65}$$

where K_{y_0} has the same order as (A, B), $\|K_{y_0}\| \leqslant \varepsilon$, and T_{y_0} is of lower order. We then define

$$\left.\begin{aligned} R_l(y_0) &:= (I + R(y_0) \circ K_{y_0})^{-1} \circ R(y_0), \\ R_r(y_0) &:= R(y_0) \circ (I + K_{y_0} \circ R(y_0))^{-1}, \end{aligned}\right\} \tag{66}$$

Choose a neighbourhood $W \ni y_0$ such that $\bar{W} \subset U$, and choose a function $\varphi_0 \in C_0^\infty(U)$ such that $\varphi_0 = 1$ on W. If $\psi \in C_0^\infty$ has support in W then

$$\mathfrak{L}\psi = \varphi_0 \mathfrak{L}\psi + (1 - \varphi_0)\mathfrak{L}\psi$$

where the supports of $1 - \varphi_0$ and ψ are disjoint, so by quasi-locality we have $(1 - \varphi_0)\mathfrak{L}\psi \in OP^{-\infty}$. Let us use the notation $\equiv$ to denote equality of operators modulo OP^{-1}. We have by (65) for all $\varphi, \psi \in C_0^\infty$ with support in W,

$$\begin{aligned} \varphi R_l(y_0) \circ \mathfrak{L}\psi &\equiv \varphi R_l(y_0) \circ \varphi_0 \mathfrak{L}\psi \\ &\equiv \varphi R_l(y_0) \circ [(A(y_0, D), B(y_0, D)) + K_{y_0}]\psi \end{aligned}$$

and using (66)

$$\begin{aligned} &\equiv \varphi(I + R(y_0) \circ K_{y_0})^{-1}[R(y_0) \circ (A(y_0, D), B(y_0, D)) + R(y_0) \circ K_{y_0}]\psi \\ &\equiv \varphi(I + R(y_0) \circ K_{y_0})^{-1}[I + R(y_0) \circ K_{y_0}]\psi \\ &= \varphi I \psi. \end{aligned}$$

Note that the function φ_0 was introduced so that we could use (65), but then it was dropped in the second line by quasi-locality of $(A(y_0, D), B(y_0, D))$ which is justified by multiplication of B by a cut-off function and Corollary 7.30. Thus we have shown that

$$\varphi R_l(y_0) \circ \mathfrak{L}\psi = \varphi I \psi + \varphi T_l \psi \qquad \forall \varphi, \psi \in C_0^\infty(W), \tag{67}$$

where $T_l \in OP^{-1}$, an operator of order -1. This is the local left regularizing property for $\mathfrak{L}$. The verification of the local right regularizing property is somewhat easier; we have

$$\begin{aligned} \varphi\mathfrak{L} \circ R_r(y_0)\psi &\equiv \varphi[\mathfrak{L}(y_0, D) + K_{y_0}]R_r(y_0)\psi \\ &\equiv \varphi[\mathfrak{L}(y_0, D) \circ R(y_0) + K_{y_0} \circ R(y_0)](I + K_{y_0} \circ R(y_0)))^{-1}\psi \\ &\equiv \varphi[I + K_{y_0} \circ R(y_0)] \cdot (I + K_{y_0} \circ R(y_0))^{-1}\psi \\ &= \varphi I \psi, \end{aligned}$$

and we have shown that

$$\varphi\mathfrak{L} \circ R_r(y_0)\psi = \varphi I \psi + \varphi T_r \psi \tag{68}$$

where $T_r \in OP^{-1}$. The equations (67) and (68) constitute the local regularizing property for $\mathfrak{L}$.

Step 4 We conclude the proof by constructing regularizers for the original operator

$$\mathfrak{L} = (\mathscr{A}(x, D), \mathscr{B}(y, D))$$

The compact set $\bar{\Omega}$ can be covered with a finite number of open sets $U(x_j)$, $x_j \in \bar{\Omega}$, $j = 1, \ldots, N$, such that if $x_j \in \Omega$ then $U(x_j) \subset \Omega$, and if $x_j \in \partial\Omega$ then there is an admissible coordinate system $(U(x_j), \kappa_j)$ such that $U(x_j) \cap \partial\Omega$ is a coordinate patch on $\partial\Omega$. Further, by virtue of Step 3, the open cover $\bigcup_{j=1}^{N} U(x_j)$ can be taken sufficiently fine that there exist bounded operators $R_l(x_j)\colon \mathscr{W}^{l-\mathbf{s}}(\Omega) \times \mathscr{W}^{l-\mathbf{m}-1/2}(\partial\Omega) \to \mathscr{W}^{l+\mathbf{t}}(\Omega)$ and $R_r(x_j)\colon \mathscr{W}^{l-\mathbf{s}}(\Omega) \times \mathscr{W}^{l-\mathbf{m}-1/2}(\partial\Omega) \to \mathscr{W}^{l+\mathbf{t}}(\Omega)$ such that for all $\varphi, \psi \in C_0^\infty(U(x_j))$ the following holds:

$$\left.\begin{aligned} \varphi\mathfrak{L} \circ R_r(x_j)\psi &= \varphi I\psi + T_r(x_j), \\ \varphi R_l(x_j) \circ \mathfrak{L}\psi &= \varphi I\psi + T_l(x_j), \end{aligned}\right\} \tag{69}$$

where

$$T_r(x_j)\colon \mathscr{W}^{l-\mathbf{s}}(\Omega) \times \mathscr{W}^{l-\mathbf{m}-1/2}(\partial\Omega) \to \mathscr{W}^{l+1-\mathbf{s}}(\Omega) \times \mathscr{W}^{l+1-\mathbf{m}-1/2}(\partial\Omega)$$

and $T_l(x_j)\colon \mathscr{W}^{l+\mathbf{t}}(\Omega) \to \mathscr{W}^{l+\mathbf{t}+1}(\Omega)$ are operators of order -1.

Let $\{\varphi_j\}_{j=1}^N$ be a partition of unity subordinate to the cover $\bigcup_{j=1}^N U(x_j)$, and then let ψ_j be a function from $C_0^\infty(U(y_j))$ which equals 1 on the support of φ_j. We can then define the left and right regularizers for the operator $\mathfrak{L}$ by

$$\mathfrak{R}_l = \sum_{j=1}^{N} \varphi_j R_l(x_j)\psi_j, \qquad \mathfrak{R}_r = \sum_{j=1}^{N} \varphi_j R_r(x_j)\psi_j,$$

Let us check the properties of $\mathfrak{R}_l$ and $\mathfrak{R}_r$. As before we let $\equiv$ denote equality modulo operators of order -1. By commuting ψ_j and $\mathfrak{L}$ using Corollaries 7.13 and 8.26, and using (69), we obtain

$$\begin{aligned} \mathfrak{R}_l \circ \mathfrak{L} &\equiv \sum \varphi_j R_l(x_j) \circ \mathfrak{L}\psi_j, \\ &\equiv \sum \varphi_j I \psi_j \\ &= I \end{aligned}$$

since $\psi_j\varphi_j = \varphi_j$ and $\sum \varphi_j = 1$. Thus we have shown that

$$\mathfrak{R}_l \circ \mathfrak{L} = I + T_l, \qquad \text{where } T_l \in OP^{-1}. \tag{70}$$

Once more by commuting $\mathfrak{L}$ and φ_j and using (69) we have similarly

$$\begin{aligned} \mathfrak{L} \circ \mathfrak{R}_r &\equiv \sum \varphi_j \mathfrak{L} \circ R_r(x_j)\psi_j \\ &\equiv \sum \varphi_j I \psi_j \\ &= I, \end{aligned}$$

which means that

$$\mathfrak{L} \circ \mathfrak{R}_r = I + T_r, \qquad \text{where } T_r \in OP^{-1}. \tag{71}$$

The equations (70) and (71) tell us that the operator $\mathfrak{L}$ is smoothable.

To get rid of the restriction (*R*) in the proof of the implication (a) → (b), we choose $\Lambda^q \in OClS^q(\partial\Omega)$ from Lemma 8.67 with principal part $\pi\Lambda^q = |\xi'|^q$, $\xi' \in T^*(\partial\Omega)\backslash 0$. Let us fix q such that

$$q \geqslant -\min_{\kappa, k, j} \operatorname{ord} b^{\kappa}_{kj}(y, D'), \qquad y \in \partial\Omega. \tag{72}$$

Now, suppose that $(\mathscr{A}, \mathscr{B})$ is L-elliptic, i.e. $\mathscr{A}$ is elliptic and $(\mathscr{A}, \mathscr{B})$ satisfies the L-condition in Definition 9.28I. Since this L-condition is invariant under multiplication of the boundary operator $\pi\mathscr{B}(y; (\xi', (1/i)(d/dt)))|_{t=0}$ with $|\xi'|^q \neq 0$, we see that $(\mathscr{A}, \Lambda^q \circ \mathscr{B})$ is L-elliptic too, where we denote by the same notation Λ^q the diagonal $r \times r$ matrix with Λ^q in the diagonal. In view of (72), the boundary operator $\Lambda^q \circ \mathscr{B}$ satisfies the restriction

$$q + \operatorname{ord} b^{\kappa}_{kj} \geqslant q + \min_{\kappa, k, j} \operatorname{ord} b^{\kappa}_{kj} \geqslant 0, \tag{R}$$

and we obtain by our previous reasoning that the map $(\mathscr{A}, \Lambda^q \circ \mathscr{B})$ is smoothable:

$$\mathfrak{R}_l \circ (\mathscr{A}, \Lambda^q \circ \mathscr{B}) = I + T_l, \qquad T_l \in OP^{-1} \tag{70'}$$

$$(\mathscr{A}, \Lambda^q \circ \mathscr{B}) \circ \mathfrak{R}_r = I + T_r, \qquad T_r \in OP^{-1}. \tag{71'}$$

Rewriting (70′) we obtain

$$(\mathfrak{R}_l \circ (I, \Lambda^q)) \circ (\mathscr{A}, \mathscr{B}) = I + T_l,$$

whence $\mathfrak{R}_l \circ (I, \Lambda^q)$ is a left smoothing regularizer of $(\mathscr{A}, \mathscr{B})$ with smoothing operator $T_l \in OP^{-1}$. Also, the operator

$$\Lambda^q\colon \mathscr{W}^{l-\mathbf{m}-1/2}(\partial\Omega) \to \mathscr{W}^{l-\mathbf{m}-q-1/2}(\partial\Omega)$$

in (71′) is Fredholm since Λ^q is elliptic, so it has a smoothing regularizer Λ^{-q}. If we multiply (71′) on the right by (I, Λ^q) and on the left by (I, Λ^{-q}) it follows that

$$(\mathscr{A}, \mathscr{B}) \circ (\mathfrak{R}_r \circ (I, \Lambda^q)) = I + T'_r,$$

where $T'_r = (I, \Lambda^{-q}) \circ T_r \circ (I, \Lambda^q) + (0, (I - \Lambda^{-q}\Lambda^q)\mathscr{B}) \circ \mathfrak{R}_r$. Hence $\mathfrak{R}_r \circ (I, \Lambda^q)$ is a right smoothing regularizer with smoothing operator $T'_r \in OP^{-1}$. Thus we have shown that $(\mathscr{A}, \mathscr{B})$ has both a right and a left smoothing regularizer, i.e. it is smoothable.

Remark In the proof of the implication (a) → (b), we see that the regularizers constructed there have definitions which are independent of l, i.e. they have the same values on common domains of definition. Indeed, the Sobolev spaces W^m_2 are function spaces and all operators in question either act directly on the argument or on the function value, or, as in the case of the extension operator β, can be taken independent of m for m varying in a bounded interval.

In conclusion, we mention that the proof of Theorem 9.32 applies without change to pseudo-differential operators, $\mathscr{A}(x, D)$, of the form

$$\mathscr{A}(x, D) = \mathscr{A}^i(x, D) + \mathscr{A}^b(x, D),$$

where $\mathscr{A}^i$ is a p.d.o. with compact support inside Ω, and $\mathscr{A}^b(x, D)$ is a polynomial in D with support in a small neighbourhood of the boundary $\partial\Omega$. (For the definition of support see Definition 8.4.) In §§10.7 and 14.6 we permit $\mathscr{A}^b$ to be a more general operator which is polynomial in the normal derivative D_n but not necessarily in the tangential derivatives.

Exercises

1. Let $T \in \Phi(X, X)$ be a Fredholm operator with ind $T = 0$. Show that there is a representation $T = S + K$ where S is an isomorphism and K is a compact operator.

2. Let $T \in \Phi(X, X)$ be a Fredholm operator. Show that there exists some $\rho > 0$ such that the d-numbers $\alpha(T + \lambda I)$ and $\beta(T + \lambda I)$ are independent of λ for $0 < |\lambda| < \rho$.

3. Prove that a compact operator $K \in \mathscr{K}(X, Y)$ is a Fredholm operator if and only if X and Y are finite dimensional spaces.

4. Consider the Laplace operator in Ω

$$\Delta = \sum_{j=1}^{n} \frac{\partial^2}{\partial x_j^2} \qquad \text{ord } \Delta = 2,\ r = 1$$

with the boundary operator on $\partial\Omega$

$$\mathscr{B}(y, D) = b_0(y) + \sum_{j=1}^{n} b_j(y)\frac{\partial}{\partial x_j} \qquad \text{ord } \mathscr{B} \leqslant 1.$$

Express the L-condition 9.28 in terms of the coefficients of $\mathscr{B}(y, D)$. Hint: Find the characteristic polynomial of Δ, then find $\mathfrak{M}^+$. The L-condition amounts to establishing uniqueness in $\mathfrak{M}^+$ for the operator

$$\pi\mathscr{B}\left(y, \xi', \frac{1}{i}\frac{d}{dt}\right)\bigg|_{t=0}$$

5. Let Δ^2 be the biharmonic operator. Since ord $\Delta^2 = 4$, then $r = 2$ and two boundary conditions are required. Consider the following boundary operators

$$\mathscr{B}_1 = 1,\ \mathscr{B}_2 = \frac{\partial}{\partial n},\ \mathscr{B}_3 = \Delta,\ \mathscr{B}_4 = \frac{\partial\Delta}{\partial n}.$$

Which combinations of two of them satisfy the L-condition?

6. In $\Omega \subset \mathbb{R}^2$, let

$$\frac{\partial^2 u_1}{\partial x^2} - \frac{\partial^2 u_1}{\partial y^2} - 2\frac{\partial^2 u_2}{\partial x\,\partial y} = 0, \qquad \frac{2\partial^2 u_1}{\partial x\,\partial y} + \frac{\partial^2 u_2}{\partial x^2} - \frac{\partial^2 u_2}{\partial y^2} = 0 \qquad (\dagger)$$

be the Bitsadze system, see [Bi 1], with the Dirichlet boundary condition

$$u_1|_{\partial\Omega} = 0, \qquad u_2|_{\partial\Omega} = 0$$

Does the L-condition hold?

7. A second example from Bisadze [Bi 1] is

$$\frac{\partial^2 u_1}{\partial x^2} - \frac{\partial^2 u_1}{\partial y^2} + \sqrt{2}\frac{\partial^2 u_2}{\partial x\,\partial y} = 0, \qquad \sqrt{2}\frac{\partial^2 u_1}{\partial x\,\partial y} + \frac{\partial^2 u_2}{\partial x^2} - \frac{\partial^2 u_2}{\partial y^2} = 0$$

again with the Dirichlet boundary condition

$$u_1|_{\partial\Omega} = 0, \qquad u_2|_{\partial\Omega} = 0.$$

Discuss the L-condition for this problem.

8. Consider in $\Omega \subset \mathbb{R}^2$ the Bitsadze system ($\dagger$) in Exercise 6 with the Poincaré boundary operator

$$u_1|_{\partial\Omega}, \qquad \left(\frac{\partial u_1}{\partial x} - \frac{\partial u_2}{\partial y}\right)\bigg|_{\partial\Omega}.$$

Where does it satisfy the L-condition?

For Exercises 9 to 11, see the examples in §9.2 and also [Vo 1].

9. Consider the system $\Delta u_1 = 0$, $(\partial^m u_1/\partial x_1^m) + \Delta u_2 = 0$, with the Dirichlet *boundary condition* $u_1|_{\partial\Omega}$, $u_2|_{\partial\Omega}$. Is it L-elliptic?

10. Recall that the system

$$\frac{\partial^p u_1}{\partial x_1{}^p} - \frac{\partial^q u_2}{\partial x_2^q} = 0, \qquad \frac{\partial^r u_1}{\partial x_2^r} + \frac{\partial^s u_2}{\partial x_1^s} = 0$$

is DN elliptic if

$$p + s = r + q = \text{even},$$

with the characteristic polynomial $\chi(\xi) = \xi_1^{p+s} + \xi_2^{r+q}$. Consequently, $(r+q)/2$ boundary conditions are needed. Find examples of boundary conditions that satisfy the L-condition.

11. The linearized system of Navier-Stokes is

$$\Delta u_i - \frac{\partial u_4}{\partial x_i} = 0, \qquad i = 1, 2, 3, \qquad \frac{\partial u_1}{\partial x_1} + \frac{\partial u_2}{\partial x_2} + \frac{\partial u_3}{\partial x_3} = 0.$$

Since the characteristic polynomial is of degree 6, we may consider 3 boundary conditions, for instance

$$u_1|_{\partial\Omega}, \qquad u_2|_{\partial\Omega}, \qquad u_3|_{\partial\Omega}.$$

What about L-ellipticity? Give other examples of boundary conditions and study their L-ellipticity.

12. Show that the Bitsadze system [Bi] on the disc $\Omega = \{(x, y); (x - x_0)^2 + (y - y_0)^2 < \varepsilon^2\}$,

$$\frac{\partial^2 u_1}{\partial x^2} - \frac{\partial^2 u_1}{\partial y^2} - 2\frac{\partial^2 u_2}{\partial x\, \partial y} = 0, \qquad 2\frac{\partial^2 u_1}{\partial x\, \partial y} + \frac{\partial^2 u_2}{\partial x^2} - \frac{\partial^2 u_2}{\partial y^2} = 0,$$

with the Dirichlet boundary value problem

$$u_1|_{\partial\Omega} = 0, \qquad u_2|_{\partial\Omega} = 0$$

on the circle $\partial\Omega = \{(x, y); (x - x_0)^2 + (y - y_0)^2 = \varepsilon^2\}$, admits an infinite set of linearly independent solutions.
(Hint: Introduce the complex notation $w(z) = u_1(x, y) + iu_2(x, y)$, $z = x + iy$, and write the system in the form $\partial^2 w/\partial \bar{z}^2 = 0$. Evidently all solutions of this equation have the form $w(z) = \bar{z}\varphi(z) + \psi(z)$, where $\varphi(z)$ and $\psi(z)$ are arbitrary holomorphic functions on Ω.)

13. Let $\mathscr{A}$ be a scalar ($p = 1$) differential operator of order $m = 2r$. As boundary values we consider the Dirichlet problem $\mathscr{D}$,

$$u|_{\partial\Omega}, \left.\frac{\partial u}{\partial n}\right|_{\partial\Omega}, \ldots, \left.\frac{\partial^{r-1}}{\partial n^{r-1}}\right|_{\partial\Omega}$$

where $\partial^j/\partial n^j$ are derivatives taken say in the inner normal direction to $\partial\Omega$. Show that if $\mathscr{A}$ is properly elliptic then $(\mathscr{A}, \mathscr{D})$ is always L-elliptic. Show also that the index is $\operatorname{ind}(\mathscr{A}, \mathscr{D}) = 0$.
(Hint: let $\mathscr{A}^*$ be the adjoint, the let $\bar{\mathscr{A}}$ be the operator which one obtains from $\mathscr{A}$ by replacing all the coefficients $a_s(x)$ by their complex conjugates $\overline{a_s(x)}$. Since $\mathscr{A}^* - \bar{\mathscr{A}}$ is an operator of order $m - 1$, then $\operatorname{ind}(\mathscr{A}^*, \mathscr{D}) = \operatorname{ind}(\bar{\mathscr{A}}, \mathscr{D})$.)

14. Consider the 2×2 operator in the plane

$$\mathscr{A}(x, D) = \begin{bmatrix} D_1^2 + D_2^2 & D_1^4 - D_2^4 \\ 1 & D_1^2 - D_2^2 + 1 \end{bmatrix}.$$

Show that $\det \mathscr{A}(x, \xi) = \xi_1^2 + \xi_2^2$. Since $m = 4$, then $\chi = \pi_m \det \mathscr{A} \equiv 0$. Conclude that $\mathscr{A}$ is *not* elliptic (Definition 9.21).

Part IV

Reduction of a Boundary Value Problem to an Elliptic System on the Boundary

10
Understanding the L-condition

Given a $p \times p$ elliptic operator $\mathscr{A}$ on the domain Ω, there is an associated family of matrix polynomials

$$L(\lambda) = \pi\mathscr{A}(y, (\xi', \lambda))$$

that depends on the parameters $(y, \xi') \in T^*(\partial\Omega)\backslash 0$. Note that the space $\mathfrak{M}^+$ used in the statement of the L-condition (Definition 9.28) is simply the space of solutions of

$$L\left(\frac{1}{i}\frac{d}{dt}\right)u = 0$$

corresponding to the eigenvalues of $L(\lambda)$ in the upper half-plane $\operatorname{Im}\lambda > 0$. In §10.1 we use the spectral theory of matrix polynomials to reformulate the L-condition in various equivalent forms, e.g. the Lopatinskii condition, the complementing condition of Agmon, Douglis, Nirenberg, the Δ-condition and other conditions (see Theorem 10.3). The L-condition for the Dirichlet problem is also discussed in detail in §10.2. In §§10.3 and 10.4 we prove some technical results concerning families of matrix polynomials. These results will be used in several places throughout the book, for instance, in §10.5 in order to prove a result due to Agranovič and Dynin on comparing the index of two elliptic boundary problems, $(\mathscr{A}, \mathscr{B}_1)$ and $(\mathscr{A}, \mathscr{B}_2)$, having the same elliptic operator $\mathscr{A}$ (see Theorem 10.19).

For a fixed p (the size of the system), we let

$$\mathrm{Ell}^{\mathbf{s},\mathbf{t}}$$

denote the set of elliptic operators, $\mathscr{A}$, with DN numbers $s_1, \ldots, s_p, t_1, \ldots, t_p$ and

$$\mathrm{BE}^{\mathbf{s},\mathbf{t},\mathbf{m}}$$

the set of L-elliptic boundary problems, $(\mathscr{A},\mathscr{B})$, with DN numbers $s_1, \ldots, s_p$, $t_1, \ldots, t_p, m_1, \ldots, m_r$. Then we have a natural map defined by

$$\mathrm{BE}^{\mathbf{s},\mathbf{t},\mathbf{m}} \ni (\mathscr{A}, \mathscr{B}) \mapsto \mathscr{A} \in \mathrm{Ell}^{\mathbf{s},\mathbf{t}}.$$

If we have a homotopy $\mathscr{A}_\tau$, $0 \leqslant \tau \leqslant 1$, in the class $\mathrm{Ell}^{\mathbf{s},\mathbf{t}}$ and a boundary operator $\mathscr{B}_0$ for $\mathscr{A}_0$ such that $(\mathscr{A}_0, \mathscr{B}_0) \in \mathrm{BE}^{\mathbf{s},\mathbf{t},\mathbf{m}}$ it is quite simple using the spectral theory of matrix polynomials to lift the homotopy $\mathscr{A}_\tau$ of elliptic operators to a homotopy of L-elliptic boundary problems $(\mathscr{A}_\tau, \mathscr{B}_\tau)$ as we show in §10.6. In general, however, it is difficult to construct the homotopy $\mathscr{A}_\tau$ in the first place unless we enlarge the class of elliptic operators under consideration. This is the purpose of §10.7, where for simplicity we consider only the case when the principal part of the elliptic operator is homogeneous of degree l (Definition 9.25). To extend the class Ell^l, we define the set

$$\mathfrak{E}^l$$

of elliptic operators of the form $\mathscr{A} = \mathscr{A}^b + \mathscr{A}^i$. Here $\mathscr{A}^b$ is an operator with support in a small neighbourhood of $\partial\Omega$ and a differential operator in the normal variables but pseudo-differential in the tangential variables, while $\mathscr{A}^i$ is a p.d.o. in all variables but with support in a compact subset of Ω. Note that $L(\lambda)$ is still a matrix polynomial in λ, so we can formulate the L-condition for $\mathscr{A} \in \mathfrak{E}^l$ as before. We denote by

$$\mathfrak{B}\mathfrak{E}^{l,\mathbf{m}}$$

the set of boundary value problems $(\mathscr{A}, \mathscr{B})$ where $\mathscr{A} \in \mathfrak{E}^l$ and $\mathscr{B}$ is a boundary operator with order m_k in the kth row $(k = 1, \ldots, r)$ and satisfying the L-condition relative to $\mathscr{A}$.

In §10.8 we prove another result due to Agranovič and Dynin on comparing the index of two elliptic boundary problems with the same boundary operator. Finally, in §10.9 we consider the composition of two elliptic boundary problems.

10.1 Alternative versions of the L-condition

Let $\mathscr{A}(x,D)$, $x \in \Omega$, be a $p \times p$ elliptic differential operator with the Douglis–Nirenberg numbers $s_1, \ldots, s_p, t_1, \ldots, t_p$. Associated with the DN principal part of $\mathscr{A}$ on the boundary $\partial\Omega$, there is the $p \times p$ matrix polynomial

$$L_{y,\xi'}(\lambda) := \pi\mathscr{A}(y, (\xi', \lambda)) = \sum_{j=0}^{l} A_j(y, \xi')\lambda^j,$$

$y \in \partial\Omega$, $0 \neq \xi' \in T_y^*(\partial\Omega)$, where $l = \max(s_i + t_j)$. We suppose that $\mathscr{A}$ is properly elliptic, so that $\det L_{y,\xi'}(\lambda)$ has r roots in the upper half-plane $\operatorname{Im}\lambda > 0$ and r in the lower half-plane.

Remark As in §9.2 the notation (ξ', λ) is the short notation for $\xi' + \lambda \cdot n(\lambda)$, where $n(y)$ is the inward-pointing unit conormal to $\partial\Omega$ at y.

Let $\mathscr{B}$ be a boundary operator of the type considered in §9.3 with DN numbers $m_1, \ldots, m_r, t_1, \ldots, t_p$. We can write it in the form

$$\mathscr{B}(y, D) = \sum_{\kappa=0}^{\mu} \mathscr{B}_\kappa(y, D')D_n^\kappa \tag{1}$$

where the coefficient-operators $\mathscr{B}_\kappa = [b_{kj}^\kappa]$ are $r \times p$ matrices of (-1) classical p.d.o.'s on $\partial\Omega$. Note that the order of b_{kj}^κ is $m_k + t_j - \kappa$. The integer $\mu \geqslant 0$ is the *transversal order* of the boundary operator and is the maximum of the orders of the normal derivatives that occur in the entries of $\mathscr{B}$. Associated with the DN principal part, $\pi\mathscr{B}$, there is the $r \times p$ matrix polynomial

$$B_{y,\xi'}(\lambda) := \pi\mathscr{B}(y, (\xi', \lambda)) = \sum_{j=0}^{\mu} B_j(y, \xi')\lambda^j,$$

$y \in \partial\Omega$, $0 \neq \xi' \in T_y^*(\partial\Omega)$, where $B_j = \pi\mathscr{B}_j$ is the principal symbol of the jth coefficient-operator $\mathscr{B}_j$.

Remark To simplify the notation, we often suppress the variables y, ξ' and write just $L(\lambda)$ rather than $L_{y,\xi'}(\lambda)$. The same is done for $B(\lambda)$.

The goal of this section is to reformulate the L-condition in several equivalent versions. We start with an abstract, quite general formulation, and use the matrix notation of Chapter 2.

Let $L(\lambda) = \sum_{j=0}^{l} A_j\lambda^j$ be a $p \times p$ matrix polynomial. Suppose that $\det L(\lambda) \neq 0$ for $\lambda \in \mathbb{R}$ and let γ^+ be a simple, closed contour in the upper half-plane $\operatorname{Im}\lambda > 0$ that contains all the eigenvalues of $L(\lambda)$ there. By Chapter 2 there exists a γ^+-spectral triple (X_+, T_+, Y_+) for $L(\lambda)$, and the base space of the γ^+-spectral pair (X_+, T_+) has dimension r, where r is the number of roots of $\det L(\lambda) = 0$ inside γ^+. We let $\mathfrak{M}^+ = \mathfrak{M}^+_{L(1/i\, d/dt)}$ denote the subspace of $C^\infty(\mathbb{R}, \mathbb{C}^p)$ consisting of the solutions of

$$L\left(\frac{1}{i}\frac{d}{dt}\right)u = 0$$

such that $u(t) \to 0$ as $t \to +\infty$. Recall that $\dim \mathfrak{M}^+ = r$ and by Corollary 2.9 every $u \in \mathfrak{M}^+$ admits a representation of the form

$$u(t) = X_+ e^{itT_+}c \tag{2}$$

for a unique $c \in \mathbb{C}^r$.

Remark By Theorem 2.21, the matrix polynomial $L(i^{-1}\lambda)$ has the spectral pair (X_+, iT_+) with respect to the eigenvalues in the half-plane $\operatorname{Re}\lambda < 0$.

The (right) Calderón projector is

$$P_+ = P_{\gamma^+} = \frac{1}{2\pi i}\int_{\gamma^+} \begin{pmatrix} I \\ \vdots \\ \lambda^{l-1}I \end{pmatrix} L^{-1}(\lambda)[L_1(\lambda) \ldots L_l(\lambda)]\, d\lambda$$

where $L_j(\lambda) = A_j + A_{j+1}\lambda + \cdots + A_\ell\lambda^{l-j}, j = 0, \ldots, l$, and we know that the image of P_{γ^+} is equal to the set of initial conditions

$$\operatorname{col}\left(\left(\frac{1}{i}\frac{d}{dt}\right)^j u(0)\right)_{j=0}^{l-1}, \qquad u \in \mathfrak{M}^+.$$

Theorem 10.1 *Let $B(\lambda) = \sum_{j=0}^{\mu} B_j\lambda^j$ be an $r \times p$ matrix polynomial of degree μ. The following statements are equivalent:*

(*a*) *For any $y \in \mathbb{C}^r$, there is a unique $u \in \mathfrak{M}^+$ such that*

$$B\left(\frac{1}{i}\frac{d}{dt}\right)u\bigg|_{t=0} = y.$$

(*b*) *If (X_+, T_+) is a γ-spectral pair for $L(\lambda)$, where X_+ is a $p \times r$ matrix and T_+ is an $r \times r$ matrix, then the $r \times r$ matrix $\Delta_B^+ = \sum_{j=0}^{\mu} B_j X_+ T_+^j$ is invertible, i.e.*

$$\det \Delta_B^+ = \det\left(\sum_{j=0}^{\mu} B_j X_+ T_+^j\right) \neq 0. \tag{3}$$

Proof As we remarked above, any $u \in \mathfrak{M}^+$ can be represented in the form $u(t) = X_+ e^{itT_+}c$ for a unique $c \in \mathbb{C}^r$. The equivalence of (*i*) and (*ii*) is clear since

$$B\left(\frac{1}{i}\frac{d}{dt}\right)u\bigg|_{t=0} = \left(\sum_{j=0}^{\mu} B_j X_+ T_+^j\right)\cdot c$$

Corollary *Further, if $\mu \leqslant l-1$ and we define the $r \times pl$ matrix*

$$\tilde{B} = [B_0 \cdots B_\mu \quad 0 \cdots 0]$$

then (a) and (b) are also equivalent to: (b′) The rank of $\tilde{B}\cdot P_{\gamma+}$ is equal to r.

Proof Let (X_+, T_+, Y_+) be a matrix γ-spectral triple for $L(\lambda)$. Then X_+ is a $p \times r$ matrix, T_+ is $r \times r$ and Y_+ is $r \times p$. Consider the $r \times pl$ matrix $\tilde{B}\cdot P_{\gamma+}$ as an operator $\mathbb{C}^{pl} \to \mathbb{C}^r$ and the $r \times r$ matrix $\sum_{j=0}^{\mu} B_j X_+ T_+^j$ as an operator $\mathbb{C}^r \to \mathbb{C}^r$. Since

$$P_{\gamma+} = \begin{pmatrix} X_+ \\ \vdots \\ X_+ T_+^{l-1} \end{pmatrix} \cdot [Y_+ \cdots T_+^{l-1} Y_+] \cdot \mathscr{L} \tag{4}$$

where $\mathscr{L}$ is defined by §2.2(5), it is clear that the image of $\tilde{B}P_{\gamma+}$ is contained in that of $\sum_{j=0}^{\mu} B_j X_+ T_+^j$. Conversely, if $y = \sum_{j=0}^{\mu} B_j X_+ T_+^j c$, $c \in \mathbb{C}^r$, then $y = \tilde{B}\,\mathrm{col}(X_+ T_+^j)_{j=0}^{l-1} c$. Since the image of $P_{\gamma+}$ is equal to that of $\mathrm{col}(X_+ T_+^j)_{j=0}^{l-1}$, we have $\mathrm{col}(X_+ T_+^j)_{j=0}^{l-1} c = P_{\gamma+} c'$ for some $c' \in \mathbb{C}^{pl}$, whence $y = \tilde{B}P_{\gamma+}c'$. This means that $\tilde{B}P_{\gamma+}$ and $\sum_{j=0}^{\mu} B_j X_+ T_+^j$ have the same image, and, therefore, the same rank. This proves the equivalence of (b) and (b′).

In the following theorem, $\tilde{L}(\lambda)$ denotes the cofactor matrix of $L(\lambda)$, i.e. the matrix polynomial such that $L(\lambda)\tilde{L}(\lambda) = \tilde{L}(\lambda)L(\lambda) = \det L(\lambda)\cdot I$. Also we let $P'_{\gamma+}$ be the left Calderón projector (Theorem 2.15)

$$P'_+ = P'_{\gamma+} = \frac{1}{2\pi i}\int_{\gamma+} \begin{pmatrix} L_1(\lambda) \\ \vdots \\ L_l(\lambda) \end{pmatrix} L^{-1}(\lambda)[I \cdots \lambda^{l-1} I]\, d\lambda$$

Theorem 10.2 *The conditions* (*a*) *and* (*b*) *of Theorem* 10.1 *are also equivalent to*:

(*c*) *The* $r \times pl$ *matrix (called the Lopatinskii matrix)*

$$G = \frac{1}{2\pi i}\int_{\gamma^+} B(\lambda)L^{-1}(\lambda)[I \cdots \lambda^{l-1}I]\, d\lambda$$

has rank r.

(*d*) *There exists a* $pl \times r$ *matrix* S *such that* $GS = I_r$ *and* $SG = P'_{\gamma^+}$, *where* P'_{γ^+} *is the left Calderón projector.*

(*e*) *The rows of the* $r \times p$ *matrix polynomial* $B(\lambda)\cdot\tilde{L}(\lambda)$ *are linearly independent modulo* $\rho^+(\lambda)$, *where we have factored the scalar polynomial* $\det L(\lambda)$ *as* $\rho^-(\lambda)\cdot\rho^+(\lambda)$, ρ^+ *containing all the roots of* $\det L(\lambda) = 0$ *inside* γ^+ *and* ρ^- *all the roots outside* γ^+.

Proof As a preliminary remark, note that by linear algebra the $r \times pl$ matrix G has rank r if and only if the corresponding linear map $G: \mathbb{C}^{pl} \to \mathbb{C}^r$ is surjective.

(b) $\Leftrightarrow$ (c) Let (X_+, T_+, Y_+) be a matrix γ^+-spectral triple for $L(\lambda)$. By the formula (ii′) in Proposition 2.2, the Lopatinskii matrix is equal to

$$G = \left(\sum_{j=0}^{\mu} B_j X_+ T_+^j\right)\cdot[Y_+ \cdots T_+^{l-1}Y_+] \tag{5}$$

and we know that $[Y_+ \cdots T_+^{l-1}Y_+]$ has rank r (see Definition 2.1(iv)). Hence G has rank r if and only if the $r \times r$ matrix $\Delta_B^+ = \sum_{j=0}^{\mu} B_j X_+ T_+^j$ is invertible.

(b) $\Leftrightarrow$ (d) The left Calderón projector is equal to

$$P'_{\gamma^+} = \mathscr{X}\cdot\mathrm{col}(X_+T_+^j)_{j=0}^{l-1}\cdot[Y_+ \cdots Y_+^{l-1}Y_+] \tag{6}$$

If (b) holds we let

$$S = \mathscr{X}\cdot\mathrm{col}(X_+T_+^j)_{j=0}^{l-1}\cdot\left(\sum_{j=0}^{\mu} B_j X_+ T_+^j\right)^{-1}, \tag{7}$$

then $SG = P'_{\gamma^+}$, and by Corollary 2.11 we have $GS = I_r$. Conversely, if (d) holds then the equation $GS = I_r$ implies that G has rank r. In view of (5) this implies that the matrix $\Delta_B^+ = \sum_{j=0}^{\mu} B_j X_+ T_+^j$ has rank r, hence it is invertible since it is a square $r \times r$ matrix.

(c) $\Rightarrow$ (e) Note that $\tilde{L}(\lambda) = \det L(\lambda)\cdot L^{-1}(\lambda)$; as remarked above, $\tilde{L}(\lambda)$ is the cofactor matrix for $L(\lambda)$ so it is a polynomial. If $x \in \mathbb{C}^r$ is such that

$$xB(\lambda)\tilde{L}(\lambda) = \rho^+(\lambda)\cdot M(\lambda) \tag{8}$$

for some $1 \times p$ matrix polynomial $M(\lambda)$, then dividing (8) by $\det L(\lambda)$ gives $xB(\lambda)L^{-1}(\lambda) = \rho^-(\lambda)^{-1}\cdot M(\lambda)$, and the right-hand side is holomorphic inside

γ whence

$$x \int_{\gamma^+} B(\lambda) L^{-1}(\lambda)[I \cdots \lambda^{l-1} I]\, d\lambda = 0. \tag{9}$$

Thus (c) implies that $x = 0$.

(e) $\Rightarrow$ (c) Conversely, let $x \in \mathbb{C}^r$ such that (9) holds. In view of (5) we see that $x \cdot \sum_{j=0}^{\mu} B_j X_+ T_+^j = 0$. Multiplying both sides of this equation by $T_+^k Y_+$ for $k = 0, 1, \ldots$, and making use of the formula (ii′) in Proposition 2.2 we obtain

$$x \int_{\gamma^+} B(\lambda) L^{-1}(\lambda) \lambda^k \, d\lambda = 0$$

for all $k = 0, 1, \ldots$. Hence $xB(\lambda)L^{-1}(\lambda)$ has an analytic continuation inside γ as a matrix function of λ. Since $\tilde{L}(\lambda) = \det L(\lambda) \cdot L^{-1}(\lambda)$, this means that $xB(\lambda)\cdot \tilde{L}(\lambda)$ vanishes at the roots of $\det L(\lambda) = 0$ inside γ^+, i.e. vanishes at the roots of $\rho^+(\lambda) = 0$. Hence (8) holds for some $1 \times p$ matrix polynomial $M(\lambda)$, and then (e) implies that $x = 0$. Thus G is surjective, i.e. has rank r.

Remark The matrix S is uniquely determined by the conditions in (d). The first equation implies that G is surjective, and if S and S_1 satisfy $SG = P'_{\gamma^+}$ and $S_1 G = P'_{\gamma^+}$ then $(S - S_1)G = 0$ whence $S - S_1 = 0$.

Remark From the proof of (c) $\Leftrightarrow$ (e) it is clear that condition (e) can be formulated in another way: *The only* $x \in \mathbb{C}^r$ such that $xB(\lambda)\cdot L^{-1}(\lambda)$ *has an analytic continuation inside* γ^+ *is* $x = 0$.

There is a natural correspondence between the conditions (a) and (c), i.e. depending only on $L(\lambda)$, not on the choice of spectral triple. First of all, note that (5), (6) and Corollary 2.11 imply that $G \cdot P'_{\gamma^+} = G$. Also, for the map $\phi\colon \mathbb{C}^{pl} \to \mathfrak{M}^+$ defined by

$$c \mapsto \phi c(t) = \frac{1}{2\pi i} \int_{\gamma^+} e^{it\lambda} L^{-1}(\lambda) \cdot \sum_{j=0}^{l-1} c_{j+1} \lambda^j \, d\lambda$$

we have

$$B\left(\frac{1}{i}\frac{d}{dt}\right)\bigg|_{t=0} \circ \phi = G,$$

and, in view of Proposition 2.5, ϕ is surjective. Further, since $\phi c = X_+ e^{itT_+}[Y_+ \cdots T_+^{l-1} Y_+]c$, it follows that $\ker \phi = \ker[Y_+ \cdots T_+^{l-1} Y_+]$, which is equal to $\ker \mathrm{P}'_{\gamma^+}$ by Corollary 2.11. Thus we have the commutative diagram

$$\begin{array}{ccccc} \mathbb{C}^{pl} & \xrightarrow{P'_{\gamma^+}} & \mathbb{C}^{pl} & \xrightarrow{\phi} & \mathfrak{M}^+ \\ & \searrow^{G} & \downarrow^{G} & \swarrow_{B\left(\frac{1}{i}\frac{d}{dt}\right)\big|_{t=0}} & \\ & & \mathbb{C}^r & & \end{array}$$

and that fact that P'_{γ^+} and ϕ have the same kernel means that we can identify $\operatorname{im} P'_{\gamma^+}$ and $\mathfrak{M}^+$ along the top row of the diagram. With this identification, the map $G|_{\mathrm{im}\,P'_{\gamma^+}}: \mathrm{im}\, P'_{\gamma^+} \to \mathbb{C}^r$ is identified with the map

$$B\left(\frac{1}{i}\frac{d}{dt}\right)\bigg|_{t=0} : \mathfrak{M}^+ \to \mathbb{C}^r.$$

Thus, condition (a) holds if and only if

$$G|_{\mathrm{im}\,P'_{\gamma^+}}: \mathrm{im} P'_{\gamma^+} \to \mathbb{C}^r \text{ is bijective.}$$

Now we apply the matrix theory to the L-condition for elliptic systems $(\mathscr{A}, \mathscr{B})$. For the sake of emphasis we gather all the equivalent conditions together in one long theorem.

From now on we usually omit mention of the contour γ^+, i.e. we write $\int_+$ instead of $\int_{\gamma^+}$ because the value of the contour integral depends only on the residues in the upper half-plane, $\mathrm{Im}\,\lambda > 0$, not on the particular contour.

In the following theorem, no assumptions are made concerning the transversal order, μ, of the boundary operator ($\mu \geqslant l$ is permitted).

Theorem 10.3 *We suppose that the operator $\mathscr{A}(x, D)$ is properly elliptic, let $\mathscr{B}(y, D)$ be a boundary operator as in* (1), *and fix $y \in \partial\Omega$, $0 \neq \xi' \in T_y^*(\partial\Omega)$. The following statements are equivalent:*

(*i*) *The initial value problem*

$$\pi\mathscr{A}\left(y, \left(\xi', \frac{1}{i}\frac{d}{dt}\right)\right)u(t) = 0, \qquad t \geqslant 0 \tag{10}$$

$$\pi\mathscr{B}\left(y, \left(\xi', \frac{1}{i}\frac{d}{dt}\right)\right)u(t)|_{t=o} = g,$$

has for every choice of $g \in \mathbb{C}^r$ a unique solution $u \in \mathfrak{M}^+_{L(1/i\, d/dt)}$. As usual, $\mathfrak{M}^+_{L(1/i\, d/dt)}$ is the space of solutions of (10) *such that* $\lim_{t\to\infty} u(t) = 0$ (*or, in other words, corresponding to the eigenvalues of $L(\lambda)$ with a positive imaginary part*).

(*ii*) *The $r \times pl$ matrix $G = G_{y,\xi'}$ defined by*

$$G = \int_+ \pi\mathscr{B}(y, (\xi', \lambda))\pi\mathscr{A}(y, (\xi', \lambda))^{-1}[I \cdots \lambda^{l-1} I]\, d\lambda \tag{11}$$

has rank r, where $\int_+$ denotes the integral along a simple, closed contour γ^+ in the upper half-plane containing all roots of $\det L_{y,\xi'}(\lambda) = 0$ *with a positive imaginary part.*

(*iii*) *If $(X_+(y, \xi'), T_+(y, \xi'), Y_+(y, \xi'))$ is a γ^+-spectral triple for $L_{y,\xi'}(\lambda)$, where $X_+(y, \xi')$ is a $p \times r$ matrix, $T_+(y, \xi')$ a $r \times r$ matrix and $Y_+(y, \xi')$ a $r \times p$ matrix (existence by Theorem 2.6), then*

$$\det \Delta^+_{\mathscr{B}}(y, \xi') = \det\left(\sum_{j=0}^{\mu} B_j(y, \xi')X_+(y, \xi')T^j_+(y, \xi')\right) \neq 0. \tag{12}$$

(*iv*) *There is a unique $pl \times r$ matrix $S = S_{y,\xi'}$ such that $GS = I_r$ and $SG = P'_+$, where $P'_+ = P'_{\gamma^+}$ is the left Calderón projector for $L_{y,\xi'}(\lambda)$.*

(v) *Let us factor the scalar polynomial,* $\det L_{y,\xi'}(\lambda)$, *as* $\rho^-(\lambda)\cdot\rho^+(\lambda)$, *where* ρ^+ *contains all the roots above the real axis, and* ρ^- *contains all roots below the real axis. If* $\tilde{L}(\lambda)$ *denotes the matrix polynomial*

$$\tilde{L}(\lambda) = \det L_{y,\xi'}(\lambda)\cdot L_{y,\xi'}^{-1}(\lambda),$$

then the rows of $B_{y,\xi'}(\lambda)\cdot\tilde{L}(\lambda)$ *are linearly independent modulo* $\rho^+(\lambda)$.

The first condition is of course the L-condition stated in §9.3. Condition (ii) is known as the *Lopatinskii condition*; the matrix (11) is the Lopatinskii matrix, and was introduced by Lopatinskii in his paper [Lo]. Fedosov used condition (iv) in a series of papers where he developed an index formula for elliptic boundary value problems [Fe]. The last condition (v) is called the "covering condition" or "complementing condition", and was introduced and used by Agmon, Douglis and Nirenberg in their fundamental paper [ADN]. We call (iii) the Δ-condition.

Example The covering condition (v) is quite difficult to apply in general. However, for the case of a scalar operator $\mathscr{A}(x, D)$ (i.e. $p = 1$) it is perhaps the simplest of the five conditions. Supposing that the operator $\mathscr{A}$ is properly elliptic, let $l = 2r$ and let $\lambda_1, \dots, \lambda_r$ be the roots of $L(\lambda)$ above the real axis, and factor the polynomial $L(\lambda)$ as

$$L(\lambda) = \rho^-(\lambda)\cdot\rho^+(\lambda),$$

where $\rho^+(\lambda) = \prod_{j=1}^{r} (\lambda - \lambda_j)$. Since $p = 1$ we have to consider only the $r \times 1$ matrix

$$B(\lambda) = \pi\mathscr{B}(y, (\xi', \lambda)), \qquad y \in \partial\Omega,\ 0 \neq \xi' \in T_y^*(\partial\Omega)$$

for linear independence modulo $\rho^+(\lambda)$. By the Euclidean algorithm, we write the rows of $B(\lambda)$ as

$$b_j(\lambda) = Q_j(\lambda)\rho^+(\lambda) + \sum_{k=0}^{r-1} b_{jk}(y, \xi')\lambda^k, \qquad j = 1, \dots, r$$

(b_j are just principal parts of the individual boundary conditions) and have:

The rows $b_j(\lambda)$ *are linearly independent modulo* $\rho^+(\lambda)$ *if and only if the* r *remainder polynomials*

$$\sum_{k=0}^{r-1} b_{jk}(y, \xi')\lambda^k, \qquad j = 1, \dots, r, \tag{13}$$

are linearly independent.

This is equivalent to

$$\det[b_{jk}(y, \xi')]_{\substack{j=1,\dots,r\\ k=0,\dots,r-1}} \neq 0, \tag{14}$$

which in turn is equivalent to the Lopatinskii conditions by Theorem 10.3. When all boundary conditions are of order $\leqslant r - 1$ in the normal derivatives, the polynomials (13) constitute the principal parts of the boundary conditions (in this case we can set $Q_j \equiv 0$) and (14) become the most simple test for checking the L-condition. For instance, the boundary operator for the

Dirichlet problem is

$$\mathscr{B} = \begin{pmatrix} 1 \\ \partial_n \\ \vdots \\ \partial_n^{r-1} \end{pmatrix},$$

where ∂_n is the operator $\partial/\partial n$, and (14) is then the determinant of the diagonal matrix $[i^k \delta_{jk}]$. Hence, the Dirichlet problem for a properly elliptic scalar operator $\mathscr{A}(x, D)$, $p = 1$, always satisfies the L-condition.

Remark When verifying the L-condition for particular examples of $(\mathscr{A}, \mathscr{B})$ it is not essential to include the factor $1/i$. The transformation $t \to it$ carries over the equations

$$\begin{aligned} &L\left(\frac{d}{dt}\right)u(t) = 0 && && L\left(\frac{1}{i}\frac{d}{dt}\right)u(it) = 0 \\ & && \text{into} && && (15) \\ &B\left(\frac{d}{dt}\right)u(t)|_{t=0} = g && && B\left(\frac{1}{i}\frac{d}{dt}\right)u(it)|_{t=0} = g \end{aligned}$$

and it gives obviously an isomorphism of the solution spaces

$$\mathfrak{M}^+_{L(d/dt)} \simeq \mathfrak{M}^+_{L(1/i\, d/dt)}.$$

where $\mathfrak{M}^+_{L(d/dt)}$ denotes the solutions of $L(d/dt)u = 0$ corresponding to the eigenvalues in the half-plane $\operatorname{Re} \lambda < 0$. It is evident that if the first initial value problem in (15) has a unique solution $u(t)$ for every $g \in \mathbb{C}^r$, then the second is also uniquely solvable and vice versa.

10.2 The Dirichlet problem

If $\mathscr{A}$ is a $p \times p$ elliptic operator such that $r = ps$ for some s, we can pose the Dirichlet problem $(\mathscr{A}, \mathscr{D})$ where

$$\mathscr{D} = \begin{pmatrix} I \\ I\,\partial_n \\ \vdots \\ I\,\partial_n^{s-1} \end{pmatrix} \qquad I \text{ is } p \times p \text{ identity},$$

but in general it will not satisfy the L-condition. In fact Theorem 3.14 implies the following result.

Proposition 10.4 *Let $\mathscr{A}(x, D)$ be a properly elliptic operator and suppose that the number r is divisible by p, i.e., $r = ps$ for some s. The Dirichlet problem $(\mathscr{A}, \mathscr{D})$ satisfies the L-condition if and only if for all $y \in \partial\Omega$ and $0 \neq \xi' \in T^*_y(\partial\Omega)$, $L_{y,\xi'}(\lambda)$ has a γ^+-spectral monic right divisor of degree s.*

Proof The Lopatinskii matrix of the Dirichlet problem is

$$\int_+ \begin{pmatrix} I \\ \vdots \\ \lambda^{s-1}I \end{pmatrix} L_{y,\xi'}^{-1}(\lambda)[I \cdots \lambda^{l-1}I]\, d\lambda$$

which is also the deciding matrix in Theorem 3.14.

Recall that $L_{y,\xi'}(\lambda) = \pi\mathscr{A}(y, (\xi', \lambda))$. Suppose now that the principal part of $\mathscr{A}$ is homogeneous of degree l, that is, with DN numbers $s_i = l$, $t_j = 0$. In this case the leading coefficient $L_{y,\xi'}(\lambda)$ is $A_l = \pi\mathscr{A}(y, n(y))$, which is invertible. Then $\det L_{y,\xi'}(\lambda)$ has degree pl. In order to pose the Dirichlet problem we need $r = ps$ for some s, hence $l = 2s$.

In this case we can strengthen the result of Proposition 10.4. Suppose there is a γ^+-spectral right factorization

$$L_{y,\xi'}(\lambda) = L_{y,\xi'}^-(\lambda)L_{y,\xi'}^+(\lambda),$$

with $L_{y,\xi'}^+(\lambda)$ monic of degree s. Then $L_{y,\xi'}^-(\lambda)$ must also have degree s with invertible leading coefficient, since A_l is invertible and $l = 2s$. Due to the homogeneity of $\pi\mathscr{A}$ we have $\pi\mathscr{A}(y, (\xi', \lambda)) = (-1)^l\pi\mathscr{A}(y, (-\xi', -\lambda))$, hence

$$L_{y,\xi'}(\lambda) = (-1)^l L_{y,-\xi'}^-(-\lambda)L_{y,-\xi'}^+(-\lambda).$$

This means that whenever $L_{y,\xi'}(\lambda)$ has a monic γ^+-spectral right divisor of degree s for all $y \in \partial\Omega$, $0 \neq \xi' \in T_y^*(\partial\Omega)$, it also has a monic γ^+-spectral *left* divisor of degree s.

In view of Theorem 3.16 we can now improve on Proposition 10.4.

Theorem 10.5 *Let $\mathscr{A}(x, D)$ be a properly elliptic operator and suppose that the principal part of $\mathscr{A}$ is homogeneous of degree $l = 2s$. The Dirichlet problem $(\mathscr{A}, \mathscr{D})$ satisfies the L-condition if and only if*

$$\det \int_+ \begin{pmatrix} I \\ \vdots \\ \lambda^{s-1}I \end{pmatrix} \pi\mathscr{A}(y, (\xi', \lambda))^{-1}[I \cdots \lambda^{s-1}I]\, d\lambda \neq 0 \qquad (16)$$

for all $y \in \partial\Omega$, $0 \neq \xi' \in T_y^*(\partial\Omega)$.

One of the hypotheses of Theorems 3.14 and 3.16 is that there should be exactly ps zeros of $\det L_{y,\xi'}(\lambda) = 0$ inside γ^+. This is where it is necessary to assume that the operator $\mathscr{A}$ is properly elliptic. However, it is not hard to show that the invertibility of the matrix in (16) for all $y \in \partial\Omega$, $0 \neq \xi' \in T_y^*(\partial\Omega)$ in fact implies the properly ellipticity of $\mathscr{A}$.

In checking the Lopatinskii condition we are confronted by the difficulty to decide which r of the lp ($l = 2s$) columns of (11) are independent. The theorem above shows that for the Dirichlet problem it is only necessary to check for linear independence in the first r columns. The same is true for any boundary operator once the Dirichlet problem is known to satisfy the Lopatinskii condition.

Theorem 10.6 *Let the principal part of the operator $\mathscr{A}$ be homogeneous of degree $l = 2s$ and suppose that the Dirichlet problem $(\mathscr{A}, \mathscr{D})$ satisfies the L-condition, i.e. the $ps \times ps$ matrix (16) is nonsingular. Then the $ps \times p$ boundary operator $\mathscr{B}(y, D)$ satisfies the L-condition relative to $\mathscr{A}$ and only if*

$$\det \int_{+} \pi\mathscr{B}(y, (\xi', \lambda))\pi\mathscr{A}(y, (\xi', \lambda))^{-1}[I \cdots \lambda^{s-1}I]\, d\lambda \neq 0 \tag{17}$$

for all $y \in \partial\Omega$, $0 \neq \xi' \in T_y^*(\partial\Omega)$.

Proof Let (X_+, T_+, Y_+) be a γ^+-spectral triple for $L_{y,\xi'}(\lambda)$. The matrix in (16) is equal to

$$\begin{pmatrix} X_+ \\ \vdots \\ X_+ T_+^{s-1} \end{pmatrix} \cdot [Y_+ \cdots T_+^{s-1} Y_+] \tag{18}$$

and the invertibility of (16) implies that both factors in (18) are invertible. Since the matrix in (17) is equal to

$$\sum_{j=0}^{\mu} B_j X_+ T_+^j \cdot [Y_+ \cdots T_+^{s-1} Y_+] \tag{19}$$

the invertibility of $[Y_+ \cdots T_+^{s-1} Y_+]$ implies that (19) has nonzero determinant. By (12) this is equivalent to the L-condition for $(\mathscr{A}, \mathscr{B})$.

As an example of an elliptic system for which the Dirichlet problems does satisfy the L-condition, we consider strongly elliptic systems. In the following definition, (c, d) denotes the usual inner product of column vectors in $\mathbb{C}^p$, i.e. if $c = [c_j]_{j=1}^p$, $d = [d_j]_{j=1}^p$ then $(c, d) = d^h c = \sum_{j=1}^p c_j \bar{d}_j$.

Definition *Let $\mathscr{A}(x, D)$, $x \in \bar{\Omega}$, be a $p \times p$ differential operator with DN numbers $t_i + t_j$ where t_j are nonnegative integers (i.e. $s_i = t_i \geqslant 0$). Then $\mathscr{A}$ is said to be* strongly elliptic *at $x \in \bar{\Omega}$ if* $\operatorname{Re}(\pi\mathscr{A}(x, \xi)c, c) \neq 0$ *for all $\xi \in \mathbb{R}^n \setminus 0$ and $c \in \mathbb{C}^p \setminus 0$.*

Clearly, if the operator $\mathscr{A}$ is strongly elliptic at x then it is elliptic.

We assume that $\bar{\Omega}$ is connected and $n \geqslant 2$ which means that $\bar{\Omega} \times \mathbb{R}^n \setminus 0 \times \mathbb{C}^p \setminus 0$ is connected. It follows that if $\mathscr{A}$ is strongly elliptic at all $x \in \bar{\Omega}$ then $\operatorname{Re}(\pi\mathscr{A}(x, \xi)c, c)$ has constant sign. Thus, either

$$\operatorname{Re}(\pi\mathscr{A}(x, \xi)c, c) > 0 \tag{20}$$

or $\operatorname{Re}(\pi\mathscr{A}(x, \xi)c, c) < 0$ for all $x \in \bar{\Omega}$, $\xi \in \mathbb{R}^n \setminus 0$ and $c \in \mathbb{C}^p \setminus 0$. Without loss of generality we assume that (20) holds. In view of the homogeneity of $\pi\mathscr{A}(x, \xi) = [a_{jk}(x, \xi)]$ and the compactness of $\bar{\Omega}$ and the unit spheres in $\mathbb{R}^n$

and $\mathbb{C}^p$, we see that

$$\operatorname{Re} \sum_{jk} a_{jk}(x, \xi) c_j \bar{c}_k \geqslant \text{const} \cdot \sum_j |\xi|^{2t_j} |c_j|^2$$

for all $x \in \bar{\Omega}$, $\xi \in \mathbb{R}^n \backslash 0$, $c \in \mathbb{C}^p \backslash 0$.

For strongly elliptic systems we may consider Dirichlet boundary conditions of the form

$$\partial_n^\nu u_j = g_j(y), \qquad y \in \partial\Omega, \tag{21}$$

where $\nu = 0, \ldots, t_j - 1$, $j = 1, \ldots, p$ (when $t_j = 0$ there are no boundary conditions).

Theorem 10.7 *Let $\mathscr{A}(x, D)$ be strongly elliptic for all $x \in \bar{\Omega}$. Then the Dirichlet problem* (21) *satisfies the L-condition relative to $\mathscr{A}$.*

Proof First we make a preliminary remark. Let $\hat{u}(\tau) = \int e^{-it\tau} u(t)\, dt$ denote the Fourier transform of a function on the real line. Let $\mathscr{S}(\bar{\mathbb{R}}_+)$ denote the set of functions $u \in \mathbb{C}^\infty(\mathbb{R}_+)$ on the half-line $t \geqslant 0$ such that u and all its derivatives satisfy $|d^j u/dt^j| \leqslant c_j(1 + t)^{-k}$, $\forall k$, i.e. are rapidly decreasing as $t \to +\infty$. Suppose that the derivatives of $u \in \mathscr{S}(\bar{\mathbb{R}}_+)$ vanish at $t = 0$ up to order $j - 1$. If we extend u by zero for $t < 0$ its derivatives are continuous on $\mathbb{R}$ up to order $j - 1$ and the jth order derivative is integrable. For any polynomial $p(\lambda)$ of degree $\leqslant j + k$, it follows that

$$\int_0^\infty p\left(\frac{1}{i}\frac{d}{dt}\right) u \cdot \bar{v}\, dt = \int_{-\infty}^\infty p(\tau) \hat{u}(\tau) \overline{\hat{v}(\tau)}\, d\tau \tag{22}$$

where $u, v \in \mathscr{S}(\bar{\mathbb{R}}_+)$ with derivatives vanishing at $t = 0$ up to order $j - 1$ and $k - 1$, respectively, and extended by zero for $t < 0$. In fact, on the right-hand side we can integrate by parts to put the integral in a form involving derivatives of u and v of order no more than j and k (because of the initial conditions for u and v no contribution occurs at the initial point $t = 0$) and then apply Parseval's Theorem.

Since the L-condition is verified pointwise we fix y, ξ' and write $L(\lambda) = \pi\mathscr{A}(y, (\xi', \lambda))$. Now $\mathscr{A}$ is strongly elliptic so we have $L = [L_{jk}]$ where L_{jk} is a polynomial of degree $\leqslant t_j + t_k$ and

$$\operatorname{Re}(L(\lambda)c, c) > 0, \qquad c \in \mathbb{C}^p \backslash 0, \quad \text{for all real } \lambda. \tag{23}$$

Let $u \in \mathfrak{M}^+$ have zero Dirichlet data, i.e. $u_j^{(\nu)}(0) = 0$ for $\nu = 0, \ldots, t_j - 1$. By definition $L(i^{-1}\, d/dt)u = 0$ and $u(t) \to 0$ exponentially as $t \to +\infty$; then multiplying by $\bar{u}(t)$ and integrating with respect to t on the half-line $t \geqslant 0$ gives

$$\sum_{j,k} \int_0^\infty L_{jk}\left(\frac{1}{i}\frac{d}{dt}\right) u_j(t) \bar{u}_k(t)\, dt = 0,$$

so we can apply (22) to each term in the sum which yields

$$\sum_{j,k} \int_{-\infty}^\infty L_{jk}(\tau) \hat{u}_j(\tau) \overline{\hat{u}_k(\tau)}\, d\tau = 0.$$

In view of (23) we have $\hat{u} = 0$, whence $u = 0$.

Remark There is a related proof which is due to Lopatinskii [Lo, p. 133] in the case where $\pi\mathscr{A}$ is homogeneous of degree $2s$, i.e. $s_i = t_i = s$. We wish to show that the $ps \times ps$ matrix in (16) is invertible. Let $c \in \mathbb{C}^{ps}$ such that

$$M_{ss}c = 0, \qquad \text{where } M_{ss} = \int_+ \begin{pmatrix} I \\ \vdots \\ \lambda^{s-1}I \end{pmatrix} L^{-1}(\lambda)[I \cdots \lambda^{s-1}I]\, d\lambda. \tag{24}$$

Then define $u \in \mathfrak{M}^+$ by

$$u(t) = \frac{1}{2\pi i} \int_+ e^{it\lambda} L^{-1}(\lambda)[I \cdots \lambda^{s-1}I]\, d\lambda \cdot c,$$

and let $u(t) = 0$ for $t < 0$. The Fourier transform of u is analytic in the upper λ-half-plane and $\hat{u}(\lambda) = L^{-1}(\lambda)[I \cdots \lambda^{s-1}I]c$. Since $L(\lambda) = O(\lambda^{-l})$ as $|\lambda| \to \infty$ it follows that the terms in M_{ss} are $O(\lambda^{-2})$ as $|\lambda| \to \infty$, so we can replace the contour integral $\int_+$ by the integral $\int_{-\infty}^{\infty}$ along the real line. Now observe that

$$\int_{-\infty}^{\infty} (L(\lambda)\hat{u}(\lambda), \hat{u}(\lambda))\, d\lambda = \int_{-\infty}^{\infty} (\hat{u}(\lambda))^* L(\lambda)\hat{u}(\lambda)\, d\lambda = c^*(M_{ss})^* \cdot c = 0.$$

By taking the real part of this equation and using strong ellipticity we find that $\hat{u}(\lambda) = 0$ whence $c = 0$ and the matrix in (16) is invertible.

It follows that Theorem 10.6 is applicable to strongly elliptic operators.

Corollary *Let $\mathscr{A}(x, D)$ be homogeneously elliptic and strongly elliptic. Then the boundary operator $\mathscr{B}(y, D)$ satisfies the L-condition relative to $\mathscr{A}$ if and only if*

$$\det \int_+ \pi\mathscr{B}(y, (\xi', \lambda))\pi\mathscr{A}(y, (\xi', \lambda))^{-1}[I \cdots \lambda^{s-1}I]\, d\lambda \neq 0$$

for all $y \in \partial\Omega$, $0 \neq \xi' \in T_y^*(\partial\Omega)$.

10.3 Matrix polynomials depending on parameters

Let M be a C^∞ manifold, and let

$$L_x(\lambda) = \sum_{j=0}^{l} A_j(x)\lambda^j, \qquad x \in M$$

be a family of $p \times p$ matrix polynomials of degree $\leqslant l$ with matrix coefficients which are C^∞ functions of $x \in M$ and such that $\det L_x(\lambda) \neq 0$ for all real λ.

We also assume that the number, r, of zeros of $\det L_x(\lambda)$ in the upper half-plane $\operatorname{Im} \lambda > 0$ is locally constant on M, multiplicities counted (i.e. constant on each component of M). This is equivalent to the following assumption: For any $x_0 \in M$ and simple, closed contour γ^+ containing the zeros of $\det L_{x_0}(\lambda)$ in $\operatorname{Im} \lambda > 0$,

there is an open set $U \ni x_0$ such that γ^+ contains the zeros of $\det L_x(\lambda)$ in $\operatorname{Im} \lambda > 0$ for all $x \in U$. (*)

Indeed, if we let $f_x(\lambda) = \det L_x(\lambda)$, then by the argument principle from complex analysis we have

$$r = \frac{1}{2\pi i}\int_{\gamma^+} \frac{f'_{x_0}(\lambda)}{f_{x_0}(\lambda)}\, d\lambda.$$

Since $f_{x_0}(\lambda) \neq 0$ for all λ on the contour γ^+, there is a neighbourhood U of x_0 such that $f_x(\lambda) \neq 0$ for all λ on the contour γ^+; by shrinking U, if necessary, it follows that

$$r = \frac{1}{2\pi i}\int_{\gamma^+} \frac{f'_x(\lambda)}{f_x(\lambda)}\, d\lambda \tag{25}$$

because the right-hand side is a continuous function of x, and integer-valued, hence locally constant. But r is the number of zeros of $f_x(\lambda)$ in $\operatorname{Im} \lambda > 0$ (by assumption), so the contour γ^+ must contain all of them, i.e. (*) holds. Conversely, if (*) holds, we conclude from the argument principle (25), and the continuity of the right-hand side, that the number of zeros of $f_x(\lambda)$ in $\operatorname{Im} \lambda > 0$ is locally constant.

We do not assume that the degree of $\det L_x(\lambda)$ is constant, although for matrix polynomials that arise from an elliptic operator this will always be true, of course.

Example $L_\varepsilon(\lambda) = (\varepsilon\lambda + i)(\lambda - i)$, $0 \leqslant \varepsilon \leqslant 1$, is permitted; it has one zero with positive imaginary part, for all ε. On the other hand, $L_\varepsilon(\lambda) = (\varepsilon\lambda - i)(\lambda - i)$, $0 \leqslant \varepsilon \leqslant 1$, is not permitted because it has two zeros with positive imaginary part when $\varepsilon \neq 0$ but only one such zero when $\varepsilon = 0$.

As above let M be a C^∞ manifold. For each $x \in M$, there is the vector space $\mathfrak{M}_x^+$ of solutions of $L_x(i^{-1}\, d/dt)u = 0$ such that $u(t) \to 0$ as $t \to +\infty$. Let γ_x^+ be a simple, closed contour containing the eigenvalues of $L_x(\lambda)$ in the upper half-plane $\operatorname{Im} \lambda > 0$. As in the proof of Theorem 2.6, $L_x(\lambda)$ has the γ_x^+-spectral triple $(X_+|_x, T_+|_x, Y_+|_x)$ where

$$\left.\begin{aligned}
&X_+|_x\colon \mathfrak{M}_x^+ \to \mathbb{C}^p \quad \text{maps} \quad u \mapsto u(0)\\
&T_+|_x\colon \mathfrak{M}_x^+ \to \mathfrak{M}_x^+ \quad \text{maps} \quad u \mapsto \frac{1}{i}\frac{d}{dt}u\\
&Y_+|_x\colon \mathbb{C}^p \to \mathfrak{M}_x^+ \quad \text{maps} \quad c \mapsto (2\pi i)^{-1}\int_{+} e^{it\lambda} L_x^{-1}(\lambda)c\, d\lambda
\end{aligned}\right\} \tag{26}$$

Lemma 10.8 *Let $L_x(\lambda)$ have the properties stated above. The family of vector spaces $\mathfrak{M}^+ = \{\mathfrak{M}_x^+\}$ has a unique vector bundle structure over M such that the maps $X_+\colon \mathfrak{M}^+ \to \mathbb{C}^p$, $T_+\colon \mathfrak{M}^+ \to \mathfrak{M}^+$ and $Y_+\colon \mathbb{C}^p \to \mathfrak{M}^+$ defined on each fibre by* (26) *are vector bundle homomorphisms.*

Proof The Calderón projector for $L_x(\lambda)$ with respect to the eigenvalues in the upper half-plane is given by

$$P_x^+ = \frac{1}{2\pi i} \int_{\gamma_x^+} \begin{pmatrix} I \\ \vdots \\ \lambda^{l-1} I \end{pmatrix} L_x^{-1}(\lambda)[I \cdots \lambda^{l-1} I] \mathscr{Z}_x \, d\lambda \tag{27}$$

where γ_x^+ contains the eigenvalues of $L_x(\lambda)$ in the upper half-plane $\operatorname{Im} \lambda > 0$. Let $x_0 \in M$. By assumption, see (*), there exists a neighbourhood U of x_0 such that the contour γ_x^+ in (27) can be replaced by $\gamma_{x_0}^+$ for all $x \in U$. Since the coefficients of $L_x(\lambda)$ depend smoothly on x it is clear that the matrix P_x^+ depends smoothly on x. Hence P_x^+ is a projection operator (Definition 5.53) for the trivial bundle $M \times \mathbb{C}^{pl}$. By virtue of Proposition 5.52, the family of vector spaces $\operatorname{im} P_+ = \{\operatorname{im} P_x^+\}$ is a vector bundle over M.

We have a map $\varphi\colon \mathfrak{M}^+ \to \operatorname{im} P_+$ defined by the Cauchy data on each fibre

$$u \in \mathfrak{M}_x^+ \mapsto \mathscr{U} = \operatorname{col}\left\{\left(\frac{1}{i}\frac{d}{dt}\right)^j u(0)\right\}_{j=0}^{l-1} \in \mathbb{C}^{pl}$$

which we know is a linear isomorphism on each fibre (Theorem 2.8). We give $\mathfrak{M}^+$ the vector bundle structure that makes φ into a vector bundle isomorphism. To prove that X_+, T_+ and Y_+ are smooth we just have to show that $X_+ \circ \varphi^{-1}$, $\varphi \circ T_+ \circ \varphi^{-1}$ and $\varphi \circ Y_+$ are smooth. Recall that for any matrix polynomial, $L(\lambda)$, every $u \in \mathfrak{M}^+$ has a representation of the form

$$u(t) = \frac{1}{2\pi i} \int_+ e^{it\lambda} L^{-1}(\lambda)[I \cdots \lambda^{l-1} I] \mathscr{Z} \cdot \mathscr{U} \, d\lambda$$

where $\mathscr{U}$ is the Cauchy data of u (see §2.2 and recall that $\int_+$ is the short notation for $\int_{\gamma^+}$). It follows that $X_+ \circ \varphi^{-1} = \mathscr{X}|_{\operatorname{im} P_+}$, $\varphi \circ T_+ \circ \varphi^{-1} = \mathscr{T}|_{\operatorname{im} P_+}$ and $\varphi \circ Y_+ = \mathscr{Y}$ where

$$\mathscr{X} = \frac{1}{2\pi i} \int_+ L^{-1}(\lambda)[I \cdots \lambda^{l-1} I] \mathscr{Z} \, d\lambda \qquad p \times pl \text{ matrix}$$

$$\mathscr{T} = \frac{1}{2\pi i} \int_+ \operatorname{col}(\lambda^j I)_{j=0}^{l-1} \lambda L^{-1}(\lambda)[I \cdots \lambda^{l-1} I] \mathscr{Z} \, d\lambda, \qquad pl \times pl \text{ matrix}$$

$$\mathscr{Y} = \frac{1}{2\pi i} \int_+ \operatorname{col}(\lambda^j I)_{j=0}^{l-1} L^{-1}(\lambda) \, d\lambda \qquad pl \times p \text{ matrix}$$

Since $\mathscr{X}$, $\mathscr{T}$ and $\mathscr{Y}$ are smooth matrix functions on M, and $\operatorname{im} P_+$ is a sub-bundle of $M \times \mathbb{C}^{pl}$, it follows that $X_+ \circ \varphi^{-1} \in \mathscr{L}(\operatorname{im} P_+, \mathbb{C}^p)$, $\varphi \circ T_+ \circ \varphi^{-1} \in \mathscr{L}(\operatorname{im} P_+)$ and $\varphi \circ Y_+ \in \mathscr{L}(\mathbb{C}^p, \operatorname{im} P_+)$ are smooth.

The definitions we had previously in Chapter 1 for admissible triples of operators carry over to families of such admissible triples. A triple of vector bundle homomorphisms (X, T, Y) is called an *admissible triple over* M if $X \in \mathscr{L}(E, \mathbb{C}^p)$, $T \in \mathscr{L}(E)$ and $Y \in \mathscr{L}(\mathbb{C}^p, E)$, where E is a vector bundle over M. As before, E is called the *base space* of (X, T, Y). Two admissible triples over M, (X, T, Y) and (X', T', Y'), with base spaces E and E', respectively, are called *similar* if there exists a vector bundle isomorphism $\varphi \in \mathscr{L}(E', E)$ such that $X' = X\varphi$, $T' = \varphi^{-1} T \varphi$ and $Y' = \varphi^{-1} Y$.

Theorem 10.9 *Let $L_x(\lambda) = \sum_{j=0}^{l} A_j(x)\lambda^j$ be a family of matrix polynomials with the properties stated above. For each $x \in M$, let γ_x^+ be a simple, closed contour containing the eigenvalues of $L_x(\lambda)$ in the upper half-plane* $\operatorname{Im} \lambda > 0$. *Then there exists an admissible triple (X_+, T_+, Y_+) over M such that $(X_+(x), T_+(x), Y_+(x))$ is a γ_x^+-spectral triple of $L_x(\lambda)$ for all $x \in M$. Any two such admissible triples are similar.*

Note: we call (X_+, T_+, Y_+) a *spectral triple with respect to the upper half-plane* for the family of matrix polynomials $L(\lambda) = L_x(\lambda)$, $x \in M$.

Proof The existence of a spectral triple (X_+, T_+, Y_+) has been proved in Lemma 10.8 (the base space is $E = \mathfrak{M}^+$). Now let (X'_+, T'_+, Y'_+) be another spectral triple for the family $L(\lambda)$, i.e. an admissible triple over M, with base space E', such that $(X_+(x), T_+(x), Y_+(x))$ is a γ_x^+-spectral triple of $L_x(\lambda)$ for all $x \in M$. By Proposition 2.12, it follows that $X'_+(x) = X_+(x) \circ \mathcal{M}_x$, $T'_+(x) = \mathcal{M}_x^{-1} \circ T_+(x) \circ \mathcal{M}_x$ and $Y'_+(x) = \mathcal{M}_x^{-1} \circ Y_+(x)$ for all $x \in M$, where

$$\mathcal{M}_x = \operatorname{row}(T_+^j(x)Y_+(x))_{j=0}^{l-1} \cdot \mathscr{L}_x \cdot \operatorname{col}(X'_+(x)T'^j_+(x))_{j=0}^{l-1}.$$

Obviously $\mathcal{M} \in \mathscr{L}(E', E)$ and is a (smooth) isomorphism. Hence (X_+, T_+, Y_+) and (X'_+, T'_+, Y'_+) are similar.

Remark If (X_+, T_+, Y_+) is any admissible triple over M satisfying the conditions of Theorem 10.9 then its base space E is necessarily isomorphic to $\mathfrak{M}^+$ due to the isomorphism $E \to \mathfrak{M}^+$ given by $E_x \ni v \mapsto X_+(x)\, e^{itT_+(x)}\, v \in \mathfrak{M}_x^+$.

Corollary 10.10 *There exist vector bundle isomorphisms*

$$\mathfrak{M}^+ \simeq \operatorname{im} P_+ \quad \text{and} \quad \mathfrak{M}^+ \simeq \operatorname{im} P'_+$$

where P_+ and P'_+ are the right and left Calderón projectors with respect to the eigenvalues in the upper half-plane.

Proof The isomorphism $\mathfrak{M}^+ \to \operatorname{im} P_+$ is given by the Cauchy data

$$u \mapsto \operatorname{col}(u^{(j)}(0))_{j=0}^{l-1}$$

on each fibre. As for the second isomorphism, consider the map $\phi\colon \mathbb{C}^{pl} \to \mathfrak{M}^+$ defined by

$$c \mapsto \phi c(t) = (2\pi i)^{-1} \int_+ e^{it\lambda}\, L^{-1}(\lambda) \cdot \sum_{j=0}^{l-1} c_{j+1} \lambda^j \, d\lambda.$$

Let (X_+, T_+, Y_+) be a spectral triple with respect to the upper half-plane for the family $L(\lambda)$. Since $\phi c = X_+ e^{itT_+}[Y_+ \cdots T_+^{l-1} Y_+]c$ it follows due to injectivity of $\operatorname{col}(X_+ T_+^j)_{j=0}^{l-1}$ that $\ker \phi = \ker[Y_+ \cdots T_+^{l-1} Y_+] = \ker P'_+$ (for the second equality, see Theorem 2.15). Since P'_+ is a smooth projector, then $\ker P'_+$ is a vector bundle over M. By surjectivity of ϕ, we have an isomorphism

$$\mathbb{C}^{pl}/\ker \phi \simeq \mathfrak{M}^+$$

But also $\mathbb{C}^{pl}/\ker \phi = \mathbb{C}^{pl}/\ker P'_+ \simeq \operatorname{im} P'_+$, so there is a natural isomorphism $\mathfrak{M}^+ \simeq \operatorname{im} P'_+$.

In §§10.4 to 10.6 and §15.1 it will be clear that the triple (X_+, T_+, Y_+) is a key step in the construction of boundary operators satisfying the *L*-condition for a given elliptic operator; this in turn is crucial for the proof of the index formula for elliptic systems in the plane in Chapter 16.

In fact we would like to have spectral triples (X_+, T_+, Y_+) which are C^∞ *matrix* functions on M. (Precisely, $X_+(x)$ is a $p \times r$ matrix, $T_+(x)$ is an $r \times r$ matrix and $Y_+(x)$ is an $r \times p$ matrix, with entries depending smoothly on $x \in M$.) One necessary condition for the existence of such a triple (X_+, T_+, Y_+) is obvious: $\mathfrak{M}^+$ must be trivial, since there is the isomorphism from $M \times \mathbb{C}^r$ to $\mathfrak{M}^+$ defined by

$$(x, v) \in \{x\} \times \mathbb{C}^r \mapsto X_+(x)\, e^{itT_+(x)}\, v \in \mathfrak{M}_x^+$$

The triviality of $\mathfrak{M}^+$ is also sufficient as we show in the next proposition.

Note that the condition that the number of eigenvalues of $L_x(\lambda)$ in $\operatorname{Im} \lambda > 0$ be locally constant in $x \in M$ just means that the number of them is constant on each component of M.

Theorem 10.11 *As usual, suppose that* $\det L_x(\lambda) \neq 0$ *for all real* λ, *and let the number, r, of roots of* $\det L_x(\lambda) = 0$ *in the upper half-plane be independent of* $x \in M$. *Then there exists a spectral triple* (X_+, T_+, Y_+) *with respect to the upper half-plane for the family* $L(\lambda) = L_x(\lambda)$ *consisting of* C^∞ matrix *functions on M if and only if the vector bundle* $\mathfrak{M}^+$ *is trivial.*

Proof First, we claim that the base space E of any spectral triple (X_+, T_+, Y_+) is isomorphic to $\mathfrak{M}^+$. Indeed, there is a map $\Phi: E \to \mathfrak{M}^+$ defined on each fibre by

$$v \in E_x \mapsto X_+(x)\, e^{itT_+(x)}\, v \in \mathfrak{M}_x^+$$

which is a linear isomorphism on each fibre (Corollary 2.9). Since Φ is smooth it follows by Lemma 5.46 that Φ is a vector bundle isomorphism.

Suppose now that $\mathfrak{M}^+$ is trivial. Then E is also trivial, i.e. there exist C^∞ sections $v_1, \ldots, v_r$ of E such that, for all $x \in M$, the vectors $v_1(x), \ldots, v_r(x)$ form a basis of E_x. Relative to this basis, the vector bundle homomorphisms X_+, T_+ and Y_+ are C^∞ matrix functions.

Remark Exercise 1 at the end of this chapter gives an example of a family of matrix polynomials $L_x(\lambda)$ for which the vector bundle $\mathfrak{M}^+$ is not trivial.

10.4 Homogeneity properties of spectral triples

Let $\mathscr{A}(x, D)$ be an elliptic differential operator in $\bar{\Omega}$ with DN numbers $s_1, \ldots, s_p, t_1, \ldots, t_p$. The coefficients of the matrix polynomial

$$L(\lambda) = \pi\mathscr{A}(y, (\xi', \lambda)) = \sum_{j=0}^{l} A_j(y, \xi')\lambda^j$$

depend smoothly on the parameters $(y, \xi') \in T^*(\partial\Omega)\backslash 0$ and have certain homogeneity properties in ξ' which follow from the homogeneity properties of $\pi\mathscr{A}$.

We consider first the case where $\mathscr{A}$ is homogeneously elliptic, that is, $s_1 = \cdots = s_p = l$ and $t_1 = \cdots = t_p = 0$ (see Definition 9.25), because the notation is simpler then and some of the results take on a more complete form. $L(\lambda)$ has degree l with invertible leading coefficient, $A_l = \pi\mathscr{A}(y, n(y))$, and det $L(\lambda)$ has degree $\alpha = pl$. A finite spectral pair (X, T) for $L(\lambda)$ is therefore a standard pair (Definition 3.1), that is, $\operatorname{col}(XT^j)_{j=0}^{l-1}$ is invertible. The homogeneity of $\pi\mathscr{A}$ implies $L_{y,c\xi'}(\lambda) = c^l L_{y,\xi'}(c^{-1}\lambda)$, whence

$$A_j(y, c\xi') = c^{l-j} \cdot A_j(y, \xi'), \qquad j = 0, \ldots, l, \forall c \in \mathbb{C} \tag{28}$$

Let $ST^*(\partial\Omega)$ denote the unit cotangent bundle to $\partial\Omega$ and $\imath$: $T^*(\partial\Omega)\backslash 0 \to ST^*(\partial\Omega)$ the map defined by $(y, \xi') \mapsto (y, \xi'/|\xi'|)$.

Lemma 10.12 *There exists a spectral triple (X_+, T_+, Y_+) with respect to the upper half-plane for the family $L(\lambda) = \pi\mathscr{A}(y, (\xi', \lambda))$, $(y, \xi') \in T^*(\partial\Omega)\backslash 0$, such that for any $c > 0$*

$$\left.\begin{aligned} X_+(y, c\xi') &= X_+(y, \xi') \\ T_+(y, c\xi') &= cT_+(y, \xi') \\ Y_+(y, c\xi') &= c^{1-l}Y_+(y, \xi') \end{aligned}\right\} \tag{29}$$

where $X_+ \in \mathscr{L}(\imath^{-1}E, \mathbb{C}^p)$, $T_+ \in \mathscr{L}(\imath^{-1}E)$, $Y_+ \in \mathscr{L}(\mathbb{C}^p, \imath^{-1}E)$, and $\imath^{-1}E$ is the pullback to $T^*(\partial\Omega)\backslash 0$ of some vector bundle E over $ST^*(\partial\Omega)$. (E is necessarily isomorphic to $\mathfrak{M}^+$.)

Proof By virtue of Theorem 10.9 there exists a γ^+-spectral triple (X_+, T_+, Y_+) for $L(\lambda)$ with $X_+ \in \mathscr{L}(E, \mathbb{C}^p)$, $T_+ \in \mathscr{L}(E)$, $Y_+ \in \mathscr{L}(\mathbb{C}^p, E)$ for some vector bundle E over $ST^*(\partial\Omega)$. Then we define

$$X_+(y, \xi') = X_+(y, \xi'/|\xi'|),$$
$$T_+(y, \xi') = |\xi'|T_+(y, \xi'/|\xi'|),$$
$$Y_+(y, \xi') = |\xi'|^{1-l}Y_+(y, \xi'/|\xi'|).$$

The triple (X_+, T_+, Y_+) satisfies the homogeneity properties listed above and it remains to show that its satisfies properties (i), (ii′), (iii) and (iv) of Definition 2.1 for all $(y, \xi') \in T^*(\partial\Omega)\backslash 0$ given that it satisfies them on $ST^*(\partial\Omega)$. Condition (i) obviously holds. As for condition (ii′) we have

$$X_+(y, c\xi')T_+^j(y, c\xi')Y_+(y, c\xi') = (2\pi i)^{-1} \int_+ \lambda^j L_{y,c\xi'}^{-1}(\lambda)\, d\lambda, \qquad |\xi'| = 1, c > 0$$

for $j = 0, 1, \ldots,$ for it holds when $c = 1$ and due to (28), (29) it then holds for any $c > 0$ (by making the substitution $\lambda \to c^{-1}\lambda$). Finally, conditions (iii),

(iv) hold since

$$\operatorname{col}(X_+(y, c\xi')T_+^j(y, c\xi'))_{j=0}^{l-1} = F(c)\cdot\operatorname{col}(X_+(y, \xi')T_+^j(y, \xi'))_{j=0}^{l-1}, \qquad c > 0 \quad (30)$$

and

$$\operatorname{row}(T_+^j(y, c\xi')Y_+(y, c\xi'))_{j=0}^{l-1} = \operatorname{row}(T_+^j(y, \xi')Y_+(y, \xi'))_{j=0}^{l-1}\cdot E(c), \qquad c > 0 \quad (31)$$

where $F(c) = \operatorname{diag}(c^j I)_{j=0}^{l-1}$ and $E(c) = c\cdot\operatorname{diag}(c^j I)_{j=0}^{l-1}$ are $pl \times pl$ diagonal matrices which are invertible.

For the matrix $\mathscr{Z}$ defined in §2.1(5), the homogeneity properties (28) imply

$$\mathscr{Z}(y, c\xi') = E^{-1}(c)\mathscr{Z}(y, \xi')F^{-1}(c) \quad (32)$$

where $E(c)$ and $F(c)$ are as defined above. Hence for the right and left Calderón projectors P_+ and P'_+, (4), (6) and (29) give

$$P_+(y, c\xi') = F(c)P_+(y, \xi')F^{-1}(c), \quad (33)$$

$$P'_+(y, c\xi') = E(c)P'_+(y, \xi')E^{-1}(c) \quad (34)$$

Remark If we write $P_+ = [P_{ki}]$, where P_{ki} are $p \times p$ blocks ($i, k = 0, \ldots, l-1$), then (33) implies $P_{ki}(y, c\xi') = c^{k-i}P_{ki}(y, \xi')$, whence P_{ki} is the principal symbol of a pseudo-differential operator on $\partial\Omega$ of order $k - i$. Similarly for P'_+.

From now on it is always assumed that the γ^+-spectral triple of $L(\lambda)$ is chosen with the properties (29). By repeating the proof of Lemma 10.12 we see that the triple

$$(X_-(y, \xi'), T_-(y, \xi'), Y_-(y, \xi')) := (X_+(y, -\xi'), -T_+(y, -\xi'), (-1)^{1-l}Y_+(y, -\xi')) \quad (35)$$

is a γ^--spectral triple of $L_{y,\xi'}(\lambda)$, and has the (positive) homogeneity properties (29). The Calderón projectors $P_+(y, -\xi')$ and $P_-(y, \xi')$ are similar (with similarity matrix $F(-1)$), and the same is true for the left Calderón projectors.

Let $\mathscr{B}$ be a boundary operator and consider the associated $r \times p$ matrix polynomial

$$B(\lambda) = \pi\mathscr{B}(y, (\xi', \lambda)) = \sum_{j=0}^{\mu} B_j(y, \xi')\lambda^j.$$

The kth row of $\pi\mathscr{B}$ is homogeneous of degree m_k in the variables $\xi = (\xi', \lambda)$, whence the kth row of the coefficient B_j is homogeneous of degree $m_k - j$. In other words,

$$B_j(y, c\xi') = M(c)B_j(y, \xi')c^{-j}, \qquad c > 0 \quad (36)$$

for $j = 0, \ldots, \mu$ where $M(c)$ is the $r \times r$ diagonal matrix $[c^{m_k}\delta_{ki}]$. Now let

$$\Delta_{\mathscr{B}}^+(y, \xi') = \sum_{j=0}^{\mu} B_j(y, \xi')X_+T_+^j \quad (37)$$

(we omit the argument (y, ξ') in X_+ and T_+ to simplify the notation). By Theorem 10.3(iii), $\mathscr{B}$ satisfies the L-condition relative to $\mathscr{A}$ if and only if $\Delta^+_{\mathscr{B}}(y, \xi')$ is invertible for all $(y, \xi') \in T^*(\partial\Omega)\backslash 0$. By virtue of (29) and (36)

$$\Delta^+_{\mathscr{B}}(y, c\xi') = M(c)\,\Delta^+_{\mathscr{B}}(y, \xi'), \qquad c > 0, \tag{38}$$

so it suffices to verify invertibility of $\Delta^+_{\mathscr{B}}(y, \xi')$ when $(y, \xi') \in ST^*(\partial\Omega)$.

Remark If the rows of $\pi\mathscr{B}(y, \xi)$ are not only positive homogeneous but also *negative* homogeneous in $\xi = (\xi', \lambda)$ then the B_j's satisfy the homogeneity property (36) for all $c \neq 0$. In such a case we also define

$$\Delta^-_{\mathscr{B}}(y, \xi') = \sum_{j=0}^{\mu} B_j(y, \xi')X_-T^j_- \tag{37'}$$

and it follows from (36) with $c = -1$ and (35) that

$$\Delta^+_{\mathscr{B}}(y, -\xi') = M(-1)\,\Delta^-_{\mathscr{B}}(y, \xi') \tag{38'}$$

The next proposition shows that if $\mu \leqslant l - 1$ and $\pi\mathscr{B}$ is both positive *and* negative homogeneous then $\Delta^+_{\mathscr{B}_1} = \Delta^+_{\mathscr{B}_2}$ if and only if $\pi\mathscr{B}_1 = \pi\mathscr{B}_2$. In other words, given the pair (X_+, T_+), *the principal part $\pi\mathscr{B}$ of $\mathscr{B}$ is uniquely determined by $\Delta^+_{\mathscr{B}}$*. Recall that μ is the transversal order; there is no restriction on the total order of $\mathscr{B}$ because the m_k's are permitted to be any real numbers.

Remark The condition that $\pi\mathscr{B}$ be both positive and negative homogeneous holds if $\mathscr{B}$ is a differential operator.

Proposition 10.13 *Let $\mathscr{B}$ be a boundary operator with transversal order $\mu \leqslant l - 1$ such that $\pi\mathscr{B}(y, \xi)$ is positive and negative homogeneous in ξ. Then we have $\pi\mathscr{B}(y; \xi', \lambda) = \sum_{j=0}^{l-1} B_j(y, \xi')\lambda^j$ where*

$$B_j = \Delta^+_{\mathscr{B}} \cdot \sum_{k=0}^{l-j-1} T^k_+ Y_+ A_{j+k+1} + \Delta^-_{\mathscr{B}} \cdot \sum_{k=0}^{l-j-1} T^k_- Y_- A_{j+k+1} \tag{39}$$

$j = 0, \ldots, l - 1$ (and $B_j = 0$ if $j > \mu$).

Proof Since $\sum_{j=0}^{l-1} B_j(y, \xi')X_\pm T^j_\pm = \Delta^\pm_{\mathscr{B}}$, this follows immediately from the fourth corollary of Theorem 2.17, because $\operatorname{col}(XT^j)_{j=0}^{l-1}$ is invertible.

As we know from §10.3 the family of vector spaces $\mathfrak{M}^+ = \{\mathfrak{M}^+_{L_{y,\xi'}}\}$ is an r-dimensional vector bundle over $ST^*(\partial\Omega)$. A boundary operator $\mathscr{B}$ that satisfies the L-condition relative to $\mathscr{A}$ defines a trivialization

$$\mathfrak{M}^+ \to ST^*(\partial\Omega) \times \mathbb{C}^r$$

$$u \in \mathfrak{M}^+_{L_{y,\xi'}} \mapsto B_{y,\xi'}\left(\frac{1}{i}\frac{d}{dt}\right)u|_{t=0}$$

This means that there is a "topological obstruction" to the existence of boundary operators satisfying the L-condition. Conversely, if it is assumed that $\mathfrak{M}^+$ is a trivial bundle then there exists a boundary operator $\mathscr{B}$ satisfying the L-condition. In fact we have the following theorem.

Theorem 10.14 *Suppose that $\mathfrak{M}^+$ is a trivial bundle. Let $\Delta\colon ST^*(\partial\Omega) \to GL_r(\mathbb{C})$ be any C^∞ matrix function (for example, $\Delta \equiv I$). Then for any $m_k \in \mathbb{R}$, $k = 1, \ldots, r$, there exsts a boundary operator $\mathscr{B}$ with tranversal order $\mu \leqslant l - 1$ such that the kth row of the principal part $\pi\mathscr{B}$ is both positive and negative homogeneous of degree m_k and*

$$\Delta^+_{\mathscr{B}} = \Delta \quad on \quad ST^*(\partial\Omega).$$

Moreover, the principal part of $\mathscr{B}$ is uniquely determined by these conditions. Since Δ is invertible, $\mathscr{B}$ satisfies the L-condition.

Proof In Theorem 10.11 we showed that if $\mathfrak{M}^+$ is trivial then there exists a γ^+-spectral triple of $L(\lambda)$ that consists of C^∞ matrix functions on $ST^*(\partial\Omega)$, i.e. X_+, T_+ and Y_+ are C^∞ matrix functions on $ST^*(\partial\Omega)$ of dimensions $p \times r$, $r \times r$ and $r \times p$, respectively. This triple of matrices can be extended by homogeneity so that it satisfies the properties (29). Now let (X_-, T_-, Y_-) be the γ^--spectral triple of $L(\lambda)$ defined as in (35). In view of the first corollary to Theorem 2.17, the triple

$$X = [X_+ \quad X_-], \quad T = \begin{pmatrix} T_+ & \\ & T_- \end{pmatrix} \quad \text{and} \quad Y = \begin{pmatrix} Y_+ \\ Y_- \end{pmatrix} \tag{40}$$

is a finite spectral triple of $L(\lambda)$.

Extend Δ to $T^*(\partial\Omega)\backslash 0$ by the formula $\Delta(y, \xi') = M(|\xi'|)\cdot\Delta(y, \xi'/|\xi'|)$, and then define matrix functions $\Delta_\pm\colon T^*(\partial\Omega)\backslash 0 \to GL_r(\mathbb{C})$ by

$$\Delta_+ = \Delta \quad \text{and} \quad \Delta_-(y, \xi') = M(-1)\,\Delta(y, -\xi')$$

Now define $r \times p$ matrix functions B_j, $j = 0, 1, \ldots, l - 1$, by the formulas (39) where $\Delta^\pm_{\mathscr{B}}$ is replaced by $\Delta_\pm$. Let $B^+_j(y, \xi')$ and $B^-_j(y, \xi')$ denote the $+$ and $-$ terms on the right-hand side of (39), then due to (29) and (28) it follows that for $c > 0$

$$\begin{aligned} B^+_j(y, c\xi') &= \Delta_+(y, c\xi')\cdot \sum_{k=0}^{l-j-1} T^k_+(y, c\xi')\, Y_+(y, c\xi') A_{j+k+1}(y, c\xi') \\ &= M(c)\,\Delta_+(y, \xi')\cdot \sum_{k=0}^{l-j-1} c^k T^k_+(y, \xi') c Y_+(y, \xi') A_{j+k+1}(y, \xi') c^{-j-k-1} \\ &= M(c) B^+_j(y, \xi') c^{-j} \end{aligned}$$

and, similarly, $B^-_j(y, c\xi') = M(c) B^-_j(y, \xi') c^{-j}$. Hence (36) holds for $c > 0$. In view of $\Delta_-(y, \xi') = M(-1)\,\Delta_+(y, \xi')$ and the equations (35), a similar calculation yields

$$B^+_j(y, -\xi') = M(-1) B^-_j(y, \xi')(-1)^j$$

and

$$B^-_j(y, -\xi') = M(-1) B^+_j(y, \xi')(-1)^{-j},$$

so that (36) holds for $c = -1$, and thus for all $c \neq 0$. Finally, let $\mathscr{B}_j$ be a $r \times p$ matrix of (-1) classical p.d.o.'s on $\partial\Omega$ with principal symbol equal to B_j (existence by Theorem 8.28), and then let

$$\mathscr{B} = \sum_{j=0}^{l-1} \mathscr{B}_j(y, D')D_n^j$$

By construction the kth rows of $\mathscr{B}$ has order m_k $(k = 1, \ldots, r)$ and $\Delta_{\mathscr{B}}^{\pm} = \Delta_{\pm}$. This completes the proof of the theorem, for the uniqueness of $\pi\mathscr{B}$ follows from Proposition 10.13.

Remark 10.15 The kth row of $\tilde{B}_j(y, \xi')$ has order $m_k - j$, where $j = 0, \ldots, l-1$. If $m_k \geqslant l - 1$ for all k, then all these entries have nonnegative order.

For elliptic operators in the plane ($n = 2$) on a simply connected domain Ω, the "topological obstruction" mentioned above is proper ellipticity; that is to say, $\mathscr{A}$ has a boundary operator satisfying the L-condition if and only if it is properly elliptic. Necessity has been proved in Remark 9.31, and sufficiency is proved in Theorem 15.9.

It should be noted that triviality of $\mathfrak{M}^+$ is the basic restriction for existence of boundary operators only because we have assumed that the image of the boundary operator lies in the sections of a trivial bundle. If the boundary operators had been permitted in the form

$$\mathscr{B}: C^\infty(\bar{\Omega}, \mathbb{C}^p) \to \bigoplus_{j=1}^{J} C^\infty(\partial\Omega, G_j),$$

where G_j are vector bundles over $\partial\Omega$ (and the jth component of $\mathscr{B}$ is of order $m_j, j = 1, \ldots, J$) then a weaker restriction on $\mathfrak{M}^+$ would have been obtained which is fundamental:

The necessary and sufficient condition for existence of a boundary operator $\mathscr{B}$ is that $\mathfrak{M}^+ \simeq p_*^{-1}V$ for some vector bundle V over $\partial\Omega$.

Here $p_*: ST^*(\partial\Omega) \to \partial\Omega$ is the cotangent bundle projection $(y, \xi') \mapsto y$, and the condition on $\mathfrak{M}^+$ means that the fibres $\mathfrak{M}^+_{y,\xi'} = V_y$ are independent of ξ'. One could, of course, allow the elliptic operator $\mathscr{A}$ to act in vector bundles too, i.e. $\mathscr{A}: C^\infty(\bar{\Omega}, E) \to C^\infty(\bar{\Omega}, F)$ where E, F are vector bundles over $\bar{\Omega}$ with the same fibre dimensions. We leave the details to the reader. See also Exercise 2.

Remark The results of Chapters 9 to 14 can be generalized quite easily to elliptic boundary problems $(\mathscr{A}, \mathscr{B})$ where the operators $\mathscr{A}$ and $\mathscr{B}$ act on sections of vector bundles. We have chosen to fix attention on systems so that the connection with the matrix theory of Part I is more readily apparent.

The following is another version of Theorem 10.14. Note that the condition $G \cdot P'_+ = G$ can always be achieved since we can replace G by $G \cdot P'_+$ (where $P'_+ = P'_{\gamma+}$ is the left Calderón projector).

Theorem 10.16 *Let G be a smooth $r \times pl$ matrix function on $ST^*(\partial\Omega)$ such that at each point $(y, \xi') \in ST^*(\partial\Omega)$ we have $G \cdot P'_+ = G$ and $G\colon \operatorname{im} P'_+ \to \mathbb{C}^r$ is invertible. Then for any $m_k \in \mathbb{R}$, $k = 1, \ldots, r$, there exists a boundary operator $\mathscr{B}$ with transversal order $\mu \leqslant l - 1$ such that the kth row of $\pi\mathscr{B}$ is positive and negative homogeneous of degree m_k and*

$$G = \frac{1}{2\pi i} \int_+ B(\lambda) L^{-1}(\lambda)[I \cdots \lambda^{l-1} I]\, d\lambda \tag{41}$$

The principal part of $\mathscr{B}$ is uniquely determined by these conditions. Since the rank of G is r, then $\mathscr{B}$ satisfies the L-condition.

Proof First note that $\mathfrak{M}^+$ is a trivial bundle, due to the isomorphism $G\colon \operatorname{im} P'_+ \to \mathbb{C}^r$ and the natural isomorphism $\mathfrak{M}^+ \simeq \operatorname{im} P'_+$ (see Corollary 10.10). Therefore we can apply Theorem 10.14. Let (X_+, T_+, Y_+) be a γ^+-spectral triple for $L(\lambda)$ consisting of smooth matrix functions on $T^*(\partial\Omega)\backslash 0$, with the usual homogeneity properties. Now, let $\Delta = G \cdot \mathscr{Z} \cdot \operatorname{col}(X_+ T_+^j)_{j=0}^{l-1}$, an $r \times r$ matrix function on $ST^*(\partial\Omega)$. In view of Corollary 2.11, we see that

$$\Delta \cdot [Y_+ \cdots T_+^{l-1} Y_+] = G \tag{42}$$

Since G is surjective, it follows that Δ is surjective at each point of $ST^*(\partial\Omega)$. However, Δ is a square matrix so $\det \Delta \neq 0$, and Δ defines a smooth matrix function $ST^*(\partial\Omega) \to GL_r(\mathbb{C})$. By Theorem 10.14 there exists a boundary operator $\mathscr{B}$ with $\Delta_{\mathscr{B}}^+ = \Delta$. The equation (41) follows from (42). The principal part of the boundary operator $\mathscr{B}$ is uniquely determined in view of Theorem 10.14.

We now return to discuss operators of general Douglis–Nirenberg type. We have $\pi\mathscr{A}(x, c\xi) = S(c)\pi\mathscr{A}(x, \xi)T(c)$, where $S(c) = [\delta_{ik} c^{s_k}]$, $T(c) = [\delta_{ik} c^{t_k}]$, so

$$L_{y, c\xi'}(\lambda) = S(c) \cdot L_{y, \xi'}(c^{-1}\lambda) \cdot T(c),$$

which implies that the coefficients, A_j, *of* $L(\lambda)$ are homogeneous of degree $s_i + t_k - j$ in the (i, k) entry (k and $i = 1, \ldots, p$), that is,

$$A_j(y, c\xi') = S(c) A_j(y, \xi') T(c) c^{-j}, \qquad j = 0, \ldots, l \tag{43}$$

(for any $c \in \mathbb{C}$). Once again we let $\imath$ be the map from $T^*(\partial\Omega)\backslash 0$ to $ST^*(\partial\Omega)$ defined by $\imath\colon (y, \xi') \mapsto (y, \xi'/|\xi'|)$.

Lemma 10.17 *There exists a γ^+-spectral triple (X_+, T_+, Y_+) for $L(\lambda)$ with the following properties for any $c > 0$*

$$\left.\begin{aligned} X_+(y, c\xi') &= T^{-1}(c) X_+(y, \xi') \\ T_+(y, c\xi') &= c T_+(y, \xi') \\ Y_+(y, c\xi') &= c Y_+(y, \xi') S^{-1}(c) \end{aligned}\right\} \tag{44}$$

where $X_+ \in \mathscr{L}(\imath^{-1}E, \mathbb{C}^p)$, $T_+ \in \mathscr{L}(\imath^{-1}E)$, $Y_+ \in \mathscr{L}(\mathbb{C}^p, \imath^{-1}E)$ and $\imath^{-1}E$ is the *pull-back of E to $T^*(\partial\Omega)\backslash 0$ of some vector bundle E over $ST^*(\partial\Omega)$. (E is necessarily isomorphic to $\mathfrak{M}^+$.)*

Proof The proof is by the same method as in Lemma 10.12, where

$$X_+(y,\xi') = T^{-1}(|\xi'|)\cdot X_+(y,\xi'/|\xi'|),$$
$$T_+(y,\xi') = |\xi'|T_+(y,\xi'/|\xi'|),$$
$$Y_+(y,\xi') = |\xi'|Y_+(y,\xi'/|\xi'|)\cdot S^{-1}(|\xi'|).$$

Remark The equations (30) to (34) still hold provided we define

$$F(c) = \operatorname{diag}(c^j T^{-1}(c))_{j=0}^{l-1} \quad \text{and} \quad E(c) = c\cdot\operatorname{diag}(c^j S^{-1}(c))_{j=0}^{l-1}.$$

From now on it is always assumed that the γ^+-spectral triple of $L(\lambda)$ is chosen with the properties (44). By the same method as in the proof of Lemma 10.12 it is clear that the triple

$$(X_-(y,\xi'), T_-(y,\xi'), Y_-(y,\xi')) := (T(-1)\cdot X_+(y,-\xi'),$$
$$-T_+(y,-\xi'), -Y_+(y,-\xi')\cdot S(-1)) \quad (45)$$

is a γ^--spectral triple of $L_{y,\xi'}(\lambda)$; it also has the homogeneity properties (44).

Let $\mathscr{B}$ be a general boundary operator. The (k,i) entry of $\pi\mathscr{B}$ is homogeneous of degree $m_k + t_i$, so that the (k,i) entry of the coefficient B_j of the corresponding matrix polynomial $B(\lambda)$ is homogeneous of degree $m_k + t_i - j$. In other words

$$B_j(y, c\xi') = M(c)B_j(y,\xi')T(c)c^{-j}, \qquad c>0 \quad (46)$$

for $j = 0, \ldots, \mu$. We define $\Delta_{\mathscr{B}}^+$ just as in (37) and then (38) holds without change. If $\pi\mathscr{B}$ is positive *and* negative homogeneous we define $\Delta_{\mathscr{B}}^-$ as in (37′) and then (38′) holds.

In general for DN operators we have $2r = \alpha < pl$, in which case the leading coefficient of $L(\lambda)$ is not invertible. We have a finite spectral pair (X, T)

$$X = [X_+ \quad X_-], \quad T = \begin{pmatrix} T_+ & \\ & T_- \end{pmatrix} \quad \text{and} \quad Y = \begin{pmatrix} Y_+ \\ Y_- \end{pmatrix}$$

but $\operatorname{col}(XT^j)_{j=0}^{l-1}$ is not invertible. A pair of equations of the form $\sum_{j=0}^{l-1} B_j X_\pm T_\pm^j = \Delta_\pm$ is still solvable for B_j since $\mathscr{X}\cdot\operatorname{row}(T^jY)_{j=0}^{l-1}$ is a left inverse of $\operatorname{col}(XT^j)_{j=0}^{l-1}$ but the solution is not unique (see §4.2). Hence Theorems 10.14 and 10.16 still hold, except for the uniqueness (and of course the condition of the homogeneity of $\pi\mathscr{B}$ is modified: $\pi\mathscr{B}$ has DN numbers $m_k + t_j$).

We end this section with the proof of a result which was needed in the proof of Lemma 9.36. Fix $y \in \partial\Omega$. In §9.4 we said that an $r \times p$ matrix $\omega(t) = \omega(\xi', t)$ satisfying the system

$$\pi\mathscr{A}\left(y, \xi', \frac{1}{i}\frac{d}{dt}\right)\omega = 0, \qquad t > 0,$$

$$\pi\mathscr{B}\left(y, \xi', \frac{1}{i}\frac{d}{dt}\right)\omega|_{t=0} = 1_r, \qquad t = 0,$$

is called a *canonical* matrix. In terms of the notation introduced in the present chapter, we have

$$\omega(\xi', t) = X_+(y, \xi')\, e^{itT_+(y,\xi')}(\Delta^+_{\mathscr{B}}(y, \xi'))^{-1}$$

The family of matrix polynomials $L_{y,\xi'}(\lambda)$, $(y, \xi') \in ST^*(\partial\Omega)$, satisfies the condition (*) at the beginning of §10.3, hence by compactness of $ST^*(\partial\Omega)$ there is a $\delta > 0$ such that the roots of $\det L_{y,\xi'}(\lambda) = 0$ satisfy $\operatorname{Im}\lambda \geqslant \delta > 0$ for all $(y, \xi') \in ST^*(\partial\Omega)$. Since the roots of $\det L_{y,\xi'}(\lambda) = 0$ are exactly the eigenvalues of $T_+(y, \xi')$ (see Proposition 2.7), it follows that

$$\|T^\nu_+(y, \xi')\, e^{iT_+(y,\xi')}\| \leqslant C_\nu\, e^{-\delta t},$$

for all $|\xi'| = 1$ and $\nu = 0, 1, \ldots,$ where C_ν depends on ν but not on ξ'. It follows that we have an estimate of the form

$$\int_0^\infty \left|\frac{d^\nu}{dt^\nu}\omega_{jk}(\xi', t)\right|^2 dt \leqslant C'_\nu < \infty, \qquad \forall|\xi'| = 1,$$

for the entries ω_{jk} of ω, a fact which was used in the proof of Lemma 9.36.

10.5 The classes $\mathrm{Ell}^{\mathbf{s},\mathbf{t}}$ and $\mathrm{BE}^{\mathbf{s},\mathbf{t},\mathbf{m}}$: two theorems of Agranovič and Dynin

In this section we prove two theorems, the first one on comparing the index of two boundary problems $(\mathscr{A}, \mathscr{B}_1)$ and $(\mathscr{A}, \mathscr{B}_2)$ having the same elliptic operator $\mathscr{A}$, and the second on reducing the transversal order of a boundary operator $\mathscr{B}$. These results are due to Agranovič and Dynin; also see §10.8 for another result of similar type.

Let $\mathbf{s} = (s_1, \ldots, s_p)$, $\mathbf{t} = (t_1, \ldots, t_p)$ and $\mathbf{m} = (m_1, \ldots, m_r)$. We let $\mathrm{Ell}^{\mathbf{s},\mathbf{t}}$ denote the set of $p \times p$ properly elliptic differential operators on $\bar{\Omega}$ with DN numbers $s_i + t_j$, and let

$$\mathrm{BE}^{\mathbf{s},\mathbf{t},\mathbf{m}}$$

be the set of pairs $(\mathscr{A}, \mathscr{B})$, where $\mathscr{A} \in \mathrm{Ell}^{\mathbf{s},\mathbf{t}}$ and $\mathscr{B}$ is a boundary operator on $\partial\Omega$ with DN numbers $m_k + t_j$ and *satisfying the L-condition.*

Notation. If we have two boundary value problems $(\mathscr{A}, \mathscr{B}^i) \in \mathrm{BE}^{\mathbf{s},\mathbf{t},\mathbf{m}}$, $i = 0, 1$, with the same elliptic operator $\mathscr{A}$ such that $\Delta^+_{\mathscr{B}^0} = \Delta^+_{\mathscr{B}^1}$, we write

$$(\mathscr{A}, \mathscr{B}^0) \approx (\mathscr{A}, \mathscr{B}^1).$$

Lemma 10.18 *If* $(\mathscr{A}, \mathscr{B}^0) \approx (\mathscr{A}, \mathscr{B}^1)$ *then* $\operatorname{ind}(\mathscr{A}, \mathscr{B}^0) = \operatorname{ind}(\mathscr{A}, \mathscr{B}^1)$.

Proof Let $\Delta^+_{\mathscr{B}^i} = \Delta$, $i = 0, 1$. Consider the boundary operators $\mathscr{B}^\tau = \tau\mathscr{B}^0 + (1-\tau)\mathscr{B}^1$, $0 \leqslant \tau \leqslant 1$. Since

$$\Delta^+_{\tau\mathscr{B}^0 + (1-\tau)\mathscr{B}^1} = \tau\cdot\Delta^+_{\mathscr{B}^0} + (1-\tau)\cdot\Delta^+_{\mathscr{B}^1} = \Delta$$

is invertible, then by Theorem 10.3(iii) the boundary operators $\mathscr{B}^\tau$ satisfy the L-condition relative to $\mathscr{A}$ for all $0 \leqslant \tau \leqslant 1$. Hence we have the linear homotopy

$$(\mathscr{A}, \mathscr{B}^\tau) = \tau\cdot(\mathscr{A}, \mathscr{B}^0) + (1-\tau)\cdot(\mathscr{A}, \mathscr{B}^1)$$

in the space of Fredholm operators $\mathscr{W}^{l+\mathbf{t}}(\Omega) \to \mathscr{W}^{l-\mathbf{s}}(\Omega) \times \mathscr{W}^{l-\mathbf{m}-1/2}(\partial\Omega)$, so the lemma follows by Theorem 9.12.

Sometimes it is convenient to modify the DN numbers of a boundary operator. Let $\Lambda^m \in OS^m(\partial\Omega)$, $m \in \mathbb{R}$, be a (-1) classical pseudo-differential operator on $\partial\Omega$ with principal symbol $|\xi'|^m$; then ind $\Lambda^m = 0$ by Lemma 8.68. For $\mathbf{m} \in \mathbb{R}^r$ let

$$\Lambda^{\mathbf{m}} = [\Lambda^{m_k}\,\delta_{kj}]^r_{k,j=1},$$

which is a diagonal matrix operator, so ind $\Lambda^{\mathbf{m}} = 0$ by Theorem 9.8. Given any boundary operator $\mathscr{B} = \sum^{\mu}_{j=0} \mathscr{B}_j(y, D')D^j_n$ with DN numbers $m_k + t_j$, the operator

$$\mathscr{B}' = \Lambda^{\tilde{\mathbf{m}}-\mathbf{m}} \circ \mathscr{B} = \sum_{j=0}^{\mu} \Lambda^{\tilde{\mathbf{m}}-\mathbf{m}}\mathscr{B}_j(y, D')D^j_n$$

has DN numbers $\tilde{m}_k + t_j$ where $\tilde{\mathbf{m}} \in \mathbb{R}^r$ is chosen arbitrarily. (In particular we may let $\tilde{m}_k = 0$, $k = 1, \ldots, r$). Clearly, $\mathscr{B}'$ satisfies the L-condition relative to $\mathscr{A}$ if and only if $\mathscr{B}$ does. It follows once again by Theorem 9.8 that $\operatorname{ind}(\mathscr{A}, \mathscr{B}') = \operatorname{ind}(I, \Lambda^{\tilde{\mathbf{m}}-\mathbf{m}}) + \operatorname{ind}(\mathscr{A}, \mathscr{B})$, whence

$$\operatorname{ind}(\mathscr{A}, \mathscr{B}') = \operatorname{ind}(\mathscr{A}, \mathscr{B}) \tag{47}$$

Theorem 10.19 *Let $(\mathscr{A}, \mathscr{B}_i)$, $i = 1, 2$, be two boundary value problems $\in \mathrm{BE}^{\mathbf{s},\mathbf{t},\mathbf{m}}$ having the same elliptic operator. Then*

$$\operatorname{ind}(\mathscr{A}, \mathscr{B}_2) = \operatorname{ind}(\mathscr{A}, \mathscr{B}_1) + \operatorname{ind} S$$

where $S \in OClS^0(\partial\Omega, r \times r)$ with principal symbol $\pi S = \Delta^+_{\mathscr{B}_2}\cdot(\Delta^+_{\mathscr{B}_1})^{-1}$ on $ST^*(\partial\Omega)$.

Note: $\Delta^+_{\mathscr{B}_1}$ and $\Delta^+_{\mathscr{B}_2}$ are defined by (37) with respect to any γ^+-spectral pair of $L(\lambda) = \pi\mathscr{A}(y, (\xi', \lambda))$.

Proof Observe that the $r \times r$ matrix function $\Delta^+_{\mathscr{B}_2}\cdot(\Delta^+_{\mathscr{B}_1})^{-1}$ is independent of the choice of spectral pair since any two such pairs are similar. In view of (47) we may assume that $\mathscr{B}_1$ and $\mathscr{B}_2$ have the same DN numbers, say $m_k = 0$. By Theorem 8.28 there exists a (-1) classical operator $S \in OS^0(\partial\Omega, r \times r)$ such that $\pi S = \Delta^+_{\mathscr{B}_1}\cdot(\Delta^+_{\mathscr{B}_2})^{-1}$. Consider the operator $(\mathscr{A}, S\mathscr{B}_1)$; since $\Delta^+_{S\mathscr{B}_1} = \pi S\cdot\Delta^+_{\mathscr{B}_1} = \Delta^+_{\mathscr{B}_2}$ then by Lemma 10.18 we have

$$\operatorname{ind}(\mathscr{A}, \mathscr{B}_2) = \operatorname{ind}(\mathscr{A}, S\mathscr{B}_1).$$

Now, since $(\mathscr{A}, S\mathscr{B}_1) = (I, S) \oplus (\mathscr{A}, \mathscr{B}_1)$ it follows by Theorem 9.8 that

$$\operatorname{ind}(\mathscr{A}, S\mathscr{B}_1) = \operatorname{ind}(I, S) + \operatorname{ind}(\mathscr{A}, \mathscr{B}_1) = \operatorname{ind} S + \operatorname{ind}(\mathscr{A}, \mathscr{B}_1),$$

so the proof is complete.

The particular values of the matrix function $\Delta_{\mathscr{B}}^+$ are of no significance because they depend on the choice of (X_+, T_+). In fact, replacing (X_+, T_+) by the similar pair $(X_+(\Delta_{\mathscr{B}}^+)^{-1}, \Delta_{\mathscr{B}}^+ T_+(\Delta_{\mathscr{B}}^+)^{-1})$, we may assume that $\Delta_{\mathscr{B}}^+ = I_r$, the $r \times r$ identity matrix. On the other hand, given two boundary operators $\mathscr{B}_1$ and $\mathscr{B}_2$ for the same elliptic operator $\mathscr{A}$ then the matrix $\Delta_{\mathscr{B}_2}^+(\Delta_{\mathscr{B}_1}^+)^{-1}$ does have a topological significance as Theorem 10.19 indicates.

It should be mentioned that for an elliptic operator $\mathscr{A}$ which is defined on a simply connected region, $\bar{\Omega}$, in the plane ($n = 2$), there is an essentially canonical choice of (X_+, T_+) that depends on the values of the coefficients of $\mathscr{A}$ on all of $\bar{\Omega}$ (not just $\partial\Omega$). In this case $\Delta_{\mathscr{B}}^+$ becomes an interesting function to study in its own right; see §15.3. In Chapter 16 we shall obtain a formula for the index of $(\mathscr{A}, \mathscr{B})$ in terms of a winding number of $\Delta_{\mathscr{B}}^+$ along $ST^*(\partial\Omega)$.

We turn now to prove a result which in view of Lemma 10.18 implies that, for the purpose of index calculations, there is no loss of generality in considering only those boundary operators where the transversal order, μ, is less than the degree, l, of the matrix polynomial $L(\lambda) = \pi\mathscr{A}(y, (\xi', \lambda))$.

Theorem 10.20 *Let* $(\mathscr{A}, \mathscr{B}) \in \mathrm{BE}^{\mathbf{s},\mathbf{t},\mathbf{m}}$. *Then for some boundary operator* $\mathscr{R}$ *with transversal order* $\mu \leqslant l - 1$ *and having the same* DN *numbers as* $\mathscr{B}$ *we have*

$$(\mathscr{A}, \mathscr{B}) \approx (\mathscr{A}, \mathscr{R})$$

i.e., $\mathscr{R}$ *can be chosen so that* $\Delta_{\mathscr{R}}^+ = \Delta_{\mathscr{B}}^+$. *Further, if* $\pi\mathscr{B}$ *is both positive and negative homogeneous then* $\mathscr{R}$ *can be chosen so that* $\Delta_{\mathscr{R}}^\pm = \Delta_{\mathscr{B}}^\pm$.

Proof Let (X_+, T_+, Y_+) be a matrix γ^+-spectral triple of $L(\lambda)$ satisfying, as usual, the homogeneity properties (44), and then (X_-, T_-, Y_-) the γ^--spectral triple defined by (45). Suppose first that $\pi\mathscr{B}$ is both positive and negative homogeneous. We wish to choose matrix functions R_j such that

$$\sum_{j=0}^{l-1} R_j(y, \xi')X_\pm T_\pm^j = \sum_{j=0}^{\mu} B_j(y, \xi')X_\pm T_\pm^j. \tag{48}$$

Thus we let R_j be defined by the expression on the right-hand side of (39), $j = 0, \ldots, l-1$, or in other words

$$[R_0 \cdots R_{l-1}] = \Delta_{\mathscr{B}}^+ \cdot [Y_+ \cdots T_+^{l-1} Y_+]\mathscr{Z} + \Delta_{\mathscr{B}}^- \cdot [Y_- \cdots T_-^{l-1} Y_-]\mathscr{Z}$$

It is clear that $R_j(y, c\xi') = M(c)R_j(y, \xi')T(c)c^{-j}$ for all $c \neq 0$ (see the proof of Theorem 10.14), so we let

$$\mathscr{R} = \sum_{j=0}^{l-1} \mathscr{R}_j(y, D')D_n^j$$

where $\mathscr{R}_j \in OS^{\mathbf{m}+\mathbf{t}-j}(\partial\Omega; r \times p)$ is a pseudo-differential operator on $\partial\Omega$ with principal symbol R_j, $j = 0, \ldots, l-1$. Then $\mathscr{R}$ has the same DN numbers as $\mathscr{B}$ and satisfies (48) by construction since

$$\Delta_{\mathscr{R}}^{+} = [R_0 \cdots R_{l-1}] \cdot \operatorname{col}(X_+ T_+^j)_{j=0}^{l-1} = \Delta_{\mathscr{B}}^{+} \cdot I + \Delta_{\mathscr{B}}^{-} \cdot 0 = \Delta_{\mathscr{B}}^{+},$$

using the equations (12) in §2.2. Similarly, $\Delta_{\mathscr{R}}^{-} = \Delta_{\mathscr{B}}^{-}$.

In the case where $\pi\mathscr{B}$ is just positive homogeneous we define R_j by the $+$ terms on the right-hand side of (39), or, in other words,

$$[R_0 \cdots R_{l-1}] = \Delta_{\mathscr{B}}^{+} \cdot [Y_+ \cdots T_+^{l-1} Y_+] \mathscr{X}$$

and then continue as before. The proof is complete.

Let $B(\lambda) = \pi\mathscr{B}(y, (\xi', \lambda))$ and $R(\lambda) = \pi\mathscr{R}(y, (\xi', \lambda))$ be the matrix polynomials associated with $\mathscr{B}$ and $\mathscr{R}$, respectively. If (48) holds it follows that

$$B(\lambda) = Q(\lambda) L(\lambda) + R(\lambda) \tag{49}$$

for some matrix polynomial $Q(\lambda)$. Indeed, $\{B(\lambda) - R(\lambda)\} L^{-1}(\lambda)$ has an analytic continuation inside γ^+:

$$\begin{aligned}\frac{1}{2\pi i} \int_{\gamma^+} \lambda^j \{B(\lambda) - R(\lambda)\} L^{-1}(\lambda)\, d\lambda &= [B_0 \cdots B_\mu] \cdot \operatorname{col}(X_+ T_+^j)_{j=0}^{\mu} \cdot T_+^j Y_+ \\ &\quad - [R_0 \cdots R_{l-1}] \cdot \operatorname{col}(X_+ T_+^k)_{j=0}^{l-1} \cdot T_+^j Y_+ \\ &= \Delta_{\mathscr{B}}^{+} \cdot T_+^j Y_+ - \Delta_{\mathscr{R}}^{+} \cdot T_+^j Y_+ = 0\end{aligned}$$

for all $j = 0, 1, \ldots$ since $\Delta_{\mathscr{R}}^{+} = \Delta_{\mathscr{B}}^{+}$. In the same way, one also shows that it has an analytic continuation inside γ^-, so it has an analytic continuation to the finite complex plane, Since it has at most a pole at ∞, it follows that $\{B(\lambda) - R(\lambda)\} L^{-1}(\lambda)$ is a polynomial, or in other words (49) holds for some $r \times p$ matrix polynomial $Q(\lambda)$ with coefficients depending smoothly on the parameters y, ξ'. (Note the connection with Theorem 3.11, except here the divisor $L(\lambda)$ in general does not have invertible leading coefficient and the quotient $Q(\lambda)$ and remainder $R(\lambda)$ are not unique.) Since $(\mathscr{A}, \mathscr{B}) \approx (\mathscr{A}, \mathscr{R})$ then as in Lemma 10.18 we have the linear homotopy

$$(\mathscr{A}, (1-t)\mathscr{R} + t\mathscr{B}), \qquad 0 \leqslant t \leqslant 1,$$

joining $(\mathscr{A}, \mathscr{B})$ and $(\mathscr{A}, \mathscr{R})$ in the space $\mathrm{BE}^{\mathbf{s},\mathbf{t},\mathbf{m}}$. This homotopy corresponds to the following homotopy of matrix polynomials:

$$\begin{aligned}B^t(\lambda) &= (1-t) \cdot R(\lambda) + t \cdot B(\lambda) \\ &= t \cdot (B(\lambda) - R(\lambda)) + R(\lambda) \\ &= tQ(\lambda)L(\lambda) + R(\lambda), \qquad 0 \leqslant t \leqslant 1.\end{aligned}$$

Remark The equation (49) could be used for checking the L-condition. If (49) holds then $(\mathscr{A}, \mathscr{B})$ satisfies the L-condition if and only if $(\mathscr{A}, \mathscr{R})$ satisfies the L-condition (since $\Delta_{\mathscr{B}}^{+} = \Delta_{\mathscr{R}}^{+}$).

10.6 Homotopies of elliptic boundary problems

Let I be the unit interval $[0, 1]$ in $\mathbb{R}$. A *homotopy of elliptic differential operators* is a family of $p \times p$ elliptic operators in $\mathrm{Ell}^{\mathbf{s},\mathbf{t}}$ (i.e. with fixed DN numbers $s_i + t_j$),

$$\mathscr{A}^\tau = \sum A_\alpha(\tau, x)D^\alpha, \qquad \tau \in I,$$

such that the coefficients $A_\alpha(t, \cdot)$ are continuous from I to $C^\infty(\bar{\Omega}, p \times p)$.

If the coefficients A_α are C^∞ functions of the variables $(\tau, x) \in I \times \bar{\Omega}$ then we say that the homotopy is a C^∞ *homotopy*.

The following fact is important for the next lemma: We can "speed up" a homotopy $\mathscr{A}^\tau$ to make it constant in a neighbourhood of $\tau = 1$ without affecting $\mathscr{A}^0$. Indeed, choose a function $\varphi \in C_0^\infty(\mathbb{R})$ such that $0 \leqslant \varphi \leqslant 1$, $\varphi(0) = 0$ and $\varphi(\tau) = 1$ when $1 - \delta \leqslant \tau \leqslant 1$. Then for the homotopy $\bar{\mathscr{A}}^\tau = \mathscr{A}^{\varphi(\tau)}$, $0 \leqslant \tau \leqslant 1$, we have

$$\bar{\mathscr{A}}^0 = \mathscr{A}^0, \qquad \bar{\mathscr{A}}^\tau = \mathscr{A}^1 \qquad \text{when } 1 - \delta \leqslant \tau \leqslant 1.$$

Similarly, we can "slow down" a homotopy to make it constant in a neighbourhood of $\tau = 0$ without affecting its value at $\tau = 1$.

Lemma 10.21 *If two elliptic operators are homotopic then they are C^∞ homotopic.*

Proof Let $\mathscr{A}^\tau$, $0 \leqslant \tau \leqslant 1$, be a (continuous) homotopy from $\mathscr{A}^0$ to $\mathscr{A}^1$. By definition, the maps $\tau \mapsto A_\alpha(\tau, \cdot)$ are continuous from I to $C^\infty(\bar{\Omega}, p \times p)$, i.e. for any multi-index γ we have

$$\sup_{x \in \bar{\Omega}} \|D_x^\gamma[A_\alpha(\tau, x) - A_\alpha(\tau_0, x)]\| < \varepsilon$$

if $|\tau - \tau_0|$ is sufficiently small. With $\gamma = 0$ this shows that A_α is continuous in τ uniformly with respect to x. Since A_α is continuous (in fact, smooth) in x for fixed τ, it follows that the matrix functions A_α are continuous on $I \times \bar{\Omega}$. By virtue of Chapter 8, Exercise 14, for any $\varepsilon > 0$ we can find $\tilde{A}_\alpha \in C^\infty(I \times \bar{\Omega})$ such that

$$\sup_{(\tau, x) \in I \times \bar{\Omega}} \|\tilde{A}_\alpha(\tau, x) - A_\alpha(\tau, x)\| < \varepsilon.$$

Then $\tilde{\mathscr{A}}^\tau := \sum \tilde{A}_\alpha(\tau, x)D^\alpha$, $0 \leqslant \tau \leqslant 1$, is a C^∞ homotopy of elliptic operators (ellipticity is preserved if ε is chosen sufficiently small), and, further, we can join $\tilde{\mathscr{A}}^1$ and $\mathscr{A}^i$, $i = 0, 1$, by a linear homotopy $\kappa\tilde{\mathscr{A}}^i + (1 - \kappa)\mathscr{A}^i$, $0 \leqslant \kappa \leqslant 1$. Finally, we can join the three homotopies to get

$$\begin{cases} \kappa\tilde{\mathscr{A}}^0 + (1 - \kappa)\mathscr{A}^0, & 0 \leqslant \kappa \leqslant 1, \\ \tilde{\mathscr{A}}^{\kappa - 1}, & 1 \leqslant \kappa \leqslant 2, \\ (3 - \kappa)\tilde{\mathscr{A}}^1 + (\kappa - 2)\mathscr{A}^1, & 2 \leqslant \kappa \leqslant 3 \end{cases}$$

which is a piecewise C^∞ homotopy from $\mathscr{A}^0$ to $\mathscr{A}^1$. It is piecewise C^∞ only because of the points $\kappa = 1$ and $\kappa = 2$, but it becomes C^∞ if we first modify the homotopies to make them constant in a neighbourhood of these points, as explained above.

Lemma 10.22 *Let $\mathscr{A}^\tau$, $\tau \in I$, be a homotopy of elliptic operators and suppose that $\mathscr{A}^0$ has a boundary operator satisfying the L-condition. Then there exists a γ^+-spectral triple*

$$(X_+(\tau, y, \xi'), T_+(\tau, y, \xi'), Y_+(\tau, y, \xi')) \tag{50}$$

of $L_{\tau, y, \xi'}(\lambda) := \pi\mathscr{A}^\tau(y, (\xi', \lambda))$ satisfying the homogeneity properties (44) *and consisting of matrices with entries that are continuous functions from I to $C^\infty(T^*(\partial\Omega)\backslash 0)$. Further, if $\mathscr{A}^\tau$ is a C^∞ homotopy then there exists a γ^+-spectral triple with entries in $C^\infty(I \times T^*(\partial\Omega)\backslash 0)$.*

Proof We will prove the lemma only for the case of C^∞ homotopy. (The general case follows in the same way once the results of §10.3 are generalized appropriately.) Let $\mathfrak{M}^+$ denote the vector bundle over $I \times ST^*(\partial\Omega)$ with fibres $\mathfrak{M}^+_{L_{\tau,y,\xi'}}$. Since the restriction of $\mathfrak{M}^+$ to $\{0\} \times ST^*(\partial\Omega)$ is trivial (due to the existence of the boundary operator for $\mathscr{A}^0$ satisfying the L-condition) then Theorem 5.62 implies that $\mathfrak{M}^+$ is also trivial. Hence there exists a γ^+-spectral triple (50) of $L_{\tau,y,\xi'}(\lambda)$ consisting of matrix functions with entries that are C^∞ functions on $I \times ST^*(\partial\Omega)$. The operators $\mathscr{A}^\tau$ have the same DN numbers for all $\tau \in I$, hence as in Lemma 10.17 we can extend this triple by homogeneity (44) so that it is a γ^+-spectral triple of $L_{\tau,y,\xi'}(\lambda)$ for all $\tau \in I$ and $(y, \xi') \in T^*(\partial\Omega)\backslash 0$.

Remark If $\mathscr{B}^0$ is a boundary operator for $\mathscr{A}^0$ satisfying the L-condition then by relacing (X_+, T_+, Y_+) with $(X_+M^{-1}, MT_+M^{-1}, MY_+)$ where $M = \Delta^+_{\mathscr{B}^0}$ we may assume that

$$\Delta^+_{\mathscr{B}^0} = r \times r \text{ identity matrix},$$

where $\Delta^+_{\mathscr{B}^0}$ is defined by (37) with respect to the γ^+-spectral pair $(X_+(0, \cdot),$ $T_+(0, \cdot))$.

A *homotopy of boundary problems* is a family $(\mathscr{A}^\tau, \mathscr{B}^\tau)$ of boundary value problems in $\mathrm{BE}^{\mathbf{s},\mathbf{t},\mathbf{m}}$ (i.e. satisfying the L-condition), where $\mathscr{A}^\tau$ is a homotopy in $\mathrm{Ell}^{\mathbf{s},\mathbf{t}}$ and, in addition,

$$\mathscr{B}^\tau = \sum \mathscr{B}_j^{(\tau)}(y, D')D_n^j, \qquad \tau \in I,$$

where each $\mathscr{B}_j^{(\tau)} \in OS^{\mathbf{m}+\mathbf{t}-j}(\partial\Omega, r \times p)$ is a (-1) classical operator with principal symbol $B_j(\tau, \cdot)$ whose restriction to the unit cotangent bundle $ST^*(\partial\Omega)$ is continuous from I to $C^\infty(ST^*(\partial\Omega), r \times p)$.

It is called a C^∞ *homotopy* if $\mathscr{A}^\tau$ is a C^∞ homotopy and the principal symbols of the operators $\mathscr{B}_j^{(\tau)}$ are C^∞ functions of the variables $(\tau, y, \xi') \in I \times ST^*(\partial\Omega)$. If two boundary value problems are homotopic then they are C^∞ homotopic. (Proof: As in Lemma 10.21 a continuous homotopy of principal symbols implies existence of a C^∞ homotopy and then we may apply Theorem 8.28 to lift the result to the operator level.)

The next theorem is one of the main results of this section. We have a map $\mathrm{BE}^{\mathbf{s},\mathbf{t},\mathbf{m}} \to \mathrm{Ell}^{\mathbf{s},\mathbf{t}}$ defined by $(\mathscr{A}, \mathscr{B}) \mapsto \mathscr{A}$, and we show that a given homotopy $\mathscr{A}^\tau$ of elliptic operators in $\mathrm{Ell}^{\mathbf{s},\mathbf{t}}$ can be "lifted" to a homotopy of boundary value problems in $\mathrm{BE}^{\mathbf{s},\mathbf{t},\mathbf{m}}$.

Theorem 10.23 *Let $\mathscr{A}^\tau$, $0 \leqslant \tau \leqslant 1$, be a homotopy (or C^∞ homotopy) of $p \times p$ elliptic differential operators in* $\mathrm{Ell}^{\mathbf{s},\mathbf{t}}$ *(i.e. with the* DN *numbers $s_i + t_j$). If $\mathscr{B}$ is a boundary operator satisfying the L-condition relative to $\mathscr{A}^0$ with* DN *numbers $m_k + t_j$ then there exists a homotopy (or C^∞ homotopy) of boundary value problems $(\mathscr{A}^\tau, \mathscr{B}^\tau) \in \mathrm{BE}^{\mathbf{s},\mathbf{t},\mathbf{m}}$ with $(\mathscr{A}^0, \mathscr{B}^0) \approx (\mathscr{A}^0, \mathscr{B})$.*

Remarks

(i) In particular, it follows that there exists a boundary operator $\mathscr{B}^1$ satisfying the L-condition relative to $\mathscr{A}^1$.

(ii) The condition $(\mathscr{A}^0, \mathscr{B}^0) \approx (\mathscr{A}^0, \mathscr{B})$ implies that the two boundary value problem operators are (linearly) homotopic, as in the proof of Lemma 10.18. Then by joining this homotopy with the homotopy $(\mathscr{A}^\tau, \mathscr{B}^\tau)$, as in the proof of Lemma 10.21, we obtain a homotopy from the given boundary value problem operator, $(\mathscr{A}^0, \mathscr{B})$, to the new one $(\mathscr{A}^1, \mathscr{B}^1)$.

Proof Again, we prove the theorem for C^∞ homotopies since that is sufficient for our purposes (and, besides, if a homotopy is given, then a C^∞ homotopy exists as shown above so there is little loss of generality). By Lemma 10.22 there exists a γ^+-spectral triple (50) with the properties indicated there. Also, we may define a γ^--spectral triple by the formula (45) and then a finite spectral triple (X, T, Y) as in the proof of Theorem 10.14 (see (40)); we need this triple in order to use §2.2(14). The matrices X, T and Y are C^∞ matrix functions on $I \times T^*(\partial\Omega)\backslash 0$. As in Theorem 10.14, there exist matrix functions B_j such that

$$\sum_{j=0}^{l-1} B_j(\tau, \cdot)X_+(\tau, \cdot)T_+^j(\tau, \cdot) = \Delta_{\mathscr{B}}^+, \qquad \forall \tau \in I, \tag{51}$$

namely,

$$B_j(\tau, \cdot) = \Delta_{\mathscr{B}}^+ \cdot \sum_{k=0}^{l-j-1} T_+^k(\tau, \cdot)Y_+(\tau, \cdot)A_{j+k+1}(\tau, \cdot)$$

(i.e. let $M_+ = \Delta_{\mathscr{B}}^+$ and $M_- = 0$ in the formula (14) of §2.2). Then $B_j \in C^\infty(I \times T^*(\partial\Omega)\backslash 0, r \times p)$, and, in view of the homogeneity properties (38), (43) and (44), it is easily seen that

$$B_j(\tau, y, c\xi') = M(c)B_j(\tau, y, \xi')T(c)c^{-j}, \qquad c > 0$$

for all $j = 0, \ldots, l-1$. Now let $\mathscr{B}_j^{(\tau)} \in OS^{\mathbf{m}+\mathbf{t}-j}(\partial\Omega, r \times p)$ be a (-1) classical pseudo-differential operator on $\partial\Omega$ with principal symbol $B_j(\tau, \cdot)$. Then the boundary operator $\mathscr{B}^\tau = \sum_{j=0}^{l-1} \mathscr{B}_j^{(\tau)} D_n^j$ has the same DN numbers as $\mathscr{B}$; further, the pair $(\mathscr{A}^\tau, \mathscr{B}^\tau)$ satisfies the L-condition for all $\tau \in I$ due to (51) and Theorem 10.3(iii), so it is a homotopy in $\mathrm{BE}^{\mathbf{s},\mathbf{t},\mathbf{m}}$. This completes the proof of the theorem because in view of (51) with $\tau = 0$ we have $\Delta^+_{\mathscr{B}_0} = \Delta^+_{\mathscr{B}}$, i.e. $(\mathscr{A}, \mathscr{B}^0) \approx (\mathscr{A}, \mathscr{B})$.

Remark Just as in Theorem 10.20 if $\pi\mathscr{B}^0$ is positive and negative homogeneous then the boundary operators $\mathscr{B}^\tau$ can also be constructed with this property. We let $M_+ = \Delta^+_{\mathscr{B}}(y, \xi')$, $M_- = \Delta^-_{\mathscr{B}}(y, \xi')$ in the formula §2.2 (14) to obtain a solution of the equations

$$\sum_{j=0}^{l-1} B_j(\tau, \cdot)X_\pm(\tau, \cdot)T^j_\pm(\tau, \cdot) = \Delta^\pm_{\mathscr{B}} \tag{51'}$$

of the form

$$B_j(\tau, \cdot) = \Delta^+_{\mathscr{B}} \cdot \sum_{k=0}^{l-j-1} T^k_+(\tau, \cdot)Y_+(\tau, \cdot)A_{j+k+1}(\tau, \cdot) + \Delta^-_{\mathscr{B}} \cdot \sum_{k=0}^{l-j-1} T^k_-(\tau, \cdot)Y_-(\tau, \cdot)A_{j+k+1}(\tau, \cdot)$$

and then continue as before.

Theorem 10.24 *Let $(\mathscr{A}_\tau, \mathscr{B}_\tau)$, $\tau \in I$, be a homotopy of boundary value problems in* $\mathrm{BE}^{\mathbf{s},\mathbf{t},\mathbf{m}}$. *Then the index of $(\mathscr{A}_\tau, \mathscr{B}_\tau)$ is independent of τ.*

Proof To simplify the notation let us consider the homogeneous case, i.e. $s_1 = \cdots = s_p = l$, $t_1 = \cdots = t_p = 0$. We have

$$\mathscr{A}_\tau = \sum A_\alpha(\tau, x)D^\alpha, \quad \mathscr{B}_\tau = \sum \mathscr{B}^{(\tau)}_j(y, D')D^j_n,$$

and when $s \geqslant l$, $s \geqslant \max(m_k + 1)$,

$$(\mathscr{A}_\tau, \mathscr{B}_\tau)\colon W^s_2(\Omega, \mathbb{C}^p) \to W^{s-l}_2(\Omega, \mathbb{C}^p) \times \mathop{\times}_{k=1}^{r} W^{s-m_k-1/2}_2(\partial\Omega),$$

is a bounded operator for each τ. If we can show that its norm depends continuously on τ then Theorem 9.12 will complete the proof.

The topology on $C^\infty(\bar{\Omega})$ is defined by the semi-norms

$$\|\varphi\|_{C^d} = \sup_{x\in\bar{\Omega},\, |\gamma|\leqslant d} |D^\gamma\varphi(x)|, \qquad d = 0, 1, 2, \ldots$$

Now, multiplication by a function $\varphi \in C^\infty(\bar{\Omega})$ defines a continuous map $W^s_2(\Omega) \to W^s_2(\Omega)$ with norm bounded by a semi-norm of φ, i.e. $\|\varphi\cdot u\|_s \leqslant C\|\varphi\|_{C^d}\|u\|_s$ for some d; we can take $d = s$ if s is an integer and $d = [s] + 1$ if $s \in \mathbb{R}_+$ (see [W1], p. 64). Since $D^\alpha\colon W^s_2(\Omega) \to W^{s-|\alpha|}_2(\Omega)$ is bounded, it follows that for any $\tau_0 \in I$

$$\|\mathscr{A}_\tau - \mathscr{A}_{\tau_0}\| \leqslant C\cdot\max_\alpha\|A_\alpha(\tau, \cdot) - A_\alpha(\tau_0, \cdot)\|_{C^d(\bar{\Omega})}, \qquad d \geqslant s - l + 1,$$

where $\|\ \|$ denotes the operator norm for maps $W^s_2(\Omega, \mathbb{C}^p) \to W^{s-l}_2(\Omega, \mathbb{C}^p)$,

$s \geqslant l$. By assumption, the matrices A_α are continuous from I to $C^\infty(\bar{\Omega}, p \times p)$, so the continuity of $\tau \mapsto \mathscr{A}_\tau$ as a map $I \to \mathscr{L}(W_2^s, W_2^{s-l})$ follows.

For the boundary operator, the entries in the kth row of the coefficient-operator $\mathscr{B}_j^{(\tau)}$ belong to $OClS^{m_k-j}(\partial\Omega)$, $k = 1, \ldots, r$, and then the continuity of $\tau \mapsto \mathscr{B}_j^{(\tau)}$ from

$$I \to \mathscr{L}\Big(W_2^{s-j-1/2}(\partial\Omega, \mathbb{C}^p), \mathop{\times}_{k=1}^{r} W_2^{s-m_k-1/2}(\partial\Omega)\Big)$$

follows by Theorem 8.81. Since D_n^j is bounded from $W_2^s(\Omega) \to W_2^{s-j-1/2}(\partial\Omega)$, this establishes the continuity of $\tau \mapsto \mathscr{B}_\tau$ as a map

$$I \to \mathscr{L}\left(W_2^s, \mathop{\times}_{k=1}^{r} W_2^{s-m_k-1/2}\right).$$

In virtue of Theorem 9.12, there exists $\delta > 0$ such that $\mathrm{ind}(\mathscr{A}_\tau, \mathscr{B}_\tau) = \mathrm{ind}(\mathscr{A}_{\tau_0}, \mathscr{B}_{\tau_0})$ when $|\tau - \tau_0| < \delta$. By connectedness of $I = [0, 1]$, it follows that $\mathrm{ind}(\mathscr{A}_1, \mathscr{B}_1) = \mathrm{ind}(\mathscr{A}_0, \mathscr{B}_0)$.

Remark A somewhat different way to see the continuity of the maps $\tau \mapsto \mathscr{B}_j^{(\tau)}$ is as follows. Let $1 = \sum \varphi_i$ be a partition of unity subordinate to an atlas (U_i, κ_i) on $\partial\Omega$, and as usual let $\psi_i \in C_0^\infty(U_i)$ be functions which are equal to 1 on supp φ_i. In the argument we may replace $\mathscr{B}_j^{(\tau)}$ by $\sum \varphi_i \mathscr{B}_j^{(\tau)} \psi_i$ since the second operator differs from the first by an element of $OP^{-\infty}(\partial\Omega)$ which can be deformed to 0 by means of a linear homotopy. In that case it follows from the estimate (86) in Chapter 8 that

$$\|\mathscr{B}_j^{(\tau)} - \mathscr{B}_j^{(\tau_0)}\| \leqslant \mathrm{const}\|B_j(\tau, \cdot) - B_j(\tau_0, \cdot)\|_{C^d(ST^*(\partial\Omega))}, \qquad \text{for some } d,$$

where the norm $\| \;\; \|$ on the left denotes the norm of operators $W_2^{s-j-1/2} \to W_2^{s-m_k-1/2}$ and $\| \;\; \|_{C^d(ST^*(\partial\Omega))}$ is some semi-norm in $C^\infty(ST^*(\partial\Omega))$ (cf. Remark 8.88).

10.7 The classes $\mathfrak{E}^l$ and $\mathfrak{BE}^{l,m}$

To apply Theorem 10.24 we must be able to construct the homotopies $\mathscr{A}_\tau$. In general, this is problematic if the operators $\mathscr{A}$ are restricted to differential operators, i.e. symbols which are polynomials in ξ. Consequently, we consider a larger class which keeps the polynomial nature in the normal direction, and in which homotopies are easily constructed. For simplicity, we consider only the case of homogeneous elliptic operators, i.e. with DN numbers $s_i = l$, $t_j = 0$.

Let a tubular neighbourhood of $\partial\Omega$ in $\mathbb{R}^n$ be identitied with $\partial\Omega \times (-1, 1)$, with $\partial\Omega \times [0, 1) \subset \bar{\Omega}$. Coordinates in $\partial\Omega \times (-1, 1)$ are denoted by (x', x_n) and those in $\partial\Omega$ by $y = (x', 0)$. We also write $\Omega_\delta = \partial\Omega \times [0, \delta)$. Let l be a positive integer; we consider operators of order l of the form

$$\mathscr{A} = \mathscr{A}^b + \mathscr{A}^i,$$

where $\mathscr{A}^i \in OClS^l(\mathbb{R}^n, p \times p)$ has compact support in $\bar{\Omega}\backslash\Omega_{1/2}$, and

$$\mathscr{A}^b = \sum_{j=0}^{l} \mathscr{A}_j D_n^j, \tag{52}$$

where the coefficients, $\mathscr{A}_j$, are functions of $x_n \in (-1, 1)$ with values in $OClS^{l-j}(\partial\Omega, p \times p)$ that vanish for $|x_n| > 2/3$, and satisfy the following condition. *For every coordinate patch U' on $\partial\Omega$ and all $\varphi, \psi \in C_0^\infty(U')$*:

$$\varphi\, \mathscr{A}_j \psi = \tilde{A}_j(x, D), \qquad \text{where } \tilde{A}_j \in S^{l-j}(\mathbb{R}^n \times \mathbb{R}^{n-1}, p \times p) \text{ is } (-1) \text{ classic.}$$

Further, the leading coefficient $\mathscr{A}_l$ is assumed to be a C^∞ matrix function on $\partial\Omega$, rather than a pseudo-differential operator.

Remark $S^{l-j}(\mathbb{R}^n \times \mathbb{R}^{n-1})$ is the set of symbols $\tilde{A}_j \in C^\infty(\mathbb{R}^n \times \mathbb{R}^{n-1})$ satisfying the estimates

$$|D_{\xi'}^\alpha D_x^\beta \tilde{A}_j(x, \xi')| \leqslant C_{\alpha\beta}(1 + |\xi'|)^{l-j-|\alpha|}, \qquad \forall x \in \mathbb{R}^n, \xi' \in \mathbb{R}^{n-1}$$

and $\tilde{A}_j$ is (-1) classic if it has a principal part $\pi\tilde{A}_j \in C^\infty(\mathbb{R}^n \times \mathbb{R}^{n-1})$ satisfying Definition 7.21.

It follows by Chapter 8, Exercise 18, that $\mathscr{A}_j := \mathscr{A}_j \otimes I$ is an operator of order $l - j$ in the tubular neighbourhood $\partial\Omega \times (-1, 1)$. Since $\mathscr{A}_j = 0$ when $x_n > 2/3$ it follows that

$$\mathscr{A}^b \colon W_2^s(\Omega, \mathbb{C}^p) \to W_2^{s-l}(\Omega, \mathbb{C}^p), \qquad s \geqslant l,$$

is continuous, i.e. $\mathscr{A}^b$ is an operator of order l. The same is true of course for the interior part, $\mathscr{A}^i$, since it is an ordinary symbol.

The principal symbol of $\mathscr{A}$ is defined to be the sum of those of $\mathscr{A}^b$ and $\mathscr{A}^i$. It is important to note that $\pi\mathscr{A}$ can be made *continuous* for all $(x, \xi) \in \bar{\Omega} \times \mathbb{R}^n\backslash 0$. We have $\pi\mathscr{A} = \pi\mathscr{A}^b + \pi\mathscr{A}^i$. By definition $\pi\mathscr{A}^i$ is already smooth for all $\xi \neq 0$, but for the boundary term

$$\pi\mathscr{A}^b(x, \xi', \lambda) = \sum_{j=0}^{l-1} \pi\mathscr{A}_j(x', \xi'; x_n)\lambda^j + \mathscr{A}_l(x', x_n)\lambda^l,$$

there is a possible difficulty at $\xi' = 0$. But the assumption on $\mathscr{A}_l$ means that we can regard it as a smooth function on $T^*(\partial\Omega) \times (-1, 1)$ (independent of ξ'), while for $j = 0, \ldots, l - 1$ the principal symbol of $\mathscr{A}_j$ is homogeneous of degree $l - j > 0$ and it is smooth for $(x', \xi'; x_n) \in T^*(\partial\Omega)\backslash 0 \times (-1, 1)$, so we may define $\pi\mathscr{A}_j = 0$ when $\xi' = 0$. In this way, *$\pi\mathscr{A}^b$ is continuously defined on $T^*(\partial\Omega) \times (-1, 1)$, including $\xi' = 0$.*

As usual, the operator $\mathscr{A}$ is said to be elliptic if $\det \pi\mathscr{A}(x, \xi) \neq 0$ for all $x \in \bar{\Omega}$, $0 \neq \xi \in \mathbb{R}^n$. Note that the assumption that $\mathscr{A}_l$ be independent of ξ' implies

$$\det \pi\mathscr{A}(y, \xi', \lambda) = a_m(x)\lambda^m + \cdots, \qquad m = pl,$$

with $a_m(x)$ independent of ξ', and $a_m(x) \neq 0$ due to ellipticity with $(\xi', \lambda) = (0, 1)$. Hence the degree of $\det \pi\mathscr{A}(y, \xi', \lambda)$ in the variable λ is $m = pl$, and

$\mathscr{A}_l$ is an invertible matrix. The continuity (and homogeneity) of $\pi\mathscr{A}$ implies

$$|\det \pi\mathscr{A}(x, \xi)| \geqslant c|\xi|^{pl} \qquad \forall \xi \neq 0, \tag{53}$$

where $c > 0$. The set of all $p \times p$ elliptic operators of order l is denoted $\mathfrak{E}^l$ (or sometimes $\mathfrak{E}^l_{p \times p}$).

The boundary operators that we consider are of the same type as before, i.e.

$$\mathscr{B} = \sum_{j=0}^{\mu} \mathscr{B}_j \cdot D_n^j,$$

where $\mathscr{B}_j$ is an $r \times p$ matrix with kth row having entries in $OClS^{m_k - j}(\partial\Omega)$, $k = 1, \ldots, r$. For convenience we assume $\mu \leqslant l - 1$. The pair $(\mathscr{A}, \mathscr{B})$ is said to be *L-elliptic* if $\mathscr{A}$ is elliptic and $\mathscr{B}$ satisfies the L-condition, i.e., we require only conditions (i) and (iii) of Definition 9.30. We denote the set of all such L-elliptic boundary value problem operators by $\mathfrak{BE}^{l,\mathbf{m}}$.

In the case that $\mathscr{A}$ is a differential operator, we showed in Remark 9.31 that proper ellipticity is a consequence of L-ellipticity (by virtue of the negative homogeneity of $\pi\mathscr{A}$) but in the present context all that we can conclude is that the number of roots of $\det \pi\mathscr{A}(y, \xi', \lambda) = 0$ in the upper half-plane $\operatorname{Im} \lambda > 0$ is independent of ξ', i.e. equal to the number of boundary conditions. (Note: $r = 0$ is possible; for instance, the operator $\mathscr{A}_+$ in Theorem 11.1 is Fredholm *without* boundary conditions.)

Now consider Theorem 9.32 for the operator $(\mathscr{A}, \mathscr{B}) \in \mathfrak{BE}^{l,\mathbf{m}}$. We claim that the theorem remains true as stated. Here we are dealing not with general DN operators, but only with operators $\mathscr{A}$ whose principal part is homogeneous of degree l, i.e. $s_i = 0$, $t_j = l$. Note: In the statement of Theorem 9.32 the letter l was used in a different sense, as an index for the Sobolev spaces.

Lemma 10.25 *Let $\mathscr{A} \in \mathfrak{E}^l$. Then for any $\varphi \in C^\infty(\bar{\Omega})$ we have*

$$[\mathscr{A}, \varphi] = \mathscr{A}\varphi - \varphi\mathscr{A} \in OP^{l-1}(\Omega),$$

i.e. an operator of order $l - 1$.

Proof First extend φ to $\mathbb{R}^n$ to obtain $\varphi \in C_0^\infty(\mathbb{R}^n)$. By definition, $\mathscr{A} = \mathscr{A}^b + \mathscr{A}^i$ where $\mathscr{A}^i \in OS^l(\mathbb{R}^n)$ has support in $\Omega \backslash \Omega_{1/2}$ and

$$\mathscr{A}^b = \sum_{j=0}^{l} \mathscr{A}_j D_n^j,$$

where $\mathscr{A}_j$, $j < l$, has the properties mentioned above and $\mathscr{A}_l$ is a C^∞ matrix function. Then $\mathscr{A}^b\varphi u = \varphi\mathscr{A}^b u + \tilde{\mathscr{A}}^b u$, where $\tilde{\mathscr{A}}^b = \sum_{j<l} [\mathscr{A}_j, \varphi] D_n^j u + \sum \mathscr{A}_j \mathscr{R}_j$ and $\mathscr{R}_j$ is a differential operator in D_n of order $<j$. (Note that we have used the fact that $\mathscr{A}_l$ is just a matrix function rather than a p.d.o.) By virtue of Chapter 8, Exercise 18, it follows that $\tilde{\mathscr{A}}^b \in OP^{l-1}(\Omega)$. Since

$$[\mathscr{A}, \varphi] = \tilde{\mathscr{A}}^b + [\mathscr{A}^i, \varphi],$$

and $[\mathscr{A}^i, \varphi] \in OS^{l-1}$ by Corollary 7.13, the proof is complete.

This lemma is all that we need to carry over the proof of Theorem 9.32 to operators in $\mathfrak{BE}^{l,\mathbf{m}}$. Consider the proof of the implication (a) → (b) in that theorem. Steps 1 and 2 obviously hold because they are just results concerning operators of constant coefficients. In the proof of Step 3 we referred to quasi-locality of p.d.o.'s. Even though quasi-locality does not hold in the present context, the fact that $[\mathscr{A}, \varphi]$ is an operator of order $l-1$ is sufficient to carry over the proof of the existence of local regularizers in Step 3 and then Step 4 follows from Step 3 as before. Note that in the present context we do not use Lemma 7.41; we use the second version of the Freezing Lemma (Lemma 7.41′) instead.

The implications (b) → (c) and (c) → (d) are functional analytic and hold once again. It is also clear that the proof of (d) → (a) is still valid due to the lemma above.

For future reference we summarize some of the properties of the boundary value problem operator $\mathfrak{L} = (\mathscr{A}, \mathscr{B})$ in the following theorem. However, we postpone until §14.4 the proof that the image is the orthogonal space of a subspace $\subset C^\infty$.

Theorem 10.26 *If $(\mathscr{A}, \mathscr{B}) \in \mathfrak{BE}^{l,\mathbf{m}}$, then the continuous operator*

$$\mathfrak{L} = (\mathscr{A}, \mathscr{B}): W_2^s(\Omega, \mathbb{C}^p) \to W_2^{s-l}(\Omega, \mathbb{C}^p) \times \mathop{\times}_{k=1}^{r} W_2^{s-m_k-1/2}(\partial\Omega),$$

$s \geqslant l$, is Fredholm. The kernel is a finite dimensional subspace of $C^\infty(\bar{\Omega}, \mathbb{C}^p)$ and the image is the orthogonal space of a finite dimensional subspace of

$$C^\infty(\bar{\Omega}, \mathbb{C}^p) \times \mathop{\times}_{k=1}^{r} C^\infty(\partial\Omega).$$

Thus, the index is independent of s. It is also independent of the lower-order terms in $\mathscr{A}$ and in $\mathscr{B}$.

If $(\mathscr{A}, \mathscr{B}) \in \mathfrak{BE}^{l,\mathbf{m}}$, the principal symbol of $\mathscr{B}$ has the form $B_{y,\xi'}(\lambda) = \pi B(y, (\xi', \lambda)) = \sum_{j=0}^{\mu} B_j(y, \xi')\lambda^j$, a matrix polynomial in λ. The results of §§10.1 to 10.3 carry over to $\mathfrak{BE}^{l,\mathbf{m}}$ since they are essentially just results concerning the associated matrix polynomials depending on parameters. In particular, by Theorem 10.3, the L-condition for $(\mathscr{A}, \mathscr{B})$ holds if only and only if

$$\det \Delta_{\mathscr{B}}^{+} = \det \sum_{j=0}^{\mu} B_j(y, \xi') X_+ T_+^j \neq 0$$

where (X_+, T_+) is a spectral pair of $L_{y,\xi'}(\lambda) = \pi\mathscr{A}(y, (\xi', \lambda))$ with respect to the eigenvalues in the upper half-plane $\operatorname{Im} \lambda > 0$.

The same is true of §10.4, except that there is no connection between spectral triples for the upper and lower half-plane as in (35) because $\pi\mathscr{A}$ is now only *positive* homogeneous in ξ.

In the following definitions, it is to be understood that we work with a fixed tubular neighbourhood of $\partial\Omega$. A homotopy of elliptic operators is a family of elliptic operators in $\mathfrak{E}^l$,

$$\mathscr{A}_\tau = \mathscr{A}_\tau^b + \mathscr{A}_\tau^i, \qquad 0 \leqslant \tau \leqslant 1,$$

in which the principal symbol of the operator $\mathscr{A}_\tau^i \in OS^l(\mathbb{R}^n)$ is a continuous function from I to $C^\infty(\bar{\Omega} \times \mathbb{R}^n, p \times p)$, and $\mathscr{A}_\tau^b = \sum \mathscr{A}_j^{(\tau)} D_n^j$ where the principal symbol of $\mathscr{A}_j^{(\tau)}$ is continuous from I to $C^\infty(ST^*(\partial\Omega) \times (-1, 1))$.

A *homotopy of boundary problems* is a family of boundary value problems $(\mathscr{A}_\tau, \mathscr{B}_\tau) \in \mathfrak{B}\mathfrak{E}^{l,\mathbf{m}}$, $0 \leqslant \tau \leqslant 1$, such that $\mathscr{A}_\tau$ is a homotopy in $\mathfrak{E}^l$ and

$$\mathscr{B}_\tau = \sum \mathscr{B}_j^{(\tau)} D_n^j,$$

where the operators $\mathscr{B}_j^{(\tau)}$ have principal symbols $B_j(\tau, \cdot)$ which are continuous from I to $C^\infty(ST^*(\partial\Omega))$.

It is called a C^∞ *homotopy* if the principal symbols of $\mathscr{A}_\tau^i$, $\mathscr{A}_j^{(\tau)}$ and $\mathscr{B}_j^{(\tau)}$ are C^∞ functions of the variables $(\tau, x, \xi) \in I \times \Omega \times \mathbb{R}^n$, $(\tau, x', \xi', x_n) \in I \times ST^*(\partial\Omega) \times (-1, 1)$ and $(\tau, y, \xi') \in I \times ST^*(\partial\Omega)$, respectively. As before, if two boundary value problems are homotopic then they are C^∞ homotopic.

The proofs of Lemma 10.22 and of Theorem 10.23 carry over to show that a homotopy of elliptic operators in $\mathfrak{E}^l$ can be lifted to a homotopy of boundary value problem operators in $\mathfrak{B}\mathfrak{E}^{l,\mathbf{m}}$, i.e. we have the following theorem.

Theorem 10.27 *Let $\mathscr{A}_\tau$, $0 \leqslant \tau \leqslant 1$, be a homotopy (or C^∞ homotopy) of $p \times p$ elliptic operators in $\mathfrak{E}^l$. If $\mathscr{B}$ is a boundary operator satisfying the L-condition relative to $\mathscr{A}^0$ with order m_k in the kth row, $k = 1, \ldots, r$, then there exists a homotopy (or C^∞ homotopy) of boundary value problems $(\mathscr{A}_\tau, \mathscr{B}_\tau) \in \mathfrak{B}\mathfrak{E}^{l,\mathbf{m}}$, $0 \leqslant \tau \leqslant 1$, with $(\mathscr{A}_0, \mathscr{B}_0) \approx (\mathscr{A}_0, \mathscr{B})$.*

Remark In particular, it follows that there exists a boundary operator $\mathscr{B}^1$ satisfying the L-condition relative to $\mathscr{A}^1$.

As mentioned above, however, we cannot use spectral triples for the lower half-plane, so the remark after Theorem 10.23 does not hold in this case. The proof of Theorem 10.19 holds without change. Finally, it will be straightforward to generalize the proof of Theorem 10.24 once we have developed the appropriate symbol calculus in §14.4 that is needed to handle the elliptic operator $\mathscr{A}$ near $\partial\Omega$. (We leave the details for the reader.)

Lemma 10.28 *Any $\mathscr{A} \in \mathfrak{E}^l$ is homotopic to an operator $\mathscr{A}' \in \mathfrak{E}^l$ which is equal to $\mathscr{A}$ on $\partial\Omega$ and which near the boundary has the form $\sum \mathscr{A}'_j D_n^j$ where the operators $\mathscr{A}'_j \in OS^{l-j}(\partial\Omega)$ do not depend on x_n.*

Proof We have $\mathscr{A} = \mathscr{A}^b + \mathscr{A}^i$, where $\mathscr{A}^b = \sum \mathscr{A}_j D_n^j$. Choose $\delta < \frac{1}{2}$ so that the support of $\mathscr{A}^i$ does not meet $\partial\Omega \times [0, \delta]$, choose $\varphi \in C_0^\infty((-\delta, \delta))$, $0 \leqslant \varphi \leqslant 1$, with $\varphi(x_n) = 1$ for $0 \leqslant x_n \leqslant \delta/2$, and put

$$\mathscr{A}_\tau = \mathscr{A}_\tau^b + \mathscr{A}^i, \qquad \text{where } \mathscr{A}_\tau^b = \sum \mathscr{A}_j(x', x_n[1 - \tau\varphi(x_n)]) D_n^j.$$

The operators $\mathscr{A}_\tau$ have principal symbol arbitrarily close to that of $\mathscr{A}$ if the support of φ is chosen sufficiently small; since $\mathscr{A}$ is elliptic, it follows that $\mathscr{A}_\tau$ is elliptic for all $0 \leqslant \tau \leqslant 1$ if δ is sufficiently small. Also note that $\mathscr{A}_\tau = \mathscr{A}$ on the boundary $x_n = 0$ for all $0 \leqslant \tau \leqslant 1$. This homotopy satisfies the conditions of the lemma since $\mathscr{A}_0 = \mathscr{A}$ and when $\tau = 1$ the coefficients of $\mathscr{A}_1$ do not depend on x_n for $x_n \leqslant \delta/2$.

Remark If $(\mathscr{A}, \mathscr{B}) \in \mathfrak{BE}^{l,\mathbf{m}}$ then $(\mathscr{A}_\tau, \mathscr{B}) \in \mathfrak{BE}^{l,\mathbf{m}}$ for all $0 \leqslant \tau \leqslant 1$, since $\mathscr{A}_\tau$ does not depend on τ at the boundary.

We can strengthen the conclusion of Lemma 10.28 as follows.

Corollary 10.29 *Any $\mathscr{A} \in \mathfrak{E}^l$ is homotopic to an operator $\tilde{\mathscr{A}} \in \mathfrak{E}^l$ which is equal to $\mathscr{A}$ on $\partial\Omega$ and which has the form*

$$\tilde{\mathscr{A}} = \varphi(x_n)\cdot\sum \tilde{\mathscr{A}}_j D_n^j + \mathscr{A}^i \tag{54}$$

where $\varphi(x_n) \in C_0^\infty((-\delta, \delta))$ is equal to 1 for $0 \leqslant x_n \leqslant \delta/2$, $\delta > 0$ is sufficiently small, the operators $\tilde{\mathscr{A}}_j \in OS^{l-j}(\partial\Omega, p \times p)$ do not depend on x_n, and $\mathscr{A}^i \in OS^l(\mathbb{R}^n, p \times p)$ has compact support in $\bar{\Omega}\backslash\Omega_{\delta/3}$.

Proof Suppose $\mathscr{A}'$ has the property stated in Lemma 10.28, i.e. $\mathscr{A}' = \mathscr{A}'^b + \mathscr{A}^i$ where $\mathscr{A}'^b = \sum \mathscr{A}'_j D_n^j$ for $0 \leqslant x_n \leqslant \delta$ and the operators $\mathscr{A}'_j$ do not depend on x_n, and the support of $\mathscr{A}^i$ does not meet $\partial\Omega \times [0, \delta]$. Then write

$$\mathscr{A}' = \varphi\cdot\mathscr{A}'^b + (1-\varphi)\cdot\mathscr{A}'^b + \mathscr{A}^i;$$

since the terms $(1-\varphi)\mathscr{A}'_j$, $j = 0, \ldots, l-1$, have compact support in $\bar{\Omega}\backslash\Omega_{\delta/3}$ and have order $l - j > 0$, then using Lemma 8.78 (and a partition of unity on $\partial\Omega$) we can find a genuine p.d.o. $\tilde{\mathscr{A}}^i$ in $OClS^l(\mathbb{R}^n)$ with compact support in $\bar{\Omega}\backslash\Omega_{\delta/3}$ and symbol so close to that of $(1-\varphi)\cdot\mathscr{A}'^b + \mathscr{A}^i$ that

$$(1-\kappa)\mathscr{A}_1 + \kappa\tilde{\mathscr{A}}$$

is elliptic for $0 \leqslant \kappa \leqslant 1$ if $\tilde{\mathscr{A}} = \varphi\cdot\mathscr{A}'^b + \tilde{\mathscr{A}}^i$.

Let $\mathfrak{MP}^l$, or $\mathfrak{MP}^l_{p\times p}$, denote the class of $p \times p$ matrix polynomials

$$L(\lambda) = L_{y,\xi'}(\lambda) = \sum_{j=0}^{l} A_j(y, \xi')\cdot\lambda^j$$

with coefficients depending smoothly on $(y, \xi') \in ST^*(\partial\Omega)$ such that A_l does not depend on ξ', $\det A_l(y) \neq 0$, and $\det L(\lambda) \neq 0$ for all $\lambda \in \mathbb{R}$. The coefficients $\tilde{A}_j$ are then extended to be homogeneous of degree $l - j$:

$$A_j(y, \xi') = A_j\left(y, \frac{\xi'}{|\xi'|}\right)\cdot|\xi'|^{l-j}, \qquad j = 0, \ldots, l-1, \tag{55}$$

or stated another way

$$L_{y,c\xi'}(\lambda) = c^l L_{y,\xi'}(\lambda/c), \qquad c > 0 \tag{55'}$$

Also we agree that $L_{y,0}(\lambda) := A_l(y)\lambda^l$ when $\xi' = 0$. A homotopy in $\mathfrak{MP}^l$ has the form $L_\tau(\lambda) = \sum_{j=0}^{l} A_j(\tau, y, \xi')\lambda^j$, $0 \leqslant \tau \leqslant 1$, where the coefficients A_j are continuous from I to $C^\infty(ST^*(\partial\Omega), p \times p)$. It is called a C^∞ *homotopy* if the coefficients A_j depend smoothly on τ, y and ξ'.

The following result can be viewed as a "homotopy extension theorem" for elliptic operators, i.e. a homotopy of $\mathscr{A}$ on the boundary can be extended to a homotopy on all of $\bar{\Omega}$.

Theorem 10.30 *Let $\mathscr{A} \in \mathfrak{E}^l$, where the coefficients of $\mathscr{A}$ do not depend on x_n near the boundary and let $L_\tau(\lambda)$, $0 \leqslant \tau \leqslant 1$, be a homotopy (or C^∞ homotopy) in $\mathfrak{MP}^l$ such that $L_0(\lambda) = \pi\mathscr{A}(y, (\xi', \lambda))$. Then there exists a homotopy (or C^∞ homotopy) $\mathscr{A}_\tau$ in $\mathfrak{E}^l$ such that $\mathscr{A}_0 = \mathscr{A}$ and $L_\tau(\lambda) = \pi\mathscr{A}_\tau(y, (\xi', \lambda))$.*

Proof Once again we focus only on the case of C^∞ homotopy. Write $\mathscr{A} = \mathscr{A}^b + \mathscr{A}^i$ and choose $\delta > 0$ so small that the support of $\mathscr{A}^i$ does not meet $\partial\Omega \times [0, \delta]$, and such that the coefficients of $\mathscr{A} = \mathscr{A}^b$ do not depend on x_n for $0 \leqslant x \leqslant \delta$. Let $\varphi \in C_0^\infty((-\delta, \delta))$ be equal to 1 in a neighbourhood of 0 and $0 \leqslant \varphi \leqslant 1$ everywhere. Write $L_\tau(\lambda) = \sum_{j=0}^{l} A_j(\tau, y, \xi')\lambda^j$, and choose operators $\mathscr{A}_j(\tau, y, D') \in OClS^{l-j}(\partial\Omega)$ depending smoothly on $\tau \in [0, 1]$ with principal part $A_j(\tau, y, \xi')$. Define $\mathscr{A}_\tau^b = \mathscr{A}^b$ when $x_n \geqslant \delta$ and

$$\mathscr{A}_\tau^b = \sum_{j=0}^{l} \mathscr{A}_j(\tau\varphi(x_n), x', D')\cdot D_n^j \qquad \text{when } x_n \leqslant \delta,$$

and set $\mathscr{A}_\tau := \mathscr{A}_\tau^b + \mathscr{A}^i$. The operator $\mathscr{A}_\tau$ is elliptic. Indeed, $\mathscr{A}_\tau = \mathscr{A}$ when $x_n \geqslant \delta$ (including points in Ω outside the collar) so it remains to verify ellipticity for $x_n \leqslant \delta$. The assumption on the support of $\mathscr{A}^i$ implies that $\pi\mathscr{A}_\tau = \pi\mathscr{A}_\tau^b$ when $x_n \leqslant \delta$. Since

$$\begin{aligned}\pi\mathscr{A}_\tau(y, (\xi', \lambda)) &= \pi\mathscr{A}_\tau^b(y, (\xi', \lambda)) \\ &= \sum_{j=0}^{l} A_j(\tau, y, \xi')\lambda^j = L_\tau(\lambda),\end{aligned}$$

and $L_\tau(\lambda) \in \mathfrak{MP}^l$, it is evident that $\mathscr{A}_\tau$ is elliptic when $x_n \leqslant \delta$. The fact that $\mathscr{A}_0 = \mathscr{A}$ is clear.

10.8 Comparing the index of two problems having the same boundary operator

If M is a compact, closed manifold then as in §8.10 we let $\mathscr{P}(M, p \times p)$ denote the closure of $OClS^0(M, p \times p)$ in the L_2 operator norm. By stereographic projection we may regard the sphere $M = S^n$ as the one-point compactification $\mathbb{R}^n \cup \infty$, which makes $\bar{\Omega}$ a submanifold of S^n.

The next theorem is valid only for boundary value problems where the elliptic operator is homogeneous of degree l.

Theorem 10.31 *Let $(\mathscr{A}_i, \mathscr{B})$, $i = 1, 2$, be boundary value problems $\in \mathfrak{BE}^{l,\mathbf{m}}$ having the same boundary operator $\mathscr{B}$. Suppose that the principal symbols $\pi\mathscr{A}_1(x, \xi)$ and $\pi\mathscr{A}_2(x, \xi)$ coincide for $x \in \partial\Omega$. Then*

$$\operatorname{ind}(\mathscr{A}_1, \mathscr{B}) = \operatorname{ind}(\mathscr{A}_2, \mathscr{B}) + \operatorname{ind} Q,$$

where $Q \in \mathscr{P}(S^n, p \times p)$ has the principal symbol

$$q(x, \xi) = \begin{cases} 1_p & \text{if } x \in S^n \backslash \Omega \\ \pi\mathscr{A}_1(x, \xi)\pi\mathscr{A}_2(x, \xi)^{-1} & \text{if } x \in \bar{\Omega} \end{cases}$$

Proof Since $\pi\mathscr{A}_i(x, \xi)$, $i = 1, 2$, is continuous for all $\xi \neq 0$ (see §10.7), then q defines a continuous map $ST^*(S^n)\backslash 0 \to GL_p(\mathbb{C})$. Hence by Theorem 8.81 there exists an operator $Q \in \mathscr{P}(S^n, p \times p)$ with principal symbol q; since q has invertible values it follows that the operator Q: $L_2(S^n, \mathbb{C}^p) \to L_2(S^n, \mathbb{C}^p)$ is Fredholm.

By virtue of a small deformation (homotopy) as in Lemma 10.28, we may assume that $\pi\mathscr{A}_1$ and $\pi\mathscr{A}_2$ coincide in a neighbourhood on $\partial\Omega$. Then there is some open set $V \subset \bar{V} \subset \Omega$ such that q is equal to 1_p on $S^n \backslash V$. Now if $\mathscr{A}_1$ and $\mathscr{A}_2$ were differential operators it would then follow immediately that q is C^∞ for $\xi \neq 0$ and homogeneous of degree 0 in ξ and hence is the principal symbol of a p.d.o. of order 0 on S^n. For general $\mathscr{A}_1$ and $\mathscr{A}_2$, however, this is not true because of the singularities at $\xi' = 0$. Nevertheless, we can apply Lemma 8.78 to find a p.d.o. $Q \in OClS^0(Sn, p \times p)$ with principal symbol arbitrarily close to q (in the sup norm). Assume that this has been done.

Now, we wish to choose $\tilde{Q} \in OClS^0(S^n, p \times p)$ so that $\pi\tilde{Q} = \pi Q$ but with the additional property that

$$\tilde{Q}u = u \quad \text{when } u = 0 \text{ in } \Omega$$

and

$$\tilde{Q}u = u \text{ in } S^n \backslash \bar{V}.$$

This can be achieved by choosing $\varphi, \psi \in C_0^\infty(\Omega)$ with $\varphi = 1$ in $\bar{V}$ and $\psi = 1$ in $\operatorname{supp} \varphi$ and then setting $\tilde{Q} = \varphi Q \psi + (1 - \varphi)I$ which has the same principal symbol as Q. The operator $\tilde{Q} \circ \mathscr{A}_2 \in \mathfrak{E}^l$ then has principal symbol arbitrarily close to that of $\mathscr{A}_1$ (by choice of Q), so we have

$$\operatorname{ind}(\mathscr{A}_1, \mathscr{B}) = \operatorname{ind}(\tilde{Q} \circ \mathscr{A}_2, \mathscr{B}).$$

On the other hand by Theorem 9.8 it follows that

$$\operatorname{ind}(\tilde{Q} \circ \mathscr{A}_2, \mathscr{B}) = \operatorname{ind}(\tilde{Q}, I) \circ (\mathscr{A}_2, \mathscr{B}) = \operatorname{ind} \tilde{Q} + \operatorname{ind}(\mathscr{A}_2, \mathscr{B}),$$

and the proof is complete once we take into account the following lemma which says that $\tilde{Q}$ has the same index whether it is regarded as an operator on Ω or on S^n.

Lemma 10.32 *Let $V \subset \bar{V} \subset \Omega$, and let T: $W_2^k(S^n) \to W_2^k(S^n)$, $k \geqslant 0$, be a continuous operator such that $Tu = u$ when $u = 0$ on Ω, and $Tu = u$ in $S^n \backslash \bar{V}$ for every u. There is a continuous operator $\tilde{T}$: $W_2^k(\Omega) \to W_2^k(\Omega)$ uniquely* *defined by equation*

$$\tilde{T}\tilde{u} = Tu|_{\Omega},$$

where $u \in W_2^k(S^n)$ *is any extension of* $\tilde{u}$, *i.e.* $u|_{\Omega} = \tilde{u}$. *Then* $\tilde{T}$ *is Fredholm if and only if* T *is Fredholm, and*

$$\operatorname{ind} \tilde{T} = \operatorname{ind} T.$$

Proof Note that $\tilde{T}$ is well-defined, for if $u_i \in W_2^k(S^n)$ are both extensions of $\tilde{u}$ then $u_1 - u_2 = 0$ in Ω, whence $T(u_1 - u_2) = u_1 - u_2$, which implies $Tu_1 = Tu_2$ in Ω. It is evident that $\tilde{T}$ is continuous because there exists a continuous extension operator β: $W_2^k(\Omega) \to W_2^k(S^n)$ and the restriction operator $\imath^+$: $W_2^k(S^n) \to W_2^k(\Omega)$ is also continuous.

Let $u \in \ker T$; then $\tilde{T}\imath^+ u = 0$ and $u = 0$ in $S^n \setminus \bar{V}$. Conversely, let $\tilde{u} \in \ker \tilde{T}$; then $\tilde{u} = 0$ in $\Omega \setminus \bar{V}$ so it can be extended by 0 on all $S^n \setminus \bar{V}$ giving us a function $u \in W_2^k(S^n)$ with $Tu = 0$ and $\imath^+ u = \tilde{u}$. Hence the restriction operator $\imath^+$ yields an isomorphism from $\ker T$ to $\ker \tilde{T}$.

Now suppose that N is a finite dimensional complement of $\operatorname{im} T$,

$$W_2^k(S^n) = \operatorname{im} T \oplus N. \tag{56}$$

It is clear that $\imath^+(\operatorname{im} T) = \operatorname{im} \tilde{T}$, and we claim that

$$W_2^k(\Omega) = \imath^+(\operatorname{im} T) \oplus \imath^+ N. \tag{56$'$}$$

Indeed, by applying the restriction operator $\imath^+$ we certainly obtain $W_2^k(\Omega) = \imath^+(\operatorname{im} T) + \imath^+ N$. Now, if $\tilde{u} \in \imath^+(\operatorname{im} T) \cap \imath^+ N$ then $\tilde{u} = \imath^+ Tw = \imath^+ u$ where $w \in W_2^k(S^n)$, $u \in N$. Hence

$$Tw - u = 0 \quad \text{in} \quad \Omega;$$

by the assumption on T this means $Tw - u \in \operatorname{im} T$, whence $u \in \operatorname{im} T \cap N = 0$, so $\tilde{u} = 0$. Further, $\imath^+$ is injective on N, for if $u \in N$ and $\imath^+ u = 0$ then $u = 0$ in Ω, so $Tu = u \in \operatorname{im} T \cap N = 0$. We conclude that $\operatorname{im} T$ and $\operatorname{im} \tilde{T}$ have the same (finite) codimension. Conversely, if we assume that $\operatorname{im} \tilde{T}$ has finite codimension then it is easily verified in the same way that $\operatorname{im} T$ must have the same finite codimension. The proof is complete.

Suppose that $l = 2s$ and the Dirichlet problem $(\mathscr{A}, \mathscr{D})$ satisfies the L-condition. According to Theorems 3.14 and 10.9, the matrix polynomial $L_{y,\xi'}(\lambda) = \pi\mathscr{A}(y, \xi' + \lambda n(y))$ has a γ^+-spectral monic right divisor $L_{y,\xi'}^+(\lambda)$ of degree s, with coefficients depending smoothly on $(y, \xi') \in T^*(\partial\Omega)\setminus 0$:

$$L_{y,\xi'}(\lambda) = L_{y,\xi'}^-(\lambda) L_{y,\xi'}^+(\lambda).$$

Of course, it then follows that $L_{y,\xi'}^-(\lambda)$ is a γ^--spectral left divisor of degree s with invertible leading coefficient $(=\pi\mathscr{A}(y, n(y)))$. Since $L_{y,c\xi'}(\lambda) = c^l L_{y,\xi'}(\lambda/c)$, it follows that

$$L_{y,\xi'}^{\pm}(\lambda) = c^s L_{y,\xi'}^{\pm}(\lambda/c), \qquad c > 0,$$

that is, the λ^j-coefficient is positive homogeneous of degree $s-j$ ($j = 0, \ldots, s-1$) and the λ^s-coefficient does not depend on ξ'. Note that $L^-_{y,0}(\lambda) = \pi\mathscr{A}(y, n(y))\lambda^s$ and $L^+_{y,0}(\lambda) = \lambda^s I$.

Consider now the following homotopy in $\mathfrak{MP}^l$:

$$L^\tau_{y,\xi'}(\lambda) = L^-_{y,(1-\tau)\xi'}(\lambda + i\tau|\xi'|)\cdot L^+_{y,(1-\tau)\xi'}(\lambda - i\tau|\xi'|), \qquad 0 \leqslant \tau \leqslant 1; \quad (57)$$

thus $L^0 = L$ and

$$L^1_{y,\xi'}(\lambda) = L^-_{y,0}(\lambda + i|\xi'|)\cdot L^+_{y,0}(\lambda - i|\xi'|) = a(y)\cdot(\lambda^2 + |\xi'|^2)^s$$

where $a(y) = \pi\mathscr{A}(y, n(y))$. By Theorem 10.30, this homotopy may be lifted to $\mathfrak{E}^l$, i.e. there exists a homotopy from $\mathscr{A}_\tau$ in $\mathfrak{E}^l$ such that $\mathscr{A}_0 = \mathscr{A}$ and

$$\pi\mathscr{A}_1 = a(y)\cdot\Delta^s \quad \text{on } \partial\Omega, \tag{58}$$

for some invertible C^∞ matrix function. By (57) the Dirichlet problem $(\mathscr{A}_\tau, \mathscr{D})$ satisfies the L-condition for all $0 \leqslant \tau \leqslant 1$. In particular, it follows that $(\mathscr{A}_1, \mathscr{D})$ satisfies the L-condition and by the stability of the index under homotopies for the class $\mathfrak{BE}^{l,\mathbf{m}}$ we have

$$\operatorname{ind}(\mathscr{A}, \mathscr{D}) = \operatorname{ind}(\mathscr{A}_1, \mathscr{D}).$$

Using (58) we now show that the index problem for $(\mathscr{A}_1, \mathscr{D})$ can be reduced to the index of an operator on the sphere S^n.

Theorem 10.33 *Suppose $\mathscr{A}_1 \in \mathfrak{E}^{2s}$, and that* (58) *holds. As in Theorem* 10.31, *we regard $\bar{\Omega} \subset S^n$ by stereographic projection. Then*

$$\operatorname{ind}(\mathscr{A}_1, \mathscr{D}) = \operatorname{ind} R,$$

where $R \in \mathscr{P}(S^n, p \times p)$ with principal symbol

$$\pi R(x, \xi) = \begin{cases} 1_p & x \in S^n \backslash \Omega \\ \pi\mathscr{A}_1(x, \xi)\cdot\pi\mathscr{A}_1(x, \xi_0)^{-1}|\xi|^{-2s}, & x \in \Omega \end{cases}$$

and $\xi_0 \in \mathbb{R}^n$ is a fixed unit vector.

Proof Note that $\pi\mathscr{A}_1(y, \xi_0) = a(y)\cdot|\xi_0|^{2s} = a(y)$. Thus, the principal part of $\mathscr{A}_1$ coincides on $\partial\Omega$ with the principal part of $c(x)\cdot\Delta^s$ where $c(x) = \pi\mathscr{A}_1(x, \xi_0)$, $x \in \bar{\Omega}$. Then by Theorem 10.31

$$\operatorname{ind}(\mathscr{A}_1, \mathscr{D}) = \operatorname{ind}(c(x)\cdot\Delta^s, \mathscr{D}) + \operatorname{ind} R = \operatorname{ind} R,$$

since $\operatorname{ind}(c(x)\cdot\Delta^s, \mathscr{D}) = \operatorname{ind}(\Delta^s, \mathscr{D}) = 0$.

Remark Let n be an *odd* number. It can be shown that if $\mathscr{A}$ is an elliptic *differential* operator of order $2s$ and the Dirichlet problem $(\mathscr{A}, \mathscr{D})$ satisfies the L-condition then the Dirichlet problem has zero index. For the proof, see [Agr], Chapter 18.

10.9 Composition of boundary problems

Given boundary value problem operators $(\mathscr{A}_\nu, \mathscr{B}_\nu)$, $\nu = 1, 2$, where $\mathscr{A}_\nu$ are differential operators and

$$\mathscr{A}_1 \text{ is } p \times p \text{ with DN numbers } -s_i' + t_j,$$

$$\mathscr{B}_1 \text{ is } r_1 \times p \text{ with DN numbers } m_k' + t_j,$$

$$\mathscr{A}_2 \text{ is } p \times p \text{ with DN numbers } -s_i'' + s_j',$$

$$\mathscr{B}_2 \text{ is } r_2 \times p \text{ with DN numbers } m_k'' + s_j',$$

the composite boundary value problem operator $(\mathscr{A}, \mathscr{B}) = (\mathscr{A}_2, \mathscr{B}_2) \circ (\mathscr{A}_1, \mathscr{B}_1)$ is defined as

$$\mathscr{A}(x, D) = \mathscr{A}_2(x, D)\mathscr{A}_1(x, D), \qquad x \in \bar{\Omega},$$

$$\mathscr{B}(y, D) = \begin{pmatrix} \mathscr{B}_1(y, D) \\ \mathscr{B}_2(y, D)\mathscr{A}_1(y, D) \end{pmatrix}, \qquad y \in \partial\Omega$$

It is not hard to verify that

$$\mathscr{A} \text{ is } p \times p \text{ with DN numbers } -s_i'' + t_j \tag{59a}$$

$$\mathscr{B} \text{ is } (r_1 + r_2) \times p \text{ with DN numbers } m_k + t_j \tag{59b}$$

where

$$\left.\begin{aligned} m_k &= m_k', \qquad && k = 1, \ldots, r_1 \\ m_k &= m_{k-r_1}'', \qquad && k = r_1 + 1, \ldots, r_1 + r_2, \end{aligned}\right\} \tag{59c}$$

The boundary operator has the required form, i.e. $\mathscr{B} = \sum \mathscr{B}_\kappa(y, D')D_n^\kappa$ where the coefficients $\mathscr{B}_\kappa$ are $(r_1 + r_2) \times p$ matrices where the (k, j) entry is a (-1) classical operator in $OS^{m_k + t_j - \kappa}(\partial\Omega)$. The transversal order of $\mathscr{B}$ is less than or equal to the sum of the transversal orders of $\mathscr{B}_1$ and $\mathscr{B}_2$. For the DN principal parts we have

$$\pi\mathscr{A}(x, \xi) = \pi\mathscr{A}_2(x, \xi)\pi\mathscr{A}_1(x, \xi), \qquad x \in \bar{\Omega}, \xi \in \mathbb{R}^n \tag{60}$$

$$\pi\mathscr{B}(y, \xi) = \begin{pmatrix} \pi\mathscr{B}_1(y, \xi) \\ \pi\mathscr{B}_2(y, \xi)\pi\mathscr{A}_1(y, \xi) \end{pmatrix}, \qquad y \in \partial\Omega, \xi \in \mathbb{R}^n \tag{61}$$

Theorem 10.34 *Let the* DN *numbers of the boundary value problem operators* $(\mathscr{A}_\nu, \mathscr{B}_\nu)$, $\nu = 1, 2$, *be as given above. Then* $(\mathscr{A}_2, \mathscr{B}_2) \circ (\mathscr{A}_1, \mathscr{B}_1)$ *is L-elliptic (relative to the* DN *numbers defined above) if and only if both* $(\mathscr{A}_1, \mathscr{B}_1)$ *and* $(\mathscr{A}_2, \mathscr{B}_2)$ *are L-elliptic. In that case, we have*

$$\operatorname{ind}(\mathscr{A}, \mathscr{B}) = \operatorname{ind}(\mathscr{A}_1, \mathscr{B}_1) + \operatorname{ind}(\mathscr{A}_2, \mathscr{B}_2). \tag{62}$$

Proof We must verify conditions (i) and (ii) of Definition 9.30. Since it is clear from (60) that the operator $\mathscr{A}$ is elliptic if and only if both $\mathscr{A}_1$ and $\mathscr{A}_2$ are elliptic, it remains to verify condition (iii).

Since the L-condition is verified pointwise, we fix $(y, \xi') \in T^*(\partial\Omega)\backslash 0$. With the obvious notation for the matrix polynomials associated with the elliptic operators $\mathscr{A}$, $\mathscr{A}_1$ and $\mathscr{A}_2$ and with the boundary operators $\mathscr{B}$, $\mathscr{B}_1$ and $\mathscr{B}_2$, we obtain from (60), (61) that

$$L(\lambda) = L_2(\lambda)L_1(\lambda)$$

$$B(\lambda) = \begin{pmatrix} B_1(\lambda) \\ B_2(\lambda)L_1(\lambda) \end{pmatrix}$$

Suppose first that $(\mathscr{A}_1, \mathscr{B}_1)$ and $(\mathscr{A}_2, \mathscr{B}_2)$ satisfy the L-condition (Definition 9.28). We show that Definition 9.28 holds for $(\mathscr{A}, \mathscr{B})$. Let $u \in \mathfrak{M}^+$ such that

$$B\left(\frac{1}{i}\frac{d}{dt}\right)u = 0.$$

If we let

$$w = L_1\left(\frac{1}{i}\frac{d}{dt}\right)u$$

then

$$L_2\left(\frac{1}{i}\frac{d}{dt}\right)w = 0, \quad B_2\left(\frac{1}{i}\frac{d}{dt}\right)w = 0 \tag{63}$$

and $w(t) \to 0$ as $t \to \infty$. Hence $w \equiv 0$, or

$$L_1\left(\frac{1}{i}\frac{d}{dt}\right)u = 0.$$

Since also

$$B_1\left(\frac{1}{i}\frac{d}{dt}\right)u = 0,$$

then $u \equiv 0$, whence $(\mathscr{A}, \mathscr{B})$ satisfies the L-condition.

Conversely suppose that $(\mathscr{A}, \mathscr{B})$ satisfies the L-condition. One verifies easily that Definition 9.28 I holds for $(\mathscr{A}_1, \mathscr{B}_1)$; then Definition 9.28 III must also hold. Now let $w \in C^\infty(\mathbb{R}, \mathbb{C}^p)$ satisfy (63) with $w(t) \to 0$ as $t \to \infty$. Thus there exists $u \in C^\infty(\mathbb{R}, \mathbb{C}^p)$ such that

$$L_1\left(\frac{1}{i}\frac{d}{dt}\right)u = w, \quad B_1\left(\frac{1}{i}\frac{d}{dt}\right)u = 0$$

and $u(t) \to 0$ as $t \to \infty$. Then

$$L\left(\frac{1}{i}\frac{d}{dt}\right)u = 0, \quad B\left(\frac{1}{i}\frac{d}{dt}\right)u = 0$$

so that $u \equiv 0$, and hence $w \equiv 0$. Hence $(\mathscr{A}_2, \mathscr{B}_2)$ satisfies the L-condition.

We turn now to the proof of (62). As in Chapter 9 we have the operators

$$\mathfrak{L} = (\mathscr{A}, \mathscr{B})\colon \mathscr{W}^{l+\mathbf{t}}(\Omega) \to \mathscr{W}^{l+\mathbf{s}''}(\Omega) \times \mathscr{W}^{l-\mathbf{m}-1/2}(\partial\Omega)$$

and, similarly,

$$\mathfrak{L}_1 = (\mathscr{A}_1, \mathscr{B}_1): \mathscr{W}^{l+\mathbf{t}}(\Omega) \to \mathscr{W}^{l+\mathbf{s}'}(\Omega) \times \mathscr{W}^{l-\mathbf{m}'-1/2}(\partial\Omega)$$

and

$$\mathfrak{L}_2 = (\mathscr{A}_2, \mathscr{B}_2): \mathscr{W}^{l+\mathbf{s}'}(\Omega) \to \mathscr{W}^{l+\mathbf{s}''}(\Omega) \times \mathscr{W}^{l-\mathbf{m}''-1/2}(\partial\Omega)$$

In view of the definition of $(\mathscr{A}, \mathscr{B}) = (\mathscr{A}_2, \mathscr{B}_2) \circ (\mathscr{A}_1, \mathscr{B}_1)$ and the decomposition $\mathscr{W}^{l-\mathbf{m}-1/2}(\partial\Omega) = \mathscr{W}^{l-\mathbf{m}'-1/2}(\partial\Omega) \times \mathscr{W}^{l-\mathbf{m}''-1/2}(\partial\Omega)$, we obtain

$$\mathfrak{L} = (\mathfrak{L}_2 \times \text{id}) \circ \mathfrak{L}_1,$$

where id denotes the identity operator on $\mathscr{W}^{l-\mathbf{m}'-1/2}(\partial\Omega)$. Then by Theorem 9.8

$$\begin{aligned} \text{ind}(\mathscr{A}, \mathscr{B}) = \text{ind } \mathfrak{L} &= \text{ind}(\mathfrak{L}_2 \times \text{id}) + \text{ind } \mathfrak{L}_1 \\ &= \text{ind}(\mathscr{A}_2, \mathscr{B}_2) + \text{ind}(\mathscr{A}_1, \mathscr{B}_1). \end{aligned}$$

Let (X_i^+, T_i^+) be a γ^+-spectral pair of $L_i(\lambda) = \pi\mathscr{A}_i(y, (\xi', \lambda))$, $i = 1, 2$, satisfying the homogeneity properties (44) relative to the DN numbers of $\mathscr{A}_i$. For the composite operator, $\mathscr{A} = \mathscr{A}_2 \circ \mathscr{A}_1$, the associated matrix polynomial

$$L(\lambda) = \pi\mathscr{A}(y, (\xi', \lambda)) = L_2(\lambda)L_1(\lambda)$$

has a γ^+-spectral pair (see Theorem 4.17)

$$X_+ = [X_1^+ \quad A], \tag{64}$$

$$T_+ = \begin{pmatrix} T_1^+ & Y_1^+ \cdot X_2^+ \\ 0 & T_2^+ \end{pmatrix}, \tag{65}$$

where

$$A = \frac{1}{2\pi i}\int_{+} L_1^{-1}(\lambda)X_2^+(I\lambda - T_2^+)^{-1}\, d\lambda.$$

Note that A is a smooth $p \times r_2$ matrix function on $T^*(\partial\Omega)\backslash 0$. Also it is easy to verify that the pair (X_+, T_+) satisfies the homogeneity properties (44) corresponding to the DN numbers of $\mathscr{A}$.

Proposition 10.35 *Let $(\mathscr{A}, \mathscr{B}) = (\mathscr{A}_2, \mathscr{B}_2) \circ (\mathscr{A}_1, \mathscr{B}_1)$, and define $\Delta_{\mathscr{B}_i}^+$ as in* (37) *with respect to the γ^+-spectral pair (X_i^+, T_i^+), and $\Delta_{\mathscr{B}}^+$ with respect to the γ^+-spectral pair (X_+, T_+) given by* (64), (65). *Then*

$$\Delta_{\mathscr{B}}^+ = \begin{pmatrix} \Delta_{\mathscr{B}_1}^+ & Z \\ 0 & \Delta_{\mathscr{B}_2}^+ \end{pmatrix},$$

where Z is an $r_1 \times r_2$ matrix function with entries that are C^∞ functions on $T^*(\partial\Omega)\backslash 0$.

Proof This follows from Corollary 4.18 since $M \equiv I$ for the spectral pair (64), (65) (see the remark after Theorem 4.17′).

Remark $\Delta^+_{\mathscr{B}}$ is invertible if and only if both $\Delta^+_{\mathscr{B}_1}$ and $\Delta^+_{\mathscr{B}_2}$ are invertible, thus indicating another proof of Theorem 10.34.

Exercises

1. For $x = (x_1, x_2, x_3) \in S^2$, the unit sphere in $\mathbb{R}^3$, let

$$L_x(\lambda) = \begin{pmatrix} \lambda^2+1 & & & x_1 \\ & \lambda^2+1 & & x_2 \\ & & \lambda^2+1 & x_3 \\ x_1 & x_2 & x_3 & 0 \end{pmatrix}$$

Since $\det L_x(\lambda) = -(\lambda^2+1)^2$, the family of matrix polynomials $L_x(\lambda)$ satisfies the hypotheses stated at the beginning of §10.3, with $r = 2$. Show that $\mathfrak{M}^+$ is isomorphic to the tangent bundle $T(S^2)$, and is, therefore, not trivial.

2. Let

$$\mathscr{A} = \begin{pmatrix} \Delta & & & x_1 \\ & \Delta & & x_2 \\ & & \Delta & x_3 \\ x_1 & x_2 & x_3 & 0 \end{pmatrix} \quad \text{in } \mathbb{R}^3 \backslash 0,$$

where Δ is the Laplace operator in $\mathbb{R}^3$.

(a) Show that with the DN numbers

$$s_1 = s_2 = s_3 = 0,\ s_4 = -2,\ t_1 = t_2 = t_3 = 2,\ t_4 = 0,$$

$\mathscr{A}$ is DN-elliptic in $\mathbb{R}^3\backslash 0$ and properly elliptic.

(b) If $\bar{\Omega} = \{x; \frac{1}{2} \leqslant |x| \leqslant 1\}$ then $\partial\Omega = S^2 \cup \{x; |x| = \frac{1}{2}\}$. Show that $\mathscr{A}$ does *not* have a boundary operator satisfying the L-condition, if we consider only matrix boundary operators. Hint: Use Exercise 1 to show that $\mathfrak{M}^+_L$ is not trivial.

(c) Suppose we allow boundary operators acting in vector bundles as discussed in the text following Theorem 10.14. Now does $\mathscr{A}$ have a boundary operator satisfying the L-condition?

3. Suppose that $r = ps$ for some s and the Dirichlet boundary operator satisfies the L-condition relative to $\mathscr{A}$. Show that if $(\mathscr{A}, \mathscr{B})$ satisfies the L-condition then

$$(\mathscr{A}, \mathscr{B}) \approx (\mathscr{A}, \mathscr{R})$$

for some boundary operator $\mathscr{R}$ with transversal order $\leqslant s-1$ and the same DN numbers as $\mathscr{B}$ (see §10.5).

4. Let $(\mathscr{A}_\nu, \mathscr{B}_\nu)$, $\nu = 1, 2$, have the DN numbers as stated at the beginning of this section. Show that for the composite operator $(\mathscr{A}, \mathscr{B})$:

(a) $\mathscr{A}$ has the DN numbers (59a) and the formula (60) holds for the DN principal part $\pi\mathscr{A}$

(b) $\mathscr{B}$ has a representation $\sum \mathscr{B}_j(y, D')D_n^j$ with DN numbers (59b), (59c), and the formula (61) holds for the DN principal part $\pi\mathscr{B}$.

5. Let $\mathscr{A}$ be homogeneous elliptic (DN numbers $s_i = l$, $t_j = 0$) and let $\mathscr{B}$ be a boundary operator with DN numbers m_k. Let

$$(\mathscr{A}', \mathscr{B}') = (\mathscr{A}, \mathscr{B}) \circ (\text{Laplace, Dirichlet})^q$$

be the composition of $(\mathscr{A}, \mathscr{B})$ and q copies of the Dirichlet problem for the Laplace operator. Let $L(\lambda)$, $L'(\lambda)$ be the $p \times p$ matrix polynomials associated with $\mathscr{A}, \mathscr{A}'$ and $B(\lambda)$, $B'(\lambda)$ the $r \times p$ and $(r + q) \times p$ matrix polynomials associated with $\mathscr{B}, \mathscr{B}'$. Also let (X_+, T_+) be a γ^+-spectral pair for $L(\lambda)$. Show that

(a) The vector bundle $\mathfrak{M}_{L'}^+$ is isomorphic to $\mathfrak{M}_L^+ \oplus \mathbb{C}^{pq}$

(b) $L'(\lambda)$ has γ^+-spectral pair $X'_+ = [X_+ \quad I_{pq}]$, $T'_+ = \begin{pmatrix} T_+ & \\ & I_{pq} \end{pmatrix}$

(c) If we define $\Delta_{\mathscr{B}'}^+$ and $\Delta_{\mathscr{B}}^+$ by (37) with respect to the spectral pairs (X_+, T_+) and (X'_+, T'_+), respectively, then $\Delta_{\mathscr{B}'}^+ = \begin{pmatrix} \Delta_{\mathscr{B}}^+ & \\ & I_{pq} \end{pmatrix}$

The next exercise is closely related to Theorem 10.23.

6. Let $(\mathscr{A}_i, \mathscr{B}_i) \in \mathrm{BE}^{\mathbf{s},\mathbf{t},\mathbf{m}}$, $i = 0, 1$, and suppose that $\mathscr{A}^\tau$, $0 \leqslant \tau \leqslant 1$, is a homotopy of elliptic operators in $\mathrm{Ell}^{\mathbf{s},\mathbf{t}}$ such that $\mathscr{A}^i = \mathscr{A}_i$, $i = 0, 1$. Now define the C^∞ matrix functions $\Delta_{\mathscr{B}_i}^+ \colon ST^*(\partial\Omega) \to GL_r(\mathbb{C})$, $i = 0, 1$, with respect to a γ^+-spectral pair (50). Prove that

> *There exist boundary operators $\mathscr{B}^\tau$ such that $(\mathscr{A}^\tau, \mathscr{B}^\tau)$ is a homotopy in $\mathrm{BE}^{\mathbf{s},\mathbf{t},\mathbf{m}}$ with $(\mathscr{A}^i, \mathscr{B}^i) = (\mathscr{A}_i, \mathscr{B}_i)$, $i = 0, 1$, if and only if $\Delta_{\mathscr{B}_0}^+$ is homotopic to $\Delta_{\mathscr{B}_1}^+$.*
>
> In general it is a difficult problem to decide when two C^∞ maps $ST^*(\partial\Omega) \to GL_r(\mathbb{C})$ are homotopic (but see the addendum below). When $n = 2$ it is quite simple though; this is the subject of the next exercise.

7. Let $\Omega \subset \mathbb{R}^2$ be an open set with smooth boundary, and assume that Ω is simply connected. The degree of a C^∞ matrix function $F\colon ST^*(\partial\Omega) \to GL_r(\mathbb{C})$ is defined to be

$$\deg F = -\frac{1}{2\pi}[\arg \det F(y, \tau(y))]_{\partial\Omega} + \frac{1}{2\pi}[\arg \det F(y, -\tau(y))]_{\partial\Omega}$$

where $\tau\colon \partial\Omega \to ST^*(\partial\Omega)$ is a positively oriented C^∞ section of the unit cotangent bundle of $\partial\Omega$. (See §15.2 for the definition of $[\ ]_{\partial\Omega}$.) Prove that

> *Two C^∞ functions F_0, F_1 from $ST^*(\partial\Omega) \to GL_r(\mathbb{C})$ have the same degree if and only if F_0 is homotopic to $M(y)F_1$ for some C^∞ matrix function $M\colon \partial\Omega \to GL_r(\mathbb{C})$.*

8. Let $(\mathscr{A}_0, \mathscr{B}_0)$ and $(\mathscr{A}_1, \mathscr{B}_1)$ be two boundary value problems in the plane $(n = 2)$ satisfying the L-condition and having the same DN numbers. Suppose that $\mathscr{A}_0$ is homotopic to $\mathscr{A}_1$ and let $\Delta^+_{\mathscr{B}_0}$ and $\Delta^+_{\mathscr{B}_1}$ be defined with respect to the γ^+-spectral triple (50). Prove that $(\mathscr{A}_0, \mathscr{B}_0)$ is homotopic to $(\mathscr{A}_1, M(y)\mathscr{B}_1)$ for some invertible $M \in C^\infty(\partial\Omega, r \times r)$ if and only if $\Delta^+_{\mathscr{B}_0}$ and $\Delta^+_{\mathscr{B}_1}$ have the same degree, that is,

$$-[\arg\det \Delta^+_{\mathscr{B}_0}(y, \tau(y))]_{\partial\Omega} + [\arg\det \Delta^+_{\mathscr{B}_0}(y, -\tau(y))]_{\partial\Omega}$$
$$= -[\arg\det \Delta^+_{\mathscr{B}_1}(y, \tau(y))]_{\partial\Omega} + [\arg\det \Delta^+_{\mathscr{B}_1}(y, -\tau(y))]_{\partial\Omega}$$

Addendum

Let M be a compact manifold. The following theorem is the first step in determining the homotopy classes of continuous maps $f: M \to GL_p(\mathbb{C})$. Note that if M is 1-dimensional then the theorem implies that f is homotopic to

$$\begin{pmatrix} \det f & \\ & 1_{p-1} \end{pmatrix},$$

a fact that is needed in §15.2. See Theorem 15.10 for another proof.

In the proof we need the following simple observation: $\mathbb{R}^{2p}$ is the realization of $\mathbb{C}^p$ as discussed in §5.1, for if $\mathbb{R}^{2p}$ is identified with $\mathbb{C}^p$ by the map $\mathbb{R}^p \ni (u_1, \ldots, u_{2p}) \mapsto (u_1 + iu_2, \ldots, u_{2p-1} + iu_{2p}) \in \mathbb{C}^p$, then

$$(u, v)_{\mathbb{R}^{2p}} = \sum_{j=1}^{2p} u_j v_j = \sum_{j=1}^{p} (u_{2j-1} + iu_{2j})(v_{2j-1} - iv_{2j}) = \operatorname{Re}(u, v)_{\mathbb{C}^p}. \quad (66)$$

Theorem 10.36 *Let M be a compact manifold of dimension $2m - 1$, and let $f: M \to GL_p(\mathbb{C})$ be a continuous map where $p > m$. Then f is homotopic to*

$$\begin{pmatrix} \tilde{f} & \\ & I_{p-m} \end{pmatrix}$$

for some continuous map $\tilde{f}: M \to GL_m(\mathbb{C})$.

Proof The first column f_1 of the matrix f defines a map $f_1: M \to \mathbb{C}^p\backslash 0$, so that $g = f_1/|f_1|$ is a map $M \to S^{2p-1}$ (where $\mathbb{C}^p$ is identified with $\mathbb{R}^{2p}$). Since $\dim M < \dim S^{2p-1}$, there exists an arbitrarily small perturbation g' of g such that g' is not surjective (see [HW]). Thus, by an arbitrarily small perturbation f' of f and by the homotopy $f + t(f' - f)$, $0 \leqslant t \leqslant 1$, we may assume that g is not surjective. Without loss of generality, the south pole is not in the image of g. Let $h: M \to S^{2p-1}$ be a continuous map with the properties stated in Lemma 10.37 below. Since $(g(x), h(x))_{\mathbb{C}^p} \neq 0$, and since $g = f_1/|f_1|$, then

$$(f_1(x), h(x))_{\mathbb{C}^p} = \sum f_{j1}(x)\overline{h_j(x)} \neq 0,$$

where we have written $f = [f_{jk}]^p_{j,k=1}$ and $h = [h_j]$ with respect to the standard basis $e_1, \ldots, e_p$ of $\mathbb{C}^p$. Now, dividing this equation by $\overline{h_1(x)} = (e_1, h(x))_{\mathbb{C}^p} \neq 0$, we obtain

$$f_{11}(x) + \sum_{j=2}^{p} \psi_j(x) \cdot f_{j1}(x) \neq 0,$$

where $\psi_j = \bar{h}_j/\bar{h}_i$ are continuous functions. We will finish the proof by induction on the size p of the matrix f. In order to proceed with the inductive step we need to apply certain row operations; it is important to note here that any elementary matrix is homotopic to the identity matrix, so if two matrix functions are row equivalent then they are also homotopic. Let

$$f' = \begin{pmatrix} 1 & \psi_2 & \cdots & \psi_p \\ & 1 & & \\ & & \ddots & \\ & & & 1 \end{pmatrix} \cdot f$$

Then $f'\colon M \to GL_p(\mathbb{C})$ is homotopic to f, and, by construction, $f'_{11}(x) \neq 0$ for all $x \in M$, so a sequence of row operations (which can be expressed in terms of homotopies) reduces f' to the form

$$f'' = \left(\begin{array}{c|ccc} 1 & f''_{12} & \cdots & f''_{1p} \\ \hline 0 & f''_{22} & \cdots & f''_{2p} \\ \vdots & & & \\ 0 & f''_{p2} & \cdots & f''_{pp} \end{array}\right)$$

Moreover, if we replace f''_{1k} by $t \cdot f''_{1k}$, $0 \leqslant t \leqslant 1$, we see that f'' is homotopic to a matrix of the same form but with $f''_{1k} = 0$. Since the matrix $[f_{jk}]_{j,k=2}^{p}$ has dimension $p - 1$, the theorem follows by induction.

Lemma 10.37 *Let $g\colon M \to S^{2p-1}$ be continuous and suppose that g is not surjective, say the south pole is not in the image of g. Then there exists a continuous map $h\colon M \to S^{2p-1}$ such that*

$$h(x) \cdot e_1 \neq 0, \qquad g(x) \cdot h(x) \neq 0 \qquad \forall x \in M,$$

where $\cdot$ denotes the standard inner product in $\mathbb{R}^{2p}$ and $e_1 = (1, 0, \ldots, 0)$.

Proof Let $\theta_1, \ldots, \theta_{2p-1}$ denote the spherical coordinates in S^{2p-1}, where θ_1 is the angle between the radius vector and e_1. Then $0 \leqslant \theta_1 \leqslant \pi$, and we let the north pole have coordinate $\theta_1 = 0$ and the south pole have the coordinate $\theta_1 = \pi$. Since M is compact the image of g is closed and there exists a neighbourhood V of the south pole not in the image of g. We may assume that V is defined by the coordinates

$$\pi - \alpha < \theta_1 \leqslant \pi, \quad \text{for some } 0 < \alpha < \pi/2.$$

Take $\beta \in (0, \alpha)$ and if $g(x)$ has coordinates $\theta_1, \ldots, \theta_{2p-1}$, then define $h(x)$,

$x \in M$, as follows:

(i) $h(x)$ is the point on S^{2p-1} with spherical coordinates

$$\left(\frac{\pi}{2} - \beta\right) \cdot \frac{2\theta_1}{\pi}, \quad \theta_2, \ldots, \theta_{2p-1}$$

if $g(x)$ belongs to the lower hemisphere $\theta_1 \leqslant \pi/2$;

(ii) $h(x)$ is the point on S^{2p-1} with spherical coordinates

$$\frac{\pi}{2} - \beta, \quad \theta_2, \ldots, \theta_{2p-1}$$

if $g(x)$ belongs to the upper hemisphere $\theta_1 \geqslant \pi/2$.

Note that h is continuous, since it is defined on the union of two closed sets and the definitions agree on the overlap (the equator, $\theta_1 = \pi/2$). Clearly $h(x) \cdot e_1 \neq 0$, i.e. $h(x)$ does not lie on the equator $\theta_1 = \pi/2$. Also $g(x) \cdot h(x) \neq 0$, i.e. the angle between $g(x)$ and $h(x)$ is $< \pi/2$, for if $\theta_1 \leqslant \pi/2$ then the angle between $g(x)$ and $h(x)$ is $\theta_1 - (\pi/2 - \beta) \leqslant \pi/2 - \alpha + \beta$ (here we used the fact that $g(x) \notin V$, so $\theta_1 \leqslant \pi - \alpha$).

11
Applications to the index

The aim of this chapter is to show how to deform an elliptic boundary problem $(\mathscr{A}, \mathscr{B})$ to a simpler form which is equivalent to an elliptic system on the double, $\tilde{\Omega}$, of $\bar{\Omega}$, which is a compact manifold without boundary.

We make use of the classes

$$\mathfrak{E}^l \quad \text{and} \quad \mathfrak{BE}^{l,\mathbf{m}}$$

defined in §10.7. For convenience it is always assumed that if $(\mathscr{A}, \mathscr{B}) \in \mathfrak{BE}^{l,\mathbf{m}}$ then the transversal order of $\mathscr{B}$ is $\mu \leqslant l-1$. Further, we assume that the coefficients of $\mathscr{A}$ do not depend on x_n near the boundary (see Lemma 10.28), so that we can make use of the class of matrix polynomials

$$\mathfrak{MP}^l$$

in Theorem 10.30. Homotopies in $\mathfrak{MP}^l$ can be lifted to homotopies in $\mathfrak{E}^l$, which are then lifted to $\mathfrak{BE}^{l,\mathbf{m}}$ by Theorem 10.27.

The main steps in the reduction of an elliptic boundary problem to the double $\tilde{\Omega}$ are as follows. Let $\mathscr{A} \in \mathfrak{E}^1$ be a first-order elliptic operator, and let $\mathscr{B}$ be a boundary operator of transversal order $\mu = 0$ satisfying the L-condition relative to $\mathscr{A}$. In §11.1, we show how the boundary operator $\mathscr{B}$ can be used to deform $\mathscr{A}$ to a form which is trivial on the boundary (Definition 11.2). Once this has been achieved it is possible to express the index of $(\mathscr{A}, \mathscr{B})$ as the index of an elliptic operator $\tilde{\mathscr{A}}$ on $\tilde{\Omega}$ (Theorem 11.5). We then show in §11.2 that any $(\mathscr{A}, \mathscr{B}) \in \mathfrak{BE}^{l,\mathbf{m}}$ of higher order $l > 1$ can be deformed to an elliptic boundary problem $(\hat{\mathscr{A}}, \hat{\mathscr{B}}) \in \mathfrak{BE}^{l,\mathbf{m}}$ which is very nearly first order (Theorem 11.12). The index problem for $(\hat{\mathscr{A}}, \hat{\mathscr{B}})$ can be reduced to the index problem for a boundary problem of the type already considered in §11.1, i.e. where the elliptic operator is of first order (Proposition 11.6).

It should be mentioned that – since the double $\tilde{\Omega}$ is a compact manifold without boundary – one could apply the Atiyah-Singer index theory to obtain a formula for the index of $\tilde{\mathscr{A}}$. In principle, this makes it possible to obtain a formula for the index of the original boundary problem $(\mathscr{A}, \mathscr{B})$, but it would take us too far afield to develop the background in algebraic topology and differential geometry required to do so.

11.1 First-order elliptic systems

Consider the 1-homogeneous functions $|\xi'|$ on $T^*(\partial\Omega)$ and $|\xi|$ on $T^*(\mathbb{R}^n)$. By Lemma 8.67 we can find p.d.o.'s $\Lambda' \in OClS^1(\partial\Omega)$ with principal part $|\xi'|$ and $\Lambda \in OClS^1(\mathbb{R}^n)$ with principal part $|\xi|$.

As usual, a tubular neighbourhood of $\partial\Omega$ in $\mathbb{R}^n$ is identified with $\partial\Omega \times (-1, 1)$, with $\partial\Omega \times [0, 1) \subset \bar{\Omega}$. Coordinates in $\partial\Omega \times (-1, 1)$ are denoted by (x', x_n) and those in $\partial\Omega$ by $y = (x', 0)$. Let $\hat{\Omega} = \Omega \cup \partial\Omega \times (-1, 0]$, which is an open neighbourhood of $\bar{\Omega}$ in $\mathbb{R}^n$. Let $\varphi \in C_0^\infty((-1, 1))$, $0 \leqslant \varphi \leqslant 1$ with $\varphi = 1$ in a neighbourhood of 0, and consider $\varphi(x_n)$ as a function on $\hat{\Omega}$ with support in the tubular neighbourhood $\partial\Omega \times (-1, 1)$. We define

$$\mathscr{A}_+ = \varphi \cdot (D_n + i\Lambda') + i(1 - \varphi)\Lambda(1 - \varphi),$$

$$\mathscr{A}_- = \varphi \cdot (-D_n + i\Lambda') + i(1 - \varphi)\Lambda(1 - \varphi), \qquad \mathscr{B}_- u = u|_{\partial\Omega}$$

where we did not indicate the obvious restriction operators to Ω for $\mathscr{A}_\pm$. Note that $\mathscr{A}_\pm \in \mathfrak{E}^1$ (ellipticity is proved below). The interior part is the same for both operators, $\mathscr{A}^i_\pm = i(1 - \varphi)\Lambda(1 - \varphi)$, while the boundary parts, $\mathscr{A}^b_\pm = \varphi \cdot (\pm D_n + i\Lambda')$, differ which affects the number of boundary conditions required.

Theorem 11.1 *The operator $\mathscr{A}_+$ is elliptic and Fredholm (without boundary conditions) from $W^l_2(\Omega)$ to $W^{l-1}_2(\Omega)$, $l \geqslant 1$. The operator $(\mathscr{A}_-, \mathscr{B}_-)$ is elliptic and Fredholm from $W^l_2(\Omega)$ to $W^{l-1}_2(\Omega) \times W^{l-1/2}_2(\partial\Omega)$, $l \geqslant 1$. Furthermore,* ind $\mathscr{A}_+ =$ ind$(A_-, \mathscr{B}_-) = 0$. *Replacing $\mathscr{A}_\pm$ by $\mathscr{A}_\pm + itI$, where t is sufficiently large, we obtain that*

$$\mathscr{A}_+ \colon W^l_2(\Omega) \to W^{l-1}_2(\Omega),$$

$$(\mathscr{A}_-, \mathscr{B}_-) \colon W^l_2(\Omega) \to W^{l-1}_2(\Omega) \times W^{l-1/2}_1(\partial\Omega)$$

are isomorphisms when $l \geqslant 1$.

Proof Consider first $\mathscr{A}_+$. If $\varphi(x_n) \neq 0$ then

$$\pi\mathscr{A}_+(x, (\xi', \xi_n)) = \varphi(x_n)(\xi_n + i|\xi'|) + i(1 - \varphi(x_n))^2|\xi| \neq 0$$

when $(\xi', \xi_n) \neq 0$, and if $\varphi(x_n) = 0$ then $\pi\mathscr{A}_+(x, \xi) = i|\xi| \neq 0$ when $\xi \neq 0$. Thus $\pi\mathscr{A}_+$ is an elliptic symbol. On the boundary $\partial\Omega$, the polynomial

$$L_+(\lambda) = \pi\mathscr{A}^b_+(y, (\xi', \lambda)) = \lambda + i|\xi'|, \qquad y \in \partial\Omega,$$

has no roots in the upper half-plane Im $\lambda > 0$, thus $r = 0$ and no boundary conditions are necessary. This means that $\mathscr{A}_+$ is a Fredholm operator by Theorem 10.26. To determine the kernel of $\mathscr{A}_+$, which we know is in $C^\infty(\bar{\Omega})$ by Weyl's Lemma, we choose a positive density (measure) in $\bar{\Omega}$ which in the collar is the product of one in $\partial\Omega$ and the Lebesgue measure, dx_n. If $u \in C^\infty(\bar{\Omega})$ then since $\mathscr{A}^b_+ = \varphi \cdot (D_n + i\Lambda')$ we have

$$\begin{aligned} 2\,\mathrm{Im}(\mathscr{A}^b_+ u, u)_\Omega &= i^{-1}((\mathscr{A}^b_+ u, u) - (u, \mathscr{A}^b_+ u)) \\ &= 2\,\mathrm{Re}(\varphi\Lambda' u, u) + i^{-1}((\varphi D_n u, u) - (u, \varphi D_n u)). \end{aligned}$$

For the first term, if we use Gårding's inequality on the operator Λ' (see Chapter 8, Exercise 10) we obtain

$$\begin{aligned}\operatorname{Re}(\varphi\Lambda' u, u)_\Omega) &= \int_0^1 \varphi(x_n)\operatorname{Re}(\Lambda' u(\cdot, x_n), u(\cdot, x_n))_{\partial\Omega}\,dx_n \\ &\geqslant \int_0^1 \varphi(x_n)[c'\|u(\cdot, x_n)\|^2_{1,\partial\Omega} - c''\|u(\cdot, x_n)\|^2_{\partial\Omega}]\,dx_n \\ &\geqslant -c''\int_0^1 \varphi(x_n)\|u(\cdot, x_n)\|^2_{\partial\Omega}\,dx_n \\ &\geqslant -c\|u\|^2_\Omega, \end{aligned} \tag{1}$$

For the second term, an integration by parts gives us

$$\begin{aligned} &i^{-1}((\varphi D_n u, u) - (u, \varphi D_n u)) \\ &\quad = -\int_0^1 \varphi(x_n)\left(\frac{\partial u}{\partial x_n}, u\right)_{\partial\Omega} dx_n - \int_0^1 \varphi(x_n)\left(u, \frac{\partial u}{\partial x_n}\right)_{\partial\Omega} dx_n \\ &\quad = -\int_0^1 \frac{\partial}{\partial x_n}(\varphi(x_n)u, u)_{\partial\Omega}\,dx_n + \int_0^1 \varphi'(x_n)(u, u)_{\partial\Omega}\,dx_n \\ &\quad \geqslant \|u\|^2_{\partial\Omega} - c\|u\|^2_\Omega \\ &\quad \geqslant -c\|u\|^2_\Omega. \end{aligned} \tag{2}$$

Adding the inequalities (1), (2), and noting that

$$\operatorname{Im}\mathscr{A}^i_+ = -\operatorname{Re}(1-\varphi)\Lambda(1-\varphi)$$

is bounded from below, it follows that

$$2\operatorname{Im}(\mathscr{A}_+ u, u) \geqslant -\tilde{c}\|u\|^2_\Omega. \tag{3}$$

Now we consider $\mathscr{A}_-$. As before the ellipticity of $\mathscr{A}_-$ is clear. Also

$$L_-(\lambda) = \pi\mathscr{A}^b_-(y, (\xi', \lambda)) = -\lambda + i|\xi'|, \qquad y \in \partial\Omega,$$

and we have exactly one characteristic root $i|\xi'|$, $\xi' \neq 0$, in the upper half-plane, so $r = 1$. The solutions of the equation

$$L_-\left(\frac{1}{i}\frac{d}{dt}\right)u = 0$$

are $u(t) = \text{const}\, e^{-t|\xi'|}$, thus $\mathfrak{M}^+_{L_-} = \operatorname{span}\{e^{-t|\xi'|}\}$ and $(\mathscr{A}_-, \mathscr{B}_-)$ obviously satisfies the L-condition. By Theorem 10.26, it follows that $(\mathscr{A}_-, \mathscr{B}_-)$ is Fredholm and $\ker(\mathscr{A}_-, \mathscr{B}_-) \subset C^\infty(\bar{\Omega})$. Analogous to the above calculations, we obtain for $u \in C^\infty(\bar{\Omega})$ that

$$2\operatorname{Im}(\mathscr{A}^b_- u, u) \geqslant -\|u\|^2_{\partial\Omega} - C\|u\|^2_\Omega,$$

(note the term $-\|u\|^2_{\partial\Omega}$ appearing with a negative sign). If $\mathscr{B}_- u = u|_{\partial\Omega} = 0$

then

$$2\,\mathrm{Im}(\mathscr{A}_- u, u) \geqslant -\tilde{c}\|u\|^2_\Omega. \tag{4}$$

In view of (3) and (4), the kernels of $\mathscr{A}_+ + itI$ and $(\mathscr{A}_- + itI, \mathscr{B}_-)$ are equal to $\{0\}$ if $2t > \tilde{c}$. Suppose that $(\mathscr{A}_+ + itI)u = 0$ and $u \neq 0$. Then we have

$$\begin{aligned} 0 &= 2\,\mathrm{Im}(\mathscr{A}_+ u + itu, u) = 2\,\mathrm{Im}(\mathscr{A}_+ u, u) + 2t(u, u) \\ &> 2\,\mathrm{Im}(\mathscr{A}_+ u, u) + \tilde{c}(u, u) \end{aligned}$$

in contradiction to (3). The same proof holds for $(\mathscr{A}_-, \mathscr{B}_-)$. Replacing $\mathscr{A}_\pm$ by $\mathscr{A}_\pm + itI$, we can assume in what follows that the kernels are equal to $\{0\}$.

The densities on Ω and on $\partial\Omega$ introduced above allow us to construct the adjoints of $\mathscr{A}_+$ and $(\mathscr{A}_-, B_-)$ with respect to the scalar products on Ω and on $\partial\Omega$. By virtue of Theorem 10.26 we know that the images of $\mathscr{A}_+$ and $(\mathscr{A}_-, \mathscr{B}_-)$ are the orthogonal space of a subspace $\subset C^\infty$. Assume now that $v \in C^\infty(\bar{\Omega})$ is orthogonal to the image of $\mathscr{A}_+$, that is

$$(v, \mathscr{A}_+ u)_\Omega = 0 \qquad \text{for } u \in C^\infty(\bar{\Omega}).$$

Integrating by parts (or using a Green's formula) we get

$$(\mathscr{A}_+^* v, u)_\Omega + (v, u)_{\partial\Omega} = 0, \qquad \forall u \in C^\infty(\bar{\Omega}).$$

For $u \in C_0^\infty(\Omega)$ we obtain $(\mathscr{A}_+^* v, u)_\Omega = 0$, and since $C_0^\infty(\Omega)$ is dense in $L^2(\Omega)$ this means that $\mathscr{A}_+^* v = 0$. Hence $(v, u)_{\partial\Omega} = 0$ for all $u \in C^\infty(\bar{\Omega})$, so $v = 0$ on $\partial\Omega$. Since $-\mathscr{A}_+^*$ is of the same form as $\mathscr{A}_-$ it follows if the number t is chosen large enough that $v = 0$. Thus $\mathscr{A}_+$ is an isomorphism if t is sufficiently large, and in any case ind $\mathscr{A}_+ = 0$.

Suppose now that $v \in C^\infty(\bar{\Omega})$, $h \in C^\infty(\partial\Omega)$ is orthogonal to the image of $(\mathscr{A}_-, \mathscr{B}_-)$, that is,

$$(v, \mathscr{A}_- u)_\Omega + (h, u)_{\partial\Omega} = 0 \qquad \text{for } u \in C^\infty(\bar{\Omega}).$$

Taking u from $C_0^\infty(\Omega)$ and integrating by parts we obtain that

$$0 = (v, \mathscr{A}_- u)_\Omega = (\mathscr{A}_-^* v, u) \qquad \forall u \in C_0^\infty(\Omega).$$

Since $C_0^\infty(\Omega)$ is dense in $L^2(\Omega)$, it follows that $\mathscr{A}_-^* v = 0$ and because $-\mathscr{A}_-^*$ is of the same form as $\mathscr{A}_+$ we get $v = 0$ and then $h = 0$. Thus $(\mathscr{A}_-, \mathscr{B}_-)$ is also an isomorphism if t is sufficiently large and in any case $\mathrm{ind}(\mathscr{A}_-, \mathscr{B}_-) = 0$.

Let p and q be any nonnegative integers. In the sequel, we refer to any direct sum $\mathscr{A}_+ \oplus \cdots \oplus \mathscr{A}_+ \oplus (\mathscr{A}_-, \mathscr{B}_-) \oplus \cdots \oplus (\mathscr{A}_-, \mathscr{B}_-)$, that is,

$$\left(\begin{pmatrix} \mathscr{A}_+ I_p & \\ & \mathscr{A}_- I_q \end{pmatrix}, \begin{pmatrix} 0 & I_q \end{pmatrix}\right) \in \mathfrak{BE}^{1,0}_{(p+q)\times(p+q)},$$

as a *trivial elliptic b.v.p. operator*. Here p is the number of copies of $\mathscr{A}_+$ and q is the number of copies of $(\mathscr{A}_-, \mathscr{B}_-)$. On the boundary $\partial\Omega$, the principal symbols of $\mathscr{A}_\pm$ are $\pi\mathscr{A}_+(y, (\xi', \lambda)) = \lambda + i|\xi'|$ and $\pi\mathscr{A}_-(y, (\xi', \lambda)) = -\lambda + i|\xi'|$, so we make the following definition.

Definition 11.2 *An elliptic operator* $\mathscr{A} \in \mathfrak{E}^1_{(p+q)\times(p+q)}$ *is said to be* trivial on the boundary *if*

$$\pi\mathscr{A}(y, (\xi', \lambda)) = a(y)\cdot\begin{pmatrix} (\lambda + i|\xi'|\, I_p & \\ & (-\lambda + i|\xi'|\, I_q \end{pmatrix}, \qquad y \in \partial\Omega,$$

where $a\colon \partial\Omega \to GL_{p+q}(\mathbb{C})$ *is an invertible* C^∞ *matrix function. Such an operator always has a boundary operator satisfying the L-condition, namely,* $\mathscr{B} = [0 \quad I_q]$. *We also say that the boundary value problem operator* $(\mathscr{A}, \mathscr{B}) \in \mathfrak{BE}^{1,0}_{(p+q)\times(p+q)}$ *is trivial on the boundary.*

Our aim is to show that the direct sum of an arbitrary b.v.p. operator $(\mathscr{A}, \mathscr{B}) \in \mathfrak{BE}^{1,0}_{p\times p}$ with a suitable *trivial* $r \times r$ elliptic b.v.p. operator (with $q = 0$) is homotopic in $\mathfrak{BE}^{1,0}_{(p+r)\times(p+r)}$ to a b.v.p. operator $(\mathscr{A}_1, \mathscr{B}_1)$ that is trivial on the boundary (Theorem 11.4). The index problem for $(\mathscr{A}_1, \mathscr{B}_1)$ can then be reduced to the index of an elliptic operator on a compact manifold without boundary (Theorem 11.5), to which the Atiyah-Singer theory can be applied.

The basic idea for the construction of this homotopy is to begin on the level of matrix polynomials, $\mathfrak{MP}^1$, and then apply the homotopy extension theorem 10.30 for elliptic operators.

Lemma 11.3 *Let E be a finite dimensional complex vector space, and let* $A \in \mathscr{L}(E)$ *with no real eigenvalues. Let* $P = (2\pi i)^{-1} \int_+ (I\lambda - A)^{-1}\, d\lambda$ *be the Riesz projector corresponding to the eigenvalues of A in the upper half-plane. Then for all* $c > 0$

$$A_\tau := (1 - \tau)A + \tau(icP - ic(I - P))$$

has no real eigenvalues for $0 \leqslant \tau \leqslant 1$ *and the Riesz projector for* A_τ *coincides with that of A,*

$$P = (2\pi i)^{-1} \int_+ (I\lambda - A_\tau)^{-1}\, d\lambda, \qquad \forall\tau.$$

Proof Since P is the Riesz projector corresponding to the eigenvalues of A in the upper half-plane then, with respect to the direct sum $E = \operatorname{im} P \oplus \ker P$, we have

$$A = \begin{pmatrix} A_+ & 0 \\ 0 & A_- \end{pmatrix}$$

where $A_+ \in \mathscr{L}(\operatorname{im} P)$, $A_- \in \mathscr{L}(\ker P)$ with spectrum contained in the upper and lower half-planes, respectively. Correspondingly, we have

$$A_\tau = \begin{pmatrix} (1-\tau)A_+ + ic\tau I_+ & 0 \\ 0 & (1-\tau)A_- - ic\tau I_- \end{pmatrix}$$

where I_+, I_- denote the identity operators in $\operatorname{im} P$ and $\ker P$. Let $A^\tau_+ = (1-\tau)A_+ + ic\tau I_+$ and $A^\tau_- = (1-\tau)A_- - ic\tau I_-$. Clearly, the spectra of A^τ_+, A^τ_- lie in the upper and lower half-plane, respectively, so that A_τ has no real eigenvalues for $0 \leqslant \tau \leqslant 1$. Moreover,

$$(2\pi i)^{-1}\int_{+}(I\lambda - A_\tau)^{-1}\,d\lambda = (2\pi i)^{-1}\int_{+}\begin{pmatrix}(I\lambda - A_+^\tau)^{-1} & 0\\ 0 & (I\lambda - A_-^\tau)^{-1}\end{pmatrix}d\lambda$$

$$= \begin{pmatrix} I_+ & \\ & 0\end{pmatrix} = P.$$

In the proof of the next theorem, one should keep in mind that if $L(\lambda) = I\lambda - A$ is of degree 1 with no real eigenvalues, then the Riesz projector with respect to the eigenvalues in the upper half-plane $\operatorname{Im}\lambda > 0$ coincides with the Caldéron projector. See, for instance, the third corollary to Theorem 2.17. Also, the right and left Calderón projectors coincide; see Chapter 10 (4), (6) with $l = 1$ and $A_l = I$.

It is important to note in the proof of Theorem 11.4 that it is the existence of the boundary oprator $\mathscr{B}$ satisfying the L-condition that makes it possible to deform the elliptic operator $\mathscr{A}$ to one which is trivial on the boundary. (See the remark after Theorem 10.27.) Once again it follows that there are "topological obstructions" to the existence of boundary operators satisfying the L-condition.

Theorem 11.4 *Let* $(\mathscr{A}, \mathscr{B}) \in \mathfrak{BE}^{1,0}_{p\times p}$ *(i.e.* $\mathscr{B}$ *is of order* 0 *in all variables, including the tangential variables). Let* $\mathscr{Q} = \mathscr{A}_+ I_r \in \mathfrak{E}^1_{r\times r}$*, where* $\mathscr{A}_+ = \varphi\cdot(D_n + i\Lambda') + i(1-\varphi)\Lambda(1-\varphi)$ *is the elliptic operator defined in Theorem* 11.1 *(i.e.* $\mathscr{Q}$ *is a trivial elliptic operator as defined just before Definition* 11.2*). Then the operator*

$$\left(\begin{bmatrix}\mathscr{A} & \\ & \mathscr{Q}\end{bmatrix}, [\mathscr{B}\quad 0]\right) \tag{5}$$

is in $\mathfrak{BE}^{1,0}_{(p+r)\times(p+r)}$ *and when the support of* φ *is sufficiently small it is homotopic in that space to a boundary value problem operator* $(\mathscr{A}_1, \mathscr{B}_1)$ *that is trivial on the boundary; in fact,* $\mathscr{A}_1 = \mathscr{A}_1^b + \mathscr{A}_1^i$ *where*

$$a^{-1}\cdot\mathscr{A}_1^b = \varphi\cdot\begin{pmatrix}(D_n + i\Lambda')I_p & \\ & (D_n - i\Lambda')I_r\end{pmatrix} \tag{6}$$

and $\mathscr{B}_1 = [0\quad I_r]$*, where* $a(y)$ *is some invertible* C^∞ *matrix function on* $\partial\Omega$.

Proof By Lemma 10.28, we may assume that the coefficients of $\mathscr{A} = \mathscr{A}^b + \mathscr{A}^i$ do not depend on x_n near the boundary. Thus, for some $\delta > 0$, we have

$$a^{-1}\cdot\mathscr{A}^b = D_n - \mathbf{A}, \qquad 0 \leqslant x_n \leqslant \delta,$$

where $a(y) = \pi\mathscr{A}(y, n(y))$ is an invertible $p\times p$ C^∞ matrix function on $\partial\Omega$, and $\mathbf{A} \in OClS^1(\partial\Omega, p\times p)$. Further, we may assume that the support of $\mathscr{A}^i$ does not meet $\partial\Omega\times[0,\delta]$.

We divide the proof into two steps. In the first step, where we make an initial deformation of the operator $\mathscr{A}$, the boundary operator is not required for the construction.

Step 1. Let A be the principal symbol of $\mathbf{A}$ and let

$$P(y,\xi') = \frac{1}{2\pi i}\int_{+} (I\lambda - A(y,\xi'))^{-1}\,d\lambda$$

be the Riesz projector for $L(\lambda) = a^{-1}\cdot\pi\mathscr{A}(y,(\xi',\lambda)) = I\lambda - A(y,\xi')$ corresponding to the eigenvalues in the upper half-plane. In view of the remark following Lemma 10.12, the matrix P is homogeneous of degree 0 in ξ'. Let $A_\tau = (1-\tau)A + \tau(i|\xi'|P - i|\xi'|\cdot(I-P))$. Since $\det L(\lambda) \neq 0$ for all $\lambda\in\mathbb{R}$, Lemma 11.3 implies that $L_\tau(\lambda) := I\lambda - A_\tau$ has no real eigenvalues when $0\leqslant\tau\leqslant 1$. Then by Theorem 10.30 applied to the homotopy $a\cdot L_\tau(\lambda)$ in $\mathfrak{M}\mathfrak{P}^1$ there exists a homotopy $\mathscr{A}_\tau$ in $\mathfrak{E}^1$ with $\mathscr{A}_0 = \mathscr{A}$ and

$$\pi\mathscr{A}_\tau(y,(\xi',\lambda)) = a\cdot L_\tau(\lambda), \qquad 0\leqslant\tau\leqslant 1.$$

We can even write out the explicit formula for $\mathscr{A}_\tau$ (from the proof of Theorem 10.30). Choose $\mathbf{P}\in OCIS^0(\partial\Omega, p\times p)$ with principal symbol P and $\Lambda'\in OCIS^1(\partial\Omega)$ with principal symbol $|\xi'|$. Set $\mathscr{A}_\tau = \mathscr{A}_\tau^b + \mathscr{A}^i$, where $\mathscr{A}_\tau^b = \mathscr{A}^b$, $x_n\geqslant\delta$, and

$$a^{-1}\cdot\mathscr{A}_\tau^b = D_n - (1-\tau\varphi)\mathbf{A} - \tau\varphi(i\Lambda'\mathbf{P} - i\Lambda'(I-\mathbf{P})) \qquad \text{where } x_n\leqslant\delta.$$

(In other words, we replace A, P, λ and τ in the definition of $L_\tau(\lambda)$ by $\mathbf{A}$, $\mathbf{P}$, D_n and $\tau\varphi$, respectively).

By Lemma 11.3, $L_\tau(\lambda)$ has the same Riesz (Calderón) projector P for all τ. Let $B = \pi\mathscr{B}$, then by the Corollary to Theorem 10.1 it follows that the matrix $B(y,\xi')\cdot P(y,\xi')$ has rank r for all (y,ξ'). Hence $(\mathscr{A}_\tau,\mathscr{B})$ satisfies the L-condition for all τ. In particular, $(\mathscr{A}_1,\mathscr{B})$ is elliptic with the same index as $(\mathscr{A},\mathscr{B})$.

We make one further deformation to cut $\mathscr{A}_1^b$ off. Note that when $\tau = 1$, $a^{-1}\,\mathscr{A}_1^b$ is equal to $D_n - i\Lambda'\mathbf{P} + i\Lambda'(I-\mathbf{P})$ in a neighbourhood of $\partial\Omega$. By the corollary to Lemma 10.28 it follows that $\mathscr{A}_1$ is homotopic to an operator $\tilde{\mathscr{A}}$ with

$$a^{-1}\tilde{\mathscr{A}}^b = \varphi\cdot(D_n - i\Lambda'\mathbf{P} + i\Lambda'(I-\mathbf{P})), \tag{7}$$

the homotopy being $(1-\kappa)\mathscr{A}_1 + k\tilde{\mathscr{A}}$, $0\leqslant\kappa\leqslant 1$. At the boundary these operators do not depend on κ, thus $(\tilde{\mathscr{A}},\mathscr{B})$ is L-elliptic with the same index as $(\mathscr{A},\mathscr{B})$.

Step 2. Suppose that we start with an elliptic operator $\mathscr{A} = \mathscr{A}^b + \mathscr{A}^i$ such that $\mathscr{A}^b$ has the form (7), i.e.

$$a^{-1}\cdot\mathscr{A}^b = \varphi\cdot(D_n + i\Lambda'(I-2\mathbf{P})),$$

where $\varphi\in C_0^\infty((-\delta,\delta))$, $0\leqslant\varphi\leqslant 1$, $\varphi(x_n) = 1$ in a neighbourhood of 0, and $\mathbf{P}\in OCIS^0(\partial\Omega, p\times p)$. Here $a(y) = \pi\mathscr{A}(y,n(y))$, an invertible matrix function. The support of $\mathscr{A}^i$ is still a compact subset of Ω, but now it can meet $\partial\Omega\times[0,\delta]$. We will use the boundary operator $\mathscr{B}\in OCIS^0(\partial\Omega, r\times p)$ in order to deform $\mathbf{P}$ to the form $\begin{pmatrix}0 & \\ & I_r\end{pmatrix}$ and at the same time deform $[\mathscr{B}\quad 0]$ to $[0\quad I_r]$ while preserving the L-condition.

Let B be the principal symbol of $\mathscr{B}$, and (X_+, T_+, Y_+) a γ^+-spectral triple for $L(\lambda) = \pi\mathscr{A}(y, (\xi', \lambda))$ consisting of smooth matrix functions on $T^*(\partial\Omega)\backslash 0$ with the usual homogeneity properties. In view of Theorem 10.3(iii) with $l = 1$, the L-condition for $(\mathscr{A}, \mathscr{B})$ is that the $r \times r$ matrix BX_+ be invertible. By Theorem 10.3(iv) we have

$$GS = I_r \quad \text{and} \quad SG = P,$$

where $G = B \cdot P$ is $r \times p$ and $S = X_+(BX_+)^{-1}$ is $p \times r$. Note that

$$L(\lambda) = I\lambda + i|\xi'|(I - 2P) = I\lambda + i|\xi'|(I - 2SG).$$

Let $(\mathscr{A}_0, \mathscr{B}_0) = \left(\begin{bmatrix} \mathscr{A} & \\ & \mathscr{Q} \end{bmatrix}, [\mathscr{B} \quad 0]\right) \in \mathfrak{B}\mathfrak{E}^{1,0}_{(p+r)\times(p+r)}$ be the operator (5); for this operator we have

$$L_0(\lambda) = \pi\mathscr{A}_0(y, (\xi', \lambda)) = I_{p+r}\lambda + i|\xi'|\cdot(I_{p+r} - 2S_0G_0),$$

where $G_0 = [G \quad 0]$, $S_0 = [S \quad 0]^T$ and I_{p+r} is the $(p + r) \times (p + r)$ identity matrix. Now consider the homotopy

$$L_\tau(\lambda) := I\lambda + i|\xi'|(I - 2S_\tau G_\tau), \qquad B_\tau := [\cos\tau \cdot B \quad \sin\tau \cdot I_r], \qquad 0 \leqslant \tau \leqslant \pi/2,$$

where $G_\tau = [\cos\tau \cdot G \quad \sin\tau \cdot I_r]$ and $S_\tau = [\cos\tau \cdot S \quad \sin\tau \cdot I_r]^T$. Since $G_\tau S_\tau = I_r$, then

$$\mathscr{P}_\tau = S_\tau G_\tau = \begin{pmatrix} \cos^2\tau\, P & \sin\tau\cos\tau\, S \\ \sin\tau\cos\tau\, G & \sin^2\tau\, I_r \end{pmatrix} \tag{8}$$

is a projector, and since $L_\tau(\lambda) = I\lambda + i|\xi'|(I - 2P_\tau)$ it is easily verified (if we choose a basis for $\ker P_\tau$ and $\operatorname{im} P_\tau$) that P_τ is the Riesz projector for $L_\tau(\lambda)$ corresponding to the eigenvalues in the upper half-plane.

By Theorem 10.30, there exists a homotopy $\mathscr{A}_\tau$ in $\mathfrak{E}^1$ starting from $\mathscr{A}_0 = \begin{pmatrix} \mathscr{A} & \\ & \mathscr{Q} \end{pmatrix}$ and such that

$$\pi\mathscr{A}_\tau(y, (\xi', \lambda)) = a \cdot L_\tau(\lambda).$$

Let us write out an explicit formula for one such homotopy. Choose $\mathbf{G} \in OClS^0(\partial\Omega, r \times p)$ and $\mathbf{S} \in OClS^0(\partial\Omega, p \times r)$ with principal symbols G and S, and choose $\theta \in C_0^\infty((-\delta, \delta))$ such that $\varphi(x_n) = 1$ in $\operatorname{supp}\theta$ and $\theta(x_n) = 1$ in a neighbourhood of 0. With a view to (8), we let

$$\mathbf{P}_\tau = \begin{pmatrix} \cos^2\tau\theta\, \mathbf{P} & \sin\tau\theta\cos\tau\theta\, \mathbf{S} \\ \sin\tau\theta\cos\tau\theta\, \mathbf{G} & \sin^2\tau\theta\, I_r \end{pmatrix}$$

and then

$$\mathscr{A}_\tau = \varphi \cdot a \cdot (D_n + i\Lambda'(I - 2\mathbf{P}_\tau)) + \begin{pmatrix} \mathscr{A}^i & \\ & \mathscr{Q}^i \end{pmatrix}$$

where $\mathscr{A}^i$ and $\mathscr{Q}^i$ are the interior pseudo-differential terms in $\mathscr{A}$ and $\mathscr{Q}$ and $a(y) = \begin{pmatrix} \pi\mathscr{A}(y, n(y)) & \\ & I_r \end{pmatrix}$. Then $\mathscr{A}_\tau$ is elliptic if $\varphi \in C_0^\infty((-\delta, \delta))$ has small enough support, i.e. δ is sufficiently small. Now let

$$\mathscr{B}_\tau = [\cos \tau \cdot \mathscr{B} \quad \sin \tau \cdot I_r]$$

Since $B_\tau \cdot P_\tau = G_\tau$ has rank r, it follows by Theorem 10.3(ii) that $(\mathscr{A}_\tau, \mathscr{B}_\tau)$ satisfies the L-condition. The index is therefore independent of τ. When $\theta(x_n) = 1$, we have

$$\mathbf{G}_{\pi/2} = [0 \quad I_r], \qquad \mathbf{S}_{\pi/2} = \begin{pmatrix} 0 \\ I_r \end{pmatrix}$$

so that in a neighbourhood of $\partial\Omega$

$$a^{-1} \cdot \mathscr{A}^b_{\pi/2} = D_n + i\Lambda'\left(I - 2\begin{pmatrix} 0 \\ I_r \end{pmatrix}[0 \quad I_r]\right)$$

$$= \begin{pmatrix} (D_n + i\Lambda')I_p & \\ & (D_n - i\Lambda')I_r \end{pmatrix}$$

and $\mathscr{B}_{\pi/2} = [0\ I_r]$. As at the end of Step 1 we can cut $\mathscr{A}^b_{\pi/2}$ off, so that it has the form (6).

Remark If $(\mathscr{A}, \mathscr{B}) \in \mathfrak{B}\mathfrak{E}^{1,\mathbf{m}}$ where $\mathbf{m} \neq 0$ then we can still apply Theorem 11.4 after multiplying $\mathscr{B}$ to the left by an elliptic operator in $\partial\Omega$ of order $-\mathbf{m}$. See the text preceding Theorem 10.19.

Let $(\mathscr{A}_1, \mathscr{B}_1)$ be a boundary value problem operator in $\mathfrak{B}\mathfrak{E}^{1,0}$ which is trivial on the boundary. For the purpose of determining the index we may assume by Theorem 11.4 that $\mathscr{A}_1 = \mathscr{A}^b_1 + \mathscr{A}^i_1$, where $a^{-1} \cdot \mathscr{A}^b_1$ has the form (6) and $\mathscr{B}_1 = [0\ I_r]$.

Now, form the boundaryless double $\tilde{\Omega}$ of $\bar{\Omega}$ consisting of two copies $\bar{\Omega}_1$ and $\bar{\Omega}_2$ of $\bar{\Omega}$ identified on $\partial\Omega$. The C^∞ structure of $\tilde{\Omega}$ is defined as follows: the collar of $\partial\Omega$ in $\bar{\Omega}_1$ is identified with $\partial\Omega \times [0, 1)$, while that in $\bar{\Omega}_2$ is identified with $\partial\Omega \times (-1, 0]$ by changing the sign of the x_n-coordinate.

To define an operator on $\tilde{\Omega}$, we keep the definition (6) of $\mathscr{A}^b_1$ in the double collar $\partial\Omega \times (-1, 1) \subset \tilde{\Omega}$, and let $\mathscr{A}^i_2 = i(1 - \varphi)\Lambda(1 - \varphi)$ be an interior p.d.o. in Ω_2 of the same form as in Theorem 11.1. Since $\mathscr{A}^i_1$ and $\mathscr{A}^i_2$ can be viewed as pseudo-differential operators in $\tilde{\Omega}$ that are compactly supported in Ω_1 and Ω_2, respectively, we can define an operator on $\tilde{\Omega}$ by

$$\tilde{\mathscr{A}} := \mathscr{A}^b_1 + \mathscr{A}^i_1 + \mathscr{A}^i_2,$$

which has an elliptic principal symbol, i.e. it is continuous on $T^*(\tilde{\Omega})\backslash 0$ and has invertible values, though it is not quite a pseudo-differential operator on $\tilde{\Omega}$.

Theorem 11.5 *The operator $\tilde{\mathscr{A}}$ defined above is a Fredholm operator $W_2^s(\tilde{\Omega}) \to W_2^{s-1}(\tilde{\Omega})$ for any $s \geqslant 1$, and* ind $\tilde{\mathscr{A}} =$ ind$(\mathscr{A}_1, \mathscr{B}_1)$. *Moreover, $\tilde{\mathscr{A}}$ is a limit of elliptic pseudo-differential operators in $OClS^1(\tilde{\Omega})$. Consequently, the index of $\tilde{\mathscr{A}}$ is equal to* ind$_s$ $\pi\tilde{\mathscr{A}}$, *and can be computed by the Atiyah-Singer index formula on the compact, closed manifold $\tilde{\Omega}$.*

Proof Note that the restriction of $\tilde{\mathscr{A}}$ to $\Omega_1 = \Omega$ is equal to $\mathscr{A}_1$, while the restriction to Ω_2 is equal to an operator $\mathscr{A}_2$ which is a direct sum of operators of the form $\mathscr{A}_+$ and $\mathscr{A}_-$ considered in Theorem 11.1. We have

$$a^{-1}\cdot\mathscr{A}_2^b = \varphi\cdot\begin{pmatrix}(-D_n + i\Lambda')I_p & \\ & -(D_n + i\Lambda')I_r\end{pmatrix} \tag{9}$$

where D_n is replaced by $-D_n$ when the sign of x_n is changed. Let $\mathscr{B}_2 = [I_p\, 0]$, then $(\mathscr{A}_2, \mathscr{B}_2)$ is a trivial elliptic b.v.p. operator (see the text preceding Definition 11.2), and ind$(\mathscr{A}_2, \mathscr{B}_2) = 0$ by Theorem 11.1.

Let us assume $s = 1$. We can rewrite the equation

$$\tilde{\mathscr{A}}\tilde{u} = \tilde{f},$$

where $\tilde{u} \in W_2^1(\tilde{\Omega})$ and $\tilde{f} \in L_2(\tilde{\Omega})$, by regarding the restrictions of $\tilde{u}$ and of $\tilde{f}$ to Ω_1 and to Ω_2 as functions $u_1, u_2 \in W_2^1(\Omega)$ and $f_1, f_2 \in L_2(\Omega)$ such that

$$\mathscr{A}_1 u_1 = f_1, \mathscr{A}_2 u_2 = f_2 \quad \text{in } \Omega; \qquad u_1^+ = u_2^+, u_1^- = u_2^- \quad \text{in } \partial\Omega. \tag{11}$$

We have written $u_1 = [u_1^+, u_1^-]$ and $u_2 = [u_2^+, u_2^-]$ corresponding to the decomposition $\mathbb{C}^{p+r} = \mathbb{C}^p \times \mathbb{C}^r$ in the expressions (6), (9) for $\mathscr{A}_1^b, \mathscr{A}_2^b$. Of course there are trace operators in the boundary conditions in (11).

Now consider the boundary value problem

$$\mathscr{A}'[u_1, u_2] = [\mathscr{A}_1 u_1, \mathscr{A}_2 u_2] = [f_1, f_2];$$

$$\mathscr{B}'_\tau[u_1, u_2] = [u_2^+ - \tau u_1^+, u_1^- - \tau u_2^-] = [0, 0]$$

Then $\tau = 1$ gives the system (11). It is easy to see that $(\mathscr{A}', \mathscr{B}'_\tau)$ satisfies the L-condition for all $0 \leqslant \tau \leqslant 1$, for if $u = [u_1, u_2]$ is a bounded solution of

$$\pi\mathscr{A}'\left(y, \left(\xi', \frac{1}{i}\frac{d}{dt}\right)\right)u = 0,$$

then $u_1^+ = 0$ and $u_2^- = 0$ (see the form of (6), (9)), and the boundary conditions $u_2^+ - \tau u_1^+ = 0$ and $u_1^- - \tau u_2^- = 0$ then give $u_1^- = 0$ and $u_2^+ = 0$, whence $u = 0$. The index of $(\mathscr{A}', \mathscr{B}'_\tau)$ is therefore independent of τ, and when $\tau = 0$ there is no coupling between u_1 and u_2 so the index is equal to ind$(\mathscr{A}_1, \mathscr{B}_1)$ + ind$(\mathscr{A}_2, \mathscr{B}_2)$ where $\mathscr{B}_1 = [0\, I_r]$, $\mathscr{B}_2 = [I_p\, 0]$. Since $(\mathscr{A}_2, \mathscr{B}_2)$ has index 0, it follows that $(\mathscr{A}', \mathscr{B}'_1)$ have the same index.

Clearly the equation $\mathscr{B}'_1[u_1, u_2] = [g_1, g_2]$ can be solved for any $[g_1, g_2] \in W_2^{1/2}(\partial\Omega, \mathbb{C}^{p+r})$; simply take $u_1^+ = 0$, $u_2^- = 0$ and let $u_2^+ \in W_2^1(\Omega, \mathbb{C}^p)$, $u_1^- \in W_2^1(\Omega, \mathbb{C}^r)$ such that $u_2^+|_{\partial\Omega} = g_1$, $u_1^-|_{\partial\Omega} = g_2$. Then by virtue of Lemma 16.12

$$\text{ind}(\mathscr{A}', \mathscr{B}'_1) = \text{ind}(\mathscr{A}'|_{\ker \mathscr{B}'_1}).$$

By Lemma 7A.12 we can return from (11) to (10), i.e., given $u_1 \in W_2^1(\Omega_1)$, $u_2 \in W_2^1(\Omega_2)$ with $u_1 = u_2$ on $\partial\Omega$, there exists $u \in W_2^1(\tilde{\Omega})$ such that $u = u_1$ on Ω_1 and $u = u_2$ on Ω_2. Hence there is a one-to-one correspondence between solutions of (11) and (10). Hence $\tilde{\mathscr{A}}$ is a Fredholm operator and

$$\operatorname{ind} \tilde{\mathscr{A}} = \operatorname{ind}(\mathscr{A}'|_{\ker \mathscr{B}_1'}) = \operatorname{ind}(\mathscr{A}', \mathscr{B}_1') = \operatorname{ind}(\mathscr{A}_1, \mathscr{B}_1).$$

By Lemma 8.78, we can regard Λ' as an operator in $OP^1(\tilde{\Omega})$, and there is a sequence $\Lambda_j \in OClS^1(\tilde{\Omega})$ with principal symbols converging uniformly on compact sets to that of Λ' such that $\Lambda_j - \Lambda' \to 0$ in $OP^1(\tilde{\Omega})$. Let $\tilde{\mathscr{A}}_j$ be the operator obtained when Λ' is replaced by Λ_j in the definition of $\tilde{\mathscr{A}}$ (in (6)); then $\tilde{\mathscr{A}}_j \to \tilde{\mathscr{A}}$ in $OP^l(\tilde{\Omega})$ and the proof of Theorem 8.76 shows that $\tilde{\mathscr{A}}_j$ is a Fredholm operator (hence elliptic) with the same index as $\tilde{\mathscr{A}}$. The Atiyah-Singer index formula for ind $\tilde{\mathscr{A}}_j$ extends by continuity to ind $\mathscr{A}$.

11.2 Higher-order elliptic systems

We first study operators $\hat{\mathscr{A}} = \hat{\mathscr{A}}^b + \hat{\mathscr{A}}^i$ of order $l > 1$ which are very nearly first-order operators in the sense that

$$a^{-1} \cdot \hat{\mathscr{A}}^b = (D_n - \mathbf{A}) \cdot (D_n + i\Lambda')^{l-1} \tag{12}$$

when x_n is small, for some $A \in OClS^1(\partial\Omega, p \times p)$. Here $a(y) = \pi\hat{\mathscr{A}}(y, n(y))$ is an invertible C^∞ matrix function on $\partial\Omega$. We also assume that $\hat{\mathscr{A}}$ is elliptic and the boundary operator $\hat{\mathscr{B}}$ factors in a similar way:

$$\hat{\mathscr{B}} = \mathbf{B} \cdot (D_n + i\Lambda')^{l-1} \tag{13}$$

where $\mathbf{B}$ is an $r \times pl$ matrix of pseudo-differential operators such that the entries in the kth row belong to $OClS^{m_k+1-l}(\partial\Omega)$, $k = 1, \ldots, r$.

In view of Theorem 11.1,

$$\mathscr{Q} := \psi(D_n + i\Lambda') + i(1 - \psi)\Lambda(1 - \psi)$$

is a Fredholm operator from $W_2^s(\Omega)$ to $W_2^{s-1}(\Omega)$, $s \geqslant 1$, with index 0 (without boundary conditions). Note that $\pi\mathscr{Q} = \psi \cdot (\xi_n + i|\xi'|) + i(1 - \psi)^2 \cdot |\xi|$.

Proposition 11.6 *Let* $(\hat{\mathscr{A}}, \hat{\mathscr{B}}) \in \mathfrak{BE}^{l,\mathbf{m}}_{p \times p}$ *such that* (12) *and* (13) *hold. Choose* $\varphi \in C_0^\infty((-1, 1))$, *with* $0 \leqslant \varphi \leqslant 1$, $\varphi = 1$ *in a neighbourhood of* 0, *and such that* (12) *is valid in* supp φ, *and choose* $\psi \in C_0^\infty((-1, 1))$ *with* $0 \leqslant \psi \leqslant 1$ *and* $\psi = 1$ *in* supp φ. *Then define* $\mathscr{A}^b$ *by*

$$a^{-1}\mathscr{A}^b = \varphi(D_n - \mathbf{A}),$$

and let $\mathscr{A} = \mathscr{A}^b + \mathscr{A}^i$ *where* $\mathscr{A}^i \in OClS^1(\mathbb{R}^n, p \times p)$ *has compact support in* Ω. *Also let* $\mathscr{B} = \mathbf{B}$. *If the principal symbol of* $\mathscr{A}$ *multiplied by* $(\psi \cdot (\xi_n + i|\xi'|) + i(1 - \psi)^2|\xi|)^{l-1}$ *is sufficiently close to that of* $\hat{\mathscr{A}}$, *then* $\mathscr{A}$ *is elliptic and* $(\mathscr{A}, \mathscr{B}) \in \mathfrak{BE}^{1,\mathbf{m}+1-l}$ *has the same index as* $(\hat{\mathscr{A}}, \hat{\mathscr{B}})$.

Remark. The condition on the principal symbol of $\mathscr{A}$, i.e. that $\pi\mathscr{A} \cdot (\pi\mathscr{Q})^{l-1}$ be close to $\hat{\mathscr{A}}$, is a condition on the interior symbol $\mathscr{A}^i$ only; it already holds near $\partial\Omega$ by the definition of $\mathscr{A}$ and (12).

Proof Given that $\pi\mathscr{A}\cdot(\pi\mathscr{Q})^{l-1}$ is sufficiently close to $\pi\hat{\mathscr{A}}$ it follows that $\det \pi\mathscr{A}(x,\xi)\neq 0$ for all $x\in\bar{\Omega}$, $0\neq\xi\in\mathbb{R}^n$, whence $\mathscr{A}$ is elliptic. Now, since $(\hat{\mathscr{A}},\hat{\mathscr{B}})$ satisfies the L-condition, the system of ordinary differential equations

$$\pi\hat{\mathscr{A}}\left(y,\xi',\frac{1}{i}\frac{d}{dt}\right)u=0,\quad \pi\hat{\mathscr{B}}\left(y,\xi',\frac{1}{i}\frac{d}{dt}\right)u=0$$

has the unique solution $u=0$. Let A, B be the principal symbols of $\mathbf{A}$ and $\mathbf{B}$. Now,

$$\pi\hat{\mathscr{A}}(y,\xi',\lambda)=(\lambda-A(y,\xi'))\cdot(\lambda+i|\xi'|)^{l-1},$$
$$\pi\hat{\mathscr{B}}(y,\xi',\lambda)=B(y,\xi')\cdot(\lambda+i|\xi'|)^{l-1}$$

and, further, for any $v\in C^\infty(\mathbb{R}_+)$ which is bounded on $\mathbb{R}_+$, the system

$$\left(\frac{1}{i}\frac{d}{dt}+i|\xi'|\right)^{l-1}u=v$$

has a unique solution u which is bounded on $\mathbb{R}_+$. It follows that the system

$$\frac{1}{i}\frac{d}{dt}v-A(y,\xi')v=0,\quad B(y,\xi')v=0$$

has the unique solution $v=0$ (see proof of Theorem 10.34) whence $(\mathscr{A},\mathscr{B})$ also satisfies the L-condition and we have $(\mathscr{A},\mathscr{B})\in\mathfrak{B}\mathfrak{E}^{1,\mathbf{m}+1-l}$.

Since $\mathscr{Q}$ has index 0, then by Theorem 9.8 the operator

$$(\mathscr{A},\mathscr{B})\circ\mathscr{Q}^{l-1}=(\mathscr{A}\mathscr{Q}^{l-1},\mathscr{B}\mathscr{Q}^{l-1})\colon W_2^s(\Omega,\mathbb{C}^p)\to$$
$$W_2^{s-l}(\Omega,\mathbb{C}^p)\times\mathop{\times}_{k=1}^{r}W_2^{s-m_k-1/2}(\partial\Omega)$$

has the same index as $(\mathscr{A},\mathscr{B})$. The boundary operator $\mathscr{B}\mathscr{Q}^{l-1}$ is equal to $\hat{\mathscr{B}}$. The operator $\mathscr{A}\mathscr{Q}^{l-1}$ does not quite belong to $\mathfrak{E}^l$ but it follows by induction using Lemma 11.7 that we can choose ψ in the definition of $\mathscr{Q}$ so that $\mathscr{A}\mathscr{Q}^{l-1}$ is the sum of a compact operator and an operator $\mathscr{A}'\in\mathfrak{E}^l$ with principal symbol close to $\pi\mathscr{A}\cdot(\pi\mathscr{Q})^{l-1}$, hence close to $\pi\hat{\mathscr{A}}$ (in the sup norm). Thus,

$$\operatorname{ind}(\mathscr{A},\mathscr{B})=\operatorname{ind}(\mathscr{A}\mathscr{Q}^{l-1},\mathscr{B}\mathscr{Q}^{l-1})=\operatorname{ind}(\mathscr{A}',\hat{\mathscr{B}})=\operatorname{ind}(\hat{\mathscr{A}},\hat{\mathscr{B}}),$$

where the second equality holds by Theorem 9.11 and the third equality holds by the stability property of the index for operators in $\mathfrak{B}\mathfrak{E}^{l,\mathbf{m}}$.

Lemma 11.7 *Let $\mathscr{A}\in\mathfrak{E}^s$. Choose $\psi\in C_0^\infty((-1,1))$ with $0\leqslant\psi\leqslant 1$ and $\psi=1$ in a neighbourhood of 0 and let $\mathscr{Q}=\psi(D_n+i\Lambda')+i(1-\psi)\Lambda(1-\psi)$. If $\psi=1$ in* supp *$\mathscr{A}^b$ then*

$$\mathscr{A}\mathscr{Q}=\mathscr{A}'+T,$$

for some $\mathscr{A}'\in\mathfrak{E}^{s+1}$ with principal symbol $\pi\mathscr{A}'=\pi\mathscr{A}\cdot\pi\mathscr{Q}$, and some operator T of lower order: $T\in OP^s$.

Proof We have

$$\mathscr{A}\mathscr{Q} = \mathscr{A}\psi(D_n + i\Lambda') + \mathscr{A}(1-\psi)\Lambda(1-\psi)$$

and by commuting $\mathscr{A}$ with ψ and with $1-\psi$ we get

$$\begin{aligned}\mathscr{A}\mathscr{Q} &= \psi\mathscr{A}(D_n + i\Lambda') + [\mathscr{A}, \psi](D_n + i\Lambda') \\ &\quad + (1-\psi)\mathscr{A}\Lambda(1-\psi) + [\mathscr{A}, 1-\psi]\Lambda(1-\psi) \qquad (14)\end{aligned}$$

Now, assuming that $\psi = 1$ on $\operatorname{supp} \mathscr{A}^b$ we get $(1-\psi)\mathscr{A}^b = 0$ so $\mathscr{A}\mathscr{Q} = \mathscr{A}' + T$, where $\mathscr{A}' = \psi\mathscr{A}(D_n + i\Lambda') + (1-\psi)\mathscr{A}^i\Lambda(1-\psi)$ and

$$T = [\mathscr{A}, \psi](D_n + i\Lambda') + [\mathscr{A}, 1-\psi]\Lambda(1-\psi).$$

It is clear that $\mathscr{A}' \in \mathfrak{E}^{s+1}$. Also, by virtue of the Lemma preceding Theorem 10.26 it follows that $[\mathscr{A}, \psi]$ and $[\mathscr{A}, 1-\psi]$ are operators of order $s-1$, so T has order s.

Remark 11.8 If $\psi \neq 1$ in $\operatorname{supp} \mathscr{A}^b$ then, in general, $(1-\psi)\mathscr{A}^b \neq 0$. However, we may use Lemma 8.78 to approximate $(1-\psi)\mathscr{A}^b\Lambda(1-\psi)$ by a genuine p.d.o. in $OS^{s+1}(\mathbb{R}^n)$ with support in some compact subset of Ω. Then from (14) it follows that

$$\mathscr{A}\mathscr{Q} = \mathscr{A}' + T + R,$$

where $T \in OP^s$ is defined as before, $\mathscr{A}' \in \mathfrak{E}^{s+1}$, and R is an operator with small norm.

Let $(\mathscr{A}, \mathscr{B}) \in \mathfrak{BE}^{l,\mathbf{m}}_{p\times p}$ be any boundary value problem operator. Our aim in this section is to show that the direct sum of $(\mathscr{A}, \mathscr{B})$ with a certain $p(l-1) \times p(l-1)$ trivial elliptic b.v.p. operator is homotopic within the space $\mathfrak{BE}^{l,\mathbf{m}}_{pl\times pl}$ to an operator $(\hat{\mathscr{A}}, \hat{\mathscr{B}})$ which is nearly first order (i.e. satisfying (12), (13)). This result is proved in Theorem 11.12 and completes the reduction of the index problem for L-elliptic boundary value problems on $\bar{\Omega}$ to the index problem on the compact, closed manifold $\tilde{\Omega}$ (the double of $\bar{\Omega}$) since we can apply Proposition 11.6 and Theorems 11.4, 11.5. We leave it as an exercise to verify that the hypotheses of Proposition 11.6 can be met, i.e. the existence of the function ψ such that *the principal symbol of* $\mathscr{A}$ *multiplied by* $(\psi\cdot(\xi_n + i|\xi'|) + i(1-\psi)^2|\xi|)^{l-1}$ *is sufficiently close to that of* $\hat{\mathscr{A}}$.

As usual, we let

$$L(\lambda) = \pi\mathscr{A}(y, (\xi', \lambda)) = \sum_{j=0}^{l} A_j(y, \xi')\lambda^j.$$

Note that the leading coefficient, $A_l(y) = \pi\mathscr{A}(y, n(y))$, is an invertible C^∞ matrix function on $\partial\Omega$ which does not depend on the parameter ξ'. Now consider the Möbius transformation $w = \varphi(\lambda) = (\lambda - i|\xi'|)/(\lambda + i|\xi'|)$ and let

$$\tilde{L}(w) = \left(\frac{1-w}{2i|\xi'|}\right)^l \cdot L\left(i|\xi'|\frac{1+w}{1-w}\right)$$

We can regard $L(\lambda)$ as the transformation of $\tilde{L}(w)$ under φ^{-1} (see §2.4), for if we write $\tilde{L}(w) = \sum_{j=0}^{l} \tilde{A}_j(y, \xi')w^j$ then

$$L(\lambda) = (\lambda + i|\xi'|)^l \tilde{L}\left(\frac{\lambda - i|\xi'|}{\lambda + i|\xi'|}\right) = \sum_{j=0}^{l} \tilde{A}_j(y, \xi')\cdot(\lambda - i|\xi'|)^j(\lambda + i|\xi'|)^{l-j} \quad (15)$$

To simplify the notation later on, let us assume that $A_l \equiv I$ (i.e. we consider the matrix polynomial $A_l^{-1}L(\lambda)$). It follows that $\tilde{A}_0 + \cdots + \tilde{A}_l \equiv I$.

Remark Recall that $\mathrm{sp}(\tilde{L}) = \varphi(\mathrm{sp}(L))$. In the case $L(\lambda) = (\lambda + i|\xi'|)^l I_p$, we have $\tilde{L}(w) = I_p$ and $\mathrm{sp}(\tilde{L}) = \{\infty\}$. See §4.2 for the definition of spectrum at infinity.

We claim that the coefficients $\tilde{A}_j$ of $\tilde{L}(w)$ are *homogeneous of degree* 0 *in* ξ'. Indeed, by Lemma 2.23, we have $(\lambda - i)^j\cdot(\lambda + i)^{l-j} = \sum_{k=0}^{l} h_{jk}\lambda^k$ where $H = [h_{jk}]$ is a unique $(l + 1) \times (l + 1)$ invertible matrix. It follows that

$$(\lambda - i|\xi'|)^j\cdot(\lambda + i|\xi'|)^{l-j} = \sum_{k=0}^{l} h_{jk}\cdot|\xi'|^{l-k}\lambda^k \quad (16)$$

Let ${}^tH^{-1} = [c_{jk}]$, then

$$L(\lambda) = \sum_{k=0}^{l} A_k(y, \xi')\lambda^k$$
$$= \sum_{j=0}^{l} \tilde{A}_j(y, \xi')\cdot(\lambda - i|\xi'|)^j\cdot(\lambda + i|\xi'|)^{l-j},$$

where

$$\tilde{A}_j(y, \xi') = \sum_{k=0}^{l} c_{jk}A_k(y, \xi')\cdot|\xi'|^{k-l}. \quad (17)$$

Since A_k is homogeneous of degree $l - k$, then $\tilde{A}_j$ is homogeneous of degree 0 in ξ'.

Let $\mathscr{Q} \in \mathfrak{E}^1$ be a trivial elliptic operator as defined in Proposition 11.6. In Theorem 11.11, we will construct a homotopy in $\mathfrak{E}^l_{pl\times pl}$ from $\begin{pmatrix} \mathscr{A} & \\ & \mathscr{Q}^l\cdot I_{p(l-1)} \end{pmatrix}$ to an elliptic operator $\hat{\mathscr{A}}$ with $\hat{\mathscr{A}}^b = (D_n - \mathbf{A})\cdot(D_n + i\Lambda')^{l-1}$ near $\partial\Omega$. To do this we begin by constructing a homotopy $L^\tau(\lambda)$ in the space $\mathfrak{M}\mathfrak{P}^l_{pl\times pl}$ such that

$$L^0(\lambda) = \begin{pmatrix} L(\lambda) & \\ & (\lambda + i|\xi'|)^l\cdot I_{p(l-1)} \end{pmatrix}, \quad L^1(\lambda) = (I\lambda - A)\cdot(\lambda + i|\xi'|)^{l-1}, \quad (18)$$

with $A = \pi\mathbf{A}$ homogeneous of degree 1 in ξ'. In view of Remark 11.8, this means that the Möbius transformation $\tilde{L}^\tau(w)$ must be a homotopy from $\tilde{L}^0(w) = \begin{pmatrix} \tilde{L}(w) & \\ & I_{p(l-1)} \end{pmatrix}$ to a matrix polynomial of degree 1.

Let $C_{\tilde{L}}(w)$ denote the companion polynomial for $\tilde{L}(w)$ (see §3.2). By Theorem 3.5 we have

$$C_{\tilde{L}}(w) = E^{-1}(w)\cdot\begin{pmatrix}\tilde{L}(w) & \\ & I_{p(l-1)}\end{pmatrix}\cdot F^{-1}(w)$$

$$= \begin{pmatrix} 0 & -I & & \\ & & \ddots & \\ & & & -I \\ I & \tilde{L}_1(w) & \cdots & \tilde{L}_{l-1}(w)\end{pmatrix}\cdot\begin{pmatrix}\tilde{L}(w) & & & \\ & I & & \\ & & \ddots & \\ & & & I\end{pmatrix}$$

$$\times\begin{pmatrix} I & & & & \\ -Iw & I & & & \\ & -Iw & \ddots & & \\ & & \ddots & \ddots & \\ & & & -Iw & I\end{pmatrix}$$

where $\tilde{L}_j(w) = \tilde{A}_j + \tilde{A}_{j+1}w + \cdots + \tilde{A}_l w^{l-j}, j = 0, 1, \ldots, l$. Actually, it will be convenient to re-order the blocks rows in this equation. If we multiply on the left by the permutation matrix on $(\mathbb{C}^p)^l$ which maps $[u_1, \ldots, u_l]$ to $[u_l, -u_1, \ldots, -u_{l-1}]$, we obtain

$$\begin{pmatrix} \tilde{A}_0 & \tilde{A}_1 & \cdots & \tilde{A}_{l-1} + wI \\ -wI & I & & \vdots \\ & \ddots & \ddots & \\ & & -wI & I\end{pmatrix}$$

$$= \begin{pmatrix} I & \tilde{L}_1(w) & \cdots & \tilde{L}_{l-1}(w) \\ & I & & \\ & & \ddots & \\ & & & I\end{pmatrix}\cdot\begin{pmatrix}\tilde{L}(w) & \\ & I_{p(l-1)}\end{pmatrix}$$

$$\times\begin{pmatrix} I & & & & \\ -wI & I & & & \\ & -wI & & & \\ & & \ddots & & \\ & & & -wI & I\end{pmatrix} \tag{19}$$

$$= (I + N_1(w))\cdot\begin{pmatrix}\tilde{L}(w) & \\ & I_{p(l-1)}\end{pmatrix}\cdot(I + N_2(w)),$$

where $N_1(w)$ and $N_2(w)$ are nilpotent matrices. Then we let

$$\tilde{L}^{\tau} = (I + \tau N_1(w))\cdot\begin{pmatrix}\tilde{L}(w) & \\ & I_{p(l-1)}\end{pmatrix}\cdot(I + \tau N_2(w)), \qquad 0 \leqslant \tau \leqslant 1, \quad (20)$$

which is a homotopy such that

$$\tilde{L}^0(w) = \begin{pmatrix}\tilde{L}(w) & \\ & I_{p(l-1)}\end{pmatrix}, \quad \tilde{L}^1(w) = \begin{pmatrix}\tilde{A}_0 & \tilde{A}_1 & \cdots & & \tilde{A}_{l-1} + wI \\ -wI & I & & & \\ & \ddots & \ddots & & \vdots \\ & & & -wI & I\end{pmatrix}$$

To transform back to the λ-variable, we let $w = \varphi(\lambda) = \dfrac{\lambda - i|\xi'|}{\lambda + i|\xi'|}$, then multiply (20) by $(\lambda + i|\xi'|)^l$. Thus,

$$L^{\tau}(\lambda) := \begin{pmatrix} I & \tau^{l-1}\tilde{L}_1(\varphi(\lambda)) & \cdots & \tau\tilde{L}_{l-1}(\varphi(\lambda)) \\ & I & & \\ & & \ddots & \vdots \\ & & & I\end{pmatrix}\begin{pmatrix}L(\lambda) & \\ & (\lambda + i|\xi'|)^l I_{p(l-1)}\end{pmatrix}$$

$$\times\begin{pmatrix} I & & & \\ -\tau\cdot\varphi(\lambda) & I & & \\ & \ddots & \ddots & \\ & & -\tau\varphi(\lambda) & I\end{pmatrix} \qquad (21)$$

for $0 \leqslant \tau \leqslant 1$. (To simplify some later calculations we have introduced powers of τ in the leftmost factor, i.e. $\tau^{l-j}\tilde{L}_j(\varphi(\lambda))$, $j = 1, \dots, l$.) We have

$$L^0(\lambda) = \begin{pmatrix}L(\lambda) & \\ & (\lambda + i|\xi'|)^l\cdot I_{p(l-1)}\end{pmatrix},$$

$$L^1(\lambda) = \begin{pmatrix}\tilde{A}_0 M_0(\lambda) & \tilde{A}_1 M_0(\lambda) & \cdots & & \tilde{A}_{l-1}M_0(\lambda) + \tilde{A}_l M_1(\lambda) \\ -M_1(\lambda) & M_0(\lambda) & & & \\ & -M_1(\lambda) & & & \vdots \\ & & \ddots & \ddots & \\ & & & -M_1(\lambda) & M_0(\lambda)\end{pmatrix}$$

where $M_0(\lambda) = (\lambda + i|\xi'|)^l$ and $M_1(\lambda) = (\lambda - i|\xi'|)(\lambda + i|\xi'|)^{l-1}$. Notice that $(\lambda + i|\xi'|)^{l-1}$ factors out of $L^1(\lambda)$, as was our goal in (18).

By definition, $\tilde{L}_j(w) - \tilde{A}_j = w\tilde{L}_{j+1}(w)$, $j = 0, \dots, l-1$, and $\tilde{L}_0(w) = \tilde{L}(w)$. Since $w = \varphi(\lambda) = M_1(\lambda)/M_0(\lambda)$, it follows that

$$\tilde{L}_j(\varphi(\lambda))\cdot M_0(\lambda) - \tilde{A}_j\cdot M_0(\lambda) = \tilde{L}_{j+1}(\varphi(\lambda))\cdot M_1(\lambda) \qquad (22)$$

and $\tilde{L}_0(\varphi(\lambda))\cdot M_0(\lambda) = L(\lambda)$, see (15).

Proposition 11.9 *Let $L^\tau(\lambda)$, $0 \leqslant \tau \leqslant 1$, be defined as in* (21). *Then $L^\tau(\lambda)$ is a $pl \times pl$ matrix polynomial of degree l, and its leading coefficient, a_τ, is invertible and homogeneous of degree* 0 *in ξ'. Then $a_\tau^{-1} \cdot L^\tau(\lambda)$ is a homotopy in the space $\mathfrak{MP}^l_{pl \times pl}$ from*

$$\begin{pmatrix} L(\lambda) & \\ & (\lambda + i|\xi'|)^l I_{p(l-1)} \end{pmatrix} \quad \textit{to} \quad (I\lambda - A)\cdot(\lambda + i|\xi'|)^{l-1},$$

where the $pl \times pl$ matrix A is homogeneous of degree 1 *in ξ' (see* (25) *below).*

Proof A priori, it is not clear that $L^\tau(\lambda)$ is a polynomial in λ. But if we multiply out the right-hand side of (21) and use (22) and we find that

$$L^\tau(\lambda) = \begin{pmatrix} (1-\tau^l)L(\lambda)+\tau^l\tilde{A}_0M_0(\lambda) & \tau^{l-1}\tilde{A}_1M_0(\lambda) & \cdots & \tau\tilde{A}_{l-1}M_0(\lambda)+\tau A_lM_1(\lambda) \\ -\tau\cdot M_1(\lambda) & M_0(\lambda) & & \\ & -\tau\cdot M_1(\lambda) & \ddots & \vdots \\ & & \ddots & \\ & & -\tau\cdot M_1(\lambda) & M_0(\lambda) \end{pmatrix} \tag{23}$$

for $0 \leqslant \tau \leqslant 1$, which is a polynomial in λ. Since $A_l \equiv I$, then $\tilde{A}_0 + \cdots + \tilde{A}_l \equiv I$ (see (15)), so it is easily seen that the leading coefficient, $a_\tau = A^\tau_l$, of $L^\tau(\lambda)$ has the factorization

$$a_\tau = \begin{pmatrix} I & \tau^{l-1}(\tilde{A}_1 + \cdots + \tilde{A}_l) & \cdots & \tau(\tilde{A}_{l-1} + \tilde{A}_l) \\ & I & & \\ & & \ddots & \\ & & & I \end{pmatrix}$$
$$\times \begin{pmatrix} I & & & \\ -\tau I & I & & \\ & -\tau I & \ddots & \\ & & \ddots & I \\ & & & -\tau I \end{pmatrix} \tag{24}$$

Thus, a_τ is invertible for all $0 \leqslant \tau \leqslant 1$. The matrix polynomials $L^\tau(\lambda)$ satisfy the homogeneity condition (55′) in §10.7 since this is true of $L(\lambda)$, $M_0(\lambda)$ and $M_1(\lambda)$, and the matrices $\tilde{A}_0, \ldots, \tilde{A}_l$ are homogeneous of degree 0 in ξ'. In view of (21) we have

$$\det L^\tau(\lambda) = \det L(\lambda)\cdot(\lambda + i|\xi'|)^{pl(l-1)} \neq 0, \qquad \forall\lambda \in \mathbb{R}.$$

Hence the matrix polynomials $a_\tau^{-1}L^\tau(\lambda)$, $0 \leqslant \tau \leqslant 1$, lie in the space $\mathfrak{MP}^l_{pl \times pl}$. (We need to multiply by a_τ^{-1} in order to make the leading coefficient independent of ξ'.) Also note that

$$a_1^{-1}\cdot L^1(\lambda) = (I\lambda - A)\cdot(\lambda + i|\xi'|)^{l-1}$$

where

$$A = a_1^{-1}\cdot\begin{pmatrix} \tilde{A}_0 & \tilde{A}_1 & \cdots & \tilde{A}_{l-2} & \tilde{A}_{l-1} - \tilde{A}_l \\ I & I & & & \\ & & & & \vdots \\ & \ddots & \ddots & & \\ & & & I & I \end{pmatrix}\cdot i|\xi'| \qquad (25)$$

is homogeneous of degree 1 in ξ'. The proof is complete.

Remark 11.10 It follows from (24) that a_τ has an inverse whose block matrix entries are just linear functions of $\tilde{A}_0, \ldots, \tilde{A}_l$. This remains true if in the definition of a_τ, i.e.

$$a_\tau = \begin{pmatrix} (1-\tau^l)I + \tau^l\tilde{A}_0 & \tau^{l-1}\tilde{A}_1 & & \tau\tilde{A}_{l-1} + \tau\tilde{A}_l \\ -\tau I & I & & \\ & -\tau I & \ddots & \\ & & \ddots & \\ & & -\tau I & I \end{pmatrix}.$$

the matrices $\tilde{A}_0, \ldots, \tilde{A}_l$ are replaced by more general operators satisfying $\tilde{A}_0 + \cdots + \tilde{A}_l = I$.

Let $\mathscr{Q} \in \mathfrak{E}^1$ be an operator as defined before Proposition 11.6. Notice that $\mathscr{Q}^l$ is not quite in $\mathfrak{E}^l$. However, by Remark 11.8, there exists an operator $\mathscr{Q}_l \in \mathfrak{E}^l$ which near the boundary keeps the form $(D_n + i\Lambda')^l$ and differs in principal symbol and in norm so little from $\mathscr{Q}^l$ that the index is not affected, i.e. ind $\mathscr{Q}_l = 0$.

Theorem 11.11 *Let $\mathscr{A} \in \mathfrak{E}^l_{p\times p}$ be a $p \times p$ elliptic operator of order l. There exists a homotopy $\mathscr{A}_\tau$, $0 \leqslant \tau \leqslant 1$, in the space $\mathfrak{E}^l_{pl\times pl}$ such that $\pi\mathscr{A}_\tau(y, (\xi', \lambda)) = a_\tau^{-1}\cdot L^\tau(\lambda)$, where $L^\tau(\lambda)$ is defined by (23) and a_τ is its leading coefficient, and*

$$\mathscr{A}_0 = \begin{pmatrix} \mathscr{A} & \\ & \mathscr{Q}_l\cdot I_{p(l-1)} \end{pmatrix}, \quad a^{-1}\mathscr{A}_1^b = (D_n - \mathbf{A})\cdot(D_n + i\Lambda')^{l-1}, \quad 0 \leqslant x_n \leqslant \delta, \quad (26)$$

for some $\mathbf{A} \in OClS^1(\partial\Omega, pl \times pl)$. *Here* $a(y) = \begin{pmatrix} \pi\mathscr{A}(y, n(y)) & \\ & I_{p(l-1)} \end{pmatrix}$.

Proof This follows from Proposition 11.9 and the Homotopy Extension Theorem 10.30. Note that we assumed earlier that $A_l = \pi\mathscr{A}(y, n(y)) = I$, but in the statement of (26) we have not made this assumption.

For the sake of completeness, we will write out a more explicit formula for $\mathscr{A}_\tau$. As usual, $\mathscr{A} = \mathscr{A}^b + \mathscr{A}^i$, where

$$\mathscr{A}^b = \sum_{j=0}^{l} \mathscr{A}_j D_n^j, \qquad 0 \leqslant x_n \leqslant \delta,$$

and $\mathscr{A}_j = \mathscr{A}_j(y, D') \in OClS^{l-j}(\partial\Omega, p \times p)$. Also, $\mathscr{A}_l = A_l = I$ is just a matrix function, not a pseudo-differential operator. Set

$$\tilde{\mathscr{A}}_j := \sum_{k=0}^{l} c_{jk} \mathscr{A}_k R^{l-k},$$

where R is a parametrix of Λ'. Note that $\tilde{\mathscr{A}}_j \in OClS^0(\partial\Omega, p \times p)$ and by (17) we see that $\tilde{\mathscr{A}}_j$ has principal symbol $\tilde{A}_j(y, \xi')$. Corresponding to (15), we have

$$\mathscr{A}^b = \sum_{j=0}^{j} \tilde{\mathscr{A}}_j \cdot (D_n - i\Lambda')^j (D_n + i\Lambda')^{l-j}, \qquad 0 \leqslant x_n \leqslant \delta, \tag{27}$$

apart from terms of order $-\infty$ *in* $\partial\Omega$ *and order* $<l$ *with respect to* x_n. First of all note that $R\Lambda' - I \in OS^{-\infty}(\partial\Omega)$, so that $R^j\Lambda'^j - I \in OS^{-\infty}(\partial\Omega)$ for all $j = 1, 2, \ldots$. Since D_n and Λ' commute, then replacing λ by D_n and $|\xi|$ by Λ' in (16) we obtain

$$(D_n - i\Lambda')^j \cdot (D_n + i\Lambda')^{l-j} = \sum_{k=0}^{l} h_{jk}(\Lambda')^{l-k} D_n^k,$$

so the right-hand side of (27) is equal to

$$\sum_{k=0}^{l} \sum_{j=0}^{l} \tilde{\mathscr{A}}_j \cdot h_{jk} \cdot (\Lambda')^{l-k} D_n^k = \sum_{k=0}^{l} \mathscr{A}_k R^{l-k} (\Lambda')^{l-k} D_n^k$$

$$\equiv \sum_{k=0}^{j} \mathscr{A}_k D_n^k = \mathscr{A}^b,$$

where in the kth term we get exactly $\mathscr{A}_l D_n^l$ when $k = l$, and when $k < l$ we get $\mathscr{A}_k D_n^k$ plus terms of order $-\infty$ in $\partial\Omega$ and order $<l$ with respect to x_n.

Further, in view of (16), we have $h_{jl} = 1, j = 0, \ldots, l$, and since ${}^tH^{-1} = [c_{jk}]$ it follows that $\sum_{j=0}^{l} c_{jk} = \delta_{kl}$. Thus

$$\tilde{\mathscr{A}}_0 + \cdots + \tilde{\mathscr{A}}_l = I,$$

since $\mathscr{A}_l = I$. It follows by Remark 11.10 that it is legitimate to replace the matrix $\tilde{A}_j$ by the p.d.o. $\tilde{\mathscr{A}}_j$.

Now, without changing the index we may assume that (27) is an exact equality. Let $\chi \in C_0^\infty((-\delta, \delta))$ be equal to 1 in a neighbourhood of 0, and $0 \leqslant \chi \leqslant 1$ everywhere. Now define $\mathscr{A}_\tau^b$ by the same formula as $a_\tau^{-1} \cdot L^\tau(\lambda)$ in (23) and (24), but with τ replaced by $\tau\chi$, $L(\lambda)$ replaced by $\mathscr{A}^b$, $\tilde{A}_j$ replaced by $\tilde{\mathscr{A}}_j$, $M_0(\lambda)$ replaced by $(D_n + i\Lambda')^l$ and $M_1(\lambda)$ replaced by $(D_n - i\Lambda')(D_n + i\Lambda')^{l-1}$. Since this is equal to $\mathscr{A}_0^b$ when $\chi(x_n) = 0$, we can continue this definition by taking $\mathscr{A}_\tau^b = \mathscr{A}_0^b$ for $x_n \geqslant \delta$. Finally, we define

$$\mathscr{A}_\tau = \mathscr{A}_\tau^b + \begin{pmatrix} \mathscr{A}^i & \\ & \mathscr{Q}_l^i \cdot I_{p(l-1)} \end{pmatrix},$$

where $\mathscr{A}^i$ and $\mathscr{Q}_l^i$ are the interior pseudo-differential terms in $\mathscr{A}$ and $\mathscr{Q}_l$, respectively. Then $\mathscr{A}_\tau \in \mathfrak{E}^l_{pl \times pl}$ and satisfies the conditions of the theorem. We have $\mathscr{A}_1^b = (D_n - \mathbf{A}) \cdot (D_n + i\Lambda')^{l-1}$ whenever $\chi(x_n) = 1$, where the operator $\mathbf{A}$ is defined by (25) with replacements indicated above.

It has been assumed throughout this chapter that the coefficients of $\mathscr{A}^b$ do not depend on x_n near the boundary (for the purpose of applying Theorem 10.30). But clearly the definition of the homotopy $\mathscr{A}_\tau$ is valid even if the coefficients of $\mathscr{A}^b$ do depend on x_n and we obtain (26) with the p.d.o. $\mathbf{A}$ depending smoothly on the parameter x_n. In order to use Proposition 11.6 we would then deform $\mathbf{A}$ so that it is independent of x_n, by using Lemma 10.28.

Note that the homotopy in Theorem 11.11 was constructed without reference to any boundary operator. Now let $\mathscr{B}$ be a boundary operator satisfying the L-condition for $\mathscr{A}_0 = \mathscr{A}$. We show in the next theorem that at the same time that $\mathscr{A}$ is deformed to an operator of type (12), $\mathscr{B}$ can be deformed to a boundary operator of the type (13) while preserving the L-condition.

Theorem 11.12 *For any* $(\mathscr{A}, \mathscr{B}) \in \mathfrak{B}\mathfrak{E}^{l,\mathrm{m}}_{p \times p}$, *the boundary operator*

$$\begin{pmatrix} \mathscr{A} & \\ & \mathscr{Q}_l \cdot I_{p(l-1)} \end{pmatrix}$$

is homotopic in the space $\mathfrak{B}\mathfrak{E}^{l,\mathrm{m}}_{pl \times pl}$ *to an operator* $(\hat{\mathscr{A}}, \hat{\mathscr{B}})$ *such that*

$$a^{-1}\hat{\mathscr{A}}^b = (D_n - \mathbf{A}) \cdot (D_n + i\Lambda')^{l-1} \tag{28}$$

when x_n *is small, for some* $A \in OClS^1_{pl \times pl}(\partial\Omega)$ *and*

$$\hat{\mathscr{B}} = \mathbf{B} \cdot (D_n + i\Lambda')^{l-1} \tag{29}$$

where $\mathbf{B}$ *is an* $r \times pl$ *matrix of pseudo-differential operators such that the entries in the* k*th row belong to* $OClS^{m_k + 1 - l}(\partial\Omega)$, $k = 1, \ldots, r$. *Here the matrix function "a" is defined as in Theorem* 11.11.

Proof Since the transversal order is $\mu \leqslant l - 1$, the boundary operator has the form $\mathscr{B} = \sum_{j=0}^{l-1} \mathscr{B}_j D_n^j$ and

$$B(\lambda) = \pi\mathscr{B}(y, (\xi', \lambda)) = \sum_{j=0}^{l-1} B_j(y, \xi')\lambda^j$$

is the associated matrix polynomial ($B_j = \pi\mathscr{B}_j$). Let $\mathscr{A}_\tau$, $0 \leqslant \tau \leqslant 1$, be the homotopy constructed in Theorem 11.11, and let (X_+, T_+) be a γ^+-spectral pair for $L(\lambda)$; we will find a γ^+-spectral pair (X^τ_+, T^τ_+) for $L^\tau(\lambda)$ such that

$$\sum_{j=0}^{l-1} [B_j(y, \xi')\, 0 \cdots 0] \cdot X^\tau_+ (T^\tau_+)^j = \sum_{j=0}^{l-1} B_j(y, \xi') \cdot X_+ T^j_+. \tag{30}$$

Once that it is done, then by Theorem 10.3 (iii) it follows that $(\mathscr{A}_\tau, [\mathscr{B}\, 0 \cdots 0])$ satisfies the L-condition for all $0 \leqslant \tau \leqslant 1$. (There are $l - 1$ blocks of the $r \times p$ zero matrix so that $[\mathscr{B}\, 0 \cdots 0]$ is an $r \times pl$ boundary operator.) After that we will show that the boundary operator $[\mathscr{B}\, 0 \cdots 0]$ can be deformed to the form (29) while keeping the same elliptic operator $\mathscr{A}_1 = \hat{\mathscr{A}}$.

But first a few preliminary observations are necessary. Let $\mathfrak{M}^+$ denote the set of $u \in C^\infty(\mathbb{R})$ of the form $u(t) = \sum p_k(t)\, e^{\lambda_k t}$, where the p_k's are polynomials and $\operatorname{Re} \lambda_k < 0$. The operator $D + i|\xi'|$, where $D = i^{-1}\, d/dt$, is invertible on $\mathfrak{M}^+$: injectivity is clear, and to prove surjectivity note that if $f \in \mathfrak{M}^+$ then $(D + i|\xi'|)u = f$, where $u(t) = \int_t^\infty e^{|\xi'|(t-s)} \cdot if(s)\, ds$. Thus, it makes sense to replace λ by D in equation (21), so that $L^\tau(D)$ is an operator on $(\mathfrak{M}^+)^{pl}$.

Let $U = [u_j]_{j=0}^{l-1} \in \mathfrak{M}^+_{L^\tau}$. Since $L^\tau(D)U = 0$, we obtain by substituting λ by D in (21) that

$$\begin{pmatrix} L(D) & \\ & (D + i|\xi'|)^l I_{p(l-1)} \end{pmatrix} V = 0 \tag{31}$$

where $V = [v_j]_{j=0}^{l-1}$ is given by

$$v_0 = u_0, \quad v_j = u_j - \tau \cdot \tilde{D} u_{j-1}, \qquad j = 1, \ldots, l-1,$$

where $\tilde{D} = (D - i|\xi'|)(D + i|\xi'|)^{-1}$. It follows from (31) that $L(D)v_0 = 0$, $v_1 = \cdots = v_{l-1} = 0$. Since (X_+, T_+) is a right γ^+-spectral pair for $L(\lambda)$, then $v_0(t) = X_+\, e^{itT_+} c$ for some c in the base space of the pair, and we obtain

$$U = \operatorname{col}(\tau^j X_+ \tilde{T}_+^j)_{j=0}^{l-1}\, e^{itT_+} c,$$

where $\tilde{T}_+ = (T_+ - i|\xi'|I) \cdot (T_+ + i|\xi'|I)^{-1}$. It follows that

$$(X_+^\tau, T_+^\tau) := (\operatorname{col}(\tau^j X_+ \tilde{T}_+^j)_{j=0}^{l-1}, T_+) \tag{32}$$

is a right *partial* γ^+ spectral pair for $L^\tau(\lambda)$ (Definition 1.13). However, in view of (21), the equations $\det L^\tau(\lambda) = 0$ and $\det L(\lambda) = 0$ have the same number of roots in the upper half-plane, so that (X_+^τ, T_+^τ) is in fact a γ^+-spectral pair for $L^\tau(\lambda)$.

With this choice of (X_+^τ, T_+^τ) the equation (30) holds.

Now consider the transformation of $B(\lambda)$ under $w = (\lambda - i|\xi'|)/(\lambda + i|\xi'|)$, i.e.

$$\tilde{B}(w) = \left(\frac{1-w}{2i|\xi'|}\right)^{l-1} \cdot B\left(i|\xi'| \frac{1+w}{1-w}\right) = \sum_{j=0}^{l-1} \tilde{B}_j(y, \xi') w^j,$$

or, equivalently, $B(\lambda) = (\lambda + i|\xi'|)^{l-1} \cdot \tilde{B}\left(\dfrac{\lambda - i|\xi'|}{\lambda + i|\xi'|}\right)$. Then

$$B(\lambda) = \sum_{j=0}^{l-1} \tilde{B}_j(y, \xi') \cdot (\lambda - i|\xi'|)^j (\lambda + i|\xi'|)^{l-1-j}, \tag{33}$$

where $\tilde{B}_j(y, \xi') = \sum_{k=0}^{l-1} c'_{jk} \cdot B_k(y, \xi') \cdot |\xi'|^{k-l+1}$ for unique constants c'_{jk}. Let

$$\tilde{\mathscr{B}}_j := \sum_{k=0}^{l-1} c'_{jk} \mathscr{B}_k R^{l-1-k}$$

where R is a parametrix of Λ'; the entries of the kth row of $\tilde{\mathscr{B}}_j$ are elements of $OClS^{m_k-l+1}(\partial\Omega)$, $k = 1, \ldots, r$. Let $\mathbf{B} = [\tilde{\mathscr{B}}_0 \cdots \tilde{\mathscr{B}}_{l-1}]$ and define $\hat{\mathscr{B}}$ by (29); we have

$$\hat{\mathscr{B}} = \sum_{j=0}^{l-1} \hat{\mathscr{B}}_j \cdot D_n^j, \qquad \text{where } \hat{\mathscr{B}}_j = \binom{l-1}{j} \cdot \mathbf{B} \cdot (i\Lambda')^{l-1-j}.$$

The entries in the kth row of $\hat{\mathscr{B}}_j$ are in $OClS^{m_k-j}(\partial\Omega)$, so $\hat{\mathscr{B}}$ and $[\mathscr{B}\, 0 \cdots 0]$ have the same DN numbers. Now let

$$(X_+^1, T_+) := (\mathrm{col}(X_+ \tilde{T}_+^j)_{j=0}^{l-1}, T_+)$$

be the γ^+-spectral pair for $L^1(\lambda) = \pi\mathscr{A}_1(y, (\xi', \lambda))$ as defined by (32) with $\tau = 1$. We claim that

$$\sum_{j=0}^{l-1} B_j^1(y, \xi') X_+^1 T_+^j = \sum_{j=0}^{l-1} B_j(y, \xi') \cdot X_+ T_+^j \tag{34}$$

where $B_j^1 = \pi\hat{\mathscr{B}}_j$ and $B_j = \pi\mathscr{B}_j$. Indeed, it follows from (33) that the right-hand side of (34) is equal to

$$\begin{aligned}\sum_{j=0}^{l-1} \tilde{B}_j(y, \xi') X_+ \cdot (T_+ - i|\xi'|I)^j \cdot (T_+ + i|\xi'|I)^{l-1-j} \\ = \sum_{j=0}^{l-1} \tilde{B}_j(y, \xi') X_+ \tilde{T}_+^j \cdot (T_+ + i|\xi'|)^{l-1} \\ = \pi\mathbf{B}(y, \xi') \cdot X_+^1 (T_+ + i|\xi'|I)^{l-1}\end{aligned}$$

which is equal to the left-hand side of (34).

In view of (34) and (30), the boundary value problem operators $(\hat{\mathscr{A}}, [\mathscr{B}\, 0 \cdots 0])$ and $(\hat{\mathscr{A}}, \hat{\mathscr{B}})$ can be connected by the homotopy

$$(\hat{\mathscr{A}}, \kappa \cdot [\mathscr{B}\, 0 \cdots 0] + (1 - \kappa)\hat{\mathscr{B}}), \quad 0 \leqslant \kappa \leqslant 1.$$

The proof is complete.

12

BVP's for ordinary differential operators and the connection with spectral triples

Let $\mathscr{A} \in \mathrm{Ell}^l$ and suppose that $\mathscr{A}$ has constant coefficients in the half-space $\Omega = \mathbb{R}^n_+$. Also let $\mathscr{B}$ be some boundary operator with constant coefficients in $\partial\Omega = \mathbb{R}^{n-1}$, fulfilling the L-condition. If f and g are C^∞ functions rapidly decreasing at ∞, then by means of a partial Fourier transform $\mathfrak{F}'$ in the variables $x_1, \ldots, x_{n-1}$ the boundary value problem

$$\mathscr{A}(D)u = f, \quad x_n > 0; \qquad \mathscr{B}(D)u = g, \quad x_n = 0^+$$

can be transformed to a boundary problem for a system of ordinary differential equations

$$\mathscr{A}\left(\xi', \frac{1}{i}\frac{d}{dx_n}\right)\tilde{u} = \tilde{f}, \quad x_n > 0; \qquad \mathscr{B}\left(\xi', \frac{1}{i}\frac{d}{dx_n}\right)\tilde{u} = \tilde{g}, \quad x_n = 0^+, \quad (1)$$

for $\tilde{u} \in {}^p\mathscr{S}(\bar{\mathbb{R}}_+)$, that is, $\tilde{u}(\xi', x_n)$ approaches 0 rapidly as $x_n \to +\infty$.

Remark. ${}^p\mathscr{S}(\bar{\mathbb{R}}_+)$ denotes the restrictions to $\bar{\mathbb{R}}_+$ of p-vector functions whose components lie in the Schwartz space $\mathscr{S}$ on the real line.

In §12.1 we prove some preliminary results on the extension of C^∞ functions defined on a half-line. The main goal of this chapter is to show that every solution of (1) which is rapidly decreasing as $x_n \to +\infty$ has a representation of the form

$$\tilde{u} = G\tilde{f} + K\tilde{g}$$

where $G: {}^p\mathscr{S}(\bar{\mathbb{R}}_+) \to {}^p\mathscr{S}(\bar{\mathbb{R}}_+)$ is the Green operator and $K: \mathbb{C}^r \to {}^p\mathscr{S}(\bar{\mathbb{R}}_+)$ is the Poisson operator.

Let $(X_+(\xi'), T_+(\xi'))$ be a spectral pair for the matrix polynomial $L_{\xi'}(\lambda) = \pi\mathscr{A}(\xi', \lambda)$ with respect to the eigenvalues in the upper half-plane $\operatorname{Im}\lambda > 0$. When $\tilde{f} = 0$ we have $\tilde{u} = X_+ e^{ix_nT_+}c$ for some $c = c(\xi')$ in the base space of the spectral pair. To satisfy the boundary conditions we must have $c = (\Delta^+_{\mathscr{B}})^{-1}\tilde{g}$. Thus the Poisson operator for (1) is given by $K\tilde{g} = X_+ e^{ix_nT_+}(\Delta^+_{\mathscr{B}})^{-1}\tilde{g}$. It is also possible to express the Green operator G in terms of the spectral pair $(X_+(\xi'), T_+(\xi'))$ as we show in §12.2.

12.1 Extension of C^∞ functions defined on a half-line

If f is a function defined on $\mathbb{R}$, we let $\imath^+ f$ denote its restriction to the half-line $\mathbb{R}_+ = \{t; t > 0\}$. The restriction map $\imath^+ : C^\infty(\mathbb{R}) \to C^\infty(\mathbb{R}_+)$ is continuous, and we define

$$C^\infty(\bar{\mathbb{R}}_+) = \imath^+ C^\infty(\mathbb{R}), \qquad \mathscr{S}(\bar{\mathbb{R}}_+) = \imath^+ \mathscr{S}(\mathbb{R}),$$

where $\mathscr{S} = \mathscr{S}(\mathbb{R})$ is the Schwartz space on $\mathbb{R}$. These spaces are easy to characterize using the following lemma.

Lemma 12.1 *Let $m_j \in \mathbb{C}, j = 0, 1, 2, \ldots$, be a given sequence of constants. Then for any $\varepsilon > 0$ there exists $f \in C_0^\infty((-\varepsilon, \varepsilon))$ such that*

$$f^{(j)}(0) = m_j \qquad \textit{for all } j.$$

Proof Choose $g \in C_0^\infty((-\varepsilon, \varepsilon))$ such that $g(0) = 1$ and

$$\frac{d^k}{dt^k} g(0) = 0 \qquad \text{for } k > 0$$

(for instance, choose g such that $g(t) = 1$ when $-\varepsilon/2 \leqslant t \leqslant \varepsilon/2$). Now, our aim is to define the modified Taylor series

$$f(t) = \sum_{j=0}^{\infty} g(t/\varepsilon_j) t^j m_j / j! \tag{2}$$

with suitably chosen ε_j. By taking t/ε_j as the new variable we find that

$$\left| \frac{d^k}{dt^k} g(t/\varepsilon_j) t^j m_j / j! \right| \leqslant C_{k,j} \varepsilon_j^{j-k},$$

and we choose $0 < \varepsilon_j < 1$ such that $C_{k,j} \varepsilon_j^{j-k} \leqslant 2^{-j}$ for $k \leqslant j - 1$. Then the kth derivative of the jth term in the series (2) has magnitude $\leqslant 2^{-j}$ when $j > k$, so the series (2) and all series obtained by termwise differentiation are uniformly convergent. Hence $f \in C^\infty(\mathbb{R})$, and its support lies in $(-\varepsilon, \varepsilon)$.

Corollary 12.2 *$C^\infty(\bar{\mathbb{R}}_+)$ is the set of all functions $u \in C^\infty(\mathbb{R}_+)$ such that the limit*

$$\lim_{t \to 0+} \frac{d^k}{dt^k} u(t) \quad \textit{exists}$$

for all $k = 0, 1, 2, \ldots$. $\mathscr{S}(\bar{\mathbb{R}}_+)$ is the set of functions $u \in C^\infty(\bar{\mathbb{R}}_+)$ such that u and all its derivatives are rapidly decreasing as $t \to +\infty$, i.e.

$$\left| \frac{d^k}{dt^k} u(t) \right| \leqslant \text{const } (1 + t)^{-j}, \qquad t > 0,$$

for all $j, k = 0, 1, 2, \ldots$.

Proof The necessity is clear. To prove sufficiency let $u \in C^\infty(\bar{\mathbb{R}}_+)$. By Lemma 12.1 there exists $f \in C_0^\infty((-1, 1))$ such that $f^{(j)}(0) = u^{(j)}(0+)$. Now let

$$U(t) = \begin{cases} u(t), & t \geqslant 0 \\ f(t), & t < 0 \end{cases}$$

Then $U \in C^\infty(\mathbb{R})$ since all the derivatives of u and f match at $t = 0$. This proves the first statement in Proposition 12.2. The second statement is also clear, for if $u \in \mathcal{S}(\bar{\mathbb{R}}_+)$ then $U \in \mathcal{S}(\mathbb{R})$ since $U(t) = u(t)$ when $t > 0$ and $U(t) = 0$ when $t < -1$.

The dependence of U on u is neither linear nor continuous because ε_j depends on the constants m_j. For the sake of completeness, we also show that there exist linear, continuous extension operators $\beta\colon C^\infty(\bar{\mathbb{R}}_+) \to C^\infty(\mathbb{R})$ and $\beta\colon \mathcal{S}(\bar{\mathbb{R}}_+) \to \mathcal{S}(\mathbb{R})$. Note that in the construction of β, the constants c_p are fixed, i.e. they do not depend on the given function u.

Theorem 12.3 *There is a continuous, linear exension operator* $\beta\colon C^\infty(\bar{\mathbb{R}}_+) \to C^\infty(\mathbb{R})$ *such that* $\imath^+\beta =$ *identity, i.e.*

$$\beta u(t) = u(t), \qquad t > 0,$$

where the topology on $C^\infty(\mathbb{R})$ *(resp.* $C^\infty(\bar{\mathbb{R}}_+)$*) is that of uniform convergence of each derivative on compact subsets of* $\mathbb{R}$ *(resp.* $\bar{\mathbb{R}}_+$*). If* f *has compact support then so does* βf*. Moreover,* β *restricts to a continuous, linear extension operator* $\beta\colon \mathcal{S}(\bar{\mathbb{R}}_+) \to \mathcal{S}(\mathbb{R})$.

Proof We choose a rapidly decreasing sequence c_p, $p = 0, 1, \ldots$, such that

$$\sum_{p=1}^{\infty} p^k c_p = (-1)^k, \qquad k = 0, 1, \ldots \tag{3}$$

(see Corollary 12.5 below). Also let φ be a C^∞ function on $\mathbb{R}$ with $\varphi(t) = 1$ for $0 \leqslant t \leqslant 1$ and $\varphi(t) = 0$ for $t \geqslant 2$. Define the extension of u by $\beta u(t) = u(t)$, $t > 0$,

$$\beta u(t) = \sum_{p=1}^{\infty} c_p \cdot \varphi(-pt)u(-pt), \qquad t < 0.$$

Then because $p \to \infty$ the sum is finite for each $t < 0$; because $\sum_{p=1}^{\infty} p^k|c_p| < \infty$, all derivatives of βu converge as $t \to 0^-$; moreover, these limits agree with those for $t \to 0^+$ in view of (3). Thus, if we define $\beta u(0) = \lim_{t \to 0^+} u(t)$, then the derivatives of βf exist and are continuous everywhere on $\mathbb{R}$, whence $\beta f \in C^\infty(\mathbb{R})$. The continuity of $\beta\colon C^\infty(\bar{\mathbb{R}}_+) \to C^\infty(\mathbb{R})$ also follows from $\sum_{p=1}^{\infty} p^k|c_p| < \infty$. This proves the first statement in the Theorem. The second statement follows immediately since $\beta u(t) = u(t)$ when $t > 0$ and $\beta u(t) = u(-t)$ when $t \leqslant -1$.

Now we turn to the construction of the rapidly decreasing sequence c_p satisfying (3). The following lemma is due to [Se 1].

Lemma 12.4 *There are sequences $\{a_p\}, \{b_p\}$ such that b_p is negative and decreases to $-\infty$ as $p \to \infty$ and*

$$\sum_{p=0}^{\infty} (b_p)^k a_p = 1, \qquad k = 0, 1, 2, \ldots,$$

and

$$\sum_{p=0}^{\infty} |b_p|^k |a_p| < \infty, \qquad k = 0, 1, 2, \ldots \tag{4}$$

Proof Set $b_p = -2^p$. Then the solutions $a_p^{(N)}$ of

$$\sum_{p=0}^{N} (b_p)^k \cdot x_p = 1, \qquad k = 0, \ldots, N,$$

are by Cramer's rule and the Vandermonde determinant equal to

$$x_p = \frac{\prod_{j>i} (b'_j - b'_i)}{\prod_{j>i} (b_j - b_i)}$$

where $b'_p = 1$ and $b'_j = b_j$ if $j \neq p$. Numerous factors cancel so we obtain

$$x_p = A_p \cdot B_p^{(N)},$$

where

$$A_p = \prod_{j=0}^{p-1} \frac{1 + 2^j}{2^j - 2^p}, \qquad B_p^{(N)} = \prod_{j=p+1}^{N} \frac{1 + 2^j}{2^j - 2^p}.$$

Then $|A_p| \leqslant \prod_{j=0}^{p-1} 2^{j+2-p} = 2^{-(p^2-3p)/2}$ and, using the fact that $\log x \leqslant x - 1$, with equality holding only at $x = 1$, we have

$$\begin{aligned}
\log B_p^{(N)} &= \sum_{j=p+1}^{N} \log \frac{1 + 2^j}{2^j - 2^p} \\
&< \sum_{j=p+1}^{N} \frac{1 + 2^p}{2^j - 2^p} \\
&= \sum_{j=1}^{N-p} \frac{2^{-p} + 1}{2^j - 1} \\
&< \sum_{j=1}^{\infty} \frac{1 + 1}{2^{j-1}} = 4.
\end{aligned}$$

Since $B_p^{(N)}$ is monotone increasing with N, it converges to some limit $B_p \leqslant e^4$ as $N \to \infty$. Setting $a_p = A_p \cdot B_p$, we have

$$|a_p| \leqslant e^4 \cdot 2^{-(p^2-3p)/2},$$

and therefore (4) holds for each k. Now, by setting

$$a_p^{(N)} = \begin{cases} A_p \cdot B_p^{(N)}, & p \leqslant N \\ 0, & p > N \end{cases}$$

we have

$$\sum_{p=0}^{\infty} (b_p)^k \cdot a_p^{(N)} = 1, \qquad k \leqslant N,$$

and since $|b_p|^k \cdot |a_p^{(N)}| \leqslant |b_p|^k |a_p|$ and $\sum |b_p|^k |a_p| < \infty$, we have

$$\sum_{p=0}^{\infty} (b_p)^k a_p = \lim_{N \to \infty} \sum_{p=0}^{\infty} (b_p)^k \cdot a_p^{(N)} = 1 \qquad \text{for all } k.$$

Corollary 12.5 *There exists a rapidly decreasing sequence c_p, $p = 0, 1, \ldots$, such that $\sum_{p=1}^{\infty} p^k c_p = (-1)^k$ for $k = 0, 1, \ldots$.*

Proof From the lemma we have $\sum_{q=0}^{\infty} (-2^q)^k \cdot a_q = 1$. Thus, we let

$$c_p = \begin{cases} a_q & \text{if } p = 2^q \\ 0 & \text{otherwise} \end{cases}$$

Since $\sum_{p=1}^{\infty} p^k \cdot |c_p| < \infty$ for every k, it follows that $|c_p| \leqslant \text{const}(1 + p)^{-k}$ for any k, i.e. c_p is rapidly decreasing.

12.2 Ordinary differential operators on a half-line

Let $D = \dfrac{1}{i}\dfrac{d}{dt}$. By taking the initial values of the kth derivative $D^k u$ we obtain a continuous map

$$\gamma_k \colon C^\infty(\bar{\mathbb{R}}_+) \to \mathbb{C}^l \qquad \text{defined by } \gamma_k u = \lim_{t \to 0^+} D^k u(t)$$

for $k = 0, \ldots, l-1$. Then let $\gamma = [\gamma_0 \cdots \gamma_{l-1}]^T \colon C^\infty(\bar{\mathbb{R}}_+) \to \mathbb{C}^l$, that is,

$$\gamma u = \lim_{t \to 0^+} \begin{pmatrix} u(t) \\ \vdots \\ D^{l-1} u(t) \end{pmatrix}. \tag{5}$$

As long as it does not lead to confusion we will use the same symbol γ for the operator $\gamma_t{}^+$ defined on $C^\infty(\mathbb{R})$ by formula (5). We denote the set of p-vector C^∞ functions by

$${}^pC^\infty(\mathbb{R}) = C^\infty(\mathbb{R}, \mathbb{C}^p),$$

that is, ${}^pC^\infty = C^\infty \times \cdots \times C^\infty$. In that case (5) defines an operator

$$\gamma \colon {}^pC^\infty(\bar{\mathbb{R}}_+) \to \mathbb{C}^{pl}.$$

Similarly, we let ${}^p\mathscr{S} = \mathscr{S} \times \cdots \times \mathscr{S}$.

Let $L(\lambda) = \sum_{j=0}^{l} A_j\lambda^j$ be a matrix polynomial. In the following lemma,

$$\mathscr{L} = \begin{pmatrix} A_1 & A_2 & \cdots & A_l \\ A_2 & & \cdot^{\cdot^{\cdot}} & \\ \vdots & & & \\ A_l & & & 0 \end{pmatrix}$$

is the matrix from §2.1 (5).

Lemma 12.6 *Let $u \in {}^pC^\infty(\mathbb{R})$, and let u^0 denote the locally integrable function equal to u for $t > 0$ and zero for $t < 0$. Then*

$$L(D)u^0 = (L(D)u)^0 + L^c\gamma u, \tag{6}$$

where $L^c := i^{-1}[\delta \cdots D^{l-1}\delta]\mathscr{L}$ and δ is the Dirac distribution at $t = 0$.

Proof We have $u^0(t) = u(t)\cdot\theta(t)$, where θ is the Heaviside function: $\theta(t) = 1$, $t > 0$ and $\theta(t) = 0$, $t < 0$. Note that

$$\begin{aligned} Du^0 &= Du\cdot\theta + u\cdot D\theta \\ &= (Du)^0 + u(0)\cdot i^{-1}\delta, \end{aligned}$$

because $D = i^{-1}\, d/dt$, and by induction on j one shows that

$$D^{j+1}u^0 = (D^j u)^0 + i^{-1}\sum_{k=0}^{j} \gamma_{j-k}u\cdot D^k\delta$$

for $j = 0, 1, \ldots$. Then

$$\begin{aligned} L(D)u^0 &= \sum_{j=0}^{l} A_j D^j u^0 \\ &= \sum_{j=0}^{l} A_j((D_n^j u)^0 + i^{-1}\sum_{k=0}^{j-1}\gamma_{j-1-k}uD^k\delta) \\ &= \left(\sum_{j=0}^{l} A_j D^j u\right)^0 + i^{-1}\sum_{j=1}^{l}\sum_{k=0}^{j-1} A_j\gamma_{j-1-k}u\cdot D^k\delta \\ &= (L(D)u)^0 + L^c\gamma u \end{aligned}$$

where

$$L^c\mathscr{U} = i^{-1}\sum_{j=0}^{l-1}\sum_{k=0}^{j} A_{j+1}\mathscr{U}_{j-k}D^k\delta = i^{-1}[\delta\cdots D^{l-1}\delta]\cdot\mathscr{L}\mathscr{U},$$

for $\mathscr{U} = [\mathscr{U}_j]_{j=0}^{l-1} \in \mathbb{C}^{pl}$.

There is a close connection between Lemma 12.6 and many of the formulas of Chapter 2. Assume that $L(\lambda)$ has no spectrum on the real axis, and let $u \in \mathfrak{M}_L^+$, that is, $u \in \mathscr{S}(\bar{\mathbb{R}}_+)$ and $L(D)u = 0$, $t > 0$. Then

$$L(D)u^0 = 0 + i^{-1}[\delta\cdots D^{l-1}\delta]\cdot\mathscr{L}\gamma u$$

and taking the Fourier transform in t $(\to \tau)$ we get

$$L(\tau)\hat{u}^0(\tau) = i^{-1}[I \cdots \tau^{l-1}I] \cdot \mathscr{X}\gamma u.$$

Hence by solving for $\hat{u}(\tau)$ we find that

$$\begin{aligned} u^0(t) &= (2\pi)^{-1} \int_{-\infty}^{\infty} e^{it\tau}\hat{u}^0(\tau)\, d\tau \\ &= (2\pi i)^{-1} \int_{-\infty}^{\infty} e^{it\tau} L^{-1}(\tau)[I \cdots \tau^{l-1}I] \cdot \mathscr{X}\gamma u\, d\tau \end{aligned} \tag{7}$$

(the integral is oscillatory). For $t > 0$ we may deform the contour of integration to the upper half-plane (by Cauchy's theorem) to obtain

$$u(t) = (2\pi i)^{-1} \int_{+} e^{it\tau} L^{-1}(\lambda)[I \cdots \lambda^{l-1}I] \cdot \mathscr{X}\mathscr{U}\, d\lambda, \qquad t > 0, \tag{8}$$

where we have set $\mathscr{U} = \gamma u$. The Calderón projectors for $L(\lambda)$ corresponding to the eigenvalues of $L(\lambda)$ in the upper and lower half-planes are the projectors in $\mathbb{C}^{pl}$ defined by

$$P^{\pm} = (2\pi i)^{-1} \int_{\pm} \begin{pmatrix} I \\ \vdots \\ \lambda^{l-1}I \end{pmatrix} L^{-1}(\lambda)[I \cdots \lambda^{l-1}I] \cdot \mathscr{X}\, d\lambda,$$

see §2.2(7).

Choose corresponding spectral triples $(X_{\pm}, T_{\pm}, Y_{\pm})$ for $L(\lambda)$, then we have

$$P^{\pm} = \operatorname{col}(X_{\pm}T^j_{\pm})^{l-1}_{j=0} \cdot [Y_{\pm} \cdots T^{l-1}_{\pm}Y_{\pm}]\mathscr{X}, \tag{9}$$

see 2.2 (8),

$$[Y_{\pm} \cdots T^{l-1}_{\pm}Y_{\pm}] \cdot \mathscr{X} \cdot \operatorname{col}(X_{\pm}T^j_{\pm})^{l-1}_{j=0} = I, \tag{10}$$

see Corollary 2.11

$$[Y_{\mp} \cdots T^{l-1}_{\mp}Y_{\mp}] \cdot \mathscr{X} \cdot \operatorname{col}(X_{\pm}T^j_{\pm})^{l-1}_{j=0} = 0, \tag{11}$$

see 2.2 (12)

and hence $(P^+)^2 = P^+$, $(P^-)^2 = P^-$ and $P^-P^+ = P^+P^- = 0$. For any $\mathscr{U} \in \mathbb{C}^{pl}$, if we define $u \in \mathfrak{M}^+_L$ as in (8) then

$$\gamma u = \lim_{t \to 0^+} [u(t) \cdots D^{l-1}u(t)] = P^+\mathscr{U}.$$

In particular, given $u \in \mathfrak{M}^+_L$ then (8) implies that $\gamma u = P^+\gamma u$. Thus, we see once again that the image of P^+ consists of the Cauchy data of solutions $u \in \mathfrak{M}^+_L$.

From now on we assume that $L(\lambda)$ has invertible leading coefficient; therefore $L(\lambda)$ has no spectrum at infinity (see §4.2). Because we are also assuming that $L(\lambda)$ has no spectrum on the real axis, then P^+ and P^- are complementary projectors, $P^+ + P^- = I$ (see the remark at the end of §3.2).

For future reference, note that $L(\lambda)$ has an invertible leading coefficient with no spectrum on the real axis if it is the matrix polynomial associated to an elliptic operator $\mathscr{A} \in \mathrm{Ell}^l$.

Proposition 12.7 *For $\mathscr{U} \in \mathbb{C}^{pl}$, let*

$$v(t) = (2\pi i)^{-1} \int_{-\infty}^{\infty} e^{it\tau} L^{-1}(\tau)[I \cdots \tau^{l-1} I] \cdot \mathscr{L}\mathscr{U} \, d\tau$$

(the integral is oscillatory). Then v coincides for $t > 0$ with an element $v_+ \in \mathfrak{M}_L^+$ and when $t < 0$ with an element $v_- \in \mathfrak{M}_L^-$. For $t = 0$ we have the jump relation

$$D_t^j v_+|_{t=0^+} - D_t^j v_-|_{t=0^-} = \mathscr{U}, \qquad j = 0, 1, \ldots, l-1.$$

Proof For $t > 0$ the path of integration can be deformed to a contour in the upper half-plane, while for $t < 0$ it can be deformed to the lower half-plane, so we obtain

$$v(t) = \frac{1}{2\pi i} \int_+ e^{it\lambda} L^{-1}(\lambda)[I \cdots \lambda^{l-1} I] \mathscr{L}\mathscr{U} \, d\lambda \cdot \theta(t)$$
$$- \frac{1}{2\pi i} \int_- e^{it\lambda} L^{-1}(\lambda)[I \cdots \lambda^{l-1} I] \mathscr{L}\mathscr{U} \, d\lambda \cdot \theta(-t)$$

where θ is the Heaviside function. This gives $v(t) = v_+(t) + v_-(t)$, where $D_t^j v_+|_{t=0^+} - D_t^j v_-|_{t=0^-} = P^+\mathscr{U} + P^-\mathscr{U} = \mathscr{U}$.

Corollary *If $\mathscr{U} \in \mathbb{C}^{pl}$ then $\mathscr{U} = \gamma u$ for some $u \in \mathfrak{M}_L^+$ if and only if $v_- = 0$, i.e. if and only if $\mathscr{U} \in \operatorname{im} P^+$. The Cauchy data of solutions in $\mathfrak{M}_L^+$ and $\mathfrak{M}_L^-$ are complementary subspaces of $\mathbb{C}^{pl}$.*

Next we consider the fundamental solution of $L(D)$.

Lemma 12.8 *We have $L(D)E = \delta$, where*

$$E(t) = \frac{1}{2\pi} \int_+ e^{it\lambda} L^{-1}(\lambda) d\lambda \cdot \theta(t) - \frac{1}{2\pi} \int_- e^{it\lambda} L^{-1}(\lambda) \, d\lambda \cdot \theta(-t) \qquad (12)$$

Also, $\gamma E L^c = P^+$ and $\gamma^- E L^c = -P^-$, the Calderón projectors. (Here γ^- denotes the Cauchy data operator taken from $t < 0$.)

Proof By Fourier transformation $t \to \tau$, we have

$$L(\tau)\hat{E}(\tau) = I,$$

so that $\hat{E}(\tau) = L^{-1}(\tau)$. Thus

$$E(t) = (2\pi)^{-1} \int_{-\infty}^{\infty} e^{it\tau} L^{-1}(\tau) \, d\tau$$

and the formula (12) follows from Cauchy's theorem as in the previous lemma. Now, since $(EL^c)\hat{} = \hat{E}\hat{L}^c$, then

$$EL^c = \frac{1}{2\pi i}\int_{-\infty}^{\infty} e^{it\tau}\, L^{-1}(\tau)[I \cdots \tau^{l-1}I]\mathscr{X}\, d\tau$$

and by Cauchy's theorem

$$EL^c = \frac{1}{2\pi i}\int_{+} e^{it\lambda}\, L^{-1}(\lambda)[I \cdots \lambda^{l-1}I]\mathscr{X}\, d\lambda \cdot \theta(t)$$
$$- \frac{1}{2\pi i}\int_{-} e^{it\lambda}\, L^{-1}(\lambda)[I \cdots \lambda^{l-1}I]\mathscr{X}\, d\lambda \cdot \theta(-t) \tag{13}$$

By taking the Cauchy data from $t > 0$ and from $t < 0$, we obtain the second statement in the lemma (see (9)).

It follows from (12) that

$$\gamma E f^0 \in \operatorname{im} P^-, \qquad \text{for all } f \in {}^p\mathscr{S}(\bar{\mathbb{R}}_+),$$

where f^0 is defined as in Lemma 12.6. Indeed, since

$$Ef^0(t) = \int_0^{\infty} E(t-s) f(s)\, ds,$$

then for $t > 0$ we have

$$i^{-1}Ef^0(t) = \frac{1}{2\pi i}\int_0^t \int_{+} e^{i(t-s)\lambda}\, L^{-1}(\lambda) f(s)\, ds\, d\lambda$$
$$- \frac{1}{2\pi i}\int_0^{\infty}\int_{-} e^{i(t-s)\lambda}\, L^{-1}(\lambda) f(s)\, ds\, d\lambda$$
$$= \int_0^t X_+\, e^{i(t-s)T_+}\, Y_+ f(s)\, ds - \int_t^{\infty} X_-\, e^{i(t-s)T_-}\, Y_- f(s)\, ds$$

so applying the Cauchy data operator γ we obtain

$$i^{-1}\gamma E f^0 = -\operatorname{col}(X_- T_-^j)_{j=0}^{l-1} \cdot \int_0^{\infty} e^{-isT_-}\, Y_- f(s)\, ds. \tag{14}$$

Thus $\gamma E f^0 \in \operatorname{im} P^-$.

Remark In deriving (14) we used the fact that

$$X_+ T_+^j Y_+ + X_- T_-^j Y_- = XT^jY = \frac{1}{2\pi i}\int_{\Gamma} \lambda^j L^{-1}(\lambda)\, d\lambda = 0$$

for $j = 0, \ldots, l-2$, since $L^{-1}(\lambda)$ is of order $-l$ as $|\lambda| \to \infty$. (Recall that $L(\lambda)$ has invertible leading coefficient.) Here Γ is a contour enclosing all the eigenvalues of $L(\lambda)$.

Theorem 12.9 *Let $f \in {}^p\mathscr{S}(\bar{\mathbb{R}}_+)$ and $\mathscr{U} \in \mathbb{C}^{pl}$. Then the system*

$$L(D)u = f, \quad t > 0, \qquad \gamma u = \mathscr{U}, \quad t = 0, \tag{15}$$

has a solution $u \in {}^p\mathscr{S}(\bar{\mathbb{R}}_+)$ if and only if $P^-\mathscr{U} = \gamma E f^0$. Moreover, the solution is unique and is given by

$$u = \imath^+ E f^0 + \imath^+ E L^c \mathscr{U}. \tag{16}$$

Proof Suppose first that the system (15) has a solution. By Lemma 12.6, we have

$$L(D)u^0 = f^0 + L^c \gamma u.$$

If we apply the fundamental solution operator E, then

$$u^0 = E f^0 + E L^c \gamma u \tag{17}$$

and, by taking the Cauchy data from $t > 0$,

$$\gamma u = \gamma E f^0 + \gamma E L^c \gamma u.$$

Since $\gamma E L^c = P^+$ by Lemma 12.8, then $\mathscr{U} = \gamma E f^0 + P^+ \mathscr{U}$, whence $P^- \mathscr{U} = \gamma E f^0$.

Conversely suppose that $P^- \mathscr{U} = \gamma E f^0$, and define

$$w = E f^0 + E L^c \mathscr{U}. \tag{18}$$

Then $L(D)w = f^0 + L^c \mathscr{U}$ and if we let $u = \imath^+ w$ then $L(D)u = f + 0$. (Note: $\imath^+ L^c = 0$ since $\imath^+ D^k \delta = 0$ for all k.) Hence

$$\gamma u = \gamma E f^0 + \gamma E L^c \mathscr{U} = P^- \mathscr{U} + P^+ \mathscr{U} = \mathscr{U}.$$

Finally, the uniqueness of the solution follows from (17). The proof is complete.

Remark If w is defined by (18) then $w(t) = 0$ for $t < 0$ since

$$L(D)w = 0, \qquad t < 0,$$

and $\gamma^- w = \gamma^- E f^0 - P^- \mathscr{U} = 0$. This shows the correspondence between equations (17) and (18), that is, $w = u^0$ where $u = \imath^+ w$.

In view of Theorem 12.9, $P^- \gamma u$ is fixed once $L(D)u$ is known. Once a basis is chosen for $\operatorname{im} P^+ \simeq \mathbb{C}^r$, the operator $P^+ \gamma$ can be regarded as a boundary operator.

Corollary *If $f \in {}^p\mathscr{S}(\bar{\mathbb{R}}_+)$ and $\mathscr{U} \in \operatorname{im} P^+$ then the system*

$$L(D)u = f, \quad t > 0, \qquad P^+ \gamma u = \mathscr{U}, \quad t = 0$$

has a unique solution $u \in {}^p\mathscr{S}(\bar{\mathbb{R}}_+)$ given by

$$u = \imath^+ E f^0 + \imath^+ E L^c \mathscr{U}. \tag{19}$$

Proof By applying the operator $\imath^+$ to (17) we obtain $u = \imath^+ E f^0 + \imath^+ E L^c \gamma u$. In view of (13) and (11) it follows that $\imath^+ E L^c P^- = 0$ and hence (19) holds.

Conversely, if u is defined by (19) then $L(D)u = f + 0$ and

$$\gamma u = \gamma E f^0 + \gamma E L^c \mathscr{U} = \gamma E f^0 + P^+ \mathscr{U} = \gamma E f^0 + \mathscr{U}.$$

Since $\gamma E f^0 \in \operatorname{im} P^-$, then $P^+ \gamma u = 0 + \mathscr{U}$.

Proposition 12.10 *Let $\gamma^- u = \lim_{t \to 0^-} [u(t) \cdots D^{l-1} u(t)]^T$ be the Cauchy data taken from $t < 0$. If $f \in {}^p\mathscr{S}(\bar{\mathbb{R}}_+)$ then $\gamma^- E f^0 = \gamma E f^0$, and $E f^0 \in {}^pC^{l-1}(\mathbb{R})$.*

Proof Since $L(D) E f^0 = f^0 = 0$ when $t < 0$, then $\gamma^- E f^0 \in \operatorname{im} P^-$. Lemma 12.8 gives

$$i^{-1} E f^0(t) = -\frac{1}{2\pi i} \int_0^\infty \int_- e^{i(t-s)\lambda} L^{-1}(\lambda)\, d\lambda \cdot f(s)\, ds,$$

when $t < 0$, and by virtue of (14) it follows that $\gamma^- E f^0 = \gamma E f^0$. The fact that $E f^0 \in {}^pC^{l-1}(\mathbb{R})$ is now clear since $E f^0 \in C^\infty$ for $t \neq 0$ and we have just shown that the Cauchy data taken from $t < 0$ and $t > 0$ coincide.

We now consider the boundary operator, $B(\lambda) = \sum_{j=0}^{\mu} B_j \lambda^j$, an $r \times p$ matrix polynomial with $\mu < l$. Let

$$\Delta_\pm = \Delta_B^\pm = \sum_{j=0}^{\mu} B_j X_\pm T_\pm^j \tag{20}$$

and assume that the Δ-condition holds, i.e. $\det \Delta_+ \neq 0$. If $g \in \mathbb{C}^r$, the system

$$L(D)u = 0, \quad t > 0, \qquad B(D)u = g, \quad t = 0,$$

has the unique solution $u = Kg$ in ${}^p\mathscr{S}(\bar{\mathbb{R}}_+)$ given by

$$Kg(t) = X_+ e^{itT_+} \Delta_+^{-1} g, \qquad t > 0. \tag{21}$$

The linear map $K\colon \mathbb{C}^r \to {}^p\mathscr{S}(\bar{\mathbb{R}}_+)$ is called the *Poisson operator* for the system.

Note that $u = Kg$ is the unique solution of $L(D)u = 0$ in $\mathscr{S}(\bar{\mathbb{R}}_+)$ such that $\gamma u = Sg$, where

$$S = \operatorname{col}(X_+ T_+^j)_{j=0}^{l-1} \cdot \Delta_+^{-1} \tag{22}$$

is a $pl \times r$ matrix. Also, we write $B(D)u = B^c \gamma u$, where $B^c = [B_0 \cdots B_\mu 0 \cdots 0]$ is an $r \times pl$ matrix, and we define the $pl \times pl$ matrix S' by the equation

$$S' + SB^c = I_{pl}. \tag{23}$$

We claim that $S'P^+ = 0$. Indeed, $S' = I - SB^c$ and it follows that

$$\begin{aligned} S' \cdot \operatorname{col}(X_+ T_+^j)_{j=0}^{l-1} &= \operatorname{col}(X_+ T_+^j)_{j=0}^{l-1} - \operatorname{col}(X_+ T_+^j)_{j=0}^{l-1} \Delta_+^{-1} \Delta_+ \\ &= 0, \end{aligned}$$

hence $S'P^+ = 0$ because P^+ and $\operatorname{col}(X_+ T_+^j)_{j=0}^{l-1}$ have the same image. For future reference we summarize the properties of S and S' in the following proposition.

Proposition 12.11 *Let P^+ denote the Calderón projector for $L(\lambda)$ with respect to the eigenvalues in the upper half-plane* $\operatorname{Im}\lambda > 0$. *There exist unique matrices S, S' of dimensions $pl \times r$ and $pl \times pl$, respectively, such that*

(*i*) $B^cS = I_r,\ P^+S = S;$
(*ii*) $S' + SB^c = I_{pl},\ S'P^+ = 0$

Proof The existence of the matrices S, S' has been shown above. Now, suppose that S is any $pl \times r$ matrix satisfying the two equations in (i). Since P^+ and $\operatorname{col}(X_+T_+^j)_{j=0}^{l-1}$ have the same image, the second equation implies that

$$S = \operatorname{col}(X_+T_+^j)_{j=0}^{l-1}\cdot M,$$

for some operator $M \in \mathscr{L}(\mathbb{C}^r, \mathfrak{M}_+)$ where $\mathfrak{M}_+$ is the base space of (X_+, T_+). Now, if we multiply this equation on the left by the matrix B^c we have $B^cS = \Delta_+M$. Since $B^cS = I_r$ it follows that $M = (\Delta_+)^{-1}$. This proves the uniqueness of S, and uniqueness of S' then follows from the first equation in (ii).

In view of (16), we can write u in terms of $L(D)u$ and the Cauchy data γu. By means of the equation (23) we will then write u in terms of $L(D)u$ and $B(D)u$.

Theorem 12.12 *Let $f \in {}^p\mathscr{S}(\bar{\mathbb{R}}_+)$ and $g \in \mathbb{C}^r$. The unique solution of $L(D)u = f$, $t > 0$, $B(D)u = g$, $t = 0$, is given by*

$$u = \imath^+(I + EL^cS'\gamma)Ef^0 + \imath^+EL^cSg. \tag{24}$$

Proof We already know that there exists a unique solution for given f and g, so it remains to verify the formula (24). Let $\mathscr{U} = \gamma u$. Since $P^-\mathscr{U} = \gamma Ef^0$, see (14), we find that $S'\mathscr{U} = S'P^-\mathscr{U} = S'\gamma Ef^0$. (Recall that $S'P^+ = 0$.) Also, $B(D)u = g$ so we have $B^c\mathscr{U} = g$. Now, in view of (23), we see that

$$\mathscr{U} = S'\mathscr{U} + Sg,$$

and (24) follows from (16).

Note that $K = EL^cS$, where K is the Poisson operator (21). Indeed, (13) with $t > 0$ gives

$$EL^cSg = \frac{1}{2\pi i}\int_+ e^{it\lambda}L^{-1}(\lambda)[I\cdots\lambda^{l-1}I]\mathscr{Z}\,d\lambda\cdot Sg, \qquad t > 0.$$

In view of (10), (22) and Proposition 2.3, it follows that

$$EL^cSg = X_+e^{itT_+}\Delta_+^{-1}g = Kg.$$

The formula in Theorem 12.12 can therefore be written in another form:

$$u = Ef^0 + K(g - B(D)Ef^0), \qquad t > 0, \tag{24'}$$

To see this, we use (23) and obtain $EL^cS'\gamma Ef^0 = EL^c\gamma Ef^0 - EL^cSB^c\gamma Ef^0$.

If we recall that $\gamma E f^0 \in \operatorname{im} P^-$ then (11) implies

$$\begin{aligned} EL^cS'\gamma Ef^0 &= 0 - EL^cS \cdot B^c\gamma \cdot Ef^0, \qquad t > 0, \\ &= -KB(D)Ef^0, \end{aligned}$$

since $B(D) = B^c\gamma$, so (24′) holds.

Definition 12.13 *The operator* ${}^p\mathcal{S}(\bar{\mathbb{R}}_+) \to {}^p\mathcal{S}(\bar{\mathbb{R}}_+)$ *defined by*

$$Gf = \imath^+(I + EL^cS'\gamma)Ef^0,$$

is called the Green operator.

Note that $u = Gf$ is the unique solution of

$$L(D)u = f, \quad t > 0, \qquad B(D)u = 0, \quad t = 0 \tag{25}$$

Theorem 12.12 says that every solution can be written in the form $u = Gf + Kg$.

Proposition 12.14 *The "singular part" of the Green operator,*

$$G_1 f = \imath^+ EL^cS'\gamma Ef^0,$$

is given by

$$G_1 f = iX_+ e^{itT_+} \Delta_+^{-1} \Delta_- \cdot \int_0^\infty e^{-isT_-} Y_- f(s)\, ds, \qquad t > 0.$$

Proof In view of the calculations above, we have $G_1 f = -KB^c\gamma Ef^0$, $t > 0$, and then by (14)

$$i^{-1}G_1 f = KB^c \operatorname{col}(X_- T_-^j)_{j=0}^{l-1} \int_0^\infty e^{-isT_-} Y_- f(s)\, ds.$$

Recalling the definitions (20) and (21), we obtain the desired formula for G_1.

For the sake of completeness we write out a few more details about the Green operator and then give an example. Recall that the equation $L(D)u = f$, $t > 0$, has the particular solution

$$u = Ef^0 = \int_0^\infty G_0(t, s) f(s)\, ds, \qquad t > 0,$$

where

$$G_0(t, s) = \begin{cases} -X_- e^{i(t-s)T_-} Y_- & t < s \\ X_+ e^{i(t-s)T_+} Y_+ & t > s \end{cases} \tag{26}$$

We call G_0 *a pre-Green* operator; it has the following properties:

(a) $L\left(\frac{1}{i}\frac{d}{dt}\right) G_0(t, s) = 0, \quad t \neq s$

(b) $\left(\frac{1}{i}\frac{d}{dt}\right)^j G_0\big|_{t=s^+} - \left(\frac{1}{i}\frac{d}{dt}\right)^j G_0\big|_{t=s^-} = \begin{cases} 0 & \text{if } j = 0, \ldots, l-2 \\ I & \text{if } j = l-1 \end{cases}$

for $s > 0, t > 0$. Note: (b) follows from

$$X_+ T_+^j Y_+ + X_- T_-^j Y_- = XT^j Y = \begin{cases} 0 & \text{if } j = 0, \ldots, l-2 \\ I & \text{if } j = l-1 \end{cases}$$

Remark 12.15 Since sp(T_+) lies in the upper half-plane and sp(T_-) in the lower half-plane, $G_0(t, s)$ and all its derivatives are bounded by a constant times $e^{-\delta|t-s|}$ for some $\delta > 0$. Consequently, $f \mapsto Ef^0$ is an operator ${}^p\mathscr{S}(\bar{\mathbb{R}}_+) \to {}^p\mathscr{S}(\bar{\mathbb{R}}_+)$. (This was implicitly assumed in the definition of the Green operator.)

We would now like to determine a function $G(t, s)$ such that

$$u = Gf = \int_0^\infty G(t, s) f(s)\, ds$$

is a solution of the system (25). In view of Proposition 12.14, we see that

$$G(t, s) = G_0(t, s) + X_+ e^{itT_+} N(s) \tag{27}$$

where $N(s) = \Delta_+^{-1} \Delta_- \cdot e^{-isT_-} Y_- \in {}^p\mathscr{S}(\bar{\mathbb{R}}_+)$. Note that G has the same properties (a), (b) as G_0 but in addition

$$\text{(c)}\quad B\left(\frac{1}{i}\frac{d}{dt}\right) G(t, s)|_{t=0^+} = 0, \qquad \text{for all } s > 0.$$

Indeed, to verify (c), we obtain by differentiation of (27)

$$\begin{aligned} B\left(\frac{1}{i}\frac{d}{dt}\right) G|_{t=0^+} &= B\left(\frac{1}{i}\frac{d}{dt}\right) G_0|_{t=0^+} + \left(\sum_{j=0}^{\mu} B_j X_+ T_+^j\right) \cdot N(s) \\ &= -\left(\sum_{j=0}^{\mu} B_j X_- T_-^j\right) \cdot e^{-isT_-} Y_- + \left(\sum_{j=0}^{\mu} B_j X_+ T_+^j\right) \cdot N(s) \\ &= 0, \end{aligned}$$

since $\Delta_\pm = \sum_{j=0}^{\mu} B_j X_\pm T_\pm^j$.

Example Suppose that the matrix polynomial

$$L(\lambda) = \pi \mathscr{A}(\xi', \lambda) = I\lambda^2 + A_1 \lambda + A_0$$

arises from a second-order elliptic operator $\mathscr{A}(D)$, $D = i^{-1}\, d/dt$, for which the Dirichlet problem satisfies the L-condition. This means that $X_\pm$ is invertible and $L(\lambda)$ has $\gamma^\pm$-spectral right divisors $I\lambda - S_\pm$, where $S_\pm = X_\pm T_\pm X_\pm^{-1}$ (see §3.7). For the system

$$L\left(\frac{1}{i}\frac{d}{dt}\right) u = f, \quad t > 0, \quad u(0) = 0, \tag{28}$$

we then have $\Delta_\pm = X_\pm$ and the Green function is

$$G(t,\tau) = G_0(t,\tau) + X_+\, e^{itT_+}\, X_+^{-1} X_-\, e^{-itT_-}\, Y_-$$
$$= G_0(t,\tau) + e^{itS_+}\, e^{-itS_-} X_- Y_-$$

Since $X_- Y_- = (2\pi i)^{-1} \int_- L^{-1}(\lambda)\, d\lambda = (S_- - S_+)^{-1}$ we obtain that the solution of (28) is

$$u(t) = \int_0^\infty G_0(t,\tau) f(\tau)\, d\tau - e^{itS_+} \int_0^\infty e^{-i\tau S_-}(S_+ - S_-)^{-1} \cdot f(\tau)\, d\tau$$

where G_0 is given by (26) and hence

$$G_0(t,\tau) = \begin{cases} e^{iS_-(t-\tau)} \cdot (S_+ - S_-)^{-1}, & t < \tau \\ e^{iS_+(t-\tau)} \cdot (S_+ - S_-)^{-1}, & t > \tau \end{cases}$$

Exercise

In addition to the hypotheses of the example above, let $\mathscr{B}(D)$ be a boundary operator that satisfies the L-condition relative to $\mathscr{A}$. Let $B(\lambda) = \mathscr{B}(\xi', \lambda) = B_1\lambda + B_0$. Show that the solution of $L(D)u = f$, $t > 0$, $B(D)u = 0$, $t = 0$, is given by

$$u(t) = \int_0^\infty G_0(t,\tau) f(\tau)\, d\tau - e^{itS_+}(B_0 + B_1 S_+)^{-1} \cdot (B_0 + B_1 S_-)$$
$$\times \int_0^\infty e^{-i\tau S_-}(S_+ - S_-)^{-1} \cdot f(\tau)\, d\tau$$

13

Behaviour of a pseudo-differential operator near a boundary

13.1 C^∞ functions defined on a half-space

We begin with a brief review of some standard distributions on $\mathbb{R}$. Let $a \in \mathbb{C}$ such that $\operatorname{Re} a > -1$. Let $x^a = e^{a \log x}$, where $\log x$ is defined to be real when $x > 0$. The function on $\mathbb{R}$ defined by

$$x_+^a = x^a \quad \text{if } x > 0, \qquad x_+^a = 0 \quad \text{if } x \leqslant 0,$$

is locally integrable so it defines a distribution in $\mathscr{D}'(\mathbb{R})$. Note that

$$\frac{d}{dx} x_+^a = a x_+^{a-1} \quad \text{if } \operatorname{Re} a > 0. \tag{1}$$

This is obvious when $x \neq 0$, but near $x = 0$ there is something to show. Let $\varphi \in C_0^\infty(\mathbb{R})$, then

$$\begin{aligned}\left\langle \frac{d}{dx} x_+^a, \varphi \right\rangle = -\langle x_+^a, \varphi' \rangle &= -\int_0^\infty x^a \varphi'(x)\, dx \\ &= \int_0^\infty a x^{a-1} \varphi(x)\, dx \\ &= \langle a x_x^{a-1}, \varphi \rangle,\end{aligned}$$

where to get the third equality we integrated by parts and used the fact that $\operatorname{Re} a > 0$ so the boundary term vanishes.

We wish to extend the definition of x_+^a to $a \in \mathbb{C}$ such that the property (1) is preserved as far as possible. For $\varphi = C_0^\infty(\mathbb{R})$ the function

$$a \mapsto \langle x_+^a, \varphi \rangle = \int_0^\infty x^a \varphi(x)\, dx$$

is analytic when $\operatorname{Re} a > -1$. The definition of $\langle x_+^a, \varphi \rangle$ is then extended by analytic continuation. In view of (1) we have

$$\langle x_+^a, \varphi' \rangle = -a \langle x_+^{a-1}, \varphi \rangle \qquad \text{if } \operatorname{Re} a > 0, \tag{2}$$

thus for $\operatorname{Re} a > -1$ and any integer $k > 0$ we have

$$\langle x_+^a, \varphi \rangle = (-1)^k \langle x_+^{a+k}, \varphi^{(k)} \rangle / ((a+1) \cdots (a+k)). \tag{3}$$

This shows that we can extend $\langle x_+^a, \varphi \rangle$ analytically to be a meromorphic function in $\mathbb{C}$ with simple poles at the integers $\leqslant 0$, such that the equation (3) holds for all $\operatorname{Re} a > -k - 1$. Thus, x_+^a is a distribution of order $\leqslant k$ if $k > -1 - \operatorname{Re} a$.

In addition to x_+^a we define the function

$$x_-^a = 0 \quad \text{if } x > 0, \qquad x_-^a = |x|^a \quad \text{if } x < 0,$$

where $\operatorname{Re} a > -1$. This is the reflection of x_+^a with respect to the origin,

$$\langle x_-^a, \varphi \rangle = \langle x_+^a, \check{\varphi} \rangle \quad \text{where } \check{\varphi}(x) = \varphi(-x).$$

The definition of x_-^a may also be extended by analytic continuation to any $a \in \mathbb{C}$ not a negative integer.

Consider now the gamma function

$$\Gamma(a) = \int_0^\infty x^{a-1} e^{-x}\, dx, \qquad \operatorname{Re} a > 0.$$

The same calculation which gave (2) for $\varphi \in C_0^\infty(\mathbb{R})$ also holds for $\varphi(x) = e^{-x}$, so we obtain

$$\Gamma(a+1) = a\Gamma(a), \qquad \operatorname{Re} a > 0. \tag{4}$$

Thus, we can extend $\Gamma(a)$ analytically to a meromorphic function in $\mathbb{C}$ with simple poles at the integers $\leqslant 0$. The residue at the integer $-k \leqslant 0$ is

$$\lim_{a \to -k} (a+k)\cdot\Gamma(a) = \lim_{a \to -k} \Gamma(a+k+1)/a(a+1)\cdots(a+k-1) = (-1)^k/k!$$

It is well-known that the gamma function has no zeros, so the quotient defined by

$$\chi_+^a = \frac{x_+^a}{\Gamma(a+1)}, \qquad \operatorname{Re} a > -1, \tag{5}$$

is analytic when $\operatorname{Re} a > -1$ (with respect to the weak topology on $\mathscr{D}'$). In view of (2) and (4) we see that $\langle \chi_+^a, \varphi' \rangle = \langle -\chi_+^{a-1}, \varphi \rangle$ when $\operatorname{Re} a > 0$. Therefore, χ_+^a can be extended analytically to *all* $a \in \mathbb{C}$ such that

$$\frac{d}{dx}\chi_+^a = \chi_+^{a-1}. \tag{6}$$

In particular, since χ_+^0 is the Heaviside function ($=1$ when $x > 0$ and $=0$ when $x < 0$), then

$$\chi_+^{-k} = \delta^{(k-1)}, \qquad k = 1, 2, \ldots.$$

In the next theorem we show that the Fourier transform of $\chi_\pm^a \in \mathscr{S}'(\mathbb{R})$ is equal to a constant times the boundary values $(\xi \mp i0)^{-a-1}$ of the analytic function $f(\zeta) = \zeta^{-a-1}$ on the real axis. The existence of these boundary values in the sense of distribution theory is proved in Theorem 13.2 below.

Theorem 13.1 *The Fourier transform of the distribution $\chi_\pm^a$ is the distribution*

$$e^{\mp i\pi(a+1)/2}(\xi \mp i0)^{-a-1}.$$

Proof Due to the reflection $x \to -x$ (see above), it suffices to find the Fourier transform of χ_+^a. Let $\operatorname{Re} a > -1$. When $\lambda > 0$ the Fourier transform of $e^{-\lambda x}\chi_+^a$ is

$$\xi \mapsto \int_0^\infty x^a e^{-x(\lambda+i\xi)}\,dx/\Gamma(a+1) = (\lambda+i\xi)^{-a-1}\int_0^\infty z^a e^{-z}\,dz/\Gamma(a+1)$$
$$= (\lambda+i\xi)^{-a-1}\Gamma(a+1)/\Gamma(a+1)$$
$$= (\lambda+i\xi)^{-a-1}$$

where the second integral is taken along the ray with direction $\lambda + i\xi$ and z^a is defined on $\mathbb{C}\setminus\mathbb{R}_-$ so that $1^a = 1$, and after that we used Cauchy's theorem to show that the integral can be taken along $\mathbb{R}_+$. Now, if we observe that

$$(\lambda+i\xi)^{-a-1} = e^{-i\pi(a+1)/2}(\xi - i\lambda)^{-a-1},$$

then let $\lambda \to 0^+$, and note that $e^{-\lambda x}\chi_+^a \to \chi_+^a$ in $\mathscr{S}'$, we find that the Fourier transform of χ_+^a is given by $e^{-i\pi(a+1)/2}(\xi - i0)^{-a-1}$, as required. This proves the theorem when $\operatorname{Re} a > -1$. The proof is then finished by analytic continuation, since we have shown that

$$\langle \chi_+^a, \hat{\varphi}\rangle = e^{-i\pi(a+1)/2}\langle (\xi - i0)^{-a-1}, \varphi\rangle, \qquad \varphi \in C_0^\infty(\mathbb{R}),$$

when $\operatorname{Re} a > -1$ and both sides are entire analytic functions of $a \in \mathbb{C}$.

In Theorem 13.1 we implicitly used a simple result on existence of boundary values on the real axis of analytic functions in the sense of distribution theory. Let I be an open interval in $\mathbb{R}$ and let

$$\Omega = \{z \in \mathbb{C};\, z = x + iy,\, x \in I,\, 0 < y < \delta\}$$

be a one-sided complex neighbourhood of I. Suppose that $f(z)$ is an analytic function on Ω. We would like to examine the behaviour of $f(x+iy)$ as $y \to 0^+$ in the distribution topology in $\mathscr{D}'(I)$. Suppose first that $f(z)$ is bounded on Ω. Then if we choose $z_0 \in \Omega$ and introduce the integral

$$F(z) = \int_{z_0}^{z} f(\zeta)\,d\zeta \tag{7}$$

along some path in Ω, it follows that F is analytic in Ω, with $F'(z) = f(z)$, and it has a continuous extension to $\bar{\Omega}$. Then the limit $\lim_{y\to 0^+} f(\cdot + iy) = \lim_{y\to 0^+} dF(\cdot + iy)/dx = dF(\cdot)/dx$ certainly exists in $\mathscr{D}'(I)$, where $F(\cdot) \in C(I)$ is the restriction of F to the real axis. However, the function that we are most interested with here is $f(z) = 1/z^{a+1}$ which is not bounded, but instead satisfies the condition of the next theorem with $N = \operatorname{Re} a + 1$.

Theorem 13.2 *Let* $\Omega = \{z \in \mathbb{C};\, z = x + iy,\, x \in I,\, 0 < y < \delta\}$. *If* f *is an analytic function in* Ω *such that for some nonnegative integer* N,

$$|f(z)| \leqslant C(\operatorname{Im} z)^{-N}, \qquad z \in \Omega,$$

then $f(\cdot + iy)$ *has a limit when* $y \to 0^+$ *as a distribution of order* $\leqslant N+1$, *that is, there exists* $f_0 \in \mathscr{D}'^{N+1}(I)$ *such that*

$$\lim_{y \to 0^+} \int f(x+iy)\varphi(x)\,dx = \langle f_0, \varphi\rangle, \qquad \varphi \in C_0^{N+1}(I).$$

We denote the distribution f_0 *by* $f(x+i0)$.

Proof We may assume that I is a bounded interval. If $N > 0$ and we define $F(z)$ as in (7) and if the path from z_0 to z is taken first horizontal, then vertical, we obtain

$$|F(z)| \leqslant C_1(\operatorname{Im} z)^{1-N} \qquad \text{if } N > 1,$$
$$|F(z)| \leqslant -C\log(\operatorname{Im} z) + C_1 \qquad \text{if } N = 1.$$

Since $\log t$ is an integrable function of t near 0, we obtain after $N+1$ integrations that $f(z) = G^{(N+1)}(z)$ where G is continuous in $\bar{\Omega}$ and analytic in Ω. This proves that

$$\lim_{y\to 0^+} f(\cdot + iy) = \lim_{y \to 0^+} d^{N+1}G(\cdot + iy)/dx^{N+1} = d^{N+1}G(\cdot)/dx^{N+1}$$

in $\mathscr{D}'^{N+1}(I)$.

Corollary 13.3 *If* $f(x+i0) \equiv 0$ *then* $f(z) \equiv 0$.

Proof Take $y > 0$ and $\varphi \in C_0^\infty(I)$ and define

$$K(w) = \int \varphi(x) f(x+wy)\,dx = \int \varphi(x - \operatorname{Re} w\cdot y) f(x + i\operatorname{Im} w\cdot y)\,dx.$$

Let d be the distance from $\operatorname{supp}\varphi$ to ∂I. Then K is an analytic function of w when $0 < \operatorname{Im} w < \delta/y$ and $|\operatorname{Re} w| < d/y$ because in that case $x + wy \in \Omega$ for all $x \in \operatorname{supp}\varphi$. When $\operatorname{Im} w \to 0^+$ we know that K and all its derivatives tend to 0 since $f(x+i0) = 0$. Hence if we extend K by 0 when $|\operatorname{Re} w| < d/y$ and $\operatorname{Im} w \leqslant 0$, then K remains analytic because it satisfies the Cauchy–Riemann equations. By the uniqueness of analytic continuation it follows that $K = 0$ identically. This implies $f = 0$ in Ω.

There is another proof of Theorem 13.2 which gives a formula for $f(x+i0)$. We use the fact that if Ω has C^1 boundary then

$$2i\int_\Omega \partial\psi/\partial\bar{z}\,dx\,dy = \int_{\partial\Omega} \psi(z)\,dz, \qquad \psi \in C_0^1(G), \Omega \subset\subset G, \tag{8}$$

which follows easily from the Gauss–Green formula (see [Hö 1], p. 60) in the plane, applied to the vector function $(\psi, i\psi)$. Set for $\varphi \in C_0^{N+1}(I)$

$$\Phi(x,y) = \sum_{j \leqslant N} \varphi^{(j)}(x)(iy)^j/j!$$

Then $\Phi(x, 0) = \varphi(x)$ and

$$2\partial\Phi/\partial\bar{z} = (\partial/\partial x + i\,\partial/\partial y)\Phi = \varphi^{(N+1)}(x)(iy)^N/N!$$

Fix Y with $0 < Y < \delta$, and note that if $0 < \eta < \delta - Y$ then $f(z + i\eta)$ is analytic in $0 < \operatorname{Im} z < Y$. Now, if $\psi(z) = \Phi(z) f(z + i\eta)$ then $\partial\psi/\partial\bar{z} = \partial\Phi/\partial\bar{z} \cdot f(z + i\eta) + 0$ and we obtain from (8) that

$$\int \Phi(x, 0) f(x + i\eta)\, dx - \int \Phi(x, Y) f(x + iY + i\eta)\, dx = 2i \iint_{0 < \operatorname{Im} z < Y} f(z + i\eta)\, \partial\Phi/\partial\bar{z}\, d\mu(z)$$

where $d\mu$ is the Lebesgue measure in $\mathbb{C}$. Here we used the fact that $\Phi(x, y) = 0$ when x is in a neighbourhood of ∂I. Writing $z = x + itY$, $0 < t < 1$, gives

$$\int \varphi(x) f(x + i\eta)\, \mathrm{d}x = \int \Phi(x, Y) f(x + iY + i\eta)\, dx + 2i \iint_0^1 f(x + itY + i\eta) \varphi^{(N+1)}(x)(iY)^{N+1} t^N/N!\, dx\, dt.$$

Now, there is a uniform bound for the integrand in the double integral, since $|f(x + itY + i\eta)| \leqslant C(tY + \eta)^{-N} \leqslant C' t^{-N}$ where C' depends only on Y, so by letting $\eta \to 0$ we find that

$$\int \varphi(x) f(x + i0)\, dx = \int \Phi(x, Y) f(x + iY)\, dx + 2i \iint_0^1 f(x + itY) \varphi^{(N+1)}(x)(iY)^{N+1} t^N/N!\, dx\, dt. \quad (9)$$

Extension of C^∞ functions defined on a half-space

Let $\mathbb{R}^n_+$ be the half-space defined by $x_n > 0$ and $\bar{\mathbb{R}}^n_+$ the closed half-space $x_n \geqslant 0$. We let $C^\infty(\bar{\mathbb{R}}^n_+)$ denote the set of functions $f \in C^\infty(\mathbb{R}^n_+)$ such that f and all its derivatives can be extended continuously to $\bar{\mathbb{R}}^n_+$.

Note: In §5.9 we had another definition of smoothness of a function defined on a closed set. Theorem 13.4 implies that the two definitions coincide for $\bar{\mathbb{R}}^n_+$. Also see Exercise 3 at the end of the chapter.

Theorem 13.4 *There is a continuous, linear extension operator β: $C^\infty(\bar{\mathbb{R}}^n_+) \to C^\infty(\mathbb{R}^n)$ such that $\imath^+\beta =$ identity, i.e.*

$$\beta f(x', x_n) = f(x', x_n), \qquad x_n > 0,$$

where the topology on $C^\infty(\mathbb{R}^n)$ (resp. $C^\infty(\bar{\mathbb{R}}^n_+)$) is that of uniform convergence of each derivative on compact subsets of $\mathbb{R}^n$ (resp. $\bar{\mathbb{R}}^n_+$).

Proof We just have to modify the proof of Theorem 12.3 in the obvious way to take into account the parameters $x' \in \mathbb{R}^{n-1}$. Choose a rapidly decreasing sequence c_p, $p = 0, 1, \ldots$, such that

$$\sum_{p=1}^{\infty} p^k c_p = (-1)^k, \qquad k = 0, 1, \ldots \tag{10}$$

(see Lemma 12.4). Also let φ be a C^∞ function on $\mathbb{R}$ with $\varphi(t) = 1$ for $0 \leqslant t \leqslant 1$ and $\varphi(t) = 0$ for $t \geqslant 2$. Define the extension of f by $\beta f(x', x_n) = f(x', x_n)$, $x_n > 0$,

$$\beta f(x', x_n) = \sum_{p=1}^{\infty} c_p \cdot \varphi(-px_n) f(x', -px_n), \qquad x_n < 0.$$

Then because $p \to -\infty$ the sum is finite for each $x_n < 0$; because $\sum_{p=1}^{\infty} p^k |c_p| < \infty$, all derivatives of βf converge as $x_n \to 0^-$, uniformly in each bounded set $|x'| \leqslant c$; moreover, these limits agree with those for $x_n \to 0^+$ in view of (10). Thus, if we define $\beta f(x', 0) = \lim_{x_n \to 0^+} f(x', x_n)$, then all partial derivatives of βf exist and are continuous everywhere on $\mathbb{R}^n$, whence $\beta f \in C^\infty(\mathbb{R}^n)$. The continuity of β: $C^\infty(\bar{\mathbb{R}}^n_+) \to C^\infty(\mathbb{R}^n)$ also follows from $\sum_{p=1}^{\infty} p^k |c_p| < \infty$.

Remark Note that $\beta f(x', x_n) = f(x', -x_n)$ when $x_n \leqslant -1$.

Let $v \in \mathscr{E}'(\mathbb{R}^n)$ have the representation

$$v(x) = \int e^{ix_n\xi_n}\, b(x', \xi_n)\, d\xi_n$$

where $b \in S^m(\mathbb{R}^{n-1} \times \mathbb{R}^1)$. (This means that $b(x', \xi_n)$ is a symbol of order m in the sense of Definition 7.1 for the single variable ξ_n.) Then v is C^∞ when $x_n \neq 0$, for by regularizing the oscillatory integral (12) we get

$$v(x) = x_n^{-M} \int e^{ix_n\xi_n}\, (-D_{\xi_n})^M b(x', \xi_n)\, d\xi_n,$$

and, since $|D_{\xi_n}^M b(x', \xi_n)| \leqslant \text{const}(1 + |\xi_n|)^{m-M}$, it follows that $v \in C^k$ for $x_n \neq 0$ when $M > m + k + 1$.

In the next theorem we characterize when v has a C^∞ extension to the half-space $x_n \geqslant 0$. This result is important for §13.2 where we study the transmission property for pseudo-differential operators.

Note that if b has an asymptotic expansion $b \sim \sum b_j$, then, by definition,

$$b = \sum_{j<N} \chi(\xi_n) \cdot b_j + r_N,$$

where $r_N \in S^{m-N}$ and χ is a cut-off function. Then we have

$$v(x) = \sum_{j<N} \int e^{ix_n\xi_n} b_j(x', \xi_n)\chi(\xi_n)\, d\xi_n + \int e^{ix_n\xi_n} r_N(x', \xi_n)\, d\xi_n \tag{11}$$

Theorem 13.5 *Suppose that $v \in \mathscr{E}'(\mathbb{R}^n)$ has the representation*

$$v(x) = \int e^{ix_n\xi_n} b(x', \xi_n)\, d\xi_n, \qquad x = (x', x_n), \tag{12}$$

with $b \in S^m(\mathbb{R}^{n-1} \times \mathbb{R}^1)$, $m \in \mathbb{R}$, having the asymptotic expansion

$$b \sim \sum_{j=0}^{\infty} b_j(x', \xi_n),$$

where b_j is homogeneous of degree $m - j$ in ξ_n (i.e. b is $(-\infty)$ classic; see remark after Definition 7.21). Then $v|_{x_n>0}$ has a C^∞ extension to the closed half-space $x_n \geqslant 0$ if and only if

$$b_j(x', -1) = b_j(x, 1)\, e^{i\pi(m-j)}. \tag{13}$$

Proof (a) Sufficiency of (13). The condition on b means that it is the restriction to $\mathbb{R}\backslash 0$ of an analytic function defined on the upper half-plane, namely,

$$b_j(x', \zeta) = b_j(x', 1)\zeta^{m-j}, \qquad \operatorname{Im} \zeta > 0,$$

with ζ^{m-j} defined to be 1 at $\zeta = 1$. Let Γ be the curve in $\mathbb{C}$ consisting of the real axis with $(-1, 1)$ replaced by the half unit circle in the upper half-plane. Since $b - \sum_{j<N} \chi(\xi_n)\cdot b_j \in S^{m-N}(\mathbb{R}^{n-1} \times \mathbb{R}^1)$ where χ is a cut-off function, and $b_j(x', \xi_n) = b_j(x', 1)\cdot \xi_n^{m-j}$, $\xi_n \neq 0$, it follows that

$$v(x) - \sum_{j<N} \int_\Gamma e^{ix_n\zeta}\, b_j(x', 1)\zeta^{m-j}\, d\zeta \in C_b^\nu(\mathbb{R}^n) \tag{*}$$

if $N > m + \nu + 1$. Indeed, this follows from the equation (11) because the integral involving the remainder, r_N, is in C_b^ν since $r_N \in S^{m-N}$ and $m - N < -\nu - 1$. On the other hand, writing the terms in the sum in (*) in the form

$$D_n^k \int_\Gamma e^{ix_n\zeta}\, b_j(x', 1)\zeta^{m-j-k}\, d\zeta,$$

with $m - j - k < -1$, we find that they vanish when $x_n > 0$ by Cauchy's theorem because the integral along the semi-circle $|\zeta| = R$, $\operatorname{Im} \zeta \geqslant 0$, has a limit of 0 as $R \to \infty$. Hence it follows from (11) that all derivatives of v are bounded when $x_n > 0$. Hence $v|_{x_n>0}$ and its derivatives have continuous extensions to $x_n \geqslant 0$.

(b) Necessity of (13). Suppose that $v|_{x_n>0}$ has a C^∞ extension to $x_n \geqslant 0$. By Theorem 13.4 (or just Exercise 3 below) we can find $w \in C_0^\infty(\mathbb{R}^n)$ equal to v when $x_n > 0$. Then $v - w$ is a distribution with compact support contained in the lower half-space $x_n \leqslant 0$, so

$$B_\varphi(\zeta) := (2\pi)^{-1}\langle v - w, e^{-ix_n\zeta}\, \varphi(x')\rangle, \qquad \varphi \in C_0^\infty(\mathbb{R}^{n-1}),$$

is an entire function and

$$|B_\varphi(\zeta)| \leqslant C(1 + |\zeta|)^M, \qquad \operatorname{Im} \zeta \geqslant 0, \tag{14}$$

for some M by the Paley–Wiener–Schwartz theorem; see [Hö 1, p. 181]. On the real axis, we have $t^{-m}B_\varphi(t\xi_n) \to b_\varphi^0(\xi_n)$ as $t \to +\infty$, where

$$b_\varphi^0(\xi_n) := \int b_0(x', \xi_n)\varphi(x')\, dx'. \tag{15}$$

To see this, note that by the asymptotic expansion for b we can write the oscillatory integral (12) as

$$v(x) = \int e^{ix_n\xi_n} b_0(x', \xi_n)\, d\xi_n + \int e^{ix_n\xi_n} r(x', \xi_n)\, d\xi_n$$

where $r = b - b_0 \in S^{m-1}(\mathbb{R}^{n-1} \times \mathbb{R}^1)$. Now,

$$t^{-m}B_\varphi(t\xi_n) = t^{-m}(2\pi)^{-1}\langle v(x), e^{-itx_n\xi_n}\varphi(x')\rangle - t^{-m}(2\pi)^{-1}\langle w(x), e^{-itx_n\xi_n}\varphi(x')\rangle$$

is a difference of two terms involving v and w. For the first term, we make a change of variables and find that

$$\begin{aligned} t^{-m}(2\pi)^{-1}\langle v(x), e^{-itx_n\xi_n}\varphi(x')\rangle &= t^{-m}(2\pi)^{-1}\langle t^{-1}v(x', t^{-1}x_n), \quad e^{-ix_n\xi_n}\varphi(x')\rangle \\ &= b_\varphi^0(\xi_n) + O(t^{-1}), \end{aligned}$$

where we used homogeneity $b_0(x', t\xi_n) = t^m b_0(x', \xi_n)$ for $t > 0$ and Fourier inversion in x_n. As for the second term, involving $w \in C_0^\infty(\mathbb{R}^n)$, if we integrate by parts with respect to x_n we see that it is rapidly decreasing as a function of t. This proves that $t^{-n}B_\varphi(t\xi_n) \to b_\varphi^0(\xi_n)$ uniformly in ξ_n as $t \to +\infty$.

Let Γ be the curve defined above. In view of (14), (15) it follows that

$$B_\varphi(\zeta)\zeta^{-m}(i + \varepsilon\zeta)^{-M-|m|-1}$$

tends to 0 at ∞ in the upper half-plane and has a bound on Γ which is independent of ε. By the maximum principle it follows that there is also a fixed bound above Γ. Thus,

$$|B_\varphi(\zeta)| \leqslant C|\zeta|^m, \qquad |\zeta| > 1, \operatorname{Im}\zeta \geqslant 0. \tag{16}$$

It follows that the functions $t^{-m}B_\varphi(t\zeta)$, $t > 1$, are uniformly bounded on compact sets in the upper half-plane $\operatorname{Im}\zeta > 0$. It can then be shown by Cauchy's formula that this family of functions is equicontinuous, hence there is a sequence $t_k \to \infty$ such that the limit

$$F(\zeta) = \lim_{k\to\infty} t_k^{-m}B_\varphi(t_k\zeta)$$

converges uniformly on compact sets in the upper half-plane, and, therefore, $F(\zeta)$ is analytic there. We claim that the boundary values of F on the real axis are

$$F(\xi_n + i0) = b_\varphi^0(\xi_n).$$

Let $I \subset \mathbb{R}$ be an open interval not containing the origin. In view of (9) with

$N = 0$, we have for $\varphi \in C_0^\infty(I)$

$$\int \varphi(x)F(x+i0)\,dx = \int \Phi(x, Y)F(x+iY)\,dx$$

$$+ 2i \int\int_0^1 F(x+itY)\varphi'(x)iY\,dx\,dt.$$

For each k the boundary values of $t_k^{-m}B_\varphi(t_k\zeta)$ also satisfy a similar equation. By virtue of the estimates (16) we can then let $t_k \to \infty$ and then use (15) to obtain that $\int \varphi(x)F(x+i0)\,dx = \int \varphi(x)b_\varphi^0(x)\,dx$, as desired.

Since $b_\varphi^0(\xi_n) = b_\varphi^0(1)\cdot\xi_n^m$ for $\xi_n > 0$ it follows by Corollary 13.3 that $F(\zeta) \equiv b_\varphi^0(1)\cdot\zeta^m$ in a one-sided complex neighbourhood of I. This proves that b_0 satisfies (13). By part (a) of the proof we can now subtract a distribution corresponding to b_0 and vanishing when $x_n > 0$ and then conclude that b_1 satisfies (13) and so on. Thus, (13) holds for all $j = 0, 1, 2, \ldots$.

13.2 The transmission property

Before considering the transmission property (Definition 13.7), it is illuminating to consider first a related situation involving the kernel of a pseudodifferential operator. The Schwartz kernel of an operator $b(x, D) \in OS^m$ is the oscillatory integral

$$K(x, y) = (2\pi)^{-n}\int e^{i(x-y,\xi)}\,b(x,\xi)\,d\xi, \tag{17}$$

where $b \in S^m(\mathbb{R}^n \times \mathbb{R}^n)$, and it is singular only on the diagonal $x = y$. If we introduce new variables $x' = x$, $x'' = x - y$, $\xi'' = \xi$, we obtain

$$(2\pi)^{-n}\int e^{i(x'',\xi'')}\,b(x',\xi'')\,d\xi'' \tag{18}$$

and now the singularity occurs at $x'' = 0$.

We now allow the possibility that the x' and x'' variables have different dimensions, and change notation so that $\mathbb{R}^{2n}$ becomes $\mathbb{R}^n$ with the variables $x = (x_1, \ldots, x_n)$ split into two groups $x' = (x_1, \ldots, x_{n-k})$ and $x'' = (x_{n-k+1}, \ldots, x_n)$. Note that in Definition 7.1 and for many properties of symbols it is irrelevant that there be as many x variables as ξ variables.

Lemma 13.6 *If $b \in S^m(\mathbb{R}^n \times \mathbb{R}^k)$ and v is defined by the oscillatory integral*

$$v(x) = \int e^{i(x'',\xi'')}\,b(x,\xi'')\,d\xi'', \qquad x = (x', x''), \tag{19}$$

then we also have

$$v(x) = \int e^{i(x'',\xi'')}\tilde{b}(x',\xi'')\,d\xi'' \tag{19'}$$

where $\tilde{b} \in S^m(\mathbb{R}^{n-k} \times \mathbb{R}^k)$ *is defined by*

$$\tilde{b}(x', \xi'') = (2\pi)^{-k} \iint e^{i(x'', \theta'' - \xi'')} b(x, \theta'')\, d\theta''\, dx'' \tag{20}$$

and has the asymptotic expansion

$$\tilde{b}(x', \xi'') \sim \sum \frac{(-i)^{|\alpha|}}{\alpha!} D^{\alpha}_{\xi''} D^{\alpha}_{x''} b(x, \xi'')|_{x''=0}$$

Proof This is essentially Propositions 7.5 and 7.8 in another form, in the same way as (17) is related to (18) (for the case $n - k = k$). Therefore we retrace the steps in the proof only briefly using the current notation. Assume first that $b \in \mathscr{S}$ in the variables x, ξ''. Then $v \in \mathscr{S}$ and (19′) means exactly that $(2\pi)^k \tilde{b}$ is the Fourier transform of v with respect to x'',

$$\begin{aligned}\tilde{b}(x', \xi'') &= (2\pi)^{-k} \int e^{-i(x'', \xi'')} v(x', x'')\, dx'' \\ &= (2\pi)^{-k} \int e^{i(x'', \theta'' - \xi'')} b(x, \theta'')\, d\theta''\, dx'',\end{aligned}$$

i.e. (20) holds when $b \in \mathscr{S}$. Now, for any $b \in S^m$, the formula (20) makes sense as an oscillatory integral and defines $\tilde{b} \in S^m(\mathbb{R}^{n-k} \times \mathbb{R}^k)$ such that each semi-norm of $\tilde{b}$ is bounded by a semi-norm of b in S^m. It follows as in the proof of Proposition 7.5 that the relation between (19) and (19′) holds for all $b \in S^m$.

To obtain the asymptotic expansion we follow the second proof of Proposition 7.8. Substituting $\eta'' = \theta'' - \xi''$, the equation (20) takes the form

$$\tilde{b}(x', \xi'') = (2\pi)^{-k} \iint e^{i(x'', \eta'')} b(x, \xi'' + \eta'')\, d\eta''\, dx''. \tag{21}$$

Expanding $b(x, \xi'' + \eta'')$ by Taylor's formula in powers of η'' we obtain

$$\tilde{b}(x', \xi'') = \sum_{|\alpha| \leq N-1} \partial^{\alpha}_{\xi''} b(x, \xi'') \cdot \frac{(\eta'')^{\alpha}}{\alpha!} + r_N(x, \xi'', \eta'') \tag{22}$$

where

$$r_N(x, \xi'', \eta'') = \sum_{|\alpha|=N} N \frac{(\eta'')^{\alpha}}{\alpha!} \int_0^1 (1-t)^{N-1} D^{\alpha}_{x''} \partial^{\alpha}_{\xi''} b(x, \xi'' + t\eta'')\, dt.$$

By the Fourier inversion formula

$$(2\pi)^{-k} \iint e^{i(x'', \eta'')} \partial^{\alpha}_{\xi''} b(x, \xi'')(\eta'')^{\alpha}\, d\eta''\, dx'' = (-1)^{|\alpha|} \partial^{\alpha}_{\xi''} b(x', x'', \xi'')|_{x''=0},$$

and then by substituting (22) in (21) we get the asymptotic expansion for $\tilde{b}$, once the appropriate estimates for the remainder terms are verified as in the second proof of Proposition 7.8.

Remark When $n - k = k$, Lemma 13.6 can be obtained from Propositions 7.5 and 7.8 by the same change of variables that transformed (17) into (18). Letting $x'' = x' - y'$, the formula (20) becomes

$$\tilde{b}(x', \xi'') = (2\pi)^{-k} \int\int e^{i(x'-y', \theta''-\xi'')} b(x', x'-y', \theta'')\, d\theta''\, dy',$$

which is the ordinary symbol $\tilde{b}$ corresponding to the amplitude function $a(x', y', \theta'') = b(x', x' - y', \theta'')$.

Remark Later on, we apply Lemma 13.6 with $k = 1$ and $x = (x', x'') = (x', x_n)$. Note that if a function v has the representation (19′) with $k = 1$, then Theorem 13.5 enables us to decide if $v \in C^\infty(\bar{\mathbb{R}}^n_+)$.

We turn now to discuss the transmission property. Let $A \in S^m(\mathbb{R}^n)$ and let $\Omega \subset \mathbb{R}^n$ be a bounded open set with C^∞ boundary. For any $u \in C^\infty(\bar{\Omega})$ we define $u^0 = L_2(\mathbb{R}^n)$ by $u^0 = u$ in Ω, $u^0 = 0$ in $\mathbb{R}^n \backslash \Omega$. Note that

$$\text{sing supp}\, A(x, D)u^0 \subset \text{sing supp}\, u^0 \subset \partial\Omega.$$

Definition 13.7 *Let $\imath^+$ denote the restriction map $\mathscr{D}'(\mathbb{R}^n) \to \mathscr{D}'(\Omega)$. An operator $A \in OS^m(\mathbb{R}^n)$ is said to satisfy the* transmission property *with respect to $\partial\Omega$ if for every $u \in C^\infty(\bar{\Omega})$, we have*

$$\imath^+ A(x, D)u^0 \in C^\infty(\bar{\Omega}),$$

that is, $A(x, D)u^0$ and all its derivatives have continuous extensions from Ω to $\bar{\Omega}$.

In general $A_\Omega u = \imath^+ Au^0$ defines a map $C^\infty(\bar{\Omega}) \to \mathscr{D}'(\Omega)$, but if A satisfies the transmission property then we get a map

$$A_\Omega\colon C^\infty(\bar{\Omega}) \to C^\infty(\bar{\Omega}).$$

It is clear that if Q is any tangential differential operator and P is any differential operator then PAQ also satisfies the transmission property. We could also formulate an apparently weaker transmission property: $\imath^+ Au^0 \in C^\infty(\bar{\Omega})$ if u vanishes to some fixed order on $\partial\Omega$. But it will be clear from the proof of Theorem 13.8 that this weaker version actually implies Definition 13.7. Thus, if A satisfies the transmission property then PAQ satisfies a weakened transmission property, and hence the full transmission property, for *any* differential operators P and Q.

The question of whether or not A has the transmission property is obviously local (by virtue of a partition of unity and quasi-locality of A) so we need only consider the case when Ω is defined by $x_n \geqslant 0$ and $\partial\Omega$ is defined by $x_n = 0$. Thus, the problem of reformulated as follows: If $A(x, D) \in OS^m(\mathbb{R}^n)$ has compact support (Definition 8.6) and if $u \in C^\infty(\bar{\mathbb{R}}^n_+)$, under what conditions is it true that

$$A(x, D)u^0|_{x_n > 0} \in C^\infty(\bar{\mathbb{R}}^n_+)? \tag{23}$$

Theorem 13.8 *Let $A \in S^m$ be a $(-\infty)$ classical symbol (see the remark after Definition 7.21) with asymptotic expansion $A \sim \sum A_j$, where A_j is homogeneous of degree $m-j$, and let the operator $A(x, D)$ have compact support. Then $A(x, D)$ satisfies the transmission property (23) with respect to $\partial\mathbb{R}^n_+ = \mathbb{R}^{n-1}$ if and only if*

$$A^{(\alpha)}_{j(\beta)}(x', 0; 0, -1) = e^{i\pi(m-j-|\alpha|)} A^{(\alpha)}_{j(\beta)}(x', 0; 0, 1) \tag{24}$$

for all j, α and β. Here we have used the notation $A^{(\alpha)}_{j(\beta)} = \partial^\alpha_\xi \partial^\beta_x A_j$.

Proof First we prove sufficiency of the transmission conditions (24). Let $u \in C^\infty(\bar{\mathbb{R}}^n_+)$; since A has compact support, we may assume that u has compact support. Then we can write u^0 as the inverse Fourier transform in the x_n variable of the "symbol" $p(x', \xi_n)$,

$$u^0(x) = \frac{1}{2\pi} \int e^{ix_n\xi_n} p(x', \xi_n)\, d\xi_n, \qquad x = (x', x_n), \tag{25}$$

where

$$p(x', \xi_n) = \int_0^\infty e^{-ix_n\xi_n} u(x)\, dx_n \tag{26}$$

is the Fourier transform of u^0 in the x_n variable. For $\xi_n \neq 0$ we may integrate by parts to obtain

$$p(x', \xi_n) = -i \sum_{k=0}^{N-1} \xi_n^{-1-k} D_n^k u(x', 0) + r_N(x', \xi_n),$$

where

$$r_N(x', \xi_n) = \xi_n^{-N} \int_0^\infty e^{-ix_n\xi_n} D_n^N u(x)\, dx_n.$$

It follows that $p \in S^{-1}(\mathbb{R}^{n-1} \times \mathbb{R}^1)$, that is,

$$|D^{\alpha_n}_{\xi_n} D^\beta_{x'} p(x', \xi_n)| \leqslant C_{\alpha,\beta}(1 + |\xi_n|)^{-1-\alpha_n} \tag{27}$$

and there is the asymptotic expansion

$$p(x', \xi_n) \sim -i \sum_{j=0}^\infty \xi_n^{-1-k} D_n^k u(x', 0) \tag{28}$$

(where it is tacitly assumed that the terms on the right-hand side are multiplied by a cut-off function $\chi(\xi_n)$).

Now, when we calculate $A(x, D)u^0$ we must take the Fourier transform of (25). By the Fourier inversion formula in the x_n variable we obtain

$$\hat{u}^0(\xi) = \int e^{-i\langle y', \xi'\rangle} p(y', \xi_n)\, dy'$$

and, therefore,

$$A(x, D)u^0 = (2\pi)^{-n} \iint e^{i\langle x, \xi\rangle} A(x, \xi) p(y', \xi_n)\, e^{-i\langle y', \xi'\rangle}\, d\xi'\, dy'.$$

It follows that

$$A(x, D)u^0 = \frac{1}{2\pi}\int e^{ix_n\xi_n}\, q(x, \xi_n)\, d\xi_n, \tag{29}$$

where

$$q(x, \xi_n) = (2\pi)^{1-n}\iint e^{i(x'-y', \xi')}\, A(x, \xi)p(y', \xi_n)\, d\xi'\, dy'.$$

Substituting $z' = (y' - x')\cdot\langle\xi_n\rangle$ and $\eta' = \xi'/\langle\xi_n\rangle$, where $\langle\xi_n\rangle = (1 + \xi_n^2)^{1/2}$, we get

$$q(x, \xi_n) = (2\pi)^{1-n}\iint e^{-i(z', \eta')}\, A(x; \langle\xi_n\rangle\eta', \xi_n)p(x' + z'\cdot\langle\xi_n\rangle^{-1}, \xi_n)\, d\eta'\, dy',$$

and, since $A \in S^m$ and $p \in S^{-1}$, it is easily verified by regularizing the integral that $q \in S^{m-1}(\mathbb{R}^{n-1} \times \mathbb{R}^1)$. We claim that q has the asymptotic expansion

$$q(x, \xi_n) \sim \sum \frac{i^{|\alpha|}}{\alpha!} D_{\xi'}^{\alpha} A(x, \xi)\cdot D_{x'}^{\alpha} p(x', \xi_n)|_{\xi'=0} \tag{30}$$

Indeed, by Taylor's formula,

$$p(x' + z'\langle\xi_n\rangle^{-1}, \xi_n) = \sum_{|\alpha|<N} \partial_{x'}^{\alpha} p(x', \xi_n)\cdot\frac{z'^{\alpha}}{\alpha!}\cdot\langle\xi_n\rangle^{-|\alpha|} + r_N(x', z', \xi_n)\cdot\langle\xi_n\rangle^{-N},$$

where

$$r_N(x', z', \xi_n) = \sum_{|\alpha|=N} N\frac{z'^{\alpha}}{\alpha!}\int_0^1 (1-t)^{N-1}\, \partial_{x'}^{\alpha} p(x' + tz'\langle\xi_n\rangle^{-1}, \xi_n)\, dt$$

Now, Fourier's inversion formula implies

$$(2\pi)^{1-n}\iint e^{-i(z', \eta')}\, A(x; \langle\xi_n\rangle\eta', \xi_n)z'^{\alpha}\langle\xi_n\rangle^{-|\alpha|}\, dz'\, d\eta' = D_{\xi'}^{\alpha} A(x, \xi)|_{\xi'=0}$$

which gives us the finite terms in the asymptotic expansion (30), and the remainder after subtracting the terms for $|\alpha| < N$ belongs to $S^{m-1-N}(\mathbb{R}^n \times \mathbb{R}^1)$.

Finally, we apply Lemma 13.6 to put (29) in the form

$$A(x, D)u^0 = \int e^{ix_n\xi_n}\, \tilde{q}(x', \xi_n)\, d\xi_n, \tag{29'}$$

where $\tilde{q} \in S^{m-1}(\mathbb{R}^{n-1} \times \mathbb{R}^1)$ is given by

$$\begin{aligned}\tilde{q}(x', \xi_n) &= \frac{1}{2\pi}\iint e^{i(x_n, \theta_n - \xi_n)}\, q(x, \theta_n)\, d\theta_n\, dx_n\, . \\ &\sim \sum \frac{(-i)^{|\alpha|}}{\alpha!} D_{x_n}^{\alpha} D_{\xi_n}^{\alpha}\, q(x, \xi_n)|_{x_n=0} \\ &\sim \sum\sum \frac{(-i)^{|\alpha|}}{\alpha!}\frac{i^{|\beta|}}{\beta!} D_{x_n}^{\alpha} D_{\xi_n}^{\alpha}\, D_{\xi'}^{\beta} A(x, \xi)\cdot D_{x'}^{\beta} p(x', \xi_n)|_{\xi'=0, x_n=0}\end{aligned} \tag{30'}$$

and p has the asymptotic expansion (28). Now, consider the operator with symbol A_0; in view of (30′) (with A replaced by A_0) and the transmission conditions (24) it follows by Theorem 13.5 that $A_0(x, D)u^0 \in C^\infty(\bar{\mathbb{R}}^n_+)$, i.e. $A_0(x, D)$ has the transmission property. The operators with symbol A_1, A_2 and so on have the same property. Hence $A(x, D)$ itself has the same property, for if we substract off a large number of A_j's then $A(x, D) - \sum_{j<N} A_j(x, D) \in OS^{m-N}$, and, therefore, $A(x, D)u^0 \in C^k(\bar{\mathbb{R}}^n_+)$ if $m - N + k < -1$.

Conversely, suppose that $A(x, D)$ has the transmission property. We want to show that the conditions (24) must be fulfilled. First of all, it follows from Theorem 13.5 that the principal part of (30′), namely

$$A_0(x', 0; 0, \xi_n)\cdot u(x', 0)/i\xi_n,$$

must satisfy (13) with m replaced by $m - 1$, so

$$A_0(x', 0; 0, -1) = e^{i\pi m} A_0(x', 0; 0, 1)$$

because $u(x', 0)$ is arbitrary. Note that this condition would still have followed if we had weakened the transmission property by demanding in Definition 13.7 that u vanish of some fixed order $k - 1$ on $x_n = 0$, because then the principal part of (30′) would be $A_0(x', 0; 0, \xi_n)\cdot D_n^k u(x', 0)\cdot \xi_n^{-1-k}/i$ and must satisfy (13) with m replaced by $m - 1 - k$.

Note that if A satisfies the transmission property then $A\cdot D_n^k$ has the weakened transmission property, for if u vanishes to order $k - 1$ on the plane $x_n = 0$ then the kth derivative of u^0 does not contain any terms supported on $x_n = 0$, and thus $D_n^k u^0 = (D_n^k u)^0 \in C^\infty(\bar{\mathbb{R}}^n_+)$. It follows that a weakened transmission property remains valid if A is multiplied to the right or to the left by *any* differential operator. In particular, the commutators of A with D_j and x_j any number of times satisfy this weakened transmission property, so it follows in view of Proposition 7.24 that for arbitrary α, β we have

$$A^{(\alpha)}_{0(\beta)}(x', 0; 0, -1) = e^{i\pi(m-|\alpha|)} A^{(\alpha)}_{0(\beta)}(x', 0; 0, 1)$$

Now, as shown above in the proof of sufficiency, these conditions imply that the operator $A_0(x, D)$ has the transmission property. Subtracting it from $A(x, D)$ we conclude that $A_1(x, D)$ has the same property and so on. Hence

$$A^{(\alpha)}_{j(\beta)}(x', 0; 0, -1) = e^{i\pi(m-j-|\alpha|)} A^{(\alpha)}_{j(\beta)}(x', 0; 0, 1)$$

for all $j = 0, 1, \ldots,$ which are the conditions (24).

Remark 13.9 Let $u \in C^\infty(\bar{\mathbb{R}}^n_+)$ have compact support, and note that

$$D_n u^0 = (D_n u)^0 + i^{-1}u(x', 0) \otimes \delta(x_n).$$

If A satisfies the transmission conditions (24) then the equation

$$A(x, D)D_n u^0 = A(x, D)(D_n u)^0 + i^{-1}A(x, D)(u(x', 0) \otimes \delta(x_n))$$

implies that $A(x, D)(u(x', 0) \otimes \delta(x_n)) \in C^\infty(\bar{\mathbb{R}}^n_+)$ since the other terms belong to $C^\infty(\bar{\mathbb{R}}^n_+)$. (Note that $A(x, D)D_n$ also satisfies the conditions (24) since its symbol is $A(x, \xi)\xi_n$.) Thus, $A(x, D)w \in C^\infty(\bar{\mathbb{R}}^n_+)$ for any single layer

$w = v(x') \otimes \delta(x_n)$. It is clear that this is also true for multiple-layers $A(x, D)w$, where $w = v(x') \otimes \delta^{(j)}(x_n)$.

Let $\mathscr{A}$ be an elliptic differential operator in $\mathbb{R}^n$, homogeneous of order l. Let E be a parametrix of $\mathscr{A}$. It is important for Chapter 14 to remark here that E satisfies the transmission conditions (24) with respect to any boundary $\partial\Omega$. In fact we can choose a parametrix E having an asymptotic expansion

$$E(x, \xi) \sim E_{-l}(x, \xi) + E_{-l-1}(x, \xi) + \cdots, \tag{31}$$

where E_j is homogeneous of degree j and is a rational function of ξ. To see this we recall that the operator $E(x, D)$ is defined by

$$E(x, D)f = (2\pi)^{-n} \int e^{i(x,\xi)} E(x, \xi)\hat{f}(\xi)\, d\xi, \qquad f \in \mathscr{S}$$

and differentiation under the integral sign gives

$$\mathscr{A}(x, D)E(x, D)f = (2\pi)^{-n} \int e^{i(x,\xi)} \mathscr{A}(x, \xi + D_x)E(x, \xi)\hat{f}(\xi)\, d\xi.$$

Thus a parametrix for $\mathscr{A}$ can be constructed by choosing a function E such that

$$\mathscr{A}(x, \xi + D_x)E(x, \xi) = 1 + K(x, \xi) \tag{32}$$

with K rapidly decreasing as $\xi \to \infty$. Let

$$\mathscr{A}(x, \xi) = \mathscr{A}_l(x, \xi) + \mathscr{A}_{l-1}(x, \xi) + \cdots + \mathscr{A}_0(x, \xi)$$

be the decomposition of $\mathscr{A}(x, \xi)$ in homogeneous terms with respect to ξ. Then if we look for an asymptotic expansion (31) for $E(x, \xi)$ and equate terms of equal homogeneity in the equation (32) we obtain

$$\mathscr{A}_l(x, \xi)E_{-l}(x, \xi) = 1,$$

$$\mathscr{A}_l(x, \xi)E_{-l-1}(x, \xi) + \mathscr{A}_{l-1}(x, \xi)E_{-l}(x, \xi) + \sum \partial_{\xi_j}\mathscr{A}_l(x, \xi) \cdot D_{x_j}E_{-l}(x, \xi) = 0$$

and a sequence of equations expressing $\mathscr{A}_l(x, \xi)E_{-l-k}(x, \xi)$ for any $k > 0$ in terms of $E_{-l}, \ldots, E_{-l-k+1}$. Since $\mathscr{A}_l(x, \xi)$ is invertible if $\mathscr{A}$ is elliptic, the sequence $E_{-l}, E_{-l-1}, \ldots$ is uniquely determined and gives a solution of (32) with K decreasing as rapidly as we please. It is clear that each term E_j is a rational function of ξ; thus E satisfies the transmission conditions with respect to any boundary $\partial\Omega$. Moreover, any other parametrix E' of $\mathscr{A}$ must also satisfy (24) because $E - E' \in OS^{-\infty}$ by virtue of Lemma 7.33.

Remark Note that any operator $A \in OS^{-\infty}$ has the transmission property, because $A \sim \sum A_j$ where $A_j = 0$, which obviously fulfills the transmission conditions (24).

13.3 Boundary values of a single-layer potential

Let $A \in OS^m(\mathbb{R}^n)$. We are interested in the boundary values as $x_n \to 0^+$ of the single-layer potential

$$A(v(x') \otimes \delta(x_n)), \qquad v \in C_0^\infty(\mathbb{R}^{n-1}) \tag{33}$$

because the resolution of this problem is one of the main elements in the construction of a parametrix of an elliptic boundary value problem in Chapter 14; see the section on the Calderón operator.

We know from Remark 13.9 that if A has the transmission property then these boundary values exist in the sense of the C^∞ topology, i.e. uniform convergence of all derivatives on compact subsets of $\mathbb{R}^{n-1}$ as $x_n \to 0^+$. The purpose of this section is to show that if A is any p.d.o. satisfying the transmission conditions (24) then the boundary values of (33) take the form

$$\lim_{x_n \to 0^+} A(v(x') \otimes \delta(x_n)) = Q(x', D')v \tag{34}$$

for some pseudo-differential operator $Q \in OS^{m+1}(\mathbb{R}^{n-1})$; see Theorem 13.12.

To compute the value of the single-layer $A(x, D)w$, where $w = v(x') \otimes \delta(x_n)$, we take $\varphi \in C_0^\infty((-1, 1))$ with $\int \varphi(t)\, dt = 1$ and let $w_\varepsilon(x) = v(x') \cdot \varepsilon^{-1}\varphi(\varepsilon^{-1}x_n)$. Since $w_\varepsilon \to w$ in $\mathscr{S}'$, then $A(x, D)w = \lim_{\varepsilon \to 0} A(x, D)w_\varepsilon$, where

$$A(x, D)w_\varepsilon = (2\pi)^{-n} \int e^{i(x,\xi)} A(x, \xi)\hat{\varphi}(\varepsilon\xi_n)\hat{v}(\xi')\, d\xi.$$

Now, if $A(x, \xi)$ happened to be a rational function of ξ_n and of order ξ_n^{-2} as $\xi_n \to \infty$ in $\mathbb{C}$ then for $x_n > 0$ we could replace the integral $\int_{-\infty}^{\infty}$ with respect to ξ_n by the integral $\int_{\gamma^+}$ along a contour γ^+ containing the poles of the integrand in the upper half-plane, so that

$$A(x, D)w_\varepsilon = (2\pi)^{1-n} \int e^{i(x',\xi')}(2\pi)^{-1}\hat{v}(\xi') \int_{\gamma^+} e^{ix_n\lambda} Q(x', x_n, \lambda)\hat{\varphi}(\varepsilon\lambda)\, d\lambda\, d\xi'.$$

Letting $\varepsilon \to 0$ and then $x_n \to 0^+$ we would obtain

$$\lim_{x_n \to 0^+} A(x, D)w = (2\pi)^{1-n} \int e^{i(x',\xi')} Q(x', \xi')\hat{v}(\xi')\, d\xi'$$

where $Q(x', \xi') = (2\pi)^{-1} \int_{\gamma^+} e^{ix_n\lambda} Q(x', 0, \lambda)\, d\lambda$.

The above condition on A is of course very restrictive. The main point here is to find a suitable generalization of the operator $\int_+ = \int_{\gamma^+}$ which can be applied under much less restrictive conditions.

If $h(z)$ is a rational function with no real pole, let $H(z)$ equal the sum of terms of degree $\geqslant -1$ in the Laurent expansion as infinity. Then $h(z) - H(z) = O(z^{-2})$ as $z \to \infty$ in $\mathbb{C}$, so

$$\int_{\gamma^+} [h(z) - H(z)]\, dz = \int_{-\infty}^{\infty} [h(t) - H(t)]\, dt,$$

whence if we take γ^+ to consist of a large semi-circle $|z| = R$, $\operatorname{Im} z \geqslant 0$, including the diameter $-R \leqslant t \leqslant R$ on the real axis

$$\int_{\gamma^+} h(z)\,dz = \int_{|t|>R} [h(t) - H(t)]\,dt + \int_{-R}^{R} h(t)\,dt - \int_{\pi}^{0} H(R\,e^{i\theta})\cdot Ri\,e^{i\theta}\,d\theta.$$

We take this result as the definition of $\int_+$ when $h(t)$ is just continuous on $\mathbb{R}$, provided of course that a suitable analogue of $H(z)$ can be found.

Lemma 3.10 *Let $h(t)$, $t \in \mathbb{R}$, be a continuous function and assume that there is an analytic function $H(z)$ in $\Pi_R = \{z \in \mathbb{C};\ \mathrm{Im}\, z \geqslant 0, |z| \geqslant R\}$, for some R, such that $H(z) = O(z^N)$ for some N when $z \to \infty$ in Π_R, and $h(t) - H(t) = O(t^{-2})$ when $t \to \infty$ on $\mathbb{R}$. Then*

$$\int_{+} h(t)\,dt := \int_{|t|>R} [h(t) - H(t)]\,dt + \int_{-R}^{R} h(t)\,dt - \int_{\pi}^{0} H(R\,e^{i\theta})\cdot Ri\,e^{i\theta}\,d\theta \tag{35}$$

is independent of the choice of $H(z)$ (or the number R).

Proof If $H_1(z)$ and $H_2(z)$ are two such functions then $H = H_1 - H_2$ satisfies the conditions of the lemma with $h = 0$, i.e.

$H(z)$ is analytic in Π_R, $H(z) = O(z^N)$ when $z \to \infty$ in Π_R and $H(t) = O(t^{-2})$ when $t \to \infty$ in $\mathbb{R}$.

Now consider the function

$$\varphi(z) = z^2 H(z)(1 - i\varepsilon z)^{-N-3} \qquad \text{on } \Pi_R.$$

Since $\varphi(z) \to 0$ uniformly on $\{z;\ \mathrm{Im}\, z \geqslant 0, |z| \geqslant \rho\}$ as $\rho \to \infty$ it follows from the maximum principle that φ attains its maximum on $\partial\Pi_R$. Since $|1 - i\varepsilon z| \geqslant 1 + \varepsilon\, \mathrm{Im}\, z \geqslant 1$ on $\partial\Pi_R$, we obtain

$$\sup_{\Pi_R} |z^2 H(z)(1 - i\varepsilon z)^{-N-3}| \leqslant M,$$

where $M = \sup_{\partial\Pi_R} |z^2 H(z)| < \infty$. Letting $\varepsilon \to 0$ we conclude that $|z^2 H(z)| \leqslant M$ on Π_R, i.e. $H(z) = O(z^{-2})$ when $z \to \infty$ in Π_R. By Cauchy's theorem, the integral of $H(z)$ along the boundary of the region $\mathrm{Im}\, z \geqslant 0$, $R \leqslant |z| \leqslant \rho$, vanishes. Hence by letting $\rho \to \infty$ we obtain

$$\int_{\partial\Pi_R} H(z)\,dz = 0.$$

This proves that (35) gives a unique definition of $\int_+ h(t)\,dt$.

Remarks

(i) If $h(t)$, $t \in \mathbb{R}$, is continuous and $h(t) = O(t^{-2})$ when $t \to \infty$, then we can choose $H = 0$ in Lemma 13.10, thus

$$\int_{+} h(t)\,\mathrm{dt} = \int_{-\infty}^{\infty} h(t)\,dt.$$

(ii) If h is a rational function with no real pole then $\int_+ h(t)\,dt$ is equal to $2\pi i$ times the sum of the residues of h in the upper half-plane.

The definition (35) of the operator $\int_+$ is a good definition in the sense that it depends continuously on parameters:

Corollary 13.11 *Same hypotheses as for Lemma* 13.10. *If $F(z, s)$ is an analytic function of z when* $\operatorname{Im} z \geqslant 0$, *for* $0 \leqslant s \leqslant 1$, *and F is a bounded, continuous of (z, s) then $\int_+ h(t)F(t, s)\,dt$ is a continuous function of s.*

Proof $\int_+ h(t)F(t, s)\,dt$ is defined by (35) with h replaced by $hF(\cdot, s)$ and H by $HF(\cdot, s)$. Hence the corollary follows by the dominated convergence theorem.

Before stating Theorem 13.12 we must determine what conditions on the symbol $A(x, \xi) \in S^m$ ensure that $A(x; \xi', \xi_n)$ satisfies the conditions of Lemma 13.10 with respect to ξ_n.

Suppose first for simplicity that A is homogeneous of degree m, and does not depend on x:

$$A(c\xi) = c^m A(\xi), \qquad |\xi| \geqslant 1, c \geqslant 1. \tag{36}$$

To apply Lemma 13.10 we need an asymptotic expansion for A as $\xi_n \to \infty$. Since $A(\xi', \xi_n) = |\xi_n|^m A(\xi'/|\xi_n|, \pm 1)$, then by Taylor's formula about $\xi'/|\xi_n| = 0$ we obtain

$$A(\xi', \xi_n) = |\xi_n|^{m-|\alpha|} \sum_{|\alpha| \leqslant N} \partial^\alpha_{\xi'} A(0, \pm 1)\cdot(\xi')^\alpha/\alpha! + R_{N+1}(\xi', \xi_n). \tag{37}$$

The remainder in Lagrange's form is

$$R_{N+1}(\xi', \xi_n) = |\xi_n|^{m-|\alpha|}(N+1) \sum_{|\alpha| = N+1} \frac{\xi'^\alpha}{\alpha!} \int_0^1 (1-t)^N \partial^\alpha_{\xi'} A(t\xi'/|\xi_n|, \pm 1)\,dt$$

and hence

$$|R_{N+1}(\xi', \xi_n)| \leqslant C_N \cdot |\xi_N|^{m-N-1} |\xi'|^{N+1} \tag{38}$$

for $|\xi_n| \geqslant |\xi'|$.

The expansion (37) is not adequate for the purpose of applying Lemma 13.10; the terms $|\xi_n|^{m-|\alpha|}$ are not analytic functions. However, the function $\xi_n^{m-|\alpha|}$ has an analytic continuation to the upper half-plane if we define it to be 1 at $\xi_n = 1$ and $e^{i\pi(m-|\alpha|)}$ at $\xi_n = -1$. Thus if it happens that

$$\partial^\alpha_\xi A(0, -1) = e^{i\pi(m-|\alpha|)}\, \partial^\alpha_\xi A(0, 1) \qquad \text{when } \alpha_n = 0, \tag{39}$$

then we obtain an asymptotic expansion for A:

$$A(\xi', \xi_n) = \sum_{|\alpha| \leqslant N} \xi_n^{m-|\alpha|} \cdot \partial^\alpha_{\xi'} A(0, 1)\xi'^\alpha/\alpha! + R_{N+1}(\xi', \xi_n) \tag{40}$$

where $R_{N+1}(\xi', \xi_n)$ is of order $m - N - 1$ as $\xi_n \to \infty$ (when ξ' is fixed, or at least bounded). If we choose $N = m + 1$, so that $R_{N+1} = O(\xi_n^{-2})$, then in view of Lemma 13.10 we can now define

$$\int_+ A(\xi', \xi_n)\, d\xi_n,$$

with $H(z) = \sum_{|\alpha| \leq N} z^{m-|\alpha|} \cdot \partial^{\alpha}_{\xi'} A(0, 1)\xi'^{\alpha}/\alpha!$ and $R = |\xi'|$.

Note: To be precise in (40) we should write $(\xi_n + i0)^{m-|\alpha|}$ instead of $\xi_n^{m-|\alpha|}$ to indicate that we are dealing with the boundary values on the real line of a function analytic in the upper half-plane.

Remark Differentiating (36) with respect to ξ', we find that $\partial^{\alpha}_{\xi'} A$ is homogeneous of degree $m - |\alpha'|$. But then (39) implies that

$$\partial^{\alpha}_{\xi'} A(0, \xi_n) = \xi_n^{m-|\alpha|}\, \partial^{\alpha}_{\xi'} A(0, 1).$$

Now, if we differentiate this equation with respect to ξ_n, it follows that (39) holds for *all* α, not just when $\alpha_n = 0$.

Recall that a $(-\infty)$ classical symbol $A(x, \xi) \in S^m$ with asymptotic expansion $A \sim \sum A_j$, where A_j is homogeneous of degree $m - j$, is said to satisfy the *transmission conditions* if all derivatives $A_{j(\beta)}$ satisfy the conditions (39) at $x_n = 0$, that is

$$A^{(\alpha)}_{j(\beta)}(x', 0; 0, -1) = e^{i\pi(m-j-|\alpha|)} A^{(\alpha)}_{j(\beta)}(x', 0; 0, 1) \tag{24}$$

for all j, α and β. As usual, we have used the notation $A^{(\alpha)}_{j(\beta)} = \partial^{\alpha}_{\xi} \partial^{\beta}_{x} A_j$. Note that there is some redundancy here: (24) holds for all α, β if it holds when $\alpha_n = 0$, $\beta' = 0$.

Theorem 13.12 *Suppose that $A \in S^m(\mathbb{R}^n)$ is $(-\infty)$ classic and has an asymptotic expansion $A \sim \sum A_j$, where A_j is homogeneous of degree $m - j$, such that the transmission conditions* (24) *are satisfied. If $v \in C_0^{\infty}(\mathbb{R}^{n-1})$ and $u = v(x') \otimes \delta(x_n)$ then $A(x, D)u|_{x_n > 0}$ has a C^{∞} extension to the closed half-space $x_n \geq 0$, and*

$$\lim_{x_n \to 0^+} A(x, D)u = Q(x', D')v, \tag{41}$$

where $Q(x', \xi') = \dfrac{1}{2\pi} \displaystyle\int_+ A(x', 0, \xi', \xi_n)\, d\xi_n$, and $Q \in S^{m+1}(\mathbb{R}^{n-1})$ has the asymptotic expansion $Q \sim \sum Q_j$, with $Q_j(x', \xi') = \dfrac{1}{2\pi} \displaystyle\int_+ A_j(x', 0, \xi', \xi_n)\, d\xi_n$ being homogeneous of degree $m + 1 - j$.

Proof As mentioned earlier, the fact that $A(x, D)u|_{x_n > 0}$ has a C^{∞} extension to the closed half-space follows from §13.2, so it remains to verify (41). Since we already know that $A(x, D)u$ and all its derivatives converge uniformly on compact sets $x' \in K \subset \mathbb{R}^{n-1}$ as $x_n \to 0^+$, it suffices to verify that (41) holds pointwise.

The conditions (24) ensure that we can define $\int_+ A(x', 0, \xi', \xi_n)\, d\xi_n$. Indeed, since we may apply (40) to each term A_j in the asymptotic expansion we obtain

$$\left| A(x', 0, \xi', \xi_n) - \sum_{\substack{j+|\alpha| \leq m+1 \\ \alpha_n = 0}} \xi_n^{m-j-|\alpha|} \cdot A_j^{(\alpha)}(x', 0; 0, 1)\xi'^{\alpha}/\alpha! \right| \leq C|\xi'|^{m+2} \cdot \xi_n^{-2} \tag{42}$$

when $|\xi_n| \geq |\xi'|$. Hence Lemma 13.10 is applicable with

$$H(z) = \sum_{\substack{j+|\alpha| \leq m+1 \\ \alpha_n = 0}} z^{m-j-|\alpha|} \cdot A_j^{(\alpha)}(x', 0; 0, 1)\xi'^{\alpha}/\alpha!$$

and $R = |\xi'|$.

To compute $A(x, D)u$ we take $\varphi \in C_0^\infty((-1, 1))$ with $\int \varphi(t)\, dt = 1$ and let

$$u_\varepsilon(x) = v(x') \cdot \varepsilon^{-1} \varphi(\varepsilon^{-1} x_n).$$

Since $u_\varepsilon \to u$ in $\mathscr{S}'$, then $A(x, D)u = \lim_{\varepsilon \to 0} A(x, D)u_\varepsilon$, where

$$A(x, D)u_\varepsilon = (2\pi)^{-n} \int e^{i\langle x, \xi\rangle} A(x, \xi)\hat{\varphi}(\varepsilon\xi_n)\hat{v}(\xi')\, d\xi$$

We have

$$|e^{ix_n\xi_n}\hat{\varphi}(\varepsilon\xi_n)| \leq \int |\varphi(t)|\, dt, \qquad \operatorname{Im} \xi_n \geq 0,$$

provided that $\varepsilon \leq x_n$ (recall that $\operatorname{supp} \varphi \subset (-1, 1)$). By virtue of Corollary 13.11 we have

$$\int e^{ix_n\xi_n} A(x, \xi)\hat{\varphi}(\varepsilon\xi_n)\, d\xi_n \to \int_+ e^{ix_n\xi_n} A(x, \xi)\, d\xi_n \quad \text{when } \varepsilon \to 0. \tag{43}$$

Moreover, if we let $R = |\xi'|$ in the definition (35) of $\int_+$ it is clear from (42) and the estimate $|A(x, \xi)| \leq c(1 + |\xi|)^m$ that the integrals in (43) are bounded by a power of $(1 + |\xi'|)$, independently of ε and x_n. Therefore, by dominated convergence, it follows that

$$A(x, D)u = (2\pi)^{1-n} \int e^{i\langle x', \xi'\rangle}\hat{v}(\xi') \cdot (2\pi)^{-1} \int_+ e^{ix_n\xi_n} A(x, \xi)\, d\xi_n\, d\xi',$$

and letting $x_n \to 0$ we conclude that $A(x, D)u \to Q(x', D')v$ pointwise, with Q defined by (41). It is easily verified that Q_j is homogeneous of degree $m + 1 - j$. To prove that $Q \sim \sum Q_j$, we choose $N > m + 1$ and set

$$A = \sum_{j<N} A_j + R_N.$$

Then we have $R_N \in S^{m-N}$ when $|\xi'| > 1$, and

$$Q(x', \xi') = \sum_{j<N} Q_j(x', \xi') + \int R_N(x', 0; \xi', \xi_n)\, d\xi_n,$$

where the integral exists in the usual sense (since $m - N < -1$) and then

$$\left| D^{\beta}_{x'} D^{\alpha}_{\xi'} \int R_N(x', 0; \xi', \xi_n)\, d\xi_n \right| \leqslant C_{\alpha\beta} \int (1 + |\xi'| + |\xi_n|)^{m-N-|\alpha|}\, d\xi_n$$

$$\leqslant C'_{\alpha\beta}(1 + |\xi'|)^{m+1-N-|\alpha|}$$

The last equality follows by taking $t = \xi_n/(1 + |\xi'|)$ as the new variable instead of ξ_n. Thus $Q \sim \sum Q_j$.

Remark Let $A \in S^m(\mathbb{R}^n)$ and let $\Omega \subset \mathbb{R}^n$ be a bounded open set with C^∞ boundary. Also let a tubular neighbourhood of $\partial\Omega$ be identified with $\partial\Omega \times (-1, 1)$, and denote the points in this neighbourhood by (x', x_n). Suppose that A satisfies the transmission conditions (24) with respect to $\partial\Omega$ (in local coordinates) and consider the single-layer $A(v(x') \otimes \delta(x_n))$, where $v \in C^\infty(\partial\Omega)$. The existence of the boundary values

$$\lim_{x_n \to 0+} A(v(x') \otimes \delta(x_n))$$

is a local question, for by means of a partition of unity and quasi-locality of A it suffices to consider the boundary values of $\varphi A \psi(v(x') \otimes \delta(x_n))$, $\varphi, \psi \in C_0^\infty(\mathbb{R}^n)$, where the supports of φ and ψ lie in the tubular neighbourhood $-1 < x_n < 1$. We can then assume that $\bar{\Omega}$ is given by $x_n \geqslant 0$ and $\partial\Omega$ by $x_n = 0$, and apply Theorem 13.12.

The results of this section apply to a multiple-layer $A(v(x') \otimes \delta^{(j)}(x_n))$ since this can be viewed as a single-layer with respect to the operator $A(x, D) \circ D_n^j$ (which also satisfies the transmission conditions (24)). In fact by applying Theorem 13.12 to the operator $D_n^k \circ A(x, D) \circ D_n^j$ we obtain the following result concerning the boundary values of normal derivatives of multiple-layers.

Corollary 13.13 *Same hypotheses as for Theorem* 13.12. *If* $v \in C_0^\infty(\mathbb{R}^{n-1})$ *and* $u = v(x') \otimes \delta(x_n)$ *then* $D_n^k \circ A(x, D) \circ D_n^j u|_{x_n > 0}$ *has a* C^∞ *extension to* $x_n \geqslant 0$, *and*

$$\lim_{x_n \to 0+} D_n^k \circ A(x, D) \circ D_n^j u = Q_{jk}(x', D')v,$$

where $Q_{jk} \in S^{m+j+k+1}(\mathbb{R}^{n-1})$ *has the principal symbol*

$$\frac{1}{2\pi} \int_+ \xi_n^{j+k} A_0(x', 0, \xi', \xi_n)\, d\xi_n.$$

The symbol calculus also permits us to write down the full asymptotic expansion, but the principal symbol is sufficient for our purposes in Chapter 14.

Exercises

1. Show that

$$\frac{1}{2}\left(\frac{1}{x+i0}+\frac{1}{x-i0}\right)=\operatorname{pv}\frac{1}{x},$$

where the distribution on the right-hand side is the principal value of $1/x$, defined by

$$\left\langle \operatorname{pv}\frac{1}{x},\varphi\right\rangle = \lim_{\varepsilon\to 0}\int_{|x|>\varepsilon}\frac{\varphi(x)}{x}\,dx$$

2. Let k be a positive integer. Show that

$$(a+k)\cdot x_+^a \to (-1)^{k-1}\,\delta^{(k-1)}/(k-1)! \quad \text{in } \mathscr{D}'(\mathbb{R}),$$

as $a\to -k$, where δ is the Dirac distribution at $x=0$.

If M is any subset of $\mathbb{R}^n$, we let $C_0^\infty(M)$ denote the set of functions $f\in C_0^\infty(\mathbb{R}^n)$ with $\operatorname{supp} f\subset M$.

3. For $j=0,1,\ldots,$ let $f_j\in C_0^\infty(K)$, where K is a compact subset of $\mathbb{R}^{n-1}$. If $I\subset\mathbb{R}$ is an open interval containing 0, show that one can find $f\in C_0^\infty(K\times I)$ such that

$$\frac{\partial^j}{\partial t^j}f(x',t)=f_j(x'),\quad t=0,\quad j=0,1,\ldots$$

Hint: Choose the function g as in the proof of Lemma 12.1, then let $f(x',t)=\sum_0^\infty g(t/\varepsilon_j)t^j f_j(x')/j!$ for suitably chosen ε_j.

4. Denote by $C_{(0)}^\infty(\mathbb{R}_+^n)$ the space of restrictions to $\mathbb{R}_+^n$ of functions $f\in C_0^\infty(\mathbb{R}^n)$. Show that $C_{(0)}^\infty(\mathbb{R}_+^n)$ coincides with the set of functions in $C^\infty(\bar{\mathbb{R}}_+^n)$ that vanish outside a compact set.

5. Let $f\in C^\infty(\mathbb{R}_+^n)$. Show that $f\in C^\infty(\bar{\mathbb{R}}_+^n)$ if and only if all derivatives $D^\alpha f(x',x_n)$ converge uniformly on compact sets $|x'|\leqslant c$ as $x_n\to 0^+$.

6. Suppose that $v\in\mathscr{D}'(\mathbb{R}^n)$ has the representation (12) for some $b\in S^m(\mathbb{R}^{n-1}\times\mathbb{R}^1)$ with $m<-1$. Show that $v\in C^k(\mathbb{R}^n)$ if $m+k<-1$.

7. Suppose that $v\in\mathscr{E}'(\mathbb{R}^n)$ has the representation (12) for some $b\in S^m(\mathbb{R}^{n-1}\times\mathbb{R}^1)$, where $b\sim\sum_{j=0}^\infty b_j(x',\xi_n)$ and b_j is homogeneous of degree $m-j$ in ξ_n. If $\operatorname{supp} v\subset\bar{\mathbb{R}}_+^n$ show that:

(a) When m is an integer, v has an asymptotic expansion

$$v(x)\sim\sum_0^\infty v_j(x')\otimes x_{n+}^{j-m-1},\qquad v_j\in C^\infty(\mathbb{R}^{n-1}),$$

where x_{n+}^{j-m-1} is the homogeneous distribution on $\mathbb{R}$ defined at the beginning of this chapter. Here the expansion means that the difference between v and a partial sum of sufficiently high order is as smooth as we please.

(b) When m is an integer $\leqslant -1$, then v is C^∞ and $O(x_n^{-m-1})$ when $x_n \geqslant 0$, and vanishes when $x_n < 0$.

(c) When m is an integer $\geqslant 0$, then v is the sum of a function U which is in C^∞ when $x_n \geqslant 0$ and vanishes when $x_n < 0$, and a multiple-layer, i.e.

$$v = U + \sum_{j \leqslant m} v_j(x') \otimes \delta^{(j)}(x_n),$$

where $v_j \in C^\infty(\mathbb{R}^{n-1})$.

Hint: Since $v = 0$ for $x_n < 0$, then it certainly has a C^∞ extension to $x_n \leqslant 0$. If we change the signs of x_n and of ξ_n, it follows from Theorem 13.5 that $b_j(x', \xi_n)$ can be extended to a homogeneous analytic functions of ξ_n in the lower half-plane $\operatorname{Im} \xi_n < 0$. Now apply Theorem 13.1.

8. Let $u \in C^\infty(\bar{\Omega})$, and let $A \in OS^m(\mathbb{R}^n)$, $m < 0$. Show that $A(x, D)u^0 \in C^k(\mathbb{R}^n)$ if $m + k < 0$.

9. Suppose that $A \in OS^m(\mathbb{R}^n)$ has the transmission property with respect to $\partial\Omega$. Show that the map $A_\Omega\colon C^\infty(\bar{\Omega}) \to C^\infty(\bar{\Omega})$ given by $u \mapsto \imath^+ A(x, D)u^0$ is continuous.

14
The main theorem revisited

The goal of this chapter is to show how an elliptic boundary problem $(\mathscr{A}, \mathscr{B})$ on a bounded domain $\bar{\Omega}$ can be reduced to a system of elliptic pseudo-differential equations on $\partial\Omega$. The idea which lies behind this reduction to the boundary is based essentially on the classical representation of the solutions of the equation $\mathscr{A}u = f$ as a sum of multiple-layer potentials.

In order to explain this approach, first let us look at the case when $\mathscr{A}$ is the Laplace operator Δ. Let E be a fundamental solution of Δ, for instance

$$E(x) = \begin{cases} \dfrac{1}{2\pi}\log|x| & \text{when } n = 2 \\[2ex] -\dfrac{1}{(n-2)|S^{n-1}|}\dfrac{1}{|x|^{n-2}} & \text{when } n > 2 \end{cases}$$

and let χ_Ω denote the characteristic function of Ω. If $u \in C^\infty(\bar{\Omega})$ then by the Leibniz formula we have

$$\Delta(\chi_\Omega u) = \chi_\Omega f + 2\sum_{j=1}^{n} \frac{\partial u}{\partial x_j}\frac{\partial \chi_\Omega}{\partial x_j} + u(\Delta\chi_\Omega),$$

where $\Delta u = f$. Since both sides of this equation have compact support, we may convolve them with E; thus

$$\chi_\Omega u = E * \chi_\Omega f + 2E * \sum_{j=1}^{n} \frac{\partial u}{\partial x_j}\frac{\partial \chi_\Omega}{\partial x_j} + E * u(\Delta\chi_\Omega). \tag{1}$$

Let us work out concrete expressions for the distributions

$$B = \sum_{j=1}^{n} \frac{\partial u}{\partial x_j}\frac{\partial \chi_\Omega}{\partial x_j} \quad \text{and} \quad C = u(\Delta\chi_\Omega)$$

in the second and third terms on the right-hand side of (1). Extend u to be a C^∞ function on an open set $\mathscr{O} \supset \bar{\Omega}$ and let $\varphi \in C_0^\infty(\mathbb{R}^n)$ be a test function with support in $\mathscr{O}$. Then

$$\langle B, \varphi \rangle = -\sum_{j=1}^{n} \int_\Omega \frac{\partial}{\partial x_j}\left(\varphi \frac{\partial u}{\partial x_j}\right) dx = -\int_{\partial\Omega} \varphi \frac{\partial u}{\partial n}\, d\sigma$$

by the Divergence Theorem in $\mathbb{R}^n$; in other words, B is the measure supported on $\partial\Omega$ with density $-\partial u/\partial n$ with respect to $d\sigma$. Also,

$$\langle C, \varphi\rangle = \int_\Omega \Delta(\varphi u)\, dx = \int_{\partial\Omega} \left(\varphi \frac{\partial u}{\partial n} + u \frac{\partial \varphi}{\partial n}\right) d\sigma,$$

so that $C = -B + C_1$, where $\langle C_1, \varphi\rangle = \int_{\partial\Omega} u\, \partial\varphi/\partial n\, d\sigma$. C_1 is a distribution supported by $\partial\Omega$, called a "double-layer".

We substitute these expressions for B and C into (1), and restrict to Ω to obtained the values of $u(x)$ on Ω. Notice that, when $x \in \Omega$, $E(x - z)$ is a C^∞ function of z in a neighbourhood of $\partial\Omega$; consequently, since B and C are supported on $\partial\Omega$, we may evaluate them (as distributions in the variable z) on this function. Furthermore, E is a locally integrable function so $E * (\chi_\Omega f)$ can be computed by integration over Ω. Thus, if $x \in \Omega$, we obtain

$$u(x) = \int_\Omega E(x-z) f(z)\, dz - \int_{\partial\Omega} E(x-z) \frac{\partial u}{\partial n}(z)\, d\sigma + \int_{\partial\Omega} u(z) \frac{\partial}{\partial n} E(x-z)\, d\sigma.$$

When $\Delta u = f = 0$ we obtain Green's formula

$$u(x) = -\int_{\partial\Omega} E(x - z) \frac{\partial u}{\partial n}(z)\, d\sigma + \int_{\partial\Omega} u(z) \frac{\partial}{\partial n} E(x - z)\, d\sigma, \qquad x \in \Omega, \quad (2)$$

which is the familiar combination of single- and double-layer potentials whereby a harmonic function is represented in terms of its Cauchy data on the boundary.

The formula (2) suggests that we might try to solve the Cauchy problem

$$\Delta u = 0 \quad \text{in } \Omega, \qquad u = g_0 \quad \text{and} \quad \partial u/\partial n = g_1 \text{ on } \partial\Omega \qquad (3)$$

by the formula

$$u(x) = -\int_{\partial\Omega} E(x - z) g_1(z)\, d\sigma + \int_{\partial\Omega} g_0(z) \frac{\partial}{\partial n} E(x - z)\, d\sigma. \qquad (4)$$

This will not work in general, for we know by uniqueness for the Dirichlet problem that the solution of (3) (if it exists) is determined by g_0 alone. The function u defined by (4) will be harmonic in Ω, since $E(x - z)$ and $\partial E(x - z)/\partial n$ are harmonic functions of $x \in \Omega$ when $z \in \partial\Omega$, but it will not have the right boundary values unless g_0 and g_1 are related by a certain pseudo-differential equation on $\partial\Omega$.

Before examining this point in more detail, let us look at a special case: the Laplace equation in a half-space,

$$\Delta u = 0 \quad \text{in } \mathbb{R}^n_+, \qquad u(x', 0) = g(x'), \qquad x' \in \mathbb{R}^{n-1}$$

for $g \in \mathscr{S}(\mathbb{R}^{n-1})$, say. Applying the partial Fourier transform in x' we obtain (with $t = x_n$)

$$-|\xi'|^2 \tilde{u}(\xi', t) + \frac{d^2}{dt^2} \tilde{u}(\xi', t) = 0, \qquad \tilde{u}(\xi', 0) = \tilde{g}(\xi').$$

This is a second-order differential equation in t with one initial condition, but the only solution which is bounded as $t \to +\infty$ is

$$\tilde{u}(\xi', t) = \tilde{g}(\xi')\, e^{-|\xi'|t}.$$

Thus $u(x', t) = K_t * g(x')$, where $\tilde{K}_t(\xi') = e^{-|\xi'|t}$; the function K_t is the Poisson kernel for the Laplace operator in $\mathbb{R}^n_+$. In this context, we can see clearly the relation between u and its normal derivative $-\partial u/\partial t$ on the boundary $\mathbb{R}^{n-1}$. Indeed,

$$-\frac{\partial}{\partial t}\tilde{u}(x', t) = |\xi'|\tilde{g}(\xi')$$

and, since $(\Delta_{x'}g)\tilde{\ }(\xi') = -|\xi'|^2\tilde{g}(\xi')$, we obtain

$$-\frac{\partial}{\partial t}u(x', t)|_{t=0} = (-\Delta_{x'})^{1/2}g.$$

For a general domain Ω, if $\Delta u = 0$ on Ω and $u = g$ on $\partial\Omega$, then it can be shown that

$$\partial u/\partial n = Sg \tag{5}$$

where $S \in OClS^1(\partial\Omega)$ is a pseudo-differential operator of order 1 with principal part the same as that of $(-\Delta_{\partial\Omega})^{1/2}$. Here $\Delta_{\partial\Omega}$ is the Laplace–Beltrami operator on $\partial\Omega$ with respect to the usual Riemannian metric induced from $\mathbb{R}^n$. The existence of the operator S in equation (5) will be evident from the following discussion.

Let $g_0, g_1 \in C^\infty(\partial\Omega)$. As mentioned above, although (4) defines a harmonic function u it may not have the Cauchy data g_0, g_1. To see when this is true we let x approach $\partial\Omega$ and obtain by Theorem 13.12 that

$$g_0 = S_0 g_0 + S_1 g_1 \tag{6}$$

where S_0 and S_1 are p.d.o.'s on $\partial\Omega$ of orders -2 and -1 respectively, because convolution by E is a pseudo-differential operator of order -2 satisfying the transmission condition. Further, the principal part of the asymptotic expansion for E is $|\xi|^{-2}$, so that Theorem 13.12 also gives that the principal part of S_1 is

$$\pi S_1 = \frac{1}{2\pi}\int_+ (\lambda^2 + |\xi'|^2)^{-1}\, d\lambda = i|\xi'|^{-1}.$$

Incidentally, it follows that S_1 is an elliptic operator on $\partial\Omega$ and – as in the proof of Lemma 8.68 – it has zero index.

Conversely, if $n > 2$ then (6) implies that there is a harmonic function u with g_0 and g_1 as its Cauchy data. Indeed, if u is defined by (4) then it follows from (6) that $u = g_0$ on $\partial\Omega$. Now consider Green's formula (2) for the function u; if we subtract it from (4) we obtain

$$\int E(x - z)[g_1(z) - \partial u/\partial n]\, d\sigma(z) = 0, \qquad x \in \Omega.$$

This integral is a continuous function of x which is harmonic on $\mathbb{R}^n\backslash\bar{\Omega}$ and vanishes on $\partial\Omega$ and at infinity (since $n > 2$). Hence it is identically 0 which implies $g_1 = \partial u/\partial n$. To solve the Cauchy problem (3) is, therefore, equivalent to solving the equation (6). In particular, it follows that S_1 is injective, for if $S_1 g_1 = 0$ for some g_1 then $0 = S_0 0 + S_1 g_1$ implies that the Cauchy problem (3) with $g_0 = 0$ has a solution u. But uniqueness for the Dirichlet problem implies that $u \equiv 0$ and hence $g_1 = 0$. Note that since S_1 is injective and has index 0 it must in fact be *invertible* (which amounts to surjectivity of the Dirichlet problem).

Now consider a general boundary operator of transversal order $\leqslant 1$ for the Laplace equation.

$$\Delta u = 0 \quad \text{in } \Omega, \qquad b_0 u + b_1 \partial u/\partial n = h \quad \text{on } \partial\Omega, \tag{7}$$

where b_0 and b_1 are differential (or pseudo-differential) operators in $\partial\Omega$.

Let S_1^{-1} denote the inverse of S_1, which is a pseudo-differential operator. Due to the equivalence of (3) and (6), the Dirichlet problem $\Delta u = 0$ in Ω, $u = g$ on $\partial\Omega$, is solved by taking $g_1 = S_1^{-1}(1 - S_0)g$. (Thus the operator in (5) is $S = S_1^{-1}(1 - S_0)$.) To solve the boundary problem (7), therefore, is equivalent to solving the system of pseudo-differential equations

$$(b_0 + b_1 S_1^{-1}(1 - S_0))g = h.$$

It is not hard to see that this is an elliptic p.d.o. on $\partial\Omega$ exactly when the boundary operator in (7) satisfies the L-condition.

This example is the prototype for the procedure in Chapter 14: knowing that a boundary problem $(\mathscr{A}, \mathscr{B}_1)$ (in this case the Dirichlet problem) satisfies the L-condition then the Fredholm property for $(\mathscr{A}, \mathscr{B}_2)$ (i.e. the problem (7)) can be reduced to the Fredholm property for a p.d.o. on $\partial\Omega$. This was also the idea behind the formula of Agranovič–Dynin in §10.5. It is important to be aware, however, that – in contrast to the case of a single equation – the Dirichlet problem for $p \times p$ elliptic *systems* $(p > 1)$ does not necessarily satisfy the L-condition (see §10.2). Nevertheless, there still exists a sort of "reduction to the boundary" using the Calderón operator which we now wish to define.

Let us look at the problem locally and suppose that $\mathscr{A}$ has constant coefficients in the half-space $\Omega = \mathbb{R}^n_+$. If $u \in C^\infty(\bar{\mathbb{R}}^n_+, \mathbb{C}^p)$ we write the Cauchy data

$$\gamma u = [\gamma_j u]_{j=0}^{l-1}, \qquad \text{where } \gamma_j u = \lim_{x_n \to 0^+} D_n^j u(x', x_n)$$

(and recall that $D_n = i^{-1}\,\partial/\partial x_n$). The elliptic differential operator $\mathscr{A}$ can be written in the form

$$\mathscr{A} = \sum_{j=0}^{l} \mathscr{A}_j D_n^j$$

where $\mathscr{A}_j$ is a differential operator of order $l - j$ on $\partial\Omega = \mathbb{R}^{n-1}$ and $\mathscr{A}_l$ is an invertible matrix. If $u \in C^\infty(\bar{\mathbb{R}}^n_+, \mathbb{C}^p)$, we let $u^0 = \chi_\Omega u$, that is, u^0 denotes the locally integrable function on $\mathbb{R}^n$ which is equal to u when $x_n > 0$ and equal to 0 in the lower half-space $x_n < 0$. Then we obtain in $\mathbb{R}^n$

$$\mathscr{A} u^0 = (\mathscr{A} u)^0 + \mathscr{A}^c \gamma u, \tag{8}$$

where

$$\mathscr{A}^c\mathscr{U} := i^{-1} \sum_{j=0}^{l-1} \sum_{k=0}^{j} \mathscr{A}_{j+1}\mathscr{U}_{j-k} \otimes D_n^k\delta, \qquad \mathscr{U} = [\mathscr{U}_j]_{j=0}^{l-1} \in C^\infty(\mathbb{R}^{n-1}, \mathbb{C}^{pl}),$$

where δ is the Dirac distribution in x_n. Since $\mathscr{A}$ has constant coefficients, it has a fundamental solution $E(x - z)$; applying it to both sides of (8) we obtain the following: if $\mathscr{A}u = 0$ then

$$u = E\mathscr{A}^c\gamma u \qquad \text{when } x_n > 0.$$

This is the desired generalization of the Green's formula (2). Now, by applying the Cauchy data operator γ, we obtain

$$\mathscr{U} = P\mathscr{U}, \qquad \text{where } P = \gamma E\mathscr{A}^c, \tag{9}$$

and $\mathscr{U} = \gamma u$ denotes the Cauchy data. The operator P is a pseudo-differential operator on $\mathbb{R}^{n-1}$, called the *Calderón operator*. (Since $\mathscr{A}$ has constant coefficients it is not hard to show that P is a projector, with image being the space of Cauchy data of the solutions of the homogeneous equation $\mathscr{A}u = 0$. The principal symbol of P is equal to the Calderón projector P^+ which was introduced in §2.2 for the matrix polynomial $L(\lambda) = \pi\mathscr{A}(y, \xi', \lambda))$ with respect to the eigenvalues in the upper half-plane $\operatorname{Im} \lambda > 0$). Now let $\mathscr{B}$ be a boundary operator,

$$\mathscr{B} = \sum_{j=0}^{l-1} \mathscr{B}_j D_n^j.$$

The equation $\mathscr{B}u = g$ can be written in the form $\mathscr{B}^c\gamma u = g$ where $\mathscr{B}^c = [\mathscr{B}_0 \cdots \mathscr{B}_{l-1}]$. By applying the operator γE once again to equation (8), it is clear that the boundary problem

$$\mathscr{A}u = f \quad \text{when } x_n > 0, \qquad \mathscr{B}u = g \quad \text{when } x_n = 0$$

is equivalent to a system of p.d.o.'s on $\mathbb{R}^{n-1}$ of the form

$$\mathscr{U} = v + P\mathscr{U}, \qquad \mathscr{B}^c\mathscr{U} = g, \tag{10}$$

where $v = \gamma E f^0$ and $\mathscr{U} = \gamma u$. This might seem like too many equations, for instance, the Laplace equation with Dirichlet boundary conditions leads to three equations for two unknowns. But, as it turns out, v is not arbitrary since $Pv = 0$ (see Chapter 12 (13)).

In this chapter we treat in detail the general case when $\mathscr{A}$ and $\mathscr{B}$ have variable coefficients and Ω is any bounded domain with C^∞ boundary. Roughly speaking, the same kind of reduction to the boundary indicated by (10) is still valid, modulo operators of order $-\infty$.

Some of the specific results that will be proved in this chapter are as follows. Let $(\mathscr{A}, \mathscr{B}) \in BE^{l,\mathbf{m}}$. It was shown in Chapter 9 that the operator

$$\mathfrak{L}\colon W_2^s(\Omega, \mathbb{C}^p) \to W_2^{s-l}(\Omega, \mathbb{C}^p) \times \mathop{\times}_{k=1}^{r} W_2^{s-m_k-1/2}(\partial\Omega), \qquad s \geqslant l,$$

defined by $\mathfrak{L}u = (\mathscr{A}u, \mathscr{B}u)$, has a smoothing regularizer, i.e. an operator which is a right and left inverse of $\mathfrak{L}$ modulo operators of order -1. In this chapter we prove a stronger property: there exists a *parametrix* for $\mathfrak{L}$, i.e. an operator

$$\mathfrak{P}: W_2^{s-l}(\Omega, \mathbb{C}^p) \times \mathop{\times}_{k=1}^{r} W_2^{s-m_k-1/2}(\partial\Omega) \to W_2^s(\Omega, \mathbb{C}^p)$$

which is continuous for every $s \geqslant l$ and is a right and left inverse of $\mathfrak{L}$, modulo operators of order $-\infty$. The essential elements in the construction of this parametrix are the Green formula (35) which expresses the solutions of $\mathscr{A}u = f$ as a sum of multiple-layer potentials and the Calderón operator (36). We will then use this parametrix to show that the image of $\mathfrak{L}$ is defined by C^∞ relations, i.e. it is the orthogonal space of a (finite dimensional) subspace of $C^\infty(\bar{\Omega}, \mathbb{C}^p) \times \mathop{\times}_{k=1}^{r} C^\infty(\partial\Omega)$. In the first two sections, §§14.1 and 14.2, we study some spaces of distributions in $\mathbb{R}^n_+$ which are needed later on in the chapter. The Calderón operator is defined in §14.3 and the construction of the parametrix $\mathfrak{P}$ is given in §14.4 when $\mathscr{A}$ is a differential operator, and then extended later on in §14.6 to the wider class of elliptic boundary value problems that was introduced in §10.7. In §14.5 we apply the formula for the parametrix $\mathfrak{P}$ to obtain an interesting result on the index.

14.1 Some spaces of distributions on $\mathbb{R}^n_+$

The following notation is used throughout this section. Let $\mathbb{R}^n_+$ denote the open half-space, $x_n > 0$, and $\bar{\mathbb{R}}^n_+$ the closure, $x_n \geqslant 0$. If F is a space of distributions in $\mathbb{R}^n$, we let

$$F(\mathbb{R}^n_+) \tag{11}$$

denote the space of restrictions to $\mathbb{R}^n_+$ of elements in F, and

$$\mathring{F}(\bar{\mathbb{R}}^n_+)$$

the set of distributions in F supported by $\bar{\mathbb{R}}^n_+$ (similarly for $\mathring{F}(\bar{\mathbb{R}}^n_+)$ of course). We use this notation for the spaces $F = \mathscr{S}, \mathscr{S}', W_2^s$, and so on. An exception to the notation is made for the cases $F = \mathscr{D}'$ and $F = C_0^\infty$. $\mathscr{D}'(\mathbb{R}^n_+)$ already has a meaning as the space of distributions in $\mathbb{R}^n_+$, so instead we let $\bar{\mathscr{D}}(\mathbb{R}^n_+)$ denote the space (11) of extendible distributions. Similarly, $C_0^\infty(\mathbb{R}^n_+)$ already denotes the set of C^∞ functions with compact support in $\mathbb{R}^n_+$, so we let $\bar{C}_0^\infty(\mathbb{R}^n_+)$ or $C_{(0)}^\infty(\mathbb{R}^n_+)$ denote the space (11) of restrictions to $\mathbb{R}^n_+$ of functions in $C_0^\infty(\mathbb{R}^n)$.

Remark $\mathring{F}(\bar{\mathbb{R}}^n_+)$ is a subspace of $\mathscr{D}'(\mathbb{R}^n)$, while $F(\mathbb{R}^n_+)$ is a subspace of $\mathscr{D}'(\mathbb{R}^n_+)$ which may be identified with the quotient space $F(\mathbb{R}^n_+) = F/\mathring{F}(\bar{\mathbb{R}}^n_-)$.

Recall that $W_2^s = W_2^s(\mathbb{R}^n)$ is the set of distributions $u \in \mathscr{S}'(\mathbb{R}^n)$ such that the Fourier transform $\hat{u}$ is locally integrable and

$$\|u\|_s^2 = \int |\hat{u}(\xi)|^2(1 + |\xi|^2)^s \, d\xi < \infty. \tag{12}$$

As we know, $C_0^\infty(\mathbb{R}^n)$ is a dense subspace in the Hilbert space $W_2^s(\mathbb{R}^n)$ and the spaces $W_2^s(\mathbb{R}^n)$ and $W_2^{-s}(\mathbb{R}^n)$ are dual with respect to an extension of the sesquilinear form

$$(u, v)_0 = \int u\bar{v}\, dx, \qquad u, v \in C_0^\infty(\mathbb{R}^n).$$

For all $s \in \mathbb{R}$, we have $\mathscr{S} \subset W_2^s \subset \mathscr{S}'$, the inclusions being continuous.

Let $W_2^s(\mathbb{R}_+^n)$ denote the space of restrictions to $\mathbb{R}_+^n$ of elements of $W_2^s(\mathbb{R}^n)$, with norm

$$\|u\|_s = \inf \|U\|_s, \tag{13}$$

the infimum being taken over all $U \in W_2^s$ which are equal to u on $\mathbb{R}_+^n$. When $s \geqslant 0$, this definition of $W_2^s(\mathbb{R}_+^n)$ agrees with the definition in the appendix to Chapter 7, and (13) is an equivalent norm, due to the existence of a continuous extension operator, Theorem 7A.7.

Also, let $\mathring{W}_2^s(\bar{\mathbb{R}}_+^n)$ denote the closed subspace of $W_2^s(\mathbb{R}^n)$ consisting of elements supported by $\bar{\mathbb{R}}_+^n$.

Our first aim is to show that $C_0^\infty(\mathbb{R}_+^n)$ *is dense in* $\mathring{W}_2^s(\bar{\mathbb{R}}_+^n)$. (Thus the definition of $\mathring{W}_2^s$ is equivalent to the one given in the appendix to Chapter 7.) The following lemma and corollary are needed for the proof.

Lemma 14.1 *For* $\varphi, u \in \mathscr{S}$

$$\|\varphi \cdot u\|_s \leqslant \|\varphi\|_{1,|s|} \|u\|_s \tag{14}$$

where $\|\varphi\|_{1,|s|} = \int |\hat{\varphi}(\xi)|(1 + |\xi|)^{|s|}\, d\xi$.

Proof Since $(\varphi u)\hat{} = \hat{\varphi} * \hat{u}$, we have

$$\|\varphi u\|_s = \left[\int (1 + |\xi|^2)^s \left|\int \hat{\varphi}(\xi - \eta)\hat{u}(\eta)\, d\eta\right|^2 d\xi\right]^{1/2}.$$

But, as easily seen,

$$(1 + |\xi|^2)^{s/2} \leqslant (1 + |\xi - \eta|)^{|s|}(1 + |\eta|^2)^{s/2}, \qquad s \in \mathbb{R},$$

and, therefore,

$$\|\varphi u\|_s \leqslant \left[\int \left|\int (1 + |\xi - \eta|)^{|s|}|\hat{\varphi}(\xi - \eta)|\,|\hat{u}(\eta)|(1 + |\eta|^2)^{s/2}\, d\eta\right|^2 d\xi\right]^{1/2},$$

or, in other words, $\|\varphi u\|_s$ is bounded by the L_2 norm (in ξ) of the function $\int K(\xi, \eta)v(\eta)\, d\eta$, where $v(\eta) = |\hat{u}(\eta)|(1 + |\eta|^2)^{s/2}$ and

$$K(\xi, \eta) = (1 + |\xi - \eta|)^{|s|} \cdot |\hat{\varphi}(\xi - \eta)|$$

Since $\int K(\xi, \eta)\, d\xi$ and $\int K(\xi, \eta)\, d\eta$ are bounded by $\|\varphi\|_{1,|s|}$, we see that (14) follows by Schur's Lemma 7.26 applied to $\int K(\xi, \eta)v(\eta)\, d\eta$.

Corollary 14.2 *Choose* $\varphi \in C_0^\infty(\mathbb{R}^n)$ such that $\varphi(0) = 1$, *and set* $\varphi_\varepsilon(x) = \varphi(\varepsilon x)$. *Then for all* $u \in W_2^s$ *we have*

$$\|\varphi_\varepsilon \cdot u - u\|_s \to 0 \qquad \textit{as } \varepsilon \to 0. \tag{15}$$

Proof The norm

$$\|\varphi_\varepsilon\|_{1,|s|} = \int \varepsilon^{-n}|\hat{\varphi}(\xi/\varepsilon)|(1 + |\xi|)^{|s|}\, d\xi$$

$$= \int |\hat{\varphi}(\xi)|(1 + |\varepsilon\xi|)^{|s|}\, d\xi$$

is bounded by a constant independent of ε for $0 < \varepsilon \leqslant 1$. By Lemma 14.1 the mappings $u \mapsto \varphi_\varepsilon u$ form an equicontinuous set of linear mappings; it suffices, therefore, to verify (15) on the dense subset $\mathscr{S} \subset W_2^s$. So we may assume that $u \in \mathscr{S}$. Then it is not hard to show that $\varphi_\varepsilon \to u$ in $\mathscr{S}$; for the details see [Don, p. 135]. By the Fourier transformation (which is an isomorphism on $\mathscr{S}$) we also have $(\varphi_\varepsilon u)^\wedge \to \hat{u}$ in $\mathscr{S}$, and a priori the convergence holds pointwise. Hence (15) follows from the dominated convergence theorem.

In the proof of Theorem 14.3, it is also necessary to use the convolution of a distribution $T \in \mathscr{D}'$ with a test function $\varphi \in C_0^\infty$ defined by

$$(T * \varphi)(x) = \langle T, \varphi(x - \cdot)\rangle,$$

where the notation on the right side means that T operates on $\varphi(x - y)$ as a function of y. It can be shown that $T * \varphi \in C^\infty$, see [Hö 1] or [Wl]. Now, since the support of $u(x - \cdot)$ is $x - \operatorname{supp} u$, we have

$$\operatorname{supp}(T * \varphi) \subset \operatorname{supp} T + \operatorname{supp} \varphi,$$

because $T * \varphi(x) = 0$ unless $(x - \operatorname{supp} \varphi) \cap \operatorname{supp} T \neq \emptyset$, i.e. $x \in \operatorname{supp} T + \operatorname{supp} \varphi$. (Note that $\operatorname{supp} T + \operatorname{supp} \varphi$ is closed due to the compactness of $\operatorname{supp} \varphi$). If $T \in \mathscr{E}'$, i.e. a distribution with compact support, then we have $T * \varphi \in C_0^\infty$.

Theorem 14.3 *$C_0^\infty(\mathbb{R}^n_+)$ is dense in $\mathring{W}_2^s(\bar{\mathbb{R}}^n_+)$.*

Proof By Corollary 14.2, $\mathscr{E}' \cap \mathring{W}_2^s(\bar{\mathbb{R}}^n_+)$ is dense in $\mathring{W}_2^s(\bar{\mathbb{R}}^n_+)$. Thus, it suffices to show that $C_0^\infty(\mathbb{R}^n_+)$ is dense in $\mathscr{E}' \cap \mathring{W}_2^s(\bar{\mathbb{R}}^n_+)$. Let $u \in \mathscr{E}' \cap \mathring{W}_2^s(\bar{\mathbb{R}}^n_+)$. Then we can approximate u by an element in $C_0^\infty(\mathbb{R}^n_+)$ using the usual regularization technique. Choose $\varphi \in C_0^\infty$ with support in $\mathbb{R}^n_+$ such that $\int \varphi\, dx = 1$, and set $\varphi_\varepsilon(x) = \varepsilon^{-n}\varphi(x/\varepsilon)$, $\varepsilon > 0$. Then $u * \varphi_\varepsilon \in C_0^\infty$ has support in $\mathbb{R}^n_+$ and its Fourier transform is $\hat{u}(\xi)\hat{\varphi}(\varepsilon\xi)$. For the norm (12) we have

$$\|u * \varphi_\varepsilon - u\|_s^2 = \int |\hat{\varphi}(\varepsilon\xi) - 1|^2 |\hat{u}(\xi)|^2 (1 + |\xi|^2)^s\, d\xi,$$

which approaches 0 as $\varepsilon \to 0$ by dominated convergence since $\hat{\varphi}(\varepsilon\xi) \to \hat{\varphi}(0) = 1$ and $|\hat{\varphi}(\varepsilon\xi) - 1| \leqslant 2$.

It follows from Theorem 14.3 that a continuous linear form on $\mathring{W}_2^{-s}(\bar{\mathbb{R}}^n_+)$ is uniquely determined by its restriction to $C_0^\infty(\mathbb{R}^n_+)$, so we may identify it with an element of $\mathscr{D}'(\mathbb{R}^n_+)$. The following theorem characterizes the dual space of $\mathring{W}_2^{-s}(\bar{\mathbb{R}}^n_+)$.

Theorem 14.4 *For any $s \in \mathbb{R}$, the space $C^{\infty}_{(0)}(\mathbb{R}^n_+)$ is dense in $W^s_2(\mathbb{R}^n_+)$, and $C^{\infty}_0(\mathbb{R}^n_+)$ is dense in $\mathring{W}^{-s}_2(\bar{\mathbb{R}}^n_+)$. The spaces $W^s_2(\mathbb{R}^n_+)$ and $\mathring{W}^{-s}_2(\bar{\mathbb{R}}^n_+)$ are dual with respect to an extension of the sesquilinear form*

$$(u, v)_0 = \int u\bar{v}\, dx, \qquad u \in C^{\infty}_{(0)}(\mathbb{R}^n_+), v \in C^{\infty}_0(\mathbb{R}^n_+).$$

Proof Since $C^{\infty}_0(\mathbb{R}^n)$ is dense in $W^s_2(\mathbb{R}^n)$, it is clear that $C^{\infty}_{(0)}(\mathbb{R}^n_+)$ is dense in $W^s_2(\mathbb{R}^n_+)$. The fact that $C^{\infty}_0(\mathbb{R}^n_+)$ is dense in $\mathring{W}^{-s}_2(\bar{\mathbb{R}}^n_+)$ was proved in Theorem 14.3.

Now, if u is the restriction to $\mathbb{R}^n_+$ of a function $U \in C^{\infty}_0(\mathbb{R}^n)$, and we regard $v \in C^{\infty}_0(\mathbb{R}^n_+)$ as a C^{∞} function on $\mathbb{R}^n$ with $v = 0$ for $x_n \leqslant 0$ then by Parseval's theorem and the Cauchy–Schwarz inequality:

$$\left|\int u\bar{v}\, dx\right|^2 = \left|\int U\bar{v}\, dx\right|^2$$
$$= \int |\hat{U}(\xi)|^2(1 + |\xi|^2)^s\, d\xi \cdot \int |\hat{v}(\xi)|^2(1 + |\xi|^2)^{-s}\, d\xi$$

and by taking the infimum over all $U \in C^{\infty}_0(\mathbb{R}^n)$ with $U = u$ in $\mathbb{R}^n_+$ we obtain

$$\left|\int u\bar{v}\, dx\right| \leqslant \|u\|_s \cdot \|v\|_{-s}.$$

It follows that the bilinear form (u, v_0) extends by continuity such that

$$|(u, v)_0| \leqslant (2\pi)^{-n}\|u\|_s \cdot \|v\|_{-s}$$

for all $u \in W^s_2(\mathbb{R}^n_+)$ and $v \in \mathring{W}^{-s}_2(\bar{\mathbb{R}}^n_+)$. Since $W^s_2(\mathbb{R}^n_+)$ can be identified with the quotient of W^s_2 by the closed subspace $\mathring{W}^s_2(\bar{\mathbb{R}}^n_+)$, then its dual space is the annihilator of $\mathring{W}^s_2(\bar{\mathbb{R}}^n_-)$ in $(W^s_2)' = W^{-s}_2$, i.e. the set of $u \in W^{-s}_2$ such that

$$(u, v)_0 = 0 \qquad \text{for all } v \in \mathring{W}^s_2(\bar{\mathbb{R}}^n_-).$$

But since $C^{\infty}_0(\mathbb{R}^n_-)$ is dense in $\mathring{W}^s_2(\bar{\mathbb{R}}^n_-)$, this is precisely the set of all $u \in W^{-s}_2$ such that

$$\langle u, \bar{v}\rangle = 0 \qquad \text{for all } v \in C^{\infty}_0(\mathbb{R}^n_-),$$

that is, $u \in \mathring{W}^{-s}_2(\bar{\mathbb{R}}^n_+)$.

The convolution of a tempered distribution $T \in \mathcal{S}'$ and a function $u \in \mathcal{S}$, is defined by

$$(T * u)(x) = \langle T, u(x - \cdot)\rangle,$$

where the notation on the right side again means that T operates on $u(x - y)$ as a function of y. The convolution is well-defined, since $u(x - \cdot) \in \mathcal{S}$ for every $x \in \mathbb{R}^n$. Some of the properties of the convolution were given as exercises at the end of Chapter 7. For further details, see [Ru 2] and [Es].

Lemma 14.5 *Let $T \in \mathscr{S}'$ and $u \in \mathscr{S}$. Then $T * u \in C^\infty(\mathbb{R}^n)$ and there exists a number N such that $|D^\alpha(T * u)(x)| \leqslant C_\alpha(1 + |x|)^N$ for all α.*

In particular, $T * u \in \mathscr{S}'$, i.e. a tempered distribution, and one can then show that $\langle T * u, \varphi\rangle = \langle T, \check{u} * \varphi\rangle$, $\varphi \in \mathscr{S}$, where $\check{u}(x) = u(-x)$. It follows that

$$(T * u)\hat{} = \hat{u}\hat{T},$$

since $\langle T * u, \hat{\varphi}\rangle = \langle T, \check{u} * \hat{\varphi}\rangle = \langle T, (\hat{u}\varphi)\hat{}\,\rangle = \langle \hat{T}, \hat{u}\varphi\rangle$ for all $\varphi \in \mathscr{S}$. (This result was already used in the proof of Theorem 14.3.)

Before continuing on, let us take another look at Theorem 13.1. Let $\lambda > 0$. Then $(\xi_n - i\lambda)^{-1}$ is the Fourier transform with respect to x_n of the function which is equal to $i\, e^{-\lambda x_n}$ when $x_n > 0$ and vanishes when $x_n < 0$. By differentiating with respect to λ we obtain

$$(\xi_n - i\lambda)^{-k} = i^k \int_0^\infty e^{-ix_n\xi_n}\, e^{-\lambda x_n}(x_n)^{k-1}\, dx_n/(k-1)! \qquad k = 1, 2, \ldots.$$

Similarly, $(\xi_n + i\lambda)^{-1}$ is the Fourier transform with respect to x_n of the function which is equal to $-i\, e^{\lambda x_n}$ when $x_n < 0$ and vanishes when $x_n > 0$, and we obtain

$$(\xi_n + i\lambda)^{-k} = i^{-k} \int_{-\infty}^0 e^{-ix_n\xi_n}\, e^{\lambda x_n}(x_n)^{k-1}\, dx_n/(k-1)! \qquad k = 1, 2, \ldots.$$

In fact, it is clear from the proof of Theorem 13.1 that the following lemma holds.

Lemma 14.6 *Let $\lambda > 0$. For any $s \in \mathbb{C}$,*

$$(\xi_n \pm i\lambda)^s = e^{\pm i\pi s/2}\, \mathfrak{F}_n(e^{\pm\lambda x_n}\, \chi_\mp^{-s-1}(x_n)), \tag{16}$$

where $\mathfrak{F}_n$ denotes the Fourier transform in the x_n variable.

Now set $\Lambda_\pm = \xi_n \pm i\lambda$, where $\lambda = \langle\xi'\rangle$, i.e.

$$\Lambda_\pm = \xi_n \pm i\langle\xi'\rangle,$$

where $\langle\xi'\rangle = (1 + |\xi'|^2)^{1/2}$. Let $s \in \mathbb{R}$. Then $\Lambda_\pm(D)^s\colon \mathscr{S} \to \mathscr{S}$ is the pseudo-differential operator

$$\Lambda_\pm(D)^s u = \mathfrak{F}^{-1}((\xi_n \pm i\langle\xi'\rangle)^s\hat{u}(\xi)), \qquad u \in \mathscr{S}, \tag{17}$$

which we often express in the form $\Lambda_\pm(D)^s = (D_n \pm i\langle D'\rangle)^s$. Here $\mathfrak{F}$ is the Fourier transform on $\mathbb{R}^n$ and is an isomorphism $\mathfrak{F}\colon \mathscr{S} \to \mathscr{S}$ and $\mathfrak{F}\colon \mathscr{S}' \to \mathscr{S}'$.

Note that $\Lambda_\pm(D)^s\colon \mathscr{S} \to \mathscr{S}$ is a continuous map and it has the inverse $\Lambda_\pm(D)^{-s}$, so it is an isomorphism on $\mathscr{S}$. The equation (17) also defines an operator $\Lambda_\pm(D)^s\colon \mathscr{S}' \to \mathscr{S}'$, where $(\xi_n \pm i\langle\xi'\rangle)^s\hat{u}$ is the distribution in $\mathscr{S}'$ defined by (18).

Remark Recall that if $f \in C^\infty$ is slowly increasing in the sense that an estimate of the form $|D^\alpha f(x)| \leqslant C_\alpha(1 + |x|)^{N_\alpha}$ holds for all α then the product

$f \cdot T$ is defined for all $T \in \mathscr{S}'$ by

$$\langle fT, \varphi \rangle = \langle T, f\varphi \rangle \qquad \forall \varphi \in \mathscr{S}. \tag{18}$$

Let $\mathfrak{F}'$ denote the partial Fourier transform in the variables $x' = (x_1, \ldots, x_{n-1})$ and $\mathfrak{F}_n$ the Fourier transform in the x_n variable, i.e.

$$\mathfrak{F}': u \mapsto \int e^{-i(x', \xi')} u(x', x_n)\, dx' =: \tilde{u}(\xi', x_n)$$

$$\mathfrak{F}_n: u \mapsto \int e^{-ix_n \xi_n} u(x', x_n)\, dx_n.$$

Since $\mathfrak{F} = \mathfrak{F}_n \circ \mathfrak{F}'$, it follows by Lemma 14.6 that

$$\Lambda_\pm(D)^s u = e^{\pm i\pi s/2}\, \mathfrak{F}'^{-1}((e^{\pm\langle \xi' \rangle x_n} \chi_\mp^{-s-1}(x_n)) * \tilde{u}(\xi', x_n)), \qquad u \in \mathscr{S}, \tag{19}$$

where $\tilde{u} = \mathfrak{F}'u$, the convolution being taken with respect to the x_n variable.

In the next lemma, we let $\mathring{\mathscr{S}}(\bar{\mathbb{R}}^n_+)$ be the set of functions $f \in \mathscr{S}$ with support in $\bar{\mathbb{R}}^n_+$, and $\mathring{\mathscr{S}}'(\bar{\mathbb{R}}^n_+)$ the set of distributions $u \in \mathscr{S}'$ with support in $\bar{\mathbb{R}}^n_+$.

Lemma 14.7 *The operator $\Lambda_-(D)^s = (D_n - i\langle D' \rangle)^s: \mathscr{S} \to \mathscr{S}$ is an isomorphism and extends to an isomorphism $\Lambda_-(D)^s: \mathscr{S}' \to \mathscr{S}'$. Further, we have isomorphisms*

$$\Lambda_-(D)^s: \mathring{\mathscr{S}}(\bar{\mathbb{R}}^n_+) \to \mathring{\mathscr{S}}(\bar{\mathbb{R}}^n_+) \qquad \textit{and} \qquad \Lambda_-(D)^s: \mathring{\mathscr{S}}'(\bar{\mathbb{R}}^n_+) \to \mathring{\mathscr{S}}'(\bar{\mathbb{R}}^n_+).$$

Proof The fact that $\Lambda_-(D)^s$ is an isomorphism on $\mathscr{S}$ and $\mathscr{S}'$ is obvious from the remarks above. We could also extend $\Lambda_-(D)^s$ from $\mathscr{S}$ to $\mathscr{S}'$ by duality. Plancherel's theorem gives us

$$\begin{aligned}(\Lambda_-(D)^s u, v_0) &= (2\pi)^{-n}(\Lambda_- \hat{u}, \hat{v})_0 \\ &= (2\pi)^{-n}(\hat{u}, \Lambda_+ \hat{v})_0, \qquad \text{since } \bar{\Lambda}_- = \Lambda_+, \\ &= (u, \Lambda_+(D)^s v)_0\end{aligned}$$

for all $u, v \in \mathscr{S}$, where $(\ ,\ )_0$ denotes the L_2 inner product. Then $\Lambda_-(D)^s$ extends to a continuous map $\Lambda_-(D)^s: \mathscr{S}' \to \mathscr{S}'$ defined by

$$\langle \Lambda_-(D)^s u, \bar{v} \rangle = \langle u, \overline{\Lambda_+(D)^s v} \rangle, \qquad v \in \mathscr{S}.$$

Once again this map is an isomorphism because it has the inverse $\Lambda_-(D)^{-s}$.

To complete the proof for the spaces $\mathring{\mathscr{S}}(\bar{\mathbb{R}}^n_+)$ and $\mathring{\mathscr{S}}'(\bar{\mathbb{R}}^n_+)$, we just have to show that if $u \in \mathscr{S}'$ has support in $\bar{\mathbb{R}}^n_+$ then so does $\Lambda_-(D)^s u$. When $u \in \mathscr{S}$ this follows immediately from (19) because the support property of convolutions implies that

$$\operatorname{supp} \Lambda_-(D)^s u \subset \bar{\mathbb{R}}^n_+ + \bar{\mathbb{R}}^n_+ = \bar{\mathbb{R}}^n_+. \tag{20}$$

If $u \in \mathscr{S}^{\circ\prime}(\bar{\mathbb{R}}^n_+) \subset \mathscr{S}'$ we argue by duality. Note that if $v \in C_0^\infty(\bar{\mathbb{R}}^n_-)$ then $\operatorname{supp} \Lambda_+(D)^s v \subset \{x; x_n \leqslant -\varepsilon\}$ by virtue of (19) since $\operatorname{supp} v \subset \{x; x_n \leqslant -\varepsilon\}$ for some $\varepsilon > 0$. Hence

$$(\Lambda_-(D)^s u, v)_0 = (u, \Lambda_+(D)^s v)_0 = 0, \qquad \text{for all } v \in C_0^\infty(\bar{\mathbb{R}}^n_-),$$

because $\operatorname{supp} u \subset \bar{\mathbb{R}}^n_+$ and $\Lambda_+(D)^s v = 0$ in a neighbourhood of $\operatorname{supp} u$. By definition of the support of a distribution, this means that $\operatorname{supp} \Lambda_-(D)^s u \subset \bar{\mathbb{R}}^n_+$.

It is evident that

$$\Lambda_\pm(D)^s\colon W_2^l(\mathbb{R}^n) \to W_2^{l-s}(\mathbb{R}^n), \qquad \text{for all } l \in \mathbb{R},$$

is an isometric isomorphism (with inverse $\Lambda_\pm(D)^{-s}$) since

$$|\Lambda_\pm| = (\xi_n^2 + \langle\xi'\rangle^2)^{1/2} = (1 + |\xi|^2)^{1/2}.$$

It also follows that $\Lambda_-(D)^s$ is an isometric isomorphism

$$\Lambda_-(D)^s\colon \mathring{W}_2^l(\bar{\mathbb{R}}^n_+) \to \mathring{W}_2^{l-s}(\bar{\mathbb{R}}^n_+), \qquad \forall l. \tag{21}$$

In the following theorem, we work with the adjoint of $\Lambda_-(D)^s$, i.e. $\Lambda_+(D)^s$. As an aside, note that if $u \in \mathscr{S}$ then the values of $\Lambda_+(D)^s u$ in $x_n > 0$ depend only on the values in $x_n > 0$ by virtue of (19) since χ_- has support in $x_n \leqslant 0$.

Theorem 14.8 *The isomorphism* $\Lambda_-(D)^s = (D_n - i\langle D'\rangle)^s\colon \mathring{\mathscr{S}}(\bar{\mathbb{R}}^n_+) \to \mathring{\mathscr{S}}(\bar{\mathbb{R}}^n_+)$ *in the preceding Lemma extends by continuity to an isometric isomorphism* $\Lambda_-(D)^s\colon L_2(\mathbb{R}^n_+) \to \mathring{W}_2^{-s}(\bar{\mathbb{R}}^n_+)$. *The adjoint,* $\Lambda_+(D)^s$, extends by continuity from $C^\infty_{(0)}(\mathbb{R}^n_+)$ *to an isometric isomorphism*

$$\Lambda_+(D)^s = (D_n + i\langle D'\rangle)^s\colon W_2^s(\mathbb{R}^n_+) \to L_2(\mathbb{R}^n_+).$$

Thus, for all $u \in W_2^s(\mathbb{R}^n_+)$ *we have* $\|u\|_s = \|\Lambda_+(D)^s u\|_{L^2(\mathbb{R}^n_+)}$, *and*

$$W_2^s(\mathbb{R}^n_+) = \{\imath^+ U \mid U \in \mathscr{S}' \text{ such that } \imath^+\Lambda_+(D)^s U \in L_2(\mathbb{R}^n_+)\}.$$

Here $\imath^+$ *denotes the restriction of distrubitions to* $\mathbb{R}^n_+$.

Proof The first statement follows from (21) with $l = 0$. To compute the adjoint of $\Lambda_-(D)^s\colon L_2(\mathbb{R}^n_+) \to \mathring{W}_2^{-s}(\bar{\mathbb{R}}^n_+)$, we use Theorem 14.4. Let $v \in C_0^\infty(\mathbb{R}^n_+)$ and let u be the restriction to $\mathbb{R}^n_+$ of a function $U \in C_0^\infty(\mathbb{R}^n)$. Then as in the proof of Lemma 14.7

$$(U, \Lambda_-(D)^s v)_0 = (\Lambda_+(D)^s U, v)_0$$

and since the supports of v and $\Lambda_-(D)^s v$ lie in $\bar{\mathbb{R}}^n_+$ we have

$$(\imath^+ U, \Lambda_-(D)^s v)_0 = (\imath^+\Lambda_+(D)^s U, v)_0.$$

This implies that $\imath^+\Lambda_+(D)^s U$ depends only on $u = \imath^+ U$. Further, by Theorem 14.4 the equation holds for all $U \in W_2^s(\mathbb{R}^n)$ and $v \in \mathring{W}_2^{-s}(\bar{\mathbb{R}}^n_+)$, where $(\ ,\ )_0$ now denotes the duality bracket between $W_2^s(\mathbb{R}^n)$ and $\mathring{W}_2^{-s}(\bar{\mathbb{R}}^n_+)$. Thus,

$$(\Lambda_-(D)^s)^* \imath^+ U = \imath^+\Lambda_+(D)^s U, \qquad U \in W_2^s(\mathbb{R}^n).$$

For any $u \in W_2^s(\mathbb{R}^n_+)$, we define $\Lambda_+(D)^s u := \imath^+\Lambda_+(D)^s U$, where $U \in W_2^s(\mathbb{R}^n)$ is any extension of u, and it follows that $(\Lambda_-(D)^s)^* u = \Lambda_+(D)^s u$, i.e.

$$(\Lambda_-(D)^s)^* = \Lambda_+(D)^s.$$

Since $\Lambda_-(D)^s$ is an isometric isomorphism, the same is true of the adjoint. The proof is complete.

For any function $v \in L_2(\mathbb{R}^n_+)$, we define $v^0 \in L_2(\mathbb{R}^n)$ by

$$v^0 = v \quad \text{in } \mathbb{R}^n_+, \qquad v^0 = 0 \quad \text{in } \mathbb{R}^n \backslash \mathbb{R}^n_+.$$

This leads to the following corollary.

Corollary 14.9 *For every $u \in W^s_2(\mathbb{R}^n_+)$ there exists $U \in W^s_2(\mathbb{R}^n)$ such that $U = u$ in $\mathbb{R}^n_+$ and $\|U\|_s = \|u\|_s$.*

Proof With $v = \Lambda_+(D)^s u$, we set $U = \Lambda_+(D)^{-s} v^0 \in W^s_2(\mathbb{R}^n)$. Then $\Lambda_+(D)^s U = v^0$ is an extension of $\Lambda_+(D)^s u$, and by definition we have

$$\Lambda_+(D)^s \imath^+ U = \imath^+ \Lambda_+(D)^s U = \Lambda_+(D)^s u.$$

Since $\Lambda_+(D)^s$: $W^s_2(\mathbb{R}^n_+) \to L_2(\mathbb{R}^n_+)$ is an isomorphism by Theorem 14.8, it follows that $\imath^+ U = u$, i.e. U is an extension of u. Moreover,

$$\|U\|_s = \|\Lambda_+(D)^s U\|_0 = \|v^0\|_0 = \|\Lambda_+(D)^s u\|_0 = \|u\|_s,$$

where the last equality again follows from Theorem 14.8.

Remark Recall that $\|u\|_s$ in Corollary 14.7 refers to the norm (3), rather than the (equivalent) norm given in the appendix to Chapter 7.

Let us consider the case when s is an integer.

Proposition 14.10 *When s is a nonnegative integer, $s = k$, $k = 0, 1, 2, \ldots$, then*

$$\Lambda_+(D)^k = i^k \sum_{j=0}^{k} \binom{k}{j} (1 + |D'|^2)^{(k-j)/2} (-1)^j \frac{\partial^j}{\partial x_n^j}.$$

When s is a negative integer, $s = -k$, $k = 1, 2, \ldots$, then

$$\Lambda_+(D)^{-k} u = i^{-k} \int_0^\infty e^{-t\langle D' \rangle} t^{k-1} u(x', x_n + t)\, dt/(k-1)!, \qquad u \in \mathscr{S}. \quad (22)$$

Proof The first statement is obvious. Now, when $s = -k$ is a negative integer, we have

$$\begin{aligned}\Lambda_+(D)^{-k} u(x) &= (2\pi)^{-n} \int e^{i(x,\xi)} (\xi_n + i\langle \xi' \rangle)^{-k} \hat{u}(\xi)\, d\xi \\ &= (2\pi)^{1-n} \int e^{i(x',\xi')} \tilde{K}(\xi', x_n) * \tilde{u}(\xi', x_n)\, d\xi',\end{aligned}$$

where $\tilde{K} = \mathfrak{F}_n^{-1}((\xi_n + i\langle \xi' \rangle)^s)$, and $\tilde{u} = \mathfrak{F}_n^{-1} \hat{u}$ are inverse Fourier transforms in the ξ_n variable. Since $\tilde{K}(\xi', x_n) = i^{-k} e^{x_n\langle\xi'\rangle} (x_{n-})^{k-1}/(k-1)!$ then

$$\Lambda_+(D)^{-k}u(x) = i^{-k}(2\pi)^{1-n} \int\int_{-\infty}^{0} e^{i(x',\xi')} e^{t\langle\xi'\rangle} t^{k-1}\tilde{u}(\xi', x_n - t)\, dt\, d\xi'/(k-1)!$$

and (22) follows if we make the change of variables $t \to -t$.

We end this section by determining the dual space of $C^\infty_{(0)}(\mathbb{R}^n_+)$ and of $\mathring{C}^\infty_0(\bar{\mathbb{R}}^n_+)$. The topology on the space $C^\infty_{(0)}(\mathbb{R}^n_+)$ is defined by the semi-norms which are restrictions to $\bar{\mathbb{R}}^n_+$ of the semi-norms in $C^\infty_0(\mathbb{R}^n)$, while $\mathring{C}^\infty_0(\bar{\mathbb{R}}^n_+)$ inherits a topology as a subspace of $C^\infty_0(\mathbb{R}^n)$. Recall that $\mathring{C}^\infty_0(\bar{\mathbb{R}}^n_+)$ denotes the set of functions $f \in C^\infty_0(\mathbb{R}^n)$ with $\operatorname{supp} f \subset \bar{\mathbb{R}}^n_+$; it can be identified with the set of functions in $C^\infty \bar{\mathbb{R}}^n_+)$ with compact support which vanish of infinite order when $x_n = 0$, because any such function can be extended by 0 when $x_n < 0$, thus giving a function in $C^\infty_0(\mathbb{R}^n)$.

If $T \in \mathscr{D}'(\mathbb{R}^n)$ then by means of the definition of the support of a distribution and a partition of unity one can show that

$$\langle T, \varphi\rangle = 0 \qquad \text{if } \operatorname{supp} T \cap \operatorname{supp} \varphi = \varnothing,$$

i.e. $\langle T, \varphi\rangle = 0$ for every $\varphi \in C^\infty_0(\mathbb{R}^n)$ which is equal to 0 in a neighbourhood of supp T. In fact, a stronger result is true (see [Hö 1], p. 46): If $T \in \mathscr{E}'$ and is of order $\leqslant k$, then

$\langle T, \varphi\rangle = 0$ *for every* $\varphi \in C^\infty$ *such that* $D^\alpha\varphi(x) = 0$ *when* $x \in \operatorname{supp} T$ *and* $|\alpha| \leqslant k$.

We use this result in the proof of the following theorem. If Ω is an open set in $\mathbb{R}^n$, the same result is obviously true if $T \in \mathscr{D}'(\Omega)$ and $\varphi \in C^\infty_0(\Omega)$.

Theorem 14.11 *The dual space of* $C^\infty_{(0)}(\mathbb{R}^n_+)$ *is equal to* $\mathring{\mathscr{D}}'(\bar{\mathbb{R}}^n_+)$, *the space of distributions with support in* $\bar{\mathbb{R}}^n_+$. *The dual space of* $\mathring{C}^\infty_0(\bar{\mathbb{R}}^n_+)$ *is equal to* $\bar{\mathscr{D}}'(\mathbb{R}^n_+)$, *the space of extendible distributions (i.e., restrictions to* $\mathbb{R}^n_+$ *of distributions in* $\mathbb{R}^n$).

Proof By definition of the space $C^\infty_{(0)}(\mathbb{R}^n_+)$ and its topology, the restriction map

$$\imath^+ : C^\infty_0(\mathbb{R}^n) \to C^\infty_{(0)}(\mathbb{R}^n_+)$$

is a continuous surjection; transposing it we get an injective map from the dual of $C^\infty_{(0)}(\mathbb{R}^n_+)$ to $\mathscr{D}'(\mathbb{R}^n)$ defined by $T \mapsto T \circ \imath^+$. Note that $\tilde{T} = T \circ \imath^+$ is a distribution in $\mathbb{R}^n$ with support $\subset \bar{\mathbb{R}}^n_+$, for if $\varphi \in C^\infty_0(\bar{\mathbb{R}}^n_-)$ then $\langle\tilde{T}, \varphi\rangle = \langle T, \imath^+\varphi\rangle = \langle T, 0\rangle = 0$. Conversely, if $\tilde{T} \in \mathscr{D}'(\mathbb{R}^n)$ has support in $\bar{\mathbb{R}}^n_+$ then for $\varphi \in C^\infty_{(0)}(\mathbb{R}^n_+)$ we define

$$\langle T, \varphi\rangle = \langle\tilde{T}, \Phi\rangle, \qquad \text{where } \Phi \in C^\infty_0(\mathbb{R}^n),\ \imath^+\Phi = \varphi.$$

Note that the definition of T does not depend on the choice of extension Φ, for if Φ_1 is another such extension then $\Phi - \Phi_1 = 0$ on $\mathbb{R}^n_+$, hence $D^\alpha(\Phi - \Phi_1) = 0$ on $\bar{\mathbb{R}}^n_+$, which contains the support of $\tilde{T}$, so $\langle\tilde{T}, \Phi - \Phi_1\rangle = 0$ as required. Now let β: $C^\infty_{(0)}(\mathbb{R}^n_+) \to C^\infty_0(\mathbb{R}^n)$ be the continuous extension operator from

Theorem 13.4. Since

$$T = \tilde{T} \circ \beta,$$

then T is continuous from $C^\infty_{(0)}(\mathbb{R}^n_+) \to \mathbb{C}$, that is, $T \in (C^\infty_{(0)}(\mathbb{R}^n_+))'$. This proves the first statement in the theorem.

For the second part, let $T \in (\mathring{C}^\infty_0(\bar{\mathbb{R}}^n_+))'$. Choosing a locally finite partition of unity $1 = \sum \psi_j$ in $\mathbb{R}^n$, we may write $T = \sum \psi_j T$. The linear functionals $\psi_j T$ are bounded by a semi-norm in $C^\infty_0(\mathbb{R}^n)$; by the Hahn-Banach theorem, they can be extended from $\mathring{C}^\infty_0(\bar{\mathbb{R}}^n_+)$ to a continuous linear functional on $C^\infty_0(\mathbb{R}^n)$. In other words, $\psi_j T \in \bar{\mathscr{D}}'(\mathbb{R}^n_+)$. With these extensions, we have $T \in \bar{\mathscr{D}}'(\mathbb{R}^n_+)$ because any compact set intersects only finitely many of the supports of the ψ_j's. Conversely, if $T \in \bar{\mathscr{D}}'(\mathbb{R}^n_+)$ then there exists $\tilde{T} \in \mathscr{D}'(\mathbb{R}^n)$ such that $\tilde{T}|_{\mathbb{R}^n_+} = T$. Then we may consider T as an element of the dual $(\mathring{C}^\infty_0(\bar{\mathbb{R}}^n_+))'$ by the definition

$$\langle T, \varphi \rangle = \langle \tilde{T}, \varphi \rangle, \qquad \varphi \in \mathring{C}^\infty_0(\bar{\mathbb{R}}^n_+).$$

This definition is independent of the choice of $\tilde{T}$, for if $\tilde{T}_1$ is another distribution in $\mathscr{D}'(\mathbb{R}^n)$ with $\tilde{T}_1|_{\mathbb{R}^n_+} = T$ then $\operatorname{supp}(\tilde{T} - \tilde{T}_1) \subset \bar{\mathbb{R}}^n_-$, so that $\langle \tilde{T} - \tilde{T}_1, \varphi \rangle = 0$ for all $\varphi \in \mathring{C}^\infty_0(\bar{\mathbb{R}}^n_+)$, since $D^\alpha \varphi = 0$ on $\bar{\mathbb{R}}^n_-$ which contains the support of $\tilde{T} - \tilde{T}_1$.

Remark Similarly, one can show that the dual space of $\mathscr{S}(\mathbb{R}^n_+)$ is $\mathring{\mathscr{S}}'(\bar{\mathbb{R}}^n_+)$, while that of $\mathring{\mathscr{S}}(\bar{\mathbb{R}}^n_+)$ is $\mathscr{S}'(\mathbb{R}^n_+)$.

In §14.3 we will need a corresponding result when Ω is a bounded open set in $\mathbb{R}^n$ with C^∞ boundary:

The dual of $C^\infty(\bar{\Omega})$ is the space $\mathring{\mathscr{D}}'(\bar{\Omega})$ of distributions in $\mathbb{R}^n$ with support $\subset \bar{\Omega}$.

The proof follows as in the first part of Theorem 14.11 by transposing the continuous surjection $C^\infty_0(\mathbb{R}^n) \to C^\infty(\bar{\Omega})$, and then using a continuous extension operator β: $C^\infty(\bar{\Omega}) \to C^\infty_0(\mathbb{R}^n)$ which can be constructed using a partition of unity and Theorem 13.4 in local coordinates near $\partial\Omega$.

14.2 The spaces $H^{s,t}$

We will need some auxiliary spaces. Note that $H^{s,0} = W^s_2$.

Definition 14.12 *For any real numbers s and t, let $H^{s,t} = H^{s,t}(\mathbb{R}^n)$ denote the set of all $u \in \mathscr{S}'(\mathbb{R}^n)$ with $\hat{u} \in L^2_{\text{loc}}$ and*

$$\|u\|_{(s,t)} = \left((2\pi)^{-n} \int |\hat{u}(\xi)|^2 (1 + |\xi|^2)^s (1 + |\xi'|^2)^t \, d\xi \right)^{1/2} < \infty$$

Since $\|u\|_{(s-m,t+m)} \leqslant \|u\|_{(s,t)}$, then

$$H^{s,t} \subset H^{s-m,t+m} \qquad \text{when } m \geqslant 0. \tag{23}$$

Another way to state (23) is the following:

$$H^{s_1,t_1} \subset H^{s_2,t_2} \qquad \text{when } s_2 \leqslant s_1 \quad \text{and} \quad s_2 + t_2 \leqslant s_1 + t_1.$$

It is not hard to show that $H^{s,t}$ is complete and C_0^∞ is dense in $H^{s,t}$. We also define $H^{s,t}(\mathbb{R}^n_+)$ as the space of restrictions $u = \imath^+ U$ of distributions $U \in H^{s,t}(\mathbb{R}^n)$ with norm

$$\|u\|_{(s,t)} = \inf \|U\|_{(s,t)}, \tag{24}$$

the infimum being taken over all $U \in H^{s,t}(\mathbb{R}^n)$ which are equal to u in $\mathbb{R}^n_+$. Further, $\mathring{H}^{s,t}(\bar{\mathbb{R}}^n_+)$ is defined to be the set of $u \in H^{s,t}(\mathbb{R}^n)$ with support in $\bar{\mathbb{R}}^n_+$. Then analogues of Theorem 14.4 holds for these spaces with the same proof.

Lemma 14.7 remains true with Λ^s replaced by $\Lambda^{s,t} = (\xi_n + i\langle\xi'\rangle)^s\langle\xi'\rangle^t$, that is,

$$\Lambda_-^{s,t}(D)\colon \mathring{\mathscr{S}}(\bar{\mathbb{R}}^n_+) \to \mathring{\mathscr{S}}(\bar{\mathbb{R}}^n_+) \qquad \text{and} \qquad \Lambda_-^{s,t}(D)\colon \mathring{\mathscr{S}}'(\bar{\mathbb{R}}^n_+) \to \mathring{\mathscr{S}}'(\bar{\mathbb{R}}^n_+)$$

are isomorphisms. The counterpart to Theorem 14.8 is as follows.

Theorem 14.8′ *The isomorphism $\Lambda_-(D)^{s,t} = (D_n - i\langle D'\rangle)^s\langle D'\rangle^t\colon \mathring{\mathscr{S}}(\bar{\mathbb{R}}^n_+) \to \mathring{\mathscr{S}}(\bar{\mathbb{R}}^n_+)$ extends by continuity to an isometric isomorphism $\Lambda_-(D)^{s,t}\colon L_2(\mathbb{R}^n_+) \to \mathring{H}^{-s-t}(\bar{\mathbb{R}}^n_+)$. The adjoint, $\Lambda_+(D)^{s,t}$, extends by continuity from $C_{(0)}^\infty(\mathbb{R}^n_+)$ to an isometric isomorphism*

$$\Lambda_+(D)^{s,t} = (D_n + i\langle D'\rangle)^s\langle D'\rangle^t\colon H^{s,t}(\mathbb{R}^n_+) \to L_2(\mathbb{R}^n_+).$$

Thus, for all $u \in H^{s,t}(\mathbb{R}^n_+)$ we have $\|u\|_{(s,t)} = \|\Lambda_+(D)^{s,t}u\|_{L_2(\mathbb{R}^n_+)}$, and

$$H^{s,t}(\mathbb{R}^n_+) = \{\imath^+ U \mid U \in \mathscr{S}' \quad \textit{such that } \imath^+\Lambda_+(D)^{s,t}U \in L_2(\mathbb{R}^n_+)\}.$$

Here $\imath^+$ denotes the restriction of distributions to $\mathbb{R}^n_+$.

Writing $\Lambda_-^{s,t}(D) = (D_n - i\langle D'\rangle)\Lambda_-^{s-1,t}(D)$, we see that every $u \in \mathring{H}^{-s,-t}(\bar{\mathbb{R}}^n_+)$ can be written in the form

$$u = u_0 + D_n u_n. \tag{25}$$

where $u_0 \in \mathring{H}^{1-s,-t-1}(\bar{\mathbb{R}}^n_+)$, $u_n \in \mathring{H}^{1-s,-t}(\bar{\mathbb{R}}^n_+)$ and the norms of u_0 and u_n are bounded by the norm of u. In fact, with $u_0 = -i\langle D'\rangle\Lambda_-^{s-1,t}(D)u$ and $u_n = \Lambda_-^{s-1,t}(D)u$, we have

$$\|u_0\|_{(1-s,-t-1)} = \|u\|_{(-s,-t)} = \|u_n\|_{(1-s,-t)}.$$

Lemma 14.13 *In order that $u \in H^{s,t}(\mathbb{R}^n_+)$ it is necessary and sufficient that $u \in H^{s-1,t+1}(\mathbb{R}^n_+)$ and $D_n u \in H^{s-1,t}(\mathbb{R}^n_+)$. We have*

$$\tfrac{1}{2}\|u\|^2_{(s,t)} \leqslant \|D_n u\|^2_{(s-1,t)} + \|u\|^2_{(s-1,t+1)} \leqslant \|u\|^2_{(s,t)}. \tag{26}$$

Proof If $u \in H^{s,t}(\mathbb{R}^n_+)$ then for any $\varepsilon > 0$ there exists $U \in H^{s,t}(\mathbb{R}^n)$ such that $U = u$ on $\mathbb{R}^n_+$ and $\|U\|^2_{(s,t)} \leqslant \|u\|^2_{(s,t)} + \varepsilon$. (In fact, just as in Corollary 14.9 we could choose U such that $\|U\|_{(s,t)} = \|u\|_{(s,t)}$.) Since $|\xi|^2 = |\xi'|^2 + \xi_n^2$, then

$$\|U\|^2_{(s-1,t+1)} + \|D_n U|^2_{(s-1,t)} = \|U\|^2_{(s,t)}$$

and it follows that

$$\|u\|^2_{(s-1,t+1)} + \|D_n u\|^2_{(s-1,t)} \leqslant \|u\|^2_{(s,t)} + \varepsilon$$

which proves that $u \in H^{s-1,t+1}(\mathbb{R}^n_+)$ and $D_n u \in H^{s-1,t}(\mathbb{R}^n_+)$. Also, the first inequality in (26) follows since ε is arbitrary.

Conversely, suppose that $u \in H^{s-1,t+1}(\mathbb{R}^n_+)$ and $D_n u \in H^{s-1,t}(\mathbb{R}^n_+)$ and let $\varphi \in \mathring{H}_{(-s,-t)}(\bar{\mathbb{R}}^n_+)$. Then write $\varphi = \varphi_0 + D_n \varphi_n$ as in (25). It follows that

$$\begin{aligned} |(u, \varphi)_0| &\leqslant \|u\|_{(s-1,t+1)} \|\varphi_0\|_{(-s+1,-t-1)} + \|D_n u\|_{(s-1,t)} \|\varphi_n\|_{(-s+1,-t)} \\ &\leqslant (\|u\|_{(s-1,t+1)} + \|D_n u\|_{(s-1,t)}) \cdot \|\varphi\|_{(-s,-t)} \end{aligned}$$

Hence $u \in H^{s,t}(\mathbb{R}^n_+)$ and $\|u\|_{(s,t)} \leqslant \|u\|_{(s-1,t+1)} + \|D_n u\|_{(s-1,t)}$. Using the fact that $\frac{1}{2}(a+b)^2 \leqslant a^2 + b^2$, the first inequality in (26) follows.

When $s = 0$, the elements in $H^{0,t}(\mathbb{R}^n)$ can be considered as L_2 functions of x_n with values in $W^t_2(\mathbb{R}^{n-1})$. It follows that

$$H^{0,t}(\mathbb{R}^n_+) = L_2(\mathbb{R}_+, W^t_2(\mathbb{R}^{n-1})). \tag{27}$$

In fact, the inclusion $\subset$ is obvious, and any $u \in L_2(\mathbb{R}_+, W^t_2(\mathbb{R}^{n-1}))$ can be extended by 0 on $x_n \leqslant 0$ to obtain $U \in L_2(\mathbb{R}, W^t_2(\mathbb{R}^{n-1})) = H^{0,t}(\mathbb{R}^n)$, so we obtain the other inclusion.

Corollary 14.14 *If s is a nonnegative integer, then $H^{s,t}(\mathbb{R}^n_+)$ is the set of all $u \in \mathscr{S}'(\mathbb{R}^n_+)$ such that $D^j_n u$ is an L_2 function of $x_n \in \mathbb{R}_+$ with values in $W^{s+t-j}_2(\mathbb{R}^{n-1})$ when $0 \leqslant j \leqslant s$. The norm $\|u\|_{(s,t)}$ is equivalent to*

$$\left(\sum_{j=0}^{s} \int_0^\infty \|D^j_n u(\cdot, x_n)\|^2_{s+t-j}\, dx_n \right)^{1/2}.$$

Proof Let $u \in H^{s,t}(\mathbb{R}^n_+)$. Consider the following statement for an integer v such that $0 \leqslant v \leqslant s$:

$$P_v\colon D^j_n u \in H^{s-v,v+t-j}(\mathbb{R}^n_+), \qquad j = 0, 1, \ldots, v.$$

It follows by Lemma 14.13 that the statement P_v is equivalent to P_{v+1} for $v = 0, 1, \ldots, s-1$. Hence the statements when $v = 0$ and $v = s$ are equivalent which proves the first part of the corollary. The second part also follows easily, since (26) implies that the norms

$$\left(\sum_{j=0}^{v} \|D^j_n u\|^2_{(s-v,v+t-j)} \right)^{1/2}$$

are equivalent for $v = 0, 1, \ldots, s$.

For further details on the spaces $H^{s,t}$, see [Hö 2] and [Hö 3, Appendix B]. These spaces are useful in conjunction with the next lemma which often allows a precise analysis of regularity at the boundary.

Lemma 14.15 *Let $\mathscr{A}$ be a differential operator of order l,*

$$\mathscr{A} = \sum_{|\alpha| \leqslant l} a_\alpha(x) D^\alpha,$$

where $a_\alpha \in C^\infty_{(0)}(\mathbb{R}^n_+)$ when $\alpha_n < l$ and such that the coefficient of D^l_n is equal to 1. If $m_1 + s_1 = m_2 + s_2$ and

$$u \in H^{m_1, s_1}(\mathbb{R}^n_+), \qquad \mathscr{A}u \in H^{m_2 - l, s_2}(\mathbb{R}^n_+),$$

it follows that $u \in H^{m_2, s_2}(\mathbb{R}^n_+)$.

Proof We break the proof into two cases: $m_2 \leqslant m_1 + 1$ and $m_2 > m_1 + 1$. (Note: When $m_2 \leqslant m_1$ the proof is trivial since $H^{m_1, s_1} \subset H^{m_2, s_2}$ by virtue of (23).)

First case: $m_2 \leqslant m_1 + 1$. We will prove by induction that

$$D^j_n u \in H^{m_2 - j, s_2}, \qquad j = 0, 1, \ldots, l. \tag{28}$$

Let us first prove (28) for $j = l$. Note that

$$D^\alpha u \in H^{m_1 - \alpha_n, s_1 - |\alpha'|}$$

since $u \in H^{m_1, s_1}$, which implies $D^\alpha u \in H^{m_1 + 1 - l, s_1}$ if $\alpha_n < l$ and $|\alpha| \leqslant l$. Moreover, $H^{m_1 + 1 - l, s_1} \subset H^{m_2 - l, s_2}$ by virtue of (23) since $m_2 \leqslant m_1 + 1$. It then follows from $\mathscr{A}u \in H^{m_2 - l, s_2}$ that $D^l_n u \in H^{m_2 - l, s_2}$ since we have just shown this is true for all the other terms in $\mathscr{A}u$. Now we argue by backwards induction. As our inductive step, suppose that (28) has been proved for some $j + 1 \leqslant l$; then we must show that it is true for j. By Lemma 14.13 this is equivalent to showing that $D^j_n u \in H^{m_2 - j - 1, s_2 + 1}$ and $D^{j+1}_n u \in H^{m_2 - j - 1, s_2}$. But the first inclusion follows since $u \in H^{m_1, s_1} \subset H^{m_2 - 1, s_2 + 1}$, while the second inclusion is just the inductive hypothesis. Hence (28) holds for all j; in particular it holds for $j = 0$.

Second case: $m_2 > m_1 + 1$. We have $\mathscr{A}u \in H^{m_2 - l, s_2} \subset H^{m_1 + 1 - l, s_1 - 1}$, so the argument just given with m_2 replaced by $m_1 + 1$ shows that

$$u \in H^{m_1 + 1, s_1 - 1}.$$

There is a positive integer k such that $m_1 + k \leqslant m_2 \leqslant m_1 + k + 1$ and we can repeat the argument up to k times to get

$$u \in H^{m_1 + k, s_1 - k},$$

so we are back in the first case, and we conclude that $u \in H^{m_2, s_2}$.

We can also put the hypotheses in a form which is suitable for localizing on manifolds.

Corollary 14.16 *Let the hypotheses be as in the Lemma. If $m_1 + s_1 = m_2 + s_2$ and for some $\varphi \in C^\infty_{(0)}(\mathbb{R}^n_+)$ we have*

$$\varphi \cdot u \in H^{m_1, s_1}(\mathbb{R}^n_+) \quad \textit{and} \quad \varphi \cdot \mathscr{A}u \in H^{m_2 - l, s_2}(\mathbb{R}^n_+),$$

then it follows that $\varphi u \in H^{m_2, s_2}(\mathbb{R}^n_+)$.

Proof The commutator $\varphi\mathscr{A} - \mathscr{A}\varphi$ is of order $l-1$ so $\varphi\mathscr{A}u - \mathscr{A}\varphi u \in H^{m_1-l+1,s_1} \subset H^{m_1-l+1,s_1-1}$. If $m_2 \leqslant m_1+1$ then $\mathscr{A}\varphi u \in H^{m_2-l,s_2}$, and the inductive step in the first case above still works with u replaced by $v = \varphi \cdot u$. If $m_2 > m_1 + 1$ then $\mathscr{A}\varphi u \in H^{m_1-l+1,s_1-1}$ and the second case proceeds as before.

We denote by $\imath'$ the restriction map

$$(\imath' u)(x') = u(x', 0), \qquad u \in C^\infty_{(0)}(\mathbb{R}^n_+).$$

Proposition 14.17 *If $m > 1/2$ there exists a continuous extension of $\imath'$ to a map*

$$\imath' \colon H^{m,s}(\mathbb{R}^n_+) \to W_2^{m+s-1/2}(\mathbb{R}^{n-1}).$$

Proof First let us verify this fact for $\mathbb{R}^n$ in place of $\mathbb{R}^n_+$. Let $U \in C_0^\infty(\mathbb{R}^n)$. As in the proof of the Lemma before Theorem 7A.9, one shows that if $v = \imath' U$ then $\hat{v}(\xi') = (2\pi)^{-1}\int \hat{U}(\xi)\, d\xi_n$, whence

$$|\hat{v}(\xi')|^2 \leqslant C_m^2 \int |\hat{U}(\xi)|^2(1+|\xi|^2)^m(1+|\xi'|^2)^{1/2-m}\, d\xi_n.$$

Now if we multiply by $(1+|\xi'|^2)^{s+m-1/2}$ and integrate, it follows that

$$\|\imath' U\|_{s+m-1/2} \leqslant C_m \|U\|_{m,s}, \qquad U \in C_0^\infty(\mathbb{R}^n).$$

Now, if $u \in C^\infty_{(0)}(\mathbb{R}^n_+)$ there exists $U \in C_0^\infty(\mathbb{R}^n)$ such that $U = u$ on $\mathbb{R}^n_+$. Then $\imath' u = \imath' U$ so that $\|\imath' u\|_{s+m-1/2} \leqslant C_m \|U\|_{m,s}$ and taking the infimum over all such U's we obtain

$$\|\imath' u\|_{s+m-1/2} \leqslant C_m \|u\|_{m,s}, \qquad u \in C^\infty_{(0)}(\mathbb{R}^n_+).$$

By the density of $C^\infty_{(0)}(\mathbb{R}^n_+)$ in $H^{m,s}(\mathbb{R}^n_+)$, the map $\imath'$ extends to a continuous operator $H^{m,s}(\mathbb{R}^n_+) \to W_2^{m+2-1/2}(\mathbb{R}^{n-1})$.

Corollary 14.18 *It follows that the trace operator γ defined by*

$$\gamma u = [\gamma_j u]_{j=0}^{l-1}, \qquad \textit{where } \gamma_j u = \lim_{x_n \to 0^+} D_n^j u(x', x_n)$$

extends by continuity from $C^\infty_{(0)}(\mathbb{R}^n_+)$ to a bounded operator

$$\gamma \colon H^{m,s}(\mathbb{R}^n_+) \to \mathop{\times}_{j=0}^{l-1} W_2^{m+s-j-1/2}(\mathbb{R}^{n-1})$$

when $m > l - \frac{1}{2}$.

The following lemma will be needed in the proof of Theorem 14.24. Here $\mathscr{A}$ is an elliptic differential operator with principal part homogeneous of degree l. Recall the definition of the operator $\mathscr{A}^c$ in the introduction to this chapter (see also (34) below).

Lemma 14.19 *Let $\mathscr{A}$ be a differential operator of order l in $\bar{\Omega}$. If $\mathfrak{U}' \subset \mathbb{R}^{n-1}$ is a coordinate patch on $\partial\Omega$ then*

$$\|\mathscr{A}^c\mathscr{U}\|_{(-l,s)} \leqslant c \cdot \sum_{j=0}^{l-1} \|\mathscr{U}_j\|_{s-j-1/2}, \tag{29}$$

for every $\mathscr{U} = [\mathscr{U}_j]_{j=0}^{l-1} \in C_0^\infty(\mathfrak{U}', \mathbb{C}^{pl})$.

Proof Recall that

$$\mathscr{A}^c\mathscr{U} = \sum_{k=0}^{l-1} v_k \otimes D_n^k\delta, \qquad \text{where } v_k = i^{-1} \sum_{j=0}^{l-k-1} \mathscr{A}_{j+k+1}\mathscr{U}_j.$$

Because $\mathscr{A}_{j+k+1}$ are differential operators (hence local operators), it follows that $v_k \in C_0^\infty(\mathfrak{U}') \subset C_0^\infty(\mathbb{R}^{n-1})$, and then

$$\|v_k \otimes D_n^k\delta\|_{(-l,s)}^2 = (2\pi)^{-n} \iint (1 + |\xi'|^2 + \xi_n^2)^{-l}(1 + |\xi'|^2)^s|\hat{v}_j(\xi')|^2\xi_n^{2k}\, d\xi'\, d\xi_n$$

$$\leqslant c_1^2 \int (1 + |\xi'|^2)^{s-l+k+1/2}|\hat{v}_j(\xi')|^2\, d\xi'$$

since $k \leqslant l - 1$, where the integral with respect to ξ_n is estimated by taking $t = \xi_n/(1 + |\xi'|2)^{1/2}$ as new integration variable. Hence

$$\|v_k \otimes D_n^k\delta\|_{(-l,s)} \leqslant c_1\|v_k\|_{s-l+k+1/2} \leqslant c_1c_2 \sum_j \|\mathscr{U}_j\|_{s-j-1/2}$$

where the second inequality holds because $\mathscr{A}_{j+k+1}$ has order $l - j - k - 1$.

14.3 The Calderón operator for an elliptic operator

For convenience we identify a tubular neighbourhood of $\partial\Omega$ in $\mathbb{R}^n$ with $\partial\Omega \times (-1, 1)$, such that $\partial\Omega \times [0, 1) \subset \bar{\Omega}$. Coordinates in $\partial\Omega \times (-1, 1)$ are denoted by (x', x_n) and those in $\partial\Omega$ by $y = (x', 0)$.

Let $\mathscr{A}$ be an elliptic differential operator in $\bar{\Omega}$ with principal part homogeneous of degree l (Definition 9.25). As usual we assume that the coefficients of $\mathscr{A}$ are in $C^\infty(\bar{\Omega})$ and have been extended to be smooth functions on $\mathbb{R}^n$. Then $\mathscr{A}$ is elliptic in a small neighbourhood of $\bar{\Omega}$ in $\mathbb{R}^n$, and by shrinking the tubular neighbourhood, if necessary, we may assume that $\mathscr{A}$ is elliptic in $\hat{\Omega} = \bar{\Omega} \cup \partial\Omega \times (-1, 1)$. By Proposition 7.37, there is a parametrix $E \in OS^{-l}(\mathbb{R}^n, p \times p)$ such that $(I - \mathscr{A}E)\psi \in OS^{-\infty}$ and $\psi(I - E\mathscr{A}) \in OS^{-\infty}$ for any $\psi \in C_0^\infty(\hat{\Omega})$, that is,

$$E\mathscr{A} = I + T_l, \qquad \mathscr{A}E = I + T_r, \tag{30}$$

where $\psi T_l \in OS^{-\infty}$ and $T_r\psi \in OS^{-\infty}$. By virtue of the lemma that follows Proposition 7.37 we have $T_l\psi \in OS^{-\infty}$ and $\psi T_r \in OS^{-\infty}$ too.

If $u \in C^\infty(\bar{\Omega}, \mathbb{C}^p)$ we write

$$\gamma u = [\gamma_j u]_{j=0}^{l-1}, \qquad \text{where } \gamma_j u = \lim_{x_n \to 0^+} D_n^j u(x', x_n). \tag{31}$$

for the Cauchy data. In $\partial\Omega \times (-1, 1)$ the elliptic differential operator $\mathscr{A}$ can be written in the form

$$\mathscr{A} = \sum_{j=0}^{l} \mathscr{A}_j D_n^j \tag{32}$$

where $\mathscr{A}_j = A_j(x', x_n, D')$ is a differential operator of order $l-j$ on $\partial\Omega$ depending on the parameter x_n, and $\mathscr{A}_l = \mathscr{A}_l(x', x_n)$ is an invertible matrix function. If $u \in C^\infty(\bar{\Omega}, \mathbb{C}^p)$, we let u^0 denote the locally integrable function on $\mathbb{R}^n$ which is equal to u in Ω and 0 in $\mathbb{R}^n \backslash \Omega$. Then we obtain in $\mathbb{R}^n$

$$\mathscr{A} u^0 = (\mathscr{A} u)^0 + \mathscr{A}^c \gamma u, \tag{33}$$

where

$$\mathscr{A}^c \mathscr{U} = i^{-1} \sum_{j=0}^{l-1} \sum_{k=0}^{j} \mathscr{A}_{j+1} \mathscr{U}_{j-k} \otimes D_n^k \delta, \qquad \mathscr{U} = [\mathscr{U}_j]_{j=0}^{l-1} \in C^\infty(\partial\Omega, \mathbb{C}^{pl}), \tag{34}$$

and δ denotes the Dirac measure in x_n. Note that $\mathscr{A}^c \mathscr{U}$ is a distribution in $\mathbb{R}^n$ supported on $\partial\Omega$. Now we apply the parametrix E to the equation (33) to get a type of Green's formula:

$$u^0 + T_l u^0 = E(\mathscr{A} u)^0 + E\mathscr{A}^c \gamma u. \tag{35}$$

The Calderón operator P for the elliptic operator $\mathscr{A}$ is defined as follows. For $\mathscr{U} \in C^\infty(\partial\Omega, \mathbb{C}^{pl})$, we let

$$P\mathscr{U} = \gamma E \mathscr{A}^c \mathscr{U} \tag{36}$$

where the boundary values are taken from inside Ω, see (31). Thus $P\mathscr{U}$ is the Cauchy data of the Green potential $E\mathscr{A}^c\mathscr{U}$. The boundary values in the definition of P exist because $E\mathscr{A}^c$ is a sum of multiple-layer potentials and E satisfies the transmission conditions. (See the text following Remark 13.9, or Theorem 14.24.) By virtue of (34),

$$(P\mathscr{U})_j = \sum_{k=0}^{l-1} P_{jk} \mathscr{U}_k, \qquad j = 0, \ldots, l-1,$$

where

$$P_{jk} \mathscr{U}_k = \lim_{x_n \to 0^+} \sum_{\nu=0}^{l-1-k} i^{-i} D_n^j E \mathscr{A}_{\nu+k+1} D_n^\nu (\mathscr{U}_k(x') \otimes \delta(x_n)).$$

Since $\mathscr{A}_{\nu+k+1}$ is a differential operator of order $l-\nu-k-1$ then

$$i^{-1} D_n^j E \mathscr{A}_{\nu+k+1} D_n^\nu \in OS^{j-k-1}(\mathbb{R}^n)$$

and it follows from Theorem 13.12 and Corollary 13.13 that

$$P_{jk} \in OClS^{j-k}(\partial\Omega, p \times p)$$

with principal symbol

$$\pi P_{jk}(x', \xi') = (2\pi i)^{-1} \int_+ \sum_{\nu=0}^{l-1-k} \lambda^{j+\nu} [\pi\mathscr{A}(x', 0, \xi', \lambda)]^{-1} \pi\mathscr{A}_{\nu+k+1}(x', 0, \xi')\, d\lambda. \tag{37}$$

Since the integrand is a rational function of λ, the operator $\int_+$ can be taken as the integral along a simple, closed contour in the upper half-plane $\operatorname{Im}\lambda > 0$ containing all the poles of the integrand there.

The operator $P = [P_{jk}]$ is, therefore, a Douglis–Nirenberg operator on $\partial\Omega$ with principal symbol πP equal to the spectral projector, P^+, for $L(\lambda) = \pi\mathscr{A}(x', 0, \xi', \lambda)$ corresponding to the eigenvalues of $L(\lambda)$ in $\operatorname{Im}\lambda > 0$. The projector P^+ is the "Calderón projector" from §2.2.

In the proof of the next theorem we use the dual space of $C^\infty(\bar{\Omega})$. By transposing the continuous surjection $C_0^\infty(\mathbb{R}^n) \to C^\infty(\bar{\Omega})$ we obtain an injection $\mathscr{D}'(\bar{\Omega}) \hookrightarrow \mathscr{D}'(\mathbb{R}^n)$. The dual space of $C^\infty(\bar{\Omega})$ is the space, $\mathring{\mathscr{D}}'(\bar{\Omega})$, of distributions in $\mathbb{R}^n$ with support in $\bar{\Omega}$.

We also need the fact that if $T \in OS^{-\infty}(\mathbb{R}^n)$ then it has a C^∞ kernel and the map $T: \mathscr{E}'(\mathbb{R}^n) \to C^\infty(\mathbb{R}^n)$ is continuous; see Theorem 7.6 and [Hö 1, p. 132].

Theorem 14.20 *The Calderón operator P is an approximate projector in the sense that* $P - P^2 \in OC^\infty(\partial\Omega, pl \times pl)$, *i.e. it has a* C^∞ *kernel on* $\partial\Omega$.

Proof If we let

$$u = E\mathscr{A}^c\mathscr{U}$$

then in virtue of Theorem 13.12 (or Theorem 14.24) the restriction of u to Ω is in $C^\infty(\bar{\Omega}, \mathbb{C}^p)$. Note that (30) gives $\mathscr{A}u = (I + T_l)\mathscr{A}^c\mathscr{U}$. Since $\mathscr{A}^c\mathscr{U}$ is supported on $\partial\Omega$, then, by restriction to Ω, we have

$$\mathscr{A}u = T_l\mathscr{A}^c\mathscr{U} \qquad \text{in } \Omega. \tag{38}$$

Also, by definition, $\gamma u = P\mathscr{U}$, so if we apply (35) to u and take the Cauchy data from Ω as in (31), we find that

$$P\mathscr{U} + \gamma T_r(E\mathscr{A}^c\mathscr{U})^0 = \gamma E(T_l\mathscr{A}^c\mathscr{U})^0 + P^2\mathscr{U}. \tag{39}$$

Let $\psi \in C_0^\infty(\hat{\Omega})$ be equal to 1 in a neighbourhood of $\bar{\Omega}$. Since we can replace T_l and T_r by $T_l\psi$ and $T_r\psi$, we may assume in (38) and in (39) that T_l, $T_r \in OS^{-\infty}$.

We know that $\mathscr{D}'(\partial\Omega) = \bigcup_s W_2^s(\partial\Omega)$ and $C^\infty(\bar{\Omega}) = \bigcap_s W_2^s(\Omega)$. Now let $\imath^+$ denote the restriction operator $C^\infty(\mathbb{R}^n) \to C^\infty(\bar{\Omega})$. Since $T_l \in OS^{-\infty}$ and $\mathscr{A}^c\mathscr{U} \in \mathscr{E}'$, the map $\mathscr{U} \mapsto \imath^+ T_l\mathscr{A}^c\mathscr{U}$ is continuous from $\mathscr{D}'(\partial\Omega, \mathbb{C}^{pl})$ to $C^\infty(\bar{\Omega}, \mathbb{C}^p)$. The transmission property for E and the continuity of the trace operator γ then imply that the map $\mathscr{U} \mapsto \gamma E(T_l\mathscr{A}^c\mathscr{U})^0$ is continuous from $\mathscr{D}'$ to C^∞, hence it has a C^∞ kernel on $\partial\Omega$.

To prove $P^2 - P$ has a C^∞ kernel on $\partial\Omega$, it remains to show that $\mathscr{U} \mapsto \gamma T_r(E\mathscr{A}^c\mathscr{U})^0$ has a C^∞ kernel on $\partial\Omega$. If $\varphi \in C^\infty(\bar{\Omega}, \mathbb{C}^p)$ and $\mathscr{U} \in C^\infty(\partial\Omega, \mathbb{C}^{pl})$, we claim that

$$((E\mathscr{A}^c\mathscr{U})^0, \varphi)_\Omega = (\mathscr{A}^c\mathscr{U}, E^*\varphi^0)_\Omega. \tag{40}$$

The right-hand side of (40) is well-defined because $E^*\varphi^0 \in C^{l-1}(\hat{\Omega}, \mathbb{C}^p)$ by Chapter 13, Exercise 8, and $\mathscr{A}^c\mathscr{U}$ is a distribution in $\mathbb{R}^n$ of order $\leqslant l - 1$. To verify (40), suppose first that φ has compact support in Ω, i.e. $\varphi \in C_0^\infty(\Omega, \mathbb{C}^p)$.

Then

$$((E\mathscr{A}^c\mathscr{U})^0, \varphi)_\Omega = (E\mathscr{A}^c\mathscr{U}, \varphi)_{\mathbb{R}^n} = (\mathscr{A}^c\mathscr{U}, E^*\varphi)_{\mathbb{R}^n} = (\mathscr{A}^c\mathscr{U}, E^*\varphi^0)_\Omega$$

since $\varphi^0 = \varphi$. But we can approximate any $\varphi \in C^\infty(\bar{\Omega}, \mathbb{C}^p)$ in the L_2 norm by functions in $C_0^\infty(\Omega, \mathbb{C}^p)$, and E^* is certainly continuous from L_2 to L_2 because it is a p.d.o. of order $-l < 0$, so (40) must hold. Since E^* satisfies the transmission conditions, then $E^*\varphi^0 \in C^\infty(\bar{\Omega}, \mathbb{C}^p)$, whence (40) implies that $\mathscr{U} \mapsto (E\mathscr{A}^c\mathscr{U})^0$ can be extended to a continuous map from $\mathscr{D}'(\partial\Omega, \mathbb{C}^{pl})$ to $\mathring{\mathscr{D}}'(\bar{\Omega}, \mathbb{C}^p)$. But $\mathring{\mathscr{D}}'(\bar{\Omega}, \mathbb{C}^p)$ is a space of distributions on $\mathbb{R}^n$ with compact support and $T_r \in OS^{-\infty}$, so the map $\mathscr{U} \mapsto \gamma T_r(E\mathscr{A}^c\mathscr{U})^0$ is continuous from $\mathscr{D}'$ to C^∞, and, therefore, it has a C^∞ kernel on $\partial\Omega$. The proof is complete.

The next result is also important.

Proposition 14.21 *The map $f \mapsto P\gamma Ef^0$ from $C^\infty(\bar{\Omega}, \mathbb{C}^p)$ to $C^\infty(\partial\Omega, \mathbb{C}^{pl})$ extends continuously to a map from $L^2(\Omega, \mathbb{C}^p)$ to $C^\infty(\partial\Omega, \mathbb{C}^{pl})$.*

Proof If $f \in C^\infty(\bar{\Omega}, \mathbb{C}^p)$ then the restriction of Ef^0 to Ω belongs to $C^\infty(\bar{\Omega}, \mathbb{C}^p)$ since E has the transmission property. If $u = Ef^0|_\Omega$, it follows from (30) that $\mathscr{A}u = f + T_r f^0$ in Ω. Replacing u by $Ef^0|_\Omega$ in (35) and taking the Cauchy data γ from inside Ω we find that

$$\gamma Ef^0 + \gamma T_l(Ef^0)^0 = \gamma Ef^0 + \gamma E(T_r f^0)^0 + P\gamma Ef^0$$

and hence

$$P\gamma Ef^0 = \gamma T_l(Ef^0)^0 - \gamma E(T_r f^0)^0 \tag{41}$$

As above, we may assume that $T_l, T_r \in OS^{-\infty}$. The map $f \mapsto T_r f^0$ is continuous from L_2 to C^∞ so $f \mapsto \gamma E(T_r f^0)^0$ is continuous from L_2 to C^∞ by the transmission property for E and the continuity of γ. Also, $f \mapsto Ef^0$ is continuous from L_2 to L_2, so the map $f \mapsto \gamma T_l(Ef^0)^0$ is continuous from L_2 to C^∞.

The following proposition will be needed shortly. Let M be a compact, closed manifold.

Proposition 14.22 *Let $Q \in OClS^0(M, p \times p)$ and assume that $Q^2 - Q \equiv 0$, i.e. $Q^2 - Q \in OS^{-\infty}$. Let $B \in OClS^\mu(M, r \times p)$ and suppose that $\pi B(y, \eta)$ restricted to the image of $\pi Q(y, \eta)$ is bijective for all $(y, \eta) \in T^*M\backslash 0$. Then there exist $S \in OClS^{-\mu}(M, p \times p)$ and $S' \in OClS^0(M, p \times r)$ such that*

(*i*) $BS \equiv I_r$, $QS \equiv S$,
(*ii*) $S' + SB \equiv I_p$, $S'Q \equiv 0$.

The operators S, S' are uniquely determined modulo $OS^{-\infty}$ by these conditions.

Proof Since πB restricted to the image of πQ is surjective then by Theorem 8.65 applied to BQ we can find a right parametrix $R \in OClS^{-\mu}(M, p \times r)$ such that $BQR \equiv I_r$. Then $S = QR$ satisfies (i), where $QS \equiv S$ because $Q^2 \equiv Q$. On the other hand, consider the $(r + p) \times p$ system $B \oplus (I_p - Q)$;

its principal symbol

$$\begin{pmatrix} \pi B(y, \eta) \\ I_p - \pi Q(y, \eta) \end{pmatrix}$$

is injective for all $(y, \eta) \in T^*M\backslash 0$ by virtue of the injectivity of πB when restricted to the image of πQ. By Theorem 8.64 we can find a left parametrix for this DN system, i.e. operators

$$T'' \in OClS^{-\mu}(M, p \times r) \quad \text{and} \quad T' \in OClS^{0}(M, p \times p)$$

such that

$$[T'' \quad T']\begin{pmatrix} B \\ I_p - Q \end{pmatrix} \equiv I_p.$$

Then $T''B + T'(I_p - Q) \equiv I_p$, so that $S' = T'(I_p - Q)$, $S'' = T''$ satisfy the equations

$$S' + S''B \equiv I_p, \quad S'Q \equiv 0. \tag{ii'}$$

If we multiply the first equation, $S' + S''B \equiv I_p$, on the right by S, and use $BS \equiv I_r$, it follows that $S'S + S'' \equiv S$. Also, if we multiply the second equation, $S'Q \equiv 0$, on the right by S and use $QS \equiv S$ it follows that $S'S \equiv 0$, whence

$$S'' \equiv S. \tag{*}$$

Now (ii) follows from (ii′).

To prove uniqueness, suppose that we have two pairs of operators S_i, S_i', $i = 1, 2$, satisfying (i) and (ii). It follows that $S_1 \equiv S_2$ in the same way as we showed (*), and then (ii) implies $S_1' \equiv S_2'$.

Now let Q and B denote any pair of Douglis–Nirenberg operators on M satisfying the following conditions:

$$\begin{aligned} Q &= [Q_{ij}]_{p\times p}, \quad Q_{ij} \text{ is a classic p.d.o. on } M \text{ of order} \leqslant s_i + t_j \\ B &= [B_{kj}]_{r\times p}, \quad B_{kj} \text{ is a classic p.d.o. on } M \text{ of order} \leqslant m_k + t_j \end{aligned}$$

and $Q^2 - Q \in OS^{-\infty}$.

Corollary 14.23 *Let Q and B satisfy the conditions just mentioned. If $\pi B(y, \eta)$ restricted to the image of the projector $\pi Q(y, \eta)$ is bijective for all $(y, \eta) \in T^*M \backslash 0$ then there exist* DN *operators*

$$\begin{aligned} S &= [S_{ik}]_{p\times r}, \quad S_{ik} \text{ is a classic p.d.o. on } M \text{ of order} \leqslant -t_i - m_k \\ S' &= [S'_{ij}]_{p\times p}, \quad S'_{ij} \text{ is a classic p.d.o. on } M \text{ of order} \leqslant -t_1 + t_j \end{aligned}$$

such that $S' + SB \equiv I_p$ and $BS \equiv I_r$. Moreover, if $s_i = -t_i$ for all i then we may assume in addition that $S'Q \equiv 0$ and $QS \equiv S$.

By the method used to prove Theorem 8.69 we may assume that $s_1 = \cdots = s_p = 0$, $t_1 = \cdots = t_p = 0$ and $m_1 = \cdots = m_r = \mu$ so the corollary follows immediately from Proposition 14.22.

Note that Corollary 14.23 remains valid if the elements Q_{ij} and B_{kj} are themselves matrices (or "blocks") of p.d.o.'s of order $\leqslant s_i + t_j$ and $\leqslant m_k + t_j$, respectively. In this sense we refer to Q and B as block DN operators. The (i, j) block of Q is Q_{ij} and the (k, j) block of B is B_{kj}.

14.4 Parametrix for an elliptic boundary problem

To make use of the Green formula (35), we must investigate how the pseudo-differential operator E behaves at the boundary $\partial\Omega$ ($x_n = 0$). First we prove Theorem 14.24 which is closely related to the transmission property for E. Then we use (35) and the Calderón operator P to construct a parametrix $\mathfrak{P}$ for the elliptic boundary problem $(\mathscr{A}, \mathscr{B}) \in BE^{l,\mathbf{m}}$.

Let $\imath^+$ denote the restriction operator $\imath^+ : \mathscr{D}'(\mathbb{R}^n) \to \mathscr{D}'(\Omega)$. Then we have $\imath^+ : W_2^s(\mathbb{R}^n) \to W_2^s(\Omega)$ and $\imath^+ : C_0^\infty(\mathbb{R}^n) \to C^\infty(\bar{\Omega})$.

If $f \in L_2(\mathbb{R}^n)$ then f^0 denotes the locally integrable function on $\mathbb{R}^n$ which is equal to f in Ω and equal to 0 in $\mathbb{R}^n \backslash \Omega$. The map $f \mapsto f^0$ is continuous from $L_2(\mathbb{R}^n)$ to $L_2(\mathbb{R}^n)$. Also note that if $s \geqslant 0$ then for all $f \in \mathscr{S}$

$$\| f^0 \|_{(0,s)} = \| f \|_{(0,s)} \leqslant \| f \|_s \tag{42}$$

Theorem 14.24 *Let E be a parametrix for the elliptic operator $\mathscr{A}$ as in* (30). *If $f \in C^\infty(\bar{\Omega}, \mathbb{C}^p)$ then*

$$\| \imath^+ E f^0 \|_s \leqslant C \| f \|_{s-l}, \qquad \forall s \geqslant l. \tag{43}$$

(*In particular, the transmission property holds: if $f \in C^\infty(\bar{\Omega}, \mathbb{C}^p)$ then $\imath^+ E f^0 \in C^\infty(\bar{\Omega}, \mathbb{C}^p)$.*) *Further, if $\mathscr{U} = [\mathscr{U}_j]_{j=0}^{l-1} \in C^\infty(\partial\Omega, \mathbb{C}^{pl})$, then*

$$\| \imath^+ E \mathscr{A}^c \mathscr{U} \|_s \leqslant C \sum_{j=0}^{l-1} \| \mathscr{U}_j \|_{s-j-1/2} \qquad \textit{for any } s. \tag{44}$$

Remark These estimates are in fact valid for any symbol $E \in S^{-l}$ satisfying the transmission conditions: for instance, see §14.6 where we obtain the estimates using an asymptotic expansion for E.

Proof First of all, we have $\| Ef \|_s \leqslant C \| f \|_{s-l}$ for all $f \in C_0^\infty$ because $E \in OS^{-l}(\mathbb{R}^n, p \times p)$. If $\operatorname{supp} f \subset \Omega$ then $f^0 = f$ and it follows that (43) holds in this case. By virtue of a partition of unity argument, it remains to verify (43) when $f \in C_0^\infty$ has support in a compact subset K of a local coordinate patch $\mathfrak{U}' \times (-\varepsilon, \varepsilon)$ near the boundary, $\mathfrak{U}' \subset \mathbb{R}^{n-1}$. Choose

$$\varphi \in C_0^\infty(\mathfrak{U}' \times (-\varepsilon, \varepsilon))$$

equal to 1 in a neighbourhood of K. We will show that

$$\| \varphi \cdot E f^0 \|_{(s-k,k)} \leqslant C \| f \|_{s-l}, \qquad \text{for all integers } k \geqslant s - l, \tag{45}$$

and then use Lemma 14.15 to strengthen this estimate. By Lemma 14.25 with $S = \varphi E$ it follows that

$$\| \varphi \cdot E f^0 \|_{(s-k,k)} \leqslant C \| f^0 \|_{(s-l-k,k)}.$$

Since $k \geqslant s - l \geqslant 0$, we have by (23) that $\| f^0 \|_{(s-l-k,k)} \leqslant \| f^0 \|_{(0,s-l)} \leqslant \| f \|_{s-l}$.

This proves (45). By virtue of (30)

$$\mathscr{A}Ef^0 = f + T_r f^0 \qquad \text{in } \Omega; \tag{46}$$

and we may assume that $T_r \in OS^{-\infty}$ by choosing $\psi \in C_0^\infty(\hat{\Omega})$ equal to 1 on K and replacing T_r by $T_r\psi$, since $\psi f^0 = f^0$. Therefore, Corollary 14.16 with $(m_1, s_1) = (s - k, k)$ and $(m_2, s_2) = (s, 0)$ allows us to conclude that $\varphi \cdot \imath^+ Ef^0 \in W_2^s(\mathbb{R}^n_+)$. The proof of Lemma 14.15 even shows that we have the estimate

$$\|\varphi \cdot \imath^+ Ef^0\|_s \leqslant C'(\|\varphi \cdot \imath^+ Ef^0\|_{(s-k,k)} + \|\varphi \cdot \mathscr{A} \imath^+ Ef^0\|_{s-l}) \leqslant C''\|f\|_{s-l}$$

where the second inequality is due to (45), (46) and the fact that $T_r \in OS^{-\infty}$. Since $\partial\Omega$ is compact, it can be covered by a finite number of coordinate patches $\mathfrak{U}'$, so we get only finitely many such constants C''. Also, by quasi-locality, $(1 - \varphi)Ef^0$ is a continuous function of $f \in L_2(\mathbb{R}^n, \mathbb{C}^p)$ with values in $W_2^N(\mathbb{R}^n, \mathbb{C}^p)$ for any N when $\operatorname{supp} f \subset K$. Since $Ef^0 = \varphi Ef^0 + (1 - \varphi)Ef^0$ then by restricting Ef^0 to Ω we obtain (43).

In view of Lemma 14.19, to prove (44) it suffices (again by a partition of unity argument and quasi-locality of E) to show that when $f \in C_0^\infty$ has compact support in $\mathfrak{U}' \times (-\varepsilon, 0]$, where $\mathfrak{U}' \subset \mathbb{R}^{n-1}$ is a coordinate patch on $\partial\Omega$, then

$$\|\imath^+ Ef\|_s \leqslant C\|f\|_{(-l,s)} \qquad \text{for any } s. \tag{47}$$

If k is an integer $\geqslant \max(s, 0)$ then by Lemma 14.25 we obtain

$$\|\varphi \cdot Ef\|_{(s-k,k)} \leqslant C\|f\|_{(s-l-k,k)} \leqslant C\|f\|_{(-l,s)},$$

where we used (23) in the second inequality (since $k \geqslant s$). By virtue of (30) we have $\mathscr{A}Ef = T_r f$ in Ω, since $\operatorname{supp} f$ does not meet Ω. Once again we may assume that $T_r \in OS^{-\infty}$ and then improve the estimate to obtain (47) in the same way as in the proof of (43).

Remark The proof of (47) shows more generally that when $f \in C_0^\infty$ has support in $\mathfrak{U}' \times (-\varepsilon, 0]$ then

$$\|\imath^+ Ef\|_{l+m+s} \leqslant C\|f\|_{(m,s)} \qquad \text{for any } m \text{ and } s. \tag{47'}$$

Using this result we could give another proof of (43). Indeed, let β: $W_2^s(\Omega) \to W_2^s(\mathbb{R}^n)$ be a continuous extension operator and let θ be the characteristic function of Ω. Since $f^0 = \theta \cdot \beta f = \beta f - (1 - \theta)\beta f$, then

$$\imath^+ Ef^0 = \imath^+ E\beta f - \imath^+ E(1 - \theta)\beta f,$$

and $1 - \theta$ is the characteristic function of $\complement\Omega$. Since $E \in OS^{-l}$, the first term is continuous $W_2^s(\Omega, \mathbb{C}^p) \to W_2^{s+l}(\Omega, \mathbb{C}^p)$, $s \geqslant 0$, and the same is true of the second term due to (47') with $m = 0$.

Lemma 14.25 *Let $S \in OS^{-l}(\mathbb{R}^n)$. For any multi-index α we can write*

$$D^\alpha S = \sum_{|\beta| \leqslant |\alpha|} S_\beta D^\beta \tag{48}$$

where S_β is a p.d.o. of order $\leqslant -l$, and $\beta_n = 0$ if $\alpha_n = 0$. It follows that S is *continuous $H^{t-l,k}(\mathbb{R}^n) \to H^{t,k}(\mathbb{R}^n)$ for every integer $k \geqslant 0$, and all real numbers t, l. (Also, see Theorem 14.34 below.)*

Proof By commuting S with the derivative D_j we have

$$D_jS = SD_j + S_0$$

where $S_0 = [D_jS] \in OS^{-l-1}$. This shows that (48) holds when $|\alpha| = 1$. The general case follows by induction on $|\alpha|$. If k is a nonnegative integer then to estimate the norm $\|Su\|_{t,k}$ we just have to estimate $\|D^\alpha Su\|_t$ for all $|\alpha| \leqslant k$, $\alpha_n = 0$. In view of (48) we obtain

$$\|D^\alpha Su\|_t \leqslant \sum_{|\beta| \leqslant |\alpha|} \|S_\beta D^\beta u\|_t \leqslant \sum_{|\beta| \leqslant |\alpha|} \|D^\beta u\|_{t-l} \leqslant C\|u\|_{t-l,k},$$

where the last inequality holds because $|\beta| \leqslant k$ and $\beta_n = 0$.

Until this point we have worked only with the elliptic operator $\mathscr{A}$ and have not used the boundary operator $\mathscr{B}$. With the aid of the Calderón operator (36) and Corollary 14.23, we now reduce the solvability of $(\mathscr{A}, \mathscr{B}) \in BE^{l,\mathbf{m}}$ to the solvability of a system of pseudo-differential operators on the boundary $\partial\Omega$. Our aim in this section is to extend the types of formulas of Chapter 12 to the context of partial differential operators.

Consider the boundary value problem

$$\mathscr{A}(x, D)u = f(x) \quad \text{in } \Omega, \qquad \mathscr{B}(y, D)u = g(y) \quad \text{on } \partial\Omega \tag{49}$$

It is no restriction to assume that the transversal order of $\mathscr{B}$ is less than l. Indeed, in a neighbourhood of $\partial\Omega$, the elliptic operator has the form $\mathscr{A} = \sum_{j=0}^{l} \mathscr{A}_j D_n^j$ and the coefficient of D_n^l is invertible, so from the equation $\mathscr{A}u = f$ we obtain

$$D_n^l u = A_l^{-1}\left(f - \sum_{j=0}^{l-1} \mathscr{A}_j D_n^j u\right).$$

If the transversal order of $\mathscr{B}$ exceeds $l - 1$, we can use this equation to write $\mathscr{B} = \tilde{\mathscr{B}} + \mathscr{C}\mathscr{A}$, where $\tilde{\mathscr{B}}$ has transversal order $<l$ and $\mathscr{C}$ is an $r \times p$ boundary operator. Then the problem (49) is equivalent to $\mathscr{A}u = f$ in Ω, $\tilde{\mathscr{B}}u = g - \mathscr{C}f$ on $\partial\Omega$.

From now on we always assume that the transversal order of $\mathscr{B}$ is $\mu \leqslant l - 1$ (also see Theorem 10.21). However, we put no restriction on the total order of $\mathscr{B}$.

Recall that the trace operator γ defined by (31) extends continuously to a bounded operator

$$\gamma\colon W_2^s(\Omega) \to \mathop{\times}_{j=0}^{l-1} W_2^{s-j-1/2}(\partial\Omega)$$

for $s > l - \frac{1}{2}$; see Theorem 7A.10 (where different notation was used there). The boundary operator has the form

$$\mathscr{B}(y, D) = \sum_{j=0}^{l-1} \mathscr{B}_j(y, D')D_n^j$$

where $\mathscr{B}_j = [b_{ki}^j]_{r\times p}$ and $b_{kl}^j \in OClS^{m_k - j}(\partial\Omega)$, and the boundary conditions $\mathscr{B}(y, D)u = g$ can be written as

$$\mathscr{B}^c \gamma u = g, \qquad \text{where } \mathscr{B}^c = [\mathscr{B}_0 \cdots \mathscr{B}_{l-1}].$$

The matrix p.d.o. $\mathscr{B}^c$ is a block DN operator on $\partial\Omega$,

$$\mathscr{B}^c = [\mathscr{B}_{kj}^c],$$

with the $1 \times p$ matrix $\mathscr{B}_{kj}^c = [b_{ki}^j]_{i=1,\ldots,p}$ in the kth row and jth block ($k = 1, \ldots, r$, $j = 0, \ldots, l-1$) having order $\leqslant m_k - j$. Also, the Calderón operator

$$P = [P_{ij}]$$

is a block DN operator with the $p \times p$ matrix P_{ij} in the ith block-row and jth block-column ($i, j = 0, \ldots, l-1$) having order $\leqslant i - j$.

Proposition 14.26 *Let $\mathscr{A} \in \mathrm{Ell}^l$. Then the boundary value problem $(\mathscr{A}, \mathscr{B})$ satisfies the L-condition if and only if the principal part, $\pi\mathscr{B}^c$, restricted to the image of $\pi P(y, \xi')$ is bijective for all $(y, \xi') \in T^*(\partial\Omega)\backslash 0$.*

Proof This has already been proved in §10.1, see the corollary to Theorem 10.1. For convenience let us recall the proof of sufficiency here. Notice that

$$\pi\mathscr{B}^c = [B_0 \cdots B_{l-1}]$$

where $B_j = \pi\mathscr{B}_j$ are the coefficients of the matrix polynomial $B(\lambda) = \pi\mathscr{B}(y, (\xi', \lambda))$. Let $L(\lambda) = \pi\mathscr{A}(y, (\xi', \lambda))$ and choose a spectral triple (X_+, T_+, Y_+) for $L(\lambda)$ with respect to the eigenvalues in the upper half-plane $\operatorname{Im}\lambda > 0$. Since πP is the "Calderón projector" from §2.2 we have

$$\pi P = P^+ = \operatorname{col}(X_+ T_+^j)_{j=0}^{l-1} \cdot [Y_+ \cdots T_+^{l-1} Y_+]\mathscr{Z}, \qquad \text{see §2.2 (8).}$$

Now it follows that

$$\pi\mathscr{B}^c \cdot \pi P = \Delta_{\mathscr{B}}^+ \cdot [Y_+ \cdots T_+^{l-1} Y_+]\mathscr{Z},$$

where as usual $\Delta_{\mathscr{B}}^+ = \sum_{j=0}^{l-1} B_j X_+ T_+^j$. Since $\Delta_{\mathscr{B}}^+$ is invertible the surjectivity of $\pi\mathscr{B}^c$ when restricted to the image of πP follows immediately from surjectivity of $[Y_+ \cdots T_+^{l-1} Y_+]\mathscr{Z}$. On the other hand, if $\pi\mathscr{B}^c v = 0$ for some v in the image of πP then $[Y_+ \cdots T_+^{l-1} Y_+]\mathscr{Z} v = 0$ and hence $v = \pi P v = 0$. This completes the proof of sufficiency.

In virtue of Theorem 14.20 we have $P^2 \equiv P$, and then by Corollary 14.23 (with p replaced by pl) there exist DN operators on $M = \partial\Omega$,

$$S = [S_{ik}], \qquad S_{ik} \in OClS^{i - m_k}(\partial\Omega, p \times 1), \qquad i = 0, \ldots, l-1, k = 1, \ldots, r,$$

$$S' = [S'_{ij}], \qquad S'_{ij} \in OClS^{i-j}(\partial\Omega, p \times p), \qquad i, j = 0, \ldots, l-1$$

such that

$$\mathscr{B}^c S \equiv I_r, \qquad PS \equiv S; \tag{50}$$

$$S' + S\mathscr{B}^c \equiv I_{pl}, \qquad S'P \equiv 0. \tag{51}$$

In the next theorem we use the operators S, S' to construct a parametrix $\mathfrak{P}$ for $(\mathscr{A}, \mathscr{B})$. This construction is of course similar to the formula for the solution in Theorem 12.12. Note that the conclusion of the theorem is that

$$\mathfrak{P}\mathfrak{L} = I + K, \qquad \mathfrak{L}\mathfrak{P} = I + K'$$

where the operator $K\colon W_2^s(\Omega, \mathbb{C}^p) \to W_2^t(\Omega, \mathbb{C}^p)$ is continuous for any t $(s \geqslant l)$ and the operator

$$K'\colon W_2^0(\Omega, \mathbb{C}^p) \times W_2^m(\partial\Omega, \mathbb{C}^r) \to W_2^t(\Omega, \mathbb{C}^p) \times W_2^t(\partial\Omega, \mathbb{C}^r)$$

is continuous for any m and t.

Theorem 14.27 *If the boundary value problem operator $\mathfrak{L} = (\mathscr{A}, \mathscr{B})$ is L-elliptic then the operator*

$$\mathfrak{P}\colon W_2^{s-l}(\Omega, \mathbb{C}^p) \times \mathop{\times}_{k=1}^{r} W_2^{s-m_k-1/2}(\partial\Omega) \to W_2^s(\Omega, \mathbb{C}^p)$$

defined by $\mathfrak{P}(f, g) = \imath^+(I + E\mathscr{A}^c S'\gamma)Ef^0 + \imath^+ E\mathscr{A}^c Sg$ is continuous for every $s \geqslant l$; $\mathfrak{P}$ is a parametrix for $\mathfrak{L}$, that is,

$$\mathfrak{P}\mathfrak{L} = I + K, \qquad \mathfrak{L}\mathfrak{P} = I + K',$$

where $K\colon W_2^s(\Omega, \mathbb{C}^p) \to C^\infty(\bar{\Omega}, \mathbb{C}^p)$ is continuous and

$$K' = \begin{pmatrix} K_1 & K_2 \\ K_3 & K_4 \end{pmatrix}\colon W_2^0(\Omega, \mathbb{C}^p) \times \mathscr{D}'(\partial\Omega, \mathbb{C}^r) \to C^\infty(\bar{\Omega}, \mathbb{C}^p) \times C^\infty(\partial\Omega, \mathbb{C}^r)$$

is continuous.

Proof The continuity of $\mathfrak{P}$ is an easy consequence of Theorem 14.24, taking into account the orders of the blocks S_{ik} and S'_{ij} in the matrix operators S and S'. From (51) it follows that $S'(I - P) + S\mathscr{B}^c = I - R$, where $R \in OS^{-\infty}(\partial\Omega, pl \times pl)$. Then, substituting this equation in the Green formula (35), we obtain

$$u^0 + T_l u^0 = Ef^0 + E\mathscr{A}^c(S'(I - P)\gamma u + SB^c\gamma u + R\gamma u). \tag{52}$$

Further, by applying the trace operator γ to (35), it also follows that $\gamma u + T_l u^0 = \gamma Ef^0 + P\gamma u$, that is,

$$(I - P)\gamma u = \gamma Ef^0 - \gamma T_l u^0,$$

and substituting this result in (52) and then applying the restriction operator $\imath^+\colon \mathscr{D}'(\mathbb{R}^n) \to \mathscr{D}'(\Omega)$, we obtain

$$u + Ku = \imath^+(I + E\mathscr{A}^c S'\gamma)Ef^0 + \imath^+ E\mathscr{A}^c Sg, \tag{53}$$

where $Ku = \imath^+ E\mathscr{A}^c(S'\gamma T_l u^0 - R\gamma u) + \imath^+ T_l u^0$. As usual, we may assume that $T_l \in OS^{-\infty}$ since T_l can be replaced by $T_l\psi$ where $\psi \in C_0^\infty(\hat{\Omega})$ and $\psi = 1$ on $\bar{\Omega}$. It follows that K is a continuous map $W_2^l(\Omega, \mathbb{C}^p) \to C^\infty(\bar{\Omega}, \mathbb{C}^p)$ because $R \in OS^{-\infty}$ and $T_l \in OS^{-\infty}$.

The equation (53) has the form $u + Ku = \mathfrak{P}(f, g)$, which shows that $\mathfrak{P}$ is a left inverse of $\mathscr{L}$ modulo the operator K, i.e. $\mathfrak{P}\mathfrak{L} = I + K$.

It is also an approximate right inverse. In fact, with the notation $u = \mathfrak{P}(f, g)$ it follows by (30), and the fact that $\mathscr{A}$ is a local operator, that

$$\mathscr{A}u = (I + T_r)(f^0 + \mathscr{A}^c S'\gamma E f^0) + (I + T_r)\mathscr{A}^c Sg \quad \text{in } \Omega,$$

Since $\imath^+\mathscr{A}^c = 0$ then

$$\mathscr{A}u = f + K_1 f + K_2 g \qquad \text{in } \Omega, \tag{54}$$

where $K_1 f = \imath^+ T_r f^0 + \imath^+ T_r \mathscr{A}^c S'\gamma E f^0$ and $K_2 g = \imath^+ T_r \mathscr{A}^c Sg$. Since we can replace T_r by $T_r\psi \in OS^{-\infty}$, then K_1 is a continuous map $W_2^0(\Omega, \mathbb{C}^p) \to C^\infty(\bar{\Omega}, \mathbb{C}^p)$ and K_2 is a continuous map $\mathscr{D}'(\partial\Omega, \mathbb{C}^r) \to C^\infty(\bar{\Omega}, \mathbb{C}^p)$. With u still equal to $\mathfrak{P}(f, g)$, we have

$$\begin{aligned}\gamma u &= \gamma E f^0 + PS'\gamma E f^0 + PSg\\ &= (I + PS')\gamma E f^0 + PSg,\end{aligned}$$

whence

$$\mathscr{B}u = \mathscr{B}^c \gamma u = g + K_3 f + K_4 g, \tag{55}$$

where $K_3 f = \mathscr{B}^c(I + PS')\gamma E f^0$ and $K_4 g = (\mathscr{B}^c PS - I)g$. In view of (50), it follows that $\mathscr{B}^c PS \equiv I$, so K_4 is of order $-\infty$ and is a continuous map from $\mathscr{D}'(\partial\Omega, \mathbb{C}^r)$ to $C^\infty(\partial\Omega, \mathbb{C}^r)$.

It remains to prove the continuity of K_3. To do so, first observe the following fact which follows from the equations (50), (51). If we multiply the equation $S\mathscr{B}^c + S' \equiv I$ by $\mathscr{B}^c P$, and recall that $\mathscr{B}^c PS \equiv \mathscr{B}^c S \equiv I$, we obtain $\mathscr{B}^c + \mathscr{B}^c PS' \equiv \mathscr{B}^c P$, that is,

$$\mathscr{B}^c(I + PS') \equiv \mathscr{B}^c P. \tag{56}$$

By adding and subtracting $\mathscr{B}^c \mathscr{P}\gamma E f^0$ we can write $K_3 f$ in the form

$$K_3 f = (\mathscr{B}^c(I + PS') - \mathscr{B}^c P)\cdot\gamma E f^0 + \mathscr{B}^c \cdot P\gamma E f^0,$$

and then Proposition 14.21 implies that K_3 is a continuous map from $W_2^0(\Omega, \mathbb{C}^p)$ to $C^\infty(\partial\Omega, \mathbb{C}^r)$. The equations (54), (55) show that $\mathfrak{L}\mathfrak{P} = I + K'$.

Remark 14.28 We can write down a formula for the DN principal part of S in terms of a spectral pair, (X_+, T_+), with respect to the eigenvalues in the upper half-plane $\operatorname{Im}\lambda > 0$ for the matrix polynomial $L(\lambda) = \pi\mathscr{A}(y, (\xi', \lambda))$. By taking the DN principal parts of the two equations in (50) it follows from the uniqueness in Proposition 12.11 that

$$\pi S = \operatorname{col}(X_+ T_+^j)_{j=0}^{l-1}\cdot(\Delta_{\mathscr{B}}^+)^{-1},$$

where $\Delta_{\mathscr{B}}^+$ is defined as in Chapter 10 (36).

In view of Theorem 14.27 we have another proof of the Fredholm property for elliptic boundary problems. This follows from Theorem 9.11, since the embeddings $C^\infty(\bar{\Omega}) \subset W_2^m(\Omega)$ and $C^\infty(\partial\Omega) \subset W_2^t(\partial\Omega)$ are continuous for all m, t, and $W_2^{m+1}(\Omega) \subset W_2^m(\Omega)$ and $W_2^{t+1}(\partial\Omega) \subset W_2^t(\partial\Omega)$ are compact. Since $u = \mathfrak{P}\mathfrak{L}u - Ku$, the continuity properties of $\mathfrak{P}$ and K show that the following

Regularity Theorem holds: *If* $u \in W_2^l(\Omega, \mathbb{C}^p)$ *and*

$$\mathscr{A}u \in W_2^{s-l}(\Omega, \mathbb{C}^p), \qquad \mathscr{B}u \in \bigtimes_{k=1}^{r} W_2^{s-m_k-1/2}(\partial\Omega), \qquad s \geq l,$$

then $u \in W_2^s(\Omega, \mathbb{C}^p)$. In particular, the kernel of $\mathfrak{L}$ is contained in $C^\infty(\bar{\Omega}, \mathbb{C}^p)$.

As mentioned earlier, the properties of the parametrix, $\mathfrak{P}$, make it possible to show a further result, namely, that the image of $\mathfrak{L}$ is the orthogonal space of a finite dimensional subspace $\subset C^\infty$.

For the proof we need to use (33), (34) which when written out more fully becomes

$$(u, \mathscr{A}^*v)_\Omega = (\mathscr{A}u, v)_\Omega + i^{-1} \sum_{j=0}^{l-1} \sum_{k=0}^{j} (\mathscr{A}_{j+1}\gamma_{j-k}u, \gamma_k v)_{\partial\Omega} \tag{57}$$

for all $u, v \in C^\infty(\bar{\Omega}, \mathbb{C}^p)$. Here we have used the notation

$$(f_1, f_2)_\Omega := \int_\Omega f_1 \cdot {}^h f_2 \, dx, \qquad (g_1, g_2)_{\partial\Omega} := \int_{\partial\Omega} g_1 \cdot {}^h g_2 \, d\sigma$$

where dx is Lebesgue measure in $\mathbb{R}^n$ and $d\sigma$ is the measure on $\partial\Omega$ induced by the standard Riemannian structure in $\mathbb{R}^n$. As usual, the superscript ^{h} indicates the Hermitian adjoint, i.e. ${}^h f_2$ is the conjugate transpose of the vector function f_2.

Recall that $\mathscr{A}_j$ is a differential operator of order $l - j$ on $\partial\Omega$. By introducing the $pl \times pl$ differential operator

$$\mathscr{L} = \begin{pmatrix} \mathscr{A}_1 & \mathscr{A}_2 & \cdots & \mathscr{A}_l \\ \mathscr{A}_2 & & & \\ \vdots & & \iddots & 0 \\ \mathscr{A}_l & & & \end{pmatrix},$$

we can write the equation (57) in the form

$$(u, \mathscr{A}^*v)_\Omega = (\mathscr{A}u, v)_\Omega + i^{-1}(\mathscr{L}\gamma u, \gamma v)_{\partial\Omega} \tag{58}$$

and since $\mathscr{A}_l$ is an invertible matrix function, rather than a differential operator, the operator $\mathscr{L}$ has an inverse $\mathscr{L}^{-1}$ which is a differential operator.

In the proof of the following theorem, one should recall that as usual we identify the dual space of the Hilbert space $L_2(\Omega)$ with itself by means of the inner product $(\ ,\ )_\Omega$, and the dual space of $W_2^t(\partial\Omega)$ with $W_2^{-t}(\partial\Omega)$, $t \in \mathbb{R}$, by means of an extension of the inner product $(\ ,\ )_{\partial\Omega}$.

Theorem 14.29 *If the boundary value problem* (49) *is L-elliptic then the boundary value problem operator* $\mathfrak{L}$ *is a Fredholm operator for every* $s \geq l$. *The kernel of* $\mathfrak{L}$ *is in* $C^\infty(\bar{\Omega}, \mathbb{C}^p)$ *and the image is the orthogonal space of a finite dimensional subspace of* $C^\infty(\bar{\Omega}, \mathbb{C}^p) \times \bigtimes_{k=1}^{r} C^\infty(\partial\Omega)$, *that is, there exist a finite number of functions* $f_1, \ldots, f_\beta \in C^\infty(\bar{\Omega}, \mathbb{C}^p)$ *and* $g_1, \ldots, g_\beta \in C^\infty(\partial\Omega, \mathbb{C}^r)$ *such that if* $(f, g) \in W_2^{s-l}(\Omega, \mathbb{C}^p) \times \bigtimes_{k=1}^{r} W_2^{s-m_k-1/2}(\partial\Omega)$ *then* $f = \mathscr{A}u$, $g = \mathscr{B}u$

for some $u \in W_2^s(\Omega, \mathbb{C}^p)$ *if and only if*

$$\int_\Omega f\cdot{}^h f_i\,dx + \int_{\partial\Omega} g\cdot{}^h g_i\,d\sigma = 0 \qquad i = 1, \ldots, \beta.$$

Thus, the index is independent of s.

Proof As mentioned above we have already shown in §10.7 that $\mathfrak{L}$ is a Fredholm operator. To complete the proof, it suffices to show that the image of $\mathfrak{L}$ is defined by C^∞ relations when $s = l$, because then it holds for any $s \geqslant l$ by virtue of the Regularity Theorem mentioned above. Let (v, w) be an element of the annihilator of im $\mathfrak{L}$ in the dual space of $L_2(\Omega, \mathbb{C}^p) \times \mathsf{X}_{k=1}^r W_2^{l-m_k-1/2}(\partial\Omega)$, that is,

$$(\mathscr{A}u, v)_\Omega + (\mathscr{B}u, w)_{\partial\Omega} = 0 \qquad \text{for all } u \in C^\infty(\bar{\Omega}, \mathbb{C}^p), \tag{59}$$

where $v \in L_2(\Omega, \mathbb{C}^p)$ and $w \in \mathsf{X}_{k=1}^r W_2^{-l+m_k+1/2}(\partial\Omega)$. In particular, with $u = \mathfrak{P}(f, g)$, $f = 0$ and $g \in C^\infty(\partial\Omega, \mathbb{C}^r)$, we have $\mathscr{A}u = K_2 g$, $\mathscr{B}u = g + K_4 g$, hence

$$(g, w)_{\partial\Omega} = -(K_2 g, v)_\Omega - (K_4 g, w)_{\partial\Omega}. \tag{60}$$

Recall that the distribution topology on $\mathscr{D}'(\partial\Omega)$ is the weakest topology making the linear forms $g \mapsto (g, \varphi)_{\partial\Omega}$ continuous for every $\varphi \in C^\infty(\partial\Omega)$. Since K_2 and K_4 are continuous maps from $\mathscr{D}'$ to C^∞, then (60) implies that w is a continuous linear form on $g \in \mathscr{D}'(\partial\Omega, \mathbb{C}^r)$ in the distribution topology. By an elementary fact on weak topologies it follows that $w \in C^\infty(\partial\Omega, \mathbb{C}^r)$.

It follows from (59) that

$$\mathscr{A}^* v = 0 \qquad \text{in } \Omega,$$

by taking u equal to 0 near $\partial\Omega$. Now let $\varphi \in C_0^\infty(\mathfrak{U}' \times (-\varepsilon, \varepsilon))$, then by Corollary 14.16 we have $\varphi \cdot v \in H^{s, -s}(\mathbb{R}_+^n)$ for all s (in local coordinates), and if we choose $s \geqslant l$ then by Proposition 14.17 the Cauchy data

$$\gamma v = \lim_{x_n \to 0^+} [D_n^j v]_{j=0}^{l-1}$$

are well-defined.

We claim that these Cauchy data must be in C^∞. Indeed, by (58) and (59)

$$(\mathscr{B}u, w)_{\partial\Omega} = i^{-1}(\mathscr{L}\gamma u, \gamma v)_{\partial\Omega}, \qquad \forall u \in C^\infty(\bar{\Omega}, \mathbb{C}^p).$$

Since $\mathscr{B}u = \mathscr{B}^c \gamma u$, then

$$(\mathscr{B}^c \mathscr{L}^{-1} \mathscr{U}, w)_{\partial\Omega} = i^{-1}(\mathscr{U}, \gamma v)_{\partial\Omega}$$

for all $\mathscr{U} \in C^\infty(\partial\Omega, \mathbb{C}^{pl})$. By transposing the pseudo-differential operator $\mathscr{B}^c$ and the differential operator $\mathscr{L}^{-1}$, we would obtain an expression for γv in terms of various pseudo-differential operators acting on w. Since w is C^∞, then so is γv.

Now, because $\mathscr{A}_l$ is an invertible matrix, we can use the equation $\mathscr{A}^* v = 0$ to solve for the higher-order normal derivatives $D_n^j v$, $j \geqslant l$, in terms of the Cauchy data; hence all normal derivatives of v are C^∞ on $\partial\Omega$. If we apply Exercise 3 of Chapter 13 locally, then use a partition of unity on $\partial\Omega$, it follows that there exists $\tilde{v} \in C_0^\infty(\hat{\Omega}\backslash\Omega)$ such that $D_n^j \tilde{v} = D_n^j v$ on $\partial\Omega$ for all j. Then $\mathscr{A}^*\tilde{v}$ vanishes to infinite order on $\partial\Omega$ and we extend v to $\hat{\Omega}$ by setting $v = \tilde{v}$ on $\hat{\Omega}\backslash\Omega$. After this extension we have $v \in L_2(\mathbb{R}^n)$, supp v is a compact subset of $\hat{\Omega}$, and $\mathscr{A}^*v \in C_0^\infty(\hat{\Omega})$. Since $\mathscr{A}^*$ is elliptic on $\hat{\Omega}$, it follows that $v \in C_0^\infty(\hat{\Omega})$. Indeed, the adjoint of the equation $\mathscr{A}E = I + T_r$ implies that

$$v = E^*\mathscr{A}^*v - T_r^*v \in C^\infty,$$

since $E^*\mathscr{A}^*v \in C^\infty$ and $T_r^*v = T_r^*\psi v \in C^\infty$ (where $\psi \in C_0^\infty(\hat{\Omega})$ is equal to 1 on supp v). By restriction to $\bar{\Omega}$, we obtain $v \in C^\infty(\bar{\Omega}, \mathbb{C}^p)$.

Thus, we have show that the annihilator, $N = (\text{im } \mathfrak{L})^\circ$, of the image of $\mathfrak{L}$ is contained in $C^\infty\bar{\Omega}, \mathbb{C}^p) \times \times_{k=1}^r C^\infty(\partial\Omega)$. We already know from §10.7 that $\mathfrak{L}$ is a Fredholm operator so N has finite dimension. One could also argue as follows. An application of the closed graph theorem implies that the inclusion is a continuous embedding, i.e. the topology on N as a subspace of the dual of $L_2(\Omega, \mathbb{C}^p) \times \times_{k=1}^r W_2^{l-m_k-1/2}(\partial\Omega)$ coincides with the topology on $C^\infty(\bar{\Omega}, \mathbb{C}^p) \times \times_{k=1}^r C^\infty(\partial\Omega)$. Since the latter space is a Montel space (i.e. closed, bounded sets are compact in the C^∞ topology) and the former space is a Banach space, it follows that N is a locally compact Banach space, hence it has finite dimension.

Since im $\mathfrak{L}$ is closed it follows that im $\mathfrak{L}$ is the annihilator of N, i.e. $(f, g) \in \text{im } \mathfrak{L}$ if and only if

$$\int_\Omega f \cdot {}^h f' \, dx + \int_{\partial\Omega} g \cdot {}^h g' \, d\sigma = 0$$

for all $(f', g') \in N$. By taking a basis for N, the proof of the theorem is complete.

14.5 An application to the index

Let M be a compact manifold without boundary and suppose M is a union

$$M = \bar{\Omega}_+ \cup \bar{\Omega}_-,$$

with $\Omega_+ = \Omega$ and common boundary $\partial\Omega = \bar{\Omega}_+ \cap \bar{\Omega}_-$. For example, we can identify $\mathbb{R}^n$ as a submanifold of $M = S^n$ by means of stereographic projection from the south pole, and then $M = \bar{\Omega}_+ \cup \bar{\Omega}_-$ where $\Omega_+ = \Omega$ and $\Omega_- = S^n\backslash\bar{\Omega}$. Another possibility is to let M be the boundaryless double of $\bar{\Omega}$.

Let n be the unit normal along $\partial\Omega$ that points into Ω_+, then $-n$ points into Ω_-. Choose a tubular neighbourhood of $\partial\Omega$ such that $\partial\Omega \times [0, 1) \subset \bar{\Omega}_+$ and $\partial\Omega \times (-1, 0] \subset \bar{\Omega}_-$, i.e. $x_n > 0$ in Ω_+ and $x_n < 0$ in Ω_-.

Theorem 14.30 *Suppose that $A: C^\infty(M, \mathbb{C}^p) \to C^\infty(M, \mathbb{C}^p)$ is a $p \times p$ elliptic differential operator on M of order l, and denote by $\mathscr{A}_\pm = \imath^+ A$ the induced operator $C^\infty(\bar{\Omega}_\pm, \mathbb{C}^p) \to C^\infty(\bar{\Omega}_\pm, \mathbb{C}^p)$. Let $\mathscr{B}_\pm = \sum_{j=0}^{l-1} \mathscr{B}_j^\pm D_n^j$ be boundary operators such that $(\mathscr{A}_\pm, \mathscr{B}_\pm)$ satisfies the L-condition. Then*

$$\text{ind}(\mathscr{A}_+, \mathscr{B}_+) + \text{ind}(\mathscr{A}_-, \mathscr{B}_-) = \text{ind } A - \text{ind } R \tag{61}$$

where $R \in OClS^0(\partial\Omega, pl \times pl)$ *is an elliptic operator of order* 0 *with principal symbol* πR: $ST^*(\partial\Omega) \to GL_{pl}(\mathbb{C})$ *given by (recall that* $pl = 2r$*)*

$$\pi R(y, \xi') = [\mathrm{col}(X_+T_+^j)_{j=0}^{l-1} \cdot (\Delta_{\mathscr{B}_+}^+)^{-1} \quad \mathrm{col}(X_-T_-^j)_{j=0}^{l-1} \cdot (\Delta_{\mathscr{B}_-}^+)^{-1}] \quad (62)$$

where $\Delta_{\mathscr{B}_\pm}^+ = \sum_{j=0}^{l-1} B_j^\pm X_\pm T_\pm^j$ *and* $B_j^\pm = \pi\mathscr{B}_j^\pm$. *Here* $(X_\pm, T_\pm)$ *are spectral pairs for* $L(\lambda) = \pi A(y, \xi' + \lambda n(y))$ *with respect to the eigenvalues in the upper and lower half-planes, respectively, depending smoothly on* $(y, \xi') \in T^*(\partial\Omega)\backslash 0$. *(As above, n is the unit normal along* $\partial\Omega$ *that points into* Ω_+.*)*

Proof Without loss of generality, the order of $\mathscr{B}_\pm$ in each row is equal to $l - \frac{1}{2}$ so that the boundary operators are continuous maps $\mathscr{B}_\pm : W_2^l(\Omega_\pm, \mathbb{C}^p) \to L_2(\partial\Omega, \mathbb{C}^r)$. Denote by $\tilde{M}$ the disjoint union of $\bar{\Omega}_+$ and $\bar{\Omega}_-$. By taking the direct sum, $\mathfrak{L} = \mathfrak{L}_+ \oplus \mathfrak{L}_-$, of the operators $\mathfrak{L}_\pm = (\mathscr{A}_\pm, \mathscr{B}_\pm)$: $W_2^l(\Omega_\pm, \mathbb{C}^p) \to L_2(\Omega_\pm, \mathbb{C}^p) \times L_2(\partial\Omega, \mathbb{C}^r)$ we obtain

$$\mathfrak{L}: W_2^l(\tilde{M}, \mathbb{C}^p) \to L_2(M, \mathbb{C}^p) \times L_2(\partial\Omega, \mathbb{C}^r \oplus \mathbb{C}^r)$$

$$(u_+, u_-) \mapsto (\mathscr{A}_+u_+, \mathscr{A}_-u_-; \mathscr{B}_+u_+, \mathscr{B}_-u_-),$$

where $W_2^l(\tilde{M}, \mathbb{C}^p) = W_2^l(\Omega_+, \mathbb{C}^p) \times W_2^l(\Omega_-, \mathbb{C}^p)$. Denote by J: $W_2^l(M, \mathbb{C}^p) \to W_2^l(\tilde{M}, \mathbb{C}^p)$ the natural embedding $u \mapsto (u|_{\Omega_+}, u|_{\Omega_-})$. Then obviously

$$A = \jmath \circ \mathfrak{L} \circ J, \quad (63)$$

where $\jmath$: $L_2(M, \mathbb{C}^p) \times L_2(\partial\Omega, \mathbb{C}^r \oplus \mathbb{C}^r) \to L_2(M, \mathbb{C}^p)$ is the canonical projection. Here we have identified $L_2(\tilde{M}, \mathbb{C}^p) \simeq L_2(M, \mathbb{C}^p)$, i.e. $(u_+, u_-) \in L_2(\tilde{M}, \mathbb{C}^p)$ is identified with $u \in L_2(M, \mathbb{C}^p)$ such that $u|_{\Omega_\pm} = u_\pm$. In that way, if $u \in W_2^l(M, \mathbb{C}^p)$ then $(\mathscr{A}_+u_+, \mathscr{A}_-u_-)$ is identified with Au and (63) follows.

Let $E \in OS^{-l}(M, p \times p)$ be a parametrix for A, and let $P_+ = \gamma_+ E\mathscr{A}_+^c$, $P_- = -\gamma_- E\mathscr{A}_-^c$ be the Calderón operators for $\mathscr{A}_+, \mathscr{A}_-$, respectively. Here $\gamma_\pm u$ denotes the Cauchy data operator taken from inside $\Omega_\pm$, that is, $\gamma_\pm u = \lim_{x_n \to 0^\pm} [D_n^j u]_{j=0}^{l-1}$. Then we may construct parametrices $\mathfrak{P}_\pm$: $L_2(\Omega_\pm, \mathbb{C}^p) \times L_2(\partial\Omega, \mathbb{C}^r) \to W_2^l(\Omega_\pm, \mathbb{C}^p)$ of $(\mathscr{A}_\pm, \mathscr{B}_\pm)$ as in Theorem 14.27,

$$\mathfrak{P}_\pm(f, g) := \imath^\pm(I + E\mathscr{A}_\pm^c S'_\pm \gamma_\pm)Ef^0 + \imath^\pm E\mathscr{A}_\pm^c S_\pm g, \quad (64)$$

where $S'_\pm, S_\pm$ are defined by (50), (51) for the boundary operator $\mathscr{B}_\pm$. It follows that the direct sum $\mathfrak{P} = \mathfrak{P}_+ \oplus \mathfrak{P}_-$ is a parametrix of $\mathfrak{L}$. Now define an operator

$$G: W_2^l(\tilde{M}, \mathbb{C}^p) \to \mathop{\times}_{j=0}^{l-1} W_2^{l-j-1/2}(\partial\Omega, \mathbb{C}^p)$$

$$(u_+, u_-) \mapsto \gamma_+u_+ - \gamma_-u_-$$

By virtue of Lemma 7A.11 we have $\ker G = \mathrm{im}\, J$, i.e. $\gamma_+u_+ = \gamma_-u_-$ if and only if there exists $u \in W_2^l(M, \mathbb{C}^p)$ such that $u|_{\Omega^\pm} = u_\pm$. Let

$$\jmath: L_2(\partial\Omega, \mathbb{C}^r \oplus \mathbb{C}^r) \to L_2(M, \mathbb{C}^p) \times L_2(\partial\Omega, \mathbb{C}^r \oplus \mathbb{C}^r)$$

be the canonical embedding $(g_1, g_2) \mapsto (0, 0; g_1, g_2)$.

We claim that $G \circ \mathfrak{P} \circ \jmath$ is a DN elliptic p.d.o. on $\partial\Omega$. Indeed, by (64)

$$G \circ \mathfrak{P} \circ \jmath(g_1, g_2) = \gamma_+ E\mathscr{A}_+^c S_+ g_1 - \gamma_- E\mathscr{A}_-^c S_- g_2 = P_+S_+g_1 + P_-S_-g_2,$$

that is,

$$G \circ \mathfrak{P} \circ \mathscr{j} = [P_+S_+ \quad P_-S_-] \tag{65}$$

a $pl \times pl$ matrix operator. By Remark 14.28 the DN principal part of $P_\pm S_\pm$ is

$$\pi(P_\pm S_\pm) = \pi P_\pm \operatorname{col}(X_\pm T_\pm^j)_{j=0}^{l-1} \cdot (\Delta_{\mathscr{B}_\pm}^+)^{-1} = \operatorname{col}(X_\pm T_\pm^j)_{j=0}^{l-1} \cdot (\Delta_{\mathscr{B}_\pm}^+)^{-1},$$

(Note that $\pi P_\pm$ is the Calderón projector corresponding to the eigenvalues in the upper and lower half-planes, respectively, so $\operatorname{im} \pi P_\pm = \operatorname{im} \operatorname{col}(X_\pm T_\pm^j)_{j=0}^{l-1}$.) In view of (65) we see that $G \circ \mathfrak{P} \circ \mathscr{j}$ is a DN operator, with DN principal part $\pi(G \circ \mathfrak{P} \circ \mathscr{j})$ equal to the right-hand side of (62) on $ST^*(\partial\Omega)$. Since

$$\operatorname{im} \pi P_+ \oplus \operatorname{im} \pi P_- = \mathbb{C}^{pl},$$

it follows that $\pi(G \circ \mathfrak{P} \circ \mathscr{j})$ has invertible values, that is, $G \circ \mathfrak{P} \circ \mathscr{j}$ is DN elliptic.

Now choose $\Lambda^k \in OClS^k(\partial\Omega)$ such that $\Lambda^k: W_2^s(\partial\Omega) \to W_2^{s-k}(\partial\Omega)$ is an isomorphism for all s, let $\Lambda = [\delta_{ij}\Lambda^{-l+j+1/2}]_{i,j=0}^{l-1}$ and

$$R := (\Lambda \circ G) \circ \mathfrak{P} \circ \mathscr{j}. \tag{66}$$

Then R is a $2r \times 2r$ classic p.d.o. of order 0 on $\partial\Omega$, with principal symbol (62). We now have the following commutative diagram:

$$\begin{array}{ccccccccc}
0 & \longrightarrow & W_2^l(M, \mathbb{C}^p) & \xrightarrow{J} & W_2^l(\tilde{M}, \mathbb{C}^p) & \xrightarrow{\Lambda \circ G} & \displaystyle\mathop{\times}_{0}^{l-1} L_2(\partial\Omega, \mathbb{C}^p) & \longrightarrow & 0 \\
 & & \Big\downarrow A & & \mathfrak{L}\Big\uparrow\Big\downarrow\mathfrak{P} & & \Big\uparrow R & & \\
0 & \longleftarrow & L_2(M, \mathbb{C}^p) & \xrightarrow{\mathscr{s}} & \begin{matrix} L_2(M, \mathbb{C}^p) \\ \oplus \\ L_2(\partial\Omega, \mathbb{C}^r \oplus \mathbb{C}^r) \end{matrix} & \xrightarrow{\mathscr{j}} & L_2(\partial\Omega, \mathbb{C}^r \oplus \mathbb{C}^r) & \longleftarrow & 0
\end{array}$$

where the rows are exact and the identities (63) and (66) hold. Thus, the formula (61) is a consequence of the following proposition.

Proposition 14.31 *All species E, F, M, N, U and V in the diagram below are Banach spaces and the maps are continuous, linear operators. Let $\mathfrak{L} \in \mathscr{L}(V, U)$ be a Fredholm operator and $\mathfrak{P} \in \mathscr{L}(U, V)$ be a regularizer for $\mathfrak{L}$, and define $A := \mathscr{s} \circ \mathfrak{L} \circ J$, $R := S \circ \mathfrak{P} \circ \mathscr{j}$, so that the diagram is commutative:*

$$\begin{array}{ccccccccc}
0 & \longrightarrow & E & \xrightarrow{J} & V & \xrightarrow{S} & F & \longrightarrow & 0 \\
 & & \Big\downarrow A & & \mathfrak{L}\Big\uparrow\Big\downarrow\mathfrak{P} & & \Big\uparrow R & & \\
0 & \longleftarrow & M & \xleftarrow{\mathscr{s}} & U & \xleftarrow{\mathscr{j}} & N & \longleftarrow & 0
\end{array}$$

Suppose that the top row is exact, and the bottom row is split-exact (i.e. it is exact, and $\operatorname{im} \mathscr{j} = \ker \mathscr{s}$ is complemented in U). Then A is a Fredholm operator if and only if R is a Fredholm operator. If A is Fredholm then

$$\operatorname{ind} A = \operatorname{ind} \mathfrak{L} + \operatorname{ind} R. \tag{67}$$

Proof Before proving (67) we show that it is possible to make some simplifying assumptions. First, we claim that without loss of generality ind $\mathfrak{L} = 0$. Indeed, suppose that $\mathfrak{L}' \in \mathscr{L}(V', U')$ is another Fredholm operator, with regularizer $\mathfrak{P}' \in \mathscr{L}(U', V')$. By introducing the operators

$$\tilde{\mathfrak{L}} = \begin{pmatrix} \mathfrak{L} & \\ & \mathfrak{L}' \end{pmatrix} \in \mathscr{L}(V \oplus V', U \oplus U'), \qquad \mathfrak{P} = \begin{pmatrix} \mathfrak{P} & \\ & \mathfrak{P}' \end{pmatrix},$$

we obtain the following diagram:

$$\begin{array}{ccccccccc} 0 & \longrightarrow & E & \xrightarrow{\tilde{J}} & V \oplus V' & \xrightarrow{\tilde{S}} & F \oplus V' & \longrightarrow & 0 \\ & & \downarrow{\scriptstyle \tilde{A}} & & {\scriptstyle \tilde{\mathfrak{L}}}\downarrow\uparrow{\scriptstyle \tilde{\mathfrak{P}}} & & \uparrow{\scriptstyle \tilde{R}} & & \\ 0 & \longleftarrow & M & \xleftarrow{\tilde{\imath}} & U \oplus U' & \xleftarrow{\tilde{\jmath}} & N \oplus U' & \longleftarrow & 0 \end{array}$$

where $\tilde{S} = S \oplus I_{V'}$ and $\tilde{\jmath} = \jmath \oplus I_U$ are the extensions of S and $\jmath$ by direct sum with the identity on V' and on U', respectively, and $\tilde{J}$ is equal to J composed with the inclusion $V \to V \oplus V'$, and $\tilde{\imath}$ is the extension of $\imath$ which is equal to 0 on U'. Further, $\tilde{A}$ and $\tilde{R}$ are defined as before to make the diagram commutative, i.e. $\tilde{A} = \tilde{\imath} \circ \tilde{\mathfrak{L}} \circ \tilde{J}$ and $\tilde{R} = \tilde{S} \circ \tilde{\mathfrak{P}} \circ \tilde{\jmath}$. Note that $\tilde{A} = A$ and $\tilde{R} = R \oplus \mathfrak{P}'$. Now ind $\tilde{\mathfrak{L}} =$ ind $\mathfrak{L}$ + ind $\mathfrak{L}'$ and ind $\tilde{R} =$ ind R + ind $\mathfrak{P}' =$ ind $R -$ ind $\mathfrak{L}'$ so that

$$\text{ind } \tilde{\mathfrak{L}} + \text{ind } \tilde{R} = \text{ind } \mathfrak{L} + \text{ind } R.$$

Thus, (67) holds for the original diagram if and only if it holds for the new diagram. To make ind $A = 0$ it suffices then to choose a Fredholm operator $\mathfrak{L}'$ with ind $\mathfrak{L}' = -$ind $\mathfrak{L}$. For instance, if ind $\mathfrak{L} = k - l$ we just take any linear operator $\mathfrak{L}' \in \mathscr{L}(\mathbb{C}^l, \mathbb{C}^k)$ because ind $\mathfrak{L}' = l - k$ (which follows from the isomorphism $\mathbb{C}^l/\ker \mathfrak{L}' \simeq \operatorname{im} \mathfrak{L}'$).

The second point to notice is that the validity of (67) does not depend on the choice of $\mathfrak{P}$ in the definition of R, for if $\mathfrak{P}_1$ is another regularizer of $\mathfrak{L}$ then $\mathfrak{P} - \mathfrak{P}_1$ is a compact operator, whence $R - R_1 = S \circ (\mathfrak{P} - \mathfrak{P}_1) \circ \jmath$ is also compact, so ind $R =$ ind R_1.

Furthermore, since addition to $\mathfrak{L}$ of any compact operator does not affect (67), the two preceding observations show that we can assume that $\mathfrak{L}$ is invertible and $\mathfrak{P} = \mathfrak{L}^{-1}$. Then by using $\mathfrak{L}$ we may identify U and V, so $\mathfrak{L}$ and $\mathfrak{P}$ become the identity maps, which we assume from now on. Further, the bottom row of the diagram being split-exact, we may assume that $U = M \oplus N$, that $\imath$ is the canonical projection $(m, n) \mapsto m$ and that $\jmath$ is the canonical inclusion $n \mapsto (0, n)$. Then we have $A = \imath \circ J$ and $R = S \circ \jmath$.

$$\begin{array}{ccccccccc} 0 & \longrightarrow & E & & & & F & \longrightarrow & 0 \\ & & \downarrow{\scriptstyle A} & \searrow{\scriptstyle J} & & \swarrow{\scriptstyle S} & \uparrow{\scriptstyle R} & & \\ 0 & \longleftarrow & M & \xleftarrow{\imath} & M \oplus N & \xleftarrow{\jmath} & N & \longleftarrow & 0 \end{array}$$

With these assumptions, we now claim that

$$\jmath(\ker R) = J(\ker A). \tag{68}$$

If $n \in \ker R$ then, by definition of R, we have $\mathscr{j}x \in \ker S = \operatorname{im} J$, so

$$\mathscr{j}n = Je, \qquad \text{for some } e \in E.$$

Then $0 = \mathscr{s}\mathscr{j}n = \mathscr{a}Je = Ae$, so $e \in \ker A$, This proves $\mathscr{j}(\ker R) \subset J(\ker A)$. Conversely, if $e \in \ker A$ then $0 = Ae = \mathscr{a}Je$, whence $Je = \mathscr{j}n$ for some $n \in N$, which implies

$$Rn = S\mathscr{j}n = SJe = 0,$$

and this proves the other inclusion in (68).

Now, since $\mathscr{j}$, J are injective, (68) implies that

$$\dim \ker R = \dim \ker A \tag{69}$$

in the sense that either both dimensions are infinite or both are finite and equal.

It remains to show the cokernels of A and R have the same dimension (either both infinite, or both finite and equal). Suppose $\operatorname{im} A$ has finite codimension, i.e. $M = \operatorname{im} A \oplus M_0$ with $\dim M_0 < \infty$. Since S is surjective, then

$$F = S(M \oplus N) = S(\operatorname{im} A \oplus 0) + S(M_0 \oplus 0) + S(0 \oplus N).$$

By definition of R we have $S(0 \oplus N) = \operatorname{im} R$, and it is easily verified that $S(\operatorname{im} A \oplus 0) \subset \operatorname{im} R$ because $\operatorname{im} J \subset \ker S$. Therefore $F = \operatorname{im} R + S(M_0 \oplus 0)$. In fact,

$$F = \operatorname{im} R \oplus S(M_0 \oplus 0), \tag{70}$$

for if $f \in S(M_0 \oplus 0) \cap \operatorname{im} R$, then

$$f = S(m_0, 0) = Rn = S(0, n), \qquad \text{for some } m_0 \in M_0, n \in N,$$

whence $(m_0, -n) \in \ker S = \operatorname{im} J$, and $(m_0, -n) = Je$ for some $e \in E$. Therefore, $m_0 = \mathscr{a}Je = Ae$, and since $M_0 \cap \operatorname{im} A = 0$ it follows that $m_0 = 0$. Hence $f = 0$, so (70) holds and im R has finite codimension $\leqslant \dim M_0$. In fact, it is clear that S is injective on $M_0 \oplus 0$ (again because $\ker S \subset \operatorname{im} J$), so

$$\operatorname{codim} \operatorname{im} R = \dim S(M_0 \oplus 0) = \dim M_0 = \operatorname{codim} \operatorname{im} A. \tag{71}$$

Conversely, suppose $\operatorname{im} R$ has finite codimension, i.e. $F = \operatorname{im} R \oplus F_0$ with $\dim F_0 < \infty$. Since S is surjective, there exists a finite dimensional subspace $V_0 \subset V$ such that S is an isomorphism from V_0 to F_0. We claim that

$$M = \operatorname{im} A \oplus M_0, \qquad \text{where } M_0 = \mathscr{a}(V_0). \tag{72}$$

First we show that the sum is direct. If $m_0 \in \operatorname{im} A \cap M_0$ then $m_0 = Ae = \mathscr{a}x$ for some $e \in E$, $x \in V_0$. Since $A = \mathscr{a}J$ then we have

$$\mathscr{a}Je = \mathscr{a}x,$$

$$\mathscr{a}(Je - x) = 0,$$

$$Je - x \in \ker \mathscr{a} = \operatorname{im} \mathscr{j},$$

so $Je - x = \mathscr{j}n$, for some $n \in N$, and then, by applying the operator S and using the fact that $SJ = 0$ and $R = S\mathscr{j}$, we obtain

$$-Sx = Rn.$$

Thus, $Sx \in F_0 \cap \operatorname{im} R = 0$, so $x = 0$. Hence $m_0 = 0$. Finally, we show that $M \subset \operatorname{im} A \oplus M_0$. Let $m \in M$. Since $F = \operatorname{im} R \oplus F_0$ then in view of the definition of V_0 we have

$$S(m, 0) = Rn' + Sx, \qquad \text{for some } n' \in N, x \in V_0.$$

Since $R = S\mathscr{j}$ and $\ker S = \operatorname{im} J$, then $(m, -n') - x = Je$, for some $e \in E$. Applying the operator $\mathscr{a}$ gives

$$m = \mathscr{a}x + \mathscr{a}Je = \mathscr{a}x + Ae$$

so $m \in M_0 + \operatorname{im} A$, and the proof of (72) is complete, whence $\operatorname{im} A$ has finite codimension.

It follows that A is a Fredholm operator if and only if R is Fredholm, and (69), (71) show that $\operatorname{ind} A = \operatorname{ind} R$.

14.6 The main theorem for operators in $\mathfrak{BE}^{l,m}$, and the classes $S^{m,m'}$

Denote by $S^{m,m'} = S^{m,m'}(\mathbb{R}^n \times \mathbb{R}^n)$ the set of all $A \in C^\infty(\mathbb{R}^n \times \mathbb{R}^n)$ such that

$$|D_\xi^\alpha D_x^\beta A(x, \xi)| \leqslant C_{\alpha,\beta}(1 + |\xi|)^{m-\alpha_n}(1 + |\xi'|)^{m'-|\alpha'|}. \tag{73}$$

$S^{m,m'}$ is a Fréchet space with semi-norms defined by the smallest constants which can be used in (73). As before we define

$$A(x, D)u(x) = \frac{1}{(2\pi)^n}\int e^{i(x,\xi)} A(x, \xi)\hat{u}(\xi)\, d\xi, \tag{74}$$

which is a continuous operator $A(x, D)\colon \mathscr{S} \to \mathscr{S}$. The operator $A(x, D)$ is sometimes denoted by $\operatorname{Op}(A)$.

For $m_1 \leqslant m_2$ we have $S^{m_1,m'} \subset S^{m_2,m'}$ and for $m_1' \leqslant m_2'$ we have $S^{m,m_1'} \subset S^{m,m_2'}$. If $A \in S^{m,m'}$ then $D_\xi^\alpha D_x^\beta A \in S^{m-\alpha_n, m'-|\alpha'|}$. If $A(x, \xi') \in S^m(\mathbb{R}^n \times \mathbb{R}^{n-1})$ is independent of ξ_n then $A \in S^{0,m}(\mathbb{R}^n \times \mathbb{R}^n)$.

The following relation is also important:

$$S^{m,m'} \subset S^{m+p,m'-p} \qquad p \geqslant 0. \tag{75}$$

General symbols or amplitudes are defined as follows. We say that $a(x, y, \xi) \in S^{m,m'}(\mathbb{R}^n \times \mathbb{R}^n \times \mathbb{R}^n)$ if the derivatives have the bound

$$|D_\xi^\alpha D_x^\beta D_y^\gamma a(x, y, \xi)| \leqslant C_{\alpha,\beta,\gamma}(1 + |\xi|)^{m-\alpha_n}(1 + |\xi'|)^{m'-|\alpha'|}, \qquad x, y, \xi \in \mathbb{R}^n,$$

for all α, β and γ. Then we define the operator

$$(Au)(x) = \iint e^{i(x-y,\xi)}\, a(x, y, \xi)u(y)\, dy\, d\xi, \qquad u \in \mathscr{S}, \tag{76}$$

(the integral is oscillatory) and as in Proposition 7.4 we have a continuous map $A\colon \mathscr{S} \to \mathscr{S}$. The operator A is sometimes denoted $\operatorname{Op}(a)$. The following proposition is the analogue of Proposition 7.5.

Proposition 14.32 *Let $a(x, y, \xi) \in S^{m,m'}$ be a general symbol. If we define the ordinary symbol by*

$$A(x, \xi) := \frac{1}{(2\pi)^n} \iint e^{i(x-y,\theta)}\, e^{-i(x-y,\xi)}\, a(x, y, \theta)\, dy\, d\theta, \tag{77}$$

then $A(x, \xi) \in S^{m,m'}$, with each semi-norm of A bounded by a semi-norm of the symbol a. Further, both definitions (74) *and* (76) *of the operator Au agree, that is,* $\mathrm{Op}(A) = \mathrm{Op}(a)$.

Proof Substituting $z = y - x$, $\eta = \theta - \xi$ and regularizing, we rewrite (77) in the form

$$A(x, \xi) = \frac{1}{(2\pi)^n} \iint e^{-i(z,\eta)} \langle z\rangle^{-m} \langle D_\eta\rangle^M \langle D_z\rangle^N [\langle\eta\rangle^{-N} \cdot a(x, x+z, \xi+\eta)]\, dz\, d\eta$$

where M, N are even and nonnegative. Taking into account the inequality $\langle \xi + \eta\rangle^{\pm 1} \leqslant 2\langle\xi\rangle^{\pm 1}\langle\eta\rangle$, it follows that $\langle\xi+\eta\rangle^s \leqslant 2^{|s|}\langle\xi\rangle^s\langle\eta\rangle^{|s|}$ for all $s \in \mathbb{R}$, so we obtain when $|\alpha| + |\beta| \leqslant k$ that

$$|D_\xi^\alpha D_x^\beta A(x, \xi)| \leqslant \text{const} \iint \langle z\rangle^{-M} \langle \xi + \eta > ^{m-\alpha_n} \langle \xi' + \eta'\rangle^{m'-\alpha'} \langle\eta\rangle^{-N}\, dz\, d\eta$$

$$\leqslant \text{const}\langle\xi\rangle^{m-|\alpha|} \cdot \langle\xi'\rangle^{m'-\alpha'} \iint \langle z\rangle^{-M}$$

$$\times \langle\eta\rangle^{|m-\alpha_n|-N_1} \langle\eta'\rangle^{|m'-\alpha'|-N_2}\, dz\, d\eta$$

$$\leqslant \text{const}(1 + |\xi|)^{m-\alpha_n}(1 + |\xi'|)^{m'-\alpha'}$$

if we take $M > n$ and $N = N_1 + N_2$ where $N_1 > n + |m - \alpha_n|$, $N_2 > |m' - \alpha'|$. It follows that $A(x, \xi) \in S^{m,m'}$, and the estimates for the semi-norms of A are easily verified.

The verification that both definitions (74) and (76) of the operator Au agree is the same as in Proposition 7.5.

If $A \in S^{m,m'}$ and $B \in S^{\mu,\mu'}$ then $A \cdot B \in S^{m+\mu, m'+\mu'}$. Our first aim in this section is to prove the following analogue of Theorems 7.10 and 7.12. It is possible to obtain the full asymptotic expansion, but, for our application in this section to the main theorem in $\mathfrak{BC}^{l,\mathbf{m}}$, it is sufficient to look at the highest order terms.

Theorem 14.33 *If $A \in S^{m,m'}$ then $A^*(x, D) \in S^{m,m'}$ and each semi-norm of A^* is bounded by a semi-norm of A. If $B \in S^{\mu,\mu'}$ then $A(x, D) \circ B(x, D) = C(x, D)$ where $C \in S^{m+\mu, m'+\mu'}$, in fact,*

$$C - AB \in S^{m+\mu, m'+\mu'-1}, \tag{78}$$

and each semi-norm of C is bounded by a product of semi-norms of A and of B. Moreover, if $D_{\xi_j} A \in S^{m-1,m'}$ for every j then $C - AB \in S^{m+\mu-1, m'+\mu'}$.

Proof As in the proof of Theorem 7.10 we see that $A^*(x, D)$ has the amplitude representation (76) with amplitude $a(x, y, \xi) = \overline{A(y, \xi)}$. Since $\overline{A(y, \xi)} \in S^{m,m'}$ then $A^*(x, D) \in OS^{m,m'}$ by Proposition 14.32. Then as in the proof of Theorem 7.12 we find that $A(x, D) \circ B(x, D)$ has the representation (76) with amplitude $a(x, y, \xi) = A(x, \xi)\overline{B^*(y, \xi)}$, so once again by Proposition 14.32 we have $A(x, D) \circ B(x, D) \in OS^{m+\mu, m'+\mu'}$. The assertions concerning the semi-norms are clear from the proof of Proposition 14.32.

To prove (78) we write

$$B(x, \xi) - B(y, \xi) = \sum (x_j - y_j) B_j(x, y, \xi)$$

where

$$B_j(x, y, \xi) = \int_0^1 \partial_{x_j} B(y + t(x - y), \xi)\, dt.$$

When $y \in \mathbb{R}^n$ is regarded as a parameter it is clear that B_j is uniformly bounded on $S^{\mu,\mu'}$, and so are the y derivatives. Now by applying the operator $A(x, D)$ and using the fact that $[A(x, D), x_j] = D_{\xi_l} A(x, D)$ (see Proposition 7.24) we get

$$A(x, D) B(x, D) = A(x, D) B(y, D) + \sum (x_j - y_j) A(x, D) B_j(x, y, \xi) + R(x, y, D)$$

where

$$R(x, y, D) = \sum D_{\xi_j} A(x, D) B_j(x, y, D).$$

Since $D_{\xi_j} A \in S^{m,m'-1}$ and $B_j \in S^{\mu,\mu'}$, then by the first part of the theorem we have that $R(x, y, \xi)$ and its y derivatives are uniformly bounded in $S^{m+\mu, m'+\mu'-1}$. Taking $y = x$ we obtain

$$C(x, \xi) = A(x, \xi) B(x, \xi) + R(x, x, \xi)$$

which proves that $C - AB \in S^{m+\mu, m'+\mu'-1}$. Moreover, if $D_{\xi_j} A \in S^{m-1,m'}$ for every j then $R(x, y, \xi)$ and its y derivatives are uniformly bounded in $S^{m+\mu-1, m'+\mu'}$ so we obtain $C - AB \in S^{m+\mu-1, m'+\mu'}$.

Theorem 14.34 *If $A \in S^{m,m'}$ then $A(x, D)$ is continuous from $H^{s+m, t+m'}(\mathbb{R}^n)$ to $H^{s,t}(\mathbb{R}^n)$ for all $s, t \in \mathbb{R}$.*

Proof By Theorem 14.33 the operator

$$A_{s,t}(x, D) = (1 + |D|^2)^{s/2}(1 + |D'|^2)^{t/2} A(x, D)(1 + |D|^2)^{-(s+m)/2}(1 + |D'|^2)^{-(t+m')/2}$$

belongs to $OS^{0,0}$. Therefore, we may assume $s = t = 0$, that is, it suffices to show that if $A \in S^{0,0}$ then $A(x, D)$ is continuous from $L_2(\mathbb{R}^n)$ to $L_2(\mathbb{R}^n)$. This is a consequence of the L_2 continuity that is proved for the general symbol spaces in [Hö 3], Theorem 18.6.3.

Remark If $A \in S^{0,0}$ and is independent of x for large x, the L_2 continuity of $A(x, D)$ is also a consequence of Theorem 7.28′.

Our interest in the classes $S^{m,m'}$ is motivated by the fact that we wish to consider symbols which are polynomials of degree l in ξ_n of the form

$$A(x, \xi) = \sum_{j=0}^{l} A_j(x, \xi')\xi_n^j, \tag{79}$$

such that the coefficient A_j of ξ_n^j is a symbol of degree $l-j$ in the other variables, that is, $A_j(x, \xi') \in S^{l-j}(\mathbb{R}^n \times \mathbb{R}^{n-1})$. Then we have

$$A \in S^{l,0}(\mathbb{R}^n \times \mathbb{R}^n)$$

since $A_j(x, \xi')\xi_n^j \in S^{j,l-j} \subset S^{l,0}$. We also assume that *the leading coefficient, A_l, is independent of ξ'*. In that case the first time we differentiate A with respect to ξ' the term ξ_n^l drops out, so $D_{\xi_j}A \in S^{l-1,0}$ for *every* j and not only for $j = n$.

If $\varphi(x) \in C_0^\infty(\mathbb{R}^n) \subset S^{0,0}(\mathbb{R}^n \times \mathbb{R}^n)$ then

$$[A, \varphi] = A\varphi - \varphi A \in S^{l-1,0}. \tag{80}$$

This follows from the last part of Theorem 14.33 since $D_{\xi_j}A \in S^{l-1,0}$ and $D_{\xi_j}\varphi = 0$. Note that (80) would *not* be true if the leading coefficient $A_l(x, \xi')$ were to depend on ξ'. (For instance, if $l = 0$ then $A \in S^{0,0}$ and we could only conclude that $[A, \varphi] \in S^{0,-1}$, which is not contained in $S^{-1,0}$.)

It is also easy to verify (80) directly using the special form of A. Indeed,

$$\begin{aligned} A(x, D)\varphi u &= \sum A_j(x, D')D_n^j\varphi u \\ &= \sum A_j(x, D')\varphi D_n^j u + \sum A_j(x, D')R_j(x, D_n)u \end{aligned}$$

where R_j is a differential operator in D_n of order $<j$, with coefficients in C_0^∞. By commuting $A_j(x, D')$ with φ we obtain that

$$A(x, D)\varphi u = \varphi A(x, D)u + \tilde{A}(x, D)u$$

where $\tilde{A} = \sum_{j<l}[A_j, \varphi]D_n^j u + \sum A_j R_j$. It follows from (75) that $\tilde{A} \in S^{l-1,0}$; for instance, the symbol of $[A_j, \varphi]D_n^j$ belongs to $S^{j,l-j-1} \subset S^{l-1,0}$.

In the next lemma we assume that the symbol $A(x, \xi)$ satisfies an ellipticity condition for large ξ. The lemma is stated for a single operator but it is obvious that the proof remains valid if A is $p \times p$ matrix valued and the matrix norm of $A(x, \xi)^{-1}$ can be estimated by $(1 + |\xi|)^{-l}$.

Lemma 14.35 *Let $\chi \in C_0^\infty(\mathbb{R}^n)$, $A \in S^{l,0}$, and assume that*

$$|A(x, \xi)| \geqslant c(1 + |\xi|)^l \qquad \textit{if } \xi \in \operatorname{supp}(1 - \chi). \tag{81}$$

Then $E(x, \xi) = (1 - \chi(\xi))/A(x, \xi)$ belongs to $S^{-l,0}$. Further, if $D_{\xi_j}A \in S^{l-1,0}$ then $D_{\xi_j}E \in S^{-l-1,0}$.

Proof Assume that estimates of the form

$$|D_\xi^\alpha D_x^\beta E(x, \xi)| \leqslant C_{\alpha\beta}(1 + |\xi|)^{-l-\alpha_n}(1 + |\xi'|)^{-|\alpha'|} \tag{82}$$

have already been proved for all $|\alpha + \beta| < k$. Then we obtain by differentiating the equation $AE = 1 - \chi$ and using the symbol estimates for $A \in S^{l,0}$ that

$$|A(x, \xi) \cdot D_\xi^\alpha D_x^\beta E(x, \xi)| \leqslant C'_{\alpha\beta}(1 + |\xi|)^{-\alpha_n}(1 + |\xi'|)^{-|\alpha'|}$$

when $|\alpha + \beta| = k$ and this implies that (82) also holds for $|\alpha + \beta| = k$. By induction the estimates (82) hold for all α, β so that $E \in S^{-l,0}$. Since

$$D_{\xi_j} E = -D_{\xi_j} A(x, \xi)/A(x, \xi)^2$$

the second statement follows from the first.

Now we use Lemma 14.35 to establish the interior regularity of an operator $\mathscr{A} \in \mathfrak{E}^l$.

Proposition 14.36 *Let $\mathscr{A} \in \mathfrak{E}^l$. If $u \in W_2^{l+s-1}(\mathbb{R}^n, \mathbb{C}^p)$ and $\mathscr{A}u \in W_2^s(\mathbb{R}^n, \mathbb{C}^p)$ with u and $\mathscr{A}u$ having compact support on Ω, then $u \in W_2^{l+s}(\mathbb{R}^n, \mathbb{C}^p)$. For every compact subset K of Ω we have if* supp $u \subset K$

$$\|u\|_{s+l} \leqslant C(\|\mathscr{A}u\|_s + \|u\|_{s+l-1}). \tag{83}$$

Proof In virtue of the lemma preceding Theorem 10.26, for every $\varphi \in C_0^\infty$

$$\|\mathscr{A}\varphi u\|_s \leqslant \|\varphi\mathscr{A}u\|_s + \text{const}\|u\|_{s+l-1}, \tag{84}$$

so it is sufficient to prove the statement when u has support in an arbitrarily small open ball. If this ball lies outside some tubular neighbourhood $x_n \leqslant \delta$, then $\mathscr{A}^b u = 0$ and the statement follows by ellipticity of $\mathscr{A}^i \in OS^l(\mathbb{R}^n, \mathbb{C}^p)$ in $x_n > \delta$ (see Proposition 7.37). On the other hand, suppose that u has support in a small open ball lying in $x_n < \delta$. We may assume that supp $u \subset \mathfrak{U}' \times (-\delta, \delta)$ where $\mathfrak{U}'$ is a coordinate patch on $\partial\Omega$, such that if $\mathfrak{U}'$ is identified with a subset of $\mathbb{R}^{n-1}$ then

$$\mathfrak{U}' \times (-\delta, \delta) \subset \{x \in \mathbb{R}^n; |x| < 2\}, \qquad \operatorname{supp} u \subset \{x \in \mathbb{R}^n; |x| < \tfrac{1}{2}\}.$$

Also, choose $\psi \in C^\infty(\mathbb{R}^n)$ which is a decreasing function of the radius so that

$$\psi(x) = 1 \quad \text{when } |x| < 1, \qquad \psi(x) = 1/|x| \quad \text{when } |x| > 2.$$

Now let $\mathscr{A}(x, \xi)$ denote the symbol matrix of $\mathscr{A}$ in the local coordinates, and set

$$A(x, \xi) = \mathscr{A}(\psi(x)x, \xi). \tag{85}$$

Then $A \in S^{l,0}$, $A(x, \xi) = \mathscr{A}(x, \xi)$ in a neighbourhood of supp u and $|\xi|^l A(x, \xi)^{-1}$ is bounded for large $|\xi|$. Since $D_{\xi_j} A \in S^{l-1,0}$ for every j, then by Lemma 14.35 we can choose $E \in S^{-l,0}$ such that $D_{\xi_j} E \in S^{-l-1,0}$ for every j and $E(x, \xi)A(x, \xi)$ is the identity for large ξ. By Theorem 14.33 it follows that $E(x, D)A(x, D) = I + R(x, D)$ where $R \in S^{-1,0}$, hence R is continuous from W_2^{s+l-1} to W_2^{s+l}. Thus

$$\begin{aligned} \|u\|_{s+l} &\leqslant \|EAu\|_{s+l} + \|Ru\|_{s+l} \\ &\leqslant C'(\|Au\|_s + \|u\|_{s+l-1}) \end{aligned}$$

Choose $\varphi \in C_0^\infty$ equal to 1 on supp u. It follows from (84) that $\|Au\|_s =$

$\|A\varphi u\|_s \leqslant \|\varphi Au\|_s + C\|u\|_{s+l-1}$ and then, since $\varphi Au = \varphi\mathscr{A}u$, we obtain the desired estimate (83).

We now wish to construct a regularizer $\mathfrak{P}$ in local coordinates near $\partial\Omega$ for the boundary value problem operator $(\mathscr{A}, \mathscr{B}) \in \mathfrak{BE}^{l,\mathbf{m}}$. The definition of $\mathfrak{P}$ will be the same as in §14.4 but its properties will be weaker. (Compare Theorem 14.40 with 14.27.) Before we can look at the properties of $\mathfrak{P}$, however, we need to establish the transmission property in the present context.

Consider a local coordinate patch where $\mathscr{A} = \mathscr{A}^b$. As in (85) we extend the symbol of $\mathscr{A}$ to the whole space to get a symbol of the form

$$A(x, \xi) = \sum_{j=0}^{l} A_j(x, \xi')\xi_n^j$$

where the coefficients have the properties stated after (79) and $|\xi|^l A(x, \xi)^{-1}$ is bounded for large $|\xi|$. As in Lemma 14.35 we define $E(x, \xi) = (1 - \chi(\xi))A(x, \xi)^{-1}$ where $\chi \in C_0^\infty$ is equal to 1 in such a large set that $E \in S^{-l,0}$ and $D_{\xi_j}E \in S^{-l-1,0}$ for every j. In view of Theorem 14.33 it follows that

$$E(x, D)A(x, D) = I + T_r(x, D), \qquad A(x, D)E(x, D) = I + T_l(x, D) \quad (86)$$

where $T_r, T_l \in S^{-1,0}$. Thus the operator $E(x, D)$ is continuous from $H^{s,t}(\mathbb{R}^n)$ to $H^{s+l,t}(\mathbb{R}^n)$ and $T_r(x, D)$ and $T_l(x, D)$ are continuous from $H^{s,t}(\mathbb{R}^n)$ to $H^{s+1,t}(\mathbb{R}^n)$ for all s, t.

The following lemma shows that the symbol $E(x, \xi)$ has a nice asymptotic expansion as $\xi_n \to \infty$ which makes it straightforward to establish the transmission property for $E(x, D)$.

Lemma 14.37 *Let*

$$\Xi = \xi_n + i(|\xi'|^2 + 1)^{1/2} \qquad (= \Lambda_+ \textit{ in the notation of } \S 14.1).$$

Then $\Xi = S^{1,0}$ *and by Lemma* 14.35 *it follows that* $\Xi^{-1} \in S^{-1,0}$. *Let the hypotheses on the symbol* $A(x, \xi)$ *be as above. Then for any* N *there is an expansion of the form*

$$E(x, \xi) = \sum_{j=0}^{N-1} E_j(x, \xi')\Xi^{-l-j} + R_N(x, \xi) \quad (87)$$

where $E_j \in S^j(\mathbb{R}^n \times \mathbb{R}^{n-1})$ *and* $R_N \in S^{-l-N,N}(\mathbb{R}^n \times \mathbb{R}^n)$.

Proof Without loss of generality $A_l \equiv I$. Then we expand A in powers of Ξ,

$$A(x, \xi) = \sum_{j=0}^{l} \tilde{A}_j(x, \xi')\Xi^j$$

where $\tilde{A}_l = I$ and $\tilde{A}_j \in S^{l-j}(\mathbb{R}^n \times \mathbb{R}^{n-1})$. We claim that it is possible to choose E_j such that for $N = 1, 2, \ldots$

$$A(x, \xi) \sum_{j=0}^{N-1} E_j(x, \xi')\Xi^{-l-j} = I + \sum_{j=0}^{l-1} R_{N,j}(x, \xi')\Xi^{-j-N} \quad (88)$$

where $R_{N,j} \in S^{j+N}(\mathbb{R}^n \times \mathbb{R}^{n-1})$. For $N = 1$ this means that $E_0 = I$ and $R_{1,j} = A_{l-1-j}$. If the identity has been established for one value of N, then for $N + 1$ we have

$$A(x, \xi) \sum_{j=0}^{N} E_j(x, \xi')\Xi^{-l-j} = I + \sum_{j=0}^{l-1} R_{N,j}(x, \xi')\Xi^{-j-N}$$
$$+ \sum_{j=0}^{l} \tilde{A}_j(x, \xi')E_N(x, \xi')\Xi^{j-l-N},$$

and to put this in the desired form we choose $E_N := -R_{N,0}$ so that the term $j = 0$ in the first sum cancels with the term $j = l$ in the second sum, and then set

$$R_{N+1,j} := R_{N,j+1} + \tilde{A}_{l-1-j}E_N$$

for $j = 0, \ldots, l - 1$ (and $R_{N,l} = 0$). Thus (88) holds for N replaced by $N + 1$.

Now, multiplying the identity (88) by $E(x, \xi) = (1 - \chi(\xi))A(x, \xi)^{-1}$, we obtain (87) where

$$R_N(x, \xi) = -\chi(\xi) \sum_{j=0}^{N-1} E_j(x, \xi')\Xi^{-l-j} - (1 - \chi(\xi))A(x, \xi)^{-1} \sum_{j=0}^{l-1} R_{N,j}(x, \xi')\Xi^{-j-N}$$

The second term belongs to $S^{-l-N,N}$ since $(1 - \chi(\xi))A(x, \xi)^{-1} \in S^{-l,0}$, $R_{N,j} \in S^{0,j+N}$ and $\Xi^{-N-j} \in S^{-N-j,0} \subset S^{-N,-j}$. The first term lies in $C_0^\infty \subset S^{-\infty}$, so it follows that $R_N \in S^{-l-N,N}$.

As usual, if $f \in L_2(\mathbb{R}^n)$ we denote by f^0 the function which is equal to f in $\mathbb{R}^n_+$ and 0 elsewhere. Also $\imath^+$ denotes the operator of restriction to $x_n > 0$.

Corollary 14.38 (*Transmission Property*) *Let the hypotheses of Lemma* 14.37 *hold. If $f \in \mathscr{S}$ then*

$$\|\imath^+ E(x, D)f^0\|_{(s+l,t)} \leqslant C_{s,t}\|\imath^+ f\|_{(s,t)}, \qquad s \geqslant 0. \tag{89}$$

We also have (see (87))

$$\|\imath^+ T_r(x, D)f^0\|_{(s+1,t)} + \|\imath^+ T_l(x, D)f^0\|_{(s+1,t)} \leqslant C_{s,t}\|\imath^+ f\|_{(s,t)} \tag{90}$$

when $s \geqslant 0$.

Proof With the expansion (87) the proof is quite simple. We have

$$E(x, D)f^0 = \sum_{j=0}^{N-1} E_j(x, D')\Xi^{-l-j}(D)f^0 + R_N(x, D)f^0.$$

It is clear from (22) that the restriction of $\Xi^{-k}(D)u$, $u \in \mathscr{S}$, to the half-space $x_n > 0$ is determined by the restriction of u to that half-space. The same is, therefore, true of $E_j(x, D')\Xi^{-l-j}(D)u$. Since $E_j(x, D')\Xi^{-l-j}(D)$ is continuous from $H^{s+l,t}(\mathbb{R}^n)$ to $H^{s,t}(\mathbb{R}^n)$, it is clear from the definition of the norm (24) that we also have continuity from $H^{s+l,t}(\mathbb{R}^n_+)$ to $H^{s,t}(\mathbb{R}^n_+)$. To estimate the remainder term we choose $N \geqslant s$, and then

$$\| R_N(x, D) f^0 \|_{(s+l,t)} \leqslant C \| f^0 \|_{(s-N,t+N)} \leqslant C \| f^0 \|_{(0,s+t)}$$

where the first inequality is due to Theorem 14.34 since $R_N \in S^{-l-N,N}$ and the second inequality is due to (23). Since the elements in $H^{0,s+t}(\mathbb{R}^n)$ can be considered as L_2 functions of x_n with values in $W_2^{s+t}(\mathbb{R}^{n-1})$ it follows that $\| f^0 \|_{(0,s+t)} \leqslant \| r^+ f \|_{(0,s+t)}$ and the latter norm is $\leqslant \| r^+ f \|_{(s,t)}$ since $s \geqslant 0$, so the proof of (89) is complete.

Note that

$$(EA)(x, D) = I - \chi(D), \qquad (AE)(x, D) = I - \chi(D) \tag{91}$$

where $\chi(D) \in OS^{-\infty}$. To prove the estimates for $T_r(x, D) f^0$ and $T_l(x, D) f^0$ it suffices to examine $E(x, D) A(x, D) - (EA)(x, D)$ and $A(x, D) E(x, D) - (AE)(x, D)$ using the same expansion (87). First, observe that the symbols of $R_N(x, D) A(x, D) - (R_N A)(x, D)$ and $A(x, D) R_N(x, D) - (AR_N)(x,D)$ are in $S^{-N,N-1}$, so for large enough N we can argue in the same way as above to obtain

$$\| R_N(x, D) A(x, D) f^0 - (R_N A)(x,D) f^0 \|_{(s+1,t)} \leqslant C \| r^+ f \|_{(s,t)}$$

and the same is of course true with the order of the factors interchanged. The symbols of

$$E_j(x, D') \Xi^{-l-j}(D) A(x, D) - (E_j \Xi^{-l-j} A)(x, D),$$

and the corresponding difference with the order of the factors interchanged, are in $S^{-1,0}$. Thus the operator is continuous from $H^{s,t}(\mathbb{R}^n)$ to $H^{s+1,t}(\mathbb{R}^n)$ for all s, t which again implies continuity in the corresponding restriction spaces. Now, since

$$\begin{aligned} &\| r^+ E(x, D) A(x, D) f^0 - r^+ (EA)(x, D) f^0 \|_{(s+1,t)} \\ &\qquad \leqslant \sum_{j=0}^{N-1} \| E_j(x, D') \Xi^{-l-j}(D) A(x, D) f^0 - (E_j \Xi^{-l-j} A)(x, D) f^0 \|_{(s+1,t)} \\ &\qquad + \| R_N(x, D) A(x, D) f^0 - (R_N A)(x, D) f^0 \|_{(s+1,t)} \end{aligned}$$

then in view of (86) and (91) it follows that

$$\| r^+ T_r(x, D) f^0 \|_{(s+1,t)} \leqslant C \| r^+ f \|_{(s,t)}.$$

Similarly, the same argument with the order of the factors interchanged implies

$$\| r^+ T_l(x, D) f^0 \|_{(s+1,t)} \leqslant C \| r^+ f \|_{(s,t)}.$$

Corollary 14.39 *Let the hypotheses of Lemma* 14.37 *hold. If $f \in H^{-l,t}(\mathbb{R}^n)$ and the support of f is contained in $x_n \leqslant 0$ then we have*

$$\| r^+ E(x, D) f \|_{(v,t-v)} \leqslant C_{t,v} \| f \|_{(-l,t)} \tag{92}$$

and

$$\| r^+ T_r(x, D) f \|_{(v+1-l,t-v)} + \| r^+ T_l(x, D) f \|_{(v+1-l,t-v)} \leqslant C_{t,v} \| f \|_{(-l,t)} \tag{93}$$

for every real number v (and all t).

Proof We may assume that $f \in C_0^\infty(\mathbb{R}^n_-)$ because f can be approximated arbitrarily closely in the $H^{-l,t}$ norm. Since $\Xi^{-k}(D)f = 0$ in $x_n > 0$ it follows from (87) that

$$\imath^+ E(x, D)f = \imath^+ R_N(x, D)f.$$

Since $R_N \in S^{-l-N,N}$ then by choosing $N \geqslant \nu$ we obtain

$$\|R_N(x, D)f\|_{(\nu,t-\nu)} \leqslant C\|f\|_{(-l-N+\nu,t-\nu+N)} \leqslant C\|f\|_{(-l,t)},$$

and (92) follows. The estimate (93) is proved in a similar manner.

Having established the transmission property for $E(x, D)$, our aim now is to construct a regularizer $\mathfrak{P}$ in local coordinates near $\partial\Omega$ for the boundary value problem operator $(\mathscr{A}, \mathscr{B})$ as in §14.4. After extending the symbol of $\mathscr{B}$ in local coordinates to the whole space (see (85)) we obtain a symbol of the form

$$B(x', \xi) = \sum_{j=0}^{l-1} B_j(x', \xi')\xi_n^j$$

where $B_j = [b_k^j]_{k=1}^r$ and $b_k^j \in S^{m_k-j}(\mathbb{R}^{n-1} \times \mathbb{R}^{n-1}, 1 \times p)$. We also use the notation b_k for the kth row of B, i.e. $B = [b_k]_{k=1}^r$ where $b_k = \sum_{j=0}^{l-1} b_k^j(x', \xi')\xi_n^j$ has order m_k, $k = 1, \ldots r$.

In virtue of Proposition 14.17, the trace operator

$$\gamma u = [\gamma_j u]_{j=0}^{l-1}, \qquad \gamma_j u = \lim_{x_n \to 0^+} D_n^j u(x', x_n)$$

is continuous

$$\gamma\colon H^{s,t}(\mathbb{R}^n_+) \to \mathop{\times}_{j=0}^{l-1} W_2^{s+t-j-1/2}(\mathbb{R}^{n-1}) \qquad \text{when } s > l - \tfrac{1}{2}. \tag{94}$$

Since $B(x', D)u = B^c\gamma u$, where $B^c = [B_0 \cdots B_{l-1}]$, it follows that the boundary problem operator

$$\mathfrak{L} = (A, B)\colon H^{s+l,t}(\mathbb{R}^n_+, \mathbb{C}^p) \to H^{s,t}(\mathbb{R}^n_+, \mathbb{C}^p) \times \mathop{\times}_{k=1}^{r} W_2^{s+t+l-m_k-1/2}(\mathbb{R}^{n-1})$$

is continuous when $s \geqslant 0$. (Also recall that $A \in S^{l,0}$ so we may use Theorem 14.34.)

With $\mathscr{A}$ replaced by $A(x, D)$, we will repeat the steps which led to Theorem 14.27 except that we will just work locally with $\Omega = \mathbb{R}^n_+$ and $\partial\Omega = \mathbb{R}^{n-1}$. Let A^c be defined as in (34). The Calderón operator is defined by (36), that is,

$$P = \gamma E(x, D)A^c,$$

and we leave it as an exercise to verify that $P = [P_{jk}]$, where $P_{jk} \in OClS^{j-k}(\mathbb{R}^{n-1}, p \times p)$, $j, k = 0, \ldots, l-1$, with principal symbol of P_{jk} given by (37). Only the principal symbol of $P^2 - P$ is now equal to 0, which means that $P^2 - P$ improves differentiability by one unit.

If the hypothesis on Q in Proposition 14.22 is fulfilled only on the principal symbol level, that is, $Q^2 - Q \in OS^{-1}$, then the conclusions remain valid with

$\equiv$ interpreted as equality of principal symbols. We can also obtain $BQS - I \in OS^{-\infty}$ by taking $S = QT$ where $BQ^2T - I \in OS^{-\infty}$.

Now, let B be L-elliptic with respect to A. Then for the DN system

$$B^c = [B_0 \cdots B_{l-1}]$$

we may fulfil (50), (51) as before; here the notation $\equiv$ means equality modulo lower-order terms. For instance, it follows from (51) that $S'(I-P)+SB^c \equiv I$, which means that

$$S'(I - P) + SB^c \equiv I - R, \tag{95}$$

where $R = [R_{ij}]$ and $R_{ij} \in OClS^{i-j-1}(\mathbb{R}^{n-1}, p \times p)$, $i, j = 0, \ldots, l-1$. However, as mentioned above, we may assume that $B^cPS - I \in OS^{-\infty}$.

Theorem 14.40 *Let A and B satisfy the conditions mentioned above and suppose that B is L-elliptic with respect to A. The operator*

$$\mathfrak{P}\colon H^{s,t}(\mathbb{R}^n_+, \mathbb{C}^p) \times \mathop{\times}_{k=1}^{r} W_2^{s+t+l-m_k-1/2}(\mathbb{R}^{n-1}) \to H^{s+l,t}(\mathbb{R}^n_+, \mathbb{C}^p)$$

defined by $\mathfrak{P}(f, g) = \imath^+(I + EA^cS'\gamma)Ef^0 + \imath^+EA^cSg$ is continuous for every $s \geqslant 0$. The operator $\mathfrak{P}$ is an approximate left and right inverse of $\mathfrak{L}$ in the sense that

$$\mathfrak{P}\mathfrak{L} = I + K, \qquad \mathfrak{L}\mathfrak{P} = I + K',$$

where the error term, K, is continuous from $H^{s,t}(\mathbb{R}^n_+, \mathbb{C}^p)$ to $H^{s+1,t}(\mathbb{R}^n, \mathbb{C}^p)$ when $s \geqslant 0$ and $K' = \begin{pmatrix} K_1 & K_2 \\ K_3 & K_4 \end{pmatrix}$ where the error terms $K_1, \ldots, K_4$ have the following properties:

K_1 is continuous from $H^{s,t}(\mathbb{R}^n_+, \mathbb{C}^p)$ to $H^{s+1,t}(\mathbb{R}^n_+, \mathbb{C}^p)$ when $s \geqslant 0$;

K_2 is continuous from $\mathop{\times}_{k=1}^{r} W_2^{s+t+l-m_k-1/2}(\mathbb{R}^{n-1})$ to $H^{s+1,t}(\mathbb{R}^n_+, \mathbb{C}^p)$ for arbitrary s and t;

K_3 is continuous from $H^{s,t}(\mathbb{R}^n_+, \mathbb{C}^p)$ to $\times_{k=1}^{r} W_2^{s+t+l-m_k+1/2}(\mathbb{R}^{n-1})$ when $s \geqslant 0$;

K_4 has a C^∞ kernel on $\mathbb{R}^{n-1}$.

Proof The estimate (92) implies that $\|\imath^+Ef\|_{(s,t)} \leqslant C\|f\|_{(-l,s+t)}$, and then, since the estimate (29) holds, we obtain

$$\|\imath^+E(x, D)A^c\mathscr{U}\|_{(s,t)} \leqslant \text{const} \sum_{j=0}^{l-1} \|\mathscr{U}_j\|_{s+t-j-1/2}, \qquad \forall s, t, \tag{96}$$

for every $\mathscr{U} = [\mathscr{U}_j]_{j=0}^{l-1} \in C_0^\infty(\mathbb{R}^{n-1}, \mathbb{C}^{pl})$.

The continuity of $\mathfrak{P}$ now follows easily from (89) and (96), taking into account the orders of the blocks in the matrix operators S and S'.

We have, by definition

$$Ku = \imath^+EA^c(S'\gamma T_l u^0 - R\gamma u) + \imath^+T_l u^0,$$

where T_r and T_l are defined by (86) and R is defined by (95). The continuity of the map

$$u \mapsto \imath^+ T_l u^0$$

from $H^{s,t}(\mathbb{R}^n_+, \mathbb{C}^p)$ to $H^{s+1,t}(\mathbb{R}^n_+, \mathbb{C}^p)$ follows from (90). Recalling the continuity (94) of the trace operator γ, and the fact that $S' = [S'_{ij}]$, where S'_{ij} is a $p \times p$ operator of order $i - j$, then we obtain the continuity of

$$u \mapsto S'\gamma T_l u^0$$

from $H^{s,t}(\mathbb{R}^n_+, \mathbb{C}^p)$ to $\times_{j=o}^{l-1} W_2^{s+t-j+1/2}(\mathbb{R}^{n-1}, \mathbb{C}^p)$. The same is true of $u \mapsto R\gamma u$, since $R = [R_{ij}]$ where the $p \times p$ operator R_{ij} has order $i - j - 1$. By composing these maps with $\imath^+ EA^c$ and recalling (96), we obtain the continuity of

$$u \mapsto \imath^+ EA^c(R\gamma u - S'\gamma T_l u^0)$$

from $H^{s,t}(\mathbb{R}^n_+, \mathbb{C}^p)$ to $H^{s+1,t}(\mathbb{R}^n_+, \mathbb{C}^p)$, which proves the desired continuity of K.

For the other operators, we have

$$K_1 f = \imath^+ T_r f^0 + \imath^+ T_r A^c S' \cdot \gamma E f^0, \qquad K_2 g = \imath^+ T_r A^c S g$$

$$K_3 f = B^c(I + PS') \cdot \gamma E f^0, \qquad K_4 g = (B^c PS - I) g.$$

Note that $K_4 = B^c PS - I \in OS^{-\infty}$ by choice of S. The continuity of K_2 follows in the same way as (96), using (93) instead of (92) and taking into account the fact that $S = [S_{ik}]$, where S_{ik} is a $p \times 1$ operator of order $i - m_k$. The continuity of K_1 also follows easily from Corollaries 14.38 and 14.39.

Before proving the continuity of K_3 we need the analogue of Proposition 14.21: *For $s \geqslant 0$, the map $f \mapsto P \cdot \gamma E(x, D) f^0$ is continuous from*

$$H^{s,t}(\mathbb{R}^n_+, \mathbb{C}^p) \quad to \quad \mathop{\times}_{j=0}^{l-1} W_2^{s+t+l-j+1/2}(\mathbb{R}^{n-1}, \mathbb{C}^p). \tag{97}$$

Indeed, as in the proof of Proposition 14.21, we have

$$P\gamma E f^0 = \gamma T_l (E f^0)^0 - \gamma E(T_r f^0)^0$$

and by Corollary 14.38

$$\|\imath^+ T_l (E f^0)^0\|_{(s+l+1,t)} \leqslant \text{const} \|\imath^+ E f^0\|_{(s+l,t)} \leqslant \text{const} \|\imath^+ f\|_{(s,t)}$$

and

$$\|\imath^+ E(T_r f^0)^0\|_{(s+l+1,t)} \leqslant \text{const} \|\imath^+ T_r f^0\|_{(s+1,t)} \leqslant \text{const} \|\imath^+ f\|_{(s,t)}.$$

In virtue of (94), the trace operator γ is continuous from $H^{s+l+1,t}(\mathbb{R}^n_+)$ to $\times_{j=0}^{l-1} W_2^{s+t+l-j+1/2}(\mathbb{R}^{n-1})$ and then (97) follows.

Now, by adding and subtracting the term $B^c \cdot P\gamma E f^0$, we can write K_3 in the form

$$K_3 f = (B^c(I + PS') - B^c P) \cdot \gamma E f^0 + B^c \cdot P\gamma E f^0.$$

Note that $B^c = [B_0 \cdots B_{l-1}]$ where the kth row of the jth block, B_j, has order $m_k - j$ and

$$B^c(I + PS') - B^c P$$

has the same form except that the kth row of the jth block has order $m_k - j - 1$. Since $f \mapsto \imath^+ E f^0$ is continuous from $H^{s,t}(\mathbb{R}^n_+, \mathbb{C}^p)$ to $H^{s+l,t}(\mathbb{R}^n_+, \mathbb{C}^p)$, then due to (94) the map $f \mapsto \gamma E f^0$ is continuous from $H^{s,t}(\mathbb{R}^n_+, \mathbb{C}^p)$ to $\times_{j=0}^{l-1} W_2^{s+t+l-j-1/2}(\mathbb{R}^{n-1}, \mathbb{C}^p)$. The continuity of K_3 follows at once.

Lemma 14.41 *If $v \in L_2(\mathbb{R}^n_+, \mathbb{C}^p)$ and $w_k \in W_2^{-l+m_k+1/2}(\mathbb{R}^{n-1})$, $k = 1 \cdots r$, and we have a relation*

$$0 = \int_{x_n > 0} A(x, D)u \cdot {}^h v(x)\, dx + \int B(x', D)u \cdot {}^h w(x')\, dx', \qquad \forall u \in C_0^\infty(\bar{\mathbb{R}}^n_+, \mathbb{C}^p), \tag{98}$$

then $v \in H^{0,t}(\mathbb{R}^n_+, \mathbb{C}^p)$ and $w_k \in W_2^{t-l+m_k+1/2}(\mathbb{R}^{n-1})$ for all t. (In particular, it follows that $w_k \in C^\infty$.)

Proof Assume that $v \in H^{0,t}(\mathbb{R}^n_+, \mathbb{C}^p)$ and $w_k \in W_2^{t-l+m_k+1/2}(\mathbb{R}^{n-1})$ for some t; we shall prove that this must remain true with t replaced by $t+1$, hence for all t. Take $f \in C_0^\infty(\mathbb{R}^n, \mathbb{C}^p)$, $g_k \in C_0^\infty(\mathbb{R}^{n-1})$, $k = 1, \ldots, r$ and let $u = \mathfrak{P}(f, g)$. By definition, we have

$$A(x, D)u = f + K_1 f + K_2 g, \qquad B(x, D)u = g + K_3 f + K_4 g,$$

and by virtue of the continuity of the operators K_1, K_2, K_3 and K_4 in Theorem 14.40 it follows that

$$\|A(x, D)u - f\|_{(0,-t)} \leqslant C\Big(\|f\|_{(0,-t-1)} + \sum_k \|g_k\|_{-t+l-m_k-3/2}\Big) \tag{99}$$

and

$$\sum \|b_k(x, D)u - g_k\|_{-t+l-m_k-1/2} \leqslant C'\Big(\|f\|_{(0,-t-1)} + \sum \|g_k\|_{-t+l-m_k-3/2}\Big) \tag{100}$$

where b_k denotes the kth row of B, $k = 1, \ldots, r$. (In the proof of (99) it is necessary to use the fact that $\|K_1 f\|_{(0,-t)} \leqslant \|K_1 f\|_{(1,-t-1)}$, see (23), and similarly for $K_2 g$.) In view of (98), (99) and (100) it follows that

$$\left|\int_{x_n > 0} f \cdot {}^h v\, dx + \sum \int g_k \bar{w}_k\, dx'\right| \leqslant C\Big(\|f\|_{(0,-t-1)} + \sum \|g_k\|_{-t+l-m_k-3/2}\Big)$$

(where the constant C depends on t) which proves that $v \in H^{0,t+1}(\mathbb{R}^n_+, \mathbb{C}^p)$ and $w_k \in W_2^{t-l+m_k+3/2}(\mathbb{R}^{n-1})$. By induction on t, the proof is complete.

In the proof of the next theorem we will need a localized version of (99), (100).

Lemma 14.42 *Take $f \in C_0^\infty(\mathbb{R}^n, \mathbb{C}^p)$, $g_k \in C_0^\infty(\mathbb{R}^{n-1})$, $k = 1, \ldots, r$, and set*

$u = \mathfrak{P}(f, g)$. Let $\varphi \in C_0^\infty(\mathbb{R}^n)$ be fixed. Then for any t

$$\|A(x, D)\varphi u - \varphi f\|_{(0,-t)} \leqslant C\left(\|f\|_{(0,-t-1)} + \sum_k \|g_k\|_{-t+l-m_k-3/2}\right) \tag{101}$$

and

$$\sum \|b_k(x, D)\varphi u - \varphi g_k\|_{-t+l-m_k-1/2} \leqslant C'\left(\|f\|_{(0,-t-1)} + \sum \|g_k\|_{-t+l-m_k-3/2}\right) \tag{102}$$

where b_k denotes the kth row of B, $k = 1, \ldots, r$.

Proof First let us consider the elliptic operator A. Let $Z = A\varphi - \varphi A$, then $A(x, D)\varphi u = \varphi \cdot A(x, D)u + Zu$, whence

$$A(x, D)\varphi u = \varphi \cdot f + \varphi \cdot K_1 f + \varphi \cdot K_2 g + Z\mathfrak{P}(f, g).$$

Now, the last part of Theorem 14.33 implies that the commutator, Z, belongs to $S^{l-1,0}$, hence

$$\|Z\mathfrak{P}(f, g)\|_{(0,-t)} \leqslant \|Z\mathfrak{P}(f, g)\|_{(1,-t-1)} \leqslant \text{const}\|\mathfrak{P}(f, g)\|_{(l,-t-1)}.$$

This result, together with the continuity of the operators K_1, K_2 and $\mathfrak{P}$ in Theorem 14.40, implies (101).

Next consider the boundary operator B. We may write

$$\gamma\varphi u = \varphi\gamma u + \delta u,$$

where $\delta = [\delta_j]_{j=0}^{l-1}$ and δ_j is a polynomial in the normal derivative operator D_n of order $\leqslant j - 1$. Since $B = B^c\gamma$, then

$$B(x', D)\varphi u = \varphi \cdot g + \varphi \cdot K_3 f + \varphi \cdot K_4 g + \tilde{Z}\mathfrak{P}(f, g)$$

where $W = B^c\varphi - \varphi B^c$ and $\tilde{Z} = W\gamma + B^c\delta$. Note that the commutator, W, is a p.d.o. on $\mathbb{R}^{n-1}$ of order one less than B^c in each block: $B^c = [B^c_{kj}]$ where B^c_{kj} has order $\leqslant m_k - j$ and $W = [W_{kj}]$ where W_{kj} has order $\leqslant m_k - j - 1$. Now let $\tilde{Z}_k$ denote the kth row of $\tilde{Z}$, $k = 1, \ldots, r$. Since

$$\begin{aligned}\|\tilde{Z}_k\mathfrak{P}(f, g)\|_{-t+l-m_k-1/2} &\leqslant \|\textstyle\sum W_{kj}\gamma_j\mathfrak{P}(f, g)\|_{-t+l-m_k-1/2}\\ &\quad + \|\textstyle\sum B^c_{kj}\delta_j\mathfrak{P}(f, g)\|_{-t+l-m_k-1/2}\\ &\leqslant \|\textstyle\sum \gamma_j\mathfrak{P}(f, g)\|_{-t+l-j-3/2}\\ &\quad + \|\textstyle\sum \delta_j\mathfrak{P}(f, g)\|_{-t+l-j-1/2}\\ &\leqslant \text{const}\|\mathfrak{P}(f, g)\|_{(l,-t-1)},\end{aligned}$$

then (102) holds by the continuity of the operators K_3, K_4 and $\mathfrak{P}$.

Theorem 14.43 *Assume that $(\mathscr{A}, \mathscr{B}) \in \mathfrak{BE}^{l,\mathbf{m}}$, i.e. the elliptic operator, $\mathscr{A}$, has the form stated in §10.7, and the boundary operator, $\mathscr{B}$, satisfies the L-condition with respect to $\mathscr{A}$. Then the conclusions of Theorem 14.29 remain valid.*

Proof We have already shown the Fredholm property in Theorem 10.26, so it remains to verify that the image of $\mathfrak{L}$ is defined by C^∞ relations. Suppose that

$$(\mathscr{A}u, v)_\Omega + (\mathscr{B}u, w)_{\partial\Omega} = 0 \qquad \text{for all } u \in C^\infty(\bar{\Omega}, \mathbb{C}^p)$$

where $v \in L_2(\Omega, \mathbb{C}^p)$, $w \in \times_{k=1}^r W_2^{-l+m_k+1/2}(\partial\Omega)$. First of all, by taking u with support in Ω it follows that $(\mathscr{A}u, v)_\Omega = 0$, whence $\mathscr{A}^* v = 0$ where $\mathscr{A}^* \in \mathfrak{E}^l$ is the adjoint of $\mathscr{A}$. It follows by Proposition 14.36 that $v \in C^\infty(\Omega, \mathbb{C}^p)$. Now we wish to show that v is smooth at the boundary $\partial\Omega$, and that w is smooth on $\partial\Omega$. Take $u \in C_0^\infty(\mathbb{R}^n, \mathbb{C}^p)$ with support in a coordinate patch near $\partial\Omega$. Now, for any $\psi \in C_0^\infty$ we have

$$(\psi\mathscr{A}u, v)_\Omega + (\psi\mathscr{B}u, w)_{\partial\Omega} = F(u)$$

where $F(u) = (\psi\mathscr{A}u - \mathscr{A}\psi u, v)_\Omega + (\psi\mathscr{B}u - \mathscr{B}\psi u, w)_{\partial\Omega}$. After extending the symbols of $\mathscr{A}$ and of $\mathscr{B}$ to the whole space as in the proof of Proposition 14.36 we have (taking supp ψ in the coordinate patch)

$$\int_{\mathbb{R}^n_+} A(x, D)u \cdot {}^h\tilde{v}\, dx + \int_{\mathbb{R}^{n-1}} B(x', D)u \cdot {}^h\tilde{w}\, dx' = F(u) \tag{103}$$

where $\tilde{v} = \psi v$, $\tilde{w} = \psi w \sigma(x')$ and $\sigma(x') > 0$ is the density on $\partial\Omega$ in local coordinates. Now replace u by

$$u = \varphi \cdot \mathfrak{P}(f, g)$$

in (103), where $\varphi \in C_0^\infty$ has support in the same coordinate patch. Note that $|F(u)| \leqslant C\|u\|_{(l-1,-t)} \leqslant C\|u\|_{(l,-t-1)}$ and then by the continuity of $\mathfrak{P}$ (see Theorem 14.40) it follows that

$$|F(u)| \leqslant \text{const}(\|f\|_{(0,-t-1)} + \sum_k \|g_k\|_{-t+l-m_k-3/2}).$$

By taking $\varphi = 1$ on supp ψ and using Lemma 14.42 instead of (99), (100), we see that the inductive step from t to $t+1$ in Lemma 14.41 remains valid. Hence $\tilde{w}$ is C^∞; since ψ is arbitrary and σ is C^∞, it follows that w is C^∞ on $\partial\Omega$. Then it follows as in the proof of Theorem 14.29 that $\tilde{v} \in C^\infty(\bar{\mathbb{R}}^n_+, \mathbb{C}^p)$, whence $v \in C^\infty(\bar{\Omega}, \mathbb{C}^p)$.

Part V

An Index Formula for Elliptic Boundary Problems in the Plane

In general, the matrix function $\Delta^+_{\mathscr{B}}$ does not provide any topological information about the boundary problem $(\mathscr{A}, \mathscr{B})$ because its definition depends on the choice of γ^+ spectral pair (X_+, T_+); see the remarks after Theorem 10.19. For elliptic systems in the plane, however, there is a choice which is essentially canonical, and the winding number of $\det \Delta^+_{\mathscr{B}}$ along $ST^*(\partial\Omega)$ is then a topological invariant of $(\mathscr{A}, \mathscr{B})$.

In Chapter 15 the topological index (or symbol index) of $(\mathscr{A}, \mathscr{B})$ is defined in terms of this winding number. In Chapter 16 we prove the index formula, i.e. that the analytical index coincides with the topological index. Chapter 17 gives some examples of the use of this index formula as well as a complete homotopy classification of elliptic systems in the plane with 2×2 real coefficients.

For simplicity we treat only the case when the region Ω is simply connected.

15

Further results on the Lopatinskii condition

In §15.1 we examine the Lopatinskii condition and the results of Chapter 10 in further detail for differential operators in the plane. In §15.2 we review the definition and properties of the winding number of a continuous nonvanishing function on the unit circle. The topological index of an elliptic boundary problem $(\mathscr{A}, \mathscr{B})$ is then defined in §15.3 under the assumption that the region Ω is simply connected (i.e. conformally equivalent to the unit disc by the Riemann mapping theorem), and we show that the topological index is additive under the composition of boundary value problems. In §15.4 we show that there is no loss of generality in assuming that $\mathscr{A}$ and $\mathscr{B}$ have real matrix coefficients, which turns out to be an important simplification in the proof of the index formula.

15.1 Some preliminaries

By a *region* we shall mean a nonempty connected open set in the complex plane. Let $\Omega \subset \mathbb{R}^2$ be a bounded region with C^∞ boundary. We denote by Ell^l the set of properly elliptic differential operators

$$\mathscr{A}(x, \partial) = \sum_{|\alpha| \leqslant l} A_\alpha(x)\, \partial^\alpha, \qquad x = (x_1, x_2) \in \bar{\Omega}, \tag{1a}$$

with principal part homogeneous of degree l. As usual, the coefficients A_α are smooth $p \times p$ matrix functions on $\bar{\Omega}$. Note that we have expressed $\mathscr{A}$ in terms of ∂^α rather than D^α; for the examples in this part of the book it is convenient to omit the factor $1/i$. In contrast with §10.5 we let $\mathrm{BE}^{l,\mathbf{m}}$ denote the set of all boundary problems $(\mathscr{A}, \mathscr{B})$ where $\mathscr{A} \in \mathrm{Ell}^l$ and $\mathscr{B}$ is a *differential* boundary operator of the form

$$\mathscr{B}(y, \partial) = \left[\sum_{|\alpha| \leqslant m_k} b_k^{(\alpha)}(y)\, \partial^\alpha \right]_{k=1}^{r}, \qquad y \in \partial\Omega \tag{1b}$$

satisfying the L-condition relative to $\mathscr{A}$.

Remark More general (pseudo-differential) boundary operators had been permitted in the previous chapters for the purpose of being able to construct homotopies of elliptic boundary problems. For systems in the plane, however, this turns out to be unnecessary (see Proposition 15.2).

Let $n(y)$ be the inward-point unit normal at $y \in \partial\Omega$, and $\tau(y)$ the unit tangent vector to $\partial\Omega$ such that $\tau(y), n(y)$ is a positively oriented basis of $\mathbb{R}^2$. We identify $ST^*(\partial\Omega)$ with the unit tangent bundle $ST(\partial\Omega)$. If $(y, \xi') \in ST^*(\partial\Omega)$ then

$$|\xi'| = 1, \qquad \text{so } \xi' = \pm\tau(y),$$

and $ST^*(\partial\Omega)$ is the disjoint union of two copies of $\partial\Omega$.

Let $(\mathscr{A}, \mathscr{B}) \in \mathrm{BE}^{l,\mathbf{m}}$, i.e. $\mathscr{A}$ is an elliptic operator of the form (1a) and $\mathscr{B}$ is a boundary operator (1b) having order m_k in the kth row, $k = 1, \ldots, r$. In the notation of §10.1, we let $\tilde{L}_y(\lambda) = L_{y,\tau(y)}(\lambda)$, that is,

$$\tilde{L}_y(\lambda) = \pi\mathscr{A}(y, \tau(y) + \lambda n(y)) = \sum_{j=0}^{l} \tilde{A}_j(y)\lambda^j, \qquad y \in \partial\Omega, \tag{2}$$

where $\tilde{A}_j(y) = A_j(y, \tau(y))$. If $(\tilde{X}_\pm(y), \tilde{T}_\pm(y), \tilde{Y}_\pm(y))$ is a $\gamma^\pm$-spectral triple of $\tilde{L}_y(\lambda)$ then we may define a γ^+-spectral triple of $L_{y,\xi'}(\lambda)$ as follows (see Theorem 2.21)

$$(X_+(y, \xi'), T_+(y, \xi'), Y_+(y, \xi'))$$
$$= \begin{cases} (\tilde{X}_+(y), \tilde{T}_+(y), \tilde{Y}_+(y)) & \xi' = \tau(y) \\ (\tilde{X}_-(y), -\tilde{T}_-(y), (-1)^{1-l}\tilde{Y}_-(y)) & \xi' = -\tau(y) \end{cases}$$

We also define $(X_-(y, \xi'), T_-(y, \xi'), Y_-(y, \xi'))$ by the equation (35) in §10.4. Associated with the boundary operator $\mathscr{B}$ there is the matrix polynomial

$$\tilde{B}_y(\lambda) = \pi\mathscr{B}(y, \tau(y) + \lambda n(y)) = \sum_{j=0}^{\mu} \tilde{B}_j(y)\lambda^j, \qquad y \in \partial\Omega \tag{3}$$

where $\tilde{B}_j(y) = B_j(y, \tau(y))$ and $\mu \leqslant \max m_k$, and we let

$$\tilde{\Delta}^\pm_{\mathscr{B}}(y) = \sum_{j=0}^{\mu} \tilde{B}_j(y)\tilde{X}_\pm(y)\tilde{T}^j_\pm(y) \tag{4}$$

and also

$$\Delta^+_{\mathscr{B}}(y, \xi') = \sum_{j=0}^{\mu} B_j(y, \xi')X_+(y, \xi')T^j_+(y, \xi') \tag{5}$$

Note the following relationship between the matrix functions (4) and (5):

$$\Delta^+_{\mathscr{B}}(y, \tau(y)) = \tilde{\Delta}^+_{\mathscr{B}}(y), \qquad \Delta^+_{\mathscr{B}}(y, -\tau(y)) = M(-1)\,\tilde{\Delta}^-_{\mathscr{B}}(y), \tag{6}$$

the second equality being valid for a *differential* operator $\mathscr{B}$, because $\pi\mathscr{B}(y, c\xi) = M(c)\cdot\pi\mathscr{B}(y, \xi)$ when $c = -1$. We know from Theorem 10.3 and (38) of §10.4 that $(\mathscr{A}, \mathscr{B})$ satisfies the Lopatinskii condition if and only if $\det \Delta^+_{\mathscr{B}}(y, \xi') \neq 0$ when $\xi' = \pm\tau(y)$, which we now see is equivalent to the condition

$$\det \tilde{\Delta}^\pm_{\mathscr{B}}(y) \neq 0 \qquad \text{for all } y \in \partial\Omega. \tag{7}$$

Remark 15.1 If the coefficients of $\tilde{L}_y(\lambda)$ are real matrices then we may assume that $(\tilde{X}_-(y), \tilde{T}_-(y)) = \overline{(\tilde{X}_+(y), \tilde{T}_+(y))}$. Further, if the coefficients of $\tilde{B}_y(\lambda)$ are also real matrices then we have $\tilde{\Delta}^-_{\mathscr{B}}(y) = \overline{\tilde{\Delta}^+_{\mathscr{B}}(y)}$; in that case $(\mathscr{A}, \mathscr{B})$ satisfies the L-condition if and only if

$$\det \tilde{\Delta}^+_{\mathscr{B}}(y) \neq 0 \qquad \text{for all } y \in \partial\Omega.$$

We now specialize the results of §§10.5 and 10.6 to elliptic systems in the plane. Denote by

$$\mathrm{BE}^{l,\mu}$$

the set of elliptic boundary problems $(\mathscr{A}, \mathscr{B})$ fulfilling the L-condition such that $\mathscr{B}$ has the same order in all rows, $m_k = \mu$ $(k = 1, \ldots, r)$.

Proposition 15.2 *Let $\mathscr{A}_t$, $t \in I$, be a homotopy of elliptic operators in* Ell^l. *If $\mathscr{B}$ is a (differential) boundary operator satisfying the L-condition relative to $\mathscr{A} = \mathscr{A}_0$ such that the principal part of $\mathscr{B}$ is homogeneous of degree $l - 1$ (in all rows) then there exists a homotopy of boundary value problems $(\mathscr{A}_t, \mathscr{B}_t) \in \mathrm{BE}^{l,l-1}$, $t \in I$, such that $(\mathscr{A}, \mathscr{B}_0) \approx (\mathscr{A}, \mathscr{B})$.*

Proof This is the same proof as in Theorem 10.23, except that it is necessary to verify that the boundary operators can be chosen to be differential operators. By the remark after Theorem 10.23, there exist (unique) $r \times p$ matrix functions $B_j(t, \cdot)$ on $T^*(\partial\Omega)\backslash 0$ such that

$$\sum_{j=0}^{l-1} B_j(t, \cdot)X_\pm(t, \cdot)T^j_\pm(t, \cdot) = \Delta^\pm_{\mathscr{B}}, \tag{8}$$

Since the DN numbers of $\mathscr{B}$ are $m_1 = \cdots = m_r = l - 1$ and $t_1 = \cdots = t_p = 0$, then $B_j(t, y, c\xi') = c^{l-1-j}B_j(t, y, \xi')$, $j = 0, \ldots, l - 1$, for all c (positive and negative). We let

$$\mathscr{B}_t(y, D) = \sum_{j=0}^{l-1} B_j(t, y, \tau(y)) \frac{\partial^{l-1-j}}{\partial\tau^{l-1-j}} \frac{\partial^j}{\partial n^j},$$

which is a differential operator. Since $\Delta^\pm_{\mathscr{B}_t} = \Delta^\pm_{\mathscr{B}}$ then $(\mathscr{A}_t, \mathscr{B}_t)$ satisfies the L-condition and in this way we have constructed a homotopy $(\mathscr{A}_t, \mathscr{B}_t) \in \mathrm{BE}^{l,l-1}$. Also, by construction, the principal parts of $\mathscr{B}_0$ and $\mathscr{B}$ are the same (see Proposition 10.13), thus $(\mathscr{A}, \mathscr{B}_0) \approx (\mathscr{A}, \mathscr{B})$. (The notation $\approx$ is defined in the text preceding Lemma 10.18.)

Remark 15.3 If the operators $\mathscr{A}_t$ and $\mathscr{B}$ have real matrix coefficients then we can find $\mathscr{B}_t$ also with real matrix coefficients such that $(\mathscr{A}_t, \mathscr{B}_t)$ satisfies the L-condition for all $t \in I$. Indeed,

$$(X_-(t, \cdot), T_-(t, \cdot), Y_-(t, \cdot)) = (\overline{X_+(t, \cdot)}, \overline{T_+(t, \cdot)}, \overline{Y_+(t, \cdot)}),$$

is a γ^--spectral triple of $L_{t,\cdot}(\lambda)$ and the solution of (8) (which does not depend on the choice of $(X_\pm, T_\pm)$) can be written in the form

$$B_j(t,\cdot) = 2\,\mathrm{Re}\left\{\Delta^+_{\mathscr{B}}\cdot \sum_{k=0}^{l-j-1} T^k_+(t,\cdot)\,Y_+(t,\cdot)\,A_{j+k+1}(t,\cdot)\right\} \tag{9}$$

which is a real matrix function.

The same remark applies to the following theorem.

Theorem 15.4 *Let $(\mathscr{A},\mathscr{B}) \in \mathrm{BE}^{l,\mu}$ with $\mu \geqslant l$. Then*

$$(\mathscr{A},\mathscr{B}) \approx (\mathscr{A},(\partial/\partial\tau)^{\mu-l+1}\mathscr{R}) \tag{10}$$

for some (differential) boundary operator $\mathscr{R}$ such that $(\mathscr{A},\mathscr{R}) \in \mathrm{BE}^{l,l-1}$. In fact, since $\mathscr{B}$ is a differential operator there exists $\mathscr{R}$ with $\tilde{\Delta}^\pm_{\mathscr{R}} = \tilde{\Delta}^\pm_{\mathscr{B}}$.

Proof This is just a restatement of Theorem 10.20 for systems in the plane. There exist unique $r \times p$ matrix functions $\tilde{R}_j$ on $\partial\Omega$ such that

$$\sum_{j=0}^{l-1} \tilde{R}_j(y)\,\tilde{X}_\pm\,\tilde{T}^j_\pm = \sum_{j=0}^{\mu} \tilde{B}_j(y)\tilde{X}_\pm\,\tilde{T}^j_\pm, \qquad y \in \partial\Omega \tag{11}$$

that is,

$$[\tilde{R}_0\cdots\tilde{R}_{l-1}] = \tilde{\Delta}^+_{\mathscr{B}}\cdot[\tilde{Y}_+\cdots\tilde{T}^{l-1}_+\tilde{Y}_+]\tilde{\mathscr{Z}} + \tilde{\Delta}^-_{\mathscr{B}}\cdot[\tilde{Y}_-\cdots\tilde{T}^{l-1}_-\tilde{Y}_-]\tilde{\mathscr{Z}}$$

It follows that the differential operator

$$\mathscr{R} := \sum_{j=0}^{l-1} \tilde{R}_j(y)\,\frac{\partial^{l-1-j}}{\partial\tau^{l-1-j}}\,\frac{\partial^j}{\partial n^j}$$

satisfies (11), that is

$$\tilde{\Delta}^\pm_{\mathscr{R}} = \tilde{\Delta}^\pm_{\mathscr{B}}.$$

Since $\mathscr{R}$ is a differential operator and $\pi\mathscr{R}$ is homogeneous of degree $l-1$ we have $(\mathscr{A},\mathscr{R}) \in \mathrm{BE}^{l,l-1}$. The relation (10) follows since $\tilde{\Delta}^\pm_{(\partial/\partial\tau)^{\mu-l+1}\mathscr{R}} = \tilde{\Delta}^\pm_{\mathscr{R}} = \tilde{\Delta}^\pm_{\mathscr{B}}$.

Remark From the proof of Lemma 10.18, we see that if (10) holds then $(\mathscr{A},\mathscr{B})$ is homotopic to $(\mathscr{A},(\partial/\partial\tau)^{\mu-l+1}\mathscr{R})$ in $\mathrm{BE}^{l,l-1}$.

15.2 The degree or winding number on the unit circle

Let S^1 denote the unit circle, and let f be a continuous nonvanishing complex-valued function on S^1, i.e. $f \in C(S^1, \mathbb{C}^*)$ where $\mathbb{C}^* = \mathbb{C}\backslash 0$. Then

$$\frac{f(e^{it})}{|f(e^{it})|} = e^{i\theta(t)}$$

for some real-valued $\theta \in C([0, 2\pi])$, and the degree (or winding number) of f is defined to be

$$\deg f = [\arg f(z)]_{S^1} = (\theta(2\pi) - \theta(0))/2\pi. \tag{12}$$

For the sake of brevity we use the notation $\tilde{f}/|\tilde{f}| = e^{i\theta}$ where $\tilde{f}(t) = f(e^{it})$. Note that θ defines a continuous function on S^1 only when $\deg f = 0$, i.e. $\theta(2\pi) = \theta(0)$, and in that case we may write $f/|f| = e^{i\theta}$ where θ is periodic.

The following result is evident.

Lemma 15.5 *The degree has the following multiplicative property*

$$\deg(f \cdot g) = \deg f + \deg g,$$

In particular, $\deg(1/f) = -\deg f$.

Proof If $\tilde{f}/|\tilde{f}| = e^{i\theta}$ and $\tilde{g}/|\tilde{g}| = e^{i\alpha}$ then

$$fg/|fg| = e^{i(\theta+\alpha)},$$

whence

$$\deg(fg) = (\theta(2\pi) + \alpha(2\pi) - (\theta(0) + \alpha(0))/2\pi = \deg f + \deg g.$$

Definition 15.6 *Let $f, g \in C(S^1, \mathbb{C}^*)$. Then f and g are said to be homotopic if there exists a continuous function $F\colon S^1 \times I \to \mathbb{C}^*$ such that $F(\cdot, 0) = f$ and $F(\cdot, 1) = g$.*

If f and g are homotopic we write $f \sim g$. It is evident that if $f_1 \sim f_2$ and $g_1 \sim g_2$ then $f_1 \cdot g_1 \sim f_2 \cdot g_2$.

Definition 15.7 *A function $f \in C(S^1, \mathbb{C}^*)$ is said to be null-homotopic if $f \sim 1$, i.e. if it is homotopic to the constant function* 1.

Any nonzero constant function is obviously null-homotopic.

Also, any *real-valued* function $f \in C(S^1, \mathbb{R}^*)$ is null-homotopic. Indeed, S^1 is connected so the image of f is connected, hence either $f > 0$ or $f < 0$. If we assume that $f > 0$ then the linear homotopy $\kappa f + (1 - \kappa)1$, $0 \leqslant \kappa \leqslant 1$, shows that $f \sim 1$. If $f < 0$ then we get $f \sim -1$, and since -1 is homotopic to 1 we get $f \sim 1$ once again.

It follows that any function $f \in C(S^1, \mathbb{C}^*)$ is homotopic to a function with modulus 1 (Proof: $f = |f| u$ where $u \in C(S^1)$, $|u| = 1$; and $|f| > 0$ is homotopic to 1, hence f is homotopic to u.)

Theorem 15.8 *The degree of a function is locally constant; in fact if*

$$\max_{z \in S^1} |g(z) - f(z)| < \max_{z \in S^1} |f(z)|$$

then $\deg g = \deg f$.

Proof Dividing through by f and using Lemma 15.5 we may assume that $f = 1$, i.e. we have to show that if

$$\max_{z \in S^1} |g(z) - 1| < 1$$

then $\deg g = 0$. We may write $g = 1 + h$ where the function h satisfies $\max_{z \in S^1} |h(z)| < 1$, and then

$$g = e^{i\theta}$$

where $i\theta = \log(1 + h) = h - h^2/2 + h^3/3 - \cdots$. The function θ is continuous on S^1 since the series converges uniformly on S^1. The homotopy

$$g_\kappa = e^{i\kappa\theta}, \qquad 0 \leqslant \kappa \leqslant 1,$$

shows that $g \sim 1$.

Corollary 15.9 *A function $f \in C(S^1, \mathbb{C}^*)$ is null-homotopic if and only if it has degree* 0.

Proof Let $f \sim 1$, i.e. there exists a continuous function $F: S^1 \times [0, 1] \to \mathbb{C}^*$ such that $F_0 = f$ and $F_1 = 1$. Here F_κ, $0 \leqslant \kappa \leqslant 1$, is the function defined by $F_\kappa(z) = F(z, \kappa)$. By Theorem 15.8 the degree, $\deg F_\kappa$, is locally constant for $0 \leqslant \kappa \leqslant 1$. Since $[0, 1]$ is connected, it follows that $\deg F_\kappa$ is constant, whence

$$\deg f = \deg F_0 = \deg F_1 = 0.$$

Conversely, if $\deg f = 0$ then, by definition, $f/|f| = e^{i\theta}$ where θ is a continuous function on S^1 (see (12)). Then $e^{i\theta} \sim 1$ due to the homotopy $e^{i\kappa\theta}$, $0 \leqslant \kappa \leqslant 1$. Since $|f| > 0$ is real-valued we also have $|f| \sim 1$, so f is null-homotopic.

Now we define the degree of a continuous matrix function, $M: S^1 \to GL_r(\mathbb{C})$. Here $GL_r(\mathbb{C})$ denotes the group of invertible $r \times r$ matrices with complex entries. The degree of M is defined to be the degree of the complex-valued function, $\det M$,

$$\deg M = [\arg \det M(z)]_{S^1}.$$

The following proposition shows that two continuous $r \times r$ matrix functions on S^1 with invertible values are homotopic if and only if their degrees coincide.

Theorem 15.10 *Two continuous matrix functions, $M: S^1 \to GL_r(\mathbb{C})$ and $N: S^1 \to GL_r(\mathbb{C})$, are homotopic if and only if* $\deg M = \deg N$.

Proof For scalar functions ($r = 1$), this follows immediately from Corollary 15.9 since $f \sim g$ if and only if $f \cdot g^{-1} \sim 1$ and

$$\deg f = \deg g \Leftrightarrow \deg(f \cdot g^{-1}) = 0.$$

Thus, we assume $r > 1$. The degree of a continuous nonvanishing function is locally constant (Theorem 15.8), so it is clear that if M and N are homotopic to each other then $\deg M = \deg N$. To prove the converse, it suffices to show that M is homotopic to the matrix function

$$N(z) = \begin{pmatrix} \det M(z) & & & \\ & 1 & & \\ & & \ddots & \\ & & & 1 \end{pmatrix}.$$

Without loss of generality (by approximation), M is a rational matrix function of the form $M(z) = \sum_{j=-k}^{k} A_j z^j$, and then it may be assumed that $M(z)$ is a polynomial in z (by consideration of $z^k M(z)$). Now, $M(z)$ has the Smith canonical form, $E(z)D(z)F(z)$, where $D(z)$ is a diagonal matrix polynomial and $E(z)$ and $F(z)$ are matrix polynomials with constant nonzero determinant. By means of the homotopy $E(tz)$, $0 \leqslant t \leqslant 1$, $E(z)$ is homotopic to the constant matrix function $E(0) \in GL_r(\mathbb{C})$, and, since $GL_r(\mathbb{C})$ is path-connected, it follows that $E(z)$ (and similarly $F(z)$) is homotopic to the constant matrix function I (the identity matrix). Hence $M(z)$ is homotopic to the diagonal matrix $D(z) = \operatorname{diag}(d_i(z))_{i=1}^{r}$. Then by a succession of rotations we see that $D(z)$ is homotopic to $N(z)$. For example, if $r = 2$, the matrix polynomial

$$\begin{pmatrix} d_1(z) & \\ & d_2(z) \end{pmatrix} \text{ is homotopic to } \begin{pmatrix} d_1(z) \cdot d_2(z) & \\ & 1 \end{pmatrix}$$

by means of the homotopy

$$\begin{pmatrix} d_1(z) & \\ & 1 \end{pmatrix} \begin{pmatrix} \cos t & -\sin t \\ \sin t & \cos t \end{pmatrix} \begin{pmatrix} d_2(z) & \\ & 1 \end{pmatrix} \begin{pmatrix} \cos t & \sin t \\ -\sin t & \cos t \end{pmatrix},$$

for $0 \leqslant t \leqslant \pi/2$.

15.3 The topological index

From now on we always assume that the region Ω is bounded and simply connected. In this section we shall define the topological index of an elliptic boundary problem (Definition 15.15) but first we need to make a special choice of γ^+-spectral pair for the matrix polynomial $\tilde{L}_y(\lambda)$ of (2). Let $\mathscr{A} \in \mathrm{Ell}^l$ be an elliptic operator with principal part $\pi\mathscr{A}(x, \xi) = \sum_{|\alpha|=l} A_\alpha(x)\xi^\alpha$, and consider the matrix polynomial

$$L_x(\lambda) = \pi\mathscr{A}(x, 1, \lambda)) = \sum_{j=0}^{l} A_j(x)\lambda^j, \tag{13}$$

which has coefficients $A_j(x) = A_{l-j,j}(x)$ defined for all $x \in \bar{\Omega}$ (not just the boundary). Due to ellipticity, $\det L_x(\lambda) \neq 0$ for real λ and the leading coefficient, $A_l(x) = \pi\mathscr{A}(x, (0, 1))$, is invertible. The solution space of

$$L_x\left(\frac{1}{i}\frac{d}{dt}\right)u = 0$$

is a direct sum $\mathfrak{M}_{L_x}^+ \oplus \mathfrak{M}_{L_x}^-$, corresponding to the eigenvalues of $L_x(\lambda)$ with $\operatorname{Im} \lambda > 0$ and $\operatorname{Im} \lambda < 0$, and we let $\mathfrak{M}^\pm$ denote the vector bundle over $\bar{\Omega}$ with fibres $\mathfrak{M}_{L_x}^\pm$.

Proposition 15.11 *Let Ω be a bounded and simply connected region with C^∞ boundary. Then the vector bundles $\mathfrak{M}^+$ and $\mathfrak{M}^-$ are trivial. Hence there exists admissible pairs $(X_\pm, T_\pm)$ over $\bar{\Omega}$ such that $(X_\pm(x), T_\pm(x))$ is a $\gamma^\pm$-spectral pair of $L_x(\lambda)$ consisting of matrix functions with entries in $C^\infty(\bar{\Omega})$.*

Proof The coefficients of $\mathscr{A}$ can be extended smoothly to a neighbourhood $\Omega' \supset \bar{\Omega}$. In view of Exercise 1 at the end of the chapter, we may assume that Ω' is also a simply connected region. Also, by continuity, we may assume that $\det \pi\mathscr{A}(x, \xi) \neq 0$ for all $x \in \Omega'$, $\xi \neq 0$. In this way we obtain a vector bundle $\mathfrak{M}^\pm$ over Ω' which is trivial since Ω' is contractible. (In fact, by the Riemann mapping theorem Ω' is conformally equivalent to the open unit disc, see [Ru 1], p. 283.) Hence its restriction to $\bar{\Omega}$ is also trivial. The last statement in the proposition follows as in Theorem 10.11.

Note that $L_x(\lambda)$ is defined in terms of the basis (1, 0), (0, 1) for $\mathbb{R}^2$ whereas the matrix polynomial $\tilde{L}_y(\lambda)$ of (2) is defined in terms of the basis $\tau(y)$, $n(y)$. Writing $\tau(y) = (\tau_1(y), \tau_2(y))$ and $n(y) = (-\tau_2(y), \tau_1(y))$ we have

$$\begin{aligned}\tilde{L}_y(\lambda) &= \pi\mathscr{A}(y, \tau(y) + \lambda n(y))\\ &= \sum_{j=0}^{l} A_j(y)(\tau_1(y) - \lambda\tau_2(y))^{l-j}(\tau_2(y) + \lambda\tau_1(y))^j\\ &= (\tau_1(y) - \lambda\tau_2(y))^l \cdot L_y(\varphi^{-1}(\lambda)),\end{aligned}$$

where $\varphi^{-1}(\lambda) = (\tau_2 + \lambda\tau_1)/(\tau_1 - \lambda\tau_2)$. In other words. $\tilde{L}_y(\lambda)$ is the transformation of $L_y(\lambda)$ under $\varphi(\lambda) = (-\tau_2 + \lambda\tau_1)/(\tau_1 + \lambda\tau_2)$ as defined in §2.4. Let $(X_\pm(y), T_\pm(y))$ be a $\gamma^\pm$-spectral pair of $L_y(\lambda)$. Since φ maps the upper half-plane to itself, then by Theorem 2.21

$$\left.\begin{aligned}\tilde{X}_\pm(y) &= X_\pm(y),\\ \tilde{T}_\pm(y) &= (\tau_1(y)T_\pm(y) - \tau_2(y)I)(\tau_2(y)T_\pm(y) + \tau_1(y)I)^{-1}.\end{aligned}\right\} \tag{14}$$

is a $\gamma^\pm$-spectral pair of $\tilde{L}_y(\lambda)$.

In the next proposition we assume that $(\mathscr{A}, \mathscr{B}) \in \mathrm{BE}^{l,\mu}$, that is, the principal part of the boundary operator $\mathscr{B}$ is homogeneous of degree μ in each row, $\pi\mathscr{B}(y, \xi) = \sum_{|\alpha|=\mu} B_\alpha(y)\xi^\alpha$. Analogous to the considerations for $L_x(\lambda)$, there is the following $r \times p$ matrix polynomial

$$B_y(\lambda) = \pi\mathscr{B}(y, (1, \lambda)) = \sum_{j=0}^{\mu} B_j(y)\lambda^j, \qquad y \in \partial\Omega,$$

where $B_j(y) = B_{\mu-j,j}(y)$, and we see that $\tilde{B}_y(\lambda)$ defined by (3) is the transformation of the rectangular $r \times p$ matrix polynomial $B_y(\lambda)$ under $\varphi(\lambda) = (-\tau_2 + \lambda\tau_1)/(\tau_1 + \lambda\tau_2)$. We let

$$\Delta^\pm_{\mathscr{B}}(y) = \sum_{j=0}^{\mu} B_j(y)X_\pm(y)T^j_\pm(y) \tag{15}$$

where $(X_\pm(y), T_\pm(y))$ is a $\gamma^\pm$-spectral pair of $L_y(\lambda)$, $y \in \partial\Omega$. Note that the matrix functions $\Delta_{\mathscr{B}}^{\pm}(y)$ should be distinguished from $\Delta_{\mathscr{B}}^{+}(y, \xi')$ which was defined in (5).

Proposition 15.12 *Let $(X_{\pm}(y), T_{\pm}(y))$ be $\gamma^{\pm}$-spectral pairs of $L_y(\lambda)$, and $(\tilde{X}_{\pm}(y), \tilde{T}_{\pm}(y))$ the corresponding $\gamma^{\pm}$-spectral pairs of $\tilde{L}_y(\lambda)$ as defined by the equations (14). Let $\mathscr{B}$ be a boundary operator whose principal part is homogeneous of degree μ. Then*

$$\tilde{\Delta}_{\mathscr{B}}^{\pm}(y) = \Delta_{\mathscr{B}}^{\pm}(y)\cdot(\tau_1(y)I + \tau_2(y)T_{\pm}(y))^{-\mu}, \qquad y \in \partial\Omega$$

Hence $\det \Delta_{\mathscr{B}}^{\pm}(y) \neq 0$ *if and only if* $\det \tilde{\Delta}_{\mathscr{B}}^{\pm}(y) \neq 0$.

Proof This follows from Lemma 2.22.

Corollary 15.13 *Let the hypotheses be as in Proposition* 15.12. *Then the following are equivalent for all $y \in \partial\Omega$:*

(*i*) *The map* $u \mapsto \tilde{B}_y\left(\frac{1}{i}\frac{d}{dt}\right)u|_{t=0}$ *from* $\mathfrak{M}_{\tilde{L}_y}^{\pm}$ *to* $\mathbb{C}^r$ *is invertible;*

(*ii*) *The map* $u \mapsto B_y\left(\frac{1}{i}\frac{d}{dt}\right)u|_{t=0}$ *from* $\mathfrak{M}_{L_y}^{\pm}$ *to* $\mathbb{C}^r$ *is invertible.*

Proof In view of Theorem 10.1, (i) and (ii) are equivalent to

$$(i')\ \det \tilde{\Delta}_{\mathscr{B}}^{\pm}(y) \neq 0 \qquad (ii')\ \det \Delta_{\mathscr{B}}^{\pm}(y) \neq 0$$

and (i′) and (ii′) are equivalent by Proposition 15.12.

Proposition 15.11 has the following important consequence.

Theorem 15.14 *Let $\Omega \in \mathbb{R}^2$ be a bounded and simply connected region with C^∞ boundary. A properly elliptic operator $\mathscr{A}$ in $\bar{\Omega}$ always has a differential boundary operator satisfying the L-condition.*

Proof Let $\mathscr{A} \in \mathrm{Ell}^l$, choose $(X_{\pm}(x), T_{\pm}(x))$ as in Proposition 15.11, and then define the spectral pair (14). There is a natural differential boundary operator associated with the spectral pair $(\tilde{X}_{\pm}(y), \tilde{T}_{\pm}(y))$ since the equations

$$\sum_{j=0}^{l-1} \tilde{B}_j(y)\tilde{X}_{\pm}\tilde{T}_{\pm}^j = I_r$$

have the (unique) solution

$$[\tilde{B}_0 \cdots \tilde{B}_{l-1}] = [\tilde{Y}_+ \cdots \tilde{T}_+^{l-1}\tilde{Y}_+]\tilde{\mathscr{Z}} + [\tilde{Y}_- \cdots \tilde{T}_+^{l-1}\tilde{Y}_-]\tilde{\mathscr{Z}}.$$

The boundary operator

$$\mathscr{B}^\circ = \sum_{j=0}^{l-1} \tilde{B}_j(y)\frac{\partial^{l-1-j}}{\partial\tau^{l-1-j}}\frac{\partial^j}{\partial n^j}$$

therefore satisfies $\tilde{\Delta}_{\mathscr{B}^\circ}^{\pm} \equiv I_r$ and we have $(\mathscr{A}, \mathscr{B}^\circ) \in \mathrm{BE}^{l,l-1}$.

Since Ω is a bounded, simply connected region in $\mathbb{R}^2$, the boundary $\partial\Omega$ is diffeomorphic to S^1; see [GP], p. 64. Let φ be an orientation-preserving diffeomorphism from S^1 to $\partial\Omega$. Then the degree of a nonvanishing continuous function $f \in C(\partial\Omega)$ is defined to be $\deg(f \circ \varphi)$, where the degree on S^1 was defined in §15.2. In the sequel we also use the notation

$$\deg(f \circ \varphi) =: [\arg f(y)]_{\partial\Omega}.$$

If M is a continuous $r \times r$ matrix function on $\partial\Omega$ with nonvanishing determinant then the degree of M is defined to be the degree of the complex-valued function $\det M$, i.e. $[\arg \det M(y)]_{\partial\Omega}$.

In order to write down the definition of the topological index, we choose a C^∞ matrix $\gamma^\pm$-spectral pair $(X_\pm(x), T_\pm(x))$ of $L_x(\lambda)$, $x \in \bar{\Omega}$, as in Proposition 15.11. The definition, however, does not actually depend on the choice of such a pair (see Lemma 15.17).

Definition 15.15 *The topological index of a boundary value problem* $(\mathscr{A}, \mathscr{B})$ *is defined as follows*:

$$\operatorname{ind}_s(\mathscr{A}, \mathscr{B}) = r(2 - l) - \frac{1}{2\pi}[\arg \det \tilde{\Delta}^+_{\mathscr{B}}(y)]_{\partial\Omega} + \frac{1}{2\pi}[\arg \det \tilde{\Delta}^-_{\mathscr{B}}(y)]_{\partial\Omega}$$

where $(X_\pm(x), T_\pm(x))$, $x \in \bar{\Omega}$, *is chosen as in Proposition* 15.11, *and then* $(\tilde{X}_\pm(y), \tilde{T}_\pm(y))$, $y \in \partial\Omega$, *is defined by the transformation* (14) *and* $\tilde{\Delta}^\pm_{\mathscr{B}}(y)$ *is defined by* (4).

At first sight it might appear that the definition of $\operatorname{ind}_s(\mathscr{A}, \mathscr{B})$ depends only on the value of the coefficients of $\mathscr{A}$ on the boundary $\partial\Omega$, but in fact the choice of (X_+, T_+) and hence of $(\tilde{X}_+, \tilde{T}_+)$ depends on the coefficients through the region $\bar{\Omega}$.

If the operators $\mathscr{A}$ and $\mathscr{B}$ have *real* coefficients, we may assume $\tilde{\Delta}_-(y) = \overline{\tilde{\Delta}_+(y)}$ (see Remark 15.1) and then we have

$$\operatorname{ind}_s(\mathscr{A}, \mathscr{B}) = r(2 - l) - \frac{1}{\pi}[\arg \det \tilde{\Delta}^+_{\mathscr{B}}(y)]_{\partial\Omega} \tag{16}$$

The next two lemmas show that $\operatorname{ind}_s(\mathscr{A}, \mathscr{B})$ is well-defined, independent of the choice of $\gamma^\pm$-spectral pairs.

Lemma 15.16 *If* $(\mathscr{A}, \mathscr{B}) \in \mathrm{BE}^{l,\mu}$, *i.e. the principal part of the boundary operator* $\mathscr{B}$ *is homogeneous of degree* μ *in all rows, then the quantity*

$$[\arg \det \Delta^\pm_{\mathscr{B}}(y)]_{\partial\Omega}$$

is independent of the choice of a C^∞ *matrix* $\gamma^\pm$*-spectral pair* $(X_\pm(x), T_\pm(x))$ *of* $L_x(\lambda)$, $x \in \bar{\Omega}$, *where* $\Delta^\pm_{\mathscr{B}}(y)$ *is defined by* (15).

Proof Let $(X^\circ_\pm(x), T^\circ_\pm(x))$ be another $\gamma^\pm$-spectral pair of $L_x(\lambda)$, $x \in \bar{\Omega}$, consisting of matrix functions with entries in $C^\infty(\bar{\Omega})$. By Proposition 2.12 we have

$$X^\circ_\pm(x) = X_\pm(x)M(x), \qquad T^\circ_\pm(x) = M^{-1}(x)T_\pm(x)M(x)$$

for some matrix function M with entries in $C^\infty(\bar{\Omega})$ such that $\det M(x) \neq 0$ for all $x \in \bar{\Omega}$. For short we write $\Delta_\pm = \Delta_{\mathscr{B}}^{\pm}$. Then

$$\Delta_\pm^\circ(y) = \Delta_\pm(y) M(y) \qquad \text{for all } y \in \partial\Omega,$$

and it follows that

$$[\arg \det \Delta_\pm^\circ(y)]_{\partial\Omega} = [\arg \det \Delta_\pm(y)]_{\partial\Omega} + [\arg \det M(y)]_{\partial\Omega}.$$

As in the proof of Proposition 15.11, $\bar{\Omega}$ is contractible to a point $x_0 \in \bar{\Omega}$, whence the function $\det M(x)$, $x \in \bar{\Omega}$, is homotopic to a constant. Its restriction to $\partial\Omega$ is, therefore, homotopic to a constant function, thus Corollary 15.9 implies

$$[\arg \det M(y)]_{\partial\Omega} = 0,$$

as was to be shown.

The following lemma is another version of Lemma 15.16 for which there is no special assumption on the orders of the rows of the boundary operator.

Lemma 15.17 *Let the $\gamma^\pm$-spectral pair $(\tilde{X}_\pm(y), \tilde{T}_\pm(y))$ of $\tilde{L}_y(\lambda)$ be defined by the transformation* (14), *and then define $\tilde{\Delta}_{\mathscr{B}}^{\pm}(y)$ by* (4). *Then the quantity*

$$[\arg \det \tilde{\Delta}_{\mathscr{B}}^{\pm}(y)]_{\partial\Omega}$$

is independent of the choice of the C^∞ matrix $\gamma^\pm$-spectral pair $(X_\pm(x), T_\pm(x))$ of $L_x(\lambda)$, $x \in \bar{\Omega}$.

Proof The boundary operator has the form (1b). Now we operate on $\mathscr{B}$ with tangential derivatives to make the order the same in each row:

$$\mathscr{B}' = \left[(\partial/\partial\tau)^{\mu - m_k} \sum_{|\alpha| \leqslant m_k} b_k^{(\alpha)}(y)\, \partial^\alpha \right]_{k=1}^{r}, \tag{17}$$

where $\mu = \max m_k$ and $\partial/\partial\tau = \tau_1(y)\cdot\partial/\partial x_1 + \tau_2(y)\cdot\partial/\partial x_2$. Let us write $\tilde{\Delta}_\pm = \tilde{\Delta}_{\mathscr{B}}^{\pm}$ and $\tilde{\Delta}'_\pm = \tilde{\Delta}_{\mathscr{B}'}^{\pm}$. For the matrix polynomials associated with the boundary operators $\mathscr{B}'$ and $\mathscr{B}$, we clearly have $\tilde{B}'_y(\lambda) = \tilde{B}_y(\lambda)$. Thus $\tilde{\Delta}'_\pm(y) = \tilde{\Delta}_\pm(y)$, which has nonzero determinant for all $y \in \partial\Omega$, so $\mathscr{B}'$ also satisfies the L-condition. Due to Propositions 15.12 and 15.5 we have

$$[\arg \det \tilde{\Delta}_\pm(y)]_{\partial\Omega} = [\arg \det \Delta'_\pm(y)]_{\partial\Omega} - \mu[\arg \det(\tau_1(y) I + \tau_2(y) T_\pm(y))]_{\partial\Omega}$$

Since the spectrum of T_+ lies in the upper half-plane, the same is true of $tiI + (1-t)T_+$ for $0 \leqslant t \leqslant 1$, and, therefore,

$$H(y, t) = \tau_1(y) I + \tau_2(y)\{tiI + (1-t)T_+(y)\}$$

defines a homotopy $\partial\Omega \times I \to GL_r(\mathbb{C})$. The winding number of a continuous function is locally constant (Theorem 15.8) whence

$$\frac{1}{2\pi}[\arg \det(\tau_1(y) I + \tau_2(y) T_+(y))]_{\partial\Omega} = \frac{1}{2\pi}[\arg \det(\tau_1(y) I + \tau_2(y) iI)]_{\partial\Omega} = r.$$

Similarly, since the spectrum of T_- lies in the lower half-plane, we find that

$$\frac{1}{2\pi}[\arg\det(\tau_1(y)I + \tau_2(y)T_-(y))]_{\partial\Omega} = -r.$$

Thus

$$\frac{1}{2\pi}[\arg\det\tilde{\Delta}_+(y)]_{\partial\Omega} = \frac{1}{2\pi}[\arg\det\Delta'_+(y)]_{\partial\Omega} - r\mu$$

and

$$\frac{1}{2\pi}[\arg\det\tilde{\Delta}_-(y)]_{\partial\Omega} = \frac{1}{2\pi}[\arg\det\Delta'_-(y)]_{\partial\Omega} + r\mu.$$

According to Lemma 15.16, the quantity $[\arg\det\Delta'_\pm(y)]_{\partial\Omega}$ is independent of the choice of $(X_\pm, T_\pm)$; hence the same is true of $[\arg\det\tilde{\Delta}_\pm(y)]_{\partial\Omega}$.

In Chapter 16 it will be shown that the index of a boundary value problem is equal to its topological index, i.e. $\operatorname{ind}(\mathscr{A},\mathscr{B}) = \operatorname{ind}_s(\mathscr{A},\mathscr{B})$. For the time being, however, we just prove the following property.

Theorem 15.18 *Let $(\mathscr{A},\mathscr{B})$ be the composite of two boundary value problems $(\mathscr{A}_1,\mathscr{B}_1)$ and $(\mathscr{A}_2,\mathscr{B}_2)$ satisfying the L-condition; see §10.9. Then we have*

$$\operatorname{ind}_s(\mathscr{A},\mathscr{B}) = \operatorname{ind}_s(\mathscr{A}_1,\mathscr{B}_1) + \operatorname{ind}_s(\mathscr{A}_2,\mathscr{B}_2) \tag{18}$$

Proof First, notice that

$$\begin{aligned}\tilde{L}_y(\lambda) &= \pi\mathscr{A}(y, \tau(y) + \lambda n(y))\\ &= \pi\mathscr{A}_2(y, \tau(y) + \lambda n(y))\cdot\pi\mathscr{A}_1(y, \tau(y) + \lambda n(y))\\ &= \tilde{L}_{y,1}(\lambda)\cdot\tilde{L}_{y,2}(\lambda)\end{aligned}$$

and

$$\begin{aligned}\tilde{B}_y(\lambda) &= \pi\mathscr{B}(y, \tau(y) + \lambda n(y))\\ &= \begin{pmatrix}\pi\mathscr{B}_1(y, \tau(y) + \lambda n(y))\\ \pi\mathscr{B}_2(y, \tau(y) + \lambda \nu(y))\cdot\pi\mathscr{A}_1(y, \tau(y) + \lambda n(y))\end{pmatrix}\\ &= \begin{pmatrix}\tilde{B}_{y,1}(\lambda)\\ \tilde{B}_{y,2}(\lambda)\cdot\tilde{L}_{y,1}(\lambda)\end{pmatrix}.\end{aligned}$$

Choose $\gamma^\pm$-spectral pairs $(X_i^\pm(x), T_i^\pm(x))$ of $L_{x,i}(\lambda)$, $i = 1, 2$, consisting of smooth matrix functions on $\bar{\Omega}$, and then define a γ^+-spectral pair $(X_\pm(x), T_\pm(x))$ of the product $L_x(\lambda)$ as in Theorem 4.17. Now let $(\tilde{X}_i^\pm(y), \tilde{T}_i^\pm(y))$ and $(\tilde{X}_\pm(y), \tilde{T}_\pm(y))$, $y \in \partial\Omega$, denote the corresponding transformations under $\varphi(\lambda) = (-\tau_2 + \lambda\tau_1)/(\tau_1 + \lambda\tau_2)$. By Definition 15.15, we have

$$\operatorname{ind}_s(\mathscr{A},\mathscr{B}) = -\frac{1}{2\pi}[\arg\det\tilde{\Delta}_+(y)]_{\partial\Omega} + \frac{1}{2\pi}[\arg\det\tilde{\Delta}_-(y)]_{\partial\Omega} + r(2-l)$$

where $r = r_1 + r_2$ and $l = l_1 + l_2$, with similar formulas for $\operatorname{ind}_s(\mathscr{A}_i,\mathscr{B}_i)$ with r, l replaced by r_i, l_i (i = 1, 2).

For short we write $\tilde{\Delta}_\pm = \tilde{\Delta}_{\mathscr{B}}^\pm$ and $\tilde{\Delta}_i^\pm = \tilde{\Delta}_{\mathscr{B}_i}^\pm$, $i = 1, 2$. By virtue of Proposition 4.24 it follows that

$$\tilde{\Delta}_\pm(y) = \begin{pmatrix} \tilde{\Delta}_1^\pm(y) & * \\ 0 & \tilde{\Delta}_2^\pm(y)\cdot(\tau_1(y)I + \tau_2(y)T_2^\pm(y))^{-l_1} \end{pmatrix}$$

where * denotes an $r_1 \times r_2$ matrix function of no significance here. (Note that $M = I$ in Proposition 4.24 by the choice of $(X_\pm(x), T_\pm(x))$; see the remark after Theorem 4.17′.) Taking the winding number along $\partial\Omega$ we obtain

$$[\arg\det \tilde{\Delta}_\pm(y)]_{\partial\Omega} = [\arg\det \tilde{\Delta}_1^\pm(y)]_{\partial\Omega} + [\arg\det \tilde{\Delta}_2^\pm(y)]_{\partial\Omega} - l_1[\arg\det(\tau_1(y)I + \tau_2(y)T_2^\pm(y))]_{\partial\Omega},$$

see Proposition 15.5. Because the spectrum of T_2^+ (resp. T_2^-) lies in the upper (resp. lower) half-plane, it follows that

$$\frac{1}{2\pi}[\arg\det(\tau_1(y)I + \tau_2(y)T_2^\pm(y)]_{\partial\Omega} = \pm r_2$$

whence

$$\frac{1}{2\pi}[\arg\det \tilde{\Delta}_+(y)]_{\partial\Omega} = \frac{1}{2\pi}[\arg\det \tilde{\Delta}_1^+(y)]_{\partial\Omega} + \frac{1}{2\pi}[\arg\det \tilde{\Delta}_2^+(y)]_{\partial\Omega} - r_2 l_1$$

$$\frac{1}{2\pi}[\arg\det \tilde{\Delta}_-(y)]_{\partial\Omega} = \frac{1}{2\pi}[\arg\det \tilde{\Delta}_1^-(y)]_{\partial\Omega} + \frac{1}{2\pi}[\arg\det \tilde{\Delta}_2^-(y)]_{\partial\Omega} + r_2 l_1$$

Thus (18) will be proved if we can show that

$$(r_1 + r_2)(2 - l_1 - l_2) + 2r_2 l_1 = r_1(2 - l_1) + r_2(2 - l_2).$$

But this is easily verified since $r_1 l_2 = r_2 l_1$ $(=pl_1 l_2/2)$.

15.4 Changing from complex to real matrix coefficients

For the purpose of this section $\mathscr{A}$ is an elliptic operator with a homogeneous principal part in $\Omega \subset \mathbb{R}^n$ and $\mathscr{B}$ is any differential boundary operator. If $\mathscr{A}(x, \partial) = \sum A_\alpha \partial^\alpha$ we let $\bar{\mathscr{A}}(x, \partial) = \sum \bar{A}_\alpha \partial^\alpha$, that is, the coefficients A_α are replaced by their complex conjugates, $\bar{A}_\alpha$, and similarly for $\bar{\mathscr{B}}(y, \partial)$. Since $\mathscr{A}u = f$, $\mathscr{B}u = g$ if and only if $\bar{\mathscr{A}}\bar{u} = \bar{f}$, $\bar{\mathscr{B}}\bar{u} = \bar{g}$, there is a one-to-one correspondence between the solution spaces of $(\mathscr{A}, \mathscr{B})$ and $(\bar{\mathscr{A}}, \bar{\mathscr{B}})$; thus one problem satisfies the L-condition if and only if the other does, and $\operatorname{ind}(\bar{\mathscr{A}}, \bar{\mathscr{B}}) = \operatorname{ind}(\mathscr{A}, \mathscr{B})$.

If $\mathscr{A} = \mathscr{A}_1 + i\mathscr{A}_2$, where the coefficients of $\mathscr{A}_j$ are real $p \times p$ matrix functions we define

$$\mathscr{A}_{\mathbb{R}}(x, \partial) = \begin{pmatrix} \mathscr{A}_1(x, \partial) & -\mathscr{A}_2(x, \partial) \\ \mathscr{A}_2(x, \partial) & \mathscr{A}_1(x, \partial) \end{pmatrix}$$

which is a differential operator with coefficients that are real $2p \times 2p$ matrix functions. Similarly, if $\mathscr{B} = \mathscr{B}_1 + i\mathscr{B}_2$, where the coefficients of $\mathscr{B}_j$ are real

$r \times p$ matrix functions we define

$$\mathscr{B}_{\mathbb{R}}(y, \partial) = \begin{pmatrix} \mathscr{B}_1(y, \partial) & -\mathscr{B}_2(y, \partial) \\ \mathscr{B}_2(y, \partial) & \mathscr{B}_1(y, \partial) \end{pmatrix}$$

which is a differential operator with coefficients that are real $2r \times 2p$ matrix functions. Due to Lemma 3.19, there is a one-to-one correspondence between solutions of

$$\mathscr{A}_{\mathbb{R}}(x, \partial)\begin{pmatrix} u_1 \\ u_2 \end{pmatrix} = \begin{pmatrix} f_1(x) \\ f_2(x) \end{pmatrix}, \qquad x \in \Omega,$$

$$\mathscr{B}_{\mathbb{R}}(y, \partial)\begin{pmatrix} u_1 \\ u_2 \end{pmatrix} = \begin{pmatrix} g_1(y) \\ g_2(y) \end{pmatrix}, \qquad y \in \partial\Omega, \tag{19}$$

and solutions of

$$\begin{pmatrix} \mathscr{A}(x, \partial) & \\ & \bar{\mathscr{A}}(x, \partial) \end{pmatrix}\begin{pmatrix} v \\ \bar{v} \end{pmatrix} = \begin{pmatrix} f(x) \\ \bar{f}(x) \end{pmatrix}, \qquad x \in \Omega,$$

$$\begin{pmatrix} \mathscr{B}(y, \partial) & \\ & \bar{\mathscr{B}}(y, \partial) \end{pmatrix}\begin{pmatrix} v \\ \bar{v} \end{pmatrix} = \begin{pmatrix} g(y) \\ \bar{g}(y) \end{pmatrix}, \qquad y \in \partial\Omega, \tag{20}$$

where the correspondence is given by $v = u_1 + iu_2$, $\bar{v} = u_1 - iu_2$ (and $f = f_1 + if_2, \bar{f} = f_1 - if_2$). Without loss of generality u_1 and u_2 can be assumed to be real-valued in (19) (by taking real and imaginary parts). This does not affect the index.

Theorem 15.19 *$(\mathscr{A}_{\mathbb{R}}, \mathscr{B}_{\mathbb{R}})$ is L-elliptic if and only if $(\mathscr{A}, \mathscr{B})$ is L-elliptic. In that case, we have*

$$\operatorname{ind}(\mathscr{A}_{\mathbb{R}}, \mathscr{B}_{\mathbb{R}}) = 2 \cdot \operatorname{ind}(\mathscr{A}, \mathscr{B}).$$

Proof By Lemma 3.19

$$\pi\mathscr{A}_{\mathbb{R}}(x, \xi) = \begin{pmatrix} I & I \\ -iI & iI \end{pmatrix}\begin{pmatrix} \pi\mathscr{A}(x, \xi) & \\ & \pi\bar{\mathscr{A}}(x, \xi) \end{pmatrix}\begin{pmatrix} I & I \\ -iI & iI \end{pmatrix}^{-1} \tag{21}$$

for all $\xi \in \mathbb{R}^n$ (or $\mathbb{C}^n$). Hence $\det \pi\mathscr{A}_{\mathbb{R}}(x, \xi) = |\det \pi\mathscr{A}(x, \xi)|^2$ for all $\xi \in \mathbb{R}^n$; thus $\mathscr{A}_{\mathbb{R}}$ is elliptic if and only if the same is true of $\mathscr{A}$.

The kth row of the boundary operator $\mathscr{B}_{\mathbb{R}}$ has order $\tilde{m}_k$ where

$$\tilde{m}_k = m_k, \quad \tilde{m}_{k+r} = m_k, \qquad k = 1, \ldots, r,$$

and we have

$$\pi\mathscr{B}_{\mathbb{R}}(y, \xi) = \begin{pmatrix} I & I \\ -iI & iI \end{pmatrix}\begin{pmatrix} \pi\mathscr{B}(y, \xi) & \\ & \pi\bar{\mathscr{B}}(y, \xi) \end{pmatrix}\begin{pmatrix} I & I \\ -iI & iI \end{pmatrix}^{-1} \tag{22}$$

The claim is that $(\mathscr{A}_{\mathbb{R}}, \mathscr{B}_{\mathbb{R}})$ satisfies the L-condition if and only if the same is true of $(\mathscr{A}, \mathscr{B})$. Indeed, working in local admissible coordinates and replacing ξ by $(\xi', i^{-1}\, d/dt)$ in (21) and (22), we obtain that

$$W = \begin{pmatrix} w_1 \\ w_2 \end{pmatrix}$$

is a solution of the system

$$\begin{aligned} \pi\mathscr{A}_{\mathbb{R}}\left(y, \left(\xi', \frac{1}{i}\frac{d}{dt}\right)\right) W(t) &= 0, \qquad t \geqslant 0 \\ \pi\mathscr{B}_{\mathbb{R}}\left(y, \left(\xi', \frac{1}{i}\frac{d}{dt}\right)\right) W(t)|_{t=0} &= 0 \end{aligned} \tag{23}$$

if and only if $w = w_1 + iw_2$, $\bar{w} = w_1 - iw_2$ is a solution of the system

$$\begin{pmatrix} \pi\mathscr{A}\left(y, \left(\xi', \frac{1}{i}\frac{d}{dt}\right)\right) & \\ & \pi\bar{\mathscr{A}}\left(y, \left(\xi', \frac{1}{i}\frac{d}{dt}\right)\right) \end{pmatrix} \begin{pmatrix} w(t) \\ \bar{w}(t) \end{pmatrix} = 0, \qquad t \geqslant 0$$

$$\begin{pmatrix} \pi\mathscr{B}\left(y, \left(\xi', \frac{1}{i}\frac{d}{dt}\right)\right) & \\ & \pi\bar{\mathscr{B}}\left(y, \left(\xi', \frac{1}{i}\frac{d}{dt}\right)\right) \end{pmatrix} \begin{pmatrix} w(t) \\ \bar{w}(t) \end{pmatrix}\Bigg|_{t=0} = 0$$

As pointed out earlier we may assume that w_1 and w_2 are real-valued. Then, by homogeneity of $\pi\mathscr{A}$, we have

$$\pi\bar{\mathscr{A}}\left(y, \left(\xi', \frac{1}{i}\frac{d}{dt}\right)\right)\bar{w}(t) = \overline{(-1)^l \pi\mathscr{A}\left(y, \left(-\xi', \frac{1}{i}\frac{d}{dt}\right)\right) w(t)}$$

Similarly

$$\pi\bar{\mathscr{B}}\left(y, \left(\xi', \frac{1}{i}\frac{d}{dt}\right)\right)\bar{w}(t) = \overline{M(-1)\pi\mathscr{B}\left(y, \left(-\xi', \frac{1}{i}\frac{d}{dt}\right)\right) w(t)}$$

It is now clear that (23) has the unique solution $W \equiv 0$ if and only if

$$\begin{aligned} \pi\mathscr{A}\left(y, \left(\xi', \frac{1}{i}\frac{d}{dt}\right)\right) w(t) &= 0, \qquad t \geqslant 0 \\ \pi\mathscr{B}\left(y, \left(\xi', \frac{1}{i}\frac{d}{dt}\right)\right) w(t)|_{t=0} &= 0 \end{aligned}$$

has the unique solution $w \equiv 0$.

Moreover, due to the one-to-one correspondence between solutions of the boundary value problems (19) and (20), it follows that

$$\begin{aligned} \operatorname{ind}(\mathscr{A}_{\mathbb{R}}, \mathscr{B}_{\mathbb{R}}) &= \operatorname{ind}(\mathscr{A} \oplus \bar{\mathscr{A}}, \mathscr{B} \oplus \bar{\mathscr{B}}) \\ &= \operatorname{ind}(\mathscr{A}, \mathscr{B}) + \operatorname{ind}(\bar{\mathscr{A}}, \bar{\mathscr{B}}) \\ &= 2 \cdot \operatorname{ind}(\mathscr{A}, \mathscr{B}). \end{aligned}$$

The following proposition will be of use in §16.15.

Proposition 15.20 *Let the matrix polynomial $L(\lambda)$ and contours γ^+ and γ^- be as in Theorem 3.20, and $B(\lambda) = \sum_{j=0}^{\mu} B_j \lambda^j$ an $r \times p$ matrix polynomial of degree μ. Let $(X_\pm, T_\pm)$ be $\gamma^\pm$-spectral pairs of $L(\lambda)$, respectively, and $(X_{\mathbb{R}}^+, T_{\mathbb{R}}^+)$ the γ^+-spectral pair of $L_{\mathbb{R}}(\lambda)$ defined in Theorem 3.20. Also, let $B_{\mathbb{R}}(\lambda) = \sum_{j=0}^{\mu} B_{\mathbb{R},j}\lambda^j$ be the real $2r \times 2p$ matrix polynomial corresponding to $B(\lambda)$. Then*

$$\Delta_{B_{\mathbb{R}}}^+ = \begin{pmatrix} I_r & I_r \\ -iI_r & iI_r \end{pmatrix} \begin{pmatrix} \Delta_B^+ & \\ & \overline{\Delta_B^-} \end{pmatrix} \tag{24}$$

where $\Delta_{B_{\mathbb{R}}}^+ = \sum_{j=0}^{\mu} B_{\mathbb{R},j} X_{\mathbb{R}}^+ (T_{\mathbb{R}}^+)^j$ and $\Delta_B^\pm = \sum_{j=0}^{\mu} B_j X_\pm T_\pm^j$, and I_r denotes the $r \times r$ identity matrix.

Proof Write $B_j = M_j + iN_j$ where M_j and N_j are real $r \times p$ matrices, $j = 0, \ldots, l-1$. Then

$$B_{\mathbb{R},j} = \begin{pmatrix} M_j & -N_j \\ N_j & M_j \end{pmatrix},$$

so with the formula for $(X_{\mathbb{R}}^+, T_{\mathbb{R}}^+)$ in Theorem 3.20, it follows that

$$\sum_{j=0}^{\mu} B_{\mathbb{R},j} X_{\mathbb{R}}^+ (T_{\mathbb{R}}^+)^j = \sum_{j=0}^{\mu} \begin{pmatrix} M_j & -N_j \\ N_j & M_j \end{pmatrix} \begin{pmatrix} I & I \\ -iI & iI \end{pmatrix} \begin{pmatrix} X_+ T_+^j & \\ & \bar{X}_- \bar{T}_-^j \end{pmatrix}$$

where I denotes the $p \times p$ identity matrix. As in Lemma 3.19, it is easily verified that

$$\begin{pmatrix} M_j & -N_j \\ N_j & M_j \end{pmatrix} \begin{pmatrix} I & I \\ -iI & iI \end{pmatrix} = \begin{pmatrix} I_r & I_r \\ -iI_r & iI_r \end{pmatrix} \begin{pmatrix} B_j & \\ & \bar{B}_j \end{pmatrix}$$

and now (24) follows immediately.

Remark Note that the matrix $\Delta_{\mathscr{B}_{\mathbb{R}}}^+$ is square ($2r \times 2r$) exactly when $2r$ is equal to the degree of $\det L(\lambda)$, which is the condition of proper ellipticity.

Exercises

1. Let Ω be a bounded and simply connected region with C^∞ boundary. Show that for any $\delta > 0$ there exists a bounded region $\Omega' \supset \bar{\Omega}$ which is also simply connected and such that Ω' lies inside a δ-neighbourhood of $\bar{\Omega}$. Hint: Let $\Omega' = \Omega \cup V_\varepsilon$ where $V_\varepsilon = \{y + tn(y);\ -\varepsilon < t < \varepsilon)$ is a tubular neighbourhood of $\partial\Omega$ and ε is sufficiently small.
2. Let $\Omega = \{(x, y);\ 1 < x^2 + y^2 < 4\} \backslash \{(x, 0);\ 1 \leqslant x \leqslant 2\}$. Show that Ω is a bounded, simply connected region. Does the conclusion of Exercise 1 hold in this case?

16

The index in the plane

The region $\Omega \subset \mathbb{R}^2$ is bounded with smooth boundary, and for simplicity we assume that it is *simply connected*. Consider the 2×2 system

$$\frac{\partial u}{\partial x_2} - \begin{pmatrix} 0 & -1 \\ 1 & 0 \end{pmatrix} \frac{\partial u}{\partial x_1} + A_0(x) \cdot u = f(x), \qquad x \in \Omega \tag{1}$$

$$u_1 \cdot \cos \varphi(y) - u_2 \cdot \sin \varphi(y) = g(y), \qquad y \in \partial\Omega,$$

where $u = [u_1 \quad u_2]^T$ is a real 2-vector function. Writing this problem in complex form with $w = u_1 + iu_2$, $z = x_1 + ix_2$, and $\partial/\partial\bar{z} = \frac{1}{2}(\partial/\partial x_1 + i\,\partial/\partial x_2)$, we obtain the Riemann–Hilbert problem:

$$\partial w/\partial\bar{z} + a(z)w + b(z)\bar{w} = f(z), \qquad z \in \Omega, \tag{1'}$$

$$\operatorname{Re} e^{i\varphi(y)} w = g(y), \qquad y \in \partial\Omega,$$

where $a, b \in C^\infty(\bar{\Omega})$ and $\varphi \in C^\infty(\partial\Omega)$. (Often this is referred to as the generalized Riemann–Hilbert problem and the "Riemann–Hilbert problem" is the case $a = b = f = 0$.) The $\mathbb{R}$-linear operator $W_2^1(\Omega, \mathbb{C}) \to L_2(\Omega, \mathbb{C}) \times W_2^{1/2}(\partial\Omega, \mathbb{R})$ defined by $w \mapsto (\partial w/\partial\bar{z} - aw - b\bar{w}, \operatorname{Re} e^{i\varphi} w|_{\partial\Omega})$ is Fredholm, its solvability properties depend on the winding number, $\chi = (1/2\pi)\,[\varphi(y)]_{\partial\Omega}$, and the index formula

$$\text{index} = 1 - 2\chi = 1 - \frac{1}{\pi}\,[\varphi(y)]_{\partial\Omega} \tag{2}$$

is well known. (See [Wen] or [Ga] and also Exercises 7, 8 and 9 at the end of this chapter.)

In this chapter we consider boundary value problems $(\mathscr{A}, \mathscr{B})$ where the principal part of the elliptic operator $\mathscr{A}$ is homogeneously elliptic (Definition 9.25) of order l. We shall show that the index of $(\mathscr{A}, \mathscr{B})$ is equal to the topological index defined in §15.3. First we prove an index formula due to A. Vol'pert for first order elliptic systems, by starting from the formula (2) for the index of the Riemann–Hilbert problem. Higher-order systems are then handled by a standard device which involves a reformulation as a first-order system. It should be pointed out that we have already shown in §15.3 that the topological index is additive under composition of boundary value problems, but this result is not used in the proofs here.

16.1 A simple form for first-order elliptic systems with real coefficients

Consider a first-order elliptic boundary value problem (BVP) in the plane

$$\frac{\partial u}{\partial x_2} - A(x)\frac{\partial u}{\partial x_1} + A_0(x)u = f(x), \qquad x = (x_1, x_2) \in \Omega, \tag{3}$$

$$B(y)u = g(y), \qquad y \in \partial\Omega,$$

where $A(x)$ and $A_0(x)$ are real $2r \times 2r$ matrix functions with entries in $C^\infty(\bar{\Omega})$, and $B(y)$ is a real $r \times 2r$ matrix function with entries in $C^\infty(\partial\Omega)$. The region $\Omega \subset \mathbb{R}^2$ is bounded and simply connected, with smooth boundary.

Any elliptic operator $\mathscr{A}$ where the principal part is homogeneous of order 1 can be written in the form (3). Indeed, the principal part $\pi\mathscr{A}(x, \xi)$ is invertible for all $x \in \bar{\Omega}$, $0 \neq \xi \in \mathbb{R}^2$, and letting $\xi = (0, 1)$ we see that the coefficient of $\partial/\partial x_2$ is invertible for all $x \in \bar{\Omega}$.

The BVP (3) is L-elliptic if $\det(I\lambda - A(x)) \neq 0$ for all $x \in \bar{\Omega}$, $\lambda \in \mathbb{R}$, and the $r \times 2r$ Lopatinskii matrix

$$B(y) \cdot \int_+ (I\lambda - A(y))^{-1}\, d\lambda$$

has rank r for all $y \in \partial\Omega$, where the integral is taken along a contour γ^+ in the half-plane $\operatorname{Im} \lambda > 0$ containing the zeros of $\det(I\lambda - A(y))$ there.

If (3) is L-elliptic then the linear operator

$$(\mathscr{A}, \mathscr{B}): W_2^1(\Omega, \mathbb{R}^{2r}) \to L_2(\Omega, \mathbb{R}^{2r}) \times W_2^{1/2}(\partial\Omega, \mathbb{R}^r),$$

defined by $u \mapsto (\partial u/\partial x_2 - A\, \partial u/\partial x_1 + A_0 u, Bu)$, is Fredholm. Since the index of Fredholm operators in Banach spaces is locally constant, $\operatorname{ind}(\mathscr{A}, \mathscr{B})$ depends only on the principal part of $\mathscr{A}$, that is,

$$\operatorname{ind}(\mathscr{A}, \mathscr{B}) = \operatorname{ind}(\pi\mathscr{A}, \mathscr{B})$$

because one can replace the coefficient $A_0(x)$ by $tA_0(x)$, $0 \leqslant t \leqslant 1$, giving a homotopy between $(\mathscr{A}, \mathscr{B})$ and $(\pi\mathscr{A}, \mathscr{B})$. Thus, for the purpose of determining the index we may assume that $A_0 = 0$.

Remark Since $\mathscr{A}$ and $\mathscr{B}$ have real matrix coefficients, there is no loss of generality in considering solutions u which are real vector functions rather than complex (just take real and imaginary parts of u). This does not affect the index; see Exercise 1.

We begin by proving a lemma which will enable us to transform the $2r \times 2r$ real system (3) into a generalized Riemann–Hilbert problem (an $r \times r$ complex system). The index for this new problem is then easy to determine since it is homotopic to a diagonal system of r (scalar) Riemann–Hilbert problems. Then in §16.2 we derive a formula for the index of $(\mathscr{A}, \mathscr{B})$.

Lemma 16.1 *Let A be a $2r \times 2r$ real matrix such that $\mathrm{sp}(A) \cap \mathbb{R} = \varnothing$. Then there exists a real invertible $2r \times 2r$ matrix Q such that*

$$Q^{-1}AQ = \begin{pmatrix} A_1 & -A_2 \\ A_2 & A_1 \end{pmatrix} \tag{4}$$

where A_1, A_2 are real $r \times r$ matrices such that $\mathrm{sp}(A_1 + iA_2)$ lies in the upper half-plane.

Proof Write $\mathbb{C}^{2r} = \mathfrak{M}_+ \oplus \mathfrak{M}_-$, where $\mathfrak{M}_\pm$ are the invariant subspaces of A corresponding to the eigenvalues in the upper and lower half-planes, respectively. Since A has real coefficients, there is a one-to-one correspondence between $\mathfrak{M}_+$ and $\mathfrak{M}_-$ given by conjugation.

Let X_+ be a $2r \times r$ matrix whose columns form a basis of $\mathfrak{M}_+$; then the columns of $\bar{X}_+$ form a basis of $\mathfrak{M}_- = \bar{\mathfrak{M}}^+$. Let T_+ be the unique $r \times r$ matrix such that $AX_+ = X_+T_+$. Then,

$$\begin{aligned} A[X_+ \quad \bar{X}_+] &= [AX_+ \quad A\bar{X}_+] \\ &= [X_+T_+ \quad \bar{X}_+\bar{T}_+] \\ &= [X_+ \quad \bar{X}_+]\begin{pmatrix} T_+ & \\ & \bar{T}_+ \end{pmatrix} \end{aligned} \tag{5}$$

Thus the columns of the $2r \times 2r$ matrix $[X_+ \quad \bar{X}_+]$ form a basis of $\mathbb{C}^{2r}$ and relative to this basis A is block diagonal.

Now write $T_+ = A_1 + iA_2$, where A_1, A_2 are real $r \times r$ matrices. By Lemma 3.19, we have

$$\begin{pmatrix} A_1 & -A_2 \\ A_2 & A_1 \end{pmatrix} = \begin{pmatrix} I & I \\ -iI & iI \end{pmatrix}\begin{pmatrix} T_+ & \\ & \bar{T}_+ \end{pmatrix}\begin{pmatrix} I & I \\ -iI & iI \end{pmatrix}^{-1}, \tag{6}$$

where I is the $r \times r$ identity matrix. Let

$$Q = [X_+ \quad \bar{X}_+]\begin{pmatrix} I & I \\ -iI & iI \end{pmatrix}^{-1}$$

then (4) follows from (5) and (6). Note that the entries of Q are real since

$$Q = [X_+ \quad \bar{X}_+]\cdot\frac{1}{2}\begin{pmatrix} I & iI \\ I & -iI \end{pmatrix} = [\mathrm{Re}\, X_+ \quad -\mathrm{Im}\, X_+]$$

where $\mathrm{Re}\, X_+ = \frac{1}{2}(X_+ + \bar{X}_+)$ and $\mathrm{Im}\, X_+ = 1/2i(X_+ - \bar{X}_+)$ are the real and imaginary parts of X_+, respectively. Also $\mathrm{sp}(T_+) = \mathrm{sp}(A|_{\mathfrak{M}_+})$ which lies in the upper half-plane.

Remark In the proof of Lemma 16.1, (X_+, T_+) is actually a γ^+-spectral pair of the matrix polynomial $L(\lambda) = I\lambda - A$. Moreover, (X, T, Y) is a standard triple of $L(\lambda)$, where

$$X = [X_+ \quad \bar{X}_+], \qquad T = \begin{pmatrix} T_+ & \\ & \bar{T}_+ \end{pmatrix}, \qquad Y = \begin{pmatrix} Y_+ \\ \bar{Y}_+ \end{pmatrix}$$

This follows from the next lemma.

Lemma 16.2 *Let A be any $n \times n$ matrix. Let X, T and Y also be $n \times n$ matrices. Then (X, T, Y) is a standard triple of $L(\lambda) = I\lambda - A$ if and only if X is invertible, $AX = XT$, and $X^{-1} = Y$.*

Proof See Definition 3.1 and Remark 3.3.

Lemma 16.3 *Let A_1, A_2 be real $r \times r$ matrices such that* $\mathrm{sp}(A_1 + iA_2)$ *lies in the upper half-plane, and consider the* $2r \times 2r$ *matrix polynomial $L(\lambda) = I\lambda - A$, where*

$$A = \begin{pmatrix} A_1 & -A_2 \\ A_2 & A_1 \end{pmatrix}.$$

Then $\mathrm{sp}(L) \cap \mathbb{R} = \varnothing$ *and $L(\lambda)$ has the following standard triple:*

$$X = \begin{pmatrix} I & I \\ -iI & iI \end{pmatrix}, \qquad T = \begin{pmatrix} A_1 + iA_2 & \\ & A_1 - iA_2 \end{pmatrix}, \qquad Y = \frac{1}{2}\begin{pmatrix} I & iI \\ I & -iI \end{pmatrix}.$$

Proof Let $T_+ = A_1 + iA_2$. In view of (6) we have

$$\det L(\lambda) = |\det(I\lambda - T_+|^2 \neq 0 \qquad \text{for real } \lambda.$$

By direct computation we also have $AX = XT$ and $X^{-1} = Y$, so (X, T, Y) is a standard triple of $L(\lambda)$.

Corollary 16.4 *Under the hypotheses of Lemma* 16.3, *$L(\lambda)$ has the following γ^+-spectral triple:*

$$(X_+, T_+, Y_+) = \left(\begin{pmatrix} I \\ -iI \end{pmatrix}, A_1 + iA_2, \tfrac{1}{2}[I \quad iI] \right).$$

We return now to the BVP (3). Let $L_x(\lambda) = I\lambda - A(x)$. By Proposition 15.11 there exists a γ^+-spectral pair $(X_+(x), T_x(x))$ of $L_x(\lambda)$ consisting of matrices with entries that are in $C^\infty(\bar{\Omega})$, where the dimensions of X_+ and T_+ are $2r \times r$ and $r \times r$, respectively. Then $A(x)X_+(x) = X_+(x)T_+(x)$ and if we let

$$Q(x) = [X_+(x) \quad \overline{X_+(x)}] \begin{pmatrix} I & I \\ -iI & iI \end{pmatrix}^{-1}, \tag{7}$$

the proof of Lemma 16.1 shows that

$$Q^{-1}(x)A(x)Q(x) = \begin{pmatrix} A_1(x) & -A_2(x) \\ A_2(x) & A_1(x) \end{pmatrix}, \qquad x \in \bar{\Omega}$$

where A_1, A_2 are C^∞ real $r \times r$ matrix functions on $\bar{\Omega}$ and the spectrum of $T_+(x) = A_1(x) + iA_2(x)$ lies in the upper half-plane for all $x \in \bar{\Omega}$. Let $u = Q \cdot v$, then the BVP (3) takes the form

$$\frac{\partial v}{\partial x_2} - \tilde{A}\frac{\partial v}{\partial x_1} + \tilde{A}_0 v = \tilde{f}(x), \qquad x \in \Omega \tag{3)$^\sim$}$$

$$\tilde{B}(y)v = g(y), \qquad y \in \partial\Omega,$$

where $\tilde{A}_0 = Q^{-1}A_0 Q + \partial Q/\partial x_2 - A\,\partial Q/\partial x_1$, $\tilde{A} = Q^{-1}A\,Q$, $\tilde{B} = BQ$ and $\tilde{f} = Q^{-1}f$. For the purpose of determining an index formula, we may of course assume that the lower order terms in (3)$^\sim$ are 0, that is, $\tilde{A}_0 = 0$.

Let $(\mathscr{A}, \mathscr{B})$ and $(\tilde{\mathscr{A}}, \tilde{\mathscr{B}})$ denote boundary value problems (3) and (3)$^\sim$, respectively. Let us compare the L-condition for $(\mathscr{A}, \mathscr{B})$ with that for $(\tilde{\mathscr{A}}, \tilde{\mathscr{B}})$. First of all,

$$\int_+ \tilde{B}(y)(I\lambda - \tilde{A}(y))^{-1}\,d\lambda = \int_+ B(y)(I\lambda - A(y))^{-1}\,d\lambda \cdot Q(y) \tag{8}$$

From this equation we see that $(\mathscr{A}, \mathscr{B})$ satisfies the L-condition if and only if $(\tilde{\mathscr{A}}, \tilde{\mathscr{B}})$ satisfies the L-condition. Further, observe that (8) may be written as

$$\tilde{B}(y)\tilde{P}_+(y) = B(y)P_+(y)Q(y),$$

where $\tilde{P}_+(y)$ and $P_+(y)$ are the Riesz projectors for $\tilde{A}(y)$ and $A(y)$ with respect to the eigenvalues in the upper half-plane. Then multiplying both sides of this equation on the right by $\begin{pmatrix} I & I \\ -iI & iI \end{pmatrix}$ and using (7) we obtain

$$\tilde{B}(y)\tilde{P}_+(y)\cdot[\tilde{X}_+(y) \quad \overline{\tilde{X}_+(y)}] = B(y)P_+(y)\cdot[X_+(y) \quad \overline{X_+(y)}],$$

where $\tilde{X}_+(y) = \begin{pmatrix} I \\ -iI \end{pmatrix}$; see Corollary 16.4. Hence

$$\tilde{B}(y)\cdot[\tilde{X}_+(y) \quad 0] = B(y)\cdot[X_+(y) \quad 0]$$

that is,

$$\tilde{B}(y)\tilde{X}_+(y) = B(y)X_+(y).$$

Writing $\tilde{B}(y) = [B_1(y) \quad B_2(y)]$, where B_1, B_2 are real $r \times r$ matrix functions, then $\tilde{B}(y)\tilde{X}_+(y)$ equals $B_1(y) - iB_2(y)$, and we have proved the following:

Lemma 16.5 *The boundary value problem $(\mathscr{A}, \mathscr{B})$ satisfies the L-condition if and only if $(\tilde{\mathscr{A}}, \tilde{\mathscr{B}})$ also satisfies the L-condition. This condition can be stated in two equivalent ways:*

$$\det B(y)X_+(y) \neq 0 \qquad \textit{for all } y \in \partial\Omega$$

or

$$\det(B_1(y) - iB_2(y)) \neq 0 \textit{ for all } y \in \partial\Omega, \textit{ where } \tilde{B}(y) = [B_1(y) \quad B_2(y)].$$

The first condition can be written out in a more explicit manner. Let $q_k(y)$, $k = 1, \ldots, r$, denote the columns of $X_+(y)$ and let $b_j(y)$, $j = 1, \ldots, r$, be the rows of $B(y)$. Then the BVP (3) satisfies the L-condition if and only if

$$\det[b_j(y) \cdot q_k(y)]_{j,k=1}^r \neq 0 \qquad \text{for all } y \in \partial\Omega.$$

By taking complex conjugates, we could also write the second condition as $\det(B_1(y) + iB_2(y)) \neq 0$.

16.2 The index formula for first-order elliptic systems with real coefficients

Let A_i and B_i $(i = 1, 2)$ be smooth $r \times r$ real matrix functions on $\bar{\Omega}$ and $\partial\Omega$, respectively, and suppose that the spectrum of $A_1(x) + iA_2(x)$ lies in the upper half-plane for all $x \in \bar{\Omega}$, and that $\det(B_1(y) + iB_2(y)) \neq 0$ for all $y \in \partial\Omega$. We call the following boundary value problem for the $\mathbb{C}^r$-valued function w,

$$\frac{\partial w}{\partial x_2} - (A_1 + iA_2)\frac{\partial w}{\partial x_1} = f(x), \quad x \in \Omega \tag{9}$$

$$\operatorname{Re}(B_1(y) - iB_2(y))w = g(y), \qquad y \in \partial\Omega,$$

where $f = f_1 + if_2$ and g is real-valued, a generalized Riemann–Hilbert problem.

Lemma 16.6 *Consider the real $2r \times 2r$ boundary value problem for $v = [v_1 \quad v_2]^T$,*

$$\frac{\partial v}{\partial x_2} - \begin{pmatrix} A_1 & -A_2 \\ A_2 & A_1 \end{pmatrix} \frac{\partial v}{\partial x_1} = \begin{pmatrix} f_1(x) \\ f_2(x) \end{pmatrix}, \qquad x \in \Omega \tag{10}$$

$$[B_1(y) \quad B_2(y)]v = g(y), \qquad y \in \partial\Omega.$$

There is a one-to-one correspondence between solutions of (10) *and* (9) *given by* $[v_1 \quad v_2]^T \mapsto w = v_1 + iv_2$.

Proof This is an easy consequence of the formula (6) and the fact that

$$\frac{1}{2}\begin{pmatrix} w \\ \bar{w} \end{pmatrix} = \begin{pmatrix} I & I \\ -iI & iI \end{pmatrix}^{-1} \begin{pmatrix} v_1 \\ v_2 \end{pmatrix}.$$

Left multiplication of the system (10) by $\begin{pmatrix} I & I \\ -iI & iI \end{pmatrix}^{-1}$ gives

$$\frac{\partial}{\partial x_2}\begin{pmatrix} w \\ \bar{w} \end{pmatrix} - \begin{pmatrix} A_1 + iA_2 & \\ & A_1 - iA_2 \end{pmatrix} \frac{\partial}{\partial x_1}\begin{pmatrix} w \\ \bar{w} \end{pmatrix} = \begin{pmatrix} f(x) \\ \bar{f}(x) \end{pmatrix}$$

which is just two copies of $\dfrac{\partial w}{\partial x_2} - (A_1 + iA_2)\dfrac{\partial w}{\partial x_1} = f(x)$, and the boundary conditions in (10) may be written as

$$\tfrac{1}{2}[B_1(y) \quad B_2(y)]\begin{pmatrix} I & I \\ -iI & iI \end{pmatrix}\begin{pmatrix} w \\ \bar{w} \end{pmatrix} = g(y),$$

that is, $\operatorname{Re}(B_1(y) - iB_2(y))w = g(y)$. Conversely, if the function w satisfies (9) then $v = [v_1 \quad v_2]^T$, where $v_1 = (w + \bar{w})/2$ and $v_2 = (w - \bar{w})/2i$, satisfies (10). This establishes the one-to-one correspondence between solutions of the two problems.

We call the boundary problem (10) the *real form* of the Riemann–Hilbert problem (9). Note that by Lemma 16.3, we have

$$\det\left(I_{2r}\lambda - \begin{pmatrix} A_1 & -A_2 \\ A_2 & A_1 \end{pmatrix}\right) = |\det(I_r\lambda - (A_1 + iA_2))|^2 \neq 0 \qquad \text{for real } \lambda,$$

so (10) is elliptic. Since $\det(B_1(y) - iB_2(y)) \neq 0$ for all $y \in \partial\Omega$, then it also satisfies the L-condition (Lemma 16.5). Hence the boundary problem (10) has a finite dimensional kernel and cokernel, and the same is therefore true of the Riemann–Hilbert problem (9). (We leave the verification of this fact to the reader.)

Let $(\mathbf{A}, \operatorname{Re}(B\ \cdot))$ denote the boundary problem (9), i.e. $\mathbf{A} = \partial/\partial x_2 - (A_1 + iA_2)\,\partial/\partial x_1$ and $B = B_1 - iB_2$. Now we construct a homotopy of elliptic operators $\mathbf{A}$. Since the spectrum of $A_1 + iA_2$ is contained in the upper half-plane, the same is true of $t(A_1 + iA_2) + (1 - t)iI$ when $0 \leqslant t \leqslant 1$. The operator

$$\mathbf{A}^{(t)} = \partial/\partial x_2 - (t(A_1 + iA_2) + (1 - t)iI)\,\partial/\partial x_1, \qquad 0 \leqslant t \leqslant 1$$

is therefore elliptic, and the index, $\operatorname{ind}(\mathbf{A}^{(t)}, \operatorname{Re}(B\ \cdot))$, is independent of t. With $t = 1$ and $t = 0$, it follows that

$$\operatorname{ind}(\mathbf{A}, \operatorname{Re}(B\ \cdot)) = \operatorname{ind}(-2i\,\partial/\partial\bar{z}, \operatorname{Re}(B\ \cdot))$$

Thus we have reduced the calculation of the index to the following boundary problem:

$$\frac{\partial w}{\partial \bar{z}} = h(z), \qquad z = x_1 + ix_2 \in \Omega \tag{11}$$

$$\operatorname{Re} B(y)w = g(y), \qquad y \in \partial\Omega,$$

where $B \in C^\infty(\partial\Omega, r \times r)$ and $\det B(y) \neq 0$ for all $y \in \partial\Omega$.

Theorem 16.7 *The Riemann–Hilbert problem* (11) *has index*

$$\operatorname{ind}(\partial/\partial\bar{z}, \operatorname{Re}(B\ \cdot)) = -\frac{1}{\pi}[\arg\det B(y)]_{\partial\Omega} + r \tag{12}$$

Proof If the matrix $B(y)$ were diagonal, then (11) would be a diagonal system of r scalar Riemann–Hilbert problems, and (12) would follow directly from the index formula (2) and Theorem 9.2(b) (by induction on the dimension

r of the system). But, in view of Theorem 15.10, the matrix function $B\colon \partial\Omega \to GL_e(\mathbb{C})$ is homotopic to

$$\begin{pmatrix} \det B(y) & & & \\ & 1 & & \\ & & \ddots & \\ & & & 1 \end{pmatrix}$$

so the theorem follows at once.

Collecting all the preceding results, in particular, the equation (10), Lemma 16.6 and Theorem 16.7, we obtain the index formula for first-order elliptic systems with real coefficients. Note that, by virtue of Lemma 15.16, the quantity $[\arg\det B(y)X_+(y)]_{\partial\Omega}$ does not depend on the choice of γ^+-spectral pair $(X_+(x), T_+(x))$, $x \in \bar{\Omega}$.

Theorem 16.8 *The index of the boundary value problem* (3) *is*

$$\operatorname{ind}(\mathscr{A}, \mathscr{B}) = -\frac{1}{\pi}[\arg\det B(y)X_+(y)]_{\partial\Omega} + r,$$

where $X_+(x)$ is a $2r \times r$ matrix function depending smoothly on $x \in \bar{\Omega}$ whose columns form a basis for the invariant subspace of $A(x)$ corresponding to the eigenvalues with positive imaginary part.

16.3 A fundamental solution for first-order elliptic systems with constant real coefficients

According to a theorem of Malgrange–Ehrenpreis, every nonzero partial differential operator $P(D) \not\equiv 0$ with constant scalar coefficients has a fundamental solution, that is, a distribution $E \in \mathscr{D}'(\mathbb{R}^n)$ such that $P(D)E = \delta$, where δ is the Dirac distribution. We need a corresponding result for systems of partial differential operators. Let $P(\xi)$ be a polynomial in n variables $\xi = (\xi_1, \ldots, \xi_n)$ with $p \times p$ matrix coefficients. A $p \times p$ matrix E with elements in $\mathscr{D}'(\mathbb{R}^n)$ is a right (left) fundamental solution for $P(D)$ if

$$P(D)E = \delta I \qquad (\text{respectively, } E * P(D)\delta I = \delta I)$$

where I is the $p \times p$ identity matrix. Denote by P^{co} the matrix formed by the cofactors in P; thus

$$P^{\text{co}}(\xi)P(\xi) = P(\xi)P^{\text{co}}(\xi) = \det P(\xi) \cdot I \tag{13}$$

and the entries of $P^{\text{co}}(\xi)$ are polynomials in ξ.

Lemma *If* $\det P(\xi) \not\equiv 0$ *then there is a two-sided fundamental solution for* $P(D)$.

Proof Let F be a fundamental solution for the differential operator $\det P(D)$. Then it follows from (13) that $E = P^{co}(D)(FI)$ is both a right and a left fundamental solution for $P(D)$.

Remark If f is a distribution with compact support then one solution of $P(D)u = f$ is given by $u = E * f$, where E is a right fundamental solution.

In particular the lemma holds if P is an elliptic system. The existence of a fundamental solution is needed in the proof of Theorem 16.11, at least for first-order elliptic operators in the plane with real coefficients. In this case it is not difficult to derive an explicit formula for a fundamental solution starting from the well-known fundamental solution of the Cauchy–Riemann operator,

$$\partial/\partial\bar{z} = \tfrac{1}{2}(\partial/\partial x + i\cdot\partial/\partial y),$$

namely, $\dfrac{1}{\pi z}$, where $z = x + iy$.

Proposition 16.9 *Let A be a real $2r \times 2r$ matrix such that* $\det(I\lambda - A) \neq 0$ *for all $\lambda \in \mathbb{R}$. Then the operator $I\dfrac{\partial}{\partial x} + A\dfrac{\partial}{\partial y}$ has the fundamental solution*

$$E = -\frac{1}{2\pi^2}\operatorname{Re}\int_{+}\frac{(I\lambda - A)^{-1}}{\lambda x - y}\,d\lambda, \tag{14}$$

Proof Assume first that A has $2r$ distinct eigenvalues. Let (X_+, T_+, Y_+) be a γ^+-spectral triple for $I\lambda - A$, where X_+ is a $2r \times r$ matrix, T_+ is an $r \times r$ matrix and Y_+ is an $r \times 2r$ matrix. The columns of X_+ form a linearly independent set of eigenvectors of A corresponding to the eigenvalues in the upper half-plane, and the columns of $\bar{X}_+$ form a linearly independent set of eigenvectors corresponding to the eigenvalues in the lower half-plane. We have

$$[X_+ \quad \bar{X}_+]^{-1}A[X_+ \quad \bar{X}_+] = \begin{pmatrix} T_+ & \\ & \bar{T}_+ \end{pmatrix}, \qquad \text{(cf. (5))}$$

where $T_+ = \operatorname{diag}(\lambda_j)_{j=1}^r$ and $\operatorname{Im}\lambda_j > 0$ for all $j = 1, \ldots, r$. Also,

$$\begin{pmatrix} Y_+ \\ \bar{Y}_+ \end{pmatrix} = [X_+ \quad \bar{X}_+]^{-1} \qquad \text{(see §3.1 (6))}$$

so that the rows of Y_+ (resp. $\bar{Y}_+$) form a linearly independent set of left eigenvectors of A corresponding to the eigenvalues in the upper (resp. lower) half-plane. Now let u be a real-valued solution of the equation

$$\frac{\partial u}{\partial x} + A\frac{\partial u}{\partial y} = f, \tag{15}$$

and let $w = Y_+ u$ and $g = Y_+ f$. Then

$$\begin{pmatrix} w \\ \bar{w} \end{pmatrix} = \begin{pmatrix} Y_+ \\ \bar{Y}_+ \end{pmatrix} u, \tag{16}$$

and the equation (15) becomes

$$\frac{\partial}{\partial x}\begin{pmatrix} w \\ \bar{w} \end{pmatrix} + \begin{pmatrix} T_+ & \\ & \bar{T}_+ \end{pmatrix} \frac{\partial}{\partial y}\begin{pmatrix} w \\ \bar{w} \end{pmatrix} = \begin{pmatrix} g \\ \bar{g} \end{pmatrix} \tag{17}$$

or

$$\frac{\partial w_j}{\partial x} + \lambda_j \frac{\partial w_j}{\partial y} = g_j, \qquad j = 1, \ldots, r. \tag{18}$$

As an aside, note that (16) is inverted as follows:

$$u = [X_+ \quad \bar{X}_+]\begin{pmatrix} w \\ \bar{w} \end{pmatrix} = 2 \operatorname{Re} \bar{X}_+ w$$

In view of (18), we are left with the problem of finding a fundamental solution for a scalar operator of the form $\partial/\partial x + \lambda_0\, \partial/\partial y$, where $\lambda_0 = a + ib$ and $b > 0$; it is easily seen that a fundamental solution is given by

$$\frac{-1}{2\pi i} \frac{1}{\lambda_0 x - y}$$

(Make a change of variables in the fundamental solution $1/\pi z$ for $\partial/\partial \bar{z}$.)

Hence the operator on the left side of (18) has fundamental solution given by $\begin{pmatrix} E_+ & \\ & \bar{E}_+ \end{pmatrix}$, where

$$E_+ = \frac{-1}{2\pi i} \operatorname{diag}\left(\frac{1}{\lambda_j x - y}\right)_{j=1}^{r}.$$

Then the original operator $I \dfrac{\partial}{\partial x} + A \dfrac{\partial}{\partial y}$ has fundamental solution

$$E = [X_+ \quad \bar{X}_+]\begin{pmatrix} E_+ & \\ & \bar{E}_+ \end{pmatrix}\begin{pmatrix} Y_+ \\ \bar{Y}_+ \end{pmatrix} = 2 \operatorname{Re} X_+ E_+ Y_+ .$$

Now write $X_+ = [X_1 \cdots X_r]$ and $Y = \begin{pmatrix} Y_1 \\ \vdots \\ Y_r \end{pmatrix}$, where X_j and Y_j are the columns and rows of X_+ and Y_+, respectively. Also, let $\gamma^+ = \gamma_1 + \cdots + \gamma_r$ where γ_j is a small circle in the upper half-plane containing λ_j but no other eigenvalues of A. Then

$$X_+E_+Y_+ = \sum_{j=1}^{r} \frac{1}{\lambda_j x - y} X_j Y_j$$
$$= \frac{1}{2\pi i} \sum_{j=1}^{r} \int_{\gamma_j} \frac{(I\lambda - A)^{-1}}{\lambda x - y} d\lambda$$
$$= \frac{1}{2\pi i} \int_{\gamma^+} \frac{(I\lambda - A)^{-1}}{\lambda x - y} d\lambda$$

(the second equality holds since λ_j is a simple pole of $(I\lambda - A)^{-1}$, i.e. λ_j is an eigenvalue of multiplicity 1). This proves (14) in the case that A has distinct eigenvalues.

Suppose now that A is any real $2r \times 2r$ matrix such that $\det(I\lambda - A) \neq 0$ for all $\lambda \in \mathbb{R}$. Note that the right side of (14) defines a function in $C^\infty(\mathbb{R}^2 \setminus 0)$ that is locally integrable near the origin, and depends continuously on the matrix A. Let $\varepsilon > 0$. By Lemma 16.10, there exists a matrix A_ε with distinct eigenvalues such that $\|A_\varepsilon - A\| < \varepsilon$. By choosing ε small enough we may assume that A_ε has real entries and $\det(I\lambda - A_\varepsilon) \neq 0$ for all $\lambda \in \mathbb{R}$. Now let E_ε denote the right side of (14) when A is replaced by A_ε. Then we have

$$\frac{\partial E_\varepsilon}{\partial x} + A_\varepsilon \frac{\partial E_\varepsilon}{\partial y} = \delta I,$$

and letting $\varepsilon \to 0$, we have $E_\varepsilon \to E$ and

$$\frac{\partial E}{\partial x} + A \frac{\partial E}{\partial y} = \delta I,$$

as required. The following proof is from [AMR].

Lemma 16.10 *Let A be an $n \times n$ matrix. For any $\varepsilon > 0$, there exists a matrix A_ε with distinct eigenvalues such that $\|A_\varepsilon - A\| < \varepsilon$.*

Proof Let $p(\lambda) = \det(I\lambda - A)$ be the characteristic polynomial of A and let $\mu_1, \ldots, \mu_{n-1}$ be the roots of the derivative $p'(\lambda)$. Then A has multiple eigenvalues if and only if

$$p(\mu_1) \cdots p(\mu_{n-1}) = 0$$

This last expression is a symmetric polynomial in $\mu_1, \ldots, \mu_{n-1}$, and so is a polynomial in the coefficients of p'; it is therefore a polynomial q in the n^2 entries of A. Then $q^{-1}(0)$ is the set of complex $n \times n$ matrices which have multiple eigenvalues. Since $q \not\equiv 0$, the complement of $q^{-1}(0)$ is dense in $\mathbb{C}^{n^2}$. (Proof by contradiction: if q vanishes on an open set in $\mathbb{C}^{n^2}$, then all its derivatives also vanish, and hence all its coefficients are zero.)

16.4 Index formulas for higher-order systems with real coefficients

In this section we suppose that $\mathscr{A}$ is an elliptic operator with principal part homogeneous of degree l, and that the boundary operator $\mathscr{B}$ is a *differential* operator. In the first part of this section we also suppose that the principal part of $\mathscr{B}$ in each row is equal to the same number μ; that is, with the notation of §15.1, $(\mathscr{A}, \mathscr{B}) \in BE^{l,\mu}$. For the moment we also assume that $\mu \leqslant l-1$.

We will derive a formula for the index of $(\mathscr{A}, \mathscr{B})$ by reformulating $(\pi\mathscr{A}, \pi\mathscr{B})$ as a first-order system and then applying Theorem 16.8. In view of §15.4 we may suppose that $\mathscr{A}$ and $\mathscr{B}$ have matrix coefficients with *real entries.*

Let $L_x(\lambda) = \pi\mathscr{A}(x, (1, \lambda)) = \sum_{j=0}^{l} A_j(x)\lambda^j$, $x \in \bar{\Omega}$. Since $\mathscr{A}$ is elliptic, then $A_l(x) = \pi\mathscr{A}(x, (0, 1))$ is invertible for all $x \in \bar{\Omega}$. By considering the operator $A_l^{-1}\mathscr{A}$, we may assume that $A_l \equiv I$, that is, $L_x(\lambda)$ is monic. Also we let $B_y(\lambda) = \pi\mathscr{B}(y, (1, \lambda)) = \sum_{j=0}^{l-1} B_j(y)\lambda^j$, $y \in \partial\Omega$. Recall from §15.1 that $(\mathscr{A}, \mathscr{B})$ satisfies the L-condition if and only if $\det \Delta_{\mathscr{B}}^{+}(y) \neq 0$, where

$$\Delta_{\mathscr{B}}^{+}(y) = \sum_{j=0}^{l-1} B_j(y) X_+(y) T_+^j(y). \tag{19}$$

As usual, we let $(X_+(x), T_+(x))$ be a γ^+-spectral pair of $L_x(\lambda)$ consisting of smooth matrix functions on $\bar{\Omega}$.

To reformulate $(\pi\mathscr{A}, \pi\mathscr{B})$ as a first-order system, let

$$v_j = \frac{\partial^{l-1}u}{\partial x_1^{l-1-j}\,\partial x_2^j}, \qquad j = 0, \ldots, l-1, \tag{20}$$

so that the equation $\pi\mathscr{A}(x, \partial/\partial x)u = f(x)$ takes the form

$$\frac{\partial v_{l-1}}{\partial x_2} + A_{l-1}(x)\frac{\partial v_{l-1}}{\partial x_1} + \cdots + A_0(x)\frac{\partial v_0}{\partial x_1} = f(x), \tag{21}$$

and there are the "compatibility conditions"

$$\frac{\partial v_j}{\partial x_2} = \frac{\partial v_{j+1}}{\partial x_1} \qquad j = 0, \ldots, l-2. \tag{22}$$

Thus, we obtain the first-order system

$$\tilde{\mathscr{A}}(x, \partial/\partial x)v = \frac{\partial v}{\partial x_2} - C_1(x)\frac{\partial v}{\partial x_1} = F(x), \qquad x \in \Omega \tag{23a}$$

where $v = [v_0 \cdots v_{l-1}]^T$ and $F = [0 \cdots 0\ f]^T$ are pl-vector functions and $C_1(x)$ is the companion matrix for $L_x(\lambda)$:

$$C_1 = \begin{pmatrix} 0 & I & & & \\ & 0 & I & & \\ \vdots & & \ddots & \ddots & \\ & & & & I \\ -A_0 & -A_1 & \cdots & \cdots & -A_{l-1} \end{pmatrix}$$

Note that $\mathscr{A}$ is elliptic if and only if $\tilde{\mathscr{A}}$ is elliptic, since $\det L_x(\lambda) = \det(I\lambda - C_1(x))$.

If $\mu = l - 1$ the boundary conditions $\pi\mathscr{B}(y, \partial/\partial x)u = g(y)$ take the form

$$\tilde{\mathscr{B}}(y)v = g(y), \qquad y \in \partial\Omega, \tag{23b}$$

where $\tilde{\mathscr{B}} = [B_0 \ B_1 \cdots B_{l-1}]$, an $r \times pl$ matrix function. Then (19) can be written in the form

$$\Delta^+_{\mathscr{B}}(y) = \tilde{\mathscr{B}} \cdot \operatorname{col}(X_+(y)T^j_+(y))^{l-1}_{j=0}. \tag{24}$$

Now we are ready to prove the first version of the index formula (for real coefficients). At first sight it might appear that (25) implies that the index of $(\mathscr{A}, \mathscr{B})$ depends only on the values of the coefficients of $\mathscr{A}$ on the boundary $\partial\Omega$, but in fact the choice of (X_+, T_+), and hence $\Delta^+_{\mathscr{B}}$, depends on the coefficients throughout the region $\bar{\Omega}$.

Theorem 16.11 *Let the hypotheses be as stated above: $(\mathscr{A}, \mathscr{B}) \in BE^{l,\mu}$ (for arbitrary μ) and the boundary value problem operators $\mathscr{A}$ and $\mathscr{B}$ have real matrix coefficients. Then*

$$\operatorname{ind}(\mathscr{A}, \mathscr{B}) = \frac{-1}{\pi}[\arg\det \Delta^+_{\mathscr{B}}(y)]_{\partial\Omega} + r(2\mu + 2 - l). \tag{25}$$

Proof Without loss of generality $(\mathscr{A}, \mathscr{B}) = (\pi\mathscr{A}, \pi\mathscr{B})$, that is, $\mathscr{A}$ contains no derivatives of order $< l$ and $\mathscr{B}$ no derivatives of order $< \mu$. We prove the theorem in four steps: (1) $\mu = l - 1$, and the operator $\mathscr{A}$ has constant coefficients; (2) $\mu = l - 1$, for general $\mathscr{A}$; (3) $\mu < l - 1$; and (4) $\mu \geqslant l$.

Step 1 We reformulate $(\mathscr{A}, \mathscr{B})$ as the first-order boundary value problem (23a), (23b), and apply Theorem 16.8 to $(\tilde{\mathscr{A}}, \tilde{\mathscr{B}})$. Note that the columns of $\operatorname{col}(X_+(x)T^j_+(x))^{l-1}_{j=0}$ form a basis for the invariant subspace of $C_1(x)$ corresponding to the eigenvalues with positive imaginary part. Thus by Theorem 16.8 and (24)

$$\begin{aligned}\operatorname{ind}(\tilde{\mathscr{A}}, \tilde{\mathscr{B}}) &= r - \frac{1}{\pi}[\arg\det \tilde{\mathscr{B}}(y)\cdot\operatorname{col}(X_+(x)T^j_+(x))^{l-1}_{j=0}]_{\partial\Omega} \\ &= r - \frac{1}{\pi}[\arg\det \Delta^+_{\mathscr{B}}(y)]_{\partial\Omega} \end{aligned}\tag{26}$$

We have the operators $(\mathscr{A}, \mathscr{B})$: $W^l_2(\Omega, \mathbb{R}^p) \to L_2(\Omega, \mathbb{R}^p) \times W^{1/2}_2(\partial\Omega, \mathbb{R}^r)$ and $(\tilde{\mathscr{A}}, \tilde{\mathscr{B}})$: $W^1_2(\Omega, \mathbb{R}^{pl}) \to L_2(\Omega, \mathbb{R}^{pl}) \times W^{1/2}_2(\partial\Omega, \mathbb{R}^r)$, and to complete the proof of (25) we must show that

$$\operatorname{ind}(\mathscr{A}, \mathscr{B}) = \operatorname{ind}(\tilde{\mathscr{A}}, \tilde{\mathscr{B}}) + r(l - 1). \tag{27}$$

There is a map Φ from $\ker(\mathscr{A}, \mathscr{B})$ to $\ker(\tilde{\mathscr{A}}, \tilde{\mathscr{B}})$ given by $u \mapsto v = [v_0 \cdots v_{l-1}]^T$, where v_j are defined by (20). Now, Φ is surjective for if $\tilde{\mathscr{A}}v = 0$ and $\tilde{\mathscr{B}}(y)v = 0$ then, by Weyl's lemma, $v \in C^\infty(\bar{\Omega}, \mathbb{R}^{pl})$; since (22) holds and $\bar{\Omega}$ is contractible, then, by Lemma 16.12, there exists $u \in C^\infty(\bar{\Omega}, \mathbb{R}^p)$ such that (20) holds. Moreover, the kernel of Φ is the subspace $\mathscr{P}$ of all $\mathbb{R}^p$-valued polynomials u of degree $\leqslant l-2$,

$$u(x_1, x_2) = \sum_{j+k=0}^{l-2} c_{jk} x_1^j x_2^k, \qquad \text{where } c_{jk} \in \mathbb{R}^p \text{ are constants.}$$

Since the number of linearly independent monomials of the form $c_{jk} x_1^j x_2^k$, $j+k \leqslant l-2$, is equal to $1+2+\cdots+(l-1) = (l-1)l/2$, it follows that

$$\dim \mathscr{P} = p \cdot \frac{(l-1)l}{2} = r(l-1).$$

Hence $\ker(\mathscr{A}, \mathscr{B})/\mathscr{P} \cong \ker(\tilde{\mathscr{A}}, \tilde{\mathscr{B}})$, so that

$$\dim \ker(\mathscr{A}, \mathscr{B}) = \dim \ker(\tilde{\mathscr{A}}, \tilde{\mathscr{B}}) + r(l-1) \tag{28}$$

At this point we wish to apply Lemma 16.13. The operator $\tilde{\mathscr{A}}$: $W_2^l(\Omega, \mathbb{R}^{pl}) \to L_2(\Omega, \mathbb{R}^{pl})$ is surjective for if $F \in C^\infty(\bar{\Omega}, \mathbb{R}^{pl})$ we can extend F to $C_0^\infty(\mathbb{R}^2, \mathbb{R}^{pl})$ and then let $u = E * F$ where E is a fundamental solution of $\tilde{\mathscr{A}}$; it follows that $\operatorname{im} \tilde{\mathscr{A}} \supset C^\infty(\bar{\Omega}, \mathbb{R}^{pl})$. By virtue of Lemma 16.14 the image of $\tilde{\mathscr{A}}$ has finite codimension – because this is true of the image of $(\tilde{\mathscr{A}}, \tilde{\mathscr{B}})$ – hence it must be closed by Proposition 9.2. Hence $\tilde{\mathscr{A}}$ is surjective because we have shown that the image of $\tilde{\mathscr{A}}$ is closed and dense in $L_2(\Omega, \mathbb{R}^{pl})$. Similarly $\mathscr{A}$: $W_2^l(\Omega, \mathbb{R}^p) \to L_2(\Omega, \mathbb{R}^p)$ is surjective. Indeed, this can be derived from the surjectivity of $\tilde{\mathscr{A}}$ for if $f \in C^\infty(\bar{\Omega}, \mathbb{R}^p)$ then there exists $v = [v_0 \cdots v_{l-1}]^T \in C^\infty(\bar{\Omega}, \mathbb{R}^{pl})$ such that $\tilde{\mathscr{A}} v = F = [0 \cdots 0\, f]^T$, then v satisfies (21), (22), and due to Lemma 16.12, there exists $u \in C^\infty(\bar{\Omega}, \mathbb{R}^p)$ such that (20) holds, whence $\mathscr{A} u = f$. Hence

$$\begin{aligned}
\operatorname{ind}(\mathscr{A}, \mathscr{B}) &= \operatorname{ind}(\mathscr{B}|_{\ker \mathscr{A}}), && \text{by Lemma 16.13} \\
&= \operatorname{ind}(\tilde{\mathscr{B}}|_{\ker \mathscr{A}_1}) + r(l-1), && \text{see (28)} \\
&= \operatorname{ind}(\tilde{\mathscr{A}}, \tilde{\mathscr{B}}) + r(l-1), && \text{by Lemma 16.13}
\end{aligned}$$

and then (26) implies

$$\operatorname{ind}(\mathscr{A}, \mathscr{B}) = \frac{-1}{\pi} [\arg \det \Delta_{\mathscr{B}}^{+}(y)]_{\partial\Omega} + rl, \tag{29}$$

which is the formula (25) for the case $\mu = l-1$.

Step 2 It follows from the proof of Proposition 15.11 that $\bar{\Omega}$ is contractible, i.e. there exists a smooth map φ: $\bar{\Omega} \times I \to \bar{\Omega}$ such that $\varphi(x, 0) = x_0$ and $\varphi(x, 1) = x$ for all $x \in \bar{\Omega}$, where x_0 is a fixed point in $\bar{\Omega}$. Consider now the following homotopy of elliptic operators:

$$\mathscr{A}_t(x, \partial/\partial x) = \mathscr{A}(\varphi(x, t), \partial/\partial x), \qquad 0 \leqslant t \leqslant 1.$$

Note that $\mathscr{A}_1 = \mathscr{A}$, and $\mathscr{A}_0$ has constant coefficients. In view of Proposition 15.2 there exists a homotopy $(\mathscr{A}_t, \mathscr{B}_t)$, $0 \leqslant t \leqslant 1$, of elliptic BVP's such that $(\mathscr{A}_1, \mathscr{B}_1) = (\mathscr{A}, \mathscr{B})$. Since the index is locally constant, we obtain $\operatorname{ind}(\mathscr{A}, \mathscr{B}) = \operatorname{ind}(\mathscr{A}_0, \mathscr{B}_0)$. By Step 1, (29) holds for $(\mathscr{A}_0, \mathscr{B}_0)$, thus it also holds for $(\mathscr{A}, \mathscr{B})$, since the winding number is locally constant and by Lemma 10.22 there exists a γ^+-spectral pair (X_+, T_+) for $L(\lambda) = \pi\mathscr{A}_t(x, (1, \lambda))$ which depends smoothly on t.

Step 3 Suppose now that $\mu < l - 1$. We then operate on the boundary operator $\mathscr{B}$ with tangential derivatives to increase its order to $l - 1$:

$$\mathscr{B}' = (\partial/\partial\tau)^{l-1-\mu} \circ \mathscr{B},$$

where $\partial/\partial\tau = \tau_1\, \partial/\partial x_1 + \tau_2\, \partial/\partial x_2$. Corresponding to (19), let

$$B'_y(\lambda) = \pi\mathscr{B}'(y, (1, \lambda)) = \sum_{j=0}^{l-1} B'_j(y)\lambda^j,$$

and $\Delta^+_{\mathscr{B}'}(y) = \sum_{j=0}^{l-1} B'_j(y)X_+(y)T^j_+(y)$. Since

$$B'_y(\lambda) = (\tau_1(y) + \tau_2(y)\lambda)^{l-1-\mu}B_r(\lambda),$$

then

$$\Delta^+_{\mathscr{B}'}(y) = \Delta^+_{\mathscr{B}}(y)(\tau_1(y)I + \tau_2(y)T_+(y))^{l-1-\mu} \tag{30}$$

(see Exercise 3). It follows that det $\Delta^+_{\mathscr{B}'}(y) \neq 0$ if and only if det $\Delta_+(y) \neq 0$ since sp$(T_+(y))$ lies in the upper half-plane. Hence $(\mathscr{A}, \mathscr{B}')$ satisfies the L-condition. From (30) we see that

$$[\arg\det \Delta^+_{\mathscr{B}'}(y)]_{\partial\Omega} = [\arg\det \Delta^+_{\mathscr{B}}(y)]_{\partial\Omega} + (l-1-\mu)[\arg\det(\tau_1 I + \tau_2 T_+)]_{\partial\Omega}$$

and, since $\dfrac{1}{2\pi}[\arg\det(\tau_1(y)I + \tau_2(y)T_+(y))]_{\partial\Omega} = r$, we obtain

$$\frac{1}{\pi}[\arg\det \Delta^+_{\mathscr{B}'}(y)]_{\partial\Omega} = \frac{1}{\pi}[\arg\det \Delta^+_{\mathscr{B}}(y)]_{\partial\Omega} + 2r(l - 1 - \mu). \tag{31}$$

Thus, to complete the proof of (25), it suffices to show that ind$(\mathscr{A}, \mathscr{B}')$ = ind$(\mathscr{A}, \mathscr{B})$. Consider the tangential derivative $\partial/\partial\tau$ as an operator from $W^{k+1}_2(\partial\Omega)$ to $W^k_2(\partial\Omega)$ for $k \geqslant 0$. This operator has index 0, because the kernel of $\partial/\partial\tau$ is one-dimensional, consisting of the constant functions on $\partial\Omega$, and the image of $\partial/\partial\tau$ has codimension 1, consisting of the functions $g \in W^k_2(\partial\Omega)$ such that $\int g = 0$; therefore, ind$(\partial/\partial\tau) = 1 - 1 = 0$, It follows that the operator

$$\Psi\colon W^l_2(\Omega, \mathbb{R}^p) \times W^{l-\mu-1/2}_2(\partial\Omega, \mathbb{R}^r) \to W^l_2(\Omega, \mathbb{R}^p) \times W^{1/2}_2(\partial\Omega, \mathbb{R}^r)$$

defined by $\Psi(f, g) = (f, (\partial/\partial\tau)^{l-1-\mu}g)$ is a composition of $l - 1 - \mu$ Fredholm operators. By Theorem 9.8, Ψ is also Fredholm of index 0. Then, since $(\mathscr{A}, \mathscr{B}') = \Psi \circ (\mathscr{A}, \mathscr{B})$, we obtain that ind$(\mathscr{A}, \mathscr{B}')$ = ind Ψ + ind$(\mathscr{A}, \mathscr{B})$ = ind$(\mathscr{A}, \mathscr{B})$. Consequently, by (29) and (31)

$$\begin{aligned} \operatorname{ind}(\mathscr{A}, \mathscr{B}) &= \operatorname{ind}(\mathscr{A}, \mathscr{B}') \\ &= \frac{-1}{\pi}[\arg\det \Delta^+_{\mathscr{B}'}(y)]_{\partial\Omega} + rl \\ &= \frac{-1}{\pi}[\arg\det \Delta^+_{\mathscr{B}}(y)]_{\partial\Omega} + r(2\mu + 2 - l) \end{aligned}$$

Step 4 Suppose that $\mu \geqslant l$. By Theorem 15.4, $(\mathscr{A}, \mathscr{B})$ is homotopic to $(\mathscr{A}, (\partial/\partial\tau)^{\mu-l+1}\mathscr{R})$ where $(\mathscr{A}, \mathscr{R}) \in BE^{l,l-1}$ and $\tilde{\Delta}^+_{\mathscr{R}} = \tilde{\Delta}^+_{\mathscr{B}}$. We then have

$$\Delta^+_{\mathscr{R}}(y) = \Delta^+_{\mathscr{B}}(y)\cdot(\tau_1(y) + \tau_2(y)T_+(y))^{l-1-\mu}, \tag{32}$$

for $y \in \partial\Omega$ (see Proposition 15.12). Hence

$$\begin{aligned}\operatorname{ind}(\mathscr{A}, \mathscr{B}) &= \operatorname{ind}(\mathscr{A}, \mathscr{R})\\ &= \frac{-1}{\pi}[\arg\det \Delta^+_{\mathscr{R}}(y)]_{\partial\Omega} + rl\\ &= \frac{-1}{\pi}[\arg\det \Delta^+_{\mathscr{B}}(y)]_{\partial\Omega} - 2r(l-1-\mu) + rl \qquad \text{by (32)}\\ &= \frac{-1}{\pi}[\arg\det \Delta^+_{\mathscr{B}}(y)]_{\partial\Omega} + r(2\mu+2-l),\end{aligned}$$

so that formula (25) holds for $(\mathscr{A}, \mathscr{B})$. This completes the proof of the theorem.

Now we turn to the proof of the three lemmas which were used in the proof of Theorem 16.11.

Lemma 16.12 *Let* $v_0, \ldots, v_{l-1} \in C^\infty(\bar{\Omega})$, $l \geqslant 2$, *such that*

$$\frac{\partial v_j}{\partial x_2} = \frac{\partial v_{j+1}}{\partial x_1} \qquad j = 0, \ldots, l-2. \tag{33}$$

Then there exists $u \in C^\infty(\bar{\Omega})$ *such that*

$$\frac{\partial^{l-1}u}{\partial x_1^{l-1-j}\,\partial x_2^j} = v_j, \qquad j = 0, \ldots, l-1. \tag{34}$$

Proof Since $\bar{\Omega}$ is contractible then by the Poincaré Lemma (Theorem 6.12) every closed form on $\bar{\Omega}$ is exact. We prove the lemma by induction on the number l. For $l = 2$, we are given $v_0, v_1 \in C^\infty(\bar{\Omega})$ such that $\partial v_0/\partial x_2 = \partial v_1/\partial x_1$. Then $\omega = v_0\,dx_1 + v_1\,dx_2$ is a closed form (i.e., $d\omega = 0$), so there exists $u \in C^\infty(\bar{\Omega})$ such that $\omega = du$, i.e., $\partial u/\partial x_1 = v_0$ and $\partial u/\partial x_2 = v_1$. Suppose now that the lemma has been proved for order $l-1$. Let $v_0, \ldots, v_{l-1} \in C^\infty(\bar{\Omega})$ such that (33) holds. Then the forms $\omega_j = v_j\,dx_1 + v_{j+1}\,dx_2$ $(j = 0, \ldots, l-2)$ are closed, so there exist $f_0, \ldots, f_{l-2} \in C^\infty(\bar{\Omega})$ such that $\omega_j = df_j$, i.e. $\partial f_j/\partial x_1 = v_j$ and $\partial f_j/\partial x_2 = v_{j+1}$ for $j = 0, \ldots, l-2$. Now, since $\partial f_j/\partial x_2 = \partial f_{j+1}/\partial x_1$ $(= v_{j+1})$ for $j = 0, \ldots, l-2$, the induction hypothesis implies that there exists $u \in C^\infty(\bar{\Omega})$ such that

$$\frac{\partial^{l-2}u}{\partial x_1^{l-2-j}\,\partial x_2^j} = f_j \qquad j = 0, \ldots, l-2.$$

Then it is easily verified that (34) holds.

Lemma 16.13 *Let E, E_1 and E_2 be Banach spaces. Let $A: E \to E_1$ and $B: E \to E_2$ be bounded linear operators. Let $A_0 = A|_{\ker B}$. If B is surjective then the following are equivalent:*

(i) *The operator (A, B) from E to $E_1 \times E_2$ is Fredholm;*
(ii) *The operator A_0 from* $\ker B$ *to E_1 is Fredholm.*

If (A, B) *and A_0 are Fredholm operators then* $\operatorname{ind}(A, B) = \operatorname{ind} A_0$. *(An analogous result is true if the roles of A and B are interchanged.)*

Proof The operators (A, B) and A_0 have the same kernel. Also note that $f \in \operatorname{im} A_0$ if and only if $(f, 0) \in \operatorname{im}(A, B)$, so the map $f \mapsto (f, 0)$ induces a map

$$E_1/\operatorname{im} A_0 \to E_1 \times E_2/\operatorname{im}(A, B) \tag{35}$$

which is injective. Hence $\operatorname{codim} \operatorname{im} A_0 \leqslant \operatorname{codim} \operatorname{im}(A, B)$. If (i) holds then $\operatorname{codim} \operatorname{im}(A, B) < \infty$ and thus $\operatorname{codim} \operatorname{im} A_0 < \infty$ also. Since $\operatorname{im} A_0$ has finite codimension it is closed and therefore (i) implies (ii). Conversely suppose (ii) holds. Let $(f, g) \notin \operatorname{im}(A, B)$. Since B is surjective there exists $u \in E$ such that $Bu = g$, then

$$(f, g) = (f - Au, 0) + (Au, Bu)$$

which implies that the map (35) is surjective. Hence $\operatorname{codim} \operatorname{im}(A, B) \leqslant \operatorname{codim} \operatorname{im} A_0$ and so (ii) implies (i). In general we see that $\operatorname{im}(A, B)$ and $\operatorname{im} A_0$ have the same codimension, that is, either both dimensions are infinite or both are finite and equal. The last statement is now clear.

We can also prove a more general result. Recall that for any operator A, we denote the dimension of its kernel by α_A and the codimension of its image by β_A, and the index is defined by $\operatorname{ind} A = \alpha_A - \beta_A$.

Lemma 16.14 *Let E, E_1 and E_2 be Banach spaces. Let $A: E \to E_1$ and $B: E \to E_2$ be bounded linear operators. Let $A_0 = A|_{\ker B}$ and $B_0 = B|_{\ker A}$. Then the following are equivalent:*

(i) *The image of the operator $\mathfrak{L} = (A, B): E \to E_1 \times E_2$ has finite codimension;*
(ii) *The images of the operator A and of B_0 have finite codimension;*
(iii) *The images of the operator A_0 and of B have finite codimension.*

If any of (i), (ii), or (iii) hold then

$$\beta_{\mathfrak{L}} = \beta_A + \beta_{B_0} = \beta_{A_0} + \beta_B. \tag{36}$$

Since $\alpha_{\mathfrak{L}} = \alpha_{A_0} = \alpha_{B_0}$ it follows that if any one of the operators $\mathfrak{L}$, A_0 and B_0 is a Fredholm operator then all three are Fredholm and

$$\operatorname{ind} \mathfrak{L} = \operatorname{ind} B_0 - \beta_A = \operatorname{ind} A_0 - \beta_B.$$

Proof There exist vector spaces $M_0 \subset \operatorname{im} A$ and $N_0 \subset \operatorname{im} B$ such that

$$\operatorname{im} A = \operatorname{im} A_0 \oplus M_0, \qquad \operatorname{im} B = \operatorname{im} B_0 \oplus N_0 \tag{37}$$

where $\oplus$ denotes a direct sum of vector spaces. Let

$$\Gamma := (M_0 \times N_0) \cap \operatorname{im} \mathfrak{L}.$$

We claim the following:

(a) Γ is the graph of an invertible operator $M_0 \to N_0$;
(b) $\operatorname{im} \mathfrak{L} = (\operatorname{im} A_0 \times \operatorname{im} B_0) \oplus \Gamma$;
(c) $\operatorname{im} A \times \operatorname{im} B = \operatorname{im} \mathfrak{L} \oplus \check{\Gamma}$, where $\check{\Gamma} = \{(f, -g); (f, g) \in \Gamma\}$.

Proof of (a). First we show that Γ is the graph of an operator $M_0 \to N_0$, i.e. we must show that for each $f \in M_0$ there is a unique $g \in N_0$ such that

$$(f, g) \in \Gamma. \tag{*}$$

Existence: Since $M_0 \subset \operatorname{im} A$, we have $f = Au$ for some $u \in E$. Then $Bu = Bu_0 + g$ for some $u_0 \in \ker A$ and $g \in N_0$. Hence

$$(f, g) = (A(u - u_0), B(u - u_0)) \in \operatorname{im} \mathfrak{L},$$

i.e. (*) holds. Uniqueness: If $(f, g) \in \Gamma$ and $(f, g') \in \Gamma$ then $(0, g - g') \in \Gamma$. This means that

$$(0, g - g') = (Au, Bu), \qquad \text{where } Au \in M_0, Bu \in N_0,$$

whence $Au = 0$, $Bu = g - g'$, so

$$g - g' \in \operatorname{im} B_0 \cap N_0 = 0$$

and therefore $g = g'$. So far we have only used the first direct sum in (37). But now if we use the second sum it follows in the same way that Γ is the graph of an operator $N_0 \to M_0$. Hence the operator defined by Γ is invertible.

Proof of (b). In view of (37) we have

$$\operatorname{im} A \times \operatorname{im} B = (\operatorname{im} A_0 \times \operatorname{im} B_0) \oplus (M_0 \times N_0). \tag{38}$$

Further,

$$\operatorname{im} A_0 \times \operatorname{im} B_0 \subset \operatorname{im} \mathfrak{L}, \tag{39}$$

for if $Bu = 0$ and $Au' = 0$ then $(Au, Bu') = (A(u + u'), B(u + u')) \in \operatorname{im} \mathfrak{L}$. Now (b) follows immediately by intersecting (38) with the subspace $\operatorname{im} \mathfrak{L}$ (recall the definition of Γ).

Proof of (c). First of all let T: $M_0 \to N_0$ denote the operator defined by Γ, i.e. $Tf = g$ if and only if (*) holds. Then $\Gamma = \Gamma(T) = \{(f, Tf); f \in M_0\}$ and it follows that

$$M_0 \times N_0 = \Gamma \oplus \check{\Gamma}, \tag{40}$$

for if $(f, g) \in M_0 \times N_0$, then $(f, g) = (f_1, Tf_1) + (f_2, -Tf_2)$, where $f_1 := \frac{1}{2}(f + T^{-1}g)$ and $f_2 := \frac{1}{2}(f - T^{-1}g)$. Now (c) is evident from (38), (40) and (b).

Now, finally, to complete the proof of the lemma, let $k = \dim M_0 = \dim N_0 = \dim \Gamma = \dim \check{\Gamma}$. From (37) we have

$$\beta_{A_0} = k + \beta_A, \qquad \beta_{B_0} = k + \beta_B.$$

Let $E_1 = \text{im } A \oplus M$, $E_2 = \text{im } B \oplus N$. Then by virtue of (c), we have

$$E_1 \times E_2 = (\text{im } A \times \text{im } B) \oplus (M \times N) = \text{im } \mathfrak{L} \oplus \check{\Gamma} \oplus (M \times N),$$

whence

$$\begin{aligned}\beta_{\mathfrak{L}} &= \dim \check{\Gamma} + \dim M + \dim N \\ &= k + \beta_A + \beta_B,\end{aligned}$$

so the equalities in (36) hold. The equivalence of (i), (ii) and (iii) is now evident. The last statement in the lemma follows immediately.

The next theorem is an index formula for the Dirichlet problem for elliptic operators of order $2s$.

Theorem 16.15 *Suppose that $\mathscr{A} \in Ell^{2s}$ has real matrix coefficients. If the Dirichlet boundary operator $\mathscr{D} = \{I, I\,\partial/\partial n, \ldots, I(\partial/\partial n)^{s-1}\}$ satisfies the L-condition then*

$$\text{ind}(\mathscr{A}, \mathscr{D}) = \frac{-1}{\pi}[\arg \det \Xi_+(y)]_{\partial\Omega} \tag{41}$$

where $\Xi_+ = \text{col}(X_+ T_+^j)_{j=0}^{s-1}$, and $(X_+(x), T_+(x))$ is a C^∞ matrix γ^+-spectral pair of $L_x(\lambda)$, $x \in \bar{\Omega}$.

Proof The Dirichlet boundary operator is

$$\mathscr{D} = \text{col}\left(I \frac{\partial^j}{\partial n^j}\right)_{j=0}^{s-1}.$$

Since $(\mathscr{A}, \mathscr{D})$ satisfies the L-condition, we have $\det \Xi_+(y) \neq 0$ for all $y \in \partial\Omega$ (see Theorem 10.5). We now increase the order of differentiation in each row (by means of tangential derivatives) so that each row has order $s - 1$:

$$\mathscr{D}' = \text{col}\left(I \frac{\partial^{s-1}}{\partial \tau^{s-1-j}\,\partial n^j}\right)_{j=0}^{s-1}. \tag{42}$$

We have

$$\pi\mathscr{D}'(y, (1, \lambda)) = \text{col}((\tau_1 + \lambda\tau_2)^{s-1-j}(-\tau_2 + \lambda\tau_1)^j)_{j=0}^{s-1} = \sum_{j=0}^{s-1} B_j'(y)\lambda^j,$$

where $\det[B_0'(y)\cdots B_{s-1}'(y)] = 1$ (see Lemma 2.23). Now, since

$$\Delta_{\mathscr{D}'}^+(y) = \sum_{j=0}^{s-1} B_j'(y) X_+(y) T_+^j(y) = [B_0'(y)\cdots B_{s-1}'(y)]\cdot\Xi_+(y),$$

then $\det \Delta_{\mathscr{D}'}^+(y) = \det \Xi_+(y)$, and, in view of Theorem 16.11 with $l = 2s$ and $\mu = s - 1$, we see that

$$\text{ind}(\mathscr{A}, \mathscr{D}') = \frac{-1}{\pi}[\arg \det \Xi_+(y)]_{\partial\Omega}$$

Since $\text{ind}(\mathscr{A}, \mathscr{D}) = \text{ind}(\mathscr{A}, \mathscr{D}')$, the formula (41) holds.

When the orders of the rows of the boundary operator are not necessarily the same, Theorem 16.11 cannot be applied directly. However, the next theorem shows that the index can be written in terms of

$$\tilde{\Delta}_{\mathscr{B}}^{+}(y) = \sum_{j=0}^{\mu} \tilde{B}_j(y)\bar{X}_+(y)\tilde{T}_+^j(y),$$

where $(\bar{X}_+(y), \bar{T}_+(y))$ is the γ^+-spectral pair defined by the transformation (14) in §15.3. In this form the index formula holds for any (differential) boundary operator $\mathscr{B}$.

Theorem 16.16 *Let $\mathscr{A} \in Ell^l$ and let $\mathscr{B}$ be a differential boundary operator satisfying the L-condition (with no restrictions on the orders m_k). If $\mathscr{A}$ and $\mathscr{B}$ have real matrix coefficients then*

$$\operatorname{ind}(\mathscr{A}, \mathscr{B}) = \frac{-1}{\pi}[\arg\det \tilde{\Delta}_{\mathscr{B}}^{+}(y)]_{\partial\Omega} + r(2 - l) \tag{43}$$

In other words, $\operatorname{ind}(\mathscr{A}, \mathscr{B}) = \operatorname{ind}_s(\mathscr{A}, \mathscr{B})$ *(see Definition* 15.15).

Proof Let m_k, $k = 1, \ldots, r$, denote the order of the kth row of $\mathscr{B}$ and let $\mu = \max(m_k)$. Now let

$$\mathscr{B}'(y, \partial/\partial x) = [\delta_{kj}(\partial/\partial\tau)^{\mu - m_k}]_{j,k=1}^{r} \circ \mathscr{B}(y, \partial/\partial x)$$

so that $\mathscr{B}'$ has the same order m in each row. Then

$$\pi\mathscr{B}'(y, \xi) = [\delta_{kj}(\tau\cdot\xi)^{\mu - m_k}]_{j,k=1}^{r} \cdot \pi\mathscr{B}(y, \xi),$$

thus $\pi\mathscr{B}'(y, \tau + \lambda n) \equiv \pi\mathscr{B}(y, \tau + \lambda n)$. It follows that $\mathscr{B}'$ satisfies the L-condition. Hence $(\mathscr{A}, \mathscr{B}') \in BE^{l,\mu}$ and by Theorem 16.11

$$\operatorname{ind}(\mathscr{A}, \mathscr{B}') = \frac{-1}{\pi}[\arg\det \Delta_{\mathscr{B}'}^{+}(y)]_{\partial\Omega} + r(2\mu + 2 - l) \tag{44}$$

By Proposition 15.12 we also have

$$\tilde{\Delta}_{\mathscr{B}'}^{+} = \Delta_{\mathscr{B}'}^{+}\cdot(\tau_1 I + \tau_2 T_+)^{-\mu}$$

and, since $\frac{1}{2\pi}[\arg\det(\tau_1 I + \tau_2 T_+)]_{\partial\Omega} = r$, it follows that

$$\frac{-1}{2\pi}[\arg\det \tilde{\Delta}_{\mathscr{B}'}^{+}]_{\partial\Omega} = \frac{-1}{2\pi}[\arg\det \Delta_{\mathscr{B}'}^{+}]_{\partial\Omega} + \mu r \tag{45}$$

In view of (44) and (45), the formula (43) holds for the boundary value problem $(\mathscr{A}, \mathscr{B}')$. Since $\tilde{\Delta}_{\mathscr{B}'}^{+} = \tilde{\Delta}_{\mathscr{B}}^{+}$ and $\operatorname{ind}(\mathscr{A}, \mathscr{B}') = \operatorname{ind}(\mathscr{A}, \mathscr{B})$, it also holds for $(\mathscr{A}, \mathscr{B})$.

16.5 The index formula for elliptic systems with complex coefficients and when the boundary operator is pseudo-differential

First we extend Theorems 16.11 and 16.16 to the case when the boundary operator is differential with complex coefficients.

Theorem 16.17 *If* $(\mathscr{A}, \mathscr{B}) \in BE^{l,\mu}$

$$\operatorname{ind}(\mathscr{A}, \mathscr{B}) = \frac{-1}{2\pi}[\arg\det \Delta^+_{\mathscr{B}}(y)]_{\partial\Omega} + \frac{1}{2\pi}[\arg\det \Delta^-_{\mathscr{B}}(y)]_{\partial\Omega} + r(2\mu + 2 - l) \tag{46}$$

Proof Let $(\mathscr{A}_{\mathbb{R}}, \mathscr{B}_{\mathbb{R}})$ and $L_{\mathbb{R}}(\lambda)$ denote the real boundary value problem operator and real matrix polynomial associated with $(\mathscr{A}, \mathscr{B})$ and $L(\lambda)$, respectively (see §15.4 and §3.8). By Theorem 3.20, we have a γ^+-spectral pair $(X^+_{\mathbb{R}}(x), T^+_{\mathbb{R}}(x))$ of $L_{\mathbb{R},x}(\lambda)$ defined in terms of $(X_\pm(x), T_\pm(x))$. Then, by virtue of Proposition 15.20,

$$\Delta^+_{\mathscr{B}_{\mathbb{R}}}(y) = \begin{pmatrix} I_r & I_r \\ -iI_r & iI_r \end{pmatrix} \begin{pmatrix} \Delta^+_{\mathscr{B}}(y) & \\ & \overline{\Delta^-_{\mathscr{B}}(y)} \end{pmatrix}$$

where I_r denotes the $r \times r$ identity matrix. By Theorem 16.11

$$\operatorname{ind}(\mathscr{A}_{\mathscr{R}}, \mathscr{B}_{\mathscr{R}}) = \frac{-1}{\pi}[\arg\det \Delta^+_{\mathscr{B}_{\mathbb{R}}}(y)]_{\partial\Omega} + 2r(2\mu + 2 - l),$$

and, since $\operatorname{ind}(\mathscr{A}_{\mathbb{R}}, \mathscr{B}_{\mathbb{R}}) = 2\cdot\operatorname{ind}(\mathscr{A}, \mathscr{B})$, the formula (46) follows immediately.

Remarks

(1) Observe that the formula (46) is not affected if the boundary operator $\mathscr{B}(y, \partial/\partial x)$ is replaced by $M(y)\mathscr{B}(y, \partial/\partial x)$, where $M(y)$ is an invertible $r \times r$ matrix function.

(2) If the operators $\mathscr{A}$ and $\mathscr{B}$ have real matrix coefficients then the formula (46) is the same as (25) because we may let $(X_-, T_-) = (\overline{X}_+, \overline{T}_+)$ and therefore $\Delta^-_{\mathscr{B}}(y) = \overline{\Delta^+_{\mathscr{B}}(y)}$.

Theorem 16.18 *If* $\mathscr{A} \in Ell^l$ *and* $\mathscr{B}$ *is any differential boundary operator satisfying the L-condition (with no restrictions on the orders* m_k*) then*

$$\operatorname{ind}(\mathscr{A}, \mathscr{B}) = \frac{-1}{\pi}[\arg\det \tilde{\Delta}^+_{\mathscr{B}}(y)]_{\partial\Omega} + \frac{1}{2\pi}[\arg\det \tilde{\Delta}^-_{\mathscr{B}}(y)]_{\partial\Omega} + r(2 - l) \tag{47}$$

Proof This is derived from Theorem 16.16 in the same manner as above.

We now wish to consider general boundary operators, i.e. $\mathscr{B} = \sum_{j=0}^{\mu} \mathscr{B}_j D_n^j$ where the coefficients $\mathscr{B}_j$ are pseudo-differential operators on $\partial\Omega$. In its present form, the formula (47) does not hold for such boundary operators;

however, if we define a γ^+-spectral triple of $L_{y,\xi'}(\lambda)$ as in §15.1:

$$(X_+(y,\xi'), T_+(y,\xi')) = \begin{cases} (\tilde{X}_+(y), \tilde{T}_+(y)), & \xi' = \tau(y) \\ (\tilde{X}_-(y), -\tilde{T}_-(y)), & \xi' = -\tau(y) \end{cases} \tag{48}$$

then for a differential operator $\mathscr{B}$ we have

$$\Delta^+_{\mathscr{B}}(y, \tau(y)) = \tilde{\Delta}^+_{\mathscr{B}}(y) \qquad \text{and} \qquad \Delta^+_{\mathscr{B}}(y, -\tau(y)) = M(-1)\,\tilde{\Delta}^-_{\mathscr{B}}(y)$$

Hence (47) can be written in the form (49).

Theorem 16.19 *Let $\mathscr{A} \in Ell^l$ and let $\mathscr{B}$ be any boundary operator satisfying the L-condition. Then with $\Delta^+_{\mathscr{B}}$ defined with respect to the γ^+-spectral pair* (48) *we have*

$$\operatorname{ind}(\mathscr{A}, \mathscr{B}) = \frac{-1}{2\pi}[\arg\det \Delta^+_{\mathscr{B}}(y, \tau(y))]_{\partial\Omega} + \frac{1}{2\pi}[\arg\det \Delta^+_{\mathscr{B}}(y, -\tau(y))]_{\partial\Omega} + r(2-l) \tag{49}$$

Proof We have shown that (49) holds for $(\mathscr{A}, \mathscr{B}^\circ)$ if $\mathscr{B}^\circ$ is a differential operator satisfying the L-condition relative to $\mathscr{A}$ (for instance, let $\mathscr{B}^\circ$ be the boundary operator constructed in Theorem 15.14), that is,

$$\operatorname{ind}(\mathscr{A}, \mathscr{B}^\circ) = \frac{-1}{2\pi}[\arg\det \Delta^+_{\mathscr{B}^\circ}(y, \tau(y))]_{\partial\Omega} + \frac{1}{2\pi}[\arg\det \Delta^+_{\mathscr{B}^\circ}(y, -\tau(y))]_{\partial\Omega} + r(2-l).$$

Now that we have the formula for one boundary problem, $(\mathscr{A}, \mathscr{B}^\circ)$, then we can reduce the index problem for $(\mathscr{A}, \mathscr{B})$ to the calculation of an index on the boundary $\partial\Omega$, to which Noether's formula can be applied since $\partial\Omega$ is diffeomorphic to S^1. By Theorem 10.19 we have

$$\operatorname{ind}(\mathscr{A}, \mathscr{B}) = \operatorname{ind}(\mathscr{A}, \mathscr{B}^\circ) + \operatorname{ind} \Delta^+_{\mathscr{B}} \cdot (\Delta_{\mathscr{B}^\circ})^{-1}$$

and $\operatorname{ind} \Delta^+_{\mathscr{B}} \cdot (\Delta_{\mathscr{B}^\circ})^{-1} = \operatorname{ind} \Delta^+_{\mathscr{B}} - \operatorname{ind} \Delta_{\mathscr{B}^\circ}$. Thus, if we apply Theorem 8.83 (Noether's formula) to p.d.o.'s with principal symbols $\Delta^+_{\mathscr{B}}$ and $\Delta^+_{\mathscr{B}^\circ}$ on $ST^*(\partial\Omega)$, respectively, then (49) follows at once.

The next theorem gives a formula for the difference between the index of a given boundary value problem for $\mathscr{A}$ and the index of the Dirichlet problem. Here $\mathscr{A}$ is a properly elliptic operator of order $l = 2s$.

In order to state this result some notation is needed. For a boundary operator $\mathscr{B}$ we let

$$G_+(y, \xi') = \frac{1}{2\pi i}\int_+ B_{y,\xi'}(\lambda) L^{-1}_{y,\xi'}(\lambda)[I \cdots \lambda^{s-1} I]\, d\lambda$$

According to Theorem 10.6, the boundary value problem $(\mathscr{A}, \mathscr{B})$ satisfies the L-condition if and only if $\det G_+(y, \xi') \neq 0$ for all $y \in \partial\Omega$, $0 \neq \xi' \in T^*_y(\partial\Omega)$. The substitution $\lambda \to c^{-1}\lambda$ implies

$$G_+(y, c\xi') = M(c)G_+(y, \xi')F(c)c^{1-l} \qquad \text{when } c > 0$$

where $M(c)$ is defined as usual and $F(c) = [c^j \delta_{kj} I]_{k,j=0}^{s-1}$. If $\mathscr{B}$ is a differential operator we also have (with G_- defined by the integral $\int_-$)

$$G_+(y, -\xi') = M(-1)G_-(y, \xi')F(-1)(-1)^{1-l} \tag{50}$$

For the Dirichlet problem the matrix G_+ takes the form

$$M^{\pm}_{ss}(y, \xi') = \frac{1}{2\pi i}\int_{\pm} \begin{pmatrix} I \\ \vdots \\ \lambda^{s-1}I \end{pmatrix} L^{-1}_{y,\xi'}(\lambda)[I \cdots \lambda^{s-1}I]\, d\lambda$$

Note that $M^+_{ss} = -M^-_{ss}$ because the integrand is $O(\lambda^{-2})$ as $|\lambda| \to \infty$. An application of Cauchy's theorem then gives $\int_{\gamma+} = \int_{-\infty}^{\infty}$ and $\int_{\gamma-} = -\int_{-\infty}^{\infty}$. As usual, γ^+ and γ^- are simple, closed contours in the upper and lower half-planes containing the eigenvalues of $L_{y,\xi'}(\lambda)$ there, oriented in the counter-clockwise direction.

The following theorem is a particular case of the formula of Agranovič–Dynin (Theorem 10.19).

Theorem 16.20 *Let $\mathscr{A} \in Ell^{2s}$ and let $\mathscr{B}$ be a boundary operator satisfying the L-condition. If the Dirichlet problem $(\mathscr{A}, \mathscr{D})$ satisfies the L-condition then*

$$\operatorname{ind}(\mathscr{A}, \mathscr{B}) = \operatorname{ind}(\mathscr{A}, \mathscr{D}) - \frac{1}{2\pi}[\arg\det G_+(y, \tau(y))]_{\partial\Omega}$$
$$+ \frac{1}{2\pi}[\arg\det G_+(y, -\tau(y))]_{\partial\Omega}$$

Hence $\operatorname{ind}(\mathscr{A}, \mathscr{B}) = \operatorname{ind}(\mathscr{A}, \mathscr{D}) + \operatorname{ind}\mathscr{G}_+$, *where $\mathscr{G}_+$ is a p.d.o. on $\partial\Omega$ with principal symbol G_+.*

Proof Let $(X_+(y, \xi'), T_+(y, \xi'))$ be the γ^+-spectral pair of $L_{y,\xi'}(\lambda)$ defined as in (48). Let $\Delta_+ = \Delta^+_{\mathscr{B}}$ and $\Xi_+ = \operatorname{col}(X_+ T^j_+)_{j=0}^{s-1}$. By Theorem 16.19 we have

$$\operatorname{ind}(\mathscr{A}, \mathscr{B}) = \frac{-1}{2\pi}[\arg\det \Delta_+(y, \tau(y))]_{\partial\Omega}$$
$$+ \frac{1}{2\pi}[\arg\det \Delta_+(y, -\tau(y))]_{\partial\Omega} + r(2 - l)$$

and

$$\text{ind}(\mathscr{A}, \mathscr{D}) = \frac{-1}{2\pi}[\arg\det \Xi_+(y, \tau(y))]_{\partial\Omega}$$

$$+\frac{1}{2\pi}[\arg\det \Xi_+(y, -\tau(y))]_{\partial\Omega} + r(2-l)$$

Hence $\text{ind}(\mathscr{A}, \mathscr{B}) = \text{ind}(\mathscr{A}, \mathscr{D}) + \kappa$, where

$$\kappa = \frac{-1}{2\pi}[\arg\det R(y, \tau(y))]_{\partial\Omega} + \frac{1}{2\pi}[\arg\det R(y, -\tau(y))]_{\partial\Omega}$$

and $R = \Delta_+ \cdot \Xi_+^{-1}$. Now, observe that

$$\begin{aligned} G_+ &= \Delta_+ \cdot [Y_+ \cdots T_+^{s-1} Y_+] \\ &= \Delta_+ \Xi_+^{-1} \Xi_+ \cdot [Y_+ \cdots T_+^{s-1} Y_+] \\ &= R \cdot M_{ss}^+ \end{aligned}$$

As mentioned above, $M_{ss}^+ = -M_{ss}^-$; thus (50) implies that $\det M_{ss}^+(y, -\tau(y))$ and $\det M_{ss}^+(y, \tau(y))$ are equal except for a constant factor ± 1. Hence they have the same winding number and

$$\kappa = \frac{-1}{2\pi}[\arg\det G_+(y, \tau(y))]_{\partial\Omega} + \frac{1}{2\pi}[\arg\det G_+(y, -\tau(y))]_{\partial\Omega}.$$

The last statement of the theorem now follows from Noether's formula for the index of p.d.o.'s on $\partial\Omega$ (Theorem 8.83).

Remark Due to §3.6, $L_y(\lambda)$ has a monic γ^+-spectral right divisor, $L_y^+(\lambda)$, of degree s. A right division by $L_y^+(\lambda)$ gives $B_y(\lambda) = Q_y(\lambda)L_y^+(\lambda) + R_y(\lambda)$ where $R_y(\lambda) = \sum_{j=0}^{s-1} R_j\lambda^j$ and

$$[R_0(y) \cdots R_{s-1}(y)] = \Delta_+(y) \cdot \{\text{col}(X_+(y)T_+^j(y))_{j=0}^{s-1}\}^{-1}$$

This is just the equation $R = \Delta_+ \Xi_+^{-1}$ in the proof above.

Corollary 16.21 *In addition to the hypotheses of Theorem* 16.20, *suppose that* $\mathscr{B}$ *is a differential operator and has order* $\leqslant s-1$. *Then*

$$\text{ind}(\mathscr{A}, \mathscr{B}) = \text{ind}(\mathscr{A}, \mathscr{D}).$$

Proof We define

$$\tilde{G}_\pm(y) = \frac{1}{2\pi i}\int_\pm \tilde{B}_y(\lambda)\tilde{L}_y^{-1}(\lambda)[I \cdots \lambda^{s-1}I]\, d\lambda$$

and note that $G_+(y, \tau(y)) = \tilde{G}_+(y)$. If $\mathscr{B}$ is a differential operator we also have $G_+(y, -\tau(y)) = M(-1)\cdot\tilde{G}_-(y)\cdot F(-1)(-1)^{1-l}$. Thus Theorem 16.20 can also be stated as follows:

$$\text{ind}(\mathscr{A}, \mathscr{B}) = \text{ind}(\mathscr{A}, \mathscr{D}) - \frac{1}{2\pi}[\arg\det \tilde{G}_+(y)]_{\partial\Omega} + \frac{1}{2\pi}[\arg\det \tilde{G}_-(y)]_{\partial\Omega}.$$

Since the order of $\mathscr{B}$ is $\leqslant s-1$, the terms in the integrand of $\tilde{G}_\pm$ are $O(\lambda^{-2})$ and we have

$$\tilde{G}_+(y) - \tilde{G}_-(y) = \frac{1}{2\pi i}\int_\Gamma \tilde{B}_y(\lambda)\tilde{L}_y^{-1}(\lambda)[I \cdots \lambda^{s-1}I]\, d\lambda = 0$$

where Γ is a simple closed contour containing all the roots of det $\tilde{L}_y(\lambda)$ in its interior. Hence $\tilde{G}_+(y) = \tilde{G}_-(y)$ which implies $\text{ind}(\mathscr{A}, \mathscr{B}) = \text{ind}(\mathscr{A}, \mathscr{D})$.

Exercises

1. Let $T: X \to Y$ be a Fredholm operator, where X and Y are Banach spaces over $\mathbb{R}$. Show that

$$\text{ind}(T_\mathbb{C}) = \text{ind}\, T$$

where $T_\mathbb{C}: X \otimes \mathbb{C} \to Y \otimes \mathbb{C}$ denotes the complexification of T defined by $v \otimes z \mapsto Tv \otimes z$, $v \in X$, $z \in \mathbb{C}$.

2. Let X and Y be Banach spaces over $\mathbb{R}$, and let $S: X \otimes \mathbb{C} \to Y \otimes \mathbb{C}$ be a ($\mathbb{C}$-linear) Fredholm operator. Show that

$$\text{ind}(S_\mathbb{R}) = 2\cdot \text{ind}\, S$$

where $S_\mathbb{R}: X \oplus X \to Y \oplus Y$ denotes the realization of S. Recall from §5.1 that the realization of S is defined as follows: If we write $S(v \otimes 1) = (S_1 v) \otimes 1 + (S_2 v) \otimes i$, where $S_i: X \to Y$, then

$$S_\mathbb{R} = \begin{pmatrix} S_1 & -S_2 \\ S_2 & S_1 \end{pmatrix}$$

3. Let $B(\lambda) = \sum_{j=0}^{\mu} B_j \lambda^j$ be an $r \times p$ matrix polynomial of degree μ and let $B'(\lambda) = (c\lambda + d)^q B(\lambda)$ where $c, d \in \mathbb{C}$ and q is any nonnegative integer. Let X be a $p \times r$ matrix and T an $r \times r$ matrix. Show that

$$\sum_{j=0}^{\mu+q} B'_j X T^j = \sum_{j=0}^{\mu} B_j X T^j \cdot (cT + dI)^q,$$

where $B'(\lambda) = \sum_{j=0}^{\mu+q} B'_j \lambda^j$.

4. Consider a system of p Laplace operators

$$\mathscr{A} = I\, \partial^2/\partial x_1^2 + I\, \partial^2/\partial x_2^2,$$

with boundary operator $\mathscr{B} = B_1(y)\, \partial/\partial x_1 + B_2(y)\, \partial/\partial x_2$.

(a) Show that the L-condition for $(\mathscr{A}, \mathscr{B})$ is $\det(B_1(y) + iB_2(y)) \neq 0$ for all $y \in \partial\Omega$.

(b) Using the fact that the Dirichlet problem for the Laplace operator has index 0, show that

$$\text{ind}(\mathscr{A}, \mathscr{B}) = \frac{-1}{\pi}[\arg \det(B_1(y) + iB_2(y))]_{\partial\Omega} + 2p.$$

The following is a generalization of the preceding exercise.

5. Let $\mathscr{A}$ be a real $p \times p$ second order elliptic operator such that

$$\det \int_{+} L_x^{-1}(\lambda)\, d\lambda \neq 0 \qquad \text{for all } x \in \bar{\Omega},$$

where $L_x(\lambda) = \pi\mathscr{A}(x, (1, \lambda))$. Show that

(a) There exists a γ^+-spectral right divisor, $I\lambda - S_x$, of $L_x(\lambda)$ for all $x \in \bar{\Omega}$ such that S_x depends smoothly on x.
(b) The pair $(X_+(x), T_+(x)) = (I, S_x)$ is a γ^+-spectral pair of $L_x(\lambda)$ and the Dirichlet problem satisfies the L-condition and has index 0.
(c) The boundary operator $\mathscr{B} = B_1(y)\, \partial/\partial x_1 + B_2(y)\, \partial/\partial x_2$ satisfies the L-condition if and only $\det(B_1(y) + B_2(y) \cdot S_y) \neq 0$ for all $y \in \partial\Omega$.
(d) If $(\mathscr{A}, \mathscr{B})$ satisfies the L-condition, then

$$\operatorname{ind}(\mathscr{A}, \mathscr{B}) = \frac{-1}{\pi} [\arg \det(B_1(y) + B_2(y) \cdot S_y)]_{\partial\Omega} + 2p$$

6. Let $\mathscr{A}$ be a (properly) elliptic operator of order $2s$. If the Dirichlet boundary operator $\mathscr{D} = \{I, I\, \partial/\partial n, \dots, I(\partial/\partial n)^{s-1}\}$ satisfies the L-condition, show that

$$\operatorname{ind}(\mathscr{A}, \mathscr{D}) = \frac{-1}{2\pi} [\arg \det \Xi_+(y)]_{\partial\Omega} + \frac{1}{2\pi} [\arg \det \Xi_-(y)]_{\partial\Omega}$$

where $\Xi_\pm = \operatorname{col}(X_\pm T_\pm^j)_{j=0}^{s-1}$, and $(X_+(x), T_+(x))$ is any smooth matrix γ^+-spectral pair of $L_x(\lambda)$, $x \in \bar{\Omega}$.

Addendum

Let U denote the unit disc $|z| < 1$ and $\mathbb{T}$ the boundary $|z| = 1$. Let $g \in C(\partial U)$ be a continuous real-valued function on $\mathbb{T}$. Consider the problem of finding an analytic function $w = u - iv$ in U such that its (harmonic) real part, u, has prescribed values,

$$\frac{\partial w}{\partial \bar{z}} = 0 \quad \text{in } U, \qquad \operatorname{Re} w = g \quad \text{on } \mathbb{T}. \tag{51}$$

It is well known that all solutions of this problem are given by

$$w(z) = \frac{1}{2\pi} \int_0^{2\pi} \frac{e^{it} + z}{e^{it} - z} g(e^{it})\, dt + ik, \tag{52}$$

where k is a real constant. Formally, this result can be obtained by looking at the Fourier expansion of g on the unit circle $\mathbb{T}$,

$$g(e^{it}) = \sum_{n \in \mathbb{Z}} \hat{g}(n)\, e^{int},$$

where the Fourier coefficients are defined by

$$\hat{g}(n) = \frac{1}{2\pi}\int_0^{2\pi} g(e^{it})\, e^{-int}\, dt$$

Since g has real values then $\overline{\hat{g}(n)} = \hat{g}(-n)$ and the Fourier series for g takes the form

$$g(e^{it}) = \operatorname{Re}\left(\hat{g}(0) + 2\sum_{n=0}^{\infty} \hat{g}(n)\, e^{int}\right).$$

Thus we let

$$w(z) = \frac{1}{2\pi}\int_0^{2\pi} S(z, e^{it})g(e^{it})\, dt, \qquad |z| \leqslant 1, \tag{53}$$

where the Schwarz kernel S is

$$S(z, e^{it}) = -1 + 2\sum_{n=0}^{\infty} z^n e^{-int} = \frac{e^{it}+z}{e^{it}-z}, \qquad |z| < 1. \tag{54}$$

Exercises (continued)

7. Let $g \in C(\partial U)$. Show that all solutions of the problem (51) have the form (52). Hint: First of all, verify that the function defined by (53) solves the problem (51). See [Ru 1], pp. 233–235.

8. As usual let Ω be a bounded and simply connected region in $\mathbb{R}^2$ with C^∞ boundary. Show that the elliptic boundary problem operator

$$\mathfrak{L}_0\colon W_2^1(\Omega) \to L_2(\Omega) \times W_2^{1/2}(\partial R, \mathbb{R}) \tag{55}$$

defined by

$$w \mapsto (\partial w/\partial \bar{z}, \operatorname{Re} w|_{\partial\Omega})$$

is surjective and has a kernel of dimension 1. Conclude that

$$\operatorname{ind} \mathfrak{L}_0 = 1.$$

Hint: By the Riemann mapping theorem (see [Ru 1], p. 283) there exists a one-to-one conformal mapping ψ of Ω onto the unit disc U. It can be shown that ψ extends to a homeomorphism of $\bar{\Omega}$ onto $\bar{U}$, and is in fact smooth up to the boundary; see [Ru 1] and [War].

The formula (2) for the index of the Riemann–Hilbert problem (1′) can now be obtained from the known index of (55) and Theorem 10.19 (Agranovič–Dynin).

9. Let $B = [\cos\varphi \quad -\sin\varphi]$ and $\varphi \in C^\infty(\partial\Omega)$. Show that the index of the operator

$$\mathfrak{L}\colon W_2^1(\Omega, \mathbb{R}^2) \to L_2(\Omega, \mathbb{R}^2) \times W_2^{1/2}(\partial\Omega, \mathbb{R})$$

defined by the left-hand side of the equations in (1), i.e.

$$u \mapsto (Au, Bu) = \left(\frac{\partial u}{\partial x_2} - \begin{pmatrix} 0 & -1 \\ 1 & 0 \end{pmatrix} \frac{\partial u}{\partial x_1} + A_0(x) \cdot u,\ Bu \right),$$

is given by the formula

$$\operatorname{ind} \mathfrak{L} = 1 - 2\chi = 1 - \frac{1}{\pi} [\varphi(y)]_{\partial\Omega}. \tag{56}$$

Hint: The matrix polynomial $L(\lambda) = I\lambda - \begin{pmatrix} 0 & -1 \\ 1 & 0 \end{pmatrix}$ associated to the elliptic operator A has a γ^+-spectral pair $X_+ = [i \quad -i]^T$, $T_+ = i$. By virtue of §15(14) the matrix polynomial $\tilde{L}(\lambda) = \pi A(y, \tau + \lambda n)$ also has γ^+-spectral pair

$$\tilde{X}_+ = \begin{pmatrix} 1 \\ -i \end{pmatrix}, \qquad \tilde{T}_+ = i.$$

Then $\tilde{\Delta}_B^+ = B\tilde{X}_+ = e^{i\varphi}$. Similarly, with an appropriate choice of γ^- spectral pair we have $\tilde{\Delta}_B^- = e^{-i\varphi}$. By virtue of the discussion at the beginning of §15.1 we obtain a γ^+-spectral pair (X_+, T_+) defined on $ST^*(\partial\Omega)$ such that

$$\Delta_B^+(y, \xi') = \begin{cases} e^{i\varphi(y)}, & \text{if } \xi' = \tau(y) \\ e^{-i\varphi(y)}, & \text{if } \xi' = -\tau(y) \end{cases}$$

Now conclude from Theorem 10.19 that

$$\operatorname{ind} \mathfrak{L} = \operatorname{ind} \mathfrak{L}_0 + \operatorname{ind} Q,$$

where Q is an elliptic p.d.o. on $\partial\Omega$ with principal symbol Δ_B^+. It follows by Noether's formula (Theorem 8.82) that $\operatorname{ind} Q = -\chi - \chi$ where $\chi = \frac{1}{2\pi} [\varphi(y)]_{\partial\Omega}$.

17
Elliptic systems with 2 × 2 real coefficients

17.1 Homotopy classification

Let $\mathscr{E} = \text{Ell}^l$ denote the set of 2 × 2 elliptic differential operators in the plane with principal parts that are homogeneous of degree l and with constant real matrix coefficients. By means of the coefficients of the operators we can regard $\mathscr{E}$ as a subset of $\mathbb{R}^N$ for some N, and thus $\mathscr{E}$ inherits a natural topology. In this section we characterize the homotopy classes of $\mathscr{E}$.

The assumption of constant coefficients is just a convenience; any elliptic operator on a simply connected region $\bar{\Omega}$ is homotopic to one with constant coefficients (see the proof of Theorem 16.11).

We first make some definitions that are needed in the proof of Theorem 17.2. If $\mathscr{A} \in \mathscr{E}$ then

$$L(\lambda) = \pi\mathscr{A}(1, \lambda) = \begin{pmatrix} a_{11}(\lambda) & a_{12}(\lambda) \\ a_{21}(\lambda) & a_{22}(\lambda) \end{pmatrix}, \tag{1}$$

is a 2 × 2 real matrix polynomial of degree l. Let

$$\alpha(\lambda) = (a_{11}(\lambda) + a_{22}(\lambda)) + i(a_{21}(\lambda) - a_{12}(\lambda)),$$
$$\beta(\lambda) = (a_{11}(\lambda) - a_{22}(\lambda)) + i(a_{21}(\lambda) + a_{12}(\lambda)),$$

and note that for all $\lambda \in \mathbb{R}$

$$|\alpha(\lambda)|^2 - |\beta(\lambda)|^2 = 4 \det L(\lambda) \tag{2}$$

Since $\det L(\lambda) \neq 0$, $\lambda \in \mathbb{R}$, then either

(a) $\det L(\lambda) > 0$, $\lambda \in \mathbb{R}$, or (b) $\det L(\lambda) < 0$, $\lambda \in \mathbb{R}$.

If (a) holds then (2) implies

$$|\alpha(\lambda)| > |\beta(\lambda)|, \qquad \lambda \in \mathbb{R} \tag{3}$$

In particular we have $\alpha(\lambda) \neq 0$ for all $\lambda \in \mathbb{R}$.

For those elliptic operators $\mathscr{A} \in \mathscr{E}$ such that $\det L(\lambda) > 0$, it is easily verified that $\alpha(\lambda)$ has degree l, and we let $\mathscr{E}_j$ ($j = 0, \ldots, l$) denote the $\mathscr{A}$'s such that the equation $\alpha(\lambda) = 0$ has exactly j roots in the upper half-plane, and $l - j$ in the lower half-plane.

Similarly, for operators $\mathscr{A} \in \mathscr{E}$ such that $\det L(\lambda) < 0$, there are $l + 1$ classes $\mathscr{E}'_j$ $(j = 0, \ldots, l)$ determined by the roots of $\beta(\lambda) = 0$. Note that case (b) can be transformed to (a) by interchanging the rows of $L(\lambda)$; this interchanges the roles of $\alpha(\lambda)$ and $\beta(\lambda)$.

Example 17.1 Let $\mathscr{A}$ be the real 2×2 operator corresponding to the Bitsadze operator $\partial^2/\partial x_1^2 + 2i\,\partial^2/\partial x_1\,\partial x_2 - \partial^2/\partial x_2^2$ (see Chapter 9, Exercise 6). Then we have

$$L(\lambda) = \pi\mathscr{A}(1, \lambda) = \begin{pmatrix} \lambda^2 - 1 & -2\lambda \\ 2\lambda & \lambda^2 - 1 \end{pmatrix}$$

and $\det L(\lambda) = (\lambda^2 + 1)^2 > 0$ for real λ. Also $\alpha(\lambda) = 2(\lambda - i)^2$ which has two roots in the upper half-plane, so that $\mathscr{A} \in \mathscr{E}_2$. If we take the conjugate of the Bitsadze operator we get an operator in $\mathscr{E}_0$.

The next theorem will show that $\mathscr{E}_j$ and $\mathscr{E}'_j$ are components (i.e. maximal connected subsets) of $\mathscr{E}$. Bojarski stated this result in [Bo 1] without proof.

Theorem 17.2 *The set* $\mathscr{E}$ *has* $2(l + 1)$ *homotopy classes, namely,* $\mathscr{E}_0, \ldots, \mathscr{E}_l$ *and* $\mathscr{E}'_0, \ldots, \mathscr{E}'_l$.

Proof It suffices to consider only those elliptic operators $\mathscr{A}$ for which $\det L(\lambda) > 0$. Also there is the homotopy $\mathscr{A} + t(\mathscr{A} - \pi\mathscr{A})$, $0 \leqslant t \leqslant 1$. Therefore we may restrict attention to the set of operators of the form $\pi\mathscr{A}$, where $\mathscr{A} \in \mathscr{E}_j$ for some $j = 0, \ldots, l$. Call this set $\mathscr{E}^+$.

Let $\mathscr{M}$ denote the set of matrix polynomials of degree l with invertible leading coefficient and $\det L(\lambda) > 0$, $\lambda \in \mathbb{R}$. Endowing $\mathscr{M}$ with the topology determined by the coefficients of the matrix polynomials, the map $\mathscr{A} \mapsto L(\lambda) = \mathscr{A}(1, \lambda)$ is a homeomorphism between the spaces $\mathscr{E}^+$ and $\mathscr{M}$, with inverse given by

$$L(\lambda) = \sum_{j=0}^{l} A_j \lambda^j \mapsto \mathscr{A}(\partial/\partial x) = \sum_{j=0}^{l} A_j \frac{\partial^l}{\partial x_1^{l-j}\,\partial x_2^j}$$

(Note that the assumption of invertible leading coefficient for $L(\lambda)$ is required in order that $\mathscr{A}(1, 0) = A_l$ be invertible.) Thus we can work within the space of matrix polynomials $\mathscr{M}$.

We prove the theorem in two steps. The first step is to show that if $\mathscr{A}^t$ is a homotopy of elliptic operators in $\mathscr{E}^+$ then $\mathscr{A}^t \in \mathscr{E}_j$ for a fixed j. Let $L^t(\lambda)$ be the corresponding homotopy of matrix polynomials in $\mathscr{M}$. In view of (3), we have $\alpha^t(\lambda) \neq 0$ for all $\lambda \in \mathbb{R}$ and $0 \leqslant t \leqslant 1$. Since $\mathscr{A}^0 \in \mathscr{E}_j$ then $\alpha^0(\lambda)$ has j roots above and $l - j$ roots below the real axis. By the argument principle in complex analysis, it follows that for sufficiently small t the number of roots of $\alpha^t(\lambda) = 0$ above the real axis is $\geqslant j$ and the number below is $\geqslant l - j$; and since $\alpha^t(\lambda)$ has degree l, equality must hold. By connectedness of $[0, 1]$ it follows that for the equation

$$\alpha^t(\lambda) = 0$$

the number of roots in the upper half-plane is constant, i.e. independent of t. This proves the first step.

The second step is to show that if $\mathscr{A}$, $\tilde{\mathscr{A}}$ are two operators belonging to the same $\mathscr{E}_j$ then they are homotopic, or, equivalently, the corresponding matrix polynomials, $L(\lambda)$ and $\tilde{L}(\lambda)$, are homotopic in $\mathscr{M}$. Let the entries of $L(\lambda)$ be denoted as in (1). Consider the homotopy

$$L^t(\lambda) = (1-t)\begin{pmatrix} a_{11}(\lambda) & a_{12}(\lambda) \\ a_{21}(\lambda) & a_{22}(\lambda) \end{pmatrix} + t\begin{pmatrix} a_{22}(\lambda) & -a_{21}(\lambda) \\ -a_{12}(\lambda) & a_{11}(\lambda) \end{pmatrix}, \qquad 0 \leqslant t \leqslant 1,$$

and note that $\alpha^t(\lambda) = \alpha(\lambda)$ and $\beta^t(\lambda) = (1-2t)\beta(\lambda)$. In view of (2), we have

$$4 \det L^t(\lambda) = |\alpha^t(\lambda)|^2 - |\beta^t(\lambda)|^2 = |\alpha(\lambda)|^2 - (1-2t)^2|\beta(\lambda)|^2 \geqslant 4 \det L(\lambda) > 0$$

for all $0 \leqslant t \leqslant 1$. When $t = \frac{1}{2}$, we have

$$L^{1/2}(\lambda) = \frac{1}{2}\begin{pmatrix} \alpha(\lambda) & -\beta(\lambda) \\ \beta(\lambda) & \alpha(\lambda) \end{pmatrix}.$$

Hence, without loss of generality, we may assume that $L(\lambda)$ and $\tilde{L}(\lambda)$ have the form

$$L(\lambda) = \begin{pmatrix} p(\lambda) & -q(\lambda) \\ q(\lambda) & p(\lambda) \end{pmatrix}, \qquad \tilde{L}(\lambda) = \begin{pmatrix} \tilde{p}(\lambda) & -\tilde{q}(\lambda) \\ \tilde{q}(\lambda) & \tilde{p}(\lambda) \end{pmatrix} \tag{4}$$

Since $\alpha = 2(p + iq)$ and $\tilde{\alpha} = \tilde{p} + i\tilde{q}$, then $p(\lambda) + iq(\lambda) \neq 0$ and $\tilde{p}(\lambda) + i\tilde{q}(\lambda) \neq 0$ for all $\lambda \in \mathbb{R}$. By Lemma 3.19, we also have

$$L(\lambda) = \begin{pmatrix} 1 & 1 \\ -i & i \end{pmatrix}\begin{pmatrix} p(\lambda) + iq(\lambda) & \\ & p(\lambda) - iq(\lambda) \end{pmatrix}\begin{pmatrix} 1 & 1 \\ -i & i \end{pmatrix}^{-1}, \tag{5}$$

with an analogous expression for $\tilde{L}(\lambda)$ where the polynomials p, q are replaced by $\tilde{p}, \tilde{q}$. By assumption, the equations $\alpha(\lambda) = 0$ and $\tilde{\alpha}(\lambda) = 0$ have the same number of roots in the upper half-plane. It is then clear that there exists a homotopy $p^t + iq^t$ connecting $p + iq$ and $\tilde{p} + i\tilde{q}$. In view of (5), this gives us a homotopy $L^t(\lambda)$ connecting $L(\lambda)$ and $\tilde{L}(\lambda)$:

$$L^t(\lambda) = \begin{pmatrix} 1 & 1 \\ -i & i \end{pmatrix}\begin{pmatrix} p^t(\lambda) + iq^t(\lambda) & \\ & p^t(\lambda) - iq^t(\lambda) \end{pmatrix}\begin{pmatrix} 1 & 1 \\ -i & i \end{pmatrix}^{-1}$$
$$= \begin{pmatrix} p^t(\lambda) & -q^t(\lambda) \\ q^t(\lambda) & p^t(\lambda) \end{pmatrix},$$

$0 \leqslant t \leqslant 1$, which completes the proof of the second step.

For second-order elliptic operators $\mathscr{A}$ with $\det \mathscr{A}(1, \lambda) > 0$, there are three homotopy classes $\mathscr{E}_0$, $\mathscr{E}_1$ and $\mathscr{E}_2$. The next theorem shows that for any $\mathscr{A} \in \mathscr{E}_1$, the Dirichlet problem $(\mathscr{A}, \mathscr{D})$ satisfies the L-condition. Since the 2×2 Laplace operator $\begin{pmatrix} \Delta & \\ & \Delta \end{pmatrix}$ also belongs to $\mathscr{E}_1$, it follows that the Dirichlet problem has index 0 for any operator in $\mathscr{E}_1$.

Theorem 17.3 *Let $\mathscr{A}$ be a second-order 2×2 elliptic operator ($l = 2$). If $\mathscr{A} \in \mathscr{E}_1$ (or $\mathscr{E}'_1$) then the Dirichlet problem for $\mathscr{A}$ satisfies the L-condition and has index 0.*

Proof To verify that the L-condition holds it is sufficient (in view of Theorem 10.5 it is also necessary) to show that

$$\det \int_{\gamma^+} [\pi\mathscr{A}(1, \lambda)]^{-1} \, d\lambda \neq 0 \tag{6}$$

For 2×2 systems, the condition (6) can be reformulated in terms of the scalar functions $\alpha(\lambda)$ and $\beta(\lambda)$. In view of (1)

$$[\pi\mathscr{A}(1, \lambda)]^{-1} = \frac{1}{D(\lambda)} \begin{pmatrix} a_{22}(\lambda) & -a_{12}(\lambda) \\ -a_{21}(\lambda) & a_{11}(\lambda) \end{pmatrix},$$

where $D(\lambda) = \det \pi A(1, \lambda) = a_{22}(\lambda)a_{11}(\lambda) - a_{12}(\lambda)a_{21}(\lambda)$, so that (6) is equivalent to

$$\int_{\gamma^+} \frac{a_{22}(\lambda)}{D(\lambda)} \, d\lambda \cdot \int_{\gamma^+} \frac{a_{11}(\lambda)}{D(\lambda)} \, d\lambda \neq \int_{\gamma^+} \frac{a_{21}(\lambda)}{D(\lambda)} \, d\lambda \cdot \int_{\gamma^+} \frac{a_{12}(\lambda)}{D(\lambda)} \, d\lambda$$

which, in turn, is equivalent to

$$\left| \int_{\gamma^+} \frac{\alpha(\lambda)}{D(\lambda)} \, d\lambda \right|^2 \neq \left| \int_{\gamma^+} \frac{\beta(\lambda)}{D(\lambda)} \, d\lambda \right|^2$$

or

$$\int_{\gamma^+} \frac{\alpha(\lambda) + \mu\beta(\lambda)}{D(\lambda)} \, d\lambda \neq 0 \tag{7}$$

for all $\mu \in \mathbb{C}$ such that $|\mu| = 1$. Since $\mathscr{A} \in \mathscr{E}_1$, the polynomial $\alpha(\lambda)$ has one root above the real axis and one below. In view of (3) and Rouché's theorem, the same is true of the λ-polynomial $\alpha(\lambda) + \mu\beta(\lambda)$; let μ_1 and μ_2 denote these roots, where $\operatorname{Im} \mu_1 > 0$ and $\operatorname{Im} \mu_2 < 0$. Also the real polynomial $D(\lambda)$ has roots λ_1, $\bar{\lambda}_1$, λ_2 and $\bar{\lambda}_2$, where $\operatorname{Im} \lambda_1 > 0$ and $\operatorname{Im} \lambda_2 > 0$. Then the condition (7) may be written in the form

$$\mathscr{I} = \frac{1}{2\pi i} \int_{\gamma^+} \frac{(\lambda - \mu_1)(\lambda - \mu_2)}{(\lambda - \lambda_1)(\lambda - \bar{\lambda}_1)(\lambda - \lambda_2)(\lambda - \bar{\lambda})} \, d\lambda \neq 0.$$

We will show that $\mathscr{I} \neq 0$ for any μ_1, μ_2, λ_1 and λ_2 such that $\operatorname{Im} \mu_1 > 0$, $\operatorname{Im} \mu_2 < 0$, $\operatorname{Im} \lambda_1 > 0$ and $\operatorname{Im} \lambda_2 > 0$.

Evaluating the residues of $\mathscr{I}$ in the upper half-plane and assuming that $\lambda_1 \neq \lambda_2$ so that the poles are simple, one obtains

$$\mathscr{I} = \frac{(\lambda_1 - \mu_1)(\lambda_1 - \mu_2)}{(\lambda_1 - \bar{\lambda}_1)(\lambda_1 - \lambda_2)(\lambda_1 - \bar{\lambda}_2)} + \frac{(\lambda_2 - \mu_1)(\lambda_2 - \mu_2)}{(\lambda_2 - \lambda_1)(\lambda_2 - \bar{\lambda}_1)(\lambda_2 - \bar{\lambda}_2)}.$$

After taking a common denominator and cancelling the factor $\lambda_1 - \lambda_2$, one obtains

$$\mathscr{I} = \frac{2i[(\mu_1 + \mu_2)\operatorname{Im}(\lambda_1\lambda_2) - \mu_1\mu_2\operatorname{Im}(\lambda_1 + \lambda_2) - |\lambda_2|^2\operatorname{Im}\lambda_1 - |\lambda_1|^2\operatorname{Im}\lambda_2]}{4\operatorname{Im}\lambda_1\cdot\operatorname{Im}\lambda_2\cdot|\lambda_1 - \bar{\lambda}_2|^2}$$

By a limiting process, it is clear that the latter formula is true even if $\lambda_1 = \lambda_2$.

Replacing λ with $\lambda + m$ in the integral $\mathscr{I}$, where m is a suitably chosen real number, we may assume that $\operatorname{Im}(\lambda_1\lambda_2) = 0$. Thus, to prove $\mathscr{I} = 0$, it is necessary to show that

$$-\mu_1\mu_2\operatorname{Im}(\lambda_1 + \lambda_2) \neq |\lambda_2|^2\operatorname{Im}\lambda_1 + |\lambda_1|^2\operatorname{Im}\lambda_2$$

or

$$-\mu_1 \neq k\bar{\mu}_2, \qquad \text{where } k = \frac{|\lambda_2|^2\operatorname{Im}\lambda_1 + |\lambda_1|^2\operatorname{Im}\lambda_2}{|\mu_2|^2\ \operatorname{Im}(\lambda_1 + \lambda_2)}.$$

But the fact that $-\mu_1 \neq k\bar{\mu}_2$ is clear since $k > 0$, $\operatorname{Im}(-\mu_1) < 0$ and $\operatorname{Im}(\bar{\mu}_2) > 0$. This proves that the Dirichlet problem for $\mathscr{A}$ satisfies the L-condition.

Now, since both $\mathscr{A}$ and the 2×2 Laplace operator are in $\mathscr{E}_1$, there is a homotopy $\mathscr{A}^t$ in $\mathscr{E}_1$ such that $\mathscr{A}^0 = \mathscr{A}$, $\mathscr{A}^1 = \begin{pmatrix}\Delta & \\ & \Delta\end{pmatrix}$. The L-condition holds for the Dirichlet problem $(\mathscr{A}^t, \mathscr{D})$ for all $0 \leqslant t \leqslant 1$. Hence by Theorem 10.25

$$\operatorname{ind}(\mathscr{A}, \mathscr{D}) = \operatorname{ind}(\mathscr{A}^1, \mathscr{D}) = 0.$$

Remarks

(1) Theorem 17,3 remains true even if the coefficients of $\mathscr{A}$ are not assumed to be constant; the same proof shows that

$$\det \int_{\gamma^+} [\pi\mathscr{A}(x, (1, \lambda))]^{-1}\, d\lambda \neq 0 \tag{8}$$

for all $x \in \bar{\Omega}$. The idea of the proof of Theorem 17.3 was sketched in the paper [Bo 2].

(2) We can also prove that $\operatorname{ind}(\mathscr{A}, \mathscr{D}) = 0$ using the formula in §16.5. Let $(X_+(x), T_+(x), Y_+(x))$ be a γ^+-spectral triple of $L_x(\lambda) = \pi\mathscr{A}(x, (1, \lambda))$, where X_+, T_+ and Y_+ are 2×2 smooth matrix functions. Since the matrix (8) is equal to $X_+(x)\cdot Y_+(x)$, it follows that $\det X_+(x) \neq 0$ for all $x \in \bar{\Omega}$. Thus

$$\operatorname{ind}(\mathscr{A}, \mathscr{D}) = -\frac{1}{\pi}[\arg\det X_+(y)]_{\partial\Omega} = 0.$$

17.2 An example: the Neumann BVP for second-order elliptic operators

Recall that if

$$\mathscr{A}(x, \partial/\partial x) = \sum_{j,k=1}^{2} A_{jk}(x)\,\partial^2/\partial x_j\,\partial x_k + \text{lower order terms}, \qquad x \in \bar{\Omega},$$

then the Neumann boundary operator is

$$\mathscr{C}(y, \partial/\partial x) = \sum_{j,k=1}^{2} A_{jk}(y)\nu_k(y)\,\partial/\partial x_j, \qquad y \in \partial\Omega,$$

where $\nu = (\nu_1, \nu_2) = (-\tau_2, \tau_1)$.

Lemma 17.4 *For the second-order elliptic operator*

$$\mathscr{A}(x, \partial/\partial x) = A(x)\frac{\partial^2}{\partial x_2^2} + 2B(x)\frac{\partial^2}{\partial x_2\,\partial x_1} + C(x)\frac{\partial^2}{\partial x_1^2} + \cdots$$

let $(X_\pm(x), T_\pm(x))$ be a $\gamma^\pm$-spectral pair of $L_x(\lambda) = \pi\mathscr{A}(x, (1, \lambda))$. If $\mathscr{C}(y, \partial/\partial x) = C_2(y)\,\partial/\partial x_2 + C_1(y)\,\partial/\partial x_1$ denotes the Neumann boundary operator then

$$C_2(y)X_\pm(y)T_\pm(y) + C_1(y)X_\pm(y)$$
$$= (A(y)X_\pm(y)T_\pm(y) + B(y)X_\pm(y))\cdot(\tau_1(y)I + \tau_2(y)T_\pm(y)) \quad (9)$$

for all $y \in \partial\Omega$.

Proof We have $\mathscr{C}(y, \partial/\partial x) = C_2(y)\,\partial/\partial x_2 + C_1(y)\,\partial/\partial x_1$ where $C_2 = \tau_1 A - \tau_2 B$ and $C_1 = \tau_1 B - \tau_2 C$. Thus

$$C_2X_\pm T_\pm + C_1X_\pm = (\tau_1 A - \tau_2 B)X_\pm T_\pm + (\tau_1 B - \tau_2 C)X_\pm$$
$$= \tau_1(AX_\pm T_\pm + BX_\pm) - \tau_2(BX_\pm T_\pm + CX_\pm)$$

Since $L(\lambda) = A\lambda^2 + 2B\lambda + C$, then $AX_\pm T_\pm^2 + 2BX_\pm T_\pm + CX_\pm = 0$. Hence $(AX_\pm T_\pm + BX_\pm)T_\pm = -(BX_\pm T_\pm + CX_\pm)$, and then (9) follows.

Corollary 17.5 *Let $\tilde{L}_y(\lambda) = \pi\mathscr{A}(y, \tau(y) + \lambda\nu(y)) = \tilde{A}(y)\lambda^2 + 2\tilde{B}(y)\lambda + C(y)$ be the transform of $L_y(\lambda)$ under $\varphi(\lambda) = (-\tau_2 + \lambda\tau_1)/(\tau_1 + \lambda\tau_2)$ with $\gamma^\pm$-spectral pair $(\tilde{X}_\pm(y), \tilde{T}_\pm(y))$ defined by §15.3 (14). Then*

$$\tilde{A}(y)\tilde{X}_\pm(y)\tilde{T}_\pm(y) + \tilde{B}(y)\tilde{X}_\pm(y) = A(y)X_\pm(y)T_\pm(y) + B(y)X_\pm(y)$$

for all $y \in \partial\Omega$.

Proof Note that $\mathscr{C}(y, \partial/\partial x) = \tilde{C}_2(y)\,\partial/\partial\nu + \tilde{C}_1(y)\,\partial/\partial\tau$, where $\tilde{C}_2(y) = \tilde{A}(y)$ and $\tilde{C}_1(y) = \tilde{B}(y)$, and the associated matrix polynomial is

$$\tilde{C}_y(\lambda) = \mathscr{C}(y, \tau(y) + \lambda\nu(y)) = \tilde{A}(y)\lambda + \tilde{B}(y).$$

By Proposition 15.12, we then have

$$\tilde{A}(y)\tilde{X}_\pm(y)\tilde{T}_\pm(y) + \tilde{B}(y)\tilde{X}_\pm(y) = (C_2(y)X_\pm(y)T_\pm(y) + C_1(y)X_\pm(y))$$
$$\times\,(\tau_1(y)I + \tau_2(y)T_\pm(y))^{-1}$$
$$= A(y)X_\pm(y)T_\pm(y) + B(y)X_\pm(y),$$

the last equality holding because of (9).

As an example, let us consider the 2 × 2 operator

$$\mathscr{A}(x, \partial/\partial x) = A(x)\frac{\partial^2}{\partial x_2^2} + 2B(x)\frac{\partial^2}{\partial x_2\,\partial x_1} + C(x)\frac{\partial^2}{\partial x_1^2} + \cdots$$

where

$$A(x) = \begin{pmatrix} \varphi_1(x) & \varphi_2(x) + 1 \\ \varphi_2(x) - 1 & -\varphi_1(x) \end{pmatrix}, \qquad B(x) = -I,\ C(x) = A(x)^T$$

and φ_1, φ_2 are real-valued functions in $C^\infty(\bar{\Omega})$ such that

$$[\varphi_1(x)]^2 + [\varphi_2(x)]^2 \neq 1$$

for all $x \in \bar{\Omega}$. It is clear that $\mathscr{A}$ is elliptic, since

$$\begin{aligned} L_x(\lambda) &= \pi\mathscr{A}(x, (1, \lambda)) \\ &= A(x)\lambda^2 + 2B(x)\lambda + C(x) \\ &= \begin{pmatrix} \varphi_1(x)(\lambda^2 + 1) - 2\lambda & \varphi_2(x)(\lambda^2 + 1) + (\lambda^2 - 1) \\ \varphi_2(x)(\lambda^2 + 1) - (\lambda^2 - 1) & -\varphi_1(x)(\lambda^2 + 1) - 2\lambda \end{pmatrix} \end{aligned}$$

and $\det L_x(\lambda) = (1 + \lambda^2)^2(1 - |\varphi(x)|^2)$ for all $\lambda \in \mathbb{R}$, where $\varphi(x) = \varphi_1(x) + i\varphi_2(x)$. In view of Example 3.17,

$$X_+(x) = \begin{pmatrix} 1 & -i\varphi_2(x) \\ -i & i\varphi_1(x) \end{pmatrix}, \qquad T_+(x) = \begin{pmatrix} i & 1 \\ 0 & i \end{pmatrix},$$

is a γ^+-spectral pair of $L_x(\lambda)$, depending smoothly on $x \in \bar{\Omega}$. The Dirichlet boundary value problem for $\mathscr{A}$ satisfies the L-condition provided that the map $u \mapsto u(0)$ from $\mathfrak{M}_{L_y}^+$ to $\mathbb{C}^p$ is invertible for all $y \in \partial\Omega$ (see Remark 15.1 and Corollary 15.13). Since $\det X_+(y) = i\overline{\varphi(y)} \neq 0$, the Dirichlet problem satisfies the L-condition if and only if

$$\varphi(y) \neq 0 \qquad \text{for all } y \in \partial\Omega,$$

and by Theorem 16.15 the index is

$$\begin{aligned} \operatorname{ind}(\mathscr{A}, \mathscr{D}) &= -\frac{1}{\pi}[\arg\det X_+(y)]_{\partial\Omega} \\ &= \frac{1}{\pi}[\arg \varphi(y)]_{\partial\Omega}. \end{aligned} \tag{10}$$

Consider now the Neumann boundary value problem for $\mathscr{A}$. By Corollary 17.5 we have

$$\begin{aligned} &\tilde{A}(y)\tilde{X}_+(y)\tilde{T}_+(y) + \tilde{B}(y)\tilde{X}_+(y) \\ &= A(y)X_+(y)T_+(y) + B(y)X_+(y) \\ &= \begin{pmatrix} \varphi_1(y) & \varphi_2(y) + 1 \\ \varphi_2(y) - 1 & -\varphi_1(y) \end{pmatrix}\begin{pmatrix} 1 & -i\varphi_2(y) \\ -i & i\varphi_1(y) \end{pmatrix}\begin{pmatrix} i & 1 \\ 0 & i \end{pmatrix} - \begin{pmatrix} 1 & -i\varphi_2(y) \\ -i & i\varphi_1(y) \end{pmatrix} \\ &= \begin{pmatrix} i\overline{\varphi(y)} & -i \\ -\overline{\varphi(y)} & |\varphi(y)|^2 - 1 \end{pmatrix}. \end{aligned} \tag{11}$$

Hence

$$\det(\tilde{A}(y)\tilde{X}_+(y)\tilde{T}_+(y) + \tilde{B}(y)\tilde{X}_+(y)) = i\overline{\varphi(y)}(|\varphi(y)|^2 - 2),$$

so the Neumann boundary value problem for $\mathscr{A}$ satisfies the L-condition if and only if $\varphi(y) \neq 0$, $|\varphi(y)| \neq 0$, $|\varphi(y)| \neq \sqrt{2}$ for all $y \in \partial\Omega$. By Theorem 16.11, the index of the Neumann problem is

$$\operatorname{ind}(\mathscr{A}, \mathscr{C}) = 4 - \frac{1}{\pi}[\arg \det(C_2 X_+ T_+ + C_1 X_+)]_{\partial\Omega}$$

(here $\mu = 1$, $l = 2$ and $r = pl/2 = 2$). By Corollary 17.4 we have

$$C_2 X_+ T_+ + C_1 X_+ = (AX_+ T_+ + BX_+)(\tau_1 I + \tau_2 T_+)$$

and by virtue of (11) we have $\det(AX_+ T_+ + BX_+) = i\bar{\varphi}(|\varphi|^2 - 2)$. Since $|\varphi|^2 - 2$ is real-valued, it has winding number equal to 0, so the winding number of $\det(AX_+ T_+ + BX_+)$ is equal to that of $\bar{\varphi}$. Also,

$$\frac{1}{2\pi}[\arg \det(\tau_1 I + \tau_2 T_+)]_{\partial\Omega} = r = 2$$

whence

$$\operatorname{ind}(\mathscr{A}, \mathscr{C}) = \frac{1}{\pi}[\arg \varphi(y)]_{\partial\Omega},$$

which is the same as the index (10) of the Dirichlet BVP.

Remark If we extend the definitions of §17.1 in the obvious way to operators with variable coefficients then

$$\alpha(\lambda) = -2i(\lambda - i)^2, \qquad \beta(\lambda) = 2(\lambda^2 + 1)\varphi$$

so that $\mathscr{A} \in \mathscr{E}_2$ if $|\varphi| < 1$ on $\bar{\Omega}$, and $\mathscr{A} \in \mathscr{E}'_1$ if $|\varphi| > 1$ on $\bar{\Omega}$. The formula (10) shows that in general the Dirichlet problem has non-zero index. However, note that $\operatorname{ind}(\mathscr{A}, \mathscr{D}) = 0$ if $\mathscr{A} \in \mathscr{E}'_1$.

17.3 An example: the elliptic system for plane elastic deformations

Consider an infinitely long, cylindrical, elastic solid with plane cross-section Ω and assume that all displacements and forces lie in this plane (and depend only on the coordinates in the plane). For an isotropic material the equilibrium equations for the displacement vector $u = [u_1 \ u_2]^T$ take the form

$$\mathscr{A}(\partial/\partial x)u = (\lambda + \mu) \operatorname{grad} \operatorname{div} u + \mu \Delta u = f(x) \qquad (12)$$

where λ, μ are Lamé's elastic constants and $f = [f_1 \ f_2]^T$ is the body force.

Let us write (12) in another form using Hooke's Law. This law, connecting stress to strain, states that

$$\text{stress} = 2\mu(\text{strain}) + \lambda(\text{trace of strain})I. \qquad (13)$$

Let the stress tensor be denoted by $\sigma = [\sigma_{ij}]$ and the strain tensor by $e = [e_{ij}]$. By definition,

$$e_{ij}(u) = \tfrac{1}{2}(\partial u_i/\partial x_j + \partial u_j/\partial x_i) \qquad i, j = 1, 2$$

so that

$$\sigma_{ij}(u) = \mu\left(\frac{\partial u_i}{\partial x_j} + \frac{\partial u_j}{\partial x_i}\right) + \lambda \operatorname{div} u\delta_{ij} \qquad i, j = 1, 2$$

and then

$$\begin{aligned}(\operatorname{div} \sigma)_i &= \sum_{j=1}^{2} \frac{\partial \sigma_{ij}}{\partial x_j} \\ &= \mu \sum_{j=1}^{2} \left(\frac{\partial^2 u_i}{\partial x_j^2} + \frac{\partial^2 u_j}{\partial x_i\, \partial x_j}\right) + \lambda \sum_{j=1}^{2} \frac{\partial}{\partial x_j} (\operatorname{div} u)\delta_{ij} \\ &= \mu\, \Delta u_i + (\lambda + \mu) \frac{\partial}{\partial x_i} \operatorname{div} u.\end{aligned}$$

Thus, $\mathscr{A}u = \operatorname{div} \sigma(u)$.

We now derive a Green formula for $\mathscr{A}$. Let n denote the outward-pointing unit normal vector field along $\partial\Omega$. By the divergence theorem we have

$$\iint_{\Omega} \operatorname{div}(\sigma \cdot v)\, dx = \int_{\partial\Omega} (\sigma \cdot v)^T n\, ds, \tag{14}$$

where $v = [v_1 \;\; v_2]^T$ is any 2-vector function and $\sigma = [\sigma_{ij}]$ is any 2×2 matrix function (not necessarily a stress matrix) with entries in $C^1(\bar{\Omega})$. Also,

$$\begin{aligned}\operatorname{div}(\sigma \cdot v) &= \sum_{i,j} \frac{\partial}{\partial x_i} (\sigma_{ij} v_j) \\ &= \sum_{i,j} \sigma_{ij} \frac{\partial v_j}{\partial x_i} + \sum_{i,j} \frac{\partial \sigma_{ij}}{\partial x_i} v_j,\end{aligned}$$

and if σ is a symmetric matrix

$$= \sum_{i,j} e_{ij}(v)\sigma_{ij} + \sum_{j} v_j(\operatorname{div} \sigma)_j.$$

Thus, we obtain

$$\operatorname{div}(\sigma \cdot v) = \operatorname{tr}(e(v)^T \sigma) + v^T \operatorname{div} \sigma,$$

where $\operatorname{tr}(e^T\sigma)$ denotes the matrix inner product $\sum_{i,j} e_{ij}\sigma_{ij}$. By (14) we then have the formula

$$\iint_{\Omega} (\operatorname{tr}(e(v)^T\sigma) + v^T \;\operatorname{div} \sigma)\, dx = \int_{\partial\Omega} v^T \sigma^T n\, ds \tag{15}$$

where v is any 2-vector function, $e = e(v)$ is the associated strain tensor and σ is any 2×2 symmetric matrix function. Substituting $\sigma = \sigma(u)$ in (15) gives us one equation, then reversing the roles of u and v gives us another; subtracting the two equations and using the fact that $\operatorname{tr}(e(v)^T\sigma(u)) =$

$\operatorname{tr}(e(u)^T\sigma(v))$, we obtain the Green formula

$$\iint_\Omega (v^T\mathscr{A}u - u^T\mathscr{A}v)\,dx = \int_{\partial\Omega} (v^T\mathscr{B}u - u^T\mathscr{B}v)\,ds,$$

for $u, v \in C^2(\bar{\Omega})$, where $\mathscr{A}u = \operatorname{div}\sigma(u)$ and $\mathscr{B}u = \sigma(u)\cdot n$. This formula suggests a boundary condition for $\mathscr{A}$, namely, that the surface traction, $\mathscr{B}u = \sigma(u)\cdot n$, be prescribed on $\partial\Omega$. For future reference note that the stress matrix is

$$\sigma(u) = \begin{pmatrix} (\lambda + 2\mu)\,\partial u_1/\partial x_1 + \lambda\,\partial u_2/\partial x_2 & \mu(\partial u_1/\partial x_2 + \partial u_2/\partial x_1) \\ \mu(\partial u_1/\partial x_2 + \partial u_2/\partial x_1) & \lambda\partial u_1/\partial x_1 + (\lambda + 2\mu)\,\partial u_2/\partial x_2 \end{pmatrix} \tag{16}$$

and then $\mathscr{B}u = \sigma(u)\cdot n$, where $n = [\tau_2 \quad -\tau_1]^T$.

Physically, there are certain restrictions on the Lamé constants λ, μ which can be determined by consideration of the strain energy. Hooke's law (13) may be written in the form $\sigma = Ce$, where $C = [c_{ijkl}]$ is a symmetric tensor of rank 4, and then the strain energy $\frac{1}{2}\operatorname{tr}(e^TCe)$ equals

$$\mu\cdot\sum_{i,j} e_{ij}^2 + \tfrac{1}{2}\lambda\cdot(\operatorname{tr} e)^2.$$

On physical grounds the strain energy should be positive definite: $\operatorname{tr}(e^TCe) \geqslant 0$ and $\operatorname{tr}(e^TCe) = 0$ if and only if $e = 0$. Given that attention is restricted to plane elastic deformations, it is not hard to show that the strain energy is positive definite and only if $\mu > 0$ and $\lambda + \mu > 0$.

However, the strain energy should be positive definite with respect to all three-dimensional elastic deformations, where the displacement vector u is $[u_1 \ u_2 \ u_3]^T$ and the u_j depend on x_1, x_2 and x_3. (All the definitions and calculations in this section remain valid in space with the obvious modifications.) This leads to a stronger restriction on λ, μ.

Exercise 1 *Show that the strain energy is positive definite with respect to the class of all three-dimensional elastic deformations if and only if* $\mu > 0$ *and* $3\lambda + 2\mu > 0$.

Under the assumption that the strain energy is positive definite, we can use the formula (15) to determine explicitly the kernel and cokernel of $(\mathscr{A}, \mathscr{B})$. Let $u \in \ker(\mathscr{A}, \mathscr{B})$. Then $\operatorname{div}\sigma(u) = 0$ in Ω and $\sigma(u)\cdot n = 0$ on $\partial\Omega$, so that (15) implies

$$0 = \iint_\Omega \operatorname{tr}(e^T\sigma)\,dx = \iint_\Omega \operatorname{tr}(e^TCe)\,dx$$

whence $\operatorname{tr}[e^TCe] = 0$. Then $e = e(u) = 0$, or

$$\frac{\partial u_i}{\partial x_j} + \frac{\partial u_j}{\partial x_i} = 0 \qquad i, j = 1, 2$$

which has solutions $u = [a_1 \ a_2]^T + b[x_2 \quad -x_1]^T$, where a_1, a_2 and $b \in \mathbb{R}$. Hence $\alpha = \dim\ker(\mathscr{A}, \mathscr{B}) = 3$.

To determine the dimension of the cokernel, we consider solvability conditions for the BVP

$$\left.\begin{aligned} \mathscr{A}(\partial/\partial x)u &= 0, & x &\in \Omega \\ \mathscr{B}(y, \partial/\partial x)u &= g(y), & y &\in \partial\Omega \end{aligned}\right\} \tag{17}$$

Suppose we let $v = [a_1 \;\; a_2]^T$ in the formula (15). Then $e = e(v) = 0$, and since a_1 and a_2 are arbitrary constants, we obtain

$$\iint_\Omega \mathscr{A}(\partial/\partial x)u \, dx = \int_{\partial\Omega} \mathscr{B}(y, \partial/\partial x)u \, ds.$$

Hence for the BVP (17) the conditions on $g = [g_1 \;\; g_2]^T$ are

$$\int_{\partial\Omega} g_1(y)\, ds = 0, \qquad \int_{\partial\Omega} g_2(y)\, ds = 0.$$

Another solvability condition is obtained by letting $v = b[x_2 \quad -x_1]^T$ in (15):

$$\int_{\partial\Omega} (y_2 g_1(y) - y_1 g_2(y))\, ds = 0.$$

These are the only solvability conditions, that is, $\beta = 3$. Since the equation $\mathscr{A}(\partial/\partial x)u = f(x)$ has a solution $u \in C^\infty(\bar{\Omega}, \mathbb{R}^2)$ for any $f \in C^\infty(\bar{\Omega}, \mathbb{R}^2)$ it follows that $\operatorname{ind}(\mathscr{A}, \mathscr{B}) = \operatorname{ind}(\mathscr{B}|_{\ker \mathscr{A}}) = 3 - 3 = 0$ (see Lemma 16.13).

Exercise 2 *Within the class of three-dimensional elastic deformations, determine the kernel and cokernel of* $(\mathscr{A}, \mathscr{B})$.

We end this chapter by applying the results of §17.1 to the operator (12), with no a priori assumptions on λ and μ (i.e. with no requirement that the strain energy be positive definite). We have

$$\begin{aligned} L(z) = \mathscr{A}(1, z) &= (\lambda + \mu)\begin{pmatrix} 1 & z \\ z & z^2 \end{pmatrix} + \mu\begin{pmatrix} z^2 + 1 & \\ & z^2 + 1 \end{pmatrix} \\ &= \begin{pmatrix} \mu z^2 + (\lambda + 2\mu) & (\lambda + \mu)z \\ (\lambda + \mu)z & (\lambda + 2\mu)z^2 + \mu \end{pmatrix}, \end{aligned}$$

so that $\det L(z) = \mu(\lambda + 2\mu)(z^2 + 1)^2$. Hence $\mathscr{A}$ is elliptic provided $\mu \neq 0$ and $\lambda + 2\mu \neq 0$ which we assume from now on. Also note that

$$\alpha(z) = (\lambda + 3\mu)(z^2 + 1), \qquad \beta(z) = -(\lambda + \mu)(z - i)^2.$$

Thus, $\mathscr{A} \in \mathscr{E}_1$ if $\mu(\lambda + 2\mu) > 0$ and $\mathscr{A} \in \mathscr{E}'_2$ if $\mu(\lambda + 2\mu) < 0$.

Since $\det L(z)$ has a single root i of multiplicity 2 above the real axis, it is easily verified that a γ^+-spectral pair for $L(z)$ is given by

$$X_+ = \begin{pmatrix} -i & -\dfrac{\lambda + 3\mu}{\lambda + \mu} \\ 1 & 0 \end{pmatrix}, \qquad T_+ = \begin{pmatrix} i & 1 \\ 0 & i \end{pmatrix}$$

provided $\lambda + \mu \neq 0$ (if $\lambda + \mu = 0$ then (12) is a system of two Laplace equations which can be dealt with separately). Since $\det X_+ = (\lambda + 3\mu)/(\lambda + \mu)$, we find that the Dirichlet problem for $\mathscr{A}$ satisfies the L-condition provided $\lambda + 3\mu \neq 0$. If this condition holds then Theorem 16.15 implies

$$\operatorname{ind}(\mathscr{A}, \mathscr{D}) = -\frac{1}{\pi}[\arg \det X_+]_{\partial\Omega} = 0,$$

since $\det X_+$ is constant. Note: When $\lambda = -3\mu$, the operator $\mathscr{A}$ is essentially the Bitsadze operator (Example 17.1).

Remark It is not hard to show that the condition $\mu(\lambda + 2\mu) > 0$ is equivalent to $\mathscr{A}$ being strongly elliptic.

Now consider the boundary operator $\mathscr{B}u = \sigma(u)\cdot n$. In view of (16)

$$B_y(z) = \mathscr{B}(y, (1, z)) = B_1(y)z + B_0(y),$$

where

$$B_1(y) = \begin{pmatrix} -\mu\tau_1 & \lambda\tau_2 \\ \mu\tau_2 & -(\lambda + 2\mu)\tau_1 \end{pmatrix}, \qquad B_0(y) = \begin{pmatrix} (\lambda + 2\mu)\tau_2 & -\mu\tau_1 \\ -\lambda\tau_1 & \mu\tau_2 \end{pmatrix}$$

After some calculation it follows that

$$\begin{aligned} \Delta_{\mathscr{B}}^{+}(y) &= B_1(y)X_+T_+ + B_0(y)X_+ \\ &= -\begin{pmatrix} 2\mu(\tau_1 + i\tau_2) & \dfrac{1}{\lambda + \mu}[-2i\mu(\lambda + 2\mu) + (4\mu\lambda + 6\mu^2)\tau_2] \\ 2\mu i(\tau_1 + i\tau_2) & \dfrac{1}{\lambda + \mu}[2\mu^2\tau_1 + 2i\mu\tau_2(\lambda + 2\mu)] \end{pmatrix}, \end{aligned}$$

and then $\det \Delta_{\mathscr{B}}^{+} = -4\mu^2(\tau_1 + i\tau_2)^2 \neq 0$. Hence $(\mathscr{A}, \mathscr{B})$ satisfies the L-condition. By Theorem 16.11,

$$\begin{aligned} \operatorname{ind}(\mathscr{A}, \mathscr{B}) &= 4 - \frac{1}{\pi}[\arg \det \Delta_{\mathscr{B}}^{+}(y)]_{\partial\Omega} \\ &= 4 - 4 = 0, \end{aligned}$$

where we have used the fact that $(1/2\pi)[\arg(\tau_1 + i\tau_2)]_{\partial\Omega} = 1$.

References

[AB] M.F. Atiyah, R. Bott, The index problem for manifolds with boundary, Bombay Colloquium on Differential Analysis, Oxford Univ. Press, Oxford (1964), 175–186.

[Ad] R.A. Adams, *Sobolev Spaces*, Academic Press, New York 1975.

[ADN] S. Agmon, A. Douglis, L. Nirenberg, Estimates near the boundary for solutions of elliptic partial differential equations satisfying general boundary conditions, Comm. Pure Appl. Math. 12 (1959), 623–727; II, Comm. Pure Appl. Math. 17 (1964), 35–92.

[Ag] S. Agmon, *Lectures on Elliptic Boundary Value Problems*, Van Nostrand Reinhold, Princeton, N.J. 1965.

[Agr] M.S. Agranovič, Elliptic singular integro-differential operators, Usp. Mat. Nauk 20, 5 (1965), 3–120 (Russian; English translation in Russian Math. Surveys 20 (1965), 1–122).

[AkG] N.I. Akheizer, I.M. Glazman, *Theory of Linear Operators in Hilbert Space*, Pitman, London 1981.

[AMR] R. Abraham, J.E. Marsden, T. Ratiu, *Manifolds, Tensor Analysis, and Applications*, Second Edition, Springer-Verlag, New York–Berlin 1988.

[Atiy] M.F. Atiyah, Algebraic topology and elliptic operators, Comm. Pure Appl. Math. 20 (1967) 237–249.

[Be] Yu.M. Berezanskii, *Expansions in Eigenfunctions of Selfadjoint Operators*, Translations of Mathematical Monographs, Vol. 17, Amer. Math. Soc., Providence, R.I. 1968.

[BGR] J.A. Ball, Gohberg, L. Rodman, *Interpolation of Rational Matrix Functions*, Operator Theory: Advances and Applications, Vol. 45, Birkhäuser Verlag, Boston–Basel 1990.

[Bi 1] A.V. Bitsadze, *Boundary Value Problems for Second-Order Elliptic Equations*. North-Holland, Amsterdam 1968.

[Bi 2] A.V. Bitsadze, *Some Classes of Partial Differential Equations*, Gordon & Breach, London 1988.

[Bo 1] B. Bojarski, On the first boundary value problem for elliptic systems of second order in the plane, Pol. Akad. Nauk 7, 9 (1959), 565–570.

[Bo 2] B. Bojarski, On the Dirichlet problem for a system of elliptic equations in space, Pol. Akad. Nauk 8, 1 (1960), 19–23.

[BT] R. Bott, L.W.Tu, *Differential Forms in Algebraic Topology*, Springer-Verlag, New York–Berlin 1982.

[Ca] H. Cartan, *Calcul Différentiel*, Hermann, Paris 1967.

[CZ] A.P. Calderón, A. Zygmund, On the existence of certain singular integrals, Acta Math. 88, 1–2 (1952), 85–139.
[DN] A. Douglis, L. Nirenberg, Interior estimates for elliptic systems of partial differential equations, Comm. Pure Appl. Math. 8 (1955), 503–538.
[Do] A. Dold, *Lectures on Algebraic Topology*, Springer-Verlag, New York–Berlin 1972.
[Don] W.F. Donoghue, *Distributions and Fourier Transforms*, Academic Press, New York 1970.
[DS 1] N. Dunford, J.T. Schwartz, *Linear Operators*, Vol. I, Wiley, New York 1958.
[DS 2] N. Dunford, J. T. Schwartz, *Linear Operators*, Vol. II, Wiley, New York 1963.
[Eg] Yu.V. Egorov, *Linear Differential Equations of the Main Types*, Nauka, Moscow 1984 (Russian; English translation in Contemp. Sov. Math., New York 1986).
[Es] G.I. Èskin, *Boundary Value Problems for Elliptic Pseudodifferential Equations*, Translations of Mathematical Monographs, Vol. 52, Amer. Math. Soc., Providence, R.I. 1981.
[Fe] B.V. Fedosov, An analytic formula for the index of an elliptic boundary-value problem, Mat. Sbornik 93, 135 (1974), No. 1 (Russian; English translation in Math. USSR Sbornik 22 (1974), No. 1, 61–90).
[Ga] F.D. Gakhov, *Boundary Value Problems*, Dover Publications, New York 1990.
[GLR] I. Gohberg, P. Lancaster, L. Rodman, *Matrix Polynomials*, Academic Press, New York 1982.
[GK] I.C. Gohberg, M.G. Krein, The basic propositions on defect numbers, root numbers and indices of linear operators, Usp. Mat. Nauk 12, 2 (1957), 43–118 (Russian; English translation in Amer. Math. Soc. Transl., Ser. 2, 13 (1960), 185–264).
[GP] V. Guillemin, A. Pollack, *Differential Topology*, Prentice-Hall, Englewood Cliffs, N.J. 1974.
[Gr] G. Grubb, *Functional Calculus of Pseudo-Differential Boundary Problems*, Birkhäuser Verlag, Boston–Basel 1986.
[Hi] M.W. Hirsch, *Differential Topology*, Springer-Verlag, New York–Berlin 1976.
[Hö 1] L. Hörmander, *The Analysis of Linear Partial Differential Operators*, Vol. I, Second Edition, Springer-Verlag, New York–Berlin 1990.
[Hö 2] L. Hörmander, *The Analysis of Linear Partial Differential Operators*, Vol. II, Springer-Verlag, New York–Berlin 1983.
[Hö 3] L. Hörmander, *The Analysis of Linear Partial Differential Operators*, Vol. III, Springer-Verlag, New York–Berlin 1985.
[HR] H. Holmann, H. Rummler, *Alternierende Differentialformen*, B.I., Mannheim, 1972.
[HW] W. Hurewicz, W. Wallman, *Dimension Theory*, Princeton University Press, Princeton 1941.
[Hus] D. Husemoller, *Fibre Bundles*, McGraw-Hill, New York 1966.
[Hu] Sze-Tsen Hu, *Homotopy Theory*, Academic Press, New York 1959.
[KN] J.J. Kohn, L. Nirenberg, An algebra of pseudo-differential operators, Comm. Pure Appl. Math. 18 (1965), 265–305.
[Ku] H. Kumano-go, *Pseudo-Differential Operators*, M.I.T. Press, Cambridge, Mass. 1974.
[La 1] S. Lang, *Algebra*, Second Edition, Addison-Wesley, Reading, Mass. 1984.

[La 2] S. Lang, *Differential Manifolds*, Springer-Verlag, New York–Berlin 1985.
[LM] J.L. Lions, E. Magenes, *Non-Homogeneous Boundary Value Problems and Applications*, Vol. I, Springer-Verlag, Berlin 1972.
[Lo] Ya. B. Lopatinskii, On a method of reducing boundary value problems for a system of differential equations of elliptic type to regular integral equations, Ukr. Mat. Zur. 5, 2 (1953), 123–152 (Russian; English translation in Amer. Math. Soc. Transl., Ser. 2, 89 (1970), 149–183).
[LS] L.H. Loomis, S. Sternberg, *Advanced Calculus*, Addison-Wesley, Reading 1968.
[Mi] S.G. Mikhlin, *Multidimensional Singular Integrals and Integral Equations*, Pergamon Press, Oxford 1965.
[MP] S.G. Mikhlin, S. Prößdorff, *Singular Integral Operators*, Springer-Verlag, New York–Berlin 1986.
[Mir] C. Miranda, *Partial Differential Equations of Elliptic Type*, Springer-Verlag, New York–Berlin 1970.
[Ne] J. Nečas, *Les méthodes directes en théorie des équations elliptiques*, Academia, Prague 1967.
[Pa] R.S. Palais, *Seminar on the Atiyah–Singer Index Theorem*, Princeton University Press, Princeton, N.J. 1965.
[Ru 1] W. Rudin, *Real and Complex Analysis*, Third Edition, McGraw-Hill, New York 1987.
[Ru 2] W. Rudin, *Functional Analysis*, McGraw-Hill, New York 1973.
[Sch] M. Schechter, *Modern Methods in Partial Differential Equations*, McGraw-Hill, New York 1977.
[Se 1] R. Seeley, Extension of C^∞ functions defined in a half space, Proc. Amer. Math. Soc. 15, 625–626 (1964).
[Se 2] R. Seeley, Integro-differential operators on vector bundles, Trans. Amer. Math. Soc. 117 (1965), 167–204.
[Se 3] R. Seeley, Singular integrals and boundary value problems, Amer. J. Math. 88 (1966), 781–809.
[Shu] M.S. Shubin, *Pseudo-Differential Operators and Spectral Theory*, Springer-Verlag, New York–Berlin 1987.
[Sp 1] M. Spivak, *A Comprehensive Introduction to Differential Geometry*, Vol. I, Publish or Perish, Inc., Boston, Mass. 1970.
[Sp 2] M. Spivak, *A Comprehensive Introduction to Differential Geometry*, Vol. V, Publish or Perish, Inc., Boston, Mass. 1975.
[Str] G. Strang, *Introduction to Applied Mathematics*, Wellesley-Cambridge Press, Wellesley, Mass. 1986.
[Tay] M.E. Taylor, *Pseudodifferential Operators*, Princeton University Press, Princeton, N.J. 1981.
[TL] A.E. Taylor, D.C. Lay, *Introduction to Functional Analysis*, Second Edition, R.E. Krieger Publ. Co., Malabar 1986.
[Tre 1] F. Trèves, *Introduction to Pseudo-Differential and Fourier Integral Operators*, Vol. I, Plenum Press, New York–London 1982.
[Tre 2] F. Trèves, *Introduction to Pseudo-Differential and Fourier Integral Operators*, Vol. II. Plenum Press, New York–London 1982.
[Tri] H. Triebel, *Interpolation Theory, Function Spaces, Differential Operators*, North-Holland, Amsterdam 1978.
[Ve] I.N. Vekua, *New Methods of Solution for Elliptic Equations*, OGIZ, Moscow–Leningrad 1948.
[Vo 1] L.R. Volevič, Solvability of boundary problems for general elliptic systems, Mat. Sbor. T. 68 (110), 3 (1965) 373–416 (Russian; English translation in Amer. Math. Soc. Transl., Ser. 2, 67 (1968), 182–225).

[Vo 2] L.R. Volevič, A problem in linear programming coming from differential equations, Usp. Mat. Nauk 18 (1963), no 3 (111), 155–162 (Russian).

[Vol'p] A.I. Vol'pert, On the index and the normal solvability of boundary value problems for elliptic systems of differential equations in the plane, Trudy Mos. Mat. Obs. 10 (1961), 41–87 (Russian).

[Wa] F.W. Warner, *Foundations of Differentiable Manifolds and Lie Groups*, Scott-Foresman, London 1971.

[War] W. Warschawski, On differentiability at the boundary in conformal mapping, Proc. Am. Math. Soc. 12 (1961), 614–620.

[Wei] J. Weidmann, *Linear Operators in Hilbert Spaces*, Springer-Verlag, New York–Berlin 1980.

[Wen] W.L. Wendland, *Elliptic Systems in the Plane*, Pitman, London 1979.

[Wl] J. Wloka, *Partial Differential Equations*, Cambridge University Press 1987.

[Wo] M.W. Wong, *An Introduction to Pseudo-Differential Operators*, World Scientific, Singapore 1991.

Index